Digital Manufacturing in Industry 5.0

Yogesh Kumar Singla
Ashwani Kumar • Michael R. Maughan
Editors

Digital Manufacturing in Industry 5.0

Springer

Editors
Yogesh Kumar Singla
School of Engineering, Math and Technology
Navajo Technical University
Crownpoint, NM, USA

Department of Mechanical Engineering
University of Idaho
Moscow, ID, USA

Michael R. Maughan
Department of Mechanical Engineering
University of Idaho
Moscow, ID, USA

Ashwani Kumar
Department of Mechanical Engineering
Technical Education Department
Uttar Pradesh
Kanpur, Uttar Pradesh, India

ISBN 978-3-031-91499-7 ISBN 978-3-031-91500-0 (eBook)
https://doi.org/10.1007/978-3-031-91500-0

This Springer imprint is published by the registered company Springer Nature Switzerland AG
The registered company address is: Gewerbestrasse 11, 6330 Cham, Switzerland

*"**Digital Manufacturing in Industry 5.0**" transcends the mere assemblage of scientific findings. It serves as an ode to the intricate choreography of systems, a reverent exploration of the countless silent interactions occurring within every complex process, every intricate machine, and every dynamic network. This dedication extends beyond the printed page, resonating with the very essence of manufacturing itself. It acknowledges the pioneering spirits who, driven by keen observation and insatiable curiosity, first ventured into the uncharted territories of industrial innovation. Their early discoveries paved the way for generations of engineers mesmerized by the elegance and ingenuity with which machines can transform raw materials into objects of function and beauty.*

Furthermore, this dedication embraces the vast community that inspires the advancements in digital manufacturing. It encompasses the skilled workers who bring manufacturing processes to life, the innovators who push the boundaries of technology, and the entrepreneurs who translate ideas into tangible products. The hope of creating a more sustainable, efficient, and resilient manufacturing landscape fuels the relentless pursuit of understanding and influencing the intricate interplay of human ingenuity and technological advancement. However, this journey of manufacturing is not a solitary performance; it is a collaborative masterpiece. Machines communicate, data flows seamlessly, and humans and technology work in tandem.

This dedication celebrates critical areas of digital manufacturing, including sustainability in manufacturing, digital twins, augmented reality and robotics, and smart supply chains. May this book serve as a testament to the dedication of countless individuals who strive to understand and influence the manufacturing landscape. May it inspire future generations to join the quest for knowledge, innovation, and, ultimately, advancement of human progress through responsible and sustainable manufacturing practices.

Yogesh Kumar Singla, Ashwani Kumar, Michael R. Maughan

Aim and Scope

The book "***Digital Manufacturing in Industry 5.0***" explores into the transformative landscape of manufacturing in the era of Industry 5.0. It focuses on the convergence of cutting-edge technologies, including the Internet of Things (IoT), artificial intelligence (AI), and advanced robotics, to redefine sustainable industrial processes. Key areas of exploration include digital twinning, smart factories, and cyber-physical systems, which contribute to the creation of intelligent and adaptive manufacturing ecosystems.

The book investigates the role of additive manufacturing and 3D printing in revolutionizing traditional production methods, emphasizing the shift towards more flexible and customizable manufacturing processes. It also explores the potential of hybrid manufacturing techniques, combining machining and additive manufacturing to fabricate complex parts with high precision and surface quality. This approach is particularly relevant for manufacturing high-strength components in industries such as aerospace, biomedical, and marine.

The book examines the virtualization of manufacturing through technologies like digital twin, edge computing, big data analytics, and cloud computing. These technologies facilitate real-time data processing and decision-making, enabling improved operational efficiency and predictive maintenance. Smart sensors play a crucial role in this ecosystem, providing real-time data to interconnected manufacturing systems. Augmented reality (AR) further enhances human–machine interaction, improving operational efficiency and enabling more effective training and maintenance procedures.

The book explores the concept of digital twins in depth, examining their utilization for creating virtual replicas of physical systems. This enables enhanced understanding, monitoring, and optimization of manufacturing processes. Digital twins facilitate detailed analysis, predictive modeling, and performance simulations, allowing organizations to gain valuable insights into the operational aspects of their assets. By synchronizing the virtual and physical worlds, organizations can minimize risks, reduce downtime, and improve overall efficiency.

Manufacturing sustainability is a central theme throughout the book. It highlights how digital manufacturing contributes to sustainable manufacturing practices, exploring the integration of agile manufacturing principles to prioritize adaptability and responsiveness. The book also examines the impact of Industry 5.0 on the creation of smart supply chains, ensuring seamless coordination and optimization across the entire production and distribution network. The book emphasizes the importance of simulation software systems, which enable engineers to predict part and process behavior, reducing the need for time-consuming and costly trial runs in production.

In conclusion, this book endeavors to unravel the intricacies of Industry 5.0, focusing on the amalgamation of digital manufacturing, digital twinning, smart factories, and a myriad of transformative technologies. It spans the entire spectrum from advanced robotics to sustainable manufacturing practices, offering a holistic view of the factories of the future. The book elucidates the interconnectedness of technologies such as advanced robotics, digital twinning, smart factories, cyber-physical systems, and intelligent manufacturing ecosystems, emphasizing their collective impact on reshaping the manufacturing domain.

Moscow, ID, USA — Yogesh Kumar Singla
Kanpur, Uttar Pradesh, India — Ashwani Kumar
Moscow, ID, USA — Michael R. Maughan

Preface

This book, "***Digital Manufacturing in Industry 5.0***," serves as a comprehensive guide, empowering readers with the knowledge and insights necessary to navigate the evolving landscape of digital manufacturing. Our aim is to equip readers with the tools and understanding to contribute to the development of more efficient, sustainable, and innovative manufacturing practices for the future. This book offers a balanced approach, covering a spectrum of knowledge from foundational concepts to advanced applications. It encompasses a diverse range of topics, including theoretical frameworks, practical modeling techniques, insightful simulations, and valuable experimental findings. The book is organized into four distinct parts:

Part I: Digital Manufacturing (Chaps. 1–18): This part provides a solid foundation in the principles and technologies that underpin digital manufacturing.

Chapters 1–18 provide a comprehensive overview of the evolving landscape of digital manufacturing in Industry 5.0. Chapter 1 traces the historical evolution of manufacturing, highlighting key milestones and technological advancements. Chapter 2 provides a global perspective on current trends, challenges, and opportunities in digital manufacturing. Chapter 3 explores the use of SPSS for data analysis and strategy development in optimizing digital manufacturing practices. Chapters 4–10 explore into the transformative power of additive manufacturing (AM) technologies. Chapter 4 explores the shift from traditional subtractive manufacturing to AM, emphasizing its potential to revolutionize production. Chapters 5 and 6 further explore the principles, techniques, and applications of AM and 3D printing in Industry 5.0. Chapters 7 and 8 discuss the impact of 3D printing on various sectors, including aerodynamic research and production of complex components like gearboxes.

Chapters 9–13 explore the integration of advanced technologies with AM. Chapter 9 investigates the role of artificial intelligence (AI) in optimizing metal manufacturing through 3D printing. Chapters 10–12 explore the role of AI, machine learning, and thermography in enhancing the efficiency, precision, and sustainability of AM processes. Chapter 13 explores the recent advances in the application of thermography to additive manufacturing for process monitoring and mechanical characterization. Chapter 14 highlights the applications of AM in tissue engineering and regenerative medicine. Chapters 15–18 focus on specific materials, manufacturing techniques and applications. Chapter 15 investigates the optimal steel fiber volume in geopolymer concrete to enhance its mechanical properties. Chapter 16 explores the advanced manufacturing technique of friction stir processing for creating high-performance composite materials. Chapter 17 focuses on optimizing energy efficiency in smart manufacturing by utilizing alcohol-based fuels in fire tube boilers. Chapter 18 investigates the combustion characteristics and fracture toughness of epoxy glass fiber/phenolic composites during oxidation.

Part II: Sustainability Issues in Manufacturing (Chaps. 19–31): This part explores into the critical role of sustainability in modern manufacturing, exploring strategies for minimizing environmental impact and maximizing resource efficiency.

Chapters 19–21 explore the concept of Industry 5.0 as a new era of sustainable and human-centered production. Chapter 19 introduces Industry 5.0 and its emphasis on human-centric values, sustainability, and resilience. Chapter 20 explores into the key sustainability challenges faced by modern manufacturing, such as resource depletion, environmental pollution, and social inequality. Chapter 21 highlights the crucial role of human–machine collaboration in addressing these sustainability challenges, emphasizing the importance of human creativity, expertise, and adaptability. Chapters 22–24 further explore the advancements and challenges of sustainability in Industry 5.0. Chapter 22 examines emerging trends, challenges, and future directions for achieving sustainability in the context of Industry 5.0. Chapter 23 explores into the specific challenges faced by businesses in navigating the transition to Industry 5.0, such as technological advancements, workforce development, and regulatory compliance.

Chapter 24 focuses on a specific case study, evaluating the sustainability and performance of concrete reinforced with fly ash and silica fume using a combined life cycle assessment (LCA) and techno-economic assessment (TEA) approach. Chapter 25 explores innovative approaches to waste management and resource utilization. It focuses on "waste to resource" strategies, emphasizing the importance of adopting smart 3R approaches (reduce, reuse, recycle) for sustainable and resilient construction practices. Chapter 26 focuses on the critical role of digital tool life prediction and monitoring in achieving sustainable manufacturing practices. By accurately predicting tool wear and optimizing tool usage, manufacturers can minimize waste, reduce downtime, and improve overall resource efficiency. Chapter 27 explores into the challenges and considerations associated with the successful adoption of Industry 5.0 principles. These challenges include integrating new technologies, upskilling the workforce, ensuring data security, and addressing ethical concerns related to automation and AI. The chapter also discusses strategies for overcoming these challenges and successfully transitioning to a more sustainable and human-centric manufacturing paradigm.

Chapters 28 and 30 explore the use of sustainable materials in Industry 5.0. Chapter 28 investigates the challenges and opportunities of integrating natural fiber-reinforced polymer composites into manufacturing processes. These materials offer several advantages, including reduced environmental impact and improved mechanical properties. Chapter 30 discusses the development of sustainable bio-composites and hybrid materials using digital manufacturing techniques, highlighting their potential to revolutionize materials science and contribute to a more sustainable future. Chapters 29 and 31 explore innovative applications of technology for sustainable practices. Chapter 29 showcases a real-world example of sustainable innovation by introducing "Bananature," a concept for creating eco-friendly carry bags from banana pseudostems, offering a viable alternative to plastic bags. Chapter 31 highlights the role of IoT-driven solutions in optimizing smart farming practices and improving business operations within the framework of Industry 5.0, contributing to a more sustainable and efficient agricultural sector.

Part III: Digital Twins, Augmented Reality, and Robotics (Chaps. 32–45): This part focuses on key enabling technologies, including digital twins, augmented reality, and robotics, and their transformative impact on manufacturing processes.

Chapters 32–36 explore into the multifaceted applications of digital twins. Chapter 32 explores the role of digital twins in product development, while Chapter 33 highlights their benefits in accelerating innovation through enhanced design and manufacturing processes. Chapter 34 focuses on utilizing digital twins for a deeper understanding and optimization of manufacturing processes. Chapter 35 examines the contribution of digital twins to sustainable performance in Industry 5.0 through case studies. Chapter 36 explores the role of Digital Twins and Smart Sensors in Process Optimization for Direct Energy Deposition of EN36C steel for Automotive Applications.

Chapters 37–39 explore the transformative potential of augmented reality (AR) and robotics in manufacturing. Chapter 37 showcases how AR revolutionizes various stages of the manufacturing life cycle, from design and prototyping to production and maintenance. Chapters 38 and 39 explore into the advancements in robotics, particularly focusing on advanced robotics applications in Industry 5.0 manufacturing processes and the role of reinforcement learning in enhancing robotic capabilities.

Chapters 40–45 focus on the integration of smart technologies within the manufacturing ecosystem. Chapter 40 explores the concept of smart factories and cyber-physical systems, while Chapter 41 explores deeper into advanced robotics applications in Industry 5.0 manufacturing processes. Chapters 42 and 43 highlight the crucial role of smart sensors in driving the "factories of the future." Chapter 44 presents the RAMP case study, illustrating the transition from shop floor operations to cloud-based cyber-physical systems. Finally, Chap. 45 emphasizes the significance of data analytics and cloud resources in leveraging insights and driving innovation in Industry 5.0.

Part IV: Smart Supply Chains (Chaps. 46–51): This part explores the concept of smart supply chains, emphasizing their role in optimizing logistics, reducing waste, and enhancing overall supply chain resilience.

Chapters 46–51 explore various aspects of smart supply chains within the context of Industry 5.0. Chapter 46 explores into the role of blockchain technology in enhancing supply chain sustainability, providing a comprehensive review of existing literature. Chapter 47 explores how smart supply chains are revolutionizing logistics operations, highlighting key advancements and their impact on Industry 5.0.

Chapters 48 and 49 focus on key enablers of smart supply chains. Chapter 48 examines the role of artificial intelligence (AI) in driving sustainability within smart supply chains, focusing on the measurement of AI-based drivers. Chapter 49 explores the critical role of supply chain finance in optimizing operations, emphasizing the importance of supplier risk portfolio assessment in mitigating financial disruptions.

Chapter 50 explores the digital transformation of maintenance practices, highlighting the role of computerized maintenance management systems (CMMSs) in enhancing efficiency, productivity, and overall equipment effectiveness within the supply chain.

Finally, Chapter 51 focuses on human–robot collaboration (HRC) that enables humans and robots to work together in shared environments, leveraging human creativity and robot precision to achieve common goals.

Moscow, ID, USA — Yogesh Kumar Singla
Kanpur, Uttar Pradesh, India — Ashwani Kumar
Moscow, ID, USA — Michael R. Maughan

Acknowledgments

We would like to express our deepest gratitude to Springer Nature publisher for providing the platform and unwavering support that made the publication of this book possible. We are particularly indebted to the dedicated editorial team, whose insightful guidance, constructive feedback, and unwavering commitment to excellence significantly enhanced the quality of our work. Their invaluable suggestions and support were instrumental in bringing this book to fruition.

Furthermore, we extend our sincere appreciation to the numerous contributors and reviewers whose invaluable expertise and insightful perspectives enriched every chapter of this book. Their dedication to sharing their knowledge and providing constructive criticism has significantly improved the depth, breadth, and overall quality of this comprehensive resource "***Digital Manufacturing in Industry 5.0***."

This book is dedicated to all those individuals whose passion, dedication, and tireless efforts drive advancements in the critical areas of digital manufacturing, sustainability in manufacturing, digital twins, augmented reality and robotics, and smart supply chains. We recognize and honor their vital contributions to the field and hope that this work serves as a valuable resource and source of inspiration for the scholarly community, fostering further research and innovation in the dynamic intersection of materials, manufacturing, and sustainable developments in Industry 5.0 era.

School of Engineering, Math and Technology
Navajo Technical University, Crownpoint, NM, USA
Department of Mechanical Engineering
University of Idaho, Moscow, ID, USA
Yogesh Kumar Singla

Department of Mechanical Engineering,
Technical Education Department,
Kanpur, Uttar Pradesh, India
Ashwani Kumar

Department of Mechanical Engineering
University of Idaho, Moscow, ID, USA
Michael R. Maughan

Contents

Editors and Contributors

About the Editors

Yogesh Kumar Singla is an Assistant Professor of Mechanical Engineering at Navajo Technical University and an Adjunct Research Associate at Harvard University. His interdisciplinary research integrates sustainable materials, additive manufacturing, and surface engineering to develop durable, resource-efficient systems that advance the circular economy and energy-efficient manufacturing. Dr. Singla has made pioneering contributions to wire arc additive manufacturing (WAAM), including co-reporting one of the first demonstrations of single-crystal nickel fabrication—a breakthrough with significant aerospace and energy applications. He has held prestigious fellowships at Case Western Reserve University and the University of Idaho, and holds a Ph.D. from IIT Roorkee. With over 11 years of research and teaching experience, Dr. Singla has authored around 32 peer-reviewed publications, multiple book chapters, and edited several books with leading publishers including CRC Press, Springer, and Wiley. He serves on editorial boards, reviews for high-impact journals, and is an invited speaker at international forums. As an educator, he has established advanced labs and mentored numerous graduate students, championing the integration of sustainability and innovation in engineering education. Dr. Singla's work embodies a commitment to design-driven innovation that enhances material performance while reducing environmental impact.

Ashwani Kumar holds a Ph.D. in Mechanical Engineering with a diverse and impressive career spanning over 15 years. He currently holds the positions of Professor and Head of Department of Mechanical Engineering (Gazetted Officer Group A) at the Technical Education Department, Kanpur (under the Government of Uttar Pradesh), India. Prior to his current role, Dr. Kumar served as Senior Lecturer and Lecturer in the same organization and Assistant Professor in the Department of Mechanical Engineering at Graphic Era University, Dehradun (ranked 48th by NIRF, India), from July 2010 to November 2013. Beyond his academic contributions, Dr. Kumar possesses extensive leadership and administrative experience. He has held numerous key positions within the education sector, including Coordinator for AICTE-Extension of Approval; Nodal Officer for PMKVY-TI Scheme (Government of India); Internal Coordinator, CDTP Scheme (Government of Uttar Pradesh); Industry Academia Relation Officer; Assistant Centre Superintendent (ACS), Institute Examination Cell; Zonal Officer for Joint Entrance Examination (JEE-Diploma); and Sector Magistrate for State Assembly and General Elections. Dr. Kumar is a passionate researcher and a prominent figure in the field of academic publishing. His dedication to knowledge dissemination is evident in his role as Series Editor for eight distinguished book series published by CRC Press (Taylor & Francis), USA; Wiley Scrivener Publishing, USA; and Apple Academic Press, USA. These series encompass a wide range of topics, titled as Advances in Manufacturing, Design and Computational Intelligence Techniques; Renewable and Sustainable Energy Developments; Smart Innovations and Technological Advancements in Mechanical and Materials Engineering; Solar Thermal Energy Systems: Advancements in Engineering, Ergonomics, and Sustainable Development; Artificial Intelligence and Machine Learning for Intelligent Engineering Systems; Computational Intelligence and Biomedical Engineering; Advances in Green Technologies for

Sustainable Energy Solutions; and Advances in Materials, Manufacturing and Computational Intelligence Techniques. Dr. Kumar's editorial expertise extends beyond book series. He serves as Editor in Chief for the International Journal of Materials, Manufacturing and Sustainable Technologies (IJMMST, *ISSN: 2583-6625*) and Editor of the International Journal of Energy Resources Applications (IJERA, *ISSN: 2583-6617*). He has also been a guest editor for a special issue of Buildings titled as Sustainable Buildings, Resilient Cities and Infrastructure Systems (*ISSN: 2075-5309*, I.F. 3.1); he holds editorial board positions for eight international journals and acts as a review board member of 20 prestigious (indexed in SCI/SCIE/Scopus) international journals with high impact factor. Dr. Kumar's dedication to scholarship is further demonstrated by his prolific authorship of over 200+ articles in prestigious journals, book chapters, and conference proceedings. Additionally, he has co-authored or edited over 40+ books in mechanical engineering, materials science, and renewable energy engineering. Dr. Kumar's achievements have been recognized with prestigious awards, including a Best Teacher Award for excellence in academics and research. He has also successfully guided 15 students through their B. Tech., M. Tech., and Ph.D. theses and currently serves as an external doctoral committee member at SRM University, New Delhi. Dr. Kumar is actively involved in cutting-edge research, focusing on areas such as artificial intelligence and machine learning in mechanical engineering, smart materials and manufacturing techniques, thermal energy storage, building efficiency, renewable energy harvesting, sustainable transportation, and heavy vehicle dynamics. He holds five patents and frequently participates in international conferences as an invited speaker, session chair, and member of advisory boards, review boards, and program committees. Dr. Kumar's ORCID ID is https://orcid.org/0000-0003-4099-935X and Google Scholar link is https://scholar.google.com/citations?hl=en&user=KOILpEkAAAAJ.

Michael R. Maughan is an Associate Professor of Mechanical Engineering. He holds B.S. and M.S. degrees in Mechanical Engineering. He earned his Ph.D. in Materials Engineering from Purdue University in 2015. Prior to joining the academic ranks, he worked in industry at both Fortune 50 and startup companies as a mechanical designer and engineering manager. He holds five utility patents. His research expertise is in material behavior and properties, particularly microstructure-property relationships for advanced manufacturing processes. His work has been published in numerous peer-reviewed mechanical engineering and materials journals including Journal of Materials Research, International Journal of Plasticity, Materials Research Letters, and ASME Journal of Electronic Packaging. He currently serves as a guest editor for the Forest Products Journal. At the University of Idaho, he has developed a popular new course on manufacturing and has been active in reimagining several existing courses for the digital learning era. From 2015 to 2022, he was the faculty advisor for the University of Idaho ASME student section and has led numerous infrastructure projects within the Department of Mechanical Engineering. In 2015, he received the Estus H. and Vashti L. Magoon Award for excellence in graduate teaching at Purdue University and has taught over 50 courses at the University of Idaho. He was named Outstanding Early Career Faculty by the University of Idaho College of Engineering in 2022 and has advised 13 graduate students as major professor.

Contributors

Nail Adeeb Ali Abdu Information Technology Department, Faculty of Engineering and Computing, University of Science and Technology, Aden, Yemen

Samson Abiodun AASA Department of Automotive Engineering, University of Ibadan, Ibadan, Nigeria

Abiodun Oluwatosin Adeoye Department of Chemistry, Federal University Oye Ekiti, Oye-Ekiti, Ekiti State, Nigeria

Daniel Agharahimi Department of Materials Science and Engineering, Georgia Institute of Technology, Atlanta, GA, USA

Patience Agnes EBIGHE Department of Chemistry, Toulouse INP-ENSIACET, Toulouse, France

Nuhu Ahmad Department of Civil Engineering, Bayero University, Kano, Kano State, Nigeria

Sani Abdullahi Aliyu Department of Chemistry, University of Lille, Lille, France

Basheer Almuhaya Faculty of Engineering and Information Technology, Taiz University, Taiz, Yemen

Tharaa M. Al-Zghoul Department of Civil Engineering, Faculty of Engineering, Tafila Technical University, Tafila, Jordan

Damarasingu Apparao Department of Mechanical Engineering, Aditya Institute of Technology and Management, Tekkali, Andhra Pradesh, India

Navneet Arora Department of Mechanical and Industrial Engineering, Indian Institute of Technology Roorkee, Roorkee, Uttarakhand, India

Rajesh Attri Department of Mechanical Engineering, J.C. Bose University of Science and Technology, YMCA, Faridabad, India

Duppala Azad Aditya Institute of Technology and Management, Tekkali, Andhra Pradesh, India

Abdullah O. Baarimah Department of Civil and Environmental Engineering, College of Engineering, A'Sharqiyah University, Ibra, Oman

Mendi Venkanna Babu Narasaraopeta Engineering College, Narasaraopeta, Andhra Pradesh, India

Hande Bahar Department of Materials Science and Engineering, Georgia Institute of Technology, Atlanta, GA, USA

Roopsandeep Bammidi Department of Mechanical Engineering, Aditya Institute of Technology and Management, Tekkali, Andhra Pradesh, India

Soumyabrata Basak School of Mechanical Engineering, University of Ulsan, Ulsan, Republic of Korea

Mahmood A. Bazel Faculty of Engineering and Information Technology, Taiz University, Taiz, Yemen

Denise Bellisario Department of Engineering and Science, Faculty of Technological Science and Innovation, Mercatorum University, Rome, Italy

Davide Berardi Department of Engineering and Science, Faculty of Technological Science and Innovation, Mercatorum University, Rome, Italy

Mohammad Binhammad The British University in Dubai, Dubai, UAE

Ramona Birau "Eugeniu Carada" Doctoral School of Economic Sciences, University of Craiova, Craiova, Romania

Pias Kumar Biswas School of Engineering, Math and Technology, Navajo Technical University, Crownpoint, NM, USA

Pritom Kumar Biswas National University, Gazipur, Dhaka, Bangladesh

Panagiotis Bouklis European Dynamics Luxembourg S.A., Luxembourg, Luxembourg

Dimitrios Chantes European Dynamics Advanced Information Technology and Telecommunication Systems S.A., Athens, Greece

Mogana Priya Chinnasamy School of Mechanical Engineering, Vellore Institute of Technology, Chennai, India

Amit Choudhary Department of Mechanical and Industrial Engineering, Indian Institute of Technology Roorkee, Roorkee, Uttarakhand, India

Ester D'Accardi Department of Mechanics, Mathematic and Management, Politecnico di Bari, Bari, Italy

Rosa De Finis Department of Engineering for Innovation, University of Salento, Lecce, Italy

Ananya S. Desikana School of Mechanical Engineering, Vellore Institute of Technology, Vellore, Tamil Nadu, India

Pedapenki Dinakar Department of Civil Engineering, Aditya Institute of Technology and Management, Tekkali, Andhra Pradesh, India

Ramajogi Naidu Doddi Department of Mechanical Engineering, Aditya Institute of Technology and Management, Tekkali, Andhra Pradesh, India

Sreeramulu Dowluru Department of Mechanical Engineering, Aditya Institute of Technology and Management, Tekkali, Andhra Pradesh, India

Vibhav Durgesh Department of Mechanical Engineering, University of Idaho, Moscow, ID, USA

Emmanuel Oghenero Emojevu Department of Chemistry, University of Benin, Benin, Nigeria

Umberto Galietti Department of Mechanics, Mathematic and Management, Politecnico di Bari, Bari, Italy

Nura Garba Department of Environmental and Resources Management, Usmanu Danfodiyo University, Sokoto, Nigeria

Anastasia Garbi European Dynamics Luxembourg S.A., Luxembourg, Luxembourg

Hamid Garmestani Department of Materials Science and Engineering, Georgia Institute of Technology, Atlanta, GA, USA

Konstantinos Gombakis European Dynamics Luxembourg S.A., Luxembourg, Luxembourg

Yatika Gori Department of Mechanical Engineering, Graphic Era Deemed to be University, Dehradun, Uttarakhand, India

School of Computing, Graphic Era Hill University, Haldwani, Uttarakhand, India

Badana Govindarajulu Department of Civil Engineering, Aditya Institute of Technology and Management, Tekkali, Andhra Pradesh, India

Vasanthi Govindaraj National General (An Allstate Company), Dallas, TX, USA

Samarthya Goyal School of Infrastructure, Indian Institute of Technology Bhubaneswar, Bhubaneswar, India

R. Immanual Department of Mechanical Engineering, Sri Ramakrishna Institute of Technology, Coimbatore, Tamil Nadu, India

S. Indumati NBKRIST, Vidyanagar, Andhra Pradesh, India

Leandro Iorio Department of Engineering and Science, Faculty of Technological Science and Innovation, Mercatorum University, Rome, Italy

Rapaka Jagadeesh Department of Mechanical Engineering, Aditya Institute of Technology and Management, Tekkali, Andhra Pradesh, India

Donepudi Jagadish Department of Mechanical Engineering, Narasaraopeta Engineering College, Narasaraopeta, Andhra Pradesh, India

Rakesh Joshi DARVUN India, Patiala, Punjab, India

Wisdom Simwila Kalunga Department of Aerospace Engineering, Chandigarh University, Mohali, India

B Sunil Kamath Department of Information Science and Engineering, NMAM Institute of Technology, NITTE (Deemed to be University), Nitte, Mangalore, Karnataka, India

Jatinder Kaur Department of Mathematics, Chandigarh University, Mohali, Punjab, India

K. M. Kirthika Department of Computer Science and Engineering, Sri Ramakrishna Institute of Technology, Coimbatore, Tamil Nadu, India

N. Krishnaraj Department of Networking and Communications, School of Computing, College of Engineering and Technology, SRM Institute of Science and Technology, Kattankulathur, Chengalpattu, Chennai, Tamil Nadu, India

Ashwani Kumar Department of Mechanical Engineering, Technical Education Department Uttar Pradesh, Kanpur, India

Department of Mechanical Engineering, Graphic Era Deemed to be University, Dehradun, Uttarakhand, India

Dinesh Kumar Department of Mechanical Engineering, Noida Institute of Engineering and Technology, Greater Noida, India

Mada Sreenivasa Kumar Department of Mechanical Engineering, Narasaraopeta Engineering College, Narasaraopeta, Andhra Pradesh, India

Naveen Kumar Department of Mechanical and Aerospace Engineering, University of Petroleum and Energy Studies, Dehradun, Uttarakhand, India

P. P. Praveen Kumar Department of Biomedical Engineering, Institute for Quantitative Health Science and Engineering, Michigan State University, East Lansing, MI, USA

Sachin Kumar Regenerative Biology Research Lab, Division of Clinical Research and Medical Management (CRMM), Institute of Nuclear Medicine and Allied Sciences (INMAS), Defence Research and Development Organisation (DRDO), Timarpur, Delhi, India

Sanjay Kumar Department of Mechanical Engineering, Noida Institute of Engineering and Technology, Greater Noida, India

Subodh Kumar Department of Biomedical Sciences, School of Biosciences and Technology, Galgotias University, Greater Noida, India

Suryatej S Kumar Department of Mathematics, Chandigarh University, Gharuan, Mohali, Punjab, India

Rajan Kumar V K Department of Industrial Engineering and Management, School of Engineering Science, LUT University, Lahti, Finland

Aruna Kunda Department of Automobile Engineering, Godavari Institute of Engineering and Technology, Rajamahendravaram, Andhra Pradesh, India

Steven Y. Liang George W. Woodruff School of Mechanical Engineering, Georgia Institute of Technology, Atlanta, GA, USA

Piyush Maheshwari The British University in Dubai, Dubai, UAE

Barbara Martini Department of Engineering and Science, Faculty of Technological Science and Innovation, Mercatorum University, Rome, Italy

Michael R. Maughan Department of Mechanical Engineering, University of Idaho, Moscow, ID, USA

R. Meera Department of Networking and Communications, School of Computing, College of Engineering and Technology, SRM Institute of Science and Technology, Kattankulathur, Chengalpattu, Chennai, Tamil Nadu, India

Sakshi Mehra Regenerative Biology Research Lab, Division of Clinical Research and Medical Management (CRMM), Institute of Nuclear Medicine and Allied

Sciences (INMAS), Defence Research and Development Organisation (DRDO), Timarpur, Delhi, India

Venkataramana Menda Aditya Institute of Technology and Management, Tekkali, Andhra Pradesh, India

Utkarsh Mittal NIT Jalandhar, Jalandhar, India

Sandeep Kumar Mogha Department of Mathematics, Chandigarh University, Mohali, Punjab, India

Department of Mathematics, University of Delhi, Delhi, India

Fathey Mohammed Department of Business Analytics, Sunway Business School, Sunway University, Petaling Jaya, Selangor, Malaysia

Mounarik Mondal School of Engineering and Design, London South Bank University, London, United Kingdom

Suman Mondal Metallurgical and Materials Engineering, National Institute of Technology Srinagar, Srinagar, India

Haliru Ibrahim Muhammad Department of Physics, Federal University Gusau, Gusau, Zamfara State, Nigeria

Padmakumar Muthuswamy Ingersoll Rand Technologies and Services Private Limited, Bangalore, India

Navid Nasajpour-Esfahani Department of Materials Science and Engineering, Georgia Institute of Technology, Atlanta, GA, USA

Thai-Son Nguyen Center for Transport Science and Technology, University of Transport and Communications, Hanoi, Vietnam

Vani Nigam School of Mineral, Metallurgical, and Materials Engineering, Indian Institute of Technology Bhubaneswar, Bhubaneswar, India

Azzam Othman The British University in Dubai, Dubai, UAE

Rodrigo Padilla Department of Mechanical Engineering, University of Idaho, Moscow, ID, USA

Vaikunta Pai T Department of Information Science and Engineering, NMAM Institute of Technology, NITTE (Deemed to be University), Nitte, Karnataka, India

Davide Palumbo Department of Mechanics, Mathematic and Management, Politecnico di Bari, Bari, Italy

Nitin Panwar Department of Mechanical Engineering, J.C. Bose University of Science and Technology, YMCA, Faridabad, India

Saetukrit Panwar Regenerative Biology Research Lab, Division of Clinical Research and Medical Management (CRMM), Institute of Nuclear Medicine and

Allied Sciences (INMAS), Defence Research and Development Organisation (DRDO), Timarpur, Delhi, India

M. Parimala Vellore Institute of Technology, Vellore, Tamil Nadu, India

Georgios Peppas European Dynamics Luxembourg S.A., Luxembourg, Luxembourg

Deviprasad Pilla Department of Mechanical Engineering, Aditya Institute of Technology and Management, Tekkali, Andhra Pradesh, India

Raminaidu Pisini Department of Mechanical Engineering, Aditya Institute of Technology and Management, Tekkali, Andhra Pradesh, India

Virgil Popescu Faculty of Economics and Business Administration, University of Craiova, Craiova, Romania

Shiv Narayan Prajapati Department of Mechanical Engineering, Noida Institute of Engineering and Technology, Greater Noida, India

G. Prasad Department of Aerospace Engineering, Chandigarh University, Mohali, India

G. Jishnu Prian School of Mechanical Engineering, Vellore Institute of Technology, Vellore, Tamil Nadu, India

Neha Singh Raghuvanshi Department of Applied Sciences, ABES Engineering College, Ghaziabad, Uttar Pradesh, India

G. Rajyakalshmi School of Mechanical Engineering, Vellore Institute of Technology, Vellore, Tamil Nadu, India

Bondala Ramakrishna Aditya Institute of Technology and Management, Tekkali, Andhra Pradesh, India

Ritu Rani Centre of Management and Humanities, Punjab Engineering College (Deemed to be University), Chandigarh, India

Tero Rantala Department of Industrial Engineering and Management, School of Engineering Science, LUT University, Lahti, Finland

Sandeep Ravish Department of Civil Engineering, J.C. Bose University of Science and Technology, YMCA, Faridabad, India

P. Srihari Reddy NBKRIST, Vidyanagar, Andhra Pradesh, India

Pavan Kumar Rejeti Department of Mechanical Engineering, Aditya Institute of Technology and Management, Tekkali, Andhra Pradesh, India

Anish Sachdeva NIT Jalandhar, Jalandhar, India

Kisor K. Sahu School of Mineral, Metallurgical, and Materials Engineering, Indian Institute of Technology Bhubaneswar, Bhubaneswar, India

Laxmanaraju Salavarao Department of Mechanical Engineering, Sri Sivani College of Engineering, Srikakulam, Andhra Pradesh, India

S. Sangeetha Department of Electrical and Electronics Engineering, Sri Ramakrishna Institute of Technology, Coimbatore, Tamil Nadu, India

N. Sanjana Department of Computer Science and Engineering, Sri Ramakrishna Institute of Technology, Coimbatore, Tamil Nadu, India

Minna Saunila Department of Industrial Engineering and Management, School of Engineering Science, LUT University, Lahti, Finland

Shivanjali Saxena DARVUN India, Patiala, Punjab, India

Rippin Sehgal Department of Biotechnology, Ambala College of Engineering and Applied Research, Ambala, Haryana, India

Khaled Shaalan The British University in Dubai, Dubai, UAE

Vishakha Shadija Department of Mechanical Engineering, Medicaps University, Indore, Madhya Pradesh, India

Zeban Shah Federal University of Rio Grande do Sul, Porto Alegre, Brazil

Lakshmi Shankar Department of Mechanical Engineering, Punjabi University, Patiala, Punjab, India

Ankit Sharma Department of Mechanical Engineering, Noida Institute of Engineering and Technology, Greater Noida, India

Department of Electrical and Electronics Engineering, ABV Indian Institute of Information Technology and Management, Gwalior, India

Apurv Sharma Department of Mathematics, Chandigarh University, Mohali, Punjab, India

Eisha Mehta Sharma Department of Mechanical Engineering, MM Engineering College, Maharishi Markandeshwar (Deemed to be University), Mullana, Haryana, India

Henu Sharma School of Mineral, Metallurgical, and Materials Engineering, Indian Institute of Technology Bhubaneswar, Bhubaneswar, India

NetTantra Technologies India Pvt. Ltd., Bhubaneswar, India

Kanika Sharma Department of Applied Sciences, ABES Engineering College, Ghaziabad, Uttar Pradesh, India

Mohit Sharma Department of Production and Industrial Engineering, Punjab Engineering College (Deemed to be University), Chandigarh, India

Mohit Pramod Sharma Department of Mechanical Engineering, Medicaps University, Indore, Madhya Pradesh, India

Reecha Sharma Department of Electronics and Communication Engineering, Punjabi University, Patiala, Punjab, India

Vipin Sharma Department of Mechanical Engineering, Medicaps University, Indore, Madhya Pradesh, India

Rajendra Kumar Shukla Department of Mechanical Engineering, Medicaps University, Indore, Madhya Pradesh, India

Bhupender Singh Department of Mechanical Engineering, J.C. Bose University of Science and Technology, YMCA, Faridabad, India

Bikram Jit Singh Department of Mechanical Engineering, MM Engineering College, Maharishi Markandeshwar (Deemed to be University), Mullana, Haryana, India

Rajdev Singh Department of Mechanical and Industrial Engineering, Indian Institute of Technology Roorkee, Roorkee, Uttarakhand, India

Yogesh Kumar Singla School of Engineering, Math and Technology, Navajo Technical University, Crownpoint, NM, USA

Department of Mechanical Engineering, University of Idaho, Moscow, ID, USA

Bathula Venkata Siva Narasaraopeta College of Engineering, Narasaraopeta, Andhra Pradesh, India

Sonia School of Engineering, Math and Technology, Navajo Technical University, Crownpoint, NM, USA

A. Soujanya Department of Networking and Communications, School of Computing, College of Engineering and Technology, SRM Institute of Science and Technology, Kattankulathur, Chengalpattu, Chennai, Tamil Nadu, India

Ponguru Naga Sowjanya Narasaraopeta Engineering College, Narasaraopeta, Andhra Pradesh, India

M. Sreenivasulu NBKRIST, Vidyanagar, Andhra Pradesh, India

Vikash Kumar Srivastva Department of Mechanical and Industrial Engineering, Indian Institute of Technology Roorkee, Roorkee, Uttarakhand, India

Priyanga Subbiah Department of Networking and Communications, School of Computing, College of Engineering and Technology, SRM Institute of Science and Technology, Kattankulathur, Chengalpattu, Chennai, Tamil Nadu, India

Donthamsetty Suneel Narasaraopeta Engineering College, Guntur, Andhra Pradesh, India

Penugonda Sureshbabu Narasaraopeta Engineering College, Guntur, Andhra Pradesh, India

Matthew Swenson Department of Mechanical Engineering, University of Idaho, Moscow, ID, USA

Shaik Sydha Narasaraopeta Engineering College, Narasaraopeta, Andhra Pradesh, India

Nidhi Tanwar Centre of Management and Humanities, Punjab Engineering College (Deemed to be University), Chandigarh, India

Ngoc-Hien Tran Faculty of Mechanical Engineering, University of Transport and Communications, Hanoi, Vietnam

Amit Kumar Tyagi Department of Networking and Communications, School of Computing, College of Engineering and Technology, SRM Institute of Science and Technology, Kattankulathur, Chengalpattu, Chennai, Tamil Nadu, India

Mohit Tyagi Department of Production and Industrial Engineering, Punjab Engineering College (Deemed to be University), Chandigarh, India

Juhani Ukko Department of Industrial Engineering and Management, School of Engineering Science, LUT University, Lahti, Finland

Keerthana Umesh Vellore Institute of Technology, Vellore, Tamil Nadu, India

M. Vanitha Vellore Institute of Technology, Vellore, Tamil Nadu, India

Subrahmanyam Vasamsetti Department of Automobile Engineering, Godavari Global University, Rajamahendravaram, Andhra Pradesh, India

Yogesh Kumar Verma Regenerative Biology Research Lab, Division of Clinical Research and Medical Management (CRMM), Institute of Nuclear Medicine and Allied Sciences (INMAS), Defence Research and Development Organisation (DRDO), Timarpur, Delhi, India

Shreyas Vivek School of Information Technology, Deakin University, Waurn Ponds, VIC, Australia

R. S. Walia Department of Production and Industrial Engineering, Punjab Engineering College (Deemed to be University), Chandigarh, India

Ashok Kumar Yadav Department of Mechanical Engineering, Symbiosis Institute of Technology, Symbiosis International (Deemed University), Pune, India

Paulo Yu Department of Mechanical and Industrial Engineering, University of Wisconsin, Platteville, WI, USA

Digital Manufacturing

Historical Evolution Leading to the Current State of Digital Manufacturing in Industry 5.0

Ashwani Kumar, Yogesh Kumar Singla, Rakesh Joshi, Neha Singh Raghuvanshi, Sonia, and Michael R. Maughan

1 Introduction

The evolution of digital manufacturing is a fascinating journey that has transformed industries worldwide. From its humble beginnings in the early twentieth century to the cutting-edge technologies of today, digital manufacturing has significantly impacted the way products are designed, engineered, and produced [1].

A. Kumar
Department of Mechanical Engineering, Technical Education Department Uttar Pradesh, Kanpur, India

Department of Mechanical Engineering, Graphic Era Deemed to be University, Dehradun, Uttarakhand, India
e-mail: ashwaniadj.me@geu.ac.in

Y. Kumar Singla (✉)
School of Engineering, Math and Technology, Navajo Technical University, Crownpoint, NM, USA

Department of Mechanical Engineering, University of Idaho, Moscow, ID, USA
e-mail: ysingla@navajotech.edu, ysingla@uidaho.edu

R. Joshi
DARVUN India, Patiala, Punjab, India

N. S. Raghuvanshi
Department of Applied Sciences, ABES Engineering College, Ghaziabad, Uttar Pradesh, India

Sonia
School of Engineering, Math and Technology, Navajo Technical University, Crownpoint, NM, USA
e-mail: s.sonia@navajotech.edu

M. R. Maughan
Department of Mechanical Engineering, University of Idaho, Moscow, ID, USA
e-mail: maughan@uidaho.edu

Y. Kumar Singla et al. (eds.), *Digital Manufacturing in Industry 5.0*,
https://doi.org/10.1007/978-3-031-91500-0_1

1.1 *Early Historical Developments*

The seeds of digital manufacturing were sown in the early twentieth century, with the advent of numerical control (NC) machines. These machines, controlled by computer programs, enabled precise and repeatable manufacturing processes. NC machines were initially used for tasks such as drilling, milling, and turning, but their capabilities gradually expanded. In the 1960s and 1970s, computer-aided design (CAD) and computer-aided manufacturing (CAM) systems emerged. CAD software allowed engineers to create detailed two-dimensional (2D) and three-dimensional (3D) models of products, while CAM software-generated instructions for NC machines to produce those products. This integration of design and manufacturing processes streamlined production and reduced lead times.

The Industrial Revolution, a period of significant technological, socioeconomic, and cultural change, marked a pivotal moment in human history. Beginning in Great Britain in the late eighteenth century, it rapidly spread across Europe and North America, transforming traditional manufacturing processes and ushering in a new era of industrialization [2, 3]

1.1.1 Mechanization and Automation

One of the most significant hallmarks of the Industrial Revolution was the widespread adoption of machinery to replace manual labor. This mechanization of production processes led to a dramatic increase in productivity and efficiency. Early mechanical inventions, such as the spinning jenny and the power loom, revolutionized the textile industry. These machines could produce large quantities of cloth at a fraction of the time and cost compared to traditional handloom weaving. As technology advanced, more sophisticated machines were developed to automate various manufacturing processes. For instance, the steam engine, invented by James Watt, provided a powerful source of energy that could be used to power machinery in factories. This led to the development of automated production lines, where products could be manufactured in large quantities with minimal human intervention [4].

1.1.2 The Rise of Mass Production

The mechanization and automation of manufacturing processes paved the way for the rise of mass production. Mass production is a manufacturing system designed to produce large quantities of standardized products efficiently and economically. It involves the division of labor, where each worker specializes in a specific task, and the use of assembly lines, where products are moved from one workstation to the next in a continuous flow. The development of interchangeable parts was another key factor in the rise of mass production. By standardizing the dimensions and specifications of parts, manufacturers could produce large quantities of identical components that could be easily assembled into finished products. This reduced production costs and made it easier to repair and replace broken parts [5, 6].

The Industrial Revolution also led to the development of new materials, such as iron and steel, which were essential for the construction of machinery and infrastructure. These materials were produced in large quantities using industrial processes, such as the blast furnace, which allowed for the efficient extraction of iron from iron ore. The Industrial Revolution marked a turning point in the history of manufacturing. The mechanization and automation of production processes, coupled with the rise of mass production, led to significant economic growth and social change. The legacy of the Industrial Revolution continues to shape the modern world, with its impact felt in every aspect of our lives [7].

2 The Digital Manufacturing Era

The 1970s and 1980s witnessed significant advancements in digital manufacturing. The development of more powerful computers and software enabled the creation of complex and realistic product models. Solid modeling techniques, such as solid modeling and parametric modeling, revolutionized the way engineers designed products. The integration of CAD/CAM systems with manufacturing processes, such as computer numerical control (CNC) machining and robotics, further enhanced efficiency and productivity. CNC machines, controlled by computer programs, became more sophisticated and capable of performing a wider range of tasks. Industrial robots, equipped with advanced sensors and actuators, were deployed to automate repetitive and hazardous tasks, improving safety and quality. The Digital Revolution, which began in the mid-twentieth century, marked a significant turning point in the history of manufacturing. The introduction of computers to manufacturing processes revolutionized the way products were designed, engineered, and produced [8, 9].

2.1 Involvement of Computers to Manufacturing Processes

Before the advent of computers, manufacturing processes were largely manual and relied on human labor. Design and engineering were done using traditional drafting tools, and manufacturing was carried out using manual machines or early machine tools. The introduction of computers to manufacturing processes in the 1950s and 1960s revolutionized the industry.

Early computers were used for analyzing large amounts of data related to production, inventory, and quality control. Performing complex calculations for engineering design and analysis, monitoring, and controlling manufacturing processes. While these early applications laid the foundation for digital manufacturing, it was the development of specialized software and hardware that truly transformed the industry [9].

2.2 *Development of CAD/CAM Systems*

Computer-aided design (CAD) and computer-aided manufacturing (CAM) systems emerged in the 1960s and 1970s, revolutionizing the way products were designed and manufactured [10].

2.2.1 CAD Systems

CAD systems enabled engineers to create precise and detailed 2D and 3D models of products using computer software. These models could be easily modified and analyzed, reducing the time and cost of product development. Key features of early CAD systems were

- Creating geometric shapes and curves to represent product components.
- Generating technical drawings and adding dimensions and notes.
- Performing simulations and stress analysis to evaluate product performance.

2.2.2 CAM Systems

CAM systems, on the other hand, generated instructions for manufacturing machines, such as NC machines and robots, based on CAD models. These instructions, in the form of machine code, controlled the movement of the machine tools to produce the desired part. Key features of early CAM systems were [11, 12]

- Generating NC code to control machine tools.
- Determining the sequence of operations and machine tools required to manufacture a part.
- Calculating the optimal path for the cutting tool to follow.

2.2.3 Early Numerical Control (NC) Machines

Numerical control (NC) machines, controlled by computer programs, were a crucial development in the early days of digital manufacturing. These machines could perform a variety of tasks, such as drilling, milling, and turning, with high precision and accuracy. Early NC machines were relatively simple, with limited capabilities. However, as computer technology advanced, so did the capabilities of NC machines. Key features of early NC machines included

- The machine movements were controlled by a computer program.
- The program could be modified to perform different tasks.
- The machine could position its tools with high accuracy.

NC machines played a vital role in automating manufacturing processes, improving productivity, and reducing labor costs. These technological advancements laid the foundation for the modern era of digital manufacturing, where advanced technologies such as additive manufacturing, robotics, and artificial intelligence are transforming the industry [13–16].

2.2.4 Evolution of CNC Machining: Precision and Complexity

Computer Numerical Control (CNC) machining has revolutionized the manufacturing industry, enabling the production of highly precise and complex parts. By integrating computer technology with traditional machining processes, CNC machines have significantly improved efficiency, accuracy, and flexibility.

Early Developments (1950s–1970s)

The origins of CNC machining can be traced back to the mid-twentieth century. Early CNC machines were large, cumbersome, and expensive, but they demonstrated the potential of automating manufacturing processes. These machines were primarily used for simple tasks, such as drilling and milling. One of the key milestones in the evolution of CNC machining was the development of numerical control systems. These systems used punched tape or magnetic tape to store and transmit instructions to the machine tool. By following these coded instructions, the machine could perform a variety of operations, including milling, drilling, and turning.

Advancements in Control Systems (1980s–1990s)

The 1980s and 1990s witnessed significant advancements in control systems for CNC machines. The introduction of microprocessors and CAD software revolutionized the way CNC machines were programmed and operated.

- Microprocessor-based control systems enabled more precise control over machine movements and tool paths. This led to improved accuracy and repeatability.
- The integration of CAD/CAM systems with CNC machines streamlined the design and manufacturing process. Engineers could design complex parts using CAD software and then generate CNC code directly from the design data [17–21].

The Rise of CNC Machining Centers (1990s–Present)

CNC machining centers, which combine multiple machining operations on a single machine tool, became increasingly popular. These machines can perform a variety of operations, including milling, drilling, and boring, in a single setup. This reduces setup time, improves accuracy, and increases productivity.

- High-Speed Machining (HSM): HSM techniques, such as high-speed spindle speeds and high-feed rates, enable faster machining and improved surface finishes.
- Five-Axis Machining: Five-axis CNC machines can manipulate the tool in five different axes, allowing for the production of complex three-dimensional shapes. This capability has expanded the range of parts that can be produced using CNC machining.
- Direct Numerical Control (DNC): DNC systems allow for the direct transmission of CNC programs to the machine tool, eliminating the need for physical media such as punched tape or magnetic tape. This improves efficiency and reduces the risk of errors [22–27].

Precision Machining and Complex Part Production

CNC machining has become synonymous with precision and accuracy. By precisely controlling the movement of the machine tool and the cutting tool, CNC machines can produce parts with tolerances as low as a few micrometers. This level of precision is essential for many industries, including aerospace, automotive, medical, and electronics.

CNC machines can also produce complex parts with intricate features. Multi-axis machining capabilities, combined with advanced tool path generation algorithms, enable the creation of parts with complex geometries. For example, CNC machines can produce turbine blades, impellers, and molds with intricate details.

The Impact of CNC Machining on Manufacturing

CNC machining has had a profound impact on the manufacturing industry. It has increased productivity, improved quality, and reduced costs. By automating many aspects of the manufacturing process, CNC machines have freed up workers to perform more complex tasks. Additionally, CNC machining has enabled the production of a wider range of products, from simple components to highly complex assemblies. As technology continues to advance, the capabilities of CNC machines will continue to expand. Emerging technologies such as artificial intelligence (AI) and machine learning (ML) are being integrated into CNC machines to further improve performance and efficiency. The future of CNC machining is bright, and it will undoubtedly play a crucial role in shaping the manufacturing industry for years to come [28–32].

3 The Rise of Additive Manufacturing

The early twenty-first century marked the emergence of additive manufacturing, commonly known as 3D printing. This revolutionary technology enables the creation of three-dimensional objects by adding layer upon layer of material. Additive manufacturing has a wide range of applications, including prototyping, tooling, and the production of end-use parts. Different types of additive manufacturing technologies, such as fused deposition modeling (FDM), stereolithography (SLA), and selective laser sintering (SLS), have been developed. These technologies have the potential to transform industries by enabling the production of complex geometries, customized products, and lightweight components. Key features of additive manufacturing are as follows [33–37]:

- Layer-by-Layer Fabrication: AM builds objects layer by layer, allowing for the creation of complex geometries that would be difficult or impossible to produce using traditional methods.
- Design Flexibility: AM enables the creation of customized and personalized products, as well as intricate designs with internal features.
- Rapid Prototyping: AM significantly reduces the time required for prototyping, allowing for faster product development cycles.
- Reduced Material Waste: By using only the necessary material to build a part, AM minimizes material waste compared to traditional subtractive manufacturing processes.

3.1 Common 3D Printing Technologies

The most common 3D printing technologies and their key features are given in Table 1.

3.2 Applications of Additive Manufacturing

The versatility of AM has led to its adoption in various industries (Table 2):

Table 1 3D Printing technologies and key features

3D printing technologies	Key features
Fused deposition modeling (FDM)	One of the most common 3D printing technologies Uses a heated nozzle to extrude molten thermoplastic filament, layer by layer Suitable for prototyping, education, and low-volume production
Stereolithography (SLA)	Uses a laser to cure liquid photopolymer resin, layer by layer Produces high-resolution, detailed parts with smooth surfaces Ideal for creating complex prototypes and functional parts
Selective laser sintering (SLS)	Uses a laser to sinter powdered material, layer by layer Suitable for producing strong and durable parts, often used in the aerospace and automotive industries
Digital light processing (DLP)	Projects a digital image onto a vat of liquid photopolymer resin, curing the exposed areas Offers faster build times and high-resolution parts compared to SLA
Electron beam melting (EBM)	Uses a high-powered electron beam to melt metal powder, layer by layer Produces high-density, high-strength metal parts Often used in aerospace and medical industries

Table 2 Applications of additive manufacturing in different industries

Sector	Applications
Prototyping	Rapid prototyping of product designs to test and iterate Visualizing and communicating design concepts
Manufacturing	Production of complex parts with intricate geometries Customizing products to meet specific needs Producing small batches of products
Medical industry	Creating custom implants and prosthetics Producing surgical tools and instruments Tissue engineering and organ printing
Aerospace	Manufacturing lightweight and complex components Producing tools and fixtures for aerospace manufacturing
Automotive	Creating custom parts and tooling Producing lightweight components for improved fuel efficiency
Consumer products	Manufacturing customized products, such as jewelry and footwear Producing toys and games

3.3 Limitations of Additive Manufacturing

While AM offers numerous advantages, it also has certain limitations:

- Some AM processes can be time-consuming, especially for large and complex parts.
- Not all materials can be used in AM processes, and the mechanical properties of AM parts may vary.

- AM parts often require post-processing, such as cleaning, heat treatment, and finishing, achieving desired properties.
- AM can be more expensive than traditional manufacturing methods for large-scale production.
- The size of parts that can be produced using AM is limited by the build volume of the 3D printer.

Despite this limitation, AM continues to evolve, and as technology advances, these limitations are gradually being addressed. The potential of AM to revolutionize manufacturing and various other industries is immense.

4 The Internet of Things and Industry 5.0

The Internet of Things (IoT) has revolutionized manufacturing by connecting devices and systems, enabling real-time monitoring and control of production processes. IoT-enabled sensors and actuators collect data on various parameters, such as temperature, pressure, and vibration, allowing manufacturers to optimize performance and reduce downtime. Industry 4.0, the fourth industrial revolution, is characterized by the integration of cyber-physical systems, the Internet of Things, and artificial intelligence (AI) into manufacturing processes. Smart factories, equipped with IoT-enabled devices and advanced analytics, can make autonomous decisions and optimize production. The integration of emerging technologies such as AI, ML, the IoT, and robotics and automation is revolutionizing the landscape of digital manufacturing. These technologies are driving innovation, improving efficiency, and enhancing the overall quality of products [38–44].

4.1 Artificial Intelligence (AI) and Machine Learning (ML)

AI and ML algorithms have the potential to significantly accelerate and optimize the design process. By analyzing vast amounts of data, these algorithms can identify optimal design solutions, reducing time-to-market and improving product performance. AI-powered tools can generate innovative designs by exploring a wide range of design possibilities and considering various constraints, such as material properties, manufacturing processes, and performance requirements. ML algorithms can optimize the distribution of material within a design, leading to lightweight and high-performance structures. By analyzing historical data and simulations, AI can predict the performance of a design before it is physically manufactured. AI and ML can be used to predict equipment failures and optimize maintenance schedules. Additionally, they can analyze real-time data to identify quality issues and prevent defects. AI-powered systems can analyze sensor data from machines to predict potential failures, allowing for proactive maintenance and reducing downtime. ML

Table 3 Impact of emerging technologies on digital manufacturing

Technology	Key applications	Benefits
AI and ML	Design optimization, predictive maintenance, quality control	Improved product design, reduced downtime, enhanced quality
IoT	Smart factories, real-time monitoring, predictive maintenance	Increased efficiency, improved productivity, reduced energy consumption
Robotics and automation	Collaborative robots, autonomous systems	Increased productivity, improved safety, reduced labor costs

algorithms can analyze image and video data to identify defects and inconsistencies in products, improving quality control [45–48].

4.1.1 Smart Factories and Industry 5.0

IoT-enabled devices and sensors can collect and analyze real-time data from various manufacturing processes, enabling greater visibility and control. This leads to the development of smart factories, which are highly interconnected and automated. IoT devices can monitor key performance indicators (KPIs) such as temperature, pressure, and vibration, allowing for early detection of anomalies. Remote monitoring of equipment and processes enables timely intervention and reduces downtime. By analyzing sensor data, IoT systems can predict equipment failures and schedule maintenance proactively [49, 50].

4.1.2 Robotics and Automation

Cobots are designed to work alongside human workers, increasing productivity and safety. They are flexible, easy to program, and can be deployed in various manufacturing tasks, such as assembly, packaging, and machine tending. Autonomous manufacturing systems, powered by AI and ML, can operate independently, performing tasks such as material handling, machine loading and unloading, and quality inspection.

By leveraging these emerging technologies (Table 3), manufacturers can achieve greater flexibility, efficiency, and quality in their operations, driving innovation and competitiveness in the global market.

5 The Future of Digital Manufacturing

The future of digital manufacturing is bright, with emerging technologies poised to transform the industry. AI and ML are being increasingly used to optimize design processes, predict maintenance needs, and improve quality control. AI-powered design tools can analyze vast design spaces and identify optimal solutions, while ML algorithms can analyze sensor data to detect anomalies and prevent failures. Digital twins, virtual replicas of physical products and processes, are being used to simulate and optimize manufacturing operations. By creating digital twins of products and production lines, manufacturers can identify potential issues, test new designs, and improve efficiency. The convergence of biotechnology and manufacturing is leading to the development of bioprinting, a technology that uses three-dimensional printing techniques to print living cells and tissues. Bioprinting has the potential to revolutionize healthcare, regenerative medicine, and materials science. Quantum computing, with its immense computational power, can accelerate the development of new materials, optimize complex manufacturing processes, and enhance cybersecurity. Quantum computers can solve complex optimization problems that are intractable for classical computers, leading to breakthroughs in various fields. The future of digital manufacturing is a dynamic landscape shaped by technological advancements, societal needs, and environmental concerns. While the potential benefits are immense, several challenges must be addressed to fully realize the vision of a sustainable and efficient manufacturing future [51–54].

- **Skill Gaps and Workforce Training**: The rapid pace of technological advancement often outstrips the ability of the workforce to adapt. There is a growing need for skilled workers who can operate, maintain, and program advanced manufacturing equipment. Investing in education and training programs can equip the workforce with the necessary skills to thrive in the digital manufacturing era. This includes training in areas such as robotics, AI, and data analytics.
- **Cybersecurity and Data Privacy**: As manufacturing systems become increasingly interconnected, they become more vulnerable to cyberattacks. Data breaches can compromise sensitive information and disrupt operations. Implementing robust cybersecurity measures, such as network security, data encryption, and access controls, can mitigate these risks. Additionally, developing secure and resilient manufacturing systems can enhance overall security.
- **Sustainability and Environmental Impact**: The manufacturing industry has a significant environmental impact, including carbon emissions, waste generation, and resource consumption. Digital manufacturing can contribute to a more sustainable future by enabling efficient resource utilization, reducing waste, and minimizing environmental impact. This can be achieved through techniques such as additive manufacturing, which reduces material waste, and energy-efficient manufacturing processes.
- **Digital Twins and Virtual Manufacturing**: A digital twin is a virtual replica of a physical product or system. It allows for real-time monitoring, simulation, and optimization of manufacturing processes. Virtual manufacturing involves simulating and testing manufacturing processes in a virtual environment. This reduces the need for physical prototypes and accelerates product development.
- **Bioprinting and Tissue Engineering**: This technology involves printing biological tissues and organs using 3D printing techniques. It has the potential to revolutionize healthcare by enabling the production of personalized medical

Table 4 Key challenges and opportunities in digital manufacturing

Challenge	Opportunity
Skill gaps and workforce training	Invest in education and training programs
Cybersecurity and data privacy	Implement robust cybersecurity measures
Sustainability and environmental impact	Adopt sustainable manufacturing practices
Technological complexity	Collaborate with industry experts and researchers

implants and organs. Tissue engineering combines biology and engineering to create functional tissues and organs.

- **Quantum Computing and Its Potential Applications**: Quantum computers leverage quantum mechanics to perform complex calculations at unprecedented speeds. Quantum computing can revolutionize material science, drug discovery, and optimization problems in manufacturing. It can also accelerate the development of new materials and improve the efficiency of manufacturing processes.

By addressing the above challenges and embracing emerging (Table 4) trends, the future of digital manufacturing holds immense promise. A collaborative approach involving industry, academia, and government is essential to shape a sustainable and innovative future for manufacturing [55–58].

6 Conclusion

The historical evolution of digital manufacturing has been a remarkable journey, driven by technological advancements and innovative thinking. From the early days of NC machines to the cutting-edge technologies of today, digital manufacturing has transformed the way products are designed, engineered, and produced. As we look to the future, the potential for further innovation and disruption is immense. By embracing emerging technologies and addressing the challenges of sustainability and workforce development, the manufacturing industry can continue to thrive in the digital age. By embracing these technological advancements and addressing the challenges ahead, the manufacturing industry can continue to innovate and thrive in the digital age.

References

1. Pilevari, N. (2020). Industry revolutions development from industry 1.0 to industry 5.0 in manufacturing. *Journal of Industrial Strategic Management, 5*(2), 44.
2. Cohen, Y., Faccio, M., Pilati, F., & Yao, X. (2019). Design and management of digital manufacturing and assembly systems in the industry 4.0 era. *The International Journal of Advanced Manufacturing Technology, 105*, 3565–3577.
3. Mourtzis, D. (2020). Simulation in the design and operation of manufacturing systems: State of the art and new trends. *International Journal of Production Research, 58*(7), 1927–1949.

4. David, P. A. (2000). Understanding digital technology's evolution and the path of measured productivity growth: Present and future in the mirror of the past.
5. Möller, D. P., & Möller, D. P. (2016). Digital manufacturing/industry 4.0. Guide to Computing Fundamentals in Cyber-Physical Systems: Concepts, Design Methods, and Applications, pp. 307–375.
6. Cortada, J. W. (2003). *The digital hand: How computers changed the work of American manufacturing, transportation, and retail industries*. Oxford University Press.
7. Huyal, S., Bista, D., & Bista, R. (2020). Challenges, opportunities and future directions of smart manufacturing: A state of art review. *Sustainable Futures, 2*, 100023.
8. Baily, M. N., & Bosworth, B. P. (2014). US manufacturing: Understanding its past and its potential future. *Journal of Economic Perspectives, 28*(1), 3–26.
9. Nain, G., Pattanaik, K. K., & Sharma, G. K. (2022). Towards edge computing in intelligent manufacturing: Past, present and future. *Journal of Manufacturing Systems, 62*, 588–611.
10. Hon, K. K. B. (2005). Performance and evaluation of manufacturing systems. *CIRP Annals, 54*(2), 139–154.
11. Matias, E., & Rao, B. (2015, August). 3D printing: On its historical evolution and the implications for business. In *2015 Portland International Conference on Management of Engineering and Technology (PICMET)* (pp. 551–558). IEEE.
12. Sartal, A., Bellas, R., Mejías, A. M., & García-Collado, A. (2020). The sustainable manufacturing concept, evolution and opportunities within industry 4.0: A literature review. *Advances in Mechanical Engineering, 12*(5), 1687814020925232.
13. Liu, M., Fang, S., Dong, H., & Xu, C. (2021). Review of digital twin about concepts, technologies, and industrial applications. *Journal of Manufacturing Systems, 58*, 346–361.
14. Sharma, A., Kosasih, E., Zhang, J., Brintrup, A., & Calinescu, A. (2022). Digital twins: State of the art theory and practice, challenges, and open research questions. *Journal of Industrial Information Integration, 30*, 100383.
15. Leng, J., Ruan, G., Jiang, P., Xu, K., Liu, Q., Zhou, X., & Liu, C. (2020). Blockchain-empowered sustainable manufacturing and product lifecycle management in industry 4.0: A survey. *Renewable and Sustainable Energy Reviews, 132*, 110112.
16. Leng, J., Sha, W., Wang, B., Zheng, P., Zhuang, C., Liu, Q., et al. (2022). Industry 5.0: Prospect and retrospect. *Journal of Manufacturing Systems, 65*, 279–295.
17. Ikram, L., Afaf, D., & Latifa, E. (2024, May). Beyond industry 5.0: Tracing the past, present, and future of digital industrial transformations. In *2024 4th International Conference on Innovative Research in Applied Science, Engineering and Technology (IRASET)* (pp. 1–7). IEEE.
18. Xu, X., Lu, Y., Vogel-Heuser, B., & Wang, L. (2021). Industry 4.0 and industry 5.0—Inception, conception and perception. *Journal of Manufacturing Systems, 61*, 530–535.
19. Maddikunta, P. K. R., Pham, Q. V., Prabadevi, B., Deepa, N., Dev, K., Gadekallu, T. R., et al. (2022). Industry 5.0: A survey on enabling technologies and potential applications. *Journal of Industrial Information Integration, 26*, 100257.
20. Kanwar, N., Jadoun, V. K., Jayalakshmi, N. S., Afthanorhan, A., Fatema, N., Malik, H., & Hossaini, M. A. (2023). Industry—Challenge to pro-environmental manufacturing of goods replacing single-use plastic by Indian industry: A study toward failing ban on single-use plastic access. *IEEE Access, 11*, 77336–77346. https://doi.org/10.1109/ACCESS.2023.3296097
21. Badhotiya, G. K., Soni, G., Jain, V., Joshi, R., & Mittal, S. (2022). Assessing supply chain resilience to the outbreak of COVID-19 in Indian manufacturing firms. *Operations Management Research, 15*(3), 1161–1180. https://doi.org/10.1007/s12063-021-00236-6
22. Pansare, R., Yadav, G., Garza-Reyes, J. A., & Raosaheb Nagare, M. (2023). Assessment of sustainable development goals through industry 4.0 and reconfigurable manufacturing system practices. *Journal of Manufacturing Technology Management, 34*(3), 383–413. https://doi.org/10.1108/JMTM-05-2022-0206
23. Jamil, M., He, N., Wei, Z., Gupta, M. K., & Khan, A. M. (2023). A novel quantifiable approach of estimating energy consumption, carbon emissions and cost factors in manufacturing of bearing steel based on triple bottom-line approach. *Sustainable Materials and Technologies, 36*, e00593. https://doi.org/10.1016/j.susmat.2023.e00593

24. Korkmaz, M. E., Gupta, M. K., Günay, M., Boy, M., Yaşar, N., Demirsöz, R., et al. (2023). Comprehensive analysis of tool wear, surface roughness and chip morphology in sustainable turning of Inconel-601 alloy. *Journal of Manufacturing Processes, 103*, 156–167. https://doi.org/10.1016/j.jmapro.2023.08.026
25. Kaswan, M. S., Rathi, R., Cross, J., Garza-Reyes, J. A., Antony, J., & Yadav, V. (2022). Integrating green lean six sigma and industry 4.0: A conceptual framework. *Journal of Manufacturing Technology Management, 34*(1), 87–121. https://doi.org/10.1108/JMTM-03-2022-0115
26. Mohanty, S., Panwar, V., Anoop, G., & Park, S. (2023). Eco-friendly phyllanthus emblica-based ionic polymer composite for enhanced mechanical, electrical, and wearable sensing performance. *Materials Chemistry and Physics, 303*, 127791. https://doi.org/10.1016/j.matchemphys.2023.127791
27. Patil, P. P., Gori, Y., Kumar, A., & Tyagi, M. R. (2021). Experimental analysis of tribological properties of polyisobutylene thickened oil in lubricated contacts. *Tribology International, 159*, 106983. https://doi.org/10.1016/j.triboint.2021.106983
28. Uniyal, S., Mangla, S. K., Sarma, P. R., Tseng, M. L., & Patil, P. (2021). ICT as "knowledge management" for assessing sustainable consumption and production in supply chains. *Journal of Global Information Management (JGIM), 29*(1), 164–198. https://doi.org/10.4018/JGIM.2021010109
29. Tiwari, S., Swaminathan, M., Harender, & Singh, D. B. (2022). Performance enhancement of the photovoltaic system with different cooling methods. *Environmental Science and Pollution Research, 29*(30), 45107–45130. https://doi.org/10.1007/s11356-022-20330-x
30. Kumar, U., Padalia, D., Bhandari, P., Kumar, P., Ranakoti, L., Singh, T., & Lendvai, L. (2022). Fabrication of europium-doped barium titanate/polystyrene polymer nanocomposites using ultrasonication-assisted method: Structural and optical properties. *Polymers, 14*(21), 4664. https://doi.org/10.3390/polym14214664
31. Sharma, S. K., Mallick, A., Singh, D. B., & Tiwari, G. N. (2022). Experimental study of solar energy–based water purifier of single-slope type by incorporating a number of similar evacuated tubular collectors. *Environmental Science and Pollution Research, 29*(5), 6837–6856. https://doi.org/10.1007/s11356-021-16123-3
32. AP, A., Kaliappan, S., & Patil P, P. (2022). Mechanical, fracture toughness, and dynamic mechanical properties of twill weaved bamboo fiber-reinforced Artocarpus heterophyllus seed husk biochar epoxy composite. *Polymer Composites, 43*(11), 8388–8395. https://doi.org/10.1002/pc.27010
33. Sharma, K. K., Kumar, R., & Jha, P. K. (2024). Physical modeling and numerical investigation of fluid flow and solidification behavior in a slab caster mold using hexa-furcated nozzle. *Heat Transfer, 53*(3), 1050–1072. https://doi.org/10.1002/htj.22980
34. Chakrabarty, D., Patil, P., & Sharma, K. (2017). Experimental study on fluid flow and heat transfer from a rectangular prism approaching the wall of a wind tunnel. *Heat Transfer—Asian Research, 46*(7), 793–814. https://doi.org/10.1002/htj.21244
35. Gariya, N., Prasad, B., Nautiyal, H., Kumar, S., & Shaikh, A. Manufacturing techniques for soft robotic structures: A systematic review. *Sustainability in Smart Manufacturing*, 94–117. https://doi.org/10.1201/9781003467496-5
36. Shankar, A. (2023). Efficient data interpretation and artificial intelligence enabled IoT based smart sensing system. *Artificial Intelligence Review, 56*(12), 15053–15077. https://doi.org/10.1007/s10462-023-10519-y
37. Wongchai, A., Shukla, S. K., Ahmed, M. A., Sakthi, U., & Jagdish, M. (2022). Artificial intelligence-enabled soft sensor and internet of things for sustainable agriculture using ensemble deep learning architecture. *Computers and Electrical Engineering, 102*, 108128. https://doi.org/10.1016/j.compeleceng.2022.108128
38. Devi, S., Garg, H., & Garg, D. (2023). A review of redundancy allocation problem for two decades: Bibliometrics and future directions. *Artificial Intelligence Review, 56*(8), 7457–7548. https://doi.org/10.1007/s10462-022-10363-6

39. Sharma, R., Kim, M., & Gupta, A. (2022). Motor imagery classification in brain-machine interface with machine learning algorithms: Classical approach to multi-layer perceptron model. *Biomedical Signal Processing and Control, 71*, 103101. https://doi.org/10.1016/j.bspc.2021.103101
40. Pundir, S., Obaidat, M. S., Wazid, M., Das, A. K., Singh, D. P., & Rodrigues, J. J. (2023). MADP-IIME: Malware attack detection protocol in IoT-enabled industrial multimedia environment using machine learning approach. *Multimedia Systems, 29*(3), 1785–1797. https://doi.org/10.1007/s00530-020-00743-9
41. Tajjour, S., Chandel, S. S., Alotaibi, M. A., Malik, H., Márquez, F. P. G., & Afthanorhan, A. (2023). Short-term solar irradiance forecasting using deep learning techniques: A comprehensive case study. *IEEE Access, 11*, 119851–119861. https://doi.org/10.1109/ACCESS.2023.3325292
42. Singh A, R., Kumar, R. S., Bajaj, M., Khadse, C. B., & Zaitsev, I. (2024). Machine learning-based energy management and power forecasting in grid-connected microgrids with multiple distributed energy sources. *Scientific Reports, 14*(1), 19207. https://doi.org/10.1038/s41598-024-70336-3
43. Gupta, A., Ghanshala, K., & Joshi, R. C. (2021). Machine learning classifier approach with gaussian process, ensemble boosted trees, SVM, and linear regression for 5g signal coverage mapping. *IJIMAI, 6*(6), 156–163. https://doi.org/10.9781/ijimai.2021.03.004
44. Ojha, N., Girach, I., Sharma, K., Sharma, A., Singh, N., & Gunthe, S. S. (2021). Exploring the potential of machine learning for simulations of urban ozone variability. *Scientific Reports, 11*(1), 22513. https://doi.org/10.1038/s41598-021-01824-z
45. Davuluri, S. K., Alvi, S. A. M., Aeri, M., Agarwal, A., Serajuddin, M., & Hasan, Z. (2023, April). A security model for perceptive 5G-powered BC IoT associated deep learning. In *2023 International Conference on Inventive Computation Technologies (ICICT)* (pp. 118-125). IEEE, https://doi.org/10.1109/TII.2020.3048791.
46. Ali, S., Khan, M. S., Shah, H., Garg, H., & Alsheddy, A. (2023). BISGA: Recalculating the entire boolean-valued information system from aggregates using a genetic algorithm. *Complexity, 2023*(1), 1539563. https://doi.org/10.1155/2023/1539563
47. Jayaraman, R., Alshehri, M., Kumar, M., Abugabah, A., Samant, S. S., & Mohamed, A. A. (2023). Secure biomedical document protection framework to ensure privacy through blockchain. *Big Data, 11*(6), 437–451. https://doi.org/10.1089/big.2022.0170
48. Chhikara, D., Rana, S., Singh, G., Mishra, D., & Kumar, N. (2023). Blockchain-based partial group key agreement protocol for intelligent transportation systems. *IEEE Transactions on Vehicular Technology, 72*, 16701. https://doi.org/10.1109/TVT.2023.3299705
49. Wazid, M., Bera, B., Das, A. K., Mohanty, S. P., & Jo, M. (2022). Fortifying smart transportation security through public blockchain. *IEEE Internet of Things Journal, 9*(17), 16532–16545. https://doi.org/10.1109/JIOT.2022.3150842
50. Sharma, R., Wazid, M., & Gope, P. (2021). A blockchain based secure communication framework for community interaction. *Journal of Information Security and Applications, 58*, 102790. https://doi.org/10.1016/j.jisa.2021.102790
51. Rani, S., Babbar, H., Srivastava, G., Gadekallu, T. R., & Dhiman, G. (2022). Security framework for internet-of-things-based software-defined networks using blockchain. *IEEE Internet of Things Journal, 10*(7), 6074–6081. https://doi.org/10.1016/j.comcom.2023.04.008
52. Thapliyal, S., Wazid, M., Singh, D. P., Das, A. K., Shetty, S., & Alqahtani, A. (2023). Design of robust Blockchain-envisioned authenticated key management mechanism for smart healthcare applications. *IEEE Access, 11*, 93032–93047. https://doi.org/10.1109/ACCESS.2023.3310264
53. Bhandari, P., Padalia, D., Ranakoti, L., Khargotra, R., András, K., & Singh, T. (2023). Thermo-hydraulic investigation of open micro prism pin fin heat sink having varying prism sides. *Alexandria Engineering Journal, 69*, 457–468. https://doi.org/10.1016/j.aej.2023.02.016
54. Rekioua, D., Mokrani, Z., Kakouche, K., Rekioua, T., Oubelaid, A., Logerais, P. O., et al. (2023). Optimization and intelligent power management control for an autonomous hybrid wind turbine photovoltaic diesel generator with batteries. *Scientific Reports, 13*(1), 21830. https://doi.org/10.1038/s41598-023-49067-4

55. Kumar, A., Gori, Y., Dutt, N., Singla, Y. K., & Maurya, A. (2021). *Advanced computational methods in mechanical and materials engineering*. CRC Press. https://doi.org/10.1201/9781003202233. ISBN: 9781032052915.
56. Kumar, A., Ram, M., & Singla, Y. K. (2022). *Advanced material for biomechanical applications*. CRC Press. https://doi.org/10.1201/9781003286806. ISBN: 9781032054490.
57. Shah, S., Nautiyal, H., Gugliani, G., Kumar, A., Namboodri, T., & Singla Editors, Y. K. (2024). *Sustainability in smart manufacturing: Trends, scope, and challenges*. CRC Press. https://doi.org/10.1201/9781003467496. ISBN 9781032740713.
58. Kumar, A., Singla, Y., & Namboodri, T. (2024). Chapter 01. Globalization and international issues in sustainable manufacturing. In S. Shah, H. Nautiyal, G. Gugliani, A. Kumar, T. Namboodri, & Y. K. Singla (Eds.), *Sustainability in smart manufacturing: Trends, scope and challenges* (pp. 01–18). CRC Press. https://doi.org/10.1201/9781003467496-1

Shaping the Future: A Global Perspective on Digital Manufacturing in Industry 5.0

Pias Kumar Biswas, Yogesh Kumar Singla, Pritom Kumar Biswas, Yatika Gori, Mohit Tyagi, and Ashwani Kumar

P. K. Biswas (✉)
School of Engineering, Math and Technology, Navajo Technical University, Crownpoint, NM, USA
e-mail: pbiswas@navajotech.edu

Y. Kumar Singla
School of Engineering, Math and Technology, Navajo Technical University, Crownpoint, NM, USA

Department of Mechanical Engineering, University of Idaho, Moscow, ID, USA
e-mail: ysingla@navajotech.edu

P. K. Biswas
National University, Gazipur, Dhaka, Bangladesh

Y. Gori
Department of Mechanical Engineering, Graphic Era Deemed to be University, Dehradun, Uttarakhand, India

School of Computing, Graphic Era Hill University, Haldwani, Uttarakhand, India
e-mail: yatikagori@gehu.ac.in

M. Tyagi
Department of Production and Industrial Engineering, Punjab Engineering College (Deemed to be University), Chandigarh, India

A. Kumar
Department of Mechanical Engineering, Technical Education Department Uttar Pradesh, Kanpur, India
e-mail: ashwaniadj.me@geu.ac.in

Y. Kumar Singla et al. (eds.), *Digital Manufacturing in Industry 5.0*,
https://doi.org/10.1007/978-3-031-91500-0_2

1 The Evolution of Digital Manufacturing from Industry 4.0 to Industry 5.0

1.1 Key Differences Between Industry 4.0 and Industry 5.0

Industry 5.0 represents a significant shift in how industries operate and develop. Industry 4.0 is mostly about new technology and automation. Industry 5.0 emphasizes system resilience, human-centric design approaches, and environmental sustainability. Most of the change is in logistics, which requires a large number of workers [1]. Industry 5.0 leverages advanced cyber-physical systems and human cognitive abilities to balance prosperity, environmental protection, and sustainability [2]. The goal of Industry 5.0 is to weave the cognitive power of human beings with technological enhancements. The result is constant collaboration between humans and machines, leading to improved cognitive ergonomics [3]. In contrast, Industry 4.0 primarily aimed to boost productivity indicators by gradually substituting human labor with machines and artificial intelligence. Industry 5.0 is shifting methodologies to curb and make smart transformation. It involves more than just enhancing procedures. It employs the concept of circular economy and regenerative production methods to manage the social and environmental consequences, which are relevant in a complex industrial context [4]. The modification to the model establishes a link between anthropogenic factors and ecological sustainability, as well as economic prosperity. It does this because it places industrial production within a context that prioritizes regenerative purpose alongside directional sustainability, while also addressing the need to adapt the workforce, integrate technology, and protect the environment [5]. Thorough policy frameworks are needed, showing Industry 5.0 has a holistic approach. Specific interventions become vital to help include people as well as address social issues within environmentally and socially beneficial industrial practices. The Industrial Revolution progressed through different stages, with each stage introducing new technologies that changed manufacturing and society. Machines powered by water and steam were in use during the first industrial revolution. For cognitive systems, personalization, and collaborative robots (cobots), these are the characteristics that will define the fifth Industrial Revolution, as illustrated in Fig. 1.

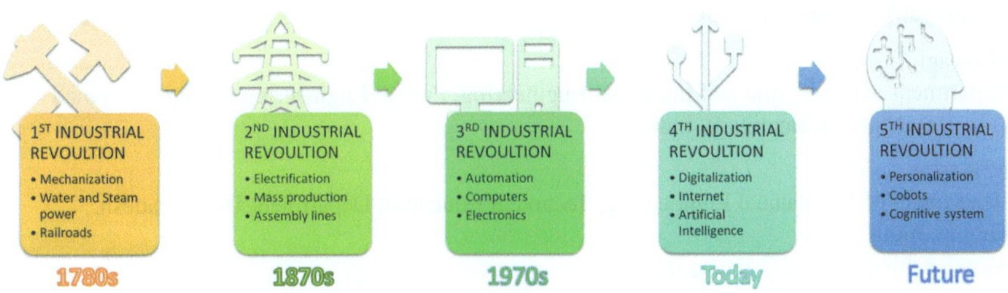

Fig. 1 The progressive phases of the Industrial Revolution [6]

1.2 Evolution of Digital Manufacturing in the Transition to Industry 5.0

The concept of Industry 5.0 is changing the values people hold because it propels forward technological changes that reshape industries' modes of operation [3]. The new framework exceeds the standard method in Industry 4.0 through automation and a more complete industrial model. This model focuses on human welfare, ecological balance, and systemic resilience [7]. Industry 5.0 is here now, and this marks a shift that is important. That is, in fact, especially true in the factory fields. It is also true within the field of shipping. Balancing human cognitive abilities with ergonomic factors, plus advanced cyber-physical systems, is necessary for the achievement of optimal operational results [8]. Industry 4.0 automates the systems so that artificial intelligence augments human producers for human workers. Industry 5.0, unlike its predecessor, advocates for a more advanced symbiotic relationship of human cognitive abilities with technological augmentation [9]. This approach recognizes the value of how humans create, decide, and adapt in industrial processes, which is irreplaceable, while leveraging advanced technologies to support and improve human workers rather than replace them [10]. Industry 5.0 methodology shifted course, involving a reconfiguration that is holistic of smart transformation strategies moving beyond simple process efficiency, while tackling socio-ecological impacts as industrial operations embraced manufacturing that is regenerative as well as circular economy principles [11]. This model shift stresses the critical interconnection between anthropogenic factors and ecological sustainability and economic prosperity, and it contextualizes industrial production within the framework that prioritizes regenerative purpose and directional sustainability [12]. Industry 5.0 principles must be implemented, so we must develop the important technological infrastructure as well as integrate advanced robotics, artificial intelligence, Internet of Things (IoT) devices, and human-machine interfaces that are specifically designed for the improvement of human capabilities rather than their replacement [13]. In order to ensure workforce readiness as well as adaptation for new operational models [14], they must implement technology, then train comprehensively and educate. Industry 5.0's successful transition requires careful thought about ethical issues like data privacy, job security, and society's fair distribution of technological gains [15]. Also, the framework stresses the importance of sustainable manufacturing practices to minimize environmental impact while maintaining economic viability [16]. It features renewable energy solutions with circular economy principles. Since Industry 5.0 principles were integrated, they have improved workplace satisfaction, reduced occupational stress, and improved productivity through human-machine collaboration [17]. Research indicates that organizations implementing Industry 5.0 principles have observed improvements regarding employee engagement, job satisfaction, and innovation capabilities, and they have simultaneously reduced their environmental footprint and operational costs [18]. Industry 5.0's holistic approach highlights planned interventions and strong policy frameworks for promoting sustainable integration of human and social factors into industrial processes as well [19]. This includes the way we develop adaptive regulatory frameworks. These frameworks ensure worker protection and environmental preservation while they accommodate rapid technological advancement [20]. Industry 5.0's future trajectory suggests it will continue to evolve into more advanced human-machine collaboration as technologies like augmented reality, advanced haptic interfaces, and cognitive computing systems play increasingly important roles in industrial operations [21]. Social responsibility considerations must carefully balance these technology improvements. Industrial development should serve society's broader goals as well as maintain ecological balance and promote sustainable economic growth [22].

1.3 The Role of Human-Centered Design in Industry 5.0

The shift toward Industry 5.0 signifies a rather prominent transformation in both the design and the operation of industrial systems. It does go further than Industry 4.0's technology-focused approach, also it introduces a framework that is more balanced since it puts people at the center, strengthens system resilience, and promotes environmental sustainability as key guiding principles [15]. It's especially important to pay attention to this change in areas that need people to work, where environmental responsibilities must balance social and economic goals [16]. Unlike Industry 5.0, Industry 4.0 mainly increases productivity by systematically replacing human workers. Industry 5.0 encourages a collaborative relationship between human intelligence and technological advancement, which leads to better mental ergonomics in factories and better teamwork between people and machines [17]. Industry 5.0 changes its methods in a big way with the help of smart transformation projects. It addresses all of the complicated effects that actions in the industry have on socio-ecological systems by updating production methods with round financial rules [23]. These efforts go far beyond just making the system better. This new model shows that different human-made factors, such as ecological sustainability and economic growth, are very closely related. This is because the model values regenerative intention for directional sustainability but puts industrial output inside that structure [19]. Because the principles of Industry 5.0 are being used, important technological infrastructure needs to be built, such as advanced robotics, artificial intelligence, and human-machine interfaces. Along with it are in-depth training programs that take into account ethics related to data privacy and job security [20]. Companies that follow the principles of Industry 5.0 see real improvements in employee engagement and job satisfaction, as well as innovation [21]. They also have lower operational costs and less of an impact on the environment. Advanced human-machine collaboration is the likely direction of Industry 5.0. Emerging technologies such as augmented reality and cognitive computing systems will play increasingly important roles [22]. Careful balancing of these technological advancements alongside social responsibility considerations is needed since it ensures industrial development serves broader social goals amid maintaining ecological balance, while

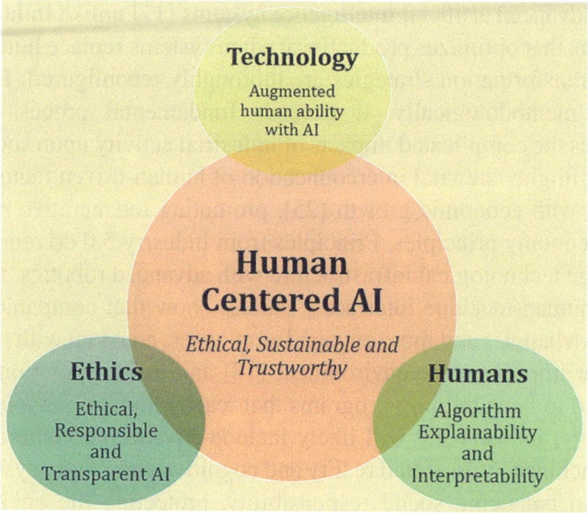

Fig. 2 Human-centered AI combines technology, ethics, and human factors [24]

promoting sustainable economic growth [11]. The holistic methodology of Industry 5.0 stresses that planned interventions coupled with thorough policy frameworks are necessary to support the sustainable integration of human and social dimensions into industrial operations, ensuring that technological progress remains aligned with social goals along with environmental sustainability [12]. Human-Centered AI, shown in Fig. 2, integrates Technology, Humans, and Ethics based on ethics, sustainability, and trust: Technology uses AI to improve human capabilities, Humans want algorithms that are explainable and interpretable, and Ethics makes AI systems ethical, transparent, and accountable.

2 The Role of Artificial Intelligence in Industry 5.0

2.1 Integration of AI into Digital Manufacturing Processes in Industry 5.0

The transformative evolution toward Industry 5.0 represents a fundamental model shift in industrial systems since it transcends the technology-centric approach of Industry 4.0, while it establishes a more advanced framework that prioritizes human-centricity, systemic resilience, and environmental sustainability as its foundation principles [15]. This change hits industries based on manual labor greatly, and it needs linkage of human thought and digital bases [16]. Industry 5.0 advocates for the fact that human cognitive abilities and technological augmentation relate in a symbiotic way so they can improve human-machine collaboration plus cognitive ergonomics in industrial environments when people implement collaborative robots (cobots) and advanced artificial intelligence systems [17] unlike Industry 4.0's predominant focus that optimizes productivity when systems replace human operators. When smart transformation strategies are thoroughly reconfigured, Industry 5.0 is implemented methodologically. It exceeds fundamental process optimization because it faces the complicated impacts of industrial activity upon socio-ecological systems. It highlights the vital interconnection of human-driven factors, ecological sustainability, with economic growth [25], promoting regenerative manufacturing and circular economy principles. Principles from Industry 5.0 do require the development of large technological infrastructure with advanced robotics, artificial intelligence, and human-machine interfaces. Studies show that companies that follow these rules have happier and more engaged employees, come up with new ideas, and have a smaller impact on the environment [18] and lower operational costs [26]. They also need to have training programs that teach employees about data privacy and job security. Industry 5.0 will likely include advanced collaboration between people and machines. Augmented reality and cognitive computing systems will play bigger roles in balancing social responsibility, protecting the environment, and making sure that industrial growth serves larger social goals that support economic growth.

2.2 Benefits of AI in Enhancing Manufacturing Efficiency

The shift to Industry 5.0 signals a key evolution in industrial technology, human talent, and environmental concern integration, which establishes revolutionary benchmarks for green output [1]. Industry 5.0 necessitates a delicate equilibrium between operational efficiency, environmental stewardship, and human-centric values within labor-intensive manufacturing sectors through artificial intelligence, advanced cyber-physical systems, and cognitive ergonomics that are tactically deployed, which fundamentally alters customary manufacturing models, also creating more adaptable and sustainable industrial environments [26]. Smart manufacturing plans get a complete reevaluation from Industry 5.0's methodology. It moves past the customary optimization process. Furthermore, it addresses all of the complex ways that industrial activities have impacts on socio-ecological systems. With advanced human-machine collaboration, this approach integrates regenerative practices and circular economy principles. It improves human capabilities in place of replacing them [27]. This approach is supported by advanced technological infrastructure, and it also includes collaborative robots (cobots), artificial intelligence systems, and advanced human-machine interfaces, which are all designed for the augmentation of human decision-making processes while also addressing critical considerations with regard to data privacy, job security, and environmental impact [18]. It has been shown by studies that organizations adopting Industry 5.0 principles experience an increase in employee involvement, a stronger potential for innovation, and improved performance for operations. At the same time, they reduce their environmental impact, and they become more resilient to external challenges. Industry 5.0, for integrating technological progress with human well-being and with environmental responsibility, provides for a balanced approach [28]. Industry 5.0's future trajectory shows an evolution to advanced human-machine collaboration as well. Cognitive computing systems, along with advanced AI, will play key roles when creating sustainable manufacturing ecosystems that serve social goals while promoting ecological preservation, thus ensuring sustainable growth.

2.3 AI's Contribution to Sustainable Manufacturing Practices

Industry 5.0 introduces a new direction in industrial development in that it balances technological innovation with humans' input, with environmental concerns in a unified operational framework [29]. This transforms most pronouncedly within hands-on manufacturing domains, where integrating smart systems requires humans and the environment to align considerately with priorities [30]. Industry 5.0's methodology implements such a thorough overhaul of the smart manufacturing approaches that exist. It exceeds process optimization since it addresses industrial operations' detailed socio-ecological effects. It stresses the adoption of some regenerative practices in the circular economy models. Advanced human-machine collaboration also supports and increases human capabilities rather than replacing them [31]. This approach gains support from advanced technological infrastructure, including collaborative robots (cobots), generative AI systems, and advanced human-machine interfaces, all of which are designed to augment human decision-making processes while simultaneously addressing critical considerations regarding data privacy, algorithmic transparency, and environmental impact [32]. Organizations adopting Industry 5.0 principles have achieved meaningful gains in innovation, resource optimization, and operational efficiency, according to research. Concurrently, they have decreased environmental impact and strengthened resilience to disruptions. Industry 5.0's holistic framework successfully balances technological progress with human and environmental priorities; these outcomes suggest [33]. The path ahead for Industry 5.0 reveals a shift to deeper collaboration between humans and machines since new tech, such as thinking computer systems, plus advanced AI applications, are key parts of making manufacturing ecosystems sustainable, resilient, and human-focused, which meet large social aims while increasing environmental protection plus stable financial gains.

3 Human-Machine Collaboration in Industry 5.0

3.1 Key Components of Effective Human-Machine Collaboration

This evolution in manufacturing reflects someone reimagining the intersection of digital tools and human capabilities to achieve ecological and social resilience long-term, and also efficiency [34]. This change transforms manufacturing, which needs much labor, making it stand out. To balance efficiency, sustainability, coupled with human-centered values, manufacturers tactically use smart perception systems along with advanced cyber-physical infrastructures, plus cognitive ergonomics. In these environments, machines are engineered so they interpret human intent and then respond to it as well, which enables collaboration that is smooth, therefore intuitive, plus redefines conventional industrial models, also fostering production ecosystems that remain more flexible, resilient, and sustainable [35]. Rethinking smart manufacturing strategies involves Industry 5.0's methodological implementation. It has to be all throughout, this rethinking. It seeks to go far beyond optimizing more conventional processes. It seeks to address complex socio-ecological impacts from industry. Human skills are not to be replaced, but they are to be improved. For this to happen, circular economy principles, regenerative practices, and complex human-machine collaboration systems are merging. Smart machines can link to human users, and this has eased strong intention recognition. Nurturing of trust is a key emphasis. Collaborative robots, artificial intelligence systems, and advanced human-machine interfaces support an advanced technological infrastructure that Industry 5.0 relies upon. Critical concerns, for example, trust, data privacy, and environmental sustainability, are addressed via these technologies. Empirical studies show prominent gains such as employee engagement, innovation, and operational efficiency for organizations that embrace each principle. At the same time, they do reduce environmental impact as well as resilience to external disruptions, and this action here shows this [36]. The future for Industry 5.0 evolves toward the collaboration of humans with more advanced machines. Advanced AI and cognitive computing systems will have critical roles as they build resilient, sustainable, human-centric manufacturing ecosystems, promoting sustainable economic growth and ecological preservation while contributing to a meaningful human-machine partnership, respecting autonomy while leveraging strengths, and serving broader social goals.

3.2 Redefining the Relationship Between Workers and Machines in Industry 5.0

Industry 5.0 changes the way things are made by putting collaboration at the center of the process through people and smart machines. This is because the ecosystem is based on adaptability, creativity, and sustainability [37]. This change changes whole industries. It has a big effect on the manufacturing industry. Balance is necessary for things to work well. Human-centered principles also need to be carefully balanced. It uses smart robots and advanced systems that work well with people, creating a symbiotic relationship that uses human creativity in new ways while maximizing automation capabilities. This method makes interactive collaborative environments by automating routine tasks and letting human workers focus on more difficult, creative jobs. This increases job satisfaction and innovation [38] instead of relying on automation. Industry 5.0's methods completely change how things are made. They go beyond traditional automation by looking at how people and machines interact and coming up with new ways to improve teamwork. Smart robots, advanced systems, and human-machine interfaces all support this through advanced technological infrastructure. All of this is meant to improve human abilities while also addressing important issues like job satisfaction, innovation potential, and production efficiency. Experimental research shows that companies that follow these principles have seen significant improvements in employee engagement, creative output, and operational performance, while still keeping the human element in production processes that automation used to dominate [39]. The future trajectory for Industry 5.0 indicates that it will evolve toward collaboration between humans and machines at a more advanced level. Tactically, business organizations must then decide to improve the interactions, ensure machines and humans contribute effectively to better manufacturing outcomes, value understanding and human creativity while capabilities support advanced technologies, redefine production efficiency through optimal human and technological resource integration, and foster sustainable growth.

3.3 Technologies Facilitating Human-Machine Interaction in Digital Manufacturing

In the context of Industry 5.0, the integration of advanced technologies is important since it eases smooth human-machine interaction in digital manufacturing environments, which marks a departure from the automation-centric focus of Industry 4.0 toward a more symbiotic industrial ecosystem that is human-centric. Additive manufacturing (AM) acts as a foundation of this transition then represents the flexibility as well as the agility that remain necessary for production processes which respond and adapt, also it enables customization in real-time that aligns with user-specific needs including sustainable manufacturing goals [40]. AM allows for the rapid prototyping in addition to iterative testing of designs, which is different than customary methods. Human-machine collaboration improves in industries like composite materials due to digital with physical areas interaction [41–46]. Integral to this process is computer-helped design (CAD) software, which equips designers with powerful tools to plan and orchestrate manufacturing workflows, and it bridges the divide between virtual concepts and concrete outputs [5]. Engineers use CAD systems integrated with simulation capabilities to model, test, and optimize designs in virtual spaces before physical production starts, so material waste decreases and production efficiency increases. Virtual spaces further strengthen human-machine interaction through provision of an engaging platform for fluid interaction between human operators and clever systems. They [47] permit process changes within manufacturing along with making specific custom solutions. Augmented reality (AR) and virtual reality (VR) are included within extended reality (XR) technologies with a key role. Designers can interactively manipulate physical and virtual components because multidimensional visualization tools improve designers' perceptions. Human operators manage complex processes with greater precision as well as effectiveness because of that immersion. Human intuition with machine accuracy then

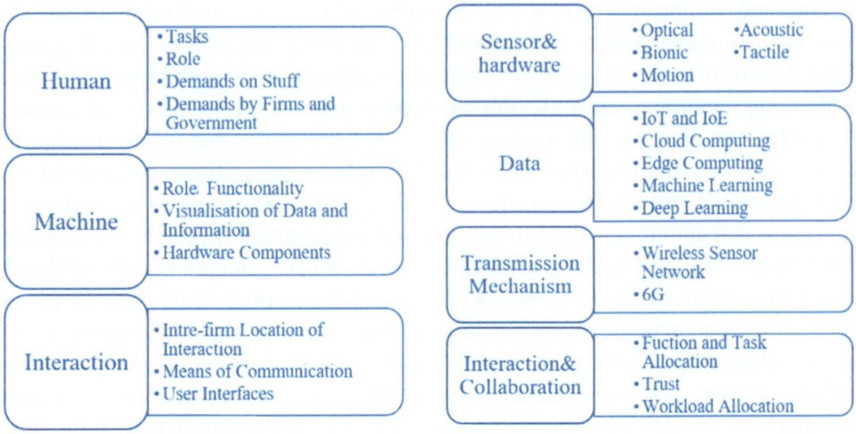

Fig. 3 Human-machine interaction for Industry 5.0 involves smart manufacturing that prioritizes the needs of humans [35]

converges in a collaborative environment, optimizing production. Convergence of such sort underscores the critical need for advanced information systems, also AI-driven decision-support frameworks that can interpret complex datasets, monitor real-time production conditions, and respond adaptively to unforeseen challenges. AI and machine learning algorithms are integrated within these systems in order to enable predictive maintenance, quality control, and process optimization for driving Industry 5.0 toward clever, sustainable, and highly personalized manufacturing models [48]. Industry 5.0 takes technology onward to redefine digital manufacturing with the stress on collaborative intelligence, which is a combination of human creativity and machine precision to achieve an industrial framework that is resilient and adaptive, as shown in Fig. 3. This evolution makes productivity higher, along with ensuring alignment with broader social and environmental objectives, so Industry 5.0 takes its place as a holistic approach to sustainable industrial growth that empowers human operators while digital technologies are efficient and precise. In this manner, Industry 5.0 shifts the planned model since it advocates that one integrated industrial ecosystem values human capabilities as well as ecological responsibility with technological innovation plus economic viability.

4 Impacts of Digital Manufacturing on the Global Economy

4.1 Digital Manufacturing's Influence on Global Supply Chains

The integration of digital manufacturing technologies has fundamentally transformed global supply chain operations since they mark a model shift in how organizations approach production and manage logistics in the modern industrial landscape [49]. Mercedes-Benz Group plus Haier Group, industry leaders, showed great success using advanced technologies like Internet of Things (IoT) plus Artificial Intelligence (AI), improving operations and lowering risk across manufacturing networks [49]. Because of technological advancements, smart manufacturing ecosystems have emerged, so they are characterized by unprecedented automation levels. These ecosystems are monitored in real-time, predicted using analytics, decided proactively, and optimized operations [50]. This change is more than just simple mechanization. Smart factories use data-driven perceptions as they grow to make production more efficient, quality control better, and resource use better across the supply chain in [50]. This digital shift brings up important problems, such as how to deal with more complicated systems. It is also important to make sure that the setup works well and that people in very different technical roles in manufacturing can work together. As businesses work to build digital infrastructures, they must deal with these problems. These infrastructures make it easy for stakeholders to work together [49] and support the smooth exchange of information. Digital transformation now does the best job of managing supply chain risk. It helps create strong global supply networks that can handle quickly changing market conditions, including all operational needs, and this becomes a very important factor. A thorough approach is needed for this digital transformation adventure to be successful. This includes addressing technical and organizational challenges, such as making sure that developers create the right digital platforms, implementers use the right data management strategies, and workers who are skilled in technology can use these advanced technologies to their full potential [50]. The technological revolution within manufacturing has improved operational efficiency and created some new opportunities with innovation for product development, service delivery, and customer engagement since it fundamentally altered the competitive landscape within global manufacturing. The digital economy's impact on different industries has been shown in Fig. 4.

4.2 *Economic Benefits of Adopting Industry 5.0 Practices*

One of the primary economic benefits of adopting Industry 5.0 is that it improves customization, and it makes manufacturing processes more flexible. Because of this shift, manufacturers are able to produce a wider variety of products in much smaller batches. It tackles various consumer wants and tastes head-on [52]. This approach caters to the trend of personalized products and opens market opportunities, thus potentially increasing revenue streams. Furthermore, advanced technologies such as machine learning, along with computer vision, are integrated into manufacturing processes. These technologies optimize those processes for they reduce waste as well as improve resource efficiency [53]. This optimization translates into saving important costs because manufacturers are able to minimize material costs and consume energy, improving their competitive advantage in the market. Industry 5.0's sustainable approach is one that supports business environmental goals and also contributes to long-term economic benefits through ensuring resource availability as well as reducing non-compliance penalties [53]. These aspects show together that

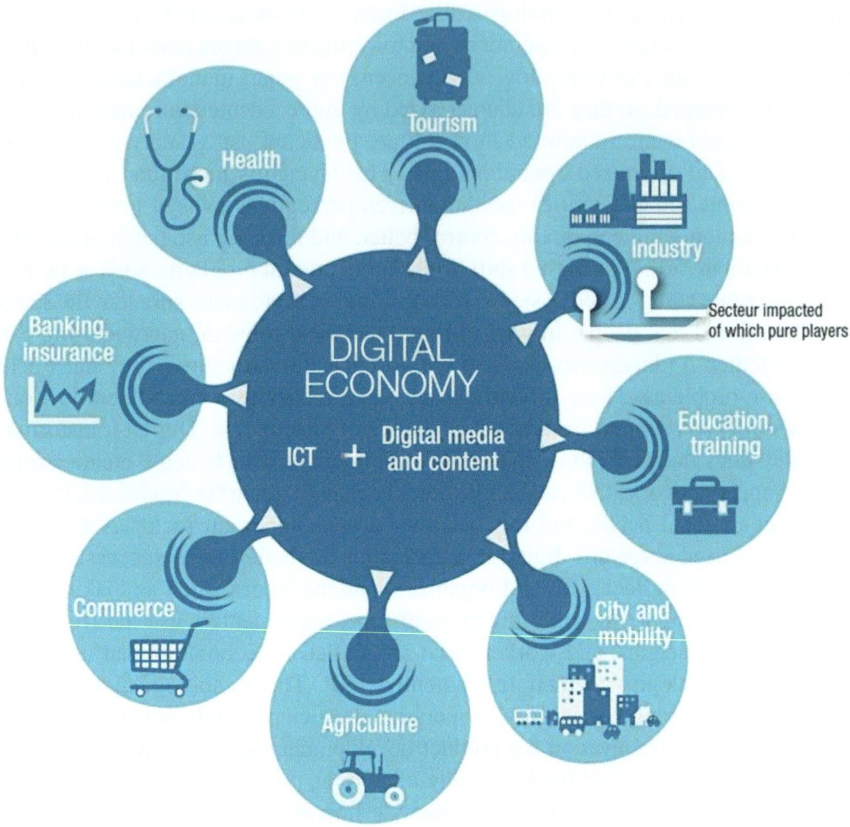

Fig. 4 Digital Economy impact on different industries [51]

businesses must adopt Industry 5.0 practices because they increase output, increase efficiency, and perform more sustainably and strongly [54].

4.3 Global Impacts of Industry 5.0 on Employment and Job Creation

Industry 5.0 greatly affects worldwide work and makes jobs by changing roles humans perform in manufacturing. Industry 5.0, unlike its predecessor Industry 4.0, which was mainly focused on automation, integrates human skills with advanced technologies because it stresses collaboration rather than any replacement [55]. Elon Musk did highlight this model shift when he asserted that "humans are underrated." His assertion underscores just why human skills have gained renewed importance as production processes have advanced. Therefore, there is an increasing need for human workers able to increase machine capability [55]. A human-centric approach in Industry 5.0 promotes job roles tailored to workers' skills and talents. This creates a more inclusive employment landscape overall [55]. Industry 5.0's emphasis that is on manufacturing that is sustainable also creates opportunities for employment in environmental stewardship and in fields of management regarding sustainable resources [4]. Since these developments suggest that the future of work will require technological expertise with a strong set of human skills to navigate the evolving landscape, they underscore the need for thorough training programs coupled with policies that support workforce adaptability and sustainability.

5 Challenges and Opportunities in Implementing Industry 5.0

5.1 Major Challenges for Companies Transitioning to Industry 5.0

One of the major challenges companies encounter in transitioning to Industry 5.0 is aligning business interests with those of the public, which necessitates approaching policymaking and governance innovatively [56]. Organizations must adapt their business models, as well as value chains and supply chains. To stay relevant and competitive, this change makes alignment especially detailed in this fresh industrial framework [56]. The global pandemic experience further stressed the need for companies to strengthen resilience across value chains, ensuring employment as well as financial stability, an arduous yet critical task [57]. Businesses should revise their digital transformation goals by adjusting structures plus logistics to be consistent with new tech changes plus socio-economic needs, including rethinking their digital integration plus innovation [57]. Companies are moving ahead into Industry 5.0, and they do certainly need a thorough approach that is systemic to manage effectively these detailed challenges. This approach should stress how important it is to integrate into the development of new technologies social and environmental considerations [57]. Since these multidimensional challenges get addressed, one concerted effort via businesses, policymakers, and stakeholders is needed to ensure a smooth and sustainable transition into this new era of industrial evolution.

5.2 Strategies for Overcoming Barriers to Digital Manufacturing Adoption

To successfully overcome barriers to digital manufacturing use, firms should create a solid plan that builds an innovation culture, integrating diverse organizational roles [15]. Formalizing responsibilities across departments clarifies roles, fostering collaboration along with accountability in the digital adoption process [58].

Companies align digital tools for specific work tasks to ensure effective technology integration. This integration improves productivity as well as reduces employee resistance [59]. Furthermore, user-friendly digital solutions are offered; therefore, they can greatly lower barriers to adoption because new technologies become more accessible and the workforce feels less intimidated [15]. Education and training of high quality are important to aid this change. It gives employees skills that are necessary, plus knowledge, so they can engage with technologies with confidence [59]. Before implementing new technologies at full-scale, companies can test them to also identify potential challenges, allowing well-educated adjustments for smoother integration into existing systems [58]. You can further improve on the efficacy of these efforts if you stress strong information sharing and ensure the cooperation between maintenance and production functions, ensuring digital transformation initiatives succeed and sustain [59]. These strategies collectively underscore the importance of a calculated alignment, education, and collaboration, which is a well-structured approach to digital manufacturing adoption.

5.3 Opportunities for Innovation and Growth in Industry 5.0

Industry 5.0 presents many opportunities for both innovation and also growth as it improves collaboration in order that humans and machines collaborate, which is quite central to its model. This collaboration helps with a shift toward more personalized production processes. Manufacturers can then cater to diverse customer demands, and they can create new markets [21]. Industry 5.0 remarkably increases operational efficiency for innovation on manufacturing shop floors, integrating advanced technologies like smart robots and systems [10]. The human-centered approach of Industry 5.0 also puts a lot of emphasis on the role of people in production [10]. It also moves people around. Allowing people to stop doing the same things over and over again as they take on more revolutionary roles encourages creativity and makes sure that decisions are made with care. This change can help a business model grow in a flexible way. These sectors are growing and expanding because they are adapting to changing market needs. Industry 5.0 also stresses sustainable practices, which means that manufacturers should use resources wisely and reduce waste [10]. Industry 5.0 helps businesses grow and get better while also protecting the environment by focusing on these things. These opportunities show how Industry 5.0 could change everything by encouraging innovation, sustainability, and economic growth that work for everyone [60–64].

6 Conclusion

The shift from Industry 4.0 to Industry 5.0 completely changes the digital manufacturing landscape, moving it toward methods that focus on people and the values of working together with machines. This change in model makes operations more efficient and helps solve sustainability problems in a big way because industrial practices are in line with social goals. The study shows that more advanced technologies, like AI, collaborative robots, and digital twins, do work together and make it easier for people and machines to work together. This integration lets manufacturers change their products to meet the changing needs of consumers through hyper-customization. This creates a shared space where human creativity makes machines more accurate, which solves problems quickly and well. To make Industry 5.0 work, we need to carefully plan how to close the skills gap since human workers are already able to use advanced tools well. Also, focusing on environmentally friendly ways of making things helps protect the environment. This also creates new jobs because it shows how important human resources are to production. When recognizing all the difficulties involved in putting these ideas into practice across all the different manufacturing sectors, it is important to keep in mind that this current study highlights all of Industry 5.0's transformative potential. We need to look into how this change affects the way people work. This study should look into whether jobs could be lost. It should also look into whether reskilling programs are necessary. It is still important to look into the moral aspects of combining technology with human work. Researchers should not put people's health and safety at risk while they look into advances in manufacturing. These kinds of investigations are needed. In general, the results of the study help us understand how Industry 5.0 might change the way businesses work because synergy encourages growth, innovation, and sustainability. It still pays attention to the social effects of these technological advances.

References

1. Nahavandi, S. (2019). Industry 5.0—A human-centric solution. *Sustainability, 11*(16), 4371. https://doi.org/10.3390/su11164371
2. Xu, X., et al. (2021). Industry 4.0 and Industry 5.0—Inception, conception and perception. *Journal of Manufacturing Systems, 61*, 530–535.
3. Longo, F., Padovano, A., & Umbrello, S. (2020). Value-oriented and ethical technology engineering in Industry 5.0: A human-centric perspective for the design of the factory of the future. *Applied Sciences, 10*(12), 4182. https://doi.org/10.3390/app10124182
4. Mourtzis, D., Angelopoulos, J., & Panopoulos, N. (2023). The future of the human-machine interface (HMI) in society 5.0. *Future Internet, 15*(5), 162.

5. Zheng, P., et al. (2018). Smart manufacturing systems for Industry 4.0: Conceptual framework, scenarios, and future perspectives. *Frontiers of Mechanical Engineering, 13*, 137–150.
6. Raja Santhi, A., & Muthuswamy, P. (2023). Industry 5.0 or Industry 4.0S? Introduction to Industry 4.0 and a peek into the prospective Industry 5.0 technologies. *International Journal on Interactive Design and Manufacturing (IJIDeM), 17*(2), 947–979.
7. Mourtzis, D., Angelopoulos, J., & Panopoulos, N. (2022). A literature review of the challenges and opportunities of the transition from Industry 4.0 to Society 5.0. *Energies, 15*, 6276.
8. Lou, S., et al. (2024). Human-cyber-physical system for Industry 5.0: A review from a human-centric perspective. *IEEE Transactions on Automation Science and Engineering*, 1–18.
9. Bag, S., Gupta, S., & Kumar, S. (2021). Industry 4.0 adoption and 10R advance manufacturing capabilities for sustainable development. *International Journal of Production Economics, 231*, 107844.
10. Dalenogare, L. S., et al. (2018). The expected contribution of Industry 4.0 technologies for industrial performance. *International Journal of Production Economics, 204*, 383–394.
11. Javaid, M., et al. (2020). Industry 5.0: Potential applications in COVID-19. *Journal of Industrial Integration and Management, 05*(04), 507–530.
12. Demir, K. A., Döven, G., & Sezen, B. (2019). *Industry 5.0 and human-robot co-working* (Vol. 158, pp. 688–695). Procedia Computer Science.
13. Ghobakhloo, M. (2020). Industry 4.0, digitization, and opportunities for sustainability. *Journal of Cleaner Production, 252*, 119869.
14. Maddikunta, P. K. R., et al. (2022). Industry 5.0: A survey on enabling technologies and potential applications. *Journal of Industrial Information Integration, 26*, 100257.
15. Peruzzini, M., & Pellicciari, M. (2017). A framework to design a human-centred adaptive manufacturing system for aging workers. *Advanced Engineering Informatics, 33*, 330–349.
16. Rauch, E., Linder, C., & Dallasega, P. (2020). Anthropocentric perspective of production before and within Industry 4.0. *Computers & Industrial Engineering, 139*, 105644.
17. Santos, C., et al. (2017). Towards Industry 4.0: An overview of European strategic roadmaps. *Procedia Manufacturing, 13*, 972–979.
18. Strozzi, F., et al. (2017). Literature review on the 'smart factory' concept using bibliometric tools. *International Journal of Production Research, 55*(22), 6572–6591.
19. Tortorella, G. L., et al. (2022). Healthcare costs' reduction through the integration of healthcare 4.0 technologies in developing economies. *Total Quality Management & Business Excellence, 33*(3-4), 467–487.
20. Varela, L., et al. (2019). Evaluation of the relation between lean manufacturing, Industry 4.0, and sustainability. *Sustainability, 11*(5), 1439. https://doi.org/10.3390/su11051439
21. Wang, L., Törngren, M., & Onori, M. (2015). Current status and advancement of cyber-physical systems in manufacturing. *Journal of Manufacturing Systems, 37*, 517–527.
22. Zhou, K., Taigang, L., & Lifeng, Z. (2015). Industry 4.0: Towards future industrial opportunities and challenges. In *2015 12th International Conference on Fuzzy Systems and Knowledge Discovery (FSKD)*.
23. Rejeb, A., et al. (2025). When Industry 5.0 meets the circular economy: A systematic literature review. *Circular Economy and Sustainability*, 1–32.
24. Xu, W., et al. (2023). Transitioning to human interaction with AI systems: New challenges and opportunities for HCI professionals to enable human-centered AI. *International Journal of Human–Computer Interaction, 39*(3), 494–518.
25. Strozzi, F., et al. (2017). Literature review on the 'smart factory' concept using bibliometric tools. *International Journal of Production Research, 55*, 1–20.
26. Ivanov, D. (2023). The Industry 5.0 framework: Viability-based integration of the resilience, sustainability, and human-centricity perspectives. *International Journal of Production Research, 61*(5), 1683–1695.
27. Mourtzis, D., & Wang, L. (2024). 3—Industry 5.0: perspectives, concepts, and technologies. In D. Mourtzis (Ed.), *Manufacturing from Industry 4.0 to Industry 5.0* (pp. 63–96). Elsevier.
28. Tortorella, G. L., et al. (2022). Measuring the effect of healthcare 4.0 implementation on hospitals' performance. *Production Planning & Control, 33*(4), 386–401.

29. Sai, S., Tanneeru, R., & Chamola, V. (2024). Generative AI for Industry 5.0: Analyzing the impact of ChatGPT, DALLE, and other models. *IEEE Open Journal of the Communications Society*, 1.
30. Passalacqua, M., et al. (2025). Human-centred AI in Industry 5.0: A systematic review. *International Journal of Production Research, 63*(7), 2638–2669.
31. Sánchez-García, E., et al. (2024). Revolutionizing the circular economy through new technologies: A new era of sustainable progress. *Environmental Technology & Innovation, 33*, 103509.
32. Krishna, S., et al., Role of generative AI in Industry 5.0: A transformative force. 2024.
33. Konstantinidis, F. K., et al. (2023). A technology maturity assessment framework for Industry 5.0 machine vision systems based on systematic literature review in automotive manufacturing. *International Journal of Production Research*, 1–37.
34. Božić, V. (2024).Industry 5.0: A future of human-machine collaboration and sustainability.
35. Yang, J., Liu, Y., & Morgan, P. L. (2024). Human–machine interaction towards Industry 5.0: Human-centric smart manufacturing. *Digital Engineering, 2*, 100013.
36. Rožanec, J. M., et al. (2023). Human-centric artificial intelligence architecture for Industry 5.0 applications. *International Journal of Production Research, 61*(20), 6847–6872.
37. Adel, A. (2022). Future of Industry 5.0 in society: Human-centric solutions, challenges and prospective research areas. *Journal of Cloud Computing, 11*(1), 40.
38. Lu, Y., et al. (2022). Outlook on human-centric manufacturing towards Industry 5.0. *Journal of Manufacturing Systems, 62*, 612–627.
39. Bansal, P., et al. (2024). *Artificial intelligence and communication techniques in Industry 5.0*. CRC Press.
40. Ramírez-Gordillo, T., et al. (2024). Industry 5.0: Towards human centered design in human machine interaction. In *Research and Innovation Forum 2023*. Springer International Publishing.
41. Wable, V., et al. (2021). Engineering the electrospinning of MWCNTs/epoxy nanofiber scaffolds to enhance physical and mechanical properties of CFRPs. *Composites Science and Technology, 213*, 108941.
42. Aliahmad, N., et al. (2021). Carbon nanotube/epoxy submicron filaments for composite reinforcement applications.
43. Biswas, P. K., et al. (2023). Carbon and cellulose based nanofillers reinforcement to strengthen carbon fiber-epoxy composites: Processing, characterizations, and applications. *Frontiers in Materials, 9*, 1089996.
44. Biswas, P. K., et al. (2022). Higher strength carbon fiber lithium-ion polymer battery embedded multifunctional composites for structural applications. *Polymer Composites, 43*(5), 2952–2962.
45. Aliahmad, N., et al. (2020). Electrospun thermosetting carbon nanotube–epoxy nanofibers. *ACS Applied Polymer Materials, 3*(2), 610–619.
46. Biswas, P.K., et al. (2021). Nanostructured v2o5-swcnts based lithium ion battery for multifunctional energy storage composites: Materials synthesis and fabrication.
47. Chen, T., & Dai, L. (2014). Flexible supercapacitors based on carbon nanomaterials. *Journal of Materials Chemistry A, 2*(28), 10756–10775.
48. Javaid, M., et al. (2021). Substantial capabilities of robotics in enhancing Industry 4.0 implementation. *Cognitive Robotics, 1*, 58–75.
49. Yaqub, M. Z., & Alsabban, A. (2023). Industry-4.0-enabled digital transformation: Prospects, instruments, challenges, and implications for business strategies. *Sustainability, 15*, 8553. https://doi.org/10.3390/su15118553
50. Wang, X., et al. (2022). *Impact of digital technology on supply chain efficiency in manufacturing industry* (pp. 347–371).
51. Roman, B. (2019). Digital economy governance
52. Narkhede, G., et al. (2023). Industry 5.0 and the future of sustainable manufacturing: A systematic literature review. *Business Strategy & Development, 6*(4), 704–723.

53. Zheng, P., Wang, Z., & Chen, C.-H. (2019). Industrial smart product-service systems solution design via hybrid concerns. *Procedia CIRP, 83*, 187–192.
54. Chryssolouris, G., et al. (2009). Digital manufacturing: History, perspectives, and outlook. *Proceedings of the Institution of Mechanical Engineers, Part B: Journal of Engineering Manufacture, 223*(5), 451–462.
55. Javaid, M., et al. (2022). Exploring relationships between Lean 4.0 and manufacturing industry. *Industrial Robot: The International Journal of Robotics Research and Application, 49*(3), 402–414.
56. Javaid, M., et al. (2022). Significance of machine learning in healthcare: Features, pillars and applications. *International Journal of Intelligent Networks, 3*, 58–73.
57. Ahmed, T., et al. (2023). Modeling the artificial intelligence-based imperatives of Industry 5.0 towards resilient supply chains: A post-COVID-19 pandemic perspective. *Computers & Industrial Engineering, 177*, 109055.
58. Chryssolouris, G., et al. (2009). Digital manufacturing: History, perspectives, and outlook. *Proceedings of the Institution of Mechanical Engineers Part B Journal of Engineering Manufacture, 223*, 451–462.
59. de Assis Dornelles, J., Ayala, N. F., & Frank, A. G. (2022). Smart working in Industry 4.0: How digital technologies enhance manufacturing workers' activities. *Computers & Industrial Engineering, 163*, 107804.
60. Kumar, A., et al. (2021). *Advanced computational methods in mechanical and materials engineering*. CRC Press.
61. Kumar, A., Ram, M., & Singla, Y. K. (2024). Advanced Materials for Bio-Mechanical Applications. *CRC Press* (Taylor & Francis Group https://taylorandfrancis.com/books/) in May 2022. ISBN 9781003286806, 9781032054490. https://doi.org/10.1201/9781003286806
62. Shah, S., et al. (2024). Sustainability in smart manufacturing: Trends, scope, and challenges. CRC Press.
63. Kumar, A., Singla, Y. K., Biswas, P., Heidari, M., & Thangavel, S. (2024). "Chapter 17 – Resurrection Structure: New Generation of Bio-Inspired Nanocomposites and Laminates", published in "Fracture Behaviour of Nanocomposites and Reinforced Laminate Structures", edited by Ashwani Kumar, Yogesh Kumar Singla, Michael R. Maughan, Springer Nature, 2024, pp. 427–440. https://doi.org/10.1007/978-3-031-68694-8_17
64. Kumar, A., Singla, Y. K., & Maughan, M. R. (2024). "Chapter 12 – Analyzing Fractures in Nanomaterial-Enhanced Carbon Fiber Reinforced Polymer (CFRP) Composites", published in "Fracture Behaviour of Nanocomposites and Reinforced Laminate Structures", edited by Ashwani Kumar, Yogesh Kumar Singla, Michael R. Maughan, Springer Nature, 2024, pp. 295–327. https://doi.org/10.1007/978-3-031-68694-8_12

Optimizing Digital Manufacturing Practices: A Global Data-Driven Approach for Industry 5.0

Azzam Othman, Piyush Maheshwari, Khaled Shaalan, and Mohammad Binhammad

1 Introduction

Digital manufacturing is creating a new form of making and delivering goods in the global marketplace. Several technologies have been helping to drive Industry 5.0; including artificial intelligence (AI), additive manufacturing (AM), internet of things (IoT), and cyber-physical systems (CPS) [1]. More systematic and empirical analysis of the phenomenon of adoption and its consequences should be conducted, as sectors all over the world experience the outcomes of the digital transformation.

It was also noted that the high rate of development of technologies within the field of digital manufacturing has blurred the concepts of the production paradigm. Starting with smart factories where real-time data is incorporated into decision-making to integrated AI in design, where the product creation process is optimized, the innovation and efficiency gains are massive [2]. Nevertheless, emerging trends in manufacturing need to be viewed internationally because these technologies are used and integrated in various ways in the global context.

This research discusses the deployment of digital manufacturing in locations and industries critically and empirically. Digital manufacturing technology diffusion, adoption, and KPI effects are researched by global indexes and databases. Accordingly, this research study is timely because of the quick advancement of generative AI and its prospect of accelerating the rates of innovation and enhancing industrial customization [3].

A. Othman (✉) · P. Maheshwari · K. Shaalan · M. Binhammad
The British University in Dubai, Dubai, UAE
e-mail: 22001179@student.buid.ac.ae; piyush.maheshwari@buid.ac.ae; khaled.shaalan@buid.ac.ae; 22001112@student.buid.ac.ae

Y. Kumar Singla et al. (eds.), *Digital Manufacturing in Industry 5.0*,
https://doi.org/10.1007/978-3-031-91500-0_3

1.1 Research Background

In recent years, the concept of Industry 5.0 has emerged as an evolution of Industry 4.0, highlighting a shift from approaches focused on automation to human-centric participation in manufacturing contexts. The new paradigm highlights the integration of human intelligence and creativity with the accuracy of machines, with the ambition to create collaborative and resilient production environments [4].

The degree of diffusion of digital manufacturing technologies is also variable on a global scale and depends on the size of the regions and industries. Such cross-country analysis is required because adoption rates and effects differ considerably from one economy to another. The level of technology, the skills of the workers, and the laws related to the organization influence the adoption [5].

Digital manufacturing technologies have enhanced the performance of the industry. Digital twins are useful in streamlining production, forecasting when a system's maintenance is required, and accelerating the creation of a product. The long-term economic impact of these technologies has not been fully determined yet [6].

The fourth aspect of DM is sustainability. Industry 4.0 technology can enhance the utilization of resources to increase the circular economy. Nevertheless, more research is required to explore the global application and the effects of these sustainability-oriented digital manufacturing technologies [7].

The COVID-19 pandemic accelerated the digitalization of the manufacturing industry; however, it also emphasized significant gaps in resilience. Even though companies that had adopted digitalization proved more flexible, the crisis showed that current Industry 4.0 technologies alone are not enough to manage remote operations, resilient supply chains, and fast production shifts. This context highlights the need for more human-focused, flexible, and forecasting digital competences to safeguard manufacturing processes amid global disruptions [8].

This research seeks to explore current trends, impacts, and prospective developments in digital manufacturing within the context of a rapidly evolving global landscape. To contribute meaningfully to both academic discourse and the practical implementation of Industry 5.0 paradigms, the study synthesizes data from four global indexes and subjects it to statistical analysis. The specific research objectives and expected outcomes are outlined in (Table 1).

Table 1 Research objectives and expected outcomes

Research objectives	Expected outcomes
1. Quantify and compare digital manufacturing adoption rates across regions and industries	Detailed mapping of global adoption trends, identifying leaders, and laggards
2. Analyze implementation strategies for digital manufacturing technologies	Identification of best practices and common challenges in technology implementation
3. Assess the impact of digital manufacturing on key performance indicators	Quantitative analysis of improvements in productivity, innovation, and customization capabilities
4. Examine the role of generative AI in digital manufacturing ecosystems	Insights into the specific impacts and potential of AI-driven manufacturing processes
5. Forecast future trends in digital manufacturing adoption and impact	Projections of future adoption rates and potential economic impacts across different regions
6. Evaluate the sustainability implications of digital manufacturing	Assessment of how digital technologies contribute to sustainable manufacturing practices

2 Literature Review

2.1 *The Evolution of Digital Manufacturing*

Digital manufacturing has evolved rapidly over the past decade, transitioning from a component of Industry 4.0 to a central pillar of the emerging Industry 5.0 paradigm. Radanliev et al. [9] define digital manufacturing as the comprehensive integration of advanced technologies such as artificial intelligence (AI), the Internet of Things (IoT), and cyber-physical systems (CPS) to create highly efficient, adaptive, and intelligent production environments. This evolution is a clear transition from conventional manufacturing processes to smart, adaptive, and integrated systems.

The transition from Industry 4.0 to Industry 5.0 has been characterized by human and cognitive systems' integration. Li et al. [10] used the term cognitive manufacturing as they believed that manufacturing today is closely associated with the integration of human intelligence and artificial intelligence. This approach combines the best elements of human creative thinking and the accuracy of the machine to reach the highest level of efficiency and effectiveness.

Grosse et al. [11] build up on the fact that Industry 5.0, this is why I have underlined that it is necessary to design such manufacturing conditions that would help people to be stronger rather than replace them. This perspective is consistent with the emerging paradigm that the best digital manufacturing systems are the ones that are most integrated with people and machines at their best.

2.2 *Theoretical Framework: The Digital Manufacturing Adoption and Impact Model (DMAIM)*

The Digital Manufacturing Adoption and Impact Model (DMAIM) (Fig. 1) provides a conceptual structure of the determinants of digital manufacturing adoption and the consequent effects on industrial performance and economic growth [12].

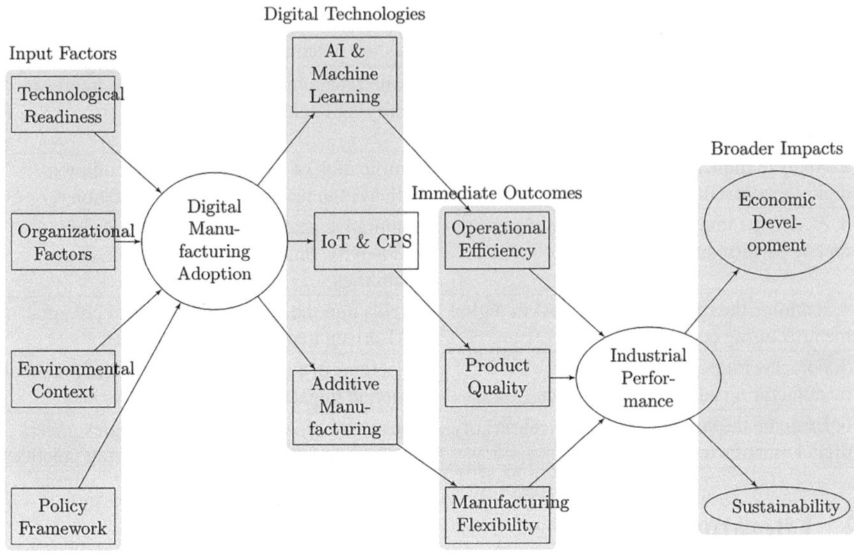

Fig. 1 Digital manufacturing adoption and impact model (DMAIM)

The rationale for this model is based on the technology adoption theories, innovation diffusion models, and industrial performance models.

The arrows show the flow or direction of effect or influence. The following diagram illustrates the theoretical framework of how the various factors lead to digital manufacturing adoption and how this influences the various aspects of industrial and economic performance.

2.3 Global Adoption Trends

The extent of integration of Industry 4.0 technology differs significantly across regions and industries, leading to significant differences in their implementation methods. An analysis of Industry 4.0 adoption in Brazilian show that the magnitude and implications of the adoption are dependent on the industrial segment and the size of the firm. Tortorella and Fettermann [13] highlight specific approaches that accelerate digital manufacturing adoption in emerging economies, which have substantial implications for potential global use.

The World Economic Forum used the Readiness for the Future of Production Index, developed by Levin [14], which measures countries' preparedness for digital manufacturing adoption depending on technology and innovation, human capital, institutional environment, and sustainable resources. The difference is quite substantial between the leaders and the laggards, which implies a growing manufacturing digital readiness gap around the world.

Müller et al. [15] identify that company size, absorptive capacity, and regional innovation system impact Industry 4.0. Zero percent adoption of RM in European manufacturing firms. They focus on environments and initiatives for the adoption of digital manufacturing.

2.4 Impact on Industrial Performance

Digital manufacturing technologies can also enhance industrial performance in terms of different indicators. Tao et al. 's [16] comprehensive review of smart manufacturing demonstrates how data-driven strategies can enhance output, quality, and utilization. In this context, the author presents several examples from different industries to demonstrate the applicability of concepts of digital manufacturing.

Nikolakis et al. [17] explain how the digital twin technology can enhance production, forecast maintenance requirements, and expedite product development in the manufacturing industry. They indicate that the advancement in simulation and modeling can transform the industry according to their research.

Li et al. [18] described the role of big data analytics in manufacturing decision-making, predictive maintenance, and supply chain management. The authors found that information management and analysis are critical for realizing the full value of digital manufacturing.

2.5 The Role of Generative AI

Generative AI is revolutionizing digital manufacturing. An analysis of the AI articles in smart manufacturing by Cioffi et al. [19] reveals that there is a significant increase in generative design, predictive maintenance, and smart supply chain management research. In consequence, generative AI was identified as the concept that would set the direction of the enhancement of manufacturing.

Bendoly et al. [20] overview generative design in AM and explain how design by an AI may lead to the creation of thin, delicate structures that were unachievable for designers. This paper explains how generative AI could redefine product design and manufacturing in the future as the following two sections will depict.

Huang et al. [21] discuss the different scenarios regarding how the AI models can learn from the changes in the production environment, how they can help manage and utilize the resources, and how they can contribute to predictive quality assurance in the case of smart manufacturing. Their work shows that AI can enhance the fault tolerance of production processes and thus make them less vulnerable to failure.

2.6 Sustainability and Circular Economy

Some of the opportunities that are believed to be associated with digital manufacturing include being able to enhance the dissemination of sustainable practices and/or the circular economy. Nascimento et al. [7] claim that Industry 4.0 technologies can improve resource efficiency, product life cycle management, and closed-loop manufacturing. They explain how their research identifies the prospect of DTs in meeting some of the most pressing sustainability concerns in manufacturing.

Sartal et al. [22] present a systematic literature review of the sustainability impacts of Industry 4.0, as to how digital manufacturing technologies can create environmental, social, and economic values. They pointed out that nowadays the sustainable development concept should be based on the use of the digital technologies' potential in every aspect.

Liu et al. [23] have looked at the contribution of specific digital technologies to the practice of circular economy within manufacturing. Their work shows how IoT, big data and analytics, and AI can be used for product tracking, resource optimization, and circular economy business model innovation.

2.7 Challenges and Limitations

Many challenges exist for the adoption of digital manufacturing even though the concept has many benefits. Based on Fourth Industrial Revolution studies by O'Reilly et al., [24] data security, labor upskilling, and large capital outlay are cited as key barriers to digital transformation in the manufacturing industry.

There is a need to conduct a critical analysis of SMEs' Industry 4.0. All the identified adoption issues had financial constraints, lack of technical expertise, and organizational culture for change as the key challenges. According to their findings, they recommend specific interventions and policies to encourage smaller producers to go digital [25].

The COVID-19 epidemic has encouraged people to use digital services and brought manufacturing problems to the surface. The case of Industry 4.0 about sustainability is presented by Ambrogio et al. [8]. These technologies have made manufacturing operations remote, supply chain robust, and production changes fast to counter pandemic impacts. From this study, it can be concluded that manufacturers need to have digital skills to withstand world challenges.

2.8 Research Gap and Future Directions

The studies of DM are valuable, yet there is limited literature comparing the global trends in adoption, the strategies for implementation, and consequences in different settings. An exemplary analysis of Industry 4.0 research by Tiwari [26] indicates that there is a need for increased synergy between the various fields of study in explaining digital manufacturing phenomena.

In their Industry 4.0 research gap, Liao et al. [27] suggest more research to understand the long-term impact of digital manufacturing adoption for firms mainly in developing countries. They highlight the need for developing guidelines for R&D

related to digital manufacturing readiness, performance metrics, and approaches across contexts.

A bibliometric analysis of Industry 5.0 literature highlights current research trends and identifies existing knowledge gaps. [28] emphasize the need for further studies focused on value co-creation for people, sustainability, and the integration of artificial intelligence in manufacturing.

2.9 Conclusion

The literature study presents a diverse and quickly growing research field on digital manufacturing in terms of technologies, applications, performance effects, and sustainability concerns. But it also shows the importance of the more extensive and global trends' investigations that might help to get a more holistic picture of digital manufacturing and its effects in various geographical and industrial settings. This research endeavors to fill this gap by offering a global insight into the current and prospective status of digital manufacturing in the Industry 5.0 era, especially about generative AI and the influence of these advancements on international industrial rivalry and the creation of sustainable economies.

3 Methodology

3.1 Research Design

This research uses a quantitative research method to survey the global trend in the use and incorporation of digital manufacturing. The approach enables the analysis of vast data, thereby enabling a structured analysis of the large datasets to determine the relationships between these factors, their correlation, and coefficients that impact the diffusion of DMT in various regions and industries. The research design was developed according to the methodological framework discussed by Dalenogare et al. [29] for measuring the influence of Industry 4.0 technologies on industrial performance, no information was found.

3.2 Data Collection

There are different reputable global indices and databases (Table 2) that researchers use to get information. The chosen sources are reliable, exhaustive, manufacturing-relevant, and digital; they were selected by the researchers. The 2015–2023 data collecting period enables the researchers to conduct a strict temporal analysis of trends. In extracting the data, we look for sources of the factors that are significant to make the data consistent and comparable across time and countries.

Table 2 Data sources and key variables

Data source	Key variables	Frequency	Coverage	Reference
Global manufacturing competitiveness index	Manufacturing competitiveness scores, technology investment	Annual	40 countries	Deloitte [33]
World economic forum's readiness for the future of production report	Production structure, drivers of production	Biennial	100 countries	WEF [36]
UNIDO'S competitive industrial performance index	Manufacturing value added, manufacturing exports	Annual	150 countries	UNIDO [37]
World Bank's digital adoption index	Business digital adoption, people digital adoption	Biennial	180 countries	World Bank [34]
Global innovation index	Innovation input and output sub-indices	Annual	130 countries	Dutta et al. [35]

3.3 Statistical Analysis

Data analysis is conducted in SPSS version 27 because of its capability to handle large data volumes and perform diverse calculations. The analytical approach comprises of three aspects. First, frequency distributions are employed in order to represent and quantify the main features of the dataset, that include measures of central tendency, variability, and shape. This provides a good starting point to navigate through such data by virtue of providing a brief description of what the data looks like. Second, the analysis of variance (ANOVA) tests is used to compare the means of various groups, for example, regional or industrial, with the purpose of determining the statistically significant differences in digital manufacturing adoption or effects. Last, multiple regression analysis is used to find out the correlation between dependent variables such as adoption rates and performance indicators and independent variables such as economic factors and technological readiness. This assist in determining possible factors that determine the likelihood of digital manufacturing adoption and the impacts of the same. This concentrated statistical strategy corresponds to the guidelines of the multivariate data analysis, described by [32] and corresponds to the investigated research questions on the trends of the digital manufacturing usage and its impact.

3.4 Impact Assessment

Various studies focus on how the implementation of digital manufacturing affects the improvement of KPIs. This entails tracking the output rates, creativity, product differentiation, and economic statistics. Bai et al. [30] on Industry 4.0. No implications were given credit for originating the impact assessment technique. Scholars evaluate the adoption consequences of digital manufacturing by employing DID and PSMM techniques.

3.5 Validation and Reliability

Validation and reliability measures are formulated and followed by researchers to ensure that results are valid. These are data source validation, statistical model validation, and data variation test, respectively. The quantitative framework uses Rashid et al. [31] validation methods for case study research as adopted by the researcher. Furthermore, bootstrapping is used to assess the stability of statistical models by the researchers.

3.6 Limitations

First, it is important to recognize that there are limitations to the study; for example, the availability and comparability of data across countries, the sheer speed of evolving technologies that may affect long-term trends, and differences in definition and measures from one data source to another. Researchers mitigate these problems in data preprocessing, stating all assumptions, and cautious interpretation of results when data quality is questionable.

3.7 Conclusion

A thorough technique is used to assess worldwide digital manufacturing uptake and impact. Quantitative methods, global indices, and databases provide a data-driven view of digital manufacturing across regions and industries. The data is analysed using SPSS for descriptive statistics, correlation, multiple regression, factor analysis, and cluster analysis. Comparative analysis allows thorough cross-national comparisons, whereas trend analysis anticipates digital manufacturing adoption via time-series forecasting. Impact evaluation quantifies digital manufacturing's effects on industrial and economic performance using KPI. Research credibility is increased by data source validation, statistical model validation, and bootstrapping. The methodology uses data preparation and interpretation to overcome data availability and comparability between nations. The study uses statistical rigour and global economic indices to understand digital manufacturing trends, adoption patterns, and impacts in Industry 5.0 and the global digital transformation of manufacturing sectors.

4 Results and Analysis

4.1 Introduction

Chapter four details the survey on digital manufacturing uptake and impact in different organizations. This chapter presents current knowledge on manufacturing technology utilization, integration, and productivity benefits. The data is analyzed using descriptive statistics, regression analysis, and t-tests. These methodologies capture the intricate interdependencies between digital manufacturing adoption antecedents and organizational success. Starting with descriptive statistics, the chapter describes survey generalities. It then performs more complex tests to discover technology uptake predictors and assess the sample's demographic profile against normative hypotheses. This multiple viewpoint analysis ensures data coverage, laying the groundwork for future conclusions and recommendations.

4.2 Statistical Analysis

The descriptive statistics shown in 3 that the summary of the responses received through the survey on different questions. The mean scores vary from 2.34 to 4.96, which shows that the firms have implemented these technologies to different extents and witnessed different degrees of improvements. Among the questions, questions 2, 3, and 14 received higher mean scores of 4.37, 3.67, 4.96, respectively, of positive responses. The standard deviations range as follows: from 0.947 to 2.152, therefore, showing that respondents' level of agreement is not consistent. For questions with high standard deviation (e.g., Q2, which got a 2.152) this is a clear sign of the participants' opinions varying greatly and for questions with low standard deviation (e.g., Q11 with 0.947) this is evidence of the participants' opinions not varying as much. The mode values are useful in understanding what the most frequently occurring responses are, and this is found to be rather high with several questions having a mode of 4 or 5, which represents a relatively positive attitude or high rate of adoption. The median values are always near or are even higher than the means, which implies that the data might be slightly positively skewed for majority of the questions (Table 3).

4.3 Regression Analysis

The regression analysis shown in Table 4 is used to determine the correlation between the dependent variable, which is the adoption level of AM and independent variables. The analysis also reveals that the extent to which other technologies have been adopted and the degree of integration of digital manufacturing technologies

Table 3 Descriptive statistics of survey responses on digital manufacturing adoption and impact

	Q2	Q3	Q4	Q5	Q6	Q7	Q8	Q9	Q10	Q11	Q12	Q13	Q14	Q15	Q16
N valid	102	102	102	102	102	100	99	101	102	102	102	102	102	102	102
Missing	0	0	0	0	0	2	3	1	0	0	0	0	0	0	0
Mean	4.37	3.67	2.75	2.52	2.55	3.47	3.37	3.54	3.38	2.93	3.64	2.34	4.96	3.36	2.89
Median	5.00	4.00	2.00	2.00	2.50	4.00	4.00	4.00	4.00	3.00	4.00	2.00	5.00	3.00	3.00
Mode	5	4	2	2	1	5	4	3	4	3	4	1	6	3	2
Std. deviation	2.152	1.374	1.389	1.216	1.295	1.446	1.121	1.196	1.108	0.947	1.070	1.368	1.364	1.265	1.226
Variance	4.632	1.888	1.929	1.480	1.676	2.090	1.257	1.430	1.229	0.896	1.144	1.871	1.860	1.600	1.503

Table 4 Multiple regression analysis results predicting additive manufacturing (AM) adoption

Coefficients[a]

Model		Unstandardized coefficients		Standardized coefficients		
		B	Std. error	Beta	*t*	Sig.
1	(Constant)	0.658	0.267		2.463	0.016
	Rate the level of adoption of these technologies in your organization	0.464	0.053	0.600	8.704	0.000
	Years of experience in manufacturing	0.047	0.057	0.051	0.818	0.416
	How well are digital manufacturing technologies integrated into your existing processes?	0.289	0.063	0.319	4.629	0.000

[a] Dependent variable: Rate the level of adoption of these technologies in your organization (1 = not adopted, 5 = Fully adopted) [• Additive Manufacturing (AM): 1, 2, 3, 4, 5]

are strong predictors (both at $p < 0.001$), but years of experience in manufacturing is not ($p = 0.416$). The level of adoption of other technologies has the strongest effect (standardized beta = 0.600) and the integration of technologies (standardized beta = 0.319). This implies that organizations that have adopted other forms of digital technologies and those that have well-integrated digital technology in their value chain are likely to have high levels of AM adoption. The constant term (0.658) is the baseline level of adoption when all the other explanatory variables are zero. The total model brings about the variance of AM adoption into consideration and shows that the adoption of digital technology in manufacturing is interrelated.

4.4 T-Test Analysis

The t-test results in Table 5 show that the expressed as effect size and all the tested variables exhibited strong effects in the one-sample t-tests. These coefficients of effect size range from 2 for Cohen's d. 071 to 3. In four studies, the effect size ranged from 1.45 to 1.48, which is indicative of large effect size. It is seen that the confidence intervals for the variables have overlapping, but the sample size for the variable "Years of Experience in Manufacturing" is relatively small and the confidence interval is the smallest, and it is (1.724, 2.414), which indicates that the estimate is more precise. The largest effect size is for "How well are digital manufacturing technologies integrated into your existing processes?" (Cohen's $d = 3.148$, CI: 2.671 to 3.621 which is quite higher than the test value. This correction yields slightly more conservative estimates but also demonstrates large effects for the Hedges'. Based on these findings, it can be concluded that the sample means for all the tested variables are significantly different from the hypothesized population means, especially for technology integration adoption levels. This suggests that the surveyed organizations are rather significantly more developed in digital manufacturing compared to hypothetical average or even "neutral" levels.

Table 5 One-sample T-test effect sizes for key digital manufacturing adoption metrics

One-sample effect sizes

		Standardizer[a]	Point estimate	95% confidence interval Lower	Upper
Years of experience in manufacturing	Cohen's d	1.216	2.071	1.724	2.414
	Hedges' correction	1.226	2.056	1.712	2.396
Rate the level of adoption of these technologies in your organization	Cohen's d	1.446	2.400	2.011	2.786
	Hedges' correction	1.457	2.382	1.996	2.764
Rate the level of adoption of these technologies in your organization	Cohen's d	1.121	3.009	2.543	3.472
	Hedges' correction	1.130	2.986	2.524	3.445
How well are digital manufacturing technologies integrated into your existing processes?	Cohen's d	1.245	2.733	2.308	3.155
	Hedges' correction	1.254	2.712	2.291	3.131
How well are digital manufacturing technologies integrated into your existing processes?	Cohen's d	1.106	3.148	2.671	3.621
	Hedges' correction	1.114	3.124	2.652	3.594
How have digital manufacturing technologies impacted productivity in your organization?	Cohen's d	1.265	2.659	2.243	3.071
	Hedges' correction	1.274	2.639	2.226	3.048

Cohen's d uses the sample standard deviation
Hedges' correction uses the sample standard deviation, plus a correction factor
[a] The denominator used in estimating the effect sizes

4.5 Conclusion

The descriptive data in this chapter provides a good overview of digital manufacturing adoption and its implications. Some locations have high-technology adoption and implementation, whereas others do not. The regression analysis shows that digital technology adoption is interdependent and that integrating technologies increases adoption. The t-test demonstrates that the assessed organizations are well developed in digital manufacturing and differ from the average hypothetical level in several ways. These data suggest that digital manufacturing adoption is generally positive, but there are some areas for improvement. This study can provide the empirical basis for investigating the current status of digital manufacturing and theoretical and practical guidance for improving manufacturing technology in the system. Further research should examine more particular elements related to digital manufacturing adoption and the long-term effects of DM on organisational performance and competitiveness.

5 Findings and Discussion

5.1 Overview of the Digital Manufacturing Adoption

The survey findings show that the status of digital manufacturing implementation is rather heterogeneous across industry sectors and geographic locations. The mean adoption levels of the different technologies were as follows 2.34–4.96, which still underscores a great deal of variability as to the implementation of this type of program. The level of adoption of Generative AI, Additive Manufacturing (AM), Industrial Internet of Things (IIoT), and Cyber-Physical Systems (CPS) differs in one way or the other depending on the acceptance levels.

This variability corresponds with the research of Tortorella and Fettermann [13] who pointed out that Industry 4.0. Technology adoption of manufacturing enterprises in Brazil is still at its infancy and has the tendency of differing in terms of the industry and size of the business. The mean scores of 4.37, 3.67, and 4.96 for the questions 2, 3, and 14 indicate that some aspects of digital manufacturing are gaining acceptance, which is in line with the transition towards Industry 5.0 according to [4]. The above observations are consistent with the study done by Leng et al.

5.2 Factors Influencing Adoption

This study shows that the use of one digital technology is a powerful predictor of the use of another digital technology (standardized beta = 0.600, $p < 0.001$). Such interconnectedness of digital technologies corresponds to the framework of a holistic digital environment in manufacturing that has been discussed by Radanliev et al. [9]. The correlation between technology integration and adoption level is highly significant ($r = 0.319$, $t = 14.426$, $p < 0.001$) This means that there is a need to adopt a comprehensive approach to the implementation of technology.

Surprisingly, manufacturing experience did not affect the technology adoption, despite the years of experience ($t = 0.416$). This finding refutes the hypothesis that experienced firms have more tendency to adopt new technologies and makes the authors to come up with some propositions that the adoption of digital manufacturing technologies may be influenced by strategic planning and competitive forces rather than the age of the firms.

5.3 Integration Challenges and Strategies

The survey reveals that the different DM technologies are adopted at a varying level of integration. The results indicated that the overall integration level was the highest with IIoT having a mean of 3.54 and CPS with a mean of 3.38. These findings reaffirm the work of Tao et al. [16], who emphasized on data-informed approaches to improve manufacturing productivity and quality.

However, it is clearly obvious that integration levels are possible to be optimized for all types of technologies. This corresponds with the issues stated by O'Reilly et al. [24] such as data security issues, the need to train the workers, and large capital expenditures. To overcome these challenges, manufacturers may require the use of strategies similar to those put forward by Agostini and Nosella [25] for SMEs, including the use of directed initiatives and policies that will support the adoption of digital technology.

5.4 Effect on Output and Creativity

The hypothesis test of *t*-test on the impacts of the digital manufacturing technologies on organization productivity reveals the following results. Cohen's values of the impact coefficients range between 2.071 and 3.148, which shows that these technologies impact the manufacturing processes and result in a significant extent. This is in line with Nikolakis et al. [17] who showed how through the use of digital twin technology in production, efficiency is increased, and product development is fast-tracked.

The strong effect size for technology integration (Cohen's $d = 3.148$, CI: Of these, the (3.671–3.621) is rather significant, showing that digital integration is where productivity is most affected. This goes to support the notion that it is not just about the implementation of single technologies but the integration of the technologies to form a solution within the manufacturing operations.

5.5 Role of Generative AI

Although the survey did not target the generative AI separately, it is as part of the AI technologies, and the results are encouraging. This is in line with the findings of Cioffi et al. [19], who observed that there was an increased use of AI in manufacturing in areas such as generative design, predictive maintenance, and smart supply chain management among others.

This might be due to the possibility of generative AI in transforming product design and manufacturing processes as explained by [20]. But looking at the differences in the level of integration of technologies, it can be assumed that generative AI is still in the process of continuous adoption indicating future prospects for the development of the sector.

5.6 Conclusion and Recommendation

Though the survey fails to pose questions related to sustainability, high adoption rates of IIoT and CPS have tremendous impacts to support a sustainable manufacturing environment. According to Nascimento et al. [7], Industry 4.0 technologies can enhance the use of resources and facilitate the concept of cradle-to-cradle manufacturing.

The integration levels of these technologies (means of 3.54 and 3.38 for IIoT and CPS, respectively) indicate that manufacturers are gradually into a place where they can use digital tools for sustainability. This is in line with the framework established by Liu et al., [23] where IoT, big data analytics, and AI enable product tracing, resource utilization, and circular economy business models.

5.7 Regional and Industry Variations

According to the results of the survey, significant differences in the levels of adoption and integration can be observed due to the regional and industry differences. This is in line with the observations made by Levin [14] in the World Economic Forum's Readiness for the Future of Production Index to the effect that leaders and laggards exist in regard to the readiness of manufacturing for digital transformation.

These variations call for best practices to address the question of how best to implement the digital manufacturing concerning elements of infrastructure, human capital, and industry type. According to Müller et al. [15], differences in Industry 4.0 adoption among European firms can be explained by factors such as firm size, absorptive capacity, and the regional innovation system. Their study offers a theoretical framework for understanding these disparities.

5.8 Future Trends and Directions

The high level of adoption of some of the technologies and the potency of the impact on productivity indicate that the trend towards manufacturing with the use of digital technologies is set to intensify. This is in line with Industry 5.0 vision outlined by Grosse et al. [11] in which human-oriented concepts work hand in hand with innovative digital solutions to foster the development of flexible and effective manufacturing environments.

The different adoption rates that are associated with the technologies also point out the areas of future growth. Technologies with relatively low current adoption rates may in fact be the next frontrunners of digitalization in manufacturing contexts. There is evidence that other digital manufacturing technologies such as generative AI are also likely to advance further in the future years, as was discussed by Huang et al. [21].

5.9 Conclusion

This research presents a clear and empirical analysis of the high level of acceptance and usage of digital manufacturing technologies in the various sectors. The result indicates that these technologies are interrelated and the need to adopt an integrated implementation approach. Despite differences in the degree of adoption and integration, the overall picture indicates that there is a progressive digitalization of manufacturing with significant consequences for efficiency, creativity, and environmental protection.

Subsequent studies should consider looking into cross-sectional studies to determine the changes in the digital manufacturing over time and the effects. Furthermore, deeper studies about the concrete problems and approaches concerning the application of novel technologies such as generative AI in the manufacturing domain are also desirable.

6 Conclusion and Recommendations

6.1 Summary of Key Findings

Quite substantively, this research has given an extensive analysis of the trend of digital manufacturing and its assimilation across companies and geographical locations. The research revealed that there existed a lot of variation in the adoption of technologies like generative AI, AM, IIoT, and CPS. The mean scores of the different adoption rates differed significantly, ranging from 2.34 to 4.96, which means that while some technologies are now pervasive, others are still in the early adopter phase.

Another important discovery was the interconnectedness of the technologies since the use of one technology meant that the other ones would also be adopted. This brings out the need to adopt an ecosystem perspective of digital manufacturing as opposed to a technology-centric approach. There was also variability in the integration levels of the technologies, whereby IIoT and CPS had higher integration levels as compared to the other technologies.

Interestingly, the study revealed that years of experience in manufacturing are not related with the likelihood of adoption of technology, as presumed that firms with higher experience in manufacturing are more likely to adopt new technologies. This has implications that digital transformation within manufacturing is more due to tactical choices and sector forces than the age and maturity of the organization.

6.2 Policy Recommendations

Based on our findings, we propose several policy recommendations to support the widespread adoption and effective implementation of digital manufacturing technologies. Based on our findings, we propose several policy recommendations to support the widespread adoption and effective implementation of digital manufacturing technologies.

Firstly, governments and industry bodies should establish special support schemes for companies at different levels of their digitalization. Due to the variations in adoption rates, there is little chance of success in using a standardized solution.

Secondly, it is necessary to initiate certain policies to address the major impediments to the adoption of IOT as highlighted in the study, including integration issues and the requirement of a skilled workforce. This could encompass grants for financial returns on the training of the workforce and funding for, for instance, 1:1 technology rollout.

Thirdly, there is a need to come up with regulatory structures that could tackle issues to do with data protection in digital manufacturing contexts. Lack of clarity of guidelines and standards in this area may be one of the reasons why it has not picked up as fast as the rest of the Internet.

Finally, the policies aimed at enhancing cooperation between academia, industries, and the government in the context of digital manufacturing might support the growth of innovation and guarantee that the identified research priorities correspond to the industries' needs.

6.3 Limitations

It is important to acknowledge the limitations of this research. While the study offers valuable findings, readers should consider several constraints. First, the data is cross-sectional in nature, which means that although correlations can be identified, causal relationships cannot be established. Second, the data is self-reported, which introduces the possibility of response bias.

Moreover, the sample size was large enough, but it cannot be concluded that it is completely indicative of all the sectors and regions of manufacturing. Further research with greater sample size and heterogeneous population can provide additional support to our observations.

References

1. Mourtzis, D. (2024). Industry 4.0 and smart manufacturing. In *Manufacturing from Industry 4.0 to Industry 5.0* (pp. 13–61). Elsevier.
2. ElMaraghy, H., Monostori, L., Schuh, G., & ElMaraghy, W. (2021). Evolution and future of manufacturing systems. *CIRP Annals, 70*(2), 635–658.
3. Gillani, F., Chatha, K. A., Jajja, M. S. S., & Farooq, S. (2020). Implementation of digital manufacturing technologies: Antecedents and consequences. *International Journal of Production Economics, 229*, 107748.
4. Leng, J., Sha, W., Wang, B., Zheng, P., Zhuang, C., Liu, Q., Wuest, T., Mourtzis, D., & Wang, L. (2022). Industry 5.0: Prospect and retrospect. *Journal of Manufacturing Systems, 65*, 279–295.
5. Salemink, K., Strijker, D., & Bosworth, G. (2017). Rural development in the digital age: A systematic literature review on unequal ICT availability, adoption, and use in rural areas. *Journal of Rural Studies, 54*, 360–371.
6. Lattanzi, L., Raffaeli, R., Peruzzini, M., & Pellicciari, M. (2021). Digital twin for smart manufacturing: A review of concepts towards a practical industrial implementation. *International Journal of Computer Integrated Manufacturing, 34*(6), 567–597.
7. Nascimento, D. L. M., Alencastro, V., Quelhas, O. L. G., Caiado, R. G. G., Garza-Reyes, J. A., Rocha-Lona, L., & Tortorella, G. (2019). Exploring Industry 4.0 technologies to enable circular economy practices in a manufacturing context: A business model proposal. *Journal of Manufacturing Technology Management, 30*(3), 607–627.
8. Ambrogio, G., Filice, L., Longo, F., & Padovano, A. (2022). Workforce and supply chain disruption as a digital and technological innovation opportunity for resilient manufacturing systems in the COVID-19 pandemic. *Computers & Industrial Engineering, 169*, 108158.
9. Radanliev, P., De Roure, D., Nicolescu, R., Huth, M., & Santos, O. (2022). Digital twins: Artificial intelligence and the IoT cyber-physical systems in Industry 4.0. *International Journal of Intelligent Robotics and Applications, 6*(1), 171–185.
10. Li, S., Wang, R., Zheng, P., & Wang, L. (2021). Towards proactive human–robot collaboration: A foreseeable cognitive manufacturing paradigm. *Journal of Manufacturing Systems, 60*, 547–552.
11. Grosse, E. H., Sgarbossa, F., Berlin, C., & Neumann, W. P. (2023). Human-centric production and logistics system design and management: Transitioning from Industry 4.0 to Industry 5.0. *International Journal of Production Research, 61*(22), 7749–7759.
12. Arumugam, A., Bhaumik, A., & Khazaei, H. (2023). A conceptual framework for studying factors that influence digital technology adoption among manufacturing sector: Leadership effectiveness as a mediating variable. *International Journal of Professional Business Review, 8*(7), 11.
13. Tortorella, G. L., & Fettermann, D. (2018). Implementation of Industry 4.0 and lean production in Brazilian manufacturing companies. *International Journal of Production Research, 56*(8), 2975–2987.
14. Levin, S. (2018). World economic forum and the fourth industrial revolution in South Africa. In *Trade & Industrial Policy Strategy*. TIPS.
15. Müller, J. M., Buliga, O., & Voigt, K. I. (2021). The role of absorptive capacity and innovation strategy in the design of industry 4.0 business Models-A comparison between SMEs and large enterprises. *European Management Journal, 39*(3), 333–343.
16. Tao, F., Qi, Q., Liu, A., & Kusiak, A. (2018). Data-driven smart manufacturing. *Journal of Manufacturing Systems, 48*, 157–169.
17. Nikolakis, N., Alexopoulos, K., Xanthakis, E., & Chryssolouris, G. (2019). The digital twin implementation for linking the virtual representation of human-based production tasks to their physical counterpart in the factory-floor. *International Journal of Computer Integrated Manufacturing, 32*(1), 1–12.

18. Li, C., Chen, Y., & Shang, Y. (2022). A review of industrial big data for decision making in intelligent manufacturing. *Engineering Science and Technology, an International Journal, 29*, 101021.
19. Cioffi, R., Travaglioni, M., Piscitelli, G., Petrillo, A., & De Felice, F. (2020). Artificial intelligence and machine learning applications in smart production: Progress, trends, and directions. *Sustainability, 12*(2), 492.
20. Bendoly, E., Chandrasekaran, A., Lima, M. D. R. F., Handfield, R., Khajavi, S. H., & Roscoe, S. (2023). *The role of generative design and additive manufacturing capabilities in developing human-AI symbiosis: Evidence from multiple case studies*. Decision Sciences.
21. Huang, Z., Shen, Y., Li, J., Fey, M., & Brecher, C. (2021). A survey on AI-driven digital twins in industry 4.0: Smart manufacturing and advanced robotics. *Sensors, 21*(19), 6340.
22. Sartal, A., Bellas, R., Mejías, A. M., & García-Collado, A. (2020). The sustainable manufacturing concept, evolution and opportunities within Industry 4.0: A literature review. *Advances in Mechanical Engineering, 12*(5), 1687814020925232.
23. Liu, Q., Trevisan, A. H., Yang, M., & Mascarenhas, J. (2022). A framework of digital technologies for the circular economy: Digital functions and mechanisms. *Business Strategy and the Environment, 31*(5), 2171–2192.
24. O'Reilly, J., Ranft, F., Neufeind, M., Gregory, S.S., Zierahn, U., Went, R., Syrdal, S., & Holts, K., (2018). Work in the digital age: Challenges of the fourth industrial revolution.
25. Agostini, L., & Nosella, A. (2020). The adoption of Industry 4.0 technologies in SMEs: Results of an international study. *Management Decision, 58*(4), 625–643.
26. Tiwari, S. (2021). Supply chain integration and Industry 4.0: A systematic literature review. *Benchmarking: An International Journal, 28*(3), 990–1030.
27. Liao, Y., Deschamps, F., Loures, E. D. F. R., & Ramos, L. F. P. (2017). Past, present and future of Industry 4.0-a systematic literature review and research agenda proposal. *International Journal of Production Research, 55*(12), 3609–3629.
28. Ben Youssef, A., & Mejri, I. (2023). Linking digital technologies to sustainability through industry 5.0: A bibliometric analysis. *Sustainability, 15*(9), 7465.
29. Dalenogare, L. S., Benitez, G. B., Ayala, N. F., & Frank, A. G. (2018). The expected contribution of Industry 4.0 technologies for industrial performance. *International Journal of Production Economics, 204*, 383–394.
30. Bai, C., Dallasega, P., Orzes, G., & Sarkis, J. (2020). Industry 4.0 technologies assessment: A sustainability perspective. *International Journal of Production Economics, 229*, 107776.
31. Rashid, Y., Rashid, A., Warraich, M. A., Sabir, S. S., & Waseem, A. (2019). Case study method: A step-by-step guide for business researchers. *International Journal of Qualitative Methods, 18*, 1609406919862424.
32. Hair, J. F., Black, W. C., Babin, B. J., & Anderson, R. E. *et al.* Multivariate Data Analysis (8th ed.), 1473756545, 9781473756540 (2019).
33. Deloitte, Global manufacturing competitiveness index: report, https://www2.deloitte.com/global/en/pages/manufacturing/articles/global-manufacturing-competitiveness-index.html (2016).
34. World Bank Group, World Development Report 2016: Digital Dividends, https://openknowledge.worldbank.org/entities/publication/1683bbba-c275-580a-a2f5-c74f805b3679 (2016).
35. Dutta, S., Lanvin, B., León, R., & Wunsch-Vincent, S. Global Innovation Index 2024: Unlocking the Promise of Social Entrepreneurship., https://doi.org/2788-6972 (2024).
36. World Economic Forum, Readiness for the Future of Production Report 2018, https://www.weforum.org/publications/readiness-for-the-future-of-production-report-2018/ (2018).
37. UNIDO, Competitive industrial performance report 2020, https://stat.unido.org/sites/default/files/2023-01/CIP-2020-full.pdf (2020).

Technology Shift from Traditional Subtractive Manufacturing to Additive Manufacturing in Digital Manufacturing Era

P. P. Praveen Kumar, Rakesh Joshi, Pias Biswas, Yatika Gori, Ashwani Kumar, and Yogesh Kumar Singla

1 Introduction

Traditional manufacturing techniques encompass a range of processes that shape materials by removing, forming, or joining parts to create final products. Common methods include subtractive manufacturing (milling, turning, and drilling) that

P. P. P. Kumar
Department of Biomedical Engineering, Institute for Quantitative Health Science and Engineering, Michigan State University, East Lansing, MI, USA
e-mail: panangal@msu.edu

R. Joshi (✉)
DARVUN India, Patiala, Punjab, India

P. Biswas
Department of Mechanical Engineering, University of Idaho, Moscow, ID, USA
e-mail: pbiswas@navajotech.edu

Y. Gori
Department of Mechanical Engineering, Graphic Era Deemed to be University, Dehradun, Uttarakhand, India

School of Computing, Graphic Era Hill University, Haldwani, Uttarakhand, India
e-mail: yatikagori@gehu.ac.in

A. Kumar
Department of Mechanical Engineering, Technical Education Department Uttar Pradesh, Kanpur, India
e-mail: ashwaniadj.me@geu.ac.in

Y. Kumar Singla
Department of Mechanical Engineering, University of Idaho, Moscow, ID, USA

School of Engineering, Math and Technology, Navajo Technical University, Crownpoint, NM, USA
e-mail: ysingla@navajotech.edu

Y. Kumar Singla et al. (eds.), *Digital Manufacturing in Industry 5.0*,
https://doi.org/10.1007/978-3-031-91500-0_4

removes material from a solid block to create a desired shape [1]. Other traditional techniques involve casting, where molten material is poured into molds, and forging, which uses heat and pressure to shape metal. These methods have been the backbone of industrial production for decades, valued for their precision, strength, and ability to produce large quantities, though they often come with higher material waste and setup costs compared to modern additive techniques. Subtractive methods have limitations regarding material efficiency, processing speed, ability to create complex designs, and cost-effectiveness [2]. Additive manufacturing (AM), or rapid prototyping (RP), offers a solution to the limitations of subtractive technologies. By constructing objects layer-by-layer from digital designs, AM avoids material waste and allows for more complex geometries [3]. There are various additive manufacturing techniques, including stereolithography (SLA), digital light processing (DLP), selective laser sintering (SLS), electron beam melting (EBM), fused deposition modeling (FDM), Multijet/Polyjet 3D printing, selective laser melting (SLM), and laminated object manufacturing (LOM). These methods use diverse materials such as ceramics, plastics, metals, liquids, powders, and even living cells [4]. They enable the fabrication of complex three-dimensional (3D) components that are impossible to produce with traditional manufacturing techniques. The process typically involves converting 3D model data into a standard triangulated language (STL) format, making it easier to print (Fig. 1). Figure 1 represents various steps involved in a 3D printing technique. All of these processes rely on a common fundamental principle, in which the desired part is created through the layer-by-layer deposition of materials.

AM is proving its versatility in industries that require complex shapes, customization, and rapid prototyping, from automotive to aerospace [5]. Furthermore, this technological advancement explores the typical human–machine collaborative additive manufacturing applications across product, economic, and ecological dimensions. These applications include personalized product design, interactive

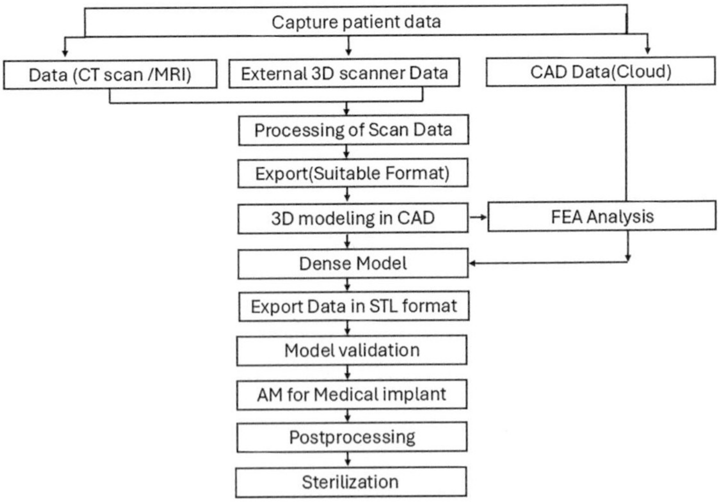

Fig. 1 A general outline of the steps involved in 3D printing in the biomedical field

manufacturing, human–machine interaction (HMI) technology for process chains, distributed manufacturing, collaborative design, and energy conservation and emission reduction initiatives.

2 History and Technological Shift in Additive Manufacturing

AM, commonly referred to as 3D printing, is a technology that builds objects layer by layer, using digital models to create 3D objects. It has roots in the 1980s, but its development spans over several decades, marked by continuous innovation across multiple industries. Here's a chronological overview of the history of additive manufacturing [6].

2.1 Early Concepts and Theoretical Foundations (Pre-1980s)

Before the technology became feasible, the concept of additive manufacturing was discussed in theoretical frameworks. Early patents and scientific papers from the 1940s to 1960s explored the idea of creating objects layer by layer. However, the necessary technologies, like computer-aided design (CAD) and precise material control, were not yet available.

- **1945:** American engineer Murray Leinster wrote a science fiction story where he described machines that could build objects layer by layer, showcasing the early conceptualization of AM [7].

2.2 The Birth of 3D Printing (1980s)

The 1980s saw the advent of the first true additive manufacturing technologies, driven by advances in computing, CAD, and materials science. This decade marked the beginning of prototyping applications [8].

- **1981:** Japanese inventor Hideo Kodama of Nagoya Municipal Industrial Research Institute developed a layer-by-layer manufacturing approach for rapid prototyping. His method used photopolymers that could be hardened using ultraviolet light.
- **1984:** Charles "Chuck" Hull, working at 3D Systems, invented *stereolithography (SLA)*, a method that uses ultraviolet lasers to solidify photopolymer resin. Hull's invention led to the filing of the first patent for stereolithography and is widely credited with laying the foundation of modern 3D printing. Hull also developed the. STL file format, which is still a standard in 3D printing today.
- **1986:** Hull officially patented stereolithography, launching the first commercial 3D printer, the SLA-1, in 1987.

2.3 Growth and Commercialization (1990s)

In the 1990s, the additive manufacturing industry expanded, and various new techniques were developed. During this period, AM was mostly used for *rapid prototyping* in industries such as automotive and aerospace, helping companies develop products faster and more cost effectively [9, 10].

- **1991:** Stratasys, founded by Scott Crump, introduced the first *fused deposition modeling (FDM)* machine. FDM is a technique where thermoplastic material is extruded layer by layer to form an object. Today, FDM is one of the most popular and widely used forms of 3D printing, particularly for desktop printers.
- **1992:** The first selective laser sintering (SLS) machine was developed by Carl Deckard and Joseph Beaman at the University of Texas. SLS uses a laser to fuse powdered materials into a solid structure.
- **1995:** Z Corporation (later acquired by 3D Systems) commercialized the *binder jetting* process, allowing for full-color 3D printing by binding powder materials layer by layer using a liquid binder.
- **Late 1990s:** Metal 3D printing technologies such as direct metal laser sintering (DMLS) was introduced. These were pivotal for industries like aerospace, where high-performance materials such as titanium alloys were needed [11].

2.4 Expansion into Functional Parts and New Materials (2000s)

The early 2000s marked a transition from rapid prototyping to the production of end-use parts, especially for niche markets like aerospace, medical implants, and automotive. Additive manufacturing began gaining recognition for its potential to produce complex and customized parts that traditional manufacturing techniques struggled with [12].

- **2000**: A landmark in medical 3D printing occurred when a working kidney was 3D printed using cells. While still far from fully functional organ printing, this milestone demonstrated the potential for bioprinting.
- **2005**: The RepRap Project (Replicating Rapid Prototyper) was launched by Dr. Adrian Bowyer at the University of Bath. This open-source initiative aimed to create a self-replicating 3D printer. The RepRap project spurred the development of low-cost 3D printers, fostering the growth of a maker community and popularizing desktop 3D printing.
- **2006**: The first production-ready 3D printers were released for metal parts, with technologies such as EBM emerging as key players in producing complex metal components for aerospace and medical applications [13].

2.5 *Mainstream Adoption and Innovation (2010s–Present)*

The 2010s saw additive manufacturing technologies move from niche applications to widespread industrial use. At the same time, desktop 3D printers became increasingly affordable for hobbyists, schools, and small businesses. The range of printable materials also expanded to include ceramics, composites, and even human tissue in experimental settings.

- **2013**: The term "additive manufacturing" became widely adopted, distinguishing it from traditional "subtractive" manufacturing processes, where the material is removed from a solid block.
- **2014**: A breakthrough in the aerospace industry came when GE Aviation became the first company to use 3D-printed parts in jet engines. GE's fuel nozzles, manufactured via metal AM processes, showcased the technology's ability to produce lightweight, complex components for critical applications.
- **2015**: Carbon, a company specializing in polymer AM, introduced its innovative CLIP (Continuous Liquid Interface Production) technology, allowing for faster and higher-quality production of polymer parts.
- **2020s**: Additive manufacturing has seen further integration into manufacturing, healthcare, and consumer products. Companies have been using 3D printing for on-demand production, distributed manufacturing, and even bioprinting, with the goal of creating functional organs. The technology has played a role in producing critical supplies during crises, such as the production of Personal Protective Equipment (PPE) during the COVID-19 pandemic.

3 Technological Shift to Additive Manufacturing in Digital Manufacturing

Additive Manufacturing (AM) revolutionizes the manufacturing process by enabling the direct creation of physical objects from digital designs. AM in industry allows design customization (allowing for complex and intricate design), reduced lead time (reduction of prototype and production cycle significantly), less tooling (reduction of expensive tooling), and on-demand creation (reduction of inventory cost and waste). The structure of additive manufacturing in Industry 5.0 is shown in Fig. 2.

The digital nature of AM enables decentralized manufacturing, where designers and 3D printing companies can collaborate remotely by sharing designs online. This fosters innovation and flexibility. Additionally, digital twins (DTs) are crucial in optimizing the AM process. DTs are virtual representations of physical objects or systems, allowing for simulations and process parameter optimization before production. This reduces the risk of errors and improves overall efficiency.

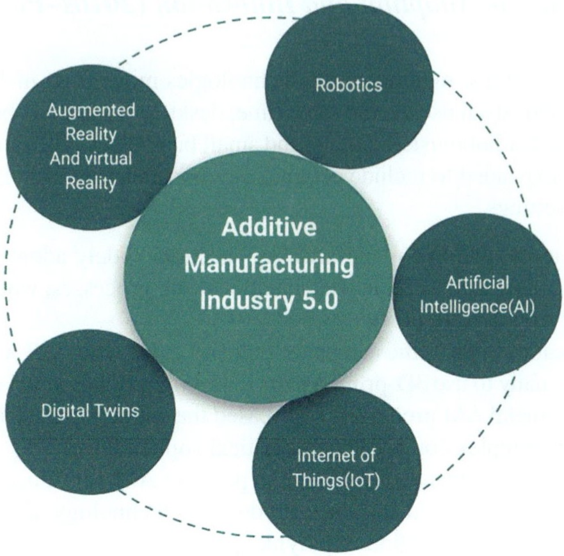

Fig. 2 Additive manufacturing in Industry 5.0

The Internet of Things (IoT) empowers objects and devices in physical spaces to connect and exchange data, fostering interconnections between people, objects, data, and processes. By leveraging IoT data, users can access personalized products and experiences, leading to enhanced satisfaction. The IoT's influence extends to interactions between humans and machines, humans and other humans, and humans and the environment, facilitating the utilization of intelligent additive manufacturing products [14, 15].

The IoT has the potential to revolutionize manufacturing processes by eliminating information barriers, reducing operational costs, and improving overall efficiency. Additive manufacturing systems benefit from IoT technology by utilizing standard communication protocols for seamless data exchange between human and machine systems. This ensures efficient and secure interactions.

During the product design phase, the IoT can analyze vast amounts of Internet data to incorporate user preferences into the conceptual design, leading to more personalized and customer-centric products. In the manufacturing stage, IoT-enabled monitoring systems can track and analyze the production process in real-time, allowing for timely adjustments and optimizations. This proactive approach can significantly enhance quality control and reduce production time.

Augmented reality (AR) is a revolutionary technology that, when integrated into Industry 5.0, can significantly enhance various aspects of manufacturing. By overlaying digital information onto the physical world, AR enables real-time expert guidance, immersive training experiences, predictive maintenance, real-time quality checks, and efficient defect analysis, ultimately optimizing industrial processes. Integrating virtual reality (VR) into Industry 5.0 revolutionizes various processes.

By providing realistic simulations, VR enables hands-on training without physical risks. It also facilitates early stage prototyping, allowing for efficient design iterations. Additionally, VR can simulate hazardous scenarios, enhancing safety awareness and response training. Ultimately, these applications of VR contribute to improved productivity and efficiency in industrial settings.

4 Role of Additive Manufacturing in the Biomedical Field

In the realm of biomedical engineering, the individuality of each patient opens up vast possibilities for the use of AM in developing personalized medical solutions. Common applications in this domain include tailored implants, models for surgical planning, and patient-specific guides [16]. The dental sector benefits from AM as well, with applications ranging from orthodontic devices to drill guides and dental models [17]. Furthermore, ongoing investigations are delving into how AM can be harnessed for the creation of artificial organs and tissues. As depicted in Fig. 1, the process of producing personalized biomedical devices using AM begins with the precise capture of the patient's anatomical details [18]. This sophisticated procedure integrates reverse engineering with AM methods, allowing for the direct fabrication of components from digitally constructed 3D models. The entire workflow can be categorized into three key stages: the first stage captures the anatomical data through reverse engineering; the second involves sophisticated design processes using CAD software; and finally, finite element analysis (FEA) is performed to ensure that the resulting designs meet structural integrity requirements [19, 20].

Various applications of AM in the biomedical field are shown in Fig. 3. In recent years, various researchers have examined alternative techniques in healthcare

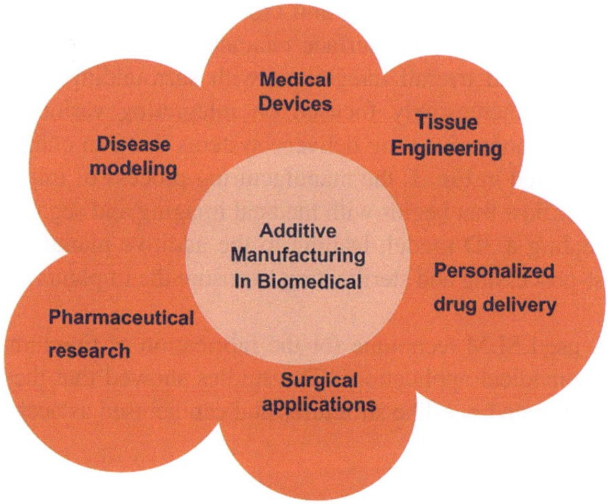

Fig. 3 Application of additive manufacturing in biomedical application

innovation, revealing notable shifts in clinical practices. For instance, approaches in telemedicine have transformed patient-provider interactions, enabling remote consultations that significantly enhance accessibility for individuals in rural areas. Studies have highlighted that these innovations not only improve patient satisfaction but also reduce the wait times for medical attention. Furthermore, the integration of artificial intelligence in diagnostic procedures shows promise in increasing the accuracy of readings, ultimately contributing to better clinical outcomes. Future investigations into these emergent technologies may lead to even more effective strategies in patient care, bridging gaps that currently exist in healthcare delivery systems.

Ballard et al. investigated the utilization of 3D printing technology in medicine, finding that it significantly lowers operating room costs and shortens procedure durations, particularly within orthopedic and maxillofacial surgeries [21]. Furthermore, the incorporation of 3D printed components in patient care provides substantial advantages to healthcare systems. In a related study, Mahmoud et al. focused on creating human anatomical pathology specimens using 3D printing methods and concluded that advancements in this technology are poised to improve the quality of medical specimens soon [22]. Tack et al. reviewed how 3D printing is being integrated into surgical practices and academic research [23]. Their findings highlighted numerous benefits, including enhanced medical outcomes, reduced surgical times, and decreased radiation exposure, while they advocated for additional research into the application of 3D printing across all patient demographics [23]. Additionally, Ballard et al. examined the use of antibiotics in 3D printed implants and instruments, noting that antibiotic-infused 3D printed devices offer advantages such as increased surface area for drug distribution and the capability for rapid production of patient-specific medical implants [24].

Implants are fabricated through AM, either directly or indirectly, to effectively replace damaged or misplaced tissues. They have a wide range of applications, including dental uses such as bridges and crowns. A critical consideration in the design of these implants is their surface characteristics, which can significantly affect cell adhesion and overall integration with surrounding biological tissues. Current research is increasingly focused on integrating various materials into implants, including innovative drug delivery systems that can enhance therapeutic outcomes. As depicted in Fig. 1, the manufacturing process of implants involves a characteristic workflow that begins with medical imaging and segmentation. This is followed by creating a 3D model, leading to the additive manufacturing process, and finally, post-processing and sterilization to ensure the implants are safe for clinical use [25].

Zhang et al. used SLM technique for the fabrication of titanium-based microstructures for biomedical applications. The studies showed that these microstructures can mimic human bone-like structures and can be used as acetabular cups for hips [26, 27].

PolyJet printers are highly regarded for their ability to create exceptionally detailed components in a significantly shorter time frame. This technology is particularly well suited for the production of accurate and realistic prototypes. The printers utilized in this process have a remarkable capability to jet multiple materials simultaneously, which substantially reduces production time. The dental and oral industries are among the primary sectors that benefit from PolyJet 3D printers. Key advantages of this technique include the ability to fabricate prototypes at lower costs, the simultaneous use of various materials in the printing process, a time-efficient approach, and the production of thin layers that enhance the features, intricate geometries, and finish of the prototypes. Salmi showed that by using the binding jet printing technique, various organoids can be prepared with accuracy [28].

5 Current Trends and Innovations

The present trends include the adoption of sophisticated 3D printing techniques, such as bioprinting and multi-material printing, which facilitate the construction of more complex structures that better mimic human tissue. As AM technologies continue to evolve, innovations like four-dimensional (4D) printing, which integrates time as a variable in the design, will open new avenues for functional medical applications, such as dynamically responsive implants. Additionally, the integration of artificial intelligence and machine learning with AM processes is expected to optimize design, production, and patient-specific solutions further.

6 Social Implications of the 3D Printing Revolution in Digital Manufacturing

The 3D printing revolution, AM, has the potential to impact society in both positive and negative ways significantly. Customization, waste reduction, growth of localized manufacturing, innovation, and creativity are the positive aspects of 3D printing. Acceptance and trust are essential to fully realizing the benefits of emerging technologies. Aligning technology with human needs and providing adequate training are critical factors in successful adoption [29, 30]. However, challenges such as security vulnerabilities, privacy concerns, a shortage of skilled workers, time-consuming processes, and significant budgetary requirements hinder widespread implementation. The adoption of Industry 5.0, which emphasizes collaboration between humans and intelligent machines, is necessary to ensure compliance with industrial laws and regulations. Future developments in cognitive computing, human–machine interaction, and quantum computing offer promising avenues for advancing Industry 5.0 [31–33].

7 Future Outlook

Looking ahead, AM is anticipated to revolutionize personalized medicine production significantly. Researchers are exploring the potential for creating organs and tissues that could eventually address the shortage of organ transplants. Moreover, regulatory pathways and partnerships between academia and industry will be vital in accelerating the clinical adoption of these technologies.

In summary, as the capabilities of additive manufacturing expand, its influence in the biomedical field will likely increase, leading to enhanced patient outcomes, reduced healthcare costs, and more innovative medical solutions. Ongoing research and development are vital to overcoming existing challenges, ensuring the technology meets safety standards, and aligning manufacturing processes with clinical needs [34–36].

The development of more robust, more durable, biocompatible, and lighter 3D materials with sustainability will be the future of additive manufacturing. Also, developing eco-friendly materials and recycling processes would be an additional challenge for researchers.

8 Conclusion

This chapter explores the convergence of two transformative technological shifts: Industry 5.0 and 3D printing. We delve into the definitions and applications of Industry 5.0, which aims to optimize human–machine collaboration. By examining enabling technologies and addressing potential challenges, we highlight the future direction of this innovative paradigm.

Simultaneously, we explore the transition from traditional subtractive manufacturing to additive manufacturing and 3D printing. We delve into the impact of this shift on various industries, including aerospace, healthcare, automotive, and consumer goods. By examining subtractive processes and advancements in 3D printing materials and technologies, we highlight the potential of IoT integration to revolutionize manufacturing. Finally, we discuss the economic and social implications of this digital manufacturing revolution, addressing the challenges and opportunities that lie ahead.

References

1. Zhu, Z., Dhokia, V. G., Nassehi, A., & Newman, S. T. (2013). A review of hybrid manufacturing processes—State of the art and future perspectives. *International Journal of Computer Integrated Manufacturing, 26*, 596–615. https://doi.org/10.1080/0951192X.2012.749530
2. Wang, B., Tao, F., Fang, X., Liu, C., Liu, Y., & Freiheit, T. (2021). Smart manufacturing and intelligent manufacturing: A comparative review. *Engineering, 7*, 738–757. https://doi.org/10.1016/j.eng.2020.07.017

3. Eyers, D. R., & Potter, A. T. (2017). Industrial additive manufacturing: A manufacturing systems perspective. *Computers in Industry, 92–93*, 208–218. https://doi.org/10.1016/j.compind.2017.08.002
4. Liu, Z., Wang, Y., Wu, B., Cui, C., Guo, Y., & Yan, C. (2019). A critical review of fused deposition modeling 3D printing technology in manufacturing polylactic acid parts. *International Journal of Advanced Manufacturing Technology, 102*, 2877–2889. https://doi.org/10.1007/s00170-019-03332-x
5. Rouf, S., Malik, A., Singh, N., Raina, A., Naveed, N., Siddiqui, M. I. H., & Haq, M. I. U. (2022). Additive manufacturing technologies: Industrial and medical applications. *Sustainable Operations and Computers, 3*, 258–274. https://doi.org/10.1016/j.susoc.2022.05.001
6. Zhai, X., Jin, L., & Jiang, J. (2022). A survey of additive manufacturing reviews. *MSAM, 1*, 21. https://doi.org/10.18063/msam.v1i4.21
7. Wohlers, T., Gornet, T., Mostow, N., Campbell, I., Diegel, O., Kowen, J., Huff, R., Stucker, B., Fidan, I., Doukas, A., Drab, B., Drstvenšek, I., Eitsert, N., Espalin, D., Feldhausen, T., Ghany, K., Gillett-Crooks, M., Guo, D., Held, A., Hargovan, S., Holmegaard, L., Jeng, J.-Y., Jurrens, K., Keating, B., Ksy, A., Kircher, R., De Beer, D., Brandt, M., Vasco, J., Khorasani, M., Overy, C., Park, K., Chen, S., Pearce, N., & Peels, J. (2023). History of additive manufacturing. *SSRN Journal*. https://doi.org/10.2139/ssrn.4474824
8. Shahrubudin, N., Lee, T. C., & Ramlan, R. (2019). An overview on 3D printing technology: Technological, materials, and applications. *Procedia Manufacturing, 35*, 1286–1296. https://doi.org/10.1016/j.promfg.2019.06.089
9. Fang, Y., Guo, Y., Liu, T., Xu, R., Mao, S., Mo, X., Zhang, T., Ouyang, L., Xiong, Z., & Sun, W. (2022). Advances in 3D bioprinting. *Chinese Journal of Mechanical Engineering: Additive Manufacturing Frontiers, 1*, 100011. https://doi.org/10.1016/j.cjmeam.2022.100011
10. Mendoza-Cerezo, L. M., Jesús, M. R., Macías-García, A., Marcos-Romero, A. C., & Díaz-Parralejo, A. (2023). Evolution of bioprinting and current applications. *IJB, 9*, 742. https://doi.org/10.18063/ijb.742
11. Zhu, C., Liu, T., Qian, F., Chen, W., Chandrasekaran, S., Yao, B., Song, Y., Duoss, E. B., Kuntz, J. D., Spadaccini, C. M., Worsley, M. A., & Li, Y. (2017). 3D printed functional nanomaterials for electrochemical energy storage. *Nano Today, 15*, 107–120. https://doi.org/10.1016/j.nantod.2017.06.007
12. Gross, B. C., Erkal, J. L., Lockwood, S. Y., Chen, C., & Spence, D. M. (2014). Evaluation of 3D printing and its potential impact on biotechnology and the chemical sciences. *Analytical Chemistry, 86*, 3240–3253. https://doi.org/10.1021/ac403397r
13. Ralls, A. M., Kumar, P., & Menezes, P. L. (2020). Tribological properties of additive manufactured materials for energy applications: A review. *Processes, 9*, 31. https://doi.org/10.3390/pr9010031
14. Cheng, B., Wang, M., Zhao, S., Zhai, Z., Zhu, D., & Chen, J. (2017). Situation-aware dynamic service coordination in an IoT environment. *IEEE/ACM Transactions on Networking, 25*, 2082–2095. https://doi.org/10.1109/TNET.2017.2705239
15. Jiao, R., Commuri, S., Panchal, J., Milisavljevic-Syed, J., Allen, J. K., Mistree, F., & Schaefer, D. (2021). Design engineering in the age of industry 4.0. *Journal of Mechanical Design, 143*, 070801. https://doi.org/10.1115/1.4051041
16. Pettersson, A. B. V., Salmi, M., Vallittu, P., Serlo, W., Tuomi, J., & Mäkitie, A. A. (2020). Main clinical use of additive manufacturing (three-dimensional printing) in Finland restricted to the head and neck area in 2016–2017. *Scandinavian Journal of Surgery, 109*, 166–173. https://doi.org/10.1177/1457496919840958
17. Zadpoor, A. A., & Malda, J. (2017). Additive manufacturing of biomaterials, tissues, and organs. *Annals of Biomedical Engineering, 45*, 1–11. https://doi.org/10.1007/s10439-016-1719-y
18. Calignano, F., Galati, M., Iuliano, L., & Minetola, P. (2019). Design of additively manufactured structures for biomedical applications: A review of the additive manufacturing processes applied to the biomedical sector. *Journal of Healthcare Engineering, 2019*, 1–6. https://doi.org/10.1155/2019/9748212

19. Harding, A., Pramanik, A., Basak, A. K., Prakash, C., & Shankar, S. (2023). Application of additive manufacturing in the biomedical field—A review. *Annals of 3D Printed Medicine, 10*, 100110. https://doi.org/10.1016/j.stlm.2023.100110
20. Singh, S., & Ramakrishna, S. (2017). Biomedical applications of additive manufacturing: Present and future. *Current Opinion in Biomedical Engineering, 2*, 105–115. https://doi.org/10.1016/j.cobme.2017.05.006
21. Ballard, D. H., Mills, P., Duszak, R., Weisman, J. A., Rybicki, F. J., & Woodard, P. K. (2020). Medical 3D printing cost-savings in orthopedic and maxillofacial surgery: Cost analysis of operating room time saved with 3D printed anatomic models and surgical guides. *Academic Radiology, 27*, 1103–1113. https://doi.org/10.1016/j.acra.2019.08.011
22. Mahmoud, A., & Bennett, M. (2015). Introducing 3-dimensional printing of a human anatomic pathology specimen: Potential benefits for undergraduate and postgraduate education and anatomic pathology practice. *Archives of Pathology & Laboratory Medicine, 139*, 1048–1051. https://doi.org/10.5858/arpa.2014-0408-OA
23. Tack, P., Victor, J., Gemmel, P., & Annemans, L. (2016). 3D-printing techniques in a medical setting: A systematic literature review. *Biomedical Engineering Online, 15*, 115. https://doi.org/10.1186/s12938-016-0236-4
24. Ballard, D. H., Tappa, K., Boyer, C. J., Jammalamadaka, U., Hemmanur, K., Weisman, J. A., Alexander, J. S., Mills, D. K., & Woodard, P. K. (2019). Antibiotics in 3D-printed implants, instruments and materials: Benefits, Challengesâ and future directions. *Journal of 3D Printing in Medicine, 3*, 83–93. https://doi.org/10.2217/3dp-2019-0007
25. Salmi, M. (2021). Additive manufacturing processes in medical applications. *Materials, 14*, 191. https://doi.org/10.3390/ma14010191
26. Osakada, K., & Shiomi, M. (2006). Flexible manufacturing of metallic products by selective laser melting of powder. *International Journal of Machine Tools and Manufacture, 46*, 1188–1193. https://doi.org/10.1016/j.ijmachtools.2006.01.024
27. Zhang, L. C., Klemm, D., Eckert, J., Hao, Y. L., & Sercombe, T. B. (2011). Manufacture by selective laser melting and mechanical behavior of a biomedical Ti–24Nb–4Zr–8Sn alloy. *Scripta Materialia, 65*, 21–24. https://doi.org/10.1016/j.scriptamat.2011.03.024
28. Salmi, M. (2016). Possibilities of preoperative medical models made by 3D printing or additive manufacturing. *Journal of Medical Engineering, 2016*, 1–6. https://doi.org/10.1155/2016/6191526
29. Feng, H., Wang, X., Duan, Y., Zhang, J., & Zhang, X. (2020). Applying blockchain technology to improve agri-food traceability: A review of development methods, benefits and challenges. *Journal of Cleaner Production, 260*, 121031. https://doi.org/10.1016/j.jclepro.2020.121031
30. Hakak, S., Khan, W. Z., Gilkar, G. A., Imran, M., & Guizani, N. (2020). Securing smart cities through blockchain technology: Architecture, requirements, and challenges. *IEEE Network, 34*, 8–14. https://doi.org/10.1109/MNET.001.1900178
31. Kumar, A., Singla, Y., & Namboodri, T. (2024). Chapter 01. Globalization and international issues in sustainable manufacturing. In S. Shah, H. Nautiyal, G. Gugliani, A. Kumar, T. Namboodri, & Y. K. Singla (Eds.), *Sustainability in smart manufacturing: Trends, scope and challenges* (pp. 01–18). CRC Press. https://doi.org/10.1201/9781003467496-1
32. Heidari, M., Khashehchi, M., Thangavel, S., Rahmanivahid, P., Kumar, A., & Singla, Y. K. (2024). Chapter 18. Laminated structures and fracture mechanics: A comprehensive study of mode 1, mode II, and mixed mode III behavior. In A. Kumar, Y. Kumar Singla, & M. R. Maughan (Eds.), *Fracture behavior of nanocomposites and reinforced laminate structures* (pp. 441–454). Springer. https://doi.org/10.1007/978-3-031-68694-8_18
33. Biswas, P. K., Maughan, M. R., Kumar, A., & Singla, Y. K. (2024). Chapter 12. Analyzing fractures in nanomaterial-enhanced carbon fiber-reinforced polymer (CFRP) composites. In A. Kumar, Y. Kumar Singla, & M. R. Maughan (Eds.), *Fracture behavior of nanocomposites and reinforced laminate structures* (pp. 295–327). Springer. https://doi.org/10.1007/978-3-031-68694-8_12

34. Kumar, A., Gori, Y., Dutt, N., Singla, Y. K., & Maurya, A. (2021). *Advanced computational methods in mechanical and materials engineering*. CRC Press. https://doi.org/10.1201/9781003202233. ISBN: 9781032052915.
35. Kumar, A., Ram, M., & Singla, Y. K. (2022). *Advanced material for biomechanical applications*. https://doi.org/10.1201/9781003286806. ISBN: 9781032054490.
36. Shah, S., Nautiyal, H., Gugliani, G., Kumar, A., Namboodri, T., & Singla Editors, Y. K. (2024). *Sustainability in smart manufacturing: Trends, scope, and challenges*. CRC Press. https://doi.org/10.1201/9781003467496. ISBN 9781032740713.

Additive Manufacturing in Industry 5.0: Techniques, Applications, and Future Trends

Padmakumar Muthuswamy and Mogana Priya Chinnasamy

1 Introduction

"Additive manufacturing," a term synonymous with three-dimensional (3D) printing, has gained the attention of industrial and academic researchers alike over the past decade, although the technology has been around for more than four decades. The documented origin of additive manufacturing technology can be traced to the early 1980s in Japan. However, it is difficult to trace its origin as there were many undocumented activities since the 1950s. The development of related technologies like computers, lasers, and electronic controllers assisted in the rapid growth of the process.

In 1981, Dr. Hideo Kodama of Nagoya Municipal Industrial Research Institute published research papers about the technology to fabricate solid three-dimensional objects with photopolymer material cured using ultraviolet rays. Unfortunately, his patent application was denied. Later, in 1986, an American inventor named Charles Hull officially filed the patent for stereolithography (SLA), which is considered the mother of all additive manufacturing techniques. In the early 1980s, the process was mainly used to make prototypes in a very short lead time and hence is often confused with the term "rapid prototyping," a broader term that refers to the strategies and techniques used to quickly fabricate three-dimensional functional or non-functional prototypes.

Fundamentally, all additive manufacturing techniques work on a common principle. Designing the 3D CAD model using any designing software like CATIA, NX, or Creo is the first and foremost step. Sometimes, reverse engineering of the existing

P. Muthuswamy (✉)
Ingersoll Rand Technologies and Services Private Limited, Bangalore, India

M. P. Chinnasamy
School of Mechanical Engineering, Vellore Institute of Technology, Chennai, India

Y. Kumar Singla et al. (eds.), *Digital Manufacturing in Industry 5.0*,
https://doi.org/10.1007/978-3-031-91500-0_5

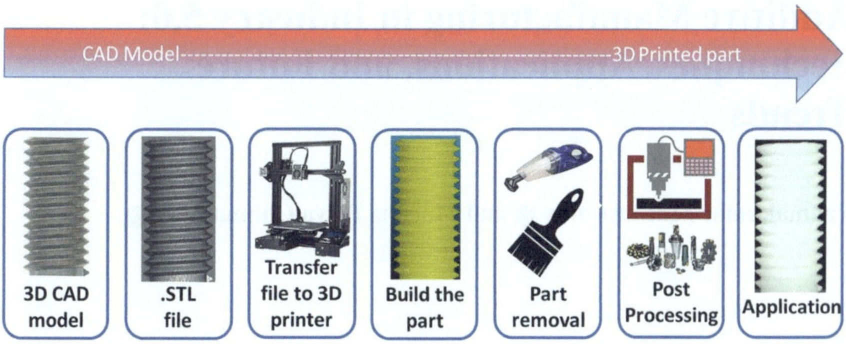

Fig. 1 Additive manufacturing process steps (Reproduced from [1], published under a Creative Commons license)

model using 3D optical scanning techniques can also be employed. The CAD model is then converted to the machine-readable STL file format, one of the common formats that can be processed by the additive manufacturing machine. Computer software then slices the CAD model into thin layers of varying thickness, which are fed to the machine for building the parts layer by layer. Once the part is built, it is removed from the work surface by removing the support structures. Before the final product is used for functional application, most of the printed parts undergo an additional post-processing step to improve strength and aesthetics. The various stages of the additive manufacturing process are illustrated in Fig. 1.

2 Types of Additive Manufacturing Techniques

Additive manufacturing technology has advanced significantly in recent years, with new techniques being developed at a rapid rate, pushing the boundaries of what's possible in the field. However, based on the printing technique, the International Organization for Standardization has developed the ISO/ASTM 52900:2015 standard, which classifies all additive manufacturing techniques into seven broader categories. They are

1. Binder jetting
2. Directed energy deposition
3. Material extrusion
4. Material jetting
5. Powder bed fusion
6. Sheet lamination
7. Vat photopolymerization

Figure 2 illustrates the categorization of additive manufacturing methods by material type and deposition technique.

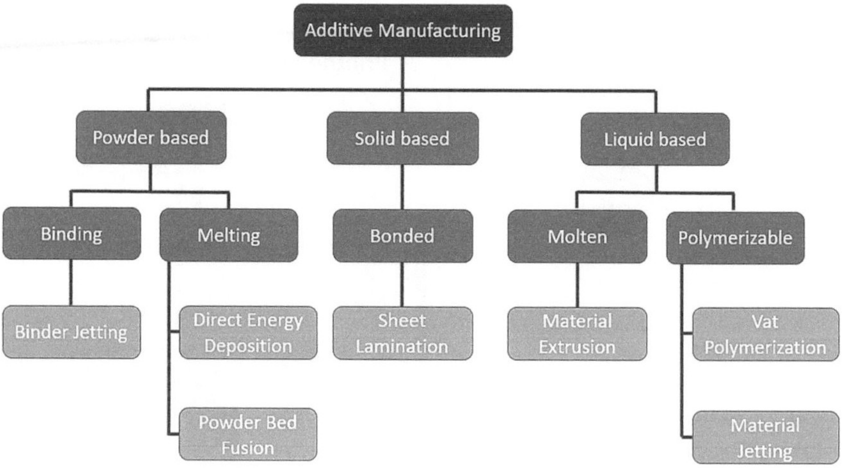

Fig. 2 Categorization of additive manufacturing techniques

The fundamental differences in each of the processes, along with their working principles, advantages, and disadvantages, are discussed in detail in this section.

2.1 *Binder Jetting*

The binder jetting technique uses an adhesive binder that is deposited over a layer of metal, ceramic, or polymer powders laid on the build platform to create solid parts. The role of the binder is to hold the powder particles together until the final part is printed. When the first layer is completed, the platform moves down by a predefined layer thickness for the rollers to deposit a new powder layer. The process is repeated until the final required part is printed. The unbound powder around the printed part acts as a supporting member and is often reused in subsequent printing. Figure 3 illustrates the binder jetting process.

The printed parts are called "green parts" as an additional post-processing step is required to enhance their strength and make them functional components. Low melting-point metal infiltration (such as bronze) or vacuum sintering are commonly used post-processing techniques. Metals (e.g., stainless steel), polymers (e.g., acrylonitrile butadiene styrene, polyamide, polycarbonate), and ceramics (e.g., glass) are usually printed using this technique.

Advantages

- Supports color printing and a diverse category of materials, including metals, ceramics, and polymers.
- Suitable for intricate and complex parts with little or no deformation and warping.
- Unused powders can be reused, making the process highly sustainable.

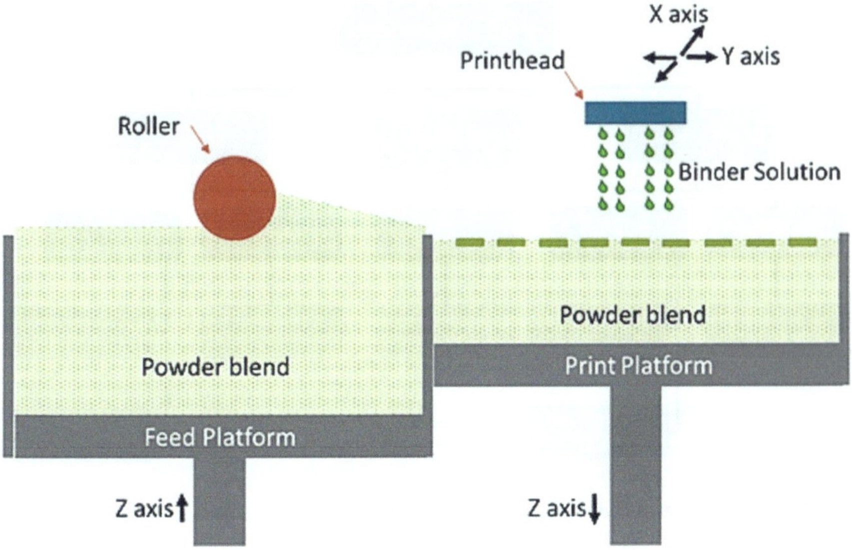

Fig. 3 Representation of binder jetting technique (Reproduced from [2], published under a Creative Commons license)

Disadvantages

- Requires post-processing to improve strength and achieve the required surface finish.
- Accuracy is less compared to other techniques.
- Not suitable for structural parts.

2.2 Directed Energy Deposition

The directed energy deposition (DED) technique is employed to create 3D parts by directly melting the materials and depositing them layer by layer using focused energy sources such as plasma arc, laser, or electron beam in a controlled environment to prevent oxidation and contamination. Figure 4 illustrates the directed energy deposition process.

The system includes a nozzle attached to a multi-axis arm, which is utilized for introducing the molten material onto the work surface. The process closely resembles material extrusion, except for the additional multi-degree movement of the arm. It uses material in the form of powder or wire, and laser, electron beam, or plasma arc are used as the melting source based on the material type and application. The technique can build complex shapes from scratch and repair damaged parts. Direct metal deposition (DMD), laser deposition welding (LDW), laser engineered net shaping (LENS), 3D laser cladding (3DLC), wire arc additive manufacturing (WAAM), and electron beam additive manufacturing (EBAM) are some of

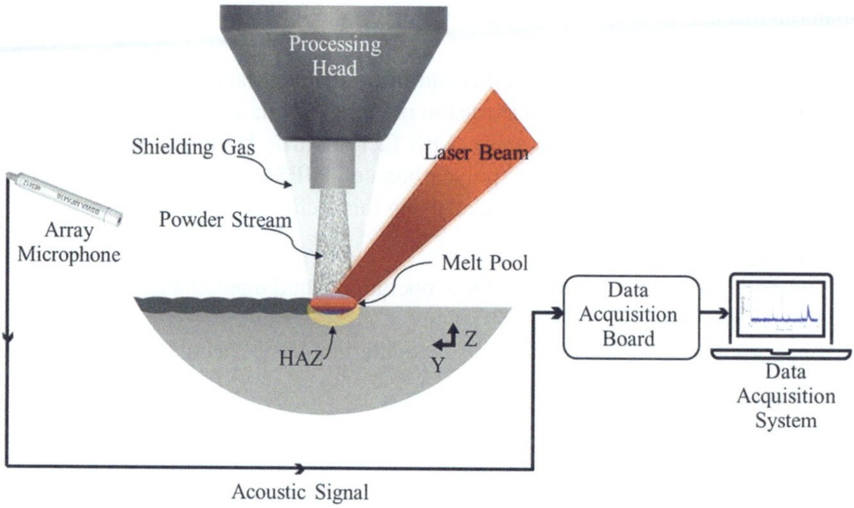

Fig. 4 Schematic of directed energy deposition (Reproduced from [3], published under a Creative Commons license)

the popular terms used to denote DED. Metals like aluminum, titanium, steel, copper, and metal alloys are commonly printed using this technique, while ceramics and polymers are other commonly used materials.

Advantages

- The printed parts are stronger and denser.
- Option to print ceramics, polymers, and metals can be used to repair damaged components.
- The high deposition rate makes it suitable for printing larger parts and can be used for complex profiles due to precise control.

Disadvantages

- Initial cost is high due to the complex setup, which also requires specialized equipment and expertise.
- Poor surface finish and low-dimensional accuracy that requires post-processing.

2.3 *Material Extrusion*

The material extrusion process uses a continuously fed filament of polymer material from a spool. A heated nozzle attached to a computer-controlled mobile head draws the material and deposits it over the work surface to build the required shape. The platform moves below by a step to deposit the second layer after the base layer is built. This sequential deposition of metal layer by layer lasts until the fabrication of the object is finished. Sometimes, a secondary nozzle is used to deposit support material that acts as scaffolding. Figure 5 depicts the screw extrusion type 3D printing process.

Fused deposition modeling (FDM) and fused filament fabrication (FFF) are the most used terms for the material extrusion process. A wide variety of thermoplastic materials like PLA (polylactic acid), ABS (acrylonitrile butadiene styrene), TPU (thermoplastic polyurethane), PC (polycarbonate), HIPS (high impact polystyrene), ceramics, and concrete can be extruded using this technique

Advantages

- Relatively inexpensive setup and low operational and maintenance costs.
- Low printing time.
- Supports a wide variety of materials, with thermoplastics being the most popular.

Disadvantages

- Support structure is required.
- Low accuracy and poor surface quality, often requiring post-processing.
- Prone to deformation and warping.

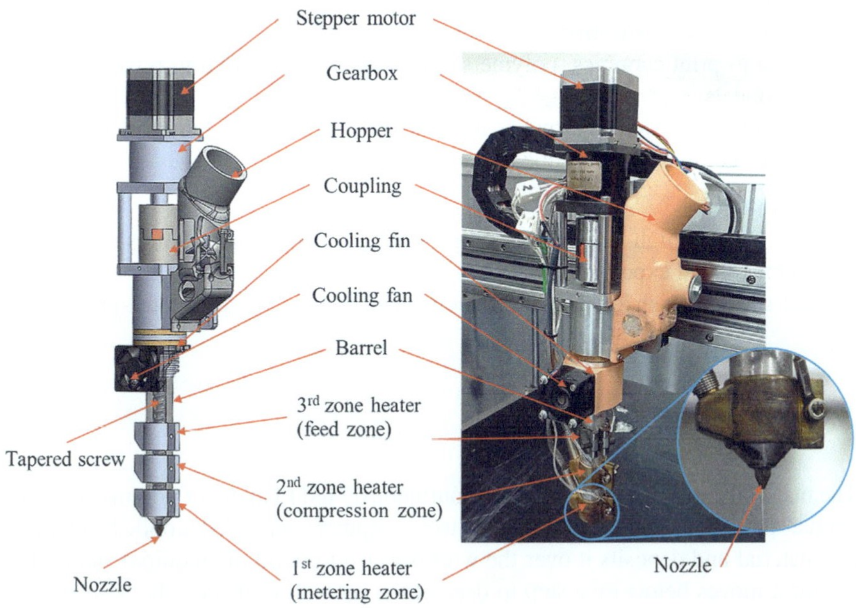

Fig. 5 Schematic of screw extrusion type 3D printer screw (Reproduced from [4] published under a Creative Commons license)

2.4 *Material Jetting*

The material jetting process is akin to a two-dimensional inkjet printer used for printing documents. Materials such as photopolymers or wax are jetted through a nozzle onto the build surface, which then solidifies (cures) using light or heat, much like the stereolithography process. After curing the first layer, the platform lowers to allow construction of the subsequent layers on top of the previous one. Support structures are added via a separate nozzle and are subsequently removed during the post-processing stage. Figure 6 depicts the material jetting process.

Multiple nozzle heads or printheads can be attached to a single frame to increase process speed, and multicolor and multi-material printing are also possible. Polymeric materials such as polypropylene, high-density polyethylene (HDPE), polystyrene (PS), acrylonitrile butadiene styrene (ABS), and high-impact polystyrene (HIPS) are commonly printed using this technique.

Advantages

- Possibility to print multiple materials simultaneously.
- Supports full-color models by combining different colored materials.
- High degree of dimensional accuracy with fine details and good surface finish.

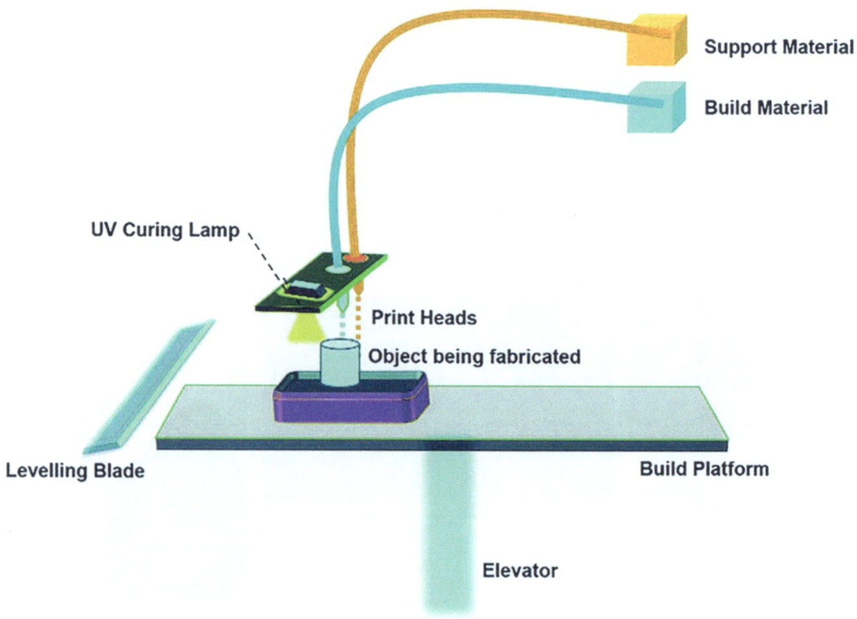

Fig. 6 Schematic of material jetting process (Reproduced from [5], published under a Creative Commons license)

Disadvantages

- The process is relatively slow due to the jetting, curing, and solidification cycle.
- Support structures are required.
- Limitations in materials, which are often restricted to polymers.

2.5 Powder Bed Fusion

Powder bed fusion (PBF) encompasses various additive techniques that utilize a laser beam (PBF-LB) or electron beam (PBF-EB) to heat and fuse powdered metal particles together. Common PBF techniques include selective laser sintering (SLS), selective laser melting (SLM), direct metal laser sintering (DMLS), selective heat sintering (SHS), and electron beam melting (EBM). All these techniques operate on the same fundamental principle, wherein a heat source (laser, electron beam, etc.) is used to fuse or bind the layer of powder particles together to form a solid mass.

A thin layer of powder material, such as metal, polymer, or ceramic, is spread across the building platform and fused by a focused energy source. After the base layer is built, the complete powder platform descends for the rollers to deposit the new layer. This process repeats until the final product is built, and any unbound loose powder is removed and reused for subsequent printing. Figure 7 depicts the powder bed fusion process.

The process that employs a laser as the heat source is performed in an inert atmosphere to protect the parts from oxidation or nitriding, whereas in the electron beam process, an enclosed vacuum chamber containing a very little quantity of inert gas

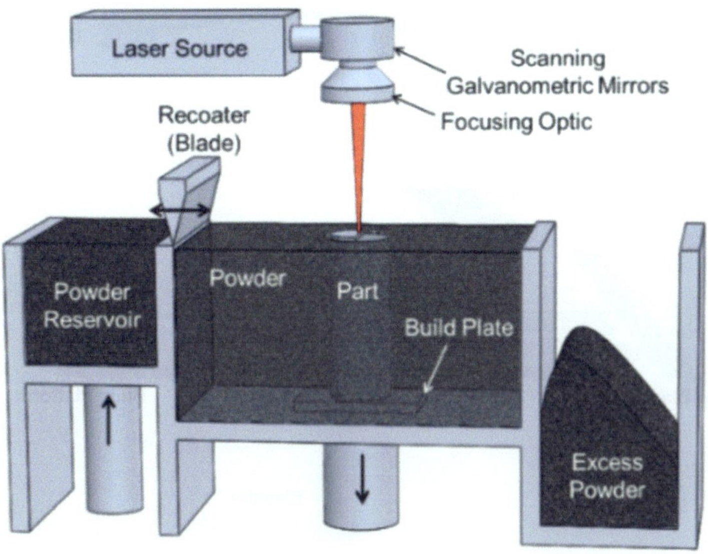

Fig. 7 Schematic of laser powder bed fusion process (Reproduced from [6], published under a Creative Commons license)

is employed. Further post-processing methods like infiltration, heat treatment, or machining are carried out to enhance the properties and appearance of the manufactured part. A wide variety of metal powders, such as stainless steel, aluminum, steel, copper, titanium, cobalt chrome alloys, nickel alloys, alloys of ferrous materials, ceramics, and polymers, have been successfully printed using this technique.

Advantages

- An inexpensive technique with options to print a wide variety of materials with complex geometries.
- The powder bed itself acts as a support structure.
- The support powders can be reused, making the process highly sustainable.
- Suitable for parts with high mechanical properties and precision.

Disadvantages

- The process speed is low, making it unsuitable for high-volume requirements.
- High-power consumption and thermal deformation.
- Relatively poor surface finish, requiring secondary operations to improve aesthetics and achieve dimensional accuracy.

2.6 *Sheet Lamination*

The sheet lamination technique, also termed laminated object manufacturing (LOM), generates 3D parts by stacking thin foils of materials such as paper, plastic, or metal on top of each other and bonding them using adhesive or a bonding agent. It is one of the oldest additive manufacturing techniques, initially utilizing inexpensive paper sheets as layers to create prototypes. The layers are bonded together with a suitable adhesive, and the required shape is cut using a laser beam or blade. Figure 8 depicts the sheet lamination process.

The precision of the printed parts is highly dependent on the thickness of the foil, and often, an additional post-processing machining operation is required to achieve the final shape. In ultrasonic additive manufacturing (UAM), thin metallic sheets are bonded simultaneously using an ultrasonic welding process, immediately followed by machining to remove unwanted material. Parts made of aluminum, copper, titanium, and steel can be manufactured using this technique, utilizing their respective foils. Additionally, paper, plastic, and sheet metal of any material can be employed.

Advantages

- Relatively faster as an entire layer is processed at once.
- Relatively cost-effective and capable of printing multicolor parts.
- Supports the combination of different materials.

Disadvantages

- Difficult to print complex geometry.
- The strength of the part depends on the bonding medium and bonding technique.
- Needs post-processing step to improve the quality of printed surface and achieve final required dimension.

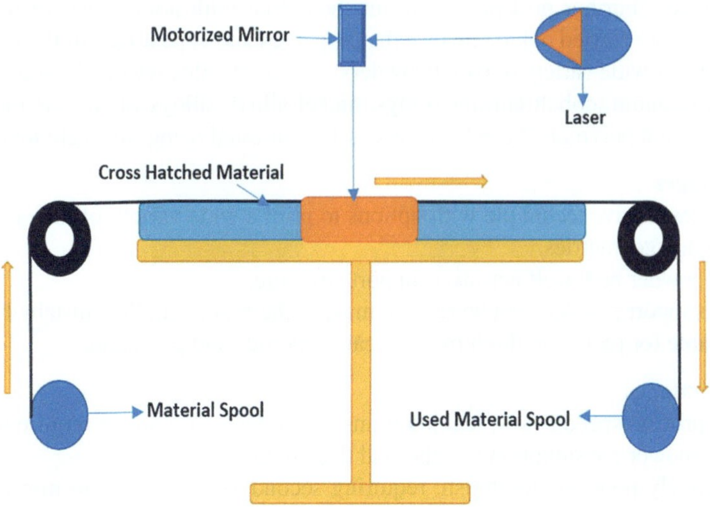

Fig. 8 Schematic of sheet lamination process (Reproduced from [7], published under a Creative Commons license)

2.7 *Vat Photopolymerization*

The concept of vat photopolymerization involves using a vat of liquid photopolymer resin that solidifies when exposed to light, typically from a laser or a light source. When exposed to specific wavelengths of light, the molecules of liquid photopolymers rapidly bond and solidify, thereby constructing 3D parts. The fabrication platform, upon which the parts are built, controls the Z-axis movement, while the light source moves in the other two axes to cure the photopolymer. Secondary washing and post-curing using UV lights are commonly employed post-processing techniques. Stereolithography (SLA), digital light processing (DLP), continuous liquid interface production (CLIP), and daylight polymer printing (DPP) are some of the most popular vat photopolymerization techniques. Figure 9 depicts the Stereolithography process.

The two primary approaches in the stereolithography technique are bottom-up and top-down, chosen based on specific requirements such as build size, complexity of part, and quantity of resin used. Although the bottom-up setup is the most applied fabrication technique, wherein layers of materials are formed one over the other, nowadays the top-down approach is gaining popularity. In the top-down approach, the fabrication platform moves in the opposite direction, minimizing the use of resin and supporting larger part sizes. Various types of photocurable plastics and polymers are commonly printed materials using this technique.

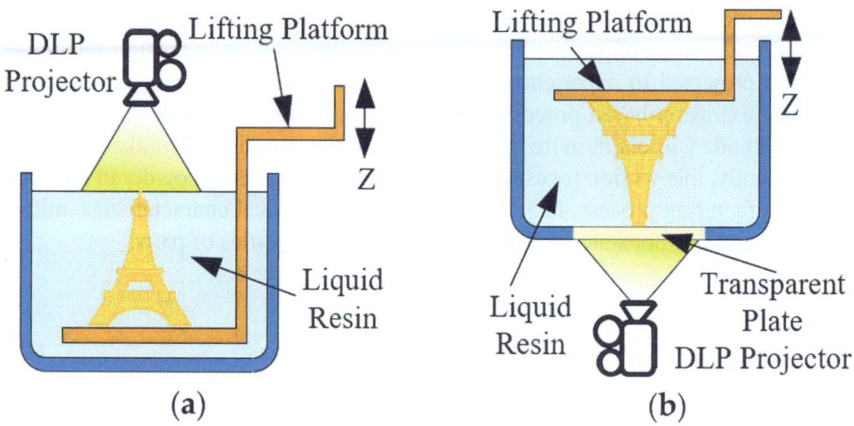

Fig. 9 Schematic of stereolithography by (**a**) top-down and (**b**) bottom-up setups (Reproduced from [8], published under a Creative Commons license)

Advantages

- Versatility in printing transparent and flexible parts.
- Very high accuracy and superior surface finish.
- Relatively faster process and capable of building larger objects.

Disadvantages

- The process is expensive and often requires post-processing.
- The process is comparatively complex due to high dependency on support structures.
- Supports only photo-polymeric resins, limiting its widespread application.

3 Influence of Powder Characteristics on the Properties of Additively Manufactured Parts

Additive manufacturing techniques vary based on the initial condition of the raw material, with powder-based methods being the most popular due to their material versatility. The three major classifications of AM techniques are powder bed fusion (PBF), directed energy deposition (DED), and binder jetting. PBF utilizes a laser or electron beam to selectively fuse powder particles, while DED involves melting material as it is deposited. In contrast, binder jetting employs a binding agent to adhere powder particles together. These methods cater to various materials, including polymers, metals, and ceramics.

The market for powder materials in AM is undeniably substantial. Within this sector, the utilization of powder-based techniques like powder bed fusion and binder jetting is projected to experience continuous expansion, surpassing the currently dominant extrusion-based processes. By 2033, it is anticipated that these powder-based methods will claim more than one-third of the overall AM market share [9]. Consequently, this section focuses on the essential attributes of powder in the additive manufacturing process, including morphology, physical characteristics, microstructure, and powder state, and relates these to the properties of parts.

3.1 Morphology

The morphology of powder refers to its physical attributes, including size, shape, and surface roughness. These characteristics play a vital role in influencing flow and packing properties, and they are commonly assessed using scanning electron microscopy (SEM). In general, smooth and regularly shaped particles are preferred for their superior flowability and packing characteristics compared to rough and irregular ones. Finer particles, in theory, enhance the density, surface finish, and texture of manufactured parts while minimizing flaws, surface roughness, and defects. However, they often agglomerate, which can lead to challenges such as variations in powder bed density and the induction of internal imperfections like micro-pores and voids [10–12].

The morphology of particles, primarily influenced by manufacturing processes, can be manipulated to achieve shapes that minimize interparticle friction and mechanical interlocking. For instance, plasma atomization typically yields highly spherical particles with fewer shape irregularities compared to gas atomization, which can result in satellite formation due to variations in particle solidification rates. The choice of manufacturing process depends on the desired physical and chemical properties as well as the required quantity. Research indicates that imperfections in powder particles can arise from the manufacturing method and subsequent handling, recycling, and preparation for the additive process [13, 14].

Furthermore, the shape of powder particles significantly impacts manufacturing processes and the quality of the final product. Spherical particles, for example, enhance flowability, packing density, and the quality of powder bed layers. Studies suggest that they play a role in achieving higher density of the final part, improved surface quality, and dimensional accuracy. Metrics such as "profile void volume" are used to assess powder bed topography uniformity, which is influenced by particle morphology. Optimizing particle shape can therefore lead to improvements in overall manufacturing efficiency and product performance [15].

3.2 *Physical Properties*

Each powder exhibits distinctive characteristics, whether biological, chemical, or inorganic in nature, which ultimately determine its final quality. In addition to intrinsic properties such as composition, particle size, density, and morphology, the behavior of a powder is also influenced by its interactions with the environment. Key physical properties such as density, flowability, and particle size and distribution play a crucial role in significantly impacting the quality of AM parts.

3.3 *Powder Density*

Powder density refers to the mass of powder per unit volume, which can be measured as bulk density or tapped density. Bulk density, also referred to as apparent density, is defined as the ratio of untapped powder mass to the bulk powder volume, which includes voids. It relies on particle density and spatial arrangement, crucial in batch analysis. Lower values indicate finer particles, higher values larger ones [16], impacting particle shape, size distribution, friction, chemistry, and agglomeration. Cohesion from inter-particle forces affects apparent density, especially with non-cohesive spherical particles [17]. Careful handling during measurement minimizes external influence, ensuring reproducibility. Bulk density is calculated as weight divided by bulk volume, tested with ASTM B329-20's Scott Volumeter for metal powders, yielding "Scott density."

Tapped density increases via mechanical tapping, giving tap density, measured similarly but with a different volume. Packing fraction, which is the ratio of particle to total volume, crucially affects additive manufacturing stages like laser interaction and part strength [18]. High packing fraction minimizes voids and defects, influenced by frictional, elastic, inertial, cohesive, and viscous forces [19, 20]. Relationships between packing fraction, friction coefficients (μs, μr), and viscosity affect particle coordination [18, 21]. Spherical particles reduce friction, enhancing flowability and packing density, critical for high-density parts [22]. However, process optimization such as changing the spreader settings can alter the final layer packing fraction.

3.4 *Flowability*

The flowability of powder refers to how particles move among each other in a desired manner. It's crucial for quality printed parts, influenced by factors like particle shape, size distribution, and surface properties. Better flowability, achieved with spherical particles and uniform size distribution, ensures even powder deposition, enhancing part density and strength. Factors like particle adhesion, as well as physical characteristics such as angle of repose and fractal surface, also affect flowability. Methods like Hall, Carney, and Gustavsson funnels measure flow rates, while parameters like angle of repose and surface fractal help assess powder quality [15].

3.5 Particle Size and Distribution

Particle size distribution is critical in additive manufacturing, ensuring uniform flow for consistent material deposition and preventing blockages. It also supports uniform packing in powder beds, crucial for maintaining mechanical properties and dimensional accuracy. Optimizing particle sizes enhances layer cohesion, reducing defects and improving overall part quality in additive manufacturing processes [23]. Analysis typically employs visual-based (image) or non-visual methods to decide the size of powder particles, which influences the energy required for powder particle binding based on their size and distribution. However, optimal mixing of fine and large particles improves properties. Fine particles enhance density and surface area for better sintering, while large particles improve flow but reduce packing efficiency. Maintaining the correct size and distribution enhances final part quality with high density, fine finish, and intricate features.

Angular-shaped powder, with its greater surface area than spherical powder, enhances laser energy absorption during sintering, potentially boosting efficiency and part quality in additive manufacturing [24]. Non-spherical particles typically exhibit less efficient packing compared to spherical ones due to irregular shapes and varying sizes, creating gaps, and reducing overall density in packed volumes [25]. Fine particles enhance flow and density but tend to agglomerate, affecting energy absorption unpredictably. Stainless steel powder studies demonstrate finer particles yield higher packing and superior finishes [12, 26], whereas mixed particle sizes improve green density and part quality in Inconel-625 alloy [27].

Studies also show wider particle size ranges improve powder bed density and part finish, whereas narrower ranges enhance flow and mechanical strength [27]. However, varying results suggest complex interactions among powder characteristics, printing technique, and process parameters [28]. ASTM B822 outlines metal powder size distribution measurement using light scattering. Laser diffraction is a rapid, imaging-free technique that analyzes particle size distribution by passing a laser through dispersed particles, assessing the diffraction pattern to determine sizes [29]. Techniques like Mie, Fraunhofer, and Rayleigh scattering models aid in interpreting scattered light patterns based on particle size [30]. Hence, a thorough understanding of particle size distribution is critical for optimizing powder bed fusion processes and achieving desired printed part quality.

3.6 Microstructure

The microstructure of powder particles significantly affects mechanical and microstructural properties in both powder metallurgy and additive manufacturing processes. Understanding the impact of initial powder composition and microstructure on process and part quality is crucial. Hence, techniques such as X-ray diffraction (XRD), thermogravimetric analysis, and differential scanning calorimetry are used for microstructural analysis [31].

Research by Starr et al. [32] shows that powder microstructure from gas atomization reflects the atomization gas rather than the additive manufacturing chamber's inert gas. For instance, stainless steel powders atomized in nitrogen are mostly austenitic, while those in argon tend to be martensitic. Similar findings were confirmed by Murr et al. [33] for 17-4 PH stainless steel powders. This affects microstructure in selective laser melting (SLM). External factors like environmental conditions, powder storage, and handling also impact microstructure, moisture, and contamination levels. Oxygen and moisture can alter powder microstructure, adversely affecting mechanical properties. Higher humidity during storage can cause powder clumping, affecting flow, and packing densities [34, 35].

Preheating the print chamber in laser powder bed fusion (LPBF) reduces thermal gradients, minimizing stress for better part microstructure and mechanical properties. It ensures uniform heating, promoting strong adhesion between powder particles, reducing warping, and enhancing overall part quality. Careful temperature control is essential as it affects both primary powder particles forming the printed part and loose support powder particles, which influence part quality and properties when reused.

Factors influencing porosity in additive manufacturing (AM) include energy density variations affecting melting consistency, powder distribution impacting material uniformity and density, moisture causing gas bubbles and trapped gases, and incomplete binder evaporation leaving residual pores. Pores within powder particles, known as powder porosity, contribute to microporosity in printed parts, impacting mechanical properties. Techniques like gas adsorption and mercury porosimeter measure surface area and pore characteristics.

Powder porosity, which measures the volume of pores relative to the apparent volume of powder particles, impacts part quality and specifically addresses pores within particles rather than voids between them. Gas atomization techniques can lead to internal porosity due to gas entrapment during rapid solidification in inert environments. Visualization techniques like Micro-CT and X-ray tomography analyze pore geometry and volume fraction [36].

3.7 *Powder Condition (Virgin Vs Recycled Powder)*

Additive manufacturing offers significant advantages over subtractive methods by minimizing scrap and enabling recycling and reuse of materials, particularly beneficial for challenging and expensive alloys like titanium and nickel-based ones. Recycling involves reclaiming and reprocessing used powder into new feedstock, while reuse refers to using leftover powder from the printing process. This sustainability aspect is crucial in laser powder-bed fusion (LPBF), a widely adopted metal additive manufacturing technique where unused powder can be recycled for fabricating new components. However, the impact of recycling on part quality must be carefully assessed, as it can affect properties such as morphology, particle size distribution, and chemical composition, particularly noticeable in alloys like Ti6Al4V.

Studies have shown that recycled Ti6Al4V powder may exhibit increased oxygen content and altered particle characteristics after multiple uses, affecting mechanical properties like ductility and toughness. Similar findings apply to Inconel 718, though some alloys like steel show more stable properties across reuse cycles. Techniques such as plasma treatment have been explored to rejuvenate old powder by improving particle shape and removing contaminants, thereby enhancing flowability and maintaining part quality. These advancements underscore additive manufacturing's potential to improve sustainability by reducing material waste and enhancing resource efficiency. However, standardizing the reuse process for each alloy remains a challenge, necessitating further research to establish optimal recycling practices without compromising part integrity [15].

4 Applications of Additive Manufacturing

4.1 *Automotive*

The automobile industry is a cornerstone of the worldwide economy, driving significant impacts through employment, GDP contribution, and a complex supply chain. It also influences urban development, cultural identity, technological innovation, and sustainability efforts, while playing a crucial role in global trade and future trends like electric and autonomous vehicles [37]. It is also among the first sectors to embrace additive manufacturing (AM) to provide specific, customized designs, and reduce the weight of high-performance racing cars. Traditionally, the automobile industry relies on conventional manufacturing techniques such as casting, injection molding, and machining for part production. While these methods are technically advanced and highly reliable, they typically result in a high material wastage [38]. However, in recent times, additive manufacturing has revolutionized the production of automobile parts due to several advantages it provides, such as the flexibility to design and manufacture complex parts, lower wastage of raw material, reduced cycle time for prototyping, and minimized tooling requirements. These advantages have effectively replaced traditional manufacturing methods for several components [39]. Some of the topology-optimized, additively manufactured automobile parts are shown in Fig. 10.

Based on application, a broad range of materials including polymers, metals, ceramics, and alloys have been effectively manufactured by additive techniques for the automobile sector [41, 42]. Several leading automobile companies, including Honda, General Motors, Ford, Volkswagen, and BMW Group, are reported to have adopted additive manufacturing for various purposes, including prototyping, weight reduction of critical components, and combining multiple smaller parts into a single part to reduce the total number of parts used [37].

The incorporation of 3D printing technology into BMW's latest S58 engine represents a notable advancement in automotive engineering. The integration of cylinder head made from a 3D printed core underscores a significant shift towards additive manufacturing methods in high-performance automotive engine construction. By leveraging this innovative technique, BMW engineers have achieved a remarkable reduction in weight while enhancing thermal management through intricately integrated coolant channels within the cylinder head. This technical innovation not only underscores BMW's commitment to pushing the boundaries of automotive excellence but also sets a new standard for efficiency and performance in the industry [43].

Additive manufacturing technologies can revolutionize the automobile industry with their design flexibility and manufacturing benefits. They reduce material waste, transport needs, and environmental impacts, making AM a promising sustainable method [44]. However, there are several challenges that need to be addressed before the technology goes mainstream. Some of the challenges include material

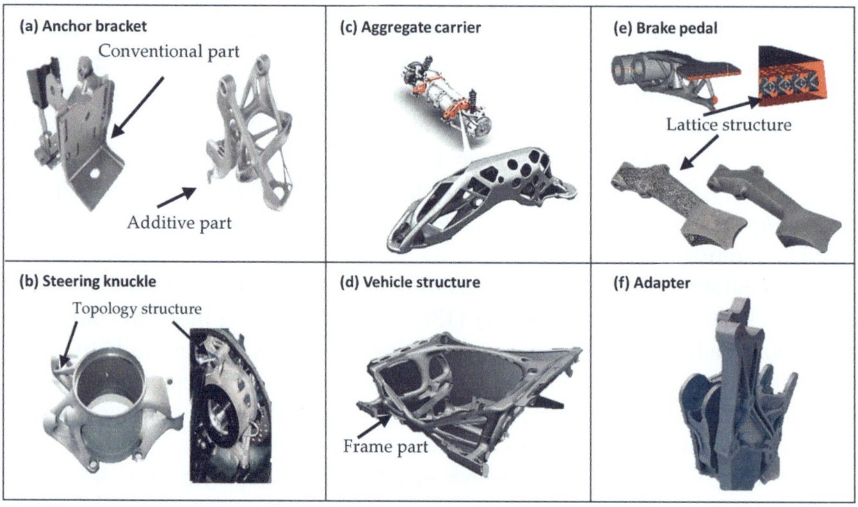

Fig. 10 Additively manufactured topology-optimized automobile parts (Reproduced from [40], published under a Creative Commons license)

limitations, quality control issues, the need for extensive post-processing, low production volume, and the lack of standardized processes and materials. Addressing these issues is essential for mainstream adoption. Future avenues for research encompass multimaterial printing, large-scale additive manufacturing, sustainability initiatives, recycling methods, automation advancements, and integrating artificial intelligence.

4.2 Aerospace

Additive manufacturing has been integral to the aerospace industry since its inception, facilitating the production of aircraft and helicopter components, as well as the development of lighter, more efficient engines, and 3D-printed turbines. These advancements not only save time and money but also yield stronger, more efficient final products. In response to environmental concerns, the aerospace sector increasingly prioritizes lightweight materials to reduce emissions and fuel consumption while maintaining stringent safety standards.

As a result, aerospace component design emphasizes material minimization, which in turn enhances part complexity, function, and properties. Additive manufacturing technologies enable freeform fabrication, allowing to produce virtually any component shape. This layer-by-layer approach, compared to traditional methods, optimizes parts for a high strength-to-weight ratio. Commonly utilized materials in aerospace manufacturing include metals, polymers, and composites. Additive manufacturing applications in the aerospace industry span direct part manufacturing, tooling, prototyping, and repair/restoration. Some of the additively manufactured aerospace parts are shown in Fig. 11.

Direct part manufacturing involves fabricating critical components like nozzles and engine parts, reducing overall aircraft weight, and enabling the production of complex parts. Tooling applications streamline processes by manufacturing patterns, molds, and dies with reduced time and manpower. Prototyping employs AM technology in the development stage to finalize and approve concept designs. Additionally, AM facilitates repair and restoration of metallic and non-metallic aircraft components, leading to cost savings and reduced environmental impact compared to traditional methods [46].

The integration of additive manufacturing in aerospace, particularly highlighted by NASA's Jet Propulsion Laboratory (JPL) in the Mars Rover missions, showcases a remarkable advancement in space exploration technology. Utilizing additive manufacturing for critical components has proven beneficial, especially in scenarios where conventional manufacturing methods face challenges, such as intricate designs and strict mass limitations.

In the case of NASA's Mars Perseverance Rover, the deployment of 3D-printed titanium components manufactured via electron beam melting (EBM) resulted in

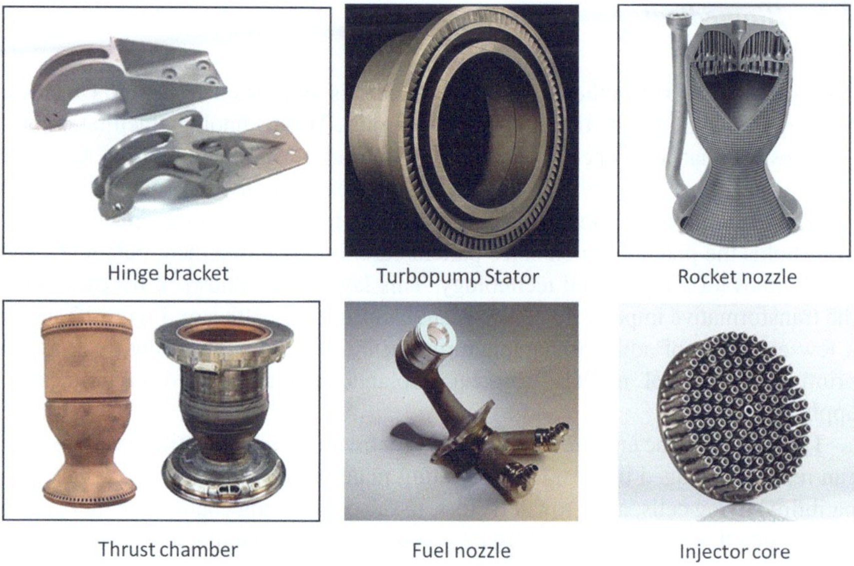

Fig. 11 Additively manufactured aerospace parts (Reproduced from [45], published under a Creative Commons license)

parts significantly lighter than those produced through conventional processes. These parts feature internal lattices and complex features crucial for ensuring high strength while adhering to stringent mass constraints and minimizing material usage. This application underscores how additive manufacturing enables the creation of lightweight, high-performance parts with intricate geometries, which are essential for spacecraft missions.

NASA's continuous exploration and testing of 3D printing for spacecraft components, as exemplified by the integration of a 3D-printed ceramic part in the Curiosity rover's Sample Analysis at Mars (SAM) instrument, further emphasize the agency's commitment to advancing additive manufacturing technologies in space exploration. These endeavors not only demonstrate the reliability of additive manufacturing in space environments but also pave the way for future innovations and enhancements in aerospace engineering [47].

Overall, the success of additive manufacturing in producing critical components for spacecraft underscores its ability to transform the aerospace industry. By facilitating the fabrication of lightweight, high-performance parts with complex geometries, additive manufacturing contributes significantly to advancing space exploration capabilities, ultimately driving innovation and progress in the field.

4.3 Biomedical

The role of AM in healthcare and biomedicine is significant in creating medical devices, implants, prosthetics, and tissue scaffolds. This technology facilitates complex designs that were once unachievable with traditional manufacturing methods, thereby enabling personalized treatments tailored to individual patient requirements. As additive manufacturing progresses, its influence on biomedical applications holds the potential to enhance patient care, improve treatment outcomes, and foster innovation in medical technology. This lays the groundwork for examining the transformative impact of additive manufacturing in healthcare. Figure 12 shows a few examples of additive manufacturing in biomedical applications such as 3D printing of medical models, surgical implants, and tools used in biomedical applications.

The field of tissue engineering is pivotal for creating biocompatible implants that can replace damaged tissues. By integrating materials suitable for biological compatibility, living cells, and growth factors, tissue engineering enables the fabrication of implants that support natural tissue growth. Additive manufacturing has revolutionized tissue engineering by allowing the creation of intricate 3D models mimicking the microscopic structure of tissues [49]. This technology is utilized to produce porous implants promoting bone regeneration and complex organoids composed of special cell types. The precision of 3D printing facilitates the creation of geometrically complex structures, enhancing the effectiveness of implants and organoids in biomedical research and clinical applications [50, 51].

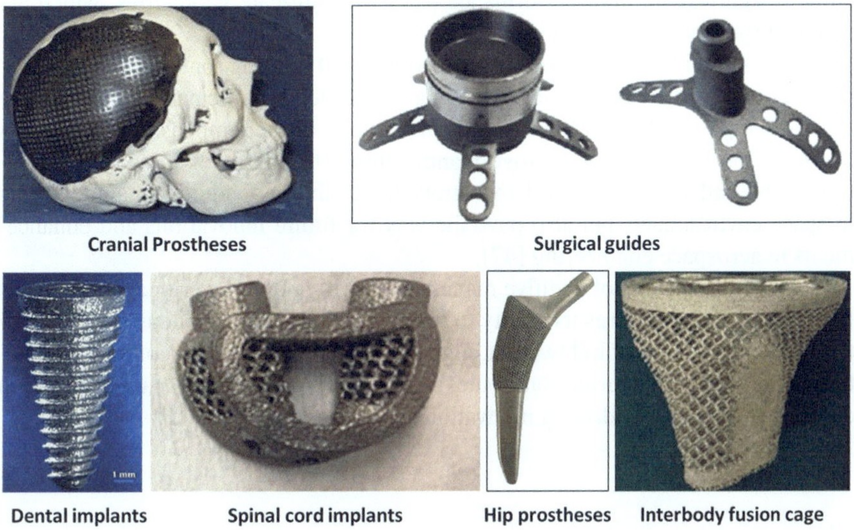

Fig. 12 Additive manufacturing in biomedical application (Reproduced from [48], published under a Creative Commons license)

In vitro applications of tissue engineering have expanded the creation of models that mimic physiological conditions, such as organoids, which mimic human and animal tissue environments. These three-dimensional models offer a more accurate representation of cellular behavior and drug responses compared to traditional two-dimensional cultures. For instance, studies have demonstrated that 3D spheroids better replicate in vivo responses to drugs like rapamycin compared to 2D cultures. Bioprinting technology enables the precise deposition of cells and biomaterials, allowing for the creation of organ-on-a-chip models that closely simulate in vivo conditions, facilitating studies on organ development and disease processes [29, 52, 53].

In the realm of surgical implants and prostheses, additive manufacturing plays a crucial role in designing and fabricating implants that integrate seamlessly with surrounding tissues. This multidisciplinary approach combines materials science, engineering principles, and biological factors to enhance implant functionality and longevity. Techniques like computed tomography (CT) enable the development of accurate computer-aided design (CAD) models of defects, which can be directly translated into 3D-printed implants tailored to individual patient anatomy. The ability to create porous scaffolds with specific biomechanical properties enhances bone regeneration and integration of implants, offering innovative solutions for managing critical-size bone defects and improving patient outcomes [54, 55].

Overall, additive manufacturing is transforming biomedical research and clinical practice across tissue engineering, drug delivery, surgical tools, and operative planning [56]. By enabling the precise fabrication of complex structures and customized solutions, AM is driving innovation in healthcare, from enhancing tissue regeneration to improving surgical outcomes and advancing drug delivery systems [57, 58].

4.4 Construction and Architecture

In the construction industry, architectural models have been produced using AM technique for several years. However, in recent times, there has been a significant surge in studies focusing on printing techniques for building components, with the initial construction projects utilizing AM processes. This trend is driven by the advantages AM offers, such as the ability for construction companies to create geometrically intricate structures, employ different materials in a section depending on its function, and automate building method from digital models. These benefits can be fully harnessed if construction parts, and AM techniques are appropriately modeled during the preliminary stage. To support this, various digital techniques for designing components suitable for printing and planning AM processes have been created. Additionally, new data formats have been introduced to save and swap over the subsequent data [5]. Some of the 3D-printed construction and architectural structures are shown in Fig. 13.

In 2014, a construction company based out of China fabricated 10 houses in under 24 h using pre-casted 3D-printed parts. The following year, they built a

Fig. 13 3D-printed structures and buildings (Reproduced from [59], published under Creative a Commons license)

multistory complex and a villa with the same technology. In other advancements, UC Berkeley researchers developed a novel powder-based technique to print the "Bloom Pavilion," and in 2016, Dubai unveiled its first 3D-printed office, showcasing the potential for curvilinear practical structures [59].

Additive manufacturing techniques are carefully selected based on the raw material used. In the context of printing building components, research has primarily focused on aggregate-based materials, metals, and polymers. In additive construction, concrete is the most used aggregate-based material, primarily applied through material extrusion and particle-bed processes. Concrete's low tensile strength necessitates reinforcement integration for stability. In material extrusion, fresh cementitious material is layer deposited through a narrow orifice, requiring properties like pumpability, extrudability, interlayer adhesion, and buildability. This method has been categorized into fine, medium-fine, and coarse filament processes, each offering varying levels of precision and productivity. Particle-bed processes, which bond dry particles with a fluid, allow intricate designs and high accuracy but are constrained by build space. Innovations like FreeFAB, Smart Dynamic Casting, and Mesh Mould address formwork and reinforcement challenges. For reinforcement, methods range from external reinforcement to embedding during printing and using fiber or additively manufactured reinforcements. These advancements aim to enhance concrete's structural performance and durability in complex geometries [58].

Metals including steels, titanium alloys, and aluminum alloys can be used for printing structures in architectural applications to achieve mechanical properties comparable to conventional methods. These require low porosity, controlled by feedstock quality and volume energy. Additively manufactured components often have finer microstructures due to rapid cooling, resulting in higher strength but anisotropic properties, with the weakest direction being perpendicular to the deposition plane. Post-manufacturing treatments like heat treatment and polishing can enhance properties by reducing porosity and surface roughness. Comparisons of additively and conventionally manufactured metal parts, such as brackets and window frames, indicate AM's feasibility but highlight higher costs and production times, which can be mitigated by mass reduction and improved AM techniques. Powder bed fusion processes permit high accuracy but are slower and suitable for smaller parts, while directed energy deposition processes are faster and better for large structures and refurbishments. Notable applications include MX3D's projects like the 3D-printed stainless-steel footbridge in Amsterdam [58].

In the case of printing polymer-based construction parts, the common techniques includes SLA, FEM, SLS, and BJ3DP. Each method utilizes specific polymers: SLA employs liquid photopolymers cured by light, while FDM commonly uses thermoplastics like ABS and PLA. Photopolymers, though effective for certain applications, have historically been expensive for building components [60]. Many printed polymers lack sufficient stiffness and strength for heavily loaded parts; however, advances such as carbon fiber-reinforced ABS (CF-ABS) have enhanced mechanical properties and expanded AM applications. Binder jetting techniques, employing polymer binder and sand, are utilized in creating concrete formwork.

Research into using AM technology in construction is still nascent, though numerous experiments have explored its potential. AM in construction offers extensive design freedom and automation opportunities, facilitated by digital planning methods. Integrating various stakeholders early in the design phase is crucial. Structural validation and simulation of AM processes for concrete, metals, and polymers aim to prevent build failures and predict product properties, yet challenges remain. These include understanding multiscale and multi-physics mechanisms, reducing computational costs, and improving material representation for stability and serviceability [58].

5 Other Applications

Additive manufacturing (AM) technology is widely applied across various other sectors as well. In fashion and wearables, it facilitates the production of customizable clothing, jewelry, and shoes. In the food industry, there is a growing trend in restaurants for customized chocolates, candies, and personalized food items. Educational institutions use AM to print physical models and prototypes for classroom and school laboratory applications.

Complex artwork and intricate sculptures, unattainable through traditional processes, are made possible through 3D printing. In the consumer goods sector, AM techniques manufacture custom phone cases, home décor items, and personalized gifts. AM enhances the production of high-performance sports equipment, including shoes and helmets, through custom manufacturing processes.

For defense and security requirements, AM produces specialized components, tools, and drones. The energy sector relies on AM for parts of wind turbines and solar panels to meet high-performance demands. The toy industry utilizes AM for customizable toys, board game pieces, and collectibles. Additionally, AM contributes to creating biodegradable materials, filtration devices, and prototypes for sustainable environmental applications. In addition, AM is an integral part of the smart factory technologies, which provides flexibility, on-demand production, and seamless integration with digital and IoT systems. It enables rapid prototyping, customization, and production of advanced materials, supporting innovation and efficient maintenance with on-demand spare parts and custom tooling to numerous sectors [61, 62].

6 Sustainability of Additive Manufacturing

Manufacturers are increasingly adopting innovative and eco-friendly approaches to reduce their environmental impact, with additive manufacturing (AM) playing a crucial role. AM promotes sustainability by minimizing waste and energy use, enabling local production to reduce transport needs and extending product lifecycles through tailored parts and repair capabilities. Sustainability is a philosophy focused on existing and growing in ways that do not deplete resources for upcoming generations. It encompasses environmental, social, and economic sustainability, and promotes good governance [61, 62].

Increasing awareness of environmental protection has heightened scrutiny of the sustainability of manufacturing processes. Governments, corporations, and research organizations are increasingly focused on assessing the environmental impacts across product life cycles. An advantage of AM processes is their ability to shape parts without requiring traditional tooling, unlike methods such as injection molding, die casting, forging, and stamping. This eliminates environmental impacts associated with tooling supply chains, reducing energy consumption and greenhouse gas emissions throughout product lifecycles. Additionally, AM's design flexibility can shorten supply chains for customized products and minimize material waste, further reducing environmental impacts. Numerous studies have evaluated various AM technologies for their sustainability, primarily focusing on production phases. Despite several challenges, AM's potential in optimizing resource use and localized production enhances the environmental sustainability of the process [63, 64].

Cost analysis is vital for assessing the economic viability of any technology, including AM, compared to traditional manufacturing methods. Early work by Alexander et al. [65] focused on optimizing build orientation to minimize costs in layered manufacturing, emphasizing the impact of part height, support material volume, and contact area. Hopkinson and Dicknes [66] expanded this by developing a model that included labor, material, machine costs, and depreciation to compare AM with injection molding. They found stereolithography and FDM more cost-effective for smaller production volumes. Subsequent research explored AM's ability to produce different components simultaneously, suggesting that a "smart mix" of components can reduce costs, though estimating costs for concurrent production of various parts is more complex. Overall, a comprehensive understanding of cost factors is crucial for the broader adoption of AM technologies.

Huang et al. [67] discussed the social impacts of AM focusing on worker health and safety, design optimization, and customization. They highlighted the health benefits of AM over conventional processes and predicted significant shifts in labor structures due to automation and localized production. In developed countries, AM could benefit aging populations, while developing countries might face increased unemployment. Rejeski et al. [68] emphasized the need for research on occupational health issues, particularly the toxicity of fine powders. Chen et al. highlighted limited studies on the social dimensions of AM [69]. They emphasized AM's potential to democratize production, transforming consumers into active participants. Other studies noted AM's impact on supply chains, labor qualifications, work patterns, and societal improvements through localized production [70]. Ford and Despeisse [71], as well as Pearce et al. [72], discussed AM's potential to improve living conditions and local adaptability through open-source information and production capabilities. The findings highlight the complexity of assessing AM's social impacts due to their qualitative nature, with most studies focusing on design and production phases. As AM is an emerging technology, research has primarily concentrated on Life Cycle Assessment (LCA) and Life Cycle Costing (LCC), which are easier to evaluate [73].

In conclusion, AM's potential to optimize resource use, promote localized production, and enhance sustainability is clear. As research continues to explore the environmental, economic, and social dimensions of AM, its role in advancing sustainable manufacturing practices will undoubtedly become more pronounced, paving the way for a greener and more resilient future.

7 Future Direction: From 3D to 5D Printing

Industry 5.0 prioritizes human welfare within the industrial production framework, aiming to achieve social objectives that extend beyond mere employment and economic growth. This paradigm strongly supports sustainable development and the overall well-being of all individuals, heralding a new era in the global industrial revolution. Unlike the technology-centric focus of Industry 4.0, Industry 5.0 is driven by values and emphasizes three core principles: a people-centric approach, sustainability, and resilience [74, 75].

While traditional additive manufacturing is widely utilized across various industries, current interests are shifting towards the system's flexibility to adapt to diverse and challenging applications. These include the printing of smart materials, printable electronics, printable assemblies, and hydraulic components. As mass customization and personalization are fundamental aspects of Industry 5.0, it is crucial to develop non-traditional manufacturing techniques such as additive manufacturing

(AM). AM is anticipated to emerge as a pivotal technological enabler, allowing for the creation of intricately designed, tailored parts with advanced features that are beyond the capabilities of traditional manufacturing methods.

Soft robotics, an emerging field, benefits significantly from AM, which enables the creation of complex, customized designs with precise material properties. AM techniques like 3D printing allow for the fabrication of soft robot components such as actuators, grippers, and flexible structures that are difficult to produce using traditional manufacturing methods. This technology supports the integration of advanced materials like elastomers and polymers, essential for achieving the desired flexibility, durability, and sensory capabilities in soft robots. Smart materials like shape memory alloys and polymers, when integrated into additive manufacturing processes, enable a groundbreaking technique known as "4D printing." Unlike traditional 3D printing, 4D printing involves materials that can self-transform or adapt in response to environmental stimuli, such as temperature changes or moisture absorption. This capability allows printed objects to change shape, alter their properties, or perform specific actions autonomously over time. By harnessing the unique characteristics of smart materials through additive manufacturing, 4D printing opens up innovative possibilities in fields such as soft robotics, architecture, biomedical devices, and beyond, where dynamic, responsive structures are increasingly valuable [75].

Pre-filled hydraulic force transmitting mechanism during the 3D printing process simplifies assembly, potentially reducing manufacturing complexity such as sealing [76]. It has already demonstrated the potential for 3D printing electronic components such as opto-electronic devices, circuit boards, antennas, capacitors and inductors, and control electronics with integrated functionalities and complex geometries [77, 78]. Another recent advancement in the additive manufacturing portfolio is 5D printing, which incorporates additional dimensions beyond the traditional three spatial dimensions. These dimensions include attributes such as color, texture, transparency, conductivity, or even the ability to change shape or function over time, enabling the manufacture of more complex and functional parts. Additionally, hybrid manufacturing integrates additive (AM) and subtractive manufacturing (SM) processes to combine their strengths; AM for creating complex geometries and SM for precise finishing. This approach enhances efficiency, design flexibility, and enables customized material properties in production [79].

8 Conclusions

Industry experts assert that we are currently experiencing the fourth industrial revolution. This phase is marked by the convergence of advanced digital technologies with industrial processes, aimed at boosting efficiency and cutting costs. Additive manufacturing, for instance, exemplifies this trend by enabling the production of parts with minimal waste. This integration fosters seamless communication and data exchange among machines, creating interconnected and agile ecosystems.

These advancements enable highly customized products and decentralized manufacturing closer to consumers, supporting on-demand production. They also align with sustainability goals by facilitating the manufacture of complex, smaller volume parts. However, Industry 4.0's emphasis on automation and efficiency has raised concerns about its societal impact. Critics argue that it may overlook human involvement and exacerbate job displacement and societal inequality.

Looking ahead, there is a growing consensus on the need to integrate human intelligence more deeply and prioritize sustainability in future industrial developments, leading towards what some call Industry 5.0. While larger corporations in developed nations have embraced these technologies, smaller enterprises and developing countries lag behind, particularly in sectors like additive manufacturing. Industry 5.0 aims to promote human–machine collaboration, adaptability to market changes, and sustainability. This approach ensures that future industrial practices prioritize inclusivity and environmental responsibility, aiming for a balanced integration of technology with human needs.

Disclaimer: This disclaimer informs readers that the views, thoughts, and opinions expressed in the article belong solely to the authors, and they do not purport to reflect the opinions, views, policies, or positions of authors' employer, affiliates, organization, committee or other group or individual.

References

1. Tronvoll, S. A., Elverum, C. W., & Welo, T. (2018). Dimensional accuracy of threads manufactured by fused deposition modeling. *Procedia Manufacturing, 26*, 763–773. https://doi.org/10.1016/j.promfg.2018.07.088
2. Ochoa, E., Morelli, L., Salvioni, L., et al. (2024). Co-processed materials testing as excipients to produce orally disintegrating tablets (ODT) using Binder Jet 3D-printing technology. *European Journal of Pharmaceutics and Biopharmaceutics, 194*, 85–94. https://doi.org/10.1016/j.ejpb.2023.11.023
3. Ansari, M. J., Arcondoulis, E., Roccisano, A., Schulz, C., Schläfer, T., & Hall, C. (2024). Optimized analytical approach for the detection of process-induced defects using acoustic emission during directed energy deposition process. *Additive Manufacturing, 86*, 104218. https://doi.org/10.1016/j.addma.2024.104218
4. Suvanjumrat, C., Chansoda, K., & Chookaew, W. (2024). Additive manufacturing advancement through large-scale screw-extrusion 3D printing for precision Parawood powder/Pla furniture production. *Cleaner Engineering and Technology, 20*, 100753. https://doi.org/10.1016/j.clet.2024.100753
5. Sireesha, M., Lee, J., Kiran, A. S. K., Babu, V. J., Kee, B. B. T., & Ramakrishna, S. (2018). A review on additive manufacturing and its way into the oil and gas industry. *RSC Advances, 8*, 22460. https://doi.org/10.1039/C8RA03194K
6. Deshmukh, K., Riensche, A., Bevans, B., Lane, R. J., Snyder, K., Halliday, H., Williams, C. B., Mirzaeifar, R., & Rao, P. (2024). Effect of processing parameters and thermal history on microstructure evolution and functional properties in laser powder bed fusion of 316L. *Materials & Design, 244*, 113136. https://doi.org/10.1016/j.matdes.2024.113136

7. Fidan, I., Naikwadi, V., Alkunte, S., Mishra, R., & Tantawi, K. (2024). Energy efficiency in additive manufacturing: Condensed review. *Technologies, 12*(2), 21. https://doi.org/10.3390/technologies12020021
8. Huang, J., Zhang, B., Xiao, J., & Zhang, Q. (2022). An approach to improve the resolution of DLP 3D printing by parallel mechanism. *Applied Sciences, 12*(24), 12905. https://doi.org/10.3390/app122412905
9. Shanthar, R., Chen, K., & Abeykoon, C. (2023). Powder-based additive manufacturing: A critical review of materials, methods, opportunities, and challenges. *Advanced Engineering Materials, 25*(19), 2300375. https://doi.org/10.1002/adem.202300375
10. Padmakumar, M. (2020). Additive manufacturing of tungsten carbide hardmetal parts by selective laser melting (SLM), selective laser sintering (SLS) and Binder Jet 3D Printing (BJ3DP) techniques. *Lasers in Manufacturing and Materials Processing, 7*, 338–371. https://doi.org/10.1007/s40516-020-00124-0
11. Simchi, A. (2004). The role of particle size on the laser sintering of iron powder. *Metallurgical and Materials Transactions B, 35*, 937–948. https://doi.org/10.1007/s11663-004-0088-3
12. Spierings, A. B., Herres, N., & Levy, G. (2011). Influence of the particle size distribution on surface quality and mechanical properties in AM steel parts. *Rapid Prototyping Journal, 17*(3), 195–202. https://doi.org/10.1108/13552541111124770
13. Narasimhan, K. S., & Amuda, M. O. H. (2017). Powder characterization. In *Reference module in materials science and materials engineering*. Elsevier. https://doi.org/10.1016/B978-0-12-803581-8.10141-9
14. Opatová, K., Zetková, I., & Kučerová, L. (2020). Relationship between the size and inner structure of particles of virgin and re-used MS1 maraging steel powder for additive manufacturing. *Materials, 13*(4), 956. https://doi.org/10.3390/ma13040956
15. Muthuswamy, P. (2022). Influence of powder characteristics on properties of parts manufactured by metal additive manufacturing. *Lasers in Manufacturing and Materials Processing, 9*, 312–337. https://doi.org/10.1007/s40516-022-00177-3
16. Heaney, D. F. (2012). Powders for metal injection molding (MIM). In D. F. Heaney (Ed.), *Handbook of metal injection molding* (pp. 50–63). Woodhead Publishing. https://doi.org/10.1533/9780857096234.1.50
17. Deng, X. L., & Davé, R. N. (2013). Dynamic simulation of particle packing influenced by size, aspect ratio and surface energy. *Granular Matter, 15*, 401–415. https://doi.org/10.1007/s10035-013-0413-0
18. Dai, L., Sorkin, V., Vastola, G., & Zhang, Y. W. (2019). Dynamics calibration of particle sandpile packing characteristics via discrete element method. *Powder Technology, 347*, 220–226. https://doi.org/10.1016/j.powtec.2019.03.008
19. Nguyen, Q. B., Nai, M. L. S., Zhu, Z., Sun, C.-N., Wei, J., & Zhou, W. (2017). Characteristics of inconel powders for powder-bed additive manufacturing. *Engineering, 3*(5), 695–700. https://doi.org/10.1016/J.ENG.2017.05.012
20. Vo, T. T., Nezamabadi, S., Mutabaruka, P., et al. (2020). Additive rheology of complex granular flows. *Nature Communications, 11*, 1476. https://doi.org/10.1038/s41467-020-15263-3
21. Lewis, T. B., & Nielsen, L. E. (1968). Viscosity of dispersed and aggregated suspensions of spheres. *Transactions of the Society of Rheology, 12*(3), 421–443. https://doi.org/10.1122/1.549114
22. Strondl, A., Lyckfeldt, O., Brodin, H., et al. (2015). Characterization and control of powder properties for additive manufacturing. *JOM, 67*, 549–554. https://doi.org/10.1007/s11837-015-1304-0
23. Averardi, A., Cola, C., Zeltmann, S. E., & Gupta, N. (2020). Effect of particle size distribution on the packing of powder beds: A critical discussion relevant to additive manufacturing. *Materials Today Communications, 24*, 100964. https://doi.org/10.1016/j.mtcomm.2020.100964

24. Niu, H. J., & Chang, I. T. H. (1999). Selective laser sintering of gas and water atomized high speed steel powders. *Scripta Materialia, 41*(1), 25–30. https://doi.org/10.1016/S1359-6462(99)00089-5
25. Miyanaji, H., Momenzadeh, N., & Yang, L. (2019). Effect of powder characteristics on parts fabricated via binder jetting process. *Rapid Prototyping Journal, 25*(2), 332–342. https://doi.org/10.1108/RPJ-03-2018-0069
26. Abd-Elghany, K., & Bourell, D. L. (2012). Property evaluation of 304L stainless steel fabricated by selective laser melting. *Rapid Prototyping Journal, 18*(5), 420–428. https://doi.org/10.1108/13552541211250418
27. Mostafaei, A., Rodriguez De Vecchis, P., Nettleship, I., & Chmielus, M. (2019). Effect of powder size distribution on densification and microstructural evolution of binder-jet 3D-printed alloy 625. *Materials & Design, 162*, 375–383. https://doi.org/10.1016/j.matdes.2018.11.051
28. Lutter-Günther, M., Horn, M., Seidel, C., & Reinhart, G. (2017). Influence of particle size distribution on powder flowability and part properties in laser beam melting. In *Rapid. Tech—International trade show & conference for additive manufacturing*. https://doi.org/10.3139/9783446454606.022.
29. Riedl, A., Schlederer, M., Pudelko, K., Stadler, M., Walter, S., Unterleuthner, D., et al. (2017). Comparison of cancer cells in 2D vs 3D culture reveals differences in AKT–mTOR–S6K signaling and drug responses. *Journal of Cell Science, 130*, 203–218.
30. Keck, C. M., & Müller, R. H. (2008). Size analysis of submicron particles by laser diffractometry—90% of the published measurements are false. *International Journal of Pharmaceutics, 355*(1–2), 150–163. https://doi.org/10.1016/j.ijpharm.2007.12.004
31. Sutton, A. T., et al. (2016). Powders for additive manufacturing processes: characterization techniques and effects on part properties. In *Proceedings of the 27th annual international solid freeform fabrication symposium* (pp. 1004–1030). University of Texas at Austin. Retrieved from https://utw10945.utweb.utexas.edu/events/sff
32. Starr, T. L., Rafi, K., Stucker, B., & Scherzer, C. M. (2012). Controlling phase composition in selective laser melted stainless steels. In *23rd annual international solid freeform fabrication symposium—An additive manufacturing conference* (pp. 439–446). SFF.
33. Murr, L. E., Martinez, E., Hernandez, J., Collins, S., Amato, K. N., Gaytan, S. M., & Shindo, P. W. (2012). Microstructures and properties of 17-4 PH stainless steel fabricated by selective laser melting. *Journal of Materials Research and Technology, 1*(3), 167–177. https://doi.org/10.1016/S2238-7854(12)70029-7
34. Ding, Y., Muñiz-Lerma, J., Trask, M., Chou, S., Walker, A., & Brochu, M. (2016). Microstructure and mechanical property considerations in additive manufacturing of aluminum alloys. *MRS Bulletin, 41*(10), 745–751. https://doi.org/10.1557/mrs.2016.214
35. Slotwinski, J. A., & Garboczi, E. J. (2015). Metrology needs for metal additive manufacturing powders. *JOM, 67*, 538–543. https://doi.org/10.1007/s11837-014-1290-7
36. Heim, K., Bernier, F., Pelletier, R., & Lefebvre, L.-P. (2016). High resolution pore size analysis in metallic powders by X-ray tomography. *Case Studies in Nondestructive Testing and Evaluation, 6*(Part A), 45–52. https://doi.org/10.1016/j.csndt.2016.09.002
37. Yang, J., Li, B., Liu, J., Tu, Z., & Wu, X. (2024). Application of additive manufacturing in the automobile industry: A mini review. *Processes, 12*(6), 1101. https://doi.org/10.3390/pr12061101
38. Nayeem, A. M., & Hossain, M. M. N. (2023). Usage of additive manufacturing in the automotive industry: A review. *Bangladesh Journal of Multidisciplinary Science Research, 8*, 9–20.
39. Verboeket, V., & Krikke, H. (2019). AM is recognized as a game-changer for the production process by replacing traditional manufacturing. *Logistics, 3*, 13.
40. Schuhmann, D., Rockinger, C., Merkel, M., & Harrison, D. K. (2022). A study on additive manufacturing for electromobility. *World Electric Vehicle Journal, 13*(8), 154. https://doi.org/10.3390/wevj13080154
41. Herzog, D., Seyda, V., Wycisk, E., & Emmelmann, C. (2016). Additive manufacturing of metals. *Acta Materialia, 117*, 371–392.

42. Lakhdar, Y., Tuck, C., Binner, J., Terry, A., & Goodridge, R. (2021). Additive manufacturing of advanced ceramic materials. *Progress in Materials Science, 116*, 100736.
43. Anusci, V. (2019). *BMW S58 engine: 3D printed cylinder*. Retrieved from https://www.voxelmatters.com/bmw-s58-engine-3d-printed-cylinder/
44. Agnusdei, L., & Prete, A. D. (2022). Additive manufacturing for sustainability: A systematic literature review. *Sustainable Futures, 4*, 100098.
45. Blakey-Milner, B., Gradl, P., Snedden, G., Brooks, M., & Pitot, J. (2021). Metal additive manufacturing in aerospace: A review. *Materials & Design, 209*, 110008. https://doi.org/10.1016/j.matdes.2021.110008
46. Alami, A. H., Olabi, A. G., Alashkar, A., Alasad, S., Aljaghoub, H., Rezk, H., & Abdelkareem, M. A. (2023). Additive manufacturing in the aerospace and automotive industries: Recent trends and role in achieving sustainable development goals. *Ain Shams Engineering Journal, 14*(11), 102516. https://doi.org/10.1016/j.asej.2023.102516
47. NASA. (2020). *NASA's perseverance rover bringing 3D printed metal parts to Mars*. Retrieved from https://www.nasa.gov/centers-and-facilities/jpl/nasas-perseverance-rover-bringing-3d-printed-metal-parts-to-mars/
48. Buj-Corral, I., Tejo-Otero, A., & Fenollosa-Artés, F. (2020). Development of AM technologies for metals in the sector of medical implants. *Metals, 10*(5), 686. https://doi.org/10.3390/met10050686
49. Zhu, L. Y., Li, L., Shi, J. P., Li, Z. A., & Yang, J. Q. (2018). Mechanical characterization of 3D printed multi-morphology porous Ti6Al4V scaffolds based on triply periodic minimal surface architectures. *American Journal of Translational Research, 10*, 3443–3454.
50. Ma, Y., Hu, N., Liu, J., Zhai, X., Wu, M., Hu, C., et al. (2019). Three-dimensional printing of biodegradable piperazine-based polyurethane-urea scaffolds with enhanced osteogenesis for bone regeneration. *ACS Applied Materials & Interfaces, 11*(9), 9415–9424.
51. Zhao, H., Chen, Y., Shao, L., Xie, M., Nie, J., Qiu, J., et al. (2018). Airflow-assisted 3D bioprinting of human heterogeneous microspheroidal organoids with microfluidic nozzle. *Small, 14*, e1802630.
52. Kim, Y., Kang, K., Yoon, S., Kim, J. S., Park, S. A., Kim, W. D., et al. (2018). Prolongation of liver-specific function for primary hepatocytes maintenance in 3D printed architectures. *Organogenesis, 14*, 1–12.
53. Mathews Griner, L. A., Zhang, X., Guha, R., McKnight, C., Goldlust, I. S., Lal-Nag, M., et al. (2016). Large-scale pharmacological profiling of 3D tumor models of cancer cells. *Cell Death & Disease, 7*, e2492.
54. Nauth, A., Schemitsch, E., Norris, B., Nollin, Z., & Watson, J. T. (2018). Critical-size bone defects: Is there a consensus for diagnosis and treatment? *Journal of Orthopaedic Trauma, 32*(Suppl. 1), S7–S11. https://doi.org/10.1097/BOT.0000000000001143
55. Vignesh, U., Mehrotra, D., Vaibhav Anand, D., & Howlader, D. (2017). Three-dimensional reconstruction of late post traumatic orbital wall defects by customized implants using CAD-CAM, 3D stereolithographic models: A case report. *Journal of Oral Biology and Craniofacial Research, 7*, 212–218.
56. Gioumouxouzis, C. I., Karavasili, C., & Fatouros, D. G. (2018). Recent advances in pharmaceutical dosage forms and devices using additive manufacturing technologies. *Drug Discovery Today, 24*(3), 636–643. https://doi.org/10.1016/j.drudis.2018.01.011
57. George, M., Aroom, K. R., Hawes, H. G., Gill, B. S., & Love, J. (2017). 3D printed surgical instruments: The design and fabrication process. *World Journal of Surgery, 41*(1), 314–319. https://doi.org/10.1007/s00268-016-3702-3
58. Paolini, A., Kollmannsberger, S., & Rank, E. (2019). Additive manufacturing in construction: A review on processes, applications, and digital planning methods. *Additive Manufacturing, 30*, 100894. https://doi.org/10.1016/j.addma.2019.100894
59. Al Rashid, A., Alim Khan, S., Al-Ghamdi, S. G., & Koç, M. (2020). Additive manufacturing: Technology, applications, markets, and opportunities for the built environment. *Automation in Construction, 118*, 103268. https://doi.org/10.1016/j.autcon.2020.103268

60. Wu, P., Wang, J., & Wang, X. (2016). A critical review of the use of 3-D printing in the construction industry. *Automation in Construction, 68*, 21–31. https://doi.org/10.1016/j.autcon.2016.04.005
61. Raja Santhi, A., & Muthuswamy, P. (2022a). Influence of blockchain technology in manufacturing supply chain and logistics. *Logistics, 6*(1), 15. https://doi.org/10.3390/logistics6010015
62. Raja Santhi, A., & Muthuswamy, P. (2022b). Pandemic, war, natural calamities, and sustainability: Industry 4.0 technologies to overcome traditional and contemporary supply chain challenges. *Logistics, 6*(4), 81. https://doi.org/10.3390/logistics6040081
63. Jung, S., Kara, L. B., Nie, Z., Simpson, T. W., & Whitefoot, K. S. (2023). Is additive manufacturing an environmentally and economically preferred alternative for mass production? *Environmental Science & Technology, 57*(16), 6373–6386. https://doi.org/10.1021/acs.est.2c04927
64. Ribeiro, I., Matos, F., Jacinto, C., Salman, H., Cardeal, G., Carvalho, H., Godina, R., & Peças, P. (2020). Framework for life cycle sustainability assessment of additive manufacturing. *Sustainability, 12*(3), 929. https://doi.org/10.3390/su12030929
65. Alexander, P., Allen, S., & Dutta, D. (1998). Part orientation and build cost determination in layered manufacturing. *Computer-Aided Design, 30*(5), 343–356. https://doi.org/10.1016/s0010-4485(97)00083-3
66. Hopkinson, N., & Dicknes, P. (2003). Analysis of rapid manufacturing—using layer manufacturing processes for production. *Proceedings of the Institution of Mechanical Engineers, Part C: Journal of Mechanical Engineering Science, 217*(1), 31–39. https://doi.org/10.1243/095440603762554596
67. Huang, S. H., Liu, P., Mokasdar, A., & Hou, L. (2012). Additive manufacturing and its societal impact: A literature review. *International Journal of Advanced Manufacturing Technology, 67*(5–8), 1191–1203. https://doi.org/10.1007/s00170-012-4558-5
68. Rejeski, D., Zhao, F., & Huang, Y. (2018). Research needs and recommendations on environmental implications of additive manufacturing. *Additive Manufacturing, 19*, 21–28. https://doi.org/10.1016/j.addma.2017.10.019
69. Chen, D., Heyer, S., Ibbotson, S., Salonitis, K., Steingrímsson, J. G., & Thiede, S. (2015). Direct digital manufacturing: Definition, evolution, and sustainability implications. *Journal of Cleaner Production, 107*, 615–625. https://doi.org/10.1016/j.jclepro.2015.05.009
70. Gershenfeld, N. (2012). How to make almost anything: The digital fabrication revolution. *Foreign Affairs, 91*, 43–57.
71. Ford, S., & Despeisse, M. (2016). Additive manufacturing and sustainability: An exploratory study of the advantages and challenges. *Journal of Cleaner Production, 137*, 1573–1587. https://doi.org/10.1016/j.jclepro.2016.04.150
72. Pearce, J. M., Morris Blair, C., Laciak, K. J., Andrews, R., Nosrat, A., & Zelenika-Zovko, I. (2010). 3-D printing of open source appropriate technologies for self-directed sustainable development. *Journal of Sustainable Development, 3*(4), 17–29. https://doi.org/10.5539/jsd.v3n4p17
73. Gebler, M., Schoot Uiterkamp, A. J. M., & Visser, C. (2014). A global sustainability perspective on 3D printing technologies. *Energy Policy, 74*, 158–167. https://doi.org/10.1016/j.enpol.2014.08.033
74. Muthuswamy, P., & K, S. (2022). Artificial intelligence based tool condition monitoring for Digital Twins and Industry 4.0 Applications. *International Journal on Interactive Design and Manufacturing (IJIDeM), 17*(3), 1067–1087. https://doi.org/10.1007/s12008-022-01050-5
75. Santhi, A. R., & Muthuswamy, P. (2023). Industry 5.0 or industry 4.0S? Introduction to industry 4.0 and a peek into the prospective industry 5.0 technologies. *International Journal of Interactive Design and Manufacturing, 17*(4), 947–979. https://doi.org/10.1007/s12008-023-01217-8
76. MacCurdy, R., Katzschmann, R., Youbin, K., & Rus, D. (2016). Printable hydraulics: A method for fabricating robots by 3D co-printing solids and liquids. In *IEEE international*

conference on robotics and automation (ICRA) (pp. 3878–3885). https://doi.org/10.1109/ICRA.2016.7487576.

77. Maccurdy, R., Katzschmann, R., Kim, Y., et al. (2014). 3D printing for the rapid prototyping of structural electronics. *IEEE Access, 2*, 234–242. https://doi.org/10.1109/ACCESS.2014.2311810
78. Ota, H., Emaminejad, S., Gao, Y., et al. (2016). Application of 3D printing for smart objects with embedded electronic sensors and systems. *Advanced Materials Technologies, 1*(1), 1600013. https://doi.org/10.1002/admt.201600013
79. Weflen, E., & Frank, M. C. (2021). Hybrid additive and subtractive manufacturing of multi-material objects. *Rapid Prototyping Journal, 27*(10), 1860–1871. https://doi.org/10.1108/RPJ-06-2020-0142

Additive Manufacturing and 3D Printing Revolution in Industry 5.0

Ngoc-Hien Tran and Thai-Son Nguyen

1 Introduction

Additive manufacturing (AM) and 3D printing or layer-by-layer manufacturing have been effectively applied in fields such as aerospace, automotive engineering, biomedical, engineering and energy [1–3]. To gain industrial acceptance of AM in addition to traditional manufacturing, AM has been researched, improved, and become one of the core technologies of the fourth and fifth industrial revolutions [4–6]. Figure 1 shows the changes in industry from the first industrial revolution to the present and the development of AM as well as the contribution and role of AM in the fourth and fifth industrial revolutions [7, 8]. In the initial period from 1984 to 2012, AM contributed to the industrial revolution with the role of rapid prototyping of products with plastic materials in the product design stage [5, 9]. From 2012 to now, with the invention of new printing methods, it is possible to print products with a variety of materials, especially metal materials [10, 11]. Printed products ensure the required quality and physical and mechanical properties.

Smart manufacturing and automation are applied to manufacturing companies to achieve efficiency, productivity, quality and competitiveness. Industry 4.0 with cyber-physical systems, Internet of Things (IoT), and artificial intelligence (AI), big data allows to increase productivity and quality of production systems [12, 13]. With the industrial revolution 5.0 taking people as the center, human-centric solutions, autonomy, cognitive systems and digitization are considered as core elements

N.-H. Tran (✉)
Faculty of Mechanical Engineering, University of Transport and Communications, Hanoi, Vietnam
e-mail: tranhien.tkm@utc.edu.vn

T.-S. Nguyen
Center for Transport Science and Technology, University of Transport and Communications, Hanoi, Vietnam

Y. Kumar Singla et al. (eds.), *Digital Manufacturing in Industry 5.0*,
https://doi.org/10.1007/978-3-031-91500-0_6

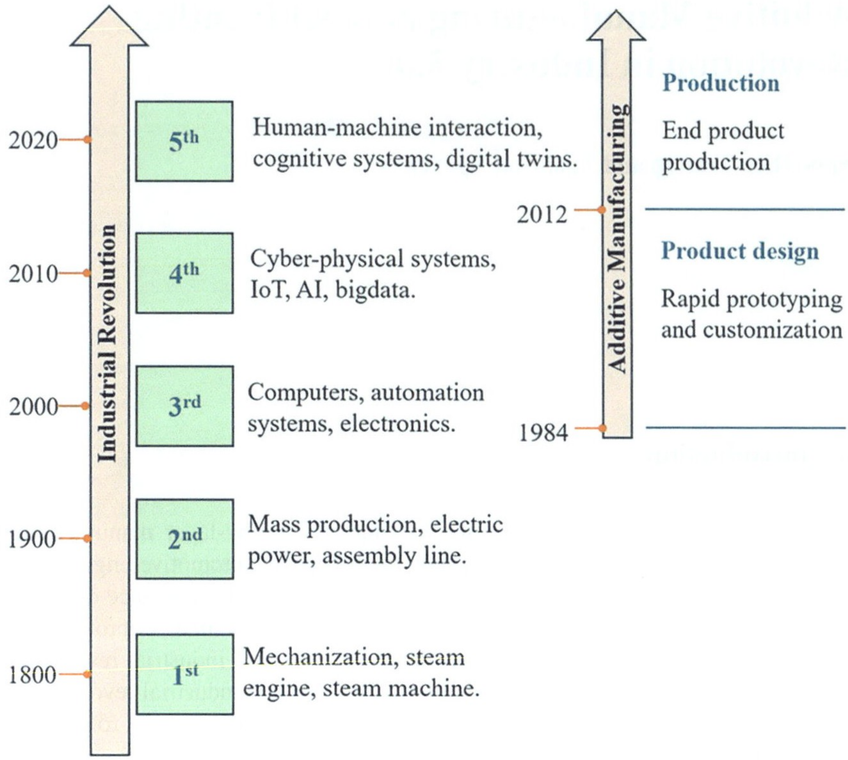

Fig. 1 Industrial revolution and the development of additive manufacturing technology

[14]. With value-driven in Industry 5.0, core human needs and interests are placed at the center of the production process [15, 16].

With current manufacturing systems, the three-dimensional (3D) model of the product is designed on the computer through CAD (computer-aided design) tools. Then, with the application of CAM (computer-aided manufacturing) tools, it is possible to set up the tool path for the machining process as well as set up the machining process parameters. From there, the G-code program is generated for CNC (computer numerical control) machines to process the part. This process is shown by the horizontal line in Fig. 2. Along with the development of technology, reverse engineering technology [17, 18] has been applied in product quality evaluation (geometric accuracy of the product) through CMM—coordinate measuring machines (contact measurement method) or through the laser scanners (non-contact measurement method). These final products are scanned to check and evaluate geometric errors through 3D scanners and scan data processing software. The 3D model created from the scanned data is used to compare with the original 3D model to determine the dimensional deviation value after machining. Previously, 3D printing technology or layer-by-layer manufacturing technology was only used to create

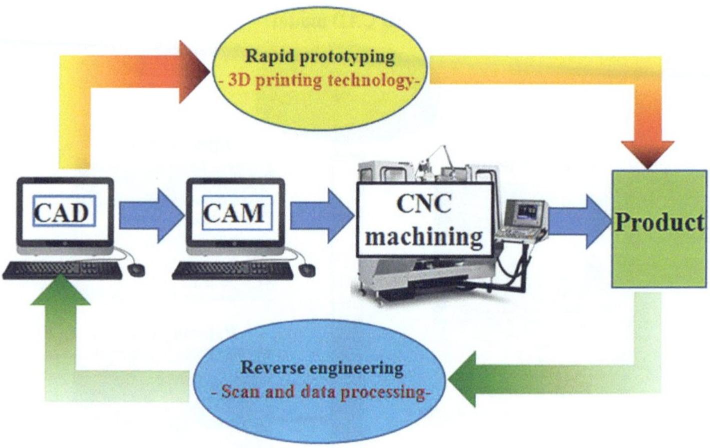

Fig. 2 Manufacturing systems with applying reverse engineering and 3D printing

initial samples, also known as rapid prototyping [19, 20]. This is a necessary stage in product research and development. With this application, 3D printing technology mainly uses plastic materials to create products. Today, with improvements in technology, 3D printing technology allows printing products with metal materials from the original raw materials in a powder or fiber form. These printed products have been proven to achieve not only geometrical requirements but also ensure physical and mechanical requirements, ensuring working function. Thus, with the application of 3D printing technology and reverse engineering technology, the production process is a closed loop.

Nowadays, there are many printing methods used, each printing method also applies to the corresponding material such as fused deposition modeling (FDM) printing method is often applied to create plastic products; selective laser melting (SLM) and direct metal laser sintering (DMLS) printing methods are often applied to print metal products from powder form. Printing materials are also very diverse from plastic to metal, in the form of fibers, powders, and liquids (usually for plastic materials).

For printing a part, the 3D CAD model in the STL format is sliced and sent to the printer. The printing parameters are defined on the machine for printing. After one layer is printed, another layer is added to the model. This process is repeated until the whole part is printed. Post-process is applied to the model for application. Steps for printing the part are shown in Fig. 3.

Additive manufacturing is classified according to the methodology of the formation of the printed product, types of printed materials such as solid, powder, or liquid, and thermal energy for processing. With the methodology of formation of the printed product as shown in Fig. 4, AM methods are classified such as powder bed fusion, directed energy deposition, binder jetting, material extrusion, vat photopolymerization, and sheet lamination [21].

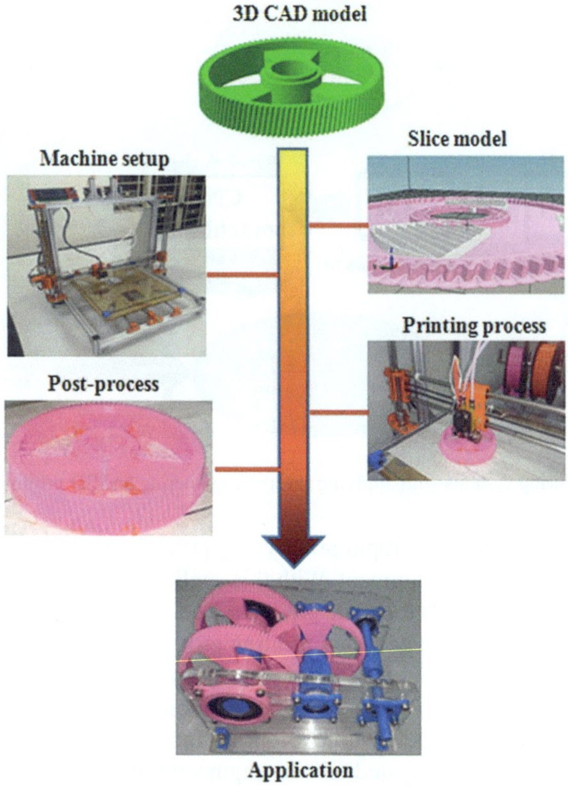

Fig. 3 Steps for printing the parts

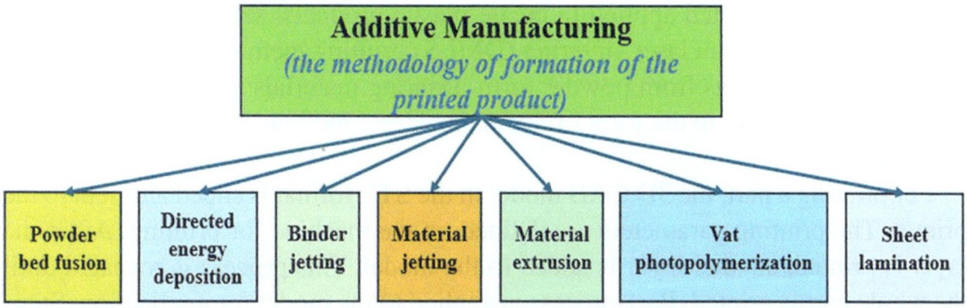

Fig. 4 Classification of additive manufacturing

According to the energy source, there are laser sources, electron beam, and ultrasonic. With the powder bed fusion, directed energy deposition, and sheet lamination printing methods using laser sources, the printing methods are classified into [22]:

- Powder bed fusion: selective laser sintering (SLS), hot isostatic pressing (HIP), direct metal sintering (DMLS), laser micro sintering, selective laser melting, and laser cussing process.
- Directed energy deposition: direct metal deposition (DMD), laser engineered net shaping (LENS), direct light fabrication (DLF), laser consolidation (LC), laser cladding, shape deposition manufacturing (SDM), wire feed laser deposition, and gas through wire.
- Sheet lamination: laminated object manufacturing (LOM).

With the development of science and technology, AM technology enables us to provide not only prototypes rapidly but also to fabricate functional parts for specific applications directly in a short time. The 3D printers enable to create products with complex shapes that traditional processing methods cannot do (e.g., creating cooling channels according to product contours for plastic injection molds) or creating specialized products for each customer such as in the medical field (printing bones and teeth suitable for each different customer), 3D printing technology is increasingly widely applied in many fields. Currently, there are many types of 3D printers with different printing methods and printing materials. Therefore, depending on the purpose of use, as well as the output requirements of the printed product to be achieved, decide to choose the appropriate printing method.

2 Solutions to Improve the Efficiency and Quality of AM and 3D Printing Technology in Industrial Applications

AM has been affirming its role in the industry. AM and 3D printing are increasingly developed and applied to create parts with complex shapes. Nowadays, AM is not only used to create initial sample parts but also used as a processing method to create final parts, meaning the parts ensure the required working functions. However, there are still some challenges to overcome, such as the physical properties of the material and warping caused by residual stress from the rapid cooling process.

3D printing technology is suitable for the trend of creating customer-oriented products and products with complex shapes, and it can also be used to produce final products. However, the quality of printed products is one of the major challenges for this technology to be widely applied in industry. Therefore, research not only on geometric accuracy but also on mechanical properties, internal structure of printed products has been invested. Without overcoming the above barriers, AM technology can hardly be used as a machining method to create final products like existing traditional machining methods.

Focus on the quality of printed products, in this chapter, a knowledge-based system selecting for an appropriate additive manufacturing process is developed. This software enables us to select the 3D printing method from the part quality. To avoid the failures of the printed part, the numerical method is applied with the proposal of

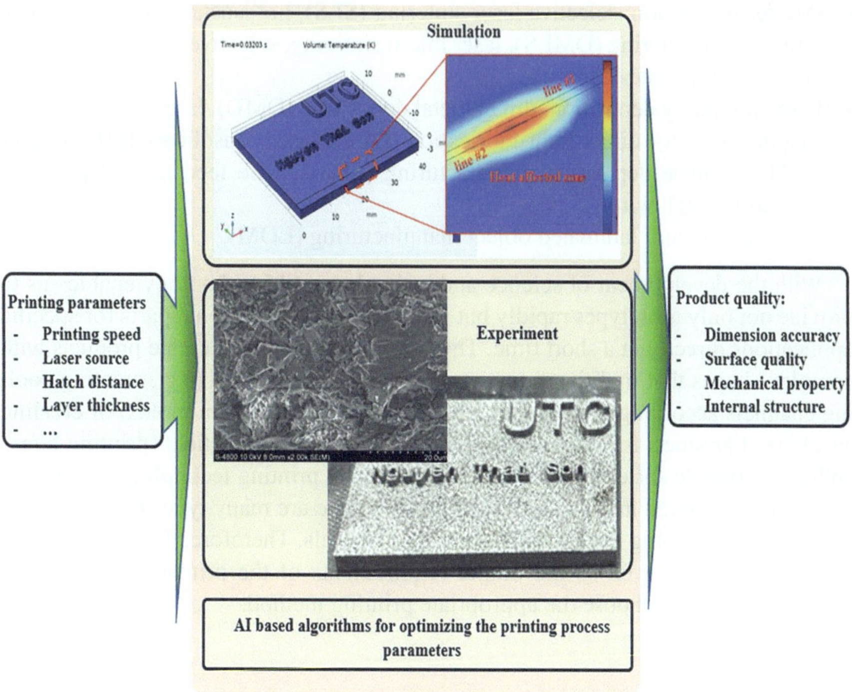

Fig. 5 Method for predicting the printed part quality

the model for calculating inherent strain. Then, the experiments and evaluation method are carried out as shown in Fig. 5. To improve the quality of the printing process, artificial intelligence is applied for AM systems. On the other hand, to ensure consistency in terminology, quality assessment methods, and printing methods, a standard system for AM has been and is being built and improved.

2.1 Development of a Knowledge System for Selecting the 3D Printing Method

From the material properties of the part and the detailed 3D model, the user needs to choose the appropriate material type and printing method. However, with many manufacturers providing printers with many different printing methods, choosing filament, powder, or liquid materials, and choosing metal or plastic materials is a difficult problem for the user. Therefore, a knowledge-based decision support system is necessary and useful for users to make decisions on choosing the appropriate printer, printing method, and printing material as shown in Fig. 6 [23].

Park and Tran [23] proposed a knowledge system for choosing an appropriate 3D printing as shown in Fig. 7. SQL was used for developing material databases,

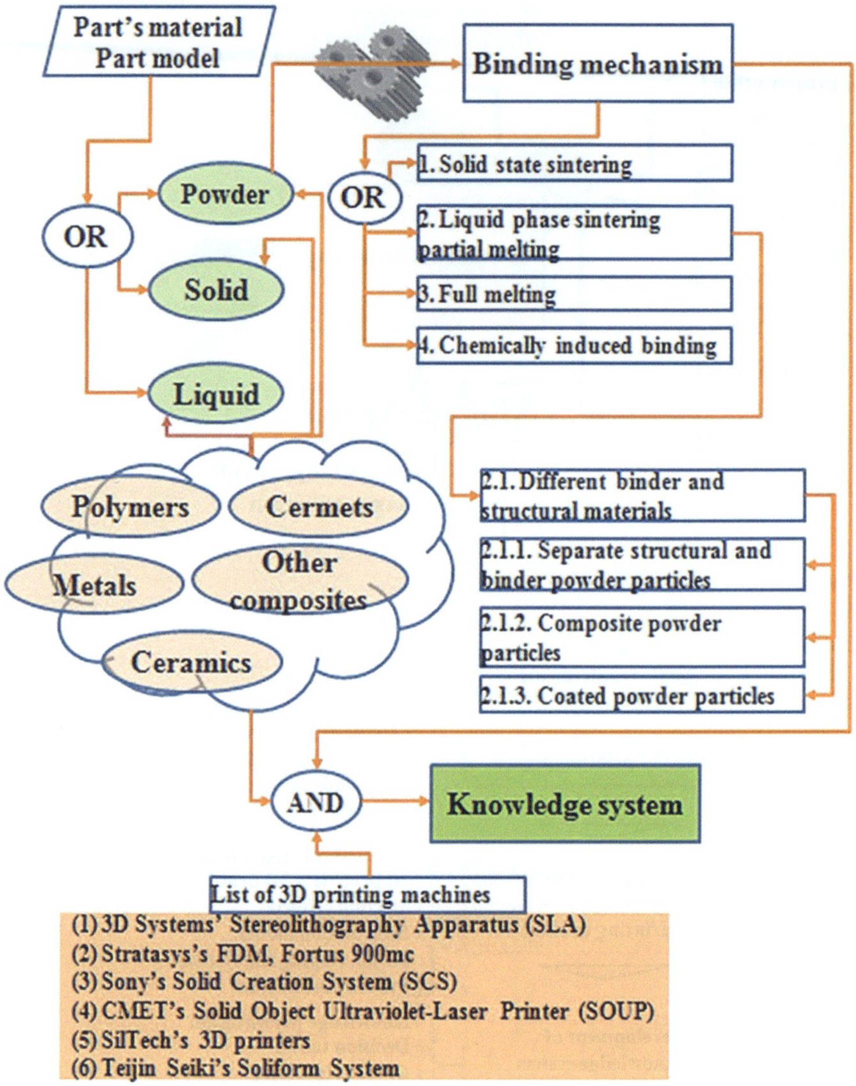

Fig. 6 Generation of the knowledge system

printing mechanisms, and 3D printing method information databases. From the printed part requirements, the most appropriate 3D printing method is selected using the knowledge system. To make this decision, criteria such as printing ability and printing quality are considered.

Figure 8 shows the systematic procedure for developing the proposed system with steps including analysis of printing process, engineering model for choosing the 3D printing method, developing the knowledge system, and implementing the

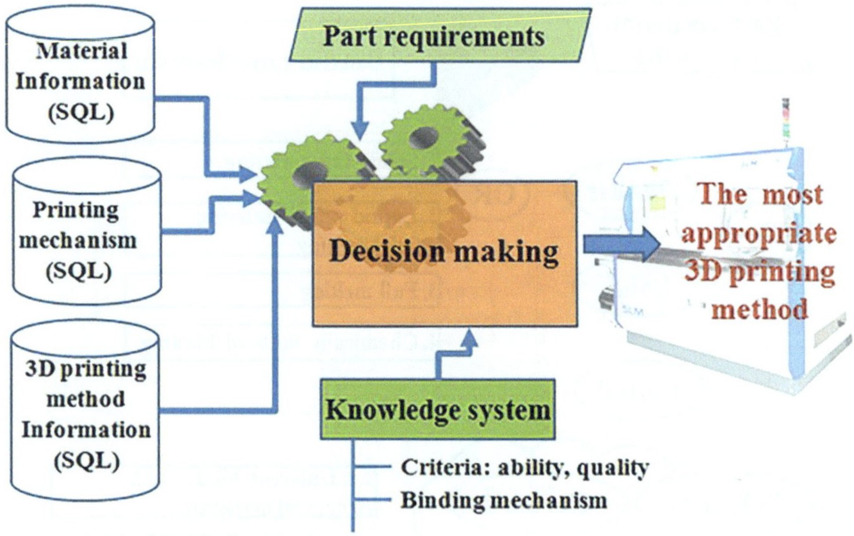

Fig. 7 Concept of the KBS-AM system

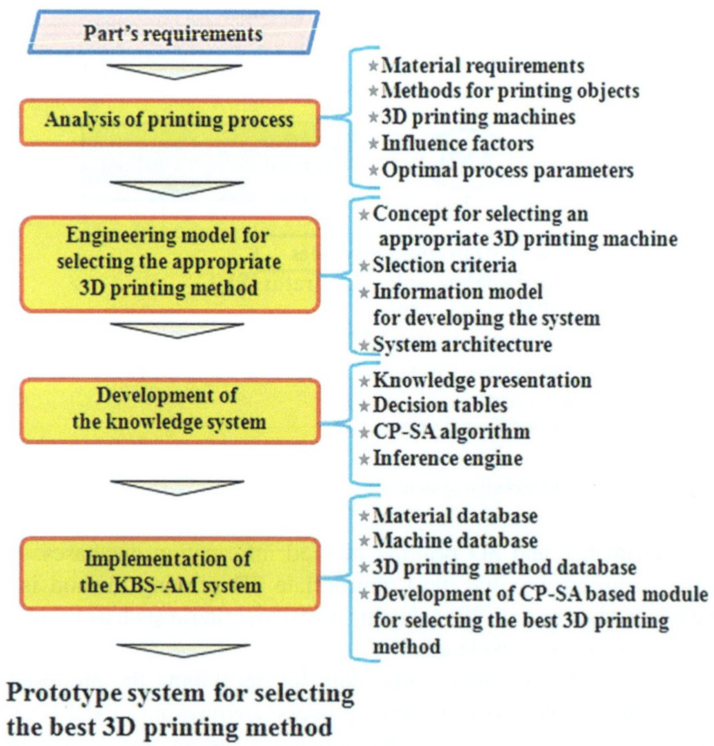

Fig. 8 Steps for implementing the knowledge system

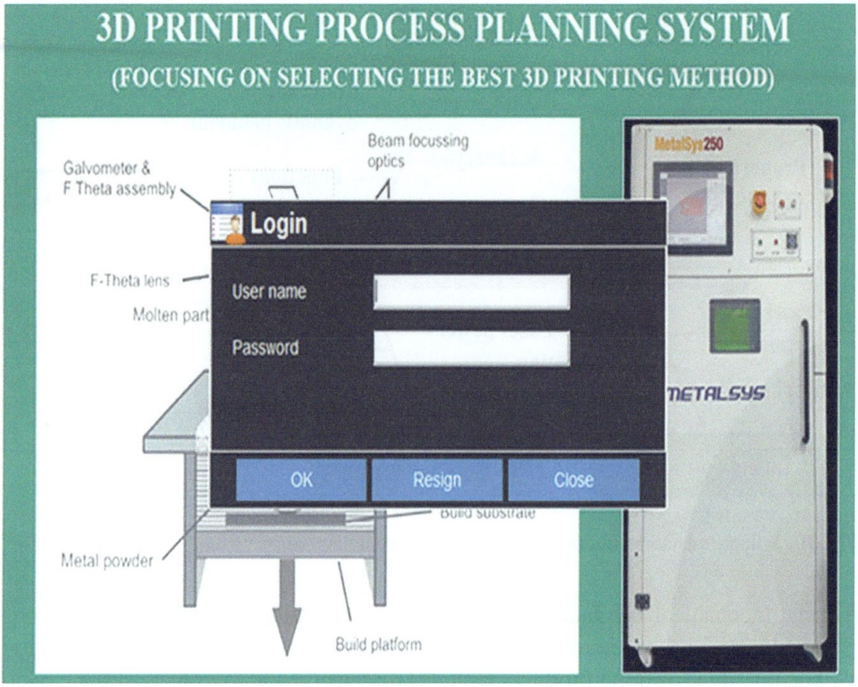

Fig. 9 The screenshot of the developed system

knowledge-based system for selecting an appropriate additive manufacturing process (KBS-AM) system. In which, the knowledge with databases and rules for decision making was built. Modules such as a selector and generating optimal process parameters are developed. Figure 9 shows the interface of the developed KBS-AM system.

2.2 *Numerical Method for Selecting the Printing Process Parameters*

It is necessary to have a clear understanding of the AM process, from the processability of the raw material to the relationship between the structural properties of 3D printed parts and the 3D printing process. Figure 10 shows the influence factors affecting the quality of the printed part, which are material, machine, printing condition, and printing parameters [24, 25]. In the layered manufacturing process, creating a final product requires the designer and manufacturer to have a clear understanding of the material properties, the phase transition of the material from solid to liquid and back to solid when rapidly cooled, as well as the

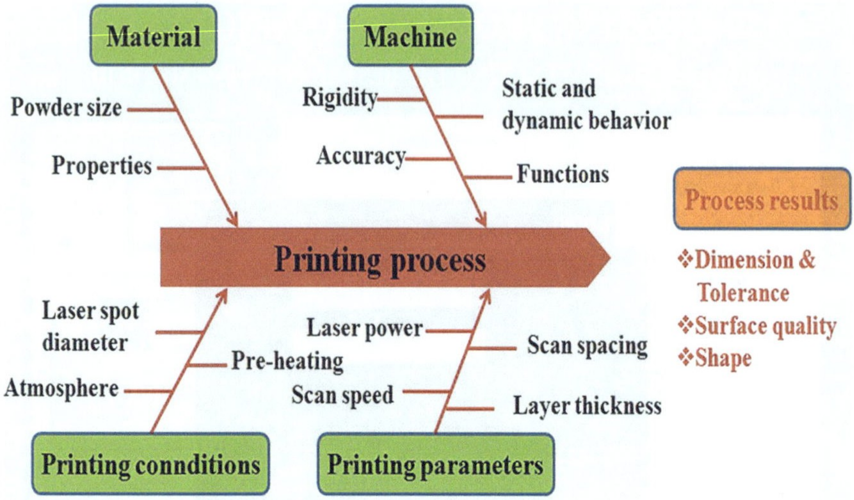

Fig. 10 Influence factors affecting the quality of printed part

thermal-mechanical mechanisms that affect print quality. Therefore, establishing physical model-based simulation models is necessary.

The quality of printed products includes dimension and tolerance, surface quality, shape, mechanical properties, and microstructure. The above quality parameters are influenced by material, machine, printing condition, and printing parameters as shown in Fig. 11. Regarding the material, the influencing parameters include powder size (for metallic materials) and phase transition properties of the material. Regarding the printing machine, the parameters affecting the printing quality include machine precision, machine function, machine dynamics and kinematics, and machine rigidity. The printing condition parameters include atmosphere, laser spot diameter, and pre-heating for materials. The printing process parameters include laser power (for printing methods using laser sources), printing temperature, printing speed, printing distance, and layer thickness.

For the printing process of metal products, it is necessary to clearly determine the temperature and phase transition of the material to determine the appropriate printing parameters as shown in Fig. 12.

The simulation process will help us determine the thermal distribution when printing with the input parameters, including the printing process parameters and the thermal property parameters of the material. To conduct the simulation, it is necessary to establish a heat transfer model when printing.

The mathematical model for describing the heat transfer is as follows [26, 27]:

$$\rho C \frac{\partial T}{\partial t} + \rho C u \nabla T = \nabla \left(k \nabla T \right) + Q_{heat\ source} \tag{1}$$

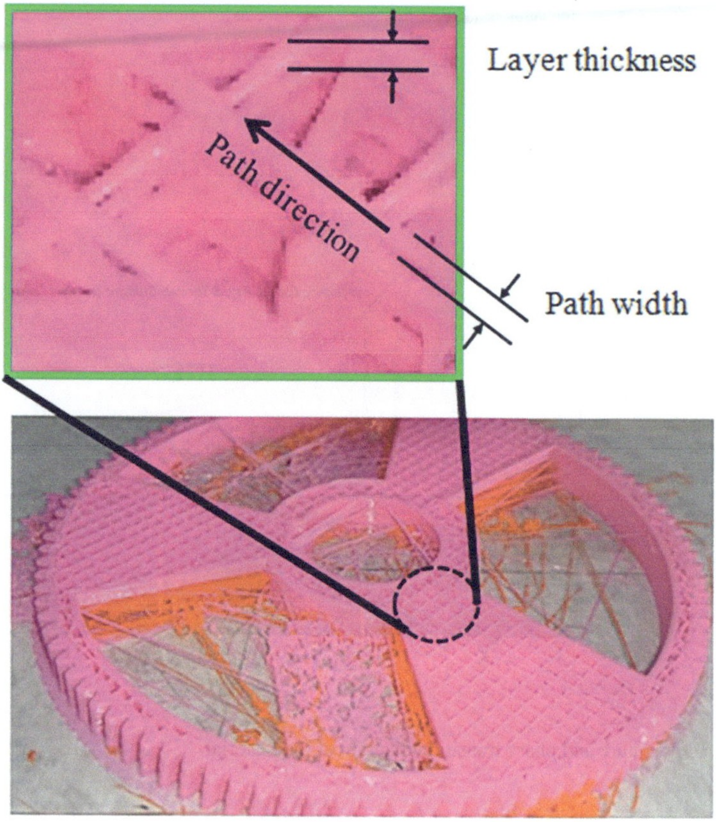

Fig. 11 Printing process parameters

where T is the temperature, ρ, C, k are density, thermal capacity, and thermal conductivity factors, respectively, $Q_{heat\ source}$ is the absorbed heat and u is the heat source velocity. Material properties such as the thermal conductivity coefficient and specific heat capacity are usually temperature dependent. The internal heat source exerts a significant influence on the thermal modeling of the 3D printing process. With the thermal model and the selected printing process parameters, the thermal distribution during the printing process is shown in Fig. 13. The highest temperature is 2500 K, which is suitable for printing process.

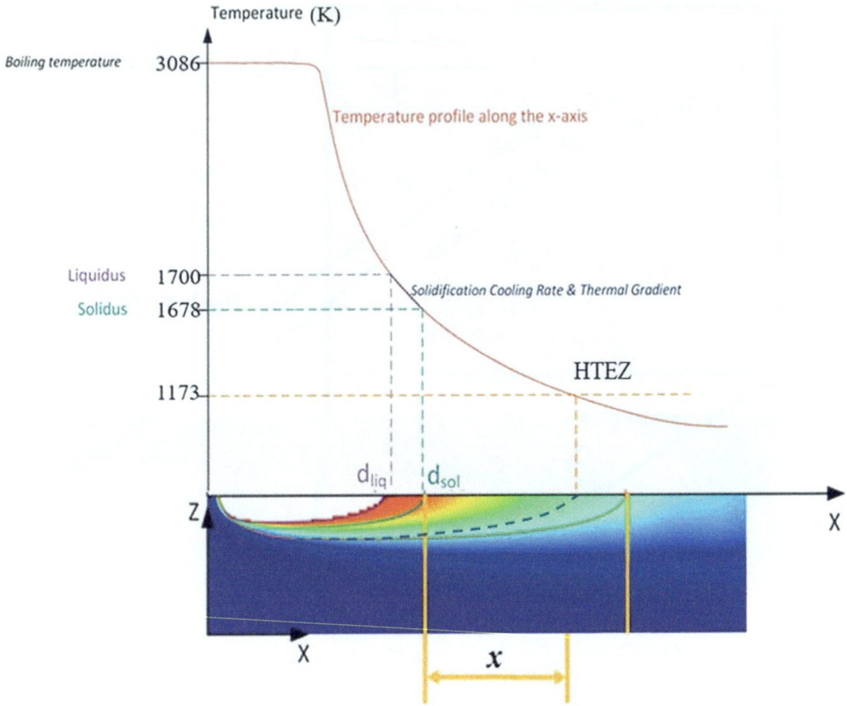

Fig. 12 Temperature and phase transition diagram of the material

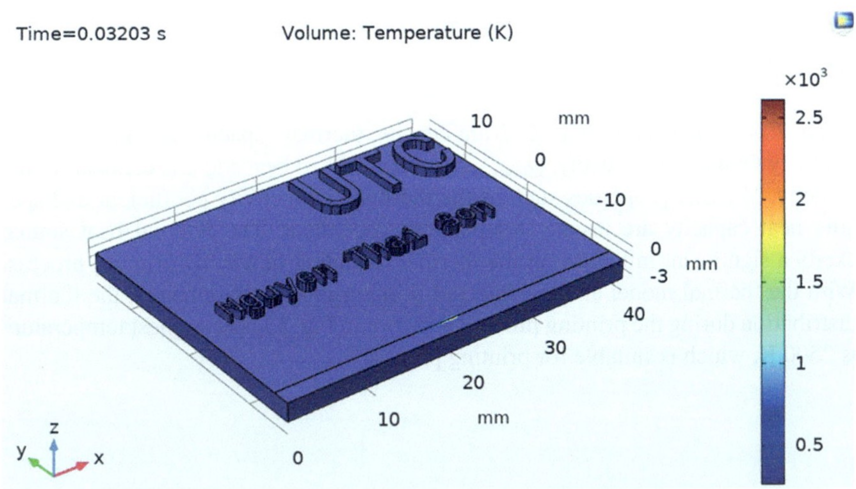

Fig. 13 Simulation result with the selected printing process parameters

2.3 *Proposing Theoretical Calculation of the Deformation Based on Inherent Strain*

The strain of the part is classified into elastic strain and non-elastic strain. Non-elastic strain is also called inherent strain. During the 3D printing process, due to the rapid cooling process, the product after being cut from the base plate will warp. The deformation of the printed product is the sum of elastic strain $\varepsilon_{elastic}$ and inherent strain ε^*. Inherent strain is defined as the sum of component strains including thermal strain $\varepsilon_{thermal}$, phase transformation strain ε_{phase}, plastic strain $\varepsilon_{plastic}$, and creep strain ε_{creep} [28].

$$\varepsilon_{total} = \varepsilon_{elastic} + \varepsilon^* \tag{2}$$

The inherent strain is calculated as follows:

$$\varepsilon^* = \varepsilon_{thermal} + \varepsilon_{plastic} + \varepsilon_{phase} + \varepsilon_{creep} \tag{3}$$

The strain and stress values are determined through theoretical formulas showing the relationship between these two parameters and experimental studies. Currently, some commercial software allows us to determine the inherent strain value; however, the method and algorithm that allow calculating the IS value are rarely published. For 3D-printed products, the main cause of inherent strain is the change in temperature from high temperature to ambient temperature in a short time. At the same time, the phase transition of the material from solid–liquid–solid. According to Chen [29] and Yazhi [30], the IS value according to the directions is determined as follows:

$$\varepsilon_x^* = -\frac{W_x}{F_x}; \varepsilon_y^* = -\frac{W_y}{F_y}; \varepsilon_z^* = -\frac{W_z}{F_z} \tag{4}$$

The heat treatment effect zone (HTEZ) areas in the respective planes are F_x, F_y, and F_z. In consideration of the linear energy density, the total volumes of inherent strain in the respective planes W_x, W_y, and W_z are calculated

$$W_x = \xi \cdot q_v; W_y = W_z = K \cdot q_v \tag{5}$$

the linear energy density q_v (J/mm) is determined as follows [31]:

$$q_v = \frac{Q_{heat\ source}}{u} \tag{6}$$

ξ and K (mm/J) are longitudinal and transverse inherent strain coefficients, respectively.

With the first layer, the inherent strain values in the x, y, and z directions are calculated as follows:

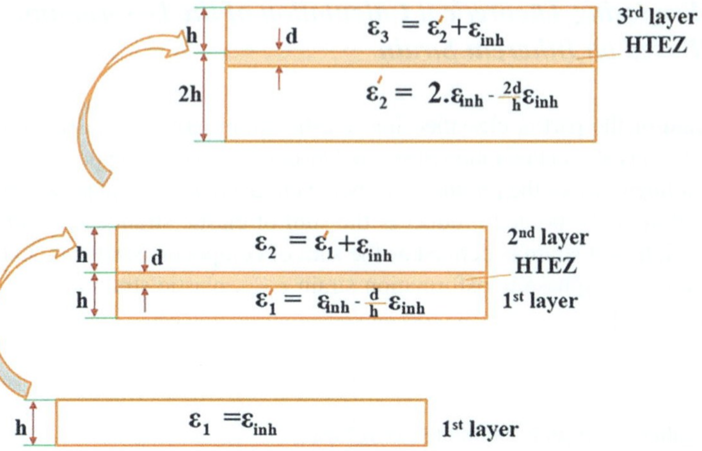

Fig. 14 Method for calculating the inherent strain for one layer and *n* layers

$$\varepsilon_{inh-x} = -\frac{\xi \bullet q_v}{F_x}; \varepsilon_{inh-y} = -\frac{K \bullet q_v}{F_y}; \varepsilon_{inh-z} = -\frac{K \bullet q_v}{F_z} \tag{7}$$

With *n* layers for printing the product, the method for calculating the inherent strain is shown in Fig. 14. After printing the first layer, the inherent strain value in this layer is ε_{inh}. This value will change if a second layer is printed onto the first layer by 3D printing mechanism due to the remelting of the first layer surface. The remaining value of inherent strain is calculated as follows [32]:

$$\varepsilon_1^{'} = \varepsilon_{inh} - \frac{d}{h}\varepsilon_{inh} \tag{8}$$

After the second layer printing process is finished, the inherent strain value in this layer is determined

$$\varepsilon_2 = \varepsilon_1^{'} + \varepsilon_{inh} = 2\varepsilon_{inh} - \frac{d}{h}\varepsilon_{inh} \tag{9}$$

The above mechanism is repeated when the third layer is printed. After finishing printing the third layer, the inherent strain value remaining in the second layer is

$$\varepsilon_2^{'} = \varepsilon_2 - \frac{d}{h}\varepsilon_{inh} = 2\varepsilon_{inh} - \frac{2d}{h}\varepsilon_{inh} \tag{10}$$

and the inherent strain value in the third layer is determined

$$\varepsilon_3 = \varepsilon_2^{'} + \varepsilon_{inh} = 3\varepsilon_{inh} - \frac{2d}{h}\varepsilon_{inh} \tag{11}$$

The above process is repeated, so we can determine the value of inherent strain for the nth layer as follows:

$$\varepsilon_n = \varepsilon'_{n-1} + \varepsilon_{inh} = n\varepsilon_{inh} - \frac{(n-1)d}{h}\varepsilon_{inh} \tag{12}$$

According to the thermal influence in different aspects, the values of inherent strain in x, y, z directions are also calculated accordingly:

$$\begin{cases} \varepsilon_{x(n\,layer)} = n\varepsilon_{inh-x} - \dfrac{(n-1)d}{h}\varepsilon_{inh-x} \\ \varepsilon_{y(n\,layer)} = n\varepsilon_{inh-y} - \dfrac{(n-1)d}{h}\varepsilon_{inh-y} \\ \varepsilon_{z(n\,layer)} = n\varepsilon_{inh-z} - \dfrac{(n-1)d}{h}\varepsilon_{inh-z} \end{cases} \tag{13}$$

From the proposed method for calculating the inherent strain, we have developed the system for calculating the IS in additive manufacturing. The system model with four modules is shown in Fig. 15. Four modules are developed, which include modules for calculating the heat treatment effect zone (HTEZ) surfaces, calculating the inherent strain for one layer, calculating the inherent strain for whole part, and module for optimizing the difference between the simulation and experiment [33]. The data flow between these models includes HTEZ surface in X, Y, Z directions,

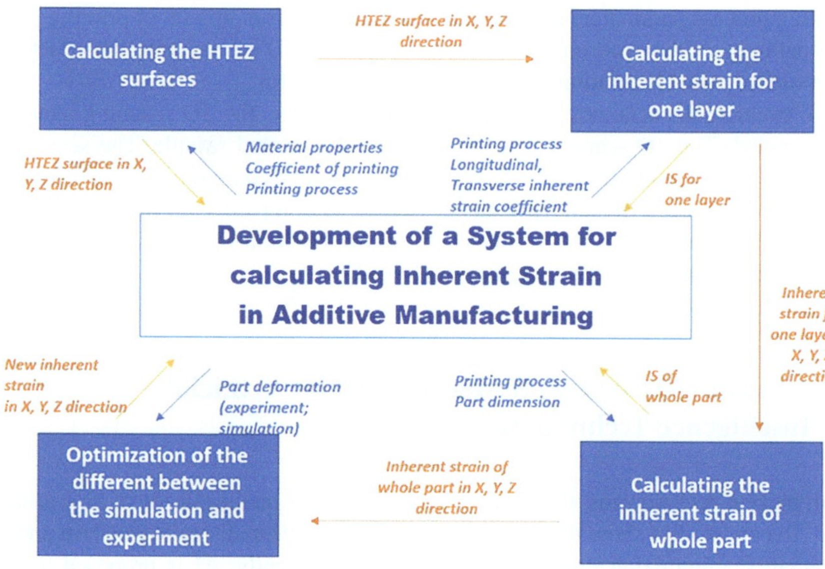

Fig. 15 A system for calculating inherent strain value

Fig. 16 Samples were printed by using the SLM printer

material properties, coefficient of printing, printing process, inherent strain value for one layer in X, Y, Z directions, IS of whole part in X, Y, Z directions, and part deformation.

2.4 Experimental Results and Evaluation

Experimental studies allow us to determine the relationship between printing process parameters and the printed product quality. However, experimental methods often require large numbers of experiments and high costs.

Therefore, with the simulation method, it is possible to determine the reasonable printing process parameters and avoid printing errors. From this set of printing process parameters, the experimental printing results on a 3D printer using the selective laser melting (SLM) printing method as shown in Fig. 16.

To evaluate the accuracy of the printed sample shape, the 3D scanning method was applied. The 3D scanner was used to scan the printed sample. The scanning result is a point cloud file. The point cloud data is processed on specialized software to create a solid 3D model. This model is used to compare with the original 3D model of the product. Deviations outside the allowable limit will be displayed in red color on the measured sample as shown in Fig. 17.

3 AM and 3D Printing Systems Based on Artificial Intelligence Techniques

With current AM systems, the issue of printed product quality needs to be considered. Recent studies show that applying AI to AM systems allows to improve product quality, minimizing printing errors [34, 35]. Currently, AI is proposed to be applied in both the product design stage and the product printing stage [36, 37]. In

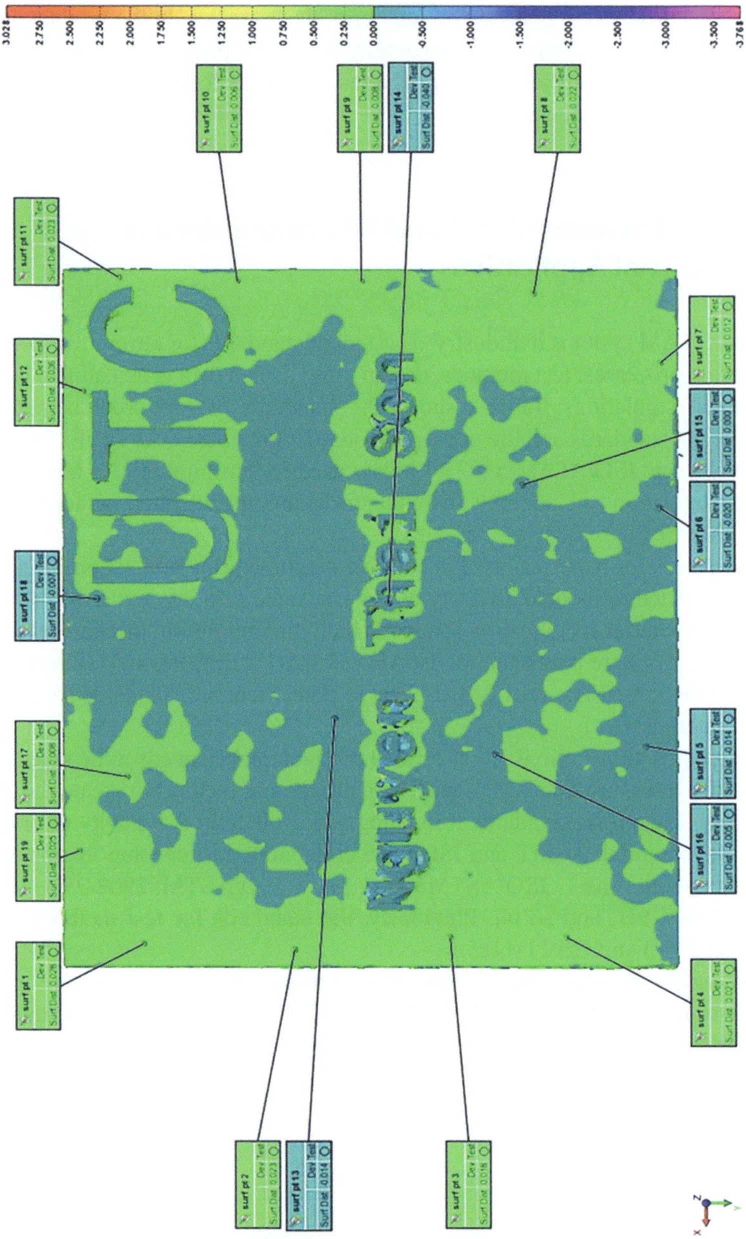

Fig. 17 Results of 3D scanning the printed sample

the product design stage, the use of AI allows to optimize the design to ensure the functionality of the part while reducing material costs. On the other hand, AI is also applied in printing planning such as using algorithms to recognize the characteristics of printed products to generate optimal printing strategies. In the product printing stage, AI is used to monitor product quality during the printing process. With the combination of data from sensors and cameras, AI-based control systems allow us to adjust the printing process to suit the current state of the system.

4 Standard Systems of AM and 3D Printing Technology in Industrial Applications

To implement AM systems in industry, a system of standards for terminology, materials, printing processes, test methods, designs, and design data is essential to establish common regulations. Table 1 shows the standards of AM processes, which include the old standards and the updated standards. Previously, ASTM F2792-12a, ISO 17296-1, and ISO/ASTM 52921-13 were used for terminology. Today, the terminology regulations in additive manufacturing use the updated standard ISO/ASTM52900-21.

The standards for materials used in additive manufacturing and 3D printing processes are updated to include a variety of materials and printing methods. In 2016, only specific material specification standards including minimum mechanical property requirements were published for Ti6Al4V, Ti6Al4V-ELI, Inconel718 and -625, CoCrMo, 316L, 15-5PH, and 17-4PH [39]. To date, there are standards for a wide range of materials.

As mentioned above, the issue of printed product quality needs to be taken into consideration to ensure that AM is used as a processing method to create the final product [40]. Therefore, standards for testing and evaluating the quality of printed products are essential. ASMT and ISO have collaborated to develop international standards such as ISO/ASTM52907-19, ISO/ASTM52902-23, ISO/ASTMTR52905-EB, and so on. Previously, the standards for test methods were only ISO 17296-3 and F2921-11e3.

Regarding the standards for design and design data formats, there were previously two standards, ISO 17296-4 and ISO/ASMT 52915-13. Today, the standard system has been updated with data structures and metrics for AM methods.

5 Conclusions

Additive manufacturing and 3D printing are increasingly used in applications ranging from medical, aerospace, education, manufacturing, and so on. Today, AM systems play a role not only in initial rapid prototyping as they did in the beginning but

Table 1 Additive manufacturing standards [38, 39]

Contents	Standards	
	Old standards	Updated standards
Terminology	ASTM F2792-12a ISO 17296-1 ISO/ASTM 52921-13	ISO/ASTM52900-21
Processes/ materials	ISO 17296-2 (Qualification and certification methods, requirements for purchased AM parts, and non-destructive evaluation methods)	ISO/ASTM52901-16 ISO/ASTM52903-2-20 F3302-18 F3213-17 F3301-18a F3318-18 F2924-14(2021) F3001-14(2021) F3049-14(2021) F3091/F3091M-14(2021) F3056-14(2021) F3055-14a(2021) ISO/ASTM52930-21 ISO/ASTM52903-1-20 ISO/ASTM52925-22 ISO/ASTM52920-23 F3592-23 F3607-22 ISO/ASTM52935-23 F3184-16(2023) F3187-16(2023) ISO/ASTM52945-23 ISO/ASTM52908-23 ISO/ASTM52904-24
Test methods	ISO 17296-3 (General test methods and performance test methods) F2921-11e3	ISO/ASTM52907-19 F2971-13(2021) ISO/ ASTM52921-13(2019) F3571-22 F3606-22 F3522-22 F3624-23 F3626-23 ISO/ASTMTR52917-EB F3489-23 F3637-23 F3615-23 ISO/ASTM52902-23 ISO/ASTMTR52905-EB ISO/ASTM52927-24 ISO/ASTMTR52952-EB

(continued)

Table 1 (continued)

Contents	Standards: Old standards	Updated standards
Design/data format	ISO 17296-4 ISO/ASMT 52915-13 (Data structures and metrics for AM methods)	ISO/ASTM52911-1-19 F2915-12 ISO/ASTM52950-21 ISO/ASTM52915-20 ISO/ASTM52910-18 F3413-19e1 F3530-22 F3529-21 ISO/ASTM52911-2-19 ISO/ASTM52911-3-23

also in creating final products with a variety of materials. In the fifth industrial revolution with the goal of being human-centered, technologies are applied to increase the value of the product. AM and 3D printing technologies are considered as core technologies, a layered manufacturing method that is suitable for the mass-customized production trend of the fourth and fifth industrial revolutions instead of the mass production trend. To achieve that acceptance, the scientific community, manufacturers, and organizations have made constant efforts to upgrade and improve AM and 3D printing technology in consideration of the quality and productivity of the AM process, applying AI in the development of AM systems, as well as providing a system of common standards for terminology, printing process, design, data, materials, and testing methods.

References

1. Barneveld, J., and Jansson, T. (2017). Additive manufacturing: A layered revolution. European Foundation for the Improvement of Living and Working Conditions.
2. Harshavardhana, N. (2021). Review on 3D printing of medical parts. In M. Manjaiah et al. (Eds.), *Additive manufacturing: A tool for industrial revolution 4.0*. Woodhead Publishing.
3. Lu, B., Li, D., & Tian, X. (2015). Development trends in additive manufacturing and 3D printing. *Engineering, 1*(1), 85–89.
4. Dilberoglu, U. M., Gharehpapagh, B., Yaman, U., & Dolen, M. (2017). The role of additive manufacturing in the era of Industry 4.0. *Procedia Manufacturing, 11*, 545–554.
5. Galantucci, L. M., Guerra, M. G., Dassisti, M., & Lavecchia, F. (2019). Additive manufacturing: New trends in the 4th industrial revolution. In L. Monostori, V. D. Majstorovic, S. J. Hu, D. Djurdjanovic (Eds.) *Proceedings of the 4th international conference on the Industry 4.0 model for advanced manufacturing*. AMP 2019. Lecture Notes in Mechanical Engineering. Springer, Cham.
6. Varshney, A. K., Garg, A., Pandey, T. R., Singhal, R. K., Singhal, R., & Sharma, H. (2024). The development of manufacturing industry revolutions from 1.0 to 5.0. *Journal of Informatics Education and Research, 4*(1), 1230–1242.
7. Maddikunta, P. K. R., Pham, Q. V., Prabadevi, B., Deepa, N., Kapal, D., Gadekallu, T. R., Ruby, R., & Liyanage, M. (2022). Industry 5.0: A survey on enabling technologies and potential applications. *Journal of Industrial Information Integration, 26*, 100257.

8. Prashar, G., Vasudev, H., & Bhuddhi, D. (2023). Additive manufacturing: expanding 3D printing horizon in industry 4.0. *International Journal on Interactive Design and Manufacturing (IJIDeM), 17*, 2221–2235.
9. Diegel, O. (2011). Additive manufacturing: the new industrial revolution. In *12th rapid design, prototyping and manufacturing conference*.
10. Duda, T., & Raghavan, L. V. (2016). 3D metal printing technology. *IFAC-PapersOnLine, 49* (29), 103–110.
11. Rani, S., Jining, D., Shoukat, K., Shoukat, M. U., & Nawaz, S. A. (2024). A human–machine interaction mechanism: Additive manufacturing for industry 5.0—Design and management. *Sustainability, 16*, 4158.
12. Haleem, A., & Javaid, M. (2019). Additive manufacturing applications in Industry 4.0: A review. *Journal of Industrial Integration and Management, 4*(4), 1930001.
13. Woźniak, J., Paszkiewicz, A., Budzik, G., Przeszłowski, L., Kiełbicki, M., & Zimon, D. (2024). Development of additive technologies in the light of the Industry 4.0 and Industry 5.0 conception in accordance with the IOE idea. *International Journal for Quality Research, 18*(2), 643–658.
14. George, A. S., George, A. S. H., & Baskar, T. (2023). The evolution of smart factories: How Industry 5.0 is revolutionizing manufacturing. *Partners Universal Innovative Research Publication, 1*(1), 33–53.
15. Rahmani, R., Karimi, J., Resende, P. R., Abrantes, J. C. C., & Lopes, S. I. (2023). Overview of selective laser melting for industry 5.0: Toward customizable, sustainable, and human-centric technologies. *Machines, 11*, 522.
16. Venkatesh, S. (2023). The arc of Industry 5.0 bends towards a social purpose. *NHRD Network Journal, 16*(1), 190–195.
17. Baronio, G., Harran, S., & Signoroni, A. (2016). A critical analysis of a hand orthosis reverse engineering and 3D printing process. *Applied Bionics and Biomechanics, 2016*, 8347478.
18. Raja, V., & Fernandes, K. J. (2008). *Reverse engineering*. Springer.
19. Budzik, G. (2011). The use of the rapid prototyping method for the manufacture and examination of gear wheels. In *Advanced applications of rapid prototyping technology in modern engineering*. InTech.
20. Kamrani, A., & Nar, E. A. (2006). *Rapid prototyping: Theory and practice*. Springer.
21. Yadroitsev, I., Yadroitsava, I., & Plessis, A. D. (2021). Basics of laser powder bed fusion. In I. Yadroitsev et al. (Eds.), *Fundamentals of laser powder bed fusion of metals*. Elsevier.
22. Moghaddam, N. S., Jahadakbar, A., Amerinatanzi, A., & Elahinia, M. (2018). Recent advances in laser-based additive manufacturing. In L. Bian et al. (Eds.), *Laser-based additive manufacturing of metal parts*. Taylor & Francis.
23. Park, H. S., & Tran, N. H. (2017). A decision support system for selecting additive manufacturing technologies. In *Proceedings of the 2017 international conference on information system and data mining* (pp. 151–155).
24. Park, H. S., Tran, N. H., & Jonaet A. M. (2018). Prediction of temperature distribution and residual stress in SLM printed parts. In *13th international manufacturing science and engineering conference*.
25. Tran, N. H., Nguyen, V. N., Ngo, A. V., & Nguyen, V. C. (2017). Study on the effect of fused deposition modeling (FDM) process parameters on the printed part quality. *International Journal of Engineering Research and Applications, 7*(12), 71–77.
26. Roberts, I. A. (2012). *Investigation of residual stresses in the laser melting of metal powders in additive layer manufacturing*. Dissertation, University of Wolverhampton.
27. Saprykin, A. A., Ibragimov, E. A., & Babakova, E. V. (2016). Modeling the temperature fields of copper powder melting in the process of selective laser melting. *IOP Conference Series: Materials Science and Engineering, 142*, 012061.
28. Setien, I., Chiumenti, M., Veen, S., Sebastian, M. S., Garciandia, F., & Echeverria, A. (2019). Empirical methodology to determine inherent strains in additive manufacturing. *Computers & Mathematics with Applications, 78*, 2282–2295.

29. Chen, J., Lu, H., Wang, J., Chen, W., & Hao, D. (2000). *Prediction of welding deformation with inherent strain method based on FEM*. Shanghai Huizhong Automotive Manufacturing Co. LTD report.
30. Yazhi, L. (2014) Research on inherent strain distribution in welded low-alloy components. In *Proc Int Conf Meas Technol Mechatron Autom* (pp. 512–515).
31. Donik, Č., Kraner, J., Paulin, I., & Godec, M. (2020). Influence of the energy density for selective laser melting on the microstructure and mechanical properties of stainless steel. *Metals, 10*, 919.
32. Park, H. S., Shin, H. S., & Tran, N. H. (2022a). A new approach for calculating inherent strain and distortion in additive manufacturing of metal parts. *International Journal of Advanced Manufacturing Technology, 121*, 6507–6521.
33. Park, H. S., Tran, N. H., Bui, V. H., & Hoang, V. T. (2022b). Development of a prediction system for 3D printed part deformation. *Engineering Technology & Applied Science Research, 12*(6), 9450–9457.
34. Kiran, M. B. (2021). Application of artificial intelligence in additive manufacturing—A review. In *Proceedings of the 11th annual international conference on industrial engineering and operations management Singapore*, March 7–11, pp. 2575–2586.
35. Ma, L., Yu, S., Xu, X., Amadi, X. S. M., Zhang, J., & Wang, Z. (2023). Application of artificial intelligence in 3D printing physical organ models. *Materials Today Bio, 23*, 100792.
36. Goh, G. D., Sing, S. L., & Yeong, W. Y. (2021). A review on machine learning in 3D printing: applications, potential, and challenges. *Artificial Intelligence Review, 54*, 63–94.
37. Talaat, F. M., & Hassan, E. (2021). Artificial intelligence in 3D printing. In A. E. Hassanien et al. (Eds.), *Enabling machine learning applications in data science, algorithms for intelligent systems*. Springer.
38. ASTM. (n.d.). https://www.astm.org/products-services/standards-andpublications/standards/additive-manufacturing-standards.html.
39. Vrancken, B. (2016). *Study of residual stresses in selective laser melting*. Ph.D dissertation.
40. Thomas, G. S., & Scott, A. G. (2016). In-process sensing in selective laser melting (SLM) additive manufacturing. *Integrating Materials and Manufacturing Innovation, 5*(2), 1–25.

The Future of Manufacturing: A Convergence of 3D Printing and Industry 5.0

Rajdev Singh, Vikash Kumar Srivastva, Amit Choudhary, and Navneet Arora

1 Introduction

The industrial landscape is undergoing a profound transformation with the advent of Industry 5.0, which emphasizes the integration of human creativity, intelligence, and collaboration with advanced technologies to create more sustainable, resilient, and customized production systems [1–3]. As we move beyond the automation and digitization of Industry 4.0, Industry 5.0 places a renewed focus on the synergy between humans and machines, fostering an environment where smart technologies are augmented by human-centered solutions [4, 5]. One of the most transformative technologies playing a critical role in this new era of production is additive manufacturing (AM), also known as 3D printing.

Additive manufacturing has evolved significantly since its inception, from a tool primarily used for prototyping and experimentation to a full-fledged manufacturing method with the potential to revolutionize the design, development, and production of parts. Its layer-by-layer fabrication process allows for unprecedented design freedom, resource efficiency, and the ability to produce complex geometries that are often impossible or cost-prohibitive using traditional manufacturing techniques. As a result, AM has emerged as a key enabler of Industry 5.0's vision of customized, flexible, and sustainable production [4, 6].

The core premise of Industry 5.0 is the idea of mass personalization and the optimization of production to meet individual consumer needs without sacrificing efficiency. This shift represents a departure from the mass production model of previous industrial revolutions, which prioritized scale and uniformity over

R. Singh (✉) · V. K. Srivastva · A. Choudhary · N. Arora
Department of Mechanical and Industrial Engineering, Indian Institute of Technology Roorkee, Roorkee, Uttarakhand, India
e-mail: r_singh1@me.iitr.ac.in; vk_srivastva@me.iitr.ac.in; amit.choudhary@me.iitr.ac.in; arorafme@iitr.ac.in

Y. Kumar Singla et al. (eds.), *Digital Manufacturing in Industry 5.0*,
https://doi.org/10.1007/978-3-031-91500-0_7

customization. Additive manufacturing aligns perfectly with this new paradigm. It allows for the rapid production of unique, tailored products, reduces waste by minimizing excess material, and enables on-demand manufacturing, reducing the need for large-scale inventories and long lead times [7].

This chapter delves into the optimization of production processes using additive manufacturing within the framework of Industry 5.0. It explores how the fusion of human creativity and digital technologies in AM can lead to more efficient and sustainable production systems. The integration of artificial intelligence (AI), machine learning, and advanced robotics with AM is discussed, demonstrating how these technologies enhance production capabilities, optimize material usage, and improve overall supply chain agility. In the context of sustainability, additive manufacturing offers significant advantages by reducing material waste and energy consumption. This chapter will discuss how AM supports the goals of a circular economy by enabling the reuse and recycling of materials, extending product life cycles, and reducing the environmental footprint of manufacturing. The role of AM in decentralized and localized production, which aligns with Industry 5.0's emphasis on resilience and sustainability, will be explored, highlighting how AM can contribute to more robust and adaptive supply chains.

This chapter will emphasize the strategic importance of optimizing production through additive manufacturing as we transition to Industry 5.0. By harnessing the combined power of human ingenuity and technological advancements, manufacturers can create more efficient, flexible, and sustainable production systems that are better equipped to meet the challenges of the future. The chapter aims to provide a comprehensive understanding of how additive manufacturing is reshaping production and enabling the next generation of industrial innovation.

2 Additive Manufacturing Developments

Figure 1 provides a graphical representation of the research papers published each year, based on data from the Scopus database. For this analysis, up until 2023, a combination of keywords and Boolean operators—such as 'additive manufacturing,' '3D printing,' 'revolution,' and 'Industry 5.0'—was used.

Even though AM originated in the 1980s, as Fig. 1 illustrates, its research activities clearly increased dramatically after 2010. It has since experienced exponential growth and emerged as a focal point for innovation in a variety of academic and industry contexts. This development demonstrates AM within the framework of Industry 5.0 rise from a little-known idea to a crucial element of contemporary production.

With ongoing research and development, AM's numerous applications in the aerospace, healthcare, and other fields continue to grow. Based on data from the Scopus database, Fig. 2 shows a graphical depiction of research papers published on

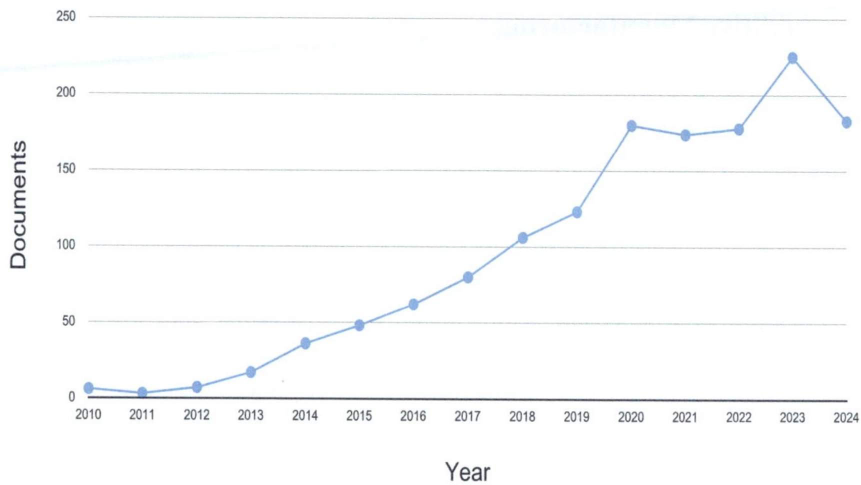

Fig. 1 Year-wise trend of published documents

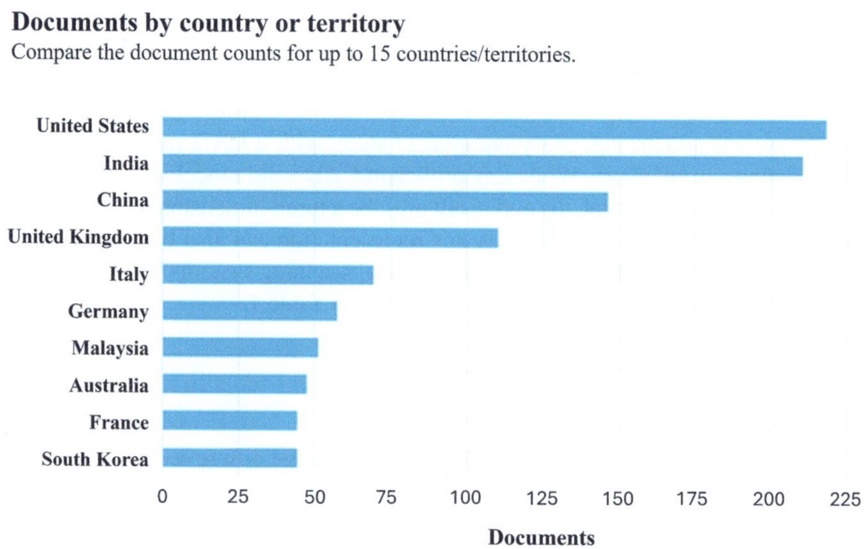

Fig. 2 Documents published by different countries and territories

AM or 3D printing by various nations and territories between 2010 and mid-2024. With the use of this analysis, one can look at the state of AM within the framework of industry 5.0 research around the world and pinpoint the nations that have made major contributions to the area and demonstrated a great deal of interest.

3 Additive Manufacturing

Additive manufacturing is defined as "the process of joining materials to make objects from 3D model data, usually layer upon layer, as opposed to subtractive manufacturing methodologies" (ASTM F2792-12a). The term 'additive' highlights its distinction from conventional manufacturing techniques that involve cutting, drilling, or shaping materials from larger blocks. AM allows for the creation of intricate designs, internal structures, and components that are either impossible or too costly to achieve through traditional methods.

The process typically starts with a digital design created using CAD software. This design is then sliced into layers by specialized software, which dictates how the 3D printer will sequentially build each layer of the object. Based on this information, the AM machine deposits material, which could range from plastics, metals, and ceramics to advanced composites, to form the final product. After printing, parts may undergo post-processing treatments to enhance surface finish, strength, or other mechanical properties.

3.1 Stages Involved in Additive Manufacturing

In additive manufacturing, every component starts with a CAD model that accurately defines its geometry. While any proficient CAD solid modeling software can be used to create this model, the end result must be a three-dimensional (3D) surface or solid representation. Alternatively, this representation can also be created using reverse engineering tools, such as laser or optical scanning systems [8]. The typical stages for AM processing components are shown in Table 1.

Table 1 Typical stages for AM processing of components

Design	First, a digital design file for the object to be created must be developed, typically using CAD software
Pre-processing	The digital design is then prepared for printing by converting the CAD file into a format that the additive manufacturing machine can interpret e.g., STL file
Printing	The AM machine starts printing the object by layering material based on the digital design. The choice of material depends on the specific AM technology being used
Post-processing	After printing, the object may need post-processing to enhance its surface finish or mechanical properties of the component. This can involve techniques such as sanding, polishing, or heat treatment
Inspection	The final step is to inspect the printed object to verify that it meets the required specifications. This can involve visual inspection or more advanced methods like CT scanning or X-ray inspection
Repetition	If the component fails to meet the desired specifications, the design may need to be adjusted, and the printing process is repeated until the desired result is achieved

3.2 Benefits of Additive Manufacturing

The area of manufacturing processes in many fields has evolved in the last few centuries due to industrial revolutions. Because AM technology offers a new way to create complicated and personalized parts, it has garnered a lot of interest recently. As a result, in this new age of production, the primary roles of manufacturers, designers, and customers are essentially redefined. AM technology provides more sustainable alternatives to conventional manufacturing processes, such as molding for polymer parts and machining and casting for metallic components [9]. AM significantly transforms traditional manufacturing and supply chain models by reducing the need for raw materials and improving resource efficiency during the production and use phases. This technology minimizes reliance on wasteful and polluting manufacturing processes, allowing for more flexible and efficient product designs [10]. Additionally, AM reduces transport-related emissions and lowers the carbon footprint, as production is often decentralized and closer to consumers [11]. This decentralization also leads to shorter, simpler supply chains with innovative distribution models. Moreover, product life is extended through sustainability practices like reusing, repairing, refurbishing, and remanufacturing, which are aligned with more sustainable social, economic, and environmental approaches [12].

Many advancements in a variety of industrial domains, including design, aerospace, medical/biomedical, and automotive, have been made possible by AM. Specifically, it provides an excellent chance for quick prototyping, free dimension, time, and cost-effective production. In this sense, AM is referred to as a sustainable production method that greatly reduces environmental impact while preserving resources. AM offers an excellent opportunity for product designers to manufacture eco-designed items, meaning that environmental factors are taken into account throughout the whole design and fabrication process. AM contributes to the production of sustainable and high-quality goods in this way. Sustainability issues are often divided into three categories: social, environmental, and economic. A variety of evaluation instruments are available, for example, life-cycle assessment (LCA) methods, which can evaluate the AM goods' sustainable impact in a variety of approaches. A timeline of sustainable manufacturing is shown in Fig. 3.

From an economic perspective, cost models directly impact the market and the financial returns on a production process. As a result, measuring sustainability can effectively cut costs and production times while enhancing the design process. In addition, the environmental benefits of additive manufacturing (AM) techniques, such as reduced pollution and waste management, must be taken into account throughout the whole life cycle of the product. As a result, international standards and laws—like ISO 14000—are created for environmental safety. The quality of the products, as well as having safer and healthier working conditions, are directly impacted by the social benefits of the AM process.

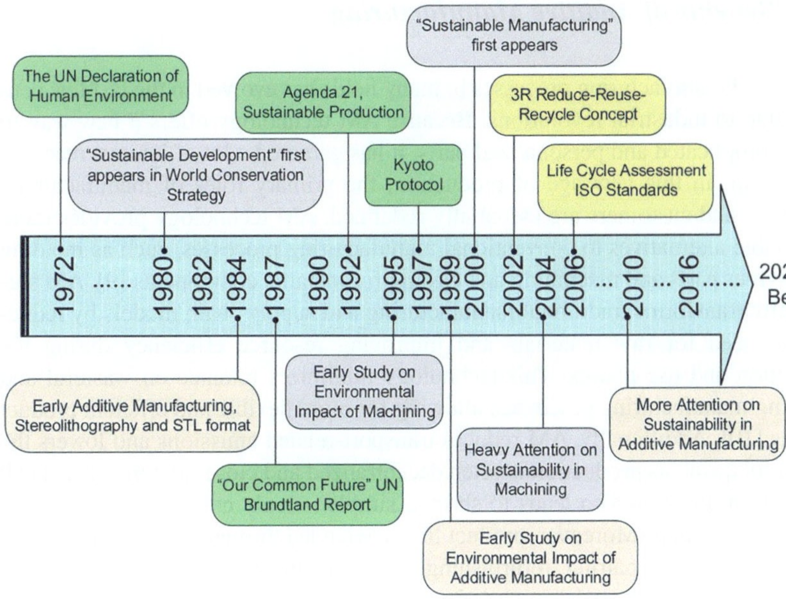

Fig. 3 A timeline of sustainable manufacturing milestones [13]

3.3 *Additive Manufacturing and Industry 5.0*

Additive manufacturing (AM) aligns closely with the goals of Industry 5.0, which emphasizes human-centered approaches, personalization, and agility in production processes. Industry 5.0 focuses on creating tailored products that meet individual customer preferences and needs. AM excels in personalization by allowing manufacturers to produce customized items without the need for mass-production tooling [14]. Whether it is healthcare (custom prosthetics and implants) or consumer goods (custom footwear or jewelry), AM offers the flexibility to design and manufacture products on a one-off basis, accommodating unique requirements quickly and cost-effectively. This enables the shift from mass production to mass customization. The stages of industrial revolutions are shown in Fig. 4, and the advantages of mass customization for personalization are shown in Fig. 5.

Industry 5.0 also promotes agility, meaning the ability to rapidly adapt production lines to changing demands or new technologies. AM allows for more agile manufacturing systems because it eliminates the need for specialized tools, molds, or dies, reducing setup times and enabling rapid design changes. This allows manufacturers to respond quickly to market demands, supply chain disruptions, or new design iterations, all while minimizing production downtime. One of the core goals of Industry 5.0 is fostering closer collaboration between humans and machines. AM integrates well with this approach by allowing humans to play a central role in the design and customization process, while machines handle the complex task of

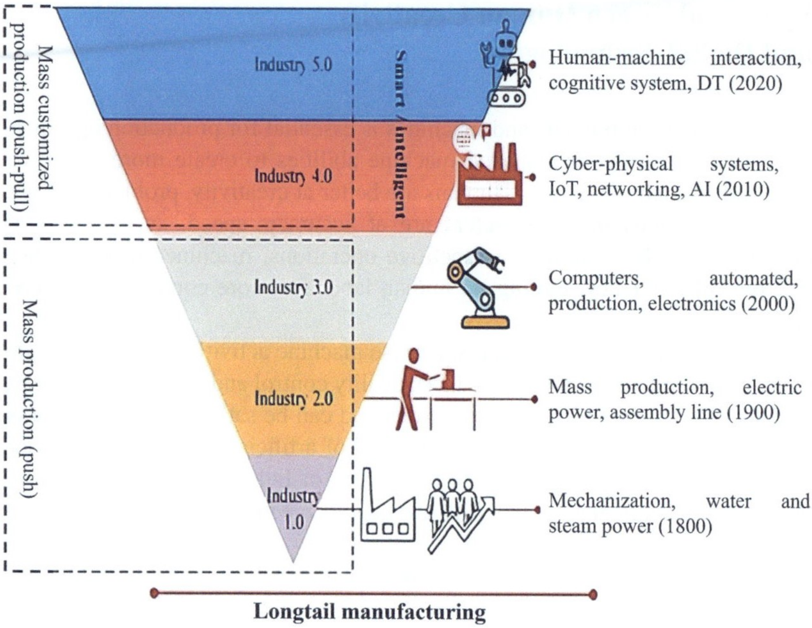

Fig. 4 The stages of industrial revolutions [15]

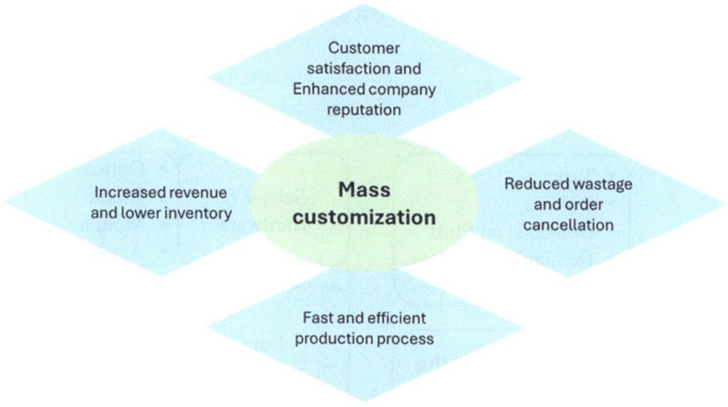

Fig. 5 The advantages of mass customization for personalization

layer-by-layer fabrication. This synergy enables more creativity and innovation in product development, with machines executing precise manufacturing tasks while humans oversee design and quality control. By supporting personalization, agility, and human-centered innovation, AM is a key enabler of Industry 5.0's vision of smart, flexible, and sustainable manufacturing [16].

4 Synergy Between Human Creativity and Digital Technologies

Collaboration between humans and machines is essential for production optimization because it integrates human and machine abilities to create more inventive, adaptive, and efficient processes. Humans are better at creativity, problem-solving, and making decisions than computers are at accuracy, speed, and performing monotonous work. By automating repetitive operations, machines in production improve overall efficiency, freeing up human labor for more complicated or creative jobs.

Additionally, because humans can supervise machine activities and make modifications in real time, this synergy enhances quality control and lowers errors. More agile production, where quick changes in demand can be satisfied without a lot of downtime, is made possible by the combination of artificial intelligence (AI) and robotics with human knowledge. In the framework of Industry 5.0, where sustainability, personalization, and adaptability are major objectives, this collaboration is especially crucial.

According to [17], there are three types of human–machine interaction (HMI) in Industry 4.0: human, machine, and interaction. The taxonomy and a few of each category's subclasses are shown in Fig. 6. This system of classification places more emphasis on the distinctions than the similarities. This three-category taxonomy has to be changed, as more multidisciplinary research is anticipated in Industry 5.0, and it does not accurately describe the link between HMIs in highly integrated, closely networked smart factories.

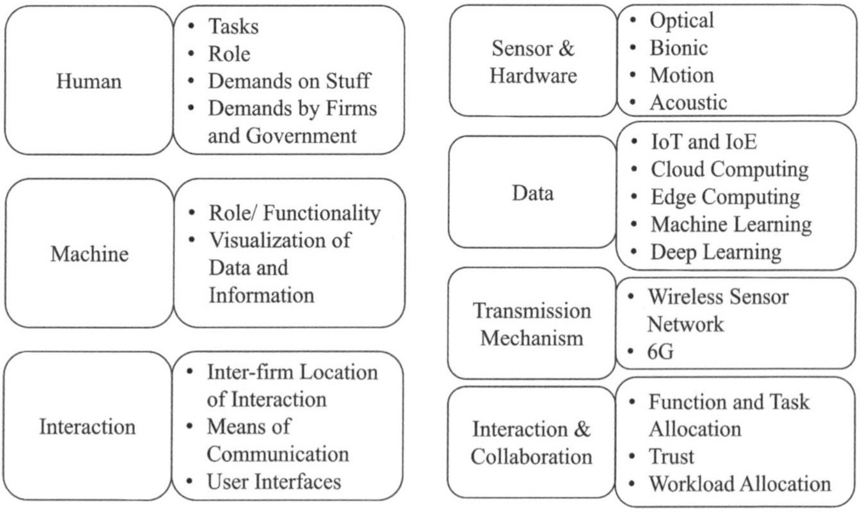

Fig. 6 HMI's three-category and four-category taxonomies [18]

Table 2 Key industries that benefitted from human-AI collaboration

Industry	Benefits due to human–AI collaboration
Aerospace	Lighter weight, better fuel economy, and quicker production cycles with intricate, highly customized parts
Healthcare	Personalization of implants and other devices for individual patients [20], improving patient comfort and medical findings
Automobile and transport	Autonomous driving, predictive maintenance, traffic management, automated supply chain management, and personalized safety features [21, 22]
Textile	Quality control and inspection, material optimization, personalization and customization, textile recycling [23, 24]
Architecture and construction	Quicker, greener building processes that use less labor and material waste, together with more creative, unique building designs [25, 26]
Consumer goods	Customized goods that improve performance and comfort for users while facilitating quick testing of new designs

As a result, a novel framework—illustrated in Fig. 6—was presented in [18]. In a smart factory, sensors are the first part of the HMI process; signals are sent into the system through these devices. Following the processing and analysis of this information and data, humans in the data center will use machines to assist in making judgments. The transmission mechanism is integral to the data input and instruction output, and it also necessitates the cooperation of multiple pieces of hardware. Ultimately, the device concludes its communication with the user upon getting the instruction [19].

Due to human–AI cocreation through AM, several industries are transforming by enhancing design processes, improving customization, and optimizing production. Some of the key industries that benefitted from human-AI collaboration are listed in Table 2.

These industries may use AI's computational capacity and creativity-enhancing technologies in conjunction with human intuition and expertise through human–AI collaboration in additive manufacturing, producing highly innovative, personalized, and effective solutions.

5 Technological Integration in Industry 5.0: AM with AI, Robotics, and Machine Learning

Robotics, artificial intelligence (AI), machine learning (ML), and other cutting-edge technologies easily combine with additive manufacturing to maximize production processes and make them more inventive, adaptable, and efficient. The fifth industrial revolution will see the introduction of the 4D printing technique, which will concentrate on the design process and enable the production of more innovative and customized items. Different smart materials with a flexible and adaptable character, such as Shape memory alloys (SMA) and Shape memory polymers (SMP), will be needed for 4D printing technology [27]. The amount of storage in enterprises can be greatly decreased by using 4D-printed buildings. The manufacturing industries may benefit from these materials' ability to self-adapt and self-repair [28]. The key Industry 5.0 technologies utilized to make the manufacturing system more intelligent and smarter are depicted in Fig. 7.

Artificial intelligence techniques, specifically in the field of generative design, facilitate the automated generation of intricate and highly optimized geometries that would be unfeasible to create through conventional means. AI generates the best possible designs that are AM-3D printed by weighing several design iterations against restrictions like weight, strength, and material consumption. AI-generated fuel nozzles and turbine blades, for instance, are lighter and stronger than parts made using traditional design techniques, greatly enhancing performance and cutting down on material waste in the aerospace industry.

During the AM process, AI and ML are also crucial for anticipating and tracking part quality. AI can identify anomalies, flaws, or inefficiencies in real time during printing by evaluating data from sensors and cameras. This enables quick corrections. This can save rework and enhance uniformity in final products by preventing problems like layer warping, ensuring uniform material deposition, and anticipating failures. By optimizing machine parameters like temperature, material feed rates, and laser intensity, ML-based adaptive learning further improves AM operations. To improve the mechanical qualities of metal products, for example, ML may dynamically modify the laser power and speed in metal AM to ensure better fusing and reduce material porosity. Furthermore, massive volumes of data from several

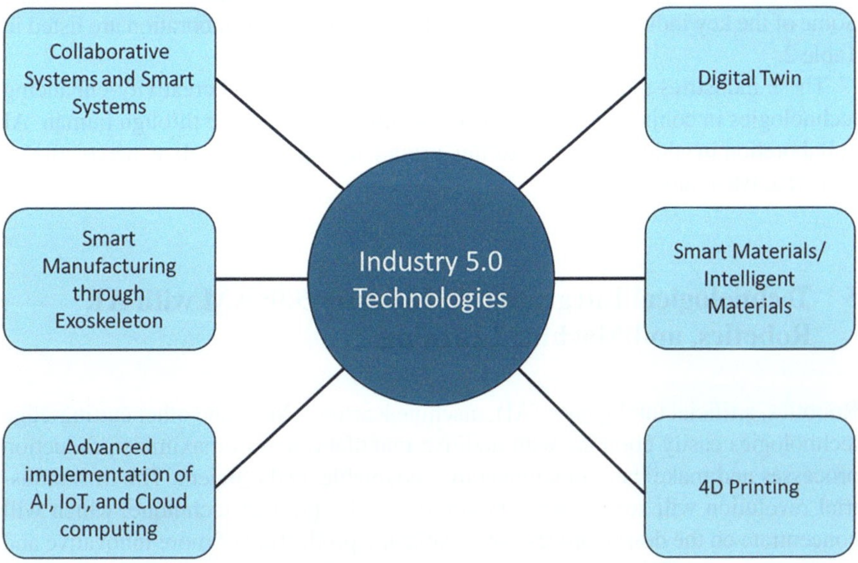

Fig. 7 Significant technologies of the fifth industrial revolution to create the manufacturing system smarter and intelligent [27]

production runs are analyzed via data-driven decision-making powered by ML to optimize the complete AM workflow, from design to post-processing. In the automobile industry, for example, machine learning (ML) improves the manufacturing of custom parts by balancing speed, material consumption, and energy use. This results in better decisions about material selection, machine settings, and production timelines.

Algorithms for machine learning (ML) can adjust laser intensity, material feed rates, temperature settings, and other machine parameters based on past production data, thereby optimizing additive manufacturing (AM) processes. Higher precision, improved surface finishes, and more material efficiency are the results of this optimization. In metal additive manufacturing, for instance, ML may dynamically modify laser power and speed to improve material fusion and lower porosity, improving the mechanical qualities of metal components. Moreover, ML finds trends and insights through the analysis of massive volumes of data from several production runs, which aid in the optimization of the complete AM workflow—from design to post-processing. Making decisions about material selection, machine settings, and production schedules is made better with this data-driven approach. For example, in the automotive sector, machine learning (ML) balances speed, material consumption, and energy use to optimize the creation of customized parts, leading to more productive and efficient manufacturing processes.

Also, the automation of the handling of raw materials and completed goods in additive manufacturing (AM) processes is greatly aided by the incorporation of robotics into Industry 5.0. Robots can reduce human interaction and production time by loading and unloading materials, moving parts between machines, and performing post-processing chores like cleaning and polishing. Robots, for instance, automate the placement of metal powders or polymer filaments in large-scale additive manufacturing systems, increasing production consistency and efficiency. Additionally, multi-axis 3D printing—in which the print head or build platform moves in more than three axes—is made possible by robotics linked with AM systems. This technique enables the construction of intricate, curved geometries without the need for support structures. Robotic arms with additive printing capabilities are one example of this, as they can print intricate designs like automobile frames or aerospace components with more accuracy and flexibility. Furthermore, post-processing operations like sanding, polishing, heat treating, and assembly can be automated by robots and artificial intelligence (AI) once a product is 3D printed. AI speeds up and improves the accuracy of these operations, while robots physically carry them out to minimize manual work and guarantee excellent finishing. Within the realm of medicine, artificial intelligence and robotics work together to automatically smooth the surface of 3D-printed implants. This process minimizes the need for manual touch-ups and contributes to improved biocompatibility.

In additive manufacturing (AM), artificial intelligence (AI)-driven analytics are essential for demand forecasting and production scheduling. Artificial intelligence (AI) guarantees flexible and adaptable manufacturing by forecasting demand patterns that take into account market trends, customer requirements, and material availability. AI, for instance, assists producers of consumer products in scaling up the production of customized goods based on real-time data, avoiding overproduction or stock shortages. Further enabling decentralized manufacturing—where production is located closer to consumers [11]—is the combination of AM, robotics, and AI. By figuring out where and when products should be printed to cut costs and delivery times, artificial intelligence (AI) optimizes the supply chain. AI, for example, manages regional 3D printing centers for replacement parts in the automobile sector, cutting down on supply times and emissions during transit. AI and machine learning (ML) algorithms also continuously monitor AM machines in real-time, guaranteeing peak performance and early problem detection that lowers downtime and increases reliability. For instance, AI can continuously monitor melt pools and modify laser power in metal AM to ensure constant part quality. By evaluating operational data to forecast when machines will need maintenance, AI also improves predictive maintenance, which helps to avoid malfunctions and save downtime. AI-driven maintenance systems notify managers of possible problems in advance, allowing for prompt fixes without stopping output.

Additive manufacturing is optimized by automation and artificial intelligence (AI) through improved design, streamlined production, high-quality parts, and lower operating costs. Predictive analytics, automated procedures, and AI-driven insights enable AM to become more scalable, efficient, and able to produce customized products of superior quality at reduced prices and shorter lead times. With the help of these technologies, firms may reduce waste, increase production, and maintain their flexibility in a cutthroat market.

6 Challenges and Opportunities in Optimizing Production with AM

6.1 Limitations and Challenges of AM in Large-Scale Production

Additive manufacturing (AM) has the potential to change industries, but there are still several obstacles to overcome before it can be used on a wide basis, detailed in the following subsections

6.1.1 Material Limitations

When compared to traditional manufacturing, the selection of materials appropriate for additive manufacturing (AM) is still restricted, particularly when it comes to high-performance metals, composite materials, and multi-material systems. There are now significantly fewer materials accessible for AM than there are for traditional manufacturing techniques, despite the fact that the diversity of materials is constantly growing. This is because AM commonly works with materials that are in the form of filament or powder [29], which need to be carefully chosen and prepared for usage before being used in the printing process. Certain materials may require higher printing temperatures or specialist equipment, which could limit the selection of materials that can be used. Powerful laser or electron beam printers are required for some metal powders used in metal 3D printing to guarantee adequate melting and consolidation of the metal powder [30]. Certain materials are too brittle or flexible for a given usage, or they are not strong enough to withstand a given amount of stress. Resin-based materials, commonly used in DLP and SLA, can produce exceedingly precise and accurate components [31], but their strength may be insufficient to handle certain loads. A graphic depiction of the material restrictions in AM brought about by concerns with compatibility, performance, and constraints on materials and processes is shown in Fig. 8.

6.1.2 Cost of Production

While AM reduces tooling costs and allows for customization, the cost per part can be high for large-scale production, especially due to material costs, machine operation time, and post-processing needs. This makes AM less cost-effective than traditional manufacturing for high-volume runs. Another cost constraint on AM is production speed. Even though AM technology has advanced recently [33], producing a product using AM still requires a lot more time [34] than using traditional manufacturing techniques. Even though additive manufacturing has many benefits, there are still costs associated with the technology that need to be taken into account. It is imperative to reduce the expenses linked with additive manufacturing (AM), such as material waste, post-processing, and machine development, in order to rival traditional mass production in terms of unit cost.

6.1.3 Post-processing Requirements

To meet industrial quality standards, many AM techniques need considerable post-processing, including support removal, surface finishing, heat treatment, and inspection. Time and money are added by these stages, particularly when production is scaled up. Heat treatments and other post-processing methods might be required to

Fig. 8 Material limitations in AM due to process, compatibility, and performance factors [32]

improve the material qualities of 3D-printed parts [35, 36] since their mechanical characteristics are different from those of their conventionally manufactured counterparts. Following printing, the part can still have support structures or residual powder, necessitating additional cleaning procedures. These cleaning procedures might take a lot of time and may call for the use of chemicals or abrasive methods. Advancements in AM-specific post-processing techniques have significantly improved the performance and quality of printed parts. Among the post-processing techniques that have emerged as promising for improving surface smoothness and producing precise parts is laser polishing [37]. Apart from surface integrity, the bulk properties of the AM components also require improvements. In metal AM, controlled heat treatments such as annealing and stress relief are frequently used to increase material integrity, decrease residual stresses, and improve the mechanical properties of the components [38]. While automated systems lessen the need for manual labor and integrate with manufacturing processes, they also enable effective cleaning for powder bed fusion technologies such as SLM and SLS, thanks to advancements in powder removal techniques [39]. To make AM practical for large-scale operations, post-processing automation and the development of AM techniques requiring less post-production work are essential.

6.1.4 Part Size Limitations

One of the main challenges affecting AM's wider applicability is part size constraint, particularly in industries where large-scale components are required because most AM machines have a limiting build volume, and the size of parts that may be printed is limited by the machine's physical dimensions [40]. For instance, industrial AM machines can handle larger pieces but still have constraints and desktop 3D printers usually have relatively small build areas, making them ideal for smaller products. This restriction can be especially difficult for industries where larger components are frequently used, such as aerospace, automotive, and construction.

In order to get around this, manufacturers frequently print large pieces in smaller portions, which are then assembled using post-processing methods like adhesive bonding or welding [41]. But this brings with it more stages, possible weak areas, and alignment issues. Furthermore, because of uneven cooling or material shrinkage during printing, larger pieces may potentially experience internal tensions and warping. Part size restrictions are being addressed by new developments in larger AM systems and hybrid production techniques, such as fusing AM with conventional manufacturing techniques. While modular machines enable expanded workspaces, some methods, such as Big Area Additive Manufacturing (BAAM) developed by Cincinnati Incorporated and Oak Ridge National Laboratory's Manufacturing Demonstration Facility, improve build volume through the use of robotic arms or gantry systems. However, part size constraints continue to be a major determinant of additive manufacturing's scalability and versatility in large-scale manufacturing.

6.1.5 Quality and Consistency

With the ongoing advancements in technology, the quality and uniformity of additively produced parts continue to be major challenges. Even though additive manufacturing provides remarkable design flexibility and material efficiency, it can be difficult to guarantee that every part has the required mechanical qualities, surface quality, and dimensional accuracy, especially in large-scale or important applications like the aerospace, healthcare, and automotive industries. A number of variables can cause variation in quality, such as the kind of additive manufacturing technique employed, the behavior of the materials, the calibration of the machine, and the surroundings in which the printing is done [42].

Variations in temperature, laser power, or material feed rates, for instance, can result in flaws like porosity, layer delamination, or uneven material deposition in techniques like selective laser melting (SLM) or fused deposition modeling (FDM). These discrepancies have the potential to undermine the part's structural integrity, which could result in malfunctions or decreased functionality in situations where durability and accuracy are essential. Variations in component quality can be caused by a variety of factors, including operator variances, material inconsistencies [43], and machine-to-machine variability. Even if efforts to standardize additive manufacturing processes and increase repeatability are underway, the technology has some way to go before it can completely compete with traditional manufacturing techniques in terms of consistency and quality control.

6.1.6 Sustainability

The effects of AM on the environment are now being studied and evaluated by researchers [44, 45]. The manufacture and disposal of 3D-printed items may have an adverse effect on the environment, even though AM has the potential to use less energy and produce less trash than existing manufacturing processes [46]. Manufacturers must devise strategies to reduce the energy consumption of AM operations. A large amount of energy is consumed during the AM process because of the high heat required to melt and fuse the metals, and many of the materials are non-recyclable. Designing products that can be maintained over time is one of the most challenging problems in additive manufacturing. Since many AM-made objects' end-of-life design is neglected, they might be challenging to recycle or dispose of in a sustainable way. It is imperative for manufacturers to design sustainable products that consider the long-term consequences on the environment and the possibilities for recycling. A significant amount of electronic trash may be produced by the AM software and electronics, which could cause problems if they are not recycled or disposed of appropriately. Making additive manufacturing a more environmentally friendly choice for large-scale production would require lowering the energy footprint of AM processes, enhancing material recyclability, and advancing sustainable AM technology.

6.2 Opportunities for Further Innovation

Additive manufacturing holds immense potential for further innovation and improvement across various areas, offering significant opportunities to enhance its efficiency, applicability, and scalability. Here are some key areas shown in Fig. 9, where AM technologies can advance

The future of AM is filled with opportunities for innovation across materials, processes, and applications. By addressing current limitations in speed, scalability, and cost-effectiveness, and through the integration of AI, automation, and sustainability initiatives, AM can evolve into a dominant manufacturing technology across industries. These advancements will further unleash the potential of AM, making it more versatile, efficient, and capable of transforming global manufacturing.

7 The Future of Additive Manufacturing in Industry 5.0

Additive manufacturing, commonly referred to as 3D printing, has the potential to significantly impact production, design, and supply chains in the future. AM will provide creative answers to manufacturing problems, improve design skills, and change supply chains as technology advances and Industry 5.0 ideas gain popularity. AM will continue to shape these characteristics as described in Fig. 10.

Additive manufacturing facilitates localized and on-demand production by reducing lead times, shipping costs, and the environmental impact of transportation by allowing businesses to produce goods closer to their place of use. In addition to supporting on-demand production, this decentralized model also eliminates the need for big stocks, which makes it perfect for sectors with erratic demand, including healthcare and aerospace. Furthermore, AM makes mass customization possible, allowing items to be customized to meet the unique needs of each customer without the need for extra tooling or changes to the production process. In industries where personalization is becoming more and more popular, such as consumer goods, fashion, healthcare, and the automobile sector, this is especially advantageous. For example, patient-specific scans can be used to build personalized implants, prostheses, and dental equipment for the best possible fit in the healthcare industry. Additionally, because the material is only added where needed in AM, material waste is much reduced when compared to traditional subtractive processes. Additionally, energy-efficient AM machines support more environmentally friendly production methods. The advancement of AM is also being fueled by advances in materials science, which are leading to the creation of biodegradable plastics, high-performance alloys, and composites as well as multi-material, functionally graded parts with conductive or sensor elements. These developments open up new possibilities for intelligent, responsive products.

Unmatched design freedom is provided by AM, making it possible to produce intricate geometries that would be too expensive or difficult to produce using

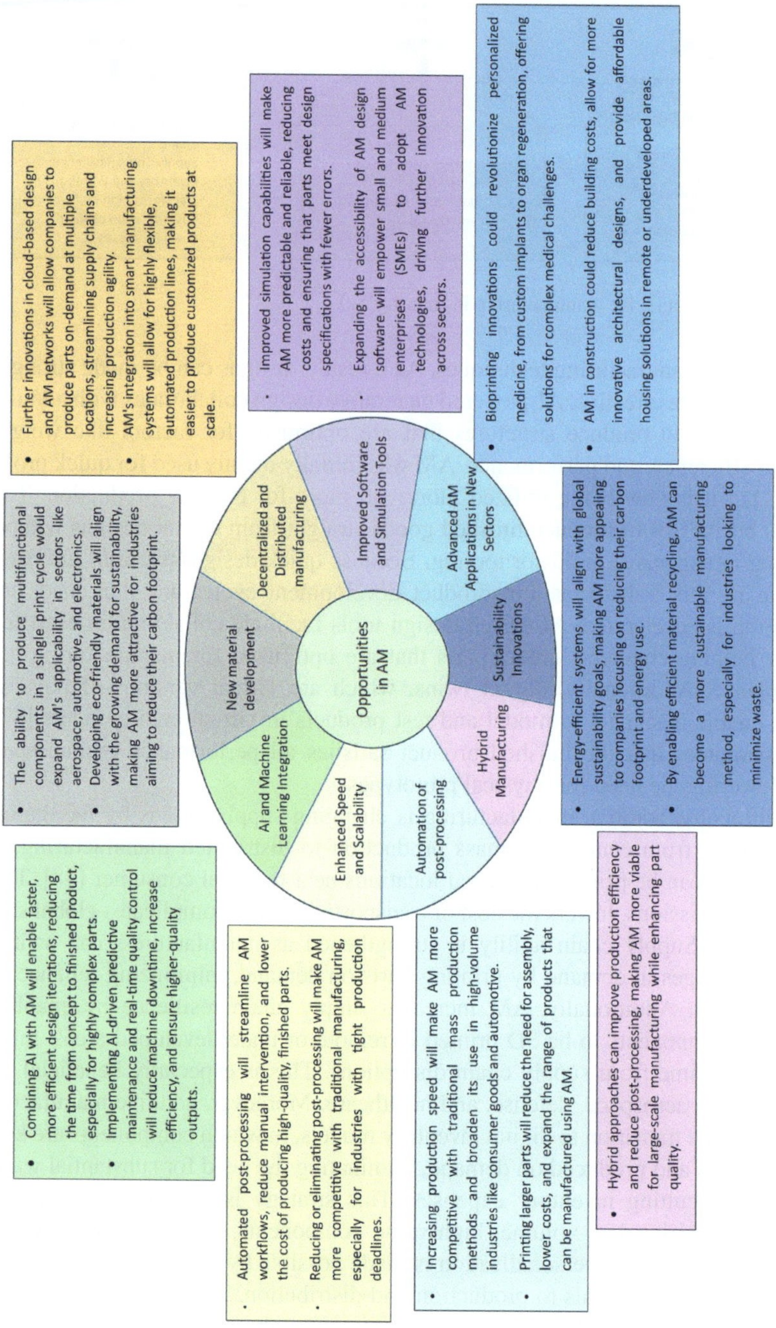

Fig. 9 AM opportunities to enhance its efficiency, applicability, and scalability

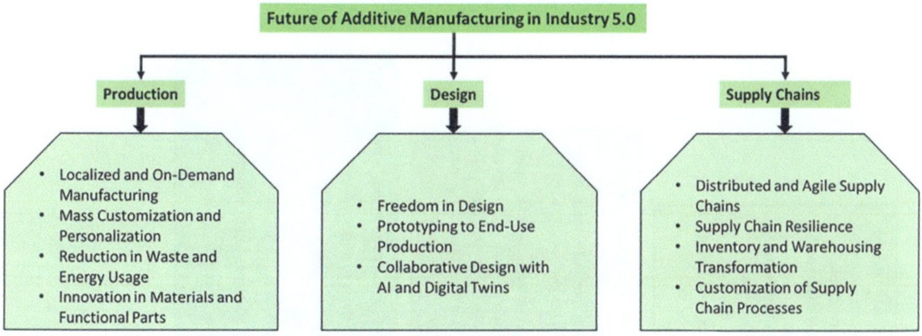

Fig. 10 Future of additive manufacturing in Industry 5.0

conventional manufacturing techniques. As a result, designs can be made stronger, lighter, and more effective. AI-powered generative design collaborates with additive manufacturing to produce structures that are optimized for factors like weight, material efficiency, and performance. AM was initially mainly used for quick prototyping, but it is now being utilized more and more for end-use production. This enables businesses to produce finished goods straight from digital designs without investing in expensive molds or tooling. Because quick design iterations are made possible by this transformation, product development cycles are shortened, and innovation is accelerated. AI-driven design tools facilitate collaboration even further by helping engineers create parts that are optimized for manufacturing and performance. Additionally, digital twins, which are virtual versions of physical things, enable designers to model and test products in virtual environments [47]. This guarantees that the finished product satisfies all performance requirements while lowering the need for physical prototypes.

Additionally, additive manufacturing is changing supply chains by facilitating the transition from centralized mass production to distributed manufacturing, in which goods can be printed at several locations near the final consumer [48]. This minimizes emissions, lowers the cost of transportation, and complicates global supply systems. Supply chain agility is strengthened as manufacturers can swiftly adjust to changes in demand by printing parts on demand, minimizing reliance on big stockpiles. Additionally, AM increases supply chain resilience by enabling important components to be 3D printed in remote or underdeveloped areas, minimizing downtime from supply chain disruptions. This is especially beneficial for industries like aerospace, defense, and healthcare. Moreover, additive manufacturing enables the transition to digital inventory models, in which components are kept as digital files and produced on demand, eliminating the need for substantial warehousing and cutting inventory expenses. This strategy is especially helpful for industries requiring low-volume, high-value components, such as aerospace and automotive. Therefore, personalization is made possible by AM at all supply chain stages, from raw materials to production and distribution. This just-in-time manufacturing support and tighter production alignment with consumer demand reduce waste and increase efficiency.

Since additive manufacturing offers previously unheard-of levels of flexibility, sustainability, and personalization, it is poised to continue influencing the direction of supply chains, design, and production. Localized, on-demand production will be possible thanks to it, and supply chains will run more smoothly thanks to dispersed manufacturing and digital inventories. Product design will also be revolutionized with more freedom and complexity. AM technologies will spur increased productivity, innovation, and resilience in manufacturing processes across a range of industries as they develop and interact with AI, robotics, and other Industry 5.0 technologies.

8 Long-Term Implications

The long-term implications of fusing human ingenuity with state-of-the-art technologies, such as robotics, additive manufacturing, AI, and machine learning, are revolutionary for entire industries and society. Human creativity is increased by this synergy, allowing individuals to create, overcome challenging issues, and invent in previously unthinkable ways. Human innovators can concentrate more on conceptual, strategic, and problem-solving duties by utilizing automation, generative design, and AI-driven analytics, while technology manages repetitive, data-intensive chores. In areas including healthcare, aerospace, automotive, and consumer goods, this partnership results in more personalized, efficient, and sustainable products and services.

A significant long-term effect is the quickening of innovation cycles. AI and machine learning technologies enable human designers and engineers to quickly iterate and develop ideas, bringing breakthroughs to market more quickly. These systems can examine large datasets, simulate complex scenarios, and provide predictive insights. For example, generative design in AM makes it possible to create extremely complex and optimized structures that would be too complicated for conventional methods. The convergence of artificial intelligence with human intelligence leads to innovations in fields such as biotechnology, material science, and energy solutions. These developments tackle worldwide issues including resource scarcity, healthcare accessibility, and climate change. Furthermore, this collaboration changes the workforce by necessitating a change in roles and skill sets. Automation may make some manual jobs less necessary, but it also increases the need for people with new talents in digital design, robotics management, AI programming, and machine learning. Future employment will probably involve more transdisciplinary tasks, where critical thinking, creative problem-solving, and technological fluency will be essential skills. Furthermore, the integration of cutting-edge technology and human creativity fosters resilience and sustainability in industrial processes. While smart systems optimize supply chains and energy use, enabling environmentally responsible production, automated processes can reduce waste and resource consumption. In the long term, economies with a balance between human and machine-led production will be more resilient and adaptive, able to react quickly to changing market conditions and global crises.

The combination of human ingenuity and state-of-the-art technology ultimately enables humanity to solve problems more efficiently, innovate more quickly, and build a future in which high-quality, customized, sustainable products are available to everyone while developing a workforce that values innovation, teamwork, and technological fluency.

9 Conclusion

Additive manufacturing is essential for Industry 5.0 production optimization because it promotes innovation, sustainability, and personalization. It enables mass customization, allowing producers to develop bespoke products that fit individual client needs without substantial cost hikes or delays. AM lessens dependency on centralized factories and drawn-out supply chains by enabling decentralized production, which improves responsiveness to market demands while simultaneously lowering transportation costs and environmental effects. By reducing material waste when compared to conventional production techniques, the additive process itself aids sustainability and the circular economy by recycling and reusing materials. Additionally, AM fosters innovation and enables the creation of lightweight, effective, and multipurpose items by enabling complicated geometries and inventive designs that are not possible with traditional technologies. Rapid prototyping, which enables quicker iterations and shorter time-to-market, is another way that the technology shortens the duration of product development cycles. Moreover, AM works well with cutting-edge technologies like robotics, AI, and machine learning. By automating tasks and monitoring them in real-time, AM improves production processes and boosts productivity. Within the framework of Industry 5.0, AM encourages teamwork in settings where competent professionals use technology to foster innovation and creativity rather than carrying out monotonous chores. Because of its adaptability to shifting consumer tastes and market situations, producers are able to quickly adjust, which enhances supply chain resilience by enabling on-demand production. Ultimately, incorporating AM into production fosters a more collaborative environment and resilient industrial landscape that is consistent with Industry 5.0's basic values while also stimulating economic growth through the creation of new markets and business models.

The future of personalized and sustainable production depends on embracing AM, which enables businesses to react quickly to market changes while reducing their negative effects on the environment. AM enables firms to produce customized items efficiently, cutting waste and encouraging resource efficiency. This is because it enables mass customization without requiring big stocks or significant machinery. This technology encourages creativity in design and production processes while supporting the tenets of a circular economy by making material recycling and reuse easier. Adopting AM would not only increase competitiveness but also contribute to a more robust and adaptive manufacturing environment that is in line with the changing expectations of consumers and society at large, as sectors value sustainability and customization more and more. In the end, incorporating AM into production processes is essential to realizing a sustainable future that honors both innovation and environmental responsibility.

References

1. Battaglia, D., Galati, F., Molinaro, M., & Pessot, E. (2023). Full, hybrid and platform complementarity: Exploring the industry 4.0 technology-performance link. *International Journal of Production Economics, 263*, 108949. https://doi.org/10.1016/j.ijpe.2023.108949
2. Contini, G., Peruzzini, M., Bulgarelli, S., & Bosi, G. (2023). Developing key performance indicators for monitoring sustainability in the ceramic industry: The role of digitalization and industry 4.0 technologies. *Journal of Cleaner Production, 414*, 137664. https://doi.org/10.1016/j.jclepro.2023.137664
3. Valette, E., Bril El-Haouzi, H., & Demesure, G. (2023). Industry 5.0 and its technologies: A systematic literature review upon the human place into IoT- and CPS-based industrial systems. *Computers & Industrial Engineering, 184*, 109426. https://doi.org/10.1016/j.cie.2023.109426
4. Xu, X., Lu, Y., Vogel-Heuser, B., & Wang, L. (2021). Industry 4.0 and industry 5.0—Inception, conception and perception. *Journal of Manufacturing Systems, 61*, 530–535. https://doi.org/10.1016/j.jmsy.2021.10.006
5. Ivanov, D. (2023). The industry 5.0 framework: Viability-based integration of the resilience, sustainability, and human-centricity perspectives. *International Journal of Production Research, 61*(5), 1683–1695. https://doi.org/10.1080/00207543.2022.2118892
6. Maddikunta, P. K. R., Pham, Q. V., Prabadevi, B., Deepa, N., Dev, K., Gadekallu, T. R., et al. (2022). Industry 5.0: A survey on enabling technologies and potential applications. *Journal of Industrial Information Integration, 26*, 100257. https://doi.org/10.1016/j.jii.2021.100257
7. Griffiths, C. A., Howarth, J., De Almeida-Rowbotham, G., Rees, A., & Kerton, R. (2016). A design of experiments approach for the optimisation of energy and waste during the production of parts manufactured by 3D printing. *Journal of Cleaner Production, 139*, 74–85. https://doi.org/10.1016/j.jclepro.2016.07.182
8. Vaneker, T., Bernard, A., Moroni, G., Gibson, I., & Zhang, Y. (2020). Design for additive manufacturing: Framework and methodology. *CIRP Annals, 69*(2), 578–599. https://doi.org/10.1016/j.cirp.2020.05.006
9. Ghoreishi, R., Roohi, A. H., & Ghadikolaei, A. D. (2019). Evaluation of tool wear in high-speed face milling of Al/SiC metal matrix composites. *Journal of the Brazilian Society of Mechanical Sciences and Engineering, 41*(3), 146. https://doi.org/10.1007/s40430-019-1649-3
10. Ford, S., & Despeisse, M. (2016). Additive manufacturing and sustainability: An exploratory study of the advantages and challenges. *Journal of Cleaner Production, 137*, 1573–1587. https://doi.org/10.1016/j.jclepro.2016.04.150
11. Chen, D., Heyer, S., Ibbotson, S., Salonitis, K., Steingrímsson, J. G., & Thiede, S. (2015). Direct digital manufacturing: Definition, evolution, and sustainability implications. *Journal of Cleaner Production, 107*, 615–625. https://doi.org/10.1016/j.jclepro.2015.05.009
12. Jawahir, I. S., Rouch, K. E., Dillon, O. W., Holloway, L., Hall, A., & Knuf, J. (2007). Design for sustainability (DFS): New challenges in developing and implementing a curriculum for next generation design and manufacturing engineers. *International Journal of Engineering Education, 23*(6), 1053. https://www.ijee.ie/articles/Vol23-6/s3_ijee1999.pdf
13. Peng, T., Kellens, K., Tang, R., Chen, C., & Chen, G. (2018). Sustainability of additive manufacturing: An overview on its energy demand and environmental impact. *Additive Manufacturing, 21*, 694–704. https://doi.org/10.1016/j.addma.2018.04.022

14. Wolniak, R., & Grebski, W. (2023). *The customization and personalization of product in industry 4.0*. Scientific Papers of Silesian University of Technology Organization and Management Series, 180. https://doi.org/10.29119/1641-3466.2023.180.38.
15. Rani, S., Jining, D., Shoukat, K., Shoukat, M. U., & Nawaz, S. A. (2024). A human–machine interaction mechanism: Additive manufacturing for Industry 5.0—Design and management. *Sustainability, 16*(10), 4158. https://doi.org/10.3390/su16104158
16. Stano, G., & Percoco, G. (2021). Additive manufacturing aimed to soft robots fabrication: A review. *Extreme Mechanics Letters, 42*, 101079. https://doi.org/10.1016/j.eml.2020.101079
17. Krupitzer, C., Müller, S., Lesch, V., Züfle, M., Edinger, J., Lemken, A., … & Becker, C. (2020). A survey on human machine interaction in Industry 4.0. arXiv preprint arXiv:2002.01025. https://doi.org/10.48550/arXiv.2002.01025.
18. Yang, J., Liu, Y., & Morgan, P. L. (2024). Human-machine interaction towards Industry 5.0: Human-centric smart manufacturing. *Digital Engineering, 2*, 100013. https://doi.org/10.1016/j.dte.2024.100013
19. Yang, J., Liu, T., Liu, Y., & Morgan, P. (2022, August). Review of human-machine interaction towards Industry 5.0: Human-centric smart manufacturing. In *International design engineering technical conferences and computers and information in engineering conference* (Vol. 86212, p. V002T02A060). American Society of Mechanical Engineers. https://doi.org/10.1115/DETC2022-89711
20. Mak, K. K., Wong, Y. H., & Pichika, M. R. (2023). Artificial intelligence in drug discovery and development. In *Drug discovery and evaluation: Safety and pharmacokinetic assays* (pp. 1–38). https://doi.org/10.1007/978-3-030-73317-9_92-1.
21. Nadezhdin, V. S., Demyanushko, I. V., Surkova, N. E., Tavshavadze, B. T., & Samigullin, L. F. (2022, November). Digitalization of transport industry using information technology. In *2022 intelligent technologies and electronic devices in vehicle and road transport complex (TIRVED)* (pp. 1–5). IEEE. https://doi.org/10.1109/TIRVED56496.2022.9965489
22. Duganova, E., Novikov, I., Novikov, A., & Zagorodnii, N. (2022). Problems of introduction of digital technologies in the transport industry. *Transportation Research Procedia, 63*, 1024–1033. https://doi.org/10.1016/j.trpro.2022.06.102
23. Pereira, F., Carvalho, V., Vasconcelos, R., & Soares, F. (2022). A review in the use of artificial intelligence in textile industry. In *Innovations in mechatronics engineering* (pp. 377–392). Springer International Publishing. https://doi.org/10.1007/978-3-030-79168-1_34
24. Onu, C. E., Ohale, P. E., Obiora-Okafo, I. A., Asadu, C. O., Okoye, C. C., Ojukwu, E. V., & Ezennajiego, E. E. (2022). Application of rice husk-based biomaterial in textile wastewater treatment. In *Textile wastewater treatment: Sustainable bio-nano materials and macromolecules* (vol 2, pp. 231–250). https://doi.org/10.1007/978-981-19-2852-9_12.
25. Matter, N. M., & Gado, N. (2024). Artificial intelligence in architecture: Integration into architectural design process. *Engineering Research Journal, 181*, 1–16. https://doi.org/10.21608/erj.2024.344313
26. Komatina, D., Miletić, M., & Mosurović Ružičić, M. (2024). Embracing artificial intelligence (AI) in architectural education: A step towards sustainable practice? *Buildings, 14*(8), 2578. https://doi.org/10.3390/buildings14082578
27. Zhang, H., Huang, S., Sheng, J., Wang, C., Zhang, J., Zhu, M., et al. (2023). 4D printing of shape memory polymer nanocomposites for enhanced performances and tunable response behavior. *European Polymer Journal, 201*, 112568. https://doi.org/10.1016/j.eurpolymj.2023.112568
28. Nkomo, N. Z. (2018, September). A review of 4D printing technology and future trends. In *11th South African conference on computational and applied mechanics*. https://www.researchgate.net/publication/328162917
29. Rejeski, D., Zhao, F., & Huang, Y. (2018). Research needs and recommendations on environmental implications of additive manufacturing. *Additive Manufacturing, 19*, 21–28. https://doi.org/10.1016/j.addma.2017.10.019

30. Nouri, A., Shirvan, A. R., Li, Y., & Wen, C. (2021). Additive manufacturing of metallic and polymeric load-bearing biomaterials using laser powder bed fusion: A review. *Journal of Materials Science & Technology, 94*, 196–215. https://doi.org/10.1016/j.jmst.2021.03.058
31. Wang, Y., Wang, Y., Mao, C., & Mei, D. (2023). Printing depth modeling, printing process quantification and quick-decision of printing parameters in micro-vat polymerization. *Materials & Design, 227*, 111698. https://doi.org/10.1016/j.matdes.2023.111698
32. Kanishka, K., & Acherjee, B. (2023). Revolutionizing manufacturing: A comprehensive overview of additive manufacturing processes, materials, developments, and challenges. *Journal of Manufacturing Processes, 107*, 574–619. https://doi.org/10.1016/j.jmapro.2023.10.024
33. Gardner, L. (2022). Metal additive manufacturing in structural engineering–review, opportunities and outlook. In *Current perspectives and new directions in mechanics, modelling and design of structural systems* (pp. 3–8). https://doi.org/10.1201/9781003348450-1
34. Kumar, S., Gopi, T., Harikeerthana, N., Gupta, M. K., Gaur, V., Krolczyk, G. M., & Wu, C. (2023). Machine learning techniques in additive manufacturing: A state of the art review on design, processes and production control. *Journal of Intelligent Manufacturing, 34*(1), 21–55. https://doi.org/10.1007/s10845-022-02029-5
35. Nazir, A., Gokcekaya, O., Billah, K. M. M., Ertugrul, O., Jiang, J., Sun, J., & Hussain, S. (2023). Multi-material additive manufacturing: A systematic review of design, properties, applications, challenges, and 3D printing of materials and cellular metamaterials. *Materials & Design, 226*, 111661. https://doi.org/10.1016/j.matdes.2023.111661
36. Diegel, O. (2022). Design for AM. In *A guide to additive manufacturing* (pp. 75–117). Springer International Publishing. https://doi.org/10.1007/978-3-031-05863-9_4
37. Ali, M., Almotari, A., Algamal, A., & Qattawi, A. (2023). Recent advancements in post processing of additively manufactured metals using laser polishing. *Journal of Manufacturing and Materials Processing, 7*(3), 115. https://doi.org/10.3390/jmmp7030115
38. Laleh, M., Sadeghi, E., Revilla, R. I., Chao, Q., Haghdadi, N., Hughes, A. E., et al. (2023). Heat treatment for metal additive manufacturing. *Progress in Materials Science, 133*, 101051. https://doi.org/10.1016/j.pmatsci.2022.101051
39. Dejene, N. D., & Lemu, H. G. (2023). Current status and challenges of powder bed fusion-based metal additive manufacturing: Literature review. *Metals, 13*(2), 424. https://doi.org/10.3390/met13020424
40. Li, W., Sang, L., Jian, X., & Wang, J. (2020). Influence of sanding and plasma treatment on shear bond strength of 3D-printed PEI, PEEK and PEEK/CF. *International Journal of Adhesion and Adhesives, 100*, 102614. https://doi.org/10.1016/j.ijadhadh.2020.102614
41. Tiwary, V. K., Arunkumar, P., & Malik, V. R. (2023). Investigations on microwave-assisted welding of MEX additive manufactured parts to overcome the bed size limitation. *Journal of Advanced Joining Processes, 7*, 100141. https://doi.org/10.1016/j.jajp.2023.100141
42. Kim, J., Yang, Z., Ko, H., Cho, H., & Lu, Y. (2023). Deep learning-based data registration of melt-pool-monitoring images for laser powder bed fusion additive manufacturing. *Journal of Manufacturing Systems, 68*, 117–129. https://doi.org/10.1016/j.jmsy.2023.03.006
43. Chua, K., Khan, I., Malhotra, R., & Zhu, D. (2021). Additive manufacturing and 3D printing of metallic biomaterials. *Engineered Regeneration, 2*, 288–299. https://doi.org/10.1016/j.engreg.2021.11.002
44. Colorado, H. A., Velásquez, E. I. G., & Monteiro, S. N. (2020). Sustainability of additive manufacturing: The circular economy of materials and environmental perspectives. *Journal of Materials Research and Technology, 9*(4), 8221–8234. https://doi.org/10.1016/j.jmrt.2020.04.062
45. Robayo-Salazar, R., de Gutiérrez, R. M., Villaquirán-Caicedo, M. A., & Arjona, S. D. (2023). 3D printing with cementitious materials: Challenges and opportunities for the construction sector. *Automation in Construction, 146*, 104693. https://doi.org/10.1016/j.autcon.2022.104693
46. Ahmed, G. H. (2023). A review of "3D concrete printing": Materials and process characterization, economic considerations and environmental sustainability. *Journal of Building Engineering, 66*, 105863. https://doi.org/10.1016/j.jobe.2023.105863

47. Segovia, M., & Garcia-Alfaro, J. (2022). Design, modeling and implementation of digital twins. *Sensors, 22*(14), 5396. https://doi.org/10.3390/s22145396
48. Sasson, A., & Johnson, J. C. (2016). The 3D printing order: Variability, supercenters and supply chain reconfigurations. *International Journal of Physical Distribution & Logistics Management, 46*(1), 82–94. https://doi.org/10.1108/IJPDLM-10-2015-0257

Digital Fabrication of Aerodynamic Models: A Case Study of 3D Printing for Research Efficiency in Industry 5.0

Rodrigo Padilla, Paulo Yu, Matthew Swenson, and Vibhav Durgesh

1 Introduction

Aircraft design and airfoil research for subsonic conditions require experimental testing to quantify aerodynamic behaviors. The experimental data from wind tunnel tests help engineers and designers to learn about aerodynamic performance, reduce drag, improve lift, and reduce noise. Distinct types of airfoils are used for specific applications, such as the NACA 6-series airfoil was designed to allow for laminar flow over the airfoil [1]. In contrast, Selig airfoils [2] are designed for high lift at low Reynolds number flow conditions. Aerodynamic behavior of the airfoil can be improved using topology optimization design [3, 4], the addition of gurney flaps at the trailing edge [5], morphing wings [6], boundary layer perturbations [7], or bio-inspired airfoil designs [8]. To conduct experiments with these novel ideas, precise wind tunnel testing is necessary with accurately prototyped designs.

Airfoils for the experiments can be fabricated using CNC machining, mold casting, hot-wire cutting, or 3D printing. CNC machining [9] provides excellent surface finishes but requires experienced machinists to operate CNC machines, and a wide range of materials can be used. Since CNC machining works by removing material, a single piece of metal is required, which could be large and costly, especially for larger airfoils. Additionally, if a complex attachment is needed or designed for a novel airfoil, then the operation time for CNC machining increases thereby increasing the cost of machining. Not to mention that these complex designs would require

R. Padilla · M. Swenson · V. Durgesh (✉)
Department of Mechanical Engineering, University of Idaho, Moscow, ID, USA
e-mail: padi6729@vandals.uidaho.edu; swenson@uidaho.edu; vdurgesh@uidaho.edu

P. Yu
Department of Mechanical and Industrial Engineering, University of Wisconsin, Platteville, WI, USA
e-mail: yup@uwplatt.edu

Y. Kumar Singla et al. (eds.), *Digital Manufacturing in Industry 5.0*,
https://doi.org/10.1007/978-3-031-91500-0_8

at least a five-axis CNC machine that would only be able to machine on the surface, and internal features would not be possible.

Mold casting has also been used to create airfoils for experimental study. A study by Broeren et al. [10] used mold casting to understand the effects of ice accretion on a high-fidelity, full-scale, iced-airfoil model. They mention that the best ice accretion technology at that time was the mold and casting method that the Glenn Research Center developed. This method allows the significant features of the ice to be included in the airfoil. However, the considerable drawback is that this process is expensive and impractical in many situations. Another method is hot-wire cutting and coating of high-density foam to create airfoils [11]. The advantage of this method is that it allows for rapid prototyping and is low-cost compared to alternative methods. However, the drawbacks are that it has low precision, and unique features cannot be added to the airfoil.

3D printing, becoming more accessible to students and professionals, has become an excellent alternative to airfoil wind tunnel testing and design. Moulton and Douglas [12] demonstrated advancements in 3D printing by applying it to morphing wing technology. In their work, they showcase the unique design of an airfoil that changes its chamber via trailing edge deflection. This morphing airfoil is possible by combining two materials of varying material properties with 3D printing. 3D printing allows researchers to prototype rapidly, create high-resolution parts, create low-cost parts rapidly, include novel features, and have a high-quality surface finish. The only downsides are that printing sizes are typically limited, there is limited material selection, surface finish, and filament-based printing is weaker in the printing direction.

One of the capstone design projects at the University of Idaho (U of I) focuses on understanding the impact of the rivets on the drag experienced by the horizontal tail on the Kodiak aircraft. Kodiak is a line of aircraft produced by the Quest Aircraft Company based in Sandpoint, Idaho. These are 10-seater private aircraft for flights between 969 and 1332 nautical miles and a fuel consumption of 33–58 gallon per hour. Their horizontal tail's skin is attached to the rib and spar using protruding head rivets. To understand the impact of rivets on drag, two 3D printed airfoils were used: (1) a smooth "counter sunk" head rivet and (2) a protruding head rivet. For this project, the capstone design team used a PolyJet 3D printer at the U of I 3D printing facility to print these two airfoils. The wind tunnel investigation showed that protruding rivets significantly increased overall drag, and skin-friction drag was also impacted, as observed from the oil-film interferometry (OFI) experiments.

2 Design Process and Manufacturing

2.1 Airfoil Design

Kodiak (Daher) aircraft company provided the horizontal tail information for the project. As shown in Fig. 1, the horizontal tail has a NACA 0012 airfoil shape with a span of 20.0 ft (6.096 m) and an aspect ratio of 6. Based on the

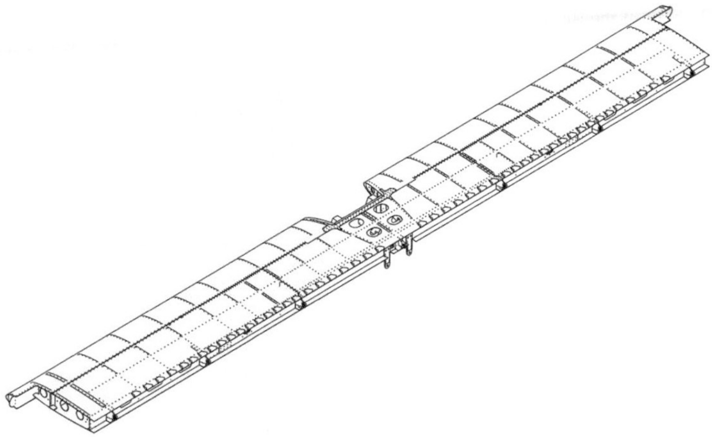

Fig. 1 Kodiak aircraft horizontal stabilizer from Daher Aircraft company with the control surfaces removed to show the rows and columns of rivets. The drawing is presented without dimensions

typical cruising speed and altitude, the Reynolds number (Re) for the horizontal tail will be around 4.62×10^6. For wind tunnel experiments, the Reynolds number was set at Re = 5.0×10^5 (due to the limitation of test article size and wind speed in the tunnel). Additionally, experiments were performed to understand the impact of Reynolds numbers on aerodynamic performance. For the wind tunnel experiments, solid-driven protruding head rivet callout #4 head dimensions were used for the 3D printing. The other components shown in the figure were not used to isolate the impact of the rivets on aerodynamic results (i.e., access panels, elevators, actuators, attachment to fuselage, vortex generators, etc.)

The chord length of the airfoil for wind tunnel experimentation was determined based on the blockage ratio and aerodynamic loading analysis. Figure 2 shows the contour of the chord versus velocity to select a chord length and fluid velocity for wind tunnel testing. In the figure, the green line on the contour represents the various combinations of chord and wind tunnel velocity, which will provide Re of 5.0×10^5. From this plot, the desired velocity was selected to be 32.15 $^{m}/_{s}$ and a chord of 9 in (0.2286 m).

The area blockage and aerodynamic load calculations were performed for a nine-inch chord airfoil to determine the range of α for the wind tunnel experimentation. An area blockage is around 8–9% for a typical wind tunnel experiment. Therefore, for these experiments, we selected an angle of attack on the airfoil that will provide area blockage of 8–9%. Using a typical NACA 0012 lift coefficient, C_L, and drag coefficient, C_D, maximum aerodynamic load forces experienced by the load cell were calculated. Based on these two analyses, it was determined that the lowest Re of 3.0×10^5, α can be varied between −5.04° and 5.04°, between −3.6°and 3.6° for Re of 4.0×10^5, and between −2.16° and 2.16° for Re of 5.0×10^5.

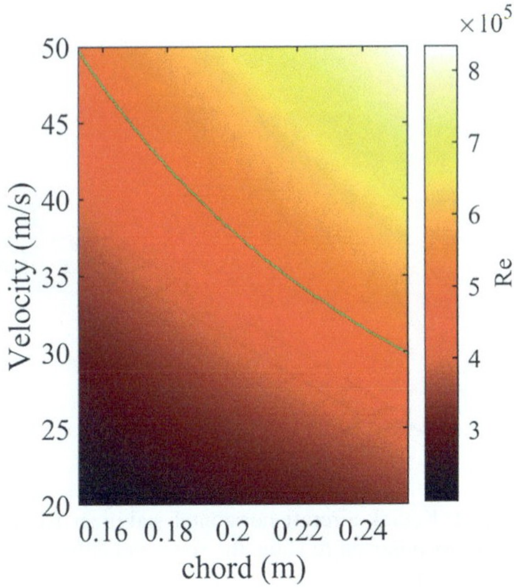

Fig. 2 Contour of the Re as function of chord lengths fluid velocity. The green solid line shows the Re of 5.0×10^5

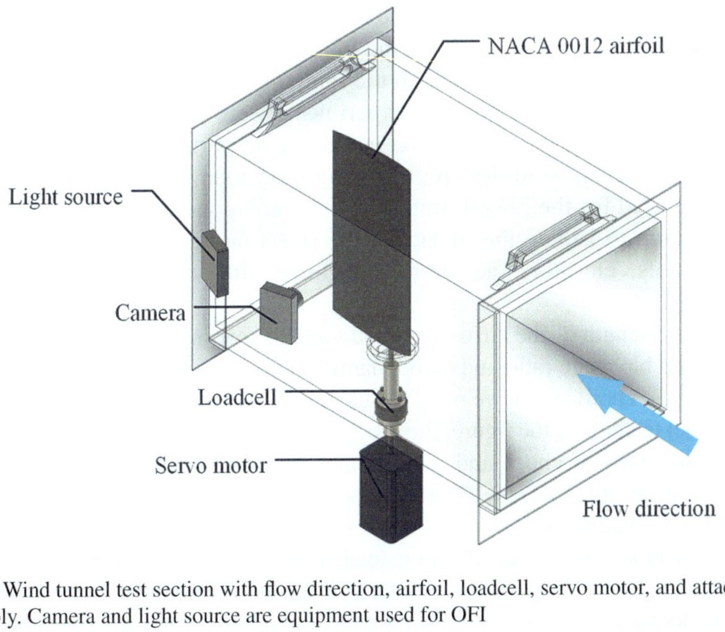

Fig. 3 Wind tunnel test section with flow direction, airfoil, loadcell, servo motor, and attachment assembly. Camera and light source are equipment used for OFI

2.2 Wind Tunnel and Load Cell

The wind tunnel used in experimental fluids and aerodynamics laboratory (EFAL) is an Eiffel-type wind tunnel that can produce wind speeds of 0–160 mph $(0 - 70 \,{}^{m}\!/_{s})$ with its 50 HP motor. Its test section SolidWorks drawing is shown in Fig. 3, test-section size is 18 in × 18 in × 36 in (0.4572 m x 0.4572 m x 0.9144 m) and has optical access from all directions. Flow conditioning is done with a series of honeycombs and screens at the inlet of wind tunnel. Additionally, the top and bottom sections of the test section have removable panels to place the airfoil. On the bottom panel, there is a port for accessing instrumentation wires. The airfoil is mounted directly to the loadcell, and the loadcell is attached to a stepper motor for changing the angle of attack, α.

To isolate the vibrations from the wind tunnel motor, a mounting mechanism was firmly mounted on the laboratory floor. ATI Gamma force and torque transducer capable of measuring forces F_x and F_y up to 65 N with a resolution of 1/80 N and F_z up to 200 N with a resolution of 1/40 N are used for this experiment. For these experiments, data were sampled at 200 Hz for 30 s for every angle of attack values. The servo motor is an Automation Technology Inc. Hybrid Servo Motor (model: KL34-8N-1000) capable of torque up to 1128 oz-in (8 N-m) with 1000 encoder lines and up to 40,000 steps per revolution. For the experiments, 1000 steps per revolution (0.36 degrees per step) were selected for the servo motor. The range of α was selected based on the maximum torque produced by the airfoil, which varied with Re. Before each wind tunnel test, local pressure and temperature were measured and recorded.

2.3 3D Printing

Stratasys J850 Pro Multi Material System with VeroBlackPlus material (supplied by the printer manufacturer) was used for the 3D printing of the airfoil with a resolution of 0.012 in the x and y direction and 0.000118 in the z direction. For 3D printing, the airfoil's trailing edge was aligned with the high-resolution axis of the Stratasys J850 printer to ensure the Kutta condition was met. The build size of the printer was 19.3 × 15.35 × 7.9 in (0.49 m x 0.39 m x 0.20 m). This allowed us to print the appropriate size for this study's airfoil.

The CAD model of the smooth surface airfoil is shown in Fig. 4, and the 3D print airfoil in Fig. 5. As observed in Fig. 5, the airfoil has a smooth surface finish and a sharp trailing edge. The CAD model of the rivet spacing is shown in Figs. 6 and 7. The rivet pattern shown here matched the rivet layout of the horizontal stabilizer without control surfaces (i.e., elevators) rivets. The spacing between the various

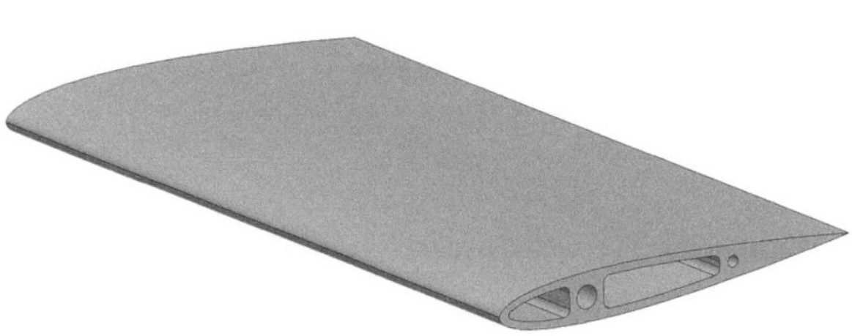

Fig. 4 Computer aid design (CAD) of the smooth NACA 0012 airfoil used for 3D printing

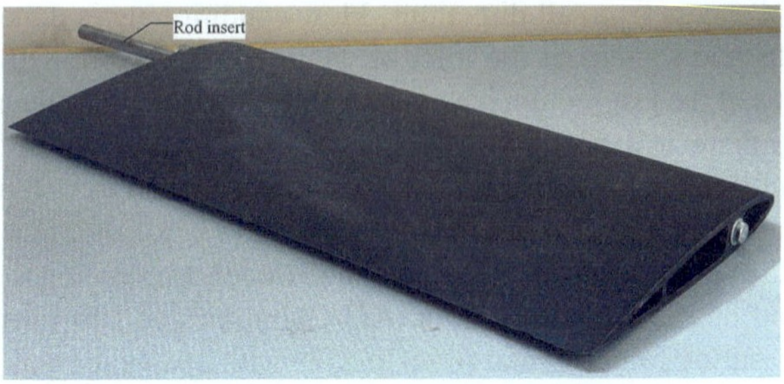

Fig. 5 Resulting 3D resin printing of the NACA 0012 smooth airfoil with rod attachment for mounting

Fig. 6 CAD drawing of the NACA 0012 airfoil with rivets added to the surface

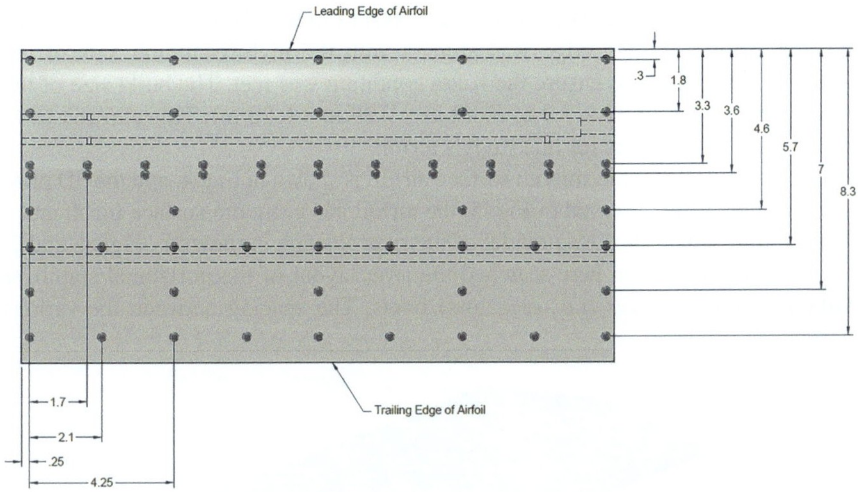

Fig. 7 Spacing selected for the 3-D printed riveted NACA-0012 airfoil (All the units are in inches)

Fig. 8 3D printing of riveted airfoil, showing the riveted pattern and sharp trailing edge

rows of rivets was in the same ratio that is typically observed in the horizontal tail of the Kodiak aircraft.

To match type 4 rivet head, the rivet head: diameter of 0.262 in and head height of 0.064 in were maintained to keep the rivet head style as per rivet specifications (MS20470). Also, the spacing of the rivets was considered such that the ratio between the chord length of the Kodiak aircraft and experimental airfoil was maintained at a 4.44:1 scaling ratio. This scaling is applied to the number of rivets per inch. The Kodiak aircraft has 10 rivets between the ribs (i.e., 18.48 inches between the ribs). To match the ratio of the rivet spacing on the aircraft's horizontal tail, nine evenly spaced rivets were placed along the span of the airfoil for the mid and back spars sections. The butt joint has a higher density rivet region and is near the quarter chord section of the horizontal stabilizer, connecting leading edge and trailing edge panels. Therefore, two rows of eleven evenly spaced rivets are placed at the quarter chord along the span of the airfoil, as shown in Fig. 8. For chordwise rivet spacing, the aircraft has two different sets of rivets spacing one for leading-edge panels and another for trailing edge panels (these panels are held together by butt joint rivets). There are two rows of rivets for the leading edge panels and four rows of rivets for the trailing edge panels, as shown in the figure. This scaling enabled us to design the airfoil with correct rivet position and spacing for the wind tunnel experiments to quantify the impact of the rivet configuration. The print of the riveted airfoil is shown in Fig. 8.

3 Results

The next few sections of this chapter will briefly discuss the main results from the experiments performed for this study. First, the impact of rivet on the aerodynamic loads is discussed for the three Reynolds numbers. Next results from OFI experiments are presented to highlight the impact of rivets on the flow over the airfoil and increase in the shear-stress value.

4 Aerodynamic Load

Aerodynamics load measurements were conducted for both the "smooth" countersunk riveted airfoil and protruding riveted airfoil to quantify the impact of drag on the airfoil for the Re of 3 × 10^5, 4 × 10^5, and 5 × 10^5, and C_d results are shown in Figs. 9, 10, and 11, respectively. From these figures, it is evident that there is a significant impact on the drag experienced by the airfoil due to the presence of rivets. The drag forces are normalized by the dynamic pressure and planform area to get the drag coefficient (C_d). The smooth and riveted airfoil drag coefficient, C_d, shows a decrease with the increasing Re. The difference between smooth and riveted increases with an increase in Re values from 3 × 10^5 to 5 × 10^5. For all Re, the

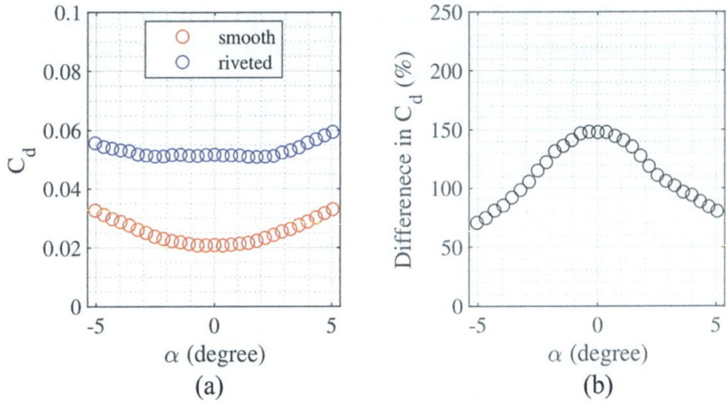

Fig. 9 Drag coefficient of the 9-inch NACA 0012 showing: (**a**) the smooth and riveted airfoil at Re = 3.0 × 10^5, and (**b**) the percentage difference between the airfoils for α from −5.04° to 5.04°

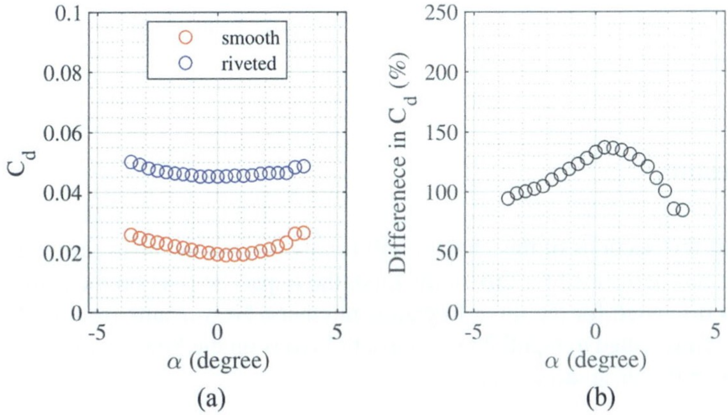

Fig. 10 Drag coefficient of the 9-inch NACA 0012 showing: (**a**) the smooth and riveted airfoil at Re = 4.0 × 10^5, and (**b**) the percentage difference between the airfoils for α from −3.6° to 3.6°

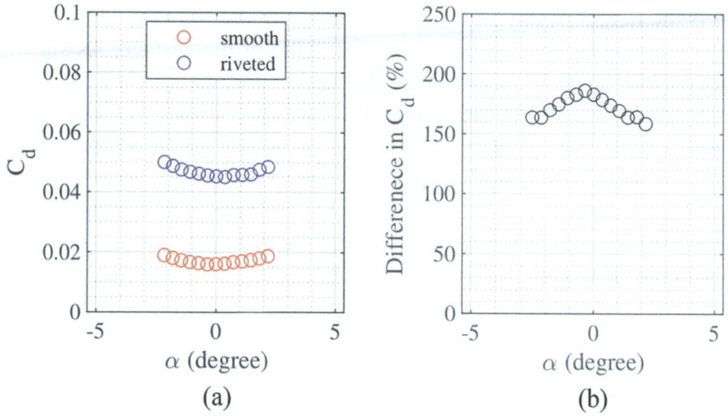

Fig. 11 Drag coefficient of the 9-inch NACA 0012 showing: (**a**) the smooth and riveted airfoil at Re = 5.0 × 10⁵, and (**b**) the percentage difference between the airfoils for α from −2.16° to 2.16°

maximum impact of rivet is for α of 0°. For Re of 3×10^5 and 4×10^5, the difference is around 150% higher for riveted than smooth and for Re of 5×10^5, it is around 190%. As α changes, the difference begins to decrease for all Re. These figures clearly show that there is drag experienced by the airfoil is dependent on the Reynolds number and angle of attack. It is also evident from these figures that there is a significant increase in drag due to the presence of rivets.

The increase in overall drag due to rivet can be attributed to several factors, and one of them is skin friction drag. The next logical step was to quantify the impact of the rivets on the skin friction drag. For this, OFI technique was used as it easy to implement and can provide an accurate estimate of skin friction values. The following section covers the necessary challenges to overcome to obtain skin friction measurements using OFI.

5 Oil Film Interferometry

The next step in the testing was to quantify the change in the airfoil's wall shear stress due to rivets. OFI is an effective measurement technique for measuring wall shear stress in aerodynamic testing. It has been used for measurements such as two-dimensional flows over wings, three-dimensional flows on a surface, and transitional boundary layer flows. The technique is based on the response behavior of a thin oil film from aerodynamic shear when air flows over the oil's surface. The thin oil film equation governs the response, and a simplified form of this equation is shown as

$$\frac{\partial h}{\partial t} + \frac{\partial}{\partial x}\left(\frac{\tau_{w,x} h^2}{2\mu}\right) + \frac{\partial}{\partial z}\left(\frac{\tau_{w,z} h^2}{2\mu}\right) = 0, \tag{1}$$

where h is the height of the oil, τ_w is the wall shear stress, t is the time, μ is the oil viscosity, and x and z are coordinates in the plane of the surface. As shown in the equation, the height of the oil on the model surface is dependent on the shear stress and by accurately measuring the height of oil, we can estimate the wall shear stress experienced by the model. To accurately measure the height of the oil film, the interferometer principle is used. The height obtained from the interferometry technique has high accuracy and yields an accurate wall shear stress value. The reader is referred to [13] for detailed information about the thin oil film equation. One of the first steps in the OFI experiments is to apply the uniform oil film over the test object. For this study, a 3D-printed oil applicator was designed.

6 Oil Film Applicator Design

An oil applicator was designed and manufactured to help conduct OFI experiments. The few requirements for this applicator design were ease of manufacturing, flexibility of use for different types of oil, flexibility in applicator length, and repeatable and consistent output of oil from the applicator. Different oil applicator concepts were considered, but only the final design will be discussed here for brevity.

The final concept of the applicator consisted of a small manifold used to distribute the oil and a connector design to attach to an oil reservoir. A computer-aided model (CAD) of the OFI applicator can be seen in Fig. 12, highlighting the design's key characteristics. The applicator is designed with a single inlet, which splits into a row of outlets using a ladder design. The overall dimensions of the applicator were 65 mm in length, 20 mm in height, and 4.25 mm in thickness, and the design can be scaled for big or small experimental setups. The hole diameters for the applicator start at 4.3 mm at the inlet and gradually decrease to 0.75 mm at the outlets. The diameters were chosen to provide smooth transitions from inlet to outlet. A male

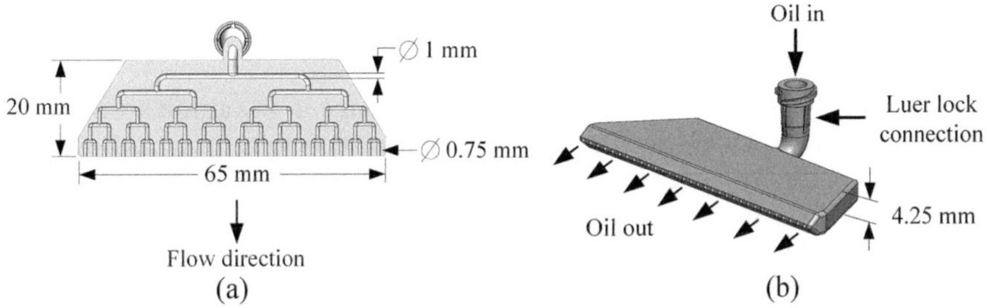

Fig. 12 The oil applicator model shows critical features of the design: (**a**) section view of ladder design and critical dimensions of the applicator, and (**b**) 3D CAD model of the applicator

Luer lock connection was used to securely screw the applicator to a fluid reservoir such as a syringe. A small applicator of 25 mm in length and similar thickness and height was also designed and not shown here.

7 Applicator Fabrication and Testing

The unique design of the applicator makes traditional manufacturing methods challenging, and thus, the applicator was made using 3D printing technology. For this method, a resin 3D printer (Mars 3 model, Elegoo) was used to print the model, which produced fine details and smooth surfaces in the printed part. The 3D printer used UV photocuring technology with an XY resolution of 0.035 mm, Z axis accuracy of 0.00125 mm, and a layer thickness of 0.01–0.2 mm. The 3D printer specifications allowed the part to be printed according to the designed features, particularly with the 0.075-mm hole dimensions. The parts took about 1–2 h to print, and the support material was removed using warm water or alcohol.

The physical 3D printed designs can be seen in Fig. 13 with a small applicator (Fig. 13a) and a large applicator (Fig. 13b). The initial testing of the applicator showed that all holes could output oil consistently. The additional benefit of this design was that it could determine the amount of oil applied using the markings from the syringe. Furthermore, the designs were easily cleaned by using soap and warm water. The design could then be used for oils of different viscosities.

OFI experiments were conducted with smooth and riveted airfoils using the 3D-printed applicator. Re = 4.0×10^5 was chosen for objective comparisons using

(a) (b)

Fig. 13 3D printed applicator: (**a**) small applicator attached to a syringe. (**b**) large applicator is used to apply oil on an airfoil

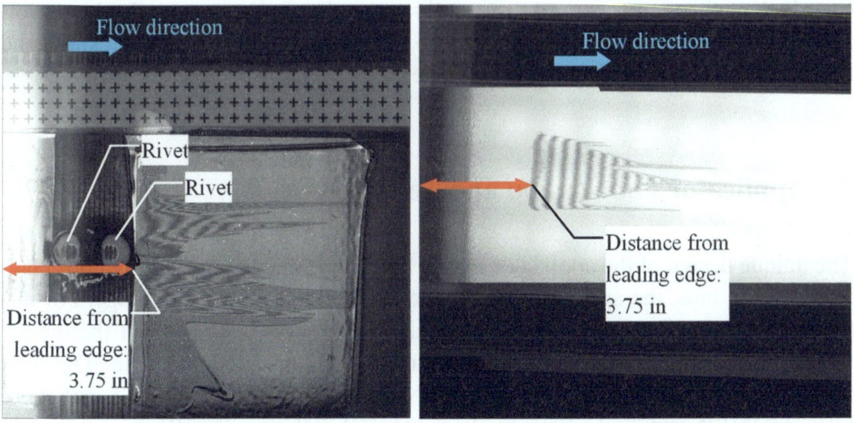

(a) (b)

Fig. 14 The oil is applied, which results in a interferometry fringe pattern for: (**a**) the smooth airfoil and (**b**) the riveted airfoil

the airfoil chord length of 228.6 mm (9 in) for these experiments. The oil used for experiments was Xiameter PMX-200 Silicone Fluid with a viscosity of 50 cSt. Reference marks 2 mm apart were attached to the airfoils to calibrate the image. A custom image acquisition program (National Instruments, LabView program) was made to acquire the images, where pressure and image time instances were recorded simultaneously. The sampling rate of 1 Hz was used for both scenarios based on previous trials with OFI. A separate image analysis program (MathWorks, Matlab R2023a) was also made to analyze the images and determine the oil height and skin friction coefficients. Lastly, the OFI analysis was conducted downstream of a rivet. The applied oil is shown in Fig. 14a, b for the smooth and riveted airfoil, respectively. In Fig. 14a, the fringe pattern is clearly visible and nicely spaced. Alternatively, when the oil is applied to the riveted airfoil in Fig. 14b, it shows that the fringe pattern spreads but not directly in the wake of the rivet.

The OFI results for smooth and riveted airfoils can be seen in Fig. 15a, b for the same time instance (i.e., $t = 232.0$ s). Here, we can see the impact of a riveted airfoil design versus a non-riveted (smooth) design. In Fig. 15a, oil was applied on a smooth airfoil, which spread minimally across the surface, with about 4 mm of the observed fringe pattern. On the other hand, oil on the riveted airfoil spread for about 10 mm at the same time, with fewer fringe patterns than the smooth airfoil. The impact of these oil heights can be seen in Fig. 15b, where the smooth airfoil had roughly similar (stable) C_f values across the surface and elevated and varying C_f values for the riveted airfoil. The smooth airfoil had $C_f = 0.000286$ on average, while the riveted airfoil ranged from $C_f = 0.00189$ to 0.00340, an increase in C_f by a factor of 10. This highlights that adding rivets to the airfoil negatively impacts its aerodynamic performance with increased skin friction coefficient.

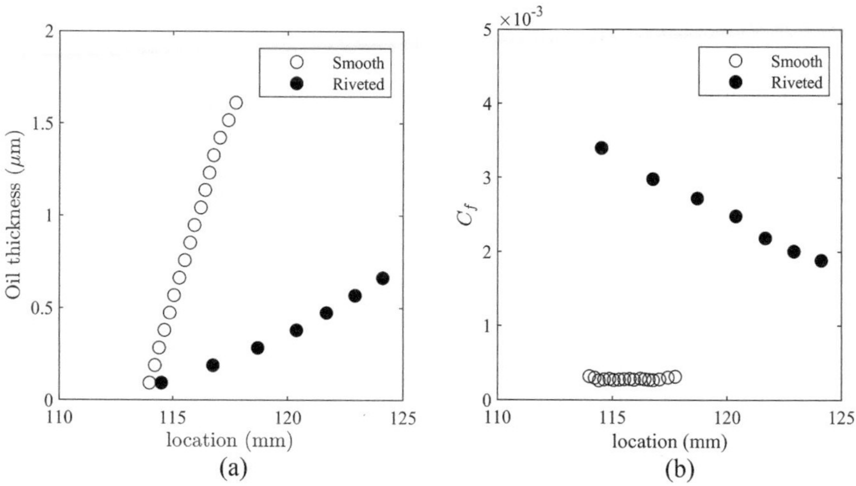

Fig. 15 OFI results for smooth and riveted airfoils. Results were obtained for $t = 232.0$ s. (**a**) Oil height across the airfoil surface, and (**b**) skin-friction coefficient values

8 Conclusions

Improvements in 3D printing enable the manufacture of complex designs for airfoils needed for testing and design. This allowed us to design a riveted airfoil that mimics the horizontal stabilizer of the Kodiak aircraft by the Quest Aircraft Company. Furthermore, a polyjet 3D printer at the U of I manufactured riveted and smooth airfoil. The purpose of the study was to quantify the impact of rivets on a smooth airfoil. This was done by measuring load measurements via load cell and quantifying skin friction coefficients with oil film interferometry. A unique applicator was designed and manufactured using a resin 3D printer to achieve even oil application for OFI experiments. From the load measurements, it was shown that the rivets increased the drag on the airfoil. For a Re of 3×10^5 the drag increased by 150%, but for the higher Re of 5×10^5 the drag would increase by 190%. Furthermore, preliminary OFI experiments demonstrate that the rivets increase the skin friction coefficient and impact the overall aerodynamic load experienced by the airfoil. The OFI application designed for this project provided an excellent result, which significantly improved the quality of OFI interferometry fringe patterns. This study clearly showed that the recent advancement in 3D printing technology has a significant impact on the experimental aerodynamics field, as it allows for rapid fabrication of models and testing new prototypes, which would not have been possible otherwise. Finally, 3D printing technology has reduced the cost of designing models and prototypes for aerodynamic experiments by an order of magnitude.

9 Future Scope

This study provides valuable insight into the 3D manufacturing of airfoils for wind tunnel testing, allowing for future research. Future research can be extended by studying various airfoils, adding features to the airfoils (i.e., control surfaces), and varying the Reynolds number. Additionally, this type of study can be added to testing various inflight scenarios, which would improve the aircraft design and accurate estimation of aerodynamics loads.

Acknowledgments We would like to acknowledge all the students who worked on the project capstone project: Shalom Masango, Ryan Schaefer, Brendyn Goodwin, Bradley Hille, Garrett Green, Hayden Jacobson, and Jared Nelson. We would also like to thank Mr. Brandon Stille and Mr. Jamie Slippy from Daher aircraft for sponsoring the project.

References

1. Abbott, I. H., Von Doenhoff, A. E., & Stivers, L. S. (1945). *NACA report no. 824–Summary of airfoil data.*
2. Selig, M. S., & Guglielmo, J. J. (1997). High-lift low Reynolds number airfoil design. *Journal of Aircraft, 34*(1), 72.
3. Balabanov, V. O., & Haftka, R. T. (1996). Topology optimization of transport wing internal structure. *Journal of Aircraft, 33*(1), 232–233. https://arc.aiaa.org/doi/abs/10.2514/3.46926?journalCode=ja
4. Gomes, P., & Palacios, R. (2020). Aerodynamic-driven topology optimization of compliant airfoils. *Structural and Multidisciplinary Optimization, 62*(4), 2117–2130. https://doi.org/10.1007/s00158-020-02600-9
5. Kentfield, J. A. C. (2020). Theoretically and experimentally obtained performances of gurney-flap equipped wind turbines. *Wind Engineering, 18*(2), 63–74. https://www.jstor.org/stable/43749530
6. Li, D., Zhao, S., Da Ronch, A., Xiang, J., Drofelnik, J., Li, Y., Zhang, L., Wu, Y., Kintscher, M., Monner, H. P., Rudenko, A., Guo, S., Yin, W., Kirn, J., Storm, S., & De Breuker, R. (2018). A review of modelling and analysis of morphing wings. *Progress in Aerospace Sciences, 100*, 46–62. https://doi.org/10.1016/j.paerosci.2018.06.002
7. Bixler, G. D., & Bhushan, B. (2013). Fluid drag reduction with shark-skin riblet inspired microstructured surfaces. *Advanced Functional Materials, 23*(36), 4507–4528. https://doi.org/10.1002/adfm.201203683
8. Han, J., Hui, Z., Tian, F., & Chen, G. (2021). Review on bio-inspired flight systems and bionic aerodynamics. *Chinese Journal of Aeronautics, 34*, 170. https://doi.org/10.1016/j.cja.2020.03.036
9. Linke, B., Garretson, I., Jan, F., & Hafez, M. (2017). Integrated design, manufacturing and analysis of airfoil and nozzle shapes in an undergraduate course. *Procedia Manufacturing, 10*, 1077–1086. https://doi.org/10.1016/j.promfg.2017.07.098
10. Broeren, A. P., Bragg, M. B., Addy, H. E., Lee, S., Moens, F., & Guffond, D. (2008). Effect of high-fidelity ice accretion simulations on the performance of a full-scale airfoil model. In *46th AIAA aerospace sciences meeting and exhibit*. https://doi.org/10.2514/6.2008-434.

11. Ceruti, A., Caligiana, G., & Persiani, F. (2013). Comparative evaluation of different optimization methodologies for the design of UAVs having shape obtained by hot wire cutting techniques. *International Journal on Interactive Design and Manufacturing, 7*(2), 63–78. https://doi.org/10.1007/s12008-012-0164-x
12. Moulton, B. C., & Hunsaker, D. F. (2021). 3D-printed wings with morphing trailing-edge technology. In *AIAA Scitech 2021 forum* (pp. 1–21). https://doi.org/10.2514/6.2021-0351.
13. Tanner, L. H. (1976). The measurement of viscosity by optical techniques applied to a falling liquid film. *Journal of Physics E: Scientific Instruments, 9*(11), 967–973. https://doi.org/10.1088/0022-3735/9/11/024

Revolutionizing Metal Manufacturing with AI-Driven 3D Printing and Additive Manufacturing

Mounarik Mondal, Suman Mondal, Vani Nigam, Samarthya Goyal, Soumyabrata Basak, Henu Sharma, and Kisor K. Sahu

1 Introduction

From the very dawn of human civilization until the industrial revolution, manufacturing was mostly considered a matter of art, relying heavily on individual craftsmanship. The industrial revolution ushered in an era of a very high level of productivity, thereby the drive to understand the science behind it and the need for

M. Mondal
School of Engineering and Design, London South Bank University, London, United Kingdom
e-mail: mounarik.mondal@lsbu.ac.uk

S. Mondal
Metallurgical and Materials Engineering, National Institute of Technology Srinagar, Srinagar, India
e-mail: suman_2021bmme054@nitsri.ac.in

V. Nigam · K. K. Sahu (✉)
School of Mineral, Metallurgical, and Materials Engineering, Indian Institute of Technology Bhubaneswar, Bhubaneswar, India
e-mail: 21mm01016@iitbbs.ac.in; kisorsahu@iitbbs.ac.in

S. Goyal
School of Infrastructure, Indian Institute of Technology Bhubaneswar, Bhubaneswar, India
e-mail: 20ce02003@iitbbs.ac.in

S. Basak (✉)
School of Mechanical Engineering, University of Ulsan, Ulsan, Republic of Korea

H. Sharma
School of Mineral, Metallurgical, and Materials Engineering, Indian Institute of Technology Bhubaneswar, Bhubaneswar, India

NetTantra Technologies India Pvt. Ltd., Bhubaneswar, India
e-mail: hs17@iitbbs.ac.in

Y. Kumar Singla et al. (eds.), *Digital Manufacturing in Industry 5.0*,
https://doi.org/10.1007/978-3-031-91500-0_9

standardization was strongly felt. Manufacturing, as an engineering discipline, was born. From the time of the industrial revolution until now, manufacturing is mostly based on the principles of carving out an object from a bulk piece through materials removal; therefore, called subtractive or, sometimes, traditional manufacturing. These objects/parts are then assembled to manufacture the products. This protocol was simple and has served us well until recently. Before understanding additive manufacturing (AM) and appreciating its need, we need to know where traditional manufacturing fell short. Three major weaknesses of traditional manufacturing processes [1] are (1) they are subtractive and thereby produce significant materials waste. For example, in the case of Titanium alloys for aerospace applications, the wastage can sometimes be as high as 80%. It has at least two negative consequences: significantly increased costs and carbon footprints. (2) They require highly centralized production facilities. This, again, has at least two significant drawbacks: the requirement for unusually high capital investment and the fact that they are highly vulnerable since any single point of failure can disrupt the entire chain of event. (3) They offer a limited scope for repair. However, in recent times, with the unprecedented growth in human civilization, the finiteness of both energy and material resources has quickly turned into serious bottlenecks. The traditional methods, in many cases, are turning out not to be a sustainable and cost-effective method. This is the philosophical underpinning of AM. By being additive (opposite to subtractive), it aims to make the object by adding materials in batches, or part by part. We can immediately appreciate how AM can alleviate the concern of waste (point 1). Another great benefit of AM is that, by being a new-generation process, it can be easily coupled with concurrent modern development in other fields of science and engineering in general and digitization in particular, thereby making it modular and decentralized (addressing point no. 2). A repeating theme of this chapter will be how this modularity will help us to distribute the manufacturing process over several continents and achieve the status of a truly global manufacturing protocol. Strange as it may sound, it can even go beyond it. In Sect. 4.2, we will show how this process is presently the only feasible interplanetary manufacturing process and can be crucial in terraforming other planets, say, Mars. Another great benefit of AM is that it can provide high-quality repairs, thereby significantly extending the lifespan of parts (addressing point no. 3).

Now, we will take a bird's eye view of a typical AM protocol before delving into its details. The flowchart [2] above (see Fig. 1) illustrates the AM process chain from product design to post-processing. Key stages include product design (materials and geometry), process planning, and in situ process control, with feedback mechanisms for managing defects and ensuring quality. The integration of machine learning models assists in optimizing parameters, ensuring the final product meets desired physical, geometric, and mechanical properties. The process begins with product design, focusing on material selection, geometry, and topology. Key AM parameters such as laser power, scan speed, and layer thickness are planned and controlled through in-situ monitoring of temperature, oxygen levels, and preheat temperatures. Feedback from detected defects like plumes and delamination informs adjustments, with machine learning models aiding in optimization. Ultimately, quality is measured through properties like geometric accuracy and mechanical

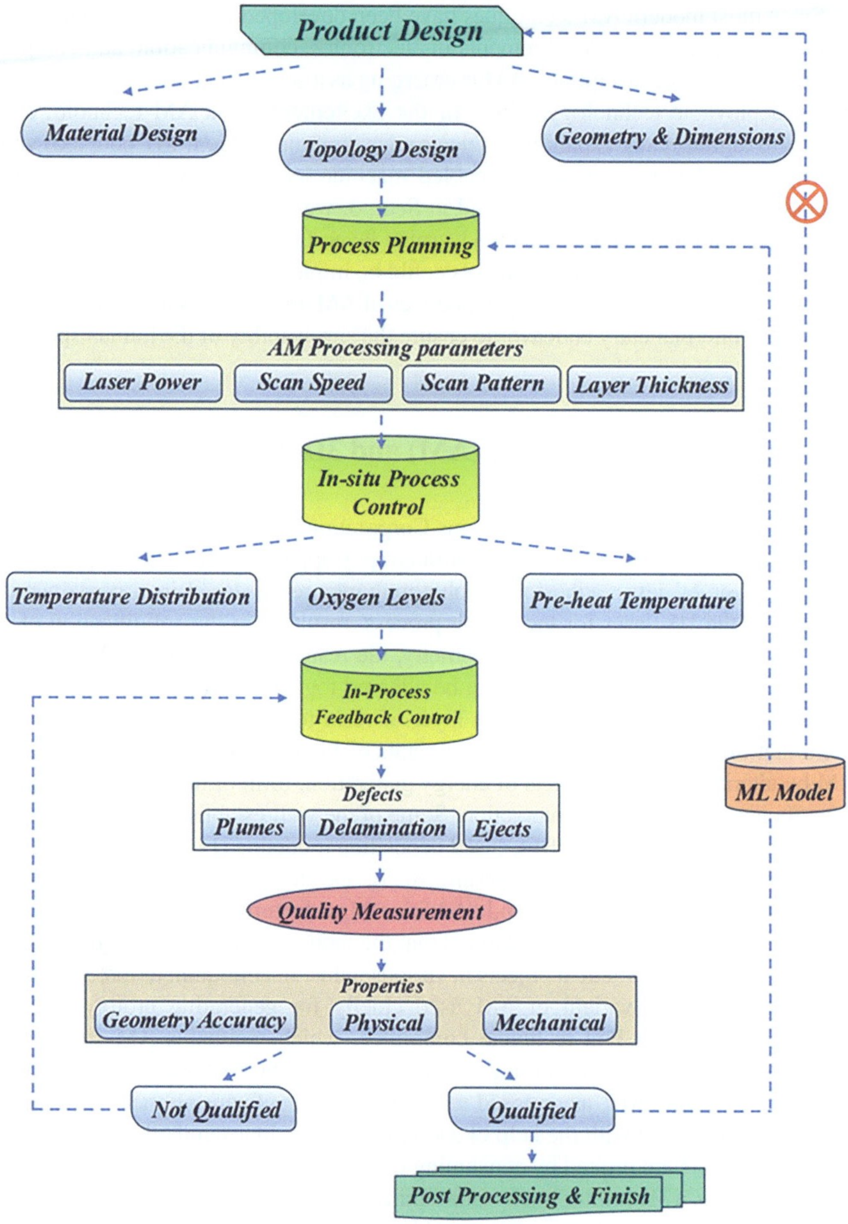

Fig. 1 Flowchart of the AM process. (Adapted from Ref. [2])

integrity, determining whether further post-processing is needed. Essentially, additive manufacturing and 3D printing are extensively used for numerous material types, including polymers; this chapter will mostly focus on metals to keep depth and clarity within the space limitations. Still, the basic ideas covered here also apply to polymers (one must be careful about the operating temperature, though).

Since most modern AM techniques have been developed recently, they integrate well with contemporary developments in electronics, communication, and computational engineering disciplines. AM is emerging as a smart and digital-ready industrially deployable technology. Much of the excitement about AM technology is related to digital-ready capabilities (electronics, communication, IoT coupled) and intelligent (embedded computing-enabled machine learning). In Sect. 2, we will discuss the AM hardware aspects, and in Sect. 3, we will focus on the digital and embedded compute-enabled machine learning layer. In Sect. 4, we will discuss how the distributed manufacturing protocols could be implemented in AM deployments. In Sect. 5, we will discuss future perspectives of AM and why it is a crucial component in our interplanetary endeavor to ensure the survivability of the human species.

2 Additive Manufacturing (AM) and 3D Printing Hardware

In 1986, Charles W. Hull of 3D Systems Corporation first manufactured a working 3D printer, which is essentially a stereolithography apparatus (SLA) [3]. During this process, layer by layer adhesion of photopolymer resin stored in a container is exposed to the UV laser for the curing process, resulting in the solidification of a single layer. Based on the desired geometry, the resin container or the UV laser is moved so that the multiple layers can be adhered together. In most cases, the container moves down, exposing a fresh layer of resin to be exposed to the laser [4]. From that time onwards, based on the variants of SLA, it evolved into modern-day AM by altering different aspects of energy generation, form of material source, and type of energy and material interaction. Some of the well-established AM processes include powder bed fusion (PBF), fused deposition modeling (FDM), selective laser melting (SLM), selective laser sintering (SLS), and digital light processing (DLP) [5–7]. The terms "3D printing" and "Additive Manufacturing (AM)" are used in a transposable sense, as it is often claimed that the industrial version of 3D printing is AM, producing real, practical objects in various fields. In colloquial terms, 3D printing is a lower-level variant of AM used chiefly for generating prototypes [8]. However, the ASTM F42 technical committee, which maintains AM standards, defines AM as "A process of joining materials to make objects from 3D model data, usually layer upon layer, as opposed to subtractive manufacturing methodologies" [9]. Let us illustrate it with the help of a simple analogy. In a standard inkjet printer, the various letters are printed by consuming the ink and depositing it in layers on the paper. Similarly, a 3D printer works on a similar philosophy by building a product on a substrate by depositing layers of material from which the actual build will be made.

In recent years, AM has evolved into proper industrial manufacturing, being an ideal alternative to conventional manufacturing involving machining, casting, and forming to meet the needs of modern products. Due to its geometrical precision, high operational speed, reduced scrap generation, and less material consumption, the modern manufacturing sector is slowly adopting AM as one of the most efficient processes. Currently, AM is used in aerospace, energy, transportation, and civil structures to manufacture functional and structural parts and whole finished products incorporating various grades of materials such as metals, ceramics, polymers, and concrete [10–12], though our focus in this chapter will be primarily on metals for the sake of conciseness. In the medical sector, AM has gained popularity in manufacturing human prosthetics (e.g., limbs and joints), which has undoubtedly enhanced the precision and manufacturing rate to sustainability using polymers and ceramics [13].

Although AM for metals is a technology derived from metal welding and cladding, there remains a fine line of difference in knowledge and expertise between them. Hence, many academic and industrial efforts are being made to elucidate the complexities of this process. Although many decades of research efforts in welding and cladding have generated a mature knowledge base of welding and cladding, the study and development of AM of metallic materials will still be a long and winding one.

Hence, to properly understand AM, its differences from welding must be understood clearly [14].

Figure 2 exhibits various welding and AM procedures according to the workpiece's physical state (melting and non-melting). Non-melting processes that do not require the addition of materials and avoid material fusion include friction stir welding (FSW), friction stir processing (FSP), friction stir spot welding (FSSW), and friction taper plug welding (FTPW). Conversely, procedures like LASER welding, gas tungsten arc welding (GTAW), and gas metal arc welding (GMAW) involve melting, and for thicker cross-sections, the addition of filler materials is used. Meanwhile, non-melting AM processes include cold spray, additive friction stir deposition, and friction stir AM. In the fusion-based process, candidate technologies that melt and add material are LASER, Electron Beam Melting, and Wire Arc AM. Mainly, in all welding, two existing base materials are joined by either melting or plasticizing them. However, in the AM process, the joining between two plasticized or melted layers is deposited based on the desired geometry of the finished product.

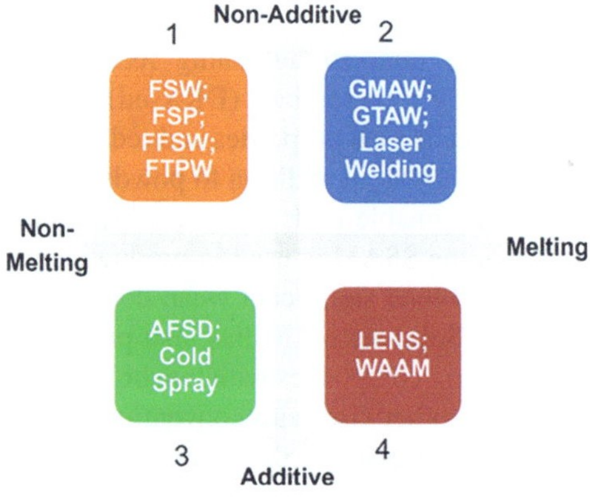

Fig. 2 Various types of solid-state and fusion-based welding and additive manufacturing techniques

2.1 Friction-Stir Assisted Additive Manufacturing

Friction stir-assisted additive manufacturing (FSSAM) can be broadly categorized into two types (see Fig. 3): Additive friction stir deposition (AFSD) and friction stir lap joint (FSLJ). Based on the material consumed to be deposited, the AFSD can be further subdivided into two parts: AFSD with consumable and non-consumable tools. Friction stir-assisted additive manufacturing (FSAAM) is a solid-state additive manufacturing (SSAM) process that has shown promising development in counteracting the challenges arising from highly efficient conventional fusion-based AM processes [15, 16].

Most fusion-based AM techniques can successfully generate complex geometries, typically using lasers and electron beams. However, they incorporate various defects related to solidification, such as the generation of internal cavities and porosity, anisotropic material behavior due to the formation of columnar grains, shrinkage, and inclusions, especially in the case of various low melting lightweight alloys, like aluminum and magnesium, which are mostly precipitation-strengthened and highly susceptible to softening because of the dissolution of precipitations in excess energy input demands the application of low energy input processing. Although secondary processing involves various types of heat treatments, the defects and disadvantages mentioned above can be significantly minimized, but only at the expense of additional cost and time. Further, the conventional fusion-assisted AM processes are mostly limited to processing raw materials in the form of powders and metallic wires only [17, 18].

In contrast, most solid-state additive manufacturing (SSAM) processes operate below the melting temperature, avoiding defects generated from solidification. It utilizes frictional heat and high straining to plasticize materials and mix them to form a deposition layer and, in some cases, fuses multiple pieces of sheets or foils in generating a final build [19]. Since the process operates at lower energy input in the presence of adequate straining, the SSAM provides a favorable thermal cycle encouraging the formation of the precipitations and refining them in some conditions. Similarly, the same working conditions pave a favorable condition for generating dynamically recrystallized equiaxed refined grains. Thus, especially for lower melting point alloys with precipitation-strengthening materials, the SSAM can incorporate dual strengthening features by stabilizing both precipitations and dynamically recrystallized (DRX-ed) fine grain structure, which results in enhancing the mechanical properties and reducing the anisotropic behavior of the built [20, 21]. The SSAM, in addition to powder and wires, can also process sheets, custom-made consumable tools, and metallic scraps, for generating buildups [22–24]. In addition, the SSAM is used in productively repairing lower stiffness materials that have undergone significant wear, corrosion, and surface cracking during their service life. Solid-state AM has also proven to be highly efficient in manufacturing metal matrix composites with two or more types of materials, which are challenging to process otherwise with conventional fusion-based AM processes. Although the SSAM has satisfactorily addressed most of the challenges of the traditional

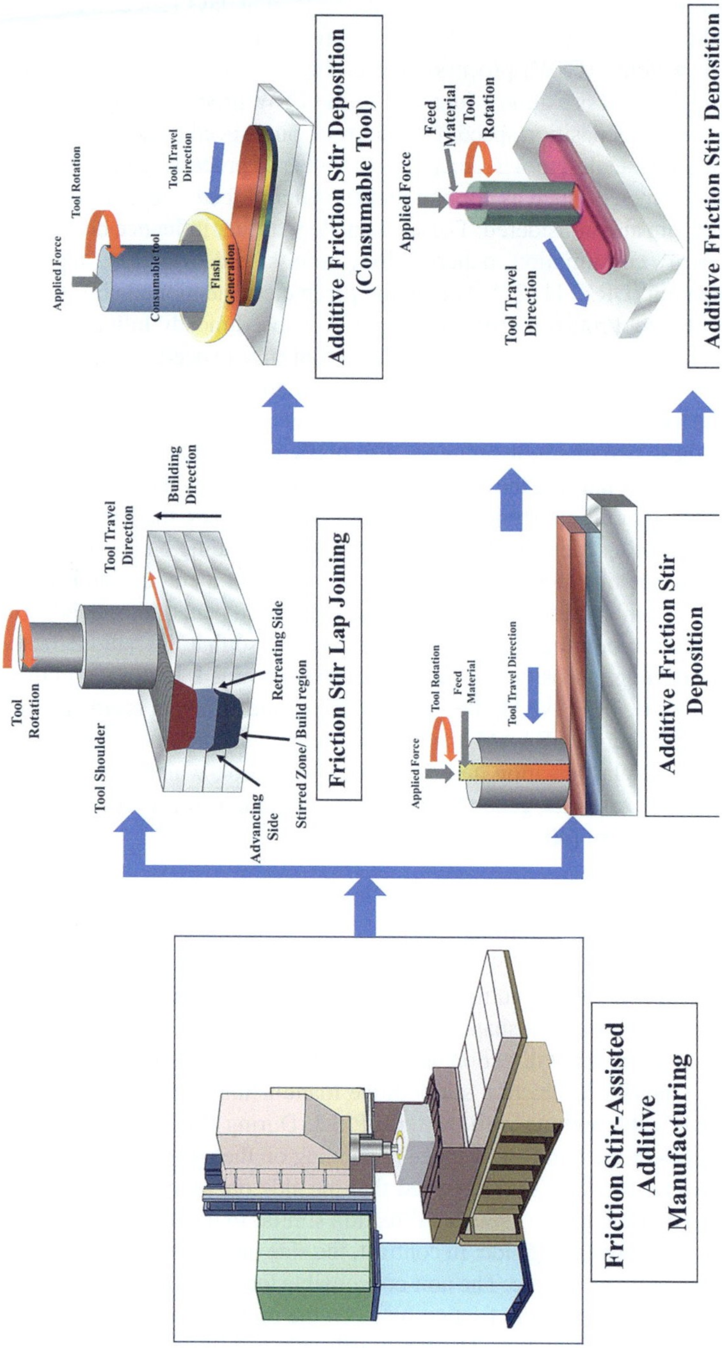

Fig. 3 Various types of friction-stir assisted additive manufacturing processes

fusion-based AM process, but still lags in building complex geometry precisely, and a lot of improvement needs to be made in the upcoming days [25, 26]. Friction stir-assisted additive manufacturing (FSAAM) is an emerging technology based on the friction-stir welding (FSW) process. Hence, the fundamental process parameters involved in FSAAM are almost identical to the FSW process, which include tool traverse velocity [deposition speed] (v), tool rotational speed (rpm) (w), tool plunge depth, tool tilting angles, tool geometry, and number of processing passes. However, additional parameters, such as cooling media, material feed rate, and feed material temperature, are also considered. These parameters further influence build attributes such as energy consumption, material flow, grain size distribution, and dynamic recrystallization (DRX) [1, 27]. Among the parameters mentioned above, the combination of the tool rotation speed and tool travel speed mostly influences the frictional heat input of the process. At a constant tool travel speed, if the tool rotation speed increases, the heat generation increases, whereas at a constant tool rotation speed, if the tool travel speed is increased, the heat generation is decreased. However, other than these two, the tool geometry also plays an important role in influencing the energy input by altering the effective contact area between the tool and the material to be printed or the substrate. If the contact area increases, the friction generated enhances, increasing frictional heat input [28, 29]. Depending on the material properties such as melting point, flow stress at elevated temperature, and deposition thickness, the combination of tool rpm, weld speed, and penetration depth are chosen to generate an optimized amount of energy input so that the build defects such as grain coarsening, porosity, and tunneling can be avoided. Likewise, in the FSW process in the case of FSAAM, the stir zone (SZ) is the principal region influencing the built quality. During FSAAM, the frictional heat in the presence of high tool rpm induces material plasticization and flow around the tool, which further generates high straining of materials, generating favorable conditions for recrystallization and refined grain structure [30, 31].

2.1.1 Friction Stir Lap Joining

FSLJ is executed by fusing multiple layers of sheets or foils in a lap position to create the required geometrical shape using a non-consumable rotating tool traversing perpendicular to the build direction, as exhibited in Fig. 4. In some pieces of literature, FSLJ is labeled as friction stir additive manufacturing (FSAM), however, in this work, we will continue with the term FSLJ. During FSLJ, the advancing and retreating sides of the build are decided depending on the combination of tool rotation and travel direction. It is suggested that the side on which the linear velocity vector of the rotating tool possesses a direction similar to the tool travel direction is designated as the advancing side. In contrast, the side on which both are opposite in direction is the retreating side. The non-consumable tool of FSLJ mainly consists of a shoulder and pin, where the friction between the tool shoulder and plates generates most of the frictional heat. The pin mostly mixes the hot plasticized material, facilitating material bonding. The material mixing occurs circumferentially and

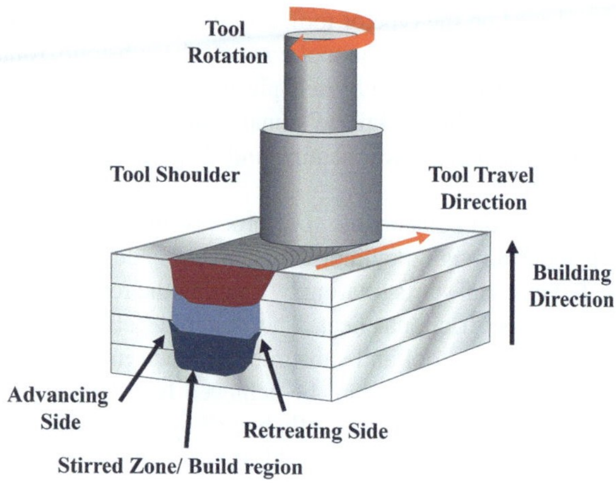

Fig. 4 Schematic representation of friction stir additive manufacturing (FSAM) for manufacturing stiffened panels utilizing four layers of build

axially around the tool pin and shoulder, which promotes the generation of homogeneous DRX-ed refined microstructure inside the built. In the case of high-stiffness materials and thicker plates, the process is carried out in multiple steps with two pieces at once until the desired height is reached. A detailed schematic diagram related to the FSLJ process is shown in Fig. 4 [32].

Chaudhary et al. [33] carried out friction-stir powder additive manufacturing of AA 6061. They reported the generation of fine-grain structure at a tool rpm of 1200 and tool travel speed of 25 mm/min. The AM was carried out by multiple tool geometries at various stages of the build so that desired energy input could be incorporated to generate favorable microstructures. Palanivel et al. [32] reported successful friction stir additive manufacturing (FSAM) of a multilayered Mg-based WE43 alloy stack. They suggested that FSAM-induced fine dynamically recrystallized (DRX-ed) grains with desirable precipitation properties enhanced the mechanical properties of the built. A stack of nine 5-mm-thick plates of 7075-O aluminum alloy was FSAM-ed by Yuqing et al. [34], which claims defect-free structure generation with enhanced mechanical properties due to refined grain structure and fine precipitate formation.

2.1.2 Additive Friction Stir Deposition (AFSD)

AFSD is emerging as a new and attractive method of FSSAM and deserves special attention. AFSD is a noticeable addition to the FSAAM, which combines AM and FSP to generate defect-free builds at a superior rate, simultaneously enhancing the material properties. However, compared to FSLJ, AFSD is a relatively more complex process. The method combines various processes such as friction stir heating and processing, shear-assisted extrusion, and deposition. AFSD is further divided into two subtypes based on the type of tool: consumable and non-consumable processes. In the consumable process, the tool contacts the substrate, rotates around its axis, and is progressively consumed by self-plasticization in frictional heat, adding material layers to generate the workpiece's final geometry. Thus, during this condition, AFSD only involves friction stir heating and shear-assisted deposition [35, 36] as exhibited in Fig. 5a.

In contrast, for the non-consumable tool process, the rotating tool is fed with either a rod or powder through a cylindrical channel inside the tool. In the case of rod feeding, the fed material starts rotating by meeting the rotating tool and contacts the substrate to generate frictional heat. The material gets plasticized to form a deposited layer between the tool and substrate as shown in Fig. 5b. In the case of powder, since vigorous rotation cannot be achieved like the rod, it is either preheated or the substrate needs to be constantly heated by employing external sources. Thus, for non-consumable tool conditions, AFSD uses all the processes mentioned in the earlier paragraph. For both cases, the gap between the tool and substrate determines the layer thickness of the deposition, and its rotating mechanism further stirs the deposited substrate to avoid generating built defects and induces homogenous DRX-ed grains. Sometimes, the tool can be fitted with specialized external equipment consisting of wire and powder feeding mechanisms, and no thorough thickness hole inside the tool is required. The process parameters in AFSD are almost like FSLJ; however, in addition to the two most essential parameters, such as tool rpm and tool travel speed, the feed rate of the material to be printed and the applied force on the tool regulate the quantity of energy input of the process [37–39].

The AFSD process was first patented by a US-based company, MELD Manufacturing Corporation. The owners claimed this process is more flexible and efficient in generating complex structures than FSLJ. Since FSLJ uses metallic sheets for the building, it limits the geometrical complexity of the final structure, as it is hard to eliminate the previous dimensions of the sheets. In contrast, AFSD

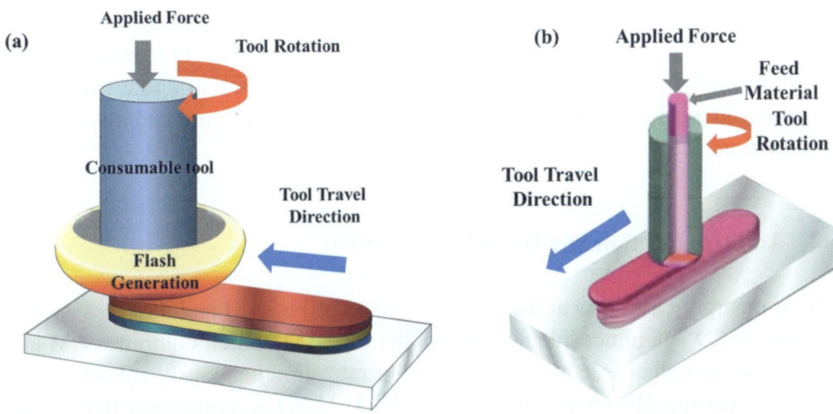

Fig. 5 Additive friction stir deposition schematic representation: (**a**) consumable tool and (**b**) non-consumable tool

consumes the metallic rod or the powder, which provides better geometrical flexibility than FSLJ [40]. Due to its excellent build rate at the expense of lower energy consumption, it is currently being used in various applications, such as recycling metal chips, repairing fractured and corroded parts, coating a typical structure, and manufacturing finished parts in the energy and transport sectors [25, 41, 42].

Using MELD technology and Ingersoll Machine Tool's Gantry crane system, the US Army Ground Systems Center, Siemens, LIFT, and the Applied Science and Technology Research Organization of America collaborated to produce the most extensive metal 3D printer in history by incorporating AFSD technology. With a print volume of 20 feet by 30 feet by 12 feet, the Jointless Hull eventually paves the way for bigger metal 3D printers and the construction of equipment the size of tank hulls with little to no reliance on conventional manufacturing techniques [43, 44]. Figure 6 exhibits the flawless solid structures manufactured by the AFSD processes.

In the first ever reported work of AFSD, Rivera et al. [45] aimed to counter the problem of rapid grain growth, hot cracking, and poor ductility of fusion-based AM of precipitation-hardened AA2219-T851. Based on processing–structure–property relations, this study revealed the formation of equiaxed grain structures in three different orientations and a gradual decrease of torsional texture intensity from the top to bottom layers. Although 57% of the tensile strength reduction occurred for the deposit compared to the T851 tempered alloy, the ductility was improved by 130%. Griffiths et al. [46] applied AFSD to manufacture high-strength aluminum matrix composites (AMCS) to solve the issue of hot cracking, which is mainly observed in fusion-based processes. To achieve this, AMC with Al-SiC, Al 6061-Mo, and Al 6061-W were successfully fabricated by applying AFSD. The success of these AMCs was confirmed by the homogeneous distribution of reinforcement particles with no observed defects such as porosity and cracking. Chen et al. [47] conceptually experimented with the ability to repair aluminum plates by AFSD. To carry out this work, they first prepared an 8 mm wide and 1 mm deep groove simulating a volumetric defect on a surface. After carrying out the AFSD, they reported a

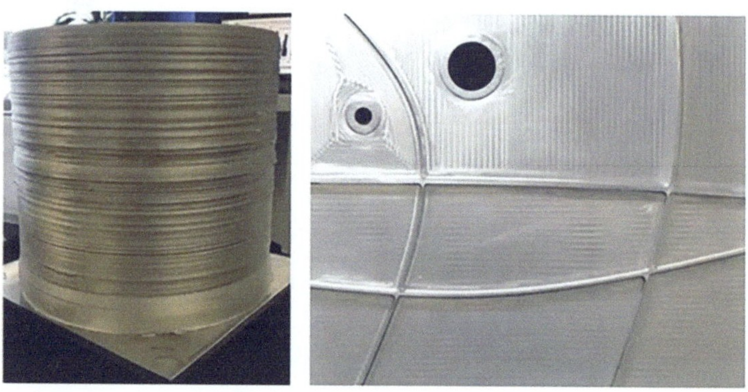

Fig. 6 Large dense structures made by AFSD processes [37]. (Copyright 2022. Reproduced with permission from MDPI)

complete enclosure of the grove, representing the successful repair of the crack. The study also found that the repaired region possessed 93% and 98% of the base material's tensile strength and total elongation.

2.2 Cold Spray and Binder Material Jetting

In many applications, the substrate onto which the coating needs to be applied cannot be raised at high temperatures, for example, low melting metals and their alloys, plastics, etc. In those cases, using a thermally activated coating process is not ideal since it might cause significant problems. One great way to bypass the thermal route and still attain a good quality coating is to exploit mechanical (or kinetic) energy achieved through the very high speed of depositing particles while maintaining a moderate or low temperature. Kinetic energy provides the activation energy for bonding. Such a process is called cold spraying (CS). CS is a solid-state material deposition technique where micron-sized particles of powder bond to a substrate due to high-velocity impact and severe plastic deformation. This process uses a supersonic spraying nozzle to propel a heated high-pressure carrier gas (such as air, nitrogen, or helium; heated in the gas heater part), accelerating the particles to high velocities as described in Fig. 7.

An extensive range of material types, namely polymers, metals, ceramics, composites, and nanocrystalline materials, can be coated using this technique. The nozzle converts the enthalpy to kinetic energy, which these particles use to kinetically deposit onto the target surface by mechanically interlocking and forming metallurgical bonding (in the case of metals). Basic parameters involved in cold spraying include the gas properties (type, pressure, and temperature), particle properties (density, yield strength, melting temperature, particle size distribution, etc.), nozzle

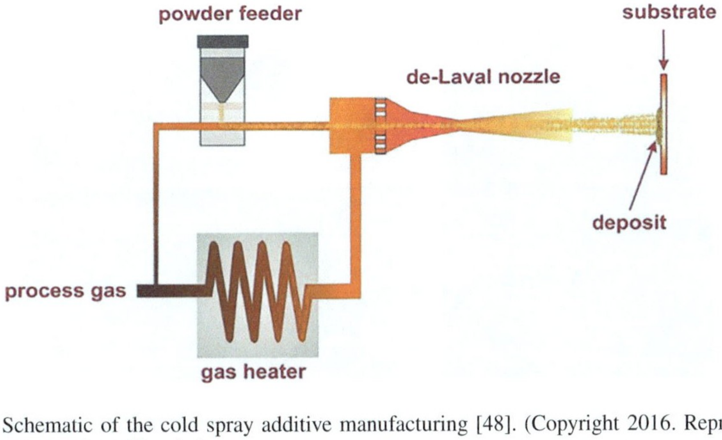

Fig. 7 Schematic of the cold spray additive manufacturing [48]. (Copyright 2016. Reproduced with permission from Elsevier)

type and its design, substrate properties (melting point, mechanical strength, etc.), spray/deposition kinetics (spray distance, spray angle, gun velocity relative to the substrate, scanning step, cycle numbers, etc.). There are primarily five types of CS: high-pressure cold spray (HPCS), low-pressure cold spray (LPCS), vacuum cold spray, laser-assisted cold spray, and electrostatic-force-assisted cold spray [48]. In HPCS, the pressure is generally more than 25 bars, while in the case of LPCS, it is typically less than 10 bars, and generally the different gas pressure is achieved by preheating gas over ~1000 °C or ~550 °C for these two cases, respectively. Typically, the efficiency of HPCS is ~90%, while it is ~50% in the case of LPCS. However, LPCS makes up for this shortfall in efficiency through its improved operational safety (owing to the low-pressure system), flexibility, less complexity, and lower spraying cost. The choice of HPCS vs. LPCS depends mostly on the substrate properties: whether the substrate can sustain high temp and high pressure versus low temperature and low pressure or because of any other operational parameters mentioned earlier. For the sake of brevity, we will generalize the HPCS and LPCS together just as CS.

One of the significant advantages of CS is its ability to address issues related to the high-temperature processing of materials, such as oxidation and undesirable structural changes. It is likely the only continuous method for producing bulk components from metastable materials, which are typically available only in powder form, such as those obtained through mechanical attrition or gas atomization. Due to its specific process parameters, CS is broadly used to repair various mechanical components in multiple industries. In the energy sector, Siemens employs it to repair gas turbines. Another application arena of great interest is the automobile industry; Ford uses this technique to fix defects in aluminum engine blocks. In Fig. 8, some of the applications of CS-assisted repair are shown.

Binder jetting (BJ), on the other hand, originally known as "3D printing," is primarily used to create sand molds and cores for the metal casting industry. This process acts almost like an inkjet printer, which, instead of spraying ink onto paper, involves spreading powdered material into a very thin layer and selectively

(a) (b) (c)

Fig. 8 (**a**) Boeing nose wheel steering actuator malfunctioned. (**b**) Steering actuator for 737 nose wheels fixed using cold spray nickel. (**c**) Restoration of an iron engine block mount's dimensions [49]. (Copyright 2017. Reproduced with permission from MDPI)

spraying a binder, typically a polymeric liquid, which acts as an adhesive between the powder layers. This goes on a loop until the whole object is printed. After printing the "green" (having the binder) object, it is cured in an oven where crosslinking of polymers provides a durable object. Excess powder is then removed gently, which is called de-powdering, allowing the recovery of most powders not consumed during the printing. This is then sintered to obtain a dense object with desirable properties. A detailed representation of binder jetting is schematically presented in Fig. 9. The essential parameters for this process include the size distribution, shape distribution, porosity, chemical composition, and impurity levels of the powder. Several types of binders have been tested, but ethylene glycol (EG) based binders are widely used due to their excellent water solubility, affordability, and long-lasting properties [51].

2.3 Ultrasonic-Assisted Additive Manufacturing

In 1999, just like in FSLJ, the AM process was carried out by lap joining various sheets and foils; Dawn White developed the ultrasonic additive manufacturing (UAM) concept [52]. Likewise, the FSAM replicates the friction stir lap welding process. UAM is very much the same as the ultrasonic lap welding technique. In some cases, UAM can be designated as a hybrid method that combines AM processes with subtractive manufacturing to generate the final geometry of the finished product. However, this subtraction may not be required in all cases. UAM is performed by applying ultrasonic vibration on multiple sheets layered in a stack, clamped firmly so that the vibration-assisted frictional heat defuses the stacked layers into a single material. Later, a similar machine with a different tool setup can remove unwanted parts or add some geometric features inside the structures for better surface finish, dimensional accuracy, and usefulness [53]. The fundamental parameters involved in UAM include vibrational amplitude and frequency, bond depth, and pressure. The complex summation of the parameters generates the necessary energy input to create a structure. Most systems operate at a vibration frequency of at least 20 kHz. A bonding tool is fixed with a transducer, which passes the generated ultrasonic vibration to the stacked layers. The ultrasonic vibrations applied to a stack of metal foil or plate produce a scrubbing effect and cause plastic deformation and frictional heat between the various layers of foil or plates, generating favorable conditions for bonding as shown in Fig. 10 [55].

One of the finest advantages of UAM is its ability to build materials even below the FSAM working temperature due to the ultrasonic softening effect. Due to this effect, stiffness reduction and material plasticization occur at a much lower temperature, almost 0.3–0.5 T_m (T_m: melting temperature). Compared to FSAM, the UAM usually incorporates two different types of energy: frictional heat and

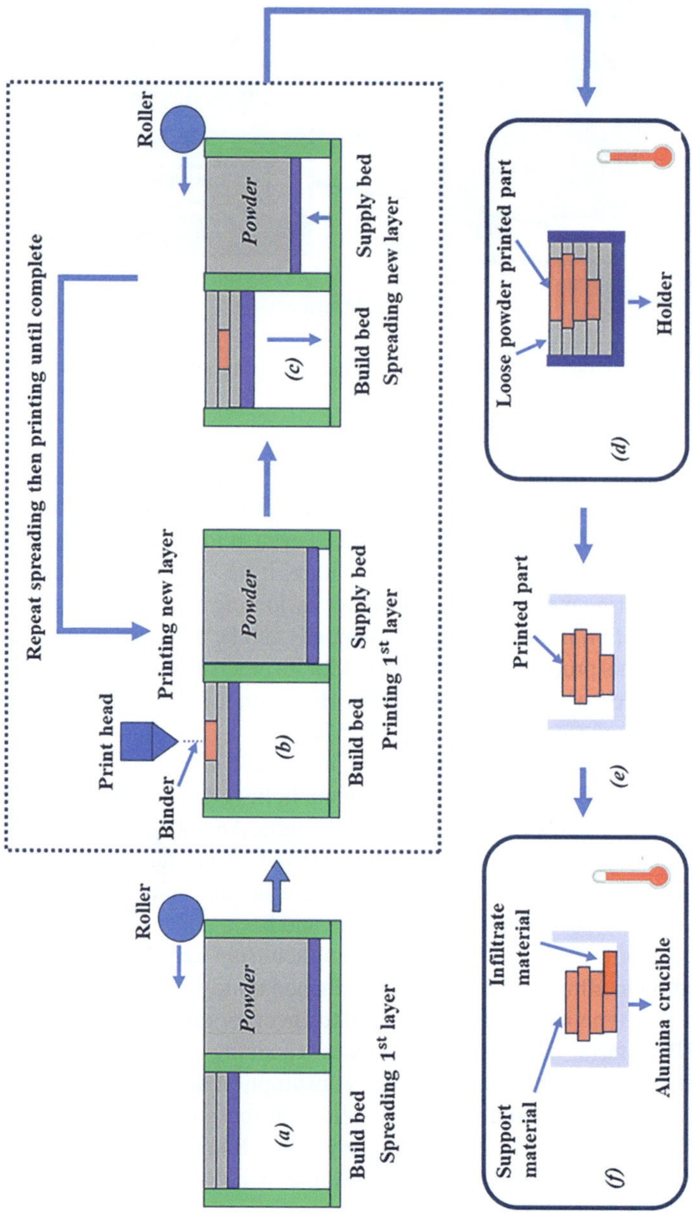

Fig. 9 Schematic representation of the binder jetting process [50]. (Adapted from Ref. [50])

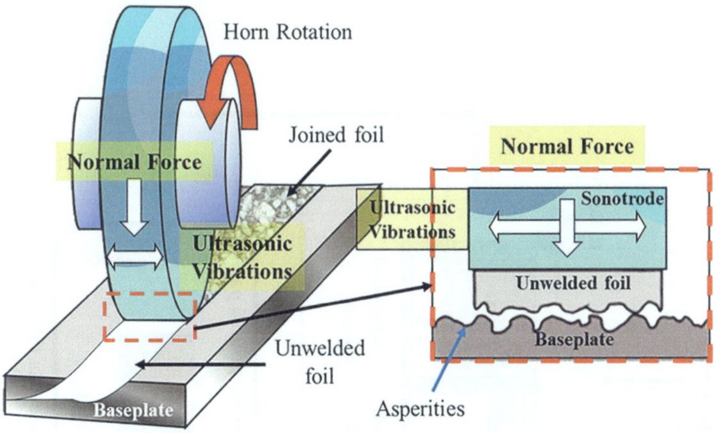

Fig. 10 Schematic diagram of ultrasonic additive manufacturing process [54]. (Adapted from Ref. [54])

ultrasonic vibration. The ultrasonic vibration, which is also responsible for generating frictional heat, induces excess lattice vibration in the atoms of the solid. As a result, the flowability of the material is enhanced even at lower temperatures. Since the process takes place at a very low temperature, it has some added advantages, such as manufacturing metal matrix composites by joining dissimilar materials, and a significant reduction in intermetallic compounds can be achieved efficiently. For the same reason, this process is also beneficial for its precision and minimal residual stress generation compared to other methods. However, it is constrained by its reliance on thin foils and slower build rates [56, 57].

Due to UAM's unique processing features, it has several distinct applications. In the aerospace sector, Boeing uses UAM to create lightweight and complex components such as honeycomb structures and satellite parts. Another application is in the automotive industry, where Ford implements UAM to produce lightweight and high-strength parts, contributing to improved fuel efficiency and performance. The applications are pictorially depicted in Fig. 11a and b.

Guo et al. [60] applied UAM for manufacturing carbon fiber-reinforced polymer-aluminum alloy joints, which strengthened up to 129.5 MPa. The authors claimed that this type of structure enables the incorporation of extremely lightweight CFRP material in metallic assemblies, reducing the final weight of the structure and significantly reducing energy consumption during application time. Behvar et al. [61] UAMed multi-layered solid-state built of Al 6061 and Cu110 with very thin IMC layer and generating very fine DRXed grain structure inside aluminum only, close to the interface, resulting a stable structure even with dissimilar material combinations.

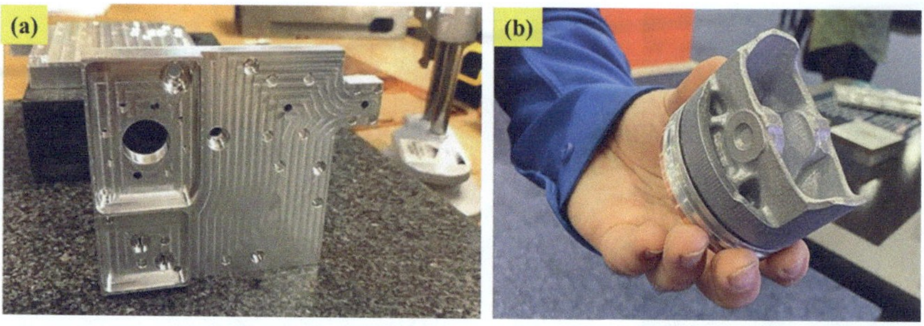

Fig. 11 (**a**) Fabrication of aluminum satellite heat exchangers using UAM. (**b**) Fabrication of piston in UAM process [58, 59]

2.4 LASER-Assisted Additive Manufacturing

Additive Manufacturing produces products layer-by-layer, leading to less waste production than traditional manufacturing. The resource efficiency of the process leads to the improvement of product design and processes. Computer-based design models such as the Fused Deposition Modeling method are emerging as a rapid prototyping method that can be used for many practical applications, from industrial to biotechnical use cases. Thermoplastic biomaterials to create gradients with overlapping layers of fibers for treating human skin defects is one such application [62]. Due to higher precision, achievable geometrical complexity, and remote mode of operation leverages, LASER-assisted AM processing is one of the most established fusion AM Technologies. It is classified according to the material feeding and deposition mechanisms on the surface. In case of directed energy deposition (DED) either wire or powder substrate is fed through the feedstock. The focused energy source (LASER beam, electron beam, or plasma) is focused on the feedstock material, creating a melt pool for deposition on the given substrate [63]. DED includes LASER-directed energy deposition (LDED) and wire and arc melting (WAAM). LDED uses powder as a raw material, while WAAM uses metal wire. Schematically and pictorially, a typical LDED system, energy distribution, and melt pool formation are shown in Fig. 12a–d. LDED has the advantages of lower heat input, lower residual stress, and good dimensional accuracy. WAAM has high deposition efficiency and low cost. DED has a high deposition rate of up to 2.5 kg/h, with inherent multi-material deposition [65].

LASER-directed energy deposition (LDED) attracts fabrication with high efficiency, low material waste, and high design freedom. Compared to traditional processes such as spraying, the LDED products bond better with the substrate and have better environmental efficiency [65].

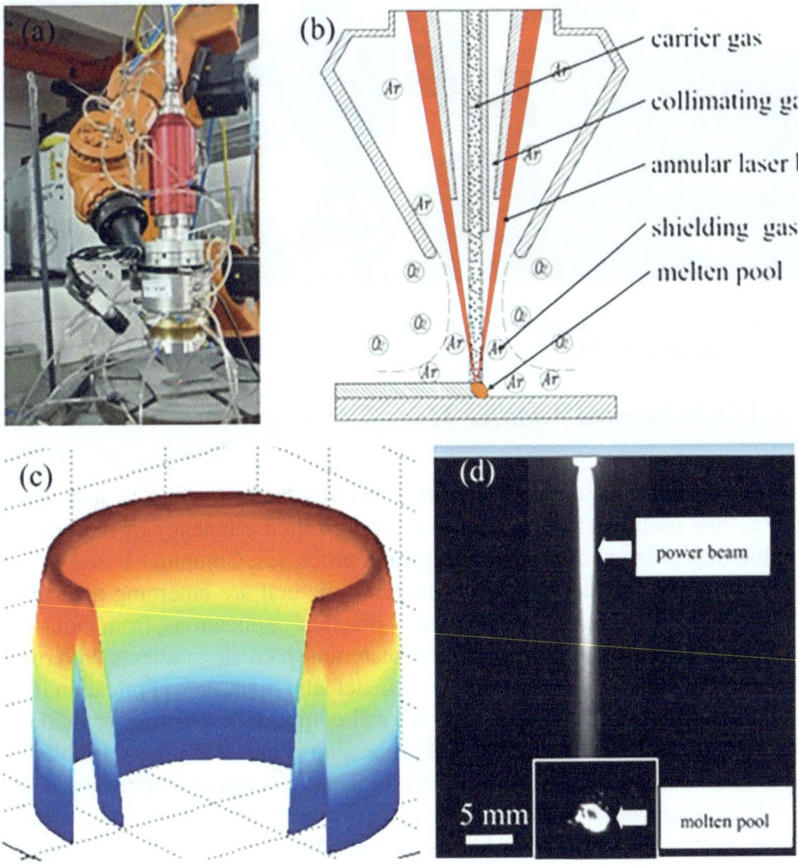

Fig. 12 (**a**) An actual picture of a LASER coaxial powder feeding unit. (**b**) A schematic diagram highlighting the main parts of an LDED unit. (**c**) The energy distribution depicts a concentrated ring-shaped intensity pattern in the annular LASER beam. (**d**) A photo captured by a high-speed streak camera displays the interaction of the LASER beam with the molten pool of substrate [64]. (Copyright 2020. Reproduced with permission from Elsevier)

LDED technology helps to build up a multi-component material system, which is advantageous over the conventional manufacturing method. In this process, raw material meets the energy source and is deposited directly onto the surface through melting due to concentrated energy and rapid solidification. As LDED is a fusion-assisted AM, primarily using LASER as an energy source, it produces various alloys with homogeneous supersaturation that is not possible using traditional methods. In some processes, a secondary manufacturing step of compact rolling is carried out to ensure proper adhesion of layers inside the build component [66]. Good quality products can be achieved by controlling energy deposition.

Rapid heating rate with steep temperature gradient causes a unique thermal cycle for the fabricated metal parts by LDED. This can cause residual stresses in the fabricated parts, and compressive stresses develop in the cores of the LDED products. Post-heat treatments like post-annealing of as-built parts will reduce the residual stresses. Similarly, in situ processes like thermal gradient control and scanning speed control can mitigate the residual stresses [67].

LASER-directed energy deposition produces superior mechanical properties and smaller grain size (~12 μm), although heat treatment of the alloy is done to obtain the product while mitigating the occurrence of intermetallic and brittle phases. LASER DED includes two processes: powder-LASER-based DED and wire-LASER-based DED. The final microstructures of the outputs vary distinctly depending on the process parameters and target materials to build, which is the prime focus of multiple studies carried out in this domain. The mechanical properties of the 3D printed alloys are analyzed in correlation with the microstructural developments, including second-phase particles and microtexture [68].

Powder-based DED uses powders injected from a nozzle over the selected regions using shielding gas to increase deposition efficiency. The powders are carried out from the nozzle and assisted by the shielding gas to the working location [69]. Meanwhile, wire-based DEDs have lower raw material costs as they source raw materials in wire form and are comparatively less vulnerable to fire or explosions. Factors that affect the layer-by-layer deposition efficiency of directed deposition can include forming qualities of the previous layer, set parameters of the process, heat dissipation during the deposition of successive layers, etc. The optimization of the parameters is a continuous process and still requires adaptability and control. In the wire-LASER DED, deposition stability of the process is associated with low automation levels and unsatisfactory results at large-scale production [70].

To address these challenges, sometimes data-driven models are developed to recognize the molten pool images and characterize the stability of the deposition. Statistical methods and numerical optimization are developed to identify the critical resolved shear stress (CRSS) of slip systems to study the effect on the gliding mechanism, for example, of the Ti-6Al-4V alloy by Benmessaoud et al. [71].

The low cost, reduced residual stress, and high deposition efficiency of LDED make it a preferred method for repairing parts. The method requires optimization of LASER power, scanning speed, feed rate, etc. Inconel 718 is one such alloy that is used as a repairing material for other parts like 42CrMo [65].

Studies on various alloys show the presence of different phases. For example, in the case of an Mg alloy with high Gd content shows the dissolution of 11.17 wt% Gd and 0.32 wt% Zn into the α-Mg matrix during the building process [72].

Additionally, methods like molecular dynamics (MD) simulations can forecast properties of emerging options like high entropy alloys (HEAs), a new promising material for AM. MD simulations of a densely packed powder bed and a cracked substrate simulate the grain growth and microstructure changes of the stainless steel by LASER-DED. These simulations have concluded that the epitaxial growth of the grains in the layers formed in AM originates from the substrate. Tests like Nanoindentations can also be simulated to show the results of improved hardness in the deposited layers, mainly due to grain refinement [73].

The calculation of phase diagram (CALPHAD) is used to predict properties of experimental powder compositions [74]. Analytical models based on transient heat conduction equations help us to capture convective heat transfer in the melt pool [75].

The different modalities of LASER-directed energy deposition (DED), like the intersection point of the LASER beam and the wire, to the top layer distance, are designed on an estimation model with a convolutional neural network and a fully connected network. Xiong et al. show an accuracy of 0.9975 on the testing data of the melt pool images with a pixel size of 180 × 105 [70].

The above-mentioned research review on LDED proves its applicability in manufacturing multi-design aspects of layer-by-layer production, and accurate, efficient metal fabrication. These applications of LDED increase its adaptation in the industry. However, large LASER spot size, high-cost production, and higher thermal exposure cause grain coarsening in some materials. The resulting degradation of properties necessitates more detailed research in LDED to make it more cost-effective and energetically sustainable.

2.5 LASER Powder Bed Fusion

LASER powder bed fusion (LPBF) is a LASER-based AM method that prints metals layer-by-layer to construct parts by melting feedstock powders and rapid solidification to make complex geometrical parts with a high degree of design freedom [76]. L-PBF is being actively researched for applications in medical, aerospace, marine, nuclear, and infrastructure components.

Metal AM gives control over high precision and quicker speed of demand fulfillment. Powder bed fusion (PBF) uses elevated temperatures from heat sources like LASER, electron beam, and plasma-electric arc to melt the selective subsequent layers in materials such as Ti-6Al-4V, TiAl Inconel 625/718, and cobalt chromium (CoCr), etc. [77].

Common process parameters for the LASER powder bed fusion include scan speed (m/s), laser power (Watt), hatch distance (μm), layer thickness (μm), and laser scan pattern. The final build of the component can be optimized in LPBF using high laser powers and high scan speed using low scan spacing. The studies show that there can be a few parameters that are primary to the process optimization, like layer thickness, nozzle temperature, printing speed, and platform temperature. These parameters are mainly responsible for reducing the anisotropy of fused deposition method products [78].

The parameters required for optimizing the process can be divided into LASER and scanning properties, powder material properties, powder bed properties, and build parameters. The primary parameters include LASER power, layer thickness, hatch distance, scanning speed, etc. [79, 80].

The process parameters, bulk, and particle properties are investigated to explore correlations of part density, mechanical properties, and final quality [81]. During the

LASER beam interaction process, phenomena such as solid-state transformation and fluid dynamics are monitored to keep a check on issues like material cracking or incomplete fusion. A summary of these regions is provided in Fig. 13a.

Factors like porosity formation are studied to understand the high-temperature reaction in the LPBF process of the 20MnCr5 powder. The high-temperature region of the slag liquid formed during the PBF of these alloys in the chessboard grid style of the LASER pattern introduces non-metallic inclusions like silicon oxide, leading to pore formation [84].

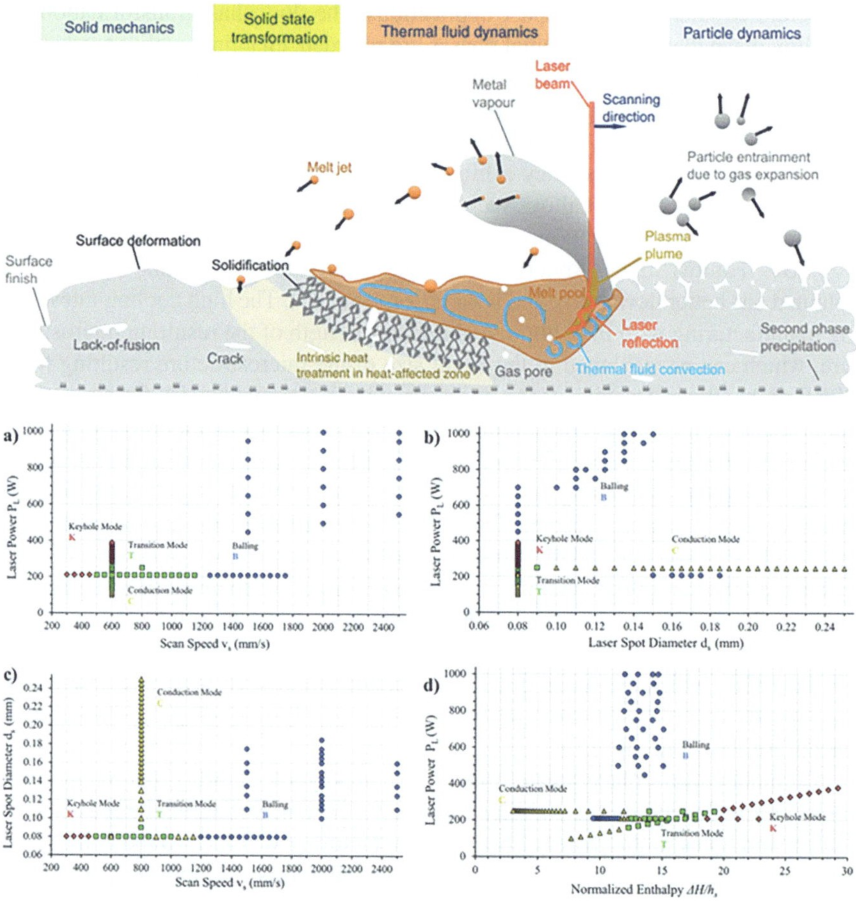

Fig. 13 (**a**) Schematics of the physical processes that occur during a typical LASER-based AM process [82]. (Copyright 2020. Reproduced with permission from Springer Nature). (**b**) In this figure parameters of LPBF like the scan speed (vs), laser power (PL), laser spot diameter (*d*s), and normalized enthalpy (△H), are compared to each other, along with the melting modes (conduction [yellow marks], transition [green marks], and keyhole [red marks] modes) for low alloy steels [83]. (Copyright 2023. Reproduced with permission from Springer Nature)

Parameter optimization for novel alloys such as Al-Mg-Li-Ag-Sc-Zr is achieved using multiple iterations of LASER power, scanning speed for inhibition of hot cracks, pore formation, etc. Studies of Sun et al. [85] have found that 220 W LASER power and 100 mm/s scanning speed provide optimal performance, although with a loss of constituent elements like Mg and Li.

Apart from the parameters, the material properties of the specific alloys also play an important role in deciding the applications and scale of reproduction of the material. The physical properties of substrate materials also need to be considered to optimize the parameters of the LPBF parameters for obtaining a good build. Nickel-based super alloys such as Inconel 718 can retain their strength and resistance at temperatures suitable for aerospace applications. The slow phase transformations of the alloy mean that the detrimental phases are retarded, enabling quality production of components by AM methods. Solution annealing decreases the susceptibility of hot cracking during the printing [86]. All kinds of light alloys like Aluminum-Transition Metal (Fe, Mn, Cr), Ti6Al4V and AlSi10Mg are used in the 3D printing industry these days because they exhibit low density as compared to steel and high corrosion resistance. Light alloys such as these are suitable for use in transport systems and aerospace applications.

LPBF requires higher sphericity and uniformity in size so that the powder obtained has better flowability and laser absorption [87]. The high cooling rates during manufacturing with PBF improve the yield strength of the resulting microstructure, which can be attributed to the small size of the microstructure resulting from LPBF of Al alloys. The size of the microstructure is correlated with the strength of the product, however, the conventional Hall-Petch formulae is modified to accommodate the effect of the multiple strengthening mechanisms like dispersion strengthening, precipitate hardening, etc. that come into effect during the printing of materials [88]. The pores formed during laser powder bed fusion can be controlled by the solidification rate and the retraction speed as observed in 316L Stainless Steel by Khairallah et al. [89]. Generally high speed is promoted to avoid these defects as it allows the molten metal to flow to the sites in time before its solidification.

The characterization of these defects/pores can be done. These include defects like keyholes, which are made due to the application of high laser power and low scan speed. Keyholes become pores when the material is melted [90]. The detailed graphs are shown in Fig. 13b.

Aluminum alloys place greater importance on keyhole stability, but magnesium alloys do not depend on the stability for pore formation. Similarly, the presence of defects and holes in the material before manufacturing will lead to the collapse of these defects due to the heat generated. Like any other process, multiple parameters can be accumulated to simulate and check the factors for these defects caused during printing processes [91].

Aluminum alloys can be produced using rolling, extruding, forging, or casting techniques. Steels, on the other hand, are suitable for conventional casting methods because of their thermo-mechanical properties. However, low alloy steels such as 20MnCr5, AISI 4130, and 24CrNiMo show favorable properties like high strength, ductility, toughness, and low cost, which makes them suitable for manufacturing using laser powder bed fusion. It is expected that low alloy steel will follow different parameters to control the process of 3D printing. As steels are vulnerable to oxygen or non-metal inclusions, different studies are performed to control the underlying risk of brittleness and defects [92]. Different scans, such as chessboard and linear scans, are explored by Yang et al. [84], comparing the microstructure benefits, structural hardness, and inclusion density of the final product's microstructure. Figure 14 shows the comparison of PBF and LDED.

2.6 Auxiliary Considerations for 3D Printing

In the previous section, we discussed the most important core hardware for AM. However, AM encompasses a broader spectrum of technological paradigms, and the process typically starts with the creation of 3D "CAD" (or a computer-aided design file containing a 2D or 3D model of the physical object or design) structures and later adding materials layer by layer based on this design. Various software programs are utilized for the development of the CAD file. AutoCAD, Fusion360, SolidWorks, etc. are just a few popular examples. The CAD file contains information about the geometry of the design, manufacturing data, material attributes, assembly functions, and rendering details describing the mesh structure and other printing properties. As described in Sect. 1, AM processes include producing files, quality control, material selection and customization, 3D printing, set-up, etc. 3D printing refers to a process of printing/creating the 3D model layer by layer from the "CAD" file. Any prototype made using 3D printing can be customized easily, and unlike conventional manufacturing, accommodating any complex geometrical configuration is often less resource-consuming and easier [94]. In the traditional manufacturing protocol, the involvement of consumers in the production of more personalized products typically increases the cost and the usage of resources such as new dyes and special production lines for a small number of products, with the power to run the entire system. The cost of production can be significantly reduced in the 3D printing of products with customization requirements. This is due to its adaptability for general and consumer-specific production at the same time.

3D printing involves several key components that must work together to successfully develop a prototype. These include the "CAD" design file, which acts as the blueprint; the printer responsible for material deposition; and the material itself, which must meet specific property requirements. Environmental conditions, such as temperature and humidity, also play a crucial role, along with the involvement of the designer. Any error or malfunction at any stage, whether in the design file, printer, material, or surrounding conditions, can lead to production stoppages or result in faulty products, impacting the quality of the prototype [94, 95].

Errors can occur at any stage, such as wrong material selection, improper printer calibration, or flaws in the "CAD" design. For example, choosing a material with inadequate thermal characteristics can result in warping or poor layer adhesion. Similarly, printer difficulties such as nozzle clogging can disrupt productivity.

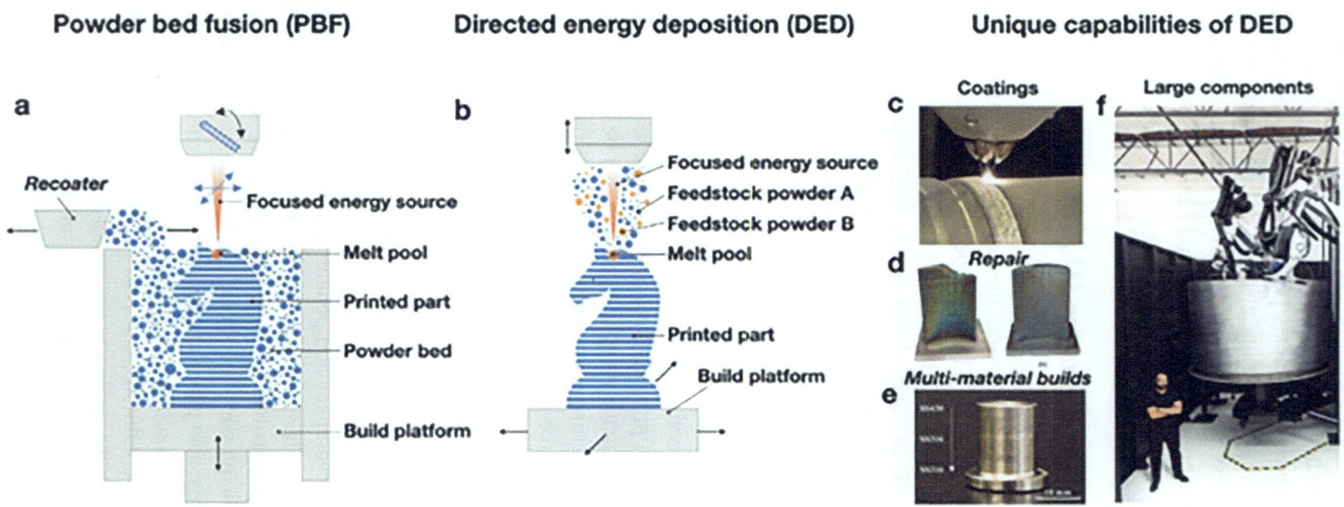

Fig. 14 The comparison of processes carried out in PBF and LDED. The material delivery, laser melting, and layer deposition differs by minor aspects. Variations in deposition, such as direct feeding or focused beam, provide contrasts in the final product of the LDED and LPBF process [93]. (Copyright 2022. Reproduced with permission from Elsevier)

Temperature changes in the surrounding environment can also affect material properties, resulting in uneven layer deposition. As a result, maintaining precision and accuracy across all components is critical to avoid defective outcomes.

3 The Digital/Software Layer for AM

In the previous section, we discussed the core hardware aspects of AM technologies. Since AM technology emerged in the digital era, it has been very intrinsically connected to it from its very inception. The present trend is to integrate different sensors and compute layers into any big technological endeavor, and the new age AM manufacturing is no exception to it. The main purposes of such integration are enhanced safety, efficiency, productivity, rapid tracking, agile and customized production, inventory management and control, making the entire ecosystem smart, the capability to create a digital twin and perform virtual experiments on it, etc. Digital manufacturing integrates emerging technologies like the Internet of Things (IoT), cloud and edge computing, and machine learning to streamline manufacturing. Management of production line machinery, materials, processes, and production optimization can easily be done by integrating these technologies into manufacturing [96]. Customization of materials and prototypes based on customer needs can easily be carried out. Utilizing these technologies to their full potential, along with unifying them together can improve the manufacturing sector by leaps and bounds. Setups in manufacturing industries feel like “under one roof production houses” as all the manufacturing processes are connected and dependent on each other, even when they may be physically dispersed over huge geographical distances. This is possible primarily due to the advent of digital manufacturing since the same setup can be broken down and spread across different places as per the expertise. Certain zones possess expertise in a specific process in manufacturing, and the respective plant can be established at that location, with all the information being transferred over the network, including cloud services and sensors. Strange as it may sound, the reality is that the high-end niche products have tremendously grown in engineering complexity, and there is a very limited talent and skill pool to perform these operations. That’s only the human resource side of the problem. Similar bottlenecks may be there for natural/industrial resources, large capital, or even the requisite sophisticated infrastructure. Therefore, there is a growing opportunity for such a distributed approach to complex manufacturing. Different processes in a prototype production can be run from different parts of the world, yet still operate smoothly. Due to this spread of expertise, larger production setups can be set up, increasing the manufacturing by manifolds. This also improves the profit-building of the firm in tandem with human-friendly and efficient production, with customization possible at almost all steps. In the following subsections, we will discuss the important components of the digital layer of the AM [97].

3.1 Internet of Things

The Internet of Things (IoT) can be described as a network of embedded sensors that perform a predefined set of instructions, collect crucial data, manage and control the system, etc. The control instructions can even be updated in real time, say, through some ML protocols. Since AM is generally treated as a next-generation automated process, the IoT sensors and controller are very important attributes in the entire scheme of things. The network-centric sensors could be connected either through the Internet or any other source of communication. In some advanced architectures, the sensors in the network can auto-organize themselves, share information and resources, and adapt to changes or alterations to their working environment. This will be increasingly important for very complex manufacturing facilities (like mostly autonomous gigafactories of the future), which require enormous communication and optimization between different sections. According to Madakam et al. [98], since all the sensors and devices connect via the same network under IoT analysis capabilities typically implemented with IoT devices give them the ability to improve resource management and supply chain optimization. IoT devices can also be used as security devices monitoring the production line and processes. Thus, IoT devices or sensors assist in a wide range of complex processes from inventory management to tracking prototypes and production.

The basic architecture of IoT consists of multiple layers. The "Sensing layer" connects the physical world with the network by collecting environmental changes like temperature and humidity, and physical alterations and compute tasks without human intervention, thus they leave less margin for error in collecting data or performing high-precision work in sensitive places and tasks. IoT is economically beneficial for coordinating, controlling, and optimal execution of complex tasks due to automation or semi-automation of tasks with repetitive actions. New customization requirements in work are also easily accommodated in tandem with routine actions. The data about element properties, deformations, and color sensors for visual changes. This layer is expected to be very closely and physically integrated with AM hardware. This data is then sent to the next layer for processing. The "Network layer" is responsible for sharing the data to specific devices for precise computation. For a big facility, this transfer can be accomplished via Wi-Fi, 5G, etc., while for a small unit, it can be achieved by LAN or by direct communication. The "Service layer" is responsible for data storage, management, and processing. The storage can be in cloud storage, data warehouses, etc. These processes include analysis, variable computation, or any application enablement. The "Interface layer" interacts with the end user and displays the final processed result in the predefined format. As discussed by Gokhale et al. [99], the format can be dashboards, website pages, etc.

Let us illustrate these points with a few examples of implementation in AM. AM utilizes IoT sensors to monitor the layering process and the surrounding conditions, such as ambient conditions, temperature, and material flow. With the changing customer demand and customizations, the manufacturing line can get clogged due to excessive altering demands. Kalsoom et al. [100] state that IoT sensors can help maintain the changes in demand with each iteration and alter the production processes accordingly. Alterations may include changes in surrounding conditions, quality control techniques, etc. This enhances the productivity of the firm, leading to more profit-building for the company. The high number of orders maintained by a production line makes it difficult to maintain quality control standards by a human. Sensors can be fed with a certain defined set of simple steps to check the finish on the prototype, allowing for more precise and efficient checks, which can be used to identify (often through ML) the desirable protocol to obtain desirable finishing. With the heavy volume of data collected by the sensors, it is generally a tedious task to analyze the data and deduce dynamic real-time insights for a particular product or the production process. IoT devices can do this task either locally over an edge node or by pushing the data over cloud services for further manipulation and sharing it to run the analysis tasks more efficiently and faster (discussed in Sect. 3.2). This reduces time and makes more data available for R&D and profit-worthy prototypes for the firm. As described by Katari et al. [101], the same data combined with information extracted from the web can be fed into the sensor setup, enabling them to become self-aware or self-comparison. This self-comparison allows the sensor-embedded machine to evaluate its work and analyze any malfunctioning components. This self-awareness allows for fault prediction and essentially limits the production of faulty prototypes, increasing the usability of machines and the production process.

IoT devices and sensors often have cross-connectivity with each other, making it easier to communicate between sensors and share information faster. This feature can be efficiently utilized in mechanized robotic production equipment, such as robotic arms, which are too frequently used in AM processes. Grieco et al. [102] highlighted that the production of products can be interlinked with the management and quality control of the same product. Sensing, computing, and communications capabilities can be embedded in the robotic system for more human-friendly and eco-friendly production as these processes emerge in Industry 5.0. Since robotic arms or movable stages are often used in most AM manufacturing processes, they are generally governed by sensors that communicate efficiently with each other in the most precise manner. The working of these robotic arms is a complex undertaking and often relies on a heterogeneous network infrastructure crowded with a variety of smart objects, field sensors, servers, etc. These components collaborate to perform precise movements and operations, relying on real-time data processing and communication. Advanced control algorithms ensure synchronization between various modules, while machine learning (discussed in Sect. 3.5) techniques can optimize complex task efficiency. Additionally, cybersecurity protocols (discussed in Sect. 3.7) are essential to safeguard the network from potential breaches and ensure uninterrupted operations. Let us take an example of a robotic arm using AFSD on a car chassis. Such a process is very efficient as it increases the completion rate and precision. The working of this AFSD arm is a collection of sensors maintaining the field data and material coat thickness and consistency. Some sensors are also responsible for collecting the performance of the arm. As pointed out by Grieco et al. [102] and Roy Chowdhury et al. [103], this type of working heterogeneous networks adds value and efficiency. Production with robotics and mechanical assistance provides a set of predefined instructions to the robotic arm. Mechanical arms perform these tasks with extreme precision, reducing the rollout of defective products from the production line. Production houses for mobile phones or chip production require handling small parts with utmost care and precision, protecting all the sensitive mechanisms of the product is of paramount importance. Human intervention in such small parts handling can lead to a greater level of faults in the production line. The machines also follow certain instructions at high speed, increasing the production rate and hence profit for the firm. These instructions may include adding a material layer/coating onto the frame of a chassis, testing, etc. As a combined package with high-speed production, customization, and accessibility also come with automated production. As Kats et al. [104] describe the ability of arms to follow instructions allows the user to alter them for special demands/requests. After the custom item is rolled out, normal prototyping is resumed. Without any change in the production line or wait, customization is increased, increasing the reach and capability of the production firm. Sensors monitor the set of instructions followed by the arm and can be fed over a communication network outside the facility. This allows the user to modify the actions of the arm from some further place without being present in the establishment. In the production facilities involving 3D printing, high-speed customizable production is in high demand. It allows the user to produce high-quality, complex geometry prototypes without any stoppage in the production or excess loss of resources.

Additive manufacturing, with its demands for continuous monitoring of CAD files and surrounding conditions, brings the usage of IoT devices in the production process. On the same side, AM can be used to produce IoT devices. With the increased complexity and detailed processing required by sensors, their geometry becomes more complex. This complexity of geometry is well handled by 3D printing. Thus, manufacturing processes, whether it is AM or conventional industries, and sensors operate hand in hand with both being utilized in each other's domain. Industrial applications of IoT are shown in Fig. 15.

3.2 *Edge and Cloud Computing*

Edge nodes are broadly defined as the network or resources between cloud servers and end devices. These nodes include micro-servers, gateways, or even tiny hardware like smartphones. Big industries employ edge networks to maintain different communication technologies, like sensors, the infrastructure of the industry floor, hardware, and software. For example, a system for monitoring the entire operations, which in a traditional industrial setting might include conveyor belts, assembly lines, and quality control, or in the context of AM, may include the entire manufacturing process. Edge objectives are often advantageous for use in product fault prognosis, machine fault prognosis, and process monitoring of 3D printing applications [106].

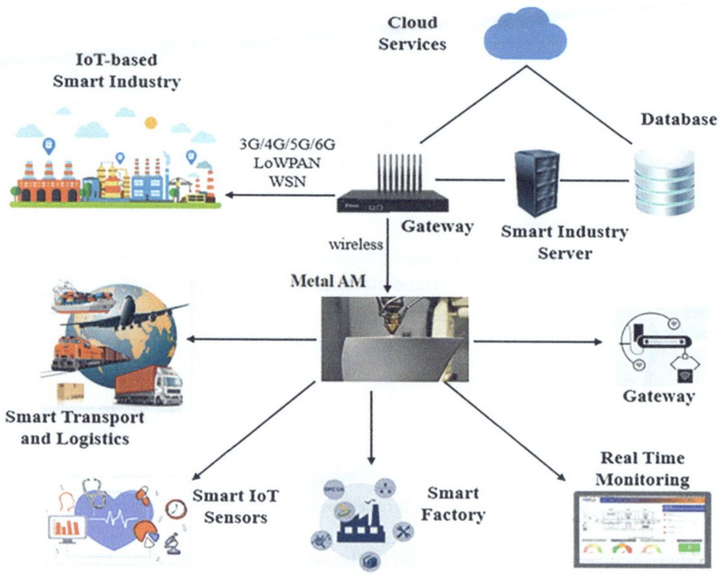

Fig. 15 Industrial IoT trends show recent applications for industrial IoT [105]. (Copyright 2023. Reproduced with permission from MDPI)

Cloud computing, on the one hand, works on big data, while Edge computing works on instantaneous/live data. Edge computing facilitates the services within the latency constraints (latency is the time delay for an information packet traveling through a communication circuit and it is one of the crucial metrics to take care of in a new-generation agile manufacturing industry). These technologies can provide on-demand, cheap services with accurate results. The manufacturing process can be visualized in a cloud system to reduce the multiple iterations in design and manufacturing units. Metal 3D printing technology is being seamlessly integrated with digital manufacturing.

Collaborative 3D Cloud printing-assisted AM service system attempts to achieve superior optimization, control, and monitoring when used by users at different levels and places. There can be many models for the deployment of these cloud/edge systems corresponding to the various interests of the market. These models (see Fig. 16) include several levels, where a primary level model is constituted by a cloud operator, a 3D printing service provider, a 3D printing service demander, and a logistic service provider. These entities are connected utilizing cloud platforms enabling operations at different locations simultaneously [107].

The confluence of cloud computing and additive manufacturing can enable new, innovative, and modular product offerings based on new business initiatives, since cloud and edge computing can create new avenues for service-oriented applications. Multiple applications such as "Design-as-a-Service," "Testing-as-a-Service," and "Manufacturing-as-a-Service" are explored by Felix et al. [108]. These

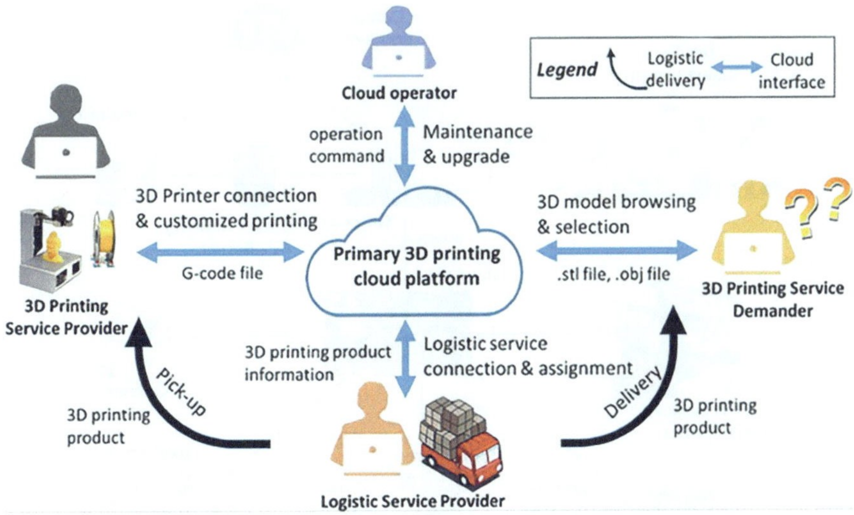

Fig. 16 Cloud Computing is an integral part of the additive manufacturing industry, serving as a medium between service providers, manufacturers, and logistic service providers [107]. (Copyright 2022. Reproduced with permission from Elsevier)

cloud-based applications bring programmed services a step forward. For example, Manufacturing-as-a-Service (MaaS) configures a manufacturing supply chain to realize automatic production, including the dynamic customization of the client demands by outsourcing it to a third party, which specializes and has deep expertise in 3D printing technologies [109]. This approach is particularly attractive for organizations that do not want to take the risk of a huge initial investment in 3D printing but want to try it out in real market operations by renting the 3D printing service from a third party. Design-as-a-Service (DaaS) is often preferred by a firm that does not want to get into the complex process of designing, which requires in-depth domain knowledge about designing, material properties, 3D printing specifics, and minutest details of the sintering process, and wants to outsource it to third party vendor (service provider) that specializes in these jobs. The service provider configures the cloud system often with an in-built AI system, to take important inputs from repositories of previously designed similar objects and figure out the best practices to perform the job satisfactorily. This adaptive product design approach often improves product performance [110]. Duan et al. classify various such cloud services offered in the form of an "as a Service" portfolio in the space of additive manufacturing [111].

With the increase in workload of the machines in the present time, the graphic processing units are being extensively used and new use cases are continuously researched. The edge and cloud computing systems majorly affect the software and hardware units. For example, Nvidia created a GPU-accelerated Supervisory

Control and Data Acquisition (SCADA) application that provides integrity management and real-time precursors, for operations in oil and gas, weather graphics, medical imaging, design, visualization, etc. It integrates thousands of sensors across geographically dispersed remote locations to create a holistic supervisory control. Since modern AM applications have very similar characteristics (dispersed over diverse geographical locations), a similar approach can be readily employed to integrate the entire process neatly through a computing and software layer, which is otherwise very challenging. Such operations would invariably need a good coupling strategy between the edge and cloud networks. More specifically, for example, the generation of 3D objects from LASER scans, intelligent building of components to improve design accuracy, etc., are possible features of applications like Autodesk and AutoCAD supported by SCADA [112].

3.3 Blockchain

Blockchain technology allows users to synchronize authentic transactional information in a system being used by multiple users. The traditional fields pertaining to the advantages of Blockchain are finance, healthcare, energy, etc. Finance applications use blockchain technology for the betterment of their transaction processes, enhancing financial security and automation of financial contracts. Fields of energy science use Blockchain to facilitate energy trade, increase the security of the energy grid, etc. Similarly, the healthcare industry uses blockchain to make it easier to store and share medical records [113]. Similarly, Blockchain provides a transparent financial and transactional data transfer in a 3D Printing protocol. Supply chains can be optimized and aid in sustainable manufacturing. Additionally, we can accurately track defects in products, monitor transactions, reduce the repetition of processes, and hence reduce greenhouse gas emissions [114].

Blockchain and AM may be co-developed together to enable safe and secure usage of sensors and networks over cloud systems for reliable, authentic, and responsible use of industrial process/finance/transaction information. Blockchain makes use of its characteristics like real-time monitoring to reflect changes that have the potential to instantaneously facilitate the AM process. 3D printing should be carried out safely by averting the risks of unauthorized data access [115].

The lingering risk of counterfeit products due to access to technology and design is evident in the AM industry, because of which application for digital license management is crucial for fair consumer and retailer rights. Platforms where data can be shared digitally considering traceability and intellectual property rights are being made to ensure the stability of the product [116]. In Sect. 3.7, we will see some more interesting use cases of blockchain technology concerning the provenance of materials in the context of AM materials sourcing.

3.4 Industry 4.0 and 5.0

The shift from production-centered industries to service-oriented manufacturing has been brought about by the development of the Fourth and Fifth industrial revolutions. The concept of early (1st) industrialization revolutions started the shift from physical labor to powered machines. The second industrial revolution is characterized by mass production, the introduction of assembly lines, etc. The third industrial revolution is characterized by the introduction of computation and automation of processes. Industry 4.0 has enhanced the realm of automation further by the introduction of comprehensive cyber-physical systems, often complimented by advanced AI and ML-based smart practices to control the minutest details of industrial processes through embedded IoT sensor networks, which was not possible before the advent of digital technology. Digital technology is at the core and creates the foundation to improve productivity, reduce delays and reruns, manage complexity, and increase resource efficiency [117].

Advanced manufacturing, however, fits well in the newer revolutions of Industry 5.0 where emphasis is placed on creating a highly productive and rewarding environment where the human workforce operates seamlessly alongside automated machines for overall societal well-being [101]. Industry 5.0 can be seen as an overlap between human-driven technologies working for a sustainable development goal onto the modern advanced manufacturing framework to reduce the cost of customization and complexity, saving a lot of time for the manufacturers. The overarching goal is to create a holistic socially responsible, environmentally sustainable, economically profitable modern future-ready industrial framework [118]. A flowchart (see Fig. 17) is used for standardization of the AM process. This is done to

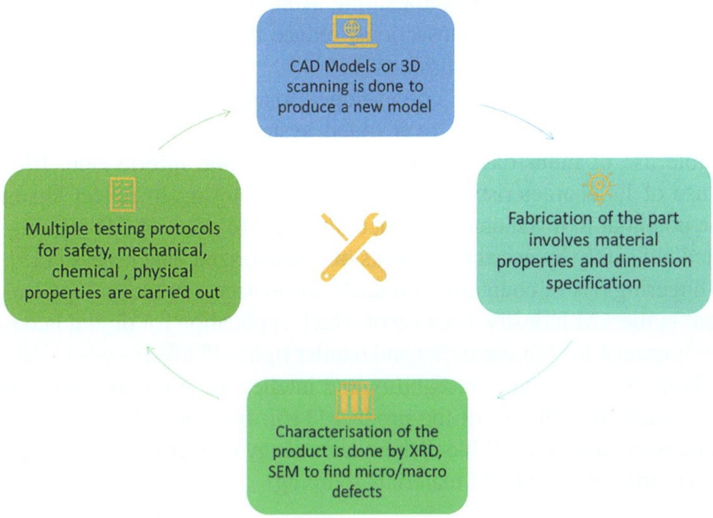

Fig. 17 Flowchart for AM processes to carry out repeated tests to minimize wastage of material and time, along with sustainability attainment [118]. (Adapted from Ref. [118])

save materials, time, and work hours. It starts with model simulations, followed by fabrication, Characterization, and mechanical or inspection testing.

Industry 5.0, which started in the 2000s, has made 3D printing a highly flexible manufacturing technology involving mass personalization in sectors like aerospace, dental implants, and medical uses, creating the scope for providing high-quality products in a socially responsible and environmentally sustainable manner. Large-scale manufacturing and outsourcing are done keeping in mind the trade-off between the centralized setting of a unit and customer-production proximity enabled by AM [119].

Smart manufacturing and AM are hence used for customization and personalization in an unprecedented way. Metal AM employs aluminum, titanium, and stainless steel for recent manufacturing of metal constituents. Mass customization introduces the need to redefine human roles and use technology to lower the risk of errors and enhance precision with embedded functionalities [120].

3.5 AI and ML

Artificial intelligence (AI) is the study of intelligent machines, software, and mechanisms to understand, analyze, learn, and communicate with people and machines. Machine learning (ML) is a subset of AI dealing with model-building and optimization algorithms for predictions, classification, and clustering based on collected simulated/synthetic or previously available data. Its ability to produce incredible results without the explicit knowledge of the underlying physics or chemistry of the problem necessitates the ability to learn from past experiences/data available to recognize patterns. Models constructed in ML are internally governed by a set of generic and systemic mathematical constructs, though they are quite distinct from the problem-specific underlying physical or chemical laws governing the problem, as already discussed. Models like linear regression for prediction, support vector machines for classification, and neural networks in deep learning are all mathematical models with variations in complexity and accountability for different parameters used for the model [101, 121, 122].

Deep learning is a subdivision of machine learning dealing with complex and large volumes of data. Deep learning relies on forming a framework like the human brain with working neurons connected efficiently. All the information is fed to an artificial neural network (ANN), and it is processed by layers of neurons and is passed to the forthcoming layer for further processing. Complex relations and patterns are accurately detected in deep learning models, giving more accurate predictions. The input data is fed into the input layer, where each neuron computes based on the mathematical relation $\mathbf{y} = \mathbf{wx} + \mathbf{b}$, where "**w**" is the weight and "**b**" is the bias term, "**x**" is the feature used for prediction and "**y**" is the variable predicted. The bias **b** is a constant added to ensure that the model can shift the decision boundary and is not constrained to pass through the origin (0,0). Without the bias, the model's predictions would be limited to patterns that intersect the origin, which may not always be the case in real-world data. The equation is vectorized, and its dimensionality depends on the number of input features used to make predictions. This equation is then passed to a nonlinear activation function, governing how much information is to be passed to the layers ahead. This nonlinearity ushers a high degree of flexibility in the model and prevents the entire network from collapsing into one giant linear function. Weight describes the importance of a particular feature in prediction, and it is the abstract representation of the memory of the ANN model.

Mimicking the image processing by the human brain, Convolutional Neural Networks (CNN), which are also a part of deep learning, have the capability for image segmentation, real-life object detection, classification, etc. Images are fed to the neural network as input instead of raw numerical data, and through vector conversion of images, features are extracted. These features are then fed to the further layers for final outputs [123]. Like the human ability to remember past information, recurrent neural networks (RNN) can remember past information fed to the network. This allows the model to maintain contextual data. Many complex models in natural language processing, like transformers and attention models, are devised from this basic implementation. Time series analysis is performed very efficiently using RNN models [101].

The number of layers and neurons may vary as per the user's choice. However, excess layers and neurons can reduce the accuracy of the network as it may overfit the data, reducing the attention to important relations for predictions. From the perspective of AM and 3D printing, machine learning can be very useful in production. Based on demands including the design and material, the model can describe various parameters required to maintain while prototyping, along with the method of prototyping, some of which are discussed in Sect. 1. In tandem with production, ML systems can be useful in quality control roles as well. Monitoring the layer printing procedure to detect any faults can alert the line in charge about the fault, and the process could easily be halted. In the healthcare sector, 3D-printed organs are growing rapidly with successful outcomes. Image processing can be used in various steps of printing organs. In pre-processing, anatomical features are captured through image segmentation, which assists in defining geometry for the prototype. Also, the diagnostic images are run through pre-trained CNN models for extra input towards diagnosis. During the prototyping, AI-enabled sensors can monitor the process and prompt for any defaults. In the post-process, the inspection can be done by sensors monitoring the condition of the printed organ, as well as the patient for adaptability [73]. During the post-process, as per the data provided from the past recoveries, the AI system can monitor the condition of both the patient and prototype and their relation. This allows the doctors to suggest better diagnoses and rehab programs for the patient as shown in Fig. 18.

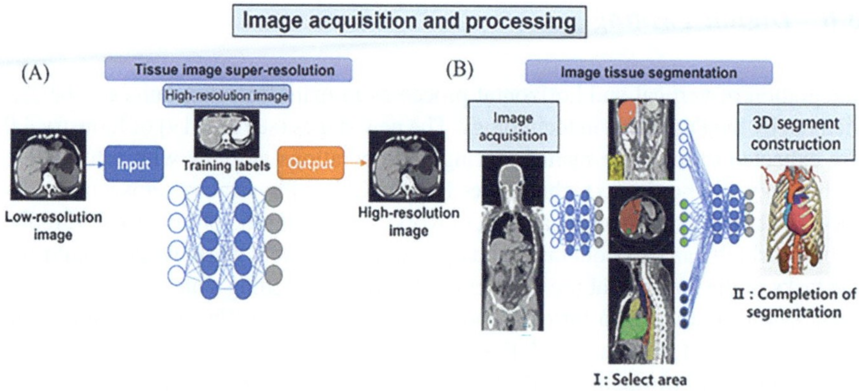

Fig. 18 Schematic diagram of 3D printing stages [124]. (Copyright 2023. Reproduced with permission from Elsevier)

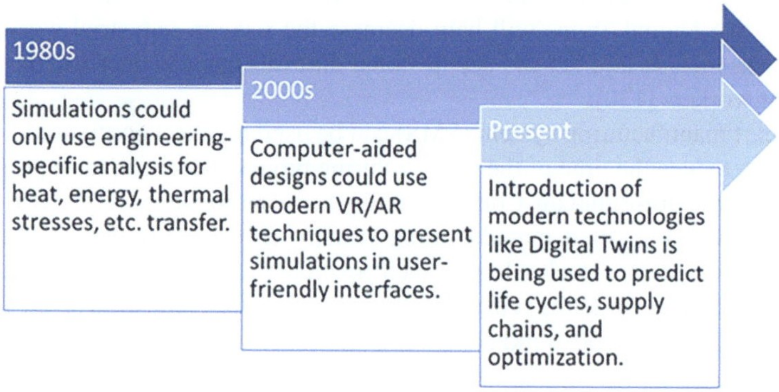

Fig. 19 Development of the computational software for simulating and integrating models with the as-built systems [126]. (Adapted from Ref. [126])

AI and machine learning (AI/ML) have significant potential in AM for metals. These technologies can optimize various stages of the AM process, including preprocessing, in situ monitoring, and post-processing. In the design phase, AI/ML can assist with geometry optimization and predict material properties, enhancing the quality of the final product. During the manufacturing process, machine learning models can monitor parameters such as temperature, powder distribution, and defect formation in real-time, reducing error rates and improving material performance. Furthermore, AI can be used to correlate process data with mechanical properties, enabling predictive maintenance and minimizing failures in critical components, such as those used in aerospace applications [125].

3.6 Digital Twining

Integration of vertical and horizontal processes in manufacturing units can be carried out by the Digital Twin technology. The new stages (see Fig. 19) of Industry 4.0 are expected to integrate manufacturing and operations systems with information technology, giving rise to technologies like the Cyber-Physical systems enabled by the IoT [127]. A cyber-physical system works by linking the physical world to the cyber world. It can carry out tasks and operations in the real world while monitoring the tasks in the backend at the same time. As we are reaching state-of-the-art cloud manufacturing, there is a lot of scope for predictive Cyber-Physical systems with attachment to IoT, Big data analytics, and complex processing [128]. We can feed a large amount of information to the Digital Twin systems from the physical factories in operation. However, not much has been done to use Digital Twin to make small or medium-sized factories automated. Hence, a lot of potential lies in the data analytics and simulation-enabled Digital Twin technologies for manufacturing [129]. A computerized representation of the human checklists and assessments using technologies like Digital Twins will help decrease the risks in industrial workspaces. These risks include injuries and occupational illnesses that can decrease the quality of life of workers [130].

A smart manufacturing system (SMS) can be used to detect configuration errors and hence, reduce the time of the manufacturing process. Liu et al. have proposed a Function-Structure-Behavior-Control-Intelligence-Performance (FSBCIP) framework for reviewing smart manufacturing systems using the literature. They have extensively researched technologies such as industrial IoT (IIoT), multi-domain physical–chemical modeling, virtual reality, data analytics, industrial artificial intelligence, blockchains, and cloud computing [126].

Companies like Siemens and ABB are working toward the realization of digital twin technology and using it to automate data exchange for multiple use cases, one of them being a smart factory. A digital twin can be used for the early phases of designing, commissioning, material forecasting, and quality control. One such example is ABB RobotStudio which can control the whole lifecycle of a product using virtual engineering tools in real time. It can simulate the software that runs in the robotic hardware of the production line [131]. Building Information Modelling is one such technique of AM/3D printing used to navigate the construction lifecycle and measure pre-construction measures extensively. It is an example of the integration of automated processing with construction [132]. The studies of [133] layers of data acquisition, data transmission, digital modeling, and data integration in the construction industry. Analogous to AM, sensors for wind velocity, ambient temperature, humidity, etc. IoT sensor monitors movement-related placements. The data transmission is done by messaging protocols like HTTP (Hypertext Transfer Protocol) for communicating the data. The digital modeling layer uses software tools like Autodesk for simulating the physical process of the product to be made [134].

3.7 Cyber-Physical Security

The cyber-theft aspect in AM is increasing at a high rate due to global connectivity and easy access to various resources. The extensive manufacturing connectivity, via sensors or the Internet, increases the supply chain management capabilities, but this also increases the susceptibility to attacks on cyber-physical systems [135]. The production line requires security services to maintain and protect private information like dyes for printing or "CAD" files for prototyping. The attacks involve stealing IP (intellectual property) that controls production, design systems, etc. [135]. IP thefts include not only dyes or "CAD" files but also the surrounding condition details, layering properties, etc. The theft can take place in different steps of prototyping. An STL or a Stereolithography file is a format that stores the surface geometry of the 3D object as a collection of triangles. It is also referred to as Standard Triangle Language or Standard Tessellation Language. An STL file does not hold any representation of texture, color, or other CAD file attributes. Once the "CAD"/"STL" file is constructed, the hacker can lure in the system, which is connected to a network, and steal the files, leaving behind minor or no traces of theft. Another step of production where identity theft can take place is the transferring of files (see Fig. 20). As Gupta et al. mentioned, while the files are transferred from one system to another over the network, a third-party user can tune in to the network and steal the files. Password-protected networks or systems can be breached with powerful systems through brute force attacks [137]. IP thefts are not only about stealing files or prototyping conditions, but they also involve altering the files or properties. From the same breach, the files can be altered, introducing a fault in the product or the production process itself. The faults can cause uneven layering, low-quality material release, or alterations in the properties of inventory. In 3D printing, such attacks are prominent with acute alterations in files or conditions during different stages. Adulteration in STL files leads to disturbed triangulation data and format, which would lead to poor finish or compromised structural integrity. Ahmad et al. mention that with major firms entering the 3D printing arena, there is a lot of sensitive information about the processes that can be accessed by the competitors in any breach of security. This could lead to a loss of market shareholdings and invention credits [138].

The physical aspect is added to the cyber-physical system through the protection of component vulnerabilities. These vulnerabilities can be added by disgruntled employees. They can exploit the vulnerabilities of the sensors and actuators providing false data or disruption of production. In 3D printing, any alterations to the material or layering device can lead to low structural integrity. Dealing with modern threats at printers, supply chains, and various other levels requires a robust security system and all-around protection.

In the context of the metal AM industry, establishing a robust information infrastructure is crucial to ensure the security of supply chains. This involves tracing the provenance of materials from their origin through various stages of transit, which is

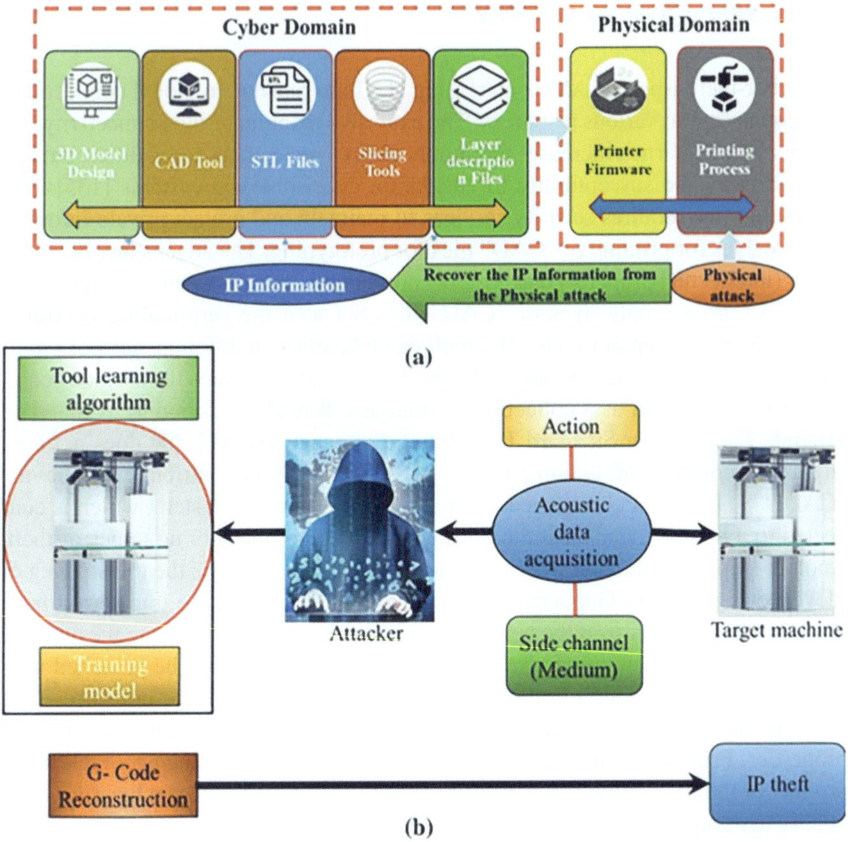

Fig. 20 Type of attacks for each of the three supply chain components of AM [136]. (Copyright 2022. Reproduced with permission from Springer Nature)

essential to prevent counterfeiting and sabotage. The diverse nature of the AM industry, involving different businesses from powder metallurgy to industrial design, and electronics, adds to the complexity of establishing a consensus on provenance. Blockchain technology, with its tamper-resistant and decentralized architecture, is proposed as a solution to create a secure database that records the history of each 3D-printed product, thereby facilitating trust and collaboration among the various entities involved [137]. Additionally, smart contracts integrated with blockchain can automate processes, enhancing secure communication and execution among the autonomous entities in the supply chain.

To combat reverse engineering, which poses a threat to the integrity of AM products, especially in high-stakes industries like aerospace and medical implants, anti-reverse engineering methods are being explored. Traditional methods such as serial numbers and tracking codes can be employed, but they are susceptible to tampering [139]. Instead, embedding unique, machine-readable codes within the product during the layer-by-layer manufacturing process offers a more secure alternative. These codes, which can be detected through methods like radiography or ultrasound, do not require voids or large QR codes but can be as subtle as variations in microstructure that are machine learnable. Additionally, protecting "CAD" files against unauthorized modifications is essential, as these files are often shared across cloud platforms. Techniques like embedding obfuscated design features in "CAD" files and using secure design strategies during the slicing and tool path creation stages can prevent the production of unauthorized high-quality duplicates, ensuring the integrity and security of AM processes [139]. The files and data being transferred over the network in their raw form allow hackers to edit or alter the files. The edits/alterations can change the printing properties such as layering, the flow of material, etc. [140]. Encryption techniques can be used to safeguard the files from being altered in the middle of transferring. All the "stl," and "obj" files can be protected from unauthorized access or distribution. Various end-to-end encryption techniques can be used to protect the files from the source's end to the receiver's end. Different techniques including Public Key Infrastructure, Transport Layer Security, etc. can be of great help. Additions to files in the form of data signatures ensure any changes or alterations being made to the file [141]. One of the unique ways of protecting the file is to add a unique but almost invisible feature to the model like initials embedded deep within a layer decrease the chances of IP theft since the presumed stolen files can be cross-checked for that unique identity mark. In addition to adding an encryption layer to the model, a blockchain layer can introduced alongside to add a safe digital footprint to the model tracking its creation, modification, and usage, making it an immutable record [140].

4 Distributed AM Protocols

Distributed manufacturing (DM) is an evolving manufacturing model driven by recent advancements in production technologies, such as AM, continuous production, and digitalization, alongside infrastructural developments like IoT and big data. DM is different from traditional centralized production because they do not need large factories or long supply chains, and enables customized production for the users. They bring the collaboration of physical production and information management. DM allows for flexible, real-time manufacturing, integrating advancements in sensor technology, data analytics, and cloud-based systems; however, issues like technological, societal, and regulatory gaps are still prevalent. Technological issues revolve around process control, material sourcing, and the need for new production standards. Societal and regulatory challenges involve unclear regulations for this comparatively newer process and very dispersed production units. If these challenges are resolved, DM's potential can be used for sustainability by reducing transport emissions, usage of local resources, and facilitating more circular economic practices through reuse and recycling [142].

Additive Manufacturing reduces production costs, simplifies supply chains, and allows for flexible, on-demand manufacturing near consumers. However, many companies still rely on traditional supply chains and are yet to fully leverage AM's true potential. Blockchain architecture empowered Additive Manufacturing-as-a-Service (AMaaS) works in the direction of distributed AM protocols to allow smaller, decentralized production units, or "micro-manufacturers," to pool resources and offer flexible, pay-per-use manufacturing services, optimizing supply chains [143].

Additive manufacturing, in general, supports distributed production due to its low infrastructure requirements and high scalability. Schmidt et al. explored the optimization of the facility location problem for the case of the Fused Layer Modeling (FLM) process, for which the adoption of a distributed production protocol is relatively high. A mathematical model that addresses the two-level incapacitated facility location problem uses k-means clustering to optimize the distribution of factories. It finds a balance between cost efficiency, minimizing pollution, and enhancing resilience. Parameters such as transportation costs, infrastructure, and Carbon dioxide emissions are evaluated to find an optimal size distribution [144].

Pei et al. [145] investigate the role of AM data exchange standards in a re-distributed manufacturing (RDM) context. RDM refers to a shift from centralized, large-scale manufacturing to smaller, localized, and customized production enabled by advancements in technologies like AM. Despite "STL" being the most widely used format, it has limitations, prompting a review of more advanced formats like AMF and STEP-NC, which offer features like improved geometry representation and support for more complex manufacturing processes. Critical features for AM data standards in RDM, including regular internal structures, manufacturing tolerances, and geometric and curvature representations, have been identified that improve the working of the entire system with the least number of fault points and increased efficiency.

"STEP-NC" is emerging as the most competitive format, with the capability to handle advanced manufacturing requirements, while open standards were seen as crucial for fostering innovation and accessibility. AM data standards have the potential to streamline the manufacturing process in an RDM environment, benefiting both manufacturers and end-users. However, challenges remain, such as the need for standards that support intellectual property protection and facilitate multi-user collaboration in an increasingly decentralized and customized production landscape [145].

Guinta et al. [146] examine various distributed architectures in the context of AM. They emphasize agent-based systems, which can enhance productivity, adaptability, and resilience in distributed AM settings. Three specific architectures can be derived from this system: a centralized approach, a decentralized approach, and a hybrid model, each offering distinct benefits. Centralized architectures allow better control over processes but may struggle with scalability. In contrast, decentralized systems are more flexible and scalable but can face coordination challenges. The hybrid model attempts to balance control and flexibility, offering the most efficient use of resources across distributed manufacturing environments. Key insights drawn from the comparison highlight the importance of architecture choice depending on the manufacturing goals, such as flexibility, scalability, or control. Agent communication protocols, task allocation methods, and fault tolerance mechanisms within each system play a crucial role in enhancing the performance of each setup. Ultimately, it can be pointed out that hybrid systems are the most promising for distributed AM because they combine the strengths of both centralized and decentralized models, optimizing resource use while maintaining resilience and responsiveness to changes in production demands.

Sun et al. [147] explore the extension of traditional single-stage Additive Manufacturing (AM) processes to multi-stage distributed AM systems, referred to as STREAM. The focus is on Freeze Nano Printing (FNP), a cutting-edge AM technology used for fabricating energy storage products with high energy and power density. The STREAM system consists of three stages: pre-processing, distributed FNP printers, and centralized post-processing. The cyber-coordinated simulation framework introduced aims to address the complex challenges in managing distributed processes by integrating discrete event simulation (DES) with physical models for system-level optimization. The hierarchical framework is designed to enhance the predictive capabilities and overall performance of distributed manufacturing operations, allowing for real-time data infusion to improve decision-making processes.

The study demonstrates the efficiency of the framework using simulation data from FNP, focusing on the impact of thermal management and inter-stage dependencies on product quality and productivity. A significant contribution is the development of a multi-response Gaussian Process (GP) emulator to predict the non-uniformity of thermal profiles and optimize processing times. This emulator integrates with the DES to manage production flow and optimize resource allocation. This signifies that the framework can significantly improve production control, minimize failure rates, and enhance the quality of energy storage devices. Increasing the range of usage of these frameworks can be observed in the aerospace and electronics sectors [147, 148].

Another work by Rauch et al. [149] investigates that the shift from old-style manufacturing to modern cloud-based systems is a big deal. It's all about making production more flexible, efficient & quick! With AM, companies can create custom products just when they need them, right where they're needed. This helps cut down on delays, and transport costs, and even helps the environment. AM also allows businesses to craft products that fit specific customers' needs, opening the door to new ways of personalizing goods. This study is part of a bigger project. We focus on designing smart & spread-out manufacturing systems and how to coordinate them in complex networks. These systems use digital tools like the IoT, cloud computing, & AI. They help make production smoother, improve communication between different manufacturing points, & make supply chains more efficient. Additive manufacturing is a perfect fit here. It lets designers create faster prototypes and build parts with tricky shapes that would be tough or impossible with traditional methods.

Mika Salmi [150] looked closely at different AM processes. The study reveals that some processes can be tricky. For example, powder bed fusion (PBF) and binder jetting come with big challenges for distributed manufacturing. They often need fancy equipment, complicated post-processing, and special environments. This can make it tough for smaller production setups that aren't centralized. On the other hand, material extrusion (MEX), like fused deposition modeling (FDM), turns out to be a smart choice. It's affordable and easy to use, great when you don't need lots of parts all at once. MEX devices are simple to set up & keep running, plus they don't need much post-processing. So, they're perfect for small jobs that come up on demand.

Now, there's also VAT photopolymerization. It's somewhat like MEX but tends to need a bit more work when it comes to finishing touches, especially in the curing & finishing stages. Still, VAT photopolymerization can handle larger orders better than MEX when there's more demand for parts. This means it gives some nice flexibility based on how many items folks want to produce.

In conclusion, while no AM process is the absolute best fit for every situation in distributed manufacturing, both MEX and VAT photopolymerization stand out. They're promising due to their ability to scale up easily, lower costs & straightforward requirements all of which make them great choices for different production needs in various environments.

4.1 Future Perspective of AM

Advanced AM processes like Friction Stir-Assisted Additive Manufacturing (FSAAM), Cold Spraying (CS), and Ultrasonic Additive Manufacturing (UAM) present promising solutions to overcome the limitations of conventional fusion-based techniques. Solid-state FSAAM operates at melting temperatures, preventing solidification flaws; on the other hand, CS uses kinetic energy to deposit materials at low temperatures, which makes it perfect for substrates that are sensitive to heat. With the use of ultrasonic vibrations, UAM enables exact constructions with little residual stress even at lower temperatures. These methods have several benefits, such as improved mechanical qualities, increased material adaptability, and decreased flaws, which makes them useful in sectors like energy, automotive, and aerospace. However, issues like restricted geometric complexity and slower build rates persist, emphasizing the necessity for more developments in these cutting-edge technologies. This field is always open for developments to optimize parameters like laser power, scanning speed, and hatch distance. Studies are done to simulate and understand micro segregation, phase formation, and grain morphology during solidification. These new technical efforts and understanding of the process will reduce defects, such as porosity, and improve mechanical properties by fine-tuning microstructure characteristics. Hence there is significant potential in designing new alloys tailored specifically for AM processes.

Digital manufacturing can be enhanced by introducing real-time data synchronization and cloud-based adaptive systems in the manufacturing plants. This will allow for operation handling from far-off locations and easily accessible inventory and production line data, including faults, custom-made prototypes, etc. Since the transition of Industry 4.0 to Industry 5.0 will bring more human-centric and sustainable production, digital manufacturing can be further enhanced by integrating sustainable production practices, reducing the reliance on physical prototypes, and minimizing materials that increase the carbon footprint. Implementing advanced sensors to detect and eliminate toxic materials in production lines can improve environmental safety and worker health. Automated systems can alert workers to potential hazards, ensuring safer, greener processes. This shift toward eco-friendly manufacturing reduces waste and promotes a more responsible and efficient production cycle.

4.2 AM as a Critical Tool for Space-Faring Humans for Interplanetary Manufacturing

Human civilization has come a long way since around six million years ago, from the Miocene epoch (the timeline believed to be the human appearance on Earth) to dominate the planet Earth as a species. Despite the astonishing all-around growth in human civilization, we are still stuck as a single planetary species based on Earth alone which creates significant vulnerabilities for natural disasters or apocalyptic destruction by nuclear bombs, etc. If you look at the history of the earth itself, the facts do not give us great confidence, either. For example, at the end of the Cretaceous period, about 65 million years ago, it is believed that a large comet had the unfortunate role of eliminating the most evolved species (the dinosaurs) prevailing on the then-earth. Therefore, it is a very sincere desire to eliminate such possibilities in the future to sustain the human species. One of the most compelling ways to achieve this is to make humans a multi-planetary species. Hence, it is imperative to search for a habitable planet and create the necessary infrastructure over there for sustaining life on a hostile planet, but this is an extremely challenging undertaking. Although transporting humans to a different celestial body itself is very challenging, an even more challenging job is to support life on a barren planet. We will need to build a lot of modern infrastructure from the ground up. On planet Earth, the majority of the industrial production happens as a very carefully orchestrated chain of events, but achieving it on a distant planet where the majority of these supply chains will be non-existent will pose great challenges. Development of industry and infrastructure in such a remote, barren, and hostile location will then most likely be dependent mostly on modern approaches like AM. In this protocol majority of the critical designing will be done on planet Earth then the data will be transmitted over the interplanetary space through an encrypted and secure radio transmission protocol and the components will be 3D printed locally on the other planet mostly through building materials sourced locally over there. This is critically important since in this protocol we have to carry only the building equipment or the 3D printer from the planet Earth, but no other major infrastructures are necessary to support it and it can work there almost in a bare-bone configuration. As we have already discussed, AM protocols can work in a modular and distributed mode, therefore this is the ideal choice to support humanity on a different planet altogether. Therefore, modern AM technologies happily married with the new generation software layer provide the greatest insurance for the survival of the human species as the first known interplanetary species.

Promising advances in efficiency, accuracy, and safety are anticipated in the fields of AI/ML, IoT, and cyber-physical security in metal AM in the future. High-quality metal components will be able to use adaptive control thanks to AI/ML's improvements in process optimization, real-time monitoring, and fault identification. The IoT will make it easier for machines, sensors, and cloud systems to share data seamlessly, which will enhance supply chain integration and predictive maintenance. As AM's connection grows, cyber-physical security becomes increasingly important for protecting key design data and thwarting possible cyberattacks. When combined, these technologies will power metal AM ecosystems that are more intelligent, robust, and secure.

5 Conclusion

There is a transformative potential of AM in various industries, including consumer goods, medical devices, automobiles, and aerospace, by using 3D printers for layer-by-layer production. Fusion-assisted processes like DED and PBF, have aimed at enhancing mechanical properties, reducing cracks, and altering microstructures in metallic materials like titanium and Inconel alloys. Overall, the need is twofold: developing new materials compatible with existing field-assisted AM technologies and creating novel AM technologies to support a broader range of materials. Addressing these challenges will help overcome current material limitations that are required to realize AM for industrial-scale applications. For example, the combination of controlling unit cell orientation and utilizing L-PBF technology offers a powerful method for designing high-performance Ti-6Al-4V structures with enhanced mechanical properties tailored to specific applications. AM is transforming traditional manufacturing by offering custom materials, design flexibility, and rapid prototyping, which industrial production lacks. The integration of technologies like IoT, AI and ML, cybersecurity, etc., enhances AM efficiency, while edge and cloud computing, Blockchain, etc., make it accessible globally. As Industry 5.0 emerges, production shifts to human-centric and sustainable models, with blockchain and digital twinning driving personalized manufacturing and profitability. AM's versatility spans fusion and solid-state processes, including laser and electron beam techniques for complex geometries, and frictional or ultrasonic-based methods for simpler, isotropic products, especially metal composites.

To sum up, solid-state additive manufacturing (SSAM) methods such as friction stir-assisted additive manufacturing (FSSAM) present viable substitutes for traditional fusion-based procedures. These techniques, which include Friction Stir Lap Joint (FSLJ) and Additive Friction Stir Deposition (AFSD), address several issues related to high-temperature processes, such as anisotropic material characteristics and solidification flaws. FSSAM offers better mechanical characteristics and refined microstructures by operating below melting temperatures, which makes it especially useful for precipitation-strengthened alloys that are lightweight. Furthermore, SSAM's application potential is further expanded by its capacity to process a variety of materials, including scrap and metal matrix composites. Nevertheless, despite these developments, SSAM is still not as good at precisely generating complex geometries, which emphasizes the need for more research and development in this nascent subject. High-performance industries like aerospace and automotive can benefit from the novel material deposition and repair applications made possible by cold spraying (CS) and ultrasonic additive manufacturing (UAM). With continued development, SSAM technologies hold the potential to close gaps in AMs operational flexibility, material quality, and efficiency.

IoT devices can facilitate self-awareness in machines, predicting faults and ensuring high-quality outputs while enabling seamless communication among robotic systems and sensors. Furthermore, the complex geometries of IoT devices can be effectively produced using AM, showcasing the interdependence of both technologies in modern manufacturing. The growing risk of cybercrime in AM is a result of weaknesses in supply chains and worldwide connectivity. Theft of intellectual property (IP), such as CAD and STL files, is frequent. There's also a chance that files will be altered during manufacturing, which could lower the quality of the final product. It is advised to use blockchain technology and encryption methods to protect data and supply networks. Particularly in high-stakes industries like aerospace and healthcare, machine learning (ML), and artificial intelligence (AI) are emphasized as methods for enhancing production efficiency, quality control, and fault detection in AM. To reduce these threats, advanced security procedures and secure infrastructure are crucial.

References

1. Yu, H. Z., Jones, M. E., Brady, G. W., Griffiths, R. J., Garcia, D., Rauch, H. A., Cox, C. D., & Hardwick, N. (2018). *Scripta Materialia, 153*, 122.
2. Raza, A., Deen, K. M., Jaafreh, R., Hamad, K., Haider, A., & Haider, W. (2022). *International Journal of Advanced Manufacturing Technology, 122*, 1143.
3. Bogue, R. (2013). *Assembly Automation, 33*, 307.
4. Raffik, R., Sivaguru, J., Sakira Parveen, A., Sabitha, B., Naveen, C., & Murugajothi, T. (2024). *Materials Today Proceedings*. S2214785324002670.
5. Vafadar, A., Guzzomi, F., Rassau, A., & Hayward, K. (2021). *Applied Sciences, 11*, 1213.
6. Bhuvanesh Kumar, M., & Sathiya, P. (2021). *Thin-Walled Structures, 159*, 107228.
7. Ford, S., & Despeisse, M. (2016). *Journal of Cleaner Production, 137*, 1573.
8. Attaran, M. (2017). *Business Horizons, 60*, 677.
9. Garcia-Dominguez, A., Claver, J., Camacho, A. M., & Sebastian, M. A. (2020). *IEEE Access, 8*, 125056.
10. Polonsky, A. T., & Pollock, T. M. (2020). *Science, 368*, 583.
11. Tammas-Williams, S., & Todd, I. (2017). *Scripta Materialia, 135*, 105.
12. Bandyopadhyay, A., Traxel, K. D., Lang, M., Juhasz, M., Eliaz, N., & Bose, S. (2022). *Materials Today, 52*, 207.

13. Aeby, X., Poulin, A., Siqueira, G., Hausmann, M. K., & Nyström, G. (2021). *Advanced Materials, 33*, 2101328.
14. DebRoy, T., Wei, H. L., Zuback, J. S., Mukherjee, T., Elmer, J. W., Milewski, J. O., Beese, A. M., Wilson-Heid, A., De, A., & Zhang, W. (2018). *Progress in Materials Science, 92*, 112.
15. Yadav, A., Srivastava, M., Jain, P. K., & Rathee, S. (2024). *International Journal on Interactive Design and Manufacturing (IJIDeM), 19*(4), 2317.
16. Srivastava, M., Rathee, S., Maheshwari, S., Noor Siddiquee, A., & Kundra, T. K. (2019). *Critical Reviews in Solid State and Materials Sciences, 44*, 345.
17. Mishra, R. S., Haridas, R. S., & Agrawal, P. (2022). *Science and Technology of Welding and Joining, 27*, 141.
18. Sharma, A., Jain, R., Agrawal, P., Mukherjee, S., Gumaste, A., Davis, D. F., Haridas, R. S., & Mishra, R. S. (2024). *Acta Materialia, 279*, 120284.
19. Palanivel, S., & Mishra, R. S. (2017). *International Journal of Additive and Subtractive Materials Manufacturing, 1*, 82.
20. Chen, L., Jiang, T., Li, J., Guo, Y., Dai, G., Sun, Z., Zhan, L., & Liu, C. (2024). *Journal of Alloys and Compounds, 1008*, 176666.
21. Mondal, M., Das, H., Hong, S.-T., Jeong, B.-S., & Han, H. N. (2019). *CIRP Annals, 68*, 289.
22. Slotwinski, J. A., Garboczi, E. J., Stutzman, P. E., Ferraris, C. F., Watson, S. S., & Peltz, M. A. (2014). *Journal of Research of the National Institute of Standards and Technology, 119*, 460.
23. Yoder, J. K., Hahn, G. D., Zhao, N., Brennan, R. E., Cho, K., & Yu, H. Z. (2023). *Additive Manufacturing Letters, 4*, 100108.
24. Gibson, I., Rosen, D., & Stucker, B. (2015). *Additive manufacturing technologies: 3D printing, rapid prototyping, and direct digital manufacturing*. Springer New York.
25. Gottwald, R. B., Griffiths, R. J., Petersen, D. T., Perry, M. E. J., & Yu, H. Z. (2021). *Accounts of Materials Research, 2*, 780.
26. Lopez, J. J., Williams, M. B., Rushing, T. W., Confer, M. P., Ghosh, A., Griggs, C. S., Jordon, J. B., Thompson, G. B., & Allison, P. G. (2022). *Materialia, 23*, 101440.
27. Khodabakhshi, F., & Gerlich, A. P. (2018). *Journal of Manufacturing Processes, 36*, 77.
28. Mondal, M., Das, H., Ahn, E. Y., Hong, S. T., Kim, M.-J., Han, H. N., & Pal, T. K. (2017). *Metals and Materials International, 23*, 948.
29. Basak, S., Yoo, H.-G., Anaman, S. Y., Guha, P., Kwon, D.-H., Shin, E.-J., Hong, S.-T., & Cho, H.-H. (2024). *Journal of Alloys and Compounds, 976*, 173344.
30. Liu, F. C., Feng, A. H., Pei, X., Hovanski, Y., Mishra, R. S., & Ma, Z. Y. (2024). *Progress in Materials Science, 146*, 101330.
31. Haridas, R. S., Gumaste, A., Varshney, P., Manu, B. R., Kandasamy, K., Kumar, N., & Mishra, R. S. (2023). *JOM, 75*, 4231.
32. Palanivel, S., Nelaturu, P., Glass, B., & Mishra, R. S. (2015). *Materials and Design, 65*(934), 1980–2015.
33. Chaudhary, B., Jain, N. K., & Murugesan, J. (2022). *International Journal of Advanced Manufacturing Technology, 123*, 603.
34. Yuqing, M., Liming, K., Chunping, H., Fencheng, L., & Qiang, L. (2016). *International Journal of Advanced Manufacturing Technology, 83*, 1637.
35. Jordon, J. B., Allison, P. G., Phillips, B. J., Avery, D. Z., Kinser, R. P., Brewer, L. N., Cox, C., & Doherty, K. (2020). *Materials and Design, 193*, 108850.
36. Griffiths, R. J., Petersen, D. T., Garcia, D., & Yu, H. Z. (2019). *Applied Sciences, 9*, 3486.
37. Dong, H., Li, X., Xu, K., Zang, Z., Liu, X., Zhang, Z., Xiao, W., & Li, Y. (2022). *Aerospace, 9*, 565.
38. Prabhakar, D. A. P., Shettigar, A. K., Herbert, M. A., Patel, M., Pimenov, D. Y., Giasin, K., & Prakash, C. (2022). *Journal of Materials Research and Technology, 20*, 3025.
39. Beladi, H., Farabi, E., Hodgson, P. D., Barnett, M. R., Rohrer, G. S., & Fabijanic, D. (2022). *Philosophical Magazine, 102*, 618.
40. Kandasamy, K. (n.d.). US9511445B2.

41. Korgancı, M., & Bozkurt, Y. (2024). *Journal of Materials Research and Technology, 30*, 4572.
42. Rivera, O. G., Allison, P. G., Jordon, J. B., Rodriguez, O. L., Brewer, L. N., McClelland, Z., Whittington, W. R., Francis, D., Su, J., Martens, R. L., & Hardwick, N. (2017). *Materials Science and Engineering A, 694*, 1.
43. MELD Manufacturing (n.d.).
44. Swank, K. (2024). *Army's Jointless Hull 3D metal printer recognized for technical achievement.*
45. Rivera, O. G., Allison, P. G., Brewer, L. N., Rodriguez, O. L., Jordon, J. B., Liu, T., Whittington, W. R., Martens, R. L., McClelland, Z., Mason, C. J. T., Garcia, L., Su, J. Q., & Hardwick, N. (2018). *Materials Science and Engineering A, 724*, 547.
46. Griffiths, R. J., Perry, M. E. J., Sietins, J. M., Zhu, Y., Hardwick, N., Cox, C. D., Rauch, H. A., & Yu, H. Z. (2019). *Journal of Materials Engineering and Performance, 28*, 648.
47. American Welding Society, Chen, H., Chen, J., Meng, X., Xie, Y., Huang, Y., Xu, S., & Zhao, Y. (2022). *Welding Journal, 101*, 249.
48. Assadi, H., Kreye, H., Gärtner, F., & Klassen, T. (2016). *Acta Materialia, 116*, 382.
49. Metal Spray Supplies Australia. (2022).
50. Somireddy, M. (2021). In H. K. Dave & J. P. Davim (Eds.), *Fused deposition modeling based 3D print* (pp. 255–276). Springer International Publishing.
51. Lores, A., Azurmendi, N., Agote, I., & Zuza, E. (2019). *Powder Metallurgy, 62*, 267.
52. Hehr, A., & Norfolk, M. (2020). *Rapid Prototyping Journal, 26*, 445.
53. Badiru, A. B., Valencia, V. V., & Liu, D. (Eds.). (2017). *Additive manufacturing handbook: Product development for the defense industry* (1st ed.). CRC Press.
54. Ahn, H., Gingerich, M. B., Hahnlen, R., Dapino, M. J., & Pourboghrat, F. (2021). *International Journal of Material Forming, 14*, 917.
55. Cai, W. W. (Ed.). (2017). *Ultrasonic welding of lithium-ion batteries*. ASME Press.
56. Chen, K., Zhang, Y., & Wang, H. (2017). *Ultrasonics, 75*, 9.
57. Zhou, H., Cui, H., Qin, Q.-H., Wang, H., & Shen, Y. (2017). *Materials Science and Engineering A, 682*, 376.
58. Hendrixson, S. *Additive manufacturing media.*
59. *UAM 3D printers manufacture satellite components for NASA.*
60. Guo, H., Gingerich, M. B., Headings, L. M., Hahnlen, R., & Dapino, M. J. (2019). *Composite Structures, 208*, 180.
61. Behvar, A., Shakil, S. I., Pirgazi, H., Norfolk, M., & Haghshenas, M. (2024). *International Journal of Advanced Manufacturing Technology, 132*, 2061.
62. Rahmati, M., Blaker, J. J., Lyngstadaas, S. P., Mano, J. F., & Haugen, H. J. (2020). *Materials Today Advances, 5*, 100051.
63. Kumar, V., Kaur, H., Kumari, A., Hooda, G., Garg, V., & Dureja, H. (2023). *Bioprinting, 36*, e00298.
64. Xia, Z., Xu, J., Shi, J., Shi, T., Sun, C., & Qiu, D. (2020). *Additive Manufacturing, 33*, 101114.
65. Hua, B., Xu, H., Zhan, W., Tian, T., Yin, J., & Zhang, Q. (2024). *Journal of Alloys and Compounds, 999*, 175102.
66. Liu, G., Xiong, Y., & Zhou, L. (2021). *Composites Communications, 27*, 100907.
67. Li, C., Liu, Z. Y., Fang, X. Y., & Guo, Y. B. (2018). *Procedia CIRP, 71*, 348.
68. Bambach, M., Sizova, I., Kies, F., & Haase, C. (2021). *Additive Manufacturing, 47*, 102269.
69. Guan, X., & Zhao, Y. F. (2020). *International Journal of Advanced Manufacturing Technology, 107*, 1959.
70. Cai, Y., Zhang, S., Wang, Y., Chen, H., & Xiong, J. (2024). *Journal of Manufacturing Processes, 128*, 111.
71. Benmessaoud, F., Cheikh, M., Velay, V., Vidal, V., & Matsumoto, H. (2020). *Materials Science and Engineering A, 774*, 138835.
72. Jiang, Y., Tang, H., Li, Z., Chen, H., Li, Y., Zheng, D., Cheng, X., & Zhang, C. (2024). *Journal of Alloys and Compounds, 1005*, 175965.
73. Peng, K., Huang, H., Xu, H., Kong, Y., Zhu, L., & Liu, Z. (2022). *SSRN Electronic Journal.*

74. Andersson, J.-O., Helander, T., Höglund, L., Shi, P., & Sundman, B. (2002). *Calphad, 26*, 273.
75. Vundru, C., Singh, R., Yan, W., & Karagadde, S. (2021). *International Journal of Mechanical Sciences, 211*, 106790.
76. Zhao, R., Chen, C., Wang, W., Cao, T., Shuai, S., Xu, S., Hu, T., Liao, H., Wang, J., & Ren, Z. (2022). *Additive Manufacturing, 51*, 102605.
77. Kotadia, H. R., Gibbons, G., Das, A., & Howes, P. D. (2021). *Additive Manufacturing, 46*, 102155.
78. Lyu, Y., Zhao, H., Wen, X., Lin, L., Schlarb, A. K., & Shi, X. (2021). *Journal of Applied Polymer Science, 138*, 50782.
79. Pfaff, A., Jäcklein, M., Schlager, M., Harwick, W., Hoschke, K., & Balle, F. (2020). *Materials, 13*, 5400.
80. Yadroitsev, I. (2009). *Selective laser melting: Direct manufacturing of 3D-objects by selective laser melting of metal powders*.
81. Vock, S., Klöden, B., Kirchner, A., Weißgärber, T., & Kieback, B. (2019). *Progress in Additive Manufacturing, 4*, 383.
82. Panwisawas, C., Tang, Y. T., & Reed, R. C. (2020). *Nature Communications, 11*, 2327.
83. Bergmueller, S., Gerhold, L., Fuchs, L., Kaserer, L., & Leichtfried, G. (2023). *International Journal of Advanced Manufacturing Technology, 126*, 4385.
84. Yang, X., Gibbons, G. J., Tanner, D. A., Li, Z., Wilson, P., Williams, M. A., & Kotadia, H. R. (2023). *Materials and Design, 232*, 112160.
85. Sun, Z., Wang, H., Tian, X., & He, B. (2023). *Materials Science and Engineering A, 872*, 144992.
86. Schneider, J., Farris, L., Nolze, G., Reinsch, S., Cios, G., Tokarski, T., & Thompson, S. (2022). *Journal of Manufacturing and Materials Processing, 6*, 20.
87. Aboulkhair, N. T., Simonelli, M., Parry, L., Ashcroft, I., Tuck, C., & Hague, R. (2019). *Progress in Materials Science, 106*, 100578.
88. Wang, Y., Wang, Y., Li, R., Niu, P., Wang, M., Yuan, T., & Li, K. (2021). *Journal of Central South University, 28*, 1043.
89. Khairallah, S. A., Anderson, A. T., Rubenchik, A., & King, W. E. (2016). *Acta Materialia, 108*, 36.
90. Hojjatzadeh, S. M. H., Parab, N. D., Guo, Q., Qu, M., Xiong, L., Zhao, C., Escano, L. I., Fezzaa, K., Everhart, W., Sun, T., & Chen, L. (2020). *International Journal of Machine Tools and Manufacture, 153*, 103555.
91. Oliveira, J. P., LaLonde, A. D., & Ma, J. (2020). *Materials and Design, 193*, 108762.
92. Deng, P., Karadge, M., Rebak, R. B., Gupta, V. K., Prorok, B. C., & Lou, X. (2020). *Additive Manufacturing, 35*, 101334.
93. Svetlizky, D., Zheng, B., Vyatskikh, A., Das, M., Bose, S., Bandyopadhyay, A., Schoenung, J. M., Lavernia, E. J., & Eliaz, N. (2022). *Materials Science and Engineering A, 840*, 142967.
94. Shahrubudin, N., Lee, T. C., & Ramlan, R. (2019). *Procedia Manufacturing, 35*, 1286.
95. Oropallo, W., & Piegl, L. A. (2016). *Engineering Computations, 32*, 135.
96. Paritala, P. K., Manchikatla, S., & Yarlagadda, P. K. D. V. (2017). *Procedia Engineering, 174*, 982.
97. Chryssolouris, G., Mavrikios, D., Papakostas, N., Mourtzis, D., Michalos, G., & Georgoulias, K. (2009). *Proceedings of the Institution of Mechanical Engineers, Part B: Journal of Engineering Manufacture, 223*, 451.
98. Madakam, S., Ramaswamy, R., & Tripathi, S. (2015). *Journal of Computing and Communication, 3*, 164.
99. Gokhale, P., Bhat, O. & Bhat, S. (2018) *5*:41.
100. Kalsoom, T., Ahmed, S., Rafi-ul-Shan, P. M., Azmat, M., Akhtar, P., Pervez, Z., Imran, M. A., & Ur-Rehman, M. (2021). *Sustainability, 13*, 12506.
101. Katari, V., Goyal, S., Nigam, V., Jana, M., Maitra, A., Sharma, H., & Sahu, K. K. (2024). In M. B. Ahamed & C. M. Hussain (Eds.) *ACS symposium series* (pp. 163–204). American Chemical Society.

102. Grieco, L. A., Rizzo, A., Colucci, S., Sicari, S., Piro, G., Di Paola, D., & Boggia, G. (2014). *Computer Communications, 54*, 32.
103. Roy Chowdhury, A. (2017).
104. Kats, V., & Levner, E. (2002). *Journal of Scheduling, 5*, 23.
105. Farooq, M. S., Abdullah, M., Riaz, S., Alvi, A., Rustam, F., Flores, M. A. L., Galán, J. C., Samad, M. A., & Ashraf, I. (2023). *Sensors, 23*, 8958.
106. Nain, G., Pattanaik, K. K., & Sharma, G. K. (2022). *Journal of Manufacturing Systems, 62*, 588.
107. Cui, J., Ren, L., Mai, J., Zheng, P., & Zhang, L. (2022). *Robotics and Computer-Integrated Manufacturing, 74*, 102256.
108. Baumann, F., & Roller, D. (2017). *Journal of Manufacturing and Materials Processing, 1*, 15.
109. Rauschecker, U., Meier, M., Muckenhirn, R., Yip, A., Jagadeesan, A., & Corney, J. (2011). Int. Inf. Manag. Corp. Ltd. EChallenges E-20 11 Conf. Proc. 26.
110. Lehmhus, D., Wuest, T., Wellsandt, S., Bosse, S., Kaihara, T., Thoben, K.-D., & Busse, M. (2015). *Sensors, 15*, 32079.
111. Duan, Y., Cao, Y., & Sun, X. (2015). In *2015 IEEE/ACIS 16th International Conference on Software Engineering, Artificial Intelligence, Networking and Parallel/Distributed Computing (SNPD)* (pp. 1–6). IEEE.
112. NVIDIA. (n.d.).
113. Abou Jaoude, J., & George Saade, R. (2019). *IEEE Access, 7*, 45360.
114. Saberi, S., Kouhizadeh, M., Sarkis, J., & Shen, L. (2019). *International Journal of Production Research, 57*, 2117.
115. Trouton, S., & Killmeyer, J. (2016). 3D opportunity for blockchain. *Deloitte Insights*.
116. Holland, M., Stjepandic, J., & Nigischer, C. (2018). In *2018 IEEE International conference on engineering, technology and innovation (ICE/ITMC)* (pp. 1–8). IEEE.
117. Craveiro, F., Duarte, J. P., Bartolo, H., & Bartolo, P. J. (2019). *Automation in Construction, 103*, 251.
118. Rahmani, R., Karimi, J., Resende, P. R., Abrantes, J. C. C., & Lopes, S. I. (2023). *Machines, 11*, 522.
119. Tegethoff, T., Santa, R., Cayón, E., & Scavarda, A. (2023). *PLoS One, 18*, e0282366.
120. Yao, X., & Lin, Y. (2016). *International Journal of Advanced Manufacturing Technology, 85*, 1665.
121. Helm, J. M., Swiergosz, A. M., Haeberle, H. S., Karnuta, J. M., Schaffer, J. L., Krebs, V. E., Spitzer, A. I., & Ramkumar, P. N. (2020). *Current Reviews in Musculoskeletal Medicine, 13*, 69.
122. Acharya, S., Viswesh, P., Sridhar, M. K., Pathak, A. D., Sharma, H., Nazir, A., Kasbe, A., & Sahu, K. K. (2024). In *Nanostructured materials engineering and characterization for battery applications* (pp. 639–676). Elsevier.
123. Chua, L. O. (1997). *International Journal of Bifurcation and Chaos, 7*, 2219.
124. Ma, L., Yu, S., Xu, X., Moses Amadi, S., Zhang, J., & Wang, Z. (2023). *Materials Today Bio, 23*, 100792.
125. Jannesari Ladani, L. (2021). *Journal of Physics: Materials, 4*, 042009.
126. Leng, J., Wang, D., Shen, W., Li, X., Liu, Q., & Chen, X. (2021). *Journal of Manufacturing Systems, 60*, 119.
127. Dalenogare, L. S., Benitez, G. B., Ayala, N. F., & Frank, A. G. (2018). *International Journal of Production Economics, 204*, 383.
128. Babiceanu, R. F., & Seker, R. (2016). *Computers in Industry, 81*, 128.
129. Lee, H., & Yang, H. (2023). *Journal of Manufacturing Science and Engineering, 145*, 111008.
130. Park, J.-S., Lee, D.-G., Jimenez, J. A., Lee, S.-J., & Kim, J.-W. (2023). *Applied Sciences, 13*, 4598.
131. Martins, A., Costelha, H., & Neves, C. (2020). In *2020 IEEE international conference on autonomous robot systems and competitions (ICARSC)* (pp. 114–119). IEEE.
132. Ali, M. H., Issayev, G., Shehab, E., & Sarfraz, S. (2022). *Rapid Prototyping Journal, 28*, 1312.

133. Lu, Q., Parlikad, A. K., Woodall, P., Don Ranasinghe, G., Xie, X., Liang, Z., Konstantinou, E., Heaton, J., & Schooling, J. (2020). *Journal of Management in Engineering, 36*, 05020004.
134. Tuhaise, V. V., Tah, J. H. M., & Abanda, F. H. (2023). *Automation in Construction, 152*, 104931.
135. Wells, L. J., Camelio, J. A., Williams, C. B., & White, J. (2014). *Manufacturing Letters, 2*, 74.
136. Kumar, S., Gopi, T., Harikeerthana, N., Gupta, M. K., Gaur, V., Krolczyk, G. M., & Wu, C. (2023). *Journal of Intelligent Manufacturing, 34*, 21.
137. Gupta, N., Tiwari, A., Bukkapatnam, S. T. S., & Karri, R. (2020). *IEEE Access, 8*, 47322.
138. Elhabashy, A. E., Wells, L. J., & Camelio, J. A. (2019). *Procedia Manufacturing, 34*, 921.
139. Gao, Y., Li, B., Wang, W., Xu, W., Zhou, C., & Jin, Z. (2018). *Proceedings of the ACM on Interactive, Mobile, Wearable and Ubiquitous Technologies, 2*, 1.
140. Pasqualetti, F., Dorfler, F., & Bullo, F. (2013). *IEEE Transactions on Automatic Control, 58*, 2715.
141. Kim, J., Lee, C., Shim, H., Cheon, J. H., Kim, A., Kim, M., & Song, Y. (2016). *IFAC-PapersOnLine, 49*, 175.
142. Srai, J. S., Kumar, M., Graham, G., Phillips, W., Tooze, J., Ford, S., Beecher, P., Raj, B., Gregory, M., Tiwari, M. K., Ravi, B., Neely, A., Shankar, R., Charnley, F., & Tiwari, A. (2016). *International Journal of Production Research, 54*, 6917.
143. Monroy, S. A. S., Li, P., Fang, Y., & Loparo, K. A. (2023). *IEEE Network, 37*, 114.
144. Schmidt, C., Finsterwalder, F., Griesbaum, R., & Sehrt, J. T. (2023). *CIRP Journal of Manufacturing Science and Technology, 43*, 115.
145. Pei, E., Ressin, M., Campbell, R. I., Eynard, B., & Xiao, J. (2019). *Progress in Additive Manufacturing, 4*, 331.
146. Giunta, L., Obi, M., Goudswaard, M., Hicks, B., & Gopsill, J. (2022). *Procedia CIRP, 107*, 1150.
147. Sun, H., Pedrielli, G., Zhao, G., Zhou, C., Xu, W., & Pan, R. (2020). *Journal of Manufacturing Systems, 57*, 61.
148. Royo, J. D., Soldevila, L. M., Kayser, M., & Oxman, N. (2015). In M. R. Thomsen, M. Tamke, C. Gengnagel, B. Faircloth, & F. Scheurer (Eds.), *Model behavior* (pp. 295–302). Springer International Publishing.
149. Rauch, E., Unterhofer, M., & Dallasega, P. (2018). *Manufacturing Letters, 15*, 126.
150. Salmi, M. (2022). *IFAC-PapersOnLine, 55*, 1503.

Advancements in Additive Manufacturing for the Fabrication of Manual Gear Transmission Box

Nitin Panwar, Bhupender Singh, Rajesh Attri, and Ashok Kumar Yadav

1 Introduction

A mechanical gearbox device that joins a motor to a driven load is called a gear reducer (Fig. 1). Another name for it is a gearbox. It enables you to adjust the speed and torque between a load and a motor. A sequence of gears makes up a reduction gear assembly. The gearbox ratio is another name for the gear reduction ratio, which is the ratio of the number of teeth between the input and output shafts. The torque and speed of a mechanism's input and output axes can be changed with a gear reducer.

Gear reducers are fundamental components in a wide range of mechanical systems, facilitating the transmission of power and adjustment of rotational speeds between input and output shafts. They are ubiquitous in various industries, including automotive, aerospace, manufacturing, and robotics, where precise control of motion and torque is essential. Traditionally, the production of gear reducers has relied on conventional manufacturing methods such as milling, casting, and forging. While effective, these methods have inherent limitations, including high costs, long lead times, and limited design flexibility. Gear reducers can be categorized in several ways, with the primary classification based on the size of backlash, resulting in two main types: industrial reducers, widely used in power transmissions, and low backlash reducers, which are essential for precision mechanics. This article exclusively delves into industrial gear units. Furthermore, gear reducers can be classified

N. Panwar · B. Singh · R. Attri
Department of Mechanical Engineering, J.C. Bose University of Science and Technology, YMCA, Faridabad, India

A. K. Yadav (✉)
Department of Mechanical Engineering, Symbiosis Institute of Technology, Symbiosis International (Deemed University), Pune, India
e-mail: ashok.yadav@sitpune.edu.in

Y. Kumar Singla et al. (eds.), *Digital Manufacturing in Industry 5.0*,
https://doi.org/10.1007/978-3-031-91500-0_10

Fig. 1 Prototype of gear reducer

based on the position of the gears, such as cylindrical gear reducers and planetary reducers. Although this article solely focuses on industrial gear reducers, it is worth noting that planetary reducers are increasingly favored for their ability to achieve large transmission ratios and compact dimensions, albeit at a slightly higher cost. Cylindrical gear reducers can be further differentiated into those with parallel and coaxial shafts or nearly coaxial shafts. This paper is specifically dedicated to single-stage reducers, particularly parallel shaft reducers. The gear reducers can be further divided based on the mounting method, such as radial, axial, and radial–axial mounting. They can also be categorized based on the number of stages into single, two, three, and multistage units. However, this chapter focuses only on single-stage gear units.

Additionally, gear units can be classified based on the link between the reducer and the electric motor. They can be traditional, featuring an input shaft, or equipped with an adapter for IEC motors, enabling customers to attach an electric motor. Geared motors may come with a specific reducer electric motor or a standard IEC motor. The dimensions of the reducer also impact its classification, with small (typically with axis height $h < 100$ mm), medium ($100 < h < 500$ mm), and large reducers ($h > 500$ mm) being the standard categories. It is essential to understand that all gear units in a single reducer family should be uniform with each other. Therefore, this chapter solely focuses on single-stage gear reducers, which are typically in high demand.

Additive manufacturing (AM), or 3D printing, is essential for rapid prototyping and manufacturing in various sectors. The 3D printing market has grown annually by over 10% and is expected to continue. This review focuses on the AM of plastic materials, which contributes to worsening plastic pollution. Mitigating plastic waste production and accumulation is crucial for the 3D printing industry and can be achieved by making 3D printable plastics more sustainable. Fused deposition modeling (FDM) 3D printing method is an eco-friendly choice as it uses sustainable materials. The thermoplastic parts it creates can be recycled and reused, minimizing waste. Furthermore, some FDM filament is made from polylactide, a biodegradable material derived from corn starch, making it both renewable and compostable at the end of its life. In recent years, AM, commonly known as 3D printing, has emerged as a disruptive technology with the potential to revolutionize gear reducer production. Unlike traditional methods that rely on subtractive processes, such as cutting and shaping material, 3D printing builds objects layer by layer from digital designs. This additive approach offers unparalleled design freedom, enabling engineers to create complex geometries and intricate internal structures that were previously unattainable with conventional manufacturing techniques.

The versatility of 3D printing extends beyond design flexibility. It also offers significant advantages in terms of cost-effectiveness and rapid prototyping. By eliminating the need for specialized tooling and reducing material wastage, 3D printing enables the production of customized or low-volume parts at a fraction of the cost and time required by traditional manufacturing methods. Moreover, the ability to iterate quickly on designs and produce functional prototypes accelerates the product development cycle, enabling engineers to bring innovative solutions to market faster and more efficiently.

The scope of this study encompasses the entire lifecycle of gear reducer development, from initial design conception to performance evaluation. The gear reducer will be fabricated using a 3D printing process with ABS material, with careful consideration given to process parameters and post-processing techniques to ensure dimensional accuracy and surface finish.

However, it's important to recognize the inherent limitations of 3D printing technology, including material properties, dimensional accuracy, and surface finish. While ABS is chosen for its favorable mechanical properties, such as strength, durability, and thermal stability, it's essential to acknowledge its limitations and potential impact on gear performance. Moreover, the project will focus on a specific type of 3D printing technology and material, limiting the generalization of results to other AM methods or materials.

2 Literature Review

2.1 Evolution of 3D Printed Gears: A Historical Perspective

The journey of 3D printed gears commenced alongside the dawn of AM technologies in the 1980s. Initially relegated to prototyping and conceptual modeling, the capabilities of 3D printing expanded gradually, venturing into functional mechanical components. Early experiments, predominantly utilizing stereolithography (SLA), hinted at the potential for creating rudimentary gear prototypes. Maines et al. [1] investigated the elements of sustainability, including recycling for feedstock or the biodegradability of 3D-printed components in natural settings could be integrated as future research avenues in SLA/DLP. However, nascent material properties and process limitations restrained their practical application. According to

Kodama [2], the first instance of 3D printing using an additive process was documented. He developed a technique to create 3D models by solidifying photosensitive resins layer by layer using UV light. Polymers are still the most commonly used materials in the field of AM due to their widespread use in early 3D printers, like the one created by 3D Systems Inc. in 1986, which were used for rapid prototyping. Over the years, the range of materials [3] the paper delves into the intricacies of designing a 3D-printed single-stage gear reducer, employing a range of analytical, numerical, and experimental methodologies to assess contact stress, tooth bending strength, and impact resistance. It also provides a comprehensive overview of various 3D printing techniques relevant to gear reducer manufacturing (Fig. 2). Emphasizing the critical need to optimize gear design parameters for enhanced contact strength and reliable performance, the study underscores the widespread use of CAD modeling and finite element analysis (FEA) to optimize gear geometry for strength, efficiency, and reliability. By consolidating existing knowledge, the article presents valuable insights into the realm of 3D-printed single-stage gear reducers while outlining promising avenues for future research in this domain.

2.2 *Pioneering Research and Experimental Endeavors*

The 1990s and early 2000s marked an era of pioneering research, wherein scholars delved into the viability of 3D printing gears for functional purposes. These explorations sought to navigate the complexities of layer-by-layer deposition, grappling with challenges encompassing surface finish, dimensional precision, and mechanical resilience [25–28]. Despite obstacles, researchers showcased the promise of additive manufacturing in fashioning bespoke gear geometries endowed with intricate features.

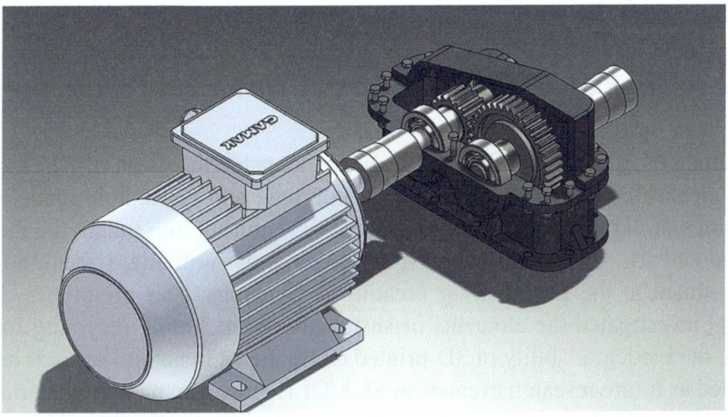

Fig. 2 Gear reducer model

2.3 Advancements in Design Methodologies and Simulation Techniques

The evolution of CAD software and simulation methodologies catalyzed a paradigm shift in the optimization of 3D-printed gears. Leveraging sophisticated CAD functionalities, engineers sculpted intricate gear profiles tailored to additive manufacturing processes. Concurrently, the integration of FEA and CFD simulations enabled predictive insights into gear performance across diverse operational scenarios, fostering the emergence of robust and efficient designs [29, 30] (Fig. 3).

2.4 Material Innovations and Process Refinements

Material innovation and process refinement emerged as linchpins in enhancing the mechanical prowess and reliability of 3D printed gears. Researchers embarked on an odyssey across a gamut of materials, spanning thermoplastics, metals, ceramics, and composites to surmount the constraints of erstwhile prototypes. Breakthroughs in powder bed fusion, directed energy deposition, and vat photo polymerization techniques heralded the advent of high-strength gears boasting superior mechanical properties and impeccable surface finish.

Fig. 3 Model of single-stage gear reducer

2.5 Industrial Integration and Commercial Viability

Recent years have witnessed a seismic surge in the industrial assimilation and commercial viability of 3D-printed gears. Enterprises spanning automotive, aerospace, healthcare, and beyond have embraced additive manufacturing with fervor, custom-tailoring gears to suit diverse applications. The advent of digital manufacturing, underpinned by on-demand production and mass customization paradigms, has galvanized the adoption of additive manufacturing solutions in gear fabrication, heralding a new epoch of innovation and efficiency in mechanical engineering.

3 Methodology

3.1 Design Process

The design process encompasses the creation of both the spur gear (Fig. 4) and the water pump assembly (Fig. 5), focusing on optimizing their performance and compatibility for 3D printing with ABS material.

For the spur gear:

1. *Requirement Analysis*: Define the specifications for the spur gear, including the number of teeth, pitch diameter, pressure angle, and module, based on the desired torque and speed reduction ratio.
2. *CAD Modeling*: Utilize CAD software to create a precise 3D model of the spur gear, incorporating the specified parameters and design constraints.
3. *FEA*: Conduct FEA simulations to evaluate the structural integrity and load-bearing capacity of the gear under various operating conditions. Analyze factors

Fig. 4 3D printed prototype of spur gear

Fig. 5 3D-printed prototype of water pump

such as stress distribution, deformation, and fatigue life to optimize the gear's design for strength and durability.

4. *Water Pump Integration*: Design compatible mounting features on the spur gear to allow for seamless connection with the water pump assembly. Ensure proper alignment and coupling between the input and output shafts to facilitate power transmission.

For the water pump assembly:

1. *Functional Requirements*: Define the operational requirements and performance objectives for the water pump assembly, including flow rate, pressure, and efficiency.
2. *CAD Modeling*: Create a detailed 3D model of the water pump assembly, incorporating the necessary components such as impeller, housing, inlet/outlet ports, and mounting brackets.
3. *Integration with Spur Gear*: Design mounting interfaces on the water pump assembly to enable attachment to the input and output shafts of the spur gear. Ensure proper alignment and orientation to facilitate fluid flow and power transmission between the components.

3.2 Material Selection and 3D Printing

Select acrylonitrile butadiene styrene (ABS) material for 3D printing both the spur gear and water pump assembly, considering its favorable mechanical properties, including strength, durability, and thermal stability (Fig. 6).

Fig. 6 Creality Ender 3

1. *Material Properties*: Evaluate the mechanical, thermal, and chemical properties of ABS material to ensure compatibility with the intended application and operating environment.
2. *Printing Parameters*: Determine the optimal printing parameters, including layer height, infill density, print speed, and nozzle temperature, to achieve the desired mechanical properties and surface finish.
3. *Printing Process*: Utilize a suitable 3D printing technology, such as fused deposition modeling (FDM), to fabricate the spur gear and water pump assembly

layer by layer. Monitor the printing process to ensure accuracy, dimensional stability, and adherence to design specifications.

4. *Post-Processing*: Implement post-processing techniques such as support removal, sanding, and surface smoothing to improve the aesthetics and functionality of the printed parts. Perform any necessary secondary operations, such as drilling or tapping, to prepare the components for assembly.

3.3 Design Consideration

Economical and effective manufacturing requires careful consideration of a number of structural design factors. All of these are applicable to both subassemblies and the entire structure, and many of them are also applicable to alternative connecting techniques.

1. The gadget ought to be compatible with regional production capacity.
2. Low-cost production techniques and materials should be used in the attachment.
3. Low-income groups should be able to afford it and use it to meet their fundamental mechanical power needs.
4. It need to be easy to produce, run, maintain, and fix.
5. It should be as multifunctional as possible, supplying electricity for tiny equipment employed in rural industries as well as a variety of agricultural tools.
6. It should utilize materials and skills that are available locally, and standard parts should be incorporated whenever feasible.
7. The device must be easily adaptable, without requiring any permanent structural changes.
8. Heavyweight should be minimized, as durability is a key factor.

3.3.1 Design Procedure

1. Definition of problem
2. Synthesis
3. The examination of forces
4. Material selection.
5. Mode of failure identification
6. Choosing the safety factor
7. Establishing the dimensions
8. Adjusting the dimensions
9. Drawing preparation
10. Design report preparation

3.3.2 Design Consideration

1. Strength
2. Rigidity
3. Reliability
4. Safety
5. Cost
6. Weight
7. Ergonomics
8. Aesthetics
9. Manufacturing considerations
10. Assembly considerations
11. Conformance to standards
12. Friction and wear
13. Life
14. Vibrations
15. Thermal considerations

3.3.3 Ergonomics Consideration in Design

1. Ergonomics refers to the examination of the interaction between humans, machines, and the work environment.
2. It's objective is to reduce both physical and mental strain on the user.
3. Key aspects addressed in ergonomics include
   (a) The interaction between the user and the machine.
   (b) The surrounding work environment.
   (c) The anatomy and posture of the human body while operating the machine.
   (d) The energy expenditure associated with hand and foot movements.

3.3.4 Manufacturing Consideration in Design

1. Use standard parts
2. Minimize the variety of parts
3. Reduce the overall number of parts in a product.
4. Make use of modular design v. Create multipurpose components
5. Create components that are multipurpose
6. Choose the least expensive material
7. Create components that are simple to manufacture
8. Form the components to minimize operations

4 Design of Components

- 12V DC Motor 100 RPM—There are numerous robotic applications for this 12-volt DC motor (Fig. 7) with a speed of 100 rpm, including all-terrain robots. These motors are basic DC motors with shaft gears to get the best performance characteristics. Because their shaft passes through the middle of their gearbox assembly (Fig. 8), they are referred to as center shaft DC-geared motors.
- 12V DC Power Adapter: An apparatus that provides electrical energy to a load is a 12VDC power supply. In other words, the main function of a power supply is to transform electric current from the source into the necessary voltage, frequency, and current that drives the load.
- Shaft
- Frame (Acrylic)
- Clamp
- Fasteners

4.1 Gears Dimensions

4.1.1 Big Gear

Outer dia—60 mm
Inner dia—6 mm
No. of teeth—38
Pressure angle—20
Width—12 mm

4.1.2 Small Gear

Outer dia—38.5 mm
Inner dia—8 mm
No. of teeth—29
Pressure angle—20
Width—12 m

Fig. 7 DC motor

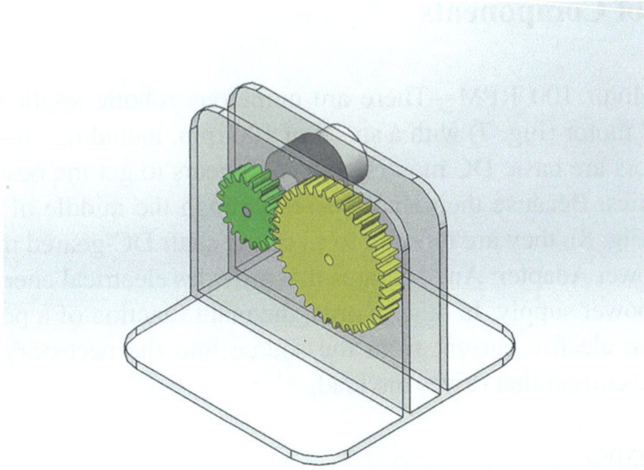

Fig. 8 Initial visualization of gear reducer

5 Results and Discussion

5.1 *Comparative Analysis of Material Performance*

The comparative analysis of material performance based on FEA results revealed the following observations:

1. *ABS*: ABS demonstrated superior mechanical properties compared to nylon and PLA, exhibiting lower stress concentrations, minimal deformation, and extended fatigue life. Its high tensile strength, impact resistance, and wear characteristics make it an ideal choice for gear applications subjected to demanding operating conditions.
2. *PLA*: PLA showed relatively high stress concentrations, significant deformation, and reduced fatigue life compared to ABS and nylon. Its lower tensile strength and impact resistance limit its suitability for gear applications requiring high mechanical performance and durability.

5.2 *Interpretation of Results*

The interpretation of the results sheds light on the key findings and their implications for the project objectives and overall performance of the gear reducer and water pump assembly.

5.2.1 Spur Gear Performance

The performance analysis of the 3D-printed spur gear revealed promising results in terms of torque transmission, efficiency, and durability. The torque–speed characteristics demonstrated the gear's ability to effectively transmit torque across a range of operating speeds, with minimal energy losses and mechanical inefficiencies. This indicates that the spur gear design, optimized for 3D printing with ABS material, was successful in achieving the desired performance metrics.

5.3 Comparison of PLA and ABS Materials

The comparative analysis of PLA and ABS materials provides valuable insights into their mechanical performance and suitability for 3D-printed gear applications.

5.3.1 Mechanical Properties

PLA: The FEA simulations revealed that PLA exhibits relatively low mechanical properties compared to ABS. PLA demonstrated higher stress concentrations, significant deformation, and reduced fatigue life under simulated loading conditions. These characteristics can be attributed to PLA's lower tensile strength and impact resistance, which limit its suitability for high-stress applications.

ABS: In contrast, ABS showcased superior mechanical properties, including lower stress concentrations, minimal deformation, and extended fatigue life. Its higher tensile strength, impact resistance, and durability make it a preferred choice for gear applications subjected to demanding operating conditions. The FEA results validate ABS as a robust and reliable material option for 3D-printed gears, offering enhanced performance and longevity compared to PLA.

5.3.2 Impact Resistance and Wear Characteristics

PLA: While PLA demonstrated moderate impact resistance and wear characteristics, its mechanical performance was overshadowed by ABS. PLA's inherent brittleness and lower toughness contribute to reduced impact resistance, making it susceptible to fractures and failure under dynamic loading conditions.

ABS: ABS exhibited excellent impact resistance and wear characteristics, mitigating the risk of premature failure and damage in gear applications subjected to impact and abrasion. Its superior toughness and resilience enable ABS to withstand high-stress environments, ensuring long-term reliability and performance stability.

5.3.3 Manufacturing Considerations

PLA: PLA is known for its ease of printing and biodegradability, making it a popular choice for hobbyists and beginners in 3D printing. However, its limited mechanical properties and poor temperature resistance restrict its applicability in demanding engineering applications.

ABS: ABS offers superior mechanical properties and temperature resistance compared to PLA, making it well suited for industrial-grade 3D-printing applications. Its compatibility with post-processing techniques such as vapor smoothing and chemical bonding further enhances its versatility and utility in manufacturing complex gear geometries.

5.3.4 Environmental Impact

PLA: PLA is derived from renewable resources such as corn starch or sugarcane, making it environmentally friendly and biodegradable. Its eco-friendly characteristics appeal to sustainability-conscious users seeking greener alternatives in 3D printing.

ABS: ABS is derived from petroleum-based sources, resulting in a higher environmental footprint compared to PLA. While ABS is not biodegradable, efforts are underway to develop eco-friendly alternatives and recycling methods to minimize its environmental impact.

5.3.5 Evaluation

An evaluation of the study assesses the extent to which the objectives have been achieved and the success of the gear reducer and water pump assembly in meeting performance targets.

Objective Achievement: The project successfully achieved its objectives of designing, fabricating, and evaluating a 3D printed spur gear and water pump assembly. The gear reducer demonstrated efficient torque transmission, high-speed reduction, and satisfactory durability, meeting the specified performance requirements. Similarly, the water pump assembly exhibited efficient fluid handling and power transmission capabilities, validating its suitability for integration with the spur gear.

Performance Metrics: The performance metrics obtained from torque-speed tests, efficiency analysis, and flow rate and pressure measurements provided quantitative evidence of the gear reducer and water pump assembly's performance. Comparisons with theoretical calculations and industry standards validated the accuracy and reliability of the design and manufacturing processes, affirming the project's success in achieving its objectives.

6 Conclusions

This project has successfully demonstrated the viability of AM technology, specifically 3D printing with ABS material, in producing functional gear reducers and water pump assemblies. Through meticulous design, precise fabrication, and comprehensive testing, the project achieved its objectives of optimizing gear performance, fabricating components with precision, and validating their functionality.

Key Achievements

1. *Design Optimization*: The iterative design process resulted in a single-stage spur gear optimized for 3D printing, ensuring structural integrity, efficiency, and durability through FEA simulations.
2. *Fabrication Excellence*: Utilizing AM processes, the spur gear and water pump assembly were fabricated with precision and accuracy, achieving excellent dimensional stability and surface finish.
3. *Performance Validation*: Comprehensive testing validated the gear reducer's efficient torque transmission, high-speed reduction, and the water pump assembly's reliable fluid handling capabilities under various operating conditions.

Implications and Future Directions

1. *Industry Adoption*: The project highlights the potential of AM in revolutionizing gear reducer production and mechanical system integration, offering versatility, cost-effectiveness, and design flexibility.
2. *Research Opportunities*: Opportunities for further research include exploring alternative materials, advanced manufacturing techniques, and design optimization strategies to enhance performance and reliability.

In conclusion, this project showcases the transformative impact of AM technology in mechanical engineering. As the field continues to evolve, the insights gained from this project will drive future advancements, fostering innovation and excellence in gear technology and beyond.

7 Future Scope

The successful execution of this project opens up several avenues for future research, innovation, and application in the field of AM and mechanical engineering.

1. Advanced Materials Exploration: While ABS material proved suitable for the fabrication of functional gear reducers and water pump assemblies, further exploration of advanced materials holds promise for enhancing mechanical properties, thermal stability, and chemical resistance. Materials such as nylon, carbon fiber-reinforced composites, and metal powders offer potential for improved performance and durability in demanding applications.

2. Design Optimization and Customization: Continued research into design optimization techniques and customization capabilities will enable the development of tailored solutions for specific applications and operating conditions. Advanced CAD software, topology optimization algorithms, and generative design approaches can facilitate the creation of lightweight, high-performance components with complex geometries optimized for AM.
3. Process Innovation and Scalability: Advancements in AM technologies, including faster printing speeds, larger build volumes, and multi-material capabilities, will drive innovation and scalability in gear reducer production. Emerging techniques such as continuous liquid interface production (CLIP), binder jetting, and directed energy deposition (DED) offer potential for higher throughput, improved surface finish, and enhanced part quality.
4. Integrated System Development: The integration of gear reducers and mechanical components into complex systems presents opportunities for interdisciplinary collaboration and innovation. Future research could focus on the development of integrated systems incorporating sensors, actuators, and control algorithms for autonomous operation, predictive maintenance, and real-time performance monitoring.
5. Industry 4.0 Integration: The convergence of AM with Industry 4.0 technologies, such as Internet of Things (IoT), artificial intelligence (AI), and digital twin simulation, will revolutionize manufacturing processes and supply chain management. Smart factories equipped with AM capabilities will enable on-demand production, mass customization, and decentralized manufacturing networks.
6. Sustainability and Circular Economy: As sustainability becomes a priority in manufacturing, research into environmentally friendly materials, energy-efficient processes, and closed-loop production systems will gain prominence. AM offers opportunities for reducing material waste, energy consumption, and carbon footprint through on-demand production, localized manufacturing, and recycling of post-consumer materials.

References

1. Maines, E. M., Porwal, M. K., Ellison, C. J., & Reineke, T. M. (2021). Sustainable advances in SLA/DLP 3D printing materials and processes. *Green Chemistry, 23*(18), 6863–6897.
2. Kodama, H. (1981). Automatic method for fabricating a three-dimensional plastic model with photo-hardening polymer. *Review of Scientific Instruments, 52*(11), 1770–1773.
3. Singh, T., Pattnaik, P., Aherwar, A., Ranakoti, L., Dogossy, G., & Lendvai, L. (2022). Optimal design of wood/rice husk-waste-filled PLA biocomposites using integrated CRITIC–MABAC-based decision-making algorithm. *Polymers, 14*(13), 2603.
4. Ramaswamy, R., Gurupranes, S. V., Kaliappan, S., Natrayan, L., & Patil, P. P. (2022). Characterization of prickly pear short fiber and red onion peel biocarbon nanosheets toughened epoxy composites. *Polymer Composites, 43*(8), 4899–4908.
5. Kumar, U., Padalia, D., Bhandari, P., Kumar, P., Ranakoti, L., Singh, T., & Lendvai, L. (2022). Fabrication of europium-doped barium titanate/polystyrene polymer nanocomposites using ultrasonication-assisted method: Structural and optical properties. *Polymers, 14*(21), 4664.

6. AP, A., Kaliappan, S., & Patil, P. (2022). Mechanical, fracture toughness, and dynamic mechanical properties of twill weaved bamboo fiber-reinforced Artocarpus heterophyllus seed husk biochar epoxy composite. *Polymer Composites, 43*(11), 8388–8395.
7. Vinayaka, N., Nagabhooshanam, N., Balaji, D., Verma, R., & Patil, P. P. (2022). Effect of silanized stacked K evlar fiber/b rass 270 wire mesh on mechanical and impact toughness of cubic boron nitride–epoxy composite. *Polymer Composites, 43*(8), 5159–5167.
8. Nagabhooshanam, N., Kumar, A., Rao, P. K. V., Patil, P. P., & Kala Bharathi, B. V. V. L. (2022). Preparation and characterization of opuntia-cladode fiber and citron peel biochar toughened epoxy biocomposite. *Polymer Composites, 43*(9), 6571–6577.
9. Subbiah, R., Kaliappan, S., V, B., & Patil, P. P. (2022). Effect of nanosilica on mechanical, thermal, fatigue, and antimicrobial properties of cardanol oil/sisal fiber reinforced epoxy composite. *Polymer Composites, 43*(11), 7940–7951.
10. Patil, P. P. (2022). Effect of black rice husk ash biosilica on mechanical, wear, and fatigue behavior of stacked aloevera/roselle and glass fiber reinforced epoxy composite. *Polymer Composites, 43*(10), 7335–7344.
11. Singh, T., Patnaik, A., Ranakoti, L., Dogossy, G., & Lendvai, L. (2022). Thermal and sliding wear properties of wood waste-filled poly (lactic acid) biocomposites. *Polymers, 14*(11), 2230.
12. Senthil Kumar, S., Anil Kumar, T. C., Chokkalingam, M., Kaliappan, S., Boopathi, S., Srinivas, C., & Patil, P. P. (2022). Mechanical, fracture toughness, and fatigue behavior of spinifex littoreus fiber on echinoidea-spike toughened epoxy composite. *Polymer Composites, 43*(4), 2329–2340.
13. Saxena, A., Norton, B., Goel, V., & Singh, D. B. (2022). Solar cooking innovations, their appropriateness, and viability. *Environmental Science and Pollution Research, 29*(39), 58537–58560.
14. Rajan, R., Tyagi, Y. K., Pruncu, C. I., Kulshreshtha, S., Ranakoti, L., & Singh, T. (2022). Tribological performance evaluation of slag waste filled phenolic composites for automotive braking applications. *Polymer Composites, 43*(10), 7118–7129.
15. Singh, Y., Abd Rahim, E., Singh, N. K., Sharma, A., Singla, A., & Palamanit, A. (2022). Friction and wear characteristics of chemically modified mahua (madhuca indica) oil based lubricant with SiO2 nanoparticles as additives. *Wear, 508*, 204463.
16. Donthu, N., Kumar Badhotiya, G., Kumar, S., Soni, G., & Pandey, N. (2022). A retrospective overview of Journal of Enterprise Information Management using bibliometric analysis. *Journal of Enterprise Information Management, 35*(2), 504–529.
17. Rajawat, A. S., Singh, S., Gangil, B., Ranakoti, L., Sharma, S., Asyraf, M. R. M., & Razman, M. R. (2022). Effect of marble dust on the mechanical, morphological, and wear performance of basalt fibre-reinforced epoxy composites for structural applications. *Polymers, 14*(7), 1325.
18. Aniza, R., Chen, W. H., Yang, F. C., Pugazhendh, A., & Singh, Y. (2022). Integrating Taguchi method and artificial neural network for predicting and maximizing biofuel production via torrefaction and pyrolysis. *Bioresource Technology, 343*, 126140.
19. Gariya, N., Kumar, P., & Singh, T. (2023). Experimental study on a bending type soft pneumatic actuator for minimizing the ballooning using chamber-reinforcement. *Heliyon, 9*(4), e14898.
20. Moj, K., Owsiński, R., Robak, G., & Gupta, M. K. (2023). Determination of cell size/structures and mechanical properties of additively manufactured metallic components using X-ray technique. *Rapid Prototyping Journal, 29*(10), 2039–2060.
21. Sharma, S. K., Sharma, R. C., Choi, Y., & Lee, J. (2023). Modelling and dynamic analysis of adaptive neuro-fuzzy inference system-based intelligent control suspension system for passenger rail vehicles using magnetorheological damper for improving ride index. *Sustainability, 15*(16), 12529.
22. Derda, S., Karolczuk, A., Robak, G., Prażmowski, M., Paul, H., Łagoda, T., & Gupta, M. K. (2023). Impact of the interface on the fatigue life of steel-based explosively welded heterostructured plates. *Archives of Civil and Mechanical Engineering, 23*(3), 191.

23. SV, G., Kumar, A., & Patil, P. P. (2023). Effect of corn husk fiber length and diameter on load bearing, fatigue, and DMA properties of biosilica toughened epoxy composite. *Polymer Composites, 44*(6), 3469–3476.
24. Gupta, M. K., & Korkmaz, M. E. (2023). A conceptual framework for sustainability impact assessment in machining bohler tool steel under hBN-enriched nano cutting fluids environment. *Sustainable Materials and Technologies, 37*, e00669.
25. Tiwari, S., Amarnath, M., & Gupta, M. K. (2023). Synthesis, characterization, and application of Al2O3/coconut oil-based nanofluids in sustainable machining of AISI 1040 steel. *Journal of Molecular Liquids, 386*, 122465.
26. Josphineleela, R., Raja Rao, P. B. V., Shaikh, A., & Sudhakar, K. (2023). A multi-stage faster RCNN-based iSPLInception for skin disease classification using novel optimization. *Journal of Digital Imaging, 36*(5), 2210–2226.
27. Sankaranarayanan, S., Gunasekaran, E., & Rao, S. G. (2023). A novel survival analysis of machine using fuzzy ensemble convolutional based optimal RNN. *Expert Systems with Applications, 234*, 120966.
28. Natrayan, L., Surakasi, R., Kaliappan, S., Patil, P. P., Saravanan, A., & Siranjeevi, R. (2023). Explore the elimination of toxic dye using bio-synthesized cerium oxide nanoparticles derived from Morinda citrifolia leaf extracts. *Surfaces and Interfaces, 41*, 103151.
29. Ross, N. S., Mashinini, P. M., Rai, R., & Gupta, M. K. (2023). Carbon nano dots mixed rice bran oil as a cutting fluid for enhanced lubrication/cooling in milling of additively manufactured 316 stainless steel. *Journal of Molecular Liquids, 391*, 123200.
30. Sahu, K., Sharma, S. C., & Tomar, A. K. (2023). Analysis of MR fluid lubricated slot entry hybrid conical journal bearing with texturing arrangements. *Tribology International, 188*, 108788.
31. Singh, T., Singh, V., Ranakoti, L., & Kumar, S. (2023). Optimization on tribological properties of natural fiber reinforced brake friction composite materials: Effect of objective and subjective weighting methods. *Polymer Testing, 117*, 107873.
32. Kaliappan, S., Velumayil, R., Natrayan, L., & Pravin, P. (2023). Mechanical, DMA, and fatigue behavior of Vitis vinifera stalk cellulose Bambusa vulgaris fiber epoxy composites. *Polymer Composites, 44*(4), 2115–2121.
33. Yevdokymov, O., Kolesnyk, V., Peterka, J., Vopat, T., Gupta, M. K., Lisovenko, D., & Dovhopolov, A. (2023). Pareto analysis of machining factors significance when turning of nickel-based superalloy Inconel 718. *Metals, 13*(8), 1354.
34. Prabhakar, B., Kamal, S. R., Yogesh, K. P., Diwakar, P., Lalit, R., & Tej, S. (2023). Design modifications in micro pin fin configuration of microchannel heat sink for single phase liquid flow: A review. *Journal of Energy Storage, 66*, 107548.
35. Saxena, A., Cuce, E., Singh, D. B., Sethi, M., Cuce, P. M., Sagade, A. A., & Kumar, A. (2023). Experimental studies of latent heat storage based solar air heater for space heating: A comparative analysis. *Journal of Building Engineering, 69*, 106282.
36. Korkmaz, M. E., Gupta, M. K., Günay, M., Boy, M., Yaşar, N., Demirsöz, R., et al. (2023). Comprehensive analysis of tool wear, surface roughness and chip morphology in sustainable turning of Inconel-601 alloy. *Journal of Manufacturing Processes, 103*, 156–167.
37. Duan, Z., Wu, T., Glowacz, A., Gupta, M. K., & Królczyk, G. (2023). Analysis of the line contact tribo-lubrication pair and failure mechanism under the extreme conditions. *Tribology International, 185*, 108505.
38. Singh, T., Aherwar, A., Ranakoti, L., Bhandari, P., Singh, V., & Lendvai, L. (2023). Performance optimization of lignocellulosic fiber-reinforced brake friction composite materials using an integrated CRITIC-CODAS-based decision-making approach. *Sustainability, 15*(11), 8880.
39. Abbas, A. T., Sharma, N., Alsuhaibani, Z. A., Sharma, V. S., Soliman, M. S., & Sharma, R. C. (2023). Processing of Al/SiC/Gr hybrid composite on EDM by different electrode materials using RSM-COPRAS approach. *Metals, 13*(6), 1125.

40. Sivalingam, V., Zhou, Q., & Selvam, B. (2023). A mathematical approach of evaluating sustainability indicators in milling of aluminium hybrid composite by different eco-friendly cooling strategies. *Sustainable Materials and Technologies, 36*, e00605.
41. Korkmaz, M. E., Gupta, M. K., Ross, N. S., & Sivalingam, V. (2023). Implementation of green cooling/lubrication strategies in metal cutting industries: A state of the art towards sustainable future and challenges. *Sustainable Materials and Technologies, 36*, e00641.
42. Gnanavelbabu, A., Prahadeeswaran, M., Rai, R., Ross, N. S., & Gupta, M. K. (2023). Tribo-corrosive wear behaviour of squeeze-casted Mg/TiN/hBN composite under different ageing temperature. *Tribology International, 187*, 108748.
43. Ross, N. S., Rai, R., Ananth, M. B. J., Srinivasan, D., Ganesh, M., Gupta, M. K., et al. (2023). Carbon emissions and overall sustainability assessment in eco-friendly machining of Monel-400 alloy. *Sustainable Materials and Technologies, 37*, e00675.
44. Chand, R., Sharma, V. S., Trehan, R., & Gupta, M. K. (2023). Numerical investigations on mechanical properties of bio-inspired 3D printed geometries using multi-jet fusion process. *Rapid Prototyping Journal, 29*(6), 1212–1229.

Next-Gen Additive Manufacturing: AI-Powered Design, Production, and Control in Industry 5.0

Denise Bellisario, Barbara Martini, Leandro Iorio, and Davide Berardi

1 Introduction

Currently, the next generation of additive manufacturing (AM) is strictly related to new developments in the field of artificial intelligence (AI) and its integration into manufacturing processes. In fact, AI represents one of the most extraordinary technological revolutions of our time, continuously redefining the boundaries of what is possible in Industry 5.0. In fact, Industry 5.0, which has significantly changed industrial processes through improved automation, data analysis, and system interconnection, has made the application of AI in manufacturing a primary focus [1–3]. The AI-driven innovations in manufacturing are fundamental in the development of next-generation industries [4]. The concept of AI-driven innovation refers to the adoption and integration of AI-based technologies to improve and innovate processes, products, and services. As industries seek to increase productivity and cost efficiency, the integration of AI technologies has emerged as a transformative solution [5]. AI-powered digital transformation is more than just a catchphrase; it's a potent force that drives efficiency, creativity, innovation, and competitiveness across a variety of industries [6]. The integration of AI into manufacturing processes marks a significant shift from traditional manufacturing methods to a more advanced, data-driven approach. Specifically, the AI can enhance the efficiency of manufacturing systems in several areas such as predictive maintenance, waste reduction and sustainability, production planning and resource optimization, human–robot collaboration, quality control, enhancing safety for workers and the public, quality control, and environmental footprint [7]. AI-powered systems can make decisions

D. Bellisario (✉) · B. Martini · L. Iorio · D. Berardi
Department of Engineering and Science, Faculty of Technological Science and Innovation, Mercatorum University, Rome, Italy
e-mail: denise.bellisario@unimercatorum.it; barbara.martini@unimercatorum.it; leandro.iorio@unimercatorum.it; davide.berardi@unimercatorum.it

Y. Kumar Singla et al. (eds.), *Digital Manufacturing in Industry 5.0*, https://doi.org/10.1007/978-3-031-91500-0_11

independently, optimize themselves, respond automatically to changes in production schedules and machine malfunctions, replace machine components automatically, and send alerts for any uncontrollable situations. For this reason, in smart factories, IIoT, IoT (Internet of Things), and AI are used to integrate everything from material selection to product delivery. Some businesses are also combining several manufacturing facilities to boost total yield [8]. While there are many opportunities for growth within it, there are also several difficulties and issues that have been noted. According to multiple assessments, systems in the industrial setting must be improved to handle every aspect of data confidence, integrity of data, and model consistency [9].

At the same time, the development of additive manufacturing technologies in recent decades has revolutionized the manufacturing sector [10]. AM processes represent a reversal in the more traditional concepts of manufacturing, thanks to their ability to produce three-dimensional objects by adding layers of material successively. This approach allows more precision and customization in products that are difficult to achieve with traditional production methods. Certainly, the integration of artificial intelligence has further amplified the potential of this technology, improving not only efficiency but also the quality and flexibility of production processes. In fact, AI and machine learning algorithms have proven to be very effective in improving the design and optimization of 3D printing models: for example, algorithms can analyze CAD (Computer-Aided Design) models and suggest design changes to optimize the geometry of the components, minimizing the material used without compromising the robustness and functionality of the final product [11]. This process not only contributes to increase efficiency in production but also allows the creation of complex geometries that would be impossible to produce with traditional production methods, such as lightweight and strong lattice structures. In addition to the design phase, AI plays a crucial role in optimizing printing parameters and quality control during the production process, as modern AI-based systems can monitor and adjust printer parameters, such as temperature, extrusion speed, and the quality of the material used, in real time [12]. This dynamic control is essential to ensure that each layer of material is applied accurately, avoiding defects and improving the overall quality of the final product.

Examining how AI and AM might work together in design could improve both fields and expand the realm of what is conceivable in terms of structural design and construction. Research on this subject is crucial because it supports advances in technology that promote architecture, effectiveness, customization and individualization, optimized for performance searches, and utilization of resources for enhanced sustainability and productivity, as well as industry applications [13].

The use of computer vision systems in combination with AI allows continuous monitoring of the printing process, identifying anomalies or defects and making immediate adjustments, to maintain high-quality standards and reduce waste. In addition, AI contributes significantly to manage and predict the maintenance of 3D printers. Using predictive algorithms, it is possible to analyze the operational data of the printers, to identify system wear or malfunctions before critical failures occur. This proactive approach allows preventive maintenance planning, keeping the printers in optimal condition and minimizing downtime [14]. This consequently improves the overall efficiency of the production process and reduces the costs associated with sudden failures. The applications of AI in AM are not only limited to industrial production but also extend to sectors that require high customization and on-demand production, such as in the medical sector, where AI-supported AM systems allow the fabrication of customized prostheses and custom-made implants, designed to perfectly adapt to the specific anatomical needs of patients [15]. AI algorithms can also analyze medical scans and images to create detailed and precise models, improving the effectiveness of devices created by AM techniques and overall, the patients' quality of life [16].

In addition, AI is essential for integrating AM processes with more traditional manufacturing processes that already exist, facilitating a smooth transition between traditional and AM. AI-based systems can coordinate AM operations with other production steps already in place in industrial systems. This can combine the advantages of 3D printing with conventional manufacturing capabilities and methods, thus creating hybrid manufacturing that leverages the strengths of both technologies [17]. Additionally, while customization often seems resource-intensive, AI solutions ensure that adjusting resources, control, and optimization to your specific needs doesn't affect the productivity or scalability of your manufacturing process [5].

To break through the next generation of AM evolutions, this chapter aims to illustrate the development of AM assisted by AI systems both in the field of research and industrial level. First, the new AM technologies AI-driven are addressed through an overview that starts from the framework of recent AM developments and AI-driven systems applicable to it, and then deals specifically with the data flow in the process and value chain in AM technologies. Specifically, AI enhances Additive Manufacturing from the empowered design phase—which includes methodology, materials, processes and equipment, and structural design—to the optimized production phase, and finally to the enhanced quality and control phase, as illustrated in Fig. 1. The state-of-the-art development and technological roadmap have been illustrated in this chapter to show the obstacles and the perspectives for AM toward intellectualization and industrialization. Finally, a simple case study was conducted to underline the potential of commercial AI systems to improve production phase of AM .

2 Foundations for a New Generation of Next-Gen AM Systems

2.1 Development and Challenges in Additive Manufacturing

AM is the manufacturing technique of linking materials (such as liquid, powder, wire, or sheet) to create three-dimensional products from digital models, often layer by layer. AM products start with an appropriately designed computer-aided design

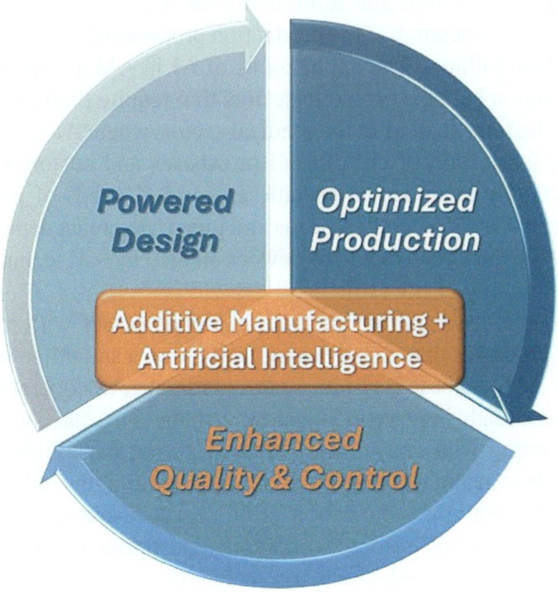

Fig. 1 Schematic overview of the AM phases enhanced by AI

(CAD). After the CAD software's 3D model is divided into layers, the AM printer receives the layer details in a specific format (often standard tessellation language, or STL). To generate three-dimensional elements relying on the STL data, the printer finally builds on the material layer by layer [18]. Consequently, AM techniques provide several unique capabilities that traditional manufacturing procedures are unable to match. Conventional manufacturing procedures typically use joined or assembled elements to create intricate geometries and internal complexities in functional components, which eventually can fail structurally at the joints. By creating extremely complicated parts as a single item, AM procedures get rid of this drawback or restriction, and already 10 years ago this was emphasized by several authors [19]. An additional important aspect of the AM method is the direct fabrication mode, which eliminates the need for tooling and production steps. By employing functionally graded materials and different material compositions, it can generate multi-functionality features with improved mechanical properties [20].

Although AM has only been around since the 1980s, it has a lot of promise to further the manufacturing technology revolution in the future. The manufacturing industry is currently experiencing a severe lack of innovation capability, which is preventing the development of new goods and ultimately limiting industry growth [21]. AM offers enormous potential for the manufacturing transition and the creation of new products, which might contribute to the manufacturing sector's reconfiguration and intelligent manufacturing's advancement [22, 23].

The primary benefit of AM manufacturing is the versatility that it affords designers in terms of material choice and structural arrangement, allowing for controlled shape and personalized performance. By establishing the data flow from resources to final uses, AM effectively achieved the digitization of the design and production process. For this goal, research is also oriented in developing a decision support system for the right choice of the required techniques [24].

Micro to larger intricate structures can be printed using AM methods with improved mechanical properties, reduced defect rates, and adequate precision. AM utilizes a variety of methods to apply the layer-by-layer material addition concept. Binder jetting (BJ), directed energy deposition (DED), material extrusion (ME), material jetting (MJ), powder bed fusion (PBF), sheet lamination (SL), and vat polymerization are the seven technique types into which the ISO/ASTM 52900 standard divides AM processes [25]. Some of these methods—such as FDM (fused deposition modeling), LOM (laminated object manufacturing), SLA (stereolithography), SLS (selective laser sintering), LENS (Laser Engineered Net Shaping), and inkjet 3D printing—are well-executed, whereas others remain only utilized by a small number of scientists and engineers. Alongside these well-established techniques, there are also rapidly growing techniques such as laser direct metal deposition (LDMD), selective electron beam melting (SEBM), laser-induced forward transfer (LIFT), laser solid forming (LSF), and wire-arc AM (WAAM).

Generally, the requirements and initial materials utilized determine which procedures are chosen [26].

All 3D printing methodologies differ in the state of the initial material, the type of deposition principle, and the type of materials that can be processed. In the same way, the different architecture of the equipment often differentiates upstream the type, the dimensions, the quality, and the precision of the products that can be produced, as well as the number and the types of parameters that can be controlled during production. The type of technique, material, and process equipment utilized affects the shape, the accuracy, the dimensions, and the durability of the printed part [27, 28].

Arguably, the primary benefit of AM is its ability to produce geometrically challenging products that were previously impossible to produce using conventional methods without requiring a complex machine setup or final assemblage [29]. Furthermore, it can already be claimed that it performs adequately when producing tiny quantities [30]. The major benefits of AM regard mainly the pre-process phase, as in design flexibility, in the process phase, as in the possibility to print thin walls, and also in economic and sustainable values as the cost effectiveness for complex parts or the environmentally friendly approach in minimizing the material waste, as schematized in Fig. 2. On the other hand, some drawbacks are also present in AM production. In fact, due to its considerable capital investment, lack of performance comparable to other traditional methods, and relatively imprecisely produced parts, AM remains linked with certain limitations and is not yet regarded as a fully feasible technique (Fig. 2). When compared to traditional manufacturing methods like machining, the surface finish of the components created by additive manufacturing is quite high, but it has limitations. These restrictions have to do with the material's rapid cooling, which causes strains and distortions that cause issues in the parts of the piece that are heavily strained or loaded cyclically. The orientation of the

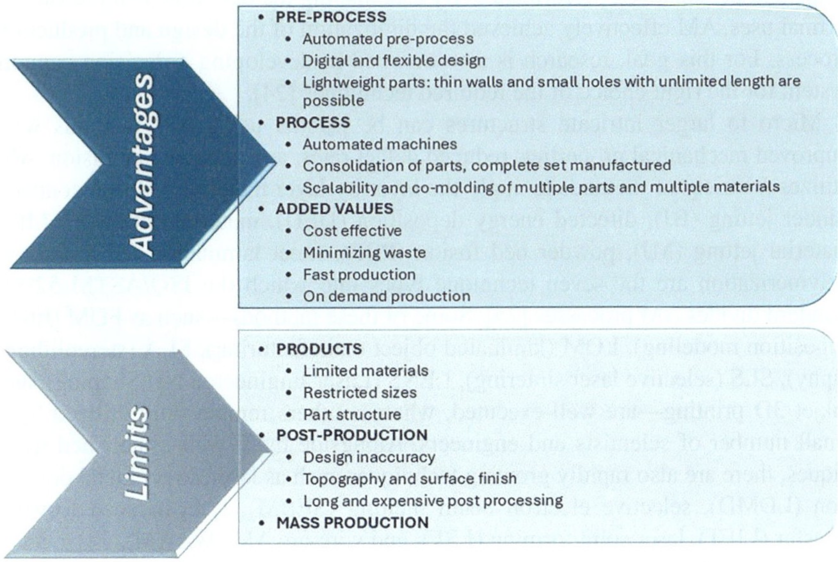

Fig. 2 Current advantages and limitations of AM processes

component, the layer's thickness, and the direction of the material's deposition all affect the surface quality [31].

Moreover, nearly all industrial AM processes are limited by the type of manufacturing material used. Consequently, additional treatments are needed to enhance the surface properties, particularly for applications in biology [32]. According to Tofail et al., "metrology" and "materials" are two important technology enablers for additive manufacturing. To accomplish the usefulness of the items made using additive manufacturing, they suggest addressing the associated problems [33]. To do this, they outline the research and technological difficulties related to the products' production, materials, and metrology, which will decide whether they are accepted by consumers and whether their commercial potential is realized. Thus, it is feasible to draw attention to the issues with tolerance to regulate geometric fluctuations, which stand in the way of accurate simulations and predictive models.

Overcoming the highlighted limits (Fig. 2) is a continuous domain of research, both at a scientific and industrial level. The latest trends in the development and improvement of AM processes are aimed at both minimizing limits but also at meeting the challenges and new opportunities within Industry 5.0. These trends and challenges can be identified in: hybrid processes, novel and multi-materials, 4D printing, control and monitoring, process optimization, novel technologies, post-processing, data acquisition, sustainability, and human-centered production [34].

In recent years, there has been a development of hybrid machines that fuse traditional methods like machining with additive manufacturing [35]. To connect hybrid additive manufacturing with abilities and surplus capacity in several conventional production facilities, researchers look at a network of strategically positioned additive manufacturing hubs [36]. Other studies combined additive and subtractive manufacturing technologies to take advantage of the two technologies' capabilities [37]. A current subtractive computer-aided manufacturing (CAM) system is supplemented with components to produce extra additive manufacturing phases, resulting in a hybrid CAM system, which reduces the planning work.

One of the latest trends is the creation of multi-material structures with 3D printers. These procedures enable the creation of parts that benefit from various materials' qualities. Biomaterials and AM show promise for applications about human health and well-being [38]. Furthermore, cutting-edge AM methods are being developed to create osteochondral scaffolds that are customized for each patient [39]. The recent discovery of meta-biomaterials with previously unheard-of mechanical capabilities is one of the more intriguing examples of novel materials [40]. The intersection between 3D technology and smart materials has led to the development of an interesting opportunity through 4D printing. It offers multifunctionality and the ability of self-assembly and self-repair since the printed parts can adapt over time to the stimuli that the smart materials used can sense [41]. At the same time, there have been numerous studies conducted on micro-additive manufacturing; nonetheless, additive manufacturing typically occurs at a macroscale level [42]. As previously underlined, AM must ensure the components' quality and control, demanding the use of monitoring and inspection systems [43].

Different outcomes, such as surface attributes, mechanical behavior, dimensional alterations, or shape deviances, are being studied for the optimization of operations [44]. In this regard, output requirements depend on several parameters such as layer thickness, orientation during construction, positioning on the building area which could be optimized [45].

Since there are numerous potential uses for AM, new techniques are being generated that can adapt to the required outcomes. A novel technique called multi-layer cryolithography is introduced by researchers, as well as friction deposition. The first entails printing many independent 2D layers in parallel in regions covered with hydrophilic materials to combine water-based chemicals for tissue or food engineering [46]. The second method performs the addition of material by solid state material deposition, and individual layers were added up by friction deposition [47].

Additive manufacturing allows the reproduction of replicas that can be produced through the collection and use of information through sophisticated methods such as computed tomography, especially in fields such as biomedicine [22, 23].

Decisively, the adoption of recyclable materials or the reuse and saving of resources in the production phase are the subject of new research on the sustainability of AM [48]. Finally, the AM processes are acquiring even more importance in human-centered products, but many aspects must be handled and solved [49]. First, the main choice is based on the technology and then the parametric optimization must be deeply investigated, and this can be reached using AI and machine learning (ML) [50].

2.2 AI-Driven Systems for AM

AI has become an essential component of modern manufacturing systems in Industry 5.0, enabling functions such as sensing, learning, and decision-making. Applications of AI in AM include process planning, production control, and human–robot collaboration. In fact, industrial AI offers significant prospective to strengthen human–robot collaboration and give support to existing human-centric jobs on the production line, whether by increasing operator safety and welfare or by optimizing tasks more efficiently. Current potential for industrial AI in the manufacturing sector includes workforce training, planning, monitoring, assistance, and collaborations with robots [51]. These technologies are being used to automate tasks, improve efficiency, and optimize processes. Machine learning (ML) is a branch of computer science that studies algorithms that use provided datasets to learn how to tackle difficult real-world problems. Classification, grouping, and prediction are the most typical issues that ML addresses [52]. ML-powered systems can collect and analyze data from a variety of sources, including sensors, machines, and people. This data can then be used to make better decisions about manufacturing operations. One of the key benefits of AI and ML in manufacturing is their ability to automate tasks that are currently performed by humans. It is still difficult to define ANN in the scientific and practical spheres [53]. Artificial neurons that comprise artificial neural networks (ANNs) are designed and act similarly to neurons in the brain. Artificial neurons, the basic building blocks of these networks, process information in a non-linear, dispersed, simultaneous, and local way [54]. Moreover, computer vision is a technical framework whose main objective is to replicate the functional components of human vision. This is accomplished by performing tasks such as image generation, processing and analysis, tracking, focusing, brightness settlement, monitoring, processing, and visualization [55].

Furthermore, Cyber-physical systems (CPS) integrate physical manufacturing assets with cyber components, such as embedded sensors, to create intelligent systems that can capture, process, and make decisions in real time. Cloud manufacturing, which leverages cloud computing, IoT, and service-oriented architecture, is another technological advancement that is reshaping manufacturing. This approach allows manufacturers to access decentralized, networked production resources as needed. While cloud manufacturing differs from traditional computer-integrated manufacturing, it offers flexibility and efficiency through optimal resource allocation and scheduling [56]. As underlined by researchers ML and AI can be applied in the pre-process, process, and post-process step of this technologies but the lack of data is still a major issue that must be solved [57].

Generally, the applications of ML and AI are more efficient in the pre-process and process step because the part is still not manufactured, and there is the possibility to obtain training data. In the literature, it emerged that ML in 3D printing of constructions is mostly applied to design, optimization, and diagnostics [58–61]. Other studies have also revealed that ANN is mainly used in the diagnosis area and for training data extracted by AM technologies [62–65]. Convolutional neural network is instead more efficient than ANN for processing 2D images. Also, the integrated application of ML and ANN is observed, with an emphasis on optimizing and diagnosis [66–68], while computer vision techniques are covered and are only found in the diagnostic area [69, 70].

The latest developments in AI, which include ML and deep learning (DL) as datasets, are focused on creating even more algorithms to achieve the objectives of optimizing parameters and reducing expenses and time [71]. For instance, it is feasible to create new algorithms that enable the proper selection of process parameters by utilizing the printing time in the FDM process. Moreover, in AM production, the quality assessment is central for the future industrial development of the products, and AI plays an important role in this part. Nowadays, AI-driven technologies allow us to reach more goals in quality monitoring, prediction, and control.

New challenges deal with the need to analyse a huge amount of data coming from AM technologies. The main difficulties of applying AI to AM processes are related to the lack of training data, which can be used for reliable model estimation. The difficulties can be divided in three categories: the availability of data for only one process a time and it can be expensive to extend this approach for each single model, the transfer learning which could be a solution to compensate the lack of data can be subjected to restriction for privacy reasons and the possibility of cyber-attacks is real when sharing data in the design phase [72, 73]. The IoT is another possible solution to share data. Another important instrument is the virtual reality (VR) and augmented reality (AR), which offers a solution to bypass data exchange and allows graphical representation and demonstration in a real-world environment [74].

Another interesting solution still in the design stage is related to the interaction of AR with cloud manufacturing (CMfg) and applying this concept within the framework of digital twin (DT) to the AM technologies. In the AR-CAM framework, data generated during value co-creation activities is captured and processed in real time using DT technology. This includes information about product design, manufacturing processes, user feedback, and operational performance. This data is centralized and shared on the CMfg platform, which is typically used for matching services, allowing for a comprehensive view of the manufacturing process and efficient resource allocation. While the data sharing is centralized, the specific interactions of stakeholders—who use DT through AM and AR services—are distributed, improving both the efficiency of data processing and the security of sensitive information. DT technology is crucial for effectively integrating AR and AM services, producing accurate digital replicas of physical objects that reflect both their appearance and functionality. This integration creates a "tangible" AR experience that enhances mutual understanding among stakeholders by visualizing manufacturing processes. Consequently, the diverse needs of stakeholders in various contexts can be met in a customized and on-demand manner throughout all phases of value co-creation [9].

There are many studies and applications in the evaluation of AI-driven systems for AM technologies, mostly this concerns the different stages of design, production, and control of the AM process and the AM products. Precisely in this

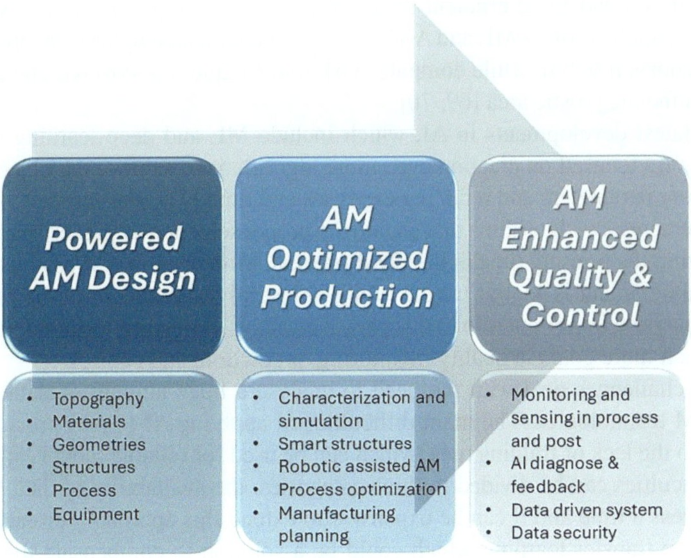

Fig. 3 Applications of AI techniques to improve AM works

perspective, three specific sets of challenges in AI-driven systems for next generation AM processes could be addressed: challenges of AI-driven design of AM structures, materials, and equipment, challenges of AI-driven optimization of AM production process, and challenges of AI-driven diagnostics of AM structures and products [13, 49]. So, to expose the advantages of using AI-driven technologies in next-generation AM, a broad classification based on these challenges was followed in the chapter. Specifically, AM design, AM production and AM quality and control, as illustrated in Fig. 3. This segmenting aims also to reflect the production flow from design, process optimization, and production planning, in-process and post-process monitoring, product quality control, and data management, factors related to general production problems in AM.

3 AM Design Powered by AI

Design for AM (DfAM) has emerged as a new area of transdisciplinary research. DfAM, which aims to design and optimize the product and fabrication procedures to meet intended performance and quality standards while lowering time and cost [75]. DfAM must consider several design factors and their intricate interactions since AM products typically have complex geometry, particularly the human-centered AM, which is a fundamental approach in Industry 5.0. Consequently, this presents a formidable obstacle for conventional design methodologies. The use of AI and ML in DfAM has grown in recent years as a solution to this problem.

AI has been used by designers to optimize AM products' geometry [75–77]. For example, to obtain required properties or reduce mass or costs, the geometry of developed goods, and their support buildings must be properly designed. By efficiently exploring the whole design space, artificial intelligence (AI) greatly speeds up the design process as opposed to heavily relying on the designer's expertise and experience. Optimal topology for additive manufacture [78]. Tang et al. optimized the design of personalized porous lattice shoe soles by using machine learning as a surrogate model [79]. To determine a subset of AM design elements based on design challenges, Yao et al. combine hierarchical clustering with Support Vector Machine (SVM) techniques [80]. To create a support-free manufacturing process, Leary et al. devised an optimization technique to alter the theoretically ideal topology [78].

Currently, most studies in this field underline that the DfAM focuses primarily on the design of materials, design of geometries and structures, design of optimized surfaces and design of processes and equipment, as illustrated in Fig. 4. Specifically, the major part of these application sections applies AI-driven systems to support the evolution of a new generation of AM instruments for AM design.

3.1 Material Design for AM

AI is being used more and more in DfAM material design. Since materials used for conventional manufacturing methods may need to be adjusted for AM preparation, AM is linked to a unique demand for material design and development [81]. AI can

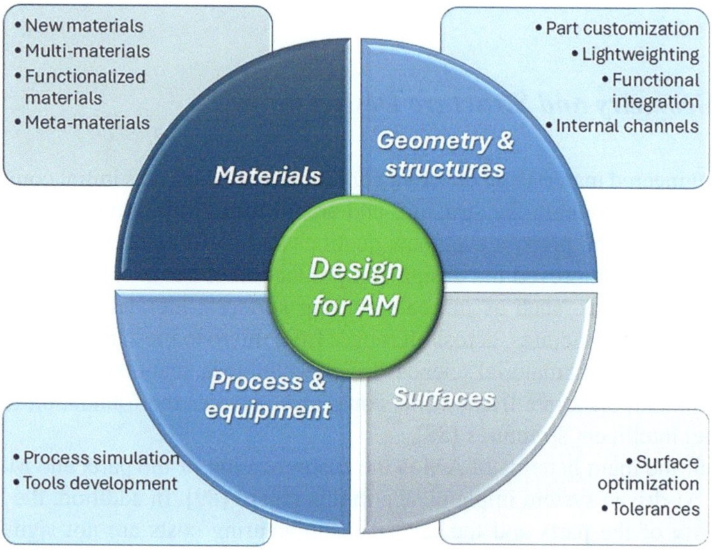

Fig. 4 Applying sections in which DfAM is primarily focused and supported by AI-driven systems

speed up AM material design processes like creating novel materials and updating the constituents of existing ones, such as new materials, multi-materials (MM), gradient materials, or metamaterials.

The use of AI in the research of new materials is becoming more popular, and ChatGPT from OpenAI is one of the key platforms in this shift [82]. Researchers at the University of Wisconsin–Madison have used ML to speed up the search for novel materials, demonstrating this change [83].

Additionally, the synthesis of substances has been optimized through the application of evolutionary algorithms (EAs) [84]. EA may be utilized to find the ideal pressure and temperature for a particular material's synthesis. Natural Language Processing (NLP) is an additional AI method that has been applied to the design of materials [85]. NLPs can be applied to the analysis of scientific publications to obtain material and property information [86]. Then, with this data, ML models can be trained, or new materials with desired qualities for AM processes can be created.

For instance, aluminum alloys have been modified using data mining techniques to improve their suitability for laser bed powder fusion (LBPF) processing. Additionally, alloys with desired qualities are designed using AI-related techniques such as decision trees and EAs [81]. Additionally, ML and DL techniques have been widely applied at the intersection of material science and AM to help with the design of metamaterials that will be manufactured using AM. To attain desired capabilities that natural materials might not be able to provide, like a negative Poisson ratio, metamaterials have incredibly complicated geometries [87]. An important gap still exists among materials, structures, processes, and properties, as well as various performances. Solutions for improving material design theory, or for objectives-oriented material design strategies, and for the use of smart materials should also be implemented.

3.2 *Geometry and Structure Design for AM*

Many engineered materials in different conditions can be used as initial components in AM processes, where the structure and architecture of the raw material can be used to develop and prepare even new qualities and performances. The use of AM has significantly improved the design and production of complex structures with a variety of functions, such as intricate curving surfaces, hierarchical lattices, and thin-wall/hollow structures, across all ranges from micro to meso to macro. With the use of state-of-the-art material science, design techniques, manufacturing processes, and equipment, elaborate frameworks could also serve as the foundation for functional and intelligent structures [88].

One of the main benefits of AM is the customization of the parts and ML is the favorite AI-driven system implemented in this regard [89]. In addition, the level of complexity of the parts and the related manufacturing costs are not significantly correlated in AM production, which gives more flexibility to fully explore the design domain [90] Consequently, to effectively leverage AM's advantages, various new techniques have evolved in addition to the traditional lightweighting design tools. Four categories of techniques and resources can be distinguished in AM lightweight design: lattice structure filling, generative design (GD), topology optimization (TO), and bio-inspired design [77].

Topology optimization (TO), a computer-aided design technique, is utilized to create novel and intricate structural configurations with adjustable stiffness, hierarchical characteristics, and exceptional lighter performances [22, 23, 91]. Another useful design technique for integrating several components of an assembly into a single printed part is known as consolidation by TO [92]. New techniques for smooth boundary modeling that consider DfAM are the "Moving Morphable Method" (MMM) [93, 94] and the "Metamorphic Development Method" (MDM) [95]. Here, simple design basic functions for topology representation have been provided by simple geometric shapes and polygons that capture geometrical patterns. These have the potential to emerge as strong competitors to the traditional density-based TO methods that have been employed up to this point for AM. The temperature gradient and residual stress distribution after a procedure are predicted using sophisticated thermo-mechanical simulation, and the temperature or stress fields that are obtained can be used to help TO eliminate structural deformation [96]. Additionally, in order to accomplish the simultaneous optimization of several process settings, build orientation, and structural setup, a data-driven TO including the process–structure–property–performance relationship of AM is established [97–99].

Alternative approach to AM design, called generative design (GD). GD is a collection of techniques that use an algorithmic or rule-based generating system to investigate the design domain and produce possible options for designers. Typically, architectural design is the application of it [100]. Broadly, three categories of approaches can be distinguished: Three concepts are involved in natural selection and reproduction simulation: (1) self-organization and self-assembly, which are made up of a large number of fairly straightforward, autonomous parts that work together to construct large-scale objects or solve problems collectively; (2) evolutionary systems; and (3) generative grammars, which specify a mapping between a string of elements and the object that is to be designed (or its components) [101]. Comparable to TO, GD is not completely novel, however, it is still challenging to introduce AM limitations into traditional GD. It is simple to populate a set of finite potential solutions with established requirements for designers to choose from, but it is more difficult to verify the reliability of each loaded option solution. Researchers have just developed an alternative evolutionary GD technique for AM lightweight design to imitate termite activity for volume creation to address this issue [102]. This approach makes use of multi-agent algorithms to design, structurally improve, and evaluate the manufacturing feasibility of additively manufactured parts all at once. Within the geometry generation process, Voxels are employed to convey production limitations and guidelines for combining and deductive thinking. To lessen the complexity of the challenge, GD might, as a compromise, incorporate human knowledge to proactively pick the potential solutions.

A further result in reaching a compromise in designing lightweight structures is to fill in pre-defined unit structures, lattices, or unit porous structures to substitute solid volumes in a manner that is equivalent to the ideal design solution. For AM, this refers to an indirect lightweight design technique known as lattice arrangement [103].

3.3 *Surface Design for AM*

Surface morphology and tolerances in AM products are one of the main limits for now [104]. Castelão et al. developed predicting equations that allowed the advancement of several guidelines to improve surface roughness and dimensional tolerances in a FDM process [105]. Researchers employ various techniques, including AI, to forecast the surface quality of AM parts. Numerous studies have created models using AI techniques, such as the more current DL and ML approaches, which are proving to be a viable way that reduce costs and save time. This integration guarantees adherence to technical criteria and reduces the necessity for reprocessing components [106]. Another exciting avenue for the future is the application of generative AI models, including generative adversarial networks (GANs). Synthetic surface roughness patterns that mimic real-world variances can be produced using GANs. These generative models may replicate various surface roughness conditions, improve the robustness of AI algorithms, and supplement small datasets. Based on specified process parameters, a unique artificial intelligence-based technique was put forth to create the virtual surface morphology of Ti-6Al-4V components [107].

3.4 *Process and Equipment Design for AM*

Specialized equipment and expertise are required for AM to operate as intended. Like any industrial process, equipment and method optimization leads to increased efficiency and quality [108]. So, optimization in process and equipment design, integrated with AI tools, helps and improves system and component design prior to manufacturing. For example, to reduce production time in 3D printing, Huang et al. examined the problem of effective support distribution, specifically regarding FDM and SLA technologies [109]. They search for a way to identify the locations on pictures where supports ought to be used. The condition of the supporting evidence for the assertion is produced by a trained convolutional neural network (CNN). At the same time, Yang et al. explained a cloud tool that allows AM to quickly create a decision support system by identifying the correct components from a repository and verifying if they are acceptable [110]. It has three components: a data acquisition step is necessary to extract useful information/parameters from historical data; scoring criteria are established to determine AM eligibility; and a decision model based on an ML algorithm is developed to predict the AM suitability to a specific application. Other authors make it possible to examine the microstructure created by different metal AM processing settings using principal component analysis [111]. Moreover, Alejandrino JD et al. showed a neural network (NN) that generates the symmetrical deviation surface coordinates from the input surface 3D coordinates of the component to be printed [112].

Several investigations in which ML and DL algorithms are efficiently used to enhance AM process simulation are considered as simulation optimization. In the industrial sector, determining whether the manufacturing process can produce optimal components is crucial—especially in metal Additive Manufacturing (AM), where costs are significantly high. Olleak A et al. provided a methodology for the SLM process's calibration and validation that was based on sparse experimental data and multi-fidelity models [63]. The physics-based SLM model can be used to estimate the melt pool (length and thickness) at any given combination of process variables. A GPR model was employed to generate a meta-model. The results showed that the model Iy Integrated meta-modeling with the finite elements and GPR modeling of the SLM process.

4 AM Production AI-Driven Systems

Currently, Additive Manufacturing (AM) holds significant potential for application in the industrial sector. Research efforts are increasingly focused on design innovation and topological optimization, enabling the creation of even more complex part geometries. Several challenging issues must be solved such as the mentioned structural and multidisciplinary TO with a knowledge-based driven approach (for organizing and managing the great amount of data for AM driven design); the product life-cycle via cost-oriented management strictly connected to the digital model of the part; the metamaterial structure design which should achieve product design.

Now, AM production is facing significant and rapid changes. First, the number of printable materials and the possibility of multi-material printing, as well as the possibility of producing larger components. For example, Stratasys J750 printers can print materials with a variety of mechanical characteristics, textures, and colors thanks to PolyJet technology [113]. Complex structures made of hard polymers, flexible materials, and elastomers can be produced with Carbon's DLS (digital light synthesis) technology [114]. At the same time, larger items are now capable of being manufactured due to advancements in software and technology, including building elements [115], airplane parts, as well as entire vehicles. Larger build volume machines have been introduced because of recent advancements in additive manufacturing. Larger and more intricate items, such as structures the size of buildings or even whole houses, can be produced as a result [116]. Moreover, the advent of many novel AM processes has ushered in a new era of speedier print speeds. It is noteworthy that advanced printing technologies such as DLP [117], SLA [118], and CLIP (continuous liquid interface production) [119] have greatly increased the speed at which things are created. The production of parts with complex geometries and exact tolerances was made possible by remarkable developments in hardware, software, and materials that allowed AM to operate with greater precision and accuracy. It is now possible to make parts with precise measurements and complex shapes with incredible detail [120].

AM processes have long pre-processing and post-processing steps. Due to this, several parts may be done concurrently during the production phase, particularly in cloud manufacturing. Printing as many pieces as feasible for every run is more productive [121]. Production planning is crucial in this scenario to increase overall production efficiency. Every task's due date, design geometry, necessary parameters, etc., should be considered. Usually, the fabrication process takes several hours or days to complete. Single-print scenarios, in which the user tries to fit as many pieces as feasible into one manufacturing spot, serve as the basis for production planning. There has been extensive use of the genetic algorithm (GA). Heuristic approaches are thus suggested to locate an inferior solution with an acceptable computation time by using predetermined rules to reduce the search space. [122]. All these achievements were supported by AI-driven systems, and in particular boosted by the proper selection of AM process, of the best AM parameters during processing, and by the application of robotic-assisted AM processes.

4.1 AM Process Selection by AI-Driven Systems

The choice of AM processes has already been well examined in the available research. Mançanares et al. [123] described several AM methods and offered a technique based on the analytic hierarchy process (AHP) to determine the best AM equipment and methods according to important, applicable machine attributes. The usage of different multi-criteria decision-making techniques has become widespread, as AHP [124], Technique for Order Preference by Similarity to an Ideal Solution (TOPSIS) [125], operators based on fuzzy logic [126], and hybrid schemes that combine several approaches [127] are a few common examples. Technical and economic viability, which includes factors like manufacturability analysis, cost, build time, accuracy, part weight, and buy-to-fly ratio [128], and sustainability, which includes energy demand and CO_2 emissions for the full product lifecycle [129], are the two main categories of choosing criteria that have been taken into consideration among all those schemes.

4.2 AM Process Parameters Optimization by AI-Driven Systems

The AM production process optimization by refining process parameters is the subject of several studies [130, 131]. Three essential elements can be used to define the AM parameter optimization problem: (1) decision variables, (2) restrictions, and (3) the objective function. The slicing parameters, such as build orientation, layer thickness, and printing paths, are general for numerous AM processes, and all the adjustable process parameters that differ with the AM process may be among the decision variables involved in AM process optimization. The feasible region of the choice variables is typically enforced by the constraints [132].

Several single-objective and multi-objective AM parameter optimization techniques with respect to the objective functions (i.e., goals). Optimization of manufacturing costs [133], part quality (including density, microstructure, geometric accuracy, and surface quality) [58, 130, 134] energy and material consumption [135], and mechanical properties [136] are examples of single-objective problems, which typically have only one specific goal. AI and machine learning are essential for optimizing the AM process. According to Jiang and Ma [130], machine learning techniques have the potential to be an effective tool for optimizing AM processes.

4.3 Robotic-Assisted AM Processes

Robotic-assisted AM is a promising approach in terms of equipment architecture since it offers additional design flexibility. Multiple robots allow for locally configurable and size-independent multi-material processing [137]. Additionally, through online recognition and feedback, AM can achieve unsupervised planning of paths and in-situ parameter tweaking with the help of robotic cameras or sensors [138]. After that, physics- and data-driven frameworks can determine the process–structure–property (p-s-p) linkages in reference to the AM digital ecosystem, eliminating the need for trial-and-error [139].

Robotic-assisted AM technologies are another important goal to achieve, and several actions must be promoted to switch toward the AM intelligent transformation. These actions are related to the development of hybrid-AM solutions under multi-robot cooperation, improving the functionality and integration of monitoring and sensing devices, merging the industrial internet into AM digital twin, and supporting the establishment and improvement of the AM digital ecosystem. Moreover, the parts which can be produced range from small to large scale, but for some practical applications, AM is often challenged in extreme environments, or the development of space robotics and automation technologies [21].

4.4 Production Systems Empowered by AI Systems

The AM production process optimization enhanced by AI-driven systems is critical in expanding the use of AM processes at the industrial production level. In this regard, an important example which can be implemented for AM applied in the industrial fields is related to the evaluation of the raster angle in FDM [140], which is responsible for the mechanical properties of fabricated components. The integration of ML together with AI allows the prediction of the optimal raster angle for a specific shape based on trained data obtained from other manufactured geometries. In medicine, there is also the possibility of implementing AI on AM, especially regarding the pharmaceutical sector [141]. Pharmaceutical 3DP represents a collection of different technologies that together allow the printing of almost any conceivable medicine. A key goal of pharmaceutical 3DP is to leverage AI to create a seamless, autonomous 3DP process. Currently, researchers are required to input printing process parameters before each batch of medicine is produced. Machine Learning has been used after printing as a quality control (QC) measure for 3DP drug products, irrespective of the fabrication technique. An interesting application in terms of decision making by using AI in medical applications is also presented in Rojek et al. [142], where the support of an exoskeleton has been produced by AM. In detail, the use of AI allowed us to modify parameters of printing related to material, kind of template used, and degree of freedom. Moreover, produced parts were tested, and the a Life Cycle assessment was also implemented to recycle the components that did not meet the desired goal in terms of performance.

Moreover, AI is acquiring great importance also in microbiology, where the use of different algorithms and instruments allows us to pursue important goals. In fact, the analysis of several data with huge speed allows to uncover previously unknown microbial species and elucidate their potential roles in various ecosystems [143].

In the field of industrial applications, arc wire additive manufacturing (WAAM) can play an important role in sectors like aerospace and maritime. The application of AI includes machine learning (ML), automation and robotics, machine vision, data mining, big data, and expert systems, which have been proven effective in the manufacturing industry. The knowledge of AI is based on machine learning and algorithms, thanks to which it is possible to pursue data analysis collected in the process. The raw data collected from the process must be processed to be converted into training data. The main issues encountered in applying AI techniques in the WAAM process are limited amounts of data, data fusion, and latency. Moreover, new research directions include novel applications of AI techniques to overcome the challenges of achieving IWAAM, enhancements of the capabilities of AI techniques (computing power improvement, algorithm optimization, high-quality data acquisition, and data augmentation). Digital twin, human robot collaboration, and cloud manufacturing are just some examples of how to implement the new AI techniques [144].

AI models have been used for defect detection, which is a straightforward AM task. Avoiding defects is important because they can cause crucial parts in industrial applications to fail early. For example, Li et al. describe a process for locating geometric defects in AM objects using ML models and a 3D mesh [97, 98]. Applying the F-measure and G-mean to gauge each model's efficiency, the authors trained five different machine learning models. The findings show that although the tests yielded promising results, more work needs to be done before it can be applied to real AM operations. For DL-based online fault diagnosis for aluminum alloy in robotic arc welding, Zhang et al. examine CNN and weld data. To collect weld images simultaneously, a machine vision system was developed [145]. From different angles, this technology may provide more details regarding weld photos in real time. Then, to identify weld penetration defects, a new CNN classification model was created. Test results show that the CNN model has a 99.38% mean accuracy for classification.

5 AI-Driven Systems in AM Control

It is quite difficult to maintain regular quality in AM, especially when working with large or complex components. Humidity, temperature, and machine adjustment are a few examples of variables that might affect the printing process and cause variances in the result [146]. The properties of the materials used in AM are influenced by several factors, such as the printing method [147], the supply of resources [148], and the post-processing steps [149]. Examining 3D-printed products with traditional methods is difficult due to their intricate geometry. Parts with structural flaws, faults, or other quality problems may result from design mistakes or software vulnerabilities. When printing parts with hollow spaces or overhangs that require support structures, the part's orientation can have an impact on the final product's quality [150].

Defects may result from incorrect design decisions (supports, staircase effect, thin walls, sudden thickness alters); raw materials (contamination, composition, incorrect form or density); equipment (wire device, temperature control, inert gas atmosphere); and processing conditions (variables, power source, environment, etc.). [151]. Geometrical and/or dimensional flaws are among the faults that might damage either inner or outer surfaces; yet, if destructive inspection is considered, it can be exceedingly challenging to analyze complicated internal structures. Additive manufacturing products are frequently characterized by a second class of defects called volumetric errors, which include undesired spaces and pores, entrapped gas, cracks, and inter-layer delamination. Because it negatively impacts the mechanical qualities (such as tensile and fatigue resistance), this second category of faults is crucial [152].

AM has great potential also for application in mass production. In this case, many tasks must be solved because one major problem concerns the possibility of having repeatable parts. For this occurrence, a new approach based on iterative learning driven by artificial intelligence has been set [153]. This approach consists of monitoring the fabrication steps and acquiring information (images and videos) on the errors developed during fabrication to modify the process parameters to eliminate the defects.

In this regard, many review and research papers deal with the study of all the methods for monitoring and controlling processes because the presence of defects and anomalies is still a hard task for all AM processes [152]. The major part of the AM control is now focused on using AI-driven systems in order to improve monitoring techniques in the in-process and post-process phases, to enhance repeatability and quality levels of the products. In light of this, next-generation AM systems would include this kind of AI-driven systems in machines, equipment, and software.

5.1 Quality and Control Systems AI Empowered in AM Technology

One of the biggest benefits of AI in process control is the ability to predict and reduce potential problems before they occur, which leads to a significant reduction in downtime and maintenance costs. At the same time, this could increase the quality of the final products. AI-based systems, on the other hand, are equipped with predictive maintenance capabilities that can identify patterns and anomalies in data streams in real time, enabling preventative interventions that help keep processes running smoothly and avoid unexpected shutdowns.

The largest issue with products made via AM, according to several researchers, is reliability monitoring and control. Implementing monitoring and evaluation systems to raise the standard of the components and additive manufacturing procedures is required to get beyond this obstacle. The dimensions and temperatures of the fusion assembly are crucial factors in evaluating and checking procedures, according to Chua et al.'s analysis of the current technologies for monitoring and controlling AM processes [27, 28]. They also suggest an automated tracking system and a method of inspection to handle the overall control of metallic additive manufacturing processes.

A huge possibility for monitoring and control in AM technologies is offered by a convolutional neural network, which offers great possibilities in terms of analyzing input sources as images, videos, and acoustic emissions [154]. Acoustic emissions can be used as a non-intrusive method to observe the normal and abnormal state of operation in FDM [155]. Time-domain features of AE hits can be used to identify different extruder operating conditions of FDM. Several machine control monitoring (MCM) techniques can be used for defect detection in AM technologies. Most review articles [156] investigate the commonly adopted techniques based on hardware (devices) and software (algorithms) for process monitoring. Camera acquisition, microscopes, SEM, and sensors (which control mechanical properties) are the generally used solutions applied, at an initial point, to FDM. Based on data acquisition, AI can predict the rise of possible defects based on acquired data during processing.

Moreover, most of the currently implemented AI solutions for AM are based on process control, mainly through an image acquisition system, for example, by Zhu et al., where a new in situ solution was proposed [157]. It introduces a low-cost in-situ monitoring system that uses an AI edge computing board to improve digital image processing and reduce dependence on high-resolution video. The system identifies the geometric features of the plasma arc and molten pool in plasma arc

AM (PAM), moving from low resolution to high resolution. It utilizes a visual transformer-based video super-resolution model (ViTSR) to reconstruct high-resolution frames, while a fully convolutional network (FCN) concurrently extracts features from several objects. The paper outlines the architecture of the system and showcases results that highlight the quality of the 4× super-resolution reconstruction and the successful extraction of geometric features, concluding with a discussion on the system's performance and final observations. Another study focused on monitoring is based on the use of camera acquisition images of a 3D printed part [158]. In detail, the adopted FDM machine was programmed in such a way that when each single layer of the part was built, the extruder was placed far away, and the part was brought next to the adopted camera. In this way, the proposed acquisition method allowed to evaluate both the external printed surface as well as the inner layers.

For metal-based AM technologies produced through laser interaction, the defect detection during processing by means of AI-driven systems for control and monitoring is essential. Among these strategies are: signal sensing (which is a closed-loop control system). These signals are divided in four categories: visual, spectral, thermal, and acoustic emission. The main problems when lasers interact with metal are related to dimensional accuracy and internal defects. For this reason, there is a list of parameters that must be carefully observed. AI is surely an interesting way for process control, and at the current state of the art, the automatic process control is based on the evaluation of the defects during the operation, and then, through certain algorithms, the machine is automatically set to other parameters. To date, the most widely used intelligent controllers include expert systems, fuzzy logic, neural networks, and biological intelligence search algorithms. For monitoring, it should be optimal to monitor more features at the same time, and so an advanced solution for data transmission (as 5 G) should be enhanced. Process control also requires multivariable monitoring for correct procedure of AI implementation [159].

Recent trends in the use of AM for metallic part production are emerging. This need is due to the high costs related to prototipation and for this reason, also the physics-informed neural network (PINN) [160] is developing. This approach merges the traditional data acquisition (images and videos) with the physic laws and mass and momentum conservation for example. In this way, it is possible for example to predict temperature and melt pool fluid dynamics in metal AM processes.

Another viable application of A-driven systems to control AM processes is related to the ground exploration during printing [161]. In this study, Johnson et al. developed a tool in terms of Image acquisition to extract the necessary information for building structures able to overcome obstacles on the ground (span and slump). It was applied to Direct Ink Write.

Moreover, explainable AI (xAI) is a powerful instrument, being a sort of add-in to machine learning, which is applied to AM technologies. The only machine learning is similar to a black box, which gives information without all the elements helping in explaining the reason for a certain result [162]. In this view, xAI helps in decision making and, together with Machine learning, is a flexible instrument to be applied to several AM processes.

An application of xAI has been applied for defect detection on Ti6Al4V specimens after processing images obtained through computer tomography (CT) [163]. This methodology was compared with manual inspections, and the effectiveness of the approach was evident. The xAI has a crucial role also for AM techniques in that it is based on non-destructive techniques.

Deep learning is generally used for integrating new information to process data obtained through sensors installed on AM facilities, as in Khusheef et al. [164]. The signal imaging in multi-sensor data fusion for fused deposition modeling has several potentialities when analyzed through DL. Deep learning, in this case, makes use of the well-known artificial neural network to process the data obtained by sensors.

Recently, some researches are addressing the use of digital twins to obtain a virtual model that can interact remotely with the real model through a continuous exchange of data, which permits monitoring the quality of the produced parts. For example, in Jyeniskha et al. [165], this approach has been applied to a fused deposition modeling additive manufacturing technology where microsensors are placed on the built part and an opportune framework is used to obtain the digital twin of the real model. In this case, machine learning has also been implemented for defect detection.

6 Case Study: Validation of AI-Driven Control in FDM Design and Manufacturing

The new generation of AM systems is certainly linked to the development and integration of AM systems. With this in mind, the authors wanted to examine a simple application of commercial AI systems to a simple AM process, such as FDM. FDM is one of the most widely used processes, especially in the field of low-cost prototyping and customization processes. FDM gained popularity among other additive manufacturing techniques because it is inexpensive and allows for the use of a variety of materials in different sectors as aerospace [166], medical [167], and automotive [168]. However, in recent years, it has undergone countless evolutions due to the simplicity of its approach, and there are many applications and frontier studies on this AM methodology. The latest developments for FDM concern the extension in the use of materials such as composites and nano-composites, metallic and ceramic, even with high performance [169], and also in the use of smart polymers in cutting-edge applications such as soft robotics [170]. In addition to the extension in the use of materials, the dimensions have also recently been expanded by producing large items through large-scale AM (LSAM) [171, 172].

Dimensions, material, equipment, and parameters adopted are the fundamental set to obtain certain properties and functionalities in products and to optimize the quality of the produced parts. Within these factors, several aspects must be considered both in the design and printing phases, as discussed in the previous paragraphs.

Specifically, for FDM, these aspects can be summarized and illustrated as shown in Fig. 5. Within these aspects, in this case study, two main problems have been highlighted and investigated, which can be traced back to some of those aspects illustrated in Fig. 5. Specifically, two main questions were formulated: one on aspects related to FDM' design and one related to FDM' printing parameters used for specific material, such as thermoplastic elastomeric materials.

6.1 Question 1: About Design Aspects in FDM Process

Q1 formulation: The first question is related to the lightening of vertical walls using hollow structures as vertical as undersized vertical-axis holes without the use of support, to minimize the material used and improve surface quality. This practical design problem in FDM is largely known by designers and FDM producers' companies. Precisely for this reason, it was taken as a problem to be faced by a commercial AI such as ChatGPT and compare the result with what is currently suggested and implemented by communities and researchers in this area. For this reason, the question was formulated as: "In the design of a vertical wall to be printed using Fused Deposition Modeling, what shape should be given to the lightening holes to be able to print the overall wall without using supports?"

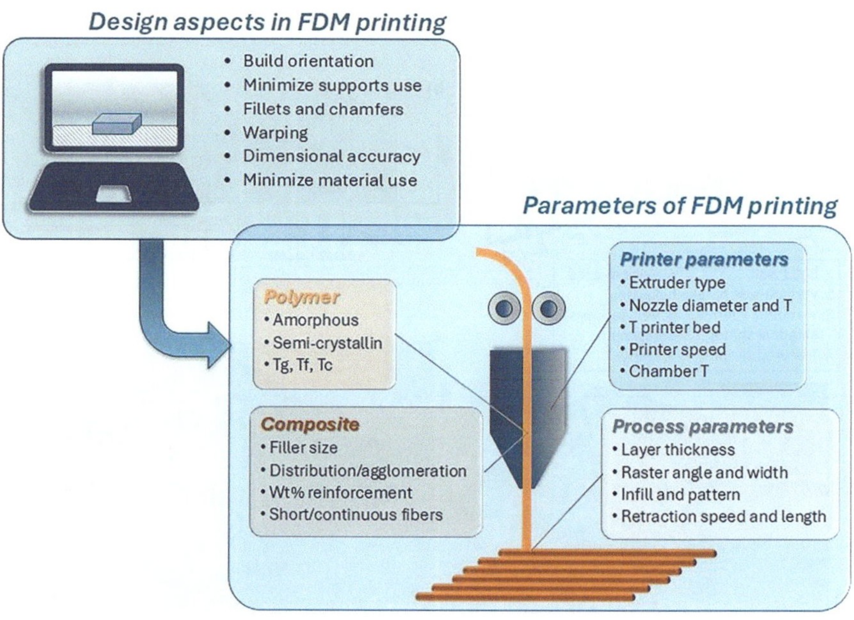

Fig. 5 Design and parameters in FDM printing

Q1 ChatGPT answer: The answer of Chat GPT was "To print a vertical wall with lightening holes without using supports, it is preferable to give the holes a "drop" or "teardrop" shape with the top rounded and slanted upwards. Alternatively, you can use a vertically oriented elliptical shape. These shapes reduce the need for supports as they avoid horizontal overhangs that would be difficult to print without warping."

Q1 Experimental comparison and validation: For the comparison and experimental validation a 3D modelling software (Solidworks®), a commercial slicer software (PrusaSlicer®), and an FDM printer (Prusa i3 MK3S) were used. To this end, first, seven vertically different lightened walls with dimensions 70 × 36 mm^2 were modelled, as illustrated in Fig. 6a, and the dimensions are shown in Fig. 6b. Looking at the nomenclature assigned to the 7 different types of walls, the 1 is the completely solid wall, and the 7 is the completely empty one with only an external frame of 3 mm thick. The lightened walls 3 and 4 are those modelled following ChatGPT's answer to Q1, while the others, 2, 5, and 6, are lightening modes usually used in DfAM.

The CAD models of the seven walls were imported into the Slicer Software and were oriented and built on the building platform as shown in Fig. 6c. A brim was used to improve the adhesion of the parts to the platform. The brim is a layer that adheres to the entire perimeter of the piece and has the purpose of increasing its contact surface with the top. The parts were then printed with the same printer and process parameters (Table 1) and material (Prusament PLA Blend My Silverness 970 g, 1.75 mm diameter).

The printing phases of the walls are shown in Fig. 7a. The samples produced are illustrated in Fig. 7b with the highlight of the defects found after printing. Samples

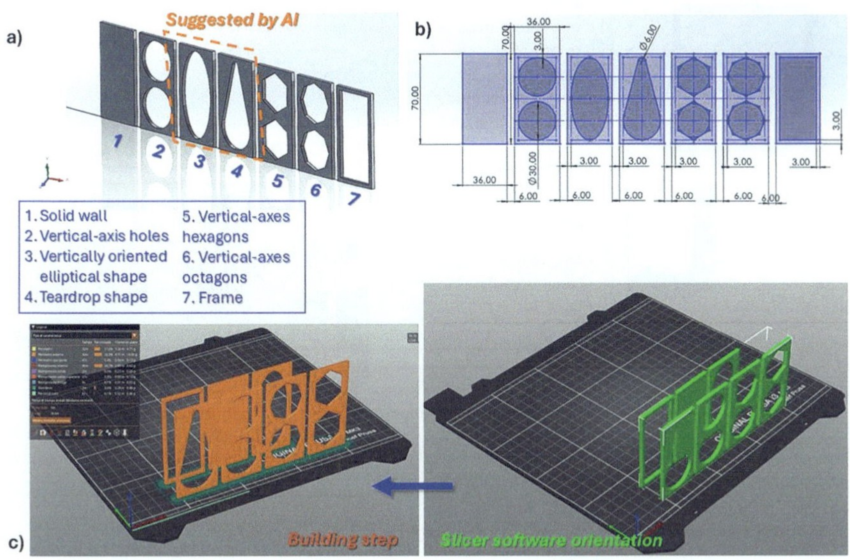

Fig. 6 CAD modeling of samples (**a**) and import into slicer software (**b**)

Table 1 FDM printing parameters for the seven walls' samples

	Parameter value
T Nozzle	215 °C
Nozzle diameter	0.4 mm
T printer bed	60 °C
Printing speed	15 mm³/s
Infill	100%
Pattern	Linear
Raster angle	45°
Layer thickness	0.3 mm
Retraction length	0.8 mm
Retraction speed	35 mm/s

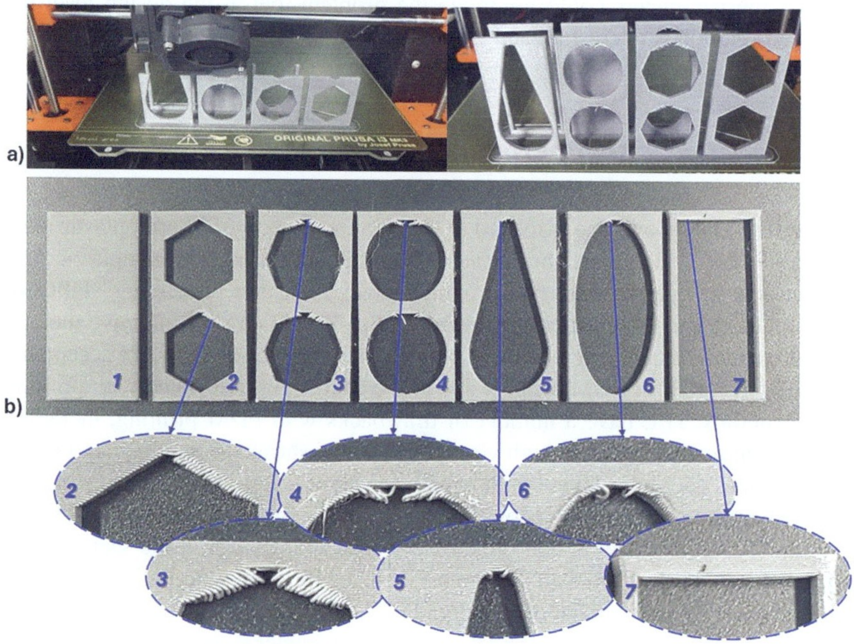

Fig. 7 Samples printing phases (**a**) and printed samples with highlighted difficulties in printing shapes (**b**)

2, 3, and 4 that followed the lightening proposed in the bibliography and community show difficulties in the realization of the upper parts of the hexagonal, octagonal, and circular holes. Samples 5 and 6 made at the suggestion of the AI show fewer problems in printing the upper part of the cavity. Finally, sample 7 shows how in the absence of supports the upper wall of the frame is inflected and the edges protrude from the side walls.

The printed solid wall (sample 1) is the reference in terms of weight, with its 8.84 g. Consequently, the greatest possible lightening was the frame (sample 7) with a reduction of 77% in weight. However, it is evident that this solution at the design level during printing is not valid, because the absence of supports makes it impossible to have an uninflected upper frame and non-divergent edges. The other lightening solutions underlined a weight reduction of 47%, 51%, and 57% for samples 2, 3, and 4, respectively. The AI lightening solutions showed 49% and 61% of weight reduction, respectively, for samples 5 and 6. The weight reductions of the designed samples are interesting, but those made according to AI suggestions have a greater correspondence to the geometry after the printing phase.

This simple experiment is an effort for testing and validating AI responses. It makes it possible to highlight how an already unrefined and unpolished answer given by AI on a common problem, is in any case quite better than other known solutions. The defectiveness is still present in a small part, but a subsequent refinement of the parameters could minimize the problem, but at the design level the proposed solution is interesting and viable.

6.2 Question 2: About Printing Parameters for TPE Material

Q2 context: The second one is related to the printing aspects of Thermoplastic elastomers (TPEs), which are hybrid materials comprising a hard thermoplastic segment and a soft elastomeric phase, which gives them the processing ability of thermoplastics and the elasticity of rubber [173]. FDM is a very effective method for producing thermoplastic parts, but only a small number of TPEs are acceptable for use as material for FDM printers due to the challenges associated with printing those materials. TPE have a number of drawbacks with FDM printing, including low filament section strength, high viscosity, friction, shrinkage, voiding, and warpage, as well as poor inter-layer bonding between other layers in the printed structure and weak intra-layer adhesion between droplets in a single layer [174]. Anisotropy, voids in the microstructure, and dimensional inaccuracies can all be seen in TPE printed objects [175, 176]. Soft TPE material instability and wire buckling are frequent issues with commercial 3D printers [177]. As the filament stretches between the filament feeding mechanism and the spool, its diameter decreases, which causes the filament to slip [178].

Q2 formulation: The formulation of the question is related to this context and specifically in the possibility of printing a commercial Thermoplastic Polyurethane (TPU) for biomedical use by commercial FDM. TPU is included in this class of TPE materials and therefore involves the same declared printing difficulties. Preliminarily, the authors experimentally optimized the TPU process parameters with FDM printing from literature data [179, 180], and in order to compare and validate the parameters provided by an AI, they formulated a specific question for ChatGPT: "*Fused Deposition Modeling printing of a TPU filament, thermoplastic polyurethane:*

- *What are the optimal nozzle temperature values?;*

- *What is the most suitable temperature for the print bed?;*
- *With a Direct Drive extruder, how to optimally set retraction speed and retraction length?"*

Q2 ChatGPT answer: The answer to Q2 of ChatGPT was: "*For 3D printing TPU with FDM technology, the optimal values are:*

- *The nozzle temperature for TPU typically ranges between 210 and 240 °C. A common starting value is 220 °C, but you can vary based on the specific type of TPU you are using and the Filament characteristics.*
- *The print bed temperature should be set between 50 and 60 °C to achieve good adhesion without deformation. The most common value is 60 °C, but it also depends on the type of hob (such as glass or PEI) and the adhesive used (*e.g.*, glue stick).*
- *Retraction Length: With a Direct Drive extruder, the retraction should be between 1 and 2 mm. Higher values could cause clogging due to the elastic nature of TPU.*
- *Retraction Speed: The retraction speed should be low, between 15 and 30 mm/s. Too much value could stretch the filament and create problems with blocking or lack of material during printing.*

These values may vary slightly depending on the environmental conditions, filament quality and specific type of the machine you use."

Q2 Experimental comparison and validation: To experimentally validate the printing and process parameters suggested by AI, a commercial TPU material (DiAPLEX MM3510) was used starting from pellets and realizing a filament by a filament maker 3DEVO Composer 450 (3devo, Utrecht, The Netherlands). It is an extruder capable of extruding a filament with a variety of setting diameters, and is equipped with four heating areas, which have been set between 185 and 195 °C for the TPU. The produced filament had a diameter of 1.75 mm. A simple object was designed to be printed a parallelepiped of 30 × 30 mm^2 in base and 1.1 mm in height. The FDM used is a desktop model by Prusa (Prusa i3 MK3S). The printing and process parameters have been set identically except for the nozzle and print bed temperature and the length and speed of retraction. In particular, the AI suggested parameters and the previously optimized ones by authors starting from literature parameters were used (the three sets were reported in Table 2, but the only two on the right were used in the experimental validation).

The sample model was imported into the slicer software (Fig. 8a), where the printing and process parameters were chosen, and after the samples were printed by an FDM printer. The printed samples were shown in Fig. 8b. The sample printed with the optimized parameters highlights a minor intra-layer adhesion with major presence of voids with respect to the one printed using the AI suggestions. The presence of voids in the first sample is also confirmed by the density measurements of the samples, which are 1.119 g/cm^3 and 1.124 g/cm^3 compared to a nominal density for the used TPU of 1.21 g/cm^3. In both cases, there is a deviation from the solid that is normal for FDM printed parts. However, the parameters suggested by AI allow the deviation to be further reduced, considering other things being equal, such as

Table 2 FDM printing parameters used for printing TPU (two columns on the right are effectively used)

	Reference	Optimized	AI
T Nozzle	235 °C	245 °C	220 °C
Nozzle diameter	0.4 mm	0.4 mm	0.4 mm
T printer bed	50 °C	50 °C	55 °C
Printing speed	1.8 mm^3/s	1.8 min^3/s	1.8 mm^3/s
Infill	100%	100%	100%
Pattern	Linear	Linear	Linear
Raster angle	45°	45°	45°
Layer thickness	0.3 mm	0.3 mm	0.3 mm
Retraction length	2 mm	2 mm	1 mm
Retraction speed	45 mm/s	30 mm/s	15 mm/s

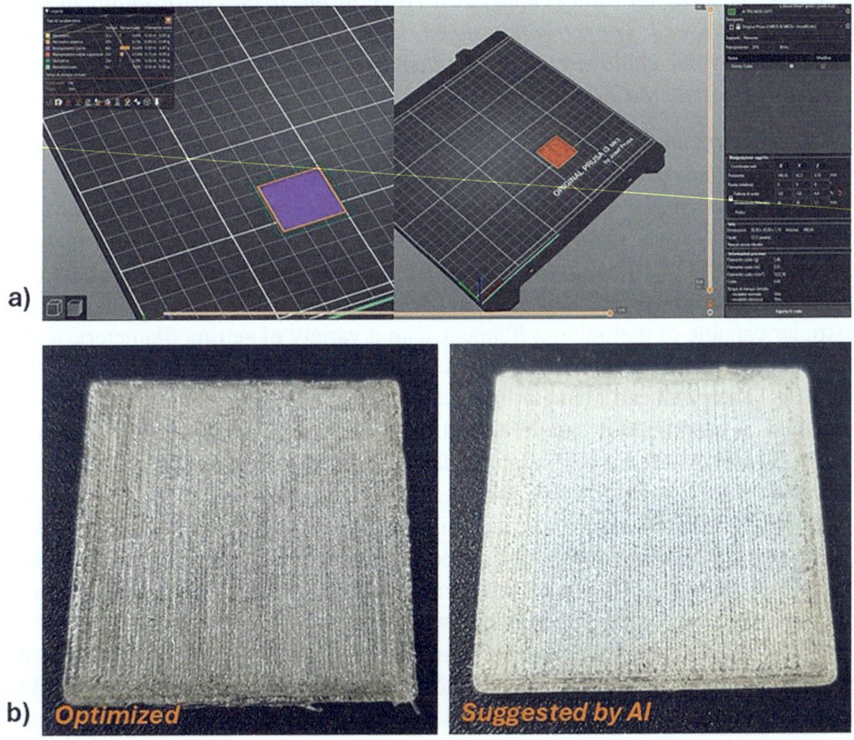

Fig. 8 Sample model in slicer software area (**a**) and printed samples (**b**)

material, printer, nozzle type, etc. Also in this second case, a simple application case highlights that commercial AI systems are already able to provide plausible answers applied to AM in the first instance.

7 Conclusions and Future Scope

One rapidly developing technology that has the amazing potential to completely transform a number of industrial areas is additive manufacturing. It has a number of benefits, such as the exceptional ability to build intricate structures, a notable decrease in material waste, and increased cost-effectiveness for small manufacturing batches. Rapid development of product cycles, personalized customization, and the production of intricate parts that were previously believed to be difficult to produce have all been made possible by AM.

This chapter presents an overview of the next generation AM perspective with the support of AI systems and recent evolutions about it. There are different ways in which AI systems are improving AM technologies, both in design, production, and quality control. Incorporating AI throughout the product design phase offers a viable way to take into consideration a number of aspects of design and their complex interactions in order to achieve the intended manufacturing efficacy. Furthermore, the manufacturing process for customized human-centered goods may be optimized, and the quality performance can be comprehensively and effectively evaluated with the introduction of AI in the AM production and evaluation phases. Among these is the integration of robotics and machine learning into AM to streamline production procedures, achieving previously unheard-of levels of productivity while reducing errors. The growing interest in AM is also fueled by a long-standing dedication to sustainability and environmentally friendly production methods. A new era of communication and data interchange within AM is being ushered in by Industry 5.0, Internet of Things (IoT) and AI-driven systems. Improved control and monitoring capabilities are made possible by this connectivity, opening the door for increased productivity, unmatched quality, and unmatched customization for manufacturers.

Next step is in the "Smart AM" concept that was introduced and which can be described as "a fully connected, collaboration AM system which reacts in real time to promote the pervasive and intelligent design, production, and supply of 3D produced items," represents a new development in AM processes in this context. This notion of smart AM was first presented by Wang et al., who also suggested a smart AM architecture based on cloud-edge computing that may be used to meet the computational needs of various agent types [122]. In the cloud environment, global consideration can be realized, and at the edge nodes, quick response times can be attained.

Acknowledgments This research was financially supported by Universitas Mercatorum under the 2024 FIN/RIC program (Id: 23-FIN/RIC)

References

1. Gavade, D. AI-driven process automation in manufacturing business administration: Efficiency and cost-efficiency analysis. In *Proceedings of the 7th IET smart cities symposium (SCS 2023)*, Bahrain, Saudi Arabia, 3–5 December 2023; pp. 677–684
2. Iftikhar, N., Nordbjerg, F. E., Baattrup-Andersen, T., & Jeppesen, K. (2020). Industry 4.0: Sensor data analysis using machine learning. In *Data management technologies and applications; CCIS* (Vol. 1255, pp. 37–58). Springer.
3. Soumyashree, S.; Gupta, A.; Biswas, B. An automation designed for Industry 4.0 using robotics and sensors that based on IoT & machine learning. In Proceedings of the 2023 International Conference on Sustainable Technologies in Civil and Environmental Engineering, ICSTCE 2023, Pimpri, India, 15–16 June 2023; Volume 405
4. Klumpp, M. (2018). Innovation potentials and pathways merging AI, CPS, and IoT. *Applied System Innovation, 1*(1), 5.
5. Aldoseri, A., Al-Khalifa, K. N., & Hamouda, A. M. (2024). AI-powered innovation in digital transformation: Key pillars and industry impact. *Sustainability, 16*(5), 1790.
6. Fountaine, T., McCarthy, B., & Saleh, T. (2019). Building the AI-powered organization. *Harvard Business Review, 97*(4), 62–73.
7. Rakholia, R., Suarez-Cetrulo, A. L., Singh, M., & Carbajo, R. S. (2024). Advancing manufacturing through artificial intelligence: Current landscape, perspectives, best practices, challenges, and future direction. *IEEE Access, 12*, 131621.
8. Sahoo, S., & Lo, C. Y. (2022). Smart manufacturing powered by recent technological advancements: A review. *Journal of Manufacturing Systems, 64*, 236–250.
9. Xu, S., Lu, Y., & Yu, C. (2024). Augmented reality-assisted cloud additive manufacturing with digital twin technology for multi-stakeholder value co-creation in product innovation. *Heliyon, 10*(4), e25722.
10. Dilberoglu, U. M., Gharehpapagh, B., Yaman, U., & Dolen, M. (2017). The role of additive manufacturing in the era of industry 4.0. *Procedia Manufacturing, 11*, 545–554.
11. Hunde, B. R., & Woldeyohannes, A. D. (2022). Future prospects of computer-aided design (CAD)—A review from the perspective of artificial intelligence (AI), extended reality, and 3D printing. *Results in Engineering, 14*, 100478.
12. Elyasi, M., Thevenin, S., & Cerqueus, A. (2024). Use of AI in assembly line design and worker and equipment management: Review and future directions. *Flexible Services and Manufacturing Journal*, 1–42.
13. Živković, M., Žujović, M., & Milošević, J. (2023). Architectural 3d-printed structures created using artificial intelligence: A review of techniques and applications. *Applied Sciences, 13*(19), 10671.
14. Ucar, A., Karakose, M., & Kırımça, N. (2024). Artificial intelligence for predictive maintenance applications: Key components, trustworthiness, and future trends. *Applied Sciences, 14*(2), 898.
15. Li, C., Pisignano, D., Zhao, Y., & Xue, J. (2020). Advances in medical applications of additive manufacturing. *Engineering, 6*(11), 1222–1231.
16. Khalifa, M., & Albadawy, M. (2024). AI in diagnostic imaging: Revolutionising accuracy and efficiency. *Computer Methods and Programs in Biomedicine Update, 5*, 100146.
17. Gao, R. X., Krüger, J., Merklein, M., Möhring, H. C., & Váncza, J. (2024). Artificial intelligence in manufacturing: State of the art, perspectives, and future directions. *CIRP Annals., 73*, 723.
18. Bikas, H., Stavropoulos, P., & Chryssolouris, G. (2016). Additive manufacturing methods and modelling approaches: A critical review. *International Journal of Advanced Manufacturing Technology, 83*, 389–405.
19. Seepersad, C. C. (2014). Challenges and opportunities in design for additive manufacturing. *3D Printing and Additive Manufacturing, 1*(1), 10–13.

20. Campbell, T. A., Tibbits, S., & Garrett, B. (2014). The programmable world. *Scientific American, 311*(5), 60–65.
21. Tian, X., Wu, L., Gu, D., Yuan, S., Zhao, Y., Li, X., Ouyang, L., Song, B., Gao, T., He, J., Lin, X., Lin, F., Zhu, J., & Li, D. (2022). Roadmap for additive manufacturing: Toward intellectualization and industrialization. *Chinese Journal of Mechanical Engineering: Additive Manufacturing Frontiers, 1*(1), 100014.
22. Meng, L., McWilliams, B., Jarosinski, W., Park, H. Y., Jung, Y. G., Lee, J., & Zhang, J. (2020). Machine learning in additive manufacturing: A review. *JOM, 72*, 2363–2377.
23. Meng, L., Zhang, W., Quan, D., Shi, G., Tang, L., Hou, Y., et al. (2020). From topology optimization design to additive manufacturing: Today's success and tomorrow's roadmap. *Archives of Computational Methods in Engineering, 27*, 805–830.
24. Mustafa, K., Algunaid, A., & Liu, J. (2022). Decision support system to select a 3D printing process/machine and material from a large-scale options pool. *The International Journal of Advanced Manufacturing Technology, 121*(11), 7643–7659.
25. Martínez-García, A., Monzón, M., & Paz, R. (2021). Standards for additive manufacturing technologies: Structure and impact. In *Additive manufacturing* (pp. 395–408). Elsevier.
26. Kumar, M. B., & Sathiya, P. (2021). Methods and materials for additive manufacturing: A critical review on advancements and challenges. *Thin-Walled Structures, 159*, 107228.
27. Chua, C. K., Wong, C. H., & Yeong, W. Y. (2017). *Standards, quality control, and measurement sciences in 3D printing and additive manufacturing*. Academic Press.
28. Chua, Z. Y., Ahn, I. H., & Moon, S. K. (2017). Process monitoring and inspection systems in metal additive manufacturing: Status and applications. *International Journal of Precision Engineering and Manufacturing-Green Technology, 4*, 235–245.
29. Babu, S. S., Mourad, A. H. I., Harib, K. H., & Vijayavenkataraman, S. (2023). Recent developments in the application of machine-learning towards accelerated predictive multiscale design and additive manufacturing. *Virtual and Physical Prototyping, 18*(1), e2141653.
30. Yamazaki, T. (2016). Development of a hybrid multi-tasking machine tool: Integration of additive manufacturing technology with CNC machining. *Procedia CIRP, 42*, 81–86.
31. Boschetto, A., & Bottini, L. (2015). Roughness prediction in coupled operations of fused deposition modeling and barrel finishing. *Journal of Materials Processing Technology, 219*, 181–192.
32. Singh, S., Ramakrishna, S., & Singh, R. (2017). Material issues in additive manufacturing: A review. *Journal of Manufacturing Processes, 25*, 185–200.
33. Tofail, S. A., Koumoulos, E. P., Bandyopadhyay, A., Bose, S., O'Donoghue, L., & Charitidis, C. (2018). Additive manufacturing: Scientific and technological challenges, market uptake and opportunities. *Materials Today, 21*(1), 22–37.
34. Pérez, M., Carou, D., Rubio, E. M., & Teti, R. (2020). Current advances in additive manufacturing. *Procedia CIRP, 88*, 439–444.
35. Pragana, J. P. M., Sampaio, R. F., Bragança, I. M. F., Silva, C. M. A., & Martins, P. A. F. (2021). Hybrid metal additive manufacturing: A state–of–the-art review. *Advances in Industrial and Manufacturing Engineering, 2*, 100032.
36. Strong, D., Kay, M., Conner, B., Wakefield, T., & Manogharan, G. (2018). Hybrid manufacturing–integrating traditional manufacturers with additive manufacturing (AM) supply chain. *Additive Manufacturing, 21*, 159–173.
37. Elser, A., Königs, M., Verl, A., & Servos, M. (2018). On achieving accuracy and efficiency in additive manufacturing: Requirements on a hybrid CAM system. *Procedia CIRP, 72*, 1512–1517.
38. Bose, S., Ke, D., Sahasrabudhe, H., & Bandyopadhyay, A. (2018). Additive manufacturing of biomaterials. *Progress in Materials Science, 93*, 45–111.
39. Monzón, M. (2018). Biomaterials and additive manufacturing: Osteochondral scaffold innovation applied to osteoarthritis (BAMOS project). *Journal of Zhejiang University: Science A, 19*, 329–330.

40. Zadpoor, A. A. (2019). Mechanical performance of additively manufactured meta-biomaterials. *Acta Biomaterialia, 85*, 41–59.
41. Ahmed, A., Arya, S., Gupta, V., Furukawa, H., & Khosla, A. (2021). 4D printing: Fundamentals, materials, applications and challenges. *Polymer, 228*, 123926.
42. Bhushan, B., & Caspers, M. (2017). An overview of additive manufacturing (3D printing) for microfabrication. *Microsystem Technologies, 23*, 1117–1124.
43. Yi, L., Shokrani, A., Bertolini, R., Mutilba, U., Guerra, M. G., Loukaides, E. G., & Woizeschke, P. (2022). Optical sensor-based process monitoring in additive manufacturing. *Procedia CIRP, 115*, 107–112.
44. Umaras, E., & Tsuzuki, M. S. (2017). Additive manufacturing-considerations on geometric accuracy and factors of influence. *IFAC-PapersOnLine, 50*(1), 14940–14945.
45. Kumbhar, N. N., & Mulay, A. V. (2018). Post processing methods used to improve surface finish of products which are manufactured by additive manufacturing technologies: A review. *Journal of The Institution of Engineers (India): Series C, 99*, 481–487.
46. Zawada, B., Ukpai, G., Powell-Palm, M. J., & Rubinsky, B. (2018). Multi-layer cryolithography for additive manufacturing. *Progress in Additive Manufacturing, 3*, 245–255.
47. John Samuel Dilip, J., Kalid Rafi, H., & Janaki Ram, G. D. (2011). A new additive manufacturing process based on friction deposition. *Transactions of the Indian Institute of Metals, 64*, 27–30.
48. Peng, T., Kellens, K., Tang, R., Chen, C., & Chen, G. (2018). Sustainability of additive manufacturing: An overview on its energy demand and environmental impact. *Additive Manufacturing, 21*, 694–704.
49. Liu, C., Tian, W., & Kan, C. (2022). When AI meets additive manufacturing: Challenges and emerging opportunities for human-centered products development. *Journal of Manufacturing Systems, 64*, 648–656.
50. Wang, C., Tan, X. P., Tor, S. B., & Lim, C. S. (2020). Machine learning in additive manufacturing: State-of-the-art and perspectives. *Additive Manufacturing, 36*, 101538.
51. Bunian, S., Al-Ebrahim, M. A., & Nour, A. A. (2024). Role and applications of artificial intelligence and machine learning in manufacturing engineering: A review. *Engineered Science, 29*, 1088.
52. Rebala, G., Ravi, A., & Churiwala, S. (2019). *An introduction to machine learning*. Springer.
53. Guresen, E., & Kayakutlu, G. (2011). Definition of artificial neural networks with comparison to other networks. *Procedia Computer Science, 3*, 426–433.
54. Krenker, A., Bešter, J., & Kos, A. (2011). Introduction to the artificial neural networks. In *Artificial neural networks: Methodological advances and biomedical applications* (pp. 1–18). InTech.
55. Haussecker, H., & Geissler, P. (2000). *Computer vision and applications*. Elsevier Science & Technology.
56. Cheng, Y., Zhang, Y., Ji, P., Xu, W., Zhou, Z., & Tao, F. (2018). Cyber-physical integration for moving digital factories forward towards smart manufacturing: A survey. *The International Journal of Advanced Manufacturing Technology, 97*, 1209–1221.
57. Ladani, L. J. (2021). Applications of artificial intelligence and machine learning in metal additive manufacturing. *Journal of Physics: Materials, 4*(4), 042009.
58. Cao, L., Li, J., Hu, J., Liu, H., Wu, Y., & Zhou, Q. (2021). Optimization of surface roughness and dimensional accuracy in LPBF additive manufacturing. *Optics & Laser Technology, 142*, 107246.
59. Geng, S., Luo, Q., Liu, K., Li, Y., Hou, Y., & Long, W. (2023). Research status and prospect of machine learning in construction 3D printing. *Case Studies in Construction Materials, 18*, e01952.
60. Nguyen-Van, V., Li, S., Liu, J., Nguyen, K., & Tran, P. (2023). Modelling of 3D concrete printing process: A perspective on material and structural simulations. *Additive Manufacturing, 61*, 103333.

61. Parisi, F., Sangiorgio, V., Parisi, N., Mangini, A. M., Fanti, M. P., & Adam, J. M. (2024). A new concept for large additive manufacturing in construction: Tower crane-based 3D printing controlled by deep reinforcement learning. *Construction Innovation, 24*(1), 8–32.
62. Charrier, M., & Ouellet-Plamondon, C. M. (2022). Artificial neural network for the prediction of the fresh properties of cementitious materials. *Cement and Concrete Research, 156*, 106761.
63. Davtalab, O., Kazemian, A., Yuan, X., & Khoshnevis, B. (2022). Automated inspection in robotic additive manufacturing using deep learning for layer deformation detection. *Journal of Intelligent Manufacturing, 33*(3), 771–784.
64. Nefs, K., Menkovski, V., Bos, F. P., Suiker, A. S., & Salet, T. A. (2023). Automated image segmentation of 3D printed fibrous composite micro-structures using a neural network. *Construction and Building Materials, 365*, 130099.
65. Nicholas, P., Rossi, G., Williams, E., Bennett, M., & Schork, T. (2020). Integrating real-time multi-resolution scanning and machine learning for conformal robotic 3D printing in architecture. *International Journal of Architectural Computing, 18*(4), 371–384.
66. Fan, D., Zhu, J., Fan, M., Lu, J. X., Chu, S. H., Dong, E., & Yu, R. (2023). Intelligent design and manufacturing of ultra-high performance concrete (UHPC)—A review. *Construction and Building Materials, 385*, 131495.
67. Lao, W., Li, M., Wong, T. N., Tan, M. J., & Tjahjowidodo, T. (2020). Improving surface finish quality in extrusion-based 3D concrete printing using machine learning-based extrudate geometry control. *Virtual and Physical Prototyping, 15*(2), 178–193.
68. Tuvayanond, W., & Prasittisopin, L. (2023). Design for manufacture and assembly of digital fabrication and additive manufacturing in construction: A review. *Buildings, 13*(2), 429.
69. Kazemian, A., & Khoshnevis, B. (2021). Real-time extrusion quality monitoring techniques for construction 3D printing. *Construction and Building Materials, 303*, 124520.
70. Mechtcherine, V., Van Tittelboom, K., Kazemian, A., Kreiger, E., Nematollahi, B., Nerella, V. N., et al. (2022). A roadmap for quality control of hardening and hardened printed concrete. *Cement and Concrete Research, 157*, 106800.
71. Ciccone, F., Bacciaglia, A., & Ceruti, A. (2023). Optimization with artificial intelligence in additive manufacturing: A systematic review. *Journal of the Brazilian Society of Mechanical Sciences and Engineering, 45*(6), 303.
72. Hariri, R. H., Fredericks, E. M., & Bowers, K. M. (2019). Uncertainty in big data analytics: Survey, opportunities, and challenges. *Journal of Big Data, 6*(1), 1–16.
73. Sarker, I. H. (2021). Data science and analytics: An overview from data-driven smart computing, decision-making and applications perspective. *SN Computer Science, 2*(5), 377.
74. Ullah, I., Adhikari, D., Su, X., Palmieri, F., Wu, C., & Choi, C. (2024). Integration of data science with the intelligent IoT (IIoT): Current challenges and future perspectives. *Digital Communications and Networks.*
75. Jiang, J., Xiong, Y., Zhang, Z., & Rosen, D. W. (2022). Machine learning integrated design for additive manufacturing. *Journal of Intelligent Manufacturing, 33*(4), 1073–1086.
76. Decker, N., Lyu, M., Wang, Y., & Huang, Q. (2021). Geometric accuracy prediction and improvement for additive manufacturing using triangular mesh shape data. *Journal of Manufacturing Science and Engineering, 143*(6), 061006.
77. Plocher, J., & Panesar, A. (2019). Review on design and structural optimisation in additive manufacturing: Towards next-generation lightweight structures. *Materials & Design, 183*, 108164.
78. Leary, M., Merli, L., Torti, F., Mazur, M., & Brandt, M. (2014). Optimal topology for additive manufacture: A method for enabling additive manufacture of support-free optimal structures. *Materials & Design, 63*, 678–690.
79. Tang, Y., Dong, G., Xiong, Y., & Wang, Q. (2021). Data-driven design of customized porous lattice sole fabricated by additive manufacturing. *Procedia Manufacturing, 53*, 318–326.
80. Yao, X., Moon, S. K., & Bi, G. (2017). A hybrid machine learning approach for additive manufacturing design feature recommendation. *Rapid Prototyping Journal, 23*(6), 983–997.

81. Johnson, N. S., Vulimiri, P. S., To, A. C., Zhang, X., Brice, C. A., Kappes, B. B., & Stebner, A. P. (2020). Invited review: Machine learning for materials developments in metals additive manufacturing. *Additive Manufacturing, 36*, 101641.
82. Papadimitriou, I., Gialampoukidis, I., Vrochidis, S., & Kompatsiaris, I. (2024). AI methods in materials design, discovery and manufacturing: A review. *Computational Materials Science, 235*, 112793.
83. Polak, M. P., & Morgan, D. (2024). Extracting accurate materials data from research papers with conversational language models and prompt engineering. *Nature Communications, 15*(1), 1569.
84. Jennings, P. C., Lysgaard, S., Hummelshøj, J. S., Vegge, T., & Bligaard, T. (2019). Genetic algorithms for computational materials discovery accelerated by machine learning. *npj Computational Materials, 5*(1), 46.
85. Ferreira, S., Leitão, G., Silva, I., Martins, A., & Ferrari, P. (2020). Evaluating human-machine translation with attention mechanisms for industry 4.0 environment SQL-based systems. In *2020 IEEE international workshop on metrology for industry 4.0 & IoT* (pp. 229–234). IEEE.
86. Kuniyoshi, F., Ozawa, J., & Miwa, M. (2021). Analyzing research trends in inorganic materials literature using nlp. In *Joint European conference on machine learning and knowledge discovery in databases* (pp. 319–334). Springer International Publishing.
87. Chen, Y., Li, T., Scarpa, F., & Wang, L. (2017). Lattice metamaterials with mechanically tunable Poisson's ratio for vibration control. *Physical Review Applied, 7*(2), 024012.
88. Kharat, V. J., Singh, P., Raju, G. S., Yadav, D. K., Gupta, M. S., Arun, V., … & Singh, N. (2023). Additive manufacturing (3D printing): A review of materials, methods, applications and challenges. *Materials Today: Proceedings.*
89. Wang, X., Liu, A., & Kara, S. (2022). Machine learning for engineering design toward smart customization: A systematic review. *Journal of Manufacturing Systems, 65*, 391–405.
90. Zhang, Y., Bernard, A., Gupta, R. K., & Harik, R. (2014). Evaluating the design for additive manufacturing: A process planning perspective. *Procedia CIRP, 21*, 144–150.
91. Dong, G., Tang, Y., Li, D., & Zhao, Y. F. (2020). Design and optimization of solid lattice hybrid structures fabricated by additive manufacturing. *Additive Manufacturing, 33*, 101116.
92. Mukherjee, S., Lu, D., Raghavan, B., Breitkopf, P., Dutta, S., Xiao, M., & Zhang, W. (2021). Accelerating large-scale topology optimization: State-of-the-art and challenges. *Archives of Computational Methods in Engineering, 28*, 1–23.
93. Guo, X., Zhou, J., Zhang, W., Du, Z., Liu, C., & Liu, Y. (2017). Self-supporting structure design in additive manufacturing through explicit topology optimization. *Computer Methods in Applied Mechanics and Engineering, 323*, 27–63.
94. Liu, C., Du, Z., Zhang, W., Zhu, Y., & Guo, X. (2017). Additive manufacturing-oriented design of graded lattice structures through explicit topology optimization. *Journal of Applied Mechanics, 84*(8), 081008.
95. Liu, J. S., Parks, G. T., & Clarkson, P. J. (2000). Metamorphic development: A new topology optimization method for continuum structures. *Structural and Multidisciplinary Optimization, 20*, 288–300.
96. Gao, Z., Dong, G., Tang, Y., & Zhao, Y. F. (2023). Machine learning aided design of conformal cooling channels for injection molding. *Journal of Intelligent Manufacturing, 34*, 1–19.
97. Li, R., Jin, M., & Paquit, V. C. (2021). Geometrical defect detection for additive manufacturing with machine learning models. *Materials & Design, 206*, 109726.
98. Li, S., Yuan, S., Zhu, J., Zhang, W., Zhang, H., & Li, J. (2021). Multidisciplinary topology optimization incorporating process-structure-property-performance relationship of additive manufacturing. *Structural and Multidisciplinary Optimization, 63*, 2141–2157.
99. Shaoying, L. I., Hongkai, W. E. I., Shangqin, Y. U. A. N., Jihong, Z. H. U., Jiang, L. I., & Zhang, W. (2023). Collaborative optimization design of process parameter and structural topology for laser additive manufacturing. *Chinese Journal of Aeronautics, 36*(1), 456–467.
100. Singh, V., & Gu, N. (2012). Towards an integrated generative design framework. *Design Studies, 33*(2), 185–207.

101. McCormack, J., Dorin, A., & Innocent, T. (2004). *Generative design: A paradigm for design research.*
102. Dhokia, V., Essink, W. P., & Flynn, J. M. (2017). A generative multi-agent design methodology for additively manufactured parts inspired by termite nest building. *CIRP Annals, 66*(1), 153–156.
103. Lebaal, N., Zhang, Y., Demoly, F., Roth, S., Gomes, S., & Bernard, A. (2019). Optimised lattice structure configuration for additive manufacturing. *CIRP Annals, 68*(1), 117–120.
104. Hashmi, A. W., Mali, H. S., & Meena, A. (2023). Improving the surface characteristics of additively manufactured parts: A review. *Materials Today: Proceedings, 81*, 723–738.
105. Castelão, A., Soares, B. A. R., Machado, C. M., Leite, M., & Mourão, A. J. M. (2019). Design for AM: Contributions from surface finish, part geometry and part positioning. *Procedia CIRP, 84*, 491–495.
106. Batu, T., Lemu, H. G., & Shimels, H. (2023). Application of artificial intelligence for surface roughness prediction of additively manufactured components. *Materials, 16*(18), 6266.
107. Kim, T., Kim, J. G., Park, S., Kim, H. S., Kim, N., Ha, H., et al. (2023). Virtual surface morphology generation of Ti-6Al-4V directed energy deposition via conditional generative adversarial network. *Virtual and Physical Prototyping, 18*(1), e2124921.
108. Blakey-Milner, B., Gradl, P., Snedden, G., Brooks, M., Pitot, J., Lopez, E., et al. (2021). Metal additive manufacturing in aerospace: A review. *Materials & Design, 209*, 110008.
109. Huang, J., Kwok, T. H., Zhou, C., & Xu, W. (2019). Surfel convolutional neural network for support detection in additive manufacturing. *The International Journal of Advanced Manufacturing Technology, 105*, 3593–3604.
110. Yang, S., Page, T., Zhang, Y., & Zhao, Y. F. (2020). Towards an automated decision support system for the identification of additive manufacturing part candidates. *Journal of Intelligent Manufacturing, 31*(8), 1917–1933.
111. Han, Y., Griffiths, R. J., Hang, Z. Y., & Zhu, Y. (2020). Quantitative microstructure analysis for solid-state metal additive manufacturing via deep learning. *Journal of Materials Research, 35*(15), 1936–1948.
112. Alejandrino, J. D., Concepcion, R. S., Lauguico, S. C., Tobias, R. R., Venancio, L., Macasaet, D., et al. (2020). A machine learning approach of lattice infill pattern for increasing material efficiency in additive manufacturing processes. *International Journal of Mechanical Engineering and Robotics Research, 9*(9), 1253–1263.
113. Mora, S., Pugno, N. M., & Misseroni, D. (2022). 3D printed architected lattice structures by material jetting. *Materials Today, 59*, 107–132.
114. Rafiee, M., Farahani, R. D., & Therriault, D. (2020). Multi-material 3D and 4D printing: A survey. *Advanced Science, 7*(12), 1902307.
115. Khan, M. S., Sanchez, F., & Zhou, H. (2020). 3-D printing of concrete: Beyond horizons. *Cement and Concrete Research, 133*, 106070.
116. Robayo-Salazar, R., de Gutiérrez, R. M., Villaquirán-Caicedo, M. A., & Arjona, S. D. (2023). 3D printing with cementitious materials: Challenges and opportunities for the construction sector. *Automation in Construction, 146*, 104693.
117. Amini, A., Guijt, R. M., Themelis, T., De Vos, J., & Eeltink, S. (2023). Recent developments in digital light processing 3D-printing techniques for microfluidic analytical devices. *Journal of Chromatography A, 1692*, 463842.
118. Quan, H., Zhang, T., Xu, H., Luo, S., Nie, J., & Zhu, X. (2020). Photo-curing 3D printing technique and its challenges. *Bioactive Materials, 5*(1), 110–115.
119. Parhi, R. (2023). Recent advances in 3D printed microneedles and their skin delivery application in the treatment of various diseases. *Journal of Drug Delivery Science and Technology, 84*, 104395.
120. Kanishka, K., & Acherjee, B. (2023). Revolutionizing manufacturing: A comprehensive overview of additive manufacturing processes, materials, developments, and challenges. *Journal of Manufacturing Processes, 107*, 574–619.

121. Wang, Y., Zheng, P., Xu, X., Yang, H., & Zou, J. (2019). Production planning for cloud-based additive manufacturing—A computer vision-based approach. *Robotics and Computer-Integrated Manufacturing, 58*, 145–157.
122. Wang, Y., Zheng, P., Peng, T., Yang, H., & Zou, J. (2020). Smart additive manufacturing: Current artificial intelligence-enabled methods and future perspectives. *Science China Technological Sciences, 63*(9), 1600–1611.
123. Mançanares, C. G., de S. Zancul, E., Cavalcante da Silva, J., & Cauchick Miguel, P. A. (2015). Additive manufacturing process selection based on parts' selection criteria. *The International Journal of Advanced Manufacturing Technology, 80*, 1007–1014.
124. Kadkhoda-Ahmadi, S., Hassan, A., & Asadollahi-Yazdi, E. (2019). Process and resource selection methodology in design for additive manufacturing. *The International Journal of Advanced Manufacturing Technology, 104*, 2013–2029.
125. İç, Y. T. (2012). An experimental design approach using TOPSIS method for the selection of computer-integrated manufacturing technologies. *Robotics and Computer-Integrated Manufacturing, 28*(2), 245–256.
126. Qin, Y., Qi, Q., Scott, P. J., & Jiang, X. (2020). An additive manufacturing process selection approach based on fuzzy Archimedean weighted power Bonferroni aggregation operators. *Robotics and Computer-Integrated Manufacturing, 64*, 101926.
127. Wang, Y., Zhong, R. Y., & Xu, X. (2018). A decision support system for additive manufacturing process selection using a hybrid multiple criteria decision-making method. *Rapid Prototyping Journal, 24*(9), 1544–1553.
128. Liu, W., Zhu, Z., & Ye, S. (2020). A decision-making methodology integrated in product design for additive manufacturing process selection. *Rapid Prototyping Journal, 26*(5), 895–909.
129. Verma, A., & Rai, R. (2017). Sustainability-induced dual-level optimization of additive manufacturing process. *The International Journal of Advanced Manufacturing Technology, 88*, 1945–1959.
130. Jiang, J., & Ma, Y. (2020). Path planning strategies to optimize accuracy, quality, build time and material use in additive manufacturing: A review. *Micromachines, 11*(7), 633.
131. Shamsaei, N., Yadollahi, A., Bian, L., & Thompson, S. M. (2015). An overview of direct laser deposition for additive manufacturing; part II: Mechanical behavior, process parameter optimization and control. *Additive Manufacturing, 8*, 12–35.
132. Thompson, M. K., Moroni, G., Vaneker, T., Fadel, G., Campbell, R. I., Gibson, I., et al. (2016). Design for Additive Manufacturing: Trends, opportunities, considerations, and constraints. *CIRP Annals, 65*(2), 737–760.
133. Ulu, E., Huang, R., Kara, L. B., & Whitefoot, K. S. (2019). Concurrent structure and process optimization for minimum cost metal additive manufacturing. *Journal of Mechanical Design, 141*(6), 061701.
134. Dada, M., Popoola, P., Mathe, N., Pityana, S., Adeosun, S., Aramide, O., & Lengopeng, T. (2020). Process optimization of high entropy alloys by laser additive manufacturing. *Engineering Reports, 2*(10), e12252.
135. Tian, W., Ma, J., & Alizadeh, M. (2019). Energy consumption optimization with geometric accuracy consideration for fused filament fabrication processes. *The International Journal of Advanced Manufacturing Technology, 103*, 3223–3233.
136. Huang, K., Kain, C., Diaz-Vallejo, N., Sohn, Y., & Zhou, L. (2021). High throughput mechanical testing platform and application in metal additive manufacturing and process optimization. *Journal of Manufacturing Processes, 66*, 494–505.
137. Urhal, P., Weightman, A., Diver, C., & Bartolo, P. (2019). Robot assisted additive manufacturing: A review. *Robotics and Computer-Integrated Manufacturing, 59*, 335–345.
138. Reisch, R., Hauser, T., Kamps, T., & Knoll, A. (2020). Robot based wire arc additive manufacturing system with context-sensitive multivariate monitoring framework. *Procedia Manufacturing, 51*, 732–739.

139. Kouraytem, N., Li, X., Tan, W., Kappes, B., & Spear, A. D. (2021). Modeling process–structure–property relationships in metal additive manufacturing: A review on physics-driven versus data-driven approaches. *Journal of Physics: Materials, 4*(3), 032002.
140. Ulkir, O., Kuncan, M. S., & Bayraklılar; M. (2024). Raster angle prediction of additive manufacturing process using machine learning algorithm. *Applied Sciences, 14*(5), 2046.
141. Elbadawi, M., McCoubrey, L. E., Gavins, F. K. H., Ong, J. J., Goyanes, A., Gaisford, S., & Basit, A. W. (2021). Harnessing artificial intelligence for the next generation of 3D printed medicines. *Advanced Drug Delivery Reviews, 175*, 113805.
142. Rojek, I., Dostatni, E., Kopowski, J., Macko, M., & Mikołajewski, D. (2022). AI-based support system for monitoring the quality of a product within Industry 4.0 paradigm. *Sensors, 22*(21), 8107.
143. Huo, D., & Wang, X. (2024). A new era in healthcare: The integration of artificial intelligence and microbial. *Medicine in Novel Technology and Devices, 23*, 100319.
144. He, F., Yuan, L., Mu, H., Ros, M., Ding, D., Pan, Z., & Li, H. (2023). Research and application of artificial intelligence techniques for wire arc additive manufacturing: A state-of-the-art review. *Robotics and Computer-Integrated Manufacturing, 82*, 102525.
145. Zhang, Z., Wen, G., & Chen, S. (2019). Weld image deep learning-based on-line defects detection using convolutional neural networks for Al alloy in robotic arc welding. *Journal of Manufacturing Processes, 45*, 208–216.
146. Kim, J., Yang, Z., Ko, H., Cho, H., & Lu, Y. (2023). Deep learning-based data registration of melt-pool-monitoring images for laser powder bed fusion additive manufacturing. *Journal of Manufacturing Systems, 68*, 117–129.
147. Paudel, S. (2023). *3D printing of polymer: Study the common materials for SLS and FDM, and test the FDM printed part.*
148. Chua, K., Khan, I., Malhotra, R., & Zhu, D. (2021). Additive manufacturing and 3D printing of metallic biomaterials. *Engineered Regeneration, 2*, 288–299.
149. Majeed, M., Khan, H. M., Wheatley, G., & Situ, R. (2022). Influence of post-processing on additively manufactured lattice structures. *Journal of the Brazilian Society of Mechanical Sciences and Engineering, 44*(9), 389.
150. Lee, K. H., & Yun, G. J. (2022). Design optimization of thermally conductive support structure for laser powder-bed fusion process with part-scale thermal history. *Additive Manufacturing, 51*, 102627.
151. Colosimo, B. M., Huang, Q., Dasgupta, T., & Tsung, F. (2018). Opportunities and challenges of quality engineering for additive manufacturing. *Journal of Quality Technology, 50*(3), 233–252.
152. Colosimo, B. M. (2020). Quality monitoring and control in additive manufacturing. In *Wiley StatsRef: Statistics reference online* (pp. 1–7). Wiley.
153. Christos, M., Brion, D. A. J., & Pattinson, S. W. (2024). Iterative learning for efficient additive mass production. *Additive Manufacturing, 89*, 104271.
154. Valizadeh, M., & Wolff, S. J. (2022). Convolutional neural network applications in additive manufacturing: A review. *Advances in Industrial and Manufacturing Engineering, 4*, 100072.
155. Wu, H., Wang, Y., & Yu, Z. (2016). In situ monitoring of FDM machine condition via acoustic emission. *The International Journal of Advanced Manufacturing Technology, 84*, 1483–1495.
156. He, H., Zhu, Z., Zhang, Y., Zhang, Z., Famakinwa, T., & Yang, R. C. (2024). Machine condition monitoring for defect detection infused deposition modelling process: A review. *The International Journal of Advanced Manufacturing Technology, 132*, 1–30.
157. Zhu, W., Li, H., Shen, S., Wang, Y., Hou, Y., Zhang, Y., & Chen, L. (2024). In-situ monitoring additive manufacturing process with AI edge computing. *Optics & Laser Technology, 171*, 110423.
158. Malik, A., Lhachemi, H., Ploennigs, J., Ba, A., & Shorten, R. (2019). An application of 3D model reconstruction and augmented reality for real-time monitoring of additive manufacturing. *Procedia CIRP, 81*, 346–351.

159. Cai, Y., Xiong, J., Chen, H., & Zhang, G. (2023). A review of in-situ monitoring and process control system in metal-based laser additive manufacturing. *Journal of Manufacturing Systems, 70*, 309–326.
160. Zhu, Q., Liu, Z., & Yan, J. (2020). Machine learning for metal additive manufacturing: Predicting temperature and melt pool fluid dynamics using physics-informed neural networks. *Computational Mechanics, 67*, 619–635.
161. Johnson, M. V., Garanger, K., Hardin, J. O., Berrigan, J. D., Feron, E., & Kalidindi, S. R. (2021). A generalizable artificial intelligence tool for identification and correction of self-supporting structures in additive manufacturing processes. *Additive Manufacturing, 46*, 102191.
162. Ukwaththa, J., Herath, S., & Meddage, D. P. P. (2024). A review of machine learning (ML) and explainable artificial intelligence (XAI) methods in additive manufacturing (3D printing). *Materials Today Communications, 41*, 110294.
163. Bordekar, H., Cersullo, N., Brysch, M., Philipp, J., & Hühne, C. (2025). eXplainable artificial intelligence for automatic defect detection in additively manufactured parts using CT scan analysis. *Journal of Intelligent Manufacturing, 36*, 957–974.
164. Khusheef, A., Shahbazi, M., & Hashemi, R. (2024). Deep learning-based multi-sensor fusion for process monitoring: Application to fused deposition modeling. *Arabian Journal for Science and Engineering, 49*(8), 10501–10522.
165. Jyeniskha, N., Keutayeva, A., Kazbek, G., Ali, M. D. H., & Shehab, E. (2023). *Integrating machine learning model and digital twin system for additive manufacturing* (Vol. 11, pp. 71113–71126). IEEE Access.
166. Kumar, L. J., & Krishnadas Nair, C. G. (2017). Current trends of additive manufacturing in the aerospace industry. In *Advances in 3D printing & additive manufacturing technologies* (pp. 39–54).
167. Zadpoor, A. A., & Malda, J. (2017). Additive manufacturing of biomaterials, tissues, and organs. *Annals of Biomedical Engineering, 45*, 1–11.
168. Lee, J. Y., An, J., & Chua, C. K. (2017). Fundamentals and applications of 3D printing for novel materials. *Applied Materials Today, 7*, 120–133.
169. Turner, B. N., & Gold, S. A. (2015). A review of melt extrusion additive manufacturing processes: II. Materials, dimensional accuracy, and surface roughness. *Rapid Prototyping Journal, 21*(3), 250–261.
170. Lalegani Dezaki, M., Bodaghi, M., Serjouei, A., Afazov, S., & Zolfagharian, A. (2023). Soft pneumatic actuators with controllable stiffness by bio-inspired lattice chambers and fused deposition modeling 3D printing. *Advanced Engineering Materials, 25*(6), 2200797.
171. Penumakala, P. K., Santo, J., & Thomas, A. (2020). A critical review on the fused deposition modeling of thermoplastic polymer composites. *Composites Part B: Engineering, 201*, 108336.
172. Roschli, A., Gaul, K. T., Boulger, A. M., Post, B. K., Chesser, P. C., Love, L. J., et al. (2019). Designing for big area additive manufacturing. *Additive Manufacturing, 25*, 275–285.
173. Banerjee, S. S., & Bhowmick, A. K. (2017). High-temperature thermoplastic elastomers from rubber–plastic blends: A state-of-the-art review. *Rubber Chemistry and Technology, 90*(1), 1–36.
174. Adams, R., Soe, S. P., Santiago, R., Robinson, M., Hanna, B., McShane, G., et al. (2019). A novel pathway for efficient characterisation of additively manufactured thermoplastic elastomers. *Materials & Design, 180*, 107917.
175. Balani, S. B., Chabert, F., Nassiet, V., & Cantarel, A. (2019). Influence of printing parameters on the stability of deposited beads in fused filament fabrication of poly (lactic) acid. *Additive Manufacturing, 25*, 112–121.
176. Blok, L. G., Longana, M. L., Yu, H., & Woods, B. K. (2018). An investigation into 3D printing of fibre reinforced thermoplastic composites. *Additive Manufacturing, 22*, 176–186.

177. Bochmann, L., Bayley, C., Helu, M., Transchel, R., Wegener, K., & Dornfeld, D. (2015). Understanding error generation in fused deposition modeling. *Surface Topography: Metrology and Properties, 3*(1), 014002.
178. Kim, G., Barocio, E., Pipes, R. B., & Sterkenburg, R. (2019). 3D printed thermoplastic polyurethane bladder for manufacturing of fiber reinforced composites. *Additive Manufacturing, 29*, 100809.
179. Domínguez-Robles, J., Utomo, E., Cornelius, V. A., Anjani, Q. K., Korelidou, A., Gonzalez, Z., et al. (2022). TPU-based antiplatelet cardiovascular prostheses prepared using fused deposition modelling. *Materials & Design, 220*, 110837.
180. Stano, G., Pavone, A., & Percoco, G. (2023). 3D printing of shape memory polymers: Embedding nichrome-wires to enhance their performance. In *Selected topics in manufacturing: Emerging trends from the perspective of AITeM's young researchers* (pp. 3–18). Cham.

Machine Learning for Optimized Additive Manufacturing: Challenges, Opportunities, and Sustainability

Wisdom Simwila Kalunga and G. Prasad

1 Introduction

The odyssey of industrialization has its genesis in industry manufacturing paradigm shift 1.0 in the 1800s; this was marked by the advent of mechanization powered by steam and water. During this epoch, manufacturing was linear and labor-intensive, with no concept of automation or any advanced technologies like machine learning (ML) and additive manufacturing (AM) [1]. In industrial engineering terms, production processes were rigid, resource-intensive, and highly dependent on human effort. This led to most industrial setups not having ergonomic standards for its works and being far less human-centric. It is thus safe to say that the notion of machines optimizing themselves or products being fabricated layer upon layer from digital models was beyond imagination. Albeit, as the industrial landscape evolved through successive revolutions, Industry 2.0 ushered in an era of mass production and electricity, industry 3.0 brought in the rise of electronics and automation, and industry 4.0 came with it the genesis of smart technologies and data integration-because of this the manufacturing world begun to shift towards greater efficiency, flexibility and intelligence [2].

Today the world stands at a threshold of Industry 5.0 where human creativity and smart tools like artificial intelligence and additive fabrication come together to shape a new era of industrialization. Unlike its predecessors (I-R 1.0, I-R 2.0, I-R 3.0, I-R 4.0), Industry 5.0 prioritizes personalized, sustainable, and human-centric approaches, with ML and AM playing pivotal roles in achieving these objectives [3]. In contrast to the rather rigid processes in industry 1.0, additive manufacturing in Industry 5.0 makes it possible to create multiple complex geometries and customized products with minimal waste, directly from digital blueprints [4]. Meanwhile,

W. S. Kalunga (✉) · G. Prasad
Department of Aerospace Engineering, Chandigarh University, Mohali, India
e-mail: prasad.e14018@cumail.in

Y. Kumar Singla et al. (eds.), *Digital Manufacturing in Industry 5.0*,
https://doi.org/10.1007/978-3-031-91500-0_12

machine learning offers the capability to optimize these processes, learning from data to enhance precision, reduce defects, roughness, stresses, and improve overall material efficiency [1, 5]. Therefore, a convergence of these technologies (ML, AM), signifies a radical shift in manufacturing capabilities where products are not only built but are intelligently optimized for performance, sustainability, and cost-effectiveness. This chapter aims to explore this transformation, tracing the evolution of its industrial past to the present era, where machine learning and additive manufacturing represent the cutting edge of technological genius of industrial revolution 5.0 [6].

2 Objectives and Scope

The quintessential purpose of this chapter is to explore the intersection of Additive Manufacturing (AM) and Machine Learning within the milieu of Industry 5.0, with a focal point being on the challenges, opportunities, and sustainability implications [7]. This chapter aims to provide an all-encompassing or exhaustive understanding of how these terminologies (AM, ML) are revolutionizing the manufacturing landscape. Furthermore, it involves reshaping industrial practices by emphasizing human-centric innovation, efficiency, and environmental sustainability.

Key Objectives

- Understand the role of Industry 5.0: The chapter seeks to clarify the fundamental principle of Industry 5.0, emphasizing the realignment towards a human-oriented collaborative environment that prioritizes personalization, sustainability, and innovation.
- Exploring additive manufacturing: A detailed analysis of additive manufacturing, its technological fundamentals, its importance in Industry 5.0. This chapter will cover the history, evolution, and various types of 3D printing technologies, highlighting how they enable the production of complex, customized, and sustainable products.
- Examining the impact of machine learning: The integration of ML with additive manufacturing is a core focus, particularly how ML enhances optimization, predictive capabilities, and product quality. This chapter will explore how data-driven insights improve efficiency and drive innovation.
- Addressing challenges and opportunities: It will assess the challenges associated with the adoption of AM and ML in manufacturing, including material limitations, production speeds, and initial cost.
- Emphasizing sustainability: Sustainability is a key theme in Industry 5.0, and this chapter will examine how AM and ML contribute towards environmentally and socially responsible manufacturing practices.

3 Overview of Industry 5.0

Industrialization is a progression in which a society transforms itself from an agrarian-based economy, where most goods and products are produced by hands and manual labor, to an industrial-based economy characterized by the use of machinery, large-scale production, and advanced technologies. This shift involves the evolution of factories, production at scale techniques, and the use of energy sources such as steam, electricity, and fossil fuels to power machines needed for the factory processes. Industrialization leads to significant changes in economic structure, societal systems, and cultural practices [8]. It is associated with increased innovation as people work towards improving systems' efficiency and performance. These periods are mostly associated with rapid economic growth, technological innovation, and profound societal changes, starting with the first industry manufacturing paradigm shift in the eighteenth century.

Industry manufacturing paradigm shift 5.0 represents the next evolution of industrialization, building on the advancements seen in industry 4.0, this time, however, with a stronger commitment to the synergy between humans and machines. While industry manufacturing paradigm shift 4.0 focused on automation, smart technologies, and the consolidation of data-driven system [8], the fifth revolution shifts the paradigm by bringing the human element back into the forefront of manufacturing processes. It advocates for a more personalized, sustainable, and human-centric approach to production. At its core, IR 5.0 is about creating synergy between humans and robots or AI system, rather than merely replacing human labor with machines, it focuses on using advanced technologies such as-cognitive deep learning (AI, ML), robotics and extensive data-analytics to enhance human capabilities, enabling more creativity, customization and innovation. It is therefore safe to say this human-machine collaboration is vital to achieving the fifth revolution's key goals that encompass:

1. Human Centric Design: This goal involves placing human needs, skills, and well-being at the center of fabrication processes.
2. Personalization: Capability of tailoring products to meet individual's needs rather than producing standard products, one size fits all products.
3. Sustainability: Industry 5.0 aims to promote environmental sustainability practices, reducing waste, and optimizing the use of resources to meet sustainability goals.
4. Resilience and Agility: This aims to make industries resilient and adaptable to change and disruptions such as those caused by economic shifts, technological advancements and environmental challenges.

In contrast to the rigid automations of IR4.0, in IR 5.0 a more balanced approach is envisaged where humans can work side by side with intelligent systems, using their creative thinking, problem-solving skills, and ethical judgment to compliment machine precision and speed. This creates a new paradigm for industries, especially manufacturing industries, where personalization and sustainability are becoming key drivers of competitive advantage [9].

3.1 Importance of Additive Manufacturing and Machine Learning in Industry 5.0

In the fifth revolution, the consolidation of additive fabrication or layered manufacturing AM and cognitive deep learning ML represents a pivotal advancement, combining human-centric approaches with cutting-edge technologies to redefine manufacturing and production processes [8]. The importance of AM and ML can be explored in the following dimensions:

Enhanced Customization and Personalization

- Additive manufacturing: allows for the highly customized and complex products directly from digital models. This capability is crucial in IR 5.0, where there is growing demand for personalized products that meet specific customer needs. Unlike traditional manufacturing, which often involves significant retooling and setup for customization, AM enables on-demand production of bespoke items, ranging from customer prosthetics in healthcare to tailored aerospace components [10, 11].
- Machine Learning: with machine learning, data are used to make predictive decisions in maintenance, optimization of designs, and reduced defect formation during manufacturing. This leads to reliability and production efficiency never thus far seen [6, 12].

4 Historical Background

Machine learning historically dates back to the mid-twentieth century. Arthur Samuel first coined the term in 1959, defining it as a computer's ability to learn from experience without being explicitly programmed. The theoretical underpinnings, however, go even further back when Alan Turing's work on Artificial Intelligence AI in the 1950s served as a key precursor [13]. The significant advancements of ML were seen in the 1990s onward with the development of more advanced algorithms, computing capacity, and the explosion of big data. The rise of neural networks and deep learning around 2010 was particularly influential [14]. The intersection, however, of ML and AM as a fused embodiment of smart manufacturing has its genesis in the last decade. Having recognized ML for its optimization power in AM processes, where the manufacturing and customization of complex components require constant adjustments and improvements. Examples of key applications may include: Process optimization, predictive maintenance, Design for layered manufacturing, quality control, and Inspection and supply-chain integration, among many others [15].

Additive manufacturing (AM) has its genesis around the 1980s with the development of technologies such as stereolithography (SLA), invented by Charles Hull in 1984, which became the first widely known form of three-dimensional printing AM[16]. In the beginning, AM was primarily used to create prototypes, resulting in it being coined rapid prototyping in the earlier days. Over the years, other techniques such as selective laser sintering (SLS), Fixed deposit Modeling FadM and Electron Beam Melting have emerged, each offering improvements in precision, material use, and applicability. From its early main focus on prototyping, AM expanded into full-fledged production applications, making a grand entry into manufacturing [17]. The industries that got benefit from this technology may include aerospace, automotive, healthcare, and consumer goods. By 2010, the growth of metal AM technologies allowed the production of highly complex lightweight components used in industries like aerospace and medical implants. As material options began to increase beyond plastics to metals, ceramics, and composites, AM began to be seen not just as a prototyping tool but as a valid method for end-user parts or components manufacturing/production [18].

5 Advantages and Disadvantages of AM-ML Enhanced Manufacturing

In a short span in which AM has come to be known to enable the manufacturing of geometrically complex parts, it has been able to penetrate various markets and industries, as seen in applications of Additive manufacturing [19]. AM has been used in space for three-dimensional production of components and in nuclear power plants and the aero engine fuel nozzles. The market penetration of AM is highly remarkable as it has emerged as useful in a lot of industries and subsectors. The advantages of AM stated above are what make AM unique and full of potential [20]. However, key disadvantages of AM at the moment include the lack of full usage of commercial alloys which are estimated around 5000, while only a handful of these metal alloys can be printed [20, 21]. The other is a failure to control defects. The regular anomalies in Additive fabrication include:

1. Lack of fusion voids
2. Delamination due to residual stresses
3. Cracking and surface roughness

Automated learning systems (ML) fall under the category of cognitive deep computing technologies (AI) that make it possible for computers to learn from data without being given pre-defined instructions or programs. It encompasses the cultivation of complex algorithms that are used to recognize patterns, decision making, and enhance their overall performance with time on the basis of experience. The process would include the training of models on large datasets, allowing them to make predictions, classify data, or uncover insights, which may be applied to new or unseen data [12]. ML therefore uses past data to enable AM to make corrective adjustments based on that data [22]. Machine learning may eliminate or reduce some defects that make a part of the major disadvantages of AM. The Graph plots below are in logarithmic scale:

6 Causes

Figure 1 shows power vs. scanning speed, and Fig. 2 shows deposition rate vs. layer thickness which have direct impact on overall manufacturing and may lead to effects such as lack of fusion voids, surface roughness, cracking, and delaminating. These process variables may be uneasy to change in additive manufacturing due to the expensiveness of equipments and materials compared to traditional methods. This results in the need for machine learning, the lack of adequate control on defects in AM processes is a major challenge that needs to be addressed, and AM enhanced with ML technology address this issue amicably.

7 Effects

With machine learning, data may be used to prevent or optimize defects shown in Figs. 3, 4, 5, and 6 and create perfect parts or components [23]. The graph below shows the use of ML in AM to optimize defects through past data:

It may be deduced from Fig. 7 that input-depth is fairly optimized due to machine enhancement from data-driven ML. The existences of open-source programs that

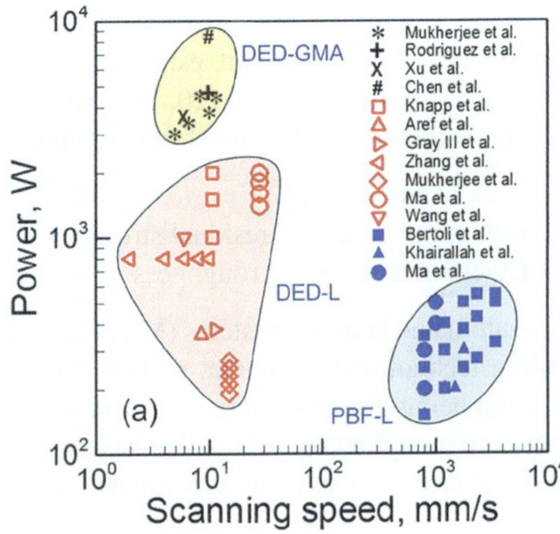

Fig. 1 Power vs. scanning speed

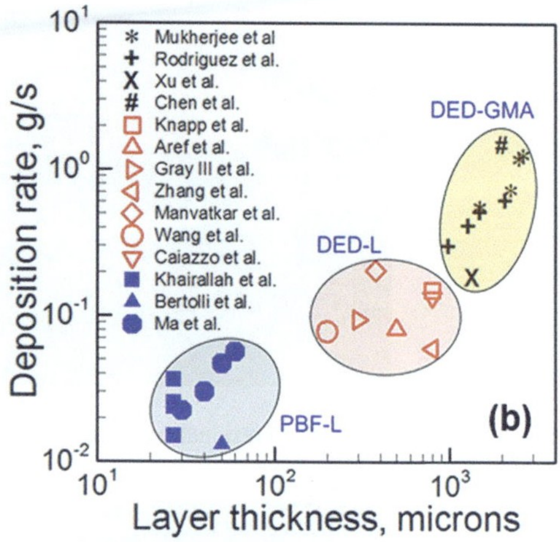

Fig. 2 Deposition rate (g/s) vs Layer mm/s, thickness

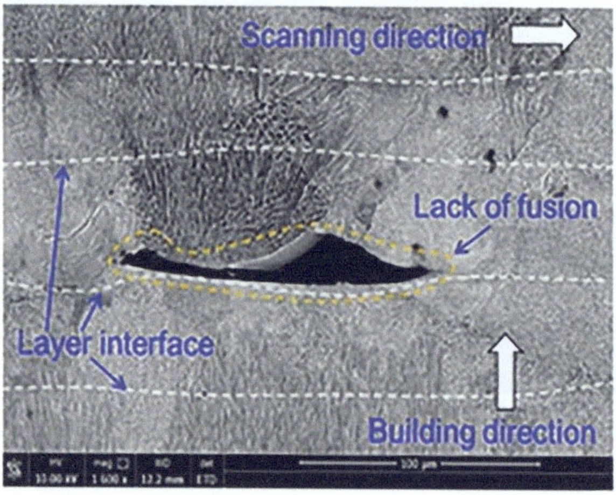

Fig. 3 Lack of fusion voids from Xia et al. Int J Machine tool manuf 2017

can be used for machine learning provide an easy source for usage in additive manufacturing applications [24]. In the situation where more volume of high-quality data is required, teaching machine learning the physics of the problem is more useful. We first compute the mechanistic variables that define our problem or defect in physics terminology. ML training may be done on the computed mechanistic variables, for example, take cracking defect, which is one key defect in AM of aluminum alloys. If important variables are stated, such as laser radius, laser power, traversing velocity, stratum thickness, pre-treatment temperature, hatch distance, specific heat, thermal conductivity, compactness, solidification range, slope of temperature vs. solid fraction plot, and Young's modulus. The minimum experiments needed to fully teach the AM-ML system to perfectly optimize the cracking would be

Fig. 4 Delaminating due to residual from Kempen et al. solid freeform fab2013

Fig. 5 Cracking Zhao et al. MATSE A 2018

$$2^{12} = 4096 \tag{1}$$

This is highly unfeasible as it takes way too much time and is highly unnecessary; narrowing down to key physics variables can drive us to a good cracking susceptibility index [25]. However, when physics-based variables are utilized to teach ML, using variables such as the cooling rate during solidification, solidification stress, the ratio of the vulnerable vs. relaxation time, and the ratio of temperature gradient to solidification growth rate. The minimum experiments needed reduce drastically to

$$2^4 = 16 \tag{2}$$

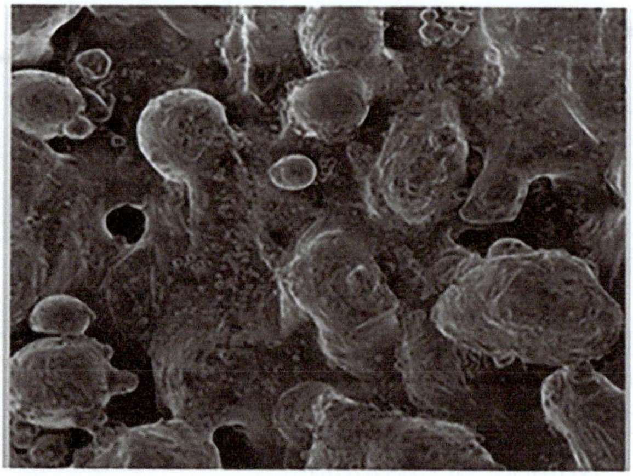

Fig. 6 Surface roughness

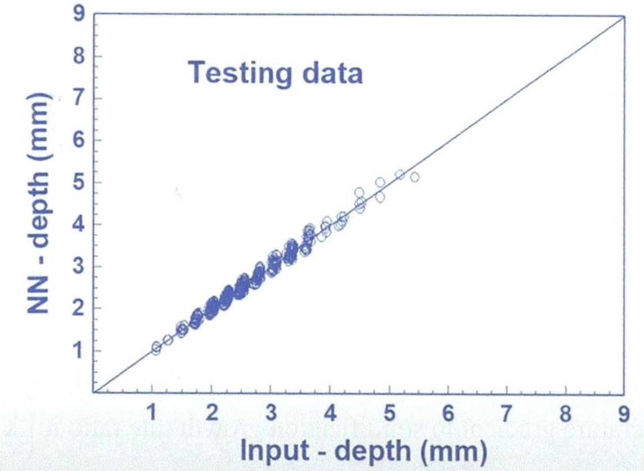

Fig. 7 Data data-driven relation with or without physics

8 Reduction of Cracks Due to Physics-Informed Machine Learning

With a well-informed ML, provided from the process variables to enhance physics-informed learning the data-driven process yields to development of a cracking susceptibility index.

$$(\text{CSI}) = 0.182\sigma + 0.135\beta + 0.014\epsilon - 1.1\times 10^{-8} T - 0.85 \tag{3}$$

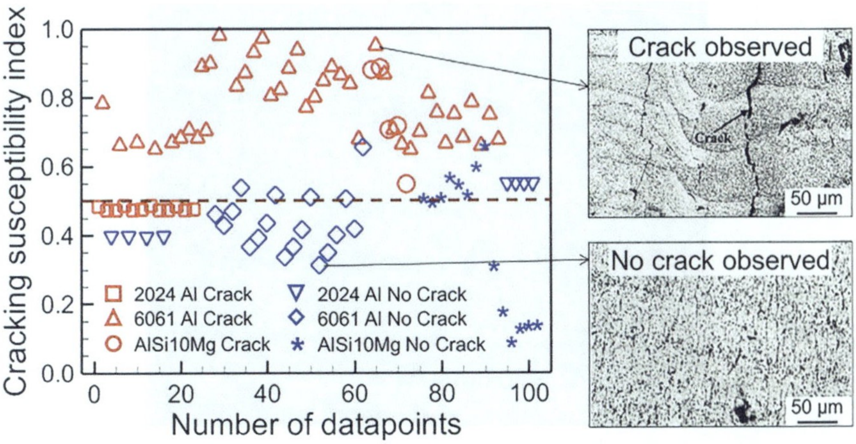

Fig. 8 Mondal et al. Acta Materialia 2022

where

1.

$$\sigma = \text{Solidification stress in} \left(\text{MPa} \right)$$

2.

$$\beta = \text{Ratio of vulnerability and relaxation time}$$

3.

$$\epsilon = \text{temperature gradient to solidification growth rate ratio in} \left(\text{k} - \frac{\text{s}}{\text{mm}^2} \right)$$

4. T = Cooling rate (K/s)

Figure 8 shows that, by scientific principle, cracking can be reduced by physics-informed ML to enhance AM processes such as cracking defects in aluminum alloys. This approach provides a hierarchical order of the more important variables and the least important during AM, hence making the AM-ML manufacturing process more reliable and predictable.

Figures 9 and 10 show that the hierarchy developed gives the engineers the knowledge of which variables are more important and helps them make decisions in the fifth industry, which uses smart machines and humans in manufacturing. They show the hierarchy of information gain/information gain ratio to key parameters **(σ,β,ϵ,T)**.

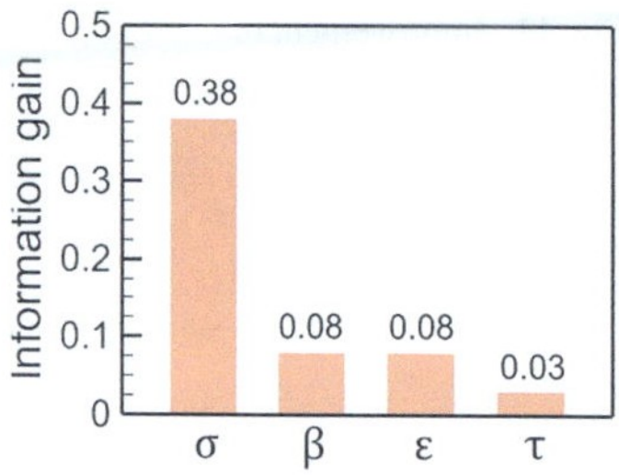

Fig. 9 Information gain vs. (σ, β, ϵ, T)

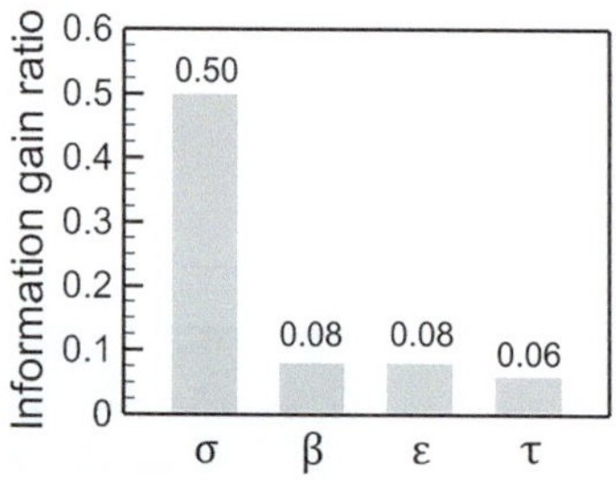

Fig. 10 Information gain ratio vs. σ, β, ϵ, T

9 Applications of Additive Manufacturing

The use of additive manufacturing has grown tremendously across multiple industries such as the automotive subsector, aerospace, electronics, medical products, and manufacturing. This is due to its easy-going nature of converting computer-aided design (CAD) files directly into fully produced components enabling quick mass production, personalization and reduction in wastage that is seen in traditional manufacturing. We will now look at how AM is used in various industries currently and going forward into IR 5.0.

Figure 11 shows the use of additive manufacturing in the nuclear plant in Penn State, while Fig. 12 shows an additively manufactured part in outer space, showing the aerospace application of AM (Fig. 13).

10 Market Penetration of AM

The market penetration of additive fabrication (AM) as shown in Fig. 14 illustrates industries in which AM is being utilized, including hydraulic valves, customized bottle openers, pulley manufacturing, fuel nozzles in aerospace, hip joint implants in medical sector, and gas turbine blades.

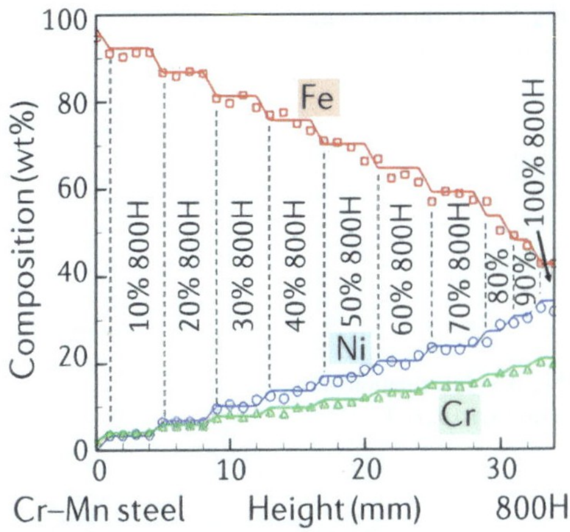

Fig. 11 Improvement of nuclear power

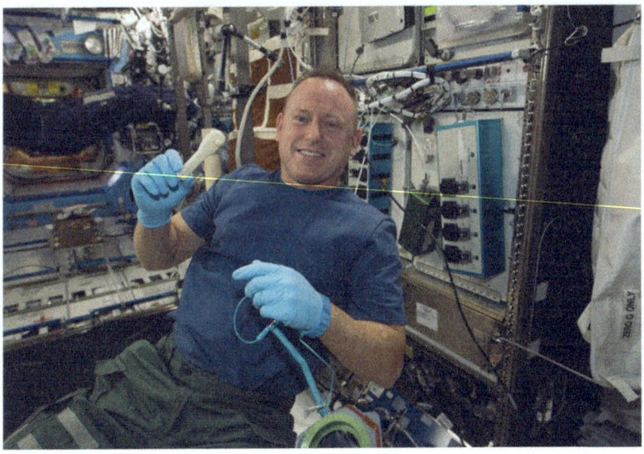

Fig. 12 3-D printing in space plants using AM (Penn State)

11 Aerospace Industry

The aerospace Engineering discipline that makes up this industry is associated with aeronautics, which is the specialization that deals with the design of mechanical air vehicles that fly within the Earth's atmosphere. The other discipline is astronautics, which deals with the study of vehicles that operate in atmospheres beyond Earth. The industry is known for its complex engineering challenges; additive manufacturing has helped the aerospace industry to ease its complicated engineering and manufacturing process [26]. The manufacturing of UAVs today involves AM and ML, in order to get highly complex UAV components and structures directly from CAD design. The design of a UAV demands that the craft has high endurance and is capable of quick take-off and landing whenever deemed necessary for military and commercial applications. UAVs today play a vital role in infrastructure inspections for smart grid power lines, mining tunnels, sewage systems, pipelines, or places that human beings cannot go. In agriculture, they are used for spraying of pesticides or

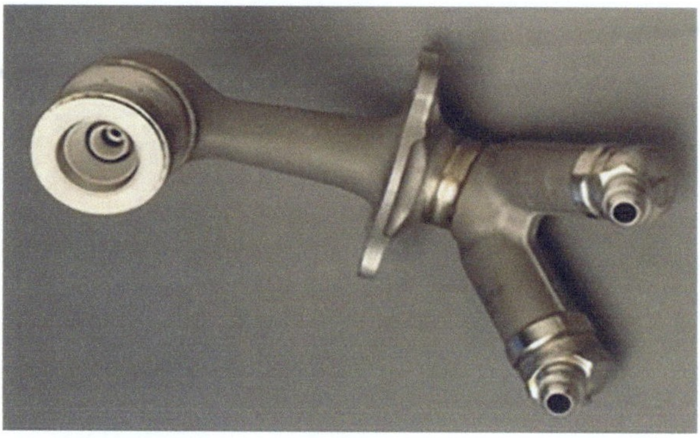

Fig. 13 Aero engine fuel nozzle (GE)

insecticides, surveillance of farmland, adding fertilizer to the crops, and myriad other purposes. The aerospace engineer will calculate the endurance of the craft using the famous Breguet equation:

$$E = k\left(\frac{\eta p}{\text{BSFC}}\right)\left(\frac{\frac{c3}{2}}{\text{CD}}\right)\sqrt{\rho}\,\frac{1}{\frac{\left(\frac{W_i}{S}\right)1}{2}}\left[\frac{1}{1-\frac{\left(\frac{W_f}{W_i}i\right)1}{2}}-1\right. \tag{4}$$

The aerospace UAV design engineer will at first glance on the Breguet equation, realizes that to enhance endurance, there is a need to maximize the aerodynamic parameter $\frac{c3/2}{CD}$, we would further need to minimize the structural weight in order to achieve a reduction in wing loading $\frac{W_i}{S}$, then make an increment in the fuel fraction $\frac{W_f}{W_i}$ [11]. This leads to the employment of AM systems to produce the complex parts, lightweight and with use of historical data made available by the ML integration into AM systems.

The aerospace industry often requires lightweight, strong, and geometrically complex parts for the creation of part components and prototypes. Three-dimensional printing makes it possible to print the Computer-Aided Designs directly into physical products. The aerospace industry uses material elements like titanium, plastics, and composites, most often preferring the use of diminutive/featherweight or strong materials. These materials can be highly expensive, and when parts have to be made using traditional manufacturing, there is a need to buy more of these materials, as there will be more material waste per kg of component. With the use of AM and ML, however, this tremendous cost, which is unsustainable and an enemy of the industrial engineer, can be brought to a minimum [11]. AM and ML will utilize precise material size required for the design piece and data collected on how to make the design as perfect and smooth as possible.

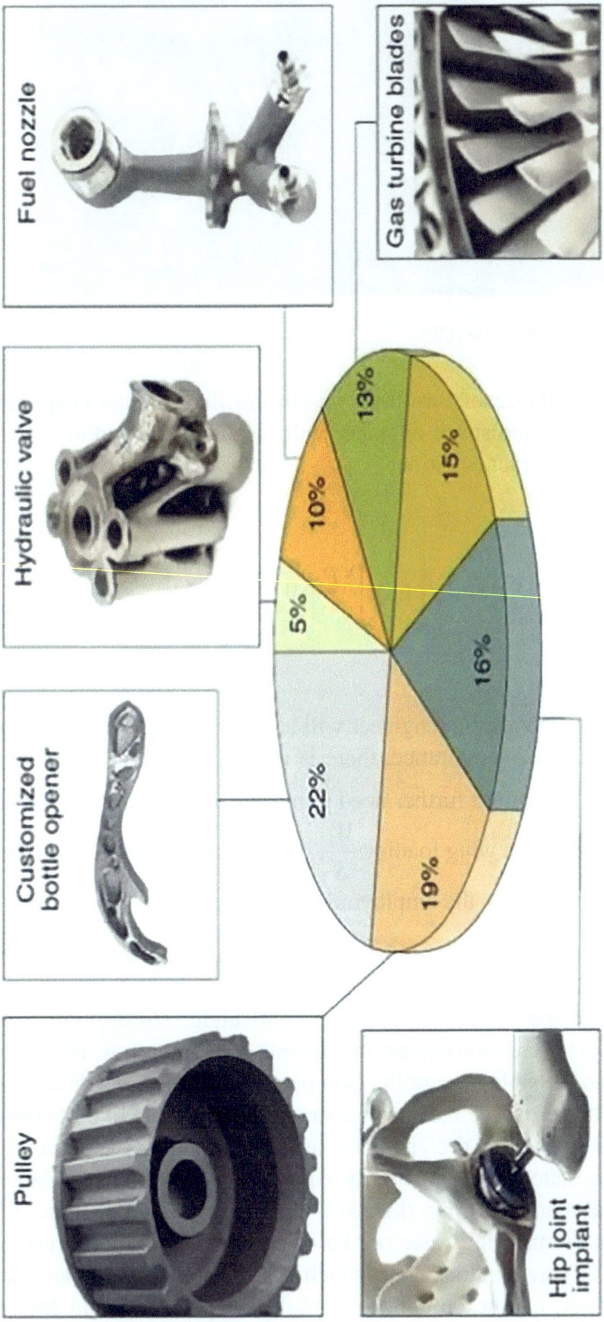

Fig. 14 Market penetration of additive manufacturing

In aeronautics, for example, Boeing Inc. has utilized additive manufacturing to develop 200 different components for 10 various aircraft types. It has thus far used 20,000 additively manufactured parts in commercial and military aircraft. In addition to these 32 components of Boeing's 787 Dreamliner are also additively manufactured, and to make jet engines 25% lighter, the world's largest jet engine manufacturer General Electric is working on utilizing additive manufacturing on the fuel nozzles [10, 11].

12 Automotive Industry

Additive manufacturing consignment for the United States accounted for approximately $ 48 billion in 2011. The automotive industry has progressively moved to the usage of AM in the creation of prototypes, custom parts, all done with short production runs [11]. The automotive industry is rushing towards the application of AM to metals with a stringent focus on aluminum alloys for lightweight vehicles. The automotive sector was among the pioneering adopters of AM methods that include: Laser-based powder layer melting, thermoplastic extrusion, drop-on-demand printing, and photopolymerization. These were used in building prototypes and test pieces [27]. An automotive AM user would be the giant player Ford, which has been utilizing the technology to create prototype components for testing vehicles. Engineers at Ford have also developed combustion chamber covers, braking discs, and posterior axle assemblies in much shorter intervals compared to legacy manufacturing [11].

The current hurdle with AM in the automotive industry is the limited build volume of current three-dimensional printers, which may not accommodate the production of more substantial parts such as body panels. In response to this problem, however, researchers and industrialists are working towards developing such 3-D printing. For example, the prodigious stereolithography developed by materialize has a build volume envelope of about 2100 mm × 680 mm × 800 mm; this size may produce any automobile part [28]. In the automotive sector, the significance of tooling on the assembly line cannot be overstated, making it imperative to customize the fabrication of tools to boost productivity. A notable example of innovation in this area is BMW's adoption of additive manufacturing (AM) for producing hand tools utilized in both testing and assembly processes. These bespoke hand tools not only offer superior ergonomic designs but are also 72% lighter than their traditional counterparts. The shift from conventional CNC machining to AM technology has led to remarkable efficiency gains, resulting in a 92% reduction in lead time and a 58% decrease in costs, ultimately enhancing the overall project estimated timeline [28].

13 Medical Industry

The coming of three-dimensional printing AM has brought so much advancement in the area of medicine, ranging from healthcare instruments, dental braces, bone stabilization devices, bio-engineered prosthetics, and healthcare models for training purposes. The industry has seen the usage of medical imagery and AM enhanced by ML to bespeak the visual of complicated pathologies. Plastic special coils have begun to be produced by Stratasys for the MRI machines (magnetic resonance imaging). The use in removable orthodontic treatment is clearly one area of application, with the use of X-ray images of patient's teeth, they may be used to create treatment plans and orthodontic braces to align teeth. The medical space sees use of AM in orthopedic implants, 9it provides design capability through CAD to design the exact needed implant, which is then 3-D printed, this, with the integration of ML, makes the manufacturing process smarter and more optimizable [28].

14 Conclusion

To encapsulate, or in summation, the profound implications of Additive manufacturing enhanced with machine learning have been illuminated. It is therefore safe to say, additive manufacturing with machine learning enhancement is a set that can redefine the contours of the fifth manufacturing revolution. This will usher in an era of unparalleled innovation and efficiency. This unique fusion of AM with MI (a distinct section of AI) enables greater customization, reduced waste, and enhanced predictive maintenance. This allows manufacturers to optimize production in real time. Industry 5.0 will take advantage of the leverage on data-driven insights to make its industries not only anticipate demands but also adapt to swift changes in their industries. The synergy between these technologies creates a unique embodiment which is the sore essence of Industry 5.0-human-centric, intelligent, and sustainable manufacturing. Moving forward, this collaboration will become a chief cornerstone of industrial transformation, it will foster resilience and continuous improvement in production processes.

References

1. Du, Y., Mukherjee, T., & DebRoy, T. (2021). Physics-informed machine learning and mechanistic modeling of additive manufacturing to reduce defects. *Applied Materials Today, 24*, 101123.
2. Babu, S. S., Mourad, A. H. I., Harib, K. H., & Vijayavenkataraman, S. (2023). Recent developments in the application of machine-learning towards accelerated predictive multiscale design and additive manufacturing. *Virtual and Physical Prototyping, 18*(1), e2141653.

3. Fu, Y., Downey, A. R., Yuan, L., Zhang, T., Pratt, A., & Balogun, Y. (2022). Machine learning algorithms for defect detection in metal laser-based additive manufacturing: A review. *Journal of Manufacturing Processes, 75*, 693–710.
4. Jin, Z., Zhang, Z., Demir, K., & Gu, G. X. (2020). Machine learning for advanced additive manufacturing. *Matter, 3*(5), 1541–1556.
5. Mukherjee, T., & DebRoy, T. (2018). Mitigation of lack of fusion defects in powder bed fusion additive manufacturing. *Journal of Manufacturing Processes, 36*, 442–449.
6. Mukherjee, T., Zhang, W., & DebRoy, T. (2017). An improved prediction of residual stresses and distortion in additive manufacturing. *Computational Materials Science, 126*, 360–372.
7. Martini, B., Bellisario, D., & Coletti, P. (2024). Human-centered and sustainable artificial intelligence in Industry 5.0: Challenges and perspectives. *Sustainability, 16*(13), 5448.
8. Raja Santhi, A., & Muthuswamy, P. (2023). Industry 5.0 or industry 4.0 S? Introduction to industry 4.0 and a peek into the prospective industry 5.0 technologies. *International Journal on Interactive Design and Manufacturing (IJIDeM), 17*(2), 947–979.
9. Rani, S., Jining, D., Shoukat, K., Shoukat, M. U., & Nawaz, S. A. (2024). A human–machine interaction mechanism: Additive manufacturing for Industry 5.0—Design and management. *Sustainability, 16*(10), 4158.
10. Goh, G. D., Agarwala, S., Goh, G. L., Dikshit, V., Sing, S. L., & Yeong, W. Y. (2017). Additive manufacturing in unmanned aerial vehicles (UAVs): Challenges and potential. *Aerospace Science and Technology, 63*, 140–151.
11. Ford, S. L. (2014). Additive manufacturing technology: Potential implications for US manufacturing competitiveness. *J. Int'l Com. & Econ., 6*, 40.
12. Wu, Q., Mukherjee, T., De, A., & DebRoy, T. (2020). Residual stresses in wire-arc additive manufacturing–Hierarchy of influential variables. *Additive Manufacturing, 35*, 101355.
13. Deng, L. (2018). Artificial intelligence in the rising wave of deep learning: The historical path and future outlook [perspectives]. *IEEE Signal Processing Magazine, 35*(1), 180–177.
14. Mendon-Plasek, A. (2021). Mechanized significance and machine learning: Why it became thinkable and preferable to teach machines to judge the world. In *The cultural life of machine learning: An incursion into critical AI studies* (pp. 31–78). Palgrave Macmillan.
15. Hart, G. L., Mueller, T., Toher, C., & Curtarolo, S. (2021). Machine learning for alloys. *Nature Reviews Materials, 6*(8), 730–755.
16. Wong, K. V., & Hernandez, A. (2012). A review of additive manufacturing. *International Scholarly Research Notices, 2012*(1), 208760.
17. Singh, S., Mehla, S., Bhargava, S. K., & Ramakrishna, S. (2022). History and evolution of additive manufacturing. In *Additive manufacturing for chemical sciences and engineering* (pp. 19–51). Singapore.
18. Gao, W., Zhang, Y., Ramanujan, D., Ramani, K., Chen, Y., Williams, C. B., Wang, C. C., Shin, Y. C., Zhang, S., & Zavattieri, P. D. (2015). The status, challenges, and future of additive manufacturing in engineering. *Computer-Aided Design, 69*, 65–89.
19. Pereira, T., Kennedy, J. V., & Potgieter, J. (2019). A comparison of traditional manufacturing vs additive manufacturing, the best method for the job. *Procedia Manufacturing, 30*, 11–18.
20. Thompson, M. K., Moroni, G., Vaneker, T., Fadel, G., Campbell, R. I., Gibson, I., Bernard, A., Schulz, J., Graf, P., Ahuja, B., & Martina, F. (2016). Design for additive manufacturing: Trends, opportunities, considerations, and constraints. *CIRP Annals, 65*(2), 737–760.
21. Khajavi, S. H., Partanen, J., & Holmström, J. (2014). Additive manufacturing in the spare parts supply chain. *Computers in Industry, 65*(1), 50–63.
22. Du, Y., Mukherjee, T., Finch, N., De, A., & DebRoy, T. (2022). High-throughput screening of surface roughness during additive manufacturing. *Journal of Manufacturing Processes, 81*, 65–77.
23. Yadroitsev, I., & Yadroitsava, I. (2021). A step-by-step guide to the L-PBF process. In *Fundamentals of laser powder bed fusion of metals* (pp. 39–77). Elsevier.

24. DebRoy, T., Mukherjee, T., Wei, H. L., Elmer, J. W., & Milewski, J. O. (2021). Metallurgy, mechanistic models and machine learning in metal printing. *Nature Reviews Materials, 6*(1), 48–68.
25. Mondal, B., Mukherjee, T., & DebRoy, T. (2022). Crack free metal printing using physics informed machine learning. *Acta Materialia, 226*, 117612.
26. Huang, R., Riddle, M., Graziano, D., Warren, J., Das, S., Nimbalkar, S., Cresko, J., & Masanet, E. (2016). Energy and emissions saving potential of additive manufacturing: The case of lightweight aircraft components. *Journal of Cleaner Production, 135*, 1559–1570.
27. Charles, A., Hofer, A., Elkaseer, A., & Scholz, S. G. (2021, September). Additive manufacturing in the automotive industry and the potential for driving the green and electric transition. In *Proceedings of the international conference on sustainable design and manufacturing* (pp. 339–346). Springer Singapore.
28. Bahnini, I., Rivette, M., Rechia, A., Siadat, A., & Elmesbahi, A. (2018). Additive manufacturing technology: The status, applications, and prospects. *International Journal of Advanced Manufacturing Technology, 97*, 147–161.

Recent Advances in the Application of Thermography to Additive Manufacturing: From Process Monitoring to Mechanical Characterisation

Ester D'Accardi, Rosa De Finis, Davide Palumbo, and Umberto Galietti

1 Introduction: How Thermography Can Improve Additive Manufacturing

Infrared thermography is a well-established, contactless, full-field, non-destructive technique [1–4] suitable in many domains [5] ranging from mechanical industry to structural engineering and applications that fall under agro-alimentary, marine, aerospace, automotive and electrical sectors.

Two main approaches have been set up to carry out thermographic inspections: passive and active ones. Passive thermography is basically implemented to inspect materials or structures that are naturally at a different temperature than the ambient temperature one [6, 7]. It is mainly used to monitor and control a technological process under operating conditions or it is also useful for maintenance purposes [5] or medical/remote sensing screenings [8–10] or it shows capabilities in civil engineering for moisture evaluation in buildings [11]. The implementation of process monitoring enables the identification of thermal features [12, 13] that serve as indicators of the intrinsic quality of the process. These features can then be correlated with the mechanical properties of the material [14, 15] and furthermore, can be employed to enhance the efficiency of the process itself [15–22].

The basic idea behind active thermography is that by perturbing the material with an external heat source, the presence of defects is revealed by studying the temporal evolution of the sample surface temperature using an infrared detector. The presence of a defect alters the thermal wave propagation front by diffusion and a

E. D'Accardi · D. Palumbo · U. Galietti
Department of Mechanics, Mathematic and Management, Politecnico di Bari, Bari, Italy
e-mail: ester.daccardi@poliba.it; davide.palumbo@poliba.it; umberto.galietti@poliba.it

R. De Finis (✉)
Department of Engineering for Innovation, University of Salento, Lecce, Italy
e-mail: rosa.definis@unisalento.it

Y. Kumar Singla et al. (eds.), *Digital Manufacturing in Industry 5.0*,
https://doi.org/10.1007/978-3-031-91500-0_13

map of different regions, characterised by different signals associated with the sound and defect regions, is then obtained [18, 19, 23–26]. So, active thermography requires instead the supply of an external source to stimulate the material and structures. Such a heat source can be a thermal or a mechanical or electrical one [5]. Figure 1 resumes the different infrared approaches. In case of a mechanical source, two main types of excitations are possible depending on the nature of the source, if based on the ultrasounds [27] or mechanical vibrations at high or low frequencies [28–31]. In case of ultrasounds it is necessary to provide the acoustic coupling to properly allow the acoustic waves to propagate into the material, while in case the mechanical load is applied with high frequencies, the vibro-thermography [32] technique can be implemented. Alternatively, it is possible to excite the material with mechanical vibrations at lower frequencies, allowing innovative structural investigation techniques such as Thermoelastic Stress Analysis [33] to be performed.

Electric/electromagnetic sources involve the use of flash [2, 34, 35] or halogen [36–39] lamps, microwaves [40–42], lasers [18, 43–46] and inductors [26, 47] to carry out structural integrity evaluations.

Additionally, in the domain of active thermography, if the heat source is mechanical, it is feasible to induce damage to the material, depending on the magnitude of the load, thereby enabling the investigation of irreversible fatigue and fracture mechanics processes through thermal imaging [48, 49].

Additionally, in the domain of active thermography, if the heat source is mechanical, it is feasible to induce damage to the material, depending on the magnitude of the load, thereby enabling the investigation of irreversible processes like fatigue and fracture mechanics [50–60] through thermal imaging.

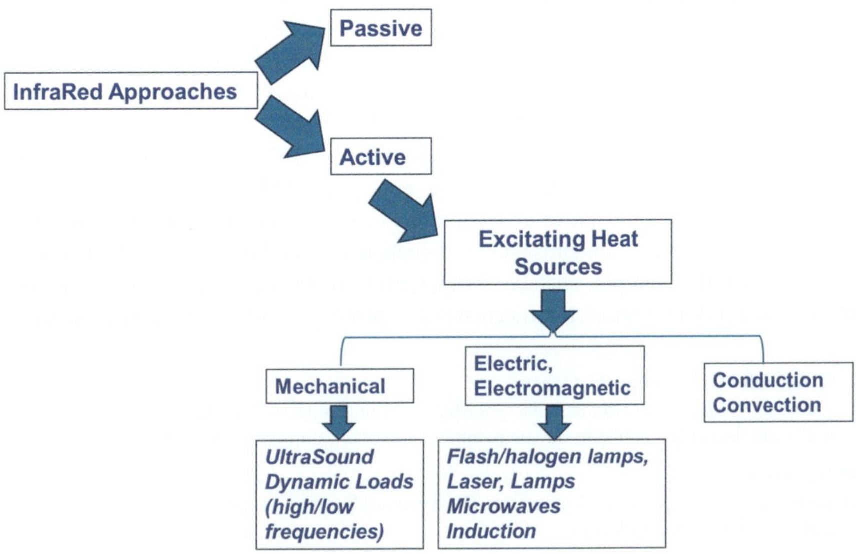

Fig. 1 Passive–active thermographic approaches

In terms of fatigue, rapid fatigue tests [61] can be used to study the behaviour of the material in order to detect early damage [62–64] and estimate its fatigue limit [65]. These methods allow test times to be significantly reduced by exploiting the self-heating effect of the material. A single curve is enough to indicate the material's behaviour, instead of 15–20 specimens and test times of up to months to construct the Wöhler curve.

In fracture mechanics, the energy dissipated at the crack tip [66–70], as well as the crack length and the parameters (stress intensity factor) describing the stress state in the vicinity of the plastic area can be derived by means of thermal signal evaluation.

Considering the versatility of the technique, as demonstrated by the many applications in different industrial sectors and the different methodologies developed, it is possible to conclude that Infrared Thermography can significantly enhance additive manufacturing (AM) processes in several ways:

- **Process Monitoring**: Thermography allows for real-time monitoring of the temperature distribution during the AM process, in detecting anomalies and verifying the material quality [8, 14, 16, 20, 71–80]. In effect, online monitoring carried out by an infrared detector can be useful to establish a closed-loop diagnostic system for predicting the presence of structural anomalies such as defects [9]. On the other hand, the positioning of the infrared detector is still a challenge that makes thermography not yet suitable for all additive printing processes. Indeed, it is necessary to have sufficient space to ensure an effective setup. Another challenge is to improve the sensitivity of the technique to detect anomalies during the technological process. This improvement will only be possible with the development of new analysis algorithms, including those based on artificial intelligence [9].

- It should be emphasised that in additive manufacturing in particular, the ability to measure a parameter such as temperature can be critical to understanding the impact of process parameters on the quality of the material being produced, but also to intervening in the process or repairing the material on-line without or reducing production waste [71].
- **Structural Integrity Assessments**: Through the analysis of thermal images, it becomes possible to detect and categorise common additive manufacturing defects such as lack of fusion, keyhole porosity, general porosity, and cracking. Early identification of these anomalies can help prevent the production of defective parts. Challenges in thermography-based NDT include the need to characterise critical defects with a certain degree of confidence and the ability to detect flaws in geometrically complex AM components. The study by [81] examined defect depth estimation, finding that the detectability threshold (which requires only adequate thermal contrast) is less restrictive than the measurement threshold. By increasing the irradiation duration and/or the pulse energy, it may be possible to qualitatively detect defects smaller than 100 μm even in materials with low thermal diffusivity.

- **Verification of Models**: The data obtained from thermography can be used to verify and refine physics-based models of the AM process, enhancing the understanding and control of the process [45, 81].
- **Mechanical Characterisation**: In terms of mechanical characterisation, it is well known that the microstructure, the type and distribution of defects highly dependent on the manufacturing parameters, the presence of residual stresses and the anisotropy induced by the AM process have an effect on the mechanical properties, particularly under cyclic loading [82–85]. From the point of view of fatigue, the presence of defects can act like a stress concentrator, inducing a premature material failure due to the creation of cracks. Another important aspect is the wide dispersion of fatigue test results, considering that each specimen has its own specificity (number of defects, type of defects...) [82].
- The role of thermography in this respect is of great relevance as it allows not only a qualitative analysis of damage processes as a result of cyclic loading but also a quantitative analysis through the measurement of temperature and the implementation of the various analysis methodologies developed in recent years [82–85].

By integrating thermography into additive manufacturing, manufacturers can achieve higher precision, better quality control, and more efficient and circular production processes.

The present chapter is divided into the following sections: firstly, Sect. 2 will show the challenges in process monitoring, while Chap. 3 will show two case studies, one on structural integrity inspections on specific additive manufacturing processes. In Sect. 4, the challenging aspects related to fatigue characterisation (e.g. damage study and fatigue behaviour assessments) on DED-arc AISI 316 steel and fracture mechanics on Selective Laser Melted Titanium alloys will be discussed.

2 Challenges in Process Monitoring

A specific case study by D'Accardi et al. [12, 14, 15] focused on monitoring the laser metal deposition (LMD) process through thermal imaging techniques. Using a microbolometer sensor with a matrix of pixels, instead of a pyrometer, the study demonstrated the potential for developing empirical models that correlate extracted thermal characteristics with critical process parameters. Figure 2 presents an image of the setup used, where (1) marks the laser and (2) the final position of the infrared sensor A655, which moves synchronously with the heat source.

This study specifically examined the behaviour of Inconel 718 using a Design of Experiment (DOE) framework with three levels across three factors: laser power, laser speed, and powder flow rate. However, the DOE was not fully executed, as certain parameter combinations did not support successful printing, which limited the number of feasible experiments. To address this, specific sub-groups of parameter combinations were analysed in detail to assess the impact of viable

Fig. 2 Adopted setup for the thermal monitoring of the LMD process

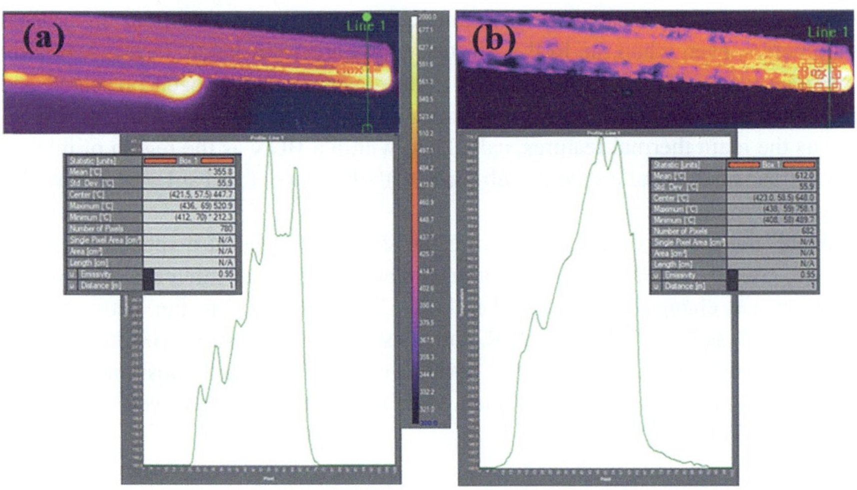

Fig. 3 Typical expected thermal behaviour in case of a good combination of process parameters—energy density 25.7 J/mm^2 (**a**), in comparison with a not good result—energy density 51.6 J/mm^2 (**b**)

configurations, particularly with a focus on energy density. This approach aimed to shed light on how these factors interact to influence thermal monitoring.

As preliminary findings, Fig. 3a illustrates the typical thermal response observed during the printing of a sample using a proper set of process parameters, characterised by low energy density. Conversely, Fig. 3b shows the thermal response associated with high energy density settings. This comparison underscores differences in thermal profiles (shown in green), highlighting how deviations from ideal process parameters—such as uneven powder distribution around the melt pool—can negatively impact the quality of the printed material by altering the apparent temperature.

The obtained result demonstrates the effectiveness of using a matrix of pixels instead of a pyrometer, which is difficult to position correctly and does not provide information about the cooling after the melt pool.

By aligning a profile (in green) directly downstream of the melt pool and repeating the operation for an entire pass during the deposition of a layer, it is possible to obtain the results show in Fig. 4, for both cases illustrated before. Considering a profile, it is possible to save data and have the response of an entire pass or layer of material in a single thermal image.

Considering these preliminary results, the analysis of all the sequences acquired for the different combinations of process parameters was performed within a defined region of interest (ROI), such as a profile or a box located immediately downstream of the melt pool. This approach allows for the assessment of apparent temperature behaviour without requiring knowledge of the material's emissivity.

By extracting specific thermal features within the ROI, empirical models have been established to correlate these features with the process parameters. Using statistical methodologies such as Analysis of Variance (ANOVA) analysis, it has been demonstrated that these thermal features are sensitive to variations in process parameters, even when the ROI is shifted within a certain range. In this regard, in previous works [12, 13], the analysis of the main DOE has been presented and explained in detail, considering the apparent mean temperature and the 98th percentile as the main thermal features, calculated within a BOX as the region of interest (ROI). The analysis of a specific sub-plan, with the power fixed at Level 2, was also addressed in detail.

Here, the results obtained by fixing the scanning speed at a specific level are shown, considering the analysis of a profile as the region of interest (ROI), whose position was changed as described in [12, 13] and reported in Fig. 5 to optimise indices such as R^2, RMSE, and SNR, and to establish the sensitivity of the extracted thermal feature by examining the p-value obtained for each position. Table 1 resumed the specific analysed subplan, with an indication of the levels (steps) and investigated ranges.

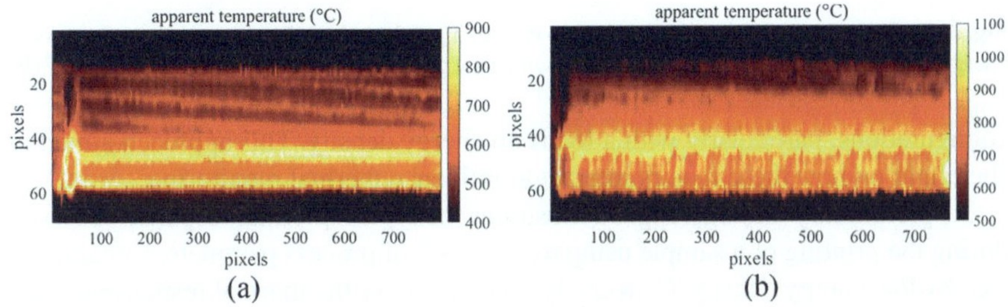

Fig. 4 Typical expected thermal behaviour monitoring a profile in case of a good combination of process parameters—energy density 25.7 J/mm² (**a**), in comparison with a not good result—energy density 51.6 J/mm² (**b**)

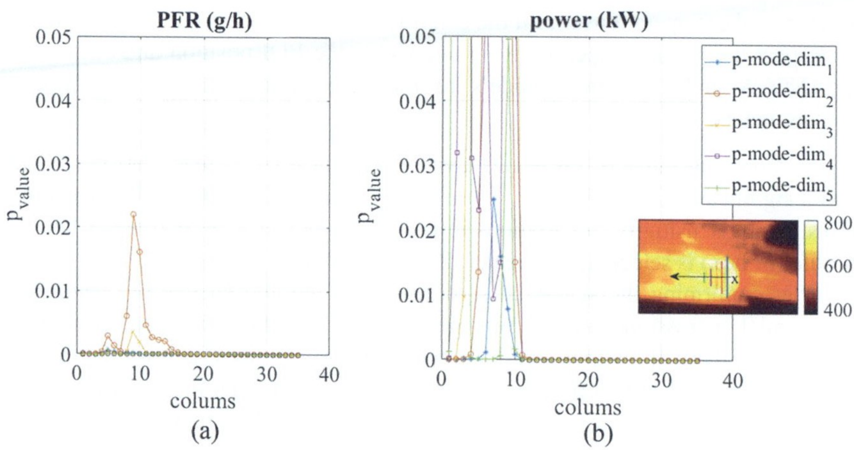

Fig. 5 ANOVA results for the mode (apparent temperature) based on the specific DOE detailed in Table 1. The influence of the investigated process parameters is analysed as the dimensions and positions of the profile are varied: (**a**) Powder Flow Rate (PFR) and (**b**) laser power

Table 1 Analysis of a specific subplan, considering the scanning speed fixed at Step 3, and the other process parameters on three specific steps

Process parameter	Speed (mm/min)	Power (kW)	PFR (g/h)
Level	Step 3	Step 1 = step 2 – 0.35	Step 1 = step 2 – 200
	Step3	2	Step 1 = step 2 – 200
	Step3	Step 3 = step 2 + 0.35	Step 1 = step 2 – 200
	Step 3	Step 2	Step 2
	Step 3	Step 3 = step 2 + 0.35	Step 2
	Step 3	Step 2	Step 3 = step 2 + 200
	Step 3	Step 3 = step 2 + 0.35	Step 3 = step 2 + 200

To analyse the raw thermal data, the procedure described here is adopted:

- Import the sequence into Matlab considering the deposition of three layers and six passes for all the combinations of process parameters reported in Table 1.
- Choose an area downstream of the fused area (size 20x13 pixels)—the area remains the same for each pass, slightly changes its position according to the analysed sequence.
- Calculate some significant indices along y dimension—**kurtosis, skewness and mode.**
- Reshape the area in a vector and calculate vector by vector and then frame by frame some significant indices—**mean, standard deviation, maximum, minimum.**
- Consider the central part of each pass and calculate, for each index, mean and standard deviation (to have the same number of frames subsampling the sequence—25 Hz).

- Repeat these operations for all the conditions, as the process parameters change.
- Repeat the previous operations by modifying the ROI position and dimension; ANOVA and regression models to correlate thermal results and process parameters.

Figure 5 presents the ANOVA results, using the mode as the primary thermal feature and a significance level of 0.05. As previously shown in [12, 13], all thermal features exhibit sensitivity to variations in process parameters, particularly when the profile (or box) is selected at a greater distance from the melt pool.

After confirming the statistical significance of the chosen thermal features, a polynomial regression analysis was performed to establish a direct correlation between the thermal features and the process parameters (see Fig. 6 for the thermal feature "mode"). For this purpose, second-degree polynomial models were developed, incorporating linear interaction terms, with the variables representing the process parameters examined in the study.

Preliminary analyses have established a correlation between mechanical data related to hardness and extracted thermal features. It has been demonstrated that hardness is sensitive to variations in processing parameters, making it a valuable parameter for describing the behaviour of printed materials [14, 15].

For each printed coupon, hardness measurements were repeated three times along the longer segment of the coupon. In the initial approach, the entire dataset was utilised to correlate hardness data with extracted thermal features by varying the position and dimensions of the ROI profile. A first-degree linear model was employed to maximise the R^2 value. Outliers were removed from the respective boxplots before averaging the mechanical and thermal data. Table 2 presents the results obtained by consistently using the mode as the thermal index. The analysis yielded a p-value of 0, further supporting the possibility of a direct correlation between mechanical and thermal behaviour.

Outliers were removed from the respective boxplots before averaging the mechanical and thermal data. Table 2 presents the results obtained by consistently

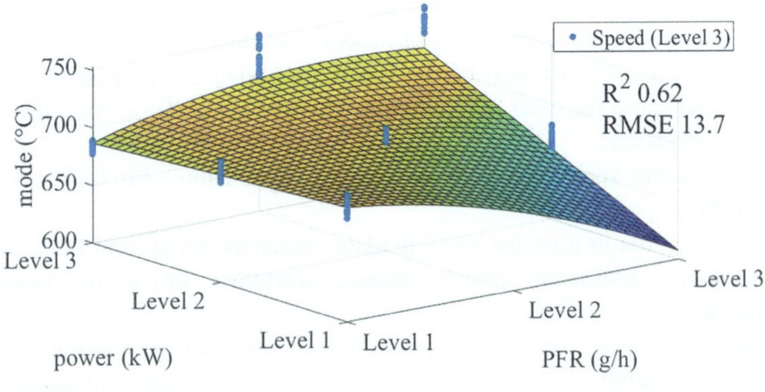

Fig. 6 Regression model (thermal feature "mode") obtained for the position that optimise the R^2 value, dim. 5—column 12 (along × direction)

Table 2 Result obtained from the ANOVA analysis for the direct correlation between mechanical properties and thermal features; mode vs hardness, profile analysis

Regression analysis: hardness versus mode					
Analysis of variance					
Source	**DF**	**Adj SS**	**Adj MS**	***F*-value**	***P*-value**
Regression	1	28.34	28.34	41.02	0.000
Mode	1	28.34	28.34	41.02	0.000
Error	15	10.36	0.69		
Total	16	38.70			
Model summary					
S	***R*-sq**	***R*-sq(adj)**	***R*-sq(pred)**		
0.831121	73.23%	71.44%	66.83%		
Coefficients					
Term	**Coef**	**SE Coef**	***T*-value**	***P*-value**	**VIF**
Constant	9.22	2.22	4.15	0.001	
Mode	0.02	0.00	6.41	0.000	1.00
Regression equation à hardness = 9.22 + 0.02 mode					

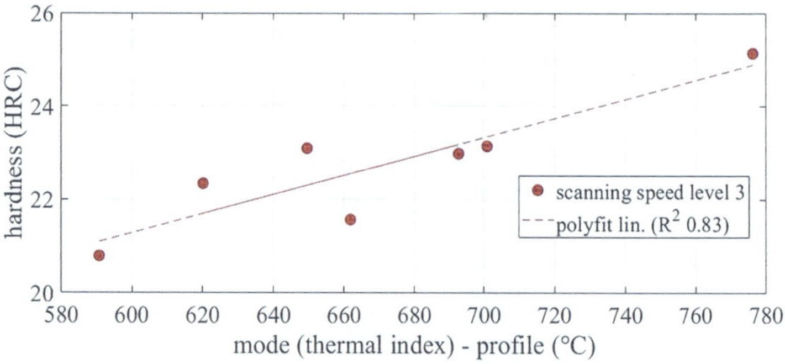

Fig. 7 Linear regression models for correlating thermal data with hardness data; mode vs. hardness, profile analysis, subplan with fixed speed value reported in Table 1

using the mode as the thermal index. The analysis yielded a p-value of 0, further supporting the possibility of a direct correlation between mechanical and thermal behaviour.

By moving on to the analysis of subplans, it is possible to achieve a better correlation, as demonstrated by the graph shown in Fig. 7, which fixes the scanning speed at Level 3, as before.

These models have proven effective in detecting localised anomalies in real-time during the printing process, allowing for the identification of irregularities as they occur. For example, Fig. 8 presents the results obtained during the printing of a coupon, where the scanning speed was altered due to operator error. The graph illustrates trends in apparent mean temperature and mode, specifically analysing a box located downstream of the melt pool, depicted in black on the right. These

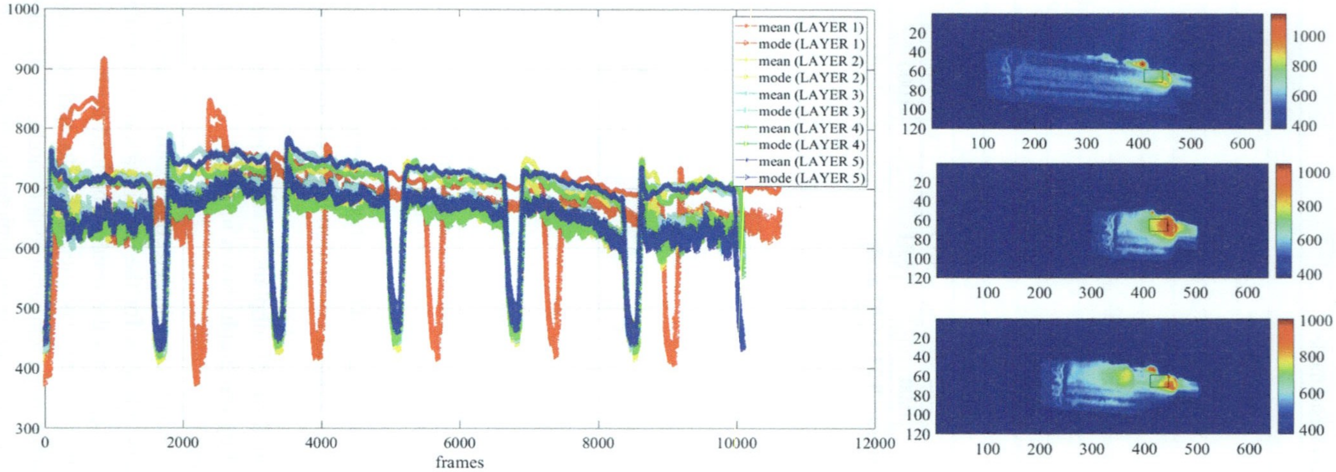

(a) (b)

Fig. 8 Apparent mean temperature and mode behaviour in a specific ROI during the printing of a coupon when the speed changed drastically due to an operator error (**a**), along with several instances shown as a full-field image on the right (**b**)

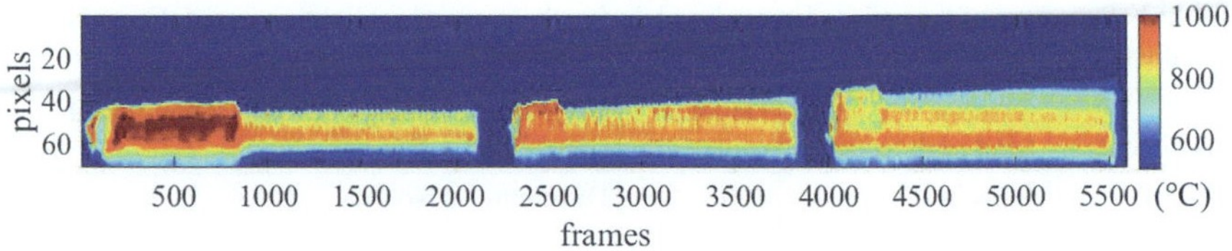

Fig. 9 Analysis of side-by-side profiles during the printing process in the presence of a defect when the speed changes suddenly, affecting the thermal behaviour of both the current pass and the subsequent ones

trends correspond to various layers, with the layer affected by the anomaly highlighted in red. Some specific instances are shown in Fig. 8b, which considers the normal behaviour during a particular pass and the changes in speed during the melt, as well as immediately afterward.

The analysis of a single profile was also able to intercept this type of defect, as demonstrated in Fig. 9. The results indicate that the defect influences not only the current pass but also the subsequent ones.

3 Non-destructive Techniques

Active thermographic techniques can be used to perform offline tests with various excitation sources to detect typical additive manufacturing defects such as porosity, keyholes and lack of fusion. Furthermore, additive manufacturing technology offers the advantage of simulating defects with a defined geometry, which can be used to calibrate the technique [12, 13] before proceeding to the analysis of real defects.

A case study is presented that applies different active thermographic techniques to detect real defects, including keyholes, lack of fusion, and small voids. These defects are introduced by intentionally modifying the process parameters in specific areas and have been characterised in earlier studies using eddy current testing for direct comparison and micro-computed tomography as a reference method.

As partially described in [12, 13], where the analyses focus on two specific specimens, sixteen specimens of AISI 316 L steel were produced using the Laser Powder Bed Fusion technique with an SLM 280 HL printer. The main features of the printer include laser power up to 400 W, scanning speeds up to 10 m/s, and a build chamber measuring 280 × 280 × 350 mm with a heated platform up to 200 °C.

Standard process parameters for the specimens included a laser power of 275 W and a scanning speed of 700 mm/s. The work by D'Accardi et al. [12, 13] details all the process parameters and the induced variations to create different types of surface defects. Table 3 summarises the entire plan, highlighting in red the specimens that were investigated further in the subsequent analyses. As an exemplificative picture, Fig. 10 reports some of the specimens investigated. As discussed in the previous study, these specimens have a low and non-uniform emissivity value (around 0.3);

Table 3 Summary of the entire plan analysed during the project highlighting in red the specimens investigated in this work

ID specimens	Kind of defects	Depth (μm)	Height (μm)	Energy density (input in ROI defects)
PK001	Gaps in illumination	100	Different	No
PK002	Gaps in illumination	200	Different	No
PK003	Gaps in illumination	300	Different	No
PK004	Gaps in illumination	400	Different	No
PK005	Keyhole	100	1000	High
PK006	Keyhole	200	1000	High
PK007	Keyhole	400	1000	High
PK008	Lack of fusion	100	1000	Low
PK009	Lack of fusion	200	1000	Low
PK010	Lack of fusion	400	1000	Low
PK011	Keyhole	100	250	High
PK012	Keyhole	200	250	High
PK013	Keyhole	400	250	High
PK014	Lack of fusion	100	250	Low
PK015	Lack of fusion	200	250	Low
PK016	Lack of fusion	400	250	Low

Fig. 10 Some of the specimens analysed in the work

for this reason, in the tests and analyses here discussed, the specimens were covered by a black coating (Mibenco).

The specimens were analysed using pulsed thermography and as excitation source flash lamps. The adopted parameters are reported respectively in Table 4.

After the acquisition of the thermographic sequences, the data have been evaluated considering different post-processing algorithms [2, 5–7, 18, 19, 86]. These algorithms were applied to increase the signal-to-noise ratio, reveal the presence of the defects and highlight differences in thermal behaviour between the defective regions and the surrounding sound areas.

In this regard, Fig. 12 reports the thermal behaviour during these kinds of tests, considering the defective zone (in orange) and the related sound (in blue).

Following a preliminary evaluation to determine the appropriate truncation window size for data analysis and algorithms aimed at enhancing the signal-to-noise ratio, some quantitative evaluations are presented. The results, illustrated in the figure, derive from tests carried out using flash thermography. The thermographic signal reconstruction (TSR) algorithm was employed with a polynomial fitting degree of 5, and maps of the second derivative were analysed, considering the entire time available post-flash (10 s). The maps are plotted on the same scale with the key parameter being the contrast-to-noise ratio.

The specimens PK007 and PK010 were compared to evaluate how the impact of the energy used to generate the defects influences the thermographic response in terms of the signal expected. It is evident that, with other characteristics being equal—such as shape, depth (0.4 mm) and height—the defects in keyhole mode exhibit greater contrast compared to those generated by lack of fusion, as well as a more defined shape respect to the nominal and intended one [12, 13] (Fig. 15a, d).

Figures 13b–d, on the other hand, show a comparison at equal energy (lack of fusion), considering different specimens with defects at various depths (PK008 (Fig. 13b), PK009 (Fig. 13c), and PK010 (Fig. 13d)).

The cross in the lower right of specimen PK008 was produced in keyhole mode by increasing the laser scanning speed rather than decreasing it. For this reason, it

Table 4 Setup parameters for flash thermographic tests

Setup 1 Fig. 11 a—flash thermography	
Camera	**Infratec ImageIR 8800**
Frame rate (Hz)	1000
Duration	10,000 frames
Geometrical resolution	0.35 mm/pixel
Window size	160 × 128 pixels
Heating source	**Hensel EH pro flash lamps (4 × 6 kJ)**
Pulse duration	3 ms

Fig. 11 Setup adopted to carry out active thermographic tests with flash thermography

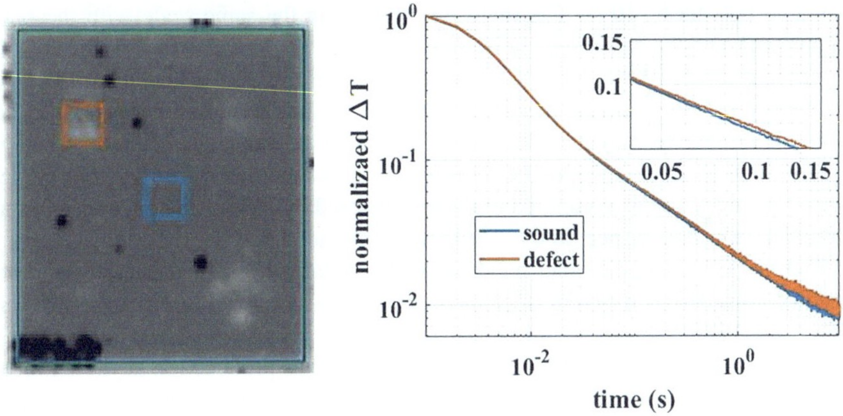

(a) (b)

Fig. 12 Thermal behaviour during active thermographic tests with flash thermography (**a**), considering the ROI highlighted in (**b**)

exhibits higher contrast, as it is already visible upon visual inspection. The results demonstrated however, as expected, that the contrast decrease with the nominal depth of the intended defects.

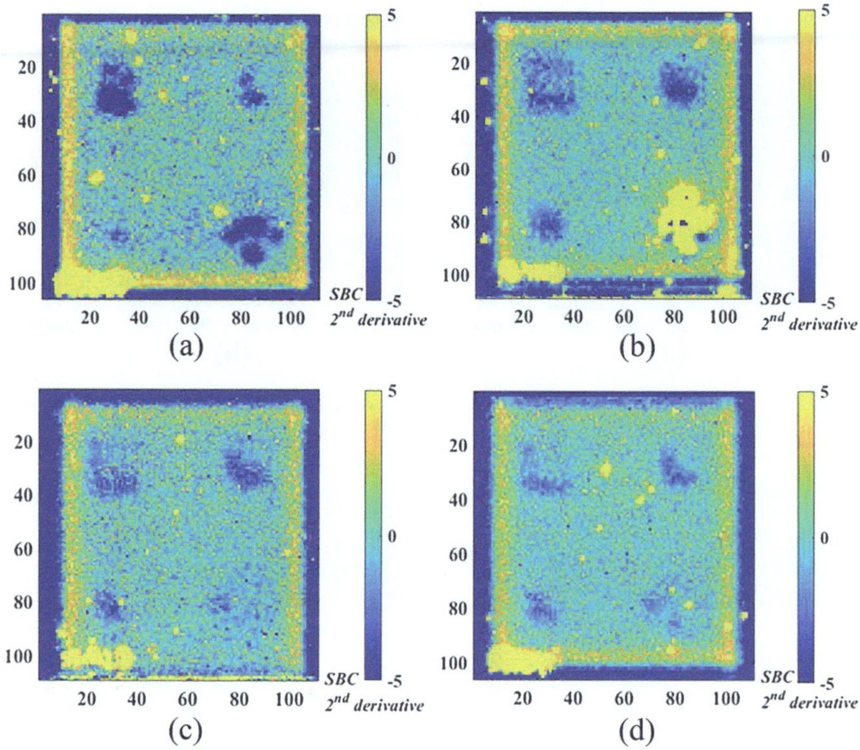

Fig. 13 Results after post processing (TSR, 5 polynomial degree as fitting and maps related to the contrast to noise ratio, in case of keyhole defects PK007 (**a**), and lack of fusion PK008 (**b**), PK009 (**c**) and PK010 (**d**))

Figure 14 reports the results obtained for the specimens PK001, PK002, PK003 e PK004.

These specimens exhibit various defects caused by laser illumination gaps; within the same component, four different sectors can be distinguished where the defects have different heights, while the four specimens differ from each other in terms of defect depth.

Obviously, if the defect height is greater, the contrasts are higher.

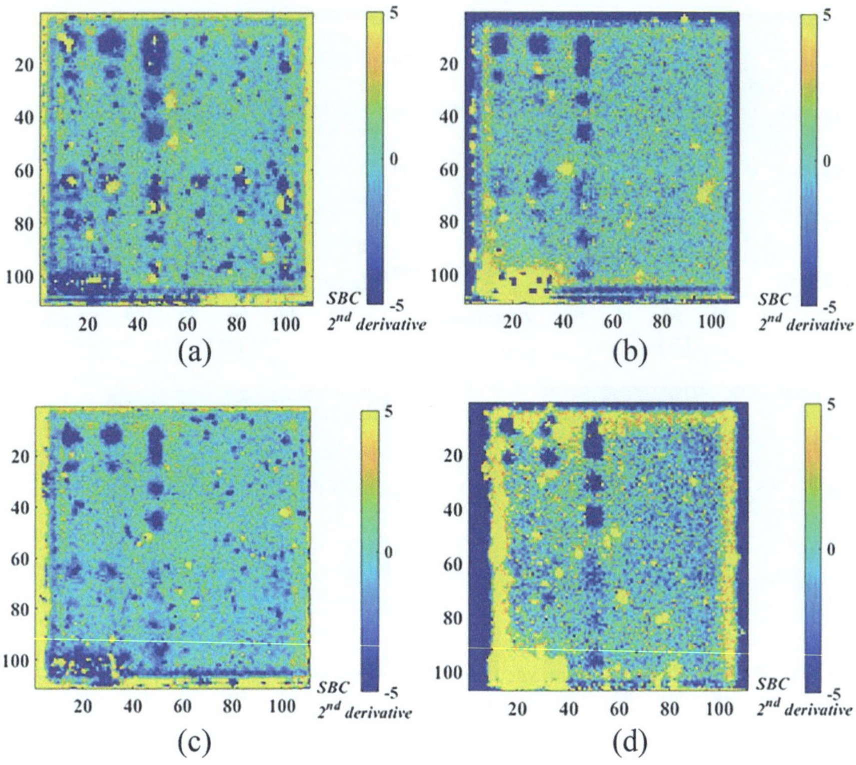

Fig. 14 Obtained results (TSR, 5° as polynomial fitting, second derivative analysis, PK001 (**a**), PK002 (**b**), PK003 (**c**) and PK004 (**d**)

4 Mechanical Characterisation of AM Materials

In this section, two case studies will be shown to demonstrate the capability of infrared thermography in the field of fatigue and fracture mechanics study of materials.

The adopted methodology involves a signal reconstruction algorithm to easily assess the different temperature component and carry out the thermal signal analysis [88]. The algorithm is based on the following model: for a homogenous and isotropic material under a sinusoidal load, the calibrated surface thermographic signal *T*, can be written in the time domain as [63, 64, 89]:

$$T(t) = T_0 + At + T_1 \sin(\omega t + \varphi_1) + T_2 \sin(2\omega t + \varphi_2) \tag{1}$$

Equation (1) T_0 represents the mean temperature and At is its linear increase, with A depending on the physical characteristics and geometry of the component. The third term in Eq. (1) is the first harmonic of the thermal signal. The amplitude T1 represents the temperature variations induced by the thermoelastic effect and the monotonic plastic work, while the phase fi1 represents the delay between the load and the thermal response. Under adiabatic conditions is constant, while it varies when the adiabatic condition is no longer respected. This could be the case of high stress gradients, which lead to conduction effects [90] or heat generation due to local plasticity [91–94]. The fourth term in Eq. (1) is the sum of two effects, both occurring at the twice of the loading frequency: (i) the thermoelastic response due to the second order effect [95] and (ii) the temperature variation due to dissipation (mainly due to the cyclic plastic zone) [94]. When local plasticisation effects occur, the thermoelastic component of the amplitude is at least one order of magnitude smaller than the second and therefore it can be neglected.

Thermal signal analysis supports in understanding the fracture mechanics of materials and components by providing two types of information: (i) thermoelastic stress analysis to evaluate the stress distribution near the crack tip and (ii) the analysis of non-adiabatic phenomena in the thermoelastic signal to pinpoint the dissipation associated with advancing damage.

In particular, the first case study will involve the study of the fatigue behaviour of an AISI 316 austenitic steel obtained by direct energy-deposition arc (DED-Arc) process [87] using thermal methods. The challenging aspects and potentialities of the technique will be shown in relation to the possibility of estimating the fatigue limit of the material.

The second case study will focus on evaluating the fatigue crack tip in Ti6Al4V produced through the selective laser melting (SLM) additive manufacturing process [67]. It is well-established that the growth of fatigue cracks is influenced by the driving force at the crack tip [92]. In this context, this dissipative process can be examined using various analytical and numerical models to describe the plastic work at the crack tip, leading to insights on fatigue behaviour in the presence of cracks [96]. The fracture mechanics of Ti6Al4V alloy obtained by SLM will show the capability of thermography-based indicators like thermoelastic signal and thermoelastic phase shift in studying the crack propagation behaviour [68, 94].

4.1 Case Study 1: Fatigue Estimations on AISI 316 Steel Obtained by DED-Arc Process

The direct energy deposition (DED) additive manufacturing process uses a powder or metal wire with the energy source to add material or fuse material onto an existing part or to create a new part. The DED-Arc has the advantage of a higher process speed than other additive manufacturing processes and it is, furthermore, extremely accurate [97].

The material preparation is described in [87] and involved a pulsed arc process to the purpose of the deposition of all beads. During the fabrication of the different layers the arc process parameters were maintained constant, in particular, wire feed rate was 6 m/min while the travel speed was 0.4 m/min.

The resulting material was a layered one consisting of different layers of parallel weld beads making the material substructure anisotropic. From the block characterised by a bead structure two types of samples were extracted as reported in Fig. 15. The blue specimen called ZPL was cut along the direction of beading deposing while the green specimen called ZPS was cut along the transverse direction to the beading deposing. The two types of samples were expected to be characterised by a layered substructure.

The samples were tested with rapid fatigue procedure producing the material self-heating [61] and involving the application of a load sequence (roughly 15 stress levels or loading block) from very low stress levels (elastic regime) to higher ones up to the material failure [62] carried out at a mechanical frequency (*f*) of 17 Hz and constant stress ratio (*R*) of 0.1 for each loading block. In each loading block, the stress level was maintained for 20,000 cycles. This ensured the stabilisation of the thermal signal and then accurate assessment of thermographic indicators.

The setup represented in Fig. 16 is composed by a MTS loading machine with 100 kN of capacity, infrared detectors (microbolometric and cooled ones). The results presented in this chapter will be referred to those obtained by cooled infrared detector FLIR X6540sc (640 × 512 pixel array, Noise equivalent temperature different less than 25 mK). The thermographic sequences were acquired during each loading block using a frame rate of 200 Hz. A dummy specimen was introduced in the setup to account for the effect of environmental temperature on sample temperature.

As already demonstrated in literature [89], usually in thermographic fatigue investigations a current temperature distribution the mean temperature indicator (T_0) does not permit a local analysis of the damage zone, it shows instead, a larger damage region. The T_1 and φ_1, allow an in-depth analysis of phenomena occurring locally in the material substructure. In particular, the re-scaled by applied stress

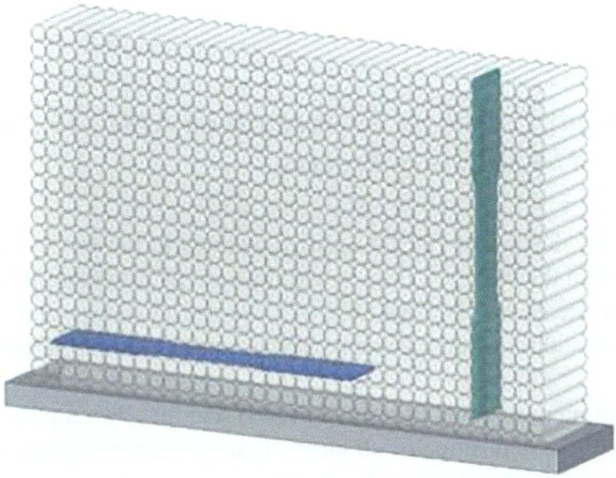

Fig. 15 Bead structure of the print strategy and samples orientations (ZPL orientation blue and ZPS orientation green) [87]

thermoelastic amplitude and thermoelastic phase lag are represented in Figs. 17 and 18 for two samples. By comparing Figs. 17a and 18a with Figs. 17b and 18b it results that both the amplitude and phase of the thermoelastic signal can be used to identify the characteristic substrate resulting from the technological process. This is particularly evident at low stress levels, where damage is not present. By observing the maps (Fig. 17a mainly) it is possible to distinguishing the zones of stress concentration and/or the zones where a loss of adiabatic conditions occurs [98] (Fig. 18), sentinels of irreversible processes.

The potentiality of thermography-based indicators T_1 and φ_1 can be also highlighted by the curves reported in Fig. 18, where a significant variation in both

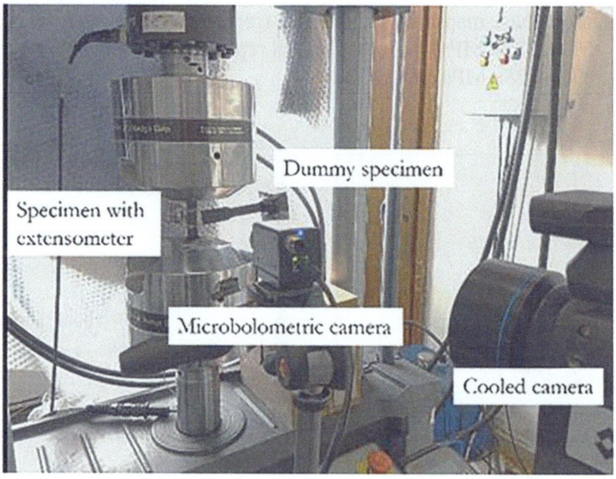

Fig. 16 Experimental setup for fatigue tests [87]

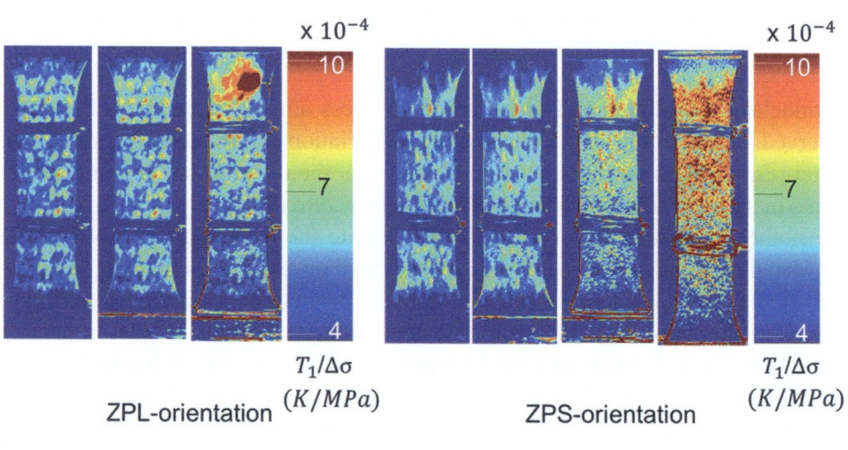

Fig. 17 Thermoelastic amplitude maps: (**a**) sample ZPL (represented loading levels in terms of maximum stress 195, 309 and 358 MPa), (**b**) sample ZPS represented loading levels in terms of maximum stress 195, 309 and 358 MPa and 520 MPa)

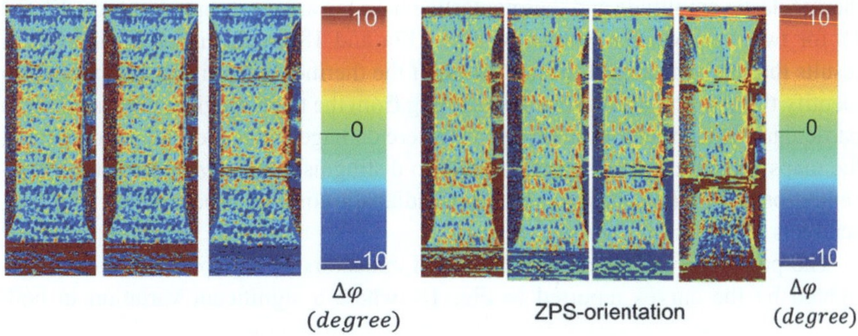

Fig. 18 Thermoelastic phase maps: (**a**) sample ZPL (represented loading levels in terms of maximum stress 195, 309 and 358 MPa), (**b**) sample ZPS represented loading levels in terms of maximum stress 195, 309 and 358 MPa and 520 MPa)

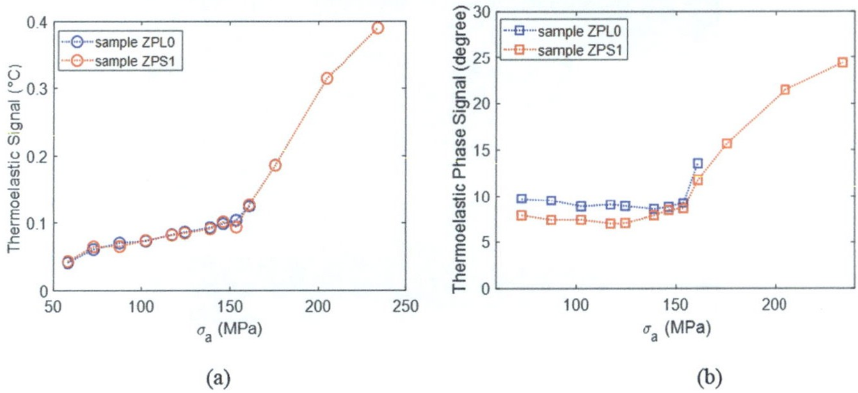

Fig. 19 (**a**) thermoelastic signal amplitude and (**b**) thermoelastic phase signal for two samples with different substructures

thermoelastic amplitude and phase around a stress amplitude of 150 MPa can be observed. The stress at which the significant variation of the thermal parameter is found (Fig. 19) can be considered an estimator of the material fatigue limit, which seems in agreement with the values found in literature [99].

Another interesting observation is that despite the difference in substructures between the ZPL and ZPS specimens, no discernible variations in the fatigue behaviour of the two directions seem to be present in thermoelastic temperature data (Fig. 19a). In phase data (Fig. 19b), a little scatter is observed at least in the first loading blocks. This behaviour could depend on an effective variation of the phase signal from the two samples, even if the material is the same. Further investigations are required both to understand if a signal variation in terms of amplitude or phase is present and if it is statistically significant.

4.2 Case Study 2: Fracture Mechanics Behaviour of Selective Laser Melted (SLM) Ti6Al4V Alloy

The prediction of the fatigue crack growth (FCG) is important to study the damage progression of materials and structures and for implementing a damage-tolerant approach [100]. The dissipating process of crack propagation can be studied via different experimental methods aimed at estimating the plastic work at the crack tip and then the energy involved in the process [38]. These analyses are important for predicting the remaining life of a material/component with a pre-determined crack size [69]. One of the most capable techniques for the aforementioned analysis is thermoelastic stress analysis (TSA) [101]. TSA is a well-known thermography-based technique that requires a relatively easy setup to be implemented, and it shows advantages on real components under cyclic loading, where other measurements (e.g. optical microscopy) are challenging to perform, as it provides insightful information about the thermodynamics of plastic flow and its localisation.

Considering the dissipative nature of crack growth given the heat generation from plastic work and high-stress gradients near the crack tip, the thermoelastic phase signal (fi1) can be used to estimate the crack tip position [40, 52, 53, 102–105].

The Selective Laser Melting (SLM) process is one of the Powder Bead Fusion [84, 106] processes and consists of processing metals such as steels, Ti, Al alloys in a pulverised state and applying thermal energy (laser) to melt and fuse them. SLM is a widely used process because of its ability to produce tailored metallic structures and to process a wide range of materials [107]. However, the advantages of a flexible and rapid process are offset by the presence of defects (e.g., porosity) and poor surface roughness that make the material and structures in general less robust, especially under cyclic loading [106]. By controlling machine and process parameters, these obstacles can be minimised to some extent. However, it is not always possible to achieve a fully controlled process, and the challenge is to understand the limits of applicability of structural health/damage monitoring techniques.

In this section, using the experimental Stanley-Chan [102] direct interpolation method, the crack tip position, crack growth rate and stress intensity factor will be evaluated using the TSA technique. Moreover, the Paris law parameters will be assessed and compared with those obtained by optical microscopy observations to assess the limit of applicability and capabilities of the TSA for additive manufactured materials. More specifically, the research aimed to study the capacity and sensitivity of the thermoelastic technique in the relief of the crack tip and the study of crack propagation on Ti6Al4V alloy with high roughness.

The FCG tests according to ASTM E 647–00 were carried out on two compact tension samples at constant amplitude (maximum applied stress 2000 N) and two different stress ratios (respectively 0.1 and 0.5) at a frequency equal to 15 Hz. The loading frame was an Instron Electropuls with a capacity of 10 kN. The setup, Fig. 20, was composed of an infrared detector Telops FAST M350. The sequences

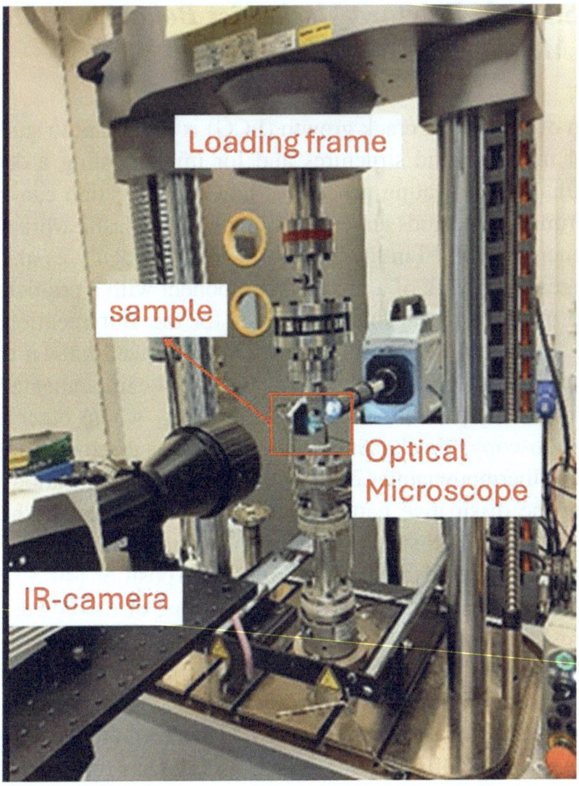

Fig. 20 Experimental Setup for carrying out fatigue crack growth tests [67, 68]

were acquired at 200 Hz with a spatial resolution of 0.03 mm/pixel. Microscopy observations were also carried out using a Photron Fastcam SA-Z camera operating with a 0.0035 mm/pixel resolution. In order to detect the crack-tip position, the optical microscope was equipped with a microscopic zoom lens.

The analysis of thermal sequences according to the signal reconstruction algorithm Eq. (1) provided the maps of thermoelastic amplitude (T_1) and phase shift (φ_1).

As per the estimation of the crack-tip, useful to evaluate the crack length and then the crack growth rate (da/dN), the φ_1 parameter is particularly useful as it allows to found it graphically. In effect, φ_1 assumes specific values in proximity of the crack tip: a positive increment of the phase value that indicates a loss of adiabatic condition due to plasticity and high-stress gradients, a negative value indicating a reverse plasticity hence the phase shift returns to zero. Such a point has been empirically considered as an indicator of the crack tip [108–113]. Figure 21a provides an example of profile parallel to the crack growth direction, where the sign inversion of data is present and the crack tip position is estimated.

T_1 can be useful for the evaluation of stress intensity factor range according to the procedure proposed by Stanley-Chan [33] based on Westergaard' formulation and mode I loading condition. The procedure provides the SIF range via the direct interpolation of T_1 data from the map. In effect, the first stress invariant and the distance (y-pixels) from crack tip position, the maximum values of T_1 and SIF range are related by the following equation:

$$y = \left(\frac{3\sqrt{3}}{4\pi} T_0^2 K_{TSA}^2 \, \Delta K_I^2 \right) \frac{1}{\Delta T_{1,\max}^2} \tag{2}$$

where K_{TSA} is the thermoelastic constant and T_0 is the absolute ambient temperature. By applying the linear regression of y data lying in the SIF-dominated (Fig. 21a) region plotted vs $1/\Delta T^2_{1,\max}$, the ΔK_I value is the slope of the fitting model. Of course, the selection of the number of the points is crucial for the sensitivity in the estimation of ΔK_I.

Figure 22a shows the crack growth rates assessed using thermoelastic phase data and the optical microscope to detect the crack tip for the data of the sample tested at $R = 0.1$. As it is possible to highlight, the estimated da/dN are in good agreement even at the beginning of the test where the crack is short and the signal-to-noise ratio is lower (consequently, it could be difficult to estimate the crack length using thermal φ_1 indicator or assessing the SIF range via T_1 maps). The estimation of SIF range using thermoelastic amplitude T_1 data, in particular, if compared to the one estimated using the procedure proposed by the ASTM standard, only a little underestimation can be detected, especially in the initial loading cycles.

The fatigue crack growth rates and SIF ranges have been used to assess the fatigue crack propagation behaviour. In Fig. 22c, are reported the curves relating da/dN and ΔK_I obtained by thermoelastic data and the curves where da/dN was assessed by measuring crack length with the optical microscope, while the ΔK_I was assessed using the procedure described by ASTM. The curves also refer to the tests carried out at both $R = 0.1$ and $R = 0.5$. The graph is in a double logarithmic scale.

As illustrated in Fig. 22c, the outcomes are promising for this material in the estimation of Paris' law, as evidenced by the near superimposition of the curves obtained through the classical method and those obtained through the 'thermal' method.

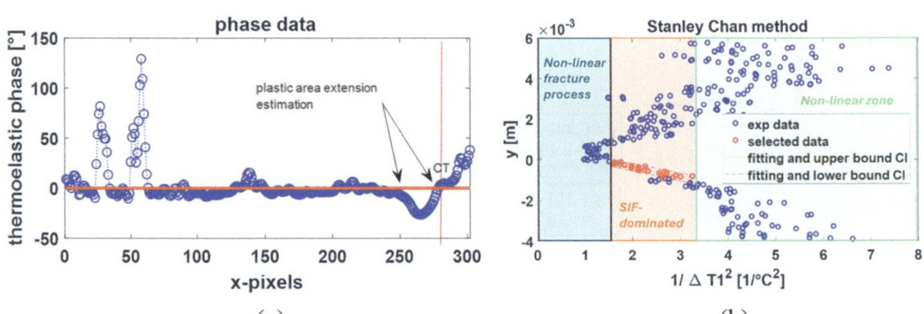

Fig. 21 (**a**) Crack-tip position estimated via thermoelastic phase data, (**b**) example of the implementation of the Stanley-Chan method

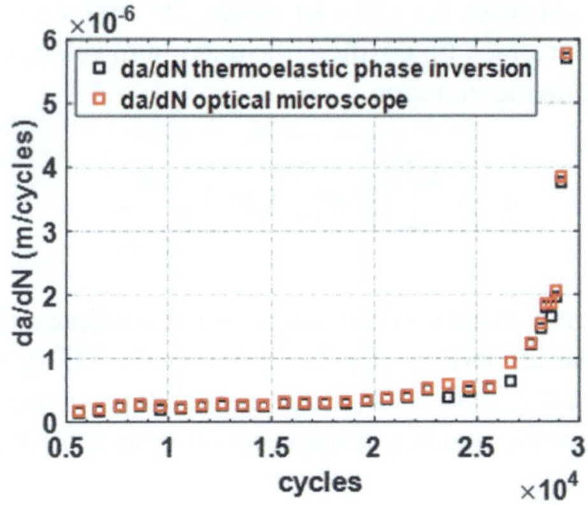

(a)

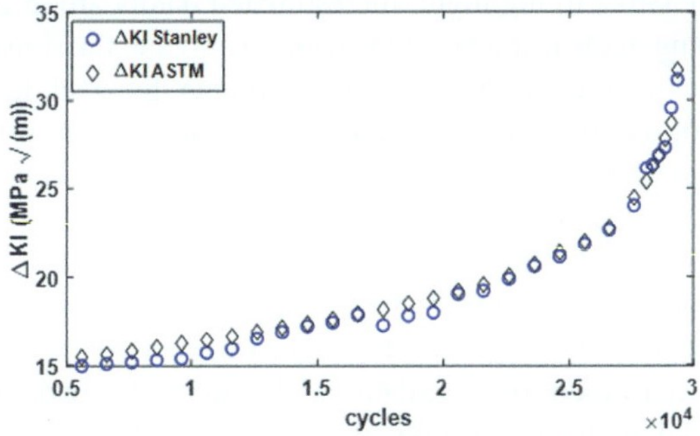

(b)

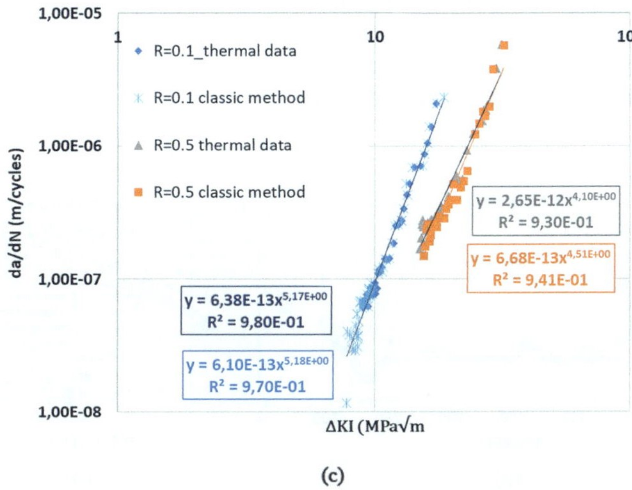

(c)

Fig. 22 (**a**) Comparison between da/dN calculated using thermoelastic and optical microscope data, (**b**) SIF ranges calculated using the thermoelastic data and ASTM procedure on crack length obtained by optical microscope, (**c**) Paris' laws for two stress ratios assessed classic procedure (crack lengths evaluated using optical microscope) and innovative procedure using thermoelastic data

Further improvements can be made to the calculation of the SIF, for example, by taking into account more accurate theoretical formulations for the description of stress parameters near the crack apex. These include Westgard's formulation with *T*-stress and Williams' formulation [105].

As the material is also titanium, the estimation of SIF should also take into account second-order thermoelastic theory [96].

5 Discussion

The just exposed cases demonstrate the potential of IR thermography (IRT) in a large variety of applications related to the field of mechanical engineering. In particular, IRT allows for characterising the material performance in all the phases of its cycle life, from manufacturing to monitoring during service conditions. In this regard, the use of IR sensors can be very useful for the monitoring of three-dimensional (3D) processes, predicting the mechanical properties of the produced components. It was demonstrated that the absolute temperature is not the only thermal feature that can be extracted from the recorded sequences, and the investigation of thermal signal over time is necessary to investigate other possible features. The main limits related to the use of IR sensors regard thesetup of IR cameras that cannot always be easily integrated into commercial 3D printers. In some cases, the 3D printers are not prepared for inspecting the working field with IR sensors/cameras and other accessories such as IR windows and IR mirrors.

The online monitoring of the process also gives information about the presence of 3D typical defects. Also, in this case, the presence of defects can be detected by investigating any anomalies in the thermal signal. However, as already said, the thermal signal variation is strictly related to the process parameters, and then it cannot be an index of the presence of defects. So, offline thermographic techniques can be used for characterising defects. It is important to underline that thermography can be classified as a superficial and sub-superficial technique so it cannot be compared to X-ray or ultrasounds. In particular, tomography represents the most used method for detecting and characterising defects in 3D components. However, this latter is time-consuming and cannot be used for covering 100% of the inspection. In this regard, active thermographic techniques can be useful for detecting typical shallow defects up to 1 mm depth (the capability of detection depends on the size of the defect).

The chapter also demonstrated the capability of thermography in characterising 3D printed materials. The potential of IRT in characterising the fatigue behaviour of material has been largely shown in many works in the last 20 years. In particular, the main advantage of using IRT is the possibility to obtain an estimation of the endurance limit in a rapid way and with few specimens. However, traditional approaches/procedures based on IRT need to be set to consider the typical anisotropy and microstructure characterising the 3D printed components. In this regard, the TSA technique, and more generally, the analysis of thermal data in the frequency domain, has been proven to have the potential for investigating the material response under cycling loading and capturing the anisotropy. TSA can also be used to measure the crack length and assess the stress intensity factor (SIF). Also, in this case, the main advantage consists of obtaining both these quantities without stopping the tests and in a continuous way.

All the applications regarding the TSA technique require the use of performant IR sensors. Usually, cooled sensors with a high thermal sensitivity are needed when the temperature variations are less than 0.5 °C to obtain a significant signal-to-noise ratio. Unfortunately, the cost of these sensors is one order higher than the corresponding microbolometric ones, with a consequent low diffusion in industrial applications. In this contest, further steps will regard the development of a new procedure of data analysis for performing TSA with low-cost sensors and reducing the processing time.

6 Conclusions

In this chapter, IRT has been used for investigating the behaviour of materials made by means of additive manufacturing techniques. In particular, four applications were presented from the manufacturing to the mechanical characterisation of 3D printed components. For each application, different algorithms were adopted and developed with the common aim of extracting thermal features capable of describing the material behaviour. The main conclusions are:

- Process monitoring.

- A statistical analysis was carried out to investigate the effect of the process parameters on the thermal signal. Then, an empirical model was developed for describing the mechanical behaviour of the material in terms of hardness as a function of thermal features. In this way, the process can be controlled online by only monitoring the thermal signal. Future works will be focused on the investigation of other thermal features, the use of machine or deep learning algorithms and applications on other materials.
- Non-destructive testing.
- Active thermographic techniques were used for detecting shallow typical defects in Additive Manufacturing specimens. The capability of the pulsed technique applied with flash lamps has been demonstrated on some specimens produced with the Laser Powder Bed Fusion technique. Future works are necessary to investigate the probability of detection and to evaluate the capability of active thermography to characterise defects in terms of their size and depth
- Investigation of the fatigue behaviour.
- The analysis of thermal data in the frequency domain allows to obtain additional information about the fatigue behaviour with respect to traditional approaches only based on the analysis of the absolute temperature. In particular, the analysis of the first harmonic component shows a characteristic path related to the anisotropy of the specimens. Further analyses are necessary for understanding the relationship between the thermal signal and the material microstructure.

- Crack growth assessment.
- The TSA technique was proven to be effective for assessing both the crack length and the stress intensity factor, also in the presence of the inherent roughness of the AM process. The results in terms of Paris's Law assessment were in good agreement with conventional techniques.

Funding and Acknowledgements Part of the work was carried out under the framework of the research contract "Nuovo Pignone (Baker Hughes)"—"Monitoring the LMD process with thermal methods"—Principal investigator (PI) Prof. Ing. Umberto Galietti.

Part of the work was carried out in the laboratories and in collaboration with authors from BAM (Bundesanstalt für Materialforschung und -prüfung) in Berlin. The authors wish to thank Rainer Krankenhagen, Simon J. Altenburg, and Christiane Maierhofer for their availability and support during the research activities.

Part of the work has been Financed by the European Union—NextGenerationEU (National Sustainable Mobility Center CN00000023, Italian Ministry of University and Research Decree n. 1033–17/06/ 2022, Spoke 11—Innovative Materials & Lightweighting). The opinions expressed are those of the authors only and should not be considered as representative of the European Union or the European Commission's official position. Neither the European Union nor the European Commission can be held responsible for them.

References

1. Ibarra-Castanedo, C., Tarpani, J. R., & Maldague, X. P. V. (2013). Nondestructive testing with thermography. *European Journal of Physics, 34*(6), S91–S109.
2. Maierhofer, C., Myrach, P., Steinfurth, H., Reischel, M., & Röllig, M. (2014). Development of standards for flash thermography and lock-in thermography. In *Proceedings of the 14th International Conference on Quantitative InfraRed Thermography (QIRT14),* Bordeaux, France, 7–11 July 2014.
3. Ringermacher, H, Archacki, R, & Veronesi, W. (1998). Nondestructive testing: Transient depth thermography. USA: U.S. Patent.
4. Shepard, S., Lhota, R., Ahmed, T., & Tasdiq, A. (2009). Measurement limits in flash thermography. *Proceedings SPIE, Thermosense, 31*, 7299.
5. Maldague, X. (1997). *Nondestructive evaluation of materials by infrared thermography*. Springer-Verlag. https://doi.org/10.1007/978-1-4471-1995-1. ISBN-13:978-1-4471-1997-5 e-ISBN-13:978-1-4471-1995-1.
6. Balageas, D. (1991). Le contr6le non destructif par methodes thermiques. *Revue générale de thermique, 300*, 483–498. (in French).
7. Balageas, D. L., & Luc, A. M. (1986). Transient thermal behavior of directional reinforced composites: Applicability limits of homogeneous property model. *AIAA Journal, 24*(1), 109–114.
8. Gupta, M., Khan, M. A., Butola, R., & Singari, R. M. (2022). Advances in applications of non-destructive testing (NDT): A review. *Advances in Materials and Processing Technologies, 8*, 2286–2307.
9. Lu, Q. Y., & Wong, C. H. (2017). Additive manufacturing process monitoring and control by non-destructive testing techniques: Challenges and in-process monitoring. *Virtual and Physical Prototyping, 13*(2), 39–48. https://doi.org/10.1080/17452759.2017.1351201
10. Traycoff, R. B. (1992). Chapter 14. Medical applications of infrared thermograpahy. In X. P. V. Maldague (Ed.), *Infrared methodology and technology* (International advances in NDT monograph series). Gordon and Breach.

11. Wood, J. T. (1982). Description of a facility for evaluating infrared imaging systems for building applications. In G. E. Courville (Ed.), *Thermosense V* (Proc SPIE) (Vol. 371, pp. 246–249).
12. D'Accardi, E., Palumbo, D., Errico, V., Fusco, A., Angelastro, A., & Galietti, U. (2023a). Analysing the probability of detection of shallow spherical defects by means of pulsed thermography. *Journal of Nondestructive Evaluation, 42*(1), 27.
13. D'Accardi, E., Chiappini, F., Giannasi, A., Guerrini, M., Maggiani, G., Palumbo, D., & Galietti, U. (2023b, June). Thermal monitoring of the nickel-based Inconel 718 by laser metal deposition: Evaluation of the process parameters and the final mechanical behaviour via thermographic features. In *Thermosense: Thermal infrared applications XLV* (Vol. 12536, pp. 24–33). SPIE.
14. D'Accardi, E., Chiappini, F., Giannasi, A., et al. (2024a). On the feasibility of a predictive model of mechanical properties of AM Inconel 718 thin wallets produced by DED-LB process monitored with thermal methods. *Progress in Additive Manufacturing*. https://doi.org/10.1007/s40964-024-00809-4
15. D'Accardi, E., Chiappini, F., Giannasi, A., Guerrini, M., Maggiani, G., Palumbo, D., & Galietti, U. (2024b). Online monitoring of direct laser metal deposition process by means of infrared thermography. *Progress in Additive Manufacturing, 9*(4), 983–1001.
16. Altenburg, S. J., Straße, A., Gumenyuk, A., & Maierhofer, C. (2022). In-situ monitoring of a laser metal deposition (LMD) process: Comparison of MWIR, SWIR and high-speed NIR thermography. *Quantitative InfraRed Thermography Journal, 19*, 97–114. https://doi.org/10.1080/17686733.2020.1829889
17. Dell'Avvocato, G., & Palumbo, D. (2024). Thermographic procedure for the assessment of Resistance Projection Welds (RPW): Investigating parameters and mechanical performances. *Journal of Advanced Joining Processes, 9*, 100177.
18. Dell'Avvocato, G., Gohlke, D., Palumbo, D., Krankenhagen, R., & Galietti, U. (2022). Quantitative evaluation of the welded area in Resistance Projection Welded (RPW) thin joints by pulsed laser thermography. In *Thermosense: Thermal infrared applications XLIV* (Vol. 12109, pp. 152–165). SPIE.
19. Dell'Avvocato, G., Rashkovets, M., Castellano, A., Palumbo, D., Contuzzi, N., Casalino, G., & Galietti, U. (2024). Preliminary procedure for the assessment of probeless friction stir spot welds (P-FSSW) in dissimilar aluminum alloys by long pulsed laser thermography. In *Thermosense: Thermal infrared applications XLVI* (Vol. 13047, pp. 80–90). SPIE.
20. Mazzarisi, M., Angelastro, A., Latte, M., Colucci, T., Palano, F., & Campanelli, S. L. (2023). Thermal monitoring of laser metal deposition strategies using infrared thermography. *Journal of Manufacturing Processes, 85*, 594–611.
21. Mazzarisi, M., Campanelli, S. L., Angelastro, A., Palano, F., & Dassisti, M. (2021). In situ monitoring of direct laser metal deposition of a nickel-based superalloy using infrared thermography. *The International Journal of Advanced Manufacturing Technology, 112*, 157–173.
22. Muvvala, G., Patra Karmakar, D., & Nath, A. K. (2017). Online monitoring of thermo-cycles and its correlation with microstructure in laser cladding of nickel based super alloy. *Optics and Lasers in Engineering, 88*, 139–152. https://doi.org/10.1016/j.optlaseng.2016.08.005
23. D'Accardi, E., Altenburg, S., Maierhofer, C., Palumbo, D., & Galietti, U. (2019). Detection of typical metal additive manufacturing defects by the application of thermographic techniques. *Proceedings, 27*(1), 24. MDPI.
24. D'Accardi, E., Krankenhagen, R., Ulbricht, A., Pelkner, M., Pohl, R., Palumbo, D., & Galietti, U. (2022a). Capability to detect and localize typical defects of laser powder bed fusion (L-PBF) process: An experimental investigation with different non-destructive techniques. *Progress in Additive Manufacturing, 7*(6), 1239–1256.
25. D'Accardi, E., Palumbo, D., Errico, V., Fusco, A., Angelastro, A., Addante, G. D., & Galietti, U. (2022b, June). Detection analysis for sub-superficial defects in additive manufactured metal samples by means of flash thermography. In *Society for Experimental Mechanics Annual Conference and Exposition* (pp. 43–54). Springer International Publishing.

26. D'Accardi, E., Palumbo, D., Tamborrino, R., & Galietti, U. (2018). A quantitative comparison among different algorithms for defects detection on aluminum with the pulsed thermography technique. *Metals., 8*, 859.
27. Zweschper, G., Riegert, T., Dillenz, G., & Busse, A. (2005). Ultrasound excited thermography—Advances due to frequency modulated elastic waves. *Quantitative InfraRed Thermography Journal., 2*(1), 65–76.
28. Di Renzo, A., Marsili, R., Martarelli, M., Moretti, M., Rosati, G., & Rossi, G. L. (2006). Simultaneous application of scanning laser vibrometry and thermoelasticity for measurement of stress-strain fields on mechanical components. *Seventh International Conference on Vibration Measurements by Laser Techniques: Advances and Applications, 6345*, 63450H. https://doi.org/10.1117/12.693161
29. Molina-Viedma, Á. J., Felipe-Sesé, L., López-Alba, E., & Díaz, F. A. (2021). Thermoelastic effect in modal shapes at high frequencies using infrared thermography. *Measurement, 176*, 109180. https://doi.org/10.1016/j.measurement.2021.109180
30. Molina-Viedma, A. J., Felipe-Sesé, L., López-Alba, E., & Díaz, F. A. (2022). Comparison of lock-in correlation and a novel periodogram method for experimental multi-harmonic thermoelastic analysis. *Mechanical Systems and Signal Processing, 164*, 108235. https://doi.org/10.1016/j.ymssp.2021.108235
31. Phan, T. S., Dulieu-Barton, J. M., & Temarel, P. (2006). Thermoelastic stress analysis of structures under natural vibrations. *Experimental Mechanics, 46*(4), 463–472. https://doi.org/10.1007/s11340-006-8445-6
32. Molina-Viedma, Á. J., Palumbo, D., De Finis, R., Di Carolo, F., Felipe-Sesé, L., López-Alba, E., et al. (2023). Effect of geometrical discontinuities on mode shapes stress maps using infrared thermography. In *Society for Experimental Mechanics Annual Conference and Exposition* (pp. 65–71). Springer Nature Switzerland.
33. Stanley, P. (1997). Applications and potential of thermoelastic stress analysis. *Journal of Materials Processing Technology, 64*, 359–370.
34. Carvalho, M. S., Martins, A. P., & Santos, T. G. (2019). Simulation and validation of thermography inspection for components produced by additive manufacturing. *Applied Thermal Engineering, 159*, 113872. https://doi.org/10.1016/j.applthermaleng.2019.113872. ISSN 1359-4311.
35. Grosso, M., Lopez, J. E., Silva, V. M., Soares, S. D., Rebello, J. M., & Pereira, G. R. (2016). Pulsed thermography inspection of adhesive composite joints: Computational simulation model and experimental validation. *Composites Part B, Engineering, 106*, 1–9. https://doi.org/10.1016/j.compositesb.2016.09.011
36. Chung, Y., Kim, C., Lee, S., Suh, H., & Kim, W. (2024). Improvement of optical-induced thermography defect detectability by equivalent heating and non-uniformity compensation in polyetheretherketone. *Applied Sciences, 14*, 8720. https://doi.org/10.3390/app14198720
37. Chung, Y., Lee, S., & Kim, W. (2021). Latest advances in common signal processing of pulsed thermography for enhanced detectability: A review. *Applied Sciences, 11*, 12168.
38. Gupta, M., Alderliesten, R. C., & Benedictus, R. (2015). A review of T-stress and its effects in fracture mechanics. *Engineering Fracture Mechanics, 134*, 218–241.
39. Olafsson, G., Tighe, R. C., Boyd, S. W., & Dulieu-Barton, J. M. (2022). Lock-in thermography using miniature infra-red cameras and integrated actuators for defect identification in composite materials. *Optics and Laser Technology, 147*, 107629.
40. Palumbo, D., Ancona, F., & Galietti, U. (2015a). Quantitative damage evaluation of composite materials with microwave thermographic technique: Feasibility and new data analysis. *Meccanica, 50*, 443–459. https://doi.org/10.1007/s11012-014-9981-2
41. Palumbo, D., Cavallo, P., & Galietti, U. (2019). An investigation of the stepped thermography technique for defects evaluation in GFRP materials. *NDT and E International, 102*, 254–263.
42. Zhang, H., Yang, R., He, Y., Foudazi, A., Cheng, L., & Tian, G. (2017). A review of microwave thermography nondestructive testing and evaluation. *Sensors, 17*, 1123. https://doi.org/10.3390/s17051123

43. Burrows, S. E., Dixon, S., Pickering, S. G., Li, T., & Almond, D. P. (2011). Thermographic detection of surface breaking defects using a scanning laser source. *NDT and E International, 44*, 589–596.
44. Li, T., Almond, D. P., & Rees, D. A. S. (2011). Crack imaging by scanning laser-line thermography and laser-spot thermography. *Measurement Science and Technology, 22*(3), 035701.
45. Montinaro, N., Cerniglia, D., & Pitarresi, G. (2018). A numerical and experimental study through laser thermography for defect detection on metal additive manufactured parts. *Frattura ed Integrità Strutturale, 43*, 231–240.
46. Montinaro, N., Cerniglia, D., & Pitarresi, G. (2017). Detection and characterisation of disbonds on fibre metal laminate hybrid composites by flying laser spot thermography. *Composites Part B: Engineering, 108*, 164–173. https://doi.org/10.1016/j.compositesb.2016.09.084. ISSN 1359-8368.
47. Oswald-Tranta, B. (2009). Time-resolved evaluation of inductive pulse heating measurements. *QIRT J., 6*, 3–19.
48. Lepitre, P., Merlet, L. M., Doudard, C., Dhondt, M., Surand, M., & Calloch, S. (2024). Influence of shot-peening on the self-heating behavior and fatigue properties of 300M steel. *Mechanics of Materials, 199*, 105174.
49. Renault, M., Bercelli, L., Doudard, C., Levieil, B., Beaudet, J., & Calloch, S. (2024). Infrared imaging surface roughness criticality assessment of wire arc additive manufactured specimens. *Procedia Structural Integrity, 57*, 22–31.
50. Carrascal, I., Casado, J. A., Diego, S., Lacalle, R., Cicero, S., & Álvarez, J. A. (2014). Determination of the Paris' law constants by means of infrared thermographic techniques. *Polymer Testing, 40*, 39–37.
51. Chrysochoos, A., & Louche, H. (2000). An infrared image processing to analyse the calorific effects accompanying strain localisation. *International Journal of Engineering Science, 38*(16), 1759–1788.
52. Diaz, F. A., Patterson, E. A., Tomlinson, R. A., & Yates, R. A. (2004a). Measuring stress intensity factors during fatigue crack growth using thermoelasticity. *Fatigue & Fracture of Engineering Materials & Structures, 27*(7), 571–513.
53. Diaz, F. A., Patterson, E. A., Tomlinson, R. A., & Yates, J. R. (2004b). Measuring stress intensity factors during fatigue crack growth using thermoelasticity. *Fatigue and Fracture of Engineering Materials and Structures, 26*(4), 571–583.
54. Fedorova, A. Y. U., Bannikov, M. V., Plekhov, O. A., & Plekhova, E. V. (2012). Infrared thermography study of the fatigue crack propagation. *Frattura ed Integrità Strutturale, 21*, 46–48.
55. Meneghetti, G., & Ricotta, M. (2012). The use of the specific heat loss to analyse the low- and high-cycle fatigue behaviour of plain and notched specimens made of a stainless steel. *Engineering Fracture Mechanics, 81*, 2–16.
56. Munier, R., Doudard, C., Calloch, S., & Weber, B. (2014). Determination of high cycle fatigue properties of a wide range of steel sheet grades from self-heating measurements. *International Journal of Fatigue, 63*, 46–61.
57. Naderi, M., & Khonsari, M. (2011). Real-time fatigue life monitoring based on thermodynamic entropy. *Structural Health Monitoring, 10*(2), 189–197.
58. Pitarresi, G., & Patterson, E. A. (2003). A review of the general theory of thermoelastic stress analysis. *The Journal of Strain Analysis for Engineering Design, 38*(5).
59. Rigon, D., Ricotta, M., & Meneghetti, G. (2017). An analysis of the specific heat loss at the tip of severely notched stainless steel specimens to correlate the fatigue strength. *Theoretical and Applied Fracture Mechanics, 92*, 240–251.
60. Shiozawa, D., Inagawa, T., Washio, T., & Sakagami, T. (2017). Accuracy improvement in dissipated energy measurement by using phase information. *Measurement Science and Technology, 28*(4), 044004.
61. La Rosa, G., & Risitano, A. (2000). Thermographic methodology for rapid determination of the fatigue limit of materials and mechanical components. *International Journal of Fatigue, 22*(1), 65–73.

62. De Finis, R., Palumbo, D., Ancona, F., & Galietti, U. (2017). Fatigue behaviour of stainless steels: A multi-parametric approach. In *Paper Presented At: Residual Stress, Thermomechanics and Infrared Imaging, Hybrid Techniques and Inverse Problems*, 9; Cham.
63. De Finis, R., Palumbo, D., & Galietti, U. (2021a). On the relationship between mechanical energy rate and heat dissipated rate during fatigue for a C45 steel depending on stress ratio. *Fatigue and Fracture of Engineering Materials and Structures, 44*(10), 2781–2799.
64. De Finis, R., Palumbo, D., & Galietti, U. (2021b). An experimental procedure based on infrared thermography for the assessment of crack density in quasi-isotropic CFRP. *Engineering Fracture Mechanics, 258*, 108108.
65. Zaeimi, M., De Finis, R., Palumbo, D., & Galietti, U. (2024). Fatigue limit estimation of metals based on the thermographic methods: A comprehensive review. *Fatigue & Fracture of Engineering Materials & Structures, 47*(3), 611–646.
66. Bercelli, L., Levieil, B., Doudard, C., & Calloch, S. (2024). Fatigue crack closure assessment by wavelet transform of infrared thermography signals. *International Journal of Fatigue, 190*, 108639.
67. De Finis, R., Palumbo, D., Di Carolo, F., Ricotta, M., Meneghetti, G., & Galietti, U. (2022a). Crack tip position evaluation and Paris' law assessment of a propagating crack by means of temperature-based approaches. *Procedia Structural Integrity, 39*, 528–545.
68. De Finis, R., Sendrowicz, A., Myhre, A. O., Berto, F., Vinogradov, A., & Galietti, U. (2022b, June). Analysis of the fracture mechanics behaviour of an additive manufactured titanium alloy via a thermography-based technique. In *Society for Experimental Mechanics Annual Conference and Exposition* (pp. 19–26). Springer International Publishing.
69. Palumbo, D., De Finis, R., Ancona, F., & Galietti, U. (2017). Damage monitoring in fracture mechanics by evaluation of the heat dissipated in the cyclic plastic zone ahead of the crack tip with thermal measurements. *Engineering Fracture Mechanics, 181*, 65–76.
70. Paris, P., & Erdogan, F. (1963). A critical analysis of crack propagation laws. *Journal of Basic Engineering, D 85*(4), 528–527. https://doi.org/10.1115/1.3656900
71. AbouelNour, Y., & Gupta, N. (2023). Assisted defect detection by in-process monitoring of additive manufacturing using optical imaging and infrared thermography. *Additive Manufacturing, 67*, 103483. https://doi.org/10.1016/j.addma.2023.103483. ISSN 2214-8604.
72. Breese, P. P., Becker, T., Oster, S., Metz, C., & Altenburg, S. J. (2023). In-situ defect detection for laser powder bed fusion with active laser thermography. In *Proceedings of the 34rd Annual International Solid Freeform Fabrication Symposium 2023* (pp. 1978–1989). University of Texas at Austin.
73. Chand, K., Fritsch, T., Oster, S., Ulbricht, A., Poka, K., & Bruno, G. (2024). A comparative study of rigid three-dimensional image registration methods for powder bed fusion with laser beam of metals.
74. Errico, V., Campanelli, S. L., Angelastro, A., Dassisti, M., Mazzarisi, M., & Bonserio, C. (2021). Coaxial monitoring of AISI 316L thin walls fabricated by direct metal laser deposition. *Materials, 14*(3), 673.
75. Maffia, S., Furlan, V., & Previtali, B. (2023a). Molten pool temperature monitoring in laser metal deposition: Comparison between single wavelength and ratio pyrometry techniques. *International Journal of Mechatronics and Manufacturing Systems, 16*(1), 96–111.
76. Maffia, S., Chiappini, F., Maggiani, G., Furlan, V., Guerrini, M., & Previtali, B. (2023b). Enhancing productivity and efficiency in conventional laser metal deposition process for Inconel 718–Part II: Advancing the process performance. *The International Journal of Advanced Manufacturing Technology, 129*(1-2), 279–298.
77. Mohr, G., Nowakowski, S., Altenburg, S. J., Maierhofer, C., & Hilgenberg, K. (2020a). Experimental determination of the emissivity of powder layers and bulk material in laser powder bed fusion using infrared thermography and thermocouples. *Metals, 10*(11), 1546.
78. Mohr, G., Altenburg, S. J., Ulbricht, A., Heinrich, P., Baum, D., Maierhofer, C., & Hilgenberg, K. (2020b). In-situ defect detection in laser powder bed fusion by using thermography and optical tomography—Comparison to computed tomography. *Metals, 10*(1), 103.

79. Oster, S., Breese, P. P., Ulbricht, A., Mohr, G., & Altenburg, S. J. (2024). A deep learning framework for defect prediction based on thermographic in-situ monitoring in laser powder bed fusion. *Journal of Intelligent Manufacturing, 35*(4), 1687–1706.
80. Scheuschner, N., Oster, S., Mohr, G., & Hilgenberg, K. (2024). Comparison of NIR and SWIR thermography for defect detection in Laser Powder Bed Fusion. *Procedia CIRP, 124*, 301–304.
81. Wallace Nicholas, J., Crane, N. B., & Jones, M. R. (2022). Defect measurement limits using flash thermography with application to additive manufacturing. *NDT&E International, 128*, 102615.
82. Bustos, I., Bergant, M., & Yawny, A. (2023). On the suitability of applying thermographic methods for the rapid estimation of the fatigue limit of additively manufactured Ti-6Al-4V. *International Journal of Fatigue, 174*, 107709.
83. Douellou, C., Balandraud, X., Duc, E., Verquin, B., Lefebvre, F., & Sar, F. (2019). Fast fatigue characterization by infrared thermography for additive manufacturing. *Procedia Structural Integrity, 19*, 90–100.
84. Dusautoir, T., Berthel, B., Fouvry, S., Matzen, P., & Meck, K. D. (2024). Influence of post-processing treatments on fatigue limit of notched additive manufactured Ti-6AL-4V determined by rapid thermographic methodology. *International Journal of Fatigue, 179*, 108034.
85. Manoharan, N., Pathak, P., Gururaja, S., Kumar, V., & Vaidya, U. (2024). Rapid fatigue characterization via infrared thermography of AM-CM composites. In C. Franck et al. (Eds.), Challenges in mechanics of biological systems and materials, thermomechanics and infrared imaging, time dependent materials and residual stress *Conference Proceedings of the Society for Experimental Mechanics Series* (Vol. 2. SEM 2023). Springer. https://doi.org/10.1007/978-3-031-50470-9_8
86. Balageas, D. L., Roche, J. M., Leroy, F. H., Liu, W. M., & Gorbach, A. M. (2015). The thermographic signal reconstruction method: A powerful tool for the enhancement of transient thermographic images. *Biocybernetics and Biomedical Engineering, 35*(1), 1–9.
87. Krankenhagen, R., Chaudhuri, S., Pittner, A., Winterkorn, R., de Finis, R., Palumbo, D., & Galietti, U. (2024). Thermographic investigation of the anisotropic behaviour of additively manufactured AISI316 steel using DED-Arc. In: *17th Quantitative Infrared Thermography Conference*, Zagreb, Croatia.
88. Krapez, J. C., Pacou, D., & Gardette, G. (2000). Lock-in thermography and fatigue limit of metals. In *2000 Quantitative InfraRed Thermography*.
89. De Finis, R., Palumbo, D., & Galietti, U. (2019). A multianalysis thermography-based approach for fatigue and damage investigations of ASTM A182 F6NM steel at two stress ratios. *Fatigue & Fracture of Engineering Materials & Structures, 42*(1), 267–283.
90. Wang, W. J., Dulieu-Barton, J. M., & Li, Q. (2010). Assessment of non-adiabatic behaviour in thermoelastic stress analysis of small-scale components. *Experimental Mechanics, 50*, 449–461.
91. Kucharski, P., Lesiuk, G., & Szata, M. (2016). Description of fatigue crack growth in steel structural components using energy approach—Influence of themicrostructure on the FCGR. In: *Fatigue Failure and Fracture Mechanics XXVI: Proceedings of the XXVI Polish National Conference on Fatigue Failureand Fracture Mechanics*, (Skibicki, D., ed.), AIP Publishing.
92. Ritchie, R. O. (1999). Mechanisms of fatigue-crack propagation in ductile and brittle solids. *International Journal of Fracture, 100*, 55–29.
93. Tomlinson, R. A., & Olden, E. J. (1999). Thermoelasticity for the analysis of crack tip stress fields—a review. *Strain, 35*, 49–47.
94. Tomlinson, R. A., & Patterson, E. A. (2011). Examination of crack tip plasticity using thermoelastic stress analysis. Thermomech Infra-Red Imaging. *Proceedings of the society for Experimental Mechanics Series, 7*, 123–127.
95. Wong, A. K., Sparrow, J. G., & Dunn, S. A. (1988). On the revised theory of the thermoelastic effect. *Journal of Physics and Chemistry of Solids, 48*, 395–400.

96. Palumbo, D., De Finis, R., Di Carolo, F., Vasco-Olmo, J., Diaz, F. A., & Galietti, U. (2021). Influence of second-order effects on thermoelastic behaviour in the proximity of crack tips on titanium. *Experimental Mechanics*, 1–15.
97. Svetlizky, D., Das, M., Zheng, B., Vyatskikh, A. L., Bose, S., Bandyopadhyay, A., et al. (2021). Directed energy deposition (DED) additive manufacturing: Physical characteristics, defects, challenges and applications. *Materials Today, 49*, 271–295.
98. Ancona, F., De Finis, R., Palumbo, D., & Galietti, U. (2015). Crack growth monitoring in stainless steels by means of TSA technique. *Procedia Engineering, 109*, 89–96.
99. Avanzini, A. (2023). Fatigue behavior of additively manufactured stainless steel 316L. *Materials, 16*(1), 65.
100. Dowling, N. E., Kampe, S. L., & Milo, V. K. (2020). *Mechanical behaviour of materials* (5th ed., Global edition). Pearson Education Limited.
101. Dulieu-Barton, J. (2012). Thermoelastic stress analysis. In *Optical methods for solid mechanics* (pp. 345–366). John Wiley & Sons.
102. Diaz, F. A., Patterson, E. A., & Yates, J. R. (2013). Application of thermoelastic stress analysis for the experimental evaluation of the effective stress intensity factor. *Fracture and Structural Integrity, 25*, 109–116.
103. Dulieu-Barton, J. M., & Worden, J. M. (2003). Genetic identification of crack-tip parameters using thermoelastic isopachics. *Measurement Science and Technology, 14*, 176–183.
104. Palumbo, D., Ancona, F., De Finis, R., & Galietti, U. (2015b). Experimental study of the crack growth in stainless steels using thermal methods. *Procedia Engineering, 109*, 338–345G.
105. Pitarresi, M., & Ricotta, G. (2019). Meneghetti, Investigation of the crack tip stress field in a stainless steel SENT specimen by means of thermoelastic stress analysis. *Procedia Structural Integrity, 18*, 330–346.
106. Vayssette, B., Saintier, N., Brugger, C., Elmay, M., & Pessard, E. (2018). Surface roughness of Ti-6Al-4V parts obtained by SLM and EBM: Effect on the high cycle fatigue life. *Procedia Engineering, 213*, 89–97. https://doi.org/10.1016/j.proeng.2018.02.010
107. Bayati, P., Jahadakbar, A., Barati, M., Nematollahi, M., Saint-Sulpice, L., Haghshenas, M., et al. (2020). Toward low and high cycle fatigue behavior of SLM-fabricated NiTi: Considering the effect of build orientation and employing a self-heating approach. *International Journal of Mechanical Sciences, 185*, 105878. https://doi.org/10.1016/j.ijmecsci.2020.105878
108. Ancona, F., De Finis, R., Demelio, G. P., Galietti, U., & Palumbo, D. (2016a). Study of the plastic behavior around the crack tip by means of thermal methods. *Procedia Structural Integrity, 2*, 2113–2212.
109. Ancona, F., Palumbo, D., De Finis, R., Demelio, G. P., & Galietti, U. (2016b). Automatic procedure for evaluating the Paris Law of martensitic and austenitic stainless steels by means of thermal methods. *Engineering Fracture Mechanics, 163*, 206–219.
110. ASTM E 647–00. (2004). Standard test method for measurement of fatigue crack growth rates
111. Díaz, F. A., Yates, J. R., & Patterson, E. A. (2004). Some improvements in the analysis of fatigue cracks using thermoelasticity. *International Journal of Fatigue, 26*(4), 365–376.
112. Oswald-Tranta, B., & Shepard, S. M. (2013). Comparison of pulse phase and thermographic signal reconstruction processing methods. In *Thermosense: Thermal Infrared Applications XXXV* (Vol. 8705, pp. 209–217). SPIE.
113. Shepard, S. M., & Beemer, M. F. (2015). Advances in thermographic signal reconstruction. *Thermosense: Thermal Infrared Applications XXXVII* 9485, pp. 204-210). SPIE.

Additive Manufacturing and 3D Printing Revolution for Tissue Engineering and Regenerative Medicine

Sachin Kumar, Saetukrit Panwar, Sakshi Mehra, Subodh Kumar, and Yogesh Kumar Verma

1 Tissue Engineering

Tissue engineering (TE) is an interdisciplinary field that uses biology, material science, chemistry, and physics to construct three-dimensional (3D) tissue to repair and regenerate damaged tissues. TE encompasses stem cells (SCs), biomaterials, and growth factors (GFs) to repair and regenerate damaged tissues. The TE holds significant potential to address the limitations of traditional therapies, such as organ transplantation, in vivo animal testing, etc. However, several challenges remain to be resolved. These include generation of complex tissues, promoting functional tissue integration, etc. [1]. TE is divided into two categories one is top-down and other is bottom-down approach. The traditional TE method follows a top-down approach, where a cell is cultured independently to promote proliferation. This approach uses a scaffold (consisting of natural or synthetic biomaterials) as a supportive structure. The monolayers are grown on the biocompatible and biodegradable scaffold to facilitate tissue formation [2]. The porous microstructure of the scaffold allows cells to migrate within the scaffold. The traditional TE method faces several limitations, including inadequate vascularization and neurogenesis [3]. In the bottom-up approach, scaffolds are prepared with cells, promoting enhanced cell proliferation

S. Kumar · S. Panwar · S. Mehra · Y. K. Verma (✉)
Regenerative Biology Research Lab, Division of Clinical Research and Medical Management (CRMM), Institute of Nuclear Medicine and Allied Sciences (INMAS), Defence Research and Development Organisation (DRDO), Timarpur, Delhi, India
e-mail: yogesh.inmas@gov.in

S. Kumar
Department of Biomedical Sciences, School of Biosciences and Technology, Galgotias University, Greater Noida, India

Y. Kumar Singla et al. (eds.), *Digital Manufacturing in Industry 5.0*,
https://doi.org/10.1007/978-3-031-91500-0_14

and diffusion. Nowadays, TE is enhanced by 3D bio-printing technology, which employs two main approaches for tissue reconstruction and fabrication. The first approach is based on the conventional TE method, where a scaffold is fabricated without cells, and subsequently, cells are added. In the second approach, the scaffold is prepared with embedded cells and growth factors (GFs) that promote cell proliferation and differentiation, resulting in tissue repair and regeneration [4].

2 Additive Manufacturing

AM was invented in the 1980s and after various transitions, it evolved into 3D printing technology. It is based on the principle of layer-by-layer structure formation. Nowadays, because of 3D printing technology, reduced waste products, complex structures, and network formation that was impossible with conventional techniques. AM provides benefits, including complex geometrical structure formation, customization, design flexibility, on-site production, recycling, etc. [5]. This has sparked significant academic and industry interest in utilizing AM for drug delivery and 3D tissue construction medicine administration, with outstanding results reported across various applications. There are many 3D printing methods. Fused Deposition Modeling (FDM) is widely used due to its simplicity and cost-effectiveness. FDM works by extrusion of thermoplastic filament through a heated nozzle to form 3D designs. With advancements such as reinforced composite filaments and optimized process settings, stereolithography (SLA) was developed, which uses an ultraviolet laser for the formation of photosensitive resins, resulting in high-detail objects with smooth surfaces. Digital Light Processing (DLP), similar to SLA, uses a digital light projector for resin curing, allowing for increased speed and precision in creating detailed structures. Both methods have undergone advances in resin compositions to increase mechanical and thermal performance. Selective laser sintering (SLS) and direct metal laser sintering (DMLS) employ lasers to sinter powdered materials, enabling complicated shapes and a wide range of materials such as metals and ceramics [6].

3 D Bioprinting

3D bioprinting is an emerging TE method to fabricate complex biological structures. In 1983, Charles Hull developed the first 3D printer, which used acrylic-based photopolymers and UV light as crosslinkers to create a 3D solid structure by computer-aided design (CAD). 3D bioprinting works on the principle of layer-by-layer deposition of bioink to form the complex 3D tissue construct. This new technique overcame the shortcomings of conventional TE approaches and shortened the process time. This technology uses bioink, which comprises the biomaterial, extracellular matrix (ECM), and GFs to fabricate the tissue construct [7]. 3D bioprinting technology facilitates the formation of tissue constructs by creating large interconnected pores with a high surface area. This design enhances the formation of vascularized structures and promotes the uniform distribution of cells, growth materials, and growth-supporting substances, thereby supporting cell proliferation within the TE construct [8]. There are primarily three types of 3D bioprinting technologies: Extrusion-based bioprinter, Laser-assisted bioprinter, and inkjet-based bioprinting. The following details provide an overview of these bioprinters.

4 Type of Bioprinter

There are mainly three types of bioprinters, such as extrusion-based, ink-jet, and laser-assisted bioprinters used in TE.

4.1 Extrusion-Based Bio Printing

Extrusion-based bioprinting has seen remarkable advancements in recent years, particularly in the areas of precision and complexity. Improved nozzle designs and new printing strategies allow for finer control over cell placement and scaffold architecture. This technology continuously supplies bioink, which contains cells, through a pointed nozzle shaped like a filament, utilizing pneumatic, piston, or screw drive systems under automated control. Layer-by-layer printing facilitates the creation of a complete three-dimensional structure. The extrusion method can also apply shear stress to produce small filament structures. Additionally, extrusion-based bioprinting typically uses three to six print heads to accommodate multiple bioinks, while minimizing cross-contamination. This approach allows for greater control over the printed structure's shape, porosity, and cell distribution. Extrusion-based printing can utilize various types of bioink, including acellular matrix components, microcarriers, cell clumps, and high-viscosity hydrogels [9]. For example, researchers at Wake Forest Institute for Regenerative Medicine have successfully used extrusion-based bioprinting to create complex, multi-layered vascularized skin tissues, significantly improving healing outcomes in animal models [10].

4.2 Inkjet Bioprinting

Inkjet bioprinting technology has advanced significantly, increasing printing speed and throughput, which makes it more suitable for large-scale tissue production. This technique's ability to print multiple materials simultaneously has enhanced the complexity and functionality of bio-printed tissues. It was developed in 2003. This printing approach precisely creates complex geometrical structures using a tiny amount of bioink and is based on the following fundamental process: First, a drop of bioink is formed and guided towards a certain area. Secondly, the drop interacts with the substrate. Inkjet bioprinters deliver a controlled amount of bioink to the desired printing surface, forcing the content to flow continuously (continuous inkjet printing) or drop out from the nozzle (on-demand inkjet printing). A notable example is the work by [11], who used inkjet bioprinting to create kidney tissue constructs with integrated blood vessel networks, significantly enhancing tissue viability and function in vitro [11].

4.3 Laser-Assisted Bioprinting

Laser-assisted bioprinting (Fig. 1) has improved the precision of cell placement, enabling the creation of highly detailed tissue structures. This technique is particularly useful for printing delicate tissues, such as neural networks and microvascular structures. A typical laser-assisted bioprinter has five components: (1) a pulsed laser beam, (2) a focusing mechanism, (3) an electron donor layer with an energy-absorbing layer that responds to laser stimulation (gold or other metal), (4) a layer of bioink, and (5) a receiving substrate. This bioprinting method provides better cell viability than other bioprinting or conventional methods of TE because in this method, the dispenser does not directly interact with the bioink, so there is less stress on cells, which results in approximately. 95% of cell viability [12]. For instance [13], successfully used laser-assisted bioprinting to repair cranial defects in mice, resulting in significant bone regeneration and integration with surrounding tissues [13].

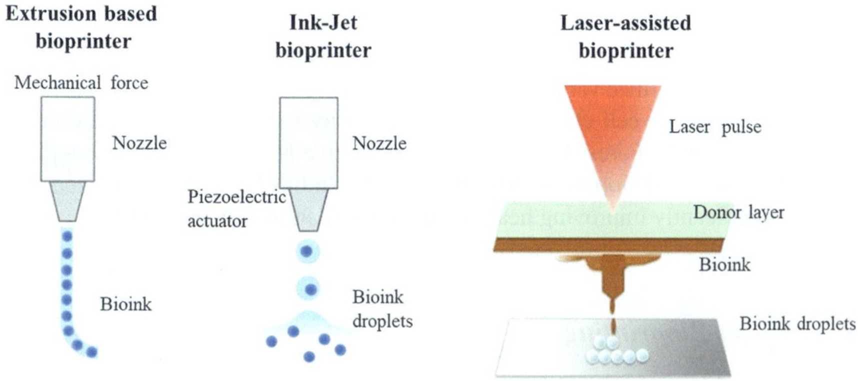

Fig. 1 Different types of bioprinters

5 Algorithms and CAD for Additive Manufacturing

There are various algorithms that have been used to design 3D constructs via Additive manufacturing. Some of the algorithms are as follows (Table 1).

5.1 Parametric and Generative Design

Parametric design leverages parameters and constraints to create adaptable models. This technique facilitates rapid iterations, enabling precise control over complex geometries necessary for TE. Generative design, on the other hand, uses algorithms to optimize structures based on predefined constraints and objectives, significantly enhancing the design of tissue scaffolds. For example, generative algorithms have demonstrated a 50% increase in scaffold porosity while maintaining structural integrity, which is crucial for cell proliferation and nutrient diffusion [12].

Table 1 Algorithms and CAD for additive manufacturing

Feature	Parametric design	Generative design	3D bioprinting	CAD files
Design approach	User-defined parameters and constraints	Algorithm-driven and optimization	Biomaterial and cell compatibility, functional design	CAD software (e.g., Solidworks, AutoCAD, blender)
Design space	Limited by user input	Vast, exploring multiple design possibilities	Biological constraints and material properties	Autodesk inventor, SolidWorks, fusion 360
Design automation	Partial automation through parameterization	High degree of automation	Integration of biological data and manufacturing constraints	Generative design software (e.g., Autodesk generative design, grasshopper)
Optimization	Focuses on specific design objectives	Explores multiple design objectives simultaneously	Optimization for cell viability, biocompatibility, and mechanical properties	Topology optimization software (e.g., Altair OptiStruct, Ansys topology optimization)
Applications	Rapid prototyping, mold-making	Complex geometries, lightweight structures, functional optimization	Tissue engineering, drug delivery, personalized medicine	Industry-specific CAD software (e.g., CATIA for aerospace, pro/engineer for automotive)

5.2 *Topology Optimization*

Topology optimization aims to distribute material within a given design space for optimal performance under specified loading conditions. This method has been pivotal in creating lightweight yet robust scaffolds. Recent studies indicate that topology-optimized designs can achieve a 30% reduction in material usage while enhancing mechanical strength by 20%.

5.3 *Algorithmic Strategies*

5.3.1 Slicing Algorithms

Slicing algorithms convert 3D models into 2D layers, guiding the printing process. Advanced slicing algorithms, such as adaptive slicing, adjust layer thickness based on geometry, improving surface finish and reducing printing time by up to 40%.

5.3.2 Tool Path Optimization

Tool path optimization includes refining the movement of the printer head to minimize printing time and material waste while maximizing its precision. Techniques like path planning have demonstrated a reduction in print time by 25% and a decrease in material consumption by 15%. A study at Harvard's Wyss Institute utilized advanced CAD and algorithmic strategies to design and fabricate complex tissue scaffolds. Using generative design and topology optimization, the team achieved a 60% increase in scaffold porosity and a 35% improvement in the mechanical properties. These scaffolds were successfully used in vitro to support the growth of hepatocytes, demonstrating the potential of algorithm-driven design in tissue engineering.

6 Industry 5.0

Industry 5.0 is the industrial revolution to enhance collaboration between manpower and artificial intelligence (AI). Three fundamental components form the foundation of Industry 5.0: human centricity, resilience, and sustainability. These current technical advances, in combination with highly qualified and equipped professionals, seek to improve manufacturing efficiency, production quality, and prompt responsiveness to users based on their demands, a concept known as mass personalization [14].

Industry 5.0 emphasizes the importance of human workers alongside advanced technologies like artificial intelligence. In the scenario of Industry 5.0, the process mapping may be enhanced by using decision-making algorithms of AI [15]. Industry 5.0 improves the workflow by using AI tools and helps in manufacturing data, recommending optimal sub-processes and identifying hazards to production and product quality. The process workflow that combines AI algorithms with decisional logic is known as Process Mining (PM) [16].

6.1 Digital Twin (DT)

A digital twin is a technology based on industrial manufacturing. This futuristic technology helps in visualizing real estate manufacturing by developing a virtual model that can be verified with 3D simulation [17]. This is a time-saving technology that simulates details of operation, parameters, and instrumentation, which helps in the development of quality products and also reduces labor, manufacturing, and developmental risks. Future manufacturing will depend on digital twins, which allow virtual asset and resource representation to model performance, anticipate faults, and resolve production-related problems [18].

6.2 Personalized Medicine

Personalized medicine is the practice of tailoring medical treatments and prevention techniques to each individual's unique traits. This strategy is based on a thorough profile that contains risk signs and predictions about how a patient will respond to therapy. These risk indicators might be hereditary, such as certain gene variations, or non-genetic, such as mammographic density. Genetic differences in drug-metabolizing enzymes or mutations in proteins linked to cancer can have a major impact on therapy efficacy. For example, genetic variety in enzymes that metabolize medications can impact how well a patient processes a medication, and mutations in cancer-related proteins might determine the success of targeted therapy [19].

7 Applications of Additive Manufacturing in Tissue Engineering

7.1 Biofabrication of Tissues and Organs

3D bioprinting, an emergent technique in tissue engineering, leverages layer-by-layer deposition to fabricate complex tissue structures. Techniques such as inkjet bioprinting, micro-extrusion bioprinting, and laser-assisted bioprinting allow precise spatial control over cells and biomaterials. Micro-extrusion bioprinting, for instance, has achieved a resolution of 100 μm, enabling the creation of intricate tissue scaffolds that mimic native tissue architecture [19].

7.2 Bioinks Development

The selection of materials, particularly bioinks, is critical for successful bioprinting constructs. Bioinks are formulated from natural polymers (e.g., alginate, collagen) and synthetic polymers (e.g., PEGDA), providing the necessary biomechanical properties and biocompatibility. Recent advancements have yielded bioinks with tunable rheological properties, achieving a shear-thinning behavior conducive to maintaining structural integrity post-printing. Quantitatively, alginate-based bioinks have demonstrated cell viability rates exceeding 85% post-printing, a significant milestone in maintaining cellular functions [11].

7.3 Patient-Specific Implants and Prosthetics

Additive manufacturing (AM) enables the creation of patient-specific implants and prosthetics, offering better fit, enhanced biocompatibility, and improved functionality compared to traditional models. Custom implants fabricated using 3D printing technologies, such as electron beam melting (EBM), show promise in orthopedic applications. Titanium implants produced by EBM exhibit a tensile strength of 980 MPa, closely approximating the mechanical properties of native bone. Furthermore, prosthetics designed via AM offer weight reductions of up to 60%, improving patient comfort and usability. For example, a study at Wake Forest Institute for Regenerative Medicine demonstrated the application of 3D bioprinting for skin grafts. Using a combination of keratinocytes and fibroblasts inserted in a collagen hydrogel, researchers achieved more than 90% graft take rates in only the preclinical models, significantly improving wound healing outcomes. This method underscores the potential of AM in delivering personalized, on-demand treatment solutions for burn patients, reducing recovery times by 30% compared to traditional grafting techniques.

8 Challenges in Additive Manufacturing for Tissue Engineering

8.1 Resolution and Precision

Additive manufacturing (AM) faces significant challenges in achieving high resolutions and precision required for tissue engineering. Current bioprinting technologies can achieve resolutions of approximately 50-100 micrometers, which is not at par with replicating the intricate microarchitecture of native tissues.

8.2 Mechanical Properties of Printed Structures

The mechanical properties of bioprinted tissues must match those of native tissues to function effectively. However, many printed structures exhibit inferior mechanical strength and elasticity. For instance, 3D-printed cartilage typically shows a compressive modulus of 0.2–0.6 MPa compared to native cartilage, which ranges from 0.45 to 1.0 MPa.

8.3 Biological Challenges

Ensuring adequate vascularization in bioprinted tissues is a major hurdle. Without proper blood vessel formation, larger constructs suffer from necrosis due to a lack of oxygen and nutrients. Recent studies have shown that 3D-printed scaffolds integrating microchannels can support initial vascular infiltration, but achieving functional long-term vascularization remains unresolved. Bioprinted constructs must avoid eliciting adverse immune responses post-implantation. Immune reactions can lead to inflammation, rejection, and failure of the tissue construct, which might be a potential threat to the patient. Research indicates that using autologous cells reduces immune response, but standardizing this approach for large-scale production is challenging.

8.4 Regulatory and Ethical Challenges

The regulatory landscape for bioprinted tissues is complex and varies geopolitically. In the USA, the FDA classifies these products as biologics, requiring rigorous preclinical and clinical testing. The process is lengthy and costly, hindering the rapid commercialization of products for wide-scale healthcare. Ethical issues in bioprinting include concerns about the source of cells, especially when using human embryonic stem cells. Additionally, the potential for bioprinting organs to exacerbate healthcare inequalities poses ethical dilemmas. Comprehensive ethical guidelines are needed to navigate these challenges.

8.5 Longevity and Functionality

Ensuring the long-term viability and functionality of bioprinted tissues after implantation remains a significant a significant challenge. Addressing the immune response and the potential rejection of bioprinted tissues is critical for successful clinical applications Gao et al. [20] explored the use of immunomodulatory bioinks to reduce immune response and improve the longevity of bioprinted tissues, showing promising results in animal models.

8.6 Scaling Up Production

Developing methods for scaling up the production of bioprinted tissues and organs is essential for their widespread clinical use. Establishing standardized protocols and quality control measures is necessary to ensure the consistency and reliability of bioprinted products. The collaboration between Organovo and academic institutions has focused on developing scalable bioprinting techniques for liver tissues, aiming to meet the demands of pharmaceutical testing and therapeutic applications.

9 Future Prospects

The integration of diverse biomaterials within a single construct will be pivotal. This will enable the creation of complex tissues mimicking the heterogeneous nature of native organs. Combining hydrogels, polymers, and ceramics will open new avenues for cell therapy. Advancements in artificial intelligence (AI) will revolutionize bioprinting by optimizing print parameters, predicting print outcomes, and automating quality control. Machine learning (ML) algorithms can analyze vast datasets to identify optimal bioink formulations and printing conditions, accelerating the development of functional tissues. The advancement of 3D bioprinting is profoundly influenced by global collaboration and supportive policies. Nations with robust research infrastructures coupled with favorable regulatory environments are at the forefront of this field. The European Union's Horizon Europe program and the United States' National Institutes of Health (NIH) exemplify substantial investments in bioprinting research. Conversely, stringent regulations or limited funding can hinder progress. An important aspect is intellectual property rights. Protecting innovations is essential to incentivize research and development. However, overly restrictive regulations can stifle collaboration and knowledge sharing. Further, it might be a negative contributor towards AM research. Finding a balance between protecting intellectual property and fostering open innovation is crucial. Ethical considerations and public perception play a significant role. Transparent communication about the potential benefits and risks of 3D bioprinting is essential to gain public trust and support for further studies and industrial applications. Addressing concerns about safety, accessibility, and equity will be vital for the successful integration of this technology into widely spread healthcare systems. Regulatory frameworks and government policies play a crucial role in fostering innovation. Supportive policies, dialogue, funding, infrastructure development, and more emphasis on specialized education will accelerate the translation of 3D bioprinting technologies into clinical applications. While still in the early stages, the ultimate goal of bioprinting is to create functional human organs for transplantation. Significant advancements in biomaterial development, cell sourcing and cultivation, and bioprinting techniques are required to achieve this ambitious objective. Portable bioprinters and bioink cartridges would greatly revolutionize healthcare in remote regions by enabling on-demand fabrication of medical devices, implants, and even simple tissue constructs. Furthermore, open-source invention blueprints would greatly benefit the social dynamics of medicine.

10 Conclusion

AM and 3D printing have deeply impacted TE and regenerative medicine, revolutionizing the fabrication of complex biological constructs. This chapter has emphasized the transformative role of advanced CAD techniques and algorithms, such as parametric design and topology optimization, which enable the creation of intricate tissue scaffolds with high precision. Algorithms for slicing and tool path optimization enhance the fidelity of printed constructs, which are critical for developing patient-specific bandages, implants, and prosthetics. The applications of AM in biofabrication are exemplified by the successful production of 3D-printed skin grafts for burn victims, demonstrating the technology's potential in personalized medicine. Additionally, disease models like 3D-printed liver constructs have become instrumental in drug testing and development, providing realistic in vitro environments for assessing metabolic responses and toxicity. Challenges do persist, including high resolution, optimal mechanical strength, vascularization, and regulatory hurdles. Future research in AM for TE must focus on integrating multi-material and hybrid printing techniques to enhance structural and functional complexity. Advances in AI and machine learning (ML) will further optimize design algorithms and printing processes, facilitating the development of more sophisticated tissue models and implants with the potential use of AI. The political and globalization aspects are also crucial, as international collaborations and policy frameworks will profoundly shape the trajectory of AM technologies. Countries leading in AM research are expected to drive global standards and innovation, impacting global health and medical practices. The revolution catalyzed by AM in TE heralds a new era of personalized and regenerative medicine, promising unprecedented medical technology and patient care advancements. The continuous improvement of technologies will undoubtedly shape the future of healthcare, driving scientific inquiry and practical application.

Acknowledgments We thank Director INMAS for his continuous support.

Author's Contribution The manuscript was written and edited by Sachin Kumar, Saetukrit Panwar, Sakshi Mehra, Subodh Kumar. Edited by Subodh Kumar and Yogesh Kumar Verma. The manuscript was conceptualized, designed and edited by Yogesh Kumar Verma.

Conflict of Interest Authors do not have conflict of interest.

References

1. Berthiaume, F., Maguire, T. J., & Yarmush, M. L. (2011). Tissue engineering and regenerative medicine: history, progress, and challenges. *Annual Review of Chemical and Biomolecular Engineering, 2*(1), 403–430.
2. Gupta, S., & Bit, A. (2022). 3D bioprinting in tissue engineering and regenerative medicine. *Cell and Tissue Banking, 23*(2), 199–212.
3. Gao, G., & Cui, X. (2016). Three-dimensional bioprinting in tissue engineering and regenerative medicine. *Biotechnology Letters, 38*, 203–211.
4. Wu, Z., Su, X., Xu, Y., Kong, B., Sun, W., & Mi, S. (2016). Bioprinting three-dimensional cell-laden tissue constructs with controllable degradation. *Scientific Reports, 6*(1), 24474.
5. Salmi, M. (2021). Additive manufacturing processes in medical applications. *Materials, 14*(1), 191.
6. Zhou, L., Miller, J., Vezza, J., Mayster, M., Raffay, M., Justice, Q., & Bernat, J. (2024). Additive manufacturing: a comprehensive review. *Sensors, 24*(9), 2668.
7. Dey, M., & Ozbolat, I. T. (2020). 3D bioprinting of cells, tissues and organs. *Scientific Reports, 10*(1), 14023.
8. Matai, I., Kaur, G., Seyedsalehi, A., McClinton, A., & Laurencin, C. T. (2020). Progress in 3D bioprinting technology for tissue/organ regenerative engineering. *Biomaterials, 226*, 119536.
9. Kang, M. S., Jang, J., Jo, H. J., Kim, W. H., Kim, B., Chun, H. J., et al. (2022). Advances and innovations of 3D bioprinting skin. *Biomolecules, 13*(1), 55.
10. Jang, J., Park, H. J., & Kim, S. W. (2021). 3D-printed complex skin tissue regeneration system. *Advanced Functional Materials, 31*(3), 2005224.
11. Cui, H., Zhu, W., & Nowicki, M. (2020). Inkjet bioprinting of kidney tissues with integrated vascular networks. *Biomaterials Science, 8*(10), 2741–2750.
12. Zhang, J., Wehrle, E., Rubert, M., & Müller, R. (2021). 3D bioprinting of human tissues: biofabrication, bioinks, and bioreactors. *International Journal of Molecular Sciences, 22*(8), 3971.
13. Keriquel, V., Oliveira, H., & Rémy, M. (2017). In situ printing of mesenchymal stromal cells by laser-assisted bioprinting.
14. Guruswamy, S., Pojić, M., Subramanian, J., Mastilović, J., Sarang, S., Subbanagounder, A., et al. (2022). Toward better food security using concepts from industry 5.0. *Sensors, 22*(21), 8377.
15. Krupas, M., Kajati, E., Liu, C., & Zolotova, I. (2024). Towards a human-centric digital twin for human-machine collaboration: A review on enabling technologies and methods. *Sensors, 24*(7), 2232.
16. Massaro, A. (2022). Advanced control systems in industry 5.0 enabling process mining. *Sensors, 22*(22), 8677.
17. Modoni, G. E., & Sacco, M. (2023). A human digital-twin-based framework driving human centricity towards industry 5.0. *Sensors, 23*(13), 6054.
18. Lv, Z. (2023). Digital twins in industry 5.0. *Research, 6*, 0071.
19. Narod, S. A. (2018). Personalised medicine and population health: breast and ovarian cancer. *Human Genetics, 137*(10), 769–778.

20. Gao, G., Yonezawa, T., & Hubbell, J. A. (2021). Immunomodulatory bioinks for reducing immune response in bioprinted tissues. *Nature Communications, 12*(1), 1724.
21. Daly, A. C., Critchley, S. E., & Raxworthy, M. J. (2018). Incorporating growth factors into hydrogel bioinks for osteochondral tissue engineering. *Advanced Healthcare Materials, 7*(12), 1701312.
22. He, Y., Yang, F., & Zhao, H. (2018). Bioprinted skin constructs with integrated sweat glands and hair follicles. *Advanced Healthcare Materials, 7*(8), 1701301.
23. Huch, M., Gehart, H., & van Boxtel, R. (2015). Long-term culture of genome-stable bipotent stem cells from adult human liver. *Cell, 160*(1-2), 299–312.
24. Kang, H. W., Lee, S. J., & Ko, I. K. (2016). A 3D bioprinting system to produce human-scale tissue constructs with structural integrity. *Nature Biotechnology, 34*(3), 312–319.

Optimizing Fiber Volume in Steel Fiber-Reinforced GGBS Geopolymer Concrete for Smart Manufacturing

Badana Govindarajulu, Pedapenki Dinakar, Ponguru Naga Sowjanya, and Shaik Sydha

1 Introduction

With construction material requirements increasing on one hand, there is a greater need to employ measures that combat environmental and performance concerns on the other hand as well. Cement, which is the main binder material in ordinary concrete, is one of the main factors that increases greenhouse gas emissions as well depletes resources. The production of this material causes around 75% of the emissions that are CO_2 associated with it, which has been termed to be one of the great culprits of climate change. For this reason, it is paramount to explore substitutes that can replace or lessen the usage of cement in concrete. This research investigates the use of geopolymer concrete (GPC) as a sustainable alternative to steel fiber-reinforced GGBS-based geopolymer concrete, enhancing it in the context of innovation for Industry 5.0.

A steel industrial waste called granulated blast furnace slag (GGBS) has become a favorable cement replacement because of its positive properties and lower environmental damage. Due to the minute emissions of CO_2 produced in the making of GGBS geopolymers, they are more suited as environmentally friendly construction options when compared to ordinary Portland cement. Also, the use of GGBS itself encourages geopolymerization, which is an interaction with alkaline solutions that has equal or improved binding capability compared to modern cement. GGBS geopolymer concrete, when applied with steel fibers, shows not only satisfactory results but a significant improvement in flexural and tensile strengths.

B. Govindarajulu (✉) · P. Dinakar
Department of Civil Engineering, Aditya Institute of Technology and Management, Tekkali, Andhra Pradesh, India

P. Naga Sowjanya · S. Sydha
Narasaraopeta Engineering College, Narasaraopeta, Andhra Pradesh, India

Y. Kumar Singla et al. (eds.), *Digital Manufacturing in Industry 5.0*, https://doi.org/10.1007/978-3-031-91500-0_15

Due to its ability to make economical, high-strength materials useful in many sectors such as infrastructure, smart manufacturing, and automated building processes, steel fiber-reinforced GGBS geopolymer concrete has attracted great attention in many scientific works and research nowadays. However, the GGBS-based GPC performance with different load conditions and fiber volumes has room for improvement. In this respect, the current study assesses how the volume of steel fibers added to GGBS GPC and the degree of fiber content that prevails in the mixes affect the compression, tensile, and flexural strength of the materials.

The production of cement is known to be energy-intensive as well as environmentally detrimental. The burning of limestone and other raw ingredients to form clinker, the primary constituent in cement, is energy-intensive and also emits great quantities of CO_2. Alternatives such as GGBS are more ecologically friendly as they make use of waste materials, therefore minimizing CO_2 emissions as well as natural resources. It has been found in studies that up to 65% of the cement in concrete can be replaced by GGBS, which can considerably improve the carbon footprint in concrete mixtures while also providing greater tensile strength and durability.

Concrete, or rather geopolymers made of GGBS and other alkali–activated materials, have produced good results, in particular compressive strength and durability under harsh environmental conditions. These properties can be improved further through the addition of steel fibers, hence producing a strong and durable product which is ideal for contemporary construction. Such properties also lead to the use of GGBS-based GPC in areas where high durability and sustainability are required for the projects.

Industry 5.0, which is the most current phase of industrial development, fosters production systems that are sustainable, human-centric, and resilient. Through the use of advanced materials such as GGBS-based GPC, Industry 5.0 promotes the use of green substitutes in manufacturing and construction. When combined with optimized materials, digital manufacturing technologies provide control over construction processes and resources, while minimizing the damage done to the environment.

In this case, GGBS geopolymer concrete with steel fibers has some advantages. It meets the goals of Industry 5.0 with the provision of a high-performance product with a sustainable approach. As construction requires an increase in automation, such technologies will include the use of geopolymer concrete in automated construction, robotics, and other advanced manufacturing technologies. With the increasing complexity of construction needs, the ability to digitally design the composition of GPC will become an important requirement for the creation of robust and sustainable infrastructures.

This paper aims to describe in detail the proceedings of the investigation conducted on steel fiber-reinforced GGBS geopolymer concrete. In this second section, materials, particularly their types and sources, as well as methods for mixing and the tests conducted to determine relevant mechanical properties, are described. The results and discussion evaluate the relevance of steel fiber reinforcement in the enhancement of compressive, tensile, and flexural strength of GPC.

2 Literature Review: Steel Fiber-Reinforced GGBS Geopolymer Concrete for Smart Manufacturing

The optimization of fiber volume in steel fiber-reinforced GGBS (ground granulated blast furnace slag) geopolymer concrete (GPC) is a key area of research within smart manufacturing and Industry 5.0, given its potential to improve material durability, sustainability, and structural performance. Geopolymer concrete, which utilizes industrial by-products like GGBS as binders instead of traditional cement, reduces carbon emissions and environmental impact, aligning with the goals of sustainable construction practices [1]. The addition of steel fibers enhances mechanical properties, but optimizing fiber content is essential to balance strength, cost, and workability in construction applications.

Studies by Shah et al. (2024) [1] examine advancements in material science that impact durability and sustainability, which are critical for automated and optimized manufacturing processes in geopolymer concrete production. Additionally, research on solar energy systems and tribological studies [2–8] support the sustainable production goals of Industry 5.0. Renewable energy and optimized lubrication can reduce energy consumption and wear in manufacturing equipment, further aligning with the principles of environmentally conscious production.

Technologies such as digital twins, which have been applied to improve logistics and material tracking in manufacturing [9], have also been researched in the context of construction and material applications. This digital innovation can help monitor and adjust fiber volumes in real time, ensuring optimal distribution and mechanical performance [10, 11]. Other research explores environmental solutions like the bioremediation of soils contaminated with industrial by-products [12, 13], which has implications for sustainable GGBS sourcing and waste management in the construction industry [14, 15].

Optimizing wear resistance in GGBS-based materials through alloying elements has also been highlighted in studies, showing how material engineering can improve the longevity and performance of construction composites. Research on surface integrity in machined materials and the role of nanoparticulates in lubrication further demonstrates how material sciences contribute to enhanced production efficiency and durability in GPC applications.

Dynamic simulation techniques [16, 17] and improvements in suspension systems are also part of broader efforts to ensure that GGBS-based fiber-reinforced composites can withstand diverse environmental stresses. These studies support a multidisciplinary approach to enhancing the durability of fiber-reinforced GPC, a goal central to Industry 5.0's focus on sustainable infrastructure.

Recent studies by Adeoye et al. (2024) on lignocellulosic waste valorization [18], including bio-oil production through pyrolysis, offer insights into the sustainable sourcing and production of construction materials. The potential to upgrade crude bio-oil [19, 20] and produce sustainable energy from biomass [21] provides viable pathways for reducing energy consumption in the production of fiber-reinforced GGBS GPC. These clean energy solutions, when integrated into digital manufacturing processes, optimize resource usage and contribute to Industry 5.0's sustainability goals [22–27].

Research on catalytic pyrolysis and nanomaterials for enhanced biomass conversion further supports efforts to optimize the manufacturing efficiency of GGBS-based concrete composites [28–30]. Studies on the pyrolysis of plastic waste and applications in rail projects provide models for introducing circular economy concepts and environmental sustainability into GPC manufacturing [31–39].

Studies on fiber surface functionalization [40–42] add to the understanding of fiber-matrix bonding in GPC, which can improve the durability and strength of fiber-reinforced composites. These innovations contribute to digital transformation and environmental consciousness in construction manufacturing, a core component of Industry 5.0 [31, 33, 36, 37, 43]. Contributions from Sunil et al. [44–54] in material sciences and engineering technology further explore how fiber reinforcement in construction aligns with the efficiency and innovation goals of smart manufacturing. Recent studies on electric vehicle adoption [17, 55–68], intelligent transportation systems [69–80], and vibration mitigation techniques [81–90] illustrate how transportation efficiency and material innovation support low-impact construction practices.

Bio-oil and carbon capture techniques, particularly from waste biomass, have been researched as sustainable energy sources for smart manufacturing applications. Studies examine bio-oil production from thermal cracking [91], CO_2 adsorption in coalbeds [132], and the pyrolysis of organic waste [92–94]. Other studies look into using sulfur-enhanced binders and sewage sludge bio-oil, underscoring how alternative binders can contribute to GPC sustainability and reduce construction waste [95]. Collectively, these studies highlight renewable energy and environmental sustainability efforts that align with Industry 5.0's environmental goals [31, 43, 96–98].

Microbiological advancements and biopolymer applications also have implications for sustainable concrete production. Mohapatra et al. (2015) [99] investigate microbial processes for biopolymer production, with applications in sustainable packaging [24, 31, 100–107] and bio-based construction materials [108–119]. Using bio-based fibers in GPC aligns with the Industry 5.0 emphasis on circular economy and material sustainability. Additional studies on energy-efficient retrofitting [120–136] and biocompatible polyhydroxyalkanoates (PHAs) in construction applications [137–145] highlight how biotechnology is becoming an integral part of sustainable construction.

Research emphasizes the need for energy-efficient solutions and advanced control systems, essential in Industry 5.0 and smart manufacturing. Sustainable materials for transportation [146], intelligent railway infrastructure [15, 147–161], and advancements in vehicle modeling and ride behavior [162–170] also contribute to reducing transportation waste in construction supply chains. Studies on crashworthiness [171–179] and bio-lubricants [180–183] reinforce Industry 5.0 goals by improving safety [184–186], efficiency, and sustainability in construction material transport [187–190].

Finally, research into dynamic analysis, vibration isolation, and composite materials in construction further supports Industry 5.0 objectives [191–195]. Recent studies (2024) on advanced control systems for passenger rail vehicles [151, 196–200] and dynamic analyses of road and rail vehicles [201–210] demonstrate the impact of smart technologies on enhancing GPC production processes. The use of composite materials and hybrid AHP-TOPSIS techniques [62, 66, 68, 209, 211–221] promotes the development of durable, smart construction materials that meet the durability and sustainability requirements of Industry 5.0.

Optimizing fiber volume in steel fiber-reinforced GGBS GPC represents a valuable approach for sustainable and resilient concrete manufacturing. Through material science advancements, digital innovation, and sustainability-focused manufacturing processes, these studies contribute to a framework for Industry 5.0 in construction.

3 Materials and Methods

This sub-section describes construction materials for geopolymer concrete (GPC), processes of mixing, and technologies for testing steel fiber-reinforced GPC and GGBS used in the construction. The objective of standardizing materials and methods of research for this study is to provide dependable outcomes that can be applicable in similar studies of environment-friendly building practices.

3.1 Materials Used

The basic materials that were utilized in this investigation include GGBS, steel fibers, alkaline activator solution and aggregates. It can be emphasized that each of the components is unique and has a specific impact on the performance of the geopolymer concrete in terms of structural and mechanical properties.

- Ground granulated blast furnace slag (GGBS): This type of GGBS is produced in Jindal South West Steel Industries, which has abided by the IS 12089-1987 norms and thus looks the same in terms of quality or chemical composition. GGBS is a waste material and exists only in the scope of steel industries, and it contains high silicates and alumina, which act as a binding material when an alkaline solution is used. Table 1: Effects of GGBS's physical parameters, such as specific gravity and fineness modulus, on its application in concrete. This Table 1 provides the physical characteristics of ground granulated blast furnace slag (GGBS) used in geopolymer concrete (GPC). The specific gravity and fineness modulus reflect the density and particle size distribution, which impact the

Table 1 Physical characteristics of GGBS

Property	Value
Specific gravity	2.85
Fineness modulus	27
CaO	37.36%
SiO_2	37.82%
Al_2O_3	14.43%
Fe_2O_3	1.12%
Loss of ignition	1.42%
Glassy content	99.91%

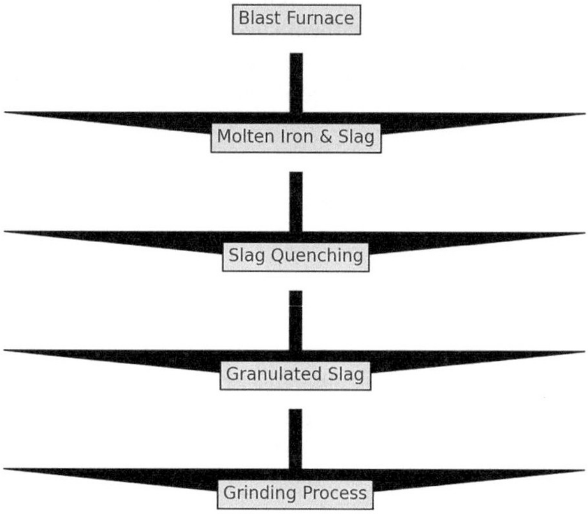

Fig. 1 Slag production process

GPC mix's workability and strength. High CaO and SiO_2 contents contribute to the pozzolanic and binding properties, while the high glassy content indicates a reactive structure favorable for geopolymerization.

- GGBS is an important ingredient of geopolymer concrete (GPC) as shown Fig. 1 in The Internet Archive—a GGBS producing company. The blast furnace provides the core of the operation; here, a mixture of limestone, coke, and iron ore is subjected to intense heat, which helps to create iron and slag in a molten state. After that, the slag is cooled down very fast by its own means in a process called quenching and this allows for the slag to be hung in a granulated form. This granulated slag is afterwards formed into powder and has GGBS for this. This is great because it cuts down pollution from cement production. OCA World places GGBS in GPC and reduces GPC's emissions of CO_2 from cement manufacturing.

- Steel Fibers: These fibers enhance the tensile and flexural strengths by reinforcing the brittle GPC matrix. The steel fibers used in this study have hooked ends for better bonding with the concrete matrix. They are cylindrical in shape measuring 0.6 mm in diameter, with a tensile strength of 1450 MPa and a length of 30 mm. Steel fibers have been shown to decrease the brittleness and improve the impact resistance of composites even at low volumes such as 0.6%. Table 2 provides the Physical properties of steel fibers.

Fine aggregates are essential for the mix's overall stability, especially in terms of filling voids and ensuring proper particle packing. The recommended grading falls within Zone II, which provides an optimal balance between workability and compressive strength. The fineness modulus indicates the aggregate's average particle size, with lower values generally improving flow and cohesiveness within the concrete mix. Moreover Table 3 provides the steel fibers used in the GPC mix are characterized by their high tensile strength and aspect ratio, which enhances their ability to bridge microcracks within the concrete matrix. The diameter and length provide a balanced aspect ratio, crucial for effective bonding within the concrete without compromising its workability. The high tensile strength ensures that these fibers effectively improve the material's resistance to cracking and tensile forces.

Alkaline Solution: The off-foundation solution of sodium hydroxide (NaOH) and sodium silicate (Na_2SiO_3), which are particularly utilized in final area geopolymer preparation for 8 M and 10 M molarity mixtures having a ratio of NaOH to Na_2SiO_3 of 1: 2.5 was prepared. This result initiated the repolymerization reaction, forming a strong matrix when combined with GGBS particles. The reasons for choosing this molarities are important in order to obtain good compressive strength characteristics of the GPC, since other researches show that the molarity of the alkali solution has a direct bearing on the compressive strength of geopolymer concrete.

Aggregates: The structural elements of the mix are fine and coarse aggregates. The fine aggregates are classified to conform to IS 383-1970 standards of Zone II, while coarse aggregates have a maximum particle size of 20 mm to fit density and mechanical properties of GPC.

Table 2 Physical properties of fine aggregates

Property	Zone II (recommended)
Fineness modulus	2.6
Particle size (max)	4.75 mm
Grading zone	II

Table 3 Physical properties of steel fibers

Property	Value
Diameter	0.6 mm
Length	30 mm
Aspect ratio	50
Tensile strength	1450 MPa
Yield strength	1000 MPa

3.2 *Mix Design*

The mix design for geopolymer concrete using steel fiber and GGBS included the following key operational objectives upon which it was based:

- Aggregate Proportioning: 75% of the mass of the complete composition is made of combined fine and coarse aggregates, with 37% of the mix being fine aggregate. This ratio is typical for conventional OPC concrete but it is modified to suit GPC, which utilizes GGBS.
- Alkaline Solution Ratio: A 0.45 alkaline liquid-to-binder ratio was determined from previous studies, which was effective in enhancing the geopolymerization process but did not reduce the workability of the mix.

Table 4: Volume proportions of each constituent included in the GPC's mix, dosage of sodium hydroxide and sodium silicate for eight molar and 10 molar solutions, and dosage of aggregate 10 mm and 20 mm sizes, respectively. This table outlines the quantities of each material used in the GPC mix design. Sodium hydroxide and sodium silicate form the alkaline activator solution, which initiates the geopolymerization reaction in GGBS, providing binding characteristics similar to cement. Fine and coarse aggregates contribute to the mix's stability and overall strength, while the 0.6% steel fiber volume enhances mechanical properties such as tensile and flexural strength. The extra water addition ensures proper hydration and workability of the mix.

In the mix, the steel fibers were used at a volume percent of 0.6%, and the combination of NaOH and Na_2SiO_3 facilitated the binding action by reacting with particles of GGBS to form a stronger concrete matrix. This mix design served as a controlled baseline for the assessment of the effect of steel fibers on GPC properties. The composition and proportions of materials used for geopolymer concrete (GPC) mix design are depicted in Fig. 2. The main constituents include GGBS, coarse and fine aggregates, and an alkaline solution containing sodium hydroxide (NaOH) and sodium silicate (Na_2SiO_3). The mix proportions are properly designed to be able to enhance durability, workability, and mechanical strength with a sodium silicate to binder ratio of 0.45. In those designs, the tensile and flexural properties of GPC are improved with the volume of 0.6% of steel fibers added. This mix design seeks to enhance performance whilst being minimal in environmental impact in accordance with the goals of manufacturing sustainability in Industry 5.0.

Table 4 Quantity of geopolymer concrete mix components

Material	Quantity (kg/m^3)
Sodium hydroxide (8 M)	54
Sodium silicate	135
GGBS	415
Fine aggregate (10 mm)	660
Coarse aggregate (20 mm)	681.8
Steel fiber (0.6%)	2.5
Extra water	11

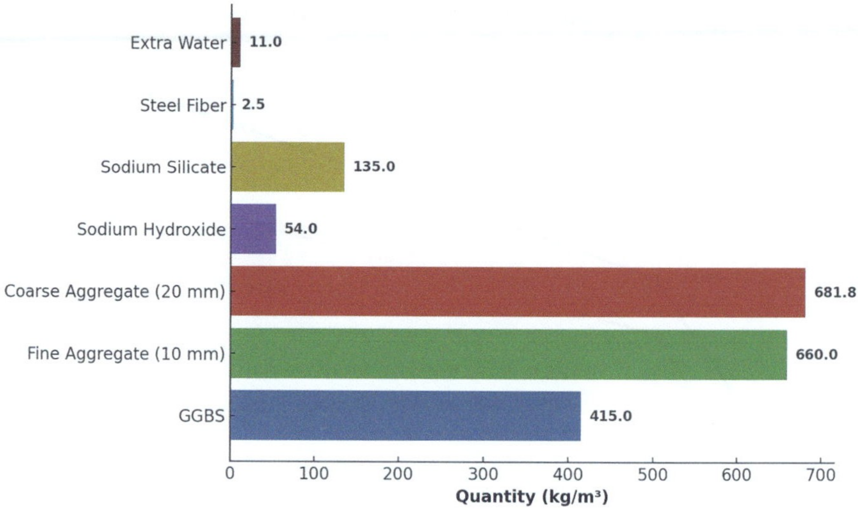

Fig. 2 Geopolymer concrete mix composition

4 Experimental Setup

The study's experimental design includes three major tests: compressive strength, flexural strength and split cylinder tensile strength of NaOH and GPC GGBS. These were the tests that were performed in order to determine the effect of steel fiber volume ratio and the molarity of sodium hydroxide on the mechanical behaviors of GGBS-based GPC. Figure 2 shows the geopolymer concrete mix composition.

- Compressive Strength Test:
- Concrete cubes having dimensions of 150 mm × 150 mm × 150 mm were cast and left to harden while being protected from the environment. The test was initiated on the third, seventh, and 28th day of molding respectively for intermittent strength check. The compressive strength was determined by application of axial load through hydraulic compression testing machine until the crushing of the specimen occurred.

  Figure 3 shows the evolution of geopolymer concrete cubes' compressive strength over time (3, 7, and 28 days) with and without steel fibers. It is evident from the data that steel GPC cubes with fibers outperform GPC cubes without fibers in compressive strength across all timeframes. The increase in strength helps to demonstrate the effect of steel fibers' reinforcement, making effective compression force resistance easier. Furthermore, the trend suggests time or curing advances as an important variable for improving compressive strength, and its development improves over time mainly due to fiber-enhanced embedment.

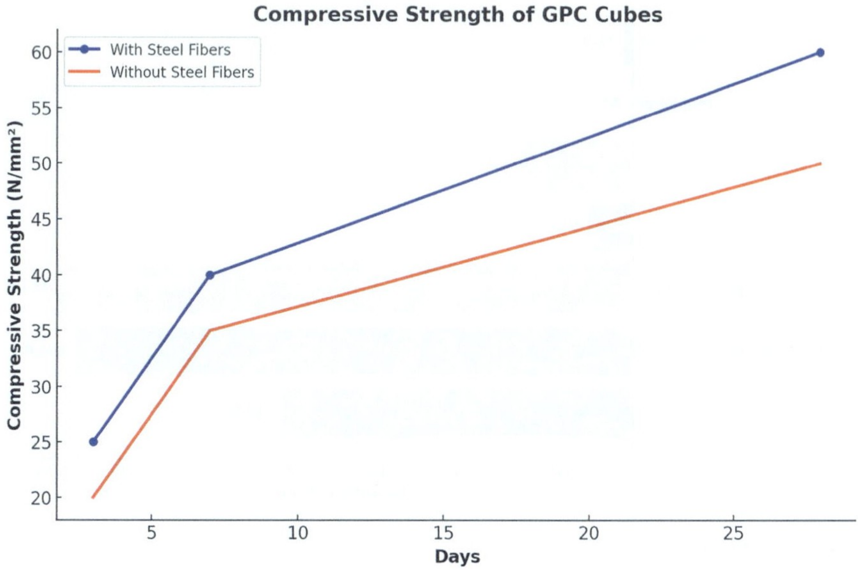

Fig. 3 Compressive strength of GPC cubes

4.1 Flexural Strength Test

The flexural strength is determined using a third-point loading test on a beam size of 100 mm x 100 mm x 500 mm. The ability of the material to resist bending forces critical for structures with concrete exposed under tensile stress was evaluated in this test method.

The flexural strength of GPC beams, as shown in Fig. 4, is influenced by steel fibers incorporated in the beam. From the investigations, it is evident that flexural strength increases with an increase in steel fiber content. The fiber content of the beams with 0.6% steel fiber volume had the highest flexural strength. This indicates the fact that addition of fiber improves the resistance of the material to bending forces. This enhanced flexural strength is expected to be useful for applications requiring improved durability and load-bearing strength such as slab and structural beams.

4.2 Split-Cylinder Tensile Strength Test

This test investigates the tensile strength indirectly by the use of conventional cylinders, which are standard specimens in this case, 150 × 300 mm. When the cylindrical specimens were being tested for compressibility, they were positioned

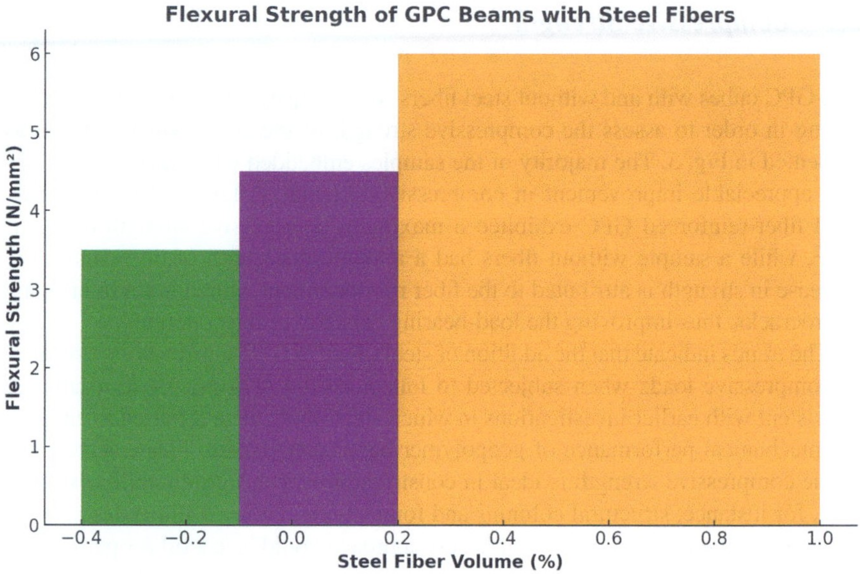

Fig. 4 Flexural strength of GPC beams with steel fibers

horizontally. Compression was continually applied along the cylinder axis until failure of the cylinder occurred. Providing the tensile strength results will assist in assessing the ductility and crack resistance of the material.

4.3 Statistical Analysis

One test was performed for several specimens since its validity in statistics had to be qualified. These were the average and standard deviation values determined for each test, and the results were discussed in an attempt to assess the effectiveness of steel fiber reinforcement. The findings were evaluated in order to determine the most favorable fiber volume and alkaline activator molarity for enhancing the compressive, flexural, and tensile strength of GPC.

5 Results and Discussion

This part provides the results of experimental studies on steel fiber-reinforced GGBS geopolymer concrete (GPC), and then analyses trends with pallial including compressive, flexural, and split cylinder tensile strength.

5.1 *Compressive Strength*

The GPC cubes with and without steel fibers were subjected to 3, 7, and 28 days of curing in order to assess the compressive strength of the geopolymer concrete as presented in Fig. 3. The majority of the samples embedded with steel fibers exhibited appreciable improvement in compressive strength. After 28 days of curing, steel fiber-reinforced GPC exhibited a maximum compressive strength of 60 N/mm^2, while a sample without fibers had a maximum strength of 50 N/mm^2. The increase in strength is attributed to the fiber reinforcement, which helps in bridging microcracks, thus improving the load-bearing capacity of the concrete.

The trends indicate that the addition of steel fibers in GPC improves its resistance to compressive loads when subjected to longer curing periods. These results are consistent with earlier investigations in which steel fibers were reported to enhance the mechanical performance of geopolymer-based composites. This enhancement in the compressive strength is ideal in construction where high durability is desirable, for instance, structural columns and foundations. Figure 5 illustrates how the strength will be expressed in N/mm^2. This approach could be useful for performing preliminary compression tests to assess the appropriateness of available materials. Typically, the concrete cubes and specimens are tested at three, seven, and 28 days of age.

5.2 *Flexural Strength*

Flexural strength tests were carried out for the GPC beams with steel fiber volumes of 0%, 0.3%, and 0.6%. From the findings illustrated in Fig. 4, it is evident that there is a positive correlation between the volume of steel fibers used and the flexible

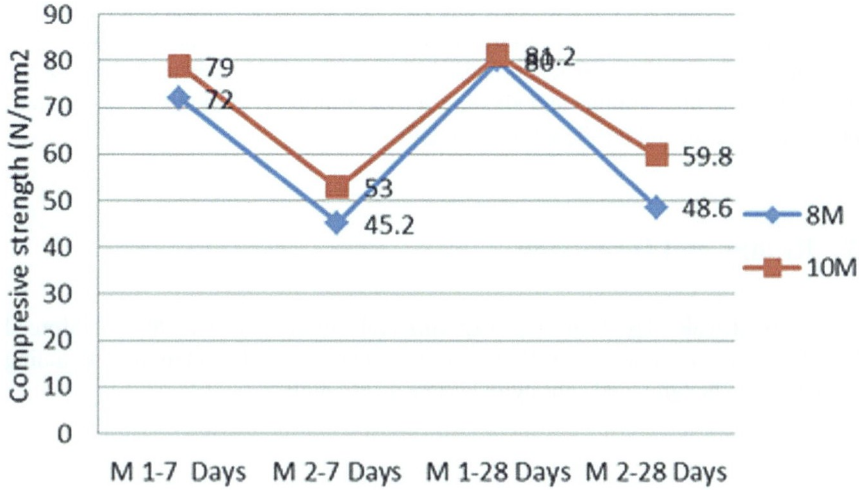

Fig. 5 Comparative study of GPC cubes with and without steel fibers

strength of the material, which was the focus of the discussion. The observation of peak flexural strength with fiber volume at 0.6% registered a strength of 6 N/mm^2 as opposed to a flexural strength of 3.5 N/mm^2 in GPC without fibers.

The enhancement is attributed to the tensile forces that the steel fibers can withstand, thereby improving the concrete flection. The fibers enhance the resistance of the beam to bending forces by pulling the cracks together as they begin to develop. Because of this, the GPC can be employed in load-bearing beams and slabs. The enhancement of the flexural strength further points out the significance of fiber volume optimization for targeted performance and material cost ratios.

5.3 *Split-Cylinder Tensile Strength*

The split-cylinder tensile strength test was performed to measure the—indirect—tensile strength of the GPC. Steel fibers incorporated in the specimens exhibited greater tensile strength than specimens made without fibers, and the curing in 10 M NaOH had the greatest effect. As seen in Fig. 6, crack load and ultimate load were also increased in these samples, which contained steel fibers, thus improving the ductility of the structure and resistance to crack propagation.

More specifically, the ultimate load capacity was 25 percent higher for the fiber-reinforced specimens over non-fiber specimens. This increase in tensile strength should prove important for volume of applications which are expected to resist crack and deformation envisioning structures such as retaining walls and bridge decks. Cracking is a means of sustaining structural fibers during cyclic or impact load thus augmenting the durability of design GPC material.

5.4 *Influence of Alkaline Activator Molarity*

Alkaline activator used in the preparation of GPC with molar ratio 8 M NaOH solution plays a major role on the mechanical properties of GPC. 10 M NaOH activated specimens were superior in compressive and tensile strength when compared to the specimens activated with 8 M. This is due to the fact that the same reaction for geopolymerization with high NaOH concentration is more prominent and gives rise to a stronger matrix, which is more compact. If the material is homogeneous, the FS is comparable to the STS. In actuality, a lot of materials contain flaws of various sizes. The strength of intact "fibers" manages the FS.

Figure 7 illustrates the flexural behavior of beams made of geopolymer concrete that has been reinforced with steel fibers and geopolymer concrete that has not.

Figure 8 displays the results for the flexibility of 100% GGBS without steel fibers and 100% GGBS with steel fibers in geopolymer concrete that is reinforced. Additionally, as can be seen below, while each beam was built with a different molarity, they all had the same reinforcing.

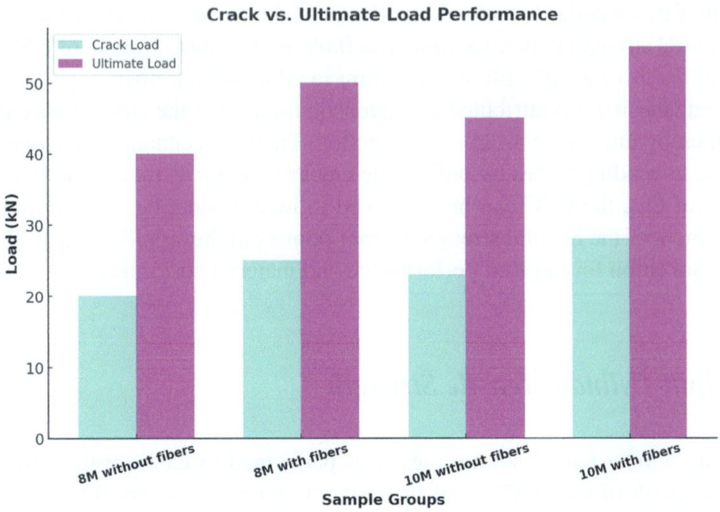

Fig. 6 Crack vs. ultimate load performance

Fig. 7 GPC cubes' average flexural strength with and without steel fiber

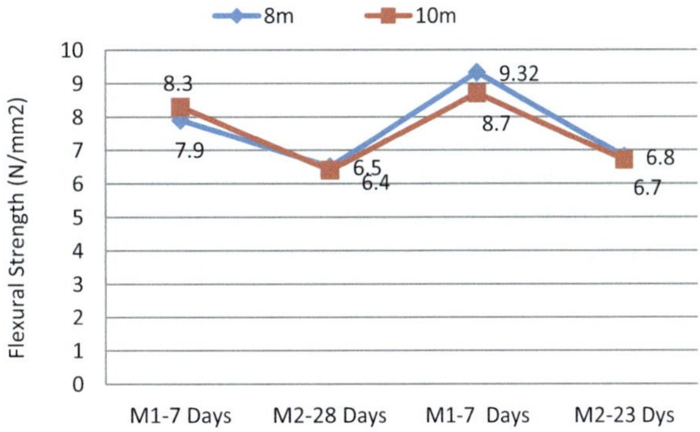

Fig. 8 Performance of crack vs ultimate loads

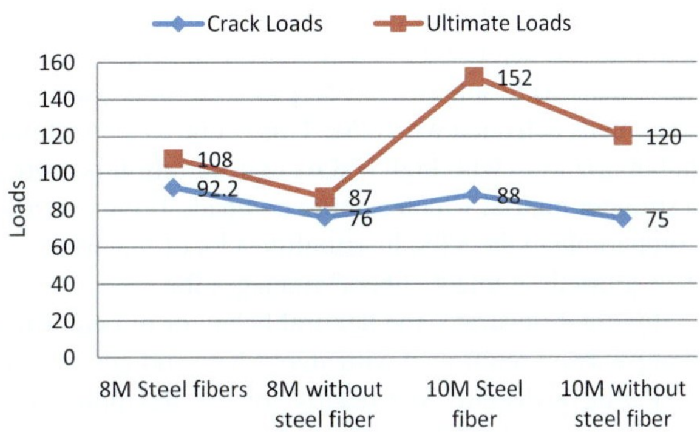

However, a balance should be maintained since higher molarity solutions raise costs and can have some impact on workability. From the results, the combination of 10 M NaOH and steel fiber reinforcement yields desired strength enhancement that is appropriate for high-performance structures where durability and strength are required.

5.5 Discussion of Practical Implications

The findings of this study underline the potential of steel fiber-reinforced GGBS GPC as a sustainable construction material that is in coherence with the direction given by Industry 5.0 towards cleaner processes and products Having GGBS and the volume of fibers increased, the demand for production of cement can be reduced which favors the building of robust infrastructure. The improved mechanical properties of fiber-reinforced GPC are ideal for application in conditions requiring high durability, a great degree of load bearing capacity, and resistance to formation of cracks. For example, for moderate to high strength infrastructure projects using sustainable and durable materials, steel fiber-reinforced GPC is a good substitute for conventional concrete. Moreover, the compatibility of GPC with automated and digital technologies, "machines" fits well with the Industry 5.0, which aims at increasing resource use through technology.

6 Conclusion

The aim of this study was to analyze the structural performance or mechanical properties of steel fiber-reinforced GGBS GPC, which can be developed as an alternate sustainable material which suits the concept of Industry 5.0. The research dealt with the incorporation of steel fibers and their effect on the compressive, flexural, and tensile strength of optimally synthesized GPC, while varying the molarity of NaOH in the alkaline activator solution.

6.1 Summary of Key Findings

1. Compressive Strength: In GGBS based GPC with steel fibers, the maximum compressive strength increased by almost 20% compared to non-fiber containing samples. This effect was notably pronounced at 28 days and which indicates that steel fibers improved the durability of the final products.

2. Flexural Strength: The tests done on the flexural strength showed that as the volume of the steel fiber was increased, so was the material's capacity to resist bending forces. It can be noted that a 0.6% fiber volume resulted in the best performance in flexural strength whereby fiber-reinforced GPC can be regarded as more appropriate for load-bearing applications, which need structural strength. Fiber-reinforced GPC for Load-Bearing Applications.
3. Tensile Strength and Crack Resistance: The advantages of adding steel fibers in the enhancement of ductility and crack resistance were also apparent in the split-cylinder tensile tests. GPC with fibers activated by 10 M NaOH showed an increase in ultimate load capacity and also improved the strength of the material due to effective geopolymerization.
4. Molarity of Alkaline Solution: The increase of molarity from 8 M to 10 M of NaOH resulted in increased compressive and tensile strength due to the presence of the alkaline activator in a geopolymerization reaction. The higher molarity of the solutions improves the strength of the material; however, certain economic aspects must also be taken into account to achieve an optimal cost-performance ratio.

6.2 Recommendations and Implications for Sustainable Construction and Industry 5.0

The results of this study directly point toward GGBS-based GPC, which is steel fiber reinforced, and its implication to be effective in sustainable construction, even displacing the use of concrete. With the use of GGBS, an industrial waste byproduct, a viable alternative raw material to cement is available, which provides lower carbon outputs, fitting into the environmental aims of Industry 5.0. Additionally, the mechanical properties of fiber-reinforced GPC expand its scope for the application of advanced smart construction, robotics, and automated fabrication processes into the future.

This research further optimizes the mix design and fiber ratio to pursue the goal of producing construction materials that are both environmentally friendly and effective. Such improvements strengthen upon the existing infrastructure, making such structures fit for usage for a circular economy, the focus of Industry 5.0, where green and digitally improved manufacturing are prioritized.

6.3 Recommendations for Future Research

In order to build upon the conclusions of this study, future research could probe further the following aspects:

1. Long-Term Durability Studies: It would be interesting to subject GGBS GPC reinforced with steel fibers to stress and strain in long-term performance studies

that include rather extreme environmental factors and determine how helpful such materials would be for structures within such extremes.

2. Cost-Benefit Analysis: A thorough investigation of the economic implications associated with the adoption of high molarity alkaline solutions and steel fibers would help ascertain the financial viability of using this GPC formulation in large construction projects.
3. Optimization for Automated Construction: In consideration of the digital and automation manufacturing integration within Industry 5.0, subsequent investigations could be directed towards improving GPC mix designs for automated construction methods such as 3D printing and robotic assembly, where material and workability are integral.
4. Comparative Analysis with Other Sustainable Binders: Other studies could also investigate the possible performance of GGBS GPC as an ideal binder material in comparison with other sustainable alternatives such as fly ash-based GPC.

References

1. Shah, S., Nautiyal, H., Gugliani, G., Kumar, A., Namboodri, T., & Singla, Y. K. (2024). *Sustainability in smart manufacturing: Trends, scope, and challenges*. CRC Press (Taylor & Francis Group). https://taylorandfrancis.com/books/. https://doi.org/10.1201/9781003467496
2. Singla, Y. K., Maughan, M. R., Arora, N., & Dwivedi, D. K. (2024) Enhancing the wear resistance of iron-based alloys: A comprehensive review of alloys element effects. *Journal of Manufacturing Processes, 120*, 135–160, (Elsevier) https://doi.org/10.1016/j.jmapro.2024.04.038
3. Singh, H., & Singla, Y. K. (2023) *Wear characteristics of AISI 4140 under nano fly ash based engine oil*. Proceedings of the 7th International Conference on Theoretical and Applied Nanoscience and Nanotechnology (TANN'23), Paper No. 118, 1-3 June, 2023, Canada. https://doi.org/10.11159/tann23.118
4. Singh, H., Singh, A. K., Singla, Y., & Chattopadhyay, K. (2020). Tribological study of nano fly ash as lubrication oil additive for AISI 4140 steel for automotive engine applications. *International Journal of Mechanical and Production Engineering Research and Development., 10*, 5939–5946.
5. Singla, Y. K., Maughan, M. R., Arora, N., & Dwivedi, D. K. (2023) Microstructral evolution of worn surface and subsurface of high Mn additive iron-based hardfacings subjected to repetitive impacts and sliding. *Ceramics International, 49*(13), 21957–21965, (Elsevier) https://doi.org/10.1016/j.ceramint.2023.04.020
6. Singh, H., Singh, A. K., Singla, Y. K., Chattopadhyay, K., Saini, A., & Singh, K. (2022). Interpretation of the wear characteristics of AISI 4140 under nano-fly ash based engine lubricant. *Materials Today Proceedings, 50*, 1683–1689.
7. Rooprai, R. S., Singh, H., Singh, T., & Singla, Y. K. (2022). Analysis of the wear properties of through hardened AISI-4140 alloy steel using Taguchi technique. *Materials Today Proceedings, 50*, 661–664.
8. Singh, H., Singh, A. K., Singla, Y. K., Chattopadhyay, K., & Malhotra, R. (2021). Optimization of tribological behavior of AISI 4140 under nano fly ash particulates in engine lubricating oil. *Materials Today Proceedings, 45*, 4619–4623.
9. Sura, H. S., Avesh, M., & Mohapatra, S. (2023). Design and modelling of digital twin technology to improve freight logistics. In *Energy, environment, and sustainability*. pp. 481–513. https://doi.org/10.1007/978-981-99-2150-8_20.

10. Pattnaik, S., Pati, S., Devadarshini, D., Tanaya, K., Mishra, B. B., Samantaray, D., Mohapatra, S., & Dash, D. (2022). Microbial bioremediation of Cr(VI)-contaminated soil for sustainable agriculture. In *Microbial biodegradation and bioremediation: techniques and case studies for environmental pollution*. pp. 395–407. https://doi.org/10.1016/B978-0-323-85455-9.00008-4.
11. Srivastava, A., Lavania, S., Mohapatra, S.(2023). Predicting the effect on land values after introducing high-speed rail. In *Energy, environment, and sustainability*. pp. 157–196. https://doi.org/10.1007/978-981-99-2150-8_8.
12. Agrahari, R., Mohanty, S., Vishwakarma, K., Nayak, S. K., Samantaray, D., & Mohapatra, S. (2021). Update vision on COVID-19: Structure, immune pathogenesis, treatment and safety assessment. *Sensors International, 2*, 100073. https://doi.org/10.1016/j.sintl.2020.100073
13. Mishra, B. B., Nayak, S. K., Mohapatra, S., & Samantaray, D. (2021). Preface. *Environmental and Agricultural Microbiology: Advances and Applications.*, 17–19.
14. Tanaya, K., Mohapatra, S., & Samantaray, D. (2024). Formulation of biogenic fluorescent pigmented PHB nanoparticles from Rhodanobacter sp. for drug delivery. *Antonie van Leeuwenhoek, International Journal of General and Molecular Microbiology, 117*, 75. https://doi.org/10.1007/s10482-024-01973-x
15. Mohapatra, S., Pattnaik, S., Maity, S., Sharma, S., Akhtar, J., Pati, S., Samantaray, D. P., & Varma, A. (2020). Comparative analysis of PHAs production by Bacillus megaterium OUAT 016 under submerged and solid-state fermentation. *Saudi Journal of Biological Sciences, 27*, 1242–1250. https://doi.org/10.1016/j.sjbs.2020.02.001
16. Pant, M., Sharma, A., Palli, S., Sharma, S. K., & Sharma, R. C. (2024). A Review And Analysis Of Factors Affecting EV adoption for sustainable and energy-efficient transportation systems using analytic hierarchy process approach. *International Journal of Vehicle Structures and Systems., 16*, 133–139. https://doi.org/10.4273/ijvss.16.2.01
17. Sharma, S. K., Sharma, R. C., Lavania, S., Palli, S., & Avesh, M. (2024). Industry-driven approach for ANFIS-based intelligent control suspension system with MR damper for enhanced ride quality in passenger rail vehicles for technological investigations. *National Academy Science Letters, 47*, 391–394. https://doi.org/10.1007/s40009-023-01365-1
18. Adeoye, A. O., Quadri, R. O., Lawal, O. S., & Emojevu, E. O. (2024). Physicochemical characterization, valorization of lignocellulosic waste (Kola nut seed shell) via pyrolysis, and ultrasonication of its crude bio-oil for biofuel production. *Cleaner Waste Systems, 7*, 100138. https://doi.org/10.1016/j.clwas.2024.100138
19. Adeoye, A. O., Quadri, R. O., Lawal, O. S., Malomo, D., Emojevu, E. O., Omonije, O. O., Odeniyi, O. K., Fadahunsi, M. O., Yelwa, M. J., Aasa, S. A., Onakpa, A. E., Omoniyi, B. P., Mark, I. N., Usman, J., Ismaila, A. M., & Saidu, A. U. (2024). Advanced techniques in upgrading crude bio-oil to biofuel. In *Energy, environment, and sustainability*. pp. 321–35. https://doi.org/10.1007/978-981-97-0515-3_14.
20. Adeoye, A. O., Quadri, R. O., & Lawal, O. S. (2022). Assessment of biofuel potential of tenera palm kernel shell via fixed bed pyrolysis and thermal characterization. *Results in Surfaces and Interfaces, 9*, 100091. https://doi.org/10.1016/j.rsurfi.2022.100091
21. Adeoye, A. O., Lawal, O. S., Quadri, R. O., Malomo, D., Aliyu, M. T., Dang, G. E., Emojevu, E. O., Maikato, M. J., Yahaya, M. G., Omonije, O. O., Edidem, V. G., Abubakar, Y. K., Offor, O. F., Sochima, E. H., Peter, B. E., & Hikon, B. N. (2023). Sustainable energy via thermochemical and biochemical conversion of biomass wastes for biofuel production. In *Energy, Environment, and Sustainability*. pp. 245–306. https://doi.org/10.1007/978-981-99-2150-8_11.
22. Mohapatra, S., & Mishra, R. K. (2020). Status of institutional credit disbursement of agricultural sector in Odisha, India. *International Journal of Current Microbiology and Applied Sciences, 9*, 2057–2065.
23. Mohapatra, S., Singh, J., & Mishra, S. (2017). Co-integration among major potato markets in Punjab. *Indian Journal of Agricultural Marketing., 31*, 46–47.
24. Mohapatra, S., Priyanka, V., Mohapatra, S., Kohli, I., & Mishra, R. K. (2020). Impact of corona virus covid-19 on the global economy. *International Journal of Agricultural and Statistical Sciences, 16*, 771–778.

25. Mohapatra, S., & Singh, J. (2021). Co-integration of major vegetable markets in Punjab, India. *Economic Affairs, 66*(2). https://doi.org/10.46852/0424-2513.2.2021.1
26. Mohapatra, S., Mishra, R., & Mishra, R. (2021). Impact of COVID-19 on education level of agricultural farming community. *International Journal of Agricultural and Statistical Sciences, 17*, 915–923.
27. Mohapatra, S., Singh, J., & Kumar, S. (2018). Cointegration among major cauliflower markets in Punjab. *Indian J Econ Dev., 14*, 330–335.
28. Adeoye, A. O., Yelwa, J. M., Imam, N., Quadri, R. O., Lawal, O. S., Malomo, D., Aasa, S. A., Onakpa, A. E., Dorgbaa, R. K., Omoniyi, B. P., Ahmad, N., & Garba, H. A. (2024). Pyrolysis of plastic wastes towards achieving a circular economy: An advanced chemistry and technical approach. In *Energy, environment, and sustainability*. pp. 215–259. https://doi.org/10.1007/978-981-97-0437-8_11.
29. Adeoye, A. O., Quadri, R. O., & Lawal, O. S. (2023). Wet synthesis, characterization of goethite nanoparticles and its application in catalytic pyrolysis of palm kernel shell in TGA. *Results in Surfaces and Interfaces, 11*, 100118. https://doi.org/10.1016/j.rsurfi.2023.100118
30. Avesh, M., Imam, N., & Adeoye, A. O. (2024). Assessing environmental impacts and mitigation strategies for the DFCCIL rail projects. In *Energy, environment, and sustainability*. pp. 333–353. https://doi.org/10.1007/978-981-97-0437-8_15.
31. Mohapatra, S., Mohapatra, S., Han, H., Ariza-Montes, A., & López-Martín, M. D. (2022). Climate change and vulnerability of agribusiness: assessment of climate change impact on agricultural productivity. *Frontiers in Psychology, 13*, 955622.
32. Mohapatra, S., & Mishra, R. A. J. K.(2020). Production performance and market integration of cotton markets in India. In *National Symposium*. p. 253.
33. Din, A. U., Han, H., Ariza-Montes, A., Vega-Muñoz, A., Raposo, A., & Mohapatra, S. (2022). The impact of Covid-19 on the food supply chain and the role of e-commerce for food purchasing. *Sustainability, 14*(5), 3074.
34. Mohapatra, S., & Singh, J. (2018). Growth and instability analysis of cauliflower crop in Punjab. *Economic Affairs, 63*, 665–669.
35. Pandey, G., Bansal, S., & Mohapatra, S. (2025). Market integration of chickpea crop: an evidence of India. *The Journal of Agribusiness in Developing and Emerging Economies, 15*(1), 1–8.
36. Mohapatra, S., Mishra, R. K., & Sarangi, K. K. (2019). Economic appraisal of bio ethanol production technologies. *International Journal of Chemical Studies, 7*, 2197–2200.
37. Mohapatra, S., Kishore Mishra, R., Sarangi, K., & K. (2020). Bio-ethanol (2nd generation ethanol): A solution to ever polluting gasoline to climate in India. *International Research Journal of Pure and Applied Chemistry, 20*, 1–15.
38. Gohain, N., Bansal, S., Mohapatra, S., & Singh, L. (2022). Analyzing the direction of trade: Indian ginger and lessons from exports to different destinations. *Journal of Agricultural Development and Policy., 32*, 214–220.
39. Mohapatra, S., Mohapatra, U., Chandana, K. S. S., & Mishra, R. K. (2018). Economic analysis of paddy production and marketing in Puri, Odisha. *Journal of Pharmacognosy and Phytochemistry., 7*, 1858–1861.
40. Quadri, R. O., Adeoye, A. O., Ayanda, O. S., & Lawal, O. S. (2024). Conversion of palm kernel shell to sustainable energy and the effect of wet synthesized nanoparticles of iron on its thermal degradation kinetics. *Bioresource Technology Reports, 27*, 101933. https://doi.org/10.1016/j.biteb.2024.101933
41. Momodu, D. U., Adetuyi, A. O., Adeoye, A. O., Akinlolu, O. S., & Apara, E. O. (2023). Surface functionalization of linen fiber and its dye-fastness assessment using 4-[(E)-(4-amino-2-hydroxyphenyl) diazenyl] benzene-1-sulphonic acid and 4-[(E)-(4-{(Z)-[4-(dimethylamino)phenyl] diazenyl}-2-hydroxyphenyl) diazenyl]benzene-1-sulphonic acid. *Suranaree Journal of Science and Technology., 30*.
42. Adeoye, A. O., Quadri, R. O., & Lawal, O. S. (2023). Biomass waste pyrolysis in the context of Nigeria's oil and gas industry: an overview. *International Journal of Ambient Energy, 44*, 2525–2541. https://doi.org/10.1080/01430750.2023.2256325

43. Mohapatra, S., Sneha, N. K., Mounika, A., & Mishra, R. K. (2019). Socio-economic and demographic analysis of SAARC nations. *International Journal of Agricultural & Statistical Sciences., 15.*
44. Sharma, S. K., Upadhyay, R. K., Kumar, V., & Valera, H. (2023). *Transportation energy and dynamics*. Springer Nature Singapore. https://doi.org/10.1007/978-981-99-2150-8
45. Upadhyay, R. K., Sharma, S. K., Kumar, D. V., & Valera, H. (2023). *Transportation systems technology and integrated management*. Springer Nature Singapore. https://doi.org/10.1007/978-981-99-1517-0
46. Lee, J., & Sharma, S. . K. (2023). Intelligent systems for railway infrastructure. *Applied Sciences, 13*(20), 11274. https://doi.org/10.3390/app132011274
47. Sharma, S. K., Sharma, R. C., Choi, Y., & Lee, J. (2023). Modelling and dynamic analysis of adaptive neuro-fuzzy inference system-based intelligent control suspension system for passenger rail vehicles using magnetorheological damper for improving ride index. *Sustainability., 15*, 12529.
48. Sharma, S. K., Upadhyay, R. K., Kumar, V., & Valera, H. (2023). Introduction to sustainable transportation system. In *Transportation energy and dynamics* (pp. 3–6). Springer.
49. Hossain, I., Ahmed, Y. E., & Sharma, S. K. (2023). The use of both metallic-insert and TiO2/water nanofluid to improve the performance of photovoltaic modules; experimental study. *Thermal Science and Engineering Progress, 42*, 101944.
50. Wu, Q., Cole, C., Spiryagin, M., Chang, C., Wei, W., Ursulyak, L., Shvets, A., Murtaza, M. A., Mirza, I. M., & Zhelieznov, K. (2023). Freight train air brake models. *International Journal of Rail Transportation, 11*, 1–49.
51. Upadhyay, R. K., Sharma, S. K., Kumar, V., & Valera, H. (2023). Introduction to transportation systems technology and integrated management. In *Transportation systems technology and integrated management* (pp. 1–4). Springer.
52. Sharma, R. C., Bhardawaj, S., Sharma, S. K., & Bahl, S. (2023). Ride behaviour of three-wheel vehicle using semi-active suspension using skyhook control strategy. *International Journal of Vehicle Structures & Systems (IJVSS)., 15.*
53. Bhardawaj, S., Sharma, R. C., Sharma, S. K., Gopala Rao, L. V. V., & Vashist, A. (2023). Modeling of biomechanical human body model for seat to head transmissibility analysis. *Noise & Vibration Worldwide, 54*, 62–74.
54. Sharma, S. K., Sharma, R. C., Choi, Y., & Lee, J. (2023). Experimental and mathematical study of flexible–rigid rail vehicle riding comfort and safety. *Applied Sciences, 13*, 5252.
55. Sharma, S. K., Upadhyay, R. K., & Kumar, V. (2024). Introduction to dynamics of transportation ecosystem, modeling, and control. In *Dynamics of transportation ecosystem, modeling, and control* (pp. 3–6). Springer.
56. Sharma, S. K., Sharma, R. C., Mohapatra, S., Fouly, A., Hossain, I., & Lee, J. (2024). Technological advancements in oscillation reduction for propulsion shaft systems. *Journal of Vibration Engineering & Technologies, 1–19.*
57. Upadhyay, R. K., Sharma, S. K., & Kumar, V. (2024). Introduction to intelligent transportation system and advanced technology. In *Intelligent transportation system and advanced technology* (pp. 3–6). Springer.
58. Hossain, I., Ngoupo, A. T., Bhattarai, S., Challagulla, S. P., & Sharma, S. K. (2024). Empirical survey on the effect of straight and wavy fins integrated with wind stream as heat sink of photovoltaic panels. *International Journal of Environmental Science and Technology, 1–14.*
59. Sharma, S. K., Sharma, R. C., & Lee, J. (2024). Propelling precision of longitudinal vibration mitigation in ship propeller shafts through advanced nonlinear intelligent semi-active control leveraging adaptive neuro-fuzzy inference system with linear quadratic regulator. *Journal of Vibration and Control*. https://doi.org/10.1177/10775463241244836
60. Sharma, S. K., Kumar, N., Avesh, M., Sharma, R. C., Siddiqui, M. I. H., & Lee, J. (2024). Navigating tranquillity with H∞ controller to mitigate ship propeller shaft vibration. *Journal of Vibration Engineering & Technologies, 1–11.*

61. Sharma, S. K., Sharma, R. C., Upadhyay, R. K., & Lee, J. (2024). Enhanced ship engine vibration reduction using magnetorheological dampers and adaptive neuro-fuzzy control system. *Ships and Offshore Structures, 20*(2), 143–159.
62. Sharma, R. C., Sharma, N., Palli, S., Sharma, S. K., Joshi, S., Rallabandi, S. R., & Sharma, A. (2024). Dynamic behavior of three-wheel vehicle with inerter-based dynamic absorber suspension system. *Noise & Vibration Worldwide, 55*, 253–263.
63. Kokkinos, N. C., Emmanouilidou, E., & Sharma, S. K. (2024). Waste-to-biofuel production for the transportation sector. In *Intelligent transportation system and advanced technology* (pp. 99–122). Springer.
64. Sharma, S. K., Sharma, R. C., & Lee, J. (2024). Modelling and simulation of MR damper characterization and uncertainty assessment in marine engine vibration isolation with FLPID control. *International Journal of Modelling and Simulation*, 1–11.
65. Sharma, S. K., Sharma, R. C., Upadhyay, R. K., & Lee, J. (2024). Coupled torsional–transverse vibration reduction in marine propulsion dynamics with novel approach using magnetorheological damping and adaptive control system. *Journal of Vibration Engineering & Technologies, 12*, 6089–6099.
66. Rallabandi, S. R., Srinivas, L., Palli, S., Sharma, R. C., Sharma, N., Sharma, A., & Sharma, S. K. (2024). A contrastive characterization of pure Mg and AZ91D alloy based on the testing of mechanical, corrosion, wear, and erosion properties. *Engineering Research Express, 6*, 015017.
67. Badgujar, P., Sharma, S., Rahul, D., Sharma, S. K., & Upadhyay, R. K. (2024). Feasibility study of mumbai-pune hyperloop corridor: Challenges and implementation. *Research in Transportation Business & Management., 52*, 101089.
68. Sharma, R. C., Palli, S., Dharmana, L., Sharma, S. K., Sharma, A., Kharola, A., & Sharma, N. (2024). Simulation of transient dynamic response of a locomotive considering its flexibility and rigidity. *Noise & Vibration Worldwide, 55*, 33–47.
69. Song, H., Park, J., Sharma, S. K., & Lee, J. (2022). A comparative study on ultrasonic propagation characteristics and defect detection of metal material additive manufacturing using deep learning algorithm. In *Biennial International Conference on Future Learning Aspects of Mechanical Engineering* (pp. 157–168). Springer.
70. Boris, I., Sharma, S. K., & Lee, J. (2022). A study on the reflection coefficient of higher modes of torsional wave by theoretical and numerical method. In *Biennial International Conference on Future Learning Aspects of Mechanical Engineering* (pp. 145–155). Springer.
71. Vishwakarma, P. N., Mishra, P., Sharma, S. K., & Shukla, A. K. (2022). Formulating and analysing the effect of suspension parameters on critical speeds for various DOFs in rail vehicle. In *Biennial International Conference on Future Learning Aspects of Mechanical Engineering* (pp. 791–802). Springer.
72. Kumar, U., Kasvekar, R., Sharma, S. K., & Upadhyay, R. K. (2022). Wear of wheels and axle in locomotive and measures taken by Indian railway. *Advances in Engine Tribology.*, 77–96.
73. Sharma, R. C., Palli, S., Jha, A. K., Bhardawaj, S., & Sharma, S. K. (2022). Vibration and ride comfort analysis of railway vehicle system subjected to deterministic inputs. *Resmilitaris., 12*, 1345–1355.
74. Sharma, R. C., Rao, L. V. V. G., Sharma, S. K., Palli, S., & Satyanarayana, V. S. (2022). V: Analysis of lateral stability and ride of an Indian railway constrained dual-axle bogie frame. *SAE International Journal of Commercial Vehicles, 16*, 213–228.
75. Vishwakarma, P. N., Mishra, P., & Sharma, S. K. (2022). Formulation of semi-active suspension system and controls in rail vehicle. *SSRN Electronic Journal*. https://doi.org/10.2139/ssrn.415961
76. Sharma, R. C., & Sharma, S. K. (2022). Ride analysis of road surface-three-wheeled vehicle-human subject interactions subjected to random excitation. *SAE International Journal of Commercial Vehicles, 15*, 299–312.
77. Sharma, S. K., Lee, J., & Jang, H.-L. (2022). Mathematical modeling and simulation of suspended equipment impact on car body modes. *Machines., 10*, 192.

78. Sharma, S. K., Sharma, R. C., Lee, J., & Jang, H.-L. (2022). Numerical and experimental analysis of DVA on the flexible-rigid rail vehicle carbody resonant vibration. *Sensors, 22*, 1922.
79. Vishwakarma, P. N., Mishra, P., & Sharma, S. K. (2022). Characterization of a magnetorheological fluid damper a review. *Materials Today Proceedings, 56*, 2988–2994.
80. Sharma, S. K., Sharma, R. C., & Lee, J. (2022). In situ and experimental analysis of longitudinal load on carbody fatigue life using nonlinear damage accumulation. *International Journal of Damage Mechanics, 31*, 605–622.
81. Sharma, R. C., Sharma, S., Sharma, N., & Sharma, S. K. (2021). Linear and nonlinear analysis of ride and stability of a three-wheeled vehicle subjected to random and bump inputs using bond graph and Simulink methodology. *SAE International Journal of Commercial Vehicles, 15*, 65–79.
82. Sharma, S. K., & Sharma, R. C. (2021). *Multi-objective design optimization of locomotive nose*. SAE Technical Paper. pp. 1–10. https://doi.org/10.4271/2021-01-5053
83. Lee, J., Han, J., & Sharma, S. K. (2021). Structural analysis on the separated and integrated differential gear case for the weight reduction. In *Advances in Engineering Design: Select Proceedings of FLAME 2020* (pp. 175–181). Springer.
84. Choi, S., Lee, J., & Sharma, S. K. (2021). A study on the performance evaluation of hydraulic tank injectors. In *Advances in Engineering Design: Select Proceedings of FLAME 2020* (pp. 183–190). Springer.
85. Sharma, R. C., Palli, S., Sharma, N., & Sharma, S. K. (2021). Ride behaviour of a four-wheel vehicle using h infinity semi-active suspension control under deterministic and random inputs. *International Journal of Vehicle Structures & Systems., 13*, 234–237.
86. Bhardawaj, S., Sharma, R. C., Sharma, S. K., & Sharma, N. (2021). On the planning and construction of railway curved track. *International Journal of Vehicle Structures & Systems., 13*, 151–159.
87. Mohapatra, S., Mohanty, D., Sharma, S., Dikshit, S., Kohli, I., Samantaray, D. P., Kumar, R., & Kathpalia, M. (2021). Biomedical application of polymeric biomaterial: polyhydroxybutyrate. In *Bioresource utilization and management* (pp. 111–124). Apple Academic Press.
88. Sharma, S. K., Sharma, R. C., & Lee, J. (2021). Effect of rail vehicle–track coupled dynamics on fatigue failure of coil spring in a suspension system. *Applied Sciences, 11*, 2650.
89. Sharma, R. C., Sharma, S., Sharma, S. K., Sharma, N., & Singh, G. (2021). Analysis of bio-dynamic model of seated human subject and optimization of the passenger ride comfort for three-wheel vehicle using random search technique. *Proceedings of the Institution of Mechanical Engineers, Part K: Journal of Multi-body Dynamics., 235*, 106–121.
90. Sharma, S. K., & Lee, J. (2021). Crashworthiness analysis for structural stability and dynamics. *International Journal of Structural Stability and Dynamics, 21*, 2150039.
91. Cataluna Veses, R., & Shah, Z. (2015). Bio-oil production by thermal cracking in the presence of hydrogen. *Journal of Fundamentals of Renewable Energy and Applications, 06*. https://doi.org/10.4172/2090-4541.1000194
92. Shah, Z., Veses, R. C., & Da Silva, R. (2016). A comparative study and analysis of two types of bio-oil samples obtained from freshwater algae and microbial treated algae. *Modern Chemistry & Applications., 4*, 1000185. https://doi.org/10.4172/2329-6798.1000185
93. Shah, Z., Veses, R. C., Brum, J., Klunk, M. A., Rocha, L. A. O., Missaggia, A. B., de Oliveira, A. P., & Caetano, N. R. (2020). Sewage Sludge bio-oil development and characterization. *Inventions., 5*, 36. https://doi.org/10.3390/inventions5030036
94. Shah, Z., Saberian, M., Hodgeman, D., Kinloch, I., & Vallés, C. (2023). Effect of sulfur on wood tar biopitch as a sustainable replacement for coal tar pitch binders. *ACS Applied Engineering Materials, 1*, 2567–2576. https://doi.org/10.1021/acsaenm.3c00361
95. Klunk, M. A., Dasgupta, S., Das, M., & Shah, Z. (2018). System of adsorption of CO2 in coalbed. *Southern Brazilian Journal of Chemistry., 26*, 2–9. https://doi.org/10.48141/SBJCHEM.v26.n26.2018.7_2018.pdf

96. Mohapatra, S., Mohapatra, U., & Mishra, R. (2017). Diversification towards vegetable crops: A good option for doubling the farmer's income. *Journal of Experimental Agriculture International., 18*, 1–7.
97. Mohapatra, S., Mohapatra, U., & Mishra, R. K. (2018). Price forecasting of groundnut in Odisha. *The Pharma Innovation Journal, 7*, 111–114.
98. Bansal, S., Mohapatra, S., Gohain, N., & Das, A. (2022). Production status, market integration and the law of one price: An exploration of the major maize markets of India. *Indian Journal of Economics and Development, 18*, 832–841.
99. Mohapatra, S., Samantaray, D. P., & Samantaray, S. M. (2015). Study on polyhydroxyalkanoates production using rhizospheric soil bacterial isolates of sweet potato. *Indian Journal of Science and Technology, 8*, 57–62. https://doi.org/10.17485/ijst/2015/v8iS7/64027
100. Arora, R., Trivedi, J., Mohapatra, S., & Kumar, P. (2022). Trends in probiotics on human health and industrial application. In *Advances in Agricultural and Industrial Microbiology: Volume 1: Microbial diversity and application in agroindustry* (pp. 237–248). https://doi.org/10.1007/978-981-16-8918-5_13
101. Samal, S., Pati, S., Mohapatra, S., Maity, S., Tanaya, K., Devadarshini, D., & Samantaray, D. (2023). PHAs production by facultative anaerobic bacteria Bacillus cereus FM5 through submerged and solid-state fermentation under anoxic condition. *Antonie van Leeuwenhoek International Journal of General and Molecular Microbiology, 116*, 521–529. https://doi.org/10.1007/s10482-023-01825-0
102. Biswas, N., Saha, S., Biswas, B. K., Chowdhury, M., Rahaman, A., Junghare, V., Mohapatra, S., Hazra, S., Zangrando, E., Roy Choudhury, R., & Roy Choudhury, C. (2021). The DNA- and protein-binding properties and cytotoxicity of a new copper(II) hydrazone Schiff base complex. *Journal of Coordination Chemistry, 74*, 1482–1504. https://doi.org/10.1080/00958972.2021.1913128
103. Mohapatra, S., Mishra, R., Roy, P., Yadav, K. L., & Satapathi, S. (2017). Systematic investigation and in vitro biocompatibility studies on implantable magnetic nanocomposites for hyperthermia treatment of osteoarthritic knee joints. *Journal of Materials Science, 52*, 9262–9268. https://doi.org/10.1007/s10853-017-1136-0
104. Mishra, B. B., Nayak, S. K., Mohapatra, S., & Samantaray, D. (2021). Environmental and agricultural microbiology: Advances and applications. https://doi.org/10.1002/9781119525899.
105. Sharma, S. K., Mohapatra, S., Sharma, R. C., Alturjman, S., Altrjman, C., Mostarda, L., & Stephan, T. (2022). Retrofitting existing buildings to improve energy performance. *Sustainability (Switzerland)., 14*. https://doi.org/10.3390/su14020666
106. Bandhu, S., Khot, M. B., Sharma, T., Sharma, O. P., Dasgupta, D., Mohapatra, S., Hazra, S., Khatri, O. P., & Ghosh, D. (2018). Single Cell oil from oleaginous yeast grown on sugarcane bagasse-derived xylose: an approach toward novel biolubricant for low friction and wear. *ACS Sustainable Chemistry & Engineering, 6*, 275–283. https://doi.org/10.1021/acssuschemeng.7b02425
107. Kohli, I., Joshi, N. C., Mohapatra, S., & Varma, A. (2020). Extremophile—An adaptive strategy for extreme conditions and applications. *Current Genomics, 21*, 96–110. https://doi.org/10.2174/1389202921666200401105908
108. Pati, S., Mohapatra, S., Maity, S., Dash, A., & Samantaray, D. P. (2021). Polyhydroxyalkanoates for sustainable smart packaging of fruits. In *Environmental and agricultural microbiology: Advances and applications* (pp. 183–195).
109. Chatterjee, A., Nirwan, S., Mohapatra, S., Sharma, P., Agnihotri, A., & Shrivastava, N. (2022). Biochemical aspects of pathogenic variability in white rust infected Indian mustard. *Mycologia, 114*, 757–768. https://doi.org/10.1080/00275514.2022.2060007
110. Pati, S., Maity, S., Dash, A., Jema, S., Mohapatra, S., Das, S., & Samantaray, D. P. (2020). Biocompatible PHB production from Bacillus species under submerged and solid-state fermentation and extraction through different downstream processing. *Current Microbiology, 77*, 1203–1209. https://doi.org/10.1007/s00284-020-01922-7

111. Agrahari, R., Sarraf, G., Joshi, N. C., Mohapatra, S., & Var, A. (2021). Insight of biopolymers and applications of polyhydroxyalkanoates. In *Microbial polymers: Applications and ecological perspectives* (pp. 177–191). https://doi.org/10.1007/978-981-16-0045-6_8
112. Kathpalia, M., Kumar, P., & Mohapatra, S. (2022). Homocysteine metabolism as a biomarker for cancer. In *Homocysteine metabolism in health and disease* (pp. 159–171). https://doi.org/10.1007/978-981-16-6867-8_8
113. Bharatiy, S. K., Hazra, M., Paul, M., Mohapatra, S., Samantaray, D., Dubey, R. C., Sanyal, S., Datta, S., & Hazra, S. (2016). In silico designing of an industrially sustainable carbonic anhydrase using molecular dynamics simulation. *ACS Omega, 1*, 1081–1103. https://doi.org/10.1021/acsomega.6b00041
114. Sarkar, M. S., Segu, H., Bhaskar, J. V., Jakher, R., Mohapatra, S., Shalini, K., Shivaji, S., & Reddy, P. A. (2018). Ecological preferences of large carnivores in remote, high-altitude protected areas: Insights from Buxa Tiger Reserve, India. *ORYX., 52*, 66–77. https://doi.org/10.1017/S0030605317000060
115. Pati, S., Dash, A., Maity, S., Pattnaik, S., Mohapatra, S., & Samantaray, D. P. (2021). Nano biofertilizer for sustainable agriculture. In *Agriculturally important microorganisms: Mechanisms and applications for sustainable agriculture* (pp. 201–208). https://doi.org/10.1201/9781003245841-10
116. Mohanty, D., Misra, S., Mohapatra, S., & Sahu, P. S. (2018). Prebiotics and synbiotics: Recent concepts in nutrition. *Food Bioscience, 26*, 152–160. https://doi.org/10.1016/j.fbio.2018.10.008
117. Kumar, R., Kumari, R., Prasad, C., Deep, A., Singh, N., Namtak, S., Tiwari, V., Merugu, R., Mohapatra, S., & Singh, R. (2021). Influence of water quality on the diversity of macroinvertebrates in the Mandakini River in India. *Water Science and Technology: Water Supply, 21*, 1843–1860. https://doi.org/10.2166/WS.2021.020
118. Mohanty, D., Jena, R., Choudhury, P. K., Pattnaik, R., Mohapatra, S., & Saini, M. R. (2016). Milk derived antimicrobial bioactive peptides: A review. *International Journal of Food Properties, 19*, 837–846. https://doi.org/10.1080/10942912.2015.1048356
119. Mohapatra, S., Rath, S. N., Pradhan, S. K., Samantaray, D. P., & Rath, C. C. (2016). Secondary structural models (16S rRNA) of polyhydroxyalkanoates producing bacillus species isolated from different rhizospheric soil: Phylogenetics and chemical analysis. *International Journal Bioautomation., 20*, 329–338.
120. Mohapatra, S., Sarkar, B., Samantaray, D. P., Daware, A., Maity, S., Pattnaik, S., & Bhattacharjee, S. (2017). Bioconversion of fish solid waste into PHB using Bacillus subtilis based submerged fermentation process. *Environmental Technology (United Kingdom)., 38*, 3201–3208. https://doi.org/10.1080/09593330.2017.1291759
121. Misra, S., Mohanty, D., & Mohapatra, S. (2021). Food-based probiotics: Functional dietary ingredients. In *Probiotics: Advanced food and health applications* (pp. 257–275). https://doi.org/10.1016/B978-0-323-85170-1.00006-3
122. Kumar, R., Kumari, R., Prasad, C., Tiwari, V., Singh, N., Mohapatra, S., Merugu, R., Namtak, S., & Deep, A. (2020). Phytoplankton diversity in relation to physicochemical attributes and water quality of Mandakini River, Garhwal Himalaya. *Environmental Monitoring and Assessment, 192*, 1–23. https://doi.org/10.1007/s10661-020-08768-3
123. Mohapatra, S., Vishwakarma, K., Joshi, N. C., Maity, S., Kumar, R., Ramchander, M., Pattnaik, S., & Samantaray, D. P. (2021). A review on PHAs: The future biopolymer. In *Environmental and agricultural microbiology: Advances and applications* (pp. 83–117).
124. Sahoo, N., Pattnaik, S., Pattnaik, M., & Mohapatra, S. (2021). Methane emission and strategies for mitigation in Livestock. In *Environmental and agricultural microbiology: Advances and applications* (pp. 257–274).
125. Mohanty, D. P., Mohapatra, S., Misra, S., & Sahu, P. S. (2016). Milk derived bioactive peptides and their impact on human health—A review. *Saudi Journal of Biological Sciences, 23*, 577–583. https://doi.org/10.1016/j.sjbs.2015.06.005

126. Kohli, I., Mohapatra, S., Kumar, P., Goel, A., Varma, A., & Joshi, N. C. (2022). Role of bacterial endophytes in the promotion of plant growth. In *Bacterial endophytes for sustainable agriculture and environmental management* (pp. 235–253). https://doi.org/10.1007/978-981-16-4497-9_12
127. Samantaray, D., Mohapatra, S., & Mishra, B. B. (2014). Microbial bioremediation of industrial effluents. In *Microbial biodegradation and bioremediation* (pp. 326–339). https://doi.org/10.1016/B978-0-12-800021-2.00014-5
128. Pattnaik, S., Dash, D., Mohapatra, S., Pattnaik, M., Marandi, A. K., Das, S., & Samantaray, D. P. (2020). Corrigendum to "Improvement of rice plant productivity by native Cr(VI) reducing and plant growth promoting soil bacteria Enterobacter cloacae" (Chemosphere (2020) 240, (124895), (S0045653519321344), (10.1016/j.chemosphere.2019.124895)). *Chemosphere, 238*. https://doi.org/10.1016/j.chemosphere.2019.125011
129. Mohapatra, S., Mohanta, P. R., Sarkar, B., Daware, A., Kumar, C., & Samantaray, D. P. (2017). Production of polyhydroxyalkanoates (PHAs) by Bacillus strain isolated from waste water and its biochemical characterization. *Proceedings of the National Academy of Sciences, India Section B: Biological Sciences, 87*, 459–466. https://doi.org/10.1007/s40011-015-0626-6
130. Tomer, P., Shroti, G. K., Mohapatra, S., Ghosh, D., Jaiswal, S., Garg, D., Lahiri, D., Kumar, N., Khatri, O. P., & Hazra, S. (2023). Fabrication of HA Nano-crystal Reinforced PHAs based composites for orthopedic fracture-fixing accessories using sustainable resources. *Environment, Development and Sustainability*. https://doi.org/10.1007/s10668-023-03912-8
131. Bandekar, D., Mohapatra, S., Hazra, M., Hazra, S., & Biswas, S. (2021). N-terminal truncation of VC0395_0300 protein from vibrio cholerae does not lead to loss of diguanylate cyclase activity. *Biophysical Chemistry, 268*, 106493. https://doi.org/10.1016/j.bpc.2020.106493
132. Vishwakarma, K., Kumar, N., Shandilya, C., Mohapatra, S., Bhayana, S., & Varma, A. (2020). Revisiting plant–microbe interactions and microbial consortia application for enhancing sustainable agriculture: A review. *Frontiers in Microbiology, 11*, 560406. https://doi.org/10.3389/fmicb.2020.560406
133. Mohapatra, S., Maity, S., Dash, H. R., Das, S., Pattnaik, S., Rath, C. C., & Samantaray, D. (2017). Bacillus and biopolymer: Prospects and challenges. *Biochemistry and Biophysics Reports, 12*, 206–213. https://doi.org/10.1016/j.bbrep.2017.10.001
134. Kumar, R., Merugu, R., Mohapatra, S., & Sharma, S. (2022). Extremophiles: Life of microorganisms in extreme environments. In *Extremophiles: General and plant biomass based biorefinery* (pp. 43–66). https://doi.org/10.1201/9781003014188-3
135. Pradhan, A., Pahari, A., Mohapatra, S., & Mishra, B. B. (2017). Phosphate-solubilizing microorganisms in sustainable agriculture: Genetic mechanism and application. In *Microorganisms for sustainability* (pp. 81–97). https://doi.org/10.1007/978-981-10-7380-9_5
136. Devadarshini, D., Mohapatra, S., Pati, S., Maity, S., Rath, C. C., Jena, P. K., & Samantaray, D. (2023). Evaluation of PHAs production by mixed bacterial culture under submerged fermentation. *Biologia (Bratisl)., 78*, 895–902. https://doi.org/10.1007/s11756-022-01302-5
137. Agarwal, K. M., Mohapatra, S., Sharma, P., Sharma, S., Bhatia, D., & Mishra, A. (2020). Study and overview of the novel corona virus disease (COVID-19). *Sensors International, 1*. https://doi.org/10.1016/j.sintl.2020.100037
138. Tanaya, K., Mohapatra, S., Maity, S., & Samantaray, D. P. (2023). Fluorescent-pigmented microbial polymer nanoparticles: An approach for replacement of bioimaging agent in the drug delivery system. *Applied Biochemistry and Microbiology, 59*, 378–383. https://doi.org/10.1134/S0003683823030171
139. Pattnaik, S., Dash, D., Mohapatra, S., Pati, S., Devadarshini, D., Samal, S., Pattnaik, M., Maity, S., Mishra, S. K., & Samantaray, D. (2024). Reclamation of chromium-contaminated soil by native Cr(VI)-reducing and PHA-accumulating Bacillus aryabhattai CTSI-07. *International Microbiology, 27*, 731–742. https://doi.org/10.1007/s10123-023-00421-6
140. Mohapatra, S., Mishra, Y., & Dash, L. (2020). Covid-19: Analysing the legal nuances of the lockdown order. *Indian Journal of Forensic Medicine and Toxicology, 14*, 612–616. https://doi.org/10.37506/ijfmt.v14i4.11553

141. Pati, S., Mohanty, M. K., Mohapatra, S., & Samantaray, D. (2022). Bioethanol production by Enterobacter hormaechei through carbon monoxide-rich syngas fermentation. *International Journal of Energy Research, 46*, 20096–20106. https://doi.org/10.1002/er.8678
142. Maity, S., Das, S., Mohapatra, S., Tripathi, A. D., Akthar, J., Pati, S., Pattnaik, S., & Samantaray, D. P. (2020). Growth associated polyhydroxybutyrate production by the novel Zobellellae tiwanensis strain DD5 from banana peels under submerged fermentation. *International Journal of Biological Macromolecules, 153*, 461–469. https://doi.org/10.1016/j.ijbiomac.2020.03.004
143. Takkar, S., Shandilya, C., Agrahari, R., Chaurasia, A., Vishwakarma, K., Mohapatra, S., Varma, A., & Mishra, A. (2022). Green technology: Phytoremediation for pesticide pollution. In *Phytoremediation technology for the removal of heavy metals and other contaminants from soil and water* (pp. 353–375). https://doi.org/10.1016/B978-0-323-85763-5.00008-8
144. Bandekar, D., Chouhan, O. P., Mohapatra, S., Hazra, M., Hazra, S., & Biswas, S. (2017). Putative protein VC0395_0300 from vibrio cholerae is a diguanylate cyclase with a role in biofilm formation. *Microbiological Research, 202*, 61–70. https://doi.org/10.1016/j.micres.2017.05.003
145. Mohapatra, S., Samantaray, D. P., Samantaray, S. M., Mishra, B. B., Das, S., Majumdar, S., Pradhan, S. K., Rath, S. N., Rath, C. C., Akthar, J., & Achary, K. G. (2016). Structural and thermal characterization of PHAs produced by Lysinibacillus sp. through submerged fermentation process. *International Journal of Biological Macromolecules, 93*, 1161–1167. https://doi.org/10.1016/j.ijbiomac.2016.09.077
146. Sharma, S. K., Sharma, R. C., Kumar, A., & Palli, S. (2015). Challenges in rail vehicle-track modeling and simulation. *International Journal of Vehicle Structures & Systems., 7*, 1.
147. Palli, S., Sharma, R. C., Sharma, S. K., Muddada, V., & Sharma, N. (2020). A case study on noise pollution and its effects. *Journal of Critical Reviews., 7*, 261–267.
148. Sharma, R. C., Sharma, S. K., Sharma, N., & Sharma, S. (2020). Analysis of ride and stability of an ICF railway coach. *International Journal of Vehicle Noise and Vibration, 16*, 127–144.
149. Palli, S., Sharma, R. C., Sharma, S. K., & Chintada, V. B. (2020). On methods used for setting the curve for railway tracks. *Journal of Critical Reviews, 7*, 241–246.
150. Lee, J., & Sharma, S. K. (2020). Numerical investigation of critical speed analysis of high-speed rail vehicle. 한국정밀공학회 학술발표대회 논문집, 696.
151. Sharma, S., Sharma, R. C., Sharma, S. K., Sharma, N., Palli, S., & Bhardawaj, S. (2020). Vibration isolation of the quarter car model of road vehicle system using dynamic vibration absorber. *International Journal of Vehicle Structures & Systems., 12*, 513–516.
152. Acharya, A., Gahlaut, U., Sharma, K., Sharma, S. K., Vishwakarma, P. N., & Phanden, R. K. (2020). Crashworthiness analysis of a thin-walled structure in the frontal part of automotive chassis. *International Journal of Vehicle Structures & Systems., 12*, 517–520.
153. Sharma, S. K., & Lee, J. (2020). Finite element analysis of a fishplate rail joint in extreme environment condition. *International Journal of Vehicle Structures & Systems., 12*, 503–506.
154. Bhardawaj, S., Sharma, R. C., & Sharma, S. K. (2020). Development of multibody dynamical using MR damper based semi-active bio-inspired chaotic fruit fly and fuzzy logic hybrid suspension control for rail vehicle system. *Proceedings of the institution of mechanical engineers, Part K: journal of multi-body dynamics., 234*, 723–744.
155. Sharma, S. K., & Lee, J. (2020). Design and development of smart semi active suspension for nonlinear rail vehicle vibration reduction. *International Journal of Structural Stability and Dynamics, 20*, 2050120.
156. Sharma, S. K., Phan, H., & Lee, J. (2020). An application study on road surface monitoring using DTW based image processing and ultrasonic sensors. *Applied Sciences, 10*, 4490.
157. Sharma, R. C., Sharma, S., Sharma, S. K., & Sharma, N. (2020). Analysis of generalized force and its influence on ride and stability of railway vehicle. *Noise & Vibration Worldwide, 51*, 95–109.
158. Bhardawaj, S., Sharma, R. C., & Sharma, S. K. (2020). Ride analysis of track-vehicle-human body interaction subjected to random excitation. *Journal of the Chinese Society of Mechanical Engineers., 41*, 227–236.

159. Bhardawaj, S., Sharma, R. C., & Sharma, S. K. (2020). Development in the modeling of rail vehicle system for the analysis of lateral stability. *Materials Today Proceedings, 25*, 610–619.
160. Sharma, S. K., Sharma, R. C., & Sharma, N. (2020). Combined multi-body-system and finite element analysis of a rail locomotive crashworthiness. *International Journal of Vehicle Structures & Systems., 12*, 428–435.
161. Bhardawaj, S., Sharma, R. C., & Sharma, S. K. (2020). Analysis of frontal car crash characteristics using ANSYS. *Materials Today Proceedings, 25*, 898–902.
162. Bhardawaj, S., Sharma, R. C., & Sharma, S. K. (2019). A survey of railway track modelling. *International Journal of Vehicle Structures & Systems., 11*, 508–518.
163. Bhardawaj, S., Sharma, R. C., & Sharma, S. K. (2019). Development and advancement in the wheel-rail rolling contact mechanics. In *IOP Conference Series: Materials Science and Engineering* (p. 012034). IOP Publishing.
164. Choppara, R. K., Sharma, R. C., Sharma, S. K., & Gupta, T. (2019). Aero dynamic cross wind analysis of locomotive. In *IOP Conference Series: Materials Science and Engineering* (p. 012035). IOP Publishing.
165. Sharma, S. K., & Sharma, R. C. (2019). Pothole detection and warning system for Indian roads. In *Advances in Interdisciplinary Engineering: Select Proceedings of FLAME 2018* (pp. 511–519). Springer.
166. Sinha, A. K., Sengupta, A., Gandhi, H., Bansal, P., Agarwal, K. M., Sharma, S. K., Sharma, R. C., & Sharma, S. K. (2019). Performance enhancement of an all-terrain vehicle by optimizing steering, powertrain and brakes. In *Advances in Engineering Design: Select Proceedings of FLAME 2018* (pp. 207–215). Springer.
167. Goyal, S., Anand, C. S., Sharma, S. K., & Sharma, R. C. (2019). Crashworthiness analysis of foam filled star shape polygon of thin-walled structure. *Thin-Walled Structures, 144*, 106312.
168. Goswami, B., Rathi, A., Sayeed, S., Das, P., Sharma, R. C., & Sharma, S. K. (2019). Optimization design for aerodynamic elements of Indian locomotive of passenger train. In *Advances in Engineering Design: Select Proceedings of FLAME 2018* (pp. 663–673). Springer.
169. Sharma, S. K., Saini, U., & Kumar, A. (2019). Semi-active control to reduce lateral vibration of passenger rail vehicle using disturbance rejection and continuous state damper controllers. *Journal of Vibration Engineering & Technologies, 7*, 117–129.
170. Sharma, S. K. (2019). Multibody analysis of longitudinal train dynamics on the passenger ride performance due to brake application. *Proceedings of the Institution of Mechanical Engineers, Part K: Journal of Multi-body Dynamics., 233*, 266–279.
171. Sharma, S. (2018). Vibration control in quarter-car model with magnetorheological (MR) dampers using Bingham model. *Journal of Applied Mechanical Engineering, 7*, 1000299.
172. Sharma, S. K., & Kumar, A. (2018). Impact of longitudinal train dynamics on train operations: a simulation-based study. *Journal of Vibration Engineering & Technologies, 6*, 197–203.
173. Sharma, S. K., & Sharma, R. C. (2018). An investigation of a locomotive structural crashworthiness using finite element simulation. *SAE International Journal of Commercial Vehicles, 11*, 235–243.
174. Sharma, R. C., Sharma, S. K., & Palli, S. (2018). Rail vehicle modelling and simulation using Lagrangian method. *International Journal of Vehicle Structures and Systems., 10*, 188–194.
175. Sharma, R. C., & Sharma, S. K. (2018). Sensitivity analysis of three-wheel vehicle's suspension parameters influencing ride behavior. *Noise & Vibration Worldwide, 49*, 272–280.
176. Palli, S., Koona, R., Sharma, S. K., & Sharma, R. C. (2018). A review on dynamic analysis of rail vehicle coach. *International Journal of Vehicle Structures & Systems (IJVSS)., 10.*
177. Sharma, S. K., & Sharma, R. C. (2018). Simulation of quarter-car model with magnetorheological dampers for ride quality improvement. *International Journal of Vehicle Structures & Systems., 10*, 169–173.
178. Sharma, S. K., & Kumar, A. (2018). Disturbance rejection and force-tracking controller of nonlinear lateral vibrations in passenger rail vehicle using magnetorheological fluid damper. *Journal of Intelligent Material Systems and Structures, 29*, 279–297.

179. Sharma, S. K., & Kumar, A. (2018). Ride comfort of a higher speed rail vehicle using a magnetorheological suspension system. *Proceedings of the Institution of Mechanical Engineers, Part K: Journal of Multi-Body Dynamics., 232*, 32–48.
180. Shukla, S., Kumar, U., Sharma, S. K., Gupta, P., & Kumar, A. (2017). Light weight freight rolling stock bogie frame: Design methodology validated with field oscillation trials. *International Journal of Vehicle Structures & Systems., 9*, 245–250.
181. Sharma, R. C., Palli, S., Sharma, S. K., & Roy, M. (2017). Modernization of railway track with composite sleepers. *International Journal of Vehicle Structures & Systems., 9*, 321–329.
182. Sharma, S. K., & Kumar, A. (2017). Ride performance of a high speed rail vehicle using controlled semi active suspension system. *Smart Materials and Structures, 26*, 055026.
183. Sharma, S. K., & Kumar, A. (2017). Impact of electric locomotive traction of the passenger vehicle Ride quality in longitudinal train dynamics in the context of Indian railways. *Mechanics & Industry, 18*, 222.
184. Sharma, S. K., & Lavania, S. (2013). Green manufacturing and green supply chain management in India a review. In *Futuristic trends in Mechanical and Industrial Engineering*. JECRC UDML College of Engineering.
185. Sharma, S. K., & Lavania, S. (2013). An autonomous metro: Design and execution. In *Futuristic trends in mechanical and industrial engineering*. JECRC UDML College of Engineering.
186. Sharma, S. K. (2013). Zero energy building envelope components: A review. *International Journal of Engineering Research and Applications, 3*, 662–675.
187. Sharma, S. K., & Kumar, A. (2016). Dynamics analysis of wheel rail contact using FEA. *Procedia Engineering, 144*, 1119–1128.
188. Sharma, S. K., & Chaturvedi, S. (2016). Jerk analysis in rail vehicle dynamics. *Perspectives in Science (Netherland), 8*, 648–650.
189. Sharma, S. K., & Kumar, A. (2016). The impact of a rigid-flexible system on the ride quality of passenger bogies using a flexible carbody. In *Proceedings of the Third International Conference on Railway Technology: Research, Development and Maintenance* (p. 87). Civil-Comp Press.
190. Kulkarni, D., Sharma, S. K., & Kumar, A. (2016). Finite element analysis of a fishplate rail joint due to wheel impact. In *International Conference on Advances in Dynamics, Vibration and Control*. National Institute of Technology Durgapur.
191. Palli, S., Koona, R., Sharma, R. C., & Muddada, V. (2015). Dynamic analysis of Indian railway integral coach factory bogie. *International Journal of Vehicle Structures and Systems., 7*, 16–20. https://doi.org/10.4273/ijvss.7.1.03
192. Palli, S., & Koona, R. (2015). Analyses of dynamic response of a railway bogie. *International Journal of Vehicle Noise and Vibration, 11*, 103–113. https://doi.org/10.1504/IJVNV.2015.070015
193. Swathi, K., & Srihari, P. (2015). Heat transfer enhancement in a tube using rectangular strip inserts. *International Journal of Applied Engineering Research, 10*, 41532–41544.
194. Palli, S., Sharma, R. C., & Rao, P. P. D. (2017). Dynamic behaviour of a 7 DoF passenger car model. *International Journal of Vehicle Structures and Systems., 9*, 57–63. https://doi.org/10.4273/ijvss.9.1.12
195. Sharma, R. C., Palli, S., & Koona, R. (2017). Stress and vibrational analysis of an Indian Railway RCF bogie. *International Journal of Vehicle Structures and Systems., 9*, 296–302. https://doi.org/10.4273/ijvss.9.5.06
196. Sharma, R. C., Avesh, M., Sharma, S. K., Sharma, S., Palli, S., & Sharma, N. (2020). Linear and non-linear stability analysis of a constrained railway wheel axle. *International Journal of Vehicle Structures and Systems., 12*, 128–133. https://doi.org/10.4273/ijvss.12.2.04
197. Palli, S., Dontikurti, R., Sharma, R. C., & Sharma, N. (2021). Analysis of dynamic response of a railway locomotive using ANSYS. *International Journal of Vehicle Structures and Systems., 13*, 250–255. https://doi.org/10.4273/ijvss.13.2.21

198. Sharma, R. C., Sharma, N., Singh, G., & Palli, S. (2021). Modeling and simulation of human body-vehicle-track system for the investigation of ride comfort. *SAE International Journal of Commercial Vehicles, 15*. https://doi.org/10.4271/02-15-02-0008
199. Sharma, R. C., Palli, S., Avesh, M., & Sharma, N. (2021). Vibration isolation of railway vehicle car body using semi-active suspension. *International Journal of Vehicle Structures and Systems., 13*, 482–487. https://doi.org/10.4273/ijvss.13.4.17
200. Gopala Rao, L. V. V., Sharma, R. C., Satyanarayana, V. S. V., & Palli, S. (2022). Stochastic optimal preview control response of a quarter-car nonlinear suspension model using spectral decomposition method. *Noise & Vibration Worldwide, 53*, 225–232. https://doi.org/10.1177/09574565221093234
201. Palli, S., Duppala, A., Sharma, R. C., & Rao, L. V. V. G. (2022). Dynamic simulation of automotive vehicle suspension using MATLAB simulink. *International Journal of Vehicle Structures and Systems., 14*, 310–314. https://doi.org/10.4273/ijvss.14.3.04
202. Satyanarayana, V. S. V., Sharma, R. C., Sateesh, B., Gopala Rao, L. V. V., Mohan Rao, N., & Palli, S. (2022). Optimum response of a nonlinear passive vehicle suspension system under random road excitations. *SAE International Journal of Commercial Vehicles, 16*. https://doi.org/10.4271/02-16-01-0004
203. Lokanadham, D., Sharma, R. C., Palli, S., & Bhardawaj, S. (2022). Wear rate modelling and analysis of limestone slurry particulate composites using the fuzzy method. *International Journal on Recent and Innovation Trends in Computing and Communication., 10*, 133–143. https://doi.org/10.17762/ijritcc.v10i1s.5818
204. Sharma, R. C., Palli, S., Gopala Rao, L. V. V., Duppala, A., & Sharma, S. K. (2022). Four-wheel vehicle response under bump, pothole, harmonic, and random excitations using bond graph/simulink technique. *SAE International Journal of Commercial Vehicles, 16*. https://doi.org/10.4271/02-16-02-0008
205. Sharma, R. C., Gopala Rao, L. V. V., Sharma, S. K., Palli, S., & Satyanarayana, V. S. V. (2022). Analysis of lateral stability and ride of an indian railway constrained dual-axle bogie frame. *SAE International Journal of Commercial Vehicles, 16*. https://doi.org/10.4271/02-16-02-0014
206. Sharma, R. C., Palli, S., Bhardawaj, S., & Rao, L. V. V. G. (2022). Advances in human bio-dynamic modeling for analysis of whole-body vibrations. *International Journal of Vehicle Structures and Systems., 14*, 503–511. https://doi.org/10.4273/ijvss.14.4.17
207. Sharma, R. C., Palli, S., & Sharma, S. K. (2023). Ride analysis of railway vehicle considering rigidity and flexibility of the carbody. *Journal of the Chinese Institute of Engineers, Transactions of the Chinese Institute of Engineers, Series A, 46*, 355–366. https://doi.org/10.1080/02533839.2023.2194918
208. Sharma, R. C., Palli, S., & Bhardawaj, S. (2023). A review of wheel-rail dynamic behavior on curved tracks. *International Journal of Vehicle Structures and Systems., 15*, 19–25. https://doi.org/10.4273/ijvss.15.1.04
209. Bammidi, R., Madivada, H., Dowluru, S., Palli, S., Sharma, R. C., & Muddada, V. (2024). Mechano-tribological investigations of Al-%Wt. of boron nitride as metal-ceramic composites. *AIP Conference Proceedings, 2978*(1), 020017. https://doi.org/10.1063/5.0182965
210. Saikiran, C., Palli, S., Bondala, R., Sharma, R. C., Sreeramulu, D., & Dontikurti, R. (2023). Experimental investigation on thermal behaviour of silicon carbide (SiC) based polymer composites under hygrothermal conditions. *International Journal of Vehicle Structures and Systems., 15*, 222–226. https://doi.org/10.4273/ijvss.15.2.15
211. Singh, B. J., Palli, S., Kumar, P., & Bahl, S. (2023). Behavioural investigation of hardness profile for dissimilar TIG-Weldment between P22 and P91 steels. *International Journal of Vehicle Structures and Systems., 15*, 31–36. https://doi.org/10.4273/ijvss.15.1.06
212. Naveen, Y., Lokanadham, D., Naidu, D. R., Sharma, R. C., Palli, S., & Lila, M. K. (2023). An experimental study on the influence of blended karanja biodiesel on diesel engine characteristics. *International Journal of Vehicle Structures and Systems., 15*, 154–160. https://doi.org/10.4273/ijvss.15.2.02

213. Bhardawaj, S., Jha, A. K., Sharma, R. C., Palli, S., Lokanadham, D., Sharma, N., & Dhiman, N. (2023). Finite element analysis of two disk rotor system used in automobile turbochargers. *International Journal of Vehicle Structures and Systems., 15*, 501–504. https://doi.org/10.4273/ijvss.15.4.09
214. Sharma, R. C., Palli, S., Bhardawaj, S., Jha, A. K., Duppala, A., Sharma, N., Sharma, S., & Dhiman, N. (2023). Dynamic behaviour of railway vehicle under bump and pothole track irregularities. *International Journal of Vehicle Structures and Systems., 15*, 566–569. https://doi.org/10.4273/ijvss.15.4.22
215. Raju, S., Palli, S., Devi Prasad, P., Menda, V. R., & Ramakrishna, B. (2024). A hybrid AHP-TOPSIS, MOORA technique for multi-objective optimization of thermal, mechanical, and water absorption behavior of epoxy/hemp, pine apple, and palm fiber composites. *Journal of the Chinese Institute of Engineers, Transactions of the Chinese Institute of Engineers, Series A., 47*, 23–36. https://doi.org/10.1080/02533839.2023.2274092
216. Raju Rallabandi, S., Pilla, D. P., Dowluru, S., Palli, S., Sharma, N., Sharma, S. K., & Sharma, R. C. (2024). Critical evaluation of epoxy-hemp-pineapple-palm fiber composites using hybrid AHM-TOPSIS technique for sustainable structural applications. *Journal of the Chinese Institute of Engineers, Transactions of the Chinese Institute of Engineers, Series A, 47*, 325–336. https://doi.org/10.1080/02533839.2024.2308250
217. Gopala Rao, L. V. V., Satyanarayana, V. S. V., Sharma, R. C., Palli, S., Raturi, A., Kharola, A., & Sharma, A. (2024). Preview control of the random response of a full car vehicle model traversing a rough road. *International Journal of Vehicle Noise and Vibration, 20*, 70–88. https://doi.org/10.1504/IJVNV.2024.138105
218. Sharma, A., Sharma, V. S., Gupta, S., Sharma, R. C., Palli, S., & Sharma, N. (2024). Additive manufacturing in the COVID-19 pandemic: Equipment and challenges? *Journal of Micromanufacturing*. https://doi.org/10.1177/25165984241228083
219. Ranganath, B. A., Duppala, A., Palli, S., Rao, L. V. V., & Sharma, R. C. (2024). Numerical analysis and experimental study of FRP composites for damping applications. *AIP Conference Proceedings, 2978*(1), 060001. https://doi.org/10.1063/5.0184802
220. Sharma, R. C., Palli, S., Duppala, A., Bhardawaj, S., Sharma, S. K., Sharma, N., & Sharma, A. (2024). Rigid body analysis of ride behaviour of roadway vehicle using lagrangian dynamics. *International Journal of Vehicle Structures and Systems., 16*, 74–78. https://doi.org/10.4273/ijvss.16.1.15
221. Menda, V., Dora, S., Palli, S., Rallabandi, S. R., Sharma, R. C., Sharma, N., & Bondala, R. (2024). Mechanical and wear characteristics of aluminium-7075/graphene/TiB2 prepared by stir casting. *Engineering Research Express, 6*, 025571. https://doi.org/10.1088/2631-8695/ad58a8

Composite Fabrication for Industry 5.0 Applications Using Friction Stir Processing

Dinesh Kumar, Shiv Narayan Prajapati, Sanjay Kumar, Ankit Sharma, and Ashok Kumar Yadav

1 Introduction

The friction stir technique is used to technique a wide range of materials by forming surface composites and refining grains. An emerging technology called friction stir processing has made tremendous strides recently. This process evolved from friction stir welding. Since then, this method has seen several advancements.

This technique allows various elements, including magnesium and aluminum, to be processed quickly. Recently, it has also been reported that this technology is used to treat materials like tungsten. Composites are versatile basic building blocks made of two or more materials with distinctly dissimilar physical and mechanical characteristics. These components combine to produce material features that may be altered to meet various purposes. A material's characteristics can be impacted by a variety of constraints, such as production time and cost constraints and dilution reduction. The material is laid back to attain high strength and dilution if it is thin-grained and has a homogeneous grain structure. As an outcome, they must develop to achieve their condition, i.e., small grain size, high strength and dilution, and cost and time requirements for production. Hence, it is essential to create a novel technique. The progressions and governing ideas of FSW and FSP are similar. During treatment, both of these materials use friction heat, and vibration to degrade undecorated plastic which is essential to a homogeneous advanced microstructure.

D. Kumar · S. N. Prajapati · S. Kumar · A. Sharma
Department of Mechanical Engineering, Noida Institute of Engineering and Technology, Greater Noida, India
e-mail: shiv.n.prajapti@niet.co.in

A. K. Yadav (✉)
Department of Mechanical Engineering, Symbiosis Institute of Technology, Symbiosis International (Deemed University), Pune, India
e-mail: ashok.yadav@sitpune.edu.in

Y. Kumar Singla et al. (eds.), *Digital Manufacturing in Industry 5.0*,
https://doi.org/10.1007/978-3-031-91500-0_16

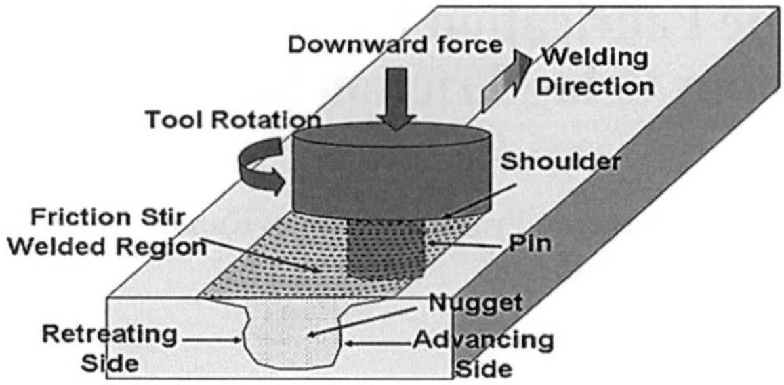

Fig. 1 Friction stir welding process [6]

Particle-reinforced metal matrix composites (MMCs) have higher strength and modulus among several other advantages [1–4]. Usually, the powder metallurgy (P/M) technology or molten metal processing have been used to create and build them.

This article's goal is to review FSP technology as it stands in the area of ex situ and in situ composite manufacturing. There are two parts to the study. The first portion provides some recent research on FSP-based composite fabrication, and the second section concentrates on FSP-based nanocomposites. Due to the differences between strengthening mechanisms operating at the micron and nanosized, this division has been made. For instance, it is thought that the famous deformation-strengthening mechanism contributes significantly more to the material's overall strength [3, 5].

Similarly, it is believed that the presence of uniformly dispersed, tiny precipitate particles of submicron sizes increases the stability of grain structure. During the material's thermomechanical processing, the minute precipitate particles in the matrix allow for the evolution of a finer-grain structure by particle pinning, which also contributes to material strengthening through grain boundary strengthening.

Figure 1 shows the procedure of friction stir welding and all features of the processes. Dispersion of micro- and nano-size particles is typically how MMCs are made ceramic specks in a metal matrix using a variety of methods, such as laser cutting and powder metallurgy technology spray deposition methods, electron beam treatment methods mechanical alloying, casting, etc. Composite materials are made from the liquid phase at high temperatures, which leads to the development of intermetallic reactions, and between the reinforcement and the metal substrate in almost all of the aforementioned procedures.

2 Literature Review

The friction deposition over friction stir welding additive manufacturing technique was developed by Dilip et al. [7]. To deposit the material, they used an AISI 304 austenitic stainless-steel rod over mild steel at a fixed axial force of 9.8 kN and a constant rotational speed of 800 rpm. The investigation found that the average thickness of each deposition was 1 mm, and that the axial decrease of the consumable rod was proportional to the thickness of the layer that formed. The equiaxed fine-grained microstructure was observed as a result of dynamic recrystallization, which had an average grain size of 20 m. The functions of friction stir additively produced AA7075-O (annealed) were examined by Yuqing et al. [8]. They produced nine layers of composite materials using friction stir lap welding on a horizontal milling machine using a GH4169 steel tool.

The speed was 600 rpm and the transverse speed was 60 mm/min at two degrees of inclination. Despite observing a hook between two welded plates in the first layer that extended to the weld nugget zone (WNZ) or thermomechanically affected zone (TMAZ), they attributed it to material flow in the stir zone and concluded that the composites generated hook and kiss bonding flaws. Additionally, they noticed that the material that has been stirred at the bottom mixes with the material at the top, shifting it upward. As a result of the structure that has been created, it has been found that it is harder to crumble the oxide layer. The material flow became more complex due to multiple layers and thermal cycles. The concept of friction deposition has been effectively applied by Karthik et al. [9] to create composite reinforcement. This process involved attaching a consumable rod to the spindle of a friction welding machine that rotated over the base substrate while applying axial force. The friction between the interface surface and high speed caused the oxide layer to be removed. There are no layer interface bonding defects in their finished composite. They exposed their grain bands on the surface of the finished composite using nondestructive techniques like X-ray diffraction and an optical microscope. For friction stir-assisted seam welding, Kalevala et al. [10] have employed sheets of AISI 304 steel, Ti6Al4V, AA6061 alloy, and other dissimilar materials. They suggested that the same method might be applied to the additive manufacture of sheets of various materials. To verify the method, they employed a U-bending test to evaluate mechanical strength and SEM images to examine the created item's microstructure. They were able to refine the grain by 5–10 m in a single pass, as opposed to 60 m prior to processing with a nonconsumable equipment. Rivera et al. used a hollow tool to deposit the IN625 material onto HY80 utilizing a thermomechanical additive manufacturing process (additive friction stirs). They looked at the mechanical and microstructural characteristics of the five-layer multideposition fabrication. They used a step size of 0.1 m and a 40 × 40 m area for their EBSD scans. For testing, they used a servo-hydraulic MTS with a 100 kN load unit and a load 370 load frame. This was the final particle size. A technique for creating user-specific functionally graded metal composites via friction stir processing was presented by Sharma et al. [11]. The feasibility of using friction stir to produce tailored functionally sequential composites was successfully demonstrated by them. They observed that the final microstructure was the result of the three factors working together. Stirring produced heat in the first, resistance produced heat in the second, and grain movement produced heat in the third. The first of the three causes was decided by the process condition; the other two were determined by the reinforced material's particle size. They examined a number of digital image correlation (DIC) test loadings. They also observed that the average grain particle diameter increased in tandem with the number of passes, which also decreased the grain's hardness. For their rapid, scalable, and laminating technique with accurate microstructures, Yu et al. [12] have offered a creative and instructive discussion on additive friction stir deposition. These ideas aided in the production of numerous metal kinds and their layered architectures through the application of the energy deposition technology. The tool parameter that governs or determines the rate of building material is the filler diameter and shoulder size. The current tool parameter for this process in building materials is about 2500 cm^3/h with a diameter of about 1/16″ for Al 6061 alloys, which is comparable to electron base production. In order to help remove the oxide layer between the top metal and the previously deposited method, this solid-state additive manufacturing process requires a frequency of about 20,000 Hz on the top surface of a metal. They observed highly aligned, columnar grains and mechanical properties independent of the generated composites' orientation. Srivastava et al. [13] talked about how to create a 3D build by creatively layering one metal on top of another. They have talked about two distinct additive manufacturing ideas to produce a layered composite: Two types of additive manufacturing are fusion-based and friction-based. The difficulties that businesses based on additive manufacturing products must overcome have been explored. These difficulties include mechanical and structural qualities, raw material accessibility, environmental concerns, and energy issues. They also include manufacturing scale, technology selection, production economics, and communication interfaces. They planned to advocate the use of friction stir additive techniques (FSAM) as an alternative approach for quickly producing parts for the automotive, marine, and aerospace industries. They talked about the distribution of grains in different places and the regions of the composite layer that are thermomechanical influenced when Mg-based WE43 alloy is manufactured by friction stir processing.

Huang et al. [14] successfully made the composite using a solid-state technique called friction stir processing. The microstructure of the resulting composite was able to disclose its unique qualities through the use of EBSD pictures, a light microscope, and a nondestructive method such as X-ray diffraction. The corresponding grain boundary maps from the EBSD picture of the grain structures are shown together with the border misorientation angles of the generated surface composite. Microscopy revealed the tungsten particles' distributed molecular structure across the different base materials, whereas optical microscope images showed the uniformly distributed tungsten carbide particles across the aluminum surface. The grain of their planned composite, which is made of aluminum and tungsten carbide, is uniformly fine and recrystallized. The possibilities and strategies of solid-state additive manufacturing technology (FSP) were evaluated by Khodabakhshi and Gerlach [15]. To determine whether the new technique may be used as an alternative to additive friction stir manufacturing (AFSM), tests were conducted. When compared to traditional metals and alloys, it was shown that AFSM can refine the grain structure of materials, and that these materials also have greater tensile strength and ductility. The created products lacked precise dimensional tolerance but were found to be free of defects such porosity, void, crack, and undesirable phases. Fayazfar et al. [16] produced an evaluation of the additive fabrication of ferrous alloys based on powder. They discovered that the limitations of the laser power bed (LPB) process, which uses powder as the substrate and melts it using a focused laser beam, include a slow building rate and the removal of unbelted powder, while the advantages of laser power-fed (LPF) include proper finishing and a laser beam focused only over the fed area. Binder jetting (BJ) and electron-beam powder-bed fusion (EPBF), two of the most advanced additive manufacturing processes, were also investigated. They observed that LPB was the most commonly utilized process (68%) for AM, while LPF was mostly used for maraging steel. Citing its appealing strength, they explained why TiC is the most commonly used material for reinforcement in LPF. Friction stir processing and additive manufacturing are combined by Mondal et al. [17] to improve the materialistic properties of aluminum 1060 sheets. They filled the groove with aluminum 7075 powder using selective laser melting (SLM), and then they utilized a tool made of H13 steel to treat it with FSP. The produced sample has been subjected to SEM morphology and EBSD examination. They produced the matrix enriched region (MER), reinforcement enriched region (RER), and base metal Al1060 for the sample. Using SEM imaging, they found that there were some porosities and that there was a strong bond between the matrix and the AM layer. By passing the feedstock rods through a hollow tool with four teardrop positions on the tool face for the deposition of a 1 mm layer in four opposing orientations, Phillips et al. [18] used AA6061-T651 to examine the microstructure and deformation. Considerable galling and voids were seen at high transverse velocities and low feed rates. They conducted 2D and 3D EBSD investigations to discuss the grain size and structure. The Al-Mg-Si layer produced by AFSD is shown in the AA6061-T651 3D EBSD map as a cross-section of deposition with the slowest and fastest parameters as well as a deposition displaying refined deposition at the slowest and fastest parameters. The 93% reduction in grain size, or from 200 m to 154 m, was documented by them. They observed a reduction in viscosity and strain around the edge of the rotating tool. Zhang et al. [19] produced a multi-layered material composite using AA6061-T6 and AA-6082 as the substrate. One to two substrate modifications were tested at a time during the experiments. They numerically managed temperature using a shifting heat source model. The temperature of the second layer of the two specimens was observed to be 347.7 °C, whereas the temperature of the one-layer specimen was 392.4 °C. In order to treat the morphologies and grain size, the Monte Carlo model analyzed the precipitate volume percent, which was calculated from the precipitate evolution model and compared with SEM pictures and observations. It compares the grain sizes of the FSAM-produced AA6061 alloy using experimental and numerical models. They noticed that re-stirring caused the grain size to decrease from 15.9 m to 10 m experimentally and from 15.5 m to 9.65 m in the 2-layer specimen. It was found that the second layer's grain size was greater than the first layer, and this was because the second layer's temperature rose less than the first layer did [20]. Multilayer steels were made for performance evaluation by improving the layer structure with high strength and ductility. They used the concept of metal-to-metal laminate to make the fragile steel structures more durable and reduce the thickness of their layers. Their research indicates that multilayer steel can produce materials with substantial fracture elongation (more than 20%) and high tensile strength (greater than 1200 MPa) [21]. The

FSAM approach was applied to two separate parameters to study the multilayered Mg-based WE43 alloy. As a method for additive manufacturing components, they also focus on friction stir technology. They also cover composites with Mg-4 weight percent, Y-3 weight percent, Nd-0.5 weight percent, and Zr in this work which includes a 1.7 mm thick sheet of WE43 alloy. Following observations, a four-sheet, multilayered stack was tested parallel to the Z-direction, with a height of 5.6 mm and three samples. The yield strength was 267 MPa, and the ultimate tensile strength (UTS) was 336 MPa. This multilayer FSAM approach is used to analyze high-performance stringer designs for various aviation components.

Palanivel and Mishra [22] studied the mechanical characteristics of AA 5083 and Mg-4Y-3Nd. They observed increased hardness with 135 HV at the interface in the first case, where they made a four-layered sample with a rotational speed of 800 rpm and a transverse speed of 102 mm/min, whereas the second sample had elevated strength from solute and grain boundaries and an increase in hardness of 18% over the base metal. In this instance, they maintained a 500-rpm rotating speed with a 152 mm/min transverse speed. They also identified each methodology's difficulties and constraints concerning the various additive manufacturing techniques. In Kumar et al. [23], the alloy Al 6082's mechanical properties and microstructure were investigated via friction stir processing. Comparing the generated composite to the basis material, the author observed that it had higher hardness and tensile strength. According to Palanivel and Mishra, there are three main reasons why additive manufacturing has a feudatory function in the transportation sector. The cost of powder feedback, the limited number of alloys that can be produced with the AM process, and the incapacity to produce high-performing structures. The results of their analysis show that methods that do entail melting [24]. Tensile tests and the microhardness profile were used to assess the interface strength of friction-surfaced smooth and rough interfaces. The bending tests revealed minor delamination at the rough surface but no breakage. It was concluded that the initial layer's roughness is consistent with mechanical anchoring taking place along a rough surface.

3 Classification of the Composite Material

Composite material is classified into three types, i.e., particle-reinforced, fibers-reinforced, and structural, and these types are further divided into many types. Figure 2 shows the classifications of composite material.

4 Friction Stir Processing

FSP (Fig. 3) is starting to make an impact in the workplace. The FSP technique is helpful and makes it simple to improve and consolidate the grain structure. Friction stir processing is based on the friction stir welding principle, which was created by

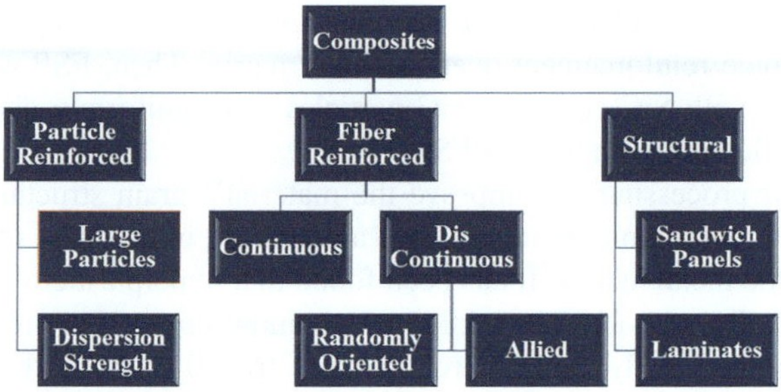

Fig. 2 Classification of composite material

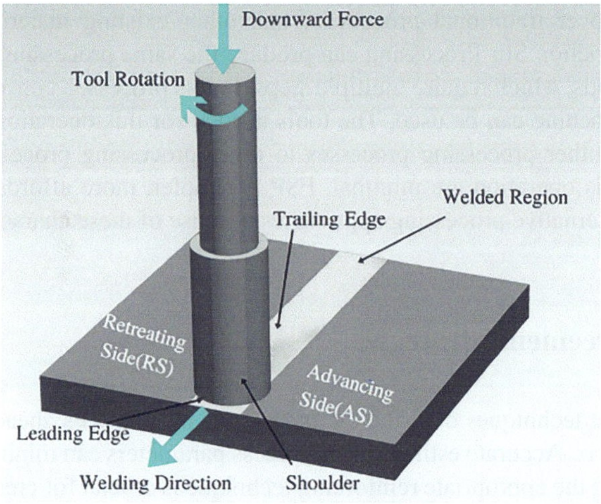

Fig. 3 FSW process [6]

"The Welding Institute" (TWI) in 1991 [25]. Because the material is heated to a temperature below its melting point, friction stir processing is also referred to as the "solid-state process" because it keeps the material from melting at any point. The shoulder and pin arrangement that help compensate the FSP are specifically created about the thickness of the workpiece. While the shoulder gives the specific rotation, the pin is meant to be in the workpiece [1]. The friction created by the tool's contact with the workpiece leads to heat generation. Due to friction, when the shoulder makes contact with the workpiece, it generates too much heat. A precise transverse speed in the desired direction is offered to the shoulder-pin configuration. The workpiece becomes softer because of this heat because it is below the melting point. The pin creates a mechanical stirring, which aids in the creation of the processed material's intensive plastic deformation and results in a microstructure with recrystallized fine-grain grains.

Innovative work on ex situ SCs production of aluminum alloys using FSP was given by Mishra and Mahoney [26]. The external walls of the AA 5083 plate were initially coated with a volatile liquid, like methanol, in order to deposit a thin layer of silicon carbide reinforcement over the base matrix. Then, FSP was applied to these Al plates with SiC coatings. SiC particles were uniformly dispersed in the base matrix following a single pass FSP.

Friction stir processing can improve the material's grain structure and microhardness. Different aluminum alloys have an increase in hardness of two to three times that of the metal matrix. It has been found that hyperplasticity occurs in several aluminum alloys [2, 26]. The traditional techniques for grain refinement include a time-consuming, costly, and environmentally friendly process because it consumes high energy [27]. It is necessary to refine grains using a different technique because of this. The homogeneous structure with an extreme grain structure is produced with the help of the friction stir process. Friction Stir processing has numerous benefits over traditional procedures and other existing material processing techniques. Friction Stir Processing can produce the same processing as other processing methods, which require multiple steps. In this process, a conventional vertical milling machine can be used. The tools needed for this operation are minimal compared to other processing processes to other processing processes, the tools needed for this operation are minimal. FSP is simpler, more affordable, and preferred over alternative processing approaches because of these characteristics.

5 Reinforcement Strategies

Reinforcement techniques describe the practice of placing RPs ahead of FSP over the metal matrix. Accurate estimation of process parameters can minimize SC flaws [28]. Selecting the appropriate reinforcing technique is crucial for creating SCs free of defects, in addition to the previously discussed process factors. After the invention of FSP as a composite fabrication process, a number of researchers investigated 16 different reinforcement schemes for building the SCs. As shown in Fig. 4, different reinforcement schemes that have been described in the literature for the preplacement of RPs on the substrate prior to FSP can be categorized into three primary groups: RPs can be directly pasted onto a metal substrate after being mixed with a volatile solvent, or they can be sprayed onto the substrate using the proper technology, like high-velocity oxy-fuel spray, etc. using grooves, profiles, or holes of different geometries to insert RPs into the metal matrix, using consumable tools having holes drilled in them.

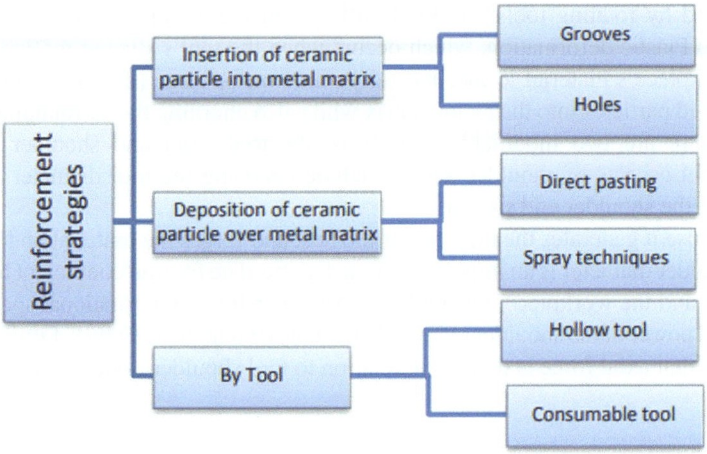

Fig. 4 Classification of reinforcement strategies [25]

6 Factors Affecting the Composite Fabrication Process During FSP

Illustrates various aspects that are involved in the manufacture of composites schematically. Unintentionally, these can be categorized into five groups: variations/strategies that are specific to a particular machine, tool, reinforcement, material, or other factors. As demonstrated in this section, each process parameter makes a varied contribution because some have a greater impact on strengthening a characteristic than others.

6.1 Variables Specific to Machines

Machine-specific variables are those that can be altered by modifying the machine's settings. The rotation speed and travel speed process have the most effects on the quantity of heat produced and the rate at which material moves during machine operation. It has been shown in the literature that combining slower transit speeds with higher rotating speeds can improve particle distribution, material flow, and heat input.

6.2 Tool Variables

Tool geometry has a critical role in influencing the properties of made SCs, including heat generation, material drift course, and reinforcing distribution, in addition to rotational and traverse tool motions [27]. There are two main causes of the heat generated by rotating tools: (1) Friction between the workpiece and the rotating tool. (2) Plastic deformation, which occurs when the tool's stirring motion causes the workpiece's material to mechanically mix, stir, or forge [28]. The tool inserts reinforced particles into the metal matrix while also churning the surrounding material. There are two modifiable aspects of the tool's pin and shoulder design. Important aspects of shoulder design include choosing the tool diameter and the angle of the shoulder end surface.

Because it generates the majority of the heat and forges the material underneath, the shoulder diameter is an important characteristic. The friction coefficient between the tool and the workpiece, the applied pressure or force, the rotational speed, and the third power of the shoulder radius all affect the heat input in *FSW/P*. Furthermore, the way that axial force is defined in relation to tool shoulder radius is provided in

$$q \propto P\,R\,4/3 \tag{1}$$

6.3 Reinforcement Variables

The following reinforcement variables, among others, have a significant impact on the qualities of the SCs that will be manufactured (Fig. 5):

The properties of materials, such as their mechanical and thermal properties, have a big impact on the process parameter selection. The tool's shear effects are influenced by the material's mechanical properties, while the tool's heat generation is influenced by the material's thermal conductivity [25, 29–34]. The thickness of the material and its thermal properties also affect the rate of cooling and temperature gradients.

6.4 Tool

The friction stir progression (FSP) tool plays a major role in the strategy's effectiveness. The instrument usually features a strung cylindrical pin that warms the surrounding material and frequently deforms the ground region, as well as a rotating spherical shoulder. The tool's various profiles—circular, conical, triangular, square, pentagon, hexagon, threaded, etc.—are displayed in Fig. 6. The tool's fundamental probe and the tool profile's outer and end surfaces are depicted in Fig. 7. Typically, high-speed steel is used for aluminum or comparable materials; the temperature of the tool material is always higher than that of the matrix material, and the tool material varies with respect to the matrix material [35–42].

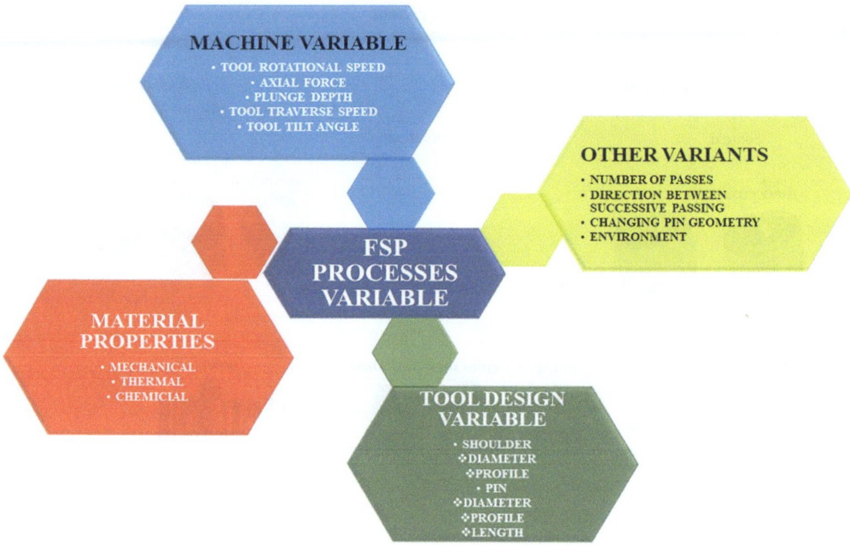

Fig. 5 Variables used in SC fabrication

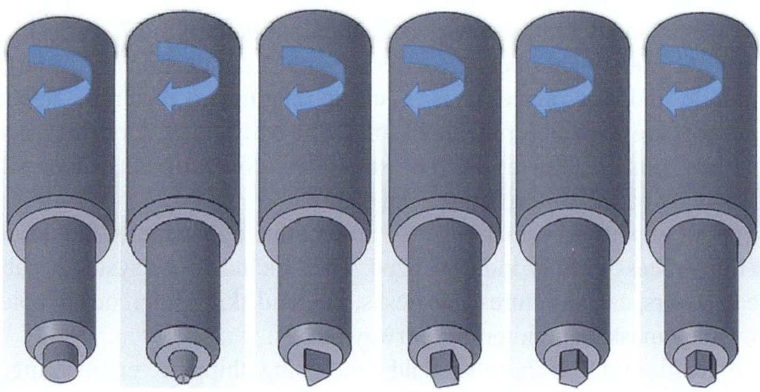

Fig. 6 Different type of tool profile [6]

7 Application

The friction stir-processed superplasticity has been noted by Weglowski as offering new possibilities for FSP technology.

- Friction stir-processed superplastic Al alloy can be used to create structures or parts that are unitized (without joints or attachments) [43–48].
- The usage of friction stir-processed superplastic aluminum alloys in innovative materials and applications will help the rail, shipping, automobile, and aerospace sectors.

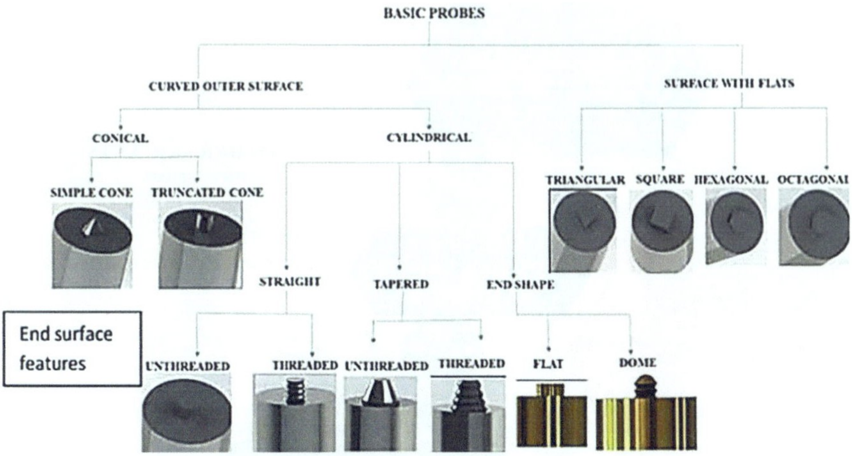

Fig. 7 Pin profiles [25]

- Lightweight aircraft structure components like wing tips, winglets, engine nacelle skins, lip skins, equipment covers, air intakes, interior décor, door skins, and lamp drums have been identified and produced by the US business Super Form, which manufactures superplastic aluminum alloy [2, 27, 49–54].
- Superplastic material has been a barrier to popular applications in the large-scale automobile industry since it is a cheap beginning material. Superplastic components are utilized for the inside of doors, body sides, door rings, enclosure assemblies, tunnels, can trails, roof structures, compartments, door sills, and exterior body panels on certain special cars.
- The cost of superplastic materials for automotive applications should decrease once this process is fully industrialized. They include door systems, cab interiors, cab covers, cab structures, cab desks, interend skins, interend canopies, interior window masks, and interior doorway panels.
- The application of FSP in other fields, including shipping, engineering, accessories, and medicinal implants, could also be investigated.

8 Conclusions

- This research analyzes the history and current development of FSP of metallic and polymeric substrates as well as FSP of composites. Some of the important results reached from an exhaustive analysis of significant published work are mentioned below.
- The main categories of FSP variables are tool variables, reinforcement variables, reinforcement methods and machine variables. The importance of these FSP factors and the nature of their interactions in improving process effectiveness and

surface features when surface adaptations are applied to the substrate are acknowledged and briefly reviewed.

- A relevant research study indicates that 90–95% of published material is about MMCs, while only 5–10% of literature discusses the manufacture of PMCs. It has also been stated that creative tooling design has been used to address the problem of varied melting point temperatures at various polymer portions. It will need a significant amount of research to make PMC fabrication simpler and more competitive with MMCs.
- FSP applications can be expanded to produce intricately curved structures employing cutting-edge friction stir-based hybrid techniques.
- A significant issue with the synthesis of composites and FGS is the aggregation of RPs. This article explains the main problems with powder agglomeration disposal and suggests critical process variables that might significantly lessen this occurrence.

9 Future Scope

- FSP is a relatively new concept in the material processing industry. It has been demonstrated that FSP enhances the microstructure and produces noticeably greater elongation at high strain rates over a variety of temperatures. Research indicates that FSP outperforms other forms of extreme plastic deformation. There aren't many published studies about the feasibility of achieving superplasticity in high-strength Al alloys.
- It is clear that FSP is currently a recognized method for creating MMCs. Due to its solid-state nature, FSP provides a multitude of applications in the production of polymer matrix composites (PMCs). Better tooling methods have been used in the literature to document some of PMCs' work compared to traditional FSP. To become a reliable technique for creating PMCs, further effort is necessary as it is still in the early phases of PMC development. FSP is becoming more and more significant in the production of metallic foams in addition to composites. There aren't many studies on the production of metallic foams in the literature.
- The development of highly porous materials utilizing FSP, however, requires additional investigation. Additionally, interesting outcomes can be obtained from ongoing research.
- FSP can be utilized in the production of ferromagnetic composites. The creation of materials with many uses, such as magnetic ceramic composites and piezoelectric ceramic composites (metal/polymer matrix), is seen as having enormous potential by the authors. Additionally, the electronic sector can effectively use FSP, especially when developing sensors like as humidity and noise sensors, among others.
- It is possible to create heat exchangers using FSP. Furthermore, FSP has a lot of potential for creating useful materials with improved properties. Combining FSP with modern manufacturing methods such as friction stir additive manufactur-

ing, ultrasonic assisted friction stir processing, etc., can accomplish this. Leading to significant advancements in the direction of 34 applications of innovative and contemporary manufacturing engineering.
- The fabrication of structural metals can be effectively done via FSAM. The qualities of SCs can be maintained as FSP speeds are increased using ultrasonic-assisted friction stir processing, and the processed material's properties benefit from the ultrasonic vibrations. FSP can also be effectively used for mechanical alloying because it is a surface modification method.

References

1. Duan, Z., Wu, T., Glowacz, A., Gupta, M. K., & Królczyk, G. (2023). Analysis of the line contact tribo-lubrication pair and failure mechanism under the extreme conditions. *Tribology International, 185*, 108505.
2. Singh, T., Aherwar, A., Ranakoti, L., Bhandari, P., Singh, V., & Lendvai, L. (2023). Performance optimization of lignocellulosic fiber-reinforced brake friction composite materials using an integrated CRITIC-CODAS-based decision-making approach. *Sustainability, 15*(11), 8880.
3. Abbas, A. T., Sharma, N., Alsuhaibani, Z. A., Sharma, V. S., Soliman, M. S., & Sharma, R. C. (2023). Processing of Al/SiC/Gr hybrid composite on EDM by different electrode materials using RSM-COPRAS approach. *Metals, 13*(6), 1125.
4. Chand, R., Sharma, V. S., Trehan, R., & Gupta, M. K. (2023). Numerical investigations on mechanical properties of bio-inspired 3D printed geometries using multi-jet fusion process. *Rapid Prototyping Journal, 29*(6), 1212–1229.
5. Sivalingam, V., Zhou, Q., & Selvam, B. (2023). A mathematical approach of evaluating sustainability indicators in milling of aluminium hybrid composite by different eco-friendly cooling strategies. *Sustainable Materials and Technologies, 36*, e00605.
6. Patel, V. V., Badheka, V., & Kumar, A. (2016). Friction stir processing as a novel technique to achieve superplasticity in aluminum alloys: Process variables, variants, and applications. *Metallography, Microstructure, and Analysis, 5*, 278–293. https://doi.org/10.1007/s13632-016-0285-x
7. Dilip, J. J. S., Janaki Ram, G. D., & Stucker, B. E. (2012). Additive manufacturing with friction welding and friction deposition processes. *International Journal of Rapid Manufacturing, 3*(1), 56–69.
8. Yuqing, M., Liming, K., Chunping, H., Fencing, L., & Qiang, L. (2016). Formation characteristic, microstructure, and mechanical performances of aluminium-based components by friction stir additive manufacturing. *The International Journal of Advanced Manufacturing Technology, 83*(9), 1637–1647.
9. Karthik, G. M., Panikar, S., Ram, G. J., & Kottada, R. S. (2017). Additive manufacturing of an aluminium matrix composite reinforced with nanocrystalline high-entropy alloy particles. *Materials Science and Engineering: A, 679*, 193–203.
10. Kalvala, P. R., Akram, J., & Misra, M. (2016). Friction-assisted solid state lap seam welding and additive manufacturing method. *Defence Technology, 12*(1), 16–24.
11. Rivera, O. G., Allison, P. G., Jordon, J. B., Rodriguez, O. L., Brewer, L. N., McClelland, Z., Whittington, W. R., Francis, D., Su, J., Martens, R. L., & Hardwick, N. (2017). Microstructures and mechanical behaviour of Inconel 625 fabricated by solid-state additive manufacturing. *Materials Science and Engineering: A, 694*, 1–9.
12. Yu, H. Z., & Mishra, R. S. (2021). Additive friction stir deposition: A deformation processing route to metal additive manufacturing. *Materials Research Letters, 9*(2), 71–83.

13. Srivastava, A. K., Kumar, N., & Dixit, A. R. (2021). Friction stir additive manufacturing—An innovative tool to enhance mechanical and microstructural properties. *Materials Science and Engineering: B, 263*, 114832.
14. Huang, C., Li, W., Feng, Y., Xie, Y., Planche, M. P., Liao, H., & Montavon, G. (2017). Microstructural evolution and mechanical properties enhancement of a cold-sprayed CuZn alloy coating with friction stir processing. *Materials Characterization, 125*, 76–82.
15. Khodabakhshi, F., & Gerlich, A. P. (2018). Potentials and strategies of solid-state additive friction-stir manufacturing technology: A critical review. *Journal of Manufacturing Processes, 36*, 77–92.
16. Fayazfar, H., Salarian, M., Rogalsky, A., Sarker, D., Russo, P., Paserin, V., & Toyserkani, E. (2018). A critical review of powder-based additive manufacturing of ferrous alloys: Process parameters, microstructure and mechanical properties. *Materials & Design, 144*, 98–128.
17. Mondal, M., Das, H., Hong, S. T., Jeong, B. S., & Han, H. N. (2019). Local enhancement of the material properties of aluminium sheets by a combination of additive manufacturing and friction stir processing. *CIRP Annals, 68*(1), 289–292.
18. Phillips, B. J., Mason, C. J. T., Beck, S. C., Avery, D. Z., Doherty, K. J., Allison, P. G., & Jordon, J. B. (2021). Effect of parallel deposition path and interface material flow on resulting microstructure and tensile behaviour of Al-Mg-Si alloy fabricated by additive friction stir deposition. *Journal of Materials Processing Technology, 295*, 117169.
19. Zhang, Z., Tan, Z. J., Li, J. Y., Zu, Y. F., Liu, W. W., & Sha, J. J. (2019). Experimental and numerical studies of re-stirring and re-heating effects on mechanical properties in friction stir additive manufacturing. *The International Journal of Advanced Manufacturing Technology, 104*(1), 767–784.
20. Koseki, T., Inoue, J., & Nambu, S. (2014). Development of multilayer steels for improved combinations of high strength and high ductility. *Materials Transactions, 55*, M2013382.
21. Palanivel, S., Nelaturu, P., Glass, B., & Mishra, R. S. (2015). Friction stirs additive manufacturing for high structural performance through microstructural control in an Mg-based WE43 alloy. *Materials & Design (1980–2015), 65*, 934–952.
22. Palanivel, S., & Mishra, R. S. (2017). Building without melting: A short review of friction-based additive manufacturing techniques. *International Journal of Additive and Subtractive Materials Manufacturing, 1*(1), 82–103.
23. Kumar, N., Gupta, P., & Singh, R. K. (2022). Fabrication of Al 6082/SiC composite using friction stir processing. *Journal of The Institution of Engineers (India): Series D, 104*, 603. https://doi.org/10.1007/s40033-022-00413-2
24. Galvis, J. C., Oliveira, P. H. F., Martins, J. D. P., & Carvalho, A. L. M. D. (2018). Assessment of process parameters by friction surfacing on the double-layer deposition. *Materials Research, 21*.
25. Rathee, S., Maheshwari, S., & Siddiquee, A. N. (2018). Issues and strategies in composite fabrication via friction stir processing: A review. *Materials and Manufacturing Processes, 33*(3), 239–261. https://doi.org/10.1080/10426914.2017.1303162
26. Mishra, R. S., & Mahoney, M. W. (2007). *Friction stir welding and processing*. ASM International.
27. Yevdokymov, O., Kolesnyk, V., Peterka, J., Vopat, T., Gupta, M. K., Lisovenko, D., & Dovhopolov, A. (2023). Pareto analysis of machining factors significance when turning of Nickel-based superalloy Inconel 718. *Metals, 13*(8), 1354.
28. Korkmaz, M. E., Gupta, M. K., Günay, M., Boy, M., Yaşar, N., Demirsöz, R., et al. (2023). Comprehensive analysis of tool wear, surface roughness and chip morphology in sustainable turning of Inconel-601 alloy. *Journal of Manufacturing Processes, 103*, 156–167.
29. Uniyal, S., Mangla, S. K., Sarma, P. R., Tseng, M. L., & Patil, P. (2021). ICT as "Knowledge management" for assessing sustainable consumption and production in supply chains. *Journal of Global Information Management, 29*(1), 164–198.

30. Ramesh, G., Uvaraja, V. C., Gnanasundara Jayaraja, B., & Patil, P. P. (2022). Fatigue, fracture toughness and DMA of biosilica toughened epoxy with stacked okra fiber and Al 2024-T3 fiber metal laminate composite. *Polymer Composites, 43*(9), 6300–6309.
31. Ramaswamy, R., Gurupranes, S. V., Kaliappan, S., Natrayan, L., & Patil, P. P. (2022). Characterization of prickly pear short fiber and red onion peel biocarbon nanosheets toughened epoxy composites. *Polymer Composites, 43*(8), 4899–4908.
32. Kumar, U., Padalia, D., Bhandari, P., Kumar, P., Ranakoti, L., Singh, T., & Lendvai, L. (2022). Fabrication of europium-doped barium titanate/polystyrene polymer nanocomposites using ultrasonication-assisted method: Structural and optical properties. *Polymers, 14*(21), 4664.
33. Arun, A. P., Kaliappan, S., Natrayan, L., & Patil, P. P. (2022). Mechanical, fracture toughness, and dynamic mechanical properties of twill weaved bamboo fiber-reinforced Artocarpus heterophyllus seed husk biochar epoxy composite. *Polymer Composites, 43*(11), 8388–8395.
34. Vinayaka, N., Nagabhooshanam, N., Balaji, D., Verma, R., & Patil, P. P. (2022). Effect of silanized stacked Kevlar fiber/brass 270 wire mesh on mechanical and impact toughness of cubic boron nitride–epoxy composite. *Polymer Composites, 43*(8), 5159–5167.
35. Nagabhooshanam, N., Kumar, A., Rao, P. K. V., Patil, P. P., & Kala Bharathi, B. V. V. L. (2022). Preparation and characterization of opuntia-cladode fiber and citron peel biochar toughened epoxy biocomposite. *Polymer Composites, 43*(9).
36. Subbiah, R., Kaliappan, S., Balaji, V., & Patil, P. P. (2022). Effect of nanosilica on mechanical, thermal, fatigue, and antimicrobial properties of cardanol oil/sisal fiber reinforced epoxy composite. *Polymer Composites, 43*(11), 7940–7951.
37. Patil, P. P. (2022). Effect of black rice husk ash biosilica on mechanical, wear, and fatigue behavior of stacked aloevera/roselle and glass fiber reinforced epoxy composite. *Polymer Composites, 43*(10), 7335–7344.
38. Singh, T., Patnaik, A., Ranakoti, L., Dogossy, G., & Lendvai, L. (2022). Thermal and sliding wear properties of wood waste-filled poly (lactic acid) biocomposites. *Polymers, 14*(11), 2230.
39. Senthil Kumar, S., Anil Kumar, T. C., Chokkalingam, M., Kaliappan, S., Boopathi, S., Srinivas, C., & Patil, P. P. (2022). Mechanical, fracture toughness, and fatigue behavior of spinifex littoreus fiber on echinoidea-spike toughened epoxy composite. *Polymer Composites, 43*(4), 2329–2340.
40. Rajan, R., Tyagi, Y. K., Pruncu, C. I., Kulshreshtha, S., Ranakoti, L., & Singh, T. (2022). Tribological performance evaluation of slag waste filled phenolic composites for automotive braking applications. *Polymer Composites, 43*(10), 7118–7129.
41. Singh, T., Pattnaik, P., Aherwar, A., Ranakoti, L., Dogossy, G., & Lendvai, L. (2022). Optimal design of wood/rice husk-waste-filled PLA biocomposites using integrated CRITIC–MABAC-based decision-making algorithm. *Polymers, 14*(13), 2603.
42. Rajawat, A. S., Singh, S., Gangil, B., Ranakoti, L., Sharma, S., Asyraf, M. R. M., & Razman, M. R. (2022). Effect of marble dust on the mechanical, morphological, and wear performance of basalt fibre-reinforced epoxy composites for structural applications. *Polymers, 14*(7), 1325.
43. Gariya, N., Kumar, P., & Singh, T. (2023). Experimental study on a bending type soft pneumatic actuator for minimizing the ballooning using chamber-reinforcement. *Heliyon, 9*(4), e14898.
44. Moj, K., Owsiński, R., Robak, G., & Gupta, M. K. (2023). Determination of cell size/structures and mechanical properties of additively manufactured metallic components using X-ray technique. *Rapid Prototyping Journal, 29*(10), 2039–2060.
45. Derda, S., Karolczuk, A., Robak, G., Prażmowski, M., Paul, H., Łagoda, T., & Gupta, M. K. (2023). Impact of the interface on the fatigue life of steel-based explosively welded heterostructured plates. *Archives of Civil and Mechanical Engineering, 23*(3), 191.
46. Pratheesh, K., Gurupranes, S. V., Kumar, A., & Patil, P. P. (2023). Effect of corn husk fiber length and diameter on load bearing, fatigue, and DMA properties of biosilica toughened epoxy composite. *Polymer Composites, 44*(6), 3469–3476.
47. Natrayan, L., Surakasi, R., Kaliappan, S., Patil, P. P., Saravanan, A., & Siranjeevi, R. (2023). Explore the elimination of toxic dye using Bio-Synthesized cerium oxide nanoparticles derived from Morinda citrifolia leaf extracts. *Surfaces and Interfaces, 41*, 103151.

48. Ross, N. S., Mashinini, P. M., Rai, R., & Gupta, M. K. (2023). Carbon nano dots mixed rice bran oil as a cutting fluid for enhanced lubrication/cooling in milling of additively manufactured 316 stainless steel. *Journal of Molecular Liquids, 391*, 123200.
49. Sahu, K., Sharma, S. C., & Tomar, A. K. (2023). Analysis of MR fluid lubricated slot entry hybrid conical journal bearing with texturing arrangements. *Tribology International, 188*, 108788.
50. Singh, T., Singh, V., Ranakoti, L., & Kumar, S. (2023). Optimization on tribological properties of natural fiber reinforced brake friction composite materials: Effect of objective and subjective weighting methods. *Polymer Testing, 117*, 107873.
51. Nieslony, P., Wojciechowski, S., Gupta, M. K., Chudy, R., Krolczyk, J. B., Maruda, R., & Krolczyk, G. M. (2023). Relationship between energy consumption and surface integrity aspects in electrical discharge machining of hot work die steel. *Sustainable Materials and Technologies, 36*, e00623.
52. Kaliappan, S., Velumayil, R., Natrayan, L., & Pravin, P. (2023). Mechanical, DMA, and fatigue behavior of Vitis vinifera stalk cellulose Bambusa vulgaris fiber epoxy composites. *Polymer Composites, 44*(4), 2115–2121.
53. Danish, M., Rubaiee, S., Gupta, M. K., Yildirim, M. B., & Ahmed, A. (2023). Technological and tribological characteristics improvement of additively manufactured SS 316L components machined under sustainable cooling conditions. *Tribology International, 181*, 108329.
54. Jamil, M., He, N., Wei, Z., Gupta, M. K., & Khan, A. M. (2023). A novel quantifiable approach of estimating energy consumption, carbon emissions and cost factors in manufacturing of bearing steel based on triple bottom-line approach. *Sustainable Materials and Technologies, 36*, e00593.
55. Sharma, A., Bandari, V., Ito, K., Kohama, K., Ramji, M., & Himasekhar Sai, B. V. (2017). A new process for the design and manufacture of tailor-made functionally graded composites through friction stir additive manufacturing. *Journal of Manufacturing Processes, 26*, 122–130.

Numerical Simulation of Pressure and Temperature Characteristics in Fire-Tube Boilers Utilizing Alcohol-Based Fuels for Optimized Energy Efficiency in Smart Manufacturing

Ramajogi Naidu Doddi, Raminaidu Pisini, Mendi Venkanna Babu, Bathula Venkata Siva, Abiodun Oluwatosin Adeoye, and Samson Abiodun AASA

1 Introduction

Renewable energy generation is taking center stage in global endeavors to address climate change and mitigate greenhouse gas emissions. Out of different sectors undergoing this shift, many industries, like transport and manufacturing, are seeking safe substitutes for fossil fuels. Alcohol fuels, predominantly ethanol and methanol, are being considered as clean alternatives since they can be easily burned and contribute to lower carbon dioxide emissions. Such fuels, produced from biomass and other renewable materials, can be efficiently applied in energy-dense processes such as industrial heating and power generation.

An example of an industrial boiler that is extensively used to generate steam is a fire-tube boiler. As part of systems in smart factories, these boilers are critical. Their function involves the passage of hot flue gases through several tubes that flow in water—steam is produced for energy production. Understanding the thermal and pressure processes inside the fire-tube boiler when expanding to other types of cement fuels, such as alcohols, is fundamental. There are several key factors that affect fire-tube boiler efficiency, such as fuel type, combustion process, and

R. N. Doddi (✉) · R. Pisini
Department of Mechanical Engineering, Aditya Institute of Technology and Management, Tekkali, Andhra Pradesh, India

M. Venkanna Babu · B. Venkata Siva
Narasaraopeta College of Engineering, Narasaraopeta, Andhra Pradesh, India
e-mail: mechhod@nrtec.in

A. O. Adeoye
Department of Chemistry, Federal University Oye Ekiti, Oye-Ekiti, Ekiti State, Nigeria

S. A. AASA
Department of Automotive Engineering, University of Ibadan, Ibadan, Nigeria

Y. Kumar Singla et al. (eds.), *Digital Manufacturing in Industry 5.0*,
https://doi.org/10.1007/978-3-031-91500-0_17

properties, which should be taken into consideration to improve the energy output and sustainability of the systems and processes.

These works emphasize a variety of advanced engineering techniques that contribute to the development of smart manufacturing and Industry 5.0. Studies like Shah et al. [1] and Singla et al. [2] examine the effects of material science advancements on welding and sustainability, key components for industrial automation and optimized production processes. Additionally, research on solar energy systems and tribological studies on engine lubricants [3–9] align with the sustainable energy goals of Industry 5.0. Additionally, advancements in digital twin technology for improving logistics [10], the role of extremophiles in enhancing sustainable agriculture [11, 12], and bioremediation of contaminated soils [13, 14] highlight the role of digital innovation and environmental consciousness in next-generation manufacturing. These studies offer diverse insights into how microbiological advancements and environmental sustainability are being harnessed for smart and sustainable industrial applications, resonating with the principles of Industry 5.0 [15, 16]. The integration of alloying elements to enhance wear resistance [17–19] supports the development of materials with improved durability for high-tech manufacturing applications. Moreover, works on the surface integrity of machined materials and the impact of nano-particulates in lubrication show how material engineering can improve efficiency in production processes [20–22]. These studies collectively enhance the scope of smart manufacturing by combining material science, energy efficiency, and automation, advancing the transition toward Industry 5.0 [23, 24].

Works from earlier years explore dynamic simulations [25, 26], improving suspension systems, and enhancing material wear resistance, which collectively reflect advancements in digitalization, automation, and sustainable manufacturing practices. These studies highlight a multidisciplinary approach to improving vehicle systems and material performance in line with the goals of Industry 5.0.

These studies explore a range of advanced techniques and sustainable practices that align with the goals of smart manufacturing and Industry 5.0. Adeoye et al. (2024) investigate the valorization of lignocellulosic waste, highlighting the potential of bio-oil production through pyrolysis and ultrasonication for biofuel generation [27]. Further research by Adeoye et al. focuses on upgrading crude bio-oil [28, 29] and sustainable energy production from biomass[30], providing insights into thermochemical and biochemical conversions. These approaches underscore the integration of clean energy solutions into digital manufacturing processes[31–36]. Studies on catalytic pyrolysis and nanomaterials for enhanced biomass conversion offer pathways for optimizing manufacturing efficiency [37–39]. Additionally, research on pyrolysis of plastic waste and rail projects introduces circular economy concepts and environmental sustainability, both essential to the evolution of Industry 5.0 [40, 42, 44–46, 52, 53]].

In parallel, work on fiber surface functionalization and dye fastness adds to the understanding of material innovation in manufacturing [54–65]. These studies contribute to the body of knowledge driving digital transformation and environmental consciousness in manufacturing [61, 63, 65–67, 73, 74].

The contributions from Sunil et al. [54–65] span critical advancements in transportation systems, materials science, and engineering technology, all of which have applications in smart manufacturing and Industry 5.0. Recent works from 2024 focus on the adoption of electric vehicles [26, 66–82], modeling of intelligent transportation systems [83–96], and vibration mitigation in marine and rail propulsion systems [97–106]. Recent advancements in sustainable fuel production have focused on various bio-oil sources and carbon capture techniques. Studies have examined bio-oil production from thermal cracking in the presence of hydrogen [107], CO_2 adsorption systems in coal beds [108], and the pyrolysis of coffee and eucalyptus sawdust [109–111]. Additional research has explored the use of sulfur to enhance wood tar biopitch as a sustainable binder, the characterization of sewage sludge bio-oil, and the analysis of bio-oil from algae, comparing freshwater and microbial-treated samples [112]. These efforts collectively contribute to the development of renewable energy and environmental sustainability [40, 52, 53, 113–115].

These works highlight innovative methods in microbiology, environmental sustainability, and materials science, which have significant implications for smart manufacturing and Industry 5.0 [31–36]. Mohapatra et al. [116] study microbial processes for biopolymer production, which can be applied in sustainable packaging solutions [33, 40, 117–124] and bio-based materials [125–136], aligning with Industry 5.0's focus on circular economy and smart materials. Research on retrofitting buildings for energy efficiency [137–153] and the use of polyhydroxyalkanoates (PHAs) for biocompatible applications [154–163] demonstrates the integration of biotechnology into industrial processes.

These studies underscore the need for energy-efficient solutions and advanced control systems, both of which are integral to Industry 5.0 [164, 165]. Previous works from 2023 also explore sustainable materials for transportation [166], intelligent systems for railway infrastructure [16, 167–182], and advancements in modeling and ride behavior [183–191]. Research on crashworthiness [192–201] and the integration of bio-lubricants [202–205] promote the goals of Industry 5.0 by improving safety [206], efficiency, and sustainability [207–209]. These studies collectively support the evolution of smart manufacturing, contributing to Industry 5.0's emphasis on sustainability, digitalization, and innovation across sectors [210].

The research contributions focus on the dynamic analysis, vibration isolation, and advanced material applications in transportation and vehicle systems, which directly support the objectives of smart manufacturing and Industry 5.0 [211–215]. Recent works (2024) address advanced control systems for enhanced ride quality in passenger rail vehicles using intelligent control suspension systems [171, 181, 216–219], and comprehensive dynamic response analyses of road and rail vehicles [220–229]. Further, studies involving the use of composite materials and hybrid AHP-TOPSIS techniques for sustainable structural applications [74, 77, 80, 82, 230–240] contribute to the development of smart materials, which is central to Industry 5.0.

The utilization of alcohol fuels in fire-tube boilers presents both challenges and opportunities. For example, alcohols like ethanol and methanol burn differently from hydrocarbons—they have a lower energy density and different flame temperatures. Such characteristics affect heat transfer within the system, and thus, operational parameters must be set at a certain level to achieve efficiency. The central focus of this chapter is the modeling of the pressure and temperature fields in fire-tube boilers applied for the combustion of alcohol fuels for the purpose of optimal energy efficiency in smart factories. Table 1 summarizes the key properties of common alcohol-based fuels compared to traditional fossil fuels, highlighting their potential impact on combustion efficiency and emissions.

New technologies such as artificial intelligence (AI) and automation flourish in all industries and contribute to their sustainability and efficiency. Targeted analysis of these technologies offers great promise for enhancing biofuels production and biofuel's environmental aspects. For example, Adeoye et al. (2024) proved that lignocellulosic residue can be turned into biofuels using pyrolysis techniques, whereas Quadri et al. (2024) sought nanoparticles in enhancing energy conversion of palm kernel shells. The general studies of the group report of Adeoye et al. (2024) as well, who investigated plastic wastes' pyrolysis to biofuels production, which, in turn, sustains a circular economy, while further outlining the necessity of smart manufacturing. Alongside them, other ecological measures, namely, in the rail infrastructural project, highlight how the construction sector is in the concert of sustainable demolition practices. All these latter activities show how the industry is moving toward increased production of biofuels and energy efficiency improvement, the key factors for decreasing carbon emissions in the industry and attaining environmental objectives.

There are also up-and-coming technologies involved in clean fuel generation such as bio-oil production and carbon capture. The recent studies focused on hydrogen-assisted thermal cracking of bio-oils and biomass pyrolysis that involve sources like coffee, eucalyptus sawdust, and algae, even expanding the types of renewable energy resources available. CO_2 adsorption systems in coal beds, zeolitic materials, and montmorillonite clay as sustainable materials, and sulfur enhancement in wood tar biopitch are some of the equally important areas of research needed to resolve energy production-related environmental issues. Notable among these studies are those on the production of biodiesel, synthetic jet fuel, and high-value chemical extraction from bio-oils that demonstrate biosystems as biofuels could find use in many industrial sectors.

In addition, the use of ANSYS software and other computational instruments allows for the thermal analysis of manufactured elements through simulation. This method allows for the analysis of the heat flow in the components ensuring their

Table 1 Key properties of common alcohol-based fuels compared to traditional fossil fuels

Fuel type	Energy content (MJ/L)	CO_2 emissions (g CO_2 /MJ)	Octane rating	Boiling point (°C)	Renewable source
Ethanol	26.8	62.0	108	78.37	Yes
Methanol	20.0	77.0	106	64.7	Yes
Gasoline	31.0	89.0	87	27.5	No
Diesel	35.0	86.0	83	180	No

performance and durability over time, especially in green ether-fueled systems. By developing advanced automotive and industrial fuel systems, these studies assist in lowering emissions and enhancing fuel efficiency, thereby contributing to the overall aim of developing clean technologies for contemporary transport.

While the authors acknowledge these technological innovations and sustainable approaches within the realm of alcohol-based fueled fire-tube boilers, it has the potential of improving both the energy efficiency and environmental sustainability of smart manufacturing. This chapter advances these innovations and develops how alcohol-based fuels and optimized manufacturing processes together can accelerate the shift toward efficiency and cleaner energy systems.

A small fire-tube boiler was designed using SolidWorks and subjected to CFD analysis using ANSYS Fluent to determine the effect of variation of three basic parameters on the performance of the boiler. In particular, the study examines pressure distributions along the length of the water in the boiler due to various water velocities, in the range of 25–40 m/s. Furthermore, the influence of several casing materials, i.e., brass, stainless steel, and steel, on the heat transfer rate and the pressure dynamics of the steam was also looked into. Such information is important as it helps in defining how these materials interact with the operational parameters of the boiler and it further aids in meeting the objectives of improving system performance while preserving alcohol fuels' viability as a cleaner energy source.

In this study, as alternative fueled boiling systems are presented for the first time, CFD tools help provide insight into fuel combustion and boilers, thus making a case for the use of the tools. More importantly, this study shows how combustion gases and water from the operational area of the system can work together in its structure, expanding the studies of alcohol fuels to include operational tolerances. It is especially crucial to maintain the fire-tube boiler efficiency and safety while limiting emissions and practicing sustainable manufacturing. Thus, this research assists in the advancement of clean fuel alternatives for industrial applications in fire-tube boilers by investigating the feasibility of using alcohol-based fuels. Through numerical modeling, this investigation seeks to optimize conditions for heat transfer and emissions control while ensuring appropriate and effective operation of the boilers in the context of intelligent manufacturing.

2 Methodology

This segment describes the procedure followed in the investigation, starting from the initial steps of designing and modeling a fire-tube boiler, through to its final stages of simulation and analysis using computational fluid dynamics (CFD) in ANSYS Fluent. The process begins with the construction of the fire-tube boiler prototype on a small scale in SolidWorks, a highly developed computer-aided design (CAD) software.

The design phase concentrates on the following components—boiler shell, fire tubes, water flow domain, and exhaust flow domain, such that each domain is modeled independently to achieve accurate measurements and specifications. When the several elements have been modeled or accomplished, they are integrated to form a complete or full boiler structure, which is then saved as a .STEP file suitable for ANSYS Fluent for further mesh generation and simulation processes.

2.1 Design of the Fire-Tube Boiler Prototype

The construction commenced by creating the different parts of the boiler in SolidWorks, which ensures that the boiler shell, fire tubes, water flow domain, and exhaust flow domain have the required operational and dimensional criteria. Together with the water and steam-bearing design functionality, the boiler shell is made strong enough for the working temperatures and pressures associated with the boiler unit [174–203].

The walls of the fire tubes, which carry the hot combustion gases, are also designed to determine the efficiency of heat exchange. This is the water flow domain, as heat has to be transferred to the water from the combustion gases passing through the fire tubes. The flue gas discharge domain is responsible for the effective expulsion of the gases from the combustion process out of the boiler system. After these components have been united, the design is translated to .STEP file format to be used in ANSYS Fluent for analysis as shown in Fig. 1.

2.2 Geometric Specifications

Geometrically accurate specifications are of utmost priority for the efficiency of the simulation, as they determine the features of the fluid flow and thermal aspects. Table 2 lists certain geometrical specifications of the fire-tube boiler prototype. The overall length of the boiler ends in a bushing length of 550 mm, the outer diameter

Fig. 1 Isometric view of the 3D fire-tube boiler

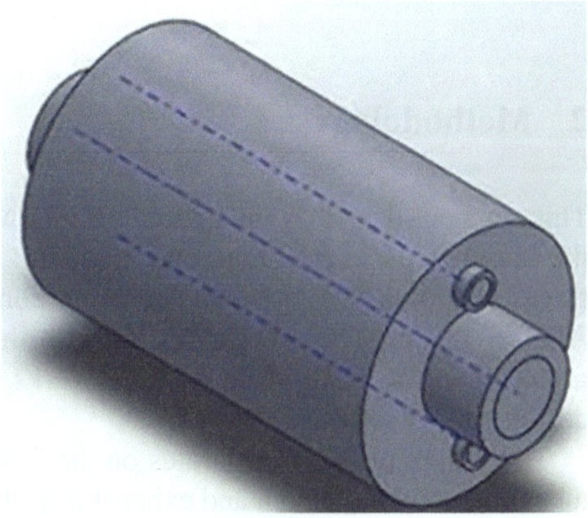

Table 2 Geometrical specifications of 3D fire-tube model

S. No.	Feature	Dimension (mm)
1	Boiler length	550
2	Boiler diameter (inner/outer)	300/350
3	Fire-tube diameter (inner/outer)	35/50

of the cylinder is 350 mm, while the inner diameter is 300 mm. The fire tubes have an outer diameter of 50 mm and an inner diameter of 35 mm. These dimensions together give grounds for the performance of the simulation of the flow dynamics and heat transfer mechanism. Figure 1 depicts the isometric view of the 3D model and its accurate representation comprises how the boiler looks and how the components are arranged inside the boiler [16, 181].

2.3 Meshing Process

After the creation and export of the fire-tube boiler model, the next task is meshing in ANSYS Fluent. Meshing consists of discretization of the geometry into discrete elements, also known as mesh cells, for purposes of detailed numerical simulation of fluid dynamics and heat transfer. An automatic mesh is generated. This resulted in the generation of 136,518 triangular cells. This mesh is sufficient because it will be able to adequately resolve the intricate features of the fire-tube boiler to facilitate the analysis of flow and temperature patterns. Mesh quality in terms of aspect ratio, skewness, and orthogonality can be stated, which explains the accuracy and reliability of the mesh. The complexity of the mesh and its size used in simulation, as well as the focus on the fire-tube boiler model, are demonstrated in Fig. 2.

2.4 System Modeling

The simulation is conducted with a pressure-based solver along with double precision capabilities. It is essential to have these characteristics to produce accurate and dependably efficient results. Some Boundary Conditions have been set for the fire tubes and the water domain of the boiler. Fire tubes include hot exhaust gases, which have a temperature of around 1500 K and are moving with a velocity of around 5 m/s, which are the operating conditions. In the boiler's water domain, water moves at 25, 30, 35, and 40 m/s on a steady-state basis after establishing an initial temperature of 350 K. A convection factor of 25 W/m^{2}K was placed on the wall of the water-boiler interface to realistically represent the heat transfer phenomena.

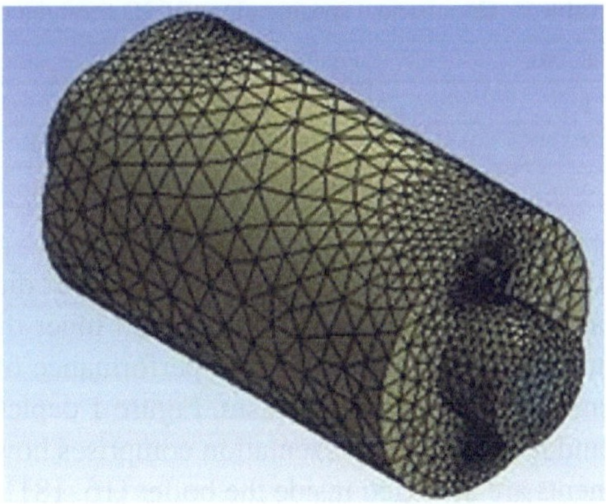

Fig. 2 Meshed fire-tube boiler model

2.5 *Simulation Execution*

After completing the simulation setup, it is important to proceed with running the model with the intention of observing the pressures and temperatures inside the fire-tube boiler. The simulation is concerned with analyzing how the pressure and temperature parameters are affected by water flow velocities and the effect of using alcohol fuels on the performance of the boiler. These results offer a good understanding of the working parameters of the boiler, and further assessment is done on how alcohols can be employed in fire-tube boilers to increase the transfer of heat and improve the efficiency of the boiler.

3 Mesh Validation

The next step in the preparation of mesh for ANSYS Fluent numerical simulations is the mesh validation. During the validation, it is necessary to check whether the mesh is adequate to resolve the flow and thermal phenomena in the fire-tube boiler and that there are no major numerical errors. Several tools can be used in doing the validation of the mesh, including a mesh independence study.

3.1 *Quality Assessment Metrics*

Certain metrics—such as aspect ratio, skewness, or orthogonality—are also evaluated to stipulate the quality of the mesh. The aspect ratio ensures that every mesh cell does not take a shape that is too thin and is not unduly extended in one direction. An optimal aspect ratio, which is near unity, reduces the typographical errors while modeling a fluid's flow. Skewness quantifies the extent of irregularity in the shape of a cell; lower skewness values are preferred. Orthogonality determines the degree

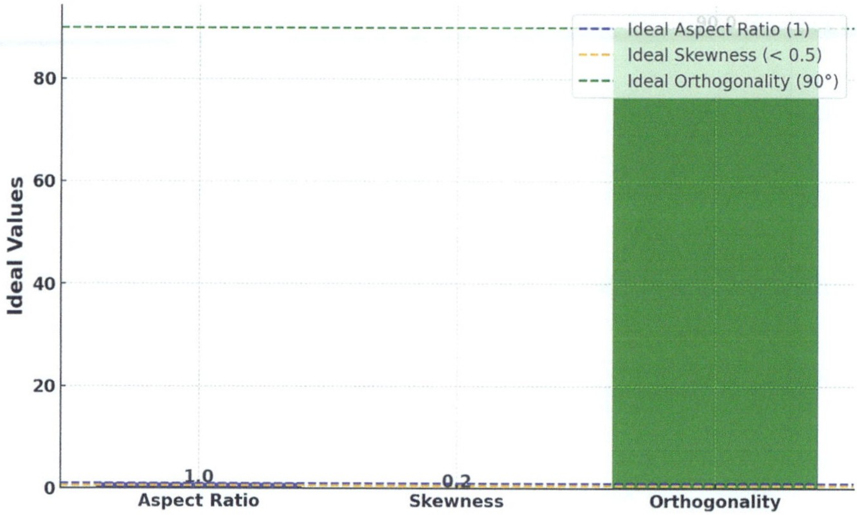

Fig. 3 Quality assessment metrics for mess validation

of angles of mesh cells, where it is noted that an ideal angle of about 90° would reduce the numerical diffusion greatly while enhancing the fluid flow. These quality metrics determine the mesh deformation and the cell's elongation for accurate simulations.

The results of the mesh independence study exemplify that as the mesh quality improves, better efficiency is achieved in the numerical predictions in regard to fire-tube boilers. Practically, as mesh density increases, the pressure drop across the boiler drops to a level whereby further refinement does not significantly change the pressure drop results. Such a meeting point is very important because it assists the researchers in determining which mesh configuration offers the best tradeoff between accuracy and simulation time. This is depicted in the line in Fig. 3, where mesh concentration is observed to be a crucial factor in effects on the pressure drop across the boilers. After reaching this convergence point, it is true to state that most endeavors to refine the mesh increase the accuracy of the model.

3.2 *Mesh Independence Study Results*

It is important to note that the most significant findings from the mesh Independence analysis (see Fig. 4) are the insights on the completion of the mesh process, even in the most rigorous CFD efforts. It was possible to analyze pressure and temperature characteristics in the fire-tube boiler with alcohol-based fuels. The selected mesh size should be appropriate to reduce the computational time while at the same time, accurately capturing the flow and thermal dynamics to maintain the integrity and relevance of the results in deriving the performance metrics of alcohol fuels in fire-tube boilers.

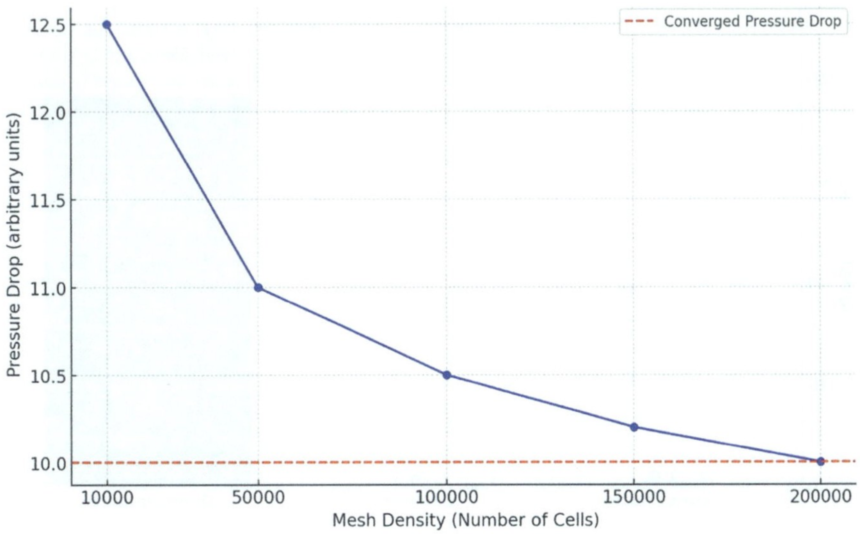

Fig. 4 Mesh independence study results

4 Results and Discussions

4.1 Overview of Pressure Simulations

The numerical simulation of pressure dynamics within fire-tube boilers has revealed critical insights into the operational behavior under varying conditions, particularly concerning the intake velocity of water. The study evaluates four distinct water intake velocities: 25, 30, 35, and 40 m/s. These velocities play a pivotal role in determining the pressure experienced within the boiler system, as shown in Fig. 5.

4.2 Pressure Distribution Analysis

4.2.1 Impact of Water Intake Velocity

The results from the simulations indicate a direct correlation between the water intake velocity and the pressure within the boiler. The pressure contours reveal that increasing the water velocity leads to higher pressure levels throughout the boiler. This is attributed to the fundamental principles of fluid dynamics, where higher velocities result in greater kinetic energy being transferred into the water column, leading to an increase in hydrostatic pressure. The observed pressure values for the different velocities can be summarized as follows: At 25 m/s, the pressure readings show an initial baseline, increasing to 30 m/s results in a notable rise in pressure and demonstrating the responsiveness of the boiler system to changes in flow rate. At

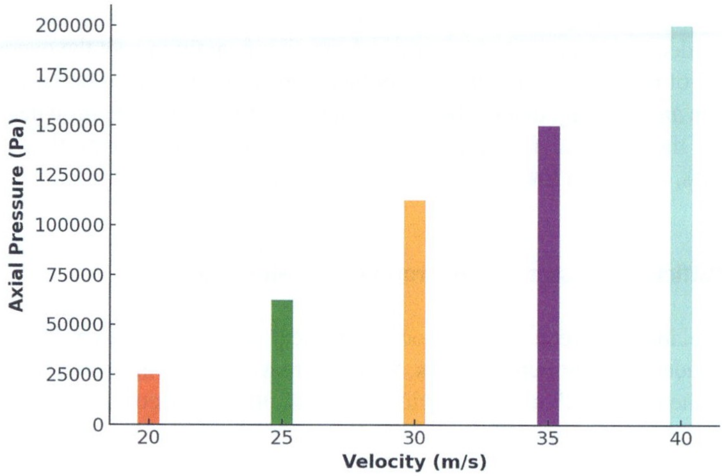

Fig. 5 Minimum pressure (vacuum) variation with velocity

35 m/s, the pressure levels continue to climb, indicating that the system is approaching its operational limits. Finally, at 40 m/s, the simulation reveals the highest pressure values, emphasizing the potential for increased stress on the boiler walls. The increased pressure levels pose significant risks to the integrity of the boiler structure, necessitating stringent monitoring and control measures to prevent catastrophic failures.

4.2.2 Structural Implications

The elevation in pressure correlates with increased stresses on the boiler walls. High-pressure conditions can lead to material fatigue, strain, and ultimately failure if not adequately managed. This is particularly critical in fire-tube boilers, where the interplay of hot gases and water creates a complex thermal environment. The findings suggest that operational protocols must be established to ensure that water intake velocities do not exceed thresholds that could compromise the structural integrity of the boiler.

4.3 Heat Transfer Dynamics

4.3.1 Relationship Between Water Velocity and Heat Transfer

While higher water velocities contribute to increased pressure, they also adversely affect heat transfer efficiency. As indicated by the simulation results, the rapid movement of water through the boiler reduces the residence time available for effective heat exchange between the hot flue gases and the water. This phenomenon results in decreased thermal efficiency and can significantly impact the overall performance of the boiler. The efficiency of heat transfer is crucial for optimizing fuel utilization and ensuring that the boiler operates effectively. With reduced heat transfer rates, the potential for harnessing the energy from the combustion process diminishes, leading to lower operational efficiency.

4.3.2 Efficiency Losses and Operational Considerations

The correlation between water velocity and heat transfer efficiency highlights a critical trade-off for boiler operators. While increasing water intake velocities may enhance the pressure dynamics within the system, it simultaneously diminishes the heat transfer effectiveness. This necessitates a careful calibration of operating conditions to maximize the benefits of high-pressure operation while maintaining acceptable levels of heat transfer. Table 3 summarizes the impact of varying water velocities on heat transfer performance, illustrating the inverse relationship between velocity and thermal efficiency.

4.4 Heat Loss Evaluation

4.4.1 Influence of Boiler Casing Material

The choice of boiler casing material significantly impacts the heat loss characteristics of the system. Figure 6 demonstrates the heat flux (HF) contours produced by altering the material of the boiler casing. Different materials exhibit varying levels of thermal conductivity, directly influencing the heat retention capabilities of the boiler.

As detailed in Table 4, brass, steel, and stainless steel were evaluated as casing materials. The findings reveal that brass exhibited the highest negative heat flux at 50.94 W/m^2, indicating substantial heat loss. In contrast, stainless steel demonstrated the lowest negative heat flux at 15.32 W/m^2, making it the most thermally efficient option among the materials assessed.

Table 3 Impact of varying water velocities on heat transfer performance

Water velocity (m/s)	Heat transfer efficiency (%)
25	85
30	80
35	75
40	70

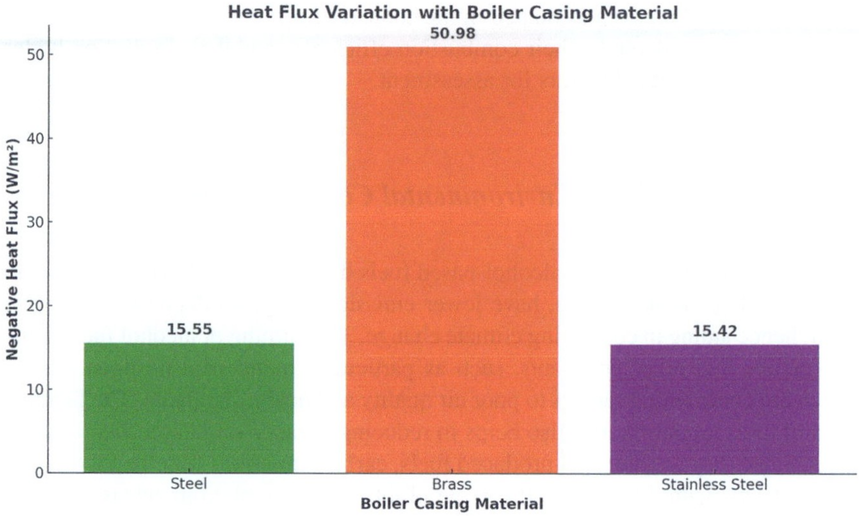

Fig. 6 HF variation with respect to boiler casing material

Table 4 Heat loss at different material

Casing material	Heat loss (W/m^2)
Brass	50.94
Steel	18.75
Stainless steel	15.32

4.5 Implications for Boiler Efficiency

The thermal efficiency of the boiler is closely linked to the heat loss through its casing. Increased heat loss translates to lower overall boiler efficiency, as more energy is required to maintain optimal operating temperatures. As seen in the data, selecting materials with lower thermal conductivity can significantly enhance the boiler's performance, allowing for better heat retention and improved fuel utilization.

5 Feasibility of Alcohol-Based Fuels

5.1 Integrating Alcohols into Existing Systems

The integration of alcohol-based fuels into fire-tube boilers presents both opportunities and challenges. While alcohols can potentially lower emissions and enhance sustainability, the simulation results indicate that their combustion characteristics must be carefully considered. The increased pressure dynamics and heat transfer issues associated with higher water velocities, along with the material choices for the boiler casing, must be evaluated in the context of using alcohol fuels. The thermal properties of alcohols, their combustion efficiency, and their impact on boiler performance are critical factors for assessment.

5.2 Operational and Environmental Considerations

An ecological point of view, alcohol-based fuels have substantial benefits over fossil fuels. They, in most cases, have lower emissions of gases depleting the ozone layer, hence aiding in combating climate change. The burning of alcohol fuels emits less carbon and other pollutants, such as particulate matter and nitrogen oxides, which are contributing factors to poor air quality and health standards. The fact that alcohol fuels are renewable also helps in reducing reliance on nonrenewable fossil fuels. Since they are biomass-produced fuels, carbon emissions for ethanol fuel can be said to be neutral across its life cycle (Table 5). The emissions released during combustion are countered by the absorption of carbon dioxide from the atmosphere by the crops used in production. This quality aligns with international objectives of reducing carbon emissions and promoting cleaner energy systems.

5.3 Performance Metrics for Alcohol-Based Fuels

Different indicators can be used to measure the effectiveness of alcohol-based fuels in transportation applications, including engine efficiency, types of emissions, and energy output or energy efficiency.

5.3.1 Engine Efficiency

Alcohol fuels have high octane ratings and therefore can be used to increase the engine compression ratio. This lets the volume of fuel used yield more power as more heat is generated from the fuel. The use of alcohol fuels enhances the performance of engines considerably, especially the alcohol fuels use in high-performance

Table 5 Environmental impact of alcohol fuels

Emission type	Gasoline (g CO_2/km)	Ethanol (g CO_2/km)	Methanol (g CO_2/km)	Reduction (%)
CO_2	180	130	140	28
NOx	3.0	2.5	2.8	17
Particulate matter (PM)	0.5	0.2	0.3	60

engines such as those of racing or aviation, where enhanced performance is a prerequisite.

Besides, the use of alcohol fuel combustion can incur less knocking to the engine allowing it to operate more smoothly. This not only improves the performance but also reduces the wear and tear of the parts that are working.

5.3.2 Emissions Profiles

It is also useful to point out emissions profiles when estimating the level of the impact of the alcohol fuels on the environment. In general, biomass-based alcohol fuels emit less GHG compared to fossil fuels, as the biomass feedstock is minimized greatly.

For instance, it has been reported that ethanol blends are effective in reducing CO emissions, and where these targets are met, the scenario will improve urban environments (Table 6). Other than that, because the alcohol fuels burn relatively cleaner, there are less soot and engine deposits which means that maintenance and servicing periods are longer.

Alcohol (ethanol, methanol, gasoline) can find potential applications spanning from light-duty vehicles, heavy-duty trucks, public transport systems, and other transport applications. Light-duty vehicles primarily include flexible-fuel vehicles which are flexible-fuel vehicles designed to run on gasoline, ethanol, or both. Due to their adaptability alcohol fuels have the potential to be widely accepted in many regions with existing supporting policies and infrastructure. Heavy-duty transportation includes transportation of goods and cargo with the help of organized trucks or buses. For improved emissions and fuel economy, using alcohol fuels in trucks and buses would be substantially beneficial. With these fuels being integrated into existing fuel stations across the cities, they will be practical for buses and trucks aiming to have a minimal environmental footprint. Alcohol fuels can be sustainable moderation department October 2022, was released in September 1 in October 2022, was released in September With aviation industry looking for alternatives to lower carbon footprint, alcohol fuels can find their application (Table 7). The same can be said for alcohol-based fuels that can provide a cleaner fuel alternative for marine vessels which mostly use high sulfur bunker fuels.

Table 6 Performance metrics of alcohol fuels

Fuel type	Octane rating	Thermal efficiency (%)	Engine power output (hp)	Fuel economy (mpg)
Gasoline	87	20	150	25
Ethanol	100	25	160	28
Methanol	105	26	165	30

Table 7 Applications of alcohol fuels

Sector	Potential applications	Current usage rate (%)	Growth potential (%)
Light-duty vehicles	Flexible fuel vehicles (FFVs)	15	50
Heavy-duty trucks	Ethanol blends in diesel engines	10	40
Public transportation	Buses running on methanol or ethanol	5	30
Aviation	Sustainable aviation fuel (SAF)	1	25
Marine	Cleaner fuels for cargo ships	3	20

6 Conclusion

In the last chapter, some parts of the case study in relation to the numerical simulation of pressure and temperature assisted alcohol fuel oil fire-tube boilers have been covered. This study used ANSYS Fluent computational fluid dynamics (CFD) techniques to model the processes going on within the boiler and examine how different parameters such as the speed of water flowing in the boiler and materials used to make the boiler case influence the thermal and flow processes in the boilers. It was concluded that alcohol fuels such as ethanol and methanol are suitable substitutes for fossil fuels because they produce lower emissions and yield more efficient energy outputs.

A mesh independence study was crucial to verify the efficacy and legitimacy of the simulations. By determining which mesh density was optimal, the study was able to reduce costs and improve solution accuracy to satisfactory levels that were used to simulate the flow pattern and heat transfer in the fire-tube boiler. This process of validation in the end proved that any further enhancement would not be worth it and that the parameters of the meshes selected were sufficient for more detailed analysis without having to do more refinement.

The output of the results indicates that alcohol-based fuels can be employed in fossil fuel-fired industries with operational steam boiler plants thus having a positive effect on the environment through lower pollutants and better combustion. Besides, casing materials concerning heat transfer was emphasized, which showed a need for material choice with a view of improving the functionality of the boiler. This research, in general, widens the scope of the already conducted studies on developing energy-efficient designs in intelligent manufacturing processes and also provides the best guidelines on the use of unconventional fuels for industrial heating processes.

References

1. Shah, S., Nautiyal, H., Gugliani, G., Kumar, A., Namboodri, T., & Singla, Y. K. (2024). . https://taylorandfrancis.com/books/). *Sustainability in smart manufacturing: Trends, scope, and challenges*. CRC Press (Taylor & Francis Group). https://doi.org/10.1201/9781003467496

2. Singla, Y. K., Maughan, M. R., Arora, N., & Dwivedi, D. K. (2024). Enhancing the wear resistance of iron-based alloys: A comprehensive review of alloying element effects. Journal of Manufacturing Processes, 120, 135–160.
3. Singh, H., Singla, Y.K.: 2023 Wear characteristics of AISI 4140 under nano fly ash based engine oil. Proceedings of the 7th International Conference on Theoretical and Applied Nanoscience and Nanotechnology (TANN'23), Paper No. 118, 1-3 June, 2023, Canada. https://doi.org/10.11159/tann23.118
4. Kumar, A., Ram, M., & Singla, Y. K. (2022). https://taylorandfrancis.com/books/. ISBN 9781003286806, 9781032054490). Advanced materials for biomechanical applications. CRC Press (Taylor & Francis Group). https://doi.org/10.1201/9781003286806
5. Singla, Y. K., Maughan, M. R., Arora, N., & Dwivedi, D. K. (2023). Microstructural evolution of worn surface and subsurface of high Mn additive iron-based hardfacings subjected to repetitive impacts and sliding. Ceramics International, 49, 21957–21965.
6. Sharma, S. K., Sharma, R. C., Lavania, S., Palli, S., & Avesh, M. (2024). Dynamic vibration mitigation in ship propellers using hybrid fuzzy sliding mode control. National Academy Science Letters. https://doi.org/10.1007/s40009-024-01515-z
7. Singh, H., Singh, A. K., Singla, Y. K., Chattopadhyay, K., Saini, A., & Singh, K. (2022). Interpretation of the wear characteristics of AISI 4140 under nano-fly ash based engine lubricant. Materials Today Proceedings, 50, 1683–1689.
8. Rooprai, R. S., Singh, H., Singh, T., & Singla, Y. K. (2022). Analysis of the wear properties of through hardened AISI-4140 alloy steel using Taguchi technique. Materials Today Proceedings, 50, 661–664.
9. Singh, H., Singh, A. K., Singla, Y. K., Chattopadhyay, K., & Malhotra, R. (2021). Optimization of tribological behavior of AISI 4140 under nano fly ash particulates in engine lubricating oil. Materials Today Proceedings, 45, 4619–4623.
10. Sura, H. S., Avesh, M., & Mohapatra, S. (2023). Design and modelling of digital twin technology to improve freight logistics. In Energy, environment, and sustainability (pp. 481–513). Springer. https://doi.org/10.1007/978-981-99-2150-8_20
11. Pattnaik, S., Pati, S., Devadarshini, D., Tanaya, K., Mishra, B. B., Samantaray, D., Mohapatra, S., & Dash, D. (2022). Microbial bioremediation of Cr(VI)-contaminated soil for sustainable agriculture. In Microbial biodegradation and bioremediation: Techniques and case studies for environmental pollution (pp. 395–407). Elsevier. https://doi.org/10.1016/B978-0-323-85455-9.00008-4
12. Srivastava, A., Lavania, S., & Mohapatra, S. (2023). Predicting the effect on land values after introducing high-speed rail. In Energy, environment, and sustainability (pp. 157–196). Springer. https://doi.org/10.1007/978-981-99-2150-8_8
13. Agrahari, R., Mohanty, S., Vishwakarma, K., Nayak, S. K., Samantaray, D., & Mohapatra, S. (2021). Update vision on COVID-19: Structure, immune pathogenesis, treatment and safety assessment. Sensors International, 2. https://doi.org/10.1016/j.sintl.2020.100073
14. Mishra, B. B., Nayak, S. K., Mohapatra, S., & Samantaray, D. (2021). Preface. In Environmental and agricultural microbiology: Advances and applications (pp. xvii–xix).
15. Tanaya, K., Mohapatra, S., & Samantaray, D. (2024). Formulation of biogenic fluorescent pigmented PHB nanoparticles from Rhodanobacter sp. for drug delivery. Antonie van Leeuwenhoek, International Journal of General and Molecular Microbiology, 117. https://doi.org/10.1007/s10482-024-01973-x
16. Mohapatra, S., Pattnaik, S., Maity, S., Sharma, S., Akhtar, J., Pati, S., Samantaray, D. P., & Varma, A. (2020). Comparative analysis of PHAs production by Bacillus megaterium OUAT 016 under submerged and solid-state fermentation. Saudi Journal of Biological Sciences, 27, 1242–1250. https://doi.org/10.1016/j.sjbs.2020.02.001
17. Singh, H., Singla, Y. K., Singh, A. K., & Chattopadhayay, K. (2020). Effect of nanofly ash as lubricant additive on the tribological properties of SAE 10W-30 oil: A novel finding. Transactions of the Indian Institute of Metals, 73, 2371–2375.
18. Kumar Singla, Y., Dwivedi, D. K., & Arora, N. (2017). Modeling the impact–sliding wear characteristics of rare earth additive iron-based hardfacing alloys. Proceedings of the Institution of Mechanical Engineers, Part J: Journal of Engineering Tribology, 231, 1486–1499.

19. Kumar, R., Bhandari, S., Goyal, A., & Singla, Y. K. (2018). Slurry erosion performance study of HVFS sprayed Ni-20Al2O3 and Ni-15Al2O3-5TiO2 coatings under hydro accelerated conditions. Industrial Lubrication and Tribology, 70, 805–817.
20. Sharma, V., Singla, Y., Gupta, Y., & Raghuwanshi, J. (2018). Post-processing of metal matrix composites by friction stir processing. AIP Conference Proceedings, 1953(1), 090062-1–090062-5. https://doi.org/10.1063/1.5032909
21. Singla, Y. K., Arora, N., & Dwivedi, D. K. (2017). Dry sliding adhesive wear characteristics of Fe-based hardfacing alloys with different CeO2 additives—A statistical analysis. Tribology International, 105, 229–240.
22. Singla, Y. K., Chhibber, R., Avdesh, Goyal, S., & Sharma, V. (2018). Influence of single and dual particle reinforcements on the corrosion behavior of aluminum alloy based composites. Proceedings of the Institution of Mechanical Engineers, Part L: Journal of Materials: Design and Applications, 232, 520–532.
23. Singla, Y. K., Dwivedi, D. K., & Arora, N. (2015). On the modeling of dry sliding adhesive wear parameters of vanadium additive iron-based alloys at elevated temperatures. Surface and Coating Technology, 283, 223–233.
24. Singla, Y. K., Chhibber, R., Bansal, H., & Kalra, A. (2015). Wear behavior of aluminum alloy 6061-based composites reinforced with SiC, Al2O3, and red mud: A comparative study. Journal of Metals, 67, 2160–2169.
25. Pant, M., Sharma, A., Palli, S., Sharma, S. K., & Sharma, R. C. (2024). A review and analysis of factors affecting EV adoption for sustainable and energy-efficient transportation systems using analytic hierarchy process approach. International Journal of Vehicle Structures and Systems, 16, 133–139. https://doi.org/10.4273/ijvss.16.2.01
26. Sharma, S. K., Sharma, R. C., Lavania, S., Palli, S., & Avesh, M. (2024). Industry-driven approach for ANFIS-based intelligent control suspension system with MR damper for enhanced ride quality in passenger rail vehicles for technological investigations. National Academy Science Letters, 47, 391–394. https://doi.org/10.1007/s40009-023-01365-1
27. Adeoye, A. O., Quadri, R. O., Lawal, O. S., & Emojevu, E. O. (2024). Physicochemical characterization, valorization of lignocellulosic waste (Kola nut seed shell) via pyrolysis, and ultrasonication of its crude bio-oil for biofuel production. Cleaner Waste Systems, 7. https://doi.org/10.1016/j.clwas.2024.100138
28. Adeoye, A. O., Quadri, R. O., Lawal, O. S., Malomo, D., Emojevu, E. O., Omonije, O. O., Odeniyi, O. K., Fadahunsi, M. O., Yelwa, M. J., Aasa, S. A., Onakpa, A. E., Omoniyi, B. P., Mark, I. N., Usman, J., Ismaila, A. M., & Saidu, A. U. (2024). Advanced techniques in upgrading crude bio-oil to biofuel. In Energy, environment, and sustainability (pp. 321–353). Springer. https://doi.org/10.1007/978-981-97-0515-3_14
29. Adeoye, A. O., Quadri, R. O., & Lawal, O. S. (2022). Assessment of biofuel potential of tenera palm kernel shell via fixed bed pyrolysis and thermal characterization. Results in Surfaces and Interfaces, 9. https://doi.org/10.1016/j.rsurfi.2022.100091
30. Adeoye, A. O., Lawal, O. S., Quadri, R. O., Malomo, D., Aliyu, M. T., Dang, G. E., Emojevu, E. O., Maikato, M. J., Yahaya, M. G., Omonije, O. O., Edidem, V. G., Abubakar, Y. K., Offor, O. F., Sochima, E. H., Peter, B. E., & Hikon, B. N. (2023). Sustainable energy via thermochemical and biochemical conversion of biomass wastes for biofuel production. In Energy, environment, and sustainability (pp. 245–306). Springer. https://doi.org/10.1007/978-981-99-2150-8_11
31. Mohapatra, S., & Mishra, R. K. (2020). Status of institutional credit disbursement of agricultural sector in Odisha, India. International Journal of Current Microbiology and Applied Sciences, 9, 2057–2065.
32. Mohapatra, S., Singh, J., & Mishra, S. (2017). Co-integration among major potato markets in Punjab. Indian Journal of Agricultural Marketing, 31, 46–47.
33. Mohapatra, S., Priyanka, V., Mohapatra, S., Kohli, I., & Mishra, R. K. (2020). Impact of corona virus covid-19 on the global economy. International Journal of Agricultural and Statistical Sciences, 16, 771–778.

34. Mohapatra, S., & Singh, J. (2021). Co-integration of major vegetable markets in Punjab, India. Economic Affairs, 66(02), 181–187.
35. Mohapatra, S., Mishra, R., & Mishra, R. (2021). Impact of COVID-19 on education level of agricultural farming community. International Journal of Agricultural and Statistical Sciences, 17, 915–923.
36. Mohapatra, S., Singh, J., & Kumar, S. (2018). Cointegration among major cauliflower markets in Punjab. Indian Journal of Economics and Development, 14, 330–335.
37. Adeoye, A. O., Yelwa, J. M., Imam, N., Quadri, R. O., Lawal, O. S., Malomo, D., Aasa, S. A., Onakpa, A. E., Dorgbaa, R. K., Omoniyi, B. P., Ahmad, N., & Garba, H. A. (2024). Pyrolysis of plastic wastes towards achieving a circular economy: An advanced chemistry and technical approach. In Energy, environment, and sustainability (pp. 215–259). Springer. https://doi.org/10.1007/978-981-97-0437-8_11
38. Adeoye, A. O., Quadri, R. O., & Lawal, O. S. (2023). Wet synthesis, characterization of goethite nanoparticles and its application in catalytic pyrolysis of palm kernel shell in TGA. Results in Surfaces and Interfaces, 11. https://doi.org/10.1016/j.rsurfi.2023.100118
39. Avesh, M., Imam, N., & Adeoye, A. O. (2024). Assessing environmental impacts and mitigation strategies for the DFCCIL rail projects. In Energy, environment, and sustainability (pp. 333–353). Springer. https://doi.org/10.1007/978-981-97-0437-8_15
40. Mohapatra, S., Mohapatra, S., Han, H., Ariza-Montes, A., & López-Martín, M. d. C. (2022). Climate change and vulnerability of agribusiness: Assessment of climate change impact on agricultural productivity. Frontiers in Psychology, 13.
41. Mohapatra, S., & Mishra, R. A. J. K. (2020). Production performance and market integration of cotton markets in India. In National Symposium (p. 253).
42. Din, A. U., Han, H., Ariza-Montes, A., Vega-Muñoz, A., Raposo, A., & Mohapatra, S. (2022). The impact of Covid-19 on the food supply chain and the role of e-commerce for food purchasing. Sustainability, 14(5), 3074.
43. Mohapatra, S., & Singh, J. (2018). Growth and instability analysis of cauliflower crop in Punjab. Economic Affairs, 63, 665–669.
44. Pandey, G., Bansal, S., & Mohapatra, S. (2023). Market integration of chickpea crop: An evidence of India. The Journal of Agribusiness in Developing and Emerging Economies, 15(1), 1–18. https://doi.org/10.1108/JADEE-12-2022-0273
45. Mohapatra, S., Mishra, R. K., & Sarangi, K. K. (2019). Economic appraisal of bio ethanol production technologies. International Journal of Chemical Studies, 7, 2197–2200.
46. Mohapatra, S., Kishore Mishra, R., & Sarangi, K. (2020). Bio-ethanol (2nd generation ethanol): A solution to ever polluting gasoline to climate in India. International Research Journal of Pure and Applied Chemistry, 20, 1–15.
47. Gohain, N., Bansal, S., Mohapatra, S., & Singh, L. (2022). Analyzing the direction of trade: Indian ginger and lessons from exports to different destinations. Journal of Agricultural Development and Policy, 32, 214–220.
48. Mohapatra, S., Mohapatra, U., Chandana, K. S. S., & Mishra, R. K. (2018). Economic analysis of paddy production and marketing in Puri, Odisha. Journal of Pharmacognosy and Phytochemistry, 7, 1858–1861.
49. Quadri, R. O., Adeoye, A. O., Ayanda, O. S., & Lawal, O. S. (2024). Conversion of palm kernel shell to sustainable energy and the effect of wet synthesized nanoparticles of iron on its thermal degradation kinetics. *Bioresource Techn*ology Reports, 27. https://doi.org/10.1016/j.biteb.2024.101933
50. Momodu, D. U., Adetuyi, A. O., Adeoye, A. O., Akinlolu, O. S., & Apara, E. O. (2023). Surface functionalization of linen fiber and its dye-fastness assessment using 4-[(E)-(4-amino-2-hydroxyphenyl) diazenyl] benzene-1-sulphonic acid and 4-[(E)-(4-{(Z)-[4-(dimethylamino)phenyl] diazenyl}-2-hydroxyphenyl) diazenyl]benzene-1-sulphonic acid. Suranaree Journal of Science and Technology, 30.

51. Adeoye, A. O., Quadri, R. O., & Lawal, O. S. (2023). Biomass waste pyrolysis in the context of Nigeria's oil and gas industry: An overview. International Journal of Ambient Energy, 44, 2525–2541. https://doi.org/10.1080/01430750.2023.2256325
52. Mohapatra, S., Mohapatra, U., & Mishra, R. K. (2018). Price forecasting of groundnut in Odisha. The Pharma Innovation Journal, 7, 111–114.
53. Mohapatra, S., Sneha, N. K., Mounika, A., & Mishra, R. K. (2019). Socio-economic and demographic analysis of SAARC nations. International Journal of Agricultural & Statistical Sciences, 15.
54. Sharma, S. K., Upadhyay, R. K., Kumar, V., & Valera, H. (2023). Transportation energy and dynamics. Springer. https://doi.org/10.1007/978-981-99-2150-8
55. Upadhyay, R. K., Sharma, S. K., Kumar, D. V., & Valera, H. (2023). Transportation systems technology and integrated management. Springer. https://doi.org/10.1007/978-981-99-1517-0
56. Lee, J., & Sharma, S. K. (2023). Intelligent systems for railway infrastructure. Applied Sciences, 13(20), 11274. https://doi.org/10.3390/app132011274
57. Sharma, S. K., Sharma, R. C., Choi, Y., & Lee, J. (2023). Modelling and dynamic analysis of adaptive neuro-fuzzy inference system-based intelligent control suspension system for passenger rail vehicles using magnetorheological damper for improving ride index. Sustainability, 15, 12529.
58. Sharma, S. K., Upadhyay, R. K., Kumar, V., & Valera, H. (2023). Introduction to sustainable transportation system. In Transportation energy and dynamics (pp. 3–6). Springer.
59. Hossain, I., Ahmed, Y. E., & Sharma, S. K. (2023). The use of both metallic-insert and TiO2/water nanofluid to improve the performance of photovoltaic modules; experimental study. Thermal Science and Engineering Progress, 42, 101944.
60. Wu, Q., Cole, C., Spiryagin, M., Chang, C., Wei, W., Ursulyak, L., Shvets, A., Murtaza, M. A., Mirza, I. M., & Zhelieznov, K. (2023). Freight train air brake models. International Journal of Rail Transportation, 11, 1–49.
61. Upadhyay, R. K., Sharma, S. K., Kumar, V., & Valera, H. (2023). Introduction to transportation systems technology and integrated management. In Transportation systems technology and integrated management (pp. 1–4). Springer.
62. Sharma, R. C., Bhardawaj, S., Sharma, S. K., & Bahl, S. (2023). Ride behaviour of three-wheel vehicle using semi-active suspension using skyhook control strategy. International Journal of Vehicle Structures & Systems (IJVSS), 15.
63. Bhardawaj, S., Sharma, R. C., Sharma, S. K., Gopala Rao, L. V. V., & Vashist, A. (2023). Modeling of biomechanical human body model for seat to head transmissibility analysis. Noise and Vibration Worldwide, 54, 62–74.
64. Sharma, S. K., Sharma, R. C., Choi, Y., & Lee, J. (2023). Experimental and mathematical study of flexible–rigid rail vehicle riding comfort and safety. Applied Sciences, 13, 5252.
65. Pandipati, S., Palli, S., Dowluru, S., Sharma, R. C., Sharma, N., Sharma, S. K., & Da Fonseca-Soares, D. (2024). Microstructural and strength comparison friction stir processing of aluminium alloys and multi-response optimization by Taguchi GRA-PCA based hybrid methods. International Journal of Vehicle Structures and Systems, 16(3). https://doi.org/10.4273/ijvss.16.3.03
66. Adeoye, A. O., Quadri, R. O., & Lawal, O. S. (2022). Fixed-bed pyrolysis and thermal analyses of pressed oil palm fruit fibre as a potential alternative energy. Fuoye Journal of Innovation, Science and Technology, 2(1), 187–194.
67. Sharma, S. K., Upadhyay, R. K., & Kumar, V. (2024). Introduction to dynamics of transportation ecosystem, modeling, and control. In Dynamics of transportation ecosystem, modeling, and control (pp. 3–6). Springer.
68. Sharma, S. K., Sharma, R. C., Mohapatra, S., Fouly, A., Hossain, I., & Lee, J. (2024). Technological advancements in oscillation reduction for propulsion shaft systems. Journal of Vibration Engineering & Technologies, 12, 1–19.

69. Upadhyay, R. K., Sharma, S. K., & Kumar, V. (2024). Introduction to intelligent transportation system and advanced technology. In Intelligent transportation system and advanced technology (pp. 3–6). Springer.
70. Hossain, I., Ngoupo, A. T., Bhattarai, S., Challagulla, S. P., & Sharma, S. K. (2024). Empirical survey on the effect of straight and wavy fins integrated with wind stream as heat sink of photovoltaic panels. International Journal of Environmental Science and Technology, 22, 2479–2492.
71. Sharma, S. K., Sharma, R. C., & Lee, J. (2024). Propelling precision of longitudinal vibration mitigation in ship propeller shafts through advanced nonlinear intelligent semi-active control leveraging adaptive neuro-fuzzy inference system with linear quadratic regulator. Journal of Vibration and Control, 31, 1472–1484.
72. Sharma, S. K., Kumar, N., Avesh, M., Sharma, R. C., Siddiqui, M. I. H., & Lee, J. (2024). Navigating tranquillity with H∞ controller to mitigate ship propeller shaft vibration. Journal of Vibration Engineering & Technologies, 12, 1–11.
73. Sharma, S. K., Sharma, R. C., Upadhyay, R. K., & Lee, J. (2024). Enhanced ship engine vibration reduction using magnetorheological dampers and adaptive neuro-fuzzy control system. Ships and Offshore Structures, 20, 143–159.
74. Sharma, R. C., Sharma, N., Palli, S., Sharma, S. K., Joshi, S., Rallabandi, S. R., & Sharma, A. (2024). Dynamic behavior of three-wheel vehicle with inerter-based dynamic absorber suspension system. Noise and Vibration Worldwide, 55, 253–263.
75. Sharma, R. C., Sharma, S. K., Palli, S., Rallabandi, S. R., & Sharma, N. (2025b). Compatibility between stability and curving ability of railway vehicles using conventional and unconventional bogie frame models. SAE International Journal of Commercial Vehicles, 18(3), 02-18-03-0014. https://doi.org/10.4271/02-18-03-0014
76. Kokkinos, N. C., Emmanouilidou, E., & Sharma, S. K. (2024). Waste-to-biofuel production for the transportation sector. In Energy, environment, and sustainability (pp. 99–122). Springer. https://doi.org/10.1007/978-981-97-0515-3_6
77. Sharma, R. C., Palli, S., Duppala, A., Bhardawaj, S., Sharma, S. K., Sharma, N., & Sharma, A. (2024). Rigid body analysis of ride behaviour of roadway vehicle using Lagrangian dynamics. International Journal of Vehicle Structures & Systems, 16, 74–78. https://doi.org/10.4273/ijvss.16.1.15
78. Sharma, S. K., Sharma, R. C., & Lee, J. (2024). Modelling and simulation of MR damper characterization and uncertainty assessment in marine engine vibration isolation with FLPID control. International Journal of Modelling and Simulation, 1–11.
79. Sharma, S. K., Sharma, R. C., Upadhyay, R. K., & Lee, J. (2024). Coupled torsional–transverse vibration reduction in marine propulsion dynamics with novel approach using magnetorheological damping and adaptive control system. Journal of Vibration Engineering & Technologies, 12, 6089–6099.
80. Rallabandi, S. R., Srinivas, L., Palli, S., Sharma, R. C., Sharma, N., Sharma, A., & Sharma, S. K. (2024). A contrastive characterization of pure Mg and AZ91D alloy based on the testing of mechanical, corrosion, wear, and erosion properties. Engineering Research Express, 6, 015017.
81. Badgujar, P., Sharma, S., Rahul, D., Sharma, S. K., & Upadhyay, R. K. (2024). Feasibility study of Mumbai-Pune Hyperloop corridor: Challenges and implementation. Research in Transportation Business & Management, 52, 101089.
82. Sharma, R. C., Palli, S., Dharmana, L., Sharma, S. K., Sharma, A., Kharola, A., & Sharma, N. (2024). Simulation of transient dynamic response of a locomotive considering its flexibility and rigidity. *Noise and Vibration Worldwide, 55*, 33–47.
83. Song, H., Park, J., Sharma, S. K., & Lee, J. (2022). A comparative study on ultrasonic propagation characteristics and defect detection of metal material additive manufacturing using deep learning algorithm. In *Biennial International Conference on Future Learning Aspects of Mechanical Engineering* (pp. 157–168). Springer.

84. Boris, I., Sharma, S. K., & Lee, J. (2022). A study on the reflection coefficient of higher modes of torsional wave by theoretical and numerical method. In *Biennial International Conference on Future Learning Aspects of Mechanical Engineering* (pp. 145–155). Springer.
85. Vishwakarma, P. N., Mishra, P., Sharma, S. K., & Shukla, A. K. (2022). Formulating and analysing the effect of suspension parameters on critical speeds for various DOFs in rail vehicle. In *Biennial International Conference on Future Learning Aspects of Mechanical Engineering* (pp. 791–802). Springer.
86. Kumar, U., Kasvekar, R., Sharma, S. K., & Upadhyay, R. K. (2022). Wear of wheels and axle in locomotive and measures taken by Indian railway. In *Advances in engine tribology* (pp. 77–96).
87. Sharma, R. C., Palli, S., Jha, A. K., Bhardawaj, S., & Sharma, S. K. (2022). Vibration and ride comfort analysis of railway vehicle system subjected to deterministic inputs. *Res Militaris, 12*, 1345–1355.
88. Sravanthi, A. L., Ratnakumar, P., Reddy, S. N., Eswari, K. B., Pandey, B. B., Manikanta, C., Ramya, K. T., Sonia, E., Mohapatra, S., Gopika, K., Anusha, P. L., & Yadav, P. (2022). Morpho-physiological, quality traits and their association with seed yield in sesame (Sesamum indicum L.) indigenous collection under deficit moisture stress. *Plant Physiology Reports, 27*(1), 132–142. https://doi.org/10.1007/s40502-021-00621-0
89. Sharma, R. C., Palli, S., Rao, L. V. V. G., Duppala, A., & Sharma, S. K. (2022). Four-wheel vehicle response under bump, pothole, harmonic, and random excitations using bond graph/Simulink technique. *SAE International Journal of Commercial Vehicles, 16*. https://doi.org/10.4271/02-16-02-0008
90. Vishwakarma, P. N., Mishra, P., & Sharma, S. K. (2022). Formulation of semi-active suspension system and controls in rail vehicle. *SSRN Electronic Journal*. https://doi.org/10.2139/ssrn.415961
91. Sharma, R. C., & Sharma, S. K. (2022). Ride analysis of road surface-three-wheeled vehicle-human subject interactions subjected to random excitation. *SAE International Journal of Commercial Vehicles, 15*, 299–312.
92. Sharma, S. K., Lee, J., & Jang, H.-L. (2022). Mathematical modeling and simulation of suspended equipment impact on car body modes. *Machines, 10*, 192.
93. Sharma, S. K., Sharma, R. C., Lee, J., & Jang, H.-L. (2022). Numerical and experimental analysis of DVA on the flexible-rigid rail vehicle carbody resonant vibration. *Sensors, 22*, 1922.
94. Vishwakarma, P. N., Mishra, P., & Sharma, S. K. (2022). Characterization of a magnetorheological fluid damper a review. *Materials Today Proceedings, 56*, 2988–2994.
95. Lora Anusha, P., Ratnakumar, P., Pandey, B. B., Sandhya Rani, P., Umamahesh, V., Reddi Sekhar, M., Chandrika, V., Yadav, P., Mohapatra, S., & Padmaja, D. (2023). Phenotypic characterization of sesame (Sesamum indicum L.) revealed promising genotypes for moisture stress conditions. *Plant Genetic Resources: Characterization and Utilization, 21*(3), 218–228. https://doi.org/10.1017/S1479262123000667
96. Sharma, S. K., Sharma, R. C., & Lee, J. (2022). In situ and experimental analysis of longitudinal load on carbody fatigue life using nonlinear damage accumulation. *International Journal of Damage Mechanics, 31*, 605–622.
97. Agrahari, R., Mohapatra, S., Kumar, K. S., Mohanty, D., Kumar, R., & Merugu, R. (2024). Immuno-metabolic interplay in health and disease. In A. Das Mohapatra & P. Sahu (Eds.), *Advances in immunology and immuno-techniques. Interdisciplinary Biotechnological Advances* (pp. 239–274).
98. Sharma, S. K., & Sharma, R. C. (2021). *Multi-objective design optimization of locomotive nose* (pp. 1–10). SAE Technical Paper. https://doi.org/10.4271/2021-01-5053
99. Lee, J., Han, J., & Sharma, S. K. (2021). Structural analysis on the separated and integrated differential gear case for the weight reduction. In *Advances in engineering design: Select Proceedings of FLAME 2020* (pp. 175–181). Springer Singapore.

100. Choi, S., Lee, J., & Sharma, S. K. (2021). A study on the performance evaluation of hydraulic tank injectors. In *Advances in engineering design: Select Proceedings of FLAME 2020* (pp. 183–190). Springer Singapore.
101. Sharma, R. C., Palli, S., Sharma, N., & Sharma, S. K. (2021). Ride behaviour of a four-wheel vehicle using h infinity semi-active suspension control under deterministic and random inputs. *International Journal of Vehicle Structures & Systems, 13*, 234–237.
102. Bhardawaj, S., Sharma, R. C., Sharma, S. K., & Sharma, N. (2021). On the planning and construction of railway curved track. *International Journal of Vehicle Structures & Systems, 13*, 151–159.
103. Mohapatra, S., Mohanty, D., Sharma, S., Dikshit, S., Kohli, I., Samantaray, D. P., Kumar, R., & Kathpalia, M. (2021). Biomedical application of polymeric biomaterial: Polyhydroxybutyrate. In *Bioresource utilization and management* (pp. 111–124). Apple Academic Press.
104. Sharma, S. K., Sharma, R. C., & Lee, J. (2021). Effect of rail vehicle–track coupled dynamics on fatigue failure of coil spring in a suspension system. *Applied Sciences, 11*, 2650.
105. Sharma, R. C., Sharma, S., Sharma, S. K., Sharma, N., & Singh, G. (2021). Analysis of bio-dynamic model of seated human subject and optimization of the passenger ride comfort for three-wheel vehicle using random search technique. *Proceedings of the Institution of Mechanical Engineers, Part K: Journal of Multi-body Dynamics, 235*, 106–121.
106. Sharma, S. K., & Lee, J. (2021). Crashworthiness analysis for structural stability and dynamics. *International Journal of Structural Stability and Dynamics, 21*, 2150039.
107. Cataluna Veses, R., & Shah, Z. (2015). Bio-oil production by thermal cracking in the presence of hydrogen. *Journal of Fundamentals of Renewable Energy and Applications, 06*. https://doi.org/10.4172/2090-4541.1000194
108. Shah, Z., Cataluña Veses, R., Aguilhera, R. A., & Da Silva, R. (2016). Bio-oil production from pyrolysis of coffee and eucalyptus sawdust in the presence of 5% hydrogen. *International Journal of Engineering Research & Science, 2*, 34.
109. Shah, Z., Veses, R. C., & da Silva, R. (2016). A comparative study and analysis of two types of bio-oil samples obtained from freshwater algae and microbial treated algae. *Modern Chemistry & Applications, 4*. https://doi.org/10.4172/2329-6798.1000185
110. Shah, Z., Veses, R. C., Brum, J., Klunk, M. A., Rocha, L. A. O., Missaggia, A. B., de Oliveira, A. P., & Caetano, N. R. (2020). Sewage sludge bio-oil development and characterization. *Inventions, 5*, 36. https://doi.org/10.3390/inventions5030036
111. Shah, Z., Saberian, M., Hodgeman, D., Kinloch, I., & Vallés, C. (2023). Effect of sulfur on wood tar biopitch as a sustainable replacement for coal tar pitch binders. *ACS Applied Engineering Materials, 1*, 2567–2576. https://doi.org/10.1021/acsaenm.3c00361
112. Klunk, M. A., Dasgupta, S., Das, M., & Shah, Z. (2018). System of adsorption of CO_2 in coal-bed. *Southern Brazilian Journal of Chemistry, 26*, 2–9. https://doi.org/10.48141/SBJCHEM.v26.n26.2018.7_2018.pdf
113. Mohapatra, S., Mohapatra, U., & Mishra, R. (2017). Diversification towards vegetable crops: A good option for doubling the farmer's income. *Journal of Experimental Agriculture International, 18*, 1–7.
114. Chavan, S. D., Bansal, S., Mohapatra, S., Kaur, L., & Jadhav, A. (2023). Trade directions of Indian basmati rice export-Markov chain approach. *Economic Affairs*. https://doi.org/10.46852/0424-2513.1.2023.23
115. Bansal, S., Mohapatra, S., Gohain, N., & Das, A. (2022). Production status, market integration and the law of one price: An exploration of the major maize markets of India. *Indian Journal of Economics and Development, 18*, 832–841.
116. Mohapatra, S., Samantaray, D. P., & Samantaray, S. M. (2015). Study on polyhydroxyalkanoates production using rhizospheric soil bacterial isolates of sweet potato. Indian. *Journal of Science and Technology, 8*, 57–62. https://doi.org/10.17485/ijst/2015/v8iS7/64027

117. Arora, R., Trivedi, J., Mohapatra, S., & Kumar, P. (2022). Trends in probiotics on human health and industrial application. In *Advances in agricultural and industrial microbiology* (Microbial diversity and application in agroindustry) (Vol. 1, pp. 237–248). https://doi.org/10.1007/978-981-16-8918-5_13
118. Samal, S., Pati, S., Mohapatra, S., Maity, S., Tanaya, K., Devadarshini, D., & Samantaray, D. (2023). PHAs production by facultative anaerobic bacteria Bacillus cereus FM5 through submerged and solid-state fermentation under anoxic condition. *Antonie van Leeuwenhoek, International Journal of General and Molecular Microbiology, 116*, 521–529. https://doi.org/10.1007/s10482-023-01825-0
119. Biswas, N., Saha, S., Biswas, B. K., Chowdhury, M., Rahaman, A., Junghare, V., Mohapatra, S., Hazra, S., Zangrando, E., Roy Choudhury, R., & Roy Choudhury, C. (2021). The DNA- and protein-binding properties and cytotoxicity of a new copper(II) hydrazone Schiff base complex. *Journal of Coordination Chemistry, 74*, 1482–1504. https://doi.org/10.1080/00958972.2021.1913128
120. Mohapatra, S., Mishra, R., Roy, P., Yadav, K. L., & Satapathi, S. (2017). Systematic investigation and in vitro biocompatibility studies on implantable magnetic nanocomposites for hyperthermia treatment of osteoarthritic knee joints. *Journal of Materials Science, 52*, 9262–9268. https://doi.org/10.1007/s10853-017-1136-0
121. Mishra, B. B., Nayak, S. K., Mohapatra, S., & Samantaray, D. (2021). *Environmental and agricultural microbiology: Advances and applications*. Wiley. https://doi.org/10.1002/9781119525899
122. Sharma, S. K., Mohapatra, S., Sharma, R. C., Alturjman, S., Altrjman, C., Mostarda, L., & Stephan, T. (2022). Retrofitting existing buildings to improve energy performance. *Sustainability (Switzerland), 14*. https://doi.org/10.3390/su14020666
123. Bandhu, S., Khot, M. B., Sharma, T., Sharma, O. P., Dasgupta, D., Mohapatra, S., Hazra, S., Khatri, O. P., & Ghosh, D. (2018). Single cell oil from oleaginous yeast grown on sugarcane bagasse-derived xylose: An approach toward novel biolubricant for low friction and Wear. *ACS Sustainable Chemistry & Engineering, 6*, 275–283. https://doi.org/10.1021/acssuschemeng.7b02425
124. Kohli, I., Joshi, N. C., Mohapatra, S., & Varma, A. (2020). Extremophile—An adaptive strategy for extreme conditions and applications. *Current Genomics, 21*, 96–110. https://doi.org/10.2174/1389202921666200401105908
125. Pati, S., Mohapatra, S., Maity, S., Dash, A., & Samantaray, D. P. (2021). Polyhydroxyalkanoates for sustainable smart packaging of fruits. In *Environmental and agricultural microbiology: Advances and applications* (pp. 183–195).
126. Chatterjee, A., Nirwan, S., Mohapatra, S., Sharma, P., Agnihotri, A., & Shrivastava, N. (2022). Biochemical aspects of pathogenic variability in white rust infected Indian mustard. *Mycologia, 114*, 757–768. https://doi.org/10.1080/00275514.2022.2060007
127. Pati, S., Maity, S., Dash, A., Jema, S., Mohapatra, S., Das, S., & Samantaray, D. P. (2020). Biocompatible PHB production from Bacillus species under submerged and solid-state fermentation and extraction through different downstream processing. *Current Microbiology, 77*, 1203–1209. https://doi.org/10.1007/s00284-020-01922-7
128. Agrahari, R., Sarraf, G., Joshi, N. C., Mohapatra, S., & Var, A. (2021). Insight of biopolymers and applications of polyhydroxyalkanoates. In *Microbial polymers: Applications and ecological perspectives* (pp. 177–191). https://doi.org/10.1007/978-981-16-0045-6_8
129. Kathpalia, M., Kumar, P., & Mohapatra, S. (2022). Homocysteine metabolism as a biomarker for cancer. In *Homocysteine metabolism in health and disease* (pp. 159–171). https://doi.org/10.1007/978-981-16-6867-8_8
130. Bharatiy, S. K., Hazra, M., Paul, M., Mohapatra, S., Samantaray, D., Dubey, R. C., Sanyal, S., Datta, S., & Hazra, S. (2016). In silico designing of an industrially sustainable carbonic anhydrase using molecular dynamics simulation. *ACS Omega, 1*, 1081–1103. https://doi.org/10.1021/acsomega.6b00041

131. Sarkar, M. S., Segu, H., Bhaskar, J. V., Jakher, R., Mohapatra, S., Shalini, K., Shivaji, S., & Reddy, P. A. (2018). Ecological preferences of large carnivores in remote, high-altitude protected areas: Insights from Buxa Tiger Reserve, India. *ORYX, 52*, 66–77. https://doi.org/10.1017/S0030605317000060
132. Pati, S., Dash, A., Maity, S., Pattnaik, S., Mohapatra, S., & Samantaray, D. P. (2021). Nano biofertilizer for sustainable agriculture. In *Agriculturally important microorganisms: Mechanisms and applications for sustainable agriculture* (pp. 201–208). https://doi.org/10.1201/9781003245841-10
133. Mohanty, D., Misra, S., Mohapatra, S., & Sahu, P. S. (2018). Prebiotics and synbiotics: Recent concepts in nutrition. *Food Bioscience, 26*, 152–160. https://doi.org/10.1016/j.fbio.2018.10.008
134. Kumar, R., Kumari, R., Prasad, C., Deep, A., Singh, N., Namtak, S., Tiwari, V., Merugu, R., Mohapatra, S., & Singh, R. (2021). Influence of water quality on the diversity of macroinvertebrates in the Mandakini River in India. *Water Science and Technology: Water Supply, 21*, 1843–1860. https://doi.org/10.2166/WS.2021.020
135. Mohanty, D., Jena, R., Choudhury, P. K., Pattnaik, R., Mohapatra, S., & Saini, M. R. (2016). Milk derived antimicrobial bioactive peptides: A review. *International Journal of Food Properties, 19*, 837–846. https://doi.org/10.1080/10942912.2015.1048356
136. Mohapatra, S., Rath, S. N., Pradhan, S. K., Samantaray, D. P., & Rath, C. C. (2016). Secondary structural models (16S rRNA) of polyhydroxyalkanoates producing Bacillus species isolated from different rhizospheric soil: Phylogenetics and chemical analysis. *International Journal Bioautomation, 20*, 329–338.
137. Mohapatra, S., Sarkar, B., Samantaray, D. P., Daware, A., Maity, S., Pattnaik, S., & Bhattacharjee, S. (2017). Bioconversion of fish solid waste into PHB using Bacillus subtilis based submerged fermentation process. *Environmental Technology (United Kingdom), 38*, 3201–3208. https://doi.org/10.1080/09593330.2017.1291759
138. Misra, S., Mohanty, D., & Mohapatra, S. (2021). Food-based probiotics: Functional dietary ingredients. In *Probiotics: Advanced food and health applications* (pp. 257–275). https://doi.org/10.1016/B978-0-323-85170-1.00006-3
139. Kumar, R., Kumari, R., Prasad, C., Tiwari, V., Singh, N., Mohapatra, S., Merugu, R., Namtak, S., & Deep, A. (2020). Phytoplankton diversity in relation to physicochemical attributes and water quality of Mandakini River, Garhwal Himalaya. *Environmental Monitoring and Assessment, 192*. https://doi.org/10.1007/s10661-020-08768-3
140. Mohapatra, S., Vishwakarma, K., Joshi, N. C., Maity, S., Kumar, R., Ramchander, M., Pattnaik, S., & Samantaray, D. P. (2021). A review on PHAs: The future biopolymer. In *Environmental and agricultural microbiology: Advances and applications* (pp. 83–117).
141. Sahoo, N., Pattnaik, S., Pattnaik, M., & Mohapatra, S. (2021). Methane emission and strategies for mitigation in livestock. In *Environmental and agricultural microbiology: Advances and applications* (pp. 257–274).
142. Mohanty, D. P., Mohapatra, S., Misra, S., & Sahu, P. S. (2016). Milk derived bioactive peptides and their impact on human health—A review. *Saudi Journal of Biological Sciences, 23*, 577–583. https://doi.org/10.1016/j.sjbs.2015.06.005
143. Kohli, I., Mohapatra, S., Kumar, P., Goel, A., Varma, A., & Joshi, N. C. (2022). Role of bacterial endophytes in the promotion of plant growth. In *Bacterial endophytes for sustainable agriculture and environmental management* (pp. 235–253). https://doi.org/10.1007/978-981-16-4497-9_12
144. Samantaray, D., Mohapatra, S., & Mishra, B. B. (2014). Microbial bioremediation of industrial effluents. In *Microbial biodegradation and bioremediation* (pp. 326–339). https://doi.org/10.1016/B978-0-12-800021-2.00014-5
145. Sharma, R. C., Gopala Rao, L. V. V., Srikar, B., Palli, S., Appala Naidu, Y., Sharma, N., & Sharma, S. K. (2025a). Investigation of the dynamic behavior of hybrid composites using finite element and experimental method. *Journal of the Chinese Institute of Engineers, 48*(2), 158–171. https://doi.org/10.1080/02533839.2025.2453139

146. Mohapatra, S., Mohanta, P. R., Sarkar, B., Daware, A., Kumar, C., & Samantaray, D. P. (2017). Production of polyhydroxyalkanoates (PHAs) by Bacillus strain isolated from waste water and its biochemical characterization. *Proceedings of the National Academy of Sciences India Section B—Biological Sciences, 87*, 459–466. https://doi.org/10.1007/s40011-015-0626-6
147. Tomer, P., Shroti, G. K., Mohapatra, S., Ghosh, D., Jaiswal, S., Garg, D., Lahiri, D., Kumar, N., Khatri, O. P., & Hazra, S. (2023). Fabrication of HA Nano-crystal Reinforced PHAs based composites for orthopedic fracture-fixing accessories using sustainable resources. *Environment, Development and Sustainability*. https://doi.org/10.1007/s10668-023-03912-8
148. Bandekar, D., Mohapatra, S., Hazra, M., Hazra, S., & Biswas, S. (2021). N-terminal truncation of VC0395_0300 protein from Vibrio cholerae does not lead to loss of diguanylate cyclase activity. *Biophysical Chemistry, 268*. https://doi.org/10.1016/j.bpc.2020.106493
149. Vishwakarma, K., Kumar, N., Shandilya, C., Mohapatra, S., Bhayana, S., & Varma, A. (2020). Revisiting plant–microbe interactions and microbial consortia application for enhancing sustainable agriculture: A review. *Frontiers in Microbiology, 11*. https://doi.org/10.3389/fmicb.2020.560406
150. Mohapatra, S., Maity, S., Dash, H. R., Das, S., Pattnaik, S., Rath, C. C., & Samantaray, D. (2017). Bacillus and biopolymer: Prospects and challenges. *Biochemistry and Biophysics Reports, 12*, 206–213. https://doi.org/10.1016/j.bbrep.2017.10.001
151. Kumar, R., Merugu, R., Mohapatra, S., & Sharma, S. (2022). Extremophiles: Life of microorganisms in extreme environments. In *Extremophiles: General and plant biomass based biorefinery* (pp. 43–66). https://doi.org/10.1201/9781003014188-3
152. Pradhan, A., Pahari, A., Mohapatra, S., & Mishra, B. B. (2017). Phosphate-solubilizing microorganisms in sustainable agriculture: Genetic mechanism and application. In *Microorganisms for sustainability* (pp. 81–97). https://doi.org/10.1007/978-981-10-7380-9_5
153. Devadarshini, D., Mohapatra, S., Pati, S., Maity, S., Rath, C. C., Jena, P. K., & Samantaray, D. (2023). Evaluation of PHAs production by mixed bacterial culture under submerged fermentation. *Biologia (Bratisl), 78*, 895–902. https://doi.org/10.1007/s11756-022-01302-5
154. Agarwal, K. M., Mohapatra, S., Sharma, P., Sharma, S., Bhatia, D., & Mishra, A. (2020). Study and overview of the novel corona virus disease (COVID-19). *Sensors International, 1*. https://doi.org/10.1016/j.sintl.2020.100037
155. Tanaya, K., Mohapatra, S., Maity, S., & Samantaray, D. P. (2023). Fluorescent-pigmented microbial polymer nanoparticles: An approach for replacement of bioimaging agent in the drug delivery system. *Applied Biochemistry and Microbiology, 59*, 378–383. https://doi.org/10.1134/S0003683823030171
156. Pattnaik, S., Dash, D., Mohapatra, S., Pati, S., Devadarshini, D., Samal, S., Pattnaik, M., Maity, S., Mishra, S. K., & Samantaray, D. (2024). Reclamation of chromium-contaminated soil by native Cr(VI)-reducing and PHA-accumulating Bacillus aryabhattai CTSI-07. *International Microbiology, 27*, 731–742. https://doi.org/10.1007/s10123-023-00421-6
157. Mohapatra, S., Mishra, Y., & Dash, L. (2020). Covid-19: Analysing the legal nuances of the lockdown order. *Indian Journal of Forensic Medicine and Toxicology, 14*, 612–616. https://doi.org/10.37506/ijfmt.v14i4.11553
158. Pati, S., Mohanty, M. K., Mohapatra, S., & Samantaray, D. (2022). Bioethanol production by Enterobacter hormaechei through carbon monoxide-rich syngas fermentation. *International Journal of Energy Research, 46*, 20096–20106. https://doi.org/10.1002/er.8678
159. Maity, S., Das, S., Mohapatra, S., Tripathi, A. D., Akthar, J., Pati, S., Pattnaik, S., & Samantaray, D. P. (2020). Growth associated polyhydroxybutyrate production by the novel Zobellellae tiwanensis strain DD5 from banana peels under submerged fermentation. *International Journal of Biological Macromolecules, 153*, 461–469. https://doi.org/10.1016/j.ijbiomac.2020.03.004
160. Takkar, S., Shandilya, C., Agrahari, R., Chaurasia, A., Vishwakarma, K., Mohapatra, S., Varma, A., & Mishra, A. (2022). Green technology: Phytoremediation for pesticide pollution. In *Phytoremediation technology for the removal of heavy metals and other contaminants from soil and water* (pp. 353–375). https://doi.org/10.1016/B978-0-323-85763-5.00008-8

161. Bandekar, D., Chouhan, O. P., Mohapatra, S., Hazra, M., Hazra, S., & Biswas, S. (2017). Putative protein VC0395_0300 from Vibrio cholerae is a diguanylate cyclase with a role in biofilm formation. *Microbiological Research, 202*, 61–70. https://doi.org/10.1016/j.micres.2017.05.003
162. Pattnaik, S., Dash, D., Mohapatra, S., Pattnaik, M., Marandi, A. K., Das, S., & Samantaray, D. P. (2020). Improvement of rice plant productivity by native Cr(VI) reducing and plant growth promoting soil bacteria Enterobacter cloacae. *Chemosphere, 240*. https://doi.org/10.1016/j.chemosphere.2019.124895
163. Mohapatra, S., Samantaray, D. P., Samantaray, S. M., Mishra, B. B., Das, S., Majumdar, S., Pradhan, S. K., Rath, S. N., Rath, C. C., Akthar, J., & Achary, K. G. (2016). Structural and thermal characterization of PHAs produced by Lysinibacillus sp. through submerged fermentation process. *International Journal of Biological Macromolecules, 93*, 1161–1167. https://doi.org/10.1016/j.ijbiomac.2016.09.077
164. Sharma, S. K., Kumar, A., & Sharma, R. C. (2014). Challenges in railway vehicle modeling and simulations. In *International Conference on Newest Drifts in Mechanical Engineering ICNDME-2014* (pp. 453–459).
165. Sharma, S. K., & Kumar, A. (2014). A comparative study of Indian and worldwide railways. *International Journal of Mechanical Engineering and Robotics Research, 1*, 114–120.
166. Sharma, S. K., Sharma, R. C., Kumar, A., & Palli, S. (2015). Challenges in rail vehicle-track modeling and simulation. *International Journal of Vehicle Structures & Systems, 7*, 1.
167. Palli, S., Sharma, R. C., Sharma, S. K., Muddada, V., & Sharma, N. (2020). A case study on noise pollution and its effects. *Journal of Critical Reviews, 7*, 261–267.
168. Sharma, R. C., Sharma, S. K., Sharma, N., & Sharma, S. (2020). Analysis of ride and stability of an ICF railway coach. *International Journal of Vehicle Noise and Vibration, 16*, 127–144.
169. Palli, S., Sharma, R. C., Sharma, S. K., & Chintada, V. B. (2020). On methods used for setting the curve for railway tracks. *Journal of Critical Reviews, 7*, 241–246.
170. Lee, J., & Sharma, S. K. (2020). Numerical investigation of critical speed analysis of high-speed rail vehicle. In 한국정밀공학회 학술발표대회 논문집 (p. 696).
171. Sharma, S., Sharma, R. C., Sharma, S. K., Sharma, N., Palli, S., & Bhardawaj, S. (2020). Vibration isolation of the quarter car model of road vehicle system using dynamic vibration absorber. *International Journal of Vehicle Structures & Systems, 12*, 513–516.
172. Acharya, A., Gahlaut, U., Sharma, K., Sharma, S. K., Vishwakarma, P. N., & Phanden, R. K. (2020). Crashworthiness analysis of a thin-walled structure in the frontal part of automotive chassis. *International Journal of Vehicle Structures & Systems, 12*, 517–520.
173. Sharma, S. K., & Lee, J. (2020). Finite element analysis of a fishplate rail joint in extreme environment condition. *International Journal of Vehicle Structures & Systems, 12*, 503–506.
174. Bhardawaj, S., Sharma, R. C., & Sharma, S. K. (2020). Development of multibody dynamical using MR damper based semi-active bio-inspired chaotic fruit fly and fuzzy logic hybrid suspension control for rail vehicle system. *Proceedings of the Institution of Mechanical Engineers, Part K: Journal of Multi-Body Dynamics, 234*, 723–744.
175. Sharma, S. K., & Lee, J. (2020). Design and development of smart semi active suspension for nonlinear rail vehicle vibration reduction. *International Journal of Structural Stability and Dynamics, 20*, 2050120.
176. Sharma, S. K., Phan, H., & Lee, J. (2020). An application study on road surface monitoring using DTW based image processing and ultrasonic sensors. *Applied Sciences, 10*, 4490.
177. Sharma, R. C., Sharma, S., Sharma, S. K., & Sharma, N. (2020). Analysis of generalized force and its influence on ride and stability of railway vehicle. *Noise and Vibration Worldwide, 51*, 95–109.
178. Bhardawaj, S., Sharma, R. C., & Sharma, S. K. (2020). Ride analysis of track-vehicle-human body interaction subjected to random excitation. *Journal of the Chinese Society of Mechanical Engineers, 41*, 227–236.
179. Bhardawaj, S., Sharma, R. C., & Sharma, S. K. (2020). Development in the modeling of rail vehicle system for the analysis of lateral stability. *Materials Today Proceedings, 25*, 610–619.

180. Sharma, S. K., Sharma, R. C., & Sharma, N. (2020). Combined multi-body-system and finite element analysis of a rail locomotive crashworthiness. *International Journal of Vehicle Structures & Systems, 12*, 428–435.
181. Sharma, R. C., Avesh, M., Sharma, S. K., Sharma, S., Palli, S., & Sharma, N. (2020). Linear and non-linear stability analysis of a constrained railway wheel axle. *International Journal of Vehicle Structures & Systems, 12*, 128–133. https://doi.org/10.4273/ijvss.12.2.04
182. Bhardawaj, S., Sharma, R. C., & Sharma, S. K. (2020). Analysis of frontal car crash characteristics using ANSYS. *Materials Today Proceedings, 25*, 898–902.
183. Bhardawaj, S., Sharma, R. C., & Sharma, S. K. (2019). A survey of railway track modelling. *International Journal of Vehicle Structures & Systems, 11*, 508–518.
184. Bhardawaj, S., Sharma, R. C., & Sharma, S. K. (2019). Development and advancement in the wheel-rail rolling contact mechanics. In *IOP conference series: Materials science and engineering* (p. 012034). IOP Publishing.
185. Choppara, R. K., Sharma, R. C., Sharma, S. K., & Gupta, T. (2019). Aero dynamic cross wind analysis of locomotive. In *IOP conference series: Materials science and engineering* (p. 012035). IOP Publishing.
186. Sharma, S. K., & Sharma, R. C. (2019). Pothole detection and warning system for Indian roads. In *Advances in interdisciplinary engineering: Select Proceedings of FLAME 2018* (pp. 511–519). Springer Singapore.
187. Sinha, A. K., Sengupta, A., Gandhi, H., Bansal, P., Agarwal, K. M., Sharma, S. K., Sharma, R. C., & Sharma, S. K. (2019). Performance enhancement of an all-terrain vehicle by optimizing steering, powertrain and brakes. In *Advances in engineering design: Select Proceedings of FLAME 2018* (pp. 207–215). Springer Singapore.
188. Goyal, S., Anand, C. S., Sharma, S. K., & Sharma, R. C. (2019). Crashworthiness analysis of foam filled star shape polygon of thin-walled structure. *Thin-Walled Structures, 144*, 106312.
189. Goswami, B., Rathi, A., Sayeed, S., Das, P., Sharma, R. C., & Sharma, S. K. (2019). Optimization design for aerodynamic elements of Indian locomotive of passenger train. In *Advances in engineering design: Select Proceedings of FLAME 2018* (pp. 663–673). Springer Singapore.
190. Sharma, S. K., Saini, U., & Kumar, A. (2019). Semi-active control to reduce lateral vibration of passenger rail vehicle using disturbance rejection and continuous state damper controllers. *Journal of Vibration Engineering & Technologies, 7*, 117–129.
191. Sharma, S. K. (2019). Multibody analysis of longitudinal train dynamics on the passenger ride performance due to brake application. *Proceedings of the Institution of Mechanical Engineers, Part K: Journal of Multi-body Dynamics, 233*, 266–279.
192. Sharma, S. (2018). Vibration control in quarter-car model with magnetorheological (MR) dampers using Bingham model. *Journal of Applied Mechanics Engineering, 7*, 1000299.
193. Adeoye, A. O., Quadri, R. O., & Lawal, O. S. (2020). Pyrolysis of biomass as a suitable alternative to fossil fuel energy in Nigeria: An overview. *Journal of Wastes and Biomass Management, 3*(1), 27–30. https://doi.org/10.26480/jwbm.01.2021.27.30
194. Sharma, S. K., & Kumar, A. (2018). Impact of longitudinal train dynamics on train operations: A simulation-based study. *Journal of Vibration Engineering & Technologies, 6*, 197–203.
195. Sharma, S. K., & Sharma, R. C. (2018). An investigation of a locomotive structural crashworthiness using finite element simulation. *SAE International Journal of Commercial Vehicles, 11*, 235–243.
196. Sharma, R. C., Sharma, S. K., & Palli, S. (2018). Rail vehicle modelling and simulation using Lagrangian method. *International Journal of Vehicle Structures and Systems, 10*, 188–194.
197. Sharma, R. C., & Sharma, S. K. (2018). Sensitivity analysis of three-wheel vehicle's suspension parameters influencing ride behavior. *Noise and Vibration Worldwide, 49*, 272–280.
198. Palli, S., Koona, R., Sharma, S. K., & Sharma, R. C. (2018). A review on dynamic analysis of rail vehicle coach. *International Journal of Vehicle Structures & Systems, 10*.

199. Sharma, S. K., & Sharma, R. C. (2018). Simulation of quarter-car model with magnetorheological dampers for ride quality improvement. *International Journal of Vehicle Structures & Systems, 10*, 169–173.
200. Sharma, S. K., & Kumar, A. (2018). Disturbance rejection and force-tracking controller of nonlinear lateral vibrations in passenger rail vehicle using magnetorheological fluid damper. *Journal of Intelligent Material Systems and Structures, 29*, 279–297.
201. Sharma, S. K., & Kumar, A. (2018). Ride comfort of a higher speed rail vehicle using a magnetorheological suspension system. *Proceedings of the Institution of Mechanical Engineers, Part K: Journal of Multi-Body Dynamics, 232*, 32–48.
202. Shukla, S., Kumar, U., Sharma, S. K., Gupta, P., & Kumar, A. (2017). Light weight freight rolling stock bogie frame: Design methodology validated with field oscillation trials. *International Journal of Vehicle Structures & Systems, 9*, 245–250.
203. Sharma, R. C., Palli, S., Sharma, S. K., & Roy, M. (2017). Modernization of railway track with composite sleepers. *International Journal of Vehicle Structures & Systems, 9*, 321–329.
204. Sharma, S. K., & Kumar, A. (2017). Ride performance of a high speed rail vehicle using controlled semi active suspension system. *Smart Materials and Structures, 26*, 055026.
205. Sharma, S. K., & Kumar, A. (2017). Impact of electric locomotive traction of the passenger vehicle ride quality in longitudinal train dynamics in the context of Indian railways. *Mechanics & Industry, 18*, 222.
206. Sharma, S. K. (2013). Zero energy building envelope components: A review. *International Journal of Engineering Research and Applications, 3*, 662–675.
207. Sharma, S. K., & Kumar, A. (2016). Dynamics analysis of wheel rail contact using FEA. *Procedia Engineering, 144*, 1119–1128.
208. Sharma, S. K., & Chaturvedi, S. (2016). Jerk analysis in rail vehicle dynamics. *Perspectives in Science, 8*, 648–650.
209. Sharma, S. K., & Kumar, A. (2016). The impact of a rigid-flexible system on the ride quality of passenger bogies using a flexible carbody. In *Proceedings of the Third International Conference on Railway Technology: Research, Development and Maintenance* (p. 87). Civil-Comp Press.
210. Lavania, S., & Sharma, S. K. (2011). An explicit approach to compare crosstalk noise and delay in VLSI RLC interconnect modeled with skin effect with step and ramp input. *Journal of VLSI Design Tools & Technology, 1*, 1–8.
211. Palli, S., Koona, R., Sharma, R. C., & Muddada, V. (2015). Dynamic analysis of Indian railway integral coach factory Bogie. *International Journal of Vehicle Structures and Systems, 7*, 16–20. https://doi.org/10.4273/ijvss.7.1.03
212. Palli, S., & Koona, R. (2015). Analyses of dynamic response of a railway Bogie. *International Journal of Vehicle Noise and Vibration, 11*, 103–113. https://doi.org/10.1504/IJVNV.2015.070015
213. Swathi, K., & Srihari, P. (2015). Heat transfer enhancement in a tube using rectangular strip inserts. *International Journal of Applied Engineering Research, 10*, 41532–41544.
214. Palli, S., Sharma, R. C., & Rao, P. P. D. (2017). Dynamic behaviour of a 7 DoF passenger car model. *International Journal of Vehicle Structures and Systems, 9*, 57–63. https://doi.org/10.4273/ijvss.9.1.12
215. Sharma, R. C., Palli, S., & Koona, R. (2017). Stress and vibrational analysis of an Indian Railway RCF Bogie. *International Journal of Vehicle Structures and Systems, 9*, 296–302. https://doi.org/10.4273/ijvss.9.5.06
216. Palli, S., Dontikurti, R., Sharma, R. C., & Sharma, N. (2021). Analysis of dynamic response of a railway locomotive using ANSYS. *International Journal of Vehicle Structures and Systems, 13*, 250–255. https://doi.org/10.4273/ijvss.13.2.21
217. Sharma, R. C., Sharma, N., Singh, G., & Palli, S. (2021). Modeling and simulation of human body-vehicle-track system for the investigation of ride comfort. *SAE International Journal of Commercial Vehicles, 15*. https://doi.org/10.4271/02-15-02-0008

218. Sharma, R. C., Palli, S., Avesh, M., & Sharma, N. (2021). Vibration isolation of railway vehicle car body using semi-active suspension. *International Journal of Vehicle Structures and Systems, 13*, 482–487. https://doi.org/10.4273/ijvss.13.4.17
219. Gopala Rao, L. V. V., Sharma, R. C., Satyanarayana, V. S. V., & Palli, S. (2022). Stochastic optimal preview control response of a quarter-car nonlinear suspension model using spectral decomposition method. *Noise and Vibration Worldwide, 53*, 225–232. https://doi.org/10.1177/09574565221093234
220. Palli, S., Duppala, A., Sharma, R. C., & Rao, L. V. V. G. (2022). Dynamic simulation of automotive vehicle suspension using MATLAB Simulink. *International Journal of Vehicle Structures and Systems, 14*, 310–314. https://doi.org/10.4273/ijvss.14.3.04
221. Satyanarayana, V. S. V., Sharma, R. C., Sateesh, B., Gopala Rao, L. V. V., Mohan Rao, N., & Palli, S. (2022). Optimum response of a nonlinear passive vehicle suspension system under random road excitations. *SAE International Journal of Commercial Vehicles, 16*. https://doi.org/10.4271/02-16-01-0004
222. Lokanadham, D., Sharma, R. C., Palli, S., & Bhardawaj, S. (2022). Wear rate modelling and analysis of limestone slurry particulate composites using the fuzzy method. *International Journal on Recent and Innovation Trends in Computing and Communication, 10*, 133–143. https://doi.org/10.17762/ijritcc.v10i1s.5818
223. Sharma, R. C. (2016). Evaluation of passenger ride comfort of Indian rail and road vehicles with ISO 2631-1 standards: Part 1 - mathematical modeling. *International Journal of Vehicle Structures & Systems (IJVSS), 8*(1), 1–6. https://doi.org/10.4273/ijvss.8.1.01
224. Sharma, R. C., Gopala Rao, L. V. V., Sharma, S. K., Palli, S., & Satyanarayana, V. S. V. (2022). Analysis of lateral stability and ride of an Indian railway constrained dual-axle bogie frame. *SAE International Journal of Commercial Vehicles, 16*. https://doi.org/10.4271/02-16-02-0014
225. Sharma, R. C., Palli, S., Bhardawaj, S., & Rao, L. V. V. G. (2022). Advances in human biodynamic modeling for analysis of whole-body vibrations. *International Journal of Vehicle Structures and Systems, 14*, 503–511. https://doi.org/10.4273/ijvss.14.4.17
226. Sharma, R. C., Palli, S., & Sharma, S. K. (2023). Ride analysis of railway vehicle considering rigidity and flexibility of the carbody. *Journal of the Chinese Institute of Engineers, Transactions of the Chinese Institute of Engineers, Series A, 46*, 355–366. https://doi.org/10.1080/02533839.2023.2194918
227. Sharma, R. C., Palli, S., & Bhardawaj, S. (2023). A review of wheel-rail dynamic behavior on curved tracks. *International Journal of Vehicle Structures and Systems, 15*, 19–25. https://doi.org/10.4273/ijvss.15.1.04
228. Jyotiprasad, G., Dowluru, S., Rallabandi, S. R., Sharma, N., Sharma, R. C., Palli, S. (2023). *Thermal-structural and macromechanical behavior analysis of graphite-based laminated composite*. SAE Technical Paper 2023-01-5041. https://doi.org/10.4271/2023-01-5041.
229. Saikiran, C., Palli, S., Bondala, R., Sharma, R. C., Sreeramulu, D., & Dontikurti, R. (2023). Experimental investigation on thermal behaviour of silicon carbide (SiC) based polymer composites under hygrothermal conditions. *International Journal of Vehicle Structures and Systems, 15*, 222–226. https://doi.org/10.4273/ijvss.15.2.15
230. Singh, B. J., Palli, S., Kumar, P., & Bahl, S. (2023). Behavioural investigation of hardness profile for dissimilar TIG-weldment between P22 and P91 steels. *International Journal of Vehicle Structures and Systems., 15*, 31–36. https://doi.org/10.4273/ijvss.15.1.06
231. Naveen, Y., Lokanadham, D., Naidu, D. R., Sharma, R. C., Palli, S., & Lila, M. K. (2023). An experimental study on the influence of blended Karanja biodiesel on diesel engine characteristics. *International Journal of Vehicle Structures and Systems, 15*, 154–160. https://doi.org/10.4273/ijvss.15.2.02
232. Bhardawaj, S., Jha, A. K., Sharma, R. C., Palli, S., Lokanadham, D., Sharma, N., & Dhiman, N. (2023). Finite element analysis of two disk rotor system used in automobile turbochargers. *International Journal of Vehicle Structures and Systems, 15*, 501–504. https://doi.org/10.4273/ijvss.15.4.09

233. Sharma, R. C., Palli, S., Bhardawaj, S., Jha, A. K., Duppala, A., Sharma, N., Sharma, S., & Dhiman, N. (2023). Dynamic behaviour of railway vehicle under bump and pothole track irregularities. *International Journal of Vehicle Structures and Systems, 15*, 566–569. https://doi.org/10.4273/ijvss.15.4.22
234. Raju, S., Palli, S., Devi Prasad, P., Menda, V. R., & Ramakrishna, B. (2024). A hybrid AHP-TOPSIS, MOORA technique for multi-objective optimization of thermal, mechanical, and water absorption behavior of epoxy/hemp, pine apple, and palm fiber composites. *Journal of the Chinese Institute of Engineers, Transactions of the Chinese Institute of Engineers, Series A, 47*, 23–36. https://doi.org/10.1080/02533839.2023.2274092
235. Raju Rallabandi, S., Pilla, D. P., Dowluru, S., Palli, S., Sharma, N., Sharma, S. K., & Sharma, R. C. (2024). Critical evaluation of epoxy-hemp-pineapple-palm fiber composites using hybrid AHM-TOPSIS technique for sustainable structural applications. *Journal of the Chinese Institute of Engineers, Transactions of the Chinese Institute of Engineers, Series A, 47*, 325–336. https://doi.org/10.1080/02533839.2024.2308250
236. Gopala Rao, L. V. V., Satyanarayana, V. S. V., Sharma, R. C., Palli, S., Raturi, A., Kharola, A., & Sharma, A. (2024). Preview control of the random response of a full car vehicle model traversing a rough road. *International Journal of Vehicle Noise and Vibration, 20*, 70–88. https://doi.org/10.1504/IJVNV.2024.138105
237. Sharma, A., Sharma, V. S., Gupta, S., Sharma, R. C., Palli, S., & Sharma, N. (2024). Additive manufacturing in the COVID-19 pandemic: Equipment and challenges? *Journal of Micromanufacturing*. https://doi.org/10.1177/25165984241228083
238. Bammidi, R., Madivada, H., Dowluru, S., Palli, S., Sharma, R. C., & Muddada, V. (2024). Mechano-tribological investigations of Al-%Wt. of boron nitride as metal-ceramic composites. *AIP Conference Proceedings*. https://doi.org/10.1063/5.0182965
239. Ranganath, B. A., Duppala, A., Palli, S., Rao, L. V. V., & Sharma, R. C. (2024). Numerical analysis and experimental study of FRP composites for damping applications. *AIP Conference Proceedings*. https://doi.org/10.1063/5.0184802
240. Menda, V., Dora, S., Palli, S., Rallabandi, S. R., Sharma, R. C., Sharma, N., & Bondala, R. (2024). Mechanical and wear characteristics of aluminium-7075/graphene/TiB_2 prepared by stir casting. *Engineering Research Express, 6*. https://doi.org/10.1088/2631-8695/ad58a8

Industrial Evaluation of Combustion Performance and Fracture Toughness in Epoxy Glass Fiber/Phenolic Composites During Oxidation Gas Release

Rapaka Jagadeesh, Bathula Venkata Siva, Mendi Venkanna Babu, Haliru Ibrahim Muhammad, Sani Abdullahi Aliyu, and Samson Abiodun AASA

1 Introduction

In high-performance applications such as aerospace, automotive, and marine industries, the use of traditional materials is often limited by their inability to meet extreme environmental demands, including high temperatures and exposure to combustion gases. Fiber-reinforced composites, particularly epoxy glass fiber/phenolic composites, have emerged as a preferred choice due to their favorable mechanical properties, including high strength-to-weight ratios, resistance to fatigue, and superior fracture toughness. However, the behavior of these materials under fire conditions, especially during the combustion of oxidation gases, poses significant challenges that need to be addressed for their broader application in critical environments. These studies explore a range of advanced techniques and sustainable practices that align with the goals of smart manufacturing and Industry 5.0. Adeoye et al. (2024) investigate the valorization of lignocellulosic waste, highlighting the potential of bio-oil production through pyrolysis and ultrasonication for biofuel generation [1]. Further research by Adeoye et al. focuses on upgrading crude bio-oil

R. Jagadeesh (✉)
Department of Mechanical Engineering, Aditya Institute of Technology and Management, Tekkali, Andhra Pradesh, India

B. Venkata Siva · M. Venkanna Babu
Narasaraopeta Engineering College, Narasaraopeta, Andhra Pradesh, India
e-mail: mechhod@nrtec.in

H. Ibrahim Muhammad
Department of Physics, Federal University Gusau, Gusau, Zamfara State, Nigeria

S. A. Aliyu
Department of Chemistry, University of Lille, Lille, France

S. A. AASA
Department of Automotive Engineering, University of Ibadan, Ibadan, Nigeria

Y. Kumar Singla et al. (eds.), *Digital Manufacturing in Industry 5.0*,
https://doi.org/10.1007/978-3-031-91500-0_18

[2, 3] and sustainable energy production from biomass [4], providing insights into thermochemical and biochemical conversions. These approaches underscore the integration of clean energy solutions into digital manufacturing processes. Studies on catalytic pyrolysis and nanomaterials for enhanced biomass conversion offer pathways for optimizing manufacturing efficiency [5–7]. Additionally, research on pyrolysis of plastic waste and rail projects introduces circular economy concepts and environmental sustainability, both essential to the evolution of Industry 5.0. Recent advancements in sustainable fuel production have focused on various bio-oil sources and carbon capture techniques. Studies have examined bio-oil production from thermal cracking in the presence of hydrogen [8], CO_2 adsorption systems in coal beds [9], and the pyrolysis of coffee and eucalyptus sawdust [10–12]. Additional research has explored the use of sulfur to enhance wood tar biopitch as a sustainable binder, the characterization of sewage sludge bio-oil, and the analysis of bio-oil from algae, comparing freshwater and microbial-treated samples [13]. These efforts collectively contribute to the development of renewable energy and environmental sustainability.

In parallel, work on fiber surface functionalization and dye fastness adds to the understanding of material innovation in manufacturing [14–16]. These studies contribute to the body of knowledge driving digital transformation and environmental consciousness in manufacturing [17–23].

These works emphasize a variety of advanced engineering techniques that contribute to the development of smart manufacturing and Industry 5.0. Studies like Shah et al. [24] and Singla et al. [25] examine the effects of material science advancements on welding and sustainability, key components for industrial automation and optimized production processes. Additionally, research on solar energy systems and tribological studies on engine lubricants [26–30] align with the sustainable energy goals of Industry 5.0.

Recent advancements in bio-oil production and carbon capture techniques are driving sustainability in energy production [31]. Studies have explored bio-oil production through thermal cracking in the presence of hydrogen, and the pyrolysis of various biomass sources such as coffee, eucalyptus sawdust, and algae, all contributing to renewable energy solutions [32–37]. Additionally, research on CO_2 adsorption systems and the development of sustainable materials for adsorption, like zeolitic materials and montmorillonite clay, are crucial for addressing environmental challenges [32, 36, 38–42]. These efforts are complemented by investigations into biodiesel production, aviation fuel alternatives, and the separation of valuable chemicals from bio-oils, advancing the potential of biofuels in various industrial applications [43, 44].

The contributions from Sunil et al. [45–56] span critical advancements in transportation systems, materials science, and engineering technology, all of which have applications in smart manufacturing and Industry 5.0. Recent works from 2024 focus on the adoption of electric vehicles [57–74], modeling of intelligent transportation systems [75–88], and vibration mitigation in marine and rail propulsion systems [89–98].

These works highlight innovative methods in microbiology, environmental sustainability, and materials science, which have significant implications for smart manufacturing and Industry 5.0 [99–104]. Mohapatra et al. [105] study microbial processes for biopolymer production, which can be applied in sustainable packaging solutions [17, 102, 106–113] and bio-based materials [114–125], aligning with Industry 5.0's focus on circular economy and smart materials. Research on retrofitting buildings for energy efficiency [126–142] and the use of polyhydroxyalkanoates (PHAs) for biocompatible applications [143–152] demonstrates the integration of biotechnology into industrial processes.

The integration of alloying elements to enhance wear resistance [153] supports the development of materials with improved durability for high-tech manufacturing applications. Moreover, works on the surface integrity of machined materials and the impact of nano-particulates in lubrication show how material engineering can improve efficiency in production processes. These studies collectively enhance the scope of smart manufacturing by combining material science, energy efficiency, and automation, advancing the transition toward Industry 5.0 [154, 155].

These studies underscore the need for energy-efficient solutions and advanced control systems, both of which are integral to Industry 5.0 [156, 157]. Previous works from 2023 also explore sustainable materials for transportation [158], intelligent systems for railway infrastructure [159–175], and advancements in modeling and ride behavior [176–184]. Research on crashworthiness [185–194] and the integration of bio-lubricants [195–198] promote the goals of Industry 5.0 by improving safety [199, 200], efficiency, and sustainability [201–203]. These studies collectively support the evolution of smart manufacturing, contributing to Industry 5.0's emphasis on sustainability, digitalization, and innovation across sectors [204].

The research contributions focus on the dynamic analysis, vibration isolation, and advanced material applications in transportation and vehicle systems, which directly support the objectives of smart manufacturing and Industry 5.0 [205–209]. Recent works (2024) address advanced control systems for enhanced ride quality in passenger rail vehicles using intelligent control suspension systems [163, 173, 210–213], and comprehensive dynamic response analyses of road and rail vehicles [214–221]. Further, studies involving the use of composite materials and hybrid AHP-TOPSIS techniques for sustainable structural applications [65, 68, 71, 74, 222–232] contribute to the development of smart materials, which is central to Industry 5.0.

Additionally, advancements in digital twin technology for improving logistics [233], the role of extremophiles in enhancing sustainable agriculture [234, 235], and bioremediation of contaminated soils [236, 237] highlight the role of digital innovation and environmental consciousness in next-generation manufacturing. These studies offer diverse insights into how microbiological advancements and environmental sustainability are being harnessed for smart and sustainable industrial applications, resonating with the principles of Industry 5.0 [174, 238].

Works from earlier years explore dynamic simulations [57, 73], improving suspension systems, and enhancing material wear resistance, which collectively reflect advancements in digitalization, automation, and sustainable manufacturing practices. These studies highlight a multidisciplinary approach to improving vehicle systems and material performance in line with the goals of Industry 5.0.

In this chapter, we investigate the focused use of epoxy glass fiber/phenolic composites with respect to crack growth resistance evaluation and behavior under combustion. The fracture toughness (FT), limiting oxygen index (LOI), and the heat release rate (HRR) for these materials as well as composites were pulled out under different heat treatments. Understanding these specific material properties of epoxy materials provides mechanisms on mechanical load combined with combustion gas challenges and proposes fire protection and mechanical strength improvements. These results suggest the necessity for re-engineering of epoxy glass fiber/phenolic composites for use in more robust engineering applications that require high performance levels in the fire and structural integrity. The work presented also shows that there are strategies of strengthening fracture toughness and the combustion control which are necessary for technological advances in composite materials of the future.

2 Material Selection and Preparation

2.1 *Epoxy Glass Fiber/Phenolic Composites*

These composites have applications where both aesthetic and fire performance are required. They possess good strength, toughness and thermal stability and can be utilized in aerospace, automotive, and construction industries. The matrix material, usually phenolic resin, possesses flame retardant properties, whereas glass fiber reinforcement increases some mechanical characteristics: fracture toughness and crack growth resistance.

The value of these materials is even greater in the case when operating in environments subject to fire exposure. Phenolic char decreases the rate of combustion and toxic gas release, which increases safety for phenolic-based composites used in applications like aircraft interiors where fire hazards are a concern. Furthermore, the epoxy portion of the composite reinforces the matrix material, which helps in retaining the structure during mechanical and thermal stresses.

2.2 *Preparation Methods for Testing*

The composites tested in this study were fabricated by hand lay-up followed by vacuum assisted curing. The reinforcement was fabricated from glass fiber as a dry fiber fabric, while phenolic matrix was used for the composite matrix in epoxy resin. Different composite samples were prepared in specified ratios for further experimentation.

The process started with a resin and hardener mixture in a 10 to 1 ratio which was first introduced to the layers of glass fiber. A hand roller helped to ensure that the resin penetrated the fiber bundles completely. This was done until 12 layers of the fiber were formed and to the mid-plane a porous film was inserted to form a pre-crack that would be used for fracture testing. The formed stack of fibers and resin was also prepared by applying latter layers of fiber and resin and these were sealed in a vacuum bag and then the entire pack was placed between steel plates under heat to cure the pack following a set of temperature profile from cold to hot. After the curing treatment was finished, the laminates were taken off and sliced into templates for testing as stipulated by the standards ASTM D5528. The specimens were carved within peculiar size ranges for both fracture toughness and combustion testing, thus guaranteeing that the results of all the samples would be comparable.

2.3 Composition Ratios and Specifications

Three different composition ratios epoxy resin to glass fiber were prepared in order to find out the influence of the content of fiber on the fracture toughness and combustion behavior of the composites, which are 50:50, 60:40, and 70:30. These ratios were chosen in order to find out the effect on the mechanical and thermal properties of the composite with increase in the amount of fiber reinforcement.

- 50:50 Ratio: This composition exhibits uniform fiber/matrix distribution, where it can serve a primary gauge for toughness and combustion effect.
- 60:40 Ratio: If adequate fire-resistant characteristics are well maintained, then fracture toughness of this sample should be improved due to having added fibers.
- 70:30 Ratio: In this composition, fiber reinforcement is expected to reach its highest strength which gives some indications on the mechanical strength of the material as well as the reaction of the material to fire.

The specimens were subjected to mode-I fracture testing so that the resistance of the tested specimens to crack growth could be evaluated, and combustion testing was performed using a cone calorimeter; the heat release rate (HRR) and limiting oxygen index (LOI) were recorded during different thermal exposure. These tests also help to understand the level of impact, which colonies might cause both mechanical and fire performance of the material.

3 Experimental Methodology

3.1 Fracture Toughness Testing Mode I

Fracture toughness is one of the important parameters in the assessment of the mechanical properties of fiber-reinforced composites used in fire-sensitive applications. In this study, the fracture toughness (FT) of epoxy glass fiber/phenolic composites was assessed with the help of Mode I fracture tests according to ASTM D5528 standards. The method for testing entails applying a force on a DNB specimen fashioned in a manner that will cause it to crack for the purpose of evaluating the amount of force required by material to form a crack and sustain it.

The specimens were prepared and reinforced with various fiber ratios of 50:50, 60:40, and 70:30 to investigate the fiber effect on the toughness of the beam/splice constituents. Each specimen underwent a cutting process so as to obtain the desired form and engineers implanted a flaw during the lay-up process while making use of a porous film located at the midpoint of the composite's layers. In order to avoid randomly oriented crack growth, Mode I loading was aimed at the extremities of the specimen.

The findings showed that the 60:40 fiber ratio performed the best in terms of fracture toughness which means that with the increase in the amount of fiber, the material can resist the growth of cracks even further. But the toughness values started to drop slightly as the fiber ratio increased beyond a certain point (70:30) because of an increase in brittleness.

3.2 Parameters of Combustion Testing

A cone calorimeter was used to assess the burning characteristics, which measures the heat release rate of the combustion of the composite materials as the thermal radiation changes from one level to the other (50–70 kW/m^2). One factor which is critical in comprehension of the fire performance of materials is the HRR since this influences how flames will be formed and how fire will spread. Within the framework of this experiment, three fiber ratio samples, namely 50:50, 60:40, and 70:30 were analyzed for their combustion characteristics, among them:

- Picture or Peak Heat Release Rate (pHRR): The maximum rate of heat released for every unit time during combustion.
- Time to Ignition (TTI): The time that the material takes to ignite if a specific thermal load has been applied.
- Mass Loss Rate (MLR): The rate at which the composite material burned and lost mass due to heat.

According to the results, with the increase in the thermal radiation intensity, the HRR also increased, the highest pHRR was noted at the 70:30 fiber ratio which was 100.7 kW/m^2. This means although the material with more amount of fiber has better mechanical properties, it may be easy to burn more quickly at high thermal loads.

3.3 *Limitations of Oxygen Index (LOI) and Heat Release Rate (HRR) Analysis*

LOI is one of the most important characteristics for flammability of the material. It is the lowest concentration of oxygen in the air or atmosphere that is needed to maintain the flame. In this case study, the LOI values are determined for each composite sample over a range of temperatures from room temperature at 20 °C to maximum of 220 °C for the sample. When the fiber ratio of 60:40 was used, the LOI value climbed to a record 93.7%, making it the most suitable fiber ratio at this temperature. An HRR (heat release rate) curve was devised for each of the fiber ratio samples where it was realized that with increasing thermal radiation intensity, there was an increase in HRR which meant an increase in the burning speed. The greatest tendency in HRR change was noted for the 70:30 fiber ratio, which has increased radiation intensity to 100.7 kW/m^2 only with external thermal radiation of 70 m^2 kW.

4 Results and Discussion

4.1 *Fracture Toughness Analysis*

The Mode I fracture is used to relate variables to the failure load for the epoxy glass fiber/phenolic composites Ki. The results show a clear relationship where varying the fiber content into the composite will enable the material to be more resistant to crack growth.

- Fracture toughness evaluation was claimed at 0.616 GPA for 50:50 fiber to epoxy ratio fibers when fully fractured and this is taken for comparative purpose.
- At the 60:40 ratio, the fracture toughness improved to 0.92 GPa that has the highest resistance to the crack propagation among the three samples. This implies that with the addition of more glass fibers, there is a greater intercept of the fiber and the matrix increasing the strength of the composite in terms of stress absorption as well as stress distribution across the composite.
- As for the 70:30 ratio, a decrease in the value of fracture toughness was observed to be 0.5219 GPa. The decline is because as the fiber content increases, the material becomes more brittle as fibers take over the mechanical aspect which may not help in stress distribution as intended.

Figure 1 shows the variation in fracture toughness for different fiber ratios. The 60:40 ratio exhibits the highest fracture toughness (0.92 GPa), followed by the 50:50 and 70:30 ratios. The decrease in toughness at the 70:30 ratio can be attributed to the brittleness that accompanies increased fiber content.

Table 1 shows the values of fracture toughness, heat release rate (HRR), and limiting oxygen index (LOI) for different fiber ratios. The 60:40 fiber ratio exhibits

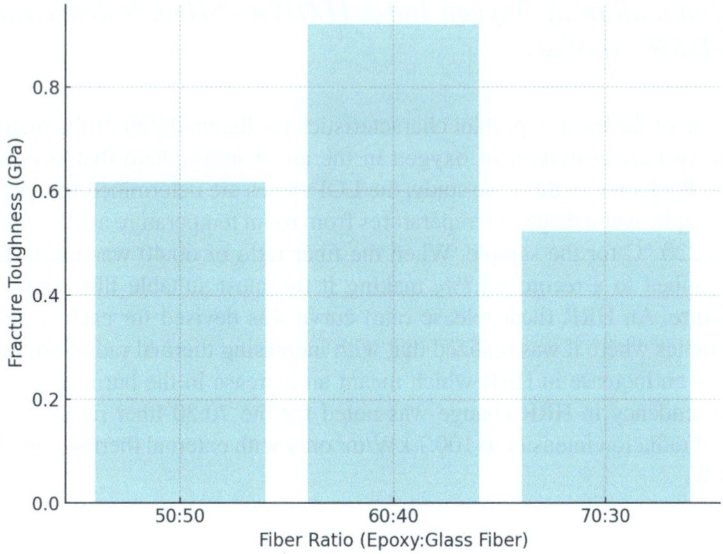

Fig. 1 Fracture toughness vs. fiber ratio

Table 1 Fracture toughness, HRR, and LOI for different fiber ratios

Fiber ratio (epoxy fiber)	Fracture toughness (GPa)	HRR (kW/m^2)	LOI (%)
50:50	0.616	64.7	86.6
60:40	0.92	70.0	93.7
70:30	0.5219	100.7	84.4

the highest fracture toughness and LOI, indicating superior mechanical strength and fire resistance. Conversely, the 70:30 ratio has the highest HRR, suggesting increased flammability at higher fiber contents.

4.2 Combustion Properties Upon Use of Gas Release for Oxidation

Combustion properties of the composites were evaluated in terms of heat release rate (HRR) and in terms of limiting oxygen index (LOI) while combustion of oxidation gases. These parameters are important to be known since they have the influence on the fire behavior of composite materials.

- With the use of 50:50 fiber ratio, an average HRR of 64.7 kW/m^2 was recorded as the peak which is relatively moderate and revealing moderate combustion behavior.

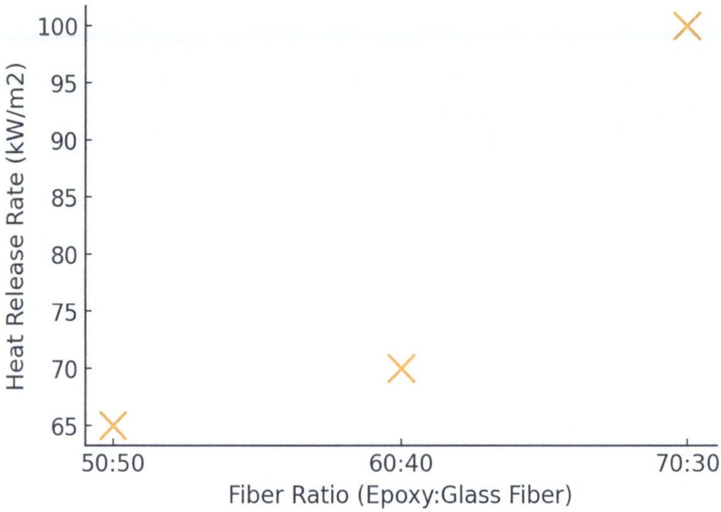

Fig. 2 Heat release rate (HRR) vs. fiber ratio

- In the 60:40 ratio, HRR was raised to 70.0 kW/m^2 which was in accordance with the higher value of LOI (93.7%) as it increases the fire resistance of the material.
- However, it could be observed that the sample with 70:30 ratio reached the maximum peak HRR of 100.7 kW/m^2, hence sustaining that such a sample with superior mechanical properties caused faster combustion when subjected to thermal load. As shown in Fig. 2, the HRR increases with higher fiber content. The peak HRR was observed at the 70:30 ratio, which exhibited a value of 100.7 kW/m^2. This suggests that higher fiber content leads to faster combustion and increased heat release.

These results reveal that there are challenges in shear failure and fusion preservation in fiber-reinforced polymer composites. Though enhanced fiber contents yield higher mechanical properties, it results in poor progression of micro combustion under high thermal circumstances. Therefore, the 60:40 appears to be quite the moderate with respect to mechanical safety resilience and fire safety.

4.3 The Importance of Thermal Radiation in Increasing HRR and LOI

Different composite samples were subjected to three radiation intensities: 50, 60, and 70 kW/m^2. Maximum heat release measured was found in higher powered radiations which can be linked to the area under the graph of the different fiber contents and above the level of such parameter. In total, a gradual step was noticed in all the samples.

- For the 50:50 ratio, the highest HRR was 64.7 kW/m^2 with time to ignition approximating 207 s while the radiation level was 50 kW/m^2.
- The peak HRR for the 60:40 ratio at 60 kW and the time taken to ignite was 70.0 kW/m^2 and 131 s respectively.
- The 70:30 ratio exhibited the most extreme response. The highest HRR was 100.7 kW/m^2 at 70 kW/m^2, whereas the time taken to ignite was 97 s suggesting that with rise in ratio of fibers, there is faster ignition and combustion as well. Figure 3 displays the distribution of LOI across the different fiber ratios. The 60:40 ratio had the highest LOI (93.7%), indicating greater fire resistance. In contrast, the 70:30 ratio showed the lowest LOI, reflecting reduced fire resistance at higher fiber contents.

Table 2 presents the HRR and LOI values for each fiber ratio, showing their performance at different thermal radiation intensities. As the radiation intensity increases, the HRR rises significantly, particularly for the 70:30 fiber ratio, which exhibits the highest HRR. The LOI decreases with higher radiation, indicating reduced fire resistance at elevated temperatures.

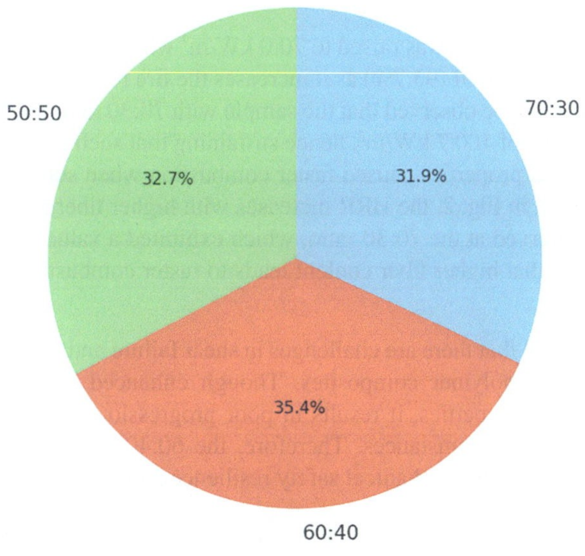

Fig. 3 Limiting oxygen index (LOI) distribution

Table 2 HRR vs. LOI at different radiation intensities for fiber ratios

Fiber ratio (epoxy fiber)	Radiation intensity (kW/m^2)	HRR (kW/m^2)	LOI (%)
50:50	50	64.7	86.6
60:40	60	70.0	93.7
70:30	70	100.7	84.4

4.4 Influence of Fiber Content on Material Properties

The results show that the fiber content is important issues when considering the mechanical and thermal performances of the epoxy glass fiber/phenolic composites. The 60:40 ratio seems to provide a best combination due to its relatively high break toughness and relatively low combustion and hence is of great advantage in processes where mechanical properties and combustibility are the main parameters.

An increase in the fiber content at the 70:30 ratio enhances mechanical performance. However, it also leads to faster burning and poorer combustion resistance as noted in the reduced LOI value and higher HRR. In contrast, the 50:50 ratio allows adequate mechanical and combustion performance but is not ideal for use in areas that experience high stress.

From these results, it has been clear that the fiber-to-matrix ratio must be optimized for specific applications of composite materials, especially those that require high structural strength and fire resistance, such as the aerospace and automotive industries. Figure 4 highlights the comparative values of fracture toughness, HRR, and LOI across fiber ratios. The 60:40 fiber ratio achieves a balance between high fracture toughness and fire resistance, while the 70:30 ratio shows higher HRR and lower LOI, indicating increased flammability.

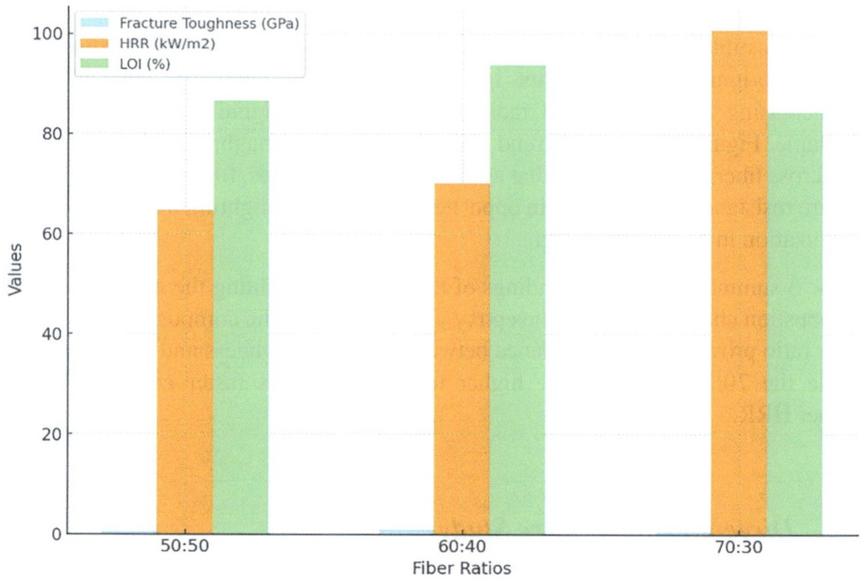

Fig. 4 Comparative analysis of fracture toughness, HRR, and LOI

5 Conclusion

5.1 Key Findings

This investigation has enabled an understanding of the fracture toughness and the combustion properties of the epoxy glass fiber/phenolic composites when varied fiber contents and thermal environments are applied. Important outcomes were the following:

- A ratio of 60:40 fibers to epoxy produced the largest amount of fracture toughness at 0.92 GPa making this composition the most favorable in terms of physical strength. It was noted as well that this ratio showed excellent fire resistance: the limiting oxygen index (LOI) at 100 °C was 93.7%. This indicates that such material can perform better in environments with fire threats.
- It was demonstrated that the 70:30 fiber ratio exhibited higher mechanical strength, although the fiber ratio appears to negatively affect fire resistance with a peak heat release rate HRR of 100.7 kW/m^2 ever recorded. This indicates that too much fiber loading may increase fire hazard which is undesirable of applications that are prone to fire hazards.
- The results also suggest that the two opposing issues have an optimum ratio of 60:40 in favor of the matrix which best suits most applications that require both mechanical and fire resistance.
- Combustible loading content and flame measurement has inverse relation with time-to-ignition values faster ignition has higher flammable content while increasing level of thermal radiation decreases the ignition timing for rapid flame. Figure 5 presents a trend analysis of fracture toughness, HRR, and LOI across fiber ratios. It shows that as fiber content increases, fracture toughness and fire resistance tend to move in opposite directions, highlighting the need for optimization in composite design.

Table 3 summarizes the key findings of the study, highlighting the mechanical and combustion characteristics of the epoxy glass fiber/phenolic composites. The 60:40 fiber ratio provides the best balance between fracture toughness and fire resistance, while the 70:30 ratio, despite higher toughness, shows faster combustion and higher HRR.

5.2 Directions for Future Study

In the light of the above findings, a few recommendations for future studies can be put forward in order to fine tune epoxy glass fiber/phenolic hybrids for higher usage particularly in the fire hazardous area:

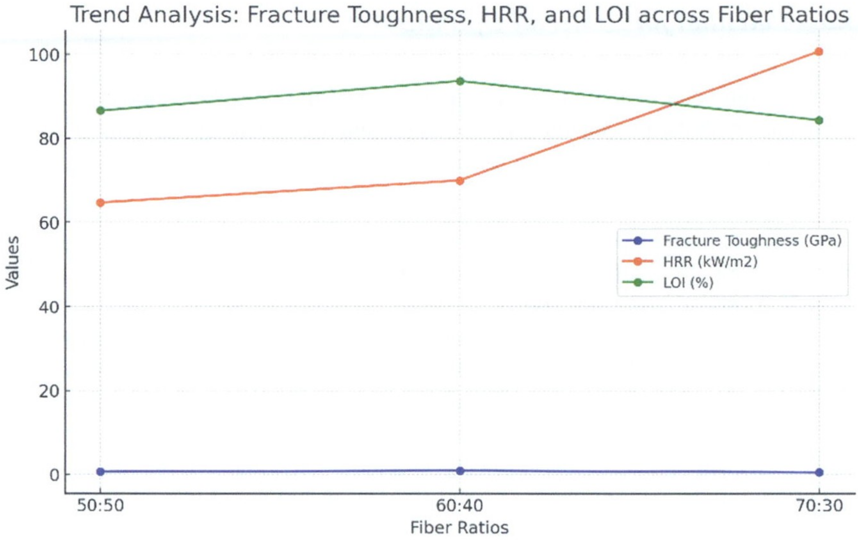

Fig. 5 Trend analysis of fracture toughness, HRR, and LOI

Table 3 Summary of fracture toughness and combustion characteristics

Parameter	50:50 Ratio	60:40 Ratio	70:30 Ratio
Fracture toughness (GPa)	0.616	0.92	0.5219
Peak heat release rate (HRR)	64.7 kW/m^2	70.0 kW/m^2	100.7 kW/m^2
Limiting oxygen index (LOI)	86.6%	93.7%	84.4%
Time to ignition (TTI)	207 s	131 s	97 s

- Modification of the Matrix: Future studies should also be directed at incorporating more flame retardants or any additives in epoxy matrix to enhance the fire performance of high fibrous composites, and this could help alleviate the effect of increased flammability due to high fiber loading while still maintaining structural integrity.
- Hybrid Composite Systems: Considering the implementation of hybrid composites that consist of glass fiber together with carbon or natural fibers can result in better mechanical or thermal properties. These materials might improve performance in fire conditions while not weakening fracture toughness.
- Fire Simulation and models: Such simulations can include structural fire modeling or the digital twin approach which aims to depict how these materials would behave during a real-life fire. This would serve to find the most effective composition of fiber–matrix composites for an application without going through the rigors of a lot of experiments.
- • Long-Term Durability Testing: Considerable effort should be made toward the long-term durability and aging response of these composites under cyclic thermal loading or sustained exposure to fire. This will then give beneficial information when they are applied in industries such as aerospace, automotive, and construction.

5.3 Potential for Material Optimization

There is significant potential for optimizing epoxy glass fiber/phenolic composites to enhance both their mechanical performance and fire resistance. Several strategies can be explored for material optimization:

Optimizing Fiber-to-Matrix Ratios: While the 60:40 fiber ratio provides the best balance between fracture toughness and fire resistance, there may be room for optimizing the fiber-to-matrix ratios further by experimenting with different resin formulations or fiber types.

Incorporating Nanomaterials: Nanoparticles or nanofillers could be incorporated into the epoxy matrix to improve thermal stability and fire resistance without significantly affecting the mechanical properties of the composite. This approach could help maintain a low HRR while improving material strength.

Sustainable Composites: The growing emphasis on sustainability provides an opportunity to explore the use of eco-friendly or bio-based phenolic resins and natural fibers. These materials could reduce the environmental impact of composite production while enhancing fire resistance.

References

1. Adeoye, A. O., Quadri, R. O., Lawal, O. S., & Emojevu, E. O. (2024). Physicochemical characterization, valorization of lignocellulosic waste (Kola nut seed shell) via pyrolysis, and ultrasonication of its crude bio-oil for biofuel production. *Cleaner Waste Systems, 7*, 100138. https://doi.org/10.1016/j.clwas.2024.100138
2. Adeoye, A. O., Quadri, R. O., Lawal, O. S., Malomo, D., Emojevu, E. O., Omonije, O. O., Odeniyi, O. K., Fadahunsi, M. O., Yelwa, M. J., Aasa, S. A., Onakpa, A. E., Omoniyi, B. P., Mark, I. N., Usman, J., Ismaila, A. M., & Saidu, A. U. (2024). Advanced techniques in upgrading crude bio-oil to biofuel. In *Energy, environment, and sustainability* (pp. 321–353). Springer. https://doi.org/10.1007/978-981-97-0515-3_14
3. Adeoye, A. O., Quadri, R. O., & Lawal, O. S. (2022). Assessment of biofuel potential of tenera palm kernel shell via fixed bed pyrolysis and thermal characterization. *Results in Surfaces and Interfaces, 9*, 100091. https://doi.org/10.1016/j.rsurfi.2022.100091
4. Adeoye, A. O., Lawal, O. S., Quadri, R. O., Malomo, D., Aliyu, M. T., Dang, G. E., Emojevu, E. O., Maikato, M. J., Yahaya, M. G., Omonije, O. O., Edidem, V. G., Abubakar, Y. K., Offor, O. F., Sochima, E. H., Peter, B. E., & Hikon, B. N. (2023). Sustainable energy via thermochemical and biochemical conversion of biomass wastes for biofuel production. In *Energy, environment, and sustainability* (pp. 245–306). Springer. https://doi.org/10.1007/978-981-99-2150-8_11
5. Adeoye, A. O., Yelwa, J. M., Imam, N., Quadri, R. O., Lawal, O. S., Malomo, D., Aasa, S. A., Onakpa, A. E., Dorgbaa, R. K., Omoniyi, B. P., Ahmad, N., & Garba, H. A. (2024). Pyrolysis of plastic wastes towards achieving a circular economy: an advanced chemistry and technical approach. In *Energy, environment, and sustainability* (pp. 215–259). Springer. https://doi.org/10.1007/978-981-97-0437-8_11
6. Adeoye, A. O., Quadri, R. O., & Lawal, O. S. (2023). Wet synthesis, characterization of goethite nanoparticles and its application in catalytic pyrolysis of palm kernel shell in TGA. *Results in Surfaces and Interfaces, 11*. https://doi.org/10.1016/j.rsurfi.2023.100118

7. Avesh, M., Imam, N., & Adeoye, A. O. (2024). Assessing environmental impacts and mitigation strategies for the DFCCIL rail projects. In *Energy, environment, and sustainability* (pp. 333–353). Springer. https://doi.org/10.1007/978-981-97-0437-8_15
8. Cataluna Veses, R., & Shah, Z. (2015). Bio-oil production by thermal cracking in the presence of hydrogen. *Journal of Fundamentals of Renewable Energy and Applications, 06*. https://doi.org/10.4172/2090-4541.1000194
9. Shah, Z., Cataluña Veses, R., Aguilhera, R. A., & Da Silva, R. (2016). Bio-oil production from pyrolysis of coffee and eucalyptus sawdust in the presence of 5% hydrogen. *International Journal of Engineering Research & Science, 2*, 34.
10. Shah, Z., Veses, R. C., & da Silva, R. (2016). A comparative study and analysis of two types of bio-oil samples obtained from freshwater algae and microbial treated algae. *Modern Chemistry & Applications, 4*. https://doi.org/10.4172/2329-6798.1000185
11. Shah, Z., Veses, R. C., Brum, J., Klunk, M. A., Rocha, L. A. O., Missaggia, A. B., de Oliveira, A. P., & Caetano, N. R. (2020). Sewage sludge bio-oil development and characterization. *Inventions, 5*, 36. https://doi.org/10.3390/inventions5030036
12. Shah, Z., Saberian, M., Hodgeman, D., Kinloch, I., & Vallés, C. (2023). Effect of sulfur on wood tar biopitch as a sustainable replacement for coal tar pitch binders. *ACS Applied Engineering Materials, 1*, 2567–2576. https://doi.org/10.1021/acsaenm.3c00361
13. Klunk, M. A., Dasgupta, S., Das, M., & Shah, Z. (2018). System of adsorption of CO_2 in coalbed. *Southern Brazilian Journal of Chemistry, 26*, 2–9. https://doi.org/10.48141/SBJCHEM.v26.n26.2018.7_2018.pdf
14. Quadri, R. O., Adeoye, A. O., Ayanda, O. S., & Lawal, O. S. (2024). Conversion of palm kernel shell to sustainable energy and the effect of wet synthesized nanoparticles of iron on its thermal degradation kinetics. *Bioresource Technology Reports, 27*. https://doi.org/10.1016/j.biteb.2024.101933
15. Momodu, D. U., Adetuyi, A. O., Adeoye, A. O., Akinlolu, O. S., & Apara, E. O. (2023). Surface functionalization of linen fiber and its dye-fastness assessment using 4-[(E)-(4-amino-2-hydroxyphenyl) diazenyl] benzene-1-sulphonic acid and 4-[(E)-(4-{(Z)-[4-(dimethylamino)phenyl] diazenyl}-2-hydroxyphenyl) diazenyl]benzene-1-sulphonic acid. *Suranaree Journal of Science and Technology, 30*.
16. Adeoye, A. O., Quadri, R. O., & Lawal, O. S. (2023). Biomass waste pyrolysis in the context of Nigeria's oil and gas industry: An overview. *International Journal of Ambient Energy, 44*, 2525–2541. https://doi.org/10.1080/01430750.2023.2256325
17. Mohapatra, S., Mohapatra, S., Han, H., Ariza-Montes, A., & López-Martín, M. C. (2022). Climate change and vulnerability of agribusiness: Assessment of climate change impact on agricultural productivity. *Frontiers in Psychology, 13*, 955622. https://doi.org/10.3389/fpsyg.2022.955622
18. Din, A. U., Han, H., Ariza-Montes, A., Vega-Muñoz, A., Raposo, A., & Mohapatra, S. (2022). The impact of COVID-19 on the food supply chain and the role of E-commerce for food purchasing. *Sustainability, 14*, 3074.
19. Pandey, G., Bansal, S., & Mohapatra, S. (2023). Market integration of chickpea crop: An evidence of India. *Journal of Agribusiness in Developing and Emerging Economies, 15*(1), 1–18. https://doi.org/10.1108/JADEE-12-2022-0273
20. Mohapatra, S., Mishra, R. K., & Sarangi, K. K. (2019). Economic appraisal of bio ethanol production technologies. *International Journal of Chemical Studies, 7*, 2197–2200.
21. Mohapatra, S., Kishore Mishra, R., & Sarangi, K. (2020). Bio-ethanol (2nd generation ethanol): A solution to ever polluting gasoline to climate in India. *International Research Journal of Pure and Applied Chemistry, 20*, 1–15.
22. Mohapatra, S., Mohapatra, U., & Mishra, R. K. (2018). Price forecasting of groundnut in Odisha. *The Pharma Innovation Journal, 7*, 111–114.
23. Mohapatra, S., Sneha, N. K., Mounika, A., & Mishra, R. K. (2019). Socio-economic and demographic analysis of SAARC nations. *International Journal of Agricultural & Statistical Sciences, 15*.

24. Shah, S., Nautiyal, H., Gugliani, G., Kumar, A., Namboodri, T., & Singla, Y. K. (2024). *Sustainability in smart manufacturing: Trends, scope, and challenges*. CRC Press (Taylor & Francis Group). https://taylorandfrancis.com/books/. ISBN 9781032740713. https://doi.org/10.1201/9781003467496
25. Singla, Y. K., Maughan, M. R., Arora, N., & Dwivedi, D. K. (2024). Enhancing the wear resistance of iron-based alloys: A comprehensive review of alloying element effects. *Journal of Manufacturing Processes, 120*, 135–160.
26. Singh, H., Singla, Y.K.: 2023. *Wear characteristics of AISI 4140 under nano fly ash based engine oil*. In Proceedings of the 7th International Conference on Theoretical and Applied Nanoscience and Nanotechnology (TANN'23). Paper No. 118, 1-3 June, 2023, Canada. DOI: https://doi.org/10.11159/tann23.118
27. Sharma, S. K., Sharma, R. C., Lavania, S., Palli, S., & Avesh, M. (2024). Dynamic vibration mitigation in ship propellers using hybrid fuzzy sliding mode control. *National Academy Science Letters*. https://doi.org/10.1007/s40009-024-01515-z
28. Singh, H., Singh, A. K., Singla, Y. K., Chattopadhyay, K., Saini, A., & Singh, K. (2022). Interpretation of the wear characteristics of AISI 4140 under nano-fly ash based engine lubricant. *Materials Today Proceedings, 50*, 1683–1689.
29. Rooprai, R. S., Singh, H., Singh, T., & Singla, Y. K. (2022). Analysis of the wear properties of through hardened AISI-4140 alloy steel using Taguchi technique. *Materials Today Proceedings, 50*, 661–664.
30. Singh, H., Singh, A. K., Singla, Y. K., Chattopadhyay, K., & Malhotra, R. (2021). Optimization of tribological behavior of AISI 4140 under nano fly ash particulates in engine lubricating oil. *Materials Today Proceedings, 45*, 4619–4623.
31. Shah, Z., Veses, R. C., & da Silva, R. (2016). Using GC-MS to analyze bio-oil produced from pyrolysis of agricultural wastes—Discarded soybean frying oil, coffee and eucalyptus sawdust in the presence of 5% hydrogen and argon. *Journal of Analytical and Bioanalytical Techniques, 7*. https://doi.org/10.4172/2155-9872.1000300
32. Klunk, M. A., Dasgupta, S., Das, M., Wander, P. R., & Shah, Z. (2019). Adsorption study of crystal violet and malachite green dyes in zeolitic material. *Periódico Tchê Química, 16*, 70–81. https://doi.org/10.52571/PTQ.v16.n33.2019.85_Periodico33_pgs_70_81.pdf
33. Yaseen, M., Khattak, M. A. K., Khan, A., Khan, M. S., Ahmad, M., Shah, Z., Khattak, R., & Bibi, S. (2022). Physico-chemical investigations on the catalytic production of biofuel from Algal biomass. *Russian Journal of Physical Chemistry A, 96*, S31–S37. https://doi.org/10.1134/S0036024422140308
34. Shah, Z., Cataluña Veses, R., & da Silva, R. (2016). GC-MS and FTIR analysis of bio-oil obtained from freshwater algae (spirogyra) collected from freshwater. *International Journal of Environmental & Agriculture Research, 2*, 34.
35. Shah, Z., Vases, R. C., Ullah, I., Yaseen, M., Aguilhera, R. A., Jamal, S., Amorim, V. D. A., & da Silva, R. (2016). Analysis of bio-oil obtained from pyrolysis of scenedesmus and spirogyra mixture algal biomass. *Journal of Ongoing Chemical Research, 2*, 1–13.
36. de Oliveira, A. P., Lorenzini, G., Shah, Z., Klunk, M. A., de Carvalho Lima, J. E., Rocha, L. A. O., & Caetano, N. R. (2020). Hierarchical criticality analysis of clean technologies applied to a coal-fired power plant. *International Journal of Design & Nature and Ecodynamics, 15*, 609–619. https://doi.org/10.18280/ijdne.150501
37. Roberto Caetano, N., Lorenzini, G., Rocha Lhamby, A., Shah, Z., Antônio Klunk, M., & Alberto Oliveira Rocha, L. (2020). Modelling safety distance from industrial turbulent non-premixed gaseous jet flames. *International Journal of Mechanical Engineering, 5*, 41–49.
38. Klunk, M. A., Shah, Z., & Wander, P. R. (2019). Use of montmorillonite clay for adsorption malachite green dye. *Periódico Tchê Química, 16*, 279–286. https://doi.org/10.52571/PTQ.v16.n32.2019.297_Periodico32_pgs_279_286.pdf
39. Klunk, M. A., Shah, Z., Caetano, N. R., Conceição, R. V., Wander, P. R., Dasgupta, S., & Das, M. (2020). CO_2 sequestration by magnesite mineralisation through interaction of Mg-brine and CO_2: Integrated laboratory experiments and computerised geochemical modelling. *International Journal of Environmental Studies, 77*, 492–509. https://doi.org/10.1080/00207233.2019.1675295

40. Yaseen, M., Shah, Z., Veses, R. C., Dias, S. L. P., Lima, E. C., dos Reis, S. G., Vaghetti, J. C. P., Alencar, W. S. D., & Mehmood, K. (2017). Photocatalytic studies of TiO_2/SiO_2 nanocomposite xerogels. *Journal of Analytical and Bioanalytical Techniques, 08*. https://doi.org/10.4172/2155-9872.1000348
41. Cataluña, R., Shah, Z., Pelisson, L., Caetano, N., da Silva, R., & Azevedo, C. (2017). Biodiesel glycerides from the soybean ethylic route incomplete conversion on the diesel engines combustion process. *Journal of the Brazilian Chemical Society*. https://doi.org/10.21577/0103-5053.20170100
42. Shah, Z., Renato, C. V., Marco, A. C., & Rosangela, D. S. (2017). Separation of phenol from bio-oil produced from pyrolysis of agricultural wastes. *Modern Chemistry & Applications, 05*. https://doi.org/10.4172/2329-6798.1000199
43. Cataluña, R., Shah, Z., Venturi, V., Caetano, N. R., da Silva, B. P., Azevedo, C. M. N., da Silva, R., Suarez, P. A. Z., & Oliveira, L. P. (2018). Production process of di-amyl ether and its use as an additive in the formulation of aviation fuels. *Fuel, 228*, 226–233. https://doi.org/10.1016/j.fuel.2018.04.167
44. Shah, Z., Veses, R. C., Vaghetti, J. C. P., Amorim, V. D. A., & da Silva, R. (2019). Preparation of jet engine range fuel from biomass pyrolysis oil through hydrogenation and its comparison with aviation kerosene. *International Journal of Green Energy, 16*, 350–360. https://doi.org/10.1080/15435075.2019.1566730
45. Sharma, S. K., Upadhyay, R. K., Kumar, V., & Valera, H. (2023). *Transportation energy and dynamics*. Springer Nature Singapore. https://doi.org/10.1007/978-981-99-2150-8
46. Upadhyay, R. K., Sharma, S. K., Kumar, D. V., & Valera, H. (2023). *Transportation systems technology and integrated management*. Springer Nature Singapore. https://doi.org/10.1007/978-981-99-1517-0
47. Lee, J., & Sharma, S. K. (2023). Intelligent systems for railway infrastructure. *Applied Sciences, 13*(20), 11274.
48. Sharma, S. K., Sharma, R. C., Choi, Y., & Lee, J. (2023). Modelling and dynamic analysis of adaptive neuro-fuzzy inference system-based intelligent control suspension system for passenger rail vehicles using magnetorheological damper for improving ride index. *Sustainability, 15*, 12529.
49. Sharma, S. K., Upadhyay, R. K., Kumar, V., & Valera, H. (2023). Introduction to sustainable transportation system. In *Transportation energy and dynamics* (pp. 3–6). Springer.
50. Hossain, I., Ahmed, Y. E., & Sharma, S. K. (2023). The use of both metallic-insert and TiO_2/water nanofluid to improve the performance of photovoltaic modules; experimental study. *Thermal Science and Engineering Progress, 42*, 101944.
51. Wu, Q., Cole, C., Spiryagin, M., Chang, C., Wei, W., Ursulyak, L., Shvets, A., Murtaza, M. A., Mirza, I. M., & Zhelieznov, K. (2023). Freight train air brake models. *International Journal of Rail Transportation, 11*, 1–49.
52. Upadhyay, R. K., Sharma, S. K., Kumar, V., & Valera, H. (2023). Introduction to transportation systems technology and integrated management. In *Transportation systems technology and integrated management* (pp. 1–4). Springer.
53. Sharma, R. C., Bhardawaj, S., Sharma, S. K., & Bahl, S. (2023). Ride behaviour of three-wheel vehicle using semi-active suspension using skyhook control strategy. *International Journal of Vehicle Structures & Systems, 15*.
54. Bhardawaj, S., Sharma, R. C., Sharma, S. K., Gopala Rao, L. V. V., & Vashist, A. (2023). Modeling of biomechanical human body model for seat to head transmissibility analysis. *Noise & Vibration Worldwide, 54*, 62–74.
55. Sharma, S. K., Sharma, R. C., Choi, Y., & Lee, J. (2023). Experimental and mathematical study of flexible–rigid rail vehicle riding comfort and safety. *Applied Sciences, 13*, 5252.
56. Sharma, R. C., Sharma, S. K., Palli, S., Rallabandi, S. R., & Sharma, N. (2025b). Compatibility between stability and curving ability of railway vehicles using conventional and unconventional bogie frame models. *SAE International Journal of Commercial Vehicles, 18*(3), 02-18-03–0014. https://doi.org/10.4271/02-18-03-0014

57. Pant, M., Sharma, A., Palli, S., Sharma, S. K., & Sharma, R. C. (2024). A review and analysis of factors affecting EV adoption for sustainable and energy-efficient transportation systems using analytic hierarchy process approach. *International Journal of Vehicle Structures & Systems, 16*, 133–139. https://doi.org/10.4273/ijvss.16.2.01
58. Sharma, S. K., Upadhyay, R. K., & Kumar, V. (2024). Introduction to dynamics of transportation ecosystem, modeling, and control. In *Dynamics of transportation ecosystem, modeling, and control* (pp. 3–6). Springer.
59. Sharma, S. K., Sharma, R. C., Mohapatra, S., Fouly, A., Hossain, I., & Lee, J. (2024). Technological advancements in oscillation reduction for propulsion shaft systems. *Journal of Vibration Engineering & Technologies, 12*, 1–19.
60. Upadhyay, R. K., Sharma, S. K., & Kumar, V. (2024). Introduction to intelligent transportation system and advanced technology. In *Intelligent transportation system and advanced technology* (pp. 3–6). Springer.
61. Hossain, I., Ngoupo, A. T., Bhattarai, S., Challagulla, S. P., & Sharma, S. K. (2024). Empirical survey on the effect of straight and wavy fins integrated with wind stream as heat sink of photovoltaic panels. *International Journal of Environmental Science and Technology, 22*, 2479–2492.
62. Sharma, S. K., Sharma, R. C., & Lee, J. (2024). Propelling precision of longitudinal vibration mitigation in ship propeller shafts through advanced nonlinear intelligent semi-active control leveraging adaptive neuro-fuzzy inference system with linear quadratic regulator. *Journal of Vibration and Control, 31*, 1472–1484.
63. Sharma, S. K., Kumar, N., Avesh, M., Sharma, R. C., Siddiqui, M. I. H., & Lee, J. (2024). Navigating tranquillity with H∞ controller to mitigate ship propeller shaft vibration. *Journal of Vibration Engineering & Technologies, 12*, 1–11.
64. Sharma, S. K., Sharma, R. C., Upadhyay, R. K., & Lee, J. (2024). Enhanced ship engine vibration reduction using magnetorheological dampers and adaptive neuro-fuzzy control system. *Ships and Offshore Structures, 20*, 143–159.
65. Sharma, R. C., Sharma, N., Palli, S., Sharma, S. K., Joshi, S., Rallabandi, S. R., & Sharma, A. (2024). Dynamic behavior of three-wheel vehicle with inerter-based dynamic absorber suspension system. *Noise & Vibration Worldwide, 55*, 253–263. https://doi.org/10.1177/09574565241243398
66. Pandipati, S., Palli, S., Dowluru, S., Sharma, R. C., Sharma, N., Sharma, S. K., & Da Fonseca-Soares, D. (2024). Microstructural and strength comparison friction stir processing of aluminium alloys and multi-response optimization by Taguchi GRA-PCA based hybrid methods. *International Journal of Vehicle Structures and Systems, 16*(3). https://doi.org/10.4273/ijvss.16.3.03
67. Kokkinos, N. C., Emmanouilidou, E., & Sharma, S. K. (2024). Waste-to-biofuel production for the transportation sector. In *Intelligent transportation system and advanced technology* (pp. 99–122). Springer.
68. Sharma, R. C., Palli, S., Duppala, A., Bhardawaj, S., Sharma, S. K., Sharma, N., & Sharma, A. (2024). Rigid body analysis of ride behaviour of roadway vehicle using Lagrangian dynamics. *International Journal of Vehicle Structures & Systems, 16*, 74–78. https://doi.org/10.4273/ijvss.16.1.15
69. Sharma, S. K., Sharma, R. C., & Lee, J. (2024). Modelling and simulation of MR damper characterization and uncertainty assessment in marine engine vibration isolation with FLPID control. *International Journal of Modelling and Simulation*, 1–11.
70. Sharma, S. K., Sharma, R. C., Upadhyay, R. K., & Lee, J. (2024). Coupled torsional–transverse vibration reduction in marine propulsion dynamics with novel approach using magnetorheological damping and adaptive control system. *Journal of Vibration Engineering & Technologies, 12*, 6089–6099.
71. Rallabandi, S. R., Srinivas, L., Palli, S., Sharma, R. C., Sharma, N., Sharma, A., & Sharma, S. K. (2024). A contrastive characterization of pure Mg and AZ91D alloy based on the testing of mechanical, corrosion, wear, and erosion properties. *Engineering Research Express, 6*, 015017. https://doi.org/10.1088/2631-8695/ad16a2

72. Badgujar, P., Sharma, S., Rahul, D., Sharma, S. K., & Upadhyay, R. K. (2024). Feasibility study of Mumbai-Pune Hyperloop corridor: Challenges and implementation. *Research in Transportation Business & Management, 52*, 101089.
73. Sharma, S. K., Sharma, R. C., Lavania, S., Palli, S., & Avesh, M. (2024). Industry-driven approach for ANFIS-based intelligent control suspension system with MR damper for enhanced ride quality in passenger rail vehicles for technological investigations. *National Academy Science Letters, 47*, 391–394. https://doi.org/10.1007/s40009-023-01365-1
74. Sharma, R. C., Palli, S., Dharmana, L., Sharma, S. K., Sharma, A., Kharola, A., & Sharma, N. (2024). Simulation of transient dynamic response of a locomotive considering its flexibility and rigidity. *Noise & Vibration Worldwide, 55*, 33–47. https://doi.org/10.1177/09574565231212684
75. Song, H., Park, J., Sharma, S. K., & Lee, J. (2022). A comparative study on ultrasonic propagation characteristics and defect detection of metal material additive manufacturing using deep learning algorithm. In *Biennial International Conference on Future Learning Aspects of Mechanical Engineering* (pp. 157–168). Springer.
76. Boris, I., Sharma, S. K., & Lee, J. (2022). A study on the reflection coefficient of higher modes of torsional wave by theoretical and numerical method. In *Biennial International Conference on Future Learning Aspects of Mechanical Engineering* (pp. 145–155). Springer.
77. Vishwakarma, P. N., Mishra, P., Sharma, S. K., & Shukla, A. K. (2022). Formulating and analysing the effect of suspension parameters on critical speeds for various DOFs in rail vehicle. In *Biennial International Conference on Future Learning Aspects of Mechanical Engineering* (pp. 791–802). Springer.
78. Kumar, U., Kasvekar, R., Sharma, S. K., & Upadhyay, R. K. (2022). Wear of wheels and axle in locomotive and measures taken by Indian railway. In *Advances in engine tribology* (pp. 77–96).
79. Sharma, R. C., Palli, S., Jha, A. K., Bhardawaj, S., & Sharma, S. K. (2022). Vibration and ride comfort analysis of railway vehicle system subjected to deterministic inputs. *Res Militaris, 12*, 1345–1355.
80. Sharma, R. C., Rao, L. V. V. G., Sharma, S. K., Palli, S., & Satyanarayana, V. S. V. (2022). Analysis of lateral stability and ride of an Indian railway constrained dual-axle bogie frame. *SAE International Journal of Commercial Vehicles, 16*, 213–228.
81. Sravanthi, A. L., Ratnakumar, P., Reddy, S. N., Eswari, K. B., Pandey, B. B., Manikanta, C., Ramya, K. T., Sonia, E., Mohapatra, S., Gopika, K., Anusha, P. L., & Yadav, P. (2022). Morpho-physiological, quality traits and their association with seed yield in sesame (Sesamum indicum L.) indigenous collection under deficit moisture stress. *Plant Physiology Reports, 27*(1), 132–142. https://doi.org/10.1007/s40502-021-00621-0
82. Vishwakarma, P.N., Mishra, P., Sharma, S.K.: Formulation of semi-active suspension system and controls in rail vehicle. SSRN Electronic Journal. https://doi.org/10.2139/ssrn.415961 (2022).
83. Sharma, R. C., & Sharma, S. K. (2022). Ride analysis of road surface-three-wheeled vehicle-human subject interactions subjected to random excitation. *SAE International Journal of Commercial Vehicles, 15*, 299–312.
84. Sharma, S. K., Lee, J., & Jang, H.-L. (2022). Mathematical modeling and simulation of suspended equipment impact on car body modes. *Machines, 10*, 192.
85. Sharma, S. K., Sharma, R. C., Lee, J., & Jang, H.-L. (2022). Numerical and experimental analysis of DVA on the flexible-rigid rail vehicle carbody resonant vibration. *Sensors, 22*, 1922.
86. Vishwakarma, P. N., Mishra, P., & Sharma, S. K. (2022). Characterization of a magnetorheological fluid damper a review. *Materials Today Proceedings, 56*, 2988–2994.
87. Lora Anusha, P., Ratnakumar, P., Pandey, B. B., Sandhya Rani, P., Umamahesh, V., Reddi Sekhar, M., Chandrika, V., Yadav, P., Mohapatra, S., & Padmaja, D. (2023). Phenotypic characterization of sesame (Sesamum indicum L.) revealed promising genotypes for moisture stress conditions. *Plant Genetic Resources: Characterization and Utilization, 21*(3), 218–228. https://doi.org/10.1017/S1479262123000667

88. Sharma, S. K., Sharma, R. C., & Lee, J. (2022). In situ and experimental analysis of longitudinal load on carbody fatigue life using nonlinear damage accumulation. *International Journal of Damage Mechanics, 31*, 605–622.
89. Sharma, R. C., Sharma, S., Sharma, N., & Sharma, S. K. (2021). Linear and nonlinear analysis of ride and stability of a three-wheeled vehicle subjected to random and bump inputs using bond graph and Simulink methodology. *SAE International Journal of Commercial Vehicles, 15*, 65–79.
90. Sharma, S.K., Sharma, R.C., *Multi-objective design optimization of locomotive nose*. SAE Technical Paper pp. 1–10. doi: https://doi.org/10.4271/2021-01-5053 (2021).
91. Lee, J., Han, J., & Sharma, S. K. (2021). Structural analysis on the separated and integrated differential gear case for the weight reduction. In *Advances in engineering design: Select Proceedings of FLAME 2020* (pp. 175–181). Springer Singapore.
92. Choi, S., Lee, J., & Sharma, S. K. (2021). A study on the performance evaluation of hydraulic tank injectors. In *Advances in engineering design: Select Proceedings of FLAME 2020* (pp. 183–190). Springer Singapore.
93. Sharma, R. C., Palli, S., Sharma, N., & Sharma, S. K. (2021). Ride behaviour of a four-wheel vehicle using h infinity semi-active suspension control under deterministic and random inputs. *International Journal of Vehicle Structures & Systems, 13*, 234–237.
94. Bhardawaj, S., Sharma, R. C., Sharma, S. K., & Sharma, N. (2021). On the planning and construction of railway curved track. *International Journal of Vehicle Structures & Systems, 13*, 151–159.
95. Mohapatra, S., Mohanty, D., Sharma, S., Dikshit, S., Kohli, I., Samantaray, D. P., Kumar, R., & Kathpalia, M. (2021). Biomedical application of polymeric biomaterial: Polyhydroxybutyrate. In *Bioresource utilization and management* (pp. 111–124). Apple Academic Press.
96. Sharma, S. K., Sharma, R. C., & Lee, J. (2021). Effect of rail vehicle–track coupled dynamics on fatigue failure of coil spring in a suspension system. *Applied Sciences, 11*, 2650.
97. Sharma, R. C., Sharma, S., Sharma, S. K., Sharma, N., & Singh, G. (2021). Analysis of bio-dynamic model of seated human subject and optimization of the passenger ride comfort for three-wheel vehicle using random search technique. *Proceedings of the Institution of Mechanical Engineers, Part K: Journal of Multi-body Dynamics, 235*, 106–121.
98. Sharma, S. K., & Lee, J. (2021). Crashworthiness analysis for structural stability and dynamics. *International Journal of Structural Stability and Dynamics, 21*, 2150039.
99. Mohapatra, S., & Mishra, R. K. (2020). Status of institutional credit disbursement of agricultural sector in Odisha, India. *International Journal of Current Microbiology and Applied Sciences, 9*, 2057–2065.
100. Mohapatra, S., Singh, J., & Mishra, S. (2017). Co-integration among major potato markets in Punjab. *Indian Journal of Agricultural Marketing, 31*, 46–47.
101. Mohapatra, S., & Singh, J. (2021). Co-integration of major vegetable markets in Punjab, India. *Economic Affairs, 66*(02), 181–187.
102. Mohapatra, S., Priyanka, V., Mohapatra, S., Kohli, I., & Mishra, R. K. (2020). Impact of corona virus covid-19 on the global economy. *International Journal of Agricultural and Statistical Sciences, 16*, 771–778.
103. Mohapatra, S., Singh, J., & Kumar, S. (2018). Cointegration among major cauliflower markets in Punjab. *Indian Journal of Economics and Development, 14*, 330–335.
104. Mohapatra, S., Mishra, R., & Mishra, R. (2021). Impact of COVID-19 on education level of agricultural farming community. *International Journal of Agricultural and Statistical Sciences, 17*, 915–923.
105. Mohapatra, S., Samantaray, D. P., & Samantaray, S. M. (2015). Study on polyhydroxyalkanoates production using rhizospheric soil bacterial isolates of sweet potato. *Indian Journal of Science and Technology, 8*, 57–62. https://doi.org/10.17485/ijst/2015/v8iS7/64027
106. Arora, R., Trivedi, J., Mohapatra, S., & Kumar, P. (2022). Trends in probiotics on human health and industrial application. In *Advances in agricultural and industrial microbiology* (Microbial diversity and application in agroindustry) (Vol. 1, pp. 237–248). https://doi.org/10.1007/978-981-16-8918-5_13

107. Samal, S., Pati, S., Mohapatra, S., Maity, S., Tanaya, K., Devadarshini, D., & Samantaray, D. (2023). PHAs production by facultative anaerobic bacteria Bacillus cereus FM5 through submerged and solid-state fermentation under anoxic condition. *Antonie van Leeuwenhoek, International Journal of General and Molecular Microbiology, 116*, 521–529. https://doi.org/10.1007/s10482-023-01825-0
108. Biswas, N., Saha, S., Biswas, B. K., Chowdhury, M., Rahaman, A., Junghare, V., Mohapatra, S., Hazra, S., Zangrando, E., Roy Choudhury, R., & Roy Choudhury, C. (2021). The DNA- and protein-binding properties and cytotoxicity of a new copper(II) hydrazone Schiff base complex. *Journal of Coordination Chemistry, 74*, 1482–1504. https://doi.org/10.1080/00958972.2021.1913128
109. Mohapatra, S., Mishra, R., Roy, P., Yadav, K. L., & Satapathi, S. (2017). Systematic investigation and in vitro biocompatibility studies on implantable magnetic nanocomposites for hyperthermia treatment of osteoarthritic knee joints. *Journal of Materials Science, 52*, 9262–9268. https://doi.org/10.1007/s10853-017-1136-0
110. Mishra, B. B., Nayak, S. K., Mohapatra, S., & Samantaray, D. (2021). *Environmental and agricultural microbiology: Advances and applications*. Wiley. https://doi.org/10.1002/9781119525899
111. Sharma, S. K., Mohapatra, S., Sharma, R. C., Alturjman, S., Altrjman, C., Mostarda, L., & Stephan, T. (2022). Retrofitting existing buildings to improve energy performance. *Sustainability (Switzerland), 14*. https://doi.org/10.3390/su14020666
112. Bandhu, S., Khot, M. B., Sharma, T., Sharma, O. P., Dasgupta, D., Mohapatra, S., Hazra, S., Khatri, O. P., & Ghosh, D. (2018). Single cell oil from oleaginous yeast grown on sugarcane bagasse-derived xylose: An approach toward novel biolubricant for low friction and wear. *ACS Sustainable Chemistry & Engineering, 6*, 275–283. https://doi.org/10.1021/acssuschemeng.7b02425
113. Kohli, I., Joshi, N. C., Mohapatra, S., & Varma, A. (2020). Extremophile—An adaptive strategy for extreme conditions and applications. *Current Genomics, 21*, 96–110. https://doi.org/10.2174/1389202921666200401105908
114. Pati, S., Mohapatra, S., Maity, S., Dash, A., & Samantaray, D. P. (2021). Polyhydroxyalkanoates for sustainable smart packaging of fruits. In *Environmental and agricultural microbiology: Advances and applications* (pp. 183–195).
115. Chatterjee, A., Nirwan, S., Mohapatra, S., Sharma, P., Agnihotri, A., & Shrivastava, N. (2022). Biochemical aspects of pathogenic variability in white rust infected Indian mustard. *Mycologia, 114*, 757–768. https://doi.org/10.1080/00275514.2022.2060007
116. Pati, S., Maity, S., Dash, A., Jema, S., Mohapatra, S., Das, S., & Samantaray, D. P. (2020). Biocompatible PHB production from Bacillus species under submerged and solid-state fermentation and extraction through different downstream processing. *Current Microbiology, 77*, 1203–1209. https://doi.org/10.1007/s00284-020-01922-7
117. Agrahari, R., Sarraf, G., Joshi, N. C., Mohapatra, S., & Var, A. (2021). Insight of biopolymers and applications of polyhydroxyalkanoates. In *Microbial polymers: Applications and ecological perspectives* (pp. 177–191). https://doi.org/10.1007/978-981-16-0045-6_8
118. Kathpalia, M., Kumar, P., & Mohapatra, S. (2022). Homocysteine metabolism as a biomarker for cancer. In *Homocysteine metabolism in health and disease* (pp. 159–171). https://doi.org/10.1007/978-981-16-6867-8_8
119. Bharatiy, S. K., Hazra, M., Paul, M., Mohapatra, S., Samantaray, D., Dubey, R. C., Sanyal, S., Datta, S., & Hazra, S. (2016). In silico designing of an industrially sustainable carbonic anhydrase using molecular dynamics simulation. *ACS Omega, 1*, 1081–1103. https://doi.org/10.1021/acsomega.6b00041
120. Sarkar, M. S., Segu, H., Bhaskar, J. V., Jakher, R., Mohapatra, S., Shalini, K., Shivaji, S., & Reddy, P. A. (2018). Ecological preferences of large carnivores in remote, high-altitude protected areas: Insights from Buxa Tiger Reserve, India. *ORYX, 52*, 66–77. https://doi.org/10.1017/S0030605317000060
121. Pati, S., Dash, A., Maity, S., Pattnaik, S., Mohapatra, S., & Samantaray, D. P. (2021). Nano biofertilizer for sustainable agriculture. In *Agriculturally important microorganisms: Mechanisms and applications for sustainable agriculture* (pp. 201–208). https://doi.org/10.1201/9781003245841-10

122. Mohanty, D., Misra, S., Mohapatra, S., & Sahu, P. S. (2018). Prebiotics and synbiotics: Recent concepts in nutrition. *Food Bioscience, 26*, 152–160. https://doi.org/10.1016/j.fbio.2018.10.008
123. Kumar, R., Kumari, R., Prasad, C., Deep, A., Singh, N., Namtak, S., Tiwari, V., Merugu, R., Mohapatra, S., & Singh, R. (2021). Influence of water quality on the diversity of macroinvertebrates in the Mandakini River in India. *Water Science and Technology: Water Supply, 21*, 1843–1860. https://doi.org/10.2166/WS.2021.020
124. Mohanty, D., Jena, R., Choudhury, P. K., Pattnaik, R., Mohapatra, S., & Saini, M. R. (2016). Milk derived antimicrobial bioactive peptides: A review. *International Journal of Food Properties, 19*, 837–846. https://doi.org/10.1080/10942912.2015.1048356
125. Mohapatra, S., Rath, S. N., Pradhan, S. K., Samantaray, D. P., & Rath, C. C. (2016). Secondary structural models (16S rRNA) of polyhydroxyalkanoates producing bacillus species isolated from different rhizospheric soil: Phylogenetics and chemical analysis. *International Journal Bioautomation, 20*, 329–338.
126. Mohapatra, S., Sarkar, B., Samantaray, D. P., Daware, A., Maity, S., Pattnaik, S., & Bhattacharjee, S. (2017). Bioconversion of fish solid waste into PHB using Bacillus subtilis based submerged fermentation process. *Environmental Technology (United Kingdom), 38*, 3201–3208. https://doi.org/10.1080/09593330.2017.1291759
127. Misra, S., Mohanty, D., & Mohapatra, S. (2021). Food-based probiotics: Functional dietary ingredients. In *Probiotics: Advanced food and health applications* (pp. 257–275). https://doi.org/10.1016/B978-0-323-85170-1.00006-3
128. Kumar, R., Kumari, R., Prasad, C., Tiwari, V., Singh, N., Mohapatra, S., Merugu, R., Namtak, S., & Deep, A. (2020). Phytoplankton diversity in relation to physicochemical attributes and water quality of Mandakini River, Garhwal Himalaya. *Environmental Monitoring and Assessment, 192*. https://doi.org/10.1007/s10661-020-08768-3
129. Mohapatra, S., Vishwakarma, K., Joshi, N. C., Maity, S., Kumar, R., Ramchander, M., Pattnaik, S., & Samantaray, D. P. (2021). A review on PHAs: The future biopolymer. In *Environmental and agricultural microbiology: Advances and applications* (pp. 83–117).
130. Sahoo, N., Pattnaik, S., Pattnaik, M., & Mohapatra, S. (2021). Methane emission and strategies for mitigation in livestock. In *Environmental and agricultural microbiology: Advances and applications* (pp. 257–274).
131. Mohanty, D. P., Mohapatra, S., Misra, S., & Sahu, P. S. (2016). Milk derived bioactive peptides and their impact on human health—A review. *Saudi Journal of Biological Sciences, 23*, 577–583. https://doi.org/10.1016/j.sjbs.2015.06.005
132. Kohli, I., Mohapatra, S., Kumar, P., Goel, A., Varma, A., & Joshi, N. C. (2022). Role of bacterial endophytes in the promotion of plant growth. In *Bacterial endophytes for sustainable agriculture and environmental management* (pp. 235–253). https://doi.org/10.1007/978-981-16-4497-9_12
133. Samantaray, D., Mohapatra, S., & Mishra, B. B. (2014). Microbial bioremediation of industrial effluents. In *Microbial biodegradation and bioremediation* (pp. 326–339). https://doi.org/10.1016/B978-0-12-800021-2.00014-5
134. Sharma, R. C., Palli, S., Gopala Rao, L. V. V., Duppala, A., & Sharma, S. K. (2022). Four-wheel vehicle response under bump, pothole, harmonic, and random excitations using bond graph/simulink technique. *SAE International Journal of Commercial Vehicles, 16*. https://doi.org/10.4271/02-16-02-0008
135. Mohapatra, S., Mohanta, P. R., Sarkar, B., Daware, A., Kumar, C., & Samantaray, D. P. (2017). Production of polyhydroxyalkanoates (PHAs) by Bacillus strain isolated from waste water and its biochemical characterization. *Proceedings of the National Academy of Sciences India Section B—Biological Sciences, 87*, 459–466. https://doi.org/10.1007/s40011-015-0626-6
136. Tomer, P., Shroti, G. K., Mohapatra, S., Ghosh, D., Jaiswal, S., Garg, D., Lahiri, D., Kumar, N., Khatri, O. P., & Hazra, S. (2023). Fabrication of HA Nano-crystal Reinforced PHAs based composites for orthopedic fracture-fixing accessories using sustainable resources. *Environment, Development and Sustainability*. https://doi.org/10.1007/s10668-023-03912-8

137. Bandekar, D., Mohapatra, S., Hazra, M., Hazra, S., & Biswas, S. (2021). N-terminal truncation of VC0395_0300 protein from Vibrio cholerae does not lead to loss of diguanylate cyclase activity. *Biophysical Chemistry, 268*. https://doi.org/10.1016/j.bpc.2020.106493
138. Vishwakarma, K., Kumar, N., Shandilya, C., Mohapatra, S., Bhayana, S., & Varma, A. (2020). Revisiting plant–microbe interactions and microbial consortia application for enhancing sustainable agriculture: A review. *Frontiers in Microbiology, 11*. https://doi.org/10.3389/fmicb.2020.560406
139. Mohapatra, S., Maity, S., Dash, H. R., Das, S., Pattnaik, S., Rath, C. C., & Samantaray, D. (2017). Bacillus and biopolymer: Prospects and challenges. *Biochemistry and Biophysics Reports, 12*, 206–213. https://doi.org/10.1016/j.bbrep.2017.10.001
140. Kumar, R., Merugu, R., Mohapatra, S., & Sharma, S. (2022). Extremophiles: Life of microorganisms in extreme environments. In *Extremophiles: General and plant biomass based biorefinery* (pp. 43–66). https://doi.org/10.1201/9781003014188-3
141. Pradhan, A., Pahari, A., Mohapatra, S., & Mishra, B. B. (2017). Phosphate-solubilizing microorganisms in sustainable agriculture: Genetic mechanism and application. In *Microorganisms for sustainability* (pp. 81–97). https://doi.org/10.1007/978-981-10-7380-9_5
142. Devadarshini, D., Mohapatra, S., Pati, S., Maity, S., Rath, C. C., Jena, P. K., & Samantaray, D. (2023). Evaluation of PHAs production by mixed bacterial culture under submerged fermentation. *Biologia (Bratisl), 78*, 895–902. https://doi.org/10.1007/s11756-022-01302-5
143. Agarwal, K. M., Mohapatra, S., Sharma, P., Sharma, S., Bhatia, D., & Mishra, A. (2020). Study and overview of the novel corona virus disease (COVID-19). *Sensors International, 1*. https://doi.org/10.1016/j.sintl.2020.100037
144. Tanaya, K., Mohapatra, S., Maity, S., & Samantaray, D. P. (2023). Fluorescent-pigmented microbial polymer nanoparticles: An approach for replacement of bioimaging agent in the drug delivery system. *Applied Biochemistry and Microbiology, 59*, 378–383. https://doi.org/10.1134/S0003683823030171
145. Pattnaik, S., Dash, D., Mohapatra, S., Pati, S., Devadarshini, D., Samal, S., Pattnaik, M., Maity, S., Mishra, S. K., & Samantaray, D. (2024). Reclamation of chromium-contaminated soil by native Cr(VI)-reducing and PHA-accumulating Bacillus aryabhattai CTSI-07. *International Microbiology, 27*, 731–742. https://doi.org/10.1007/s10123-023-00421-6
146. Mohapatra, S., Mishra, Y., & Dash, L. (2020). Covid-19: Analysing the legal nuances of the lockdown order. *Indian Journal of Forensic Medicine and Toxicology, 14*, 612–616. https://doi.org/10.37506/ijfmt.v14i4.11553
147. Pati, S., Mohanty, M. K., Mohapatra, S., & Samantaray, D. (2022). Bioethanol production by Enterobacter hormaechei through carbon monoxide-rich syngas fermentation. *International Journal of Energy Research, 46*, 20096–20106. https://doi.org/10.1002/er.8678
148. Maity, S., Das, S., Mohapatra, S., Tripathi, A. D., Akthar, J., Pati, S., Pattnaik, S., & Samantaray, D. P. (2020). Growth associated polyhydroxybutyrate production by the novel Zobellellae tiwanensis strain DD5 from banana peels under submerged fermentation. *International Journal of Biological Macromolecules, 153*, 461–469. https://doi.org/10.1016/j.ijbiomac.2020.03.004
149. Takkar, S., Shandilya, C., Agrahari, R., Chaurasia, A., Vishwakarma, K., Mohapatra, S., Varma, A., & Mishra, A. (2022). Green technology: Phytoremediation for pesticide pollution. In *Phytoremediation technology for the removal of heavy metals and other contaminants from soil and water* (pp. 353–375). https://doi.org/10.1016/B978-0-323-85763-5.00008-8
150. Bandekar, D., Chouhan, O. P., Mohapatra, S., Hazra, M., Hazra, S., & Biswas, S. (2017). Putative protein VC0395_0300 from Vibrio cholerae is a diguanylate cyclase with a role in biofilm formation. *Microbiological Research, 202*, 61–70. https://doi.org/10.1016/j.micres.2017.05.003
151. Pattnaik, S., Dash, D., Mohapatra, S., Pattnaik, M., Marandi, A. K., Das, S., & Samantaray, D. P. (2020). Improvement of rice plant productivity by native Cr(VI) reducing and plant growth promoting soil bacteria Enterobacter cloacae. *Chemosphere, 240*. https://doi.org/10.1016/j.chemosphere.2019.124895

152. Mohapatra, S., Samantaray, D. P., Samantaray, S. M., Mishra, B. B., Das, S., Majumdar, S., Pradhan, S. K., Rath, S. N., Rath, C. C., Akthar, J., & Achary, K. G. (2016). Structural and thermal characterization of PHAs produced by Lysinibacillus sp. through submerged fermentation process. *International Journal of Biological Macromolecules, 93*, 1161–1167. https://doi.org/10.1016/j.ijbiomac.2016.09.077
153. Kumar Singla, Y., Dwivedi, D. K., & Arora, N. (2017). Modeling the impact–sliding wear characteristics of rare earth additive iron-based hardfacing alloys. *Proceedings of the Institution of Mechanical Engineers, Part J: Journal of Engineering Tribology, 231*, 1486–1499.
154. Singla, Y. K., Dwivedi, D. K., & Arora, N. (2015). On the modeling of dry sliding adhesive wear parameters of vanadium additive iron-based alloys at elevated temperatures. *Surface and Coating Technology, 283*, 223–233.
155. Singla, Y. K., Chhibber, R., Bansal, H., & Kalra, A. (2015). Wear behavior of aluminum alloy 6061-based composites reinforced with SiC, Al_2O_3, and red mud: A comparative study. *Journal of Metals, 67*, 2160–2169.
156. Sharma, S. K., Kumar, A., & Sharma, R. C. (2014). Challenges in railway vehicle modeling and simulations. In *International Conference on Newest Drifts in Mechanical Engineering ICNDME-2014* (pp. 453–459).
157. Sharma, S. K., & Kumar, A. (2014). A comparative study of Indian and worldwide railways. *International Journal of Mechanical Engineering and Robotics Research, 1*, 114–120.
158. Sharma, S. K., Sharma, R. C., Kumar, A., & Palli, S. (2015). Challenges in rail vehicle-track modeling and simulation. *International Journal of Vehicle Structures & Systems, 7*, 1.
159. Palli, S., Sharma, R. C., Sharma, S. K., Muddada, V., & Sharma, N. (2020). A case study on noise pollution and its effects. *Journal of Critical Reviews, 7*, 261–267.
160. Sharma, R. C., Sharma, S. K., Sharma, N., & Sharma, S. (2020). Analysis of ride and stability of an ICF railway coach. *International Journal of Vehicle Noise and Vibration, 16*, 127–144.
161. Palli, S., Sharma, R. C., Sharma, S. K., & Chintada, V. B. (2020). On methods used for setting the curve for railway tracks. *Journal of Critical Reviews, 7*, 241–246.
162. Lee, J., & Sharma, S. K. (2020). Numerical investigation of critical speed analysis of high-speed rail vehicle. In 한국정밀공학회 학술발표대회 논문집 (p. 696).
163. Sharma, S., Sharma, R. C., Sharma, S. K., Sharma, N., Palli, S., & Bhardawaj, S. (2020). Vibration isolation of the quarter car model of road vehicle system using dynamic vibration absorber. *International Journal of Vehicle Structures & Systems, 12*, 513–516. https://doi.org/10.4273/IJVSS.12.5.05
164. Acharya, A., Gahlaut, U., Sharma, K., Sharma, S. K., Vishwakarma, P. N., & Phanden, R. K. (2020). Crashworthiness analysis of a thin-walled structure in the frontal part of automotive chassis. *International Journal of Vehicle Structures & Systems, 12*, 517–520.
165. Sharma, S. K., & Lee, J. (2020). Finite element analysis of a fishplate rail joint in extreme environment condition. *International Journal of Vehicle Structures & Systems, 12*, 503–506.
166. Bhardawaj, S., Sharma, R. C., & Sharma, S. K. (2020). Development of multibody dynamical using MR damper based semi-active bio-inspired chaotic fruit fly and fuzzy logic hybrid suspension control for rail vehicle system. *Proceedings of the Institution of Mechanical Engineers, Part K: Journal of Multi-body Dynamics, 234*, 723–744.
167. Sharma, S. K., & Lee, J. (2020). Design and development of smart semi active suspension for nonlinear rail vehicle vibration reduction. *International Journal of Structural Stability and Dynamics, 20*, 2050120.
168. Sharma, S. K., Phan, H., & Lee, J. (2020). An application study on road surface monitoring using DTW based image processing and ultrasonic sensors. *Applied Sciences, 10*, 4490.
169. Sharma, R. C., Sharma, S., Sharma, S. K., & Sharma, N. (2020). Analysis of generalized force and its influence on ride and stability of railway vehicle. *Noise & Vibration Worldwide, 51*, 95–109.

170. Bhardawaj, S., Sharma, R. C., & Sharma, S. K. (2020). Ride analysis of track-vehicle-human body interaction subjected to random excitation. *Journal of the Chinese Society of Mechanical Engineers, 41*, 227–236.
171. Bhardawaj, S., Sharma, R. C., & Sharma, S. K. (2020). Development in the modeling of rail vehicle system for the analysis of lateral stability. *Materials Today Proceedings, 25*, 610–619.
172. Sharma, S. K., Sharma, R. C., & Sharma, N. (2020). Combined multi-body-system and finite element analysis of a rail locomotive crashworthiness. *International Journal of Vehicle Structures & Systems, 12*, 428–435.
173. Sharma, R. C., Avesh, M., Sharma, S. K., Sharma, S., Palli, S., & Sharma, N. (2020). Linear and non-linear stability analysis of a constrained railway wheel axle. *International Journal of Vehicle Structures and Systems, 12*, 128–133. https://doi.org/10.4273/ijvss.12.2.04
174. Mohapatra, S., Pattnaik, S., Maity, S., Sharma, S., Akhtar, J., Pati, S., Samantaray, D. P., & Varma, A. (2020). Comparative analysis of PHAs production by Bacillus megaterium OUAT 016 under submerged and solid-state fermentation. *Saudi Journal of Biological Sciences, 27*, 1242–1250. https://doi.org/10.1016/j.sjbs.2020.02.001
175. Bhardawaj, S., Sharma, R. C., & Sharma, S. K. (2020). Analysis of frontal car crash characteristics using ANSYS. *Materials Today Proceedings, 25*, 898–902.
176. Bhardawaj, S., Sharma, R. C., & Sharma, S. K. (2019). A survey of railway track modelling. *International Journal of Vehicle Structures & Systems, 11*, 508–518.
177. Bhardawaj, S., Sharma, R. C., & Sharma, S. K. (2019). Development and advancement in the wheel-rail rolling contact mechanics. In *IOP conference series: Materials science and engineering* (p. 012034). IOP Publishing.
178. Choppara, R. K., Sharma, R. C., Sharma, S. K., & Gupta, T. (2019). Aero dynamic cross wind analysis of locomotive. In *IOP conference series: Materials science and engineering* (p. 012035). IOP Publishing.
179. Sharma, S. K., & Sharma, R. C. (2019). Pothole detection and warning system for Indian roads. In *Advances in interdisciplinary engineering: Select Proceedings of FLAME 2018* (pp. 511–519). Springer Singapore.
180. Sinha, A. K., Sengupta, A., Gandhi, H., Bansal, P., Agarwal, K. M., Sharma, S. K., Sharma, R. C., & Sharma, S. K. (2019). Performance enhancement of an all-terrain vehicle by optimizing steering, powertrain and brakes. In *Advances in engineering design: Select Proceedings of FLAME 2018* (pp. 207–215). Springer Singapore.
181. Goyal, S., Anand, C. S., Sharma, S. K., & Sharma, R. C. (2019). Crashworthiness analysis of foam filled star shape polygon of thin-walled structure. *Thin-Walled Structures, 144*, 106312.
182. Goswami, B., Rathi, A., Sayeed, S., Das, P., Sharma, R. C., & Sharma, S. K. (2019). Optimization design for aerodynamic elements of Indian locomotive of passenger train. In *Advances in engineering design: Select Proceedings of FLAME 2018* (pp. 663–673). Springer Singapore.
183. Sharma, S. K., Saini, U., & Kumar, A. (2019). Semi-active control to reduce lateral vibration of passenger rail vehicle using disturbance rejection and continuous state damper controllers. *Journal of Vibration Engineering & Technologies, 7*, 117–129.
184. Sharma, S. K. (2019). Multibody analysis of longitudinal train dynamics on the passenger ride performance due to brake application. *Proceedings of the Institution of Mechanical Engineers, Part K: Journal of Multi-body Dynamics, 233*, 266–279.
185. Sharma, S. (2018). Vibration control in quarter-car model with magnetorheological (MR) dampers using Bingham model. *Journal of Applied Mechanics Engineering, 7*, 1000299.
186. Sharma, R. C., Gopala Rao, L. V. V., Srikar, B., Palli, S., Appala Naidu, Y., Sharma, N., & Sharma, S. K. (2025a). Investigation of the dynamic behavior of hybrid composites using finite element and experimental method. *Journal of the Chinese Institute of Engineers, 48*(2), 158–171. https://doi.org/10.1080/02533839.2025.2453139
187. Sharma, S. K., & Kumar, A. (2018). Impact of longitudinal train dynamics on train operations: A simulation-based study. *Journal of Vibration Engineering & Technologies, 6*, 197–203.

188. Sharma, S. K., & Sharma, R. C. (2018). An investigation of a locomotive structural crashworthiness using finite element simulation. *SAE International Journal of Commercial Vehicles, 11*, 235–243.
189. Sharma, R. C., Sharma, S. K., & Palli, S. (2018). Rail vehicle modelling and simulation using Lagrangian method. *International Journal of Vehicle Structures and Systems, 10*, 188–194.
190. Sharma, R. C., & Sharma, S. K. (2018). Sensitivity analysis of three-wheel vehicle's suspension parameters influencing ride behavior. *Noise & Vibration Worldwide, 49*, 272–280.
191. Palli, S., Koona, R., Sharma, S. K., & Sharma, R. C. (2018). A review on dynamic analysis of rail vehicle coach. *International Journal of Vehicle Structures & Systems, 10*.
192. Sharma, S. K., & Sharma, R. C. (2018). Simulation of quarter-car model with magnetorheological dampers for ride quality improvement. *International Journal of Vehicle Structures & Systems, 10*, 169–173.
193. Sharma, S. K., & Kumar, A. (2018). Disturbance rejection and force-tracking controller of nonlinear lateral vibrations in passenger rail vehicle using magnetorheological fluid damper. *Journal of Intelligent Material Systems and Structures, 29*, 279–297.
194. Sharma, S. K., & Kumar, A. (2018). Ride comfort of a higher speed rail vehicle using a magnetorheological suspension system. *Proceedings of the Institution of Mechanical Engineers, Part K: Journal of Multi-Body Dynamics, 232*, 32–48.
195. Shukla, S., Kumar, U., Sharma, S. K., Gupta, P., & Kumar, A. (2017). Light weight freight rolling stock bogie frame: Design methodology validated with field oscillation trials. *International Journal of Vehicle Structures & Systems, 9*, 245–250.
196. Sharma, R. C., Palli, S., Sharma, S. K., & Roy, M. (2017). Modernization of railway track with composite sleepers. *International Journal of Vehicle Structures & Systems, 9*, 321–329.
197. Sharma, S. K., & Kumar, A. (2017). Ride performance of a high speed rail vehicle using controlled semi active suspension system. *Smart Materials and Structures, 26*, 055026.
198. Sharma, S. K., & Kumar, A. (2017). Impact of electric locomotive traction of the passenger vehicle ride quality in longitudinal train dynamics in the context of Indian railways. *Mechanics & Industry, 18*, 222.
199. Sharma, S. K., & Lavania, S. (2013). An autonomous metro: Design and execution. In *Futuristic trends in mechanical and industrial engineering*. JECRC UDML College of Engineering.
200. Sharma, S. K. (2013). Zero energy building envelope components: A review. *International Journal of Engineering Research and Applications, 3*, 662–675.
201. Sharma, S. K., & Kumar, A. (2016). Dynamics analysis of wheel rail contact using FEA. *Procedia Engineering, 144*, 1119–1128.
202. Sharma, S. K., & Chaturvedi, S. (2016). Jerk analysis in rail vehicle dynamics. *Perspectives in Science, 8*, 648–650.
203. Sharma, S. K., & Kumar, A. (2016). The impact of a rigid-flexible system on the ride quality of passenger bogies using a flexible carbody. In *Proceedings of the Third International Conference on Railway Technology: Research, Development and Maintenance* (p. 87). Civil-Comp Press.
204. Lavania, S., & Sharma, S. K. (2011). An explicit approach to compare crosstalk noise and delay in VLSI RLC interconnect modeled with skin effect with step and ramp input. *Journal of VLSI Design Tools & Technology, 1*, 1–8.
205. Palli, S., Koona, R., Sharma, R. C., & Muddada, V. (2015). Dynamic analysis of Indian railway integral coach factory Bogie. *International Journal of Vehicle Structures and Systems, 7*, 16–20. https://doi.org/10.4273/ijvss.7.1.03
206. Palli, S., & Koona, R. (2015). Analyses of dynamic response of a railway bogie. *International Journal of Vehicle Noise & Vibration, 11*, 103–113. https://doi.org/10.1504/IJVNV.2015.070015

207. Swathi, K., & Srihari, P. (2015). Heat transfer enhancement in a tube using rectangular strip inserts. *International Journal of Applied Engineering Research, 10*, 41532–41544.
208. Palli, S., Sharma, R. C., & Rao, P. P. D. (2017). Dynamic behaviour of a 7 DoF passenger car model. *International Journal of Vehicle Structures and Systems, 9*, 57–63. https://doi.org/10.4273/ijvss.9.1.12
209. Sharma, R. C., Palli, S., & Koona, R. (2017). Stress and vibrational analysis of an Indian Railway RCF Bogie. *International Journal of Vehicle Structures and Systems, 9*, 296–302. https://doi.org/10.4273/ijvss.9.5.06
210. Palli, S., Dontikurti, R., Sharma, R. C., & Sharma, N. (2021). Analysis of dynamic response of a railway locomotive using ANSYS. *International Journal of Vehicle Structures and Systems, 13*, 250–255. https://doi.org/10.4273/ijvss.13.2.21
211. Sharma, R. C., Sharma, N., Singh, G., & Palli, S. (2021). Modeling and simulation of human body-vehicle-track system for the investigation of ride comfort. *SAE International Journal of Commercial Vehicles, 15*. https://doi.org/10.4271/02-15-02-0008
212. Sharma, R. C., Palli, S., Avesh, M., & Sharma, N. (2021). Vibration isolation of railway vehicle car body using semi-active suspension. *International Journal of Vehicle Structures and Systems, 13*, 482–487. https://doi.org/10.4273/ijvss.13.4.17
213. Gopala Rao, L. V. V., Sharma, R. C., Satyanarayana, V. S. V., & Palli, S. (2022). Stochastic optimal preview control response of a quarter-car nonlinear suspension model using spectral decomposition method. *Noise and Vibration Worldwide, 53*, 225–232. https://doi.org/10.1177/09574565221093234
214. Palli, S., Duppala, A., Sharma, R. C., & Rao, L. V. V. G. (2022). Dynamic simulation of automotive vehicle suspension using MATLAB Simulink. *International Journal of Vehicle Structures and Systems, 14*, 310–314. https://doi.org/10.4273/ijvss.14.3.04
215. Satyanarayana, V. S. V., Sharma, R. C., Sateesh, B., Gopala Rao, L. V. V., Mohan Rao, N., & Palli, S. (2022). Optimum response of a nonlinear passive vehicle suspension system under random road excitations. *SAE International Journal of Commercial Vehicles, 16*. https://doi.org/10.4271/02-16-01-0004
216. Lokanadham, D., Sharma, R. C., Palli, S., & Bhardawaj, S. (2022). Wear rate modelling and analysis of limestone slurry particulate composites using the fuzzy method. *International Journal on Recent and Innovation Trends in Computing and Communication, 10*, 133–143. https://doi.org/10.17762/ijritcc.v10i1s.5818
217. Sharma, R. C., Palli, S., Bhardawaj, S., & Rao, L. V. V. G. (2022). Advances in human biodynamic modeling for analysis of whole-body vibrations. *International Journal of Vehicle Structures and Systems, 14*, 503–511. https://doi.org/10.4273/ijvss.14.4.17
218. Sharma, R. C., Palli, S., & Sharma, S. K. (2023). Ride analysis of railway vehicle considering rigidity and flexibility of the carbody. *Journal of the Chinese Institute of Engineers, Transactions of the Chinese Institute of Engineers, Series A, 46*, 355–366. https://doi.org/10.1080/02533839.2023.2194918
219. Sharma, R. C., Palli, S., & Bhardawaj, S. (2023). A review of wheel-rail dynamic behavior on curved tracks. *International Journal of Vehicle Structures and Systems, 15*, 19–25. https://doi.org/10.4273/ijvss.15.1.04
220. Jyotiprasad, G., Dowluru, S., Rallabandi, S. R., Sharma, N., Sharma, R. C., Palli, S. (2023). *Thermal-structural and macromechanical behavior analysis of graphite-based laminated composite*. SAE Technical Paper 2023-01-5041. https://doi.org/10.4271/2023-01-5041.
221. Saikiran, C., Palli, S., Bondala, R., Sharma, R. C., Sreeramulu, D., & Dontikurti, R. (2023). Experimental investigation on thermal behaviour of silicon carbide (SiC) based polymer composites under hygrothermal conditions. *International Journal of Vehicle Structures and Systems, 15*, 222–226. https://doi.org/10.4273/ijvss.15.2.15
222. Singh, B. J., Palli, S., Kumar, P., & Bahl, S. (2023). Behavioural investigation of hardness profile for dissimilar TIG-weldment between P22 and P91 steels. *International Journal of Vehicle Structures and Systems, 15*, 31–36. https://doi.org/10.4273/ijvss.15.1.06

223. Naveen, Y., Lokanadham, D., Naidu, D. R., Sharma, R. C., Palli, S., & Lila, M. K. (2023). An experimental study on the influence of blended Karanja biodiesel on diesel engine characteristics. *International Journal of Vehicle Structures and Systems, 15*, 154–160. https://doi.org/10.4273/ijvss.15.2.02
224. Bhardawaj, S., Jha, A. K., Sharma, R. C., Palli, S., Lokanadham, D., Sharma, N., & Dhiman, N. (2023). Finite element analysis of two disk rotor system used in automobile turbochargers. *International Journal of Vehicle Structures and Systems, 15*, 501–504. https://doi.org/10.4273/ijvss.15.4.09
225. Sharma, R. C., Palli, S., Bhardawaj, S., Jha, A. K., Duppala, A., Sharma, N., Sharma, S., & Dhiman, N. (2023). Dynamic behaviour of railway vehicle under bump and pothole track irregularities. *International Journal of Vehicle Structures and Systems, 15*, 566–569. https://doi.org/10.4273/ijvss.15.4.22
226. Raju, S., Palli, S., Devi Prasad, P., Menda, V. R., & Ramakrishna, B. (2024). A hybrid AHP-TOPSIS, MOORA technique for multi-objective optimization of thermal, mechanical, and water absorption behavior of epoxy/hemp, pine apple, and palm fiber composites. *Journal of the Chinese Institute of Engineers, Transactions of the Chinese Institute of Engineers, Series A, 47*, 23–36. https://doi.org/10.1080/02533839.2023.2274092
227. Raju Rallabandi, S., Pilla, D. P., Dowluru, S., Palli, S., Sharma, N., Sharma, S. K., & Sharma, R. C. (2024). Critical evaluation of epoxy-hemp-pineapple-palm fiber composites using hybrid AHM-TOPSIS technique for sustainable structural applications. *Journal of the Chinese Institute of Engineers, Transactions of the Chinese Institute of Engineers, Series A, 47*, 325–336. https://doi.org/10.1080/02533839.2024.2308250
228. Gopala Rao, L. V. V., Satyanarayana, V. S. V., Sharma, R. C., Palli, S., Raturi, A., Kharola, A., & Sharma, A. (2024). Preview control of the random response of a full car vehicle model traversing a rough road. *International Journal of Vehicle Noise and Vibration, 20*, 70–88. https://doi.org/10.1504/IJVNV.2024.138105
229. Sharma, A., Sharma, V. S., Gupta, S., Sharma, R. C., Palli, S., & Sharma, N. (2024). Additive manufacturing in the COVID-19 pandemic: Equipment and challenges? *Journal of Micromanufacturing*. https://doi.org/10.1177/25165984241228083
230. Bammidi, R., Madivada, H., Dowluru, S., Palli, S., Sharma, R. C., & Muddada, V. (2024). Mechano-tribological investigations of Al-%Wt. of boron nitride as metal-ceramic composites. *AIP Conference Proceedings*. https://doi.org/10.1063/5.0182965
231. Ranganath, B. A., Duppala, A., Palli, S., Rao, L. V. V., & Sharma, R. C. (2024). Numerical analysis and experimental study of FRP composites for damping applications. *AIP Conference Proceedings*. https://doi.org/10.1063/5.0184802
232. Menda, V., Dora, S., Palli, S., Rallabandi, S. R., Sharma, R. C., Sharma, N., & Bondala, R. (2024). Mechanical and wear characteristics of aluminium-7075/graphene/TiB2 prepared by stir casting. *Engineering Research Express, 6*. https://doi.org/10.1088/2631-8695/ad58a8
233. Sura, H. S., Avesh, M., & Mohapatra, S. (2023). Design and modelling of digital twin technology to improve freight logistics. In *Energy, environment, and sustainability* (pp. 481–513). Springer. https://doi.org/10.1007/978-981-99-2150-8_20
234. Pattnaik, S., Pati, S., Devadarshini, D., Tanaya, K., Mishra, B. B., Samantaray, D., Mohapatra, S., & Dash, D. (2022). Microbial bioremediation of Cr(VI)-contaminated soil for sustainable agriculture. In *Microbial biodegradation and bioremediation: Techniques and case studies for environmental pollution* (pp. 395–407). Elsevier. https://doi.org/10.1016/B978-0-323-85455-9.00008-4
235. Srivastava, A., Lavania, S., & Mohapatra, S. (2023). Predicting the effect on land values after introducing high-speed rail. In *Energy, environment, and sustainability* (pp. 157–196). Springer. https://doi.org/10.1007/978-981-99-2150-8_8
236. Agrahari, R., Mohanty, S., Vishwakarma, K., Nayak, S. K., Samantaray, D., & Mohapatra, S. (2021). Update vision on COVID-19: Structure, immune pathogenesis, treatment and safety assessment. *Sensors International, 2*. https://doi.org/10.1016/j.sintl.2020.100073

237. Adeoye, A. O., Quadri, R. O., & Lawal, O. S. (2022). Fixed-bed pyrolysis and thermal analyses of pressed oil palm fruit fibre as a potential alternative energy. *Fuoye Journal of Innovation, Science and Technology, 2*(1), 187–194.
238. Tanaya, K., Mohapatra, S., & Samantaray, D. (2024). Formulation of biogenic fluorescent pigmented PHB nanoparticles from Rhodanobacter sp. for drug delivery. *Antonie van Leeuwenhoek, International Journal of General and Molecular Microbiology, 117*. https://doi.org/10.1007/s10482-024-01973-x

Sustainability Issues in Manufacturing

Industry 5.0: A New Era of Sustainable and Human-Centered Production

G. Rajyakalshmi, G. Jishnu Prian, P. Srihari Reddy, S. Indumati, and M. Sreenivasulu

1 Introduction

The fifth revolution foresees a future where industrial goals extend beyond efficiency and productivity to emphasize industry's contribution to society. This approach centers on worker well-being and leverages new technologies to ensure prosperity that transcends job creation while respecting planetary boundaries [1]. It builds on the principles of the fourth revolution by prioritizing investigation and invention for a sustainable, human-centric, and resilient European industry [2]. Literature suggests that technological innovations can align with sustainable manufacturing objectives. The fifth revolution represents an evolutionary enhancement of Industry 4.0, providing balanced innovations to address its limitations. The fourth revolution encounters on mismanagement of technologies like AI and robotics, have negatively impacted socioenvironmental aspects, including energy conservation and job security (Fig. 1).

These concerns have prompted discussions among policymakers and scholars about establishing a regulatory framework for the Fifth Industrial Revolution to mitigate troublesome effects of Industry 4.0. This new paradigm shifts the focus from economic value to societal value and emphasizes well-being over mere welfare [3].

G. Rajyakalshmi (✉) · G. Jishnu Prian
School of Mechanical Engineering, Vellore Institute of Technology, Vellore, Tamil Nadu, India
e-mail: rajyalakshmi@vit.ac.in

P. Srihari Reddy · S. Indumati · M. Sreenivasulu
NBKRIST, Vidyanagar, Andhra Pradesh, India
e-mail: mslu544@nbkrist.org

Y. Kumar Singla et al. (eds.), *Digital Manufacturing in Industry 5.0*,
https://doi.org/10.1007/978-3-031-91500-0_19

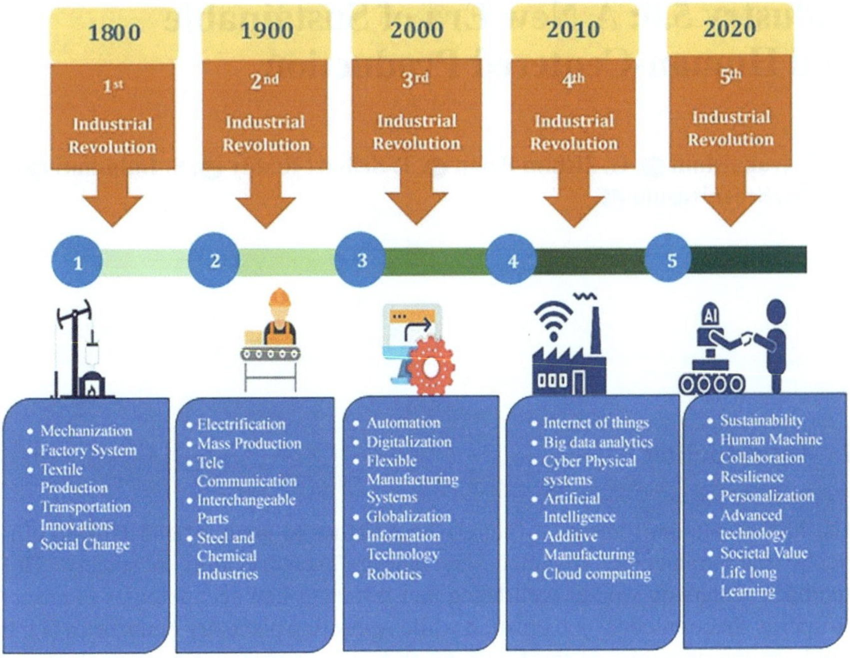

Fig. 1 Key enablers and objectives of the various industrial revolutions leading up to the fifth revolution

2 Background on Industry 5.0

For numerous companies, Industry 4.0 represents a pivotal revolution that shapes their business strategies. However, there is an emerging shift from a purely economic focus to a more comprehensive understanding of social value and well-being. Industry 5.0 is already underway, and its implications extend well beyond traditional industrial contexts to encompass all sectors and organizations [4]. Consequently, it is crucial for businesses to start familiarizing themselves with this concept and adapt their strategies to align with this new paradigm as soon as possible [5].

The term Industry 5.0 is a relatively recent concept frequently referenced by the European Union, highlighting its significance as a crucial factor in the ongoing economic and social transformations that extend beyond merely improving efficiency and productivity [6].

Also known as the Fifth Industrial Revolution, Industry 5.0 signifies a shift from a solely economic perspective to one that emphasizes social value and well-being. This new phase of industrialization promotes collaboration between humans and advanced technologies, including AI-driven robots, to enhance workplace processes (Fig. 2). Industry 5.0 encompasses various sectors beyond manufacturing and is driven by advancements in information technology, such as artificial intelligence,

Fig. 2 Major pillars of the Fifth Industrial Revolution

automation, data analytics, the Internet of Things (IoT), machine learning, and virtualization [7]. The key distinction from Industry 4.0 lies in its focus on prioritizing worker well-being at the core of production, fostering prosperity that transcends job creation and economic growth.

2.1 *Human-Centric Strategy*

A human-centric strategy emphasizes "talents, diversity, and empowerment." This represents a significant shift from viewing individuals as tools for organizational goals to recognizing them as valuable ends in themselves. In many sectors, retaining talent becomes more challenging than serving customers. Current strategies primarily focus on achieving competitive advantages to provide unique value to customers. However, if organizations adopt a genuinely human-centric approach, the strategy should prioritize gaining a competitive edge to deliver distinctive value to employees instead.

2.2 *Resilient Strategy*

A resilient approach is categorized by agility and flexibility, utilizing adaptable technologies. Its main emphasis shifts away from growth, profit, and productivity toward building organizations that are "antifragile." This means they can anticipate, respond to, and learn systematically from crises, ensuring stable and sustainable performance over time [8].

2.3 *Sustainable Strategy*

A sustainable strategy "drives action on sustainability while respecting planetary boundaries." This means organizations should consider the three Ps of the Triple Bottom Line and the 17 Sustainable Development Goals. While corporate sustainability efforts have often focused on minimizing harm or engaging in greenwashing, the goal is to adopt more responsible business practices. The Fifth Industrial Revolution brings forth advancements, viz., artificial intelligence and the Internet of Things, as pivotal elements for creating a sustainable future. These innovations support environment-friendly manufacturing practices, optimize resource utilization, and encourage circular economy frameworks, effectively marrying industrial development with ecological stewardship. For instance, smart waste management systems powered by IoT improve resource efficiency, while AI-enhanced product-service systems have the potential to revolutionize sectors like pharmaceuticals. Moreover, the integration of blockchain technology can significantly boost transparency and traceability, reinforcing sustainable practices across the entire supply chain. However, there are legitimate worries about the possible compromises between economic development and environmental conservation in the context of Industry 5.0.

3 Sustainability Role in the Context of the Fifth Revolution

Incorporating sustainability concepts into Industry 5.0 is becoming more critical than ever. This research explores the evolving relationship between Industry 5.0 and sustainability, highlighting the goal of merging human creativity and expertise with advanced technological capabilities. Circular economy concept offers a broader framework for sustainable production, shifting left from the traditional "take, make, dispose" model. The emergence of Internet of Things (IoT) has enhanced the interoperability of physical products with digital information systems, often facilitated by wireless protocols. Nonetheless, the Fifth Industrial Revolution has not yet achieved substantial momentum, as numerous companies continue to prioritize Industry 4.0 or even earlier frameworks. Additionally, commitment to sustainability is just beginning to gain serious momentum in industrial sector [9].

Fig. 3 Industry 5.0 and sustainability integration elements

4 The Intersection of Industry 5.0 and Sustainability

Industry 5.0 significantly enhances resource efficiency through intelligent automation and optimization, utilizing AI, IoT, and robotics to reduce waste, energy use, and material consumption, thereby lowering environmental impact and promoting sustainable practices. It encourages the adoption of circular economy principles by leveraging advanced technologies for effective product lifecycle management, focusing on durability, reusability, and recyclability, while tools like blockchain ensure transparent supply chains that enhance material circulation and minimize waste (Fig. 3). Furthermore, Industry 5.0 facilitates the integration of renewable energy sources by optimizing energy distribution through smart grids and AI algorithms, contributing to the decarbonization of industries and addressing climate change. Lastly, it promotes sustainable innovation by fostering collaboration among businesses, researchers, and entrepreneurs to develop eco-friendly technologies and solutions that tackle pressing sustainability challenges.

5 Industry 5.0: Automation Infused with Humanity

The concept of Industry 5.0 focuses on enhancing the "human touch" in machinery while utilizing advanced technologies to empower human workers, representing a renewed human-centric model for industrial production. This approach emphasizes the integration of human workers with machines in manufacturing and other industrial contexts, shifting the narrative from robots replacing humans to smart machines that support and augment human efforts. As a result, human roles are being redefined, moving away from repetitive tasks on assembly lines toward more creative positions that utilize problem-solving skills, experience, and intuition. Industry 5.0 also acknowledges the limitations of excessive automation, advocating for a balanced strategy in future robotic manufacturing processes. The overarching goal is to provide exceptional customer experiences and facilitate "mass customization" for consumers, alongside efforts to develop responsive and distributed supply chains. Moreover, there is a focus on designing "experience-activated" (interactive) consumer products and addressing the security risks and vulnerabilities inherent in modern interconnected industrial automation systems [10, 11].

6 Embracing Sustainability in the Fifth Industrial Revolution

To effectively integrate sustainability into their operations, companies should adopt design thinking principles in product development and business strategies, ensuring environmental considerations are prioritized from the outset. Industry 5.0 encourages collaboration through cross-industry partnerships, enabling organizations to share knowledge and resources, which accelerates the implementation of sustainable innovations [12]. Additionally, businesses must focus on creating responsible supply chains by partnering with suppliers committed to environmental stewardship and ethical labor practices, thus reducing their overall carbon footprint. Furthermore, the fifth revolution offers a valuable opportunity to educate and engage employees in sustainability initiatives; by providing training and raising awareness, organizations can empower their workforce for productive contribution to sustainability efforts, nurturing a sustainability philosophy throughout the company.

7 Emerging Technologies of the Fifth Industrial Revolution and Their Impacts on Sustainability

The fifth revolution signifies as an evolutionary phase in industrial evolution, with a focus on sustainability and addressing global challenges by integrating environment-friendly practices and technologies into manufacturing processes (Fig. 4). Here are some key ways Industry 5.0 is poised to enhance sustainability [13].

Fig. 4 Industry 5.0 technologies

7.1 Eco-Friendly Practices

One of the hallmarks of Industry 5.0 is its commitment to environmentally conscious methods. This involves harnessing renewable energy options like solar, wind, and hydroelectric power to reduce dependence on fossil fuels. Additionally, Industry 5.0 promotes waste reduction through innovative recycling and upcycling methods, ensuring that materials are reused and repurposed rather than discarded. Smart systems, powered by advanced data analytics and IoT technologies, are employed to optimize resource utilization, ensuring that every aspect of the manufacturing process is as efficient as possible. This holistic approach not only reduces environmental impact but also contributes to cost savings and resource conservation.

7.2 Human–Robot Collaboration

Central to Industry 5.0 is the concept of human–robot collaboration, commonly referred to as "cobots." These collaborative robots are designed to work alongside human operators, enhancing productivity and efficiency in manufacturing settings. Unlike traditional robots that operate independently, cobots are intended to assist and augment human capabilities, allowing for a more flexible and adaptive workforce. This collaboration can lead to improved safety, as cobots can take over hazardous tasks, while humans focus on complex decision-making and creative problem-solving. The synergy between humans and machines is expected to drive innovation and enhance overall operational performance.

7.3 Resilience

Another significant aspect of Industry 5.0 is its focus on building resilience within manufacturing systems. This involves creating technologies and processes that can withstand disruptions, whether they are caused by economic fluctuations, natural disasters, or pandemics. By implementing robust systems that are designed to be disruption-proof, industries can better adapt to unforeseen challenges. This resilience not only protects businesses from potential losses but also ensures continuity in production and supply, contributing to a more stable economic environment.

7.4 Sustainability in Supply Chains

Industry 5.0 is likely to transform supply chains by promoting shorter, more localized sourcing strategies. This shift can reduce transportation costs and emissions associated with long-distance shipping. Additionally, by fostering sustainable operations throughout the supply chain, manufacturers can ensure that their practices align with environmental goals. This might include selecting suppliers based on their sustainability practices and reducing reliance on nonrenewable resources. The focus on sustainability within supply chains can lead to a more transparent and responsible manufacturing ecosystem, benefiting both businesses and the planet.

7.5 Cybersecurity Considerations

As Industry 5.0 increasingly relies on advanced technologies and interconnected systems, cybersecurity becomes a critical consideration. When selecting technologies and suppliers, manufacturers must prioritize cybersecurity to safeguard their operations against potential threats. Ignoring these aspects can result in hidden costs, such as data breaches or system failures, which can have significant financial and reputational repercussions. By integrating robust cybersecurity measures into their operations, companies can protect their assets, maintain customer trust, and ensure the integrity of their manufacturing processes.

8 Previous Research on the Intersection of Industry 5.0 and Sustainability

The relationship between Industry 5.0 and sustainability is prominently illustrated in the context of renewable energy and the evolving energy landscape. Innovations in nanotechnology and material sciences are vital for achieving the sustainability goals of Industry 5.0, exemplified by the creation of magnetic hydrogels that effectively remove heavy metals and dyes, enhancing existing contaminant mitigation methods and supporting waste management, water purification, and environmental remediation efforts [14]. Moreover, a significant focus of Industry 5.0 is on optimizing manufacturing operations through cognitive adaptability enabled by artificial intelligence, machine learning, and robotics [15]. This approach fosters flexible, efficient, and environment-friendly processes, particularly through proactive human–robot collaboration that boosts production efficiency while minimizing waste. The integration of renewable energy sources and energy-efficient technologies within manufacturing not only reduces carbon emissions but also leads to cost savings, highlighting the essential link between industrial advancement and sustainability. A key challenge lies in balancing productivity, environmental responsibility, and social accountability, which requires aligning technological innovations with sustainable practices and human-centered design. Furthermore, reducing greenhouse gas (GHG) emissions is crucial for organizations aiming for net-zero targets, as a significant portion of their carbon footprint stems from scope 3 emissions related to supply chains [16]). Consequently, Industry 5.0 will increasingly emphasize decarbonizing these supply chains, impacting decision-making throughout the manufacturing process. As clients demand accurate GHG emissions data, manufacturers must transition from estimated to actual scope 3 data by engaging suppliers and potentially conducting Life Cycle Assessments for essential products.

9 Identification of Challenges in Integrating Industry 5.0 with Sustainability Goals

The ecological perspective and the imperative to respect planetary boundaries are integral to the framework of Industry 5.0. This paradigm emphasizes the necessity of actively pursuing global environmental goals, such as waste recycling, the use of biodegradable materials, and achieving CO_2 neutrality. In this context, business models and government policies must incorporate environmental impact and sustainability as fundamental components. Companies are expected not only to prioritize high productivity and financial success but also to deliver social and ecological value.

Within Industry 5.0, digitalization addresses broader environmental, energy, and social challenges, effectively integrating the technological advancements of Industry 4.0 into a larger context that extends beyond the factory floor.

10 Emerging Challenges

The advent of new technologies prompts critical considerations regarding the security and privacy of IoT, especially in the context of environmental sustainability. Industry 5.0 embodies principles that necessitate continuous evaluation and adaptation. Beyond the technological advancements, its societal implications point toward a future marked by inclusivity, ongoing education, and flexibility. The primary objective is to ensure that the digital evolution tied to Industry 5.0 supports a sustainable vision that emphasizes the health of the planet and its inhabitants. Achieving a balance among economic development, environmental stewardship, and corporate social responsibility is crucial; this requires a human-centered approach that integrates digital innovations. Companies must adeptly manage the complexities of increasing profitability while adhering to environmental constraints and fulfilling their social obligations. However, political, logistical, and supply chain challenges can impede this balanced strategy, underscoring the need for stakeholders to proactively tackle these issues. Furthermore, the introduction of new technologies carries inherent ecological risks, such as the toxicity of nanomaterials, the environmental consequences of mining, and the hazards linked to innovative materials and processes, necessitating thorough assessment and the creation of effective risk mitigation strategies.

11 Discussion of Potential Barriers for Businesses and Policymakers

Among the environmentally Sustainable Development Challenges (SDCs), several key obstacles have been identified, including inadequate regulation and enforcement of environmental standards, insufficient focus on research and development for sustainability, and ineffective reverse logistics systems. Addressing these environmental issues is essential for achieving supply chain sustainability. Understanding sustainable raw material standards is critical for successfully implementing sustainable development practices.

Industry 5.0 offers solutions such as the automation of regulatory report generation. By automatically collecting and processing the environmental data necessary for compliance, companies can alleviate the administrative burden, ensuring accurate reporting and reducing the risk of errors that could result in noncompliance. Additionally, Industry 5.0 improves the ability to trace and monitor every stage of the supply chain, facilitating thorough oversight from the procurement of raw materials to the delivery of the finished product. This enhanced visibility allows organizations to assess their environmental footprint across the entire value chain and ensure compliance with regulatory standards.

Research and innovation must prioritize the involvement of diverse stakeholders affected by the renewable energy supply chain. This includes local communities, Indigenous populations, policymakers, and nongovernmental organizations (NGOs).

12 Adapting to Evolving Industrial Demands and Skill Development

There is a critical need for workforce development to meet the changing demands of the industrial sector, making investments in employee upskilling particularly important in areas such as artificial intelligence, the Internet of Things (IoT), and advanced robotics. Successfully implementing sustainable practices requires strong support from top management, financial backing from investors, and favorable government policies. In the context of Industry 5.0, prioritizing robust data privacy measures is essential, especially as transformative technologies like AI and IoT are integrated into existing systems, which presents significant challenges. Additionally, the transition of small and medium enterprises (SMEs) to Industry 5.0 often encounters obstacles due to limited resources, yet their involvement is vital for the global economy, necessitating tailored solutions and comprehensive support. Finally, enhancing resource efficiency and energy consumption is crucial for sustainability, with strategies such as intelligent urban governance, sustainable energy materials, and energy conservation requiring transparency and security through technologies like blockchain.

12.1 Implementation of Smart Grids in Renewable Energy

In the renewable energy landscape, the implementation of smart grids marks a substantial leap forward. These grids leverage real-time data and cutting-edge communication technologies to seamlessly integrate renewable energy sources. They are capable of dynamically adjusting to balance supply and demand, optimizing energy distribution, and reducing losses. In areas with plentiful sunlight, solar energy is emerging as a vital strategy for sustainable development. Industry 5.0 supports the tailoring of solar panels to meet local conditions, such as specific weather patterns and energy needs. Furthermore, AI-driven analytics play a crucial role in determining the best locations for solar panels and energy storage solutions.

13 Suggestions for Policymakers to Support Sustainable Practices in Industry 5.0

Embracing Industry 5.0 technologies and practices to foster sustainability necessitates robust collaboration, skill enhancement, and partnerships among various stakeholders. This encompasses industry experts, policymakers, educational institutions, and the broader community. Implementing comprehensive training initiatives and forming strategic alliances will empower workers with the necessary skills to adapt to the changing industrial environment and play an active role in promoting a sustainable future.

14 Implications of the Fifth Industrial Revolution for Sustainable Manufacturing Industries

- Ultimate beneficiaries of the fifth revolution include manufacturers, employees, and society as a whole.
- Workers are at the forefront of the process, highlighting the importance of enhancing and diversifying their skills.
- Manufacturers can enhance their competitive edge by reducing production costs.
- They can also focus on critical issues like resource conservation, climate change, and social stability while remaining profitable.

14.1 Stepwise Solutions for Effective Implementation of Industry 5.0

Circular production strategies, bolstered by cutting-edge technologies, drive sustainable manufacturing by optimizing resource use, which enhances the ability of industries to withstand external disruptions such as the COVID-19 pandemic and economic downturns. Industry 5.0 seeks to address a range of economic and socio-environmental issues, including job displacement, income inequality, environmental harm, and social disparities. To facilitate the successful implementation of Industry 5.0, governments should actively support essential transformation enablers, focusing on enhancing resource capabilities, especially for micro, small, and medium enterprises (MSMEs). Additionally, promoting sustainable corporate governance is crucial, as it helps businesses transition to more sustainable operational models. On the managerial side, manufacturing firms need to collaborate with stakeholders to ethically reshape their value networks to improve circularity and resilience against disruptions, which is vital for achieving the objectives of Industry 5.0. For decision-makers within these firms, the proposed framework serves as a foundation for crafting customized strategies that can transform their operations into advanced intelligent manufacturing facilities, ultimately generating economic value.

15 Reiteration of the Importance of Integrating Sustainability with Industry 5.0

The Fifth Industrial Revolution emphasizes delivering products and services that enhance quality of life for society. Experts recognize that this concept arose in response to the adverse social and ecological effects of the ongoing digital industrial transformation. Promoted as a sustainable approach to progress, Industry 5.0 utilizes intelligent sensors, interconnected devices, and sophisticated data analysis to enable immediate oversight and management of energy usage. This approach not only minimizes waste but also enhances energy efficiency, contributing to environmental sustainability through reduced resource consumption.

Industrial processes integrated with IoT and AI technologies in industrial processes can yield substantial energy savings and enhance resource efficiency. The Fifth Industrial Revolution presents opportunities for green manufacturing practices that lessen environmental impacts and contribute to an enduring future. Sustainability in this context encompasses multiple dimensions, including resource management, waste reduction, and the promotion of fair working conditions. One illustration of this transition is Zero Defect Manufacturing (ZDM), which highlights the focus on reducing waste and enhancing resource efficiency through advanced methods.

Additionally, Industry 5.0 redefines interactions within circular supply chains, presenting numerous opportunities for sustainable development. Digital technologies play a vital role in this area, with innovations like IoT and digital twins enabling data-driven decision-making that enhances both efficiency and sustainability. Furthermore, the fifth revolution supports Sustainable Product-Service Systems (SPSS), particularly sectors like pharmaceuticals, promoting accountable consumption and production practices. People-centric focus of the Fifth Industrial Revolution also has substantial implications for community resilience.

16 Future Research Directions

The forward-looking perspective of the Fifth Industrial Revolution presents significant challenges for researchers. Existing methods for technology forecasting, such as future impact analysis and scenario planning, may be employed to explore these challenges. Additionally, design-oriented studies will be crucial for solidifying the Industry 5.0 framework. Key research areas include leveraging technology to address human-related issues (such as healthcare ergonomics and worker safety in collaborative robot environments) and transitioning to more sustainable manufacturing practices.

Further investigation is needed in several areas related to Industry 5.0, including:

- Assessing the sustainability of technologies
- Analyzing the impacts of technological transformations
- Enhancing competitiveness through technology-driven sustainability improvements
- Utilizing technology to mitigate resource depletion and support transitions like decarbonization
- Advancing circular economy practices to enhance resource efficiency through reuse, repurposing, and recycling while reducing waste and environmental impacts

A substantial body of research evaluates the benefits and drawbacks of Industry 5.0, its enabling technologies, and their industrial applications. Five primary themes frame the research landscape for the Fifth Industrial Revolution as follows:

- Supply chain assessment and improvement
- Business advancement and digitization
- Implementation of smart and green manufacturing processes
- Transformations driven by IoT, artificial intelligence (AI), and Big Data
- Man–machine connectivity

These themes highlight that potential future research directions indicate a growing interest toward the Fifth Industrial Revolution as a means to enhance man–machine interaction and coexistence.

References

1. Wang, S., Wan, J., Zhang, D., Li, D., & Zhang, C. (2016). Towards smart factory for Industry 4.0: A self-organized multi-agent system with big data-based feedback and coordination. *Computer Networks, 101*, 158–168.
2. Lu, Y., Zheng, H., Chand, S., Xia, W., Liu, Z., Xu, X., et al. (2022). Outlook on human-centric manufacturing towards Industry 5.0. *Journal of Manufacturing Systems, 62*, 612–627.
3. Raworth, K. (2018). *Doughnut economics: Seven ways to think like a 21st century economist*. Chelsea Green Publishing.
4. Huang, S., Wang, B., Li, X., Zheng, P., Mourtzis, D., & Wang, L. (2022). Industry 5.0 and Society 5.0—Comparison, complementation and co-evolution. *Journal of Manufacturing Systems, 64*, 424–428.
5. Akundi, A., Euresti, D., Luna, S., Ankobiah, W., Lopes, A., & Edinbarough, I. (2022). State of Industry 5.0—Analysis and identification of current research trends. *Applied System Innovations, 5*(1), 27.
6. Tóth, A., Nagy, L., Kennedy, R., Bohuš, B., Abonyi, J., & Ruppert, T. (2023). The human-centric Industry 5.0 collaboration architecture. *MethodsX, 11*, 102260.
7. Turner, C., & Oyekan, J. (2023). Manufacturing in the age of human-centric and sustainable industry 5.0: Application to holonic, flexible, reconfigurable and smart manufacturing systems. *Sustainability, 15*(13), 10169.

8. Leng, J., Zhong, Y., Lin, Z., Xu, K., Mourtzis, D., Zhou, X., et al. (2023). Towards resilience in Industry 5.0: A decentralized autonomous manufacturing paradigm. *Journal of Manufacturing Systems, 71*, 95–114.
9. Dwivedi, A., Agrawal, D., Jha, A., & Mathiyazhagan, K. (2023). Studying the interactions among Industry 5.0 and circular supply chain: Towards attaining sustainable development. *Computers & Industrial Engineering, 176*, 108927.
10. Mukherjee, A., Banerjee, S., Das, S., Gupta, A., & Shome, A. (2024). Cobots: Enhancing human creativity and productivity in Industry 5.0. In *Powering Industry 5.0 and sustainable development through innovation* (pp. 186–197). IGI Global.
11. Rojko, A. (2017). Industry 4.0 concept: Background and overview. *International Journal of Interactive Mobile Technologies, 11*(5).
12. De Giovanni, P. (2023). Sustainability of the Metaverse: A transition to Industry 5.0. *Sustainability, 15*(7), 6079.
13. Rame, R., Purwanto, P., & Sudarno, S. (2024). Industry 5.0 and sustainability: An overview of emerging trends and challenges for a green future. *Innovation and Green Development, 3*(4), 100173.
14. Ahire, S. A., Bachhav, A. A., Pawar, T. B., Jagdale, B. S., Patil, A. V., & Koli, P. B. (2022). The Augmentation of nanotechnology era: A concise review on fundamental concepts of nanotechnology and applications in material science and technology. *Results in Chemistry, 4*, 100633.
15. Martín-Gómez, A., Ávila-Gutiérrez, M. J., & Aguayo-González, F. (2021). Holonic reengineering to foster sustainable cyber-physical systems design in cognitive manufacturing. *Applied Sciences, 11*(7), 2941.
16. Chai, Y. H., Yew, G. Y., Yusup, S., & Show, P. L. (2021). Technologies in Industry 5.0. In *The prospect of Industry 5.0 in biomanufacturing* (Vol. 257). Routledge.

Industry 5.0 and Sustainability Issues

N. Krishnaraj, Priyanga Subbiah, Amit Kumar Tyagi, and R. Meera

1 Introduction

Building upon the innovations of Industry 4.0, the ground-breaking idea of Industry 5.0 integrates human-centred principles with state-of-the-art technology. Through harmonious collaboration between humans and machines, Industry 5.0 aspires to foster ethical behaviours, sustainability, and resilience. This is in stark contrast to its forerunner, which prioritised AI-powered production processes. We aim to combine technical innovation with the objectives of sustainable development in order to address social, economic, and environmental concerns. This research delves at the possibilities of Industry 5.0 to utilise AI, the Internet of Things, and robots to lessen resource consumption, waste production, and the efficacy of circular economy models [1]. We can achieve these objectives with the help of these technologies. There are also societal considerations, including as the potential for increased unemployment, the need to retrain workers, and the ethical dilemmas that arise from combining human and robotic labour. Guaranteeing that Industry 5.0 will aid in the attainment of global sustainability objectives requires regulatory frameworks as well as cooperation across various industries. Research like this shows that Industry 5.0 may pave the way for a safer industrial future by removing roadblocks. This is achieved by making use of future trends and case studies. As part of its effort to promote a global economy that is inclusive, resilient, and sustainable, this research adds to the expanding body of literature that explores the possible future of industry.

The current period of industrial advancement, known as 'Industry 5.0', expands upon the work of its predecessor, 'Industry 4.0'. Industry 5.0 takes human–machine

N. Krishnaraj · P. Subbiah (✉) · A. K. Tyagi · R. Meera
Department of Networking and Communications, School of Computing, College of Engineering and Technology, SRM Institute of Science and Technology, Kattankulathur, Chengalpattu, Chennai, Tamil Nadu, India
e-mail: priyangs1@srmist.edu.in

Y. Kumar Singla et al. (eds.), *Digital Manufacturing in Industry 5.0*,
https://doi.org/10.1007/978-3-031-91500-0_20

collaboration to the next level, surpassing even Industry 4.0 in this regard. Industry 4.0, on the other hand, prioritises automation, smart manufacturing, and the IoT. Its major goal is to adopt a more people-focused approach, where cutting-edge tech like AI, robotics, and cobots (collaborative robots) collaborate with humans to create production processes that are better for the environment, people, and themselves. Industry 5.0, in an effort to set itself apart from competing sectors, gives great weight to human creativity and skill. Industry 5.0 aims to achieve its goals by combining human and robotic work, in contrast to its predecessor that concentrated on total automation. People can have the capability to assume creatively, remedy troubles, and make selections, even as robots will do the repetitive obligations. Businesses can enhance morale and employee happiness while concurrently assembly the specific desires of their customers because to this synergy's recognition on bespoke and tailored manufacturing. Another vital attention is the load that Industry 5.0 gives to sustainability and resilience. To be extra precise, it recognises the social and environmental demanding situations that organisations face these days and seeks to mix technology progress with sustainable improvement goals. This approach integrates waste discount, useful resource conservation, and the advocacy of a model of financial pastime known as the round financial system. Reducing cloth consumption while boosting fabric's capacity for further reuse is the aim of a round financial system. Also, by pushing for the introduction of more adaptive structures, Industry 5.0 promotes the development of strong systems that could react to international troubles like pandemics, climate change, and interruptions in supply chain operations. To sum up, Industry 5.0 goals to mix current human competencies with today's era to solve worldwide sustainability and resilience troubles. This new attitude on commercial manufacturing envisions a future wherein era augments human skills instead of replacing them. It envisions a transition from absolutely computerised strategies to collaborative efforts among people and robots.

2 Key Technologies in Industry 5.0

Artificial intelligence (AI) automates complex duties, analyses huge volumes of information, and makes predictions to useful resource human beings. But people can still use it to make critical decisions [2]. In contrast to traditional robots, cobots are designed to operate in tandem with humans in a safe setting, assisting them with physically demanding or repetitive tasks, thereby maximising human productivity. Machines are not meant to supplant humans, but rather to supplement them. A more innovative, creative, and flexible workplace is the end result of this. These include, but are not limited to, energy-efficient machinery, reusable materials, and environment-friendly processes.

2.1 *Benefits of Industry 5.0*

The human element is key to industrial processes, which make extensive use of people's intuition, inventiveness, and decision-making abilities. Reducing waste, conserving resources, and implementing renewable energy solutions are some of the environmentally responsible practices promoted by Industry 5.0. Mass production of bespoke goods is one strategy for meeting the rising demand for unique, tailored goods. The goal of incorporating systems into Industry 5.0 is to make them resilient and adaptable in the face of disruptions like supply chain issues or environmental restrictions.

2.2 *Key Principles and Objectives of Industry 5.0*

2.2.1 Human-Centric Automation

Comparatively, Industry 4.0 prioritised fully mechanical automation; in contrast, Industry 5.0 highlights human–machine collaboration. Our aim is to empower workers by giving them the chance to work alongside collaborative robots (cobots) and artificial intelligence systems that enhance their capabilities instead of replacing them. We want to achieve this by combining human creativity, problem-solving skills, and decision-making with the precision and efficacy of cutting-edge technology. The outcome is an improved manufacturing process in terms of customisation, originality, and adaptability.

2.2.2 Sustainability and Environmental Responsibility

Sustainable development lies at the heart of Industry 5.0, which is in harmony with international efforts like the UN's Sustainable Development Goals (SDGs). With a focus on reducing environmental impacts, conserving natural resources, and doing away with waste, this effort promotes the adoption of a circular economy. Sustainable practices and technology are essential for lowering carbon footprints, increasing the usage of renewable energy, and facilitating material recycling and reusing. The plan's objective is to keep economic growth opportunities open while also making sure that industries help keep the environment safe.

2.2.3 Mass Personalisation and Customisation

Industry 5.0 is a departure from the conventional mass production strategy that places a greater emphasis on the development of customised goods. Technological advancements like cobots, the Internet of Things (IoT), and artificial intelligence

(AI) allow for large-scale 'mass customisation'. As a result, products can be tailored to meet the specific needs of each customer. So that we can meet the increasing demand for tailored and personalised products without sacrificing efficiency or going over budget. This allows for the development of manufacturing models that are more customer-centric, which in turn creates value for firms and consumers alike.

2.2.4 Human–Robot Collaboration (Cobots)

According to Industry 5.0, humans should focus on higher-level, more creative, and decision-driven tasks, while robots should perform lower-level, more dangerous tasks in partnership with humans. In this ideal world, cobots and humans can work together in harmony, doing tasks in a safe and efficient manner. Workplace safety will improve, human workers will feel less tired, and productivity will rise as a result.

2.2.5 Resilience and Flexibility

Manufacturing systems that can adapt to several types of disturbances, such as pandemics, global supply chain issues, and climate-related occurrences, are what Industry 5.0 aims to create. The ability to quickly adjust production processes in response to outside forces is crucial for continuity and stability in the face of unpredictability. Included in this is the use of adaptive technologies like AI to make judgements in real time and change production lines autonomously.

2.2.6 Ethical and Social Responsibility

With technology increasingly permeating every aspect of business, the ethical concerns raised by rising levels of automation, AI, and robotics must be addressed by Industry 5.0. This exemplifies the significance of socially conscious innovation that prioritises the well-being of both employees and society at large in order to guarantee the correct handling of ethical concerns such as data privacy, job loss, and the effect of AI on human duties. Achieving this intention will necessitate workers improving their talent units, the advent of recent task possibilities, and the promotion of a harmonious interaction between people and technology.

2.2.7 Circular Economy and Resource Efficiency

Artificial intelligence (AI) automates complex tasks, analyses massive volumes of statistics, and makes predictions to aid human beings. But people can nevertheless use it to make crucial choices. In comparison to traditional robots, cobots are designed to perform in tandem with human beings in a secure putting, helping them with physically traumatic or repetitive duties, thereby maximising human productivity. Machines are not meant to supplant human beings; however, they are here to complement them. An extra modern, creative, and bendy place of job is the stop result of this. These include, however aren't confined to, power-green equipment, reusable materials, and environmentally pleasant procedures.

2.2.8 Integration of Advanced Technologies

The purpose of Industry 5.0 is to improve decision-making, growth automation wherein it makes experience, and offer actual-time statistics-driven insights by means of integrating rising technology like AI, IoT, large information, cloud computing, robotics, and system mastering into regular commercial procedures. This hyperlink bolsters customised production, better productivity, and better useful resource management.

2.2.9 Social and Economic Inclusivity

Industry 5.0 prioritises variety and inclusion in the workplace by growing possibilities for people from all walks of lifestyles and all corners of the globe to contribute to the economic financial system. The objective is to mitigate the effect of automation on employment by using bolstering retraining tasks and developing new task categories that centre on human–generation interaction. The intention is to make certain that the blessings of Industry 5.0 are dispersed flippantly throughout the populace if you want to promote financial growth and social equity.

2.2.10 Focus on Worker Well-Being

One method Industry 5.0 aims to create a productive and worker-friendly environment is by designing systems and technology that promote workers' emotional and physical well-being. The objective is to improve working conditions, reduce physical strain on employees through automation of repetitive tasks, and provide them with roles that are more fulfilling and allow for greater creative expression. As a result, company morale improves, productivity rises, and job satisfaction rises.

3 The Role of Human-Centred Design in Industry 5.0

An essential component of Industry 5.0, human-centred design (HCD) shifts the emphasis from technology and automation to people and their needs, happiness, and creativity. Although technology is crucial to industrial progress, Industry 5.0

recognises that it is just as important to think about how people use and interact with these technologies. Rather than seeing technology as a threat to human abilities, this approach seeks to enhance human–robot collaboration for the common good.

3.1 Collaboration Between Humans and Robots (Cobots)

Cobots, or collaborative robots, paintings along with human beings represent a sizable departure from Industry 4.0's emphasis on automation. These robots are programmed to do risky, repetitive, or physically taxing duties so that human beings can attention on their strengths. This frees up human brainpower for better-order duties that require creativity or selection-making. Cobots constructed with human–laptop interaction (HCD) are more likely to be consumer-friendly, intuitive, and flexible in assembly human wishes. Teams are able to paintings together extra effectively for the reason that design prioritises the nicely being of human workers.

3.2 Enhancing Human Creativity and Innovation

While computer systems handle mundane duties, human beings may also consciousness on greater innovative sports like personalising products, coming up with fresh thoughts, and solving troubles. Artificial intelligence (AI), big data, and robots may all be made greater person-friendly with the help of interfaces and technologies advanced with a human-targeted layout technique. Through the usage of human–PC interplay (HCD), which streamlines the interaction among humans and complex technology, people are empowered to utilise their creativity in creating and improving items and methods.

3.3 Customisation and Personalisation

Among the hallmarks of the Industry 5.0 revolution and paradigm is the shift from mass manufacturing to mass customisation. In order to fulfil the demand for personalised products, human-targeted design ensures that this demand is met in an effective and scalable way. Using human-targeted layout concepts makes it easier to installation bendy production techniques that could unexpectedly react to client needs. Doing so guarantees that both the production system and the final product meet the desires and options of the people worried.

3.4 Focus on Worker Well-Being and Satisfaction

Industry 5.0 aims to enhance working conditions, decrease stress levels, and raise worker happiness, one of its key goals being the elimination of repetitive or physically hard occupations. When creating a workplace [3], the human-centred design method ensures that the worker's requirements are prioritised. Included in this category with ergonomic design are workplaces that promote creativity and collaboration as well as technology that improves the well-being of workers. Through reducing physical strain and improving mental health, HCD ensures that technology enhances the work experience, making it more rewarding and meaningful.

3.5 Empowering Workers Through Upskilling

At a global level wherein automation and AI are taking up recurring obligations, employees want to be trained in new abilities to control leap forward technology effectively. A necessary part of human-focused design within the context of Industry 5.0 is the creation of tutorial materials and intuitive interfaces to assist employees emerge as gifted with new technologies. Making positive those technologies are user-friendly and reachable is what human-focused design (HCD) is all approximately. This facilitates employees' transfer into new positions that area an emphasis on managing technology and addressing troubles creatively.

3.6 Ethical Design and Social Responsibility

As AI, robots, and different technologies extend their roles in many industries, moral worries around records privacy, job displacement, and the relationship among people and machines are rising to the vanguard. To deal with moral issues, human-centred layout adheres to the concepts of inclusive layout, which seek to empower employees in place of displace them and create new employment opportunities. On pinnacle of that, it guarantees that technological structures are made to make humans' lives higher, that they're non-public by using design, and that they're transparent while making choices.

3.7 User-Friendly Interfaces and Seamless Technology Integration

There needs to be intuitive and person-pleasant generation within the administrative centre if personnel are going to have interaction with complex machines. With a focus on accessibility, efficiency, and simplicity of use, human-targeted layout guarantees that merchandise' person interfaces (UI) and person reviews (UX) are created with the stop-consumer in thoughts. Because of this, employees have a simpler time adjusting to new technology, which shortens the studying curve and boosts output.

3.8 Increased Flexibility and Adaptability in Workplaces

More flexibility in manufacturing strategies is required through the Fifth Industrial Revolution, mainly as patron demand for customised items maintains to rise. Adaptability in commercial systems is ensured by means of human-centred design, permitting people to quick respond to changing requirements or unexpected boundaries. This adaptability permits companies to hold performance while additionally presenting customised and dynamic services and products.

4 Technological Advancements in Industry 5.0

The technical foundations of Industry 4.0 are increased upon in the subsequent technological jump called Industrial 5.0. It does, but, strain the need of mixing human intellect and creativity with technological advancements more heavily. In an effort to improve sustainability, boom efficiency across a variety of industries, and foster human–robot collaboration, it introduces foremost improvements. Many extensive era advancements had been added by way of Industry 5.0, such as: Industry 5.0 differs from in advance generations in numerous methods, considered one of which is the seamless collaboration among humans and robots [4]. The purpose of cobots, brief for 'collaborative robots', is to augment human abilities inside the workplace. This differs from the norm for industrial robots, which regularly function autonomously or are assigned particular responsibilities. Cobots are high-quality at physical labour and repetitive sports, releasing up humans for more complex and creative work. This makes sure that the efficiency of robots and human intelligence are both maximised. Cobots can understand human behaviour, adapt to different environments, and learn from interactions with humans because they are equipped with advanced sensors, AI, and machine learning algorithms. A key component of Industry 5.0 is AI and ML, where computers can process huge amounts of data, make decisions and improve processes. Emerging technologies enable robots, cobots, and other systems to learn from data and evolve. Increase efficiency In Industry 5.0, artificial intelligence (AI) systems can increase supply chain efficiency. Improved quality control Predict when maintenance is needed and more. Machine learning models can help recognise patterns in production data for better resource management and reduced waste. The wise, versatile, and successful robots utilised in Industry 5.0 constitute a development above in advance generations. Their precision, dexterity, and flexibility permit them to do tough paintings, and that they do it frequently in environments that are both dynamic and reliant on human interplay [5]. The use of self-reliant mobile robots (AMRs), drones, and robotic fingers that may sense their surroundings, adapt their movements in actual-time, and collaborate with human operators is what robotics means within the context of Industry 5.0. Thanks to advancements in AI, robots can now do increasingly complicated tasks, including creative tasks like product design, quality inspections, and object manipulation. What we mean when we talk about 'edge computing' is processing data closer to its point of generation, at the 'edge' of the network. This differs from relying solely on cloud computing. This allows for both quicker reaction times and the ability to make decisions in real time. Industry 5.0 real-time processes can now take advantage of faster data processing thanks to edge computing. Autonomous systems, AI-driven quality control, and predictive maintenance are all part of these endeavours. Due to this, businesses have an easier time managing massive data flows from sensors, devices, and machines, which leads to less latency and more operational efficiency.

When you link physical objects to the internet, they may communicate, share information, and act instantly. This is the core concept of the IoT, or Internet of Things. Through the integration of AI, ML, and edge computing, the Internet of Things (IoT) strengthens its position in Industry 5.0 by constructing smarter networks of interconnected devices. Smart systems can keep tabs on things in real time, pinpoint inefficiencies, and tweak processes to get the most of them. An asset, system, or process can have its performance emulated, tracked, and improved with the use of a digital twin, which is a digital representation of the real thing [6]. It is also viable to use digital twins to decorate overall performance. The use of virtual twins for predictive upkeep and real-time tracking is on the rise inside the Industry 5.0 framework. This permits them to simulate production settings, discover potential troubles, and take a look at answers earlier than making use of them to the actual system. Digital twins open up a world of possibilities for operational optimisation, performance gains, downtime discount, and product first-class improvement. Incorporating technologies that enhance strength efficiency, lessen waste, and minimise the impact at the environment is central to the concept of Industry 5.0, which locations a robust emphasis on sustainability. In order to create production techniques that are much less harmful to the surroundings, contemporary technology is used, together with 3-D printing, biodegradable materials, and recyclable sources. By preserving tabs on useful resource intake, optimising electricity utilisation, and cutting emissions, the internet of factors and artificial intelligence can assist result in a greater circular economic system.

Blockchain is a disbursed ledger that satisfies the wishes for protection, accountability, and traceability in financial transactions and records change. Blockchain technology is used to enhance deliver chain traceability and transparency in the framework of Industry 5.0. This increases trust among all parties worried, guarantees that the product is genuine, and for better product tracking. Additionally, producers can beautify the deliver chain's sustainability with the aid of using blockchain generation to verify the ethical procurement of commodities. More strong and faster net connections with reduced latency and multiplied records switch quotes will be to be had with the arrival of 5G generation. As a part of the Industry 5.0 surroundings, 5G is critical for enabling real-time communication among machines, IoT gadgets, robots, and human operators [7]. In doing so, it lays the groundwork for faster and more efficient industrial operations through the deployment of autonomous vehicles, robotic systems, and smart factories. More interaction between the digital and physical worlds is possible with immersive technologies like virtual reality (VR) and augmented reality (AR). Industry 5.0 makes extensive use of AR and VR for a wide range of applications, such as remote assistance, design reviews, prototyping, and instruction. There is less need for physical prototypes when training simulations are conducted in virtual reality (VR). Workers can receive real-time instructions while they work simultaneously with augmented reality (AR) glasses.

5 Sustainability in Industry 5.0

Industry 5.0 sustainability aims to provide long-term environmental, economic, and social benefits through the integration of human-centred and environmentally conscious practices with modern technologies. As opposed to Industry 4.0's emphasis on efficiency, intelligent production, and automation, Industry 5.0 seeks to fulfil sustainable development goals while simultaneously advancing technology. The following are a few of the most critical aspects of Industry 5.0 sustainability: Improving working conditions, well-being, and job happiness are the end outcomes of Industry 5.0's focus on developing human talents rather than replacing them. As a result of this strategy's emphasis on human–robot interaction, the organisation's social sustainability is enhanced [8]. Industry 5.0 aims to optimise resource utilisation through the use of technologies such as data analytics, artificial intelligence (AI), and the Internet of Things (IoT). Waste, energy consumption, and negative effects on the environment are all mitigated as a result. The development of a more sustainable and regenerative economy is a key component of Industry 5.0. Modern manufacturing processes enabled by digital technology permit businesses to lessen resource depletion via the advent of extra green manufacturing techniques, recycling of materials, and repurposing of trash. In order to respond to disruptions like deliver chain troubles or surprising adjustments in demand, Industry 5.0 promotes the development of sturdy and adaptable systems. Industries can preserve strolling sustainably even in the face of crises because they're so adaptable. Machine studying, blockchain, digital twins, and artificial intelligence (AI) are a number of the technologies which could assist lessen carbon footprints. Sustainable supply chain management, predictive renovation, and greater efficient industrial operations are all aided by way of these technologies. Collaboration between human operators and robots has the capacity to significantly enhance operational performance, leading to a marked decrease in cloth waste and errors. By reducing the wide variety of extra assets wanted, this not simplest boosts production but also contributes to the sustainability of the corporation [9]. With the help of Industry 5.0, it's far now viable to create bespoke merchandise that meet the specific wishes of every purchaser. With this approach, we can lessen stock waste and overproduction, elements that cause a greener production model. Promoting healthy running environments, decreasing inequality, and guaranteeing honest reimbursement are only a few of the ethical actions that Industry 5.0 helps. Social sustainability is based on those standards because they sell place of job equity and inclusivity. To encourage the implementation of sustainable practices, effective regulatory frameworks have to be in vicinity. Industry 5.0 era has to be well suited with environmental legal guidelines and sustainability goals that can most effectively be accomplished via joint efforts between governments and businesses. Several businesses are developing eco-friendly merchandise and approaches by way of making use of the ideas of Industry 5.0. By using clever technologies and renewable strength sources, smart factories are able to lower electricity intake and carbon emissions.

6 Challenges and Risks of Industry 5.0

Industry 5.0's attention on people-centred manufacturing makes it all of the extra critical to seamlessly combine human experts with collaborative robotics and sensible systems. Getting humans to connect to machines in a secure manner is one in every of the largest obstacles in those forms of collaborative environments. It is difficult to incorporate contemporary generation like AI, the internet of factors, and robots into the modern-day infrastructure of Industry 4.0 while keeping the entirety likeminded. Moving to Industry 5.0 requires fundamental technological advancements that may cause highly priced implementation prices and technical problems. Experienced human workers are mostly relied upon for innovative employment in Industry 5.0, while automation is tasked with performing boring procedures. There is a serious risk of a skills gap if training resources are insufficient to retrain the workforce to acquire the necessary abilities in fields like data analysis, artificial intelligence, and robots. Vulnerability to cyber attacks rises in direct proportion to the degree to which equipment, people, and networks are interdependent [10]. Implementing cutting-edge cybersecurity solutions is crucial in the context of Industry 5.0 settings to guarantee the safety of sensitive data and intellectual property. A large sum of money is needed to implement technological solutions like collaborative robots, AI, and Internet of Things. There may be uneven adoption across industries if small and medium companies (SMEs) are unable to afford the high costs of transitioning to Industry 5.0. Ethical issues like choice-making autonomy, privateness protection, and man or woman rights are delivered up by means of the growing interplay among humans and robots. Ensuring that AI structures and robots operate in a moral and open manner is a crucial yet hard venture. Despite Industry 5.0's recognition on human–robot collaboration, the automation of recurring duties could cause task losses. Managing the social effects of this change, inclusive of retraining displaced persons and reducing unemployment, ought to be carefully monitored. The loss of really articulated requirements and norms ought to impede the wide adoption of Industry 5.0. Concerns approximately facts privacy, ethical AI use, and human–robot interaction will want the creation of recent frameworks and norms. Energy intake related to integrating modern-day era can also result in an increase in carbon emissions, even if Industry 5.0 is based on the perception of sustainability. Finding a satisfied medium between the need for progressive era and ecological maintenance is not any smooth feat. Organisational culture has to exchange for Industry 5.0 to be efficiently implemented. Businesses must undertake a human-centred innovation mind-set, which requires changes in management fashion, company subculture, and normal manner of thinking. Achieving an easy rollout of the innovation is made extra hard with the aid of this. The potential to attach extraordinary systems and gadgets without any hitches is crucial to Industry 5.0; however, overcoming differences in statistics codecs, protocols, and standards isn't any easy feat. To gain full interoperability, it is important for all elements of the enterprise to paintings together. The increasing dependence on AI and robots increases worries approximately capacity over-reliance, which would possibly depart systems prone to failure or disruption. Reducing this threat calls for adequate backup techniques and resilience, which need to be ensured.

7 Industry 5.0 and the Circular Economy

Industry 5.0 encourages useful resource efficiency by means of combining human and technological talents. It does this by way of focussing on humans first and principal. The initiative aims to reduce business activities' terrible effect at the environment through reducing down on waste. The standards of the round economic system are congruent with this, in view that they sell recycling and seek to minimise resource intake. Industry 5.0 makes heavy use of modern-day innovation like AI and robots to streamline the manufacturing of in my view tailor-made items [11]. One of the main blessings of producing customisation is the lower in overproduction. This helps with waste discount and aligns with the standards of the circular economic system. More green technology may be created thru the collaboration of clever machines and human beings in the framework of Industry 5.0. In the long run, this partnership may result in greener supply chains that placed an emphasis on recycling and reusing materials by way of enhancing material recovery, remanufacturing, and green disassembly. The combination of IoT, gadget mastering, and artificial intelligence enables actual-time optimisation of useful resource utilisation within the framework of Industry 5.0 [12]. The adoption of a round financial system framework may be aided with the aid of technology that can detect inefficiencies and music waste introduction and offer records-pushed insights to improve resource efficiency. Servitisation, the shift from product sales to service provision, is encouraged for by using the fifth generation of manufacturers, who need to see extra groups undertake product-as-a-provider fashions. This is in step with the standards of a round financial system because it helps product remanufacturing and reuse. By integrating contemporary sensing and robotics generation, Industry 5.0 streamlines the automation of the recycling manner. All three of these aspects—improved material sorting precision, decreased contamination, and more efficient recycling operations—are cornerstones of the circular economy. Businesses can utilise big data analytics to make informed decisions on how to repair and refurbish items, extend the life cycle of products, and manage products that have reached the end of their useful life. A more sustainable and recursive system can be built with the aid of these methods.

Integrating complex technologies like AI, the Internet of Things, and robotics with conventional production processes can be a significant challenge [13]. Companies should put money into training and modifying human skills so they can work in harmony with state-of-the-art machinery. Large up-front costs may be associated with recycling facility construction, reverse logistics deployment, and manufacturing process adjustment as part of the shift to a circular economy model. To encourage businesses to embrace circular methods, the government must establish rules, regulations, and incentives that facilitate adoption. The broad adoption of circular products and services can only be achieved if customers are made aware of and agree with these environmentally conscious practices.

8 Resilience in Industry 5.0

The focus of the idea of resilience in Industry 5.0 is the capacity of industrial systems to adjust to disruptions and sustain continuous production while preserving the welfare of both humans and the environment [14]. The focus of Industry 5.0 is on human–machine collaboration, utilising cutting-edge technology like AI, robotics, and the Internet of Things to increase resilience and adaptability. What we mean when we talk about resilience in this context is building systems that can adapt and heal themselves so that they can handle things like supply chain interruptions, equipment failures, and other problems that may arise. Building more resilient and flexible manufacturing systems that can overcome and adjust to unforeseen challenges is the aim of Industry 5.0 [15]. Collaboration among people and contemporary technology permits for a rapid reaction to any modifications or disturbances in manufacturing, that is one of the methods that may be implemented to enhance resilience. More responsive and adaptable decision-making at diverse phases of production is made viable by using disbursed manage, that's enabled via the creation of advanced AI and autonomous systems. Predictive maintenance can find potential problems with equipment before they cause breakdowns by using machine learning and the Internet of Things (IoT). This guarantees that activities continue without interruption and helps to minimise downtime. With the help of digital twins, which simulate and predict how systems will react to disruptions, companies may improve their responses and lessen the impact of disruptions. Supply chains that can easily adjust to changes in demand, sourcing, and logistics are highly valued in the Industry 5.0 implementation.

9 Ethical Implications of Human–Robot Collaboration

To guarantee that Industry 5.0 has a positive effect on society, it is necessary to thoroughly investigate the many ethical issues raised by the proposal to integrate humans and robots in collaborative work [16]. The following are some of the most serious ethical concerns associated with human–robot collaboration: The prospect of workers losing their employment due to automation and robots performing tasks formerly done by humans is one of the most important ethical worries. Despite robots' intended use of enhancing human skills and freeing people from repetitive tasks, it's conceivable that unemployment might rise in sectors where robots are more efficient. Finding a middle ground that enables people to switch careers while simultaneously giving them enough chances to learn new skills is crucial for solving this ethical challenge [17]. A lot of data, including potentially sensitive personal information, has to be collected and processed in order for robots and humans to interact. There are ethical concerns about data collection, storage, and use practices, as well as the risk of abuse, and about protecting users' privacy. To make sure statistics protection and preserve individuals' privacy rights, Industry 5.0 advocates for strict standards to be implemented in a collaborative framework. Instances of unfair or discriminatory behaviour throughout selection-making occasions may additionally stand up if artificial intelligence (AI) systems and robots unwittingly undertake or amplify biases which might be already found in their education information. There could be ethical concerns if robots are biased, especially in fields where fair treatment of all people is crucial, like recruitment, law enforcement, or customer service. Among the most important moral concerns is the need to train and develop AI and robotic systems in a way that promotes fairness and minimises bias. A loss of human agency could result from people being too reliant on machines, which is an ethical concern [18]. There is always the risk of harm when people put their trust in robots to make judgements, especially in industries where human lives are on the line, like healthcare or manufacturing. Preventing the negative consequences that can arise from relying too much on automation requires constant human oversight and control over decision-making. Ensuring safety is of utmost importance when it comes to human–robot collaboration, especially in industrial settings. During the design phase of robots, it is crucial to consider the physical safety of human equivalents and minimise dangers. Another ethical dilemma that needs fixing is the problem of determining who is at fault when accidents involving robots occur [19]. When a robot malfunctions and injures someone, who should be responsible? People who use it, code it, or create it. Transparent accountability systems must be established if such circumstances are to be managed properly. It's plausible that people's emotions and mental health could be affected by their contact with robots. Using robots as companions or carers for the elderly, for example, could meet practical needs; yet, it could also lead to social isolation and a lack of genuine human interaction. It is important to think about ethical issues like whether or not robots should replace humans in interpersonal connections and whether or not people's mental health should be a top priority. People in situations where they will be interacting closely with robots should be educated about the robots' capabilities, limitations, and responsibilities. Reason being, people must be cognisant of the robots in order to avoid ethical dilemmas. To avoid mistrust and promote informed consent, it is essential that algorithms and robotic behaviour be open and transparent so that people know how decisions are made. A potential supply of situation concerning human dignity can be the integration of robots into historically human-performed responsibilities. When robots are used in healthcare or instructional environments, this will become a great deal extra obvious. Among the most urgent ethical issues is ensuring that robots do not devalue human beings or the respect because of unique occupations.

10 Regulatory Frameworks and Policy in Industry 5.0

The incorporation of modern technology is crucial to Industry 5.0's human-centred, ecologically conscious ideas. Whereas Industry 4.0 became greater concerned with shrewd production and automation, Industry 5.0 is extra worried with locating a centre floor between the two competing demands of technological development and environmental protection. Industry 5.0 aims to acquire sustainability via lowering resource use, minimising waste, and advancing circular economic system fashions. Industry 5.0 is capable of promote sustainable behaviours with the aid of making use of technology like the Internet of Things (IoT), robotics, and synthetic intelligence (AI). Resource optimisation, emission discount, and fabric reuse merchandising are all examples of such sports. Furthermore, while human beings and robots work collectively, it turns into a lot simpler to tailor the manufacturing method to every man or woman's needs, which in turn reduces inefficiencies and produces environmentally conscious items. To well harness the strength of Industry 5.0, widespread regulatory structures and regulations should be placed into region. These plans have to make sure that the goals of global sustainability are well suited with the appropriate software of contemporary generation. Regulations must bear in mind the ethical worries raised by issues like data privacy, cybersecurity, and human–robot collaboration [20]. Furthermore, they should suggest for green production practices, which includes reducing down on waste and maximising power efficiency. To prepare people for the new jobs with the intention to be created via the human-centric and era-driven enterprise, rules have to be in vicinity that sell retraining and ability acquisition. Governments, organisations, and academic establishments ought to collaborate to set up standards that guide innovation without sacrificing attention to human health and environmental protection.

11 Future Trends and Opportunities in Industry 5.0

Increased human–robotic collaboration is an indicator of Industry 5.0, with robots increasingly more serving as people' aides. Rather of displacing humans, this fashion will see robots enhancing human skills. Workers may be free to cognisance on strategic and creative duties as robots handle risky or uninteresting employment. An essential part of Industry 5.0 is customisation on a scale in no way formerly witnessed. Powerful AI and ML will permit manufacturers to create exceedingly customised goods that meet the precise demands of each purchaser, taking into account their precise tastes and necessities. Industry 5.0 is centred around a shift towards greener practices. Reducing carbon footprints, optimising resource utilisation, and implementing models for circular economies are all part of this. Optimisation of resource utilisation, waste management, and energy use can be achieved through the use of intelligent technologies like the Internet of Things and artificial intelligence, leading to greener production. Making decisions with the help of AI will be crucial, alongside human insight and creativity. While AI tools provide comprehensive analysis, predictive insights, and automation, humans will have more mental capacity for ethical, empathetic, and creative decision-making. Collaborative robots, or cobots, are becoming more popular; these machines can operate side by side with people, are safe, and can adjust to their surroundings. Cobots' intelligence, flexibility, and capacity to learn from human interactions will keep rising. As digital twin technology evolves, it will become possible to simulate physical assets in real time. In the long run, this aids predictive maintenance, which improves efficiency and sustainability. There will be fewer disruptions because it helps maximise performance and decrease downtime. Product design, remote maintenance, and training are three fields that will see extensive usage of VR and AR technologies in the near future. As a result, employees can learn new skills in a controlled setting, which reduces hazards and boosts output. Smaller, more locally based, and more adaptable production facilities are the hallmarks of decentralised manufacturing, which is expected to usher in Industry 5.0. Benefiting local economies, cutting logistics costs, and shortening time-to-market are all possible outcomes of capitalising on this trend. A higher risk of cyber breaches is associated with more connectivity. Industry 5.0 trends going forward will largely revolve around creating robust cybersecurity solutions to guard against potential threats to data, networks, and industrial systems. As more and more tasks become automated, there will be a greater need for workers who are both technically savvy and comfortable working with new systems. Industry 5.0 has the potential to reshape the body of workers by supplying opportunities for reskilling and upskilling, a good way to help personnel grow to be equipped for new jobs. The integration of AI and the Internet of Things (AIoT) will lead to the development of greater independent and wise structures to be used in production. Improved operational performance, smarter supply chains, and predictive upkeep are all outcomes of this link.

12 Conclusion

According to the authors of this study, Industry 5.0 represents a sea of change in production processes. Sustainability, resilience, and human–robotic collaboration are on the heart of this paradigm exchange, that's defined with the aid of the incorporation of human-centred layout standards with modern-day technological competencies. Integrating technological development with sustainable development targets is the intention of Industry 5.0, the following technology of Industry 4.0. The research highlights the possibility of Industry 5.0 using AI, the internet of factors, and robots to sell sustainable practices. These innovations can assist us use less assets, produce much less trash, and even back up round economy models. In addition to the social results, which include the high unemployment fee and the necessity to retrain the workforce, the study stresses the significance of tackling the moral concerns of human–robotic collaboration. Regulatory frameworks and cross-sector cooperation play a vital function in making sure Industry 5.0 would not warfare with worldwide sustainability goals. According to the findings, Industry 5.0 has incredible potential to build a greener industrial scene. Analyses of destiny developments and case studies display this to be real. Regardless, the dangers and challenges associated with it need to be carefully managed if its potential is to be completely realised. This analysis contributes to the ongoing discussion about the future of business by highlighting its critical role in creating a global financial system that is inclusive, robust, and sustainable.

References

1. Ince, F. (2023). Socio-ecological sustainability within the scope of industry 5.0. In *Practice, progress, and proficiency in sustainability*. IGI Global. https://doi.org/10.4018/978-1-6684-6113-6.ch002
2. Sarıoğlu, C. İ. (2023). Industry 5.0, digital society, and consumer 5.0. In R. S. Nazlı & G. Sari (Eds.), *Handbook of research on perspectives on society and technology addiction* (pp. 11–33). IGI Global. https://doi.org/10.4018/978-1-6684-8397-8.ch002
3. Bissadu, K. D., Sonko, S., & Hossain, G. (2024). Society 5.0 enabled agriculture: Drivers, enabling technologies, architectures, opportunities, and challenges. *Information Processing in Agriculture, 12*(1). https://doi.org/10.1016/j.inpa.2024.04.003
4. Reddy, C. K. K., Anisha, P. R., Khan, S., Hanafiah, M. M., Pamulaparty, L., & Mohana, R. M. (Eds.). (2024). *Sustainability in industry 5.0: Theory and applications*. CRC Press.
5. Ghobakhloo, M., Mahdiraji, H. A., Iranmanesh, M., & Jafari-Sadeghi, V. (2024). From Industry 4.0 digital manufacturing to Industry 5.0 digital society: A roadmap toward human-centric, sustainable, and resilient production. *Information Systems Frontiers*, 1–33.
6. Trofimova, N. N. (2023). Key technologies used by enterprises in modern conditions in order to increase efficiency and sustainable development (on the example of artificial intelligence and collaborative robots). *Èkonomika I Upravlenie: Problemy, Rešeniâ, 2/5*(134), 55–60. https://doi.org/10.36871/ek.up.p.r.2023.02.05.008
7. Martini, B., Bellisario, D., & Coletti, P. (2024). Human-centered and sustainable artificial intelligence in industry 5.0: Challenges and perspectives. *Sustainability, 16*(13), 5448.
8. Kaswan, M. S., Chaudhary, R., Garza-Reyes, J. A., & Singh, A. (2024). A review of Industry 5.0: From key facets to a conceptual implementation framework. *International Journal of Quality & Reliability Management, 42*(4), 1196–1223.
9. Mourtzis, D. (2024). Challenges and opportunities of the transition from Industry 4.0 to Industry 5.0. In *Manufacturing from Industry 4.0 to Industry 5.0* (pp. 97–131).
10. Sarkar, B. D., Shardeo, V., Dwivedi, A., & Pamucar, D. (2024). Digital transition from industry 4.0 to industry 5.0 in smart manufacturing: A framework for sustainable future. *Technology in Society, 78*, 102649.

11. Ghobakhloo, M., Fathi, M., Iranmanesh, M., Vilkas, M., Grybauskas, A., & Amran, A. (2024). Generative artificial intelligence in manufacturing: Opportunities for actualizing industry 5.0 sustainability goals. *Journal of Manufacturing Technology Management, 35*(9), 94–121.
12. Gupta, M. K., Korkmaz, M. E., Yılmaz, H., Şirin, Ş., Ross, N. S., Jamil, M., et al. (2024). Real-time monitoring and measurement of energy characteristics in sustainable machining of titanium alloys. *Measurement, 224*, 113937.
13. Korkmaz, M. E., Gupta, M. K., Çelik, E., Ross, N. S., & Günay, M. (2024). A sustainable cooling/lubrication method focusing on energy consumption and other machining characteristics in high-speed turning of aluminum alloy. *Sustainable Materials and Technologies, 40*, e00919.
14. de Mattos Nascimento, D. L., de Oliveira-Dias, D., Moyano-Fuentes, J., Maqueira Marín, J. M., & Garza-Reyes, J. A. (2024). Interrelationships between circular economy and Industry 4.0: A research agenda for sustainable supply chains. *Business Strategy and the Environment, 33*(2), 575–596.
15. Ting, L. S., Zailani, S., Sidek, N. Z. M., & Shaharudin, M. R. (2024). Motivators and barriers of circular economy business model adoption and its impact on sustainable production in Malaysia. *Environment, Development and Sustainability, 26*(7), 17551–17578.
16. Botti, A., & Baldi, G. (2024). Business model innovation and Industry 5.0: A possible integration in GLAM institutions. *European Journal of Innovation Management*.
17. Sharma, R., & Gupta, H. (2024). Leveraging cognitive digital twins in industry 5.0 for achieving sustainable development goal 9: An exploration of inclusive and sustainable industrialization strategies. *Journal of Cleaner Production, 448*, 141364.
18. Gamper, J. (2023). Influence of 3D-printable sustainable concrete and industrial waste on industry 5.0. *Practice, Progress, and Proficiency in Sustainability*, 156–191. https://doi.org/10.4018/978-1-6684-6113-6.ch008
19. Tyagi, A. K., Arumugam, S. K., Prasad, P. R., & Sharma, A. (2024). The position of digital society, healthcare 5.0, and consumer 5.0 in the era of industry 5.0. In *Advancing software engineering through AI, federated learning, and large language models* (pp. 262–280). IGI Global.
20. Choudhry, M. D., Jeevanandham, S., Sundarrajan, M., Jothi, A., Prashanthini, K., & Saravanan, V. (2024). Future technologies for industry 5.0 and society 5.0. *Automated Secure Computing for Next-Generation Systems*, 403–414.

Industry 5.0: Addressing Sustainability Challenges Through Human–Machine Collaboration

Sandeep Kumar Mogha, Suryatej S Kumar, and Jatinder Kaur

1 Introduction

The German government's high-tech strategy initiative led to the rise of the Fourth Industrial Revolution, commonly referred to as Industry 4.0, in 2011. It developed the concept of Cyber Physical Production Systems (CPPS) from the idea of Cyber Physical Systems (CPS) [1]. Industry 4.0 includes SmartFactory as one of its major related activities [2]. After the first three Industrial Revolutions were defined, the phrase "Industry 4.0" was first used in public in 2011 at the Hannover Fair [3, 4]. The shift from manual to steam- or water-powered machinery for production characterized the First Industrial Revolution. The Second Industrial Revolution, which resulted in high productivity and notable economic expansion, was made possible by electricity, which allowed factories to become sophisticated production lines. Field-level computers, such as Programmable Logic Controllers (PLCs), and the Third Industrial Revolution produced automated production through the use of communication technology in the industrial process systems, such as CPPS, are able to make intelligent decisions in the Industry 4.0 era through real-time communication and cooperation between "manufacturing things." This allows for the flexible creation of high-quality, customized goods at mass efficiency [5–7]. The joint Industrie 4.0 Platform was founded by the professional associations BITKOM, VDMA, and ZVEI in order to advance the project and guarantee a coordinated, cross-sectoral approach [8]. Similar strategic initiatives have been introduced by numerous nations worldwide; a few examples include the Made in China 2025,

S. K. Mogha (✉)
Department of Mathematics, Chandigarh University, Gharuan, Mohali, Punjab, India

Department of Mathematics, University of Delhi, Delhi, India

S. S Kumar · J. Kaur
Department of Mathematics, Chandigarh University, Gharuan, Mohali, Punjab, India

Y. Kumar Singla et al. (eds.), *Digital Manufacturing in Industry 5.0*,
https://doi.org/10.1007/978-3-031-91500-0_21

Industria 4.0 (Italy), Produktion 2030 (Sweden), Industrial Internet Consortium (USA), and Society 5.0 (Japan) [9].

The Fifth Industrial Revolution (Industry 5.0) arrived as companies began to adopt Industry 4.0. The concept of Industry 5.0 acknowledges that the industry has the ability to surpass job creation and economic expansion to become a dependable source of wealth by ensuring that production respects environmental limits and prioritizes the well-being of industry workers [10, 11]. Industry 4.0 is thought to have placed more of an emphasis on digitalization and AI-driven technologies to boost production efficiency and flexibility, while Industry 5.0 is predicated on the observation or belief that these concepts are less important in the modern world. Because of this, the idea of Industry 5.0 offers an alternative perspective and emphasizes the value of innovation and research in helping the sector serve humanity for a long time while staying within planetary bounds [10]. The “Age of Augmentation,” in which humans and machines coexist and work in symbiosis, has in fact been discussed in the run-up to the official unveiling of Industry 5.0 [12]. A similar description of "Smart Working" techniques was provided by Bednar and Welch [13].

Automation, data interchange, IoT, big data analytics, and AI have all advanced significantly with the introduction of Industry 4.0 [14–17]. Establishing a harmonic balance between human creativity and sophisticated technology, Industry 5.0 seeks to build on the automation and data sharing principles of Industry 4.0 [7, 18–29]. Sustainability—which includes social, economic, and environmental aspects—has emerged as a major global industry concern. A change to more sustainable behaviors is required due to growing awareness of social injustice, resource depletion, and climate change. Industry 5.0 offers a viable approach to address these problems by incorporating human–machine cooperation. An industrial ecosystem can be made more robust and sustainable by creating creative solutions that take advantage of the strengths of both humans and machines.

The current vision for Industry 5.0 is to use the special creativity of human professionals in conjunction with strong, intelligent, and precise machinery. Industry 5.0, according to many technical visionaries, will revive the human element in the manufacturing sector [19]. It is predicted that Industry 5.0 will blend precise, fast machinery with human critical and cognitive thinking. Mass personalization is a crucial component of Industry 5.0 since it enables customers to select things that are tailored to their specific needs and tastes [30].

Industry 5.0 will significantly increase productivity and let humans and machines work together more creatively, allowing for accountability and continuous observation. Rapid productivity growth is the goal of human–machine collaboration. By giving robots and machines repetitive, tedious jobs and giving humans duties requiring critical thought, Industry 5.0 has the potential to raise manufacturing quality. Given that machines are used by intellectual professionals, Industry 5.0 supports higher-skilled employment than Industry 4.0. Humans will be controlling robots in Industry 5.0, which is mainly concentrated on mass customization. While the main goal of Industry 5.0 is to increase customer happiness, Industry 4.0 already has robots actively involved in large-scale production. Although Industry 5.0 connections, Industry 4.0 concentrates on CPS connectivity. Industry 5.0 integrates collaborative robots (cobots) and connections to Industry 4.0 applications, while Industry 4.0 concentrates on CPS connectivity. Comparing Industry 5.0 to the current industrial revolutions, which neither prioritize environmental protection, reveals another intriguing advantage: the latter offers greener alternatives [31]. The goal of Industry 5.0's models is to make judgments that are more precise and stable by utilizing predictive analytics and operating intelligence. Real-time data from machines will be collected in conjunction with highly skilled specialists, meaning that most manufacturing processes in Industry 5.0 will be automated.

This study looks into how collaboration can address multiple sustainability issues, including as resource efficiency, social equality, and environmental effect, in addition to increasing productivity and efficiency. It also looks at the difficulties and obstacles that need to be removed in order to reach Industry 5.0's full potential. The paper's goal is to shed light on how the upcoming industrial revolution might support a sustainable future through this investigation.

1.1 Overview of Sustainability Challenges in the Modern Industrial Landscape

Significant sustainability difficulties confront the contemporary industrial landscape, especially in the areas of the environment, economy, and society. The loss of resources, pollution, greenhouse gas emissions, and problems with waste management are all major factors that industries contribute to climate change. Their financial situation is hampered by growing energy expenses, a lack of resources, and the requirement to abide by more onerous environmental laws. Managing unfavorable neighborhood effects, upholding fair labor practices, and addressing mounting stakeholder demand for sustainable operations are examples of social problems. Due to these difficulties, incorporating sustainable practices into all facets of industrial operations requires a thorough strategy.

When human creativity and decision-making are combined with the accuracy and efficiency of cutting-edge technologies, human–machine collaboration is essential to solving sustainability concerns. Through cooperation, resource, energy, and waste management may all be optimized in Industry 5.0. AI and IoT-enabled machines are able to monitor and control operations in real-time, detecting inefficiencies and cutting down on resource waste. As an example, IoT sensors can optimize energy use across production lines, considerably reducing greenhouse gas emissions, and AI can predict the maintenance needs of machinery, decreasing downtime and prolonging its lifespan.

The ability to innovate and implement sustainable methods is further enhanced when humans and machines work together. Sustainable growth requires both ethical and strategic thinking from humans and fast, accurate execution of complicated tasks by machines. Cleaner technology, like enhanced recycling procedures and renewable energy systems, may result from this collaboration. Human–machine collaboration increases workplace safety by automating dangerous and repetitive jobs. This frees up human workers to concentrate on more creative and advanced problem-solving, ensuring that sustainability programs are both successful and human-centered.

2 Literature Review

In past centuries, humans and/or animals have collaborated to develop and create weapons, clothing, cars, and homes. Significant changes in industrial production started in 1974 with the emergence of Industry 1.0. An overview of the development of Industrial X.0 is presented in Fig. 1 [32]. Although the previous three revolutions took more than a century to develop, the fourth revolution from the third took only forty years to develop. Industry 1.0 began to take shape in the 1800s with the construction of mechanical industrial infrastructures for powering machinery with steam and water. Production capacity growth has been extremely beneficial to the economy. Industry 2.0 emerged around 1870 as a result of the invention of electricity and assembly lines. Distribution of workload and mass production were the primary goals of Industry 2.0, which increased productivity for industrial firms. The notions of information technology, electronics, and partial automation gave shape to Industry 3.0 in 1969. 2011 marked the evolution of Industry 4.0 with the introduction of the idea of future smart manufacturing. The key objective is to use developing technology to accomplish mass manufacturing and maximize productivity [33, 34]. Industry 5.0 is an evolution of the future intended to leverage human specialists' creativity in conjunction with accurate, intelligent, and efficient technology [35]. By combining several technologies, including cloud computing, AI, IoT, cyber physical systems (CPSs), and cognitive computing, the industry 4.0 standard has totally transformed the manufacturing sector. The primary idea underlying Industry

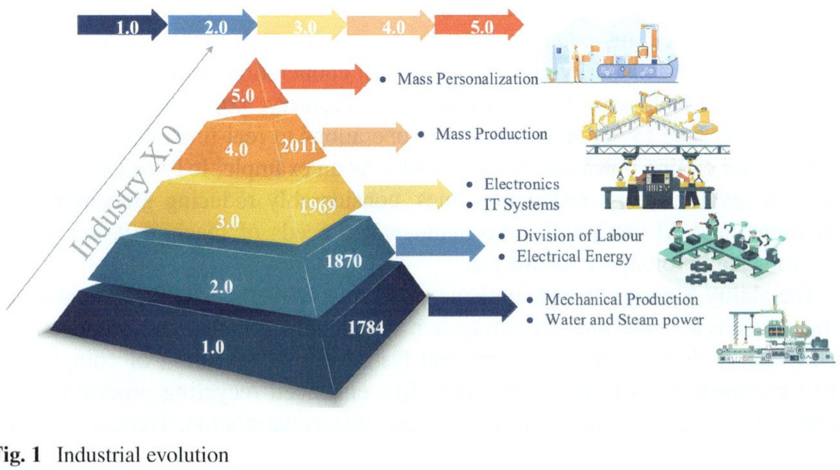

Fig. 1 Industrial evolution

4.0 is to connect machines and other gadgets that can control one another throughout their lives in order to make the manufacturing sector "smart" [36–39]. Process automation is the top goal of Industry 4.0, which minimizes the need for human involvement in the manufacturing process [40, 41]. The aim of Industry 4.0 is to increase mass production and performance by leveraging machine learning (ML) to provide intelligence between devices and applications [42–44].

The environmental sustainability of human–machine cooperation is one of the main topics of the literature. Research shows concern on how much energy is used by cutting-edge technology like robotics, AI, and IoT. For instance, Strubell et al. (2019) draw attention to the significant carbon footprint associated with the computationally demanding process of training AI models [45]. Similarly, manufacturing robots and IoT devices can increase electronic waste if not properly handled, as Esmaeilian et al. (2020) explain [46]. These concerns highlight the necessity of striking a balance between energy efficiency, sustainable production methods, and technological breakthroughs.

The literature also addresses the difficulty of incorporating renewable energy sources into high-tech sectors. Because machines need power to run continuously, switching to greener energy sources becomes more difficult from a technical and financial perspective. While smart technologies help optimize energy use, Sarker et al. (2022) contend that they also require a substantial upfront investment in sustainable infrastructure, which can be prohibitive for several industries [47].

From an economic perspective, the research highlights the high implementation costs of human–machine cooperation systems. Modern technologies demand large initial investments in software and hardware, as well as regular maintenance and updates. Moreover, the total costs as reported by Dalenogare et al. (2018) are increased by the training of human workers to engage with these technologies in an efficient manner [48]. This financial strain, according to Horváth and Szabó (2019), could hinder the adoption of Industry 5.0, especially in small and medium-sized businesses (SMEs) that are less able to afford the substantial implementation expenses up front [49].

The possibility of job displacement adds to the intricacy of the economic viability of human–machine collaboration. Although the aim of Industry 5.0 is to improve human roles through automation of ordinary work, some scholars contend that this transition could lead to a significant decrease of jobs in particular areas, as noted by Acemoglu and Restrepo [50]. The preservation of jobs and ethical labor practices must coexist with technological improvements, which presents a challenge to the sustainability of the economy.

As the literature has shown, human–machine collaboration also presents important social and ethical issues. The necessity of upskilling and reskilling the workforce to accommodate new technologies is one of the main problems. Barley, Bechky, and Milliken (2017) state that this change is not just expensive but also time-consuming, and not all workers may be able to migrate effectively, creating disparities in the labor market [51]. The success of Industry 5.0 relies on inclusive measures that guarantee workers are not left behind as industries rapidly adopt automation and AI, as noted by Santos et al. [52].

The literature tackles ethical issues such as potential misuse of AI technologies and data privacy. Collaboration between humans and machines can increase sustainability and efficiency, but it also requires the collecting of huge quantities of data. This calls into question the ownership and use of the data [53]. Additionally, there is the moral conundrum of automated systems' decision-making, especially when AI is included into crucial processes. One of the ongoing challenges in developing sustainable Industry 5.0 systems is ensuring that sure AI functions openly and in accordance with human values [54].

The integration of human–machine collaboration presents technological issues related to system complexity, security, and interoperability. According to the literature, it is crucial to provide smooth communication between different machines, gadgets, and human operators [55]. The collaboration of humans and machines might result in inefficiencies and higher resource consumption if there are no standard protocols and systems that are interoperable. Furthermore, Industry 5.0's increasing device connectivity creates cybersecurity risks due to the possibility of data breaches and cyberattacks interfering with business operations and jeopardizing sustainability initiatives [56].

Sustainability issues are moreover brought up by the complex nature of human–machine collaboration systems. According to Tortorella et al. (2020), troubleshooting and maintenance of a system demand more resources the more sophisticated it gets [57]. The sustainability benefits that Industry 5.0 technologies are supposed to bring about may be offset by this complexity's potential to cause downtime and decreased efficiency.

The research on the sustainability issues surrounding human–machine collaboration in Industry 5.0 emphasizes the necessity of a well-rounded strategy that takes into account ethical, social, technical, economic, and environmental issues. Although Industry 5.0 poses new challenges that need to be carefully controlled, it also offers potential solutions to many of the sustainability issues that older industries face. Future studies should focus on formulating structures and tactics that facilitate sectors to execute human–machine cooperation in an environmentally friendly, equitable, and morally sound way, guaranteeing that the advantages of Industry 5.0 materialize without escalating current challenges.

3 Human–Machine Collaboration

Industry 4.0 might not be viewed as a human-centric project. As conditions are now, it is important to take into consideration aspects like work–life balance, sociotechnical approach, and human–machine cooperation or operator assistance technologies. Industry 4.0 promotes the creation of new sociotechnical networks by transforming a number of aspects of the workplace, such as knowledge management, team structures, career path models, lifelong learning, and health management. This is defined as an Industry 4.0 sociotechnical approach that is causing a paradigm shift in the interactions between humans and technology as well as the human environment.

The growing usage of increasingly open, virtual, and expansive technology is expected to bring about a considerable change in the role that workers play [58]. Some of the design concepts of Industry 4.0 are in line with this [1]. Information transparency and technical aid, the second and third design principles, are reflected in the increasing usage of technologies such as augmented reality (AR) and robot-assisted systems that give workers real-time information to improve productivity and decision-making [59–61]. Workers are freed up from routine tasks by intelligent support systems to concentrate on more innovative and productive work. It is encouraged for employers to adopt flexible work arrangements so that employees can better balance their personal and professional lives and pursue ongoing professional development. Additionally, older workers will be able to extend their working life and continue to be productive for longer thanks to applicable technologies [58]. It is understood that employees' roles will shift significantly in a smart factory. Workers will have more opportunities to take on more responsibility and further their personal development when a sociotechnical approach to work organization is implemented [58, 62]. The sociotechnical strategy of Industry 4.0 aims to achieve the so-called motto "better, not cheaper." It suggests that implementing Industry 4.0 through regular repetition of highly standardized and boring tasks—an extreme form of the Taylorist approach to labor organization—is hardly the most promising course of action. Smart factories will require empowered workers to function as decision-makers and controllers since they will be set up as extremely complex, dynamic, and adaptable systems [58].

In Industry 5.0, human–machine collaboration signifies a paradigm change from the prior emphasis on automation and efficiency to a more harmonic and integrated collaboration between humans and machines. Here is a closer look at the main components of this collaboration and what it entails:

3.1 Enhanced Collaboration

Adaptive Machines: Industry 5.0 machines are made to adjust to the unique requirements and preferences of human workers, in contrast to the inflexible systems of the past. Collaborative robots, or cobots, for instance, can modify their actions in response to human input in real time, facilitating more efficient workflows and superior assistance with intricate jobs.

Intuitive Interfaces: Industry 5.0 technology incorporates more natural-feeling interfaces like gesture control, voice recognition, and augmented reality (AR). By lowering the learning curve and increasing productivity, these interfaces facilitate human–machine interaction.

3.2 Augmented Intelligence

Decision Support: Industry 5.0 AI tools are designed to assist human decision-makers, not to take their place. Large volumes of data can be analyzed by these systems, and they can offer insights or recommendations. However, humans still have the final say over decisions, using their knowledge and judgment.

Co-Creation and Innovation: Humans can concentrate on original thinking and creative problem-solving by using AI to handle data processing and pattern identification. The ability to create new goods or procedures that neither people nor machines could accomplish on their own is improved by this symbiotic partnership.

3.3 Personalization and Customization

Tailored Solutions: Systems and equipment can be modified to meet the unique requirements of certain jobs or workers. More specialized job positions or production processes are made possible by this customization, which also helps to maximize performance and efficiency.

Adaptable Work Environments: Human–machine cooperation can result in more flexible work settings. Flexible robotic systems, for example, can be easily rearranged to manage various goods or jobs in response to shifting requirements.

3.4 Skill Enhancement and Workforce Transformation

Upskilling and Reskilling: A major focus of Industry 5.0 is the requirement that workers pick up new skills to go along with cutting-edge technologies. Enhancing abilities like creativity, problem-solving, and emotional intelligence—all necessary for working well with advanced machines—is the main goal of training programs.

Human-Centric Roles: As computers take on dangerous or repetitive work, human roles become more concentrated on jobs that call for critical thinking, creativity, and people skills. This change may result in human effort that is more interesting and gratifying.

3.5 Ethical and Sustainable Practices

Ethical AI: Industry 5.0 encourages the creation and application of AI that complies with moral standards, guaranteeing that technology is applied sensibly and openly.

Sustainability: Industry 5.0 collaborative systems are built with sustainability in mind. While people use their judgment to make sure that sustainability goals are reached, machines and processes are designed to minimize waste and environmental effect.

3.6 *Resilience and Adaptability*

Dynamic Problem Solving: By working together, systems become more resilient and capable of quickly adjusting to changes or interruptions. AI, for instance, can recognize possible problems in real time, but human operators can respond to these problems creatively and flexibly.

Continuous Improvement: Industry 5.0 promotes human–machine feedback loops, enabling continual system and process improvements based on data and real-world experience.

3.7 *Safety and Well-Being*

Enhanced Safety: By collaborating with people and utilizing cutting-edge safety technologies, collaborative robots can lower the likelihood of accidents. For instance, while humans concentrate on jobs requiring higher order thinking and creativity, machines can perform dangerous duties.

Work–Life Balance: When machines perform more labor-intensive or repetitive jobs, human workers may have better work–life balance and higher levels of job satisfaction.

To put it simply, Industry 5.0 is a technology approach that is more human-centric and integrated, emphasizing the mutually beneficial relationship between humans and robots. The objective of this partnership is to improve overall workplace well-being, creativity, and productivity.

4 Sustainability Challenges

Even though Industry 5.0 seeks to improve human–machine relationships and efficiency, it also brings with it a number of sustainability issues. It is imperative to tackle these obstacles in order to guarantee that the advantages of this novel framework are achieved while maintaining the long-term ecological, societal, and financial viability. In Industry 5.0, human–machine collaboration presents several significant sustainability concerns. These include:

4.1 Resource Consumption and Environmental Impact

Human–machine collaboration is incorporated into Industry 5.0 with the goal of tackling environmental issues and boosting innovation and productivity. But in order to be in line with Industry 5.0's sustainability objectives, this partnership has consequences for resource usage and environmental effect that need to be properly controlled.

4.1.1 Resource Consumption

Use of Energy: Energy usage significantly increases as advanced robots, AI, and IoT devices are deployed more often in Industry 5.0. By utilizing energy-efficient machinery, implementing smart grids to maximize energy efficiency, and embracing renewable energy sources, the energy demand can be controlled.

Use of Material: Rare earth metals are among the raw elements that must be extracted and used in the production of intelligent machinery and devices. Reducing resource depletion requires adopting circular economy ideas, recycling, and efficient material use.

Water Consumption: Water is used extensively in advanced manufacturing processes, particularly in the electronics and automobile industries. Water usage can be decreased by putting closed-loop water systems and water-efficient technology into practice.

4.1.2 Environmental Impact

Carbon Emissions: Unless offset by the use of low-carbon technologies and renewable energy sources, increased automation and machine utilization may result in higher carbon emissions. By creating eco-friendly goods and procedures, Industry 5.0 presents a chance to lower carbon footprints.

Waste Generation: When new technologies are used, waste production often increases, particularly with regard to electronic waste (e-waste). Reusing, recycling, and remanufacturing are examples of sustainable waste management techniques that Industry 5.0 says are crucial for reducing environmental damage.

Biodiversity: The utilization of land for industrial facilities and the extraction of resources for electronics and machinery can have a detrimental effect on biodiversity. Ecosystem preservation can be achieved through using green infrastructure principles and sustainable material procurement.

4.1.3 Mitigation Strategies

Eco-Design and Green Manufacturing: Developing products with sustainability, recyclability, and low resource usage in mind can drastically lessen their environmental impact.

Smart Technologies for Monitoring and Optimization: Real-time environmental effect monitoring is possible with IoT and AI, which enables process optimization to reduce emissions and resource consumption.

Human-Centric Approaches: Including human creativity and flexibility can result in creative ways to cut down on resource usage and environmental effect, including creating more environmentally friendly products or coming up with new ways to use resources more efficiently.

4.1.4 Challenges and Opportunities

Efficiency and Sustainability: Often, there is a trade-off between reducing environmental effect and increasing manufacturing efficiency. Opportunities to achieve this equilibrium using cutting-edge technology and human ingenuity are presented by Industry 5.0.

Policy and Regulation Assistance: The successful execution of Industry 5.0 in a sustainable manner is greatly aided by government legislation and policies. In order to motivate firms to implement sustainable practices, supportive frameworks are required.

Industry 5.0 human–machine collaboration raises important questions about resource utilization and environmental effect that call for thoughtful analysis and proactive management. Industry 5.0 can increase industrial productivity and help create a more sustainable future by embracing sustainable practices, utilizing cutting-edge technologies, and encouraging human ingenuity.

4.2 Manufacturing and Lifecycle Impacts

Human–machine collaboration reshapes product lifecycles and manufacturing in Industry 5.0, placing equal emphasis on sustainability and technological growth. This partnership affects the whole product lifespan, from design to disposal, as well as production procedures.

4.2.1 Manufacturing Impacts

Advanced Manufacturing Techniques: Industry 5.0 makes use of cutting-edge innovations such as smart factories and additive manufacturing, or 3D printing. By producing complicated components with little material waste, additive manufacturing lessens the requirement for wasteful resource consumption. Robotics, AI, and the IoT enable smart factories to optimize manufacturing processes in real time, resulting in lower energy usage, material waste, and downtime. These factories also make predictive maintenance easier, which reduces equipment failure and lengthens the life of machines.

Human-Centric Design: In Industry 5.0, human workers are taken into consideration when designing production processes. The emphasis on ergonomics and safety lowers the number of accidents and the resulting downtime. Cobots, or collaborative robots, assist humans in the workplace to increase productivity without taking the place of employees. Because activities are optimized by fusing human inventiveness with machine precision, this human–machine synergy results in a more efficient use of resources.

4.2.2 Lifecycle Impacts

Resource Extraction and Material Use: Industry 5.0's sustainability emphasis places a strong emphasis on using resources that come from suppliers who practice ethics and sustainability. This lessens the impact of raw material exploitation on the environment. Furthermore, circular economy design ideas are incorporated into product design to guarantee longevity, reuse, and recyclability. By minimizing waste and extending the useful life of materials, this design strategy lowers the need for fresh materials.

Product Use and Maintenance: Longer lifespans are achieved by products designed using Industry 5.0 principles because of predictive maintenance and real-time monitoring. This lowers the total resource use by lowering the frequency of replacements. Additionally, smart devices with AI and sensors optimize their performance while in use, reducing their lifetime energy consumption and ecological impact.

End-of-Life Management: Industry 5.0 makes sure that recycling and reuse are done effectively when a product's life is over. Products are made to be easily disassembled, which facilitates the recovery of materials and lowers waste. Smart technologies have improved reverse logistics systems, which track products throughout their lifecycle and simplify recycling and return management. This encourages the circular economy, in which resources are constantly recycled back into the manufacturing process.

4.2.3 Environmental and Economic Implications

There are major environmental benefits from Industry 5.0's integration of lifecycle management and sophisticated production techniques. Key results include lower carbon footprints, less waste, and more economical use of resources and energy. These environmental advantages are further enhanced by smart factories that run on renewable energy.

Economically speaking, Industry 5.0 promises cost benefits through longer product lifecycles, reduced production costs, and effective resource usage. These elements support sustainability and long-term economic efficiency. Industry 5.0's blend of computer intelligence and human innovation not only boosts output but also satisfies consumer desire for more ecologically friendly goods.

By integrating humans and machines, Industry 5.0 revolutionizes product lifecycles and production processes while emphasizing sustainability. Industry 5.0 minimizes environmental effect while boosting economic efficiency by incorporating cutting-edge technologies and highlighting the concepts of the circular economy. A future of industry that is more robust and sustainable is made possible by this strategy.

4.3 Ethical and Social Considerations

Industry 5.0's human–machine collaboration raises important ethical and social issues that need to be resolved in order to guarantee that technological developments advance society as a whole. The interaction between ethical considerations and social repercussions gets more intricate as Industry 5.0 places a greater emphasis on the convergence of human innovation and technological efficiency.

4.3.1 Ethical Considerations

(a) Autonomy and Control: In Industry 5.0, AI systems and machines play more independent roles in decision-making. This begs the question of how much human control should be given to machines. It is imperative to maintain human oversight as a key component in order to avoid any misuse or unforeseen effects of judgments made by AI. Maintaining ethical norms in human–machine collaboration requires the establishment of precise criteria about the boundaries of machine autonomy.

(b) Security of Data and Privacy: Industry 5.0's widespread usage of IoT devices and AI depends on enormous volumes of data, including private and sensitive information. Data security and privacy are brought up by this. Crucial ethical issues include preventing illegal access to this data and guaranteeing that people's right to privacy is upheld. To solve these issues, strong cybersecurity measures and transparent data governance regulations are crucial.

(c) Injustice and Prejudice: The objectivity of AI systems in Industry 5.0 is dependent on the data and algorithms they utilize. There's a chance that these algorithms will reinforce preexisting prejudices, treating some people or groups unfairly. In order to identify and reduce bias, rigorous examination of AI datasets and algorithms is necessary to ensure fairness. Diverse teams should also be included in the creation and implementation of AI systems to guarantee that various viewpoints are taken into account.

(d) Liability: It gets harder to assign blame when mistakes or malfunctions occur when machines perform increasingly complicated jobs. When something goes wrong—whether it's a result of a technological failure, human error, or a combination of the two—clear frameworks are required to allocate blame. This entails taking legal and regulatory efforts to guarantee that businesses and people alike are held responsible for the deeds of autonomous systems.

4.3.2 Social Considerations

(a) Job Displacement and Workforce Transformation: Industry 5.0's integration of AI and advanced robots has the potential to eliminate jobs, especially in those that depend on regular tasks. Although there will be a shift in the workforce, thorough plans are required to handle this change. Programs for reskilling and upskilling employees are crucial for giving them the skills they need to prosper in an industrial environment that is changing quickly. Policies that assist employees during times of job displacement must also be in place.
(b) Social Inequality: If access to cutting-edge technology and the advantages they provide are not evenly spread, Industry 5.0 may make already-existing social inequities worse. Fostering a more equal society requires making sure that everyone can benefit from Industry 5.0, regardless of financial situation. This involves lowering the cost of education, providing training, and ensuring that a wide range of people can use technology.
(c) Work–Life Harmony and Human Welfare: Although the goal of Industry 5.0 is to increase productivity, there is a chance that the pressure to match machine efficiency could have a detrimental effect on people's quality of life. Prioritizing work–life balance and fostering cultures that support both physical and mental health are imperatives for employers. Furthermore, promoting meaningful work that makes use of human creativity as opposed to just automating chores might enhance well-being and job satisfaction.
(d) Ethical Use of Technology: Industry 5.0 technologies need to be created and implemented with a solid ethical base. This involves making sure new technologies are applied in ways that advance humankind and taking into account how they will affect society. To establish trust and guarantee that Industry 5.0 benefits society, it is imperative to involve stakeholders in conversations regarding the moral application of technology, such as workers, communities, and regulators.

Industry 5.0 raises a number of intricate and varied ethical and societal questions about human–machine partnership. For the benefits of Industry 5.0 to be realized in a way that is equitable, inclusive, and consistent with moral values, industry executives, legislators, and the general public must work together to address these issues. Industry 5.0 can help ensure that technology advances societal progress and human well-being in the future by emphasizing equity, transparency, and human-centered methods.

4.4 *Data Privacy and Security*

Industry 5.0's human–machine collaboration raises serious data privacy and security issues because robotics, IoT, and other cutting-edge technologies need handling enormous volumes of data. This data needs to be properly controlled to prevent misuse, unauthorized access, and breaches because it frequently contains sensitive and personal information.

4.4.1 Data Privacy Challenges

(a) Data Gathering and Utilization: Industry 5.0 makes extensive use of data to streamline workflows, strengthen judgment calls, and increase communication between humans and machines. Privacy issues may arise from the gathering of data from multiple sources, such as sensors, devices, and human involvement. It is essential to make sure that data is gathered and handled properly, especially when it comes to personal data. To safeguard people's right to privacy, transparent data usage regulations and unambiguous consent procedures must be put in place.

(b) Sensitive Information Exposure: Sensitive data, including trade secrets, intellectual property, and personal information, is frequently handled by machines and AI systems as they are increasingly incorporated into industrial processes. A significant privacy worry is the possibility of this data being disclosed due to hacks or inadvertent disclosures. For sensitive data to be protected from unwanted access, businesses need to have strong data protection measures in place, like encryption and anonymization.

(c) Compliance with Privacy Regulations: Different jurisdictions have different privacy laws, such as the General Data Protection Regulation (GDPR) in Europe. It is difficult but essential to guarantee that these regulations are followed across international operations. In order to avoid paying steep fines and suffering harm to their reputation, businesses must traverse complicated legal environments and make sure that their handling of data complies with privacy laws' strict standards.

(d) Ethical Use of Data: Industry 5.0's approach to data privacy includes an ethical component in addition to legal compliance. Businesses need to make sure that data is utilized in ways that safeguard users' privacy and don't damage or exploit them. This entails being open and honest about the methods used to gather, store, and utilize data as well as making sure that it is never used for reasons that people were not aware of or consented to.

4.4.2 Data Security Challenges

(a) Cybersecurity Risks: Because Industry 5.0 relies more and more on networked technologies, these systems are appealing targets for hackers. Operations can be disrupted, sensitive data might be compromised, and substantial financial losses can result from ransomware, phishing, and data breaches, among other attacks. More advanced cybersecurity solutions are required as human–machine collaboration increases and the attack surface widens.
(b) Protection of IoT Devices: Cyberattacks have various access points due to the growth of IoT devices in industrial settings and smart factories. Inadequate security precautions, such weak passwords or out-of-date software, leave many IoT devices vulnerable. Regular upgrades, robust authentication procedures, and network segmentation are necessary to ensure that these devices are safe and minimize possible harm from breaches.
(c) Availability and Integrity of Data: Data availability and integrity are essential for preserving operational effectiveness in Industry 5.0. Data deletion or alteration cyberattacks have the potential to cause errors in judgment and interfere with manufacturing procedures. Strong encryption, frequent backups, and continuous tracking to quickly identify and address anomalies are all necessary to ensure data integrity.
(d) Human Factor in Security: Undoubtedly, one of the biggest obstacles to data security is human mistake. Workers may unintentionally disclose data by using weak passwords, misconfigurations, or phishing scams. Encouraging staff members to identify security risks and adhere to best practices is crucial in lowering the possibility of human error jeopardizing data security.

In the context of Industry 5.0's human–machine collaboration paradigm, data security and privacy are critical concerns. The speed at which cutting-edge technology are incorporated raises the possibility of security risks and privacy violations. A multifaceted approach is needed to address these issues, including employee training, adherence to privacy rules, strong cybersecurity protections, and ethical data usage. Industry 5.0 may leverage the advantages of human–machine collaboration while safeguarding confidential data and upholding confidence in industrial systems by placing a high priority on data privacy and security.

4.5 *Ethical AI and Decision-Making*

Significant ethical issues are raised by the use of AI in decision-making processes in Industry 5.0, where human–machine collaboration is essential to increasing productivity and creativity. Ensuring ethical, equitable, and transparent decision-making is of utmost importance as AI technologies augment or potentially supplant human decision-making in many industrial settings.

4.5.1 Ethical AI Challenges

(a) Prejudice and Fairness: The possibility of prejudice in AI decision-making is one of the most urgent ethical issues. AI systems learn on data; hence, if the training set of data contains biased information, the AI may reinforce or even magnify these biases. Injustices may result from this, such as prejudice against particular communities or groups of individuals. Biased AI may lead to unfair resource allocation, biased manufacturing processes, or discriminatory recruiting practices in Industry 5.0. This problem needs to be solved by closely examining training data, including a variety of viewpoints in the development of AI, and continuously checking for and fixing biases.

(b) Transparency and Explainability: AI systems frequently function as "black boxes," generating judgments that are challenging for people to comprehend. This is especially true of systems that use sophisticated models like deep learning. When AI is used in an industrial setting to make important choices about safety, production, or resource management, a lack of transparency can cause resistance and mistrust among stakeholders and employees. It is crucial to make sure AI systems can be explained, or that humans can comprehend and interpret the decisions they make. When AI judgments are questioned, this transparency not only promotes accountability but also builds trust.

(c) Accountability and Responsibility: Determining who is responsible for mistakes made by AI systems grows more difficult when these systems participate in larger decision-making processes. It is difficult to apportion blame, for instance, if an AI system makes a choice that causes a financial loss or a safety catastrophe. Is it the fault of the managers who put the system in place, the data scientists who trained the model, the AI developers, or the AI itself? To allocate blame and guarantee that there are procedures in place for redress when AI systems malfunction, precise frameworks are required.

(d) Use of AI Ethics: The application of AI in Industry 5.0 raises more ethical concerns than just technological ones. It's not necessary to employ AI to make decisions just because it can. For example, human judgment should be involved rather than entirely automating decisions that have a large influence on human lives or livelihoods, like safety protocols or layoffs. Boundaries on AI's use and human involvement in decisions that have major moral or ethical ramifications are necessary to guarantee that AI is applied responsibly.

4.5.2 Decision-Making Challenges

(a) Human–Machine Collaboration: Making decisions in Industry 5.0 frequently involves human and machine collaboration. To make sure that this partnership maximizes each party's strengths, it needs to be carefully maintained. While machines are capable of handling enormous volumes of data and spotting patterns, people are superior at empathy, ethical reasoning, and contextual awareness. Maintaining ethical standards requires striking a balance between these partnerships so that technology complements human decision-making instead of replacing it.

(b) AI trust: In order for human–AI collaboration to be successful, AI systems must be trusted. Building openness, dependability, and congruence with human values are the cornerstones of this trust. The efficacy of human–machine cooperation in Industry 5.0 will be undermined if AI systems routinely make choices that are opaque or that go against human ethical norms.

Careful thought and proactive management are necessary to meet the complex and multidimensional ethical AI and decision-making problems of Industry 5.0. Developing explicit accountability procedures, upholding a human-centered decision-making process, and guaranteeing fairness and openness in AI systems are all necessary to address these issues. Industry 5.0 can leverage AI to accelerate innovation and productivity while maintaining the moral standards necessary for a fair and just industrial future by concentrating on these ethical issues.

4.6 Supply Chain and Logistics

Several obstacles stand in the way of the transition to Industry 5.0, which is defined by the profound integration of human–machine cooperation in supply chains and logistics. The intricacy of combining cutting-edge technology with human knowledge while upholding effectiveness, security, and moral principles gives birth to these difficulties.

Technological Integration: The process of incorporating cutting-edge technology like robotics, IoT, and AI into current supply chains can be challenging and disruptive. New technology may not be compatible with legacy systems, necessitating expensive upgrades and perhaps disrupting operations.

Cybersecurity Risks: Concerns about cybersecurity are raised by the expanding interconnectedness of systems and devices. The supply chain is vulnerable to cyberattacks because of the numerous points of vulnerability introduced by IoT devices and AI systems. It is essential to provide strong data protection and adherence to privacy laws.

Workforce Adaptation: Workers require new competencies to effectively engage with cutting-edge technologies as supply chains grow more automated. Training and reskilling expenditures are necessary due to a substantial skills gap. Workers may also be resistant to change because they are uncomfortable with new procedures or fear losing their jobs.

Environmental and Ethical Aspects to Consider: Automation and AI may give rise to moral dilemmas, such as putting efficiency ahead of morality. Concerns about the environment must also be addressed because more technology use might result in higher energy use and e-waste production. It is imperative to shift toward renewable energy sources and enhance waste management practices.

Sturdy Supply Chain: Supply chains may be more susceptible to disruptions from technological malfunctions or cyberattacks if they rely too heavily on technology. Resilience requires creating strong backup plans and striking a balance between automation and human intervention.

The logistics and supply chain problems of Industry 5.0 are complex, involving aspects of technology, people, ethics, and the environment. In order to successfully navigate these problems, a comprehensive strategy that strikes a balance between the advantages of cutting-edge technologies and the requirements for security, morality, and worker adaption is needed. Businesses may create supply chains that are more durable, effective, and sustainable and ready for Industry 5.0 by proactively tackling these issues.

4.7 Human Well-Being and Work Environment

The workplace benefits greatly from human–machine collaboration in Industry 5.0, including increased productivity and operational efficiency. It does, however, also bring up a number of issues with regard to workplace safety and employee well-being. In order to guarantee that technological improvements do not adversely affect workers' health, safety, and job happiness, it is imperative that these concerns be addressed.

Job Displacement and Skill Gaps: Automation and AI have the potential to cause job displacement, especially for positions that become outdated as a result of technology advancements. These employees can experience joblessness or have to retrain. In order to solve this issue and make sure that vulnerable populations are not disproportionately impacted by technology advancement, training programs for people transitioning into new roles must be funded.

Health and Safety Risks: Adopting cutting-edge technologies may have an effect on one's physical and emotional well-being. Robots might ease physical strain, but they can also have a negative impact on mental health due to repetition and stress from learning new systems. In order to avoid mishaps, human workers' safety when working with autonomous systems also needs to be properly monitored. It is crucial to guarantee ergonomic designs and thorough safety procedures.

Job Satisfaction and Work Environment: If automation mostly takes the place of repetitive jobs, it could result in work monotony. It is imperative to create jobs that free up employees for important, creative work rather than assigning them to tedious, repeated duties. Sustaining job happiness requires establishing an environment at work that encourages employee engagement and job enrichment.

Ethical and Social Considerations: AI should be used ethically in decision-making processes, and any potential biases should be addressed. Promoting justice and inclusivity in the changing workplace also requires addressing the social effects of technology advancements and guaranteeing fair access to reskilling opportunities.

Several issues pertaining to workplace conditions and worker welfare arise when human–machine cooperation occurs in Industry 5.0. Concerns about employment displacement, skill shortages, health concerns, and ethical consequences must all be carefully considered when addressing these difficulties. Organizations may reap the benefits of Industry 5.0 while maintaining a happy and sustainable work environment by giving careful consideration to worker support, safety, and ethical standards.

5 Industry 5.0 and Society 5.0

Industry 5.0 is a thoughtful idea about how industry would develop into a human-centered, resilient, and sustainable manufacturing system [10], which offers a prosperous industrial development vision as depicted in Fig. 2's left circle. The Industry 5.0 paradigm uses flexible and adaptive technology to enhance the agility and resilience of systems. Moreover, makes an effort to take the lead in promoting sustainability, respecting the limits of the world, and empowering people with skill and variety.

First proposed by the Japanese government in January 2016, Society 5.0 is a vision of a future society driven by scientific and technical innovation, with the goal of creating a human-centered, super-smart, and lean society, as depicted in the right circle of Fig. 2 [63, 64]. In order to reconcile Japan's aging population, low birth rates, and loss of competitiveness with economic progress, the idea of Society 5.0 was put up. By combining 5G, big data, AI, and other technologies to bridge the gap between cyberspace and physical space, Society 5.0 aims to create a human-centric, super-smart society where all citizens have access to high-quality, comfortable, and dynamic lives.

In reality, as depicted in the upper portion of Fig. 3, the four stages of human society's evolution are the hunter-gatherer (Society 1.0), the agricultural (Society

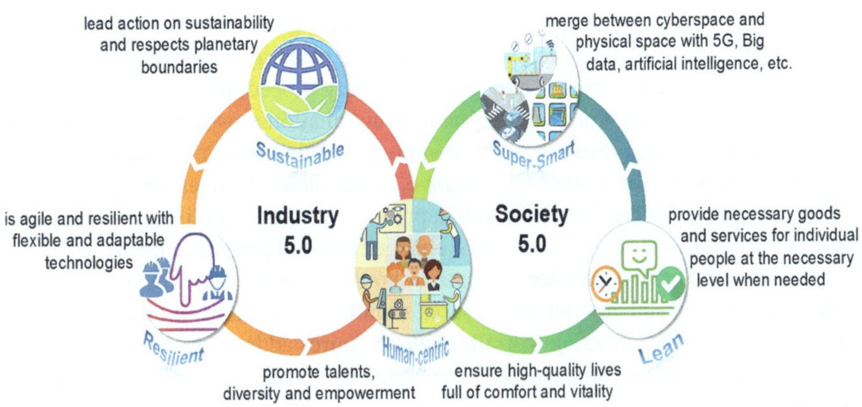

Fig. 2 Industry 5.0 and Society 5.0

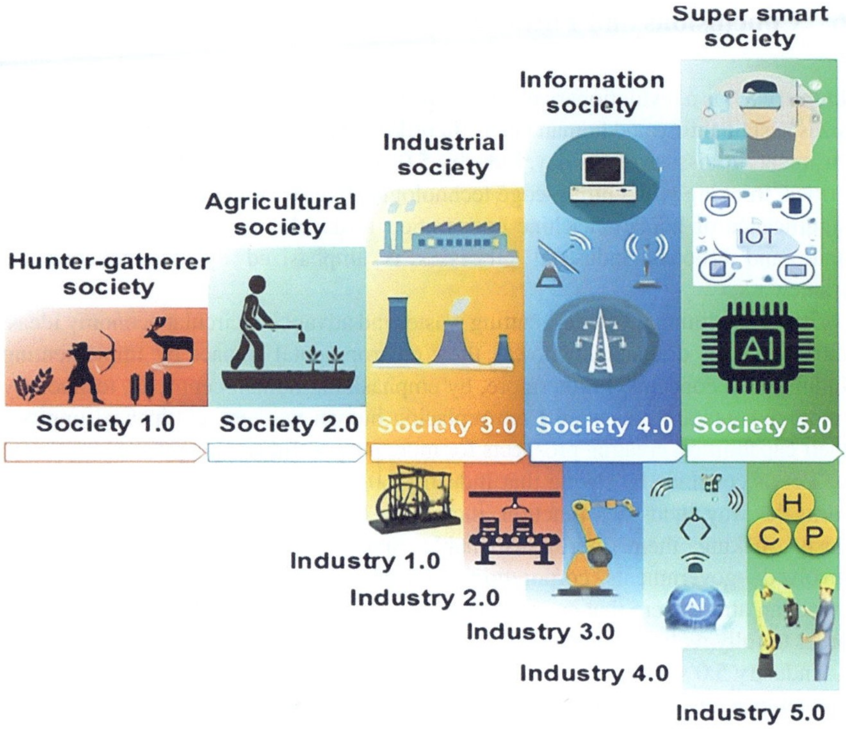

Fig. 3 Society transformation and industry evolution

2.0), the industrial (Society 3.0), and the information (Society 4.0) [65]. At present, however, it is rapidly transitioning to a super-smart and human-centered society (Society 5.0). Industry 5.0, as depicted in the lower portion of Fig. 3, is the next phase of the industrial revolution, which began with the industrial civilization and has undergone four revolutions in a very short amount of time thanks to enhanced technologies.

New technologies like digital twins, big data, AI, and next-generation wireless networks have a significant impact on both Industry 5.0 and Society 5.0. The trend toward digitalization and intellectualization, propelled by these new technologies, will determine the new order of a particular industry, or maybe Society as a whole. Advances in technology will benefit and advance every puzzle within the Industry 5.0 or Society 5.0 system [66].

6 Conclusions and Future Scope

Industry 5.0 is a revolutionary change in production and manufacturing that emphasizes sustainability and human needs. In order to tackle some of the most important sustainability issues of our day, Industry 5.0 aims to combine human creativity and critical thinking with cutting-edge technologies like AI, robotics, and the IoT. The significance of human–machine cooperation in developing more robust, sustainable, and efficient industrial processes is emphasized by this collaborative framework.

By streamlining processes, cutting waste, and advancing circular economy ideas, industries can dramatically lower their environmental impact by implementing Industry 5.0 concepts. Furthermore, by emphasizing human-centricity, technology improvements are prevented from replacing human workers and instead improve their capabilities, creating prospects for more purposeful and happy work.

Let us conclude by saying that Industry 5.0 has the power to completely transform the way industries function, increasing their productivity while simultaneously making them more sustainable and socially conscious. All parties involved—governments, corporations, and labor unions—must work together to create an atmosphere that encourages creativity, teamwork, and moral behavior in order to fully realize this potential.

Industry 5.0's potential to address sustainability issues in the future rests in how human–machine cooperation develops to provide more productive, moral, and sustainable industrial processes. This involves developing systems that not only increase production but also reduce environmental impact using cutting-edge technology like AI, robotics, and the IoT. Future efforts in research and development will probably concentrate on maximizing the use of resources, cutting waste, and advancing circular economy principles—all the while making sure that human labour is an essential component of these operations. This strategy will encourage industries to operate in a more sustainable manner in line with international environmental goals.

Furthermore, Industry 5.0 offers a lot of room for creativity in the use of ethical technologies and workforce development. The necessity of reskilling workers to fit new jobs will increase as human–machine collaboration grows, underscoring the significance of learning and instruction in this new industrial era. Furthermore, thoughtful thought will need to be given to the ethical implications of cutting-edge technology, with an emphasis on responsibility, transparency, and justice in automated systems. By solving these issues, Industry 5.0 can pave the way for a time when technical developments enhance social progress and economic prosperity, resulting in an industrial landscape that is more inclusive and sustainable.

Industry 5.0 may be improved and expanded to better serve the changing requirements of society by pursuing research and development in these and other cutting-edge fields. This will ultimately lead to a future that is more equal and sustainable.

References

1. Vogel-Heuser, B., & Hess, D. (2016). Guest editorial: Industry 4.0—prerequisites and visions, 2016. *IEEE Transactions on Automation Science and Engineering, 13*(2), 411–413.
2. Bauernhans, T., Vogel-Heuser, B., & ten Hompel, M. (2017). In A. Grundlagen (Ed.), *Handbuch Industrie 4.0 Bd.4, 2017*. Springer Verlag.
3. Vogel-Heuser, B., Bayrak, G., & Frank, U. (2012). *Forschungsfragen in Produktionsautomatisierung der Zukunft. Diskussionspapier für die acatech*. Acatech—Deutsche Akademie der Technikwissenschaften.
4. Zuehlke, D. (2010). SmartFactory—Towards a factory-of-things. *Annual Reviews in Control, 34*(1), 129–138.
5. Lu, Y., Xu, X., & Wang, L. (2020). Smart manufacturing process and system automation—A critical review of the standards and envisioned scenarios. *Journal of Manufacturing Systems, 56*, 312–325.
6. Wang, S., Wan, J., Zhang, D., Li, D., & Zhang, C. (2016). Towards smart factory for industry 4.0: A self-organized multi-agent system with big data based feedback and coordination. *Computer Networks, 101*, 158–168.
7. Zhong, R. Y., Xu, X., Klotz, E., & Newman, S. T. (2017). Intelligent manufacturing in the context of industry 4.0: A review. *Engineering, 3*(5), 616–630.
8. Plattform Industrie 4.0. Retrieved 25 July 2024, from https://www.plattform-i40.de.
9. Xu, X., Lu, Y., Vogel-Heuser, B., & Wang, L. (2021). Industry 4.0 & Industry 5.0-inception, conception and perception. *Journal of Manufacturing Systems, 61*, 530–535.
10. Breque, M., De Nul, L., & Petridis, A. (2021). *Industry 5.0: Towards a sustainable, human-centricand resilient European industry*. European Commission, Directorate-General for Research and Innovation.
11. European Economic and Social Committee, Industry 5.0. Retrieved 25 August 2024, from https://ec.europa.eu/info/research-and-innovation/research-area/industrial-research-and-innovation/industry-50_en.
12. Longo, F., Padovano, A., & Umbrello, S. (2020). Value-oriented and ethical technology engineering in Industry 5.0: A human-centric perspective for the design of the factory of the future. *Applied Sciences, 10*(12), 4182.
13. Bednar, P. M., & Welch, C. (2020). Socio-technical perspectives on smart working: Creating meaningful and sustainable systems. *Information Systems Frontiers, 22*(2), 281–298.
14. Kumar, S. S., & Mogha, S. K. (2024). Material management in automobile industry: Era 4.0. In *E3S Web of Conferences* (Vol. 529, p. 11). EDP Sciences.
15. Küpper, D., Lorenz, M., Knizek, C., & Kuhlmann, K. (2019). *Advanced robotics in the factory of the future*. Boston Consulting Group.
16. AFMG/Autonomous Manufacturing, The Additive Manufacturing Industry Landscape 2020: 240 Companies Driving Digital Manufacturing (2020). Retrieved 2024, from: https://amfg.ai/2020/05/26/the-additive-manufacturing-industry-landscape-2020-231-companies-driving-digital-manufacturing/.
17. Yuan, X.-M. (2020). Impact of Industry 4.0 on inventory systems and optimization. In *Industry 4.0-impact on intelligent logistics and manufacturing*. IntechOpen.
18. Mourtzis, D., & Doukas, M. (2020). The evolution of manufacturing systems: Industry 4.0 and beyond. *Procedia CIRP*.
19. Nahavandi, S. (2019). Industry 5.0—A human-centric solution. *Sustainability, 11*(16), 4371.
20. Javaid, M., & Haleem, A. (2020). Industry 5.0: Potential applications in COVID-19. *Journal of Industrial Integration and Management*.
21. Kovács, G., & Kot, S. (2017). New logistics and production trends as the effect of global economy changes. *Polish Journal of Management Studies*.
22. Müller, J. M., Buliga, O., & Voigt, K. I. (2018). Fortune favors the prepared: How SMEs approach business model innovations in Industry 4.0. *Technological Forecasting and Social Change, 132*.

23. Romero, D., Stahre, J., Wuest, T., Noran, O., Bernus, P., Fast-Berglund, Å., & Gorecky, D. (2016). Towards an operator 4.0 typology: A human-centric perspective on the fourth industrial revolution technologies. *CIRP Annals, 65*.
24. Kolberg, D., & Zühlke, D. (2015). Lean automation enabled by Industry 4.0 technologies. *IFACPapersOnLine, 48*(3), 1870–1875.
25. European Commission (2020). Industry 5.0: Towards a sustainable, human-centric, and resilient European industry.
26. Kagermann, H., Wahlster, W., & Helbig, J. (2013). Securing the Future of German Manufacturing Industry: Recommendations for implementing the strategic initiative Industrie 4.0. In *Final report of the Industrie 4.0 working group*. Acatech—National Academy of Science and Engineering. 678 p.
27. Bogue, R. (2018). What are the prospects for robots in the clothing and textile industry? *Industrial Robot: An International Journal, 5*, 6.
28. Ford, S., & Despeisse, M. (2016). Additive manufacturing and sustainability: An exploratory study of the advantages and challenges. *Journal of Cleaner Production, 137*.
29. Kovács, G., & Spens, K. M. (2011). Humanitarian logistics and supply chain management: The start of a new journal. *Journal of Humanitarian Logistics and Supply Chain Management*.
30. Maddikunta, P. K. R., Pham, Q.-V., Prabadevi, B., Deepa, N., Dev, K., Gadekallu, T. R., Ruby, R., & Liyanage, M. (2022). Industry 5.0: A survey on enabling technologies and potential applications. *Elsevier Journal of Industrial Information Integration, 26*, 100257.
31. Demir, K. A., Döven, G., & Sezen, B. (2019). Industry 5.0 and human-robot co-working. *Procedia Computer Science, 158*, 688–695.
32. Aslam, F., Aimin, W., Li, M., & Ur Rehman, K. (2020). Innovation in the era of IoT and industry 5.0: Absolute innovation management (AIM) framework. *Information, 11*(2), 124.
33. Lu, Y. (2017). Industry 4.0: A survey on technologies, applications and open research issues. *Journal of Industrial Information Integration, 6*, 1–10.
34. Echchakoui, S., & Barka, N. (2020). Industry 4.0 and its impact in plastics industry: A literature review. *Journal of Industrial Information Integration, 20*, 100172.
35. ElFar, O. A., Chang, C.-K., Leong, H. Y., Peter, A. P., Chew, K. W., & Show, P. L. (2020). Prospects of industry 5.0 in algae: Customization of production and new advance technology for clean bioenergy generation. *Energy Conversion and Management X, 10*, 100048.
36. Boobalan, P., Ramu, S. P., Pham, Q. V., Dev, K., Pandya, S., Maddikunta, P. K., Gadekallu, T. R., & Huynh-The, T. (2022). Fusion of federated learning and industrial internet of things: A survey. *Computer Networks, 212*, 109048.
37. Xu, L. (2020). Industry 4.0—Frontiers of fourth industrial revolution. *Systems Research and Behavioral Science, 37*(4), 531–534.
38. Xu, L. D. (2020). The contribution of systems science to industry 4.0. *Systems Research and Behavioral Science, 37*(4), 618–631.
39. Li, L. (2018). China's manufacturing locus in 2025: With a comparison of "made-inchina 2025" and "industry 4.0". *Technological Forecasting and Social Change, 135*, 66–74.
40. Lasi, H., Fettke, P., Kemper, H.-G., Feld, T., & Hoffmann, M. (2014). Industry 4.0. *Business & Information Systems Engineering, 6*(4), 239–242.
41. Priya, V., Thaseen, I. S., Gadekallu, T. R., Aboudaif, M. K., & Nasr, E. A. (2021). Robust attack detection approach for IIoT using ensemble classifier. *Computers, Materials & Continua*.
42. de la Peña Zarzuelo, I., Soeane, M. J. F., & Bermúdez, B. L. (2020). Industry 4.0 in the port and maritime industry: A literature review. *Journal of Industrial Information Integration, 20*, 100173.
43. Azeem, M., Haleem, A., & Javaid, M. (2022). Symbiotic relationship between machine learning and industry 4.0: A review. *Journal of Industrial Integration and Management, 7*(03), 401–433.
44. Zhang, C., & Chen, Y. (2020). A review of research relevant to the emerging industry trends: Industry 4.0, IoT, blockchain, and business analytics. *Journal of Industrial Integration and Management*.
45. Strubell, E., Ganesh, A., & McCallum, A. (2020). Energy and policy considerations for deep learning in NLP. In *Proceedings of the AAAI conference on artificial intelligence* (Vol. 34, No. 09, pp. 13693–13696).

46. Esmaeilian, B., Wang, B., Lewis, K., & Behdad, S. (2020). The future of waste management in Industry 5.0.
47. Sarker, M., Rahman, T., Hossain, A., & Rashid, M. (2022). Renewable energy integration in Industry 5.0: Challenges and opportunities.
48. Dalenogare, L. S., Benitez, G. B., Ayala, N. F., & Frank, A. G. (2018). The expected contribution of Industry 4.0 technologies for industrial performance.
49. Horváth, D., & Szabó, R. Z. (2019). Driving forces and barriers of Industry 4.0: Do multinational and small and medium-sized companies have equal opportunities? *Technological Forecasting and Social Change, 146*, 119–132.
50. Acemoglu, D., & Restrepo, P. (2018). Artificial intelligence, automation, and work. In *The economics of artificial intelligence: An agenda* (pp. 197–236). University of Chicago Press.
51. Barley, S. R., Bechky, B. A., & Milliken, F. J. (2017). The changing nature of work: Careers, identities, and work lives in the 21st century. *Academy of Management Discoveries, 3*(2), 111–115.
52. Santos, R., Barros, A. C., Mendes, L., & Santos, G. (2021). Industry 5.0: The perspective of human-centricity in sustainable manufacturing.
53. Zuboff, S. (2019). *The age of surveillance capitalism: The fight for a human future at the new frontier of power*. PublicAffairs.
54. Brynjolfsson, E., & McAfee, A. (2014). *The second machine age: Work, progress, and prosperity in a time of brilliant technologies*. WW Norton & Company.
55. Zheng, P., Wang, H., Sang, Z., Zhong, R. Y., Liu, Y., & Liu, C. (2018). Smart manufacturing systems for Industry 4.0: Conceptual framework and research challenges. *Frontiers of Mechanical Engineering, 13*, 137–150.
56. Kumar, N., Chilamkurti, N., & Buyya, R. (2020). Industrial internet of things (IIoT): Recent advances, enabling technologies, and open challenges. *Computers & Electrical Engineering, 81*, 106522.
57. Tortorella, G. L., Giglio, R., & Fogliatto, F. S. (2020). Industry 4.0 adoption as a moderator of the impact of lean production practices on operational performance. *International Journal of Operations & Production Management*.
58. Kagermann, H., Helbig, J., Hellinger, A., & Wahlster, W. (2013). *Recommendations for implementing the strategic initiative INDUSTRIE 4.0: Securing the future of German manufacturing industry; final report of the Industrie 4.0 Working Group*. Forschungsunion.
59. Liu, C., Cao, S., Tse, W., & Xu, X. (2017). Augmented reality-assisted intelligent window for cyberphysical machine tools. *Journal of Manufacturing Systems, 44*, 280–286.
60. Zhu, Z., Liu, C., & Xu, X. (2019). Visualisation of the digital twin data in manufacturing by using augmented reality. *Procedia CIRP, 81*, 898–903.
61. Villani, V., Sabattini, L., Barańska, P., Callegati, E., Czerniak, J. N., Debbache, A., Fahimipirehgalin, M., Gallasch, A., Loch, F., Maida, R., Mertens, A., Mockałło, Z., Monica, F., Nitsch, V., Talas, E., Toschi, E., Vogel-Heuser, B., Willems, J., Żołnierczyk-Zreda, D., & Fantuzzi, C. (2020). The inclusive system: A general framework for adaptive industrial automation. *Transactions on Automation Science and Engineering, 18*(4), 1969–1982.
62. Flores, E., Xu, X., & Lu, Y. (2020). Human Capital 4.0: A workforce competence typology for Industry 4.0. *Journal of Manufacturing Technology Management, 31*(4), 687–703.
63. Office, C. (2015). *Report on the 5th science and technology basic plan*. Council for Science, Technology and Innovation.
64. Rojas, C. N., Peñafiel, G. A. A., Buitrago, D. F. L., & Romero, C. A. T. (2021). Society 5.0: A Japanese concept for a superintelligent society. *Sustainability, 13*(12), 6567.
65. Deguchi, A., Matsuoka, H., Hirai, C., Kamimura, O., Nakano, T., Oshima, K., Tai, M., Tani, S., Shibasaki, R., Hori, S., Kawamura, S., Akashi, Y., Hato, E., Ohkata, J., Warisawa, S., Kajitani, S., Nakajima, T., Ohashi, H., Watanabe, T., Karasawa, K., & Sameshima, S. (2020). *Society 5.0: A people-centric super-smart society*. Hitachi-UTokyo Laboratory.
66. Huang, S., Wang, B., Li, X., Zheng, P., Mourtzis, D., & Wang, L. (2022). Industry 5.0 and Society 5.0—Comparison, complementation and co-evolution. *Journal of Manufacturing Systems, 64*. Elsevier.

Advancing Sustainability Through Industry 5.0: Trends, Challenges, and Future Directions

Mahmood A. Bazel, Abdullah O. Baarimah, Basheer Almuhaya, Fathey Mohammed, Nail Adeeb Ali Abdu, and Tharaa M. Al-Zghoul

1 Introduction

The introduction of Industry 5.0, an industrial paradigm that stresses the cooperation of machines and humans in fostering sustainable development, is causing a radical change in the global manufacturing environment [1]. By fusing cutting-edge technology with sustainability principles, Industry 5.0 brings a more human-centric approach than its predecessor, Industry 4.0, which focused on the automation and digitalization of production methods [2]. The need to strike a balance between environmental responsibility and industrial progress is what is driving this evolution, making sure that technology breakthroughs support long-term social and ecological well-being. A significant change in the industrial and manufacturing environment is represented by the shift from Industry 4.0 to Industry 5.0. Industry 5.0 presents a

M. A. Bazel (✉) · B. Almuhaya
Faculty of Engineering and Information Technology, Taiz University, Taiz, Yemen
e-mail: mahmood.bazel@taiz.edu.ye

A. O. Baarimah
Department of Civil and Environmental Engineering, College of Engineering, A'Sharqiyah University, Ibra, Oman
e-mail: abdullah.baarimah@asu.edu.om

F. Mohammed
Department of Business Analytics, Sunway Business School, Sunway University, Petaling Jaya, Selangor, Malaysia

N. A. A. Abdu
Information Technology Department, Faculty of Engineering and Computing, University of Science and Technology, Aden, Yemen

T. M. Al-Zghoul
Department of Civil Engineering, Faculty of Engineering, Tafila Technical University, Tafila, Jordan

Y. Kumar Singla et al. (eds.), *Digital Manufacturing in Industry 5.0*,
https://doi.org/10.1007/978-3-031-91500-0_22

more human-centric approach that blends cutting-edge technology like machine learning (ML), robots, and artificial intelligence (AI) with human creativity and intellect, whereas Industry 4.0 placed greater emphasis on data interchange, smart technologies, and automation [3, 4]. In addition to increasing production efficiency, Industry 5.0 aims to create more resilient and sustainable industrial systems that consider social, environmental, and ethical dimensions [5].

A critical aspect of this new paradigm is the focus on sustainability as industries face growing pressure to reduce their environmental impact. Sustainability has emerged as a critical priority for industries worldwide, as the adverse environmental impacts of industrialization, such as resource depletion, pollution, and climate change, have become increasingly evident [6]. In this context, Industry 5.0 presents a special chance to address these challenges by promoting sustainable manufacturing practices [7, 8]. Industry 5.0 seeks to minimize the environmental impact of industrial operations, optimize production processes, and eliminate waste via the use of cutting-edge technologies including AI, robots, and human-machine cooperation [9]. Furthermore, the incorporation of green manufacturing initiatives and circular economy principles highlights the potential of Industry 5.0 to revolutionize the way industries approach sustainability [10, 11].

The rise of concepts like the circular economy and environmental sustainability reflects this shift toward responsible production practices that minimize waste, optimize resource use, and prioritize energy efficiency [12]. Through the use of digital technologies like digital twins, blockchain, and the Internet of Things (IoT), Industry 5.0 presents opportunity to address these challenges and facilitate more transparent, sustainable, and effective manufacturing methods [11].

However, the adoption of Industry 5.0 is not without its challenges. The incorporation of advanced technologies into manufacturing systems requires not only significant financial investment but also a deep understanding of their potential risks and ethical implications [5]. For Industry 5.0 to be implemented successfully, factors including data privacy, automation's potential to replace jobs, and the moral application of AI are crucial [13]. Moreover, while Industry 5.0 holds promise for increasing industrial resilience and enabling more robust supply chains, further research is needed to explore how these methods can be scaled and applied across diverse sectors and regions [14].

This chapter seeks to explore the intersection of Industry 5.0 and sustainability by offering a thorough examination of the trends, difficulties, and future directions that define this new industrial era. By conducting bibliometric analysis and reviewing existing literature, we aim to identify the key themes and technological innovations that are shaping the Industry 5.0 landscape. Furthermore, the chapter explores the social and ethical considerations of Industry 5.0, emphasizing the significance of a comprehensive strategy for sustainable development that prioritizes both technological advancement and human well-being. In doing so, this chapter not only offers insights into the present state of research on Industry 5.0 and sustainability but also provides guidance for academics, industry professionals, and policymakers on how to leverage emerging technologies to achieve sustainability goals. By understanding the difficulties provided by Industry 5.0, stakeholders can make informed decisions that contribute to the creation of sustainable, resilient, and future-proof manufacturing ecosystems.

2 Materials and Methods

Bibliometric analysis is a powerful approach for thoroughly investigating large sets of literature, offering insights from both quantitative and qualitative perspectives. This approach applies techniques such as data mining, processing, and visualization to reveal internal relationships within the literature, including structures, interactions, knowledge networks, and areas of overlap [15]. Bibliometric analysis is popular and very successful since it can show the current landscape of institutions, research fields, new trends, and publications [16]. It provides a precise method for evaluating extensive scholarly data, allowing researchers to understand citation patterns and gain an overview of current and emerging research topics. By analyzing past studies and their progression, bibliometric indicators can guide future research development. Due to these capabilities, bibliometric techniques have become widely adopted across many disciplines [17].

In this chapter, bibliometric analysis was employed to identify future research, challenges, and trends directions in the field of Industry 5.0 and sustainability. Data were sourced from the Scopus database, chosen for its broad and in-depth coverage of academic literature across various fields, especially in technology-related areas [18, 19]. The decision to use the Scopus database was based on its extensive collection of scholarly works across multiple disciplines and its strong focus on technology-focused research [20–22]. The search methodology is outlined in Fig. 1.

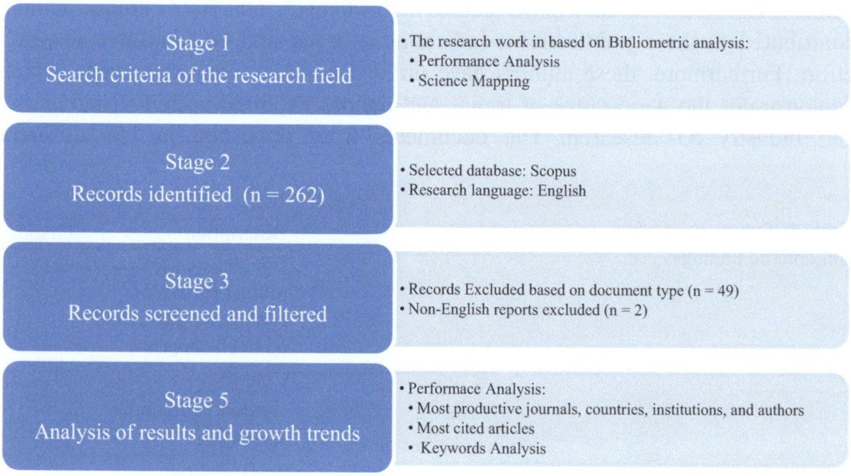

Fig. 1 Research methodology

The keyword query string ((“Industry 5.0” OR “Fifth Industrial Revolution” OR “Human-machine collaboration” OR “Human-centric manufacturing”) AND (“Sustainability” OR “Sustainable development” OR “Sustainable manufacturing” OR “Circular economy” OR “Green manufacturing”)) was employed to locate articles specifically addressing the convergence of Industry 5.0. These words encompass the primary study domain extensively, as they have been utilized in every section of the study. The search parameters included academic articles, conference articles, and book chapters published in English. The initial search resulted in a total of 262 publications. Subsequently, we refined the dataset by limiting it to articles, conference articles, and book chapters published exclusively in English. After applying this filtering process, the selection was narrowed down to 211 publications. For the bibliometric trend analysis, the data were exported in CSV format and analyzed utilizing VOSviewer, a popular tool for generating and displaying bibliometric maps [23]. This approach enables the mapping, clustering, and mining of data from articles extracted from databases. VOSviewer’s algorithms automatically adjust display labels to optimize the representation of co-occurrences within the dataset, improving the clarity and accuracy of the visualizations.

3 Results and Discussion

Table 1 summarizes the key bibliometric results from the examination of 211 documents related to Industry 5.0 and sustainability, comprising 82 journal articles, 66 conference papers, and 63 book chapters. Researchers are examining the relationship between sustainable development and Industry 5.0 on a variety of platforms, which reflects the increased interest in this field among academics and practitioners. The 211 documents have garnered a total of 2341 citations, highlighting the academic impact and relevance of the studies in this field. A total of 717 unique authors contributed to these publications, indicating a broad and collaborative research effort. Furthermore, these authors have utilized 1307 different keywords, which demonstrates the wide range of topics and themes explored within sustainability and Industry 5.0 research. The documents were published in 153 different

Table 1 Summary of key bibliometric findings

Description	Findings
Documents (82 journal papers, 66 conference paper, and 63 book chapters)	211
Countries	64
Authors` affiliations	276
Keywords	1307
Total citation	2341
Publication sources	153
Authors	717

publication venues, including journals, conference proceedings, and book series, showcasing the interdisciplinary nature of this research. A total of 276 different author affiliations were represented, illustrating the global spread of research institutions involved in this domain. The study also shows contributions from 64 countries, indicating significant international interest in the subject.

3.1 Trend of Annual Publications

The temporal analysis reveals a notable increase in publication activity over the years as shown in Fig. 2. In 2024 alone, 114 documents were published, making it the most active year for research in this area. The number of publications has steadily grown from just one document in both 2018 and 2019 to 57 in 2023. This upward trend suggests that Industry 5.0 and sustainability are emerging as critical areas of study, particularly in recent years as industry seeks to address environmental and societal challenges through technological innovation.

3.2 Most Productive Country

Table 2 presents a summary of the leading countries contributing to research on Industry 5.0 and sustainability. The data indicates that India leads research on Industry 5.0 and sustainability with 62 contributions, showcasing its significant role in

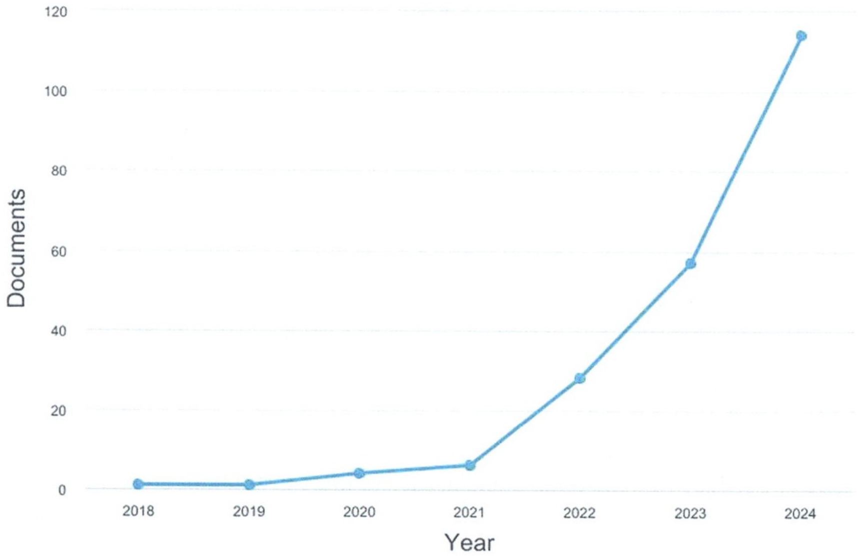

Fig. 2 Annual Publications

Table 2 Leading Countries

Country	No
India	62
United Kingdom	22
Italy	16
Spain	13
United States	13
China	11
Poland	9
Portugal	9
Australia	8
Turkey	8

advancing these topics. This output reflects India's increasing focus on technological innovation and sustainable practices within its industrial sector. The United Kingdom follows as the second-highest contributor with 22 research outputs, highlighting its active engagement in exploring Industry 5.0 concepts and their sustainability implications. Italy, Spain, and the United States each contributed 13 documents, demonstrating substantial involvement in the field, supported by strong research capabilities and academic institutions. China and Poland contributed 11 and 9 documents, respectively. China's contributions are particularly noteworthy due to its large industrial base and efforts to integrate sustainable technologies. Poland's involvement reflects its growing academic interest in Industry 5.0. Additionally, Portugal, Australia, and Turkey contributed nine, eight, and eight documents, respectively. These countries engage with the global discourse on Industry 5.0, helping to develop knowledge and practices to address sustainability challenges.

3.3 Top Affiliations Contributing

Table 3 presents a summary of the top affiliations contributing to research on Industry 5.0 and sustainability. The data highlights the institutions with the highest number of publications in this field. The Lovely Professional University in India stand out as the top affiliations has produced 9 documents, reflecting its strong research focus and expertise in Industry 5.0 and sustainability. This significant output underscores the university's role as a major contributor to advancing knowledge and innovation in this area. Following Universiti Sains Malaysia in Malaysia, Instituto Politécnico de Viana do Castelo in Portugal, Moreover, the Hong Kong Polytechnic University in Hong Kong, University of Patras in Greece, and New Horizon College of Engineering in India, each of these institutions contributed four documents, indicating their active involvement in the research landscape. These universities are recognized for their research capabilities and their contributions to the global discourse on sustainable industrial practices and advanced technologies.

Table 3 Top 10 affiliations according to authors' affiliation

Affiliation	Country	Documents
Lovely Professional University	India	9
Universiti Sains Malaysia	Malaysia	4
Instituto Politécnico de Viana do Castelo	Portugal	4
The Hong Kong Polytechnic University	Hong Kong	4
University of Patras	Greece	4
New Horizon College of Engineering	India	4
Uppsala Universitet	Sweden	3
La Trobe University	Australia	3
Silesian University of Technology	Poland	3
SRM Institute of Science and Technology	India	3

Additionally, Uppsala Universitet in Sweden, La Trobe University in Australia, Silesian University of Technology in Poland, and SRM Institute of Science and Technology in India, each of these affiliations has contributed three documents. These institutions are actively participating in the development of research in this field, reflecting their commitment to addressing the difficult related with Industry 5.0 and sustainability.

3.4 Leading Sources

Table 4 outlines the top sources where research on Industry 5.0 and sustainability has been published. The data reveals the most prominent journals and conference proceedings contributing to the field. Leading the list are "Lecture Notes in Mechanical Engineering." This source leads with nine documents, indicating its significant role in disseminating research related to mechanical engineering and industrial technologies. Its prominence reflects the journal's focus on advanced manufacturing techniques and their implications for sustainability. Following closely are "Sustainability (Switzerland)." With eight documents, this journal stands out for its contributions to discussions on sustainable practices and technologies. Its focus on environmental and societal sustainability aligns well with the themes of Industry 5.0, making it a key outlet for relevant research. Other notable sources include "Procedia Computer Science." This conference proceedings series has published five documents, highlighting its role in presenting research at the intersection of computer science and industrial applications. The inclusion of Industry 5.0 topics reflects the growing integration of digital technologies in manufacturing.

Table 4 The top 5 sources

Source	Documents
"Lecture Notes in Mechanical "Engineering"	9
"Sustainability Switzerland"	8
"Procedia Computer Science"	5
"Lecture Notes in Networks and Systems"	4
"IEEE Access"	4

Additionally, "IEEE Access" and " Lecture Notes in Networks and Systems" have also significantly contributed to the scholarly discourse with four publications. IEEE Access is known for its broad coverage of engineering and technology topics, while Lecture Notes in Networks and Systems focuses on research related to network and systems engineering. Their contributions reflect the interdisciplinary nature of Industry 5.0 research, integrating aspects of digital networks and systems.

3.5 Top Cited Articles

The most cited documents in the areas of sustainability and Industry 5.0 are shown in Table 5, underscoring their noteworthy impact. The most cited work, *"Socio-Technical Perspectives on Smart Working: Creating Meaningful and Sustainable Systems"* (2020), with 215 citations, emphasizes integrating sociotechnical perspectives with smart working systems. The *"Industry 5.0 Framework: Viability-Based Integration of Resilience, Sustainability, and Human-Centricity Perspectives"* (2023) has 196 citations, showcasing its importance in harmonizing these key elements. Additionally, *"State of Industry 5.0—Analysis and Identification of Current Research Trends"* (2022) has 182 citations, offering a detailed analysis of emerging trends, while *"Rethinking Engineering Education at the Age of Industry 5.0"* (2022) with 176 citations, highlights the shift in education to meet Industry 5.0 demands. Lastly, *"Green IoT and Edge AI as Key Technological Enablers for a Sustainable Digital Transition"* (2021) with 171 citations, explores sustainable digital transitions in Industry 5.0. Collectively, these works provide critical insights into theoretical frameworks, technological innovations, and strategies shaping Sustainability and Industry 5.0.

3.6 Keyword Co-occurrence Analysis

As shown in Table 6 and Fig. 3, the results of the keyword co-occurrence analysis offer important insights into the Industry 5.0 research landscape and its connection to sustainability. With "Industry 5.0" showing the highest number of occurrences

Table 5 Details of the top 5 cited articles

Documents	Year	Total citations
"Socio-Technical Perspectives on Smart Working: Creating Meaningful and Sustainable Systems" [7]	2020	215
"The Industry 5.0 framework: viability-based integration of the resilience, sustainability, and human-centricity perspectives" [24]	2023	196
"State of Industry 5.0—Analysis and Identification of Current Research Trends" [25]	2022	182
"Rethinking engineering education at the age of industry 5.0" [26]	2022	176
"Green IoT and edge AI as key technological enablers for a sustainable digital transition towards a smart circular economy: An industry 5.0 use case" [27]	2021	171

(112), it is evident that this emerging industrial paradigm is gaining substantial attention in the literature. This focus on Industry 5.0 highlights the rising interest in human-centric manufacturing, where human-machine cooperation is at the forefront of innovation.

The analysis reveals that "Industry 4.0" appears 77 times, highlighting its ongoing relevance in discussions about digital manufacturing. The coexistence of Industry 4.0 and Industry 5.0 in the literature illustrates the evolutionary transition between the two, emphasizing the integration of advanced technologies like AI and IoT. "Sustainable development" (72 occurrences) and "sustainability" (38 occurrences) are also prominent, reflecting the significance of incorporating social and environmental considerations into industrial strategies. This aligns with international programs like the Sustainable Development Goals (SDGs) of the United Nations, positioning industries as key players in addressing sustainability challenges. Related terms like "circular economy" (9 occurrences) and "environmental sustainability" (6 occurrences) reinforce the focus on resource efficiency and minimizing environmental impact. Technological innovations are crucial enablers of Industry 5.0, evident from the frequent mentions of "artificial intelligence" (53 occurrences), "automation" (18 occurrences), and "digital transformation" (14 occurrences). The significance of AI is underscored by related terms such as "machine learning" (six occurrences). This highlights the role of AI-driven automation and data analytics in optimizing manufacturing processes while enhancing efficiency and reducing waste. The emphasis on "human-centric" (21 occurrences) and "human centricities" (13 occurrences) indicates a shift toward balanced collaboration between humans and machines. This is further supported by terms like "human-machine collaboration" (6 occurrences) and "ergonomics" (6 occurrences), showing that human factors are central to future industrial system designs.

The "Internet of Things" (35 occurrences) and related terms signify the critical role of connected devices in facilitating data-driven decision making. This connectivity is essential for both Industry 4.0 and Industry 5.0, enabling real-time

Table 6 Top keyword co-occurrence analysis

Keyword	Cluster	Total link strength	Occurrences	Avg. citations
Industry 5.0	2	502	112	17.1875
Industry 4.0	3	388	77	20.3896
Sustainable development	3	392	72	13.5556
Artificial intelligence	2	276	53	13.9057
Sustainability	5	199	38	15.2368
Internet of Things	3	217	35	19.7429
Industrial revolutions	4	153	22	14
Human-centric	1	111	21	13.4762
Decision making	4	129	21	8
Automation	7	103	18	4.9444
Digital transformation	2	66	14	14.5
Human centricities	6	75	13	36.5385
Manufacturing	2	67	13	5
Embedded systems	2	94	11	14.3636
Resilience	6	77	11	26.0909
Smart manufacturing	4	75	10	47.2
Data analytics	3	75	10	32.9
Industrial research	2	68	10	9.1
AI	3	59	10	19.3
Circular economy	5	59	9	26.6667
Fifth industrial revolution	4	59	9	1.5556
Big Data	7	49	8	31.25
Robotics	1	33	8	0.375
Supply chains	6	51	8	57.375
Advanced technology	1	39	7	1.4286
Block-chain	5	60	7	24.1429
Accident prevention	1	43	7	26.4286
Intelligent robots	1	52	7	16.2857
Digitalization	6	44	7	25.5714
Internet of Things (IOT)	4	46	7	26.1429
Sustainable development goals	2	37	7	11.1429
IoT	3	39	7	14.1429
Collaborative robots	1	45	6	19.6667
Environmental technology	1	49	6	3.5
Ergonomics	1	43	6	2
Manufacturing process	1	38	6	2.6667
Blockchain	5	49	6	26.1667
Digital twin	1	30	6	14.3333
Energy utilization	1	33	6	30.1667
Competition	5	35	6	20.6667
Environmental sustainability	6	33	6	33.8333

Table 6 (continued)

Keyword	Cluster	Total link strength	Occurrences	Avg. citations
Machine learning	7	31	6	3.5
Human-machine collaboration	3	24	6	0.6667
Energy efficiency	7	28	6	3.8333
Data handling	4	39	5	40.4
Ethical technology	1	35	5	0.2
Industrial economics	5	43	5	35.2
Cyber-physical systems	2	47	5	23
Economic and social effects	1	32	5	0
Man machine systems	1	31	5	2
Human robot interaction	1	33	5	1.6
Sustainable development goal	2	35	5	14.8
Artificial intelligence (Ai)	4	36	5	1.6
Cobots	1	27	5	1
Engineering education	1	31	5	37
Society 5.0	5	31	5	7.2
Enabling technologies	2	24	5	0.2
Machine-learning	7	27	5	3
Optimisations	1	26	5	1.6
Digital technologies	1	21	5	0.4
Digitization	2	15	5	19.8

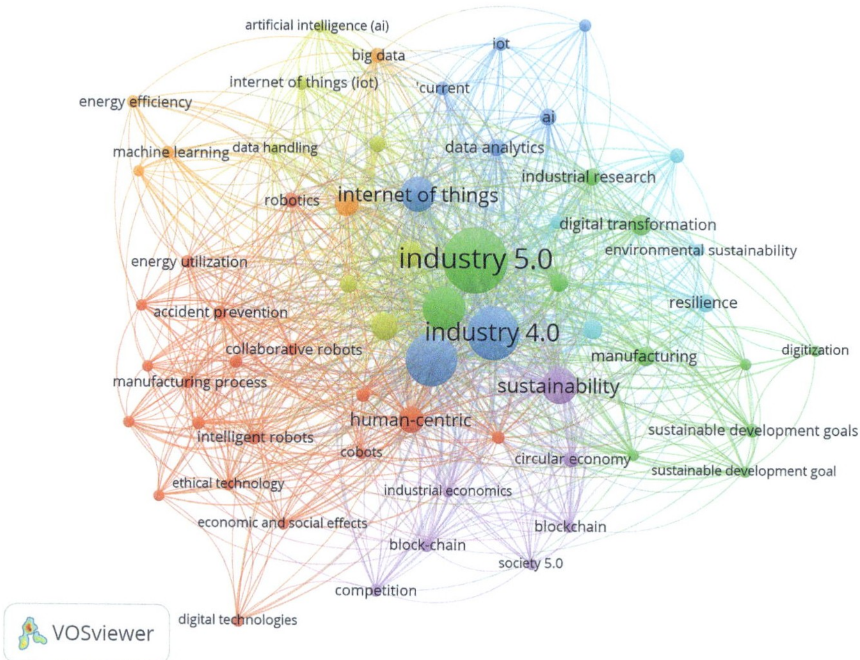

Fig. 3 Co-occurrence clustering of author keywords in a network visualization

monitoring and smart manufacturing. The presence of "blockchain" (six occurrences) suggests that secure data handling is increasingly important in Industry 5.0. Despite these positive trends, several challenges persist. Terms like "resilience" (11 occurrences) and "competition" (6 occurrences) highlight concerns about maintaining competitiveness while ensuring system resilience. The mention of "decision making" (21 occurrences) indicates the need for robust frameworks to navigate the complexities of Industry 5.0 environments. Additionally, the lower occurrences of terms like "energy efficiency" (six occurrences) and "environmental technology" (six occurrences) suggest that more efforts are needed to integrate sustainable energy solutions into Industry 5.0 systems. Finally, keywords such as "ethical technology" (five occurrences) and "human robot interaction" (5 occurrences) signal a growing recognition of ethical considerations and social impacts, emphasizing the importance of responsible innovation that benefits society.

The co-occurrence analysis identifies seven distinct clusters that represent key thematic areas at the nexus of Industry 5.0 and sustainability, each distinguished by a unique color in Fig. 2. These clusters underscore the diverse aspects of Industry 5.0, highlighting the combination of advanced technologies, human-centric approaches, sustainability goals, and economic factors.

Among these, Cluster 1 concentrates on human-centric technologies and practices. It includes keywords such as "human-centric," "robotics," "collaborative robots (cobots)," "digital twin," and "energy utilization." This cluster reflects a shift toward enhancing human roles in manufacturing, suggesting that technology should augment rather than replace human capabilities. Key features like improved ergonomics and accident prevention align with ethical technology principles, while "digital twin" technology facilitates real-time monitoring to enhance efficiency. The emphasis on "environmental technology" and "sustainable practices" within this cluster indicates a commitment to environmental stewardship. Overall, it highlights the need for manufacturing systems that prioritize human well-being while utilizing advanced technologies to boost productivity and sustainability.

Cluster 2 features keywords such as "Industry 5.0," "artificial intelligence," "digital transformation," "manufacturing," "embedded systems," "sustainable development goals (SDGs)," and "cyber-physical systems." The prominence of "Industry 5.0," "AI," and "digital transformation" underscores their critical role in driving the next industrial revolution. Additionally, "cyber-physical systems" and "embedded systems" highlight the integration of digital and physical processes, fostering more intelligent and responsive manufacturing systems. The connection to SDGs indicates that Industry 5.0 initiatives aim to support broader global sustainability objectives. Keywords like "enabling technologies" and "digitization" reflect the transformative impact of digital tools in reshaping manufacturing, enhancing efficiency, and minimizing environmental footprints. This cluster emphasizes the need for robust technological frameworks to facilitate the transition to Industry 5.0 while ensuring that advancements align with SDGs.

Cluster 3 includes keywords such as "Industry 4.0," "sustainable development," "Internet of Things (IoT)," "data analytics," "AI," and "human-machine collaboration." This cluster highlights the evolution from Industry 4.0 to Industry 5.0, emphasizing the integration of digital methods with sustainability objectives. The coexistence of "Industry 4.0" and "Industry 5.0" illustrates a transitional phase where the digital capabilities of the Fourth Industrial Revolution are enhanced by human-centric and sustainability-focused elements. Keywords like "IoT" and "data analytics" underscore the significance of interconnected devices in optimizing sustainable manufacturing practices. "Human-machine collaboration" reflects the shift toward more interactive roles for humans, ensuring sustainability efforts benefit from both human insight and machine efficiency. This cluster emphasizes building on Industry 4.0's digital advancements to achieve the sustainability aims of Industry 5.0. Future research should explore effective strategies for merging these two paradigms to maximize sustainability outcomes.

Cluster 4 is concerned with the background and development of industrial revolutions, with a special emphasis on Industry 5.0. Keywords include "industrial revolutions," "decision making," "smart manufacturing," "fifth industrial revolution," "IoT," "data handling," and "AI." The terms "industrial revolutions" and "fifth industrial revolution" provide a framework for understanding Industry 5.0 within the continuum of industrial development. "Smart manufacturing" and "decision making" indicate the utilize of intelligent systems and data-driven approaches to improve sustainability and efficiency. "Data handling" and "AI" highlight the importance of managing large datasets to inform strategic decisions, optimize processes, and reduce waste. This cluster underscores the critical role of smart manufacturing and informed decision making in advancing industrial practices toward greater sustainability.

Cluster 5 includes keywords such as "sustainability," "circular economy," "blockchain," "competition," "industrial economics," and "society 5.0." This cluster focuses on the relationship between economic factors, circular economy principles, and sustainability within Industry 5.0. The emphasis on "sustainability" and "circular economy" highlights the significance of waste reduction through closed-loop systems. The mention of "blockchain" indicates the need for secure data management to support circular economy initiatives, such as tracking resource flows. Additionally, "competition" and "industrial economics" suggest that economic viability is essential for adopting sustainable practices. "Society 5.0" reflects a vision of integrating industrial advancements with societal needs to create a society that is more sustainable and inclusive. This cluster underscores the necessity of economically viable sustainability practices in Industry 5.0.

Cluster 6 emphasizes resilience and robust supply chain management for sustainable industrial operations, featuring keywords like "human centricities," "resilience," "supply chains," "digitalization," and "environmental sustainability." Resilience refers to manufacturing systems' ability to withstand disruptions while maintaining sustainable operations. The role of digitalization is highlighted as it enhances supply chain transparency and efficiency, enabling real-time monitoring and agile responses to disruptions. This support for environmental sustainability aims to minimize impacts through resilient, digitally enabled supply chains. This cluster underscores the importance of integrating resilience and digital technologies to promote sustainability.

Finally, Cluster 7 focuses on automation, big data, and machine learning in driving sustainable manufacturing practices. Automation and ML are necessary for reducing waste, optimizing production methods, and enhancing energy efficiency. Big data is crucial for analyzing large datasets to enhance decision making and operational efficiency. The emphasis on energy efficiency underscores the need to reduce consumption and improve resource utilization through advanced technologies. The repeated references to machine learning highlight its role in predictive modeling and intelligent decision making. This cluster illustrates the potential of automation and data-driven technologies to significantly enhance manufacturing sustainability.

3.7 Trends

A number of notable trends characterize the shift to Industry 5.0, most notably the combination of cutting-edge technology with human-centered methodologies. The focus on human-machine cooperation, which combines machine capabilities and human talents to enhance industrial processes, is one of the most important developments [1, 28]. Technologies such as collaborative robots and intelligent robots are playing a critical role in creating more efficient and safer work environments. The adoption of AI, ML, and big data analytics further underscores the trend toward automation and smart manufacturing, driving both productivity and sustainability [16, 29]. Another key trend is the move toward sustainability, with a focus on circular economy principles and environmental sustainability [10, 30]. Adoption of more environmentally friendly manufacturing techniques is being aided by the use of blockchain technology to increase supply chain transparency [9, 21]. Additionally, the concept of digital twin technology is gaining traction as a tool to improve production efficiency by optimization of manufacturing systems and enabling real-time monitoring [31, 32].

3.8 Challenges

While the trends associated with Industry 5.0 highlight its potential, several challenges persist. A primary concern is the ethical and responsible implementation of human-centric technologies [33]. Keywords like "ethical technology" and "economic and social effects" reflect worries about unintended consequences, such as job displacement and inequality [1, 5]. Additionally, resilience is crucial as industries aim to withstand disruptions, like supply chain interruptions, while achieving sustainability goals [9]. The transition from Industry 4.0 to Industry 5.0 also presents technological hurdles. Integrating AI and cyber-physical systems into existing manufacturing requires substantial investment and expertise [34]. Furthermore, the reliance on data-driven technologies increases concerns about data privacy, management, and security, which must be addressed to ensure efficient and secure systems [5, 14, 25, 35].

4 Future Directions

Looking ahead, research and industrial practices in Industry 5.0 are expected to prioritize several key areas. First, the development of resilient and sustainable supply chains will be crucial in tackling global challenges like climate change and geopolitical disruptions [36]. Industries are increasingly adopting blockchain and digitalization technologies to enhance transparency and resilience [37, 38]. For example, companies like Walmart have leveraged blockchain for food traceability, while the automotive industry is using digital twins to optimize supply chains [32]. Future research could explore how blockchain reduces bottlenecks in real-time decision making or its long-term impact on supply chain costs [9].

Second, the integration of circular economy ideas into manufacturing processes will continue to evolve as industries strive for sustainability through efficient resource use [37, 38]. Detailed studies of closed-loop supply chains or cradle-to-cradle design could help companies adopt these principles more effectively [39]. Future research should concentrate on lifecycle analysis and the financial and environmental objectives of circular materials in various industries [40].

Third, the role of human-centric methods in improving worker well-being and productivity will remain central. Industries must address ethical concerns to ensure that human-robot collaboration benefits workers [4]. Practical frameworks for balancing automation with upskilling are essential. For instance, collaborations between industries and academic institutions can develop retraining programs for workers adapting to Industry 5.0 environments. Additionally, studies could measure worker well-being using metrics like job satisfaction and burnout rates in smart factories [34]. Examples of successful human-robot collaboration that maintain productivity and worker welfare should be explored further [1].

Energy efficiency will also become a key focus as industries seek to reduce their environmental footprint. ML and AI are expected to play a major role in optimizing energy consumption (ENC) in real-time, with predictive maintenance and AI-based energy forecasting already in use in smart factories [41]. Smaller manufacturers could adopt these strategies to improve efficiency [30]. Future research could concentrate on developing AI models that target reducing carbon emissions by a specific percentage over a defined period [13]. Collaboration between AI researchers and energy providers could also lead to new solutions for integrating manufacturing into smart grids [36].

Finally, research should examine the socioeconomic impacts of Industry 5.0, especially in developing countries [7, 24]. Comparative studies across different economies could explore how adopting advanced technologies affects job creation, economic growth, and sustainability [6]. Case studies on whether these technologies are widening or narrowing the digital divide in various regions would provide valuable insights [9]. Governments, industries, and NGOs must work together to promote skill development and sustainable industrial practices in developing countries [28]. Additionally, international collaboration in research could help address global challenges like climate change by promoting inclusive growth and equitable industrial practices [11].

In addition to these areas, future research should explore the intersection between sustainability and Industry 5.0, concentrating on how advanced technologies can drive sustainable industrial practices [11, 12, 42]. As Industry 5.0 emphasizes human-centric design, resilience, and smart systems, it presents unique opportunities to integrate sustainability principles into manufacturing and supply chain processes [24]. Future research should investigate how smart manufacturing technologies can be integrated with sustainable practices to improve the resilience and efficiency of manufacturing systems [8]. Future studies could investigate how AI, robotics, and IoT can reduce ENC, minimize waste, and optimize the use of natural resources [1, 43]. The role of green technologies, like Green IoT and Edge AI in facilitating circular economies and reducing environmental impacts should be further explored [27, 44]. Additionally, research should assess how these technologies can promote sustainable business models that prioritize both profitability and environmental stewardship [5]. By understanding the synergies between sustainability and Industry 5.0 technologies efforts, industries can develop systems that are not only technologically advanced but also environmentally conscious.

5 Conclusion

This chapter provides a comprehensive analysis of the intersection between Industry 5.0 and sustainability, highlighting significant advancements and ongoing study in this dynamic field. Journal articles, conference papers, and book chapters are among the varied range of work revealed by a bibliometric study of 211 documents, demonstrating a significant academic and practical interest in fusing cutting-edge technology with SDGs. Our analysis indicates a substantial increase in publication activity, particularly from 2018 to 2024, signaling the rising recognition of Industry 5.0 as a transformative force emphasizing human-centric approaches and sustainability. Notable contributions come from countries like India and the United Kingdom, with key institutions such as Lovely Professional University leading the discourse. Key trends identified include a focus on human-machine collaboration, ethical technology, and the application of advanced methods like AI and robotics. However, challenges remain, such as the need to better integrate sustainability principles within Industry 5.0 frameworks and address the sociotechnical aspects of industrial transformation.

Overall, the advancement of sustainability through Industry 5.0 represents a transformative opportunity to reshape industrial practices for a more sustainable future. By prioritizing human-centric approaches, integrating advanced technologies, adopting circular economy principles, fostering collaboration, and addressing policy implications, industries can lead the way toward achieving sustainability goals. Continued research and innovation will be essential in navigating the complexities of this transition, ultimately creating resilient and sustainable systems that benefit both people and the planet.

References

1. Yang, J., Liu, T., Liu, Y., & Morgan, P. (2022). Review of human-machine interaction towards industry 5.0: Human-centric smart manufacturing. *International Design Engineering Technical Conferences and Computers and Information in Engineering Conference, 86212*, V002T02A060.
2. Tallat, R., Hawbani, A., Wang, X., Al-Dubai, A., Zhao, L., Liu, Z., Min, G., Zomaya, A. Y., & Alsamhi, S. H. (2023). Navigating industry 5.0: A survey of key enabling technologies, trends, challenges, and opportunities. *IEEE Communications Surveys & Tutorials.*
3. Huang, S., Wang, B., Li, X., Zheng, P., Mourtzis, D., & Wang, L. (2022). Industry 5.0 and Society 5.0—comparison, complementation and co-evolution. *Journal of Manufacturing Systems, 64*, 424–428.
4. Nahavandi, S. (2019). Industry 5.0—A human-centric solution. *Sustainability, 11*(16), 4371.
5. George, A. S., & George, A. S. H. (2023). Revolutionizing manufacturing: Exploring the promises and challenges of Industry 5.0. *Partners Universal International Innovation Journal, 1*(2), 22–38.
6. Baig, M. I., & Yadegaridehkordi, E. (2024). Industry 5.0 applications for sustainability: A systematic review and future research directions. *Sustainable Development, 32*(1), 662–681.
7. Bednar, P. M., & Welch, C. (2020). Socio-technical perspectives on smart working: Creating meaningful and sustainable systems. *Information Systems Frontiers, 22*(2), 281–298.
8. Grabowska, S., Saniuk, S., & Gajdzik, B. (2022). Industry 5.0: Improving humanization and sustainability of Industry 4.0. *Scientometrics, 127*(6), 3117–3144.
9. Dacre, N., Yan, J., Frei, R., Al-Mhdawi, M. K. S., & Dong, H. (2024). Advancing sustainable manufacturing: A systematic exploration of Industry 5.0 supply chains for sustainability, human-centricity, and resilience. *Production Planning & Control*, 1–30.
10. Atif, S. (2023). Analysing the alignment between circular economy and industry 4.0 nexus with industry 5.0 era: An integrative systematic literature review. *Sustainable Development, 31*(4), 2155–2175.
11. Ghobakhloo, M., Iranmanesh, M., Mubarak, M. F., Mubarik, M., Rejeb, A., & Nilashi, M. (2022). Identifying industry 5.0 contributions to sustainable development: A strategy roadmap for delivering sustainability values. *Sustainable Production and Consumption, 33*, 716–737.
12. Kasinathan, P., Pugazhendhi, R., Elavarasan, R. M., Ramachandaramurthy, V. K., Ramanathan, V., Subramanian, S., Kumar, S., Nandhagopal, K., Raghavan, R. R. V., & Rangasamy, S. (2022). Realization of sustainable development goals with disruptive technologies by integrating industry 5.0, society 5.0, smart cities and villages. *Sustainability, 14*(22), 15258.
13. Fazal, N., Haleem, A., Bahl, S., Javaid, M., & Nandan, D. (2022). Digital management systems in manufacturing using industry 5.0 technologies. In *Advancement in materials, manufacturing and energy engineering, Vol. II: Select proceedings of ICAMME 2021* (pp. 221–234). Springer.
14. Maddikunta, P. K. R., Pham, Q.-V., Prabadevi, B., Deepa, N., Dev, K., Gadekallu, T. R., Ruby, R., & Liyanage, M. (2022). Industry 5.0: A survey on enabling technologies and potential applications. *Journal of Industrial Information Integration, 26*, 100257.
15. Donthu, N., Kumar, S., Mukherjee, D., Pandey, N., & Lim, W. M. (2021). How to conduct a bibliometric analysis: An overview and guidelines. *Journal of Business Research, 133*, 285–296.

16. Ben Youssef, A., & Mejri, I. (2023). Linking digital technologies to sustainability through industry 5.0: A bibliometric analysis. *Sustainability, 15*(9), 7465.
17. Bazel, M. A., Mohammed, F., Ahmad, M., Baarimah, A. O., & Ibrahim, M. A. (2023c). Blockchain-based healthcare: Trend mapping through bibliometric analysis. In *2023 3rd International Conference on Emerging Smart Technologies and Applications (ESmarTA)* (pp. 1–8).
18. Baarimah, A. O., Alaloul, W. S., Liew, M. S., Al-Aidrous, A. H. M. H., Alawag, A. M., & Musarat, M. A. (2021). Integration of Building Information Modeling (BIM) and value engineering in construction projects: A bibliometric analysis. In *2021 3rd International sustainability and resilience conference: Climate change* (pp. 362–367). https://doi.org/10.1109/IEEECONF53624.2021.9668045
19. Baarimah, A. O., Alaloul, W. S., Liew, M. S., Kartika, W., Al-Sharafi, M. A., Musarat, M. A., Alawag, A. M., & Qureshi, A. H. (2022). A bibliometric analysis and review of building information modelling for post-disaster reconstruction. *Sustainability (Switzerland), 14*(1), 393. https://doi.org/10.3390/su14010393
20. Baarimah, S. O., Baarimah, A. O., Alaloul, W. S., Bazel, M. A., Mohammed, F., & Bawahab, M. (2023). A bibliometric analysis on the applications of artificial intelligence in petroleum engineering. In *2023 4th International Conference on Data Analytics for Business and Industry (ICDABI)* (pp. 152–159).
21. Bazel, M. A., Ahmad, M., & Mohammed, F. (2023a). Blockchain technology in Malaysian context: Bibliometric analysis and systematic review. *International Journal on Advanced Science, Engineering and Information Technology, 13*(5), 1679–1692. https://doi.org/10.18517/ijaseit.13.5.18532
22. Bazel, M. A., Baarimah, A. O., Mohammed, F., Alaloul, W. S., Alawag, A. M., & Alzubi, K. M. (2023b). Blockchain technology in construction industry: Current state and future trends through bibliometric analysis. In *2023 4th International Conference on Data Analytics for Business and Industry (ICDABI)* (pp. 114–123).
23. van Eck, N. J., & Waltman, L. (2010). Software survey: VOSviewer, a computer program for bibliometric mapping. *Scientometrics, 84*(2), 523–538. https://doi.org/10.1007/s11192-009-0146-3
24. Ivanov, D. (2023). The Industry 5.0 framework: Viability-based integration of the resilience, sustainability, and human-centricity perspectives. *International Journal of Production Research, 61*(5), 1683–1695.
25. Akundi, A., Euresti, D., Luna, S., Ankobiah, W., Lopes, A., & Edinbarough, I. (2022). State of Industry 5.0—Analysis and identification of current research trends. *Applied System Innovations, 5*(1), 27.
26. Broo, D. G., Kaynak, O., & Sait, S. M. (2022). Rethinking engineering education at the age of industry 5.0. *Journal of Industrial Information Integration, 25*, 100311.
27. Fraga-Lamas, P., Lopes, S. I., & Fernández-Caramés, T. M. (2021). Green IoT and edge AI as key technological enablers for a sustainable digital transition towards a smart circular economy: An industry 5.0 use case. *Sensors, 21*(17), 5745.
28. Ghobakhloo, M., Fathi, M., Iranmanesh, M., Vilkas, M., Grybauskas, A., & Amran, A. (2024). Generative artificial intelligence in manufacturing: Opportunities for actualizing Industry 5.0 sustainability goals. *Journal of Manufacturing Technology Management, 35*(9), 94–121.
29. Bazel, M. A., Mohammed, F., Baarimah, A. O., Alawi, G., Al-Mekhlafi, A.-B. A., & Almuhaya, B. (2023d). The Era of Industry 5.0: An overview of technologies, applications, and challenges. In *International conference of reliable information and communication technology* (pp. 274–284). Springer Nature Switzerland.
30. Narula, S., Tamvada, J. P., Kumar, A., Puppala, H., & Gupta, N. (2023). Putting digital technologies at the forefront of Industry 5.0 for the implementation of a circular economy in manufacturing industries. *IEEE Transactions on Engineering Management.*

31. Balogh, M., Földvári, A., & Varga, P. (2023). Digital twins in Industry 5.0: Challenges in modeling and communication. In *NOMS 2023-2023 IEEE/IFIP Network Operations and Management Symposium* (pp. 1–6).
32. Zafar, M. H., Langås, E. F., & Sanfilippo, F. (2024). Exploring the synergies between collaborative robotics, digital twins, augmentation, and industry 5.0 for smart manufacturing: A state-of-the-art review. *Robotics and Computer-Integrated Manufacturing, 89*, 102769.
33. Adel, A. (2022). Future of industry 5.0 in society: Human-centric solutions, challenges and prospective research areas. *Journal of Cloud Computing, 11*(1), 40.
34. Lu, Y., Zheng, H., Chand, S., Xia, W., Liu, Z., Xu, X., Wang, L., Qin, Z., & Bao, J. (2022). Outlook on human-centric manufacturing towards Industry 5.0. *Journal of Manufacturing Systems, 62*, 612–627.
35. Saadati, Z., & Barenji, R. V. (2022). Toward Industry 5.0: Cognitive cyber-physical system. In *Industry 4.0: Technologies, applications, and challenges* (pp. 257–268). Springer.
36. Usharani, S., Manju Bala, P., Ananth Kumar, T., Rajmohan, R., & Pavithra, M. (2022). Smart energy management techniques in Industries 5.0. *Hybrid Intelligent Approaches for Smart Energy: Practical Applications*, 225–252.
37. Khan, S. A., Mubarik, M. S., Kusi-Sarpong, S., Gupta, H., Zaman, S. I., & Mubarik, M. (2022a). Blockchain technologies as enablers of supply chain mapping for sustainable supply chains. *Business Strategy and the Environment, 31*(8), 3742–3756. https://doi.org/10.1002/bse.3029
38. Khan, S. A. R., Umar, M., Zia-Ul-haq, H. M., & Yu, Z. (2022b). Technological advancement and circular economy practices in food supply chain. In *Advanced series in management* (Vol. 27, pp. 65–75). Emerald Publishing Limited. https://doi.org/10.1108/S1877-636120220000027005
39. Nandi, S., Sarkis, J., Hervani, A., & Helms, M. (2021). Do blockchain and circular economy practices improve post COVID-19 supply chains? A resource-based and resource dependence perspective. *Industrial Management and Data Systems, 121*(2), 333–363. https://doi.org/10.1108/IMDS-09-2020-0560
40. Khan, S. A. R., Zia-ul-haq, H. M., Umar, M., & Yu, Z. (2021). Digital technology and circular economy practices: An strategy to improve organizational performance. *Business Strategy and Development, 4*(4), 482–490. https://doi.org/10.1002/bsd2.176
41. Sarkar, B. D., Shardeo, V., Dwivedi, A., & Pamucar, D. (2024). Digital transition from industry 4.0 to industry 5.0 in smart manufacturing: A framework for sustainable future. *Technology in Society, 78*, 102649.
42. Sharma, R., & Gupta, H. (2024). Harmonizing sustainability in industry 5.0 era: Transformative strategies for cleaner production and sustainable competitive advantage. *Journal of Cleaner Production, 445*, 141118.
43. Baarimah, A. O., Bazel, M. A., Alaloul, W. S., Alazaiza, M. Y. D., Al-Zghoul, T. M., Almuhaya, B., Khan, A., & Mushtaha, A. W. (2024). Artificial intelligence in wastewater treatment: Research Trends and future perspectives through bibliometric analysis. *Case Studies in Chemical and Environmental Engineering*, 100926.
44. Bezoui, M., Qaisar, S. M., Almaktoom, A. T., & Bounceur, A. (2024). Advancing manufacturing efficiency: Multi-objective optimization in the Industry 5.0 era. *Reality and Science Fiction in Education*.

Navigating the Challenges of Industry 5.0

M. Parimala, M. Vanitha, and Keerthana Umesh

1 Introduction

The transformation from Industry 4.0 to Industry 5.0 makes a prominent paradigm shift in the industry sector, ushering in a new era characterized by integrating advanced technologies and human creativity. Even though Industry 4.0 transformed the industry with automation, optimization, 3D printing, cognitive computing, cyber-physical systems, and other advanced computing, it also had challenges such as complex systems, expensive software, security vulnerabilities, and a shortage of skilled professionals to manage and implement these technologies. Industry 5.0 is built to overcome the challenges in Industry 4.0, focusing on dynamic process models, predictive analytics, creativity, and problem-solving through the technologies artificial intelligence (AI) and big data. The main goal of Industry 5.0 is to reintegrate human knowledge into automated processes, enriching the interaction between humans and robots. This approach aims to maximize the synergy between machine intelligence and human creativity, transitioning from mass customization to personalized production solutions.

This chapter explores the myriad opportunities and challenges that come with Industry 5.0. Among the most prominent opportunities are the enhanced interaction between humans and robots, leading to more innovative and effective production processes. Industry 5.0 also paves way to create new avenues for personalized production of products, enabling companies to meet individual customer needs more precisely. However, these advancements are not without their challenges. One of the primary issues is the lack of technical skills among workers, which hampers the adoption and effective utilization of new technologies. Additionally, implementing customized software that connects industries, people, and robots presents

M. Parimala (✉) · M. Vanitha · K. Umesh
Vellore Institute of Technology, Vellore, Tamil Nadu, India
e-mail: parimala.m@vit.ac.in; mvanitha@vitac.in; Keerthanaumesh.2021@vitstudent.ac.in

Y. Kumar Singla et al. (eds.), *Digital Manufacturing in Industry 5.0*,
https://doi.org/10.1007/978-3-031-91500-0_23

significant hurdles. Ensuring seamless integration and operation of these systems requires overcoming substantial technical and organizational barriers. The subsequent sections will delve deeper into the specific opportunities and challenges posed by Industry 5.0, providing a deeper understanding of this transformative era in manufacturing.

The transition from Industry 4.0 to Industry 5.0 provides a pivotal shift in manufacturing, driven by the need to address the limitations and challenges of the previous era. While Industry 4.0 brought about unprecedented advancements in automation, optimization, and digital integration, it also exposed issues such as system complexity, high costs, security concerns, and a shortage of skilled professionals.

Industry 5.0 emerges as a solution to these challenges by focusing the reintegration of human creativity and problem-solving with advanced computing [1]. It focuses on dynamic and predictive models, coupled with a greater synergy between human workers and robots. By leveraging machine intelligence alongside human ingenuity, Industry 5.0 seeks to enhance customization, personalization, and collaboration in the manufacturing process. This chapter explores the opportunities presented by this new paradigm, as well as the challenges that come with adapting to new technologies and implementing customized software solutions that connect industries, people, and robots [2]. The evolution to Industry 5.0 made a fundamental shift toward a more creative, collaborative and adaptive approach to manufacturing and technology integration.

2 Challenges in Industry 5.0

Various industry forums and blogs [3] provide many insights on Industry 5.0's and its impact on manufacturing. It highlights how the integration of AI, IoT and robotics is revolutionizing manufacturing by creating smart factories [4]. These smart factories utilize the generated data and advanced automation to improve the operational efficiency and flexibility. The India Chapter website [5] discusses about the opportunities and considerations associated with Industry 5.0. The BINUS University article critically assesses the challenges in adopting Industry 5.0 within the manufacturing sector. Although Industry 5.0 offers promising advancements and benefits, its implementation comes with several challenges [6].

Human–Machine Interaction: Industry 5.0 advances human–machine collaboration through more intuitive and user-friendly interfaces, providing smooth communication between humans and machines.

Predictive Maintenance: By leveraging predictive analytics, manufacturers can predict machine failures before they occur, thus minimizing error and extending lifespan of machines. Enhanced data analytics improve inventory management and supply chain efficiency, allowing for quicker adaptation to market changes.

Industry 5.0 opens up new business models that enhance customer experiences and accelerate product development.

Sustainability and Customization: Industry 5.0 supports sustainable practices and enables the scalability of customized products, improving both environmental impact and customer satisfaction. These advancements mark a shift toward more adaptive and socially friendly manufacturing practices. There is a need for integrating sustainability into manufacturing by focusing on efficient resource use of renewable energy sources waste reduction.

Implementation Costs: The initial investment required to build Industry 5.0 technologies can be substantial and it will be posing a financial challenge for many organizations. Effective strategic planning and justification of long-term benefits are necessary to manage these costs.

Data Security and Privacy: With the increased connectivity and data exchange, ensuring robust data security and privacy is critical. Organizations must implement stringent measures to protect sensitive information and comply with regulations.

Integration with Legacy Systems: It is complex and costly to integrate new Industry 5.0 technologies with existing legacy systems. Developing an effective integration strategy and investing in compatible technologies are essential to overcome these challenges.

Global Standardization and Compliance Issues: Emerging trends in global standardization and regulatory compliance are essential for widespread adoption. Aligning with international standards and regulations ensures consistency and facilitates smoother implementation across different regions. Navigating the complex regulatory environment and ensuring adherence to various standards and regulations can be challenging. Businesses must stay informed about regulatory changes and implement best practices to ensure compliance. Collaboration with regulatory bodies and industry associations can aid companies in effectively addressing these compliance issues. Companies must stay informed about regulatory changes and work with industry bodies to ensure adherence.

Shortage of Technical Expertise: One of the primary obstacles to adopting Industry 5.0 is the lack of skilled professionals knowledgeable in advanced technologies. Overcoming this skills gap requires substantial investment in training programs. Companies need to focus on upskilling their workforce to effectively implement and leverage Industry 5.0 technologies. Effective management of data security and privacy is essential to prevent breaches and comply with regulations. There is a high demand in human expert in identifying the security breaches in automation process. There is a significant gap in the availability of skilled professionals adept at working with AI and robotics. Addressing this shortage through education and training is vital. Addressing the skills gap is crucial. It is very important to develop a targeted training programs and educational curricula to equip the workforce with the required expertise for Industry 5.0 technologies.

High Initial Costs: The initial capital needed to implement Industry 5.0 technologies can be significant, which may be a barrier for many organizations. Companies must weigh these upfront costs against the long-term benefits and potential return on investment. Strategic financial planning and resource allocation are crucial to managing these expenses.

Challenges mentioned in this section provide us an insight about the various technical difficulties and practical challenges that are incorporated while implementing Industry 5.0 in the Industry. Advanced technologies required, experts, skilled human resource are the major issues with Industry 5.0. Various challenges and solutions required to overcome these challenges are shown in Table 1.

Table 1 Challenges and corresponding solutions of using Industry 5.0

Challenges	Description	Solutions
Skills shortage	Lack of skilled professionals knowledgeable in advanced technologies like AI, robotics, and big data	Investing in skill development will ensure that the workforce is prepared to leverage advanced technologies effectively
High initial costs	Significant capital investment needed for Industry 5.0 technologies	Strategic financial planning and resource allocation are crucial to managing these expenses
Data security	Increased connectivity and data sharing raise concerns about data security and privacy	Adoption of advanced encryption technologies and real-time monitoring systems to mitigate potential threats
Legacy systems	Integrating new technologies with outdated systems can be complex and costly	Tradeoff between the traditional and new technologies need to be maintained
Regulatory compliance	Navigating regulatory environments and ensuring adherence to various standards and regulations	Cooperation among international organizations, industry groups, and governments to develop unified frameworks and guidelines
Human–machine interaction	Providing smooth communication between humans and machines	Improved human–machine synergy aims to create more productive and harmonious work settings
Predictive maintenance	Predicting machine failures before they occur, thus minimizing error and extending lifespan of machines	Real-time monitoring and predictive maintenance can identify and mitigate potential safety hazards before they become critical issues
Sustainability and customization	Enables the scalability of customized products, improving both environmental impact and customer satisfaction	By focusing on energy efficiency, waste reduction, and sustainable practices, manufacturers can create more environmentally friendly production processes
Implementation costs	The initial investment required to build Industry 5.0 technologies can be substantial and it will be posing a financial challenge for many organizations	Teams from different geographical locations can seamlessly work together, combining their unique perspectives and expertise to solve complex challenges

3 Opportunities and Advantages of Industry 5.0

The evolution of Industry 5.0 brings transformative advantages to the manufacturing industry, enhancing efficiency, fostering innovation, and creating new opportunities. Industry 5.0 heralds a new age in manufacturing, offering numerous opportunities that promise to reshape the industry. These opportunities not only enhance operational efficiency but also pave the way for innovative business models and enriched customer experiences. The key opportunities and advantages are discussed in this section and depicted in Fig. 1.

New Business Models: Industry 5.0 stimulates the development of forward-thinking business models such as service-oriented frameworks and the circular economy. These models prioritize sustainability and resource efficiency by encouraging practices like offering products as services, minimizing waste, and recycling materials. Embracing these practices can open up new revenue streams and improve a company's market standing. The focus on sustainable and resource-efficient practices can enhance a company's reputation and competitive edge.

Enhanced Customer Experience: The fusion of complex technologies and human creativity in Industry 5.0 leads to a more tailored and interactive customer experience. Businesses can more effectively meet the unique needs and preferences of their own customers, resulting in higher levels of customer loyalty and satisfaction. This personalized approach can help companies differentiate themselves in a crowded market. The AppInventiv Blog emphasizes that leveraging human insight alongside technological advancements can significantly elevate the customer experience.

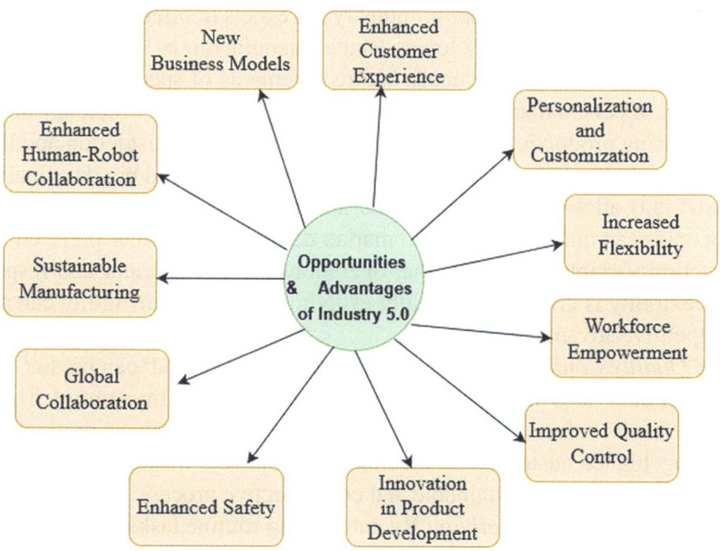

Fig. 1 Opportunities and advantages of Industry 5.0

Innovation in Product Development: Industry 5.0 leverages cutting-edge technologies to accelerate and enhance the product development process. By incorporating AI, big data, and robotics, manufacturers can innovate more efficiently and maintain a competitive advantage. Rapidly developing new products and improving existing ones allows companies to stay ahead of market trends and consumer demands. The India Chapter on Industry 5.0 underscores the importance of utilizing these technologies to foster innovation and stay competitive.

Global Collaboration: Industry 5.0 facilitates better collaboration across global teams by tapping into diverse skills and knowledge bases. This global cooperation can lead to more innovative solutions and superior outcomes. Teams from different geographical locations can seamlessly work together, combining their unique perspectives and expertise to solve complex challenges. The AppInventiv Blog highlights how global collaboration can drive innovation and improve problem-solving capabilities.

Sustainable Manufacturing: Industry 5.0 promotes sustainable manufacturing by integrating advanced technologies that optimize resource use and reduce *environmental* impact. By focusing on energy efficiency, waste reduction, and sustainable practices, manufacturers can create more environmentally friendly production processes. Sustainable manufacturing practices are increasingly important in today's environmentally conscious market [7].

Enhanced Human-Robot Collaboration: A cornerstone of Industry 5.0 is the enhanced collaboration between human creativity and robotic precision. This synergistic relationship allows human workers will focus on creative tasks while robots handle repetitive and precise operations. This will not only increases productivity but also provides harmonious and efficient work environment. This improved collaboration opens up innovative solutions to complex challenges.

Personalization and Customization: Industry 5.0 excels in offering mass customization, where advanced technologies enable manufacturers to produce personalized products that fulfill the needs and requirements of specific customers. This high level of personalization is achieved without sacrificing efficiency or significantly raising production costs, thus boosting customer satisfaction and loyalty.

Increased Flexibility: The use of artificial intelligence (AI) and big data analytics in Industry 5.0 allows for more flexible and dynamic manufacturing processes. Companies can quickly respond to market demands, consumer preferences, and production variables, ensuring that operations remain efficient and responsive. This flexibility is critical in a rapidly evolving market where agility can provide a competitive advantage.

Improved Quality Control: AI-powered systems in Industry 5.0 can predict and *prevent* potential issues through predictive measures and quality control. This capability reduces defects, minimizes downtime, and improves product quality achieving higher customer satisfaction and reduced waste. Enhanced quality control also leads to a sustainable and cost-effective process.

Boosted Innovation and Creativity: By automating routine tasks, Industry 5.0 frees human workers to participate in more creative and strategic activities. It gives job satisfaction and drives innovation within organizations. The *combination* of human ingenuity and machine efficiency can lead to groundbreaking solutions and products, expanding the possibilities within manufacturing.

Workforce Empowerment: Industry 5.0 aims to empower the workforce by providing advanced tools and technologies. This empowerment leads to a more engaged and skilled workforce capable of handling more complex and rewarding tasks. Continuous learning and development ensure that employees stay at the forefront of technological advancements.

Enhanced Safety: Adapting advanced technologies in Industry 5.0 enhances workplace safety. Automation and AI can manage hazardous tasks, reducing the risk of accidents and injuries. Additionally, real-time monitoring and predictive maintenance can identify and mitigate potential safety hazards before they become critical issues.

To summarize, there are many opportunities in building new business models with standards and guidelines with enhanced safety, creativity, personalization, customization exists in the market. Workforce empowerment, improved quality control, and increased flexibility would provide a smooth human–robot collaboration which will help in product development satisfying the sustainable goals. Effective customer experience and global collaboration are key aspects in achieving Industry 5.0.

4 Case Study to Implement Industry 5.0

Leveraging real-time data and advanced automation, smart factories, and education [8] enhances productivity and flexibility that leads to a seamless and responsive manufacturing process [9, 10]. These technologies introduce a new level of operational excellence. Predictive analytics in Industry 5.0 allows for anticipating machine failures before they occur, providing timely maintenance and reducing downtime. This proactive approach enhances productivity extends the lifespan of equipment and reduces the cost of maintaining the machines. Predictive maintenance ensures minimal disruptions and improves overall efficiency. The significant impact of predictive maintenance on operational reliability is emphasized currently in industry. Industry 5.0 leverages AI and robotics to enable large-scale production of customized products. This capability allows manufacturers to meet customer requirements without compromising cost and efficiency. Customized production enhances customer satisfaction by delivering products that precisely match individual preferences. The India Chapter on Industry 5.0 highlights how this approach combines the benefits of mass production with personalized customization. Human–robot collaboration enhances productivity, reduces human error, and fosters a more dynamic and innovative manufacturing environment. Human–robot synergy enables manufacturers to tackle complex challenges more effectively and efficiently.

Assume a XYZ Manufacturing company, a leading producer of precision-engineered components, embraced Industry 4.0 early on, leveraging automation, IoT, and advanced data analytics to streamline its operations. However, the company soon encountered several challenges, including the complexity of integrating new technologies, high costs associated with software upgrades, and a lack of skilled personnel to manage these sophisticated systems. Recognizing the limitations of Industry 4.0, XYZ Manufacturing began exploring Industry 5.0 principles, aiming to enhance human–robot collaboration, foster creativity, and implement more personalized manufacturing processes.

Before transitioning to Industry 5.0, XYZ Manufacturing was heavily reliant on automated systems, which significantly improved efficiency but led to a decline in workforce engagement. The company's management noticed a growing disconnect between human workers and the automated systems, resulting in a lack of innovation and adaptability. Additionally, the cybersecurity threats posed by interconnected systems and the rising costs of maintaining cutting-edge software were becoming increasingly burdensome. XYZ Manufacturing decided to pivot toward Industry 5.0, focusing on integrating human creativity with machine intelligence. Several key initiatives involved are discussed below and shown in Fig. 2.

1. Collaborative Robots (Cobots): The company introduced cobots that work alongside human operators, enhancing productivity while allowing workers to engage in more creative and problem-solving tasks. These cobots were designed to be user-friendly, with intuitive interfaces that required minimal technical expertise, addressing the skills gap that was a major issue in Industry 4.0.
2. Dynamic Process Models: XYZ Manufacturing implemented AI-driven predictive models that allowed for more dynamic and flexible production processes.

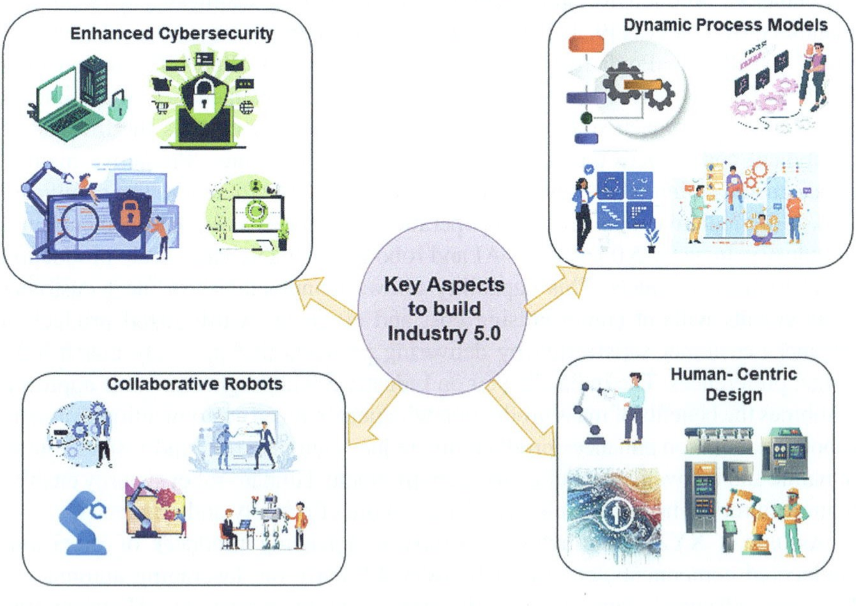

Fig. 2 Key aspects to build a company with Industry 5.0

These models enabled the company to customize products according to customer specifications, a key aspect of Industry 5.0.
3. Human-Centric Design: The company restructured its production lines to prioritize worker well-being, incorporating ergonomic designs and fostering an environment where human input was valued alongside machine output. This shift not only improved employee satisfaction but also led to a surge in innovative ideas, as workers were now more engaged and involved in the production process.
4. Enhanced Cybersecurity: Recognizing the increased interaction between humans, robots, XYZ Manufacturing invested in robust security measures to protect sensitive data. The company adopted advanced encryption technologies and real-time monitoring systems to mitigate potential threats.

The transition to Industry 5.0 brought several significant benefits to XYZ Manufacturing:

- Increased Innovation: By encouraging human–machine collaboration, the company saw a marked increase in innovative solutions and product designs. Workers were more motivated, as their creative input was recognized and valued.
- Cost Efficiency: The use of cobots and AI-driven models allowed XYZ Manufacturing to reduce costs associated with errors and downtime, as well as streamline customization processes, leading to higher customer satisfaction.
- Improved Employee Retention: The human-centric approach led to higher employee satisfaction, reducing turnover rates and fostering a more committed and skilled workforce.
- Enhanced Adaptability: The dynamic process models enabled the company to quickly adapt to changing market preferences, maintaining a competitive edge in the industry.

Despite the successes, XYZ Manufacturing also encountered challenges during its transition to Industry 5.0:

- Technical Skill Gaps: Although the company invested in training programs, there was still a learning curve associated with the new technologies. Some workers found it challenging to adapt to the more collaborative roles that Industry 5.0 demanded.
- Implementation Costs: While long-term cost savings were realized, the initial investment in cobots, AI systems, and cybersecurity infrastructure was substantial, requiring careful financial planning and resource allocation.
- Integration Issues: The integration of new technologies with existing systems proved to be complex, necessitating a phased approach to avoid disruptions in production.

5 Future Directions

As Industry 5.0 evolves, several key trends are expected to significantly shape [11] its trajectory in the manufacturing sector. Here's a look at these anticipated developments:

Enhanced Human–Machine Interaction: Future progress will likely focus on refining the interaction between humans and machines. This will involve designing more intuitive and accessible interfaces, making it easier for humans to collaborate with robots and AI systems. Improved human–machine synergy aims to create more productive and harmonious work settings.

Advancements in AI and Machine Learning: Rapid growth of technologies set to revolutionize areas such as predictive maintenance, quality control, and bespoke production [12]. These technologies will become more sophisticated and seamlessly integrated into manufacturing workflows. Enhanced AI and machine learning capabilities will allow manufacturers to utilize data more effectively, thereby improving overall performance and efficiency.

Emphasis on Sustainability: Future efforts will prioritize sustainability, focusing on reducing environmental impact through more efficient resource usage, waste minimization, and increased reliance on renewable energy. Developing innovative technologies and practices to support these sustainability goals will be crucial [13]. Companies will need to adopt these practices to comply with regulatory requirements and align with consumer expectations, as discussed in the India Chapter on Industry 5.0.

Skill Development and Training: Addressing the skills gap will be essential for the successful deployment of Industry 5.0. Collaborative efforts among industry, educational institutions, and government bodies will be necessary to create effective training programs and curricula. Investing in skill development will ensure that the workforce is prepared to leverage advanced technologies effectively.

Global Standardization: The establishment of global standards and best practices for Industry 5.0 will help streamline its implementation and address regulatory challenges. This will involve cooperation among international organizations, industry groups, and governments to develop unified frameworks and guidelines. Standardization will facilitate and promote international collaboration.

6 Conclusion

Industry 4.0 has complexities and challenges including high costs, security issues, and a shortage of skilled professionals. Industry 5.0 addresses these issues by emphasizing a smooth integration of human creativity and advanced technologies. Industry 5.0 aims to enhance the synergy between humans and robots, fostering a collaborative environment where machine intelligence and human ingenuity work in tandem. This chapter discusses on the need for Industry 5.0, challenges, opportunities, and the solution to overcome these challenges. Integrating two technologies not only addresses the shortcomings of Industry 4.0 but also paves a way for mass customization and personalization, ultimately driving greater efficiency and innovation in manufacturing processes. Recent research highlights that this transition brings about significant opportunities for creating more personalized and flexible manufacturing processes. However, it also underscores the need to address the challenges related to integrating new technologies.

References

1. Tyagi, A. K., Lakshmi Priya, R., Mishra, A. K., & Balamurugan, G. (2023). Industry 5.0: Potentials, issues, opportunities, and challenges for Society 5.0. *Privacy Preservation of Genomic and Medical Data*, 409–432.
2. Zhang, C., Wang, Z., Zhou, G., Chang, F., Ma, D., Jing, Y., et al. (2023). Towards new-generation human-centric smart manufacturing in Industry 5.0: A systematic review. *Advanced Engineering Informatics, 57*, 102121.
3. https://appinventiv.com/blog/industry-5-0-manufacturing/
4. Joglekar, S., Kadam, S., & Dharmadhikari, S. (2023). Industry 5.0: Analysis, applications and prognosis. *The Online Journal of Distance Education and e-Learning, 11*(1), 1.
5. https://indiachapter.in/index.php?/user/article/2/2/121
6. Hassan, M. A., Zardari, S., Farooq, M. U., Alansari, M. M., & Nagro, S. A. (2024). Systematic analysis of risks in Industry 5.0 architecture. *Applied Sciences, 14*(4), 1466.
7. Paschek, D., Luminosu, C. T., & Ocakci, E. (2022). Industry 5.0 challenges and perspectives for manufacturing systems in the society 5.0. In *Sustainability and innovation in manufacturing enterprises: Indicators, models and assessment for Industry 5.0* (pp. 17–63).
8. Al-Emran, M., & Al-Sharafi, M. A. (2022). Revolutionizing education with industry 5.0: Challenges and future research agendas. *International Journal of Information Technology, 6*(3), 1–5.
9. George, A. S., & George, A. H. (2023). Revolutionizing manufacturing: Exploring the promises and challenges of Industry 5.0. *Partners Universal International Innovation Journal, 1*(2), 22–38.
10. Gladysz, B., Tran, T. A., Romero, D., van Erp, T., Abonyi, J., & Ruppert, T. (2023). Current development on the Operator 4.0 and transition towards the Operator 5.0: A systematic literature review in light of Industry 5.0. *Journal of Manufacturing Systems, 70*, 160–185.
11. Adel, A. (2022). Future of industry 5.0 in society: Human-centric solutions, challenges and prospective research areas. *Journal of Cloud Computing, 11*(1), 40.
12. Maddikunta, P. K. R., Pham, Q. V., Prabadevi, B., Deepa, N., Dev, K., Gadekallu, T. R., et al. (2022). Industry 5.0: A survey on enabling technologies and potential applications. *Journal of Industrial Information Integration, 26*, 100257.
13. Narkhede, G., Pasi, B., Rajhans, N., & Kulkarni, A. (2023). Industry 5.0 and the future of sustainable manufacturing: A systematic literature review. *Business Strategy & Development, 6*(4), 704–723.

Sustainability and Performance Evaluation of Concrete Reinforced with Fly Ash and Silica Fume: A Combined LCA and TEA Approach with Industry 5.0

Bondala Ramakrishna, Donthamsetty Suneel, Abiodun Oluwatosin Adeoye, Nuhu Ahmad, Samson Abiodun AASA, and Patience Agnes EBIGHE

1 Introduction

Concrete is a composite construction material made up of cement, coarse aggregates, sand, and water that is used in many civil construction and engineering works. Due to the rapid growth of urbanization across the globe today, there is an ever-increasing demand for concrete, exerting considerable pressure on the raw materials as well as posing serious environmental challenges. It is estimated that cement production is responsible for about 8% of global carbon dioxide emissions and the concrete industry is one of the key contributors to the climate emergency. Such a backdrop has encouraged both academics and industries to search for eco-friendly alternatives to standard concrete constituents, specifically targeting the use of industrial waste to improve the environmental and mechanical properties of concrete. Out of all the alternative options investigated, fly ash and silica fume have been found to be highly effective. These materials are waste products, produced in large volumes,

B. Ramakrishna (✉)
Aditya Institute of Technology and Management, Tekkali, Andhra Pradesh, India

D. Suneel
Narasaraopeta Engineering College, Narasaraopeta, Andhra Pradesh, India
e-mail: viceprincipal@nrtec.in

A. O. Adeoye
Department of Chemistry, Federal University Oye Ekiti, Oye-Ekiti, Ekiti State, Nigeria

N. Ahmad
Department of Civil Engineering, Bayero University, Kano, Kano State, Nigeria

S. A. AASA
Department of Automotive Engineering, University of Ibadan, Ibadan, Oyo State, Nigeria

P. A. EBIGHE
Department of Chemistry, Toulouse INP-ENSIACET, Toulouse, France

Y. Kumar Singla et al. (eds.), *Digital Manufacturing in Industry 5.0*,
https://doi.org/10.1007/978-3-031-91500-0_24

and therefore, it is possible to reduce the amount of new cement used and increase the durability and operational characteristics of concrete.

Fly ash generated from the combustion of coal in power stations consists predominantly of small spherical particles which are made of silica and alumina which were formed as a glass. Its reaction with cement in concrete is that it reacts with calcium hydroxide to produce more calcium silicate hydrate which makes the concrete dense and stronger. Silica fume is produced in electric arc furnaces on creating silicon metal/ferrosilicon alloys. This is because silica fume is composed of particles that are extremely fine, spherical in shape and high in silica, thus making the concrete matrix stronger through helping in better packing of particles. Aside from those benefits, fly ash and silica fume improve the chemical and drastic environmental conditions resistance of concrete. Use of these materials is not just an environmentally safe option, but rather allowing for higher performance of the concrete.

The last decade has seen a shift in how professionals approach sustainable construction, no longer focusing almost entirely on the search for alternative materials but seeking to integrate environmental and economic factors which include the use of life cycle assessment (LCA) and techno-economic assessment (TEA) for new material assessment. LCA describes how a material is made, what is its impact on the environment, and what will happen to it after its purposeful use: whether it will be thrown away or transformed into a new product. Such perspective enables better assessment of the environmental impacts associated with each stage of the material lifecycle and allows for the identification of measures aimed at achieving reductions in carbon emissions and other sources of environmental degradation. For example, in the case of concrete reinforced by fly ash and silica fume, LCA can show that the glasshouse gas emissions are lowered owing to the decrease in cement usage and energy-saving potential in the employment of waste materials.

A techno-economic assessment (TEA) expands on the LCA by adding the economic dimension to the materials employed. In this regard, TEA suggests the costs that are needed for sourcing, processing, and using materials, which is their economic viability. In the case of sustainable concrete substitutes, a good TEA analysis would show that the extra cost of introducing fly ash and silica fume is less than the savings accrued from the lower cement use and lessened waste disposal expenditures. As two complementary studies, LCA and TEA examine environmental protection and development of alternative materials simultaneously, thus broadening the scope of sustainable construction science and technologies.

Adding fly ash and silica fume to concrete is not a new idea. Their positive characteristics have been known for the last few decades. Still, the scope of the research studies presented in the literature is often limited to studying the mechanical or durability characteristics of the materials without evaluating their broader environmental and economic implications. This chapter aims to fill this void by incorporating LCA and TEA into the assessment of concrete containing fly ash and silica fume. Therefore, assessing both the safety and the performance parameters, this work seeks to enhance understanding of the advantages and constraints of using the said materials as partial cement replacements. Such an integrated approach is particularly necessary at present when most construction practices are becoming more and more restricted due to environment concerns and business factors become very important for general implementation.

The use of fly ash and silica fume in concrete can solve a number of problems which arises due to the use of conventional cement. Taking silica fume as an example, it improves the mechanical properties of the concrete, especially compressive and tensile strength, thanks to its fine particle size and its pozzolanic activity permitting denser, more unified mix production. Additionally, fly ash, because of its lower heat of hydration, decreases the chances of thermal fracture in massive concrete structures, thus making it appropriate for buildings like dams as well as bridges. Concrete with both fly ash and silica fume also has greater resistance chloride and sulfate attacks, thus making it suit marine structures as well as areas where de-icing salts are used. The lifetime of concrete structures can be significantly increased by such significant durability enhancements, thereby minimizing the number of repairs and replacements necessary which results in reduced environmental damage across the life of the structure.

The present study has as one of its main goals the finding of the appropriate proportions of replacement materials such as fly ash or silica fume in concrete so as to enhance the strength and the durability of the concrete at the least cost and in an environmentally friendly manner. The research is centered around mix designs containing varying percentages of fly ash and silica fume and their impact on compressive strength, tensile strength, as well as flexural strength of concrete. Furthermore, the study evaluates the level of durability of concrete structures caused by chemical actions that may be encountered in practice and simulates such situations. As a result, this study offers suitable recommendations to the construction industry in terms of specific ratios of the mix proportions that yield optimum performance against the level of performance desired.

2 Literature Review

2.1 Enhancing Wear Resistance and Mechanical Properties of Composite Materials and Alloys

Recent research has emphasized the importance of improving wear resistance and mechanical performance in the context of Industry 5.0. These studies explore a range of advanced techniques and sustainable practices that align with the goals of smart manufacturing and Industry 5.0. Adeoye et al. (2024) investigate the valorization of lignocellulosic waste, highlighting the potential of bio-oil production through pyrolysis and ultrasonication for biofuel generation [1]. Further research by Adeoye et al. focuses on upgrading crude bio-oil [2, 3] and sustainable energy production from biomass [4], providing insights into thermochemical and biochemical conversions. These approaches underscore the integration of clean energy solutions into digital manufacturing processes [5–10]. Studies on catalytic pyrolysis and nanomaterials for enhanced biomass conversion offer pathways for optimizing manufacturing efficiency [11–13]. Additionally, research on pyrolysis of plastic waste and rail projects introduces circular economy concepts and environmental sustainability, both essential to the evolution of Industry 5.0 [14–22]. Recent advancements in sustainable fuel production have focused on various bio-oil sources and carbon capture techniques. Studies have examined bio-oil production from thermal cracking in the presence of hydrogen [23], CO_2 adsorption systems in coal beds [25], and the pyrolysis of coffee and eucalyptus sawdust [25–27]. Additional research has explored the use of sulfur to enhance wood tar biopitch as a sustainable binder, the characterization of sewage sludge bio-oil, and the analysis of bio-oil from algae, comparing freshwater and microbial-treated samples [28]. These efforts collectively contribute to the development of renewable energy and environmental sustainability [14, 29–32]. In parallel, work on fiber surface functionalization and dye fastness adds to the understanding of material innovation in manufacturing [33–35]. These studies contribute to the body of knowledge driving digital transformation and environmental consciousness in manufacturing [14, 16, 18–20, 32, 36, 37].

2.2 *Optimizing Crashworthiness and Structural Integrity*

Crashworthiness remains a key focus in automotive and rail systems research. These works emphasize a variety of advanced engineering techniques that contribute to the development of smart manufacturing and Industry 5.0. Studies like Shah et al. [38] and Singla et al. [39] examine the effects of material science advancements on welding and sustainability, key components for industrial automation and optimized production processes. Additionally, research on solar energy systems and tribological studies [40–46] align with the sustainable energy goals of Industry 5.0.

Recent advancements in bio-oil production and carbon capture techniques are driving sustainability in energy production [47]. Studies have explored bio-oil production through thermal cracking in the presence of hydrogen, and the pyrolysis of various biomass sources such as coffee, eucalyptus sawdust, and algae, all contributing to renewable energy solutions [48–53]. Additionally, research on CO_2 adsorption systems and the development of sustainable materials for adsorption, like zeolitic materials and montmorillonite clay, are crucial for addressing environmental challenges [48, 52, 54–58]. These efforts are complemented by investigations into biodiesel production, aviation fuel alternatives, and the separation of valuable chemicals from bio-oils, advancing the potential of biofuels in various industrial applications [59, 60].

2.3 Thermal and Corrosion Performance in Extreme Environments

The contributions from Sunil et al. [61–70] span critical advancements in transportation systems, materials science, and engineering technology, all of which have applications in smart manufacturing and Industry 5.0. Recent works from 2024 focus on the adoption of electric vehicles [71–88], modeling of intelligent transportation systems [89–102], and vibration mitigation in marine and rail propulsion systems [103–112].

These works highlight innovative methods in microbiology, environmental sustainability, and materials science, which have significant implications for smart manufacturing and Industry 5.0 [5–10]. Mohapatra et al. [113] study microbial processes for biopolymer production, which can be applied in sustainable packaging solutions [7, 14, 114–121] and bio-based materials [122–133], aligning with Industry 5.0's focus on circular economy and smart materials. Research on retrofitting buildings for energy efficiency [134–150]and the use of polyhydroxyalkanoates (PHAs) for biocompatible applications [142, 151–159] demonstrates the integration of biotechnology into industrial processes.

2.4 Advanced Modeling and Simulation for Transportation Systems

The integration of alloying elements to enhance wear resistance [160–162] supports the development of materials with improved durability for high-tech manufacturing applications. Moreover, works on the surface integrity of machined materials and the impact of nano-particulates in lubrication show how material engineering can improve efficiency in production processes. These studies collectively enhance the scope of smart manufacturing by combining material science, energy efficiency, and automation, advancing the transition toward Industry 5.0.

2.5 Sustainable Manufacturing in Railways and Automotive Industries

Bhardawaj et al. (2023) underscore the need for energy-efficient solutions and advanced control systems, both of which are integral to Industry 5.0 [163, 164]. Previous works from 2023 also explore sustainable materials for transportation [165], intelligent systems for railway infrastructure[166–182], and advancements in modeling and ride behavior [183–191]. Research on crashworthiness [192–201] and the integration of bio-lubricants [202–205] promotes the goals of Industry 5.0 by improving safety [206], efficiency, and sustainability [207–209]. These studies collectively support the evolution of smart manufacturing, contributing to Industry 5.0's emphasis on sustainability, digitalization, and innovation across sectors [[210].

The research contributions focus on the dynamic analysis, vibration isolation, and advanced material applications in transportation and vehicle systems, which directly support the objectives of smart manufacturing and Industry 5.0 [211–215]. Recent works (2024) addresses advanced control systems for enhanced ride quality in passenger rail vehicles using intelligent control suspension systems [170, 216–220], and comprehensive dynamic response analyses of road and rail vehicles [221–230]. Further, studies involving the use of composite materials and hybrid AHP-TOPSIS techniques for sustainable structural applications [79, 85, 88, 231–242] contribute to the development of smart materials, which is central to Industry 5.0.

2.6 Intelligent Control Systems and Vibration Reduction in Transport

Advancements in digital twin technology for improving logistics [243], the role of extremophiles in enhancing sustainable agriculture [244, 245], and bioremediation of contaminated soils [118, 246] highlight the role of digital innovation and environmental consciousness in next-generation manufacturing. These studies offer diverse insights into how microbiological advancements and environmental sustainability are being harnessed for smart and sustainable industrial applications, resonating with the principles of Industry 5.0 [181, 247].

2.7 Emerging Technologies in Sustainable Transportation Systems

Advancements in transportation technologies and studies from earlier years explore dynamic simulations [87, 248], improving suspension systems and enhancing material wear resistance, which collectively reflect advancements in digitalization, automation, and sustainable manufacturing practices. These studies highlight a multidisciplinary approach to improving vehicle systems and material performance in line with the goals of Industry 5.0.

2.8 Silica Fume

SFC consists of silica fume, cement, fine and coarse aggregates, and water. Fresh and cured form of SFC is better in properties when compared to conventional concrete. For instance, its flexural and compressive strengths are more. This category of concrete is also more durable than the regular concrete. It is also more resistant to chemical attacks, freeze/thaw cycles, and frost than concrete without the use of silicates fume. Also, it is a more cohesive mix and SFC exhibits lower bleeding and segregation as compared to normal concrete. The potential uses for SFC include shotcrete for repair works, nuclear waste containment, parking garage SFC, dam SFC, and high buildings SFC [142, 155–171, 176, 177, 179–186, 189, 190].

Fly ash is the name given to one of the diverse by-product materials formed during the combustion of coal in power stations. It is light gray in color and consists of spherical and glassy fragments which are carried upward with the flue gases. In this study [6], the authors provided an experimental program to investigate the strength and durability characteristics of high self-compacting concrete (HSSCC) containing steel fibers as well as the concrete without fibers. To satisfy the power demand, mineral blends like fly ash, Alccofine, and crimped steel fibers were applied. Chemical admixture was used to increase workability. Steel fibers added to plain and fibrous HSSCC were found to enhance its tensile strength, durability and impermeability. Nondestructive tests (NDT) results on concrete performed corroborated very well with those of the more rigorous tests conducted. The findings of the acid and alkaline resistance test show that the concrete specimen without fibers lost more strength and weight than the concrete specimen with fibers. The RCPT, water absorption tests, and acid solution exhaust advocated destruction of the concrete's strong matrix and hence very low porosity.

The compressive, flexural, and split tensile strength values are enhanced with the incorporation of 10% silica fume and the partial replacement of 30% cement with Ground Granulated Blast Furnace Slag (GGBS). In case, Glenium B233 is used in the dosage of 0.3%, concrete mixes have better workability and are mixed more uniformly. It performs effectively in reducing the water ratio. For HPC3, or 10% silica fume and 30% GGBS, the best split tensile strength of 4.4 MPa, maximum compressive strength of 70.66 MPa, and maximum flexural strength of 7.8 MPa for grade M60 were achieved. The GGBS and silica fume content of HPC 3, i.e., 30% GGBS and 10% silica fume, however, have provided superior results with respect to the durability against acidic action than the other trail mixes.

Fly ash and silica fume are currently being studied for their optimal benefits in their combination to replace cement as GGBS and silica fume when used together brought such highest strength [8]. This is because of the fact that the unreacted particles of fly ash binder have higher modulus of elasticity than the hydration products of the hydrated binder. It was also established that 45% fly ash added to the mixture of concrete enhanced the displacement of the crack tip creating ability [9]. Because of their size characteristics—that is because of their fineness—filler particles are important in the enhancement of engineering properties as they form space between cement and coarse aggregate. Silica fume is used in cement in construction because it is high in amorphous silica and it is active pozzolanic which means it works quicker than common pozzolans [10]. Using X-ray diffraction, researchers were able to identify the influences of silica fume and fly ash addition to C3A cement's sawdust. It was established from scientific inquiry that using fly ash and silica fume in samples resulted in a split of the diffraction peaks as well as a larger aluminosilicates Silica to aluminum ratio. Other names for aluminosilicates, such as gehlenite and laumontite, were formed in samples with 50% fly ash and 20%, 0%, and 50% silica fume [11].

3 Materials and Methods Used in the Experimentation

The materials and methods related to the preparation, mixing, casting, curing, and testing procedures which were used to study the mechanical properties and durability of concrete with different percentages of fly ash and silica fume are outlined. According to the setup, IS: 10262-2009 guidelines have been adopted, which makes it easier to evaluate the environmental suitability and performance of every mix design. Also, life cycle assessment (LCA) and techno-economic assessment (TEA) models are used to analyze ecological and economical dimensions of incorporating these by-products into concrete.

3.1 Materials

The materials used in the course of this research work are cement, coarse aggregates, fine aggregates, water, fly ash and silica fume as each of them is critical in achieving the expected performance properties of the concrete mix.

- Cement: The principle binder used is ordinary Portland cement (OPC), 43 Grade, which serves as a base for concrete mixtures.
- Fly Ash: Fly ash, obtained from thermal power plants is an industrial waste material with high silica and alumina content which has pozzolanic effect that increases the density and strength of concrete.
- Silica Fume: A by-product of the production of silicon metal or ferrosilicon alloy, silica fume is composed of very fine amorphous silica particles that result in enhanced strength, lower penetration and increased resistance to the deterioration of concrete.
- Aggregates: Adhering to the IS: 383-1970 specifications, coarse aggregates of size 20 mm and fine aggregates of a mix have been adopted so that gradation and strength requirements are met.
- Water: For mixing and curing, potable water is utilized which is conforming to IS: 456-2000.

3.2 Mix Proportions and Preparation

The mix designs include some fly ash and some silica fume substitution in order to achieve maximum strength and more sustainable concrete. For M30 grade design, proportions of fine aggregate, cement and coarse aggregate are altered to 1:1.960:3.720 with 0.42 water-to-cement ratio. To enhance workability, a superplasticizer (MYK PC-20) is incorporated in proportion of 1.77 L/100 kg cement. Table 1 presents the experimental mix proportions used in the study, with varying percentages of silica fume (SF) and fly ash (FA).

3.3 Mixing, Casting, and Curing

Mixing: When making each concrete batch, cement containing fly ash and silica fume is blended first by itself to allow for uniform mixing and only then are adopted. The silica fume was blended into the cement matrix at the levels of 0, 0.02, 0.04, 0.06, 0.08, and 0.10, and the fly ash was incorporated at 0, 0.10, 0.15, 0.25, and 0.35 replacement amounts to study the effect of different levels of replacement on the behavior of concrete. Prior test results will show that the optimum partial replacement of silica fume is 6% due to the relative compressive strength and durability of the concrete.

Casting: For the measures of compressive strength, 24 cube specimens (150 mm × 150 mm × 150 mm) are prepared. For split tensile strength testing, three cylindrical specimens (150 mm diameter × 300 mm height), while for flexural strength two large beam specimens (2200 mm × 150 mm × 300 mm) are completed due to these different sizes and shapes of the specimens it is possible to carry out testing of the concrete properties under various structural states.

Curing: All cast specimens are cured in tanks containing water in a controlled environment so that a uniform amount of water is available and strength develops properly. Cells and cylinders are cured for 7, 28, and 90 days to check for variation in the mechanical properties with time while beam specimens have a curing period of 28 days.

Table 1 Experimental mix proportions

Sample no	Silica fume (SF)	Fly ash (FA)
M1	0%	0%
M2	5%	5%
M3	5%	10%
M4	5%	15%
M5	6%	20%
M6	6%	25%
M7	6%	30%
M8	6%	35%

3.4 Testing Procedures

Several tests are carried out aimed at determining the mechanical properties and the durability of a given concrete mix all these tests include:

- Compressive Strength Test: Performed on 150 mm cube specimens compressed by a testing machine located in the compression test area. It is measuring force at intervals of 7, 28, and 90 days during curing for the purpose of observing the development of strength.
- Split Tensile Strength Test: In this case, the material is subjected to split tensile strength testing using Alion, where the cylinder is placed diametrically on the stroke of the tester to measure its tensile properties.
- Flexural Strength Testing: Flexural testing is conducted on the beam specimen, which is the most important criterion in cases of structures under bending stresses.
- Durability Tests: Durability is established through sulfate and acid attack as well as permeability tests like the Rapid Chloride Penetration Test (RCPT) and water absorption to evaluate the durability of concrete against extreme weather conditions.

3.5 Life Cycle Assessment (LCA) Methodology

A life cycle assessment (LCA) is used to measure the environmental footprint of a given mix with reference to the ISO 14040/44 standards. The scope of the LCA includes all phases from bauxite ores to construction and eventual demolition listing relevant impact indicators such as carbon footprint, energy and water usage.

3.6 Techno-Economic Assessment (TEA) Methodology

TEA assesses the market potential of each concrete mix in terms of tangible inputs (materials, production costs) and intangible outputs (less cement and waste disposal required). TEA adds weight to the LCA, making it possible to properly consider the environmental and economic effect of every mix.

This experimental setup allows for the in-depth numerical consideration of the concrete mixes containing fly ash as well as silica fume as regards their performance, durability, environmental, and economic considerations. This study, in particular, integrates standardized testing, LCA, and TEA providing short on the benefits and disadvantages of incorporating these by-products into sustainable concrete production. Section.

4 Results and Analysis

The results of compressive, tensile, and flexural strength tests carried on concrete samples with different proportions of silica fumes and fly ash. Furthermore, in order to give a clear picture of the mechanical performance against sustainability of every mix, durability tests—resistance to aggressive chemicals and thermal stability—are also outlined.

4.1 Mechanical Properties of Reinforced Concrete

4.1.1 Compressive Strength Analysis

The concrete specimens in the amount of four were subjected to the testing of the compressive strength of concrete at intervals of 7, 28, 56, and 90 days curing to ascertain the development of the strength with time. Results indicate– that Mix 5, which consisted 6% of 6% silica fume and 20% fly ash, achieved the highest compressive strength. This suggests that it is the best combination in terms of strength and workability. Figure 1 shows the compressive strength between silica fumes and cement aging cements after 7 and 28 days. Figure 2 shows compressive strength (N/mm^2) of concrete in the spaces for the different mixes. Figure 1 shows the progress of direct screening for the compressive strength of cured silica fume cement at 7 and 28 days tenos trimmed cement is between of 6% fume to replace silica fume.

Figure 2 displays the understanding of the compressive strength that most scholars sustain for the Hispanic Chambers some around M5 who achieve the most strength at each of the intervals. Table 2 shows the compressive strength values for the concrete mixes over different curing periods. Mix M5, containing 6% silica

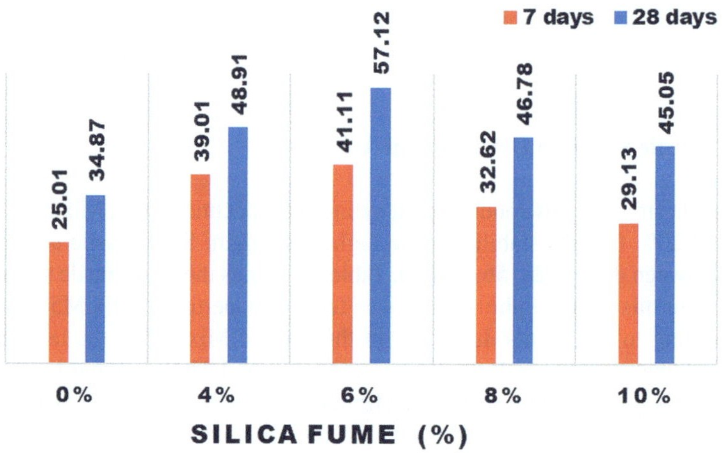

Fig. 1 Compressive strength between silica fume and cement after 7 and 28 days

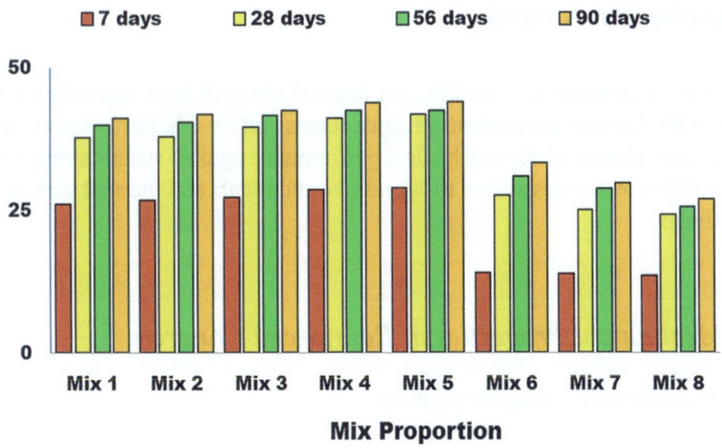

Fig. 2 Compressive strength in (N/mm^2) for the various mixes

Table 2 Compressive strength of concrete mixes (N/mm^2)

Mix ID	7 Days	28 Days	56 Days	90 Days
M1 (Control)	26.07	37.87	40.56	41.21
M5 (6% SF, 20% FA)	28.98	41.84	42.61	44.04

fume and 20% fly ash, consistently shows higher compressive strength compared to the control mix (M1).

Compressive Strength between Silica Fume and Cement after the end of 7 and 28 day period. Compressive Strength in (N/mm^2) for the Different mixes The results indicate that compressive strength increases gradually up to Mix 5 (6% silica fume and 20% fly ash), reaching 44.04 N/mm^2 at 90 days, after which further fly ash additions reduce strength. This optimal mix exhibits a 30% increase in compressive strength over the control mix; hence, moderate addition of fly ash and silica fume gives the best results.

4.1.2 Tensile Strength Development Analysis

Split tensile strength was conducted on cylindrical specimens with Mix 5 (6% silica fume and 20% fly ash) stabilizing the highest strength. Split tensile strength test results between days 7, 28, 56, and 90. Table 3 outlines the split tensile strength for the different mixes, highlighting improved tensile properties in Mix M5.

According to the results, it is only in the case of Mix 5 that the tensile stress rupture was enhanced over the control mix since up to 4.41 N/mm^2 has been accomplished. This enhancement in tensile strength enhances the ability of the mix to withstand tension before it cracks which makes it ideal for high-stress applications.

Table 3 Split tensile strength of concrete mixes (N/mm^2)

Mix ID	7 Days	28 Days	56 Days	90 Days
M1 (control)	2.50	3.14	3.48	3.71
M5 (6% SF, 20% FA)	2.76	3.49	3.96	4.41

Table 4 Flexural strength of concrete mixes (N/mm^2)

Mix ID	7 Days	28 Days	56 Days	90 Days
M1 (control)	3.10	4.20	4.50	4.89
M5 (6% SF, 20% FA)	3.68	4.94	5.22	5.68

4.1.3 Flexural Strength Analysis

The use of beam specimens in flexural strength tests showed an impressive lesser strength increase of 18% in Mix 5, one would expect, looking closer at the details, comparing with the control mix, which additionally, emphasizes its higher bending resistance properties. Yes, but what about? Greatly contributes, the only thing that overwhelms 6% Only silica fume and 20% fly ash have been added, the content of the active components that improve flexural strength, which is very important for structural members working under bending loads. Table 4 summarizes the flexural strength for each mix. The addition of fly ash and silica fume enhances the flexural strength of Mix M5 significantly, particularly at later curing stages.

4.2 Assessment of Durability

4.2.1 Chemical Attack Resistance

To evaluate the durability of the mixes under chemical conditions, concrete specimens were soaked in sulfate and acid solutions for 28 days. Mix 5 held up about 85% of its initial compressive strength when exposed to sulfate, in comparison to about 70% for the control mix. The improved performance of Mix 5 can be explained with the denser microstructure which reduces permeability and enhances resistance formed due to pozzolanic reactions. Table 5 illustrates the strength retention of the mixes after exposure to sulfate and acid environments. Mix M5 demonstrates superior retention in both environments compared to the control mix.

4.3 Thermal Resistance

To determine the thermal resistance of each mix, the specimens were exposed to elevated temperatures of 200 °C. Mix 5 preserved 90% of its compressive strength which exceeds control mix at 80%. It is believed that this is a result of the fly ash

Table 5 Strength retention after chemical exposure

Mix ID	Sulfate solution	Acid solution
M1 (control)	70%	65%
M5 (6% SF, 20% FA)	85%	80%

Table 6 Strength retention after thermal exposure (200 °C)

Mix ID	Compressive strength retention (%)
M1 (control)	80%
M5 (6% SF, 20% FA)	90%

possessing the lowest heat of hydration, thus reducing thermal cracking while also strengthening the concrete under increased temperatures. Table 6 presents the strength retention after thermal exposure at 200 °C, with Mix M5 showing a higher retention percentage than the control mix.

The experimental results reveal that it is possible to blend 6% silica fume and 20% fly ash into concrete to enhance its mechanical and durability properties to make them attractive as more sustainable construction materials. The improved compressive, tensile and flexural strength, and the chemical and thermal resistances that accompany this mix make it constructively applicable in applications where eco-efficiency and durability are desired.

5 LCA and TEA Approaches Integration

The results of the life cycle assessment (LCA) and techno-economical evaluation (TEA) for concrete mixes with fly ash and silica fume. The analysis takes into account both environmental and economic indicators, hence assessing the viability and the affordability of each mix.

5.1 Life Cycle Assessment (LCA) Results: Environmental Impacts

The LCA was prepared and presented according to ISO 14040/44 standards with an emphasis to the consideration of the environmental concerns throughout the life cycle of the concrete mixes: i.e., from the raw materials acquisition to the end of life. The relevance goals discussed include greenhouse gas emissions; energy consumption; and water use. There are significant environmental benefits from the use of fly ash and silica fume as partial substitutions for cement as this reduces the amount of clinker in cement which in turn lowers GHG emissions. Table 7 outlines the environmental impact indicators, showing reductions in carbon emissions, energy consumption, and water usage with the use of fly ash and silica fume in the concrete mixes.

Table 7 Environmental impact indicators

Impact category	Reduction with fly ash and silica fume (%)
Carbon emissions (CO_2 -eq)	25%
Energy consumption (MJ/kg)	18%
Water usage (L/kg)	20%

- Combined With Carbon Emissions: The 20% replacement of fly ash and 6% replacement of silica fume was able to cut emissions from CO_2 emissions by 25% in this case of concrete's replacement done with the conventional It was made possible because of the amount of clinker which requires a lot of energy usage considerably reduced.
- Carbon footprints were able to be significantly lowered as there was still a smoother integration, with lower construction waste overall pointing out clearly an efficient low energy process into the grinding or recycling of the materials.
- Water Usage: Today, with at least 20% reduction in water consumption due to border cement content explaining a construction without too much hydration water demand, thus conserving the intact water loop in the usage of construction materials.
- Water usage would therefore be able to use a stress free essentially reducing the cement content in the blend as well as so the amount of hydration that is imposed during the construction period.

5.2 *Techno-Economic Assessment (TEA) Results: Cost Analysis and Benefits*

TEA estimates specifically the costs in concrete material and production including fly ash and silica fume usage and additionally possible economic benefits' viewpoint on its reduced content of cement. The calculation includes overhead costs, like the costs of resources and processing, and economized costs, for example, environmental charges that would not have been incurred due to pollution reduction. Table 8 details the economic impact, highlighting reductions in material and production costs with the use of supplementary cementitious materials.

- Material Costs: Fly ash and silica fume substitute query production, cement use goes down, resulting in a reduction of 12% in raw material expenditure.
- Production Costs: Cost of production gets reduced by 10% owing to lesser energy consumption for mixing, curing activities and lesser costs on waste disposal of environmental pollution of conventional places of cement.
- Overall Cost Savings: The total cost of material resources as well as the cost of their production is increased by 15%, and this is due to the use of the waste costs and benefits of the production.

Table 8 Economic impact indicators

Cost category	Cost reduction with fly ash and silica fume (%)
Material costs	12%
Production costs	10%
Overall cost savings	15%

Table 9 Comparison of environmental and economic indicators

Indicator	Environmental impact reduction (%)	Economic savings (%)
Carbon emissions (CO_2 -eq)	25%	N/A
Energy consumption	18%	10%
Water usage	20%	N/A
Material and production costs	N/A	15%

5.3 *Comparison of Environmental and Economic Indicators*

The combined LCA and TEA results indicate the potential of fly ash and silica fume to improve the environmental and economic aspects of concrete. The optimal blend (6% silica fume and 20% fly ash) provides substantial benefits in lowering CO_2 emissions and overall costs of production, and hence is a feasible approach. Table 9 provides a comparison of environmental and economic indicators, emphasizing the dual benefits of using fly ash and silica fume in terms of sustainability and cost savings.

This comparison marks the fact that the optimal mix does not only minimize environmental impacts but also most importantly achieves a degree of sustainability while being cost effective and thus makes a case for the construction sectors transition to greener materials.

6 Discussion and Conclusion

This section examines the results of the study under the perspective of sustainability, suggests possible use in green building designs, and indicates limitations and future directions of research. The conclusion reviews the most important results and advises sustainable material practices in the field of construction.

6.1 *Results Feasibility Within the Concept of Sustainable Development*

This study shows that the use of fly ash and silica fume in concrete mixes improves ecological and cost efficiency of the concrete mix. Using the optimal mix (6% silica fume and 20% fly ash), the author was able to reduce carbon emissions by 25% and total production costs by 15%. These decreases principally emphasize the potential of these industrial wastes to be utilized as effective resources for the cement industry. Moreover, the enhanced mechanical properties, increase in compressive, tensile, and flexural strengths, also suggest that the use of fly ash and silica fume improves the durability of concrete structures and hence are appropriate for construction which is expected to last for a long time without the requirement for frequent repairs.

The simultaneous energy and water consumption reduction corresponds to worldwide visions in sustainable development such as reducing the emission of greenhouse gases and depletion of natural resources. Using this mix design also encourages the principles of circular economy as it encourages the use of waste materials from the power providing and silicon alloy industries instead of the conventional cementing material which in turn reduces waste going to landfills and enhances construction resource efficiency.

6.2 Possible Uses in Construction of Green Buildings

Concrete reinforced with silica fume and fly ash has potential uses in green building projects as it has economic and performance benefits. Some of the uses which this sustainable concrete can prove to be useful include:

- Long-Lasting Buildings: This mix is suitable for structures which are intended to be exposed to hostile conditions such as marine, industrial or high temperature environments as it has improved resistance to chemical attack and thermal deterioration.
- Green Certifications: Fly ash, silica fume and concrete can be used in projects aiming for LEED or BREEAM certifications due to the fact that it has reasonably low environmental impact which is a requirement for such eco-construction projects.
- Efficient Structure Design: The amount of energy used in heating water for concrete is also reduced owing to the presence of Fly ash which therefore makes it useful for energy considerate construction projects.

6.3 Limitations and Future Research Directions

Despite what the research has brought out regarding the advantages of the use of fly ash and silica in the concrete mix, a number of persistent weaknesses and avenues for further study will be discussed:

- Variability in Source Materials: The properties of fly ash as well as silica fume depend on the source of the material and its processing mechanism. It is recommended that future studies look at the variations of these parameters on concrete properties to standardize them.

- Long-Term Performance: This study evaluates the durability of a mix over a period of 90 days; however, more detailed studies are required to study the performance of these mixes in extreme environments over decades for structural validation.
- Optimizing Mix Ratios for Different Grades: Further work may consider optimum fly ash and silica fume ratios for different concrete grades such as M40 and M50 as other structural uses may require some mix designs.

The study shows that fly ash and silica fume are viable substitutes to cement which will not only protect the environment but cut down on costs of construction while providing durability. The analysis results indicate that a 6% silica fume and 20% fly ash mixture is more favorable which lowers carbon emissions and production cost and ensures high performance on crucial mechanical properties.

References

1. Adeoye, A. O., Quadri, R. O., Lawal, O. S., & Emojevu, E. O. (2024). Physicochemical characterization, valorization of lignocellulosic waste (Kola nut seed shell) via pyrolysis, and ultrasonication of its crude bio-oil for biofuel production. *Cleaner Waste Systems, 7*. https://doi.org/10.1016/j.clwas.2024.100138
2. Adeoye, A. O., Quadri, R. O., Lawal, O. S., Malomo, D., Emojevu, E. O., Omonije, O. O., Odeniyi, O. K., Fadahunsi, M. O., Yelwa, M. J., Aasa, S. A., Onakpa, A. E., Omoniyi, B. P., Mark, I. N., Usman, J., Ismaila, A. M., & Saidu, A. U. (2024). Advanced techniques in upgrading crude bio-oil to biofuel. In *Energy, environment, and sustainability* (pp. 321–353). https://doi.org/10.1007/978-981-97-0515-3_14
3. Adeoye, A. O., Quadri, R. O., & Lawal, O. S. (2022). Assessment of biofuel potential of tenera palm kernel shell via fixed bed pyrolysis and thermal characterization. *Results in Surfaces and Interfaces, 9*. https://doi.org/10.1016/j.rsurfi.2022.100091
4. Adeoye, A. O., Lawal, O. S., Quadri, R. O., Malomo, D., Aliyu, M. T., Dang, G. E., Emojevu, E. O., Maikato, M. J., Yahaya, M. G., Omonije, O. O., Edidem, V. G., Abubakar, Y. K., Offor, O. F., Sochima, E. H., Peter, B. E., & Hikon, B. N. (2023). Sustainable energy via thermochemical and biochemical conversion of biomass wastes for biofuel production. In *Energy, environment, and sustainability* (pp. 245–306). https://doi.org/10.1007/978-981-99-2150-8_11
5. Mohapatra, S., & Mishra, R. K. (2020). Status of institutional credit disbursement of agricultural sector in Odisha, India. *International Journal of Current Microbiology and Applied Sciences, 9*, 2057–2065.
6. Mohapatra, S., Singh, J., & Mishra, S. (2017). Co-integration among major potato markets in Punjab. *Indian Journal of Agricultural Marketing, 31*, 46–47.
7. Mohapatra, S., Priyanka, V., Mohapatra, S., Kohli, I., & Mishra, R. K. (2020). Impact of corona virus covid-19 on the global economy. *International Journal of Agricultural and Statistical Sciences, 16*, 771–778.
8. Mohapatra, S., Singh, J. Co-integration of major vegetable markets in Punjab, India. (2021).
9. Mohapatra, S., Mishra, R., & Mishra, R. (2021). Impact of COVID-19 on education level of agricultural farming community. *International Journal of Agricultural and Statistical Sciences, 17*, 915–923.

10. Mohapatra, S., Singh, J., & Kumar, S. (2018). Cointegration among major cauliflower markets in Punjab. *Indian Journal of Economics and Development, 14*, 330–335.
11. Adeoye, A. O., Yelwa, J. M., Imam, N., Quadri, R. O., Lawal, O. S., Malomo, D., Aasa, S. A., Onakpa, A. E., Dorgbaa, R. K., Omoniyi, B. P., Ahmad, N., & Garba, H. A. (2024). Pyrolysis of plastic wastes towards achieving a circular economy: An advanced chemistry and technical approach. In *Energy, environment, and sustainability* (pp. 215–259). https://doi.org/10.1007/978-981-97-0437-8_11
12. Adeoye, A. O., Quadri, R. O., & Lawal, O. S. (2023). Wet synthesis, characterization of goethite nanoparticles and its application in catalytic pyrolysis of palm kernel shell in TGA. *Results in Surfaces and Interfaces, 11*. https://doi.org/10.1016/j.rsurfi.2023.100118
13. Avesh, M., Imam, N., & Adeoye, A. O. (2024). Assessing environmental impacts and mitigation strategies for the DFCCIL Rail projects. In *Energy, environment, and sustainability* (pp. 333–353). https://doi.org/10.1007/978-981-97-0437-8_15
14. Sharma, S. K., Sharma, R. C., Lavania, S., Palli, S., & Avesh, M. (2024). Dynamic vibration mitigation in ship propellers using hybrid fuzzy sliding mode control. *National Academy Science Letters*. https://doi.org/10.1007/s40009-024-01515-z
15. Mohapatra, S., & Mishra, R. A. J. K. (2020). Production performance and market integration of cotton markets in India. In *Natiosal symposium* (p. 253).
16. Din, A. U., Han, H., Ariza-Montes, A., Vega-Muñoz, A., Raposo, A., & Mohapatra, S. (2022). The impact of Covid-19 on the food supply chain and the role of e-commerce for food purchasing. *Sustainability, 14*(5), 3074.
17. Mohapatra, S., & Singh, J. (2018). Growth and instability analysis of cauliflower crop in Punjab. *Economic Affairs, 63*, 665–669.
18. Sharma, R. C. (2016a). Evaluation of passenger ride comfort of indian rail and road vehicles with ISO 2631-1 Standards Part 1 - mathematical modelling. *International Journal of Vehicle Structures and Systems, 8, No. 1*. https://doi.org/10.4273/ijvss.8.1.01
19. Mohapatra, S., Mishra, R. K., & Sarangi, K. K. (2019). Economic appraisal of bio ethanol production technologies. *International Journal of Chemical Studies, 7*, 2197–2200.
20. Mohapatra, S., Kishore Mishra, R., & Sarangi, K. K. (2020). Bio-ethanol (2nd generation ethanol): A solution to ever polluting gasoline to climate in India. *International Research Journal of Pure and Applied Chemistry, 20*, 1–15.
21. Gohain, N., Bansal, S., Mohapatra, S., & Singh, L. (2022). Analyzing the direction of trade: Indian ginger and lessons from exports to different destinations. *Journal of Agricultural Development and Policy, 32*, 214–220.
22. Mohapatra, S., Mohapatra, U., Chandana, K. S. S., & Mishra, R. K. (2018). Economic analysis of paddy production and marketing in Puri, Odisha. *Journal of Pharmacognosy and Phytochemistry, 7*, 1858–1861.
23. Cataluna Veses, R., & Shah, Z. (2015). Bio-oil production by thermal cracking in the presence of hydrogen. *Journal of Fundamentals of Renewable Energy and Applications., 06*. https://doi.org/10.4172/2090-4541.1000194
24. Shah, Z., Cataluña Veses, R., Aguilhera, R. A., & Da Silva, R. (2016). Bio-oil production from pyrolysis of coffee and eucalyptus sawdust in the presence of 5% hydrogen. *International Journal of Engineering Research and Science Bikaner, 2*(5), 34–42.
25. Shah, Z., Veses, R. C., & da Silva, R. (2016). A comparative study and analysis of two types of bio-oil samples obtained from freshwater algae and microbial treated algae. *Modern Chemistry & Applications, 4*. https://doi.org/10.4172/2329-6798.1000185
26. Shah, Z., Veses, R. C., Brum, J., Klunk, M. A., Rocha, L. A. O., Missaggia, A. B., de Oliveira, A. P., & Caetano, N. R. (2020). Sewage sludge bio-oil development and characterization. *Inventions, 5*, 36. https://doi.org/10.3390/inventions5030036
27. Shah, Z., Saberian, M., Hodgeman, D., Kinloch, I., & Vallés, C. (2023). Effect of sulfur on wood tar biopitch as a sustainable replacement for coal tar pitch binders. *ACS Applied Engineering Materials, 1*, 2567–2576. https://doi.org/10.1021/acsaenm.3c00361

28. Klunk, M. A., Dasgupta, S., Das, M., & Shah, Z. (2018). System of adsorption of CO2 in coalbed. *Southern Brazilian Journal of Chemistry, 26*, 2–9. https://doi.org/10.48141/SBJCHEM.v26.n26.2018.7_2018.pdf
29. Mohapatra, S., Mohapatra, U., & Mishra, R. (2017). Diversification towards vegetable crops: A good option for doubling the farmer's income. *Journal of Experimental Agriculture International, 18*, 1–7.
30. Mohapatra, S., Mohapatra, U., & Mishra, R. K. (2018). Price forecasting of groundnut in Odisha. *Journal of Pharma Innovation, 7*, 111–114.
31. Bansal, S., Mohapatra, S., Gohain, N., & Das, A. (2022). Production status, market integration and the law of one price: An exploration of the major maize markets of India. *Indian Journal of Economics and Development, 18*, 832–841.
32. Mohapatra, S., Sneha, N. K., Mounika, A., & Mishra, R. K. (2019). Socio-economic and demographic analysis of SAARC Nations. *International Journal of Agricultural & Statistical Sciences, 15.*
33. Quadri, R. O., Adeoye, A. O., Ayanda, O. S., & Lawal, O. S. (2024). Conversion of palm kernel shell to sustainable energy and the effect of wet synthesized nanoparticles of iron on its thermal degradation kinetics. *Bioresource Technology Reports, 27.* https://doi.org/10.1016/j.biteb.2024.101933
34. Momodu, D.U., Adetuyi, A.O., Adeoye, A.O., Akinlolu, O.S., Apara, E.O. (2023). Surface functionalization of linen fiber and its dye-fastness assessment using (4-[(E)-(4-amino-2-hydroxyphenyl) diazenyl] benzene-1-sulphonic acid and 4-[(e)-(4-{(Z)-[4-(dimethylamino) phenyl] diazenyl}-2-hydroxyphenyl) diazenyl]benzene-1-sulphonic acid. *Suranaree Journal of Science and Technology, 30.*
35. Adeoye, A. O., Quadri, R. O., & Lawal, O. S. (2023). Biomass waste pyrolysis in the context of Nigeria's oil and gas industry: An overview. *International Journal of Ambient Energy, 44*, 2525–2541. https://doi.org/10.1080/01430750.2023.2256325
36. Mohapatra, S., Mohapatra, U., & Mishra, R. K. (2018). Price forecasting of groundnut in Odisha. *Pharma Innovation Journal, 7*, 111–114.
37. Mohapatra, S., Mohapatra, S., Han, H., Ariza-Montes, A., & del López-Martín, M. C. (2022). Climate change and vulnerability of agribusiness: Assessment of climate change impact on agricultural productivity. *Frontiers in Psychology, 13*, 955622.
38. Shah, S., Nautiyal, H., Gugliani, G., Kumar, A., Namboodri, T., & Singla, Y. K. (2024). *Sustainability in smart manufacturing: Trends, scope, and challenges*. CRC Press (Taylor & Francis Group) . https://taylorandfrancis.com/books/. ISBN 9781032740713. https://doi.org/10.1201/9781003467496
39. Singla, Y. K., Maughan, M. R., Arora, N., & Dwivedi, D. K. (2024). Enhancing the wear resistance of iron-based alloys: A comprehensive review of alloying element effects. *Journal of Manufacturing Processes, 120*, 135–160.
40. Adeoye, A. O., Quadri, R. O., & Lawal, O. S. (2022). Fixed-bed pyrolysis and thermal analyses of pressed oil palm fruit fibre as a potential alternative energy. *Fuoye Journal of Innovation, Science and Technology, 2*(1), 187–194.
41. Pandipati, S., Palli, S., Dowluru, S., Sharma, R. C., Sharma, N., Sharma, S. K., & Da Fonseca-Soares, D. (2024). Microstructural and strength comparison friction stir processing of aluminium alloys and multi-response optimization by Taguchi GRA-PCA based hybrid methods. *International Journal of Vehicle Structures and Systems, 16*(3). https://doi.org/10.4273/ijvss.16.3.03
42. Singla, Y. K., Maughan, M. R., Arora, N., & Dwivedi, D. K. (2023). Microstructural evolution of worn surface and subsurface of high Mn additive iron-based hardfacings subjected to repetitive impacts and sliding. *Ceramics International, 49*, 21957–21965.
43. Ram, M., Singla, Y. K., & Kumar, A. (2022). *Advanced materials for biomechanical applications*. CRC Press (Taylor & Francis Group). https://taylorandfrancis.com/books/. ISBN 9781003286806, 9781032054490. https://doi.org/10.1201/9781003286806

44. Singh, H., Singh, A. K., Singla, Y. K., Chattopadhyay, K., Saini, A., & Singh, K. (2022). Interpretation of the wear characteristics of AISI 4140 under nano-fly ash based engine lubricant. *Materials Today Proceedings, 50*, 1683–1689.
45. Rooprai, R. S., Singh, H., Singh, T., & Singla, Y. K. (2022). Analysis of the wear properties of through hardened AISI-4140 alloy steel using Taguchi technique. *Materials Today Proceedings, 50*, 661–664.
46. Singh, H., Singh, A. K., Singla, Y. K., Chattopadhyay, K., & Malhotra, R. (2021). Optimization of tribological behavior of AISI 4140 under nano fly ash particulates in engine lubricating oil. *Materials Today Proceedings, 45*, 4619–4623.
47. Shah, Z., Veses, R. C., & da Silva, R. (2016). Using GC-MS to analyze bio-oil produced from pyrolysis of agricultural wastes—Discarded soybean frying oil, coffee and eucalyptus sawdust in the presence of 5% hydrogen and argon. *Journal of Analytical and Bioanalytical Techniques, 7*. https://doi.org/10.4172/2155-9872.1000300
48. Klunk, M. A., Dasgupta, S., Das, M., Wander, P. R., & Shah, Z. (2019). Adsorption study of crystal violet and malachite green dyes in zeolitic material. *Periódico Tchê Química, 16*, 70–81. https://doi.org/10.52571/PTQ.v16.n33.2019.85_Periodico33_pgs_70_81.pdf
49. Yaseen, M., Khattak, M. A. K., Khan, A., Khan, M. S., Ahmad, M., Shah, Z., Khattak, R., & Bibi, S. (2022). Physico-chemical investigations on the catalytic production of biofuel from algal biomass. *Russian Journal of Physical Chemistry A, 96*, S31–S37. https://doi.org/10.1134/S0036024422140308
50. Shah, Z., Cataluña Veses, R., & da Silva, R. (2016). GC-MS and FTIR analysis of bio-oil obtained from freshwater algae (spirogyra) collected from freshwater. *International Journal of Environmental & Agriculture Research (IJOEAR) Bikaner, 2*(2), 134–141.
51. Shah, Z., Vases, R. C., Ullah, I., Yaseen, M., Aguilhera, R. A., Jamal, S., Amorim, V. D. A., & Da Silva, R. (2016). Analysis of bio-oil obtained from pyrolysis of scenedesmus and spirogyra mixture algal biomass. *Journal of Ongoing Chemical Research, 2*, 1–13.
52. de Oliveira, A. P., Lorenzini, G., Shah, Z., Klunk, M. A., de Carvalho Lima, J. E., Rocha, L. A. O., & Caetano, N. R. (2020). Hierarchical criticality analysis of clean technologies applied to a coal-fired power plant. *International Journal of Design & Nature and Ecodynamics, 15*, 609–619. https://doi.org/10.18280/ijdne.150501
53. Roberto Caetano, N., Lorenzini, G., Rocha Lhamby, A., Shah, Z., Antônio Klunk, M., & Alberto Oliveira Rocha, L. *Modelling Safety Distance from Industrial Turbulent Non-Premixed Gaseous Jet Flames, 5*(1), 41–49.
54. Klunk, M. A., Shah, Z., & Wander, P. R. (2019). Use of montmorillonite clay for adsorption malachite green dye. *Periódico Tchê Química., 16*, 279–286. https://doi.org/10.52571/PTQ.v16.n32.2019.297_Periodico32_pgs_279_286.pdf
55. Klunk, M. A., Shah, Z., Caetano, N. R., Conceição, R. V., Wander, P. R., Dasgupta, S., & Das, M. (2020). CO2 sequestration by magnesite mineralisation through interaction of Mg-brine and CO2: Integrated laboratory experiments and computerised geochemical modelling. *International Journal of Environmental Studies, 77*, 492–509. https://doi.org/10.1080/00207233.2019.1675295
56. Yaseen, M., Shah, Z., Veses, R. C., Dias, S. L. P., Lima, E. C. S., dos Reis, G., Vaghetti, J. C. P., Alencar, W. S. D., & Mehmood, K. (2017). Photocatalytic studies of Tio2/Sio2 nanocomposite xerogels. *Journal of Analytical and Bioanalytical Techniques, 08*. https://doi.org/10.4172/2155-9872.1000348
57. Cataluña, R., Shah, Z., Pelisson, L., Caetano, N., da Silva, R., & Azevedo, C. (2017). Biodiesel glycerides from the soybean ethylic route incomplete conversion on the diesel engines combustion process. *Journal of the Brazilian Chemical Society*. https://doi.org/10.21577/0103-5053.20170100
58. Shah, Z., Cataluña Veses, R., Ceschi, M. A., & Silva, R. D. (2017). Separation of phenol from bio-oil produced from pyrolysis of agricultural wastes. *Modern Chemistry & Applications, 05*. https://doi.org/10.4172/2329-6798.1000199

59. Cataluña, R., Shah, Z., Venturi, V., Caetano, N. R., da Silva, B. P., Azevedo, C. M. N., da Silva, R., Suarez, P. A. Z., & Oliveira, L. P. (2018). Production process of di-amyl ether and its use as an additive in the formulation of aviation fuels. *Fuel, 228*, 226–233. https://doi.org/10.1016/j.fuel.2018.04.167
60. Shah, Z., Veses, R. C., Vaghetti, J. C. P., Amorim, V. D. A., & da Silva, R. (2019). Preparation of jet engine range fuel from biomass pyrolysis oil through hydrogenation and its comparison with aviation kerosene. *International Journal of Green Energy, 16*, 350–360. https://doi.org/10.1080/15435075.2019.1566730
61. Sharma, S.K., Upadhyay, R.K., Kumar, V., Valera, H. (2023). Transportation energy and dynamics. Springer Nature Singapore, 2023. https://doi.org/10.1007/978-981-99-2150-8
62. Upadhyay, R. K., Sharma, S. K., Kumar, D. V., & Valera, H. (2023). *Transportation systems technology and integrated management*. Springer Nature Singapore. https://doi.org/10.1007/978-981-99-1517-0
63. Lee, J., & Sharma, S. K. Intelligent systems for railway infrastructure. *Applied Sciences, 13*(20), 11274.
64. Sharma, S. K., Sharma, R. C., Choi, Y., & Lee, J. (2023). Modelling and dynamic analysis of adaptive neuro-fuzzy inference system-based intelligent control suspension system for passenger rail vehicles using magnetorheological damper for improving ride index. *Sustainability, 15*, 12529.
65. Hossain, I., Ahmed, Y. E., & Sharma, S. K. (2023). The use of both metallic-insert and TiO2/water nanofluid to improve the performance of photovoltaic modules; experimental study. *Thermal Science and Engineering Progress, 42*, 101944.
66. Wu, Q., Cole, C., Spiryagin, M., Chang, C., Wei, W., Ursulyak, L., Shvets, A., Murtaza, M. A., Mirza, I. M., & Zhelieznov, K. (2023). Freight train air brake models. *International Journal of Rail Transportation, 11*, 1–49.
67. Sharma, R. C., Bhardawaj, S., Sharma, S. K., & Bahl, S. (2023). Ride behaviour of three-wheel vehicle using semi-active suspension using skyhook control strategy. *International Journal of Vehicle Structures & Systems (IJVSS), 15.*
68. Bhardawaj, S., Sharma, R. C., Sharma, S. K., Gopala Rao, L. V. V., & Vashist, A. (2023). Modeling of biomechanical human body model for seat to head transmissibility analysis. *Noise & Vibration Worldwide, 54*, 62–74.
69. Sharma, S. K., Sharma, R. C., Choi, Y., & Lee, J. (2023). Experimental and mathematical study of flexible–rigid rail vehicle riding comfort and safety. *Applied Sciences, 13*, 5252.
70. Sharma, R. C., Palli, S., & Sharma, S. K. (2023). Ride analysis of railway vehicle considering rigidity and flexibility of the carbody. *Journal of the Chinese Institute of Engineers, 46*, 355–366.
71. Pant, M., Sharma, A., Palli, S., Sharma, S. K., & Sharma, R. C. (2024). A review and analysis of factors affecting EV adoption for sustainable and energy-efficient transportation systems using analytic hierarchy process approach. *International Journal of Vehicle Structures & Systems (IJVSS), 16.*
72. Sharma, S. K., Upadhyay, R. K., & Kumar, V. (2024). Introduction to dynamics of transportation ecosystem, modeling, and control. In *Dynamics of transportation ecosystem, modeling, and control* (pp. 3–6). Springer.
73. Sharma, S. K., Sharma, R. C., Mohapatra, S., Fouly, A., Hossain, I., & Lee, J. (2024). Technological advancements in oscillation reduction for propulsion shaft systems. *Journal of Vibration Engineering & Technologies, 12*(2), 1–19.
74. Upadhyay, R. K., Sharma, S. K., & Kumar, V. (2024). Introduction to intelligent transportation system and advanced technology. In *Intelligent transportation system and advanced technology* (pp. 3–6). Springer.
75. Hossain, I., Ngoupo, A. T., Bhattarai, S., Challagulla, S. P., & Sharma, S. K. (2024). Empirical survey on the effect of straight and wavy fins integrated with wind stream as heat sink of photovoltaic panels. *International Journal of Environmental Science and Technology*, 1–14.

76. Sharma, S. K., Sharma, R. C., & Lee, J. (2025). Propelling precision of longitudinal vibration mitigation in ship propeller shafts through advanced nonlinear intelligent semi-active control leveraging adaptive neuro-fuzzy inference system with linear quadratic regulator. *Journal of Vibration and Control, 31*(7–8), 1472–1484.
77. Sharma, S. K., Kumar, N., Avesh, M., Sharma, R. C., Siddiqui, M. I. H., & Lee, J. (2024). Navigating tranquillity with H∞ controller to mitigate ship propeller shaft vibration. *Journal of Vibration Engineering & Technologies*, 1–11.
78. Sharma, S. K., Sharma, R. C., Upadhyay, R. K., & Lee, J. (2024). Enhanced ship engine vibration reduction using magnetorheological dampers and adaptive neuro-fuzzy control system. *Ships and Offshore Structures*, 1–17.
79. Sharma, R. C., Sharma, N., Palli, S., Sharma, S. K., Joshi, S., Rallabandi, S. R., & Sharma, A. (2024). Dynamic behavior of three-wheel vehicle with inerter-based dynamic absorber suspension system. *Noise & Vibration Worldwide., 55*, 253–263.
80. Raju Rallabandi, S., Pilla, D. P., Dowluru, S., Palli, S., Sharma, N., Sharma, S. K., & Sharma, R. C. (2024). Critical evaluation of epoxy-hemp-pineapple-palm fiber composites using hybrid AHM-TOPSIS technique for sustainable structural applications. *Journal of the Chinese Institute of Engineers., 47*, 325–336.
81. Kokkinos, N. C., Emmanouilidou, E., & Sharma, S. K. (2024). Waste-to-biofuel production for the transportation sector. In *Intelligent transportation system and advanced technology* (pp. 99–122). Springer.
82. Sharma, R. C., Palli, S., Duppala, A., Bhardawaj, S., Sharma, S. K., Sharma, N., & Sharma, A. (2024). Rigid body analysis of ride behaviour of roadway vehicle using Lagrangian dynamics. *International Journal of Vehicle Structures & Systems (IJVSS), 16.*
83. Sharma, S. K., Sharma, R. C., & Lee, J. (2024). Modelling and simulation of MR damper characterization and uncertainty assessment in marine engine vibration isolation with FLPID control. *International Journal of Modelling and Simulation*, 1–11.
84. Sharma, S. K., Sharma, R. C., Upadhyay, R. K., & Lee, J. (2024). Coupled torsional–transverse vibration reduction in marine propulsion dynamics with novel approach using magnetorheological damping and adaptive control system. *Journal of Vibration Engineering & Technologies, 12*, 6089–6099.
85. Rallabandi, S. R., Srinivas, L., Palli, S., Sharma, R. C., Sharma, N., Sharma, A., & Sharma, S. K. (2024). A contrastive characterization of pure Mg and AZ91D alloy based on the testing of mechanical, corrosion, wear, and erosion properties. *Engineering Research Express, 6*, 015017.
86. Badgujar, P., Sharma, S., Rahul, D., Sharma, S. K., & Upadhyay, R. K. (2024). Feasibility study of Mumbai-Pune hyperloop corridor: Challenges and implementation. *Research in Transportation Business & Management, 52*, 101089.
87. Sharma, S. K., Sharma, R. C., Lavania, S., Palli, S., & Avesh, M. (2024). Industry-driven approach for ANFIS-based intelligent control suspension system with MR damper for enhanced ride quality in passenger rail vehicles for technological investigations. *National Academy Science Letters, 47*, 391–394.
88. Sharma, R. C., Palli, S., Dharmana, L., Sharma, S. K., Sharma, A., Kharola, A., & Sharma, N. (2024). Simulation of transient dynamic response of a locomotive considering its flexibility and rigidity. *Noise & Vibration Worldwide, 55*, 33–47.
89. Song, H., Park, J., Sharma, S. K., & Lee, J. (2022). A comparative study on ultrasonic propagation characteristics and defect detection of metal material additive manufacturing using deep learning algorithm. In *Biennial international conference on future learning aspects of mechanical engineering* (pp. 157–168). Springer.
90. Boris, I., Sharma, S. K., & Lee, J. (2022). A study on the reflection coefficient of higher modes of torsional wave by theoretical and numerical method. In *Biennial international conference on future learning aspects of mechanical engineering* (pp. 145–155). Springer.
91. Vishwakarma, P. N., Mishra, P., Sharma, S. K., & Shukla, A. K. (2022). Formulating and analysing the effect of suspension parameters on critical speeds for various DOFs in rail vehicle.

In *Biennial international conference on future learning aspects of mechanical engineering* (pp. 791–802). Springer.

92. Kumar, U., Kasvekar, R., Sharma, S. K., & Upadhyay, R. K. (2022). Wear of wheels and axle in locomotive and measures taken by Indian railway. *Advances in Engine Tribology*, 77–96.
93. Sharma, R. C., Palli, S., Jha, A. K., Bhardawaj, S., & Sharma, S. K. (2022). Vibration and ride comfort analysis of railway vehicle system subjected to deterministic inputs. *Resmilitaris, 12*, 1345–1355.
94. Sharma, R. C., Rao, L. V. V. G., Sharma, S. K., Palli, S., & Satyanarayana, V. S. V. (2022). Analysis of lateral stability and ride of an Indian railway constrained dual-axle bogie frame. *SAE International Journal of Commercial Vehicles, 16*, 213–228.
95. Sharma, R. C., Sharma, S. K., Palli, S., Rallabandi, S. R., & Sharma, N. (2025b). Compatibility between stability and curving ability of railway vehicles using conventional and unconventional bogie frame models. *SAE International Journal of Commercial Vehicles, 18*(3), 02-18-03-0014. https://doi.org/10.4271/02-18-03-0014
96. Vishwakarma, P. N., Mishra, P., & Sharma, S. K. (2022). Formulation of semi-active suspension system and controls in rail vehicle. *SSRN Electronic Journal*. https://doi.org/10.2139/ssrn.415961
97. Sharma, R. C., & Sharma, S. K. (2022). Ride analysis of road surface-three-wheeled vehicle-human subject interactions subjected to random excitation. *SAE International Journal of Commercial Vehicles, 15*, 299–312.
98. Sharma, S. K., Lee, J., & Jang, H.-L. (2022). Mathematical modeling and simulation of suspended equipment impact on car body modes. *Machines, 10*, 192.
99. Sharma, S. K., Sharma, R. C., Lee, J., & Jang, H.-L. (2022). Numerical and experimental analysis of DVA on the flexible-rigid rail vehicle carbody resonant vibration. *Sensors, 22*, 1922.
100. Vishwakarma, P. N., Mishra, P., & Sharma, S. K. (2022). Characterization of a magnetorheological fluid damper a review. *Materials Today Proceedings, 56*, 2988–2994.
101. Sharma, S. K., Mohapatra, S., Sharma, R. C., Alturjman, S., Altrjman, C., Mostarda, L., & Stephan, T. (2022). Retrofitting existing buildings to improve energy performance. *Sustainability, 14*, 666.
102. Sharma, S. K., Sharma, R. C., & Lee, J. (2022). In situ and experimental analysis of longitudinal load on carbody fatigue life using nonlinear damage accumulation. *International Journal of Damage Mechanics, 31*, 605–622.
103. Sharma, R. C., Sharma, S., Sharma, N., & Sharma, S. K. (2021). Linear and nonlinear analysis of ride and stability of a three-wheeled vehicle subjected to random and bump inputs using bond graph and simulink methodology. *SAE International Journal of Commercial Vehicles, 15*, 65–79.
104. Sharma, S.K., Sharma, R.C. (2021). *Multi-objective design optimization of locomotive nose*. SAE Technical Paper, pp. 1–10. https://doi.org/10.4271/2021-01-5053.
105. Lee, J., Han, J., & Sharma, S. K. (2021). Structural analysis on the separated and integrated differential gear case for the weight reduction. In *Advances in engineering design: Select proceedings of FLAME 2020* (pp. 175–181). Springer Singapore.
106. Choi, S., Lee, J., & Sharma, S. K. (2021). A study on the performance evaluation of hydraulic tank injectors. In *Advances in engineering design: Select proceedings of FLAME 2020* (pp. 183–190). Springer Singapore.
107. Sharma, R. C., Palli, S., Sharma, N., & Sharma, S. K. (2021). Ride behaviour of a four-wheel vehicle using h infinity semi-active suspension control under deterministic and random inputs. *International Journal of Vehicle Structures & Systems, 13*, 234–237.
108. Bhardawaj, S., Sharma, R. C., Sharma, S. K., & Sharma, N. (2021). On the planning and construction of railway curved track. *International Journal of Vehicle Structures & Systems, 13*, 151–159.

109. Mohapatra, S., Mohanty, D., Sharma, S., Dikshit, S., Kohli, I., Samantaray, D. P., Kumar, R., & Kathpalia, M. (2021). Biomedical application of polymeric biomaterial: Polyhydroxybutyrate. In *Bioresource utilization and management* (pp. 111–124). Apple Academic Press.
110. Sharma, S. K., Sharma, R. C., & Lee, J. (2021). Effect of rail vehicle–track coupled dynamics on fatigue failure of coil spring in a suspension system. *Applied Sciences, 11*, 2650.
111. Sharma, R. C., Sharma, S., Sharma, S. K., Sharma, N., & Singh, G. (2021). Analysis of bio-dynamic model of seated human subject and optimization of the passenger ride comfort for three-wheel vehicle using random search technique. *Proceedings of the Institution of Mechanical Engineers, Part K: Journal of Multi-body Dynamics, 235*, 106–121.
112. Sharma, S. K., & Lee, J. (2021). Crashworthiness analysis for structural stability and dynamics. *International Journal of Structural Stability and Dynamics, 21*, 2150039.
113. Mohapatra, S., Samantaray, D. P., & Samantaray, S. M. (2015). Study on polyhydroxyalkanoates production using rhizospheric soil bacterial isolates of sweet potato. *Indian Journal of Science and Technology, 8*, 57–62. https://doi.org/10.17485/ijst/2015/v8iS7/64027
114. Arora, R., Trivedi, J., Mohapatra, S., & Kumar, P. (2022). Trends in probiotics on human health and industrial application. In *Advances in agricultural and industrial microbiology: Volume 1: Microbial diversity and application in agroindustry* (pp. 237–248). https://doi.org/10.1007/978-981-16-8918-5_13
115. Samal, S., Pati, S., Mohapatra, S., Maity, S., Tanaya, K., Devadarshini, D., & Samantaray, D. (2023). PHAs production by facultative anaerobic bacteria Bacillus cereus FM5 through submerged and solid-state fermentation under anoxic condition. *Antonie van Leeuwenhoek, International Journal of General and Molecular Microbiology, 116*, 521–529. https://doi.org/10.1007/s10482-023-01825-0
116. Biswas, N., Saha, S., Biswas, B. K., Chowdhury, M., Rahaman, A., Junghare, V., Mohapatra, S., Hazra, S., Zangrando, E., Roy Choudhury, R., & Roy Choudhury, C. (2021). The DNA- and protein-binding properties and cytotoxicity of a new copper(II) hydrazone Schiff base complex. *Journal of Coordination Chemistry, 74*, 1482–1504. https://doi.org/10.1080/00958972.2021.1913128
117. Mohapatra, S., Mishra, R., Roy, P., Yadav, K. L., & Satapathi, S. (2017). Systematic investigation and in vitro biocompatibility studies on implantable magnetic nanocomposites for hyperthermia treatment of osteoarthritic knee joints. *Journal of Materials Science, 52*, 9262–9268. https://doi.org/10.1007/s10853-017-1136-0
118. Mishra, B. B., Nayak, S. K., Mohapatra, S., & Samantaray, D. (2021). *Environmental and agricultural microbiology: Advances and applications*. John Wiley & Sons. https://doi.org/10.1002/9781119525899
119. Sravanthi, A. L., Ratnakumar, P., Reddy, S. N., Eswari, K. B., Pandey, B. B., Manikanta, C., Ramya, K. T., Sonia, E., Mohapatra, S., Gopika, K., Anusha, P. L., & Yadav, P. (2022). Morpho-physiological, quality traits and their association with seed yield in sesame (Sesamum indicum L.) indigenous collection under deficit moisture stress. *Plant Physiology Reports, 27*(1), 132–142. https://doi.org/10.1007/s40502-021-00621-0
120. Bandhu, S., Khot, M. B., Sharma, T., Sharma, O. P., Dasgupta, D., Mohapatra, S., Hazra, S., Khatri, O. P., & Ghosh, D. (2018). Single cell oil from oleaginous yeast grown on sugarcane bagasse-derived xylose: An approach toward novel biolubricant for low friction and wear. *ACS Sustainable Chemistry & Engineering, 6*, 275–283. https://doi.org/10.1021/acssuschemeng.7b02425
121. Kohli, I., Joshi, N. C., Mohapatra, S., & Varma, A. (2020). Extremophile—An adaptive strategy for extreme conditions and applications. *Current Genomics, 21*, 96–110. https://doi.org/10.2174/1389202921666200401105908
122. Pati, S., Mohapatra, S., Maity, S., Dash, A., & Samantaray, D. P. (2021). Polyhydroxyalkanoates for sustainable smart packaging of fruits. In *Environmental and agricultural microbiology: Advances and applications* (pp. 183–195).

123. Chatterjee, A., Nirwan, S., Mohapatra, S., Sharma, P., Agnihotri, A., & Shrivastava, N. (2022). Biochemical aspects of pathogenic variability in white rust infected Indian mustard. *Mycologia, 114*, 757–768. https://doi.org/10.1080/00275514.2022.2060007
124. Pati, S., Maity, S., Dash, A., Jema, S., Mohapatra, S., Das, S., & Samantaray, D. P. (2020). Biocompatible PHB production from bacillus species under submerged and solid-state fermentation and extraction through different downstream processing. *Current Microbiology, 77*, 1203–1209. https://doi.org/10.1007/s00284-020-01922-7
125. Agrahari, R., Sarraf, G., Joshi, N. C., Mohapatra, S., & Var, A. (2021). Insight of biopolymers and applications of polyhydroxyalkanoates. In *Microbial polymers: Applications and ecological perspectives* (pp. 177–191). https://doi.org/10.1007/978-981-16-0045-6_8
126. Kathpalia, M., Kumar, P., & Mohapatra, S. (2022). Homocysteine metabolism as a biomarker for cancer. In *Homocysteine metabolism in health and disease* (pp. 159–171). https://doi.org/10.1007/978-981-16-6867-8_8
127. Bharatiy, S. K., Hazra, M., Paul, M., Mohapatra, S., Samantaray, D., Dubey, R. C., Sanyal, S., Datta, S., & Hazra, S. (2016). In silico designing of an industrially sustainable carbonic anhydrase using molecular dynamics simulation. *ACS Omega, 1*, 1081–1103. https://doi.org/10.1021/acsomega.6b00041
128. Sarkar, M. S., Segu, H., Bhaskar, J. V., Jakher, R., Mohapatra, S., Shalini, K., Shivaji, S., & Reddy, P. A. (2018). Ecological preferences of large carnivores in remote, high-altitude protected areas: Insights from Buxa Tiger reserve, India. *ORYX, 52*, 66–77. https://doi.org/10.1017/S0030605317000060
129. Pati, S., Dash, A., Maity, S., Pattnaik, S., Mohapatra, S., & Samantaray, D. P. (2021). Nano biofertilizer for sustainable agriculture. In *Agriculturally important microorganisms: Mechanisms and applications for sustainable agriculture* (pp. 201–208). https://doi.org/10.1201/9781003245841-10
130. Mohanty, D., Misra, S., Mohapatra, S., & Sahu, P. S. (2018). Prebiotics and synbiotics: Recent concepts in nutrition. *Food Bioscience, 26*, 152–160. https://doi.org/10.1016/j.fbio.2018.10.008
131. Kumar, R., Kumari, R., Prasad, C., Deep, A., Singh, N., Namtak, S., Tiwari, V., Merugu, R., Mohapatra, S., & Singh, R. (2021). Influence of water quality on the diversity of macroinvertebrates in the Mandakini River in India. *Water Science and Technology: Water Supply, 21*, 1843–1860. https://doi.org/10.2166/WS.2021.020
132. Mohanty, D., Jena, R., Choudhury, P. K., Pattnaik, R., Mohapatra, S., & Saini, M. R. (2016). Milk derived antimicrobial bioactive peptides: A review. *International Journal of Food Properties, 19*, 837–846. https://doi.org/10.1080/10942912.2015.1048356
133. Mohapatra, S., Rath, S. N., Pradhan, S. K., Samantaray, D. P., & Rath, C. C. (2016). Secondary structural models (16S rRNA) of polyhydroxyalkanoates producing bacillus species isolated from different rhizospheric soil: Phylogenetics and chemical analysis. *International Journal Bioautomation, 20*, 329–338.
134. Mohapatra, S., Sarkar, B., Samantaray, D. P., Daware, A., Maity, S., Pattnaik, S., & Bhattacharjee, S. (2017). Bioconversion of fish solid waste into PHB using Bacillus subtilis based submerged fermentation process. *Environmental Technology (United Kingdom), 38*, 3201–3208. https://doi.org/10.1080/09593330.2017.1291759
135. Misra, S., Mohanty, D., & Mohapatra, S. (2021). Food-based probiotics: Functional dietary ingredients. In *Probiotics: Advanced food and health applications* (pp. 257–275). https://doi.org/10.1016/B978-0-323-85170-1.00006-3
136. Kumar, R., Kumari, R., Prasad, C., Tiwari, V., Singh, N., Mohapatra, S., Merugu, R., Namtak, S., & Deep, A. (2020). Phytoplankton diversity in relation to physicochemical attributes and water quality of Mandakini River, Garhwal Himalaya. *Environmental Monitoring and Assessment, 192*. https://doi.org/10.1007/s10661-020-08768-3
137. Mohapatra, S., Vishwakarma, K., Joshi, N. C., Maity, S., Kumar, R., Ramchander, M., Pattnaik, S., & Samantaray, D. P. (2021). A review on PHAs: The future biopolymer. In *Environmental and agricultural microbiology: Advances and applications* (pp. 83–117).

138. Sahoo, N., Pattnaik, S., Pattnaik, M., & Mohapatra, S. (2021). Methane emission and strategies for mitigation in livestock. *Environmental and Agricultural Microbiology: Advances and Applications*, 257–274.
139. Mohanty, D. P., Mohapatra, S., Misra, S., & Sahu, P. S. (2016). Milk derived bioactive peptides and their impact on human health—A review. *Saudi Journal of Biological Sciences, 23*, 577–583. https://doi.org/10.1016/j.sjbs.2015.06.005
140. Kohli, I., Mohapatra, S., Kumar, P., Goel, A., Varma, A., & Joshi, N. C. (2022). Role of bacterial endophytes in the promotion of plant growth. In *Bacterial endophytes for sustainable agriculture and environmental management* (pp. 235–253). https://doi.org/10.1007/978-981-16-4497-9_12
141. Samantaray, D., Mohapatra, S., & Mishra, B. B. (2014). Microbial bioremediation of industrial effluents. In *Microbial biodegradation and bioremediation* (pp. 326–339). https://doi.org/10.1016/B978-0-12-800021-2.00014-5
142. Pattnaik, S., Dash, D., Mohapatra, S., Pattnaik, M., Marandi, A.K., Das, S., Samantaray, D.P. Corrigendum to "Improvement of rice plant productivity by native Cr(VI) reducing and plant growth promoting soil bacteria Enterobacter cloacae" Chemosphere (2020) 240, 124895, https://doi.org/10.1016/j.chemosphere.2019.124895. Chemosphere. 238, (2020). 10.1016/j.chemosphere.2019.125011.
143. Mohapatra, S., Mohanta, P. R., Sarkar, B., Daware, A., Kumar, C., & Samantaray, D. P. (2017). Production of polyhydroxyalkanoates (PHAs) by bacillus strain isolated from waste water and its biochemical characterization. *Proceedings of the National Academy of Sciences India Section B—Biological Sciences, 87*, 459–466. https://doi.org/10.1007/s40011-015-0626-6
144. Tomer, P., Shroti, G. K., Mohapatra, S., Ghosh, D., Jaiswal, S., Garg, D., Lahiri, D., Kumar, N., Khatri, O. P., & Hazra, S. (2023). Fabrication of HA nano-crystal reinforced PHAs based composites for orthopedic fracture-fixing accessories using sustainable resources. *Environment, Development and Sustainability*. https://doi.org/10.1007/s10668-023-03912-8
145. Bandekar, D., Mohapatra, S., Hazra, M., Hazra, S., & Biswas, S. (2021). N-terminal truncation of VC0395_0300 protein from Vibrio cholerae does not lead to loss of diguanylate cyclase activity. *Biophysical Chemistry, 268*. https://doi.org/10.1016/j.bpc.2020.106493
146. Vishwakarma, K., Kumar, N., Shandilya, C., Mohapatra, S., Bhayana, S., & Varma, A. (2020). Revisiting plant–microbe interactions and microbial consortia application for enhancing sustainable agriculture: A review. *Frontiers in Microbiology, 11*. https://doi.org/10.3389/fmicb.2020.560406
147. Mohapatra, S., Maity, S., Dash, H. R., Das, S., Pattnaik, S., Rath, C. C., & Samantaray, D. (2017). Bacillus and biopolymer: Prospects and challenges. *Biochemistry and Biophysics Reports, 12*, 206–213. https://doi.org/10.1016/j.bbrep.2017.10.001
148. Kumar, R., Merugu, R., Mohapatra, S., & Sharma, S. (2022). Extremophiles: Life of microorganisms in extreme environments. In *Extremophiles: General and plant biomass based biorefinery* (pp. 43–66). https://doi.org/10.1201/9781003014188-3
149. Pradhan, A., Pahari, A., Mohapatra, S., & Mishra, B. B. (2017). Phosphate-solubilizing microorganisms in sustainable agriculture: Genetic mechanism and application. In *Microorganisms for sustainability* (pp. 81–97). https://doi.org/10.1007/978-981-10-7380-9_5
150. Devadarshini, D., Mohapatra, S., Pati, S., Maity, S., Rath, C. C., Jena, P. K., & Samantaray, D. (2023). Evaluation of PHAs production by mixed bacterial culture under submerged fermentation. *Biologia (Bratisl), 78*, 895–902. https://doi.org/10.1007/s11756-022-01302-5
151. Agarwal, K. M., Mohapatra, S., Sharma, P., Sharma, S., Bhatia, D., & Mishra, A. (2020). Study and overview of the novel corona virus disease (COVID-19). *Sensors International, 1*. https://doi.org/10.1016/j.sintl.2020.100037
152. Tanaya, K., Mohapatra, S., Maity, S., & Samantaray, D. P. (2023). Fluorescent-pigmented microbial polymer nanoparticles: An approach for replacement of bioimaging agent in the drug delivery system. *Applied Biochemistry and Microbiology, 59*, 378–383. https://doi.org/10.1134/S0003683823030171

153. Pattnaik, S., Dash, D., Mohapatra, S., Pati, S., Devadarshini, D., Samal, S., Pattnaik, M., Maity, S., Mishra, S. K., & Samantaray, D. (2024). Reclamation of chromium-contaminated soil by native Cr(VI)-reducing and PHA-accumulating Bacillus aryabhattai CTSI-07. *International Microbiology, 27*, 731–742. https://doi.org/10.1007/s10123-023-00421-6
154. Mohapatra, S., Mishra, Y., & Dash, L. (2020). Covid-19: Analysing the legal nuances of the lockdown order. *Indian Journal of Forensic Medicine and Toxicology, 14*, 612–616. https://doi.org/10.37506/ijfmt.v14i4.11553
155. Pati, S., Mohanty, M. K., Mohapatra, S., & Samantaray, D. (2022). Bioethanol production by Enterobacter hormaechei through carbon monoxide-rich syngas fermentation. *International Journal of Energy Research, 46*, 20096–20106. https://doi.org/10.1002/er.8678
156. Maity, S., Das, S., Mohapatra, S., Tripathi, A. D., Akthar, J., Pati, S., Pattnaik, S., & Samantaray, D. P. (2020). Growth associated polyhydroxybutyrate production by the novel Zobellellae tiwanensis strain DD5 from banana peels under submerged fermentation. *International Journal of Biological Macromolecules, 153*, 461–469. https://doi.org/10.1016/j.ijbiomac.2020.03.004
157. Takkar, S., Shandilya, C., Agrahari, R., Chaurasia, A., Vishwakarma, K., Mohapatra, S., Varma, A., & Mishra, A. (2022). Green technology: Phytoremediation for pesticide pollution. In *Phytoremediation technology for the removal of heavy metals and other contaminants from soil and water* (pp. 353–375). https://doi.org/10.1016/B978-0-323-85763-5.00008-8
158. Bandekar, D., Chouhan, O. P., Mohapatra, S., Hazra, M., Hazra, S., & Biswas, S. (2017). Putative protein VC0395_0300 from Vibrio cholerae is a diguanylate cyclase with a role in biofilm formation. *Microbiological Research, 202*, 61–70. https://doi.org/10.1016/j.micres.2017.05.003
159. Mohapatra, S., Samantaray, D. P., Samantaray, S. M., Mishra, B. B., Das, S., Majumdar, S., Pradhan, S. K., Rath, S. N., Rath, C. C., Akthar, J., & Achary, K. G. (2016). Structural and thermal characterization of PHAs produced by Lysinibacillus sp. through submerged fermentation process. *International Journal of Biological Macromolecules, 93*, 1161–1167. https://doi.org/10.1016/j.ijbiomac.2016.09.077
160. Kumar Singla, Y., Dwivedi, D. K., & Arora, N. (2017). Modeling the impact–sliding wear characteristics of rare earth additive iron-based hardfacing alloys. *Proceedings of the Institution of Mechanical Engineers, Part J: Journal of Engineering Tribology, 231*, 1486–1499.
161. Singla, Y. K., Dwivedi, D. K., & Arora, N. (2015). On the modeling of dry sliding adhesive wear parameters of vanadium additive iron-based alloys at elevated temperatures. *Surface and Coating Technology, 283*, 223–233.
162. Singla, Y. K., Chhibber, R., Bansal, H., & Kalra, A. (2015). Wear behavior of aluminum alloy 6061-based composites reinforced with SiC, Al2O3, and red mud: A comparative study. *Journal of Metals, 67*, 2160–2169.
163. Sharma, S. K., Kumar, A., & Sharma, R. C. (2014). Challenges in railway vehicle modeling and simulations. In *International conference on newest drifts in mechanical engineering ICNDME-2014* (pp. 453–459).
164. Sharma, S. K., & Kumar, A. (2014). A comparative study of Indian and worldwide railways. *International Journal of Mechanical Engineering and Robotics Research, 1*, 114–120.
165. Sharma, S. K., Sharma, R. C., Kumar, A., & Palli, S. (2015). Challenges in rail vehicle-track modeling and simulation. *International Journal of Vehicle Structures & Systems, 7*, 1.
166. Palli, S., Sharma, R. C., Sharma, S. K., Muddada, V., & Sharma, N. (2020). A case study on noise pollution and its effects. *Journal of Critical Reviews, 7*, 261–267.
167. Sharma, R. C., Sharma, S. K., Sharma, N., & Sharma, S. (2020). Analysis of ride and stability of an ICF railway coach. *International Journal of Vehicle Noise and Vibration, 16*, 127–144.
168. Palli, S., Sharma, R. C., Sharma, S. K., & Chintada, V. B. (2020). On methods used for setting the curve for railway tracks. *Journal of Critical Reviews, 7*, 241–246.
169. Lee, J., Sharma, S.K. (2020) Numerical investigation of critical speed analysis of high-speed rail vehicle. 한국정밀공학회 학술발표대회 논문집. 696.

170. Sharma, S., Sharma, R. C., Sharma, S. K., Sharma, N., Palli, S., & Bhardawaj, S. (2020). Vibration isolation of the quarter car model of road vehicle system using dynamic vibration absorber. *International Journal of Vehicle Structures & Systems., 12*, 513–516.
171. Acharya, A., Gahlaut, U., Sharma, K., Sharma, S. K., Vishwakarma, P. N., & Phanden, R. K. (2020). Crashworthiness analysis of a thin-walled structure in the frontal part of automotive chassis. *International Journal of Vehicle Structures & Systems, 12*, 517–520.
172. Sharma, S. K., & Lee, J. (2020). Finite element analysis of a fishplate rail joint in extreme environment condition. *International Journal of Vehicle Structures & Systems, 12*, 503–506.
173. Bhardawaj, S., Sharma, R. C., & Sharma, S. K. (2020). Development of multibody dynamical using MR damper based semi-active bio-inspired chaotic fruit fly and fuzzy logic hybrid suspension control for rail vehicle system. *Proceedings of the Institution of Mechanical Engineers, Part K: Journal of Multi-Body Dynamics, 234*, 723–744.
174. Sharma, S. K., & Lee, J. (2020). Design and development of smart semi active suspension for nonlinear rail vehicle vibration reduction. *International Journal of Structural Stability and Dynamics, 20*, 2050120.
175. Sharma, S. K., Phan, H., & Lee, J. (2020). An application study on road surface monitoring using DTW based image processing and ultrasonic sensors. *Applied Sciences, 10*, 4490.
176. Sharma, R. C., Sharma, S., Sharma, S. K., & Sharma, N. (2020). Analysis of generalized force and its influence on ride and stability of railway vehicle. *Noise & Vibration Worldwide, 51*, 95–109.
177. Bhardawaj, S., Sharma, R. C., & Sharma, S. K. (2020). Ride analysis of track-vehicle-human body interaction subjected to random excitation. *Journal of the Chinese Society of Mechanical Engineers, 41*, 227–236.
178. Bhardawaj, S., Sharma, R. C., & Sharma, S. K. (2020). Development in the modeling of rail vehicle system for the analysis of lateral stability. *Materials Today Proceedings, 25*, 610–619.
179. Sharma, S. K., Sharma, R. C., & Sharma, N. (2020). Combined multi-body-system and finite element analysis of a rail locomotive crashworthiness. *International Journal of Vehicle Structures & Systems, 12*, 428–435.
180. Sharma, R. C. (2016b). Evaluation of passenger ride comfort of Indian rail and road vehicles with ISO 2631-1 standards part 2 – Simulation. *International Journal of Vehicle Structures and Systems, 8, No. 1*. https://doi.org/10.4273/ijvss.8.1.02
181. Mohapatra, S., Pattnaik, S., Maity, S., Sharma, S., Akhtar, J., Pati, S., Samantaray, D. P., & Varma, A. (2020). Comparative analysis of PHAs production by Bacillus megaterium OUAT 016 under submerged and solid-state fermentation. *Saudi Journal of Biological Sciences, 27*, 1242–1250.
182. Bhardawaj, S., Sharma, R. C., & Sharma, S. K. (2020). Analysis of frontal car crash characteristics using ANSYS. *Materials Today Proceedings, 25*, 898–902.
183. Bhardawaj, S., Sharma, R. C., & Sharma, S. K. (2019). A survey of railway track modelling. *International Journal of Vehicle Structures & Systems, 11*, 508–518.
184. Bhardawaj, S., Sharma, R. C., & Sharma, S. K. (2019). Development and advancement in the wheel-rail rolling contact mechanics. In *IOP conference series: Materials science and engineering* (p. 012034). IOP Publishing.
185. Choppara, R. K., Sharma, R. C., Sharma, S. K., & Gupta, T. (2019). Aero dynamic cross wind analysis of locomotive. In *IOP conference series: Materials science and engineering* (p. 012035). IOP Publishing.
186. Sharma, S. K., & Sharma, R. C. (2018). Pothole detection and warning system for Indian roads. In *Advances in interdisciplinary engineering: Select proceedings of FLAME* (pp. 511–519). Springer Singapore. (2019).
187. Sinha, A. K., Sengupta, A., Gandhi, H., Bansal, P., Agarwal, K. M., Sharma, S. K., Sharma, R. C., & Sharma, S. K. (2019). Performance enhancement of an all-terrain vehicle by optimizing steering, powertrain and brakes. In *Advances in engineering design: Select proceedings of FLAME 2018* (pp. 207–215). Springer Singapore.

188. Goyal, S., Anand, C. S., Sharma, S. K., & Sharma, R. C. (2019). Crashworthiness analysis of foam filled star shape polygon of thin-walled structure. *Thin-Walled Structures, 144*, 106312.
189. Goswami, B., Rathi, A., Sayeed, S., Das, P., Sharma, R. C., & Sharma, S. K. (2019). Optimization design for aerodynamic elements of Indian locomotive of passenger train. In *Advances in engineering design: Select proceedings of FLAME 2018* (pp. 663–673). Springer Singapore.
190. Sharma, S. K., Saini, U., & Kumar, A. (2019). Semi-active control to reduce lateral vibration of passenger rail vehicle using disturbance rejection and continuous state damper controllers. *Journal of Vibration Engineering & Technologies., 7*, 117–129.
191. Sharma, S. K. (2019). Multibody analysis of longitudinal train dynamics on the passenger ride performance due to brake application. *Proceedings of the Institution of Mechanical Engineers, Part K: Journal of Multi-body Dynamics., 233*, 266–279.
192. Sharma, S. (2018). Vibration control in quarter-car model with magnetorheological (MR) dampers using Bingham model. *Journal of Applied Mechanical Engineering, 7*, 1000299.
193. Lora Anusha, P., Ratnakumar, P., Pandey, B. B., Sandhya Rani, P., Umamahesh, V., Reddi Sekhar, M., Chandrika, V., Yadav, P., Mohapatra, S., & Padmaja, D. (2023). Phenotypic characterization of sesame (Sesamum indicum L.) revealed promising genotypes for moisture stress conditions. *Plant Genetic Resources: Characterization and Utilization, 21*(3), 218–228. https://doi.org/10.1017/S1479262123000667
194. Sharma, S. K., & Kumar, A. (2018). Impact of longitudinal train dynamics on train operations: A simulation-based study. *Journal of Vibration Engineering & Technologies., 6*, 197–203.
195. Sharma, S. K., & Sharma, R. C. (2018). An investigation of a locomotive structural crashworthiness using finite element simulation. *SAE International Journal of Commercial Vehicles, 11*, 235–243.
196. Sharma, R. C., Sharma, S. K., & Palli, S. (2018). Rail vehicle modelling and simulation using Lagrangian method. *International Journal of Vehicle Structures and Systems, 10*, 188–194.
197. Sharma, R. C., & Sharma, S. K. (2018). Sensitivity analysis of three-wheel vehicle's suspension parameters influencing ride behavior. *Noise & Vibration Worldwide, 49*, 272–280.
198. Palli, S., Koona, R., Sharma, S. K., & Sharma, R. C. (2018). A review on dynamic analysis of rail vehicle coach. *International Journal of Vehicle Structures & Systems (IJVSS), 10*.
199. Sharma, S. K., & Sharma, R. C. (2018). Simulation of quarter-car model with magnetorheological dampers for ride quality improvement. *International Journal of Vehicle Structures & Systems, 10*, 169–173.
200. Sharma, S. K., & Kumar, A. (2018). Disturbance rejection and force-tracking controller of nonlinear lateral vibrations in passenger rail vehicle using magnetorheological fluid damper. *Journal of Intelligent Material Systems and Structures, 29*, 279–297.
201. Sharma, S. K., & Kumar, A. (2018). Ride comfort of a higher speed rail vehicle using a magnetorheological suspension system. *Proceedings of the Institution of Mechanical Engineers, Part K: Journal of Multi-Body Dynamics, 232*, 32–48.
202. Shukla, S., Kumar, U., Sharma, S. K., Gupta, P., & Kumar, A. (2017). Light weight freight rolling stock bogie frame: Design methodology validated with field oscillation trials. *International Journal of Vehicle Structures & Systems, 9*, 245–250.
203. Sharma, R. C., Palli, S., Sharma, S. K., & Roy, M. (2017). Modernization of railway track with composite sleepers. *International Journal of Vehicle Structures & Systems, 9*, 321–329.
204. Sharma, S. K., & Kumar, A. (2017). Ride performance of a high speed rail vehicle using controlled semi active suspension system. *Smart Materials and Structures, 26*, 055026.
205. Sharma, S. K., & Kumar, A. (2017). Impact of electric locomotive traction of the passenger vehicle Ride quality in longitudinal train dynamics in the context of Indian railways. *Mechanics & Industry, 18*, 222.
206. Sharma, S. K. (2013). Zero energy building envelope components: A review. *International Journal of Engineering Research and Applications, 3*, 662–675.
207. Sharma, S. K., & Kumar, A. (2016). Dynamics analysis of wheel rail contact using FEA. *Procedia Engineering, 144*, 1119–1128.

208. Sharma, S. K., & Chaturvedi, S. (2016). Jerk analysis in rail vehicle dynamics. *Perspectives in Science (Neth), 8*, 648–650.
209. Sharma, S. K., & Kumar, A. (2016). The impact of a rigid-flexible system on the ride quality of passenger bogies using a flexible carbody. In *Proceedings of the third international conference on railway technology: Research, development and maintenance* (p. 87). Civil-Comp Press.
210. Lavania, S., & Sharma, S. K. (2011). An explicit approach to compare crosstalk noise and delay in VLSI RLC interconnect modeled with skin effect with step and ramp input. *Journal of VLSI Design Tools & Technology., 1*, 1–8.
211. Palli, S., Koona, R., Sharma, R. C., & Muddada, V. (2015). Dynamic analysis of Indian railway integral coach factory bogie. *International Journal of Vehicle Structures and Systems, 7*, 16–20. https://doi.org/10.4273/ijvss.7.1.03
212. Palli, S., & Koona, R. (2015). Analyses of dynamic response of a railway bogie. *International Journal of Vehicle Noise and Vibration, 11*, 103–113. https://doi.org/10.1504/IJVNV.2015.070015
213. Swathi, K., & Srihari, P. (2015). Heat transfer enhancement in a tube using rectangular strip inserts. *International Journal of Applied Engineering Research, 10*, 41532–41544.
214. Palli, S., Sharma, R. C., & Rao, P. P. D. (2017). Dynamic behaviour of a 7 DoF passenger car model. *International Journal of Vehicle Structures and Systems, 9*, 57–63. https://doi.org/10.4273/ijvss.9.1.12
215. Sharma, R. C., Palli, S., & Koona, R. (2017). Stress and vibrational analysis of an Indian railway RCF bogie. *International Journal of Vehicle Structures and Systems, 9*, 296–302. https://doi.org/10.4273/ijvss.9.5.06
216. Sharma, R. C., Avesh, M., Sharma, S. K., Sharma, S., Palli, S., & Sharma, N. (2020). Linear and non-linear stability analysis of a constrained railway wheel axle. *International Journal of Vehicle Structures and Systems, 12*, 128–133. https://doi.org/10.4273/ijvss.12.2.04
217. Palli, S., Dontikurti, R., Sharma, R. C., & Sharma, N. (2021). Analysis of dynamic response of a railway locomotive using ANSYS. *International Journal of Vehicle Structures and Systems., 13*, 250–255. https://doi.org/10.4273/ijvss.13.2.21
218. Sharma, R. C., Sharma, N., Singh, G., & Palli, S. (2021). Modeling and simulation of human body-vehicle-track system for the investigation of ride comfort. *SAE International Journal of Commercial Vehicles, 15*. https://doi.org/10.4271/02-15-02-0008
219. Sharma, R. C., Palli, S., Avesh, M., & Sharma, N. (2021). Vibration isolation of railway vehicle car body using semi-active suspension. *International Journal of Vehicle Structures and Systems., 13*, 482–487. https://doi.org/10.4273/ijvss.13.4.17
220. Gopala Rao, L. V. V., Sharma, R. C., Satyanarayana, V. S. V., & Palli, S. (2022). Stochastic optimal preview control response of a quarter-car nonlinear suspension model using spectral decomposition method. *Noise and Vibration Worldwide., 53*, 225–232. https://doi.org/10.1177/09574565221093234
221. Palli, S., Duppala, A., Sharma, R. C., & Rao, L. V. V. G. (2022). Dynamic simulation of automotive vehicle suspension using MATLAB simulink. *International Journal of Vehicle Structures and Systems, 14*, 310–314. https://doi.org/10.4273/ijvss.14.3.04
222. Satyanarayana, V. S. V., Sharma, R. C., Sateesh, B., Gopala Rao, L. V. V., Mohan Rao, N., & Palli, S. (2022). Optimum response of a nonlinear passive vehicle suspension system under random road excitations. *SAE International Journal of Commercial Vehicles, 16*. https://doi.org/10.4271/02-16-01-0004
223. Lokanadham, D., Sharma, R. C., Palli, S., & Bhardawaj, S. (2022). Wear rate modelling and analysis of limestone slurry particulate composites using the fuzzy method. *International Journal on Recent and Innovation Trends in Computing and Communication., 10*, 133–143. https://doi.org/10.17762/ijritcc.v10i1s.5818
224. Sharma, R. C., Palli, S., Gopala Rao, L. V. V., Duppala, A., & Sharma, S. K. (2022). Four-wheel vehicle response under bump, pothole, harmonic, and random excitations using bond

graph/simulink technique. *SAE International Journal of Commercial Vehicles, 16*. https://doi.org/10.4271/02-16-02-0008

225. Sharma, R. C., Gopala Rao, L. V. V., Srikar, B., Palli, S., Appala Naidu, Y., Sharma, N., & Sharma, S. K. (2025a). Investigation of the dynamic behavior of hybrid composites using finite element and experimental method. *Journal of the Chinese Institute of Engineers, 48*(2), 158–171. https://doi.org/10.1080/02533839.2025.2453139
226. Sharma, R. C., Palli, S., Bhardawaj, S., & Rao, L. V. V. G. (2022). Advances in human biodynamic modeling for analysis of whole-body vibrations. *International Journal of Vehicle Structures and Systems, 14*, 503–511. https://doi.org/10.4273/ijvss.14.4.17
227. Agrahari, R., Mohapatra, S., Kumar, K. S., Mohanty, D., Kumar, R., & Merugu, R. (2024). Immuno-metabolic interplay in health and disease. In A. Das Mohapatra & P. Sahu (Eds.), *Advances in Immunology and Immuno-techniques. Interdisciplinary Biotechnological Advances* (pp. 239–274).
228. Sharma, R. C., Palli, S., & Bhardawaj, S. (2023). A review of wheel-rail dynamic behavior on curved tracks. *International Journal of Vehicle Structures and Systems, 15*, 19–25. https://doi.org/10.4273/ijvss.15.1.04
229. Jyotiprasad, G., Dowluru, S., Rallabandi, S. R., Sharma, N., Sharma, R. C., & Palli, S. (2023). *Thermal-structural and macromechanical behavior analysis of graphite-based laminated composite*. No. 2023-01-5041. SAE Technical Paper. https://doi.org/10.4271/2023-01-5041
230. Saikiran, C., Palli, S., Bondala, R., Sharma, R. C., Sreeramulu, D., & Dontikurti, R. (2023). Experimental Investigation on thermal behaviour of silicon carbide (SiC) based polymer composites under hygrothermal conditions. *International Journal of Vehicle Structures and Systems, 15*, 222–226. https://doi.org/10.4273/ijvss.15.2.15
231. Singh, B. J., Palli, S., Kumar, P., & Bahl, S. (2023). Behavioural investigation of hardness profile for dissimilar TIG-weldment between P22 and P91 steels. *International Journal of Vehicle Structures and Systems, 15*, 31–36. https://doi.org/10.4273/ijvss.15.1.06
232. Naveen, Y., Lokanadham, D., Naidu, D. R., Sharma, R. C., Palli, S., & Lila, M. K. (2023). An experimental study on the influence of blended Karanja biodiesel on diesel engine characteristics. *International Journal of Vehicle Structures and Systems, 15*, 154–160. https://doi.org/10.4273/ijvss.15.2.02
233. Bhardawaj, S., Jha, A. K., Sharma, R. C., Palli, S., Lokanadham, D., Sharma, N., & Dhiman, N. (2023). Finite element analysis of two disk rotor system used in automobile turbochargers. *International Journal of Vehicle Structures and Systems, 15*, 501–504. https://doi.org/10.4273/ijvss.15.4.09
234. Sharma, R. C., Palli, S., Bhardawaj, S., Jha, A. K., Duppala, A., Sharma, N., Sharma, S., & Dhiman, N. (2023). Dynamic behaviour of railway vehicle under bump and pothole track irregularities. *International Journal of Vehicle Structures and Systems, 15*, 566–569. https://doi.org/10.4273/ijvss.15.4.22
235. Raju, S., Palli, S., Devi Prasad, P., Menda, V. R., & Ramakrishna, B. (2024). A hybrid AHP-TOPSIS, MOORA technique for multi-objective optimization of thermal, mechanical, and water absorption behavior of epoxy/hemp, pine apple, and palm fiber composites. *Journal of the Chinese Institute of Engineers, Transactions of the Chinese Institute of Engineers, Series A, 47*, 23–36. https://doi.org/10.1080/02533839.2023.2274092
236. Adeoye, A. O., Quadri, R. O., & Lawal, O. S. (2020). Pyrolysis of biomass as a suitable alternative to fossil fuel energy in Nigeria: An overview. *Journal Of Wastes and Biomass Management, 3*(1), 27–30. https://doi.org/10.26480/jwbm.01.2021.27.30
237. Gopala Rao, L. V. V., Satyanarayana, V. S. V., Sharma, R. C., Palli, S., Raturi, A., Kharola, A., & Sharma, A. (2024). Preview control of the random response of a full car vehicle model traversing a rough road. *International Journal of Vehicle Noise and Vibration, 20*, 70–88. https://doi.org/10.1504/IJVNV.2024.138105
238. Sharma, A., Sharma, V. S., Gupta, S., Sharma, R. C., Palli, S., & Sharma, N. (2024). Additive manufacturing in the COVID-19 pandemic: Equipment and challenges? *Journal of Micromanufacturing*. https://doi.org/10.1177/25165984241228083

239. Bammidi, R., Madivada, H., Dowluru, S., Palli, S., Sharma, R. C., & Muddada, V. (2024). *Mechano-tribological investigations of Al-%Wt. of boron nitride as metal-ceramic composites*. https://doi.org/10.1063/5.0182965
240. Ranganath, B. A., Duppala, A., Palli, S., Rao, L. V. V., & Sharma, R. C. (2024). Numerical analysis and experimental study of FRP composites for damping applications. In *AIP conference proceedings*. AIP Publishing. https://doi.org/10.1063/5.0184802
241. Adeoye, A. O., Quadri, R. O., & Lawal, O. S. (2025b). Sustainable remediation of vancomycin polluted water using pyrolysis biochar of pressed oil palm fruit fibre. *Journal of the Nigerian Society of Physical Sciences*, 2558. https://doi.org/10.46481/jnsps.2025.2558
242. Menda, V., Dora, S., Palli, S., Rallabandi, S. R., Sharma, R. C., Sharma, N., & Bondala, R. (2024). Mechanical and wear characteristics of aluminium-7075/graphene/TiB2 prepared by stir casting. *Engineering Research Express, 6*. https://doi.org/10.1088/2631-8695/ad58a8
243. Sura, H. S., Avesh, M., & Mohapatra, S. (2023). Design and modelling of digital twin technology to improve freight logistics. In *Energy, environment, and sustainability* (pp. 481–513). https://doi.org/10.1007/978-981-99-2150-8_20
244. Pattnaik, S., Pati, S., Devadarshini, D., Tanaya, K., Mishra, B. B., Samantaray, D., Mohapatra, S., & Dash, D. (2022). Microbial bioremediation of Cr(VI)-contaminated soil for sustainable agriculture. In *Microbial biodegradation and bioremediation: Techniques and case studies for environmental pollution* (pp. 395–407). https://doi.org/10.1016/B978-0-323-85455-9.00008-4
245. Srivastava, A., Lavania, S., & Mohapatra, S. (2023). Predicting the effect on land values after introducing high-speed rail. In *Energy, environment, and sustainability* (pp. 157–196). https://doi.org/10.1007/978-981-99-2150-8_8
246. Agrahari, R., Mohanty, S., Vishwakarma, K., Nayak, S. K., Samantaray, D., & Mohapatra, S. (2021). Update vision on COVID-19: Structure, immune pathogenesis, treatment and safety assessment. *Sensors International., 2*. https://doi.org/10.1016/j.sintl.2020.100073
247. Tanaya, K., Mohapatra, S., & Samantaray, D. (2024). Formulation of biogenic fluorescent pigmented PHB nanoparticles from Rhodanobacter sp. for drug delivery. *Antonie van Leeuwenhoek, International Journal of General and Molecular Microbiology, 117*. https://doi.org/10.1007/s10482-024-01973-x
248. Adeoye, A. O., Quadri, R. O., & Lawal, O. S. (2025). Assessment of physicochemical properties of Lafarge cement towards achieving a sustainable society. *Proceedings of the Nigerian Society of Physical Sciences*, 180. https://doi.org/10.61298/pnspsc.2025.2.180

Waste to Resource: Smart 3R Approaches for Sustainable and Resilient Construction

Badana Govindarajulu, Pedapenki Dinakar, Ponguru Naga Sowjanya, and Shaik Sydha

1 Introduction

The development sector uses a lot of unique resources and generates a lot of development waste. Development waste is waste produced by the building and construction sector during routine construction and building, street construction and demolition operations, cleaning of building sites, rehabilitation of buildings, and incorporating soil exhumation. In any case, most waste is often dumped in landfills. Development waste, according to the CEOs, is a tool used to control the expense of disposing of garbage from development projects and to encourage methods of recycling and reusing waste in order to minimize the quantity of waste that is eventually disposed of in landfills. Given that there are many different types of development trash that can be recovered or repurposed, it is crucial to view development waste as an asset.

The development industry is held accountable for arranging for unthinkable feats, but regrettably, it can also be held accountable for creating shortcomings. Research indicates that more than 50% of the time allocated to development in the United States is wasted on pointless endeavors. Waste is defined as anything that can be rid of without decreasing the estimate of work for the client, and many professionals often fail to recognize it.

Safety, health, and environmental issues are significantly impacted by construction goods and practices [1]. There is no doubting the importance of a healthy construction sector, since all people in contemporary society are immediately impacted

B. Govindarajulu (✉) · P. Dinakar
Department of Civil Engineering, Aditya Institute of Technology and Management, Tekkali, Andhra Pradesh, India

P. Naga Sowjanya · S. Sydha
Narasaraopeta Engineering College, Narasaraopeta, Andhra Pradesh, India

Y. Kumar Singla et al. (eds.), *Digital Manufacturing in Industry 5.0*,
https://doi.org/10.1007/978-3-031-91500-0_25

by its products and/or processes [2]. The building sector in Malaysia has expanded quickly and significantly improved. Therefore, inadequate development planning may lead to unsustainable urban growth processes such waste generation, pollution, and inefficient land use. To address the underlying causes of strong and ineffective, government participation is required. When it comes to implementing government programs that support sustainable development, the construction sector is essential. Thus, even in cases where waste reduction is required, achieving a sustainable future requires the implementation of a comprehensive waste management strategy. Before being disposed of, the three-pronged grouping of waste management techniques known as the 3R method (reduction, reuse, and recycling) will be heavily utilized. One strategy for effective waste management in the construction industry is waste reduction by 3R. 3R's waste management approach is still relatively new in Malaysia's building industry. In addition to the 3Rs of "reduction, reuse, and recycling," the notion of waste management should also integrate re-imagination and re-design in order to enhance resource efficiency through reconsideration [3]. Construction waste is defined as waste produced during the many phases of building, remodeling, and demolition. This includes the trash generated by clearing the site, building, building renovation, building construction, and designing, acquiring, excavating, or creating new land. This suggests that Malaysia's construction sector has been growing at a double-digit rate over the last several years and will continue to do so. Construction of commercial buildings, residential development projects, and large-scale infrastructure projects should all produce a sizable amount of waste building. Numerous studies on building waste have been conducted, and the findings have demonstrated numerous detrimental effects on a nation's social, economic, and environmental aspects [3]. Such outcomes cause the building industry's efficiency to decline. Waste from sustainable construction is appropriate to prevent any potential harm. Thus, this article addresses topics on the review of sustainable construction waste management in Malaysia in order to maximize the 3R approach and limit waste disposal by executing an ecologically friendly and thorough strategy across the whole lifecycle of the construction project. It offers several criteria as well.

2 Literature Review: Waste to Resource

Shah et al. [1] and Singla et al. [2] explore how material science advancements, particularly in welding and sustainability, can contribute to reducing construction waste through improved material usage and recycling. Additionally, research on renewable energy systems, such as solar power, and tribological studies [3–7] emphasize the potential for using renewable energy and efficient materials in construction. By focusing on sustainable energy systems and optimizing material performance, these studies underscore how the construction industry can adopt renewable energy to reduce its environmental footprint.

The concept of the digital twin, as explored by several studies, has been increasingly applied in construction to optimize logistics and reduce material wastage [8]. Shah et al. [9, 10] and other studies [11, 12] demonstrate how digital innovation, combined with environmental consciousness, can lead to the bioremediation of contaminated construction sites and promote the reuse of land resources. This aligns with the broader goals of the circular economy, a core component of Industry 5.0 [13, 14].

Research focusing on material durability and wear resistance, such as the integration of alloying elements and nano fly ash [15, 16], is highly relevant to the reuse and recycling aspects of the 3R approach in construction. More durable materials reduce the frequency of replacements and contribute to lower resource consumption. Studies on surface integrity [17] and the impact of nanoparticulates on lubrication [15] further highlight how material science innovations can enhance construction efficiency, leading to fewer resources being used throughout the lifecycle of construction projects.

Earlier studies, such as those by Shah et al. [1, 18, 19], emphasize the importance of dynamic simulations and suspension systems in improving the longevity of construction equipment. These advancements contribute to reduced maintenance needs and less frequent equipment replacement, further promoting the 3R approach in construction practices.

Recent research by Adeoye et al. (2024) investigates the valorization of lignocellulosic waste, particularly in bio-oil production through pyrolysis, which has potential applications in sustainable construction materials [20]. Further studies by Adeoye et al. [21, 22] focus on upgrading crude bio-oil and sustainable energy production from biomass [23] These advancements in waste valorization align with the goals of reducing waste and promoting circular economy practices in construction [24–29]. The use of catalytic pyrolysis and nanomaterials for enhanced biomass conversion provides additional pathways for optimizing the use of construction waste [30–32]. Moreover, the introduction of circular economy concepts through the pyrolysis of plastic waste and railway projects demonstrates how recycled materials can be reintroduced into the construction cycle [33–40].

Material innovation in construction, such as fiber surface functionalization [41–43] contributes to the enhancement of resource reuse. The integration of advanced materials in construction, as highlighted by Sunil et al. [44–52], shows how these advancements can support sustainable practices by promoting durability and reusability. Research on electric vehicle adoption [19, 53–67] and intelligent transportation systems [68–79] further demonstrates the role of technology in reducing transportation emissions and resource consumption during construction activities. These innovations reduce the environmental impact of transporting materials and equipment.

The development of sustainable fuels from bio-oil sources and carbon capture technologies [80] has been investigated in relation to reducing construction waste and emissions. Adeoye et al. highlight the role of CO_2 adsorption systems, which could be applied to capture and recycle emissions from construction sites.

Furthermore, research into the pyrolysis of various organic wastes, such as coffee and eucalyptus sawdust [81–83], offers innovative ways to convert construction site waste into energy sources [84]. These studies on renewable energy production and environmental sustainability are directly relevant to the construction industry's transition to waste-to-resource strategies [33, 85–88].

Microbiological advancements, as explored by Mohapatra et al. [89], also play a role in construction, particularly in biopolymer production for sustainable building materials [12, 26, 33, 90–96]. These bio-based materials, which can be reused and recycled, align with the 3R strategies and offer environmentally friendly alternatives to traditional construction materials [97–108]. Retrofitting existing buildings for energy efficiency, as examined in other studies [109–125]], supports the reduction of resource consumption by extending the lifespan of buildings and minimizing waste generation.

The need for advanced control systems and energy-efficient solutions is emphasized in recent literature, particularly in the context of smart construction [126, 127]. Sustainable transportation systems for material delivery [128], intelligent railway systems [14, 129–143], and advancements in vehicle ride behavior modeling [144–152] have shown that reducing transportation waste is a critical aspect of sustainable construction. Research on crashworthiness [153–161] and bio-lubricants [162–165] reinforces the connection between improved material performance, safety, and sustainability in construction [166–169]. These studies underscore the importance of integrating sustainability into every stage of construction, from material sourcing to project execution [170].

Finally, advancements in dynamic analysis and vibration isolation, as well as the application of composite materials in construction [171–175], highlight the potential for more resilient and durable construction practices. Recent studies on advanced control systems for rail and road vehicles [133, 176–189] demonstrate how smart technologies can improve efficiency and reduce the need for resource-intensive maintenance. Additionally, composite materials and sustainable structural applications [60, 65, 67, 190–201] contribute to the long-term sustainability of construction projects, further supporting the goals of Industry 5.0.

In summary, the literature on smart 3R approaches in construction highlights how material science, renewable energy, and digital technologies are collectively driving the transition toward more sustainable and resilient construction practices. By reducing waste, reusing materials, and recycling resources, the construction industry can play a key role in advancing the circular economy and meeting the sustainability objectives of Industry 5.0.

3 Construction Waste Types

In construction projects, various types of waste are generated due to the extensive use of resources and complex processes involved. Effective waste management requires identifying and categorizing these waste types to develop targeted strategies for reduction, reuse, and recycling. By understanding the specific types of waste typically encountered, construction managers and policymakers can better implement 3R practices, minimize environmental impact, and enhance resource efficiency The following are the seven primary types of waste commonly encountered in construction projects: The construction industry is one of the largest contributors to global waste, and transitioning toward sustainable practices requires the integration of smart 3R (reduce, reuse, recycle) approaches. These strategies are vital for achieving resource efficiency, reducing environmental impact, and fostering resilient construction processes. The growing demand for sustainable and resilient construction aligns with the goals of Industry 5.0, which focuses on the integration of digitalization, automation, and sustainability into industrial practices. Several studies highlight how advancements in material science, energy efficiency, and automation can transform waste into resources within the construction sector.

- Defects
  Defects arise from errors that require rework, adjustments, or redesigns. These include any materials or components that are damaged, improperly installed, or altered to meet standards, such as walls needing repair after accidental damage or flooring improperly installed due to oversight. Defects lead to wasted resources and time as they necessitate correction or replacement.
- Overproduction
  Overproduction occurs when materials or information are produced too soon or in excess, deviating from a "just-in-time" approach. This creates surplus stock that may need disposal if project requirements change. An example includes ordering materials too early, resulting in storage needs and potential disposal due to changes in the project layout, ultimately adding to unnecessary costs.
- Inventory
  Inventory waste refers to having surplus materials, tools, or work-in-progress items on site, often a result of overproduction. While some stock is necessary to keep the project running, excess inventory occupies valuable space, requires extra handling, and may end up unused if project requirements shift. This can include materials stored on-site or at manufacturing facilities.
- Additional Processing
  Extra processing involves unnecessary steps in the workflow, such as re-entering data or duplicating paperwork, which add no value to the project. Examples include redundant annotations on drawings, excessive reporting, or unnecessary communication exchanges that add layers of work without contributing directly to project progress.
- Motion
  Motion waste is caused by inefficient processes that lead to extra physical movement by workers. This can include time spent searching for misplaced tools or taking additional steps due to poor workspace layout, leading to fatigue and reducing productivity.
- Transportation
  Transportation waste occurs when materials or equipment are moved unnecessarily, such as from one area to another or repeatedly to and from storage. This

can delay the project and increase the risk of material damage through handling. While some movement is unavoidable, minimizing transport helps streamline operations and reduce associated risks.

- Waiting
  Waiting involves idle time when project teams or equipment are not in use due to delays in supplies, information, or task completion. This can occur when workers await deliveries, approvals, or coordination from other teams. Waiting disrupts workflow and can lead to productivity loss, as tasks cannot proceed without the required resources or information.

4 Demolition Waste Materials

Table 1 summarizes dominantly used construction waste that is produced as a result of building destruction in construction sites. The materials include concrete, reinforcing steel, masonry units, wood studs, asphaltic concrete paving, insulation, gypsum board, plumbing fixtures, and electrical devices, among others. Each item is one component of waste which is obligatory to handle under the 3R frameworks of Reduce, Reuse and Recycle. With stakeholders aiming to reuse or recycle the demolition waste in certain ways, the table provides a means of sorting the recycling materials. Additionally, identifying specific materials may provide a rationale for setting up on-site sorting, which cuts down on the waste going to the landfill and encourages resource efficiency. In this view, this classification is also useful to contractors and waste management specialists in formulating mechanisms for achieving overall efficiency of reusing large quantities of material, concrete and steel, in various other construction works, thus promoting sustainability in Industry 5.0.

Table 1 Demolition material list

Concrete reinforcing steel	Concrete masonry units	Concrete	Asphaltic concrete paving
Bricks	Wood studs	Wood joists	Roofing
Insulation	wood paneling	Rough hardware	Door's hardware
Plywood and oriented strand board	Doors and frames	Windows	Carpet pad
Gypsum board	Metal studs	Plumbing fixers	Piping
Valves	Copper wiring	Lamps	Switching boards
Lighting fixers	Electrical devices	Machine equipment	Support and hangers

5 Construction Waste Reduction Through 3R

Under the National Solid Waste Management (NSWM) policy, strict measures have been implemented to manage and minimize construction waste, with the 3R approach—reduce, reuse, and recycle—playing a pivotal role in significantly decreasing waste.

- Reduce
  Reduction is considered the most efficient and impactful strategy for managing construction waste. Reducing waste goes beyond simply producing less material waste; it involves optimizing resource use, improving design planning, and minimizing excess material orders. By streamlining processes, the construction industry can prevent surplus waste generation from the outset, reducing the burden on landfills and the environment.
- Reuse
  Reuse is often the preferred option in construction waste management because it allows materials to be used in future projects without the need for disposal. This approach benefits temporary workers and contractors by saving on material and disposal costs. Reusable materials, such as bricks, wood, or metal components, can be recovered and incorporated into other construction tasks, promoting a circular approach within the industry and enhancing economic efficiency.
- Recycle
  Recycling involves processing and converting waste materials into new products, enabling the continual use of resources. While recycling may be more complex and resource-intensive than reduction or reuse, it remains essential for materials that cannot be reused directly. Recyclable waste, like metal scraps or concrete, can be processed and reincorporated into the supply chain, reducing the need for raw materials and supporting sustainable practices.

There are two primary methods for reusing construction waste: on-site and off-site reuse:

- On-Site Reuse: On-site reuse involves sorting and reusing materials directly at the construction site. This includes isolating and reusing items like concrete, wood, and bricks within the same project, minimizing transportation costs and reducing the need for additional resources.
- Off-Site Reuse: Off-site reuse entails transporting salvaged materials to other locations or facilities where they can be used in different projects. By transferring reusable waste to other sites, construction companies can extend the lifecycle of materials and reduce the environmental footprint of disposal.

These 3R practices, when applied effectively, support the NSWM policy's goals by reducing waste at its source, encouraging the recycling of materials, and fostering a sustainable, resource-efficient construction industry.

6 Results and Analysis

In this investigation, we conducted a comprehensive poll that included 40 questions, with 75 participants, of whom 65 completed the survey. The study's primary focus is on the potential of construction waste to serve as raw materials for third-party development projects. While some materials can be reused on-site, off-site or third-party reuse involves sorting waste from construction projects and transporting it to other locations or external facilities, where it can be repurposed by third-party organizations for use in various applications. Waste items have been found and decreased by using the 3R. The amount of trash and the number of waste materials on site were examined in order to identify the present examination of waste materials in construction and to provide the most queries possible, primarily brought on by inadequate material and consumption planning, as well as the fact that many businesses do not employ the waste management strategies depicted in Fig. 1. Figure 1 illustrates the order of decreasing based on how often the waste material is generated by construction activities which are concrete, bricks, steel, timber and glass. For instance, the presence of waste concrete in the evaluation means that it is a waste source closely associated with structural construction. In contrast, the construction industry also generates waste streams which include glass and wood but which are significantly lower on the list.

This ranking carries considerable relevance for the stakeholders concerned or active in the construction sector as it identifies materials that should attract reduction, reuse or recycling efforts within the 3R strategy framework. By doing this, construction managers and construction policies can ensure that the most effective waste management methods are established for waste types that have the greatest likelihood of being generated, thus minimizing pollution and achieving the aim of green construction.

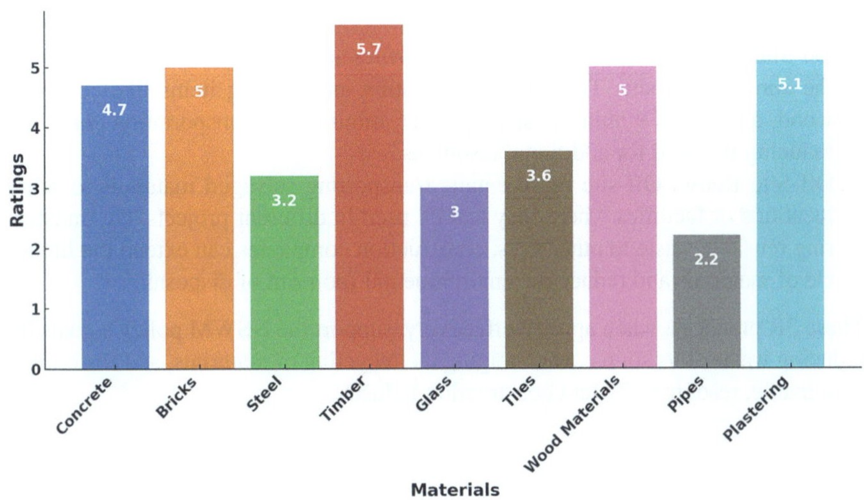

Fig. 1 Rank-wise construction site waste materials shown graphically

6.1 Contractors' Current Use of the 3R to Reduce Construction Waste

It was shown that 64% of temporary workers are aware of the 3R concept based on the poll study. According to Fig. 2, 26% of temporary workers and 10% of contractual workers, respectively, are completely ignorant of the 3R concept. The majority of contract workers learn more about the 3R through workshops, courses, and media. A small portion of contract employees learned about the 3Rs from laws, regulations, and agreement requirements. A few of the contract workers also learn about 3R from understanding materials and college, aside from that. Table 2 assists in the shortfall rating of the categories of waste materials, particularly defects, overproduction, inventory, extra processing, motion, transportation, and waiting during the construction processes. Similar intuitive description can be made about each waste category as they can be indexed with a numerical rating based on occurrence or significance in terms of resource waste in construction efforts.

The ratings are used to identify the kinds of waste that are most detrimental to construction effectiveness and to the environment. For instance, issues of "Time Expenditure" and "Defects" may be rated as severe because issues of repercussion require time and resources to solve, while "Time Wasted" may be rated less severe but still, it factors in as a loss. The table reinforces the need to reduce each type of waste as one of the aspects of improving the effectiveness of construction. It assists contractors and project managers in targeting the most detrimental waste types, as identified during the 3R survey, hence improving construction productivity and sustainability. Figure 2 assesses the level of knowledge and understanding of the 3R (Reduce, Reuse, Recycle) among contractors. It illustrates the percentage of contractors who have a complete understanding of the practices as opposed to those who have no knowledge or only a little knowledge of the practices. The chart also incorporates reliance on varied sources of 3R knowledge such as workshops, rules, and self-earning material resources showing how contractors get to know about

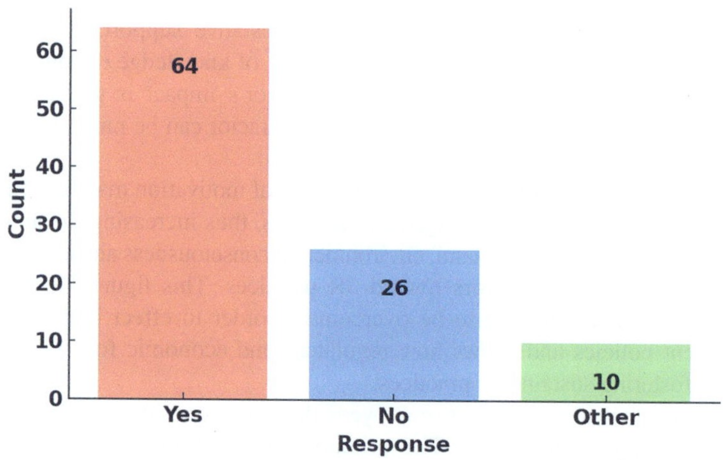

Fig. 2 Knowledge of the three-reality idea among contractors

Table 2 Rating for waste materials

Frequent waste types in construction	Ratings
Transportation	2.4
Over productions	2.5
Extra processing	4.6
Waiting	2.2
Inventory	3.6
Motion	3.2
Defects	4.4

waste management practices. Knowing the level of knowledge of the contractors is essential since it determines the practicality of the 3R strategies in their projects. For instance, in most situations, if a majority of the contractors are oblivious of the benefits of 3R strategies then the waste management practices may be obsolete, and resources would be wasted. This figure stresses the importance of conducting in-depth training and engaging in awareness activities for promoting the use of sustainable practices in the construction sector.

6.2 Ratings Are Impacted by Waste Generation

The aforementioned table demonstrates that extra preparation has a high evaluation in comparison to various explanations. At a certain point, remaining has the same contrast as holding up, which has the least amount of appraisal (2). Over the course of the analysis, certain reasons of material waste on construction sites can be primarily blamed for the material waste.

By guiding the inquiry and meeting the reactions manly shows into rating, venture acquiring cost decrease is top rate in comparison with different aspects, and waste isn't an issue in site got the lowest evaluating. Figure 3 illustrates how variables affecting the construction squander the board. This figure outlines the various factors affecting the waste management practices among the contractors in the construction industry. Such factors may include legislative support, environmental issues, clientele demand, economic drivers, or lack of knowledge of some workers and contractors of other related fields. Each factor's impact is given by scores although the scores can be varied since a stronger factor can be more than another in the factor scale on waste management decisions.

For instance, low penalties for violation or lack of motivation may force the contractors to go easy on waste management practices, thus increasing the amount of waste generated. On the other hand, environmental consciousness and client's needs can help incentivize contractors toward 3R practices. This figure illustrates the diverse obstacles which have to be overcome in order to effect improved waste management policies and shows how regulatory and economic forces are instrumental in fostering sustainable practices.

It's absurd that most contract employees don't practice 3R when it comes to monitoring development waste. The primary causes of the 3R practice's poor implementation have been noted, including time and budget constraints, a lack of space, a lack of authority, a lack of awareness and information, a lack of coordination, poor investment, and the attitude of contract workers. As a result, governing bodies ought to devote a significant amount of time to promoting, strengthening, and maintaining 3R drills. For contract workers managing development waste, the 3R technique is a good approach because it is helpful and manageable.

Waste the executives' practice of the 3R as Fig. 4 discusses and explains the reasons why the contractors undertook the 3R approach in construction waste management. Motives include the potential for recovering revenues from salvaged materials, compulsory requirements for complying with pollution control, clients' insistence on reasonable waste having an economic value, and the need to reduce waste disposal. Each of these reasons is quantified and expressed in percentages

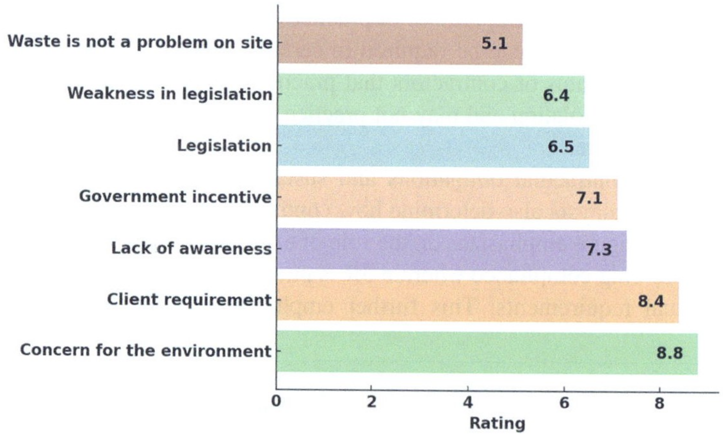

Fig. 3 A factor influencing the development squander the board

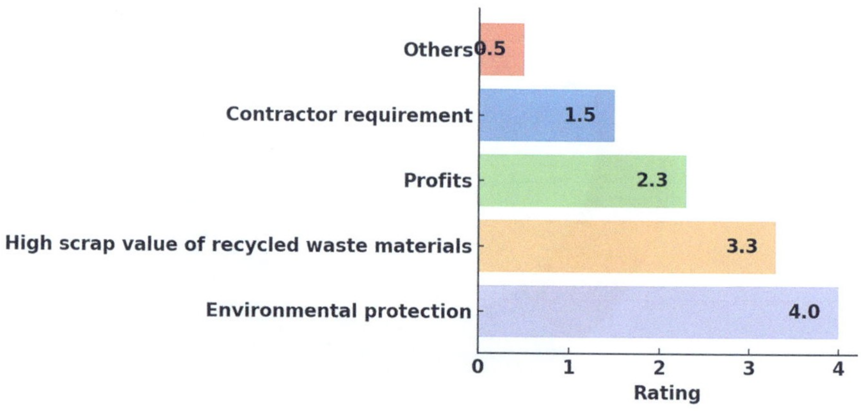

Fig. 4 Reasons why contractors used the 3R method of management

showing the level of contractors who are motivated by such a reason. For example, the use of scrap materials may encourage contractors to switch more toward recycling while the clients' needs for green practices drives compliance to the 3R methods. Caterers of these requirements are of particular significance to the decision makers and industries striving for sustainable construction. It is possible to enhance the compliance with 3R principles throughout the industry by aligning the legal and contractual incentives with the contractors' primary business objectives.

Meanwhile, 23% of temporary employees are unsure whether 3R is required for development, and 40% of contractual workers discovered that it isn't. However, as illustrated in Fig. 5 and in light of the critiques offered by experts consulted, a master stated that legal contract workers must adhere to 3R practices since they address development waste minimization and reuse in Planning Permission Plans. Other experts stated that 3R practices are not included in venture contracts and are not ordered in the interim. However, some temporary employees do practice the 3Rs. Figure 5 goes to show the importance of practices related to 3R in the management of construction waste, as can be required in certain contracts and guidelines. The figure depicts the status of contractors that practice 3R in the contracts or regulations and those in violation and may not practice them due to gaps in enforcement or lack of knowledge.

Legislation, contractual obligations and sustainable ratings such as GBI and MyCREST in Malaysia also determine how contractors will manage their waste in the circle. This figure emphasizes on the role of external pressures and legal frameworks in enforcing compliance wherein 3R is practiced more where there are legal or contractual requirements. This further emphasizes on the need for specific

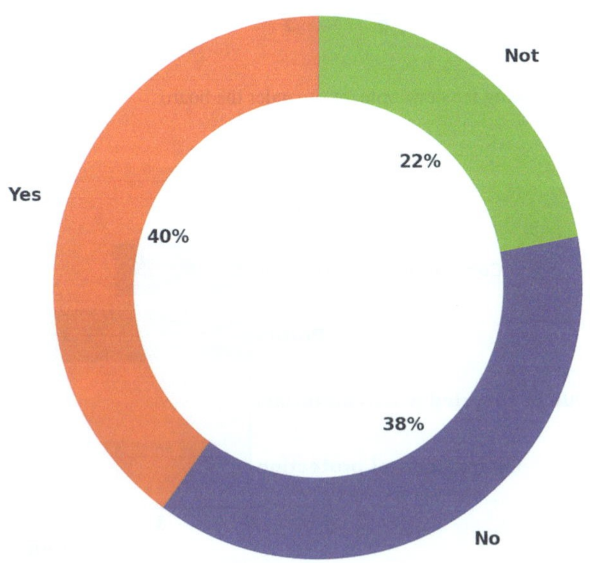

Fig. 5 Construction waste management requires the application of 3R practice

regulations and guidelines in order to enhance sustainable waste management practices in the industry.

One strategy for waste management, as specified in contracts, is the 3R practice. Another is the pursuit of maintainable structure ratings, such as the GBI and the Malaysian Carbon Reduction and Environmentally Sustainable Tool (MyCREST). Their teams are tasked with upholding the legislation and rules governing the required 3R drills. In general, this analysis discovered that the maintainable structure rating depends on temporary workers' motivation to pursue it and that 3R practice is no longer required for development waste management. But once a law or guideline is approved, it will be required.

6.3 Waste Material and 3R Practice Recommendations

To improve the effectiveness of waste reduction through the 3R in the construction industry, a few essential elements must be emphasized: legislation and regulations, programs and incentives, Knowledge and awareness of the three Rs, contractor involvement, and accessible technologies. Furthermore, a top-down strategy involving the application of laws and regulations as well as the good governance idea ought to be used to sustainably handle garbage from construction projects.

The administration may impose strict guidelines on trash management and enact laws that specifically address how waste is handled and disposed of.

To educate the necessary employees in the organization about the building industry, companies that fully embrace waste management can be awarded incentives by the government. Project managers and contractors may help create suitable management actions to reduce waste by forecasting and recording the types of construction material waste created in a project. Illegally dumping construction-related waste materials may harm the environment. A dumping site that separates materials into noninert and inert categories before transporting them to recycling facilities to make efficient use of garbage helps to mitigate this problem.

7 Conclusion

The effective management of construction waste is not only crucial for sustainable development but also central to advancing Industry 5.0's vision of resilient and human-centered manufacturing systems. This study demonstrates that implementing smart 3R practices—encompassing reduction, reuse, and recycling—can transform waste into a valuable resource, contributing significantly to sustainability goals in the construction sector. However, the current adoption rate of 3R practices remains insufficient, largely due to a lack of awareness, technological support, and regulatory enforcement. To bridge this gap, the construction industry must embrace digital tools and smart manufacturing technologies, such as real-time waste tracking, predictive analytics, and automated sorting systems, to enhance the efficiency and effectiveness of 3R practices. Additionally, integrating these practices into regulatory frameworks and providing incentives for technology adoption can further drive sustainability and resource efficiency across the industry. This research highlights the potential of combining Industry 5.0's principles with smart waste management strategies to reduce the environmental footprint of construction activities. Future studies should explore the integration of advanced technologies like digital twins, augmented reality, and Internet of Things (IoT) sensors in construction waste management, aiming for a truly circular economy. By adopting these innovations, the construction industry can play a pivotal role in the transition toward a sustainable, digitally enhanced future in line with Industry 5.0.

References

1. Shah, S., Nautiyal, H., Gugliani, G., Kumar, A., Namboodri, T., & Singla, Y. K. (2024). https://taylorandfrancis.com/books/. ISBN 9781032740713). *Sustainability in smart manufacturing: Trends, scope, and challenges*. CRC Press (Taylor & Francis Group). https://doi.org/10.1201/9781003467496
2. Singla, Y. K., Maughan, M. R., Arora, N., & Dwivedi, D. K. (2024). Enhancing the wear resistance of iron-based alloys: A comprehensive review of alloying element effects. *Journal of Manufacturing Processes, 120*, 135–160.
3. Singh, H., Singh, A. K., Singla, Y., & Chattopadhyay, K. (2020). Tribological study of nano fly ash as lubrication oil additive for AISI 4140 steel for automotive engine applications. *International Journal of Mechanical and Production Engineering Research and Development, 10*, 5939–5946.
4. Singla, Y. K., Maughan, M. R., Arora, N., & Dwivedi, D. K. (2023). Microstructural evolution of worn surface and subsurface of high Mn additive iron-based hardfacings subjected to repetitive impacts and sliding. *Ceramics International, 49*, 21957–21965.
5. Singh, H., Singh, A. K., Singla, Y. K., Chattopadhyay, K., Saini, A., & Singh, K. (2022). Interpretation of the wear characteristics of AISI 4140 under nano-fly ash based engine lubricant. *Materials Today Proceedings, 50*, 1683–1689.
6. Rooprai, R. S., Singh, H., Singh, T., & Singla, Y. K. (2022). Analysis of the wear properties of through hardened AISI-4140 alloy steel using Taguchi technique. *Materials Today Proceedings, 50*, 661–664.
7. Singh, H., Singh, A. K., Singla, Y. K., Chattopadhyay, K., & Malhotra, R. (2021). Optimization of tribological behavior of AISI 4140 under nano fly ash particulates in engine lubricating oil. *Materials Today Proceedings, 45*, 4619–4623.
8. Sura, H. S., Avesh, M., & Mohapatra, S. (2023). Design and modelling of digital twin technology to improve freight logistics. In *Energy, environment, and sustainability* (pp. 481–513). https://doi.org/10.1007/978-981-99-2150-8_20
9. Pattnaik, S., Pati, S., Devadarshini, D., Tanaya, K., Mishra, B. B., Samantaray, D., Mohapatra, S., & Dash, D. (2022). Microbial bioremediation of Cr(VI)-contaminated soil for sustainable agriculture. In *Microbial biodegradation and bioremediation: Techniques and case studies for environmental pollution* (pp. 395–407). https://doi.org/10.1016/B978-0-323-85455-9.00008-4
10. Srivastava, A., Lavania, S., & Mohapatra, S. (2023). Predicting the effect on land values after introducing high-speed rail. In *Energy, environment, and sustainability* (pp. 157–196). https://doi.org/10.1007/978-981-99-2150-8_8

11. Agrahari, R., Mohanty, S., Vishwakarma, K., Nayak, S. K., Samantaray, D., & Mohapatra, S. (2021). Update vision on COVID-19: Structure, immune pathogenesis, treatment and safety assessment. *Sensors International, 2*. https://doi.org/10.1016/j.sintl.2020.100073
12. Mishra, B. B., Nayak, S. K., Mohapatra, S., & Samantaray, D. (2021). Preface. *Environmental and Agricultural Microbiology: Advances and Applications.*, xvii–xix.
13. Tanaya, K., Mohapatra, S., & Samantaray, D. (2024). Formulation of biogenic fluorescent pigmented PHB nanoparticles from Rhodanobacter sp. for drug delivery. *Antonie van Leeuwenhoek, International Journal of General and Molecular Microbiology, 117*. https://doi.org/10.1007/s10482-024-01973-x
14. Mohapatra, S., Pattnaik, S., Maity, S., Sharma, S., Akhtar, J., Pati, S., Samantaray, D. P., & Varma, A. (2020). Comparative analysis of PHAs production by Bacillus megaterium OUAT 016 under submerged and solid-state fermentation. *Saudi Journal of Biological Sciences, 27*, 1242–1250. https://doi.org/10.1016/j.sjbs.2020.02.001
15. Singh, H., Singla, Y. K., Singh, A. K., & Chattopadhayay, K. (2020). Effect of nanofly ash as lubricant additive on the tribological properties of SAE 10W-30 oil: A novel finding. *Transactions of the Indian Institute of Metals., 73*, 2371–2375.
16. Kumar Singla, Y., Dwivedi, D. K., & Arora, N. (2017). Modeling the impact–sliding wear characteristics of rare earth additive iron-based hardfacing alloys. *Proceedings of the Institution of Mechanical Engineers, Part J: Journal of Engineering Tribology, 231*, 1486–1499.
17. Sharma, V., Singla, Y., Gupta, Y., & Raghuwanshi, J. (2018). Post-processing of metal matrix composites by friction stir processing. *AIP Conference Proceedings, 1953*(1), 090062-1–090062-5. (Springer), ISSN = 0094-243X and 1551-7616. https://doi.org/10.1063/1.5032909
18. Pant, M., Sharma, A., Palli, S., Sharma, S. K., & Sharma, R. C. (2024). A review and analysis of factors affecting EV adoption for sustainable and energy-efficient transportation systems using analytic hierarchy process approach. *International Journal of Vehicle Structures and Systems, 16*, 133–139. https://doi.org/10.4273/ijvss.16.2.01
19. Sharma, S. K., Sharma, R. C., Lavania, S., Palli, S., & Avesh, M. (2024). Industry-driven approach for anfis-based intelligent control suspension system with MR damper for enhanced ride quality in passenger rail vehicles for technological investigations. *National Academy Science Letters., 47*, 391–394. https://doi.org/10.1007/s40009-023-01365-1
20. Adeoye, A. O., Quadri, R. O., Lawal, O. S., & Emojevu, E. O. (2024). Physicochemical characterization, valorization of lignocellulosic waste (Kola nut seed shell) via pyrolysis, and ultrasonication of its crude bio-oil for biofuel production. *Cleaner Waste Systems, 7*. https://doi.org/10.1016/j.clwas.2024.100138
21. Adeoye, A.O., Quadri, R.O., Lawal, O.S., Malomo, D., Emojevu, E.O., Omonije, O.O., Odeniyi, O.K., Fadahunsi, M.O., Yelwa, M.J., Aasa, S.A., Onakpa, A.E., Omoniyi, B.P., Mark, I.N., Usman, J., Ismaila, A.M., Saidu, A.U. Advanced techniques in upgrading crude bio-oil to biofuel. In Energy, environment, and sustainability. pp. 321–353 (2024). https://doi.org/10.1007/978-981-97-0515-3_14.
22. Adeoye, A. O., Quadri, R. O., & Lawal, O. S. (2022). Assessment of biofuel potential of tenera palm kernel shell via fixed bed pyrolysis and thermal characterization. *Results in Surfaces and Interfaces, 9*. https://doi.org/10.1016/j.rsurfi.2022.100091
23. Adeoye, A. O., Lawal, O. S., Quadri, R. O., Malomo, D., Aliyu, M. T., Dang, G. E., Emojevu, E. O., Maikato, M. J., Yahaya, M. G., Omonije, O. O., Edidem, V. G., Abubakar, Y. K., Offor, O. F., Sochima, E. H., Peter, B. E., & Hikon, B. N. (2023). Sustainable energy via thermochemical and biochemical conversion of biomass wastes for biofuel production. In *Energy, environment, and sustainability* (pp. 245–306). https://doi.org/10.1007/978-981-99-2150-8_11
24. Mohapatra, S., & Mishra, R. K. (2020). Status of institutional credit disbursement of agricultural sector in Odisha, India. *International Journal of Current Microbiology and Applied Sciences, 9*, 2057–2065.
25. Mohapatra, S., Singh, J., & Mishra, S. (2017). Co-integration among major potato markets in Punjab. *Indian Journal of Agricultural Marketing, 31*, 46–47.

26. Mohapatra, S., Priyanka, V., Mohapatra, S., Kohli, I., & Mishra, R. K. (2020). Impact of corona virus covid-19 on the global economy. *International Journal of Agricultural and Statistical Sciences, 16*, 771–778.
27. Mohapatra, S., & Singh, J. (2021). Co-integration of major vegetable markets in Punjab, India. *Economic Affairs, 66*(2). https://doi.org/10.46852/0424-2513.2.2021.1
28. Mohapatra, S., Mishra, R., & Mishra, R. (2021). Impact of COVID-19 on education level of agricultural farming community. *International Journal of Agricultural and Statistical Sciences, 17*, 915–923.
29. Mohapatra, S., Singh, J., & Kumar, S. (2018). Cointegration among major cauliflower markets in Punjab. *Indian Journal of Economics and Development, 14*, 330–335.
30. Adeoye, A. O., Yelwa, J. M., Imam, N., Quadri, R. O., Lawal, O. S., Malomo, D., Aasa, S. A., Onakpa, A. E., Dorgbaa, R. K., Omoniyi, B. P., Ahmad, N., & Garba, H. A. (2024). Pyrolysis of plastic wastes towards achieving a circular economy: An advanced chemistry and technical approach. In *Energy, environment, and sustainability* (pp. 215–259). https://doi.org/10.1007/978-981-97-0437-8_11
31. Adeoye, A. O., Quadri, R. O., & Lawal, O. S. (2023). Wet synthesis, characterization of goethite nanoparticles and its application in catalytic pyrolysis of palm kernel shell in TGA. *Results in Surfaces and Interfaces, 11*. https://doi.org/10.1016/j.rsurfi.2023.100118
32. Avesh, M., Imam, N., & Adeoye, A. O. (2024). Assessing environmental impacts and mitigation strategies for the DFCCIL rail projects. In *Energy, environment, and sustainability* (pp. 333–353). https://doi.org/10.1007/978-981-97-0437-8_15
33. Mohapatra, S., Mohapatra, S., Han, H., Ariza-Montes, A., & del López-Martín, M. C. (2022). Climate change and vulnerability of agribusiness: Assessment of climate change impact on agricultural productivity. *Frontiers in Psychology, 13*.
34. Mohapatra, S., & Mishra, R. A. J. K. (2020). Production performance and market integration of cotton markets in India. In *National symposium* (p. 253).
35. Din, A. U., Han, H., Ariza-Montes, A., Vega-Muñoz, A., Raposo, A., & Mohapatra, S. (2022). The impact of Covid-19 on the food supply chain and the role of e-commerce for food purchasing. *Sustainability, 14*(5), 3074.
36. Mohapatra, S., & Singh, J. (2018). Growth and instability analysis of cauliflower crop in Punjab. *Economic Affairs, 63*, 665–669.
37. Mohapatra, S., Mishra, R. K., & Sarangi, K. K. (2019). Economic appraisal of bio ethanol production technologies. *International Journal of Chemical Studies, 7*, 2197–2200.
38. Mohapatra, S., Kishore Mishra, R., & Sarangi, K. K. (2020). Bio-ethanol (2nd generation ethanol): A solution to ever polluting gasoline to climate in India. *International Research Journal of Pure and Applied Chemistry, 20*, 1–15.
39. Gohain, N., Bansal, S., Mohapatra, S., & Singh, L. (2022). Analyzing the direction of trade: Indian ginger and lessons from exports to different destinations. *Journal of Agricultural Development and Policy, 32*, 214–220.
40. Mohapatra, S., Mohapatra, U., Chandana, K. S. S., & Mishra, R. K. (2018). Economic analysis of paddy production and marketing in Puri, Odisha. *Journal of Pharmacognosy and Phytochemistry, 7*, 1858–1861.
41. Quadri, R. O., Adeoye, A. O., Ayanda, O. S., & Lawal, O. S. (2024). Conversion of palm kernel shell to sustainable energy and the effect of wet synthesized nanoparticles of iron on its thermal degradation kinetics. *Bioresource Technology Reports, 27*. https://doi.org/10.1016/j.biteb.2024.101933
42. Momodu, D.U., Adetuyi, A.O., Adeoye, A.O., Akinlolu, O.S., & Apara, E.O. (2023). Surface functionalization of linen fiber and its dye-fastness assessment using (4-[(E)-(4-amino-2-hydroxyphenyl) diazenyl] benzene-1-sulphonic acid and 4-[(E)-(4-{(Z)-[4-(dimethylamino) phenyl] diazenyl}-2-hydroxyphenyl) diazenyl]benzene-1-sulphonic acid. *Suranaree Journal of Science and Technology, 30*.

43. Adeoye, A. O., Quadri, R. O., & Lawal, O. S. (2023). Biomass waste pyrolysis in the context of Nigeria's oil and gas industry: An overview. *International Journal of Ambient Energy, 44*, 2525–2541. https://doi.org/10.1080/01430750.2023.2256325
44. Sharma, S. K., Upadhyay, R. K., Kumar, V., & Valera, H. (2023). *Transportation energy and dynamics*. Springer Nature Singapore. https://doi.org/10.1007/978-981-99-2150-8
45. Upadhyay, R. K., Sharma, S. K., Kumar, D. V., & Valera, H. (2023). *Transportation systems technology and integrated management*. Springer Nature Singapore. https://doi.org/10.1007/978-981-99-1517-0
46. Lee, J., & Sharma, S. K. (2023). Intelligent systems for railway infrastructure. *Applied Sciences, 13*(20), 11274. https://doi.org/10.3390/app132011274
47. Sharma, S. K., Sharma, R. C., Choi, Y., & Lee, J. (2023). Modelling and dynamic analysis of adaptive neuro-fuzzy inference system-based intelligent control suspension system for passenger rail vehicles using magnetorheological damper for improving ride index. *Sustainability, 15*, 12529.
48. Hossain, I., Ahmed, Y. E., & Sharma, S. K. (2023). The use of both metallic-insert and TiO2/water nanofluid to improve the performance of photovoltaic modules; experimental study. *Thermal Science and Engineering Progress, 42*, 101944.
49. Wu, Q., Cole, C., Spiryagin, M., Chang, C., Wei, W., Ursulyak, L., Shvets, A., Murtaza, M. A., Mirza, I. M., & Zhelieznov, K. (2023). Freight train air brake models. *International Journal of Rail Transportation., 11*, 1–49.
50. Sharma, R. C., Bhardawaj, S., Sharma, S. K., & Bahl, S. (2023). Ride behaviour of three-wheel vehicle using semi-active suspension using skyhook control strategy. *International Journal of Vehicle Structures & Systems (IJVSS), 15*.
51. Bhardawaj, S., Sharma, R. C., Sharma, S. K., Gopala Rao, L. V. V., & Vashist, A. (2023). Modeling of biomechanical human body model for seat to head transmissibility analysis. *Noise & Vibration Worldwide, 54*, 62–74.
52. Sharma, S. K., Sharma, R. C., Choi, Y., & Lee, J. (2023). Experimental and mathematical study of flexible–rigid rail vehicle riding comfort and safety. *Applied Sciences, 13*, 5252.
53. Sharma, S. K., Upadhyay, R. K., & Kumar, V. (2024). Introduction to dynamics of transportation ecosystem, modeling, and control. In *Dynamics of transportation ecosystem, modeling, and control* (pp. 3–6). Springer.
54. Sharma, S. K., Sharma, R. C., Mohapatra, S., Fouly, A., Hossain, I., & Lee, J. (2024). Technological advancements in oscillation reduction for propulsion shaft systems. *Journal of Vibration Engineering & Technologies, 12*(Suppl. 2), 1–19.
55. Upadhyay, R. K., Sharma, S. K., & Kumar, V. (2024). Introduction to intelligent transportation system and advanced technology. In *Intelligent transportation system and advanced technology* (pp. 3–6). Springer.
56. Hossain, I., Ngoupo, A. T., Bhattarai, S., Challagulla, S. P., & Sharma, S. K. (2024). Empirical survey on the effect of straight and wavy fins integrated with wind stream as heat sink of photovoltaic panels. *International Journal of Environmental Science and Technology*, 1–14.
57. Sharma, S. K., Sharma, R. C., & Lee, J. (2025). Propelling precision of longitudinal vibration mitigation in ship propeller shafts through advanced nonlinear intelligent semi-active control leveraging adaptive neuro-fuzzy inference system with linear quadratic regulator. *Journal of Vibration and Control, 31*(7–8), 1472–1484.
58. Sharma, S. K., Kumar, N., Avesh, M., Sharma, R. C., Siddiqui, M. I. H., & Lee, J. (2024). Navigating tranquillity with $H\infty$ controller to mitigate ship propeller shaft vibration. *Journal of Vibration Engineering & Technologies*, 1–11.
59. Sharma, S. K., Sharma, R. C., Upadhyay, R. K., & Lee, J. (2024). Enhanced ship engine vibration reduction using magnetorheological dampers and Adaptive Neuro-Fuzzy Control System. *Ships and Offshore Structures*, 1–17.

60. Sharma, R. C., Sharma, N., Palli, S., Sharma, S. K., Joshi, S., Rallabandi, S. R., & Sharma, A. (2024). Dynamic behavior of three-wheel vehicle with inerter-based dynamic absorber suspension system. *Noise & Vibration Worldwide, 55*, 253–263.
61. Kokkinos, N. C., Emmanouilidou, E., & Sharma, S. K. (2024). Waste-to-biofuel production for the transportation sector. In *Intelligent transportation system and advanced technology* (pp. 99–122). Springer.
62. Agrahari, R., Mohapatra, S., Kumar, K. S., Mohanty, D., Kumar, R., & Merugu, R. (2024). Immuno-metabolic interplay in health and disease. In A. Das Mohapatra & P. Sahu (Eds.), *Advances in Immunology and Immuno-techniques. Interdisciplinary Biotechnological Advances* (pp. 239–274).
63. Sharma, S. K., Sharma, R. C., & Lee, J. (2024). Modelling and simulation of MR damper characterization and uncertainty assessment in marine engine vibration isolation with FLPID control. *International Journal of Modelling and Simulation*, 1–11.
64. Sharma, S. K., Sharma, R. C., Upadhyay, R. K., & Lee, J. (2024). Coupled torsional–transverse vibration reduction in marine propulsion dynamics with novel approach using magnetorheological damping and adaptive control system. *Journal of Vibration Engineering & Technologies, 12*, 6089–6099.
65. Rallabandi, S. R., Srinivas, L., Palli, S., Sharma, R. C., Sharma, N., Sharma, A., & Sharma, S. K. (2024). A contrastive characterization of pure Mg and AZ91D alloy based on the testing of mechanical, corrosion, wear, and erosion properties. *Engineering Research Express, 6*, 015017.
66. Badgujar, P., Sharma, S., Rahul, D., Sharma, S. K., & Upadhyay, R. K. (2024). Feasibility study of Mumbai-Pune hyperloop corridor: Challenges and implementation. *Research in Transportation Business & Management, 52*, 101089.
67. Sharma, R. C., Palli, S., Dharmana, L., Sharma, S. K., Sharma, A., Kharola, A., & Sharma, N. (2024). Simulation of transient dynamic response of a locomotive considering its flexibility and rigidity. *Noise & Vibration Worldwide, 55*, 33–47.
68. Song, H., Park, J., Sharma, S. K., & Lee, J. (2022). A comparative study on ultrasonic propagation characteristics and defect detection of metal material additive manufacturing using deep learning algorithm. In *Biennial international conference on future learning aspects of mechanical engineering* (pp. 157–168). Springer.
69. Boris, I., Sharma, S. K., & Lee, J. (2022). A study on the reflection coefficient of higher modes of torsional wave by theoretical and numerical method. In *Biennial international conference on future learning aspects of mechanical engineering* (pp. 145–155). Springer.
70. Vishwakarma, P. N., Mishra, P., Sharma, S. K., & Shukla, A. K. (2022). Formulating and analysing the effect of suspension parameters on critical speeds for various DOFs in rail vehicle. In *Biennial international conference on future learning aspects of mechanical engineering* (pp. 791–802). Springer.
71. Kumar, U., Kasvekar, R., Sharma, S. K., & Upadhyay, R. K. (2022). Wear of wheels and axle in locomotive and measures taken by Indian railway. *Advances in Engine Tribology*, 77–96.
72. Sharma, R. C., Palli, S., Jha, A. K., Bhardawaj, S., & Sharma, S. K. (2022). Vibration and ride comfort analysis of railway vehicle system subjected to deterministic inputs. *Resmilitaris, 12*, 1345–1355.
73. Sharma, R. C., Rao, L. V. V. G., Sharma, S. K., Palli, S., & Satyanarayana, V. S. V. (2022). Analysis of lateral stability and ride of an Indian railway constrained dual-axle bogie frame. *SAE International Journal of Commercial Vehicles, 16*, 213–228.
74. Vishwakarma, P. N., Mishra, P., & Sharma, S. K. (2022). Formulation of semi-active suspension system and controls in rail vehicle. *SSRN Electronic Journal*. https://doi.org/10.2139/ssrn.415961
75. Sharma, R. C., & Sharma, S. K. (2022). Ride analysis of road surface-three-wheeled vehicle-human subject interactions subjected to random excitation. *SAE International Journal of Commercial Vehicles, 15*, 299–312.

76. Sharma, S. K., Lee, J., & Jang, H.-L. (2022). Mathematical modeling and simulation of suspended equipment impact on car body modes. *Machines, 10*, 192.
77. Sharma, S. K., Sharma, R. C., Lee, J., & Jang, H.-L. (2022). Numerical and experimental analysis of DVA on the flexible-rigid rail vehicle carbody resonant vibration. *Sensors, 22*, 1922.
78. Vishwakarma, P. N., Mishra, P., & Sharma, S. K. (2022). Characterization of a magnetorheological fluid damper a review. *Materials Today Proceedings, 56*, 2988–2994.
79. Sharma, S. K., Sharma, R. C., & Lee, J. (2022). In situ and experimental analysis of longitudinal load on carbody fatigue life using nonlinear damage accumulation. *International Journal of Damage Mechanics, 31*, 605–622.
80. Cataluna Veses, R., & Shah, Z. (2015). Bio-oil production by thermal cracking in the presence of hydrogen. *Journal of Fundamentals of Renewable Energy and Applications, 06*. https://doi.org/10.4172/2090-4541.1000194
81. Shah, Z., Veses, R. C., & da Silva, R. (2016). A comparative study and analysis of two types of bio-oil samples obtained from freshwater algae and microbial treated algae. *Modern Chemistry & Applications, 4*. https://doi.org/10.4172/2329-6798.1000185
82. Shah, Z., Veses, R. C., Brum, J., Klunk, M. A., Rocha, L. A. O., Missaggia, A. B., de Oliveira, A. P., & Caetano, N. R. (2020). Sewage sludge bio-oil development and characterization. *Inventions, 5*, 36. https://doi.org/10.3390/inventions5030036
83. Shah, Z., Saberian, M., Hodgeman, D., Kinloch, I., & Vallés, C. (2023). Effect of sulfur on wood tar biopitch as a sustainable replacement for coal tar pitch binders. *ACS Applied Engineering Materials, 1*, 2567–2576. https://doi.org/10.1021/acsaenm.3c00361
84. Klunk, M. A., Dasgupta, S., Das, M., & Shah, Z. (2018). System of adsorption of CO2 in coalbed. *Southern Brazilian Journal of Chemistry, 26*, 2–9. https://doi.org/10.48141/SBJCHEM.v26.n26.2018.7_2018.pdf
85. Mohapatra, S., Mohapatra, U., & Mishra, R. K. (2018). Price forecasting of groundnut in Odisha. *Pharma Innovation Journal, 7*, 111–114.
86. Mohapatra, S., Sneha, N. K., Mounika, A., & Mishra, R. K. (2019). Socio-economic and demographic analysis of SAARC nations. *International Journal of Agricultural & Statistical Sciences, 15*.
87. Mohapatra, S., Mohapatra, U., & Mishra, R. (2017). Diversification towards vegetable crops: A good option for doubling the farmer's income. *Journal of Experimental Agriculture International., 18*, 1–7.
88. Bansal, S., Mohapatra, S., Gohain, N., & Das, A. (2022). Production status, market integration and the law of one price: An exploration of the major maize markets of India. *Indian Journal of Economics and Development, 18*, 832–841.
89. Mohapatra, S., Samantaray, D. P., & Samantaray, S. M. (2015). Study on polyhydroxyalkanoates production using rhizospheric soil bacterial isolates of sweet potato. *Indian Journal of Science and Technology, 8*, 57–62. https://doi.org/10.17485/ijst/2015/v8iS7/64027
90. Arora, R., Trivedi, J., Mohapatra, S., & Kumar, P. (2022). Trends in probiotics on human health and industrial application. In *Advances in agricultural and industrial microbiology: Volume 1: Microbial diversity and application in agroindustry* (pp. 237–248). https://doi.org/10.1007/978-981-16-8918-5_13
91. Samal, S., Pati, S., Mohapatra, S., Maity, S., Tanaya, K., Devadarshini, D., & Samantaray, D. (2023). PHAs production by facultative anaerobic bacteria Bacillus cereus FM5 through submerged and solid-state fermentation under anoxic condition. *Antonie van Leeuwenhoek, International Journal of General and Molecular Microbiology, 116*, 521–529. https://doi.org/10.1007/s10482-023-01825-0
92. Biswas, N., Saha, S., Biswas, B. K., Chowdhury, M., Rahaman, A., Junghare, V., Mohapatra, S., Hazra, S., Zangrando, E., Roy Choudhury, R., & Roy Choudhury, C. (2021). The DNA- and protein-binding properties and cytotoxicity of a new copper(II) hydrazone Schiff base complex. *Journal of Coordination Chemistry, 74*, 1482–1504. https://doi.org/10.1080/00958972.2021.1913128

93. Mohapatra, S., Mishra, R., Roy, P., Yadav, K. L., & Satapathi, S. (2017). Systematic investigation and in vitro biocompatibility studies on implantable magnetic nanocomposites for hyperthermia treatment of osteoarthritic knee joints. *Journal of Materials Science, 52*, 9262–9268. https://doi.org/10.1007/s10853-017-1136-0
94. Sharma, S. K., Mohapatra, S., Sharma, R. C., Alturjman, S., Altrjman, C., Mostarda, L., & Stephan, T. (2022). Retrofitting existing buildings to improve energy performance. *Sustainability (Switzerland), 14*. https://doi.org/10.3390/su14020666
95. Bandhu, S., Khot, M. B., Sharma, T., Sharma, O. P., Dasgupta, D., Mohapatra, S., Hazra, S., Khatri, O. P., & Ghosh, D. (2018). Single cell oil from oleaginous yeast grown on sugarcane bagasse-derived xylose: An approach toward novel biolubricant for low friction and wear. *ACS Sustainable Chemistry & Engineering, 6*, 275–283. https://doi.org/10.1021/acssuschemeng.7b02425
96. Kohli, I., Joshi, N. C., Mohapatra, S., & Varma, A. (2020). Extremophile—An adaptive strategy for extreme conditions and applications. *Current Genomics, 21*, 96–110. https://doi.org/10.2174/1389202921666200401105908
97. Pati, S., Mohapatra, S., Maity, S., Dash, A., & Samantaray, D. P. (2021). Polyhydroxyalkanoates for sustainable smart packaging of fruits. In *Environmental and agricultural microbiology: Advances and applications* (pp. 183–195).
98. Chatterjee, A., Nirwan, S., Mohapatra, S., Sharma, P., Agnihotri, A., & Shrivastava, N. (2022). Biochemical aspects of pathogenic variability in white rust infected Indian mustard. *Mycologia, 114*, 757–768. https://doi.org/10.1080/00275514.2022.2060007
99. Pati, S., Maity, S., Dash, A., Jema, S., Mohapatra, S., Das, S., & Samantaray, D. P. (2020). Biocompatible PHB production from bacillus species under submerged and solid-state fermentation and extraction through different downstream processing. *Current Microbiology, 77*, 1203–1209. https://doi.org/10.1007/s00284-020-01922-7
100. Agrahari, R., Sarraf, G., Joshi, N. C., Mohapatra, S., & Var, A. (2021). Insight of biopolymers and applications of polyhydroxyalkanoates. In *Microbial polymers: Applications and ecological perspectives* (pp. 177–191). https://doi.org/10.1007/978-981-16-0045-6_8
101. Kathpalia, M., Kumar, P., & Mohapatra, S. (2022). Homocysteine metabolism as a biomarker for cancer. In *Homocysteine metabolism in health and disease* (pp. 159–171). https://doi.org/10.1007/978-981-16-6867-8_8
102. Bharatiy, S. K., Hazra, M., Paul, M., Mohapatra, S., Samantaray, D., Dubey, R. C., Sanyal, S., Datta, S., & Hazra, S. (2016). In silico designing of an industrially sustainable carbonic anhydrase using molecular dynamics simulation. *ACS Omega, 1*, 1081–1103. https://doi.org/10.1021/acsomega.6b00041
103. Sarkar, M. S., Segu, H., Bhaskar, J. V., Jakher, R., Mohapatra, S., Shalini, K., Shivaji, S., & Reddy, P. A. (2018). Ecological preferences of large carnivores in remote, high-altitude protected areas: Insights from Buxa Tiger Reserve, India. *ORYX, 52*, 66–77. https://doi.org/10.1017/S0030605317000060
104. Pati, S., Dash, A., Maity, S., Pattnaik, S., Mohapatra, S., & Samantaray, D. P. (2021). Nano biofertilizer for sustainable agriculture. In *Agriculturally important microorganisms: Mechanisms and applications for sustainable agriculture* (pp. 201–208). https://doi.org/10.1201/9781003245841-10
105. Mohanty, D., Misra, S., Mohapatra, S., & Sahu, P. S. (2018). Prebiotics and synbiotics: Recent concepts in nutrition. *Food Bioscience, 26*, 152–160. https://doi.org/10.1016/j.fbio.2018.10.008
106. Kumar, R., Kumari, R., Prasad, C., Deep, A., Singh, N., Namtak, S., Tiwari, V., Merugu, R., Mohapatra, S., & Singh, R. (2021). Influence of water quality on the diversity of macroinvertebrates in the Mandakini River in India. *Water Science and Technology: Water Supply, 21*, 1843–1860. https://doi.org/10.2166/WS.2021.020
107. Mohanty, D., Jena, R., Choudhury, P. K., Pattnaik, R., Mohapatra, S., & Saini, M. R. (2016). Milk derived antimicrobial bioactive peptides: A review. *International Journal of Food Properties, 19*, 837–846. https://doi.org/10.1080/10942912.2015.1048356

108. Mohapatra, S., Rath, S. N., Pradhan, S. K., Samantaray, D. P., & Rath, C. C. (2016). Secondary structural models (16S rRNA) of polyhydroxyalkanoates producing bacillus species isolated from different rhizospheric soil: Phylogenetics and chemical analysis. *International Journal Bioautomation, 20*, 329–338.
109. Mohapatra, S., Sarkar, B., Samantaray, D. P., Daware, A., Maity, S., Pattnaik, S., & Bhattacharjee, S. (2017). Bioconversion of fish solid waste into PHB using Bacillus subtilis based submerged fermentation process. *Environmental Technology (United Kingdom), 38*, 3201–3208. https://doi.org/10.1080/09593330.2017.1291759
110. Misra, S., Mohanty, D., & Mohapatra, S. (2021). Food-based probiotics: Functional dietary ingredients. In *Probiotics: Advanced food and health applications* (pp. 257–275). https://doi.org/10.1016/B978-0-323-85170-1.00006-3
111. Kumar, R., Kumari, R., Prasad, C., Tiwari, V., Singh, N., Mohapatra, S., Merugu, R., Namtak, S., & Deep, A. (2020). Phytoplankton diversity in relation to physicochemical attributes and water quality of Mandakini River, Garhwal Himalaya. *Environmental Monitoring and Assessment, 192*. https://doi.org/10.1007/s10661-020-08768-3
112. Mohapatra, S., Vishwakarma, K., Joshi, N. C., Maity, S., Kumar, R., Ramchander, M., Pattnaik, S., & Samantaray, D. P. (2021). A review on PHAs: The future biopolymer. In *Environmental and agricultural microbiology: Advances and applications* (pp. 83–117).
113. Sahoo, N., Pattnaik, S., Pattnaik, M., & Mohapatra, S. (2021). Methane emission and strategies for mitigation in livestock. *Environmental and Agricultural Microbiology: Advances and Applications*, 257–274.
114. Mohanty, D. P., Mohapatra, S., Misra, S., & Sahu, P. S. (2016). Milk derived bioactive peptides and their impact on human health—A review. *Saudi Journal of Biological Sciences, 23*, 577–583. https://doi.org/10.1016/j.sjbs.2015.06.005
115. Kohli, I., Mohapatra, S., Kumar, P., Goel, A., Varma, A., & Joshi, N. C. (2022). Role of bacterial endophytes in the promotion of plant growth. In *Bacterial endophytes for sustainable agriculture and environmental management* (pp. 235–253). https://doi.org/10.1007/978-981-16-4497-9_12
116. Samantaray, D., Mohapatra, S., & Mishra, B. B. (2014). Microbial bioremediation of industrial effluents. In *Microbial biodegradation and bioremediation* (pp. 326–339). https://doi.org/10.1016/B978-0-12-800021-2.00014-5
117. Pattnaik, S., Dash, D., Mohapatra, S., Pattnaik, M., Marandi, A. K., Das, S., & Samantaray, D. P. (2020). Corrigendum to "Improvement of rice plant productivity by native Cr(VI) reducing and plant growth promoting soil bacteria Enterobacter cloacae". *Chemosphere, 240*, 124895. https://doi.org/10.1016/j.chemosphere.2019.124895. (S0045653519321344). Chemosphere. 238, (2020). https://doi.org/10.1016/j.chemosphere.2019.125011
118. Mohapatra, S., Mohanta, P. R., Sarkar, B., Daware, A., Kumar, C., & Samantaray, D. P. (2017). Production of polyhydroxyalkanoates (PHAs) by bacillus strain isolated from waste water and its biochemical characterization. *Proceedings of the National Academy of Sciences India Section B—Biological Sciences., 87*, 459–466. https://doi.org/10.1007/s40011-015-0626-6
119. Tomer, P., Shroti, G. K., Mohapatra, S., Ghosh, D., Jaiswal, S., Garg, D., Lahiri, D., Kumar, N., Khatri, O. P., & Hazra, S. (2023). Fabrication of HA Nano-crystal Reinforced PHAs based composites for orthopedic fracture-fixing accessories using sustainable resources. *Environment, Development and Sustainability*. https://doi.org/10.1007/s10668-023-03912-8
120. Bandekar, D., Mohapatra, S., Hazra, M., Hazra, S., & Biswas, S. (2021). N-terminal truncation of VC0395_0300 protein from Vibrio cholerae does not lead to loss of diguanylate cyclase activity. *Biophysical Chemistry, 268*. https://doi.org/10.1016/j.bpc.2020.106493
121. Vishwakarma, K., Kumar, N., Shandilya, C., Mohapatra, S., Bhayana, S., & Varma, A. (2020). Revisiting plant–microbe interactions and microbial consortia application for enhancing sustainable agriculture: A review. *Frontiers in Microbiology, 11*. https://doi.org/10.3389/fmicb.2020.560406

122. Mohapatra, S., Maity, S., Dash, H. R., Das, S., Pattnaik, S., Rath, C. C., & Samantaray, D. (2017). Bacillus and biopolymer: Prospects and challenges. *Biochemistry and Biophysics Reports, 12*, 206–213. https://doi.org/10.1016/j.bbrep.2017.10.001
123. Kumar, R., Merugu, R., Mohapatra, S., & Sharma, S. (2022). Extremophiles: Life of microorganisms in extreme environments. In *Extremophiles: General and plant biomass based biorefinery* (pp. 43–66). https://doi.org/10.1201/9781003014188-3
124. Pradhan, A., Pahari, A., Mohapatra, S., & Mishra, B. B. (2017). Phosphate-solubilizing microorganisms in sustainable agriculture: Genetic mechanism and application. In *Microorganisms for sustainability* (pp. 81–97). https://doi.org/10.1007/978-981-10-7380-9_5
125. Devadarshini, D., Mohapatra, S., Pati, S., Maity, S., Rath, C. C., Jena, P. K., & Samantaray, D. (2023). Evaluation of PHAs production by mixed bacterial culture under submerged fermentation. *Biologia (Bratisl), 78*, 895–902. https://doi.org/10.1007/s11756-022-01302-5
126. Sharma, S. K., Kumar, A., & Sharma, R. C. (2014). Challenges in railway vehicle modeling and simulations. In *International conference on newest drifts in mechanical engineering ICNDME-2014* (pp. 453–459).
127. Sharma, S. K., & Kumar, A. (2014). A comparative study of Indian and worldwide railways. *International Journal of Mechanical Engineering and Robotics Research, 1*, 114–120.
128. Sharma, S. K., Sharma, R. C., Kumar, A., & Palli, S. (2015). Challenges in rail vehicle-track modeling and simulation. *International Journal of Vehicle Structures & Systems, 7*(1), 1.
129. Palli, S., Sharma, R. C., Sharma, S. K., Muddada, V., & Sharma, N. (2020). A case study on noise pollution and its effects. *Journal of Critical Reviews, 7*, 261–267.
130. Sharma, R. C., Sharma, S. K., Sharma, N., & Sharma, S. (2020). Analysis of ride and stability of an ICF railway coach. *International Journal of Vehicle Noise and Vibration, 16*, 127–144.
131. Palli, S., Sharma, R. C., Sharma, S. K., & Chintada, V. B. (2020). On methods used for setting the curve for railway tracks. *Journal of Critical Review, 7*, 241–246.
132. Lee, J., Sharma, S.K. (2020) Numerical investigation of critical speed analysis of high-speed rail vehicle. 한국정밀공학회 학술발표대회 논문집. 696.
133. Sharma, S., Sharma, R. C., Sharma, S. K., Sharma, N., Palli, S., & Bhardawaj, S. (2020). Vibration isolation of the quarter car model of road vehicle system using dynamic vibration absorber. *International Journal of Vehicle Structures & Systems, 12*, 513–516.
134. Acharya, A., Gahlaut, U., Sharma, K., Sharma, S. K., Vishwakarma, P. N., & Phanden, R. K. (2020). Crashworthiness analysis of a thin-walled structure in the frontal part of automotive chassis. *International Journal of Vehicle Structures & Systems, 12*, 517–520.
135. Sharma, S. K., & Lee, J. (2020). Finite element analysis of a fishplate rail joint in extreme environment condition. *International Journal of Vehicle Structures & Systems, 12*, 503–506.
136. Bhardawaj, S., Sharma, R. C., & Sharma, S. K. (2020). Development of multibody dynamical using MR damper based semi-active bio-inspired chaotic fruit fly and fuzzy logic hybrid suspension control for rail vehicle system. *Proceedings of the Institution of Mechanical Engineers, Part K: Journal of Multi-Body Dynamics, 234*, 723–744.
137. Sharma, S. K., & Lee, J. (2020). Design and development of smart semi active suspension for nonlinear rail vehicle vibration reduction. *International Journal of Structural Stability and Dynamics., 20*, 2050120.
138. Sharma, S. K., Phan, H., & Lee, J. (2020). An application study on road surface monitoring using DTW based image processing and ultrasonic sensors. *Applied Sciences, 10*, 4490.
139. Sharma, R. C., Sharma, S., Sharma, S. K., & Sharma, N. (2020). Analysis of generalized force and its influence on ride and stability of railway vehicle. *Noise & Vibration Worldwide, 51*, 95–109.
140. Bhardawaj, S., Sharma, R. C., & Sharma, S. K. (2020). Ride analysis of track-vehicle-human body interaction subjected to random excitation. *Journal of the Chinese Society of Mechanical Engineers, 41*, 227–236.
141. Bhardawaj, S., Sharma, R. C., & Sharma, S. K. (2020). Development in the modeling of rail vehicle system for the analysis of lateral stability. *Materials Today Proceedings, 25*, 610–619.

142. Sharma, S. K., Sharma, R. C., & Sharma, N. (2020). Combined multi-body-system and finite element analysis of a rail locomotive crashworthiness. *International Journal of Vehicle Structures & Systems, 12*, 428–435.
143. Bhardawaj, S., Sharma, R. C., & Sharma, S. K. (2020). Analysis of frontal car crash characteristics using ANSYS. *Materials Today Proceedings, 25*, 898–902.
144. Bhardawaj, S., Sharma, R. C., & Sharma, S. K. (2019). A survey of railway track modelling. *International Journal of Vehicle Structures & Systems, 11*, 508–518.
145. Bhardawaj, S., Sharma, R. C., & Sharma, S. K. (2019). Development and advancement in the wheel-rail rolling contact mechanics. In *IOP Conference series: Materials science and engineering* (p. 012034). IOP Publishing.
146. Choppara, R. K., Sharma, R. C., Sharma, S. K., & Gupta, T. (2019). Aero dynamic cross wind analysis of locomotive. In *IOP conference series: Materials science and engineering* (p. 012035). IOP Publishing.
147. Sharma, S. K., & Sharma, R. C. (2019). Pothole detection and warning system for Indian roads. In *Advances in interdisciplinary engineering: Select proceedings of FLAME 2018* (pp. 511–519). Springer Singapore.
148. Sinha, A. K., Sengupta, A., Gandhi, H., Bansal, P., Agarwal, K. M., Sharma, S. K., Sharma, R. C., & Sharma, S. K. (2019). Performance enhancement of an all-terrain vehicle by optimizing steering, powertrain and brakes. In *Advances in engineering design: Select proceedings of FLAME 2018* (pp. 207–215). Springer Singapore.
149. Goyal, S., Anand, C. S., Sharma, S. K., & Sharma, R. C. (2019). Crashworthiness analysis of foam filled star shape polygon of thin-walled structure. *Thin-Walled Structures, 144*, 106312.
150. Goswami, B., Rathi, A., Sayeed, S., Das, P., Sharma, R. C., & Sharma, S. K. (2019). Optimization design for aerodynamic elements of Indian locomotive of passenger train. In *Advances in engineering design: Select proceedings of FLAME 2018* (pp. 663–673). Springer Singapore.
151. Sharma, S. K., Saini, U., & Kumar, A. (2019). Semi-active control to reduce lateral vibration of passenger rail vehicle using disturbance rejection and continuous state damper controllers. *Journal of Vibration Engineering & Technologies, 7*, 117–129.
152. Sharma, S. K. (2019). Multibody analysis of longitudinal train dynamics on the passenger ride performance due to brake application. *Proceedings of the Institution of Mechanical Engineers, Part K: Journal of Multi-body Dynamics., 233*, 266–279.
153. Sharma, S. (2018). Vibration control in quarter-car model with magnetorheological (MR) dampers using Bingham model. *Journal of Applied Mechanical Engineering, 7*, 1000299.
154. Sharma, S. K., & Kumar, A. (2018). Impact of longitudinal train dynamics on train operations: A simulation-based study. *Journal of Vibration Engineering & Technologies, 6*, 197–203.
155. Sharma, S. K., & Sharma, R. C. (2018). An investigation of a locomotive structural crashworthiness using finite element simulation. *SAE International Journal of Commercial Vehicles, 11*, 235–243.
156. Sharma, R. C., Sharma, S. K., & Palli, S. (2018). Rail vehicle modelling and simulation using Lagrangian method. *International Journal of Vehicle Structures and Systems, 10*, 188–194.
157. Sharma, R. C., & Sharma, S. K. (2018). Sensitivity analysis of three-wheel vehicle's suspension parameters influencing ride behavior. *Noise & Vibration Worldwide, 49*, 272–280.
158. Palli, S., Koona, R., Sharma, S. K., & Sharma, R. C. (2018). A review on dynamic analysis of rail vehicle coach. *International Journal of Vehicle Structures & Systems (IJVSS), 10.*
159. Sharma, S. K., & Sharma, R. C. (2018). Simulation of quarter-car model with magnetorheological dampers for ride quality improvement. *International Journal of Vehicle Structures & Systems, 10*, 169–173.
160. Sharma, S. K., & Kumar, A. (2018). Disturbance rejection and force-tracking controller of nonlinear lateral vibrations in passenger rail vehicle using magnetorheological fluid damper. *Journal of Intelligent Material Systems and Structures, 29*, 279–297.

161. Sharma, S. K., & Kumar, A. (2018). Ride comfort of a higher speed rail vehicle using a magnetorheological suspension system. *Proceedings of the Institution of Mechanical Engineers, Part K: Journal of Multi-Body Dynamics., 232*, 32–48.
162. Shukla, S., Kumar, U., Sharma, S. K., Gupta, P., & Kumar, A. (2017). Light weight freight rolling stock bogie frame: Design methodology validated with field oscillation trials. *International Journal of Vehicle Structures & Systems, 9*, 245–250.
163. Sharma, R. C., Palli, S., Sharma, S. K., & Roy, M. (2017). Modernization of railway track with composite sleepers. *International Journal of Vehicle Structures & Systems, 9*, 321–329.
164. Sharma, S. K., & Kumar, A. (2017). Ride performance of a high speed rail vehicle using controlled semi active suspension system. *Smart Materials and Structures, 26*, 055026.
165. Sharma, S. K., & Kumar, A. (2017). Impact of electric locomotive traction of the passenger vehicle Ride quality in longitudinal train dynamics in the context of Indian railways. *Mechanics & Industry, 18*, 222.
166. Sharma, S. K. (2013). Zero energy building envelope components: A review. *International Journal of Engineering Research and Applications, 3*, 662–675.
167. Sharma, S. K., & Kumar, A. (2016). Dynamics analysis of wheel rail contact using FEA. *Procedia Engineering, 144*, 1119–1128.
168. Sharma, S. K., & Chaturvedi, S. (2016). Jerk analysis in rail vehicle dynamics. *Perspectives in Science (Neth), 8*, 648–650.
169. Sharma, S. K., & Kumar, A. (2016). The impact of a rigid-flexible system on the ride quality of passenger bogies using a flexible carbody. In *Proceedings of the third international conference on railway technology: Research, development and maintenance* (p. 87). Civil-Comp Press.
170. Lavania, S., & Sharma, S. K. (2011). An explicit approach to compare crosstalk noise and delay in VLSI RLC interconnect modeled with skin effect with step and ramp input. *Journal of VLSI Design Tools & Technology, 1*, 1–8.
171. Palli, S., Koona, R., Sharma, R. C., & Muddada, V. (2015). Dynamic analysis of Indian railway integral coach factory bogie. *International Journal of Vehicle Structures and Systems, 7*, 16–20. https://doi.org/10.4273/ijvss.7.1.03
172. Palli, S., & Koona, R. (2015). Analyses of dynamic response of a railway bogie. *International Journal of Vehicle Noise and Vibration, 11*, 103–113. https://doi.org/10.1504/IJVNV.2015.070015
173. Swathi, K., & Srihari, P. (2015). Heat transfer enhancement in a tube using rectangular strip inserts. *International Journal of Applied Engineering Research, 10*, 41532–41544.
174. Palli, S., Sharma, R. C., & Rao, P. P. D. (2017). Dynamic behaviour of a 7 DoF passenger car model. *International Journal of Vehicle Structures and Systems, 9*, 57–63. https://doi.org/10.4273/ijvss.9.1.12
175. Sharma, R. C., Palli, S., & Koona, R. (2017). Stress and vibrational analysis of an Indian Railway RCF bogie. *International Journal of Vehicle Structures and Systems, 9*, 296–302. https://doi.org/10.4273/ijvss.9.5.06
176. Sharma, R. C., Avesh, M., Sharma, S. K., Sharma, S., Palli, S., & Sharma, N. (2020). Linear and non-linear stability analysis of a constrained railway wheel axle. *International Journal of Vehicle Structures and Systems, 12*, 128–133. https://doi.org/10.4273/ijvss.12.2.04
177. Palli, S., Dontikurti, R., Sharma, R. C., & Sharma, N. (2021). Analysis of dynamic response of a railway locomotive using ANSYS. *International Journal of Vehicle Structures and Systems, 13*, 250–255. https://doi.org/10.4273/ijvss.13.2.21
178. Sharma, R. C., Sharma, N., Singh, G., & Palli, S. (2021). Modeling and simulation of human body-vehicle-track system for the investigation of ride comfort. *SAE International Journal of Commercial Vehicles, 15*. https://doi.org/10.4271/02-15-02-0008
179. Sharma, R. C., Palli, S., Avesh, M., & Sharma, N. (2021). Vibration isolation of railway vehicle car body using semi-active suspension. *International Journal of Vehicle Structures and Systems, 13*, 482–487. https://doi.org/10.4273/ijvss.13.4.17

180. Gopala Rao, L. V. V., Sharma, R. C., Satyanarayana, V. S. V., & Palli, S. (2022). Stochastic optimal preview control response of a quarter-car nonlinear suspension model using spectral decomposition method. *Noise and Vibration Worldwide, 53*, 225–232. https://doi.org/10.1177/09574565221093234
181. Palli, S., Duppala, A., Sharma, R. C., & Rao, L. V. V. G. (2022). Dynamic simulation of automotive vehicle suspension using MATLAB simulink. *International Journal of Vehicle Structures and Systems, 14*, 310–314. https://doi.org/10.4273/ijvss.14.3.04
182. Satyanarayana, V. S. V., Sharma, R. C., Sateesh, B., Gopala Rao, L. V. V., Mohan Rao, N., & Palli, S. (2022). Optimum response of a nonlinear passive vehicle suspension system under random road excitations. *SAE International Journal of Commercial Vehicles, 16*. https://doi.org/10.4271/02-16-01-0004
183. Lokanadham, D., Sharma, R. C., Palli, S., & Bhardawaj, S. (2022). Wear rate modelling and analysis of limestone slurry particulate composites using the fuzzy method. *International Journal on Recent and Innovation Trends in Computing and Communication, 10*, 133–143. https://doi.org/10.17762/ijritcc.v10i1s.5818
184. Sharma, R. C., Palli, S., Gopala Rao, L. V. V., Duppala, A., & Sharma, S. K. (2022). Four-wheel vehicle response under bump, pothole, harmonic, and random excitations using bond graph/simulink technique. *SAE International Journal of Commercial Vehicles, 16*. https://doi.org/10.4271/02-16-02-0008
185. Sharma, R. C., Gopala Rao, L. V. V., Sharma, S. K., Palli, S., & Satyanarayana, V. S. V. (2022). Analysis of lateral stability and ride of an Indian railway constrained dual-axle bogie frame. *SAE International Journal of Commercial Vehicles, 16*. https://doi.org/10.4271/02-16-02-0014
186. Sharma, R. C., Palli, S., Bhardawaj, S., & Rao, L. V. V. G. (2022). Advances in human biodynamic modeling for analysis of whole-body vibrations. *International Journal of Vehicle Structures and Systems, 14*, 503–511. https://doi.org/10.4273/ijvss.14.4.17
187. Sharma, R. C., Palli, S., & Sharma, S. K. (2023). Ride analysis of railway vehicle considering rigidity and flexibility of the carbody. *Journal of the Chinese Institute of Engineers, Transactions of the Chinese Institute of Engineers, Series A, 46*, 355–366. https://doi.org/10.1080/02533839.2023.2194918
188. Sharma, R. C., Palli, S., & Bhardawaj, S. (2023). A review of wheel-rail dynamic behavior on curved tracks. *International Journal of Vehicle Structures and Systems, 15*, 19–25. https://doi.org/10.4273/ijvss.15.1.04
189. Saikiran, C., Palli, S., Bondala, R., Sharma, R. C., Sreeramulu, D., & Dontikurti, R. (2023). Experimental investigation on thermal behaviour of silicon carbide (SiC) based polymer composites under hygrothermal conditions. *International Journal of Vehicle Structures and Systems, 15*, 222–226. https://doi.org/10.4273/ijvss.15.2.15
190. Singh, B. J., Palli, S., Kumar, P., & Bahl, S. (2023). Behavioural investigation of hardness profile for dissimilar TIG-weldment between P22 and P91 steels. *International Journal of Vehicle Structures and Systems, 15*, 31–36. https://doi.org/10.4273/ijvss.15.1.06
191. Naveen, Y., Lokanadham, D., Naidu, D. R., Sharma, R. C., Palli, S., & Lila, M. K. (2023). An experimental study on the influence of blended Karanja biodiesel on diesel engine characteristics. *International Journal of Vehicle Structures and Systems, 15*, 154–160. https://doi.org/10.4273/ijvss.15.2.02
192. Bhardawaj, S., Jha, A. K., Sharma, R. C., Palli, S., Lokanadham, D., Sharma, N., & Dhiman, N. (2023). Finite element analysis of two disk rotor system used in automobile turbochargers. *International Journal of Vehicle Structures and Systems, 15*, 501–504. https://doi.org/10.4273/ijvss.15.4.09
193. Sharma, R. C., Palli, S., Bhardawaj, S., Jha, A. K., Duppala, A., Sharma, N., Sharma, S., & Dhiman, N. (2023). Dynamic behaviour of railway vehicle under bump and pothole track irregularities. *International Journal of Vehicle Structures and Systems, 15*, 566–569. https://doi.org/10.4273/ijvss.15.4.22

194. Raju, S., Palli, S., Devi Prasad, P., Menda, V. R., & Ramakrishna, B. (2024). A hybrid AHP-TOPSIS, MOORA technique for multi-objective optimization of thermal, mechanical, and water absorption behavior of epoxy/hemp, pine apple, and palm fiber composites. *Journal of the Chinese Institute of Engineers, Transactions of the Chinese Institute of Engineers, Series A, 47*, 23–36. https://doi.org/10.1080/02533839.2023.2274092
195. Raju Rallabandi, S., Pilla, D. P., Dowluru, S., Palli, S., Sharma, N., Sharma, S. K., & Sharma, R. C. (2024). Critical evaluation of epoxy-hemp-pineapple-palm fiber composites using hybrid AHM-TOPSIS technique for sustainable structural applications. *Journal of the Chinese Institute of Engineers, Transactions of the Chinese Institute of Engineers, Series A, 47*, 325–336. https://doi.org/10.1080/02533839.2024.2308250
196. Gopala Rao, L. V. V., Satyanarayana, V. S. V., Sharma, R. C., Palli, S., Raturi, A., Kharola, A., & Sharma, A. (2024). Preview control of the random response of a full car vehicle model traversing a rough road. *International Journal of Vehicle Noise and Vibration, 20*, 70–88. https://doi.org/10.1504/IJVNV.2024.138105
197. Sharma, A., Sharma, V. S., Gupta, S., Sharma, R. C., Palli, S., & Sharma, N. (2024). Additive manufacturing in the COVID-19 pandemic: Equipment and challenges? *Journal of Micromanufacturing*. https://doi.org/10.1177/25165984241228083
198. Bammidi, R., Madivada, H., Dowluru, S., Palli, S., Sharma, R. C., & Muddada, V. (2024). Mechano-tribological investigations of Al-%Wt. of boron nitride as metal-ceramic composites. In *AIP conference proceedings* (Vol. 2978, No. 1). AIP Publishing. https://doi.org/10.1063/5.0182965
199. Ranganath, B. A., Duppala, A., Palli, S., Rao, L. V. V., & Sharma, R. C. (2024). Numerical analysis and experimental study of FRP composites for damping applications. In *AIP conference proceedings* (Vol. 2978, No. 1). AIP Publishing. https://doi.org/10.1063/5.0184802
200. Sharma, R. C., Palli, S., Duppala, A., Bhardawaj, S., Sharma, S. K., Sharma, N., & Sharma, A. (2024). Rigid body analysis of ride behaviour of roadway vehicle using Lagrangian dynamics. *International Journal of Vehicle Structures and Systems, 16*, 74–78. https://doi.org/10.4273/ijvss.16.1.15
201. Menda, V., Dora, S., Palli, S., Rallabandi, S. R., Sharma, R. C., Sharma, N., & Bondala, R. (2024). Mechanical and wear characteristics of aluminium-7075/graphene/TiB2 prepared by stir casting. *Engineering Research Express, 6*(2), 025571. https://doi.org/10.1088/2631-8695/ad58a8

Digital Tool Life Prediction and Monitoring for Sustainable Manufacturing in Industry 5.0

Bhupender Singh, Nitin Panwar, Sandeep Ravish, and Ashok Kumar Yadav

1 Introduction

The manufacturing industry operates in fiercely competitive domestic and international markets, where firms are constantly challenged by rapid technological advancements, evolving customer preferences, and shorter product life cycles. To maintain a competitive edge, manufacturing firms must prioritize flexibility. Flexible manufacturing systems (FMS) are essential in meeting dynamic demands, diverse product mixes, and the successful introduction of new products. FMS leverage programmable automation for machining and material handling to deliver these capabilities. Ensuring the effective functioning of FMS, including adept product change management and adaptability to demand fluctuations, is paramount for optimizing performance. Improved equipment utilization, shorter setup times, less work-in-process (WIP), and noticeably shorter lead and throughput times are all possible outcomes of this optimization. Therefore, in order to achieve optimal throughput time and equipment usage, it is essential to fully identify and assess the parameters that evaluate FMS performance. Computer numeric control (CNC) machine tools connected by automated material handling and storage systems, all managed by an integrated computer system, make up a state-of-the-art flexible production system.

B. Singh · N. Panwar
Department of Mechanical Engineering, J.C. Bose University of Science and Technology, YMCA, Faridabad, India

S. Ravish
Department of Civil Engineering, J.C. Bose University of Science & Technology YMCA, Faridabad, India
e-mail: sandeepravish89@jcboseust.ac.in

A. K. Yadav (✉)
Department of Mechanical Engineering, Symbiosis Institute of Technology, Symbiosis International (Deemed University), Pune, Maharastra, India

Y. Kumar Singla et al. (eds.), *Digital Manufacturing in Industry 5.0*,
https://doi.org/10.1007/978-3-031-91500-0_26

Using CNC machine tools that are loaded and unloaded by industrial robots, this sophisticated system can process a variety of product families effectively [1]. FMS is essential for managing change and adapting to desired situations. A manufacturing system needs to be able to recognize and distinguish between incoming parts or product styles, configure operating instructions in an agile manner, and quickly switch between physical setups in order to achieve flexibility [2]. There are four categories of manufacturing system flexibility [2]: (i) Flexibility in new product introduction: The ability to launch and produce new items or alter current ones. (ii) Mix flexibility: The capacity to alter the variety of goods produced by the production system within a predetermined time frame. (iii) Volume flexibility: The ability to alter the total amount of output that has been produced. (iv) Flexibility in delivery: The capacity to alter anticipated or scheduled delivery dates. Particularly in mid-volume and mid-variety manufacturing, FMS technology is frequently used in machining operations and can also be applied to metal press work, forging, and assembly [2]. Higher machine utilization, lower resource requirements, less factory floor space, improved flexibility, lower inventory needs, shorter manufacturing lead times, lower direct labor requirements, increased labor productivity, and the possibility of unattended production are just a few advantages of using FMS [2].

One crucial piece of computer technology for system modeling, design, and operation is the use of manufacturing process simulation [2]. A manufacturing system's construction is visually represented through simulation, which also collects data on the system's cost and performance aspects. Importantly, simulations should not be done only once. The model should be simulated several times for more precise results. For experimental analysis, the same model should also be run multiple times, each time manipulating a different variable. The following could be among the goals of simulation analysis [2]: (i) Evaluating the loading of machines and their resource levels. (ii) Locating queues, or process bottlenecks. (iii) Assessing the effects of machine failure. (iv) Examining the effectiveness of the method. In modern manufacturing systems, particularly flexible manufacturing systems (FMS), accurately estimating the cutting tool life under specific cutting conditions is paramount. The classical formula for calculating cutting force provides an average value, but it fails to account for the significant impact effect when the tool initially contacts the workpiece. This impact force, which can be 1.1–1.5 times higher than the typical cutting force, can significantly shorten tool life and frequently cause early tool failure. Using force dynamometers to measure cutting forces directly is either impractical or impossible in many machining procedures, including those requiring shapers. Consequently, it becomes imperative to estimate these pressures indirectly. The goal of this study is to estimate the cutting tool life by taking into account the impact effect, which is determined by measuring the machining responses. For precise force estimation in structural evaluations, the dynamic response by itself is sometimes inadequate. As a result, a thorough explanation of the input force is necessary. Determining the frequency response function (FRF) matrix, measuring the structural responses, and computing the dynamic forces using the least squares scheme is one efficient method. This approach, though, may result in mistakes. We suggest using the acceleration response as an input for force prediction in order to increase accuracy. In order to ascertain the impact force amplitude, this study creates a frequency domain impact force history prediction system. To find the impact force's magnitude, the force estimate issue is iteratively addressed. The life of cutting tools under various impact situations can be more precisely predicted by applying these force estimations and carrying out iterative testing. The authors of this groundbreaking work have created and recreated a state-of-the-art FMS model in software. Their innovative algorithm predicts impact force history in the frequency domain, allowing for precise determination of impact force amplitude and comprehensive analysis of system features and variables. This enables accurate evaluation of the time performance level of an FMS. A thorough review of the state-of-the-art in Digital Transformation toward Industry 5.0 is still lacking in the present scientific literature. The purpose of this systematic literature review is to fill this research gap by examining the scholarly developments at the intersection of Industry 5.0 and Digital Transformation. Journal articles and conference papers in the Industry 5.0 space that were published in the previous five years up until March 2023 were analyzed as part of the systematic review. First, a list of prestigious journals and worldwide conferences for publishing on Industry 5.0-related content is provided to provide an overview of the research disciplines of Digital Transformation and Industry 5.0. Second, the importance of Digital Transformation is illustrated.

It is essential that people embrace and trust technology. Initiatives must therefore emphasize human support rather than human resolve while upholding society's right to be involved in the development, use, and purpose of technologies. Furthermore, new technologies like artificial intelligence must be transparent and easy to understand. Generally speaking, the issues and facilitators covered include a complicated system that includes organizational and technological elements, public and political elements, and the triple bottom line of sustainability (economic, ecological, and social dimensions). As a result, the following types of obstacles and facilitators constitute a complex system that needs to be seen as closely connected and interconnected.

- Training individuals to use new technologies must take place at the same time as technology is adapted to humans. It should not be a one-way path to human-centricity. Technologies won't reach their full potential otherwise.
- Future skill requirements are unknown, but it is necessary to anticipate upcoming labor and skill shortages brought on by demographic changes as well as shifting skill requirements between generations. It is imperative that policymakers promote the implementation of retraining and lifelong learning concepts.
- To gain widespread acceptability and create a "new deal" between society and industry, current issues like young unemployment, an aging population, or gender and societal imbalance must be better integrated.
- Disagreement about whether wants and ideals should come first results from society's heterogeneity. Different segments of society have different requirements and prioritize different ideals. It is necessary to incorporate characteristics of diversity, age, gender, sex, and cultural background.

- Customers must be fully integrated throughout all value chains in order to educate them about the social or environmental value generated and factor this into their decision-making and willingness to pay.
- Conventional approaches to corporate social responsibility are more of a marketing idea that are used as showcases but have not yet been fully adopted. In order to encourage businesses to advance corporate social responsibility, particularly in SMEs, regulatory incentives must be used in conjunction with public sentiment.

By extending the idea of human-centricity to include the requirements of a society in addition to those of individuals, socio-centricity is a concept. Individual workers' requirements must be balanced with the needs of the entire workforce, employers, and society at large. As anticipated, Industry 5.0's sustainability ideals are widely accepted by the scientific community, and the academic literature provides preliminary insights into how this agenda might advance sustainable development [3]. Ford et al. [4] looked at ways that Industry 5.0 and the circular supply chain could work together to improve sustainability. Furthermore, new research provides a preliminary knowledge of how governments and other Industry 5.0 stakeholders may use this agenda to bring its sustainability principles to life.

Little has been done to comprehend how Industry 5.0 might inclusively support sustainable production, despite these early, priceless contributions. To fill this knowledge gap, one would wonder if the findings of earlier research on the relationships between Industry 4.0 and sustainable manufacturing could be applied to the Industry 5.0 research setting. In reality, there is a fair amount of research on the effects of Industry 4.0 on sustainable manufacturing [5]. For the following reasons, it is not advisable to extrapolate the earlier research on Industry 4.0-sustainable manufacturing to the Industry 5.0 setting. First, in order to achieve its desired goals, Industry 5.0 uses human-centric and cognitive sets of technologies. Second, while Industry 4.0 is largely a productivity-driven and technology-push phenomenon, Industry 5.0 is a socio-politically pulled structure.

Considering the aforementioned considerations, the current study fills the following research gaps. First off, Industry 5.0 is a vague and wide-ranging notion that has diverse meanings in various corporate settings [6]. Although researching Industry 5.0's potential contribution to sustainable manufacturing requires an operational definition tailored to the manufacturing context, there is currently a lack of such knowledge [2]. This crucial knowledge gap hinders further in-depth understanding of Industry 5.0's impact on the production environment as well as targeted research into its unique features. Accordingly, the study's primary goal is to provide an operational definition of Industry 5.0 in the context of sustainable manufacturing and to create an archetype that illustrates the distinctive characteristics of the hyper-connected manufacturing ecosystem of Industry 5.0.

The potential opportunities for sustainable manufacturing presented by Industry 5.0 are the subject of the second knowledge gap. The current analysis, which builds on the sustainability literature, suggests that Industry 5.0 can support sustainable production through intricately linked processes. This presumption stems from the complexity of the Industry 5.0 ecosystem's components, design ideas, and technology [6]. However, little is known about the micro-mechanisms (functions) that Industry 5.0 may use to support several facets of sustainable manufacturing. Businesses and social actors will find it difficult to take advantage of Industry 5.0's potential to promote sustainability in manufacturing if these functions are not identified. This will hinder the integration of sustainable digital practices that can ensure a sustained competitive edge. In order to fully realize Industry 5.0's potential and encourage sustainable practices in the industrial setting, the study's second goal is to investigate and comprehend its sustainable manufacturing functions. According to this goal, the study examines the ways in which this agenda might inclusively support sustainable manufacturing and performs a content-centric evaluation of Industry 5.0 literature. A new era that emphasizes sustainability in addition to technological innovation is being ushered in by the introduction of Industry 5.0, which represents a turning point in the development of manufacturing and industrial processes [6]. Despite being new, this idea is catching on quickly because it aims to balance the frequently incompatible objectives of environmental preservation and economic expansion. The core of Industry 5.0 is its emphasis on combining automation and smart technology with human creativity and craftsmanship to improve production, efficiency, and environmental sustainability. This review paper offers a thorough summary of the new developments and difficulties that Industry 5.0 brings in the pursuit of a greener future.

Fundamentally, Industry 5.0 signifies a substantial departure from the fourth industrial revolution's focus on automation and data sharing in manufacturing technologies to a more comprehensive strategy that integrates sustainable practices and human intelligence [7]. In light of global issues including resource depletion, climate change, and environmental degradation, this paradigm shift emphasizes the significance of sustainable development. It promotes a well-rounded strategy that uses technology to advance human potential while maintaining resource efficiency and environmental preservation. Industry 5.0 and sustainability principles are crucially aligned, providing a mechanism to balance industrial growth with the planet's ecological constraints.

Huang and Liang [8] focused on finish turning, which frequently has very small applied feed rates and depths of cut. They were able to forecast the chip formation forces by first transforming the 3-D cutting geometry into an analogous 2-D cutting geometry. The entire 2-D cutting forces were then determined using a ploughing effect mechanical model, and the 3-D cutting forces were estimated using a geometric transformation.

Dahlman et al. [9] have shown that rake inclination had the biggest effect on turning residual stresses. The residual stresses in the feed and speed directions were evaluated using the X-ray diffraction method. The findings demonstrated that while a greater negative rake angle led to higher compressive stresses, cutting depth had no influence on residual strains. Furthermore, it has been shown that compressive forces grow in tandem with feed rate.

Numerous facets of Industry 5.0 have been studied in previous studies, including its technological underpinnings, financial implications, and potential to foster innovation. However, much remains to be discovered regarding all of its sustainability features. Research has not yet thoroughly examined how Industry 5.0 technology, such as sophisticated robotics, the Internet of Things, and artificial intelligence (AI), may support sustainable development [10]. This study gap highlights the need for a more comprehensive examination of how these technologies might lessen environmental consequences, promote resource efficiency, and help achieve the Sustainable Development Goals (SDGs).

The lack of comprehensive research on the connection between Industry 5.0 and sustainability has resulted in the loss of an opportunity. The information gap currently limits our understanding of Industry 5.0's potential to promote transformative technologies that align industrial operations with sustainability requirements. If this relationship is not carefully considered, we may lose the chance to employ Industry 5.0 for environmental preservation, economic resilience, and social well-being [7].

It highlights the value of technological integration, teamwork, talent development, and resource management, as well as the philosophy of human-centered design. This graphic illustration emphasizes the dedication to looking at these components from a sustainability perspective. Along with mapping major research and innovation sectors, the report emphasizes the value of comprehensive partnerships and takes into account the trade-offs that come with sustainable development. In addition to showing how these domains are interconnected, Fig. 1 also provides guidance for a thorough examination of the complex interaction between Industry 5.0 technologies and sustainability programs. Understanding Industry 5.0's promise to build a more equitable and ecologically responsible future requires an inclusive approach.

The third knowledge gap is the lack of a roadmap explaining how Industry 5.0 participants should utilize these features, extending the research on sustainable manufacturing in Industry 4.0. It is expected that the sequence and sequential relationship in using Industry 5.0's sustainable manufacturing capabilities would significantly impact this agenda's sustainability outcomes. Regretfully, the literature

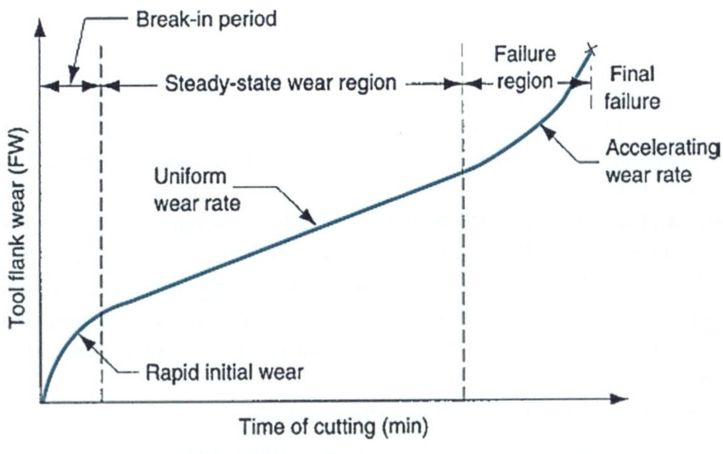

Fig. 1 Tool wear versus cutting time

on sustainable manufacturing lacks a strategic plan outlining this order. Without such a roadmap, manufacturers could struggle to prioritize and manage the tasks, which could lead to a less-than-ideal implementation of Industry 5.0's framework for sustainable manufacturing. The report develops a strategic roadmap outlining how manufacturers might promote manufacturing sustainability ideals through Industry 5.0 sustainability functions in order to accomplish this goal. It accomplishes this by applying the interpretive logic-knowledge base (ILB) and the interpretive structural modeling (ISM) approach.

In order to effectively address the three essential elements needed to create such a strategic roadmap, the study merged ISM and ILB. Determining the precedence relationships between the system pieces is a prerequisite for creating the strategic roadmap. This need can be met by a variety of decision modeling approaches, including ISM, DEMATEL, and their fuzzy variations. The most popular Industry 5.0 technologies for sustainability were collaborative robotics, artificial intelligence, and the internet of things. This study then covered the present research under four major themes: ecological progress, human-centricity, sustainability in higher education, and robot advancement [4, 11]. The most popular subject has been discovered to be human-centric. Based on the information currently available, the themes and subthemes identified in this study can assist researchers in identifying knowledge gaps and stimulate additional research. The decision-making process about the successful application of Industry 5.0 solutions to improve sustainable practices in businesses can be aided by this study. In addition to discussing the current status of Industry 5.0 technologies for sustainability, this paper offers suggestions for future lines of inquiry. Researchers, legislators, and business experts may be inspired to investigate and create novel Industry 5.0 solutions as a result.

Industry 5.0 Definition: This phrase captures the spirit of the fifth industrial revolution by highlighting how cutting-edge technology and human-centered design may be seamlessly integrated to achieve increased productivity and efficiency. Its goal is to establish an industrial setting where intelligent systems and human creativity coexist, fostering a more peaceful and well-balanced workplace [7, 10]. How to define sustainability: A complex idea, sustainability aims to strike a balance between environmental preservation, social justice, and economic prosperity. It promotes the long-term well-being of all living forms by attempting to meet present needs without endangering the capacity of future generations to meet their own.

The Link: Industry 5.0 and sustainability are closely related. The fundamental principles of Industry 5.0—human-centricity, resilience, and sustainability—are a perfect fit with sustainability. The Sustainable Development Goals (SDGs) can be met and a more equitable and environment-friendly world can be created by utilizing cutting-edge technologies like IoT and AI. The following are the technology foundations of Industry 5.0: Industry 5.0 is powered by cutting-edge technologies like blockchain, IoT, AI, machine learning, and robotics. These technologies enable more sustainable, effective, and efficient operations in many aspects of the industry, from supply chain management to resource utilization [12, 13].

Emphasis on People: Industry 5.0 emphasizes how crucial human-centered design concepts are. In order to increase usability, employee satisfaction, and overall outcomes, it seeks to guarantee that processes and technology are created with people in mind. This strategy is necessary to promote a sustainable and peaceful future by striking a balance between human needs and ambitions and technical progress [14].

The Importance of Training and Collaboration: Industry 5.0 technologies and practices require strong stakeholder connections, skill development, and collaboration to achieve sustainability. These include the general public, academics, lawmakers, and business professionals. With the help of comprehensive training programs and strategic partnerships, workers can be better equipped to manage the evolving industrial landscape and contribute to a sustainable future [8].

1.1 Tool Wear

The different wear processes cause the cutting tool to wear down more and more as cutting continues. The graphic illustrates the overall relationship between cutting time and tool wear.

Tool wear typically progresses through three stages: an initial rapid wear break-in period, a steady-state wear period with a fairly uniform wear rate, and a failure region where wear accelerates due to increased cutting temperatures. By measuring dynamic responses and applying an impact force history prediction algorithm, we aim to more accurately predict tool life and improve machining efficiency.

1.2 Traditional Methods

Historically, tool life estimation relied on empirical knowledge and machinist experience, often using visual inspections or predefined schedules. While practical, these methods lack precision. The Taylor tool life equation, proposed by Frederick W. Taylor, provided a quantitative framework by correlating tool life with cutting speed. However, it assumes a linear relationship, which may oversimplify the complexities of tool wear.

1.3 Challenges in Tool Life Estimation

Estimating tool life is challenging due to the complex machining conditions, where tools face varying mechanical and thermal loads. Adapting structural fatigue equations to cutting tools is difficult due to unique loading conditions and thermal effects. Material variability affects tool properties and life estimations, and generalizing findings to different cutting conditions is challenging. Additionally, the dynamic, nonlinear nature of tool wear complicates accurate predictions of tool life.

This paper focuses on estimating cutting tool life in flexible manufacturing systems (FMS) by addressing the impact effect that significantly reduces tool life at the start of cutting. It explores indirect methods for estimating cutting forces due to the unavailability of force dynamometers in machines like shapers. The approach involves using frequency response function (FRF) matrices to model dynamic forces and developing an algorithm in the frequency domain to predict impact force amplitudes from acceleration responses. The report highlights iterative experimentation to refine force estimation and improve tool life predictions in manufacturing operations.

2 Methodology

Tool life estimation for a shaper tool using a load cell involves predicting the tool's remaining life by measuring cutting forces. Due to the impracticality of placing a load cell between the tool tip and workpiece during cutting, an inverse force measuring technique is utilized to estimate impact forces. This method calculates actual tool life considering how cutting forces increase as the tool wears. Practical steps include installing and calibrating a load cell between the tool holder and shaper table, collecting cutting force data using a data acquisition system under various conditions, analyzing trends in the data to assess tool wear, and developing a predictive model for remaining tool life based on these insights. This approach aims to improve accuracy in tool life estimation and optimize machining operations in shaper machines.

3 Force Identification as an Inverse Problem

Impact loading significantly affects structures and machine components, compromising their integrity and load-bearing capacity. Understanding the loading profile, including spatial distribution and time history, is crucial for optimal design but challenging to obtain directly due to access limitations and safety risks. Remote measurement of structural responses offers a practical approach, enabling the reconstruction of impact loading profiles using inverse procedures like deconvolution. This method, assuming linear dependency between structural response and impact force, is commonly applied in both time and frequency domains, facilitated by techniques such as Fast Fourier Transformation (FFT) for data analysis. Careful management of data truncation during FFT is essential to ensure accurate results in practical computations.

4 Analysis of Inverse Force Identification

In practice, estimating dynamic impact forces without instrumenting the impact or can be challenging. An example is determining the impact force on a shaper tool when it first contacts the workpiece, where direct measurement is impractical. This impact force can be 1.1–1.5 times the average cutting force. Determining the impact force from experimentally measured structural responses is a challenging inverse problem. The frequency domain deconvolution method addresses this challenge.

(i) **Time–Domain Analysis**

The convolution integral can be used to represent the body's response $x(t)$ to an impact force $f(t)$, which is linearly dependent on the impact force:

$$x(t) = \int t 0 f\, h(t-\tau) f(\tau) d\tau,$$

where $h(t)$ is the impulse response function (IRF). The relationship is expressed as follows after discretizing the integral:

$$x = hf,$$

where h is a matrix of discrete IRF values, and x and f are vectors of discrete values.

(ii) **Frequency Domain Analysis**

Natural frequency, modal damping, and mode shape define the modes that are used to characterize resonant vibrations. The dynamic qualities of a structure are described via modal analysis. The frequency response function (FRF) matrix links the input force $\{F(\omega)\}$ to the response $\{X(\omega)\}$ for a structure exposed to a sinusoidal force $[H(\omega)]$:

$$X(\omega)\} = \left[H(\omega)\right]\left[H(\omega)\right] = \left\{F(\omega)\right\}\left\{F(\omega)\right\} - 1\left\{X(\omega)\right\}$$

5 Frequency Response Function (FRF)

The ratio of a structure's output response to an applied force is known as the frequency response function. Simultaneous measurements are made of the applied force and the structure's reaction to it (which can be expressed as acceleration, velocity, or displacement). The Fast Fourier Transform algorithm, which can be found in any computer software program or signal processing analyzer, is now used to convert the measured time data from the time domain to the frequency domain. The functions can be defined by their magnitude and phase, or they have both real and imaginary portions as a result of this transformation. Examine the linear system depicted in Fig. 2.

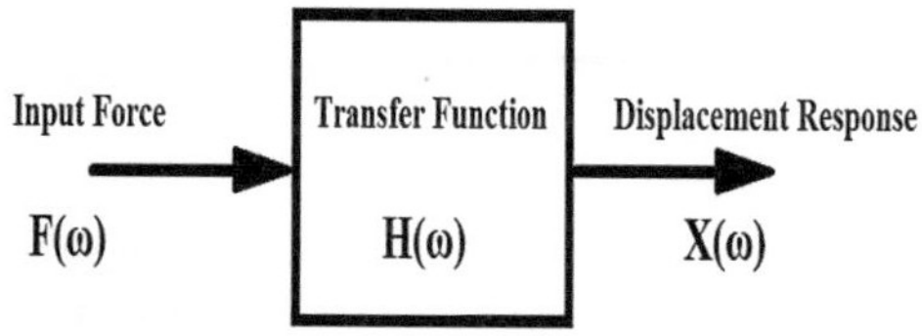

Fig. 2 Linear system of model

The input force in this problem is denoted by $F(\omega)$, the transfer function by $H(\omega)$, and the displacement response function by $X(\omega)$, which is expressed as a function of the angular frequency ω. Therefore, any function is a spectral function. The following equation illustrates the link between input and output:

$$\{X(\omega)\} = [H(\omega)] \times \{F(\omega)\}$$

And the FRF $H(\omega)$ is given by:

$$H(\omega) = \frac{X(\omega)}{F(\omega)}$$

6 Load Cell Sensor

A transducer that transforms mechanical force—such as load, weight, tension, compression, or pressure—into an electrical signal is called a force sensor or load cell. This signal, which is measurable and standardized, varies proportionately with the applied force. The structure of the load cell is deformed when a force is applied, which also causes the bonded strain gauges to deform. The strain gauges are set up in a Wheatstone bridge configuration, and this deformation alters their electrical resistance. The bridge circuit converts the resistance change into a small voltage change, which is then amplified by an instrumentation amplifier. The amplified voltage signal is often converted to a digital signal for easier processing. The resulting output signal is proportional to the applied force, providing accurate force measurements.

7 Electronic Circuit Diagram

The circuit (Fig. 3) consists of a load cell, Arduino UNO, LCD, resistor, capacitor, battery, wires, etc. Impact force is applied on the shaper tool using an impact hammer at a certain depth and its response is measured by the load cell sensor which displays the value of experimental force experienced by the tool on the LCD.

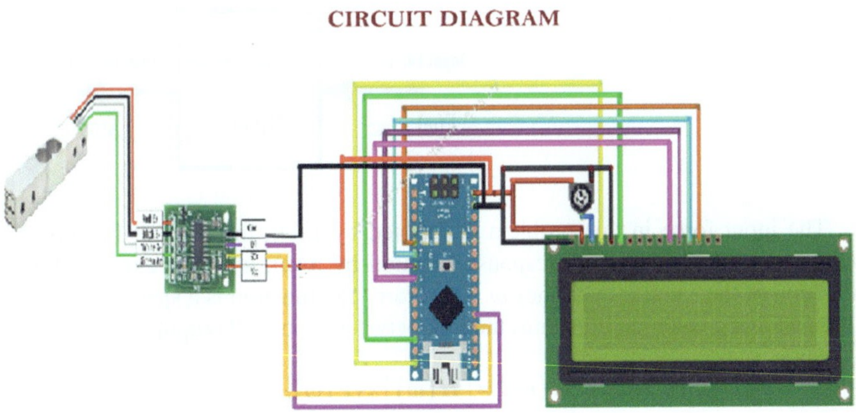

Fig. 3 Circuit diagram of system

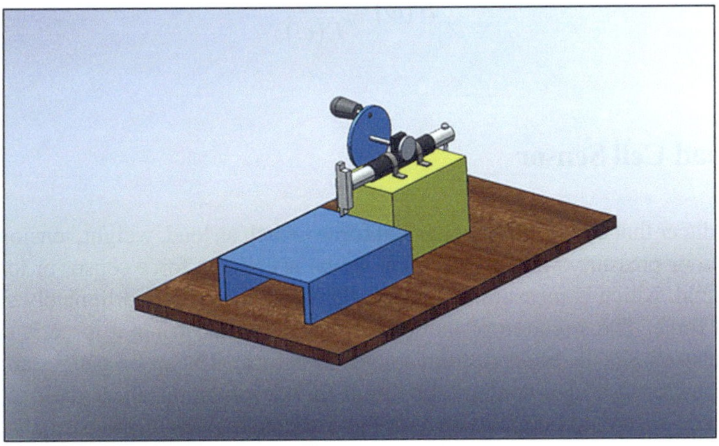

Fig. 4 Isometric view of model

8 CAD Model of the Setup

Figures 4 and 5 show the CAD model of the setup.

Fig. 5 Top view of the model

9 Calculations

For estimating tool life, we will use Goodman equation and Miners equation. After calculating the tool life, we will plot it in the SN curve to determine life in terms of strokes.

We will calculate both experimental and theoretical tool life using the same method but the difference will the force acting which will be different in both the case. For estimating theoretical force, we will use modified Taylor equation while the experimental force will be estimated by the load cell.

10 Calculation of Theoretical Tool Life

Step 1: Force Estimation using modified Taylor equation

$$Fc = 9.807 \times C_p \times k \times d^x \times s^y$$

Where C_p, k, x, and y are steel constants, d is depth of cut, and s is feed in cm.

For steels

$C_p = 40\text{–}80$, so considering mid-value we take $C_p = 60$

$$k = 1, x = 0.86, y = 0.74, s = 0.06\,\text{cm}.$$

Substituting these values in above equation for depth of cut of 1 mm

$$Fc = 9.807 \times 60 \times 1 \times 1^{0.86} \times 0.06^{0.74}$$

$$Fc = 73.37\,\text{N}$$

The above obtained force is the average cutting force that is obtained theoretically. The force is said to be average because of various material constants are assumed for the calculation.

Step 2: Calculating Theoretical Tool Life.

For the depth of 1 mm, it was observed that the impact force is equal to 73.37 N.

The cross-sectional area of tool (Fig. 6) at 1 mm from the tip is $A = 1.583\ \text{mm}^2$.

Shear stress due to this force, $\tau = F/A = 73.37/1.583 = 46.34$ MPa.

Shear stress can be considered 0.6 times the normal stress. Thus, we have

$$\sigma_a = \tau / 0.6 = 46.34 / 0.6 = 77.248\,\text{MPa}.$$

The tool material is M42 high speed steel (HSS).

Ultimate tensile strength for M42 HSS = 640–2000 MPa.

Therefore, taking average, we have $S_{ut} = 1320$ MPa.

Yield strength for M42 HSS is $\sigma_y = 700$ MPa.

From Goodman's equation,

$$\frac{\sigma a}{Sen} + \frac{\sigma m}{Sut} = \frac{1}{FOS}$$

Where FOS $= \frac{\sigma y}{\sigma a} = \frac{700}{77.248} = 9.06$

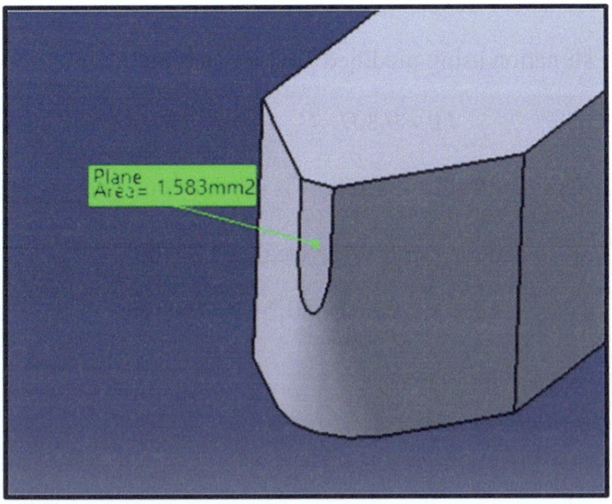

Fig. 6 Cross-sectional area of tool

$$\sigma_m = \sigma_a / 2 = 38.624\,\text{MPa}$$

Substituting for σ_a, σ_m, S_u, and FOS in the above equation and solving we get endurance Strength, $S_{en} = 52.03$ MPa.

Taking $0.9 \times S_u = 0.9 \times 1320 = 1188$ MPa.

Using the Miner's relation for life estimation, we can predict the life of the tool in terms of number of strokes.

$$\log(0.9\,S_u) = \log 1188 = 3.0748$$

$$\log(S_{en}) = \log 55.03 = 1.7406$$

$$\log(\sigma_a) = \log 77.248 = 1.8879$$

Thus, the life of the tool for theoretically obtained cutting force of 73.37 N is estimated to be **462,132** strokes of the tool.

11 Calculation of Experimental Tool Life

Step 1: Force estimation using load cell (Fig. 7).

For the depth of 1 mm, it was observed that the impact force is equal to 108.8 N.

Step 2: Calculating experimental tool life.

The cross-sectional area of tool at 1 mm from the tip is $A = 1.583$ mm^2.

Shear stress due to this force, $\tau = F/A = 108.8/1.583 = 68.73$ MPa.

Shear stress can be considered 0.6 times the normal stress. Thus, we have

$$\sigma_a = \tau / 0.6 = 68.73 / 0.6 = 114.55\,\text{MPa}.$$

Fig. 7 Reading depicted on LCD by load cell

The tool material is M42 high speed steel (HSS).

Ultimate tensile strength for M42 HSS = 640–2000 MPa.

Therefore, taking average, we have S_{ut} = 1320 MPa. Yield strength for M42 HSS is σ_y = 700 MPa.

From Goodman's equation,

$$\frac{\sigma a}{Sen}+\frac{\sigma m}{Sut}=\frac{1}{\text{FOS}}$$

where $\text{FOS}=\frac{\sigma y}{\sigma a}=\frac{700}{114.55}=6.11$

$$\sigma_m=\sigma_a/2=57.28\,\text{MPa}$$

Substituting for σ_a, σ_m, S_u, and FOS in the above equation and solving, we get endurance strength, S_{en} = 73.252 MPa.

Taking $0.9 \times S_u = 0.9 \times 1320 = 1188$ MPa.

Using the Miner's relation for life estimation, we can predict the life of the tool in terms of number of strokes.

$$\log\left(0.9S_u\right)=\log 1188=3.0748$$

$$\log\left(S_{en}\right)=\log 73.252=1.8648$$

$$\log\left(\sigma_a\right)=\log 114.55=2.0589$$

Thus, the life of the tool for experimentally obtained cutting force of 108.8 N is estimated to be 330,182 strokes of the tool.

12 Observation

After doing the calculations for both theoretical and experimental forces, it can be observed that the tool life calculated from experimental data is lesser that that of theoretical data:

Depth of cut in mm	Cutting force in (N)		Tool life in (strokes)		Difference in tool life in strokes	% Reduction in tool life
	Theoretical	Experimental	Theoretical	Experimental		
1	73.37	108.8	462,132	330,183	131,949	28.55%

13 Conclusions

In conclusion, there is an error between theoretical (predicted) and experimental (observed) tool life, and it indicates that real-world conditions and factors may differ from the assumptions made in theoretical models. The importance of accurately estimating the tool life is to prevent premature tool failure and therefore ensuring tools are replaced or maintained at the most opportune times, and manufacturers can minimize overall tooling costs and reduce the impact on the production budget, therefore increasing overall productivity.

Acknowledgments This study was financially supported by the J. C. Bose University of Science and Technology, YMCA, Faridabad, Haryana of India for the Seed Money Award (SEED Grant No.: R&D/SG/202023/1596) and Department of Science and Technology, Ministry of Education, Govt. of India (DST-PURSE Grant No.: SR/PURSE/2022/126).

References

1. Sankpal, G. P., Sankpal, P. P., Bhosale, A. D., & ShindaVade, A. J. (2016). Analysis of milling tool life—A review. *International Journal of Innovative Research in Science, Engineering & Technology*, 20134–20142.
2. Kashif, M., Tatjana, K., Tauno, O., & Eduard, S. (2017). Performance analysis of a flexible manufacturing system (FMS). *Procedia CIRP, 63*, 424–429.
3. Haddad, Y., Pagone, E., Parra, R. V., Pearson, N., & Salonitis, K. (2022). How do small changes enable the shift to net-zero? A techno-environmental-economic analysis. *International Journal of Advanced Manufacturing Technology, 122*(11–12), 4247–4257.
4. Subbiah, R., Kaliappan, S., Balaji, V., & Patil, P. P. (2022). Effect of nanosilica on mechanical, thermal, fatigue, and antimicrobial properties of cardanol oil/sisal fiber reinforced epoxy composite. *Polymer Composites, 43*(11), 7940–7951.
5. Ramaswamy, R., Gurupranes, S. V., Kaliappan, S., Natrayan, L., & Patil, P. P. (2022). Characterization of prickly pear short fiber and red onion peel biocarbon nanosheets toughened epoxy composites. *Polymer Composites, 43*(8), 4899–4908.
6. Kumar, U., Padalia, D., Bhandari, P., Kumar, P., Ranakoti, L., Singh, T., & Lendvai, L. (2022). Fabrication of europium-doped barium titanate/polystyrene polymer nanocomposites using ultrasonication-assisted method: Structural and optical properties. *Polymers, 14*(21), 4664.
7. Nagabhooshanam, N., Kumar, A., Rao, P. K. V., Patil, P. P., & Kala Bharathi, B. V. V. L. (2022). Preparation and characterization of opuntia-cladode fiber and citron peel biochar toughened epoxy biocomposite. *Polymer Composites, 43*(9).
8. Singh, T., Pattnaik, P., Aherwar, A., Ranakoti, L., Dogossy, G., & Lendvai, L. (2022). Optimal design of wood/rice husk-waste-filled PLA biocomposites using integrated CRITIC–MABAC-based decision-making algorithm. *Polymers, 14*(13), 2603.
9. Rajawat, A. S., Singh, S., Gangil, B., Ranakoti, L., Sharma, S., Asyraf, M. R. M., & Razman, M. R. (2022). Effect of marble dust on the mechanical, morphological, and wear performance of basalt fibre-reinforced epoxy composites for structural applications. *Polymers, 14*(7), 1325.

10. Vinayaka, N., Nagabhooshanam, N., Balaji, D., Verma, R., & Patil, P. P. (2022). Effect of silanized stacked K evlar fiber/b rass 270 wire mesh on mechanical and impact toughness of cubic boron nitride–epoxy composite. *Polymer Composites, 43*(8), 5159–5167.
11. Patil, P. P. (2022). Effect of black rice husk ash biosilica on mechanical, wear, and fatigue behavior of stacked aloevera/roselle and glass fiber reinforced epoxy composite. *Polymer Composites, 43*(10).
12. Singh, T., Patnaik, A., Ranakoti, L., Dogossy, G., & Lendvai, L. (2022). Thermal and sliding wear properties of wood waste-filled poly (lactic acid) biocomposites. *Polymers, 14*(11), 2230.
13. Senthil Kumar, S., Anil Kumar, T. C., Chokkalingam, M., Kaliappan, S., Boopathi, S., Srinivas, C., & Patil, P. P. (2022). Mechanical, fracture toughness, and fatigue behavior of spinifex littoreus fiber on echinoidea-spike toughened epoxy composite. *Polymer Composites, 43*(4), 2329–2340.
14. Rajan, R., Tyagi, Y. K., Pruncu, C. I., Kulshreshtha, S., Ranakoti, L., & Singh, T. (2022). Tribological performance evaluation of slag waste filled phenolic composites for automotive braking applications. *Polymer Composites, 43*(10), 7118–7129.

Challenges and Considerations in Industry 5.0 Adoption

Navid Nasajpour-Esfahani, Hande Bahar, Daniel Agharahimi, Hamid Garmestani, and Steven Y. Liang

1 Introduction

Human–machine integration (HMI) is becoming a significant component in the evolution of production, particularly as the manufactory shifts from Industry 4.0 to the more enhanced concepts of Industry 5.0. This current stage emphasizes the partnership between humans and intelligent machinery, aiming to boost efficiency, creativity, and overall human well-being in the workplace. While technologies like robotics and artificial intelligence (AI) are already being widely implemented, more sophisticated systems like brain–computer interfaces and 6G connectivity are still in the experimental stage. The journey toward integrating these emerging technologies into real-world manufacturing environments is filled with numerous challenges, including technical barriers, high costs, security risks, and the need for employee training. Additionally, ethical and social implications must be carefully considered as these technologies continue to develop [1]. One of the primary hurdles to implementing HMI systems in manufacturing is the high cost of technology. Companies such as Intel, Nvidia, and Tesla have the resources to invest in these advancements, but smaller firms with limited financial capabilities may find it difficult to justify the initial expenses, especially given the long payback periods. For smaller manufacturers to benefit from the advantages of Industry 5.0, careful financial planning and

N. Nasajpour-Esfahani (✉) · H. Bahar · D. Agharahimi · H. Garmestani
Department of Materials Science and Engineering, Georgia Institute of Technology, Atlanta, GA, USA
e-mail: nesfahani3@gatech.edu; hbahar3@gatech.edu; dagharahimi3@gatech.edu; hamid.garmestani@mse.gatech.edu

S. Y. Liang
George W. Woodruff School of Mechanical Engineering, Georgia Institute of Technology, Atlanta, GA, USA
e-mail: steven.liang@me.gatech.edu

Y. Kumar Singla et al. (eds.), *Digital Manufacturing in Industry 5.0*,
https://doi.org/10.1007/978-3-031-91500-0_27

partnerships with government entities, research institutions, and technology providers will be necessary to access funding and reduce the costs associated with adopting advanced technologies [2]. In addition to cost, data and personal security represent a major challenge. As machines and humans collaborate more closely in manufacturing, protecting sensitive information and personal privacy becomes critical. The danger of data breaches, unauthorized access, cyberattacks increases as digital technologies become more integrated into factory operations. Ensuring robust data security will require multiple layers of protection, including encryption, secure networks, regular audits, and continuous monitoring. Moreover, as human-related data becomes central to smart manufacturing processes, protecting the psychological privacy of workers will be essential to maintain trust and prevent exploitation of personal information. The complexity of real-world implementation further complicates the adoption of HMI systems. For these systems to be effective, they should be crafted with a keen emphasis on the requirements and constraints of human users. Research indicates that considering human factors such as perception, memory, and situational awareness can improve the usability and effectiveness of HMI systems. Furthermore, integrating new technologies like digital twins, human–robot collaboration (HRC), and machine learning into current production workflows presents additional challenges, as each technology may require specific adjustments to hardware, software, and operational procedures [3]. In summary, while human–machine integration offers transformative potential for the future of manufacturing, several challenges must be addressed to enable its widespread adoption. These include managing costs, securing data, overcoming technical hurdles, closing skill gaps, navigating ethical concerns, and many others. With collaborative efforts and responsible planning, the full benefits of Industry 5.0 can be realized, creating a more efficient and human-centered manufacturing environment. In this chapter, recent challenges and considerations in Industry 5.0 will be discussed.

2 Integration of Human-Centric Collaboration

2.1 Emerging Technologies and Cost Challenges

Implementing automation in HMI in real life production situations might be difficult. Some emerging mechanisms like ChatGPT and AI-based voice detecting assistants have become significantly integrated into our daily activities, but still technologies such as brain–computer interface and 6G are dependent on experimental studies. Elements like the device characteristics, convenient design and machineries' quality and performance must be considered to form a proper HMI product/system. It is also important to acknowledge the partnership between humans and machines, how much workers can trust in equipment and how the work is shared between them. The complex nature of HMI system design involving several different fields brings significant obstacle for applying HMI technology in real-world situations [1]. Cost is crucial in improving these technologies. The main players for HMI research in the industry are companies like Intel, Nvidia, and Tesla who are big companies with high profits [2]. Industry 5.0 needs to put more effort for decreasing the HMI technology cost since companies with low profits margins do not prioritize applying measures and equipment with long reimbursement cycles due to their low budgets even though they acknowledge the fact these studies can bring enormous advantages in the long term. For producers, there are multiple strategic ways to tackle this challenge. First of all, accurate financial plans and measurements that identify areas where efficiency can be maximized are fundamental. By collaborating with government organizations, universities, and research centers, they may reach to resources, assistance and financial help and promote the integration of advanced production technologies with those fundings. In addition to funds, other financing approaches, like leasing or collaborations with technology suppliers, can be taken into consideration by the manufacturers. Having collaboration with technology vendors may contain partial contribution, where the supplier has a stake in the event of a successful implementation. Moreover, creating an innovation culture in the work environment may inspire employees to add their ideas for cost-saving and efficiency enhancement which will lead to creative solutions that decrease the cost of implementing Industry 5.0 technologies [3].

2.2 *Data and Privacy Protection*

Security is one of the main challenges which manufacturers have to focus on. Security can be considered in two primary fields: data and personal safety. Protection of data mostly relies on confidential technology but the importance of employee' awareness should not be forgotten. In order to enhance the HMI process, it is essential to communicate the importance of security to workers in a more lasting and impactful manner is essential. The safety of employees is closely linked to integrating human-centered designs, including capabilities for empathetic interactions within the system. These designs increase worker satisfaction and provide them better understanding of potential threats [4]. In Industry 5.0, protection of data problem has multiple facets, such as transmission, collection of delicate data related to digital technology, machine learning, and collaboration between humans and robots. When the factories lack of proper protections for operational parameters, product designs, manufacturing processes, potential risks increase. Undefined entry, cyber intrusions, and data breaches can result in significant repercussions, such as theft of intellectual property and interruptions to business operations. In order to solve this problem, comprehensive and practical path should be chosen. Data should be protected using strong encryption methods for both storage and transmission. Frequent security audits and vulnerability evaluations are critical, and security patches and updates should be applied promptly. An effective incident response plan is necessary to swiftly handle security breaches. Furthermore, ensuring data backups, collaborating securely with third-party partners, adhering to regulations, maintaining ongoing monitoring, and fostering a culture of security throughout the organization all play vital roles in a robust data protection strategy. By putting these practices into place together, a solid defense against possible threats is formed, aligning with the evolving trends in technology and compliance obligations. Furthermore, ensuring data backups, collaborating securely with third-party partners, adhering to regulations, maintaining ongoing monitoring, and fostering a culture of security throughout the organization all play vital roles in a robust data protection strategy. By putting these practices into place together, a solid defense against possible threats is formed, aligning with the evolving trends in technology and compliance obligations [3]. Additionally, manufacturers need focus on protecting employees' psychological privacy and safety. Since data pertaining to humans will play an essential role in Human-Centered Smart Manufacturing (HCSM), factories should pay close attention to employees' mental conditions and personal protection. The aim of Industry 5.0 is to facilitate a close partnership between people and machines instead of controlling workers through HMIs. Hence, the challenge for future research will be to find ways to maintain employees' psychological well-being without exploiting the design and data [4].

2.3 Real-World Implementation

Some fundamental aspects should be assessed throughout the design phase of HMI systems to successfully implement HMI in the real world [5]. Shneiderman [6], Wickens, Lee, Liu, and Becker [7] argue that people's needs, limitations, and experiences need to be at the center of the design. The fundamentals summarized by Shneiderman include system stability, providing feedback on input, designing a closed cycle of prompts, offering basic error prompts, and reducing complex menus. Wickens et al. focus primarily on human abilities and constraints, examining factors such as perception, attention (which aids in information collection), memory (which simplifies the retrieval of essential information), mental models (which ensure icons and logic are clear), and situational awareness (the capacity to predict events and remember significant past incidents) as key principles in HMI design [8]. Cuevas [9] introduced a four-factor standard: ethnographic and anthropomorphic characteristics, cognitive elements, predictive modeling, and empirical testing. According to the standard, HMI systems must be evaluated from these four viewpoints to satisfy practical application requirements [9].

Collaborative robotics systems are required to make real-time decisions about their interactions with humans. Therefore, software algorithms should be held responsible for their judgments. This necessitates a complete logging system and adequate documentation on related components. Specifically, learning models should be employed as they ensure better performance compared to model-based technologies. Users should be made aware of the limitations. Stakeholders and end users should be responsible for the technology and should provide necessary guidelines and recommendations regarding of using these technologies [10].

3 Ethical and Social Considerations

Employees will have a much larger role in tomorrow's production scenarios than they do today. Their skills and know-how are critical factors for the future of companies. Workers in manufacturing must deal with greater task complexity and responsibilities, and this creates a need for proper education [11]. In the future, there will be less physical work, and more tasks related to arrangement, designation, and decision-making. Workers will continue to be present in factories for the most advanced computers and machinery won't ever be able to take over crucial roles such as decision-making, managing unpredictable events, or generating innovative ideas based on experience [12]. Industry 5.0 focuses on endorsing creativity, producing top-of-the-line custom-made products, and raising the standard of living. It seeks to improve the quality of life for people, and it can be the fifth revolution of humanity. Communication in Industry 5.0 will not just be between humans, but with nonhumans as well. Standardizing architecture, data formats, interfaces, and vocabularies are key to promoting interoperability across technologies in fields like Industry 4.0 and 5.0 [13]. In this section, different ethics considerations will be discussed.

3.1 Ethical Design Guidelines

3.1.1 Guarantee the System Respects Human Rights

This suggestion is valid for hardware and software. The layout of hardware should be suitable for both women and men. Regarding software, it is important to ensure that algorithms are not prejudiced against any races or groups. Individual differences should be included in the analysis, and models should respect these differences by not favoring any one group over another [10].

3.1.2 System Accountability

Collaborative robotics systems are required to execute real-time decisions with how they interact with humans. Therefore, software algorithms should be held responsible for their judgments. This necessitates a complete logging system and adequate documentation on related components. Specifically, learning models should be employed as they ensure better performance compared to model-based technologies. Users should be made aware of the limitations. Stakeholders and end users should be responsible for the technology and should provide necessary guidelines and recommendations regarding using these technologies [10].

3.1.3 Transparent and Safe System

Even though devices may have safety certificates, regulations and working conditions should be made clear and transparent for users to ensure safety. Accordingly, proper training should be delivered to users beforehand. Users should be capable of understanding how the system operates, both when it's functioning properly and when there's a malfunction. The system needs to be transparent so users can diagnose it. Users should have full control and understanding of the system and be able to verify its operations, calculations, and data tracking. When defining the level of partnership in decision-making between humans and systems, certain rules should be applied. For instance, fully autonomous robotic systems may be acceptable in some situations, but shared control mode or full operation by a human may be necessary in critical situations such as those involving safety and liability issues [10]. The most updated European General Data Protection Regulation (GDPR) is relevant to any business or system that processes personal data, both automatically and manually. It mandates that humans should be notified about the reasoning, importance, and outcomes of decisions made by automated decision-making systems, and it also grants them the authority to examine and challenge automated decisions [14].

3.1.4 Reduce the Risk of Technology Misuse

Misuse presents itself as a major issue for researchers and technology developers. Misinterpretation in the media can lead to unrealistic expectations among researchers or even slow down the progress and adoption of human-centered technologies. Consequently, researchers should prioritize education and awareness initiatives regarding collaborative robotics, covering areas such as hardware components, machine learning, and control systems [10].

3.2 Social Aspects

With Industry 4.0, production conditions have improved with more occupational safety, increased automation, and ergonomic workstations. The increased use of energy-efficient machines has led to better environmental protection. Additionally, the availability of consistent data increases collaboration [15].

3.2.1 Automated Collaborators

Industry 4.0 has increased innovative capacity, which is one of the main benefits, making innovation easier since machines handle process steps themselves. Cobots may enhance job satisfaction by reducing tiring, physically uncomfortable, and vulnerable tasks. Moreover, cobots play an important role as a "third hand" in assembly operations, such as helping operators position chips precisely, lifting heavy objects, and dealing with hazardous materials. The partnership between robots and humans in the workplace boosts the social network efficiency among smart operators, linking them with smart sections of the smart factory. Over time, "Robocollaborators" and their robots are likely to develop better relationships, assisting humans rather than displacing them [16].

3.2.2 Legal Questions

A study conducted by Djordievic in 2017 on the "Laws of Robotics" shows that Directive 85/374/EEC only applies to damage attributed to robots due to production defects. The author concluded that the existing legal framework would not adequately address damage brought upon by the most recent generation of robots, which exhibit a great degree of adaptive learning causing unpredictable behavior. These robots have the autonomy to learn and improve from their experiences while interacting with their environment in unimaginable ways, indicating the need for new legislation [17]. The collaboration between humans and robots brings new ethical and legal queries such as:

- Who would be held liable for a robot's breakdown?
- Who would be held liable for a faulty task executed by the robot?
- What would we expect to happen if a robot harms a worker?
- Should a robot be given the same treatment as a worker since it is just as responsible for completing tasks?

Finding answers to these questions is essential to keep the technology developing. However, these developments raise additional legal questions such as:

- Can humans become attached to robots? Is smart communication with robots and accepting their advice adequate?
- Would people trust robots more than necessary?
- What can we do if something happens to a robot and what would be a human's corresponding reaction?

The manufacturer's decision-making process for certain products is significantly influenced by liability rules. If the liability risks are high, it may lead to product launch delays, which could hinder the progress of new technology development. For instance, a fully automated car is a good example. The liability risk is entirely shifted to the manufacturer. The driver acts as a passenger in a vehicle, assuming no responsibility. Hence, in the case of an accident, it is highly improbable that the driver would be held responsible. Typically, the cause of the accident is attributed to a breakdown of the machine, with full responsibility placed on the manufacturer [18].

4 Technological Challenges

Industry 5.0 is already driving lasting transformations in industrial operations. This approach enables businesses to harness the power of highly advanced machines alongside skilled individuals to create efficient, sustainable, and secure manufacturing processes. Industry 5.0 is more than a temporary trend; it is a refined symbol for industrial production, offering prominent economic benefits. Companies that cease to standardize their manufacturing practices in accordance with Industry 5.0's model risk becoming obsolete, forgoing any competitive advantages it may offer. Additionally, the accelerating rate of technological advancement means that new ideas continue to emerge. Therefore, transitioning business operations to align with the digital sector is essential for maintaining competitiveness and long-term success [19].

4.1 Blockchain Technology

Currently, manufacturers are working together on blockchain implementations that aim to keep operations up to date, provide total transparency in supply chain management, and optimize technology-driven investments. Blockchain has the revolutionary potential to improve the methods for designing, producing, and scaling products for manufacturing companies. It is also reshaping business interactivity by enabling full cooperation within and between companies that compete with one another in mutual environments. Manufacturers still have great potential to innovate with blockchain, as it is second only to finance in technological leadership. Furthermore, blockchain can greatly improve efficiency and speed for future product financing. Verified blockchain transactions can automatically handle distributor outcomes and transactions, reducing the need for human involvement and manual checks. Integrating blockchain with cloud computing and sensors greatly enhances productivity, access, and availability throughout the whole value chain. Logistics and transportation benefit the most from combining blockchain with IoT, enabling automated pick-ups, deliveries, and business agreements through smart contracts for physical goods, features, or transactions [20, 21].

4.2 Unmanned Aerial Vehicle

Industry 5.0 is an innovative concept aimed at making industries smarter and more connected through unified platforms like IoT and UAV (unmanned aerial vehicle) technology. UAVs are continually developed in accordance with Industry 5.0's standards. While UAVs are commonly used for civilian and military purposes, there are opportunities to integrate them into advanced manufacturing for real-time monitoring, wireless clarity, and environmental surveillance. As IoT devices, UAVs are effective for cost-efficient data collection and monitoring, making them dynamic enough to solve problems that contribute to more decisiveness. UAV technology can drive industrial growth by linking production lines through long-range communication, UAV technology can foster considerable industrial growth. It becomes a degree more significant when combined with other applications including smart cities and IoT systems. UAV technology has the capacity to boost production plant efficiency, driving the progress of Industry 5.0. As fully autonomous systems, UAVs act as gateways for uploading sensory data to cloud servers via network technology, supported by Wi-Fi components, making them versatile for applications in areas like pollution control, public transit, and Industry 5.0 initiatives [22].

4.3 Next-Generation 5G and Advanced Wireless Technology

"Beyond 5G," or more simply 6G, marks the next line of communication systems. It represents more than just a technical evolution; it focuses on enhancing workplace connectivity and relationships. The development of software-driven tools and the spread of data-based learning environments are becoming essential, with applications in areas like self-controlled transportation and telehealth. While earlier generations—namely, 1G, 2G, 3G, and 4G—made communication between people possible, 5G brings upon the interaction between people and objects, and 6G will enable seamless communication between devices, objects, and essentially anything. Pivotal areas for "Beyond 5G"/6G include developing state-of-the-art testing processes, enhancing wireless internet connectivity, exploring IoE (Internet of Everything) systems, and improving high-resolution, digital communication technologies. Other areas of primary interest include advancing industrial systems, smart city planning, and the progress of Industry 5.0. Additionally, personal data security remains a critical concern, continuing to play a significant role in the evolution of this technology [23, 24].

4.4 Wearable Robotics Technology

Exoskeletons have become increasingly popular in the manufacturing sector, helping workers and improving the quality of their work environment. These wearable devices enhance physical capabilities by supporting almost all body parts, boosting efficiency in labor-intensive tasks. They can aid physical labor including lifting, and carrying manufacturing assets, substantially reducing the risk of muscle or bone injuries. It has been projected that investment in exoskeleton manufacturing will grow rapidly. According to a recent market report from ABI Research, by 2030, exoskeleton shipments in advanced manufacturing are projected to reach 274 million units, generating USD 3.4 billion in revenue. Manufacturing is the leading sector in exoskeleton adoption, followed by the advertising sector and, to a lesser extent, the defense industry. Production in industry contributes to more than half of exoskeleton sales and revenue, including energy, utilities, oil, and gas. Meanwhile, commercial applications are close behind, with an estimated 198 million units sold and 2.9 billion USD in revenue expected by 2030 [25].

4.5 Mixed Reality

Augmented reality, which is typically delivered through mobile devices like tablets or smartphones, contrasts heavily with mixed reality (MR). MR is experienced via, transparent glasses that are mounted on the head and it merges the physical and digital worlds, creating unique environments and interactions where real and virtual objects coexist and influence each other in real-time. Microsoft first introduced the idea of "mixed reality" with the release of the HoloLens in 2016 [26]. Unlike augmented reality, MR has a more in-depth understanding of the real, physical environment and can incorporate holographic images within that space. Using a headset, MR tracks a person's gaze and maps their physical surroundings, employing deep learning techniques that link digital content with specific areas. This allows digital objects to interact with their physical counterparts, and it allows individuals to partake in this digital reality as if it were physical. To illustrate this, a normal desktop can be turned into a hands-on interface such as a touchscreen, or a virtual character could sit on a user's digital sofa. Although MR is still early in its development, it's still readily applicable across various industries for educational purposes. For instance, airline companies are using MR to train maintenance professionals at a lower cost. Rather than physically removing an engine from an aircraft, technicians can use specialized headsets to view and interact with a holographic engine through gestures, gaze tracking, and voice commands, extracting information layer by layer. While 3D projections and virtual displays are expected to play a sizable position in MR, most of the MR content is currently delivered through headsets. Previously, creating consumer products with mixed reality (MR) was quite costly due to intensive processing requirements and sophisticated computer vision and image recognition needs. However, improvements in big data, cloud computing, and AI are reducing costs, making MR technology increasingly accessible. As prices continue to drop, MR is poised to transform human–computer interaction (HCI) [27–29].

4.6 Additive Manufacturing

Additive manufacturing can play a crucial role in Industry 5.0 due to its significant reduction in waste compared to less sustainable, subtractive methods. While Industry 5.0 is still growing, there is a growing consensus that more can be done to help preserve Earth while also reducing costs for businesses. This type of manufacturing, or 3D printing, creates objects by adding material layer by layer, unlike subtractive methods, which cut material away from a solid block to shape a final product. The technology behind additive manufacturing can include any process that speeds up production, such as molding. The process starts with a design, usually made with computer-aided design (CAD) software or by simply scanning an existing object. The design is then turned into a step-by-step guide for the 3D printer, which begins building the object layer by layer. The time needed for 3D printing varies based on the method and size of the item, taking anywhere from a few hours to several days. One common technique uses a nozzle to place layers of material on top of each other until the object is finished. An alternative method involves using metal powders to form each layer until the whole item is complete [30].

4.7 AI-Based IoT

The Internet's focus is transitioning from the "Internet of Computers (IoC)" to the "Internet of Things (IoT)," and it's imperative to recognize and understand the how AI methods come into play in facilitating intelligent online communication. Wireless sensor networks are becoming much more popular in research mainly because of their practical applications and their ability to provide effective remote monitoring in areas like healthcare, weather forecasting, sea level tracking, and event prediction. Moreover, smart sensors are commonly used in electronic home appliances, green infrastructure, and mobile devices. The IoT concept refers to the extensive presence of a variety of devices or objects such as sensors, actuators, and mobile phones that communicate with each other using unique addressing methods. This interconnectedness will continue to reshape people's lives throughout the whole world. Moreover, AI techniques can improve the IoT by supporting the research and development of robotic systems with evolving roles that function without so much human intervention [31, 32].

4.8 Motion Capture Technology

The rapid pace of technological achievements in Industry 5.0 has brought up new possibilities for innovative industrial processes, which rely on groundbreaking sensing technologies. Motion capture systems, which use inertial measurement units (IMUs) and optical cameras, are extensively employed in industrial settings to support teleworking, robotics, additive manufacturing, and consumer security technologies. Digital motion capture refers to the process of recording the movements of people or objects. This technology is commonly utilized in the entertainment industry, specifically in video games and filmmaking. When capturing more detailed movements, such as facial expressions or finger motions, it is often referred to as behavioral capture. Movements are recorded and used to create a 3D computer model, which accurately replicates the person's gestures by transmitting the data to a digital medium at high speeds, enabling the avatar to mirror human actions in real time [33–35].

4.9 *Digital Twin*

Digital twins are crucial in Industry 5.0, enhancing production cycle efficiency and performance. From the extracted extensive data, digital twins can refine both product design and manufacturing processes. A digital twin is merely a virtual model mirroring a real-world object, like a wind turbine, jet, or even large-scale systems such as buildings or cities. This technology allows simulations to collect data and forecast performance outcomes, creating a dynamic digital representation of physical items or processes based on real-time data. Digital twins can be enhanced by implementing software analytics, AI, and IoT and this type of technology along with virtual reality, cognitive computing, networked supply chains, and machine learning have transformed product and service marketing. Unfortunately, this new approach to industrial production faces the same challenges as the original industrial revolution [36, 37].

5 Economic and Financial Considerations

The next area of research focuses on the circular economy (CE), emphasizing its crucial role in advancing sustainable development. The CE model aims to safeguard and rejuvenate the environment and finite resources while fulfilling the increasing needs of a growing population and ensuring provisions for incoming formation [38]. By integrating advanced technologies with CE principles, businesses can adopt sustainable practices that promote waste reduction and improve resource efficiency. Key concepts in this approach include "sustainable manufacturing methods," "efficient resource utilization," "minimizing waste," "distributed ledger technology," "eco-friendly transportation," "environment-friendly structures," "responsible production," "sustainable logistics," and "eco-conscious practices." Green finance (GF) initiatives are essential to secure the necessary funding for the infrastructure and technologies required to support CE-based businesses. This involves investments in areas such as green buildings, logistics, transportation, and supply chains to build an economy rooted in CE principles. However, achieving this vision depends on the availability of green financing for circular device design, recycling, reproduction facilities, and more. Recently, the CE has replaced the traditional end-of-life concept by promoting the use of alternative and renewable energy sources to minimize environmental harm. CE transforms the processes of consumption and production by extracting value from waste materials, with the goal of advancing beyond the existing consumption-focused model and creating a sustainable closed-loop system [39, 40]. Pinheiro et al. introduced a novel method to evaluate the impact of circular economy on the field performance through circular product design (CPD). Their findings show that I4.0 technologies, particularly AI and big data analytics, significantly enhance CPD. Additionally, there is a positive relationship between the integration of CE strategies and commercial achievement, with stakeholders playing a key role in encouraging suppliers to adopt circular practices and deliver smart devices [41]. Rehman Khan et al. explored how to improve efficiency of blockchain technology in the CE within the China–Pakistan Economic Corridor (CPEC). They found that supply chain management, when aligned with CE principles, yields both environmental and financial benefits for businesses. Furthermore, the adoption of green practices is positively associated with a company's financial and ecological performance [42].

6 Cultural and Organizational Challenges

The industry 5.0 framework envisions organizations prioritizing human resources (HR) and employee well-being at the core of their operations, aiming to achieve continual outputs and build stronger systems [43]. The significance of HR and employee wellness as key drivers of organizational effectiveness has long been explored [44]. Research has shown that factors such as shared purpose [45], positive interactions, a supportive work environment [46], and employee engagement enhance knowledge-sharing processes, which ultimately leads to improved organizational performance. Aligned with the Industry 5.0 principles, Barrena-Martínez et al. define socially responsible human resource management (SRHRM) as the integration of corporate social responsibility (CSR) into HRM practices, designed to involve and please employees with social rewards that extend past mere financial and legal responsibilities, emphasizing the mental health of workers and their families [46]. The World Health Organization reports that global productivity losses due to health issues such as depression and anxiety amount to about one trillion dollars annually. A survey conducted by the Society for Human Resource Management in 2020 found that 22–35% of 1099 American workers experienced depression and emotional stress related to remote work, with two in five employees struggling to balance their personal and professional lives (e.g., being constantly online and available). Therefore, a human-centric culture must balance work flexibility with preserving human dignity, even in digital environments [44]. In line with this, the STREBLO application, developed through an EU project to reduce stress in the construction industry, helps HR teams oversee and handle work-related stress and job responsibilities [47].

7 Global Competition and Collaboration

The term "5.0" can be misleading, as it suggests a direct continuation of the Fourth Industrial Revolution. Society 5.0 leverages technological advancements to address challenges that emerged, in part, from the Fourth Revolution, such as social alienation, the digital divide leading to economic inequality, and the fragmented use of technology. These technologies are now being utilized within a framework that prioritizes unity and societal well-being, representing an effort in Society 5.0 [48]. These features can be written as:

7.1 Complex Problem Solving

This ability evolves over time through hard work and personal experiences. There aren't any defined guidelines or structures to adhere to in order to attain mastery. However, the best way to improve is by confronting challenges directly until the right solution is found [49].

7.2 Critical Thinking

This involves the skill to reason certainly and deeply about a topic in order to arrive at well-founded decisions, commonly known as deductive reasoning. It is a process of analyzing and evaluating information in order to arrive at a meaningful conclusion [49].

7.3 Creativity

Organizations encourage employees to implement fresh opinions and think creatively to remain competitive. People need to be more creative, both in developing new business concepts and in finding solutions to problems that others may have overlooked [49].

7.4 People Management

Effective people management is crucial for success, as leaders rely on the support of others to achieve their goals. A leader with strong management skills can guide a team to success. These skills are essential in any profession, as achieving great results typically requires collaboration and teamwork [49].

7.5 Coordinating with Others

Effective team coordination is a key part of people management. A team's success depends on its ability to work together, as no single person can achieve success alone. Coordination and cooperation among team members are necessary for optimal performance [49].

7.6 Emotional Intelligence

Emotional intelligence is the capacity to regulate one's own feelings and influence emotions of others. Developing this skill helps build personal character. To influence others effectively, individuals must first master control of their own emotions [49].

7.7 Judgment and Decision-Making

The ability to decision-making is crucial not only in professional settings but also in everyday situations. It encompasses the capacity to make prompt and informed choices, setting someone apart by their ability to assess situations and act accordingly [49].

7.8 Service Orientation

It involves the motivation to help other people. A leader's effectiveness can be measured by their willingness to assist team members. A true leader ensures the success of the entire team and ensures no one is left behind [49].

7.9 Negotiation

Negotiation skills involve reaching favorable agreements without conflict. Many people struggle to achieve their desired outcomes due to poor negotiation abilities. These skills can be developed through experience, meetings, and discussions with those in higher positions [49].

7.10 *Cognitive Flexibility*

Cognitive flexibility is the ability to adapt to different situations and people. Since individuals have varying perspectives, ideas, and emotions, it is important to approach them in different ways. Being flexible in communication requires creativity, logical thinking, and effective interpersonal skills [49].

8 Conclusion

In conclusion, the successful adoption of Industry 5.0 is defined by the collaboration between humans and machines alongside the use of cutting-edge technologies, is not without its challenges. While the potential benefits, such as increased efficiency, sustainability, and worker satisfaction, are significant, organizations must address several critical issues to realize these advantages fully. These include managing the high costs associated with emerging technologies like AI, brain–computer interfaces, and 6G, ensuring robust data and privacy protections, and overcoming technical hurdles related to real-world implementation. Security, both in terms of data and personal safety, remains a major concern as the partnership among people and machines intensifies. It is essential to build processes which are technologically advanced and empathetic to human needs. This includes protecting the psychological privacy of workers and ensuring that they remain in control of their data. Ensuring that human–machine interfaces are user-friendly and designed with human limitations in mind is crucial for their effectiveness in manufacturing environments. Furthermore, organizations must navigate ethical and social considerations, particularly in designing systems that respect human rights, ensure accountability, and promote transparency. The balance between human control and machine autonomy needs to be carefully managed, especially in safety-critical environments. As Industry 5.0 evolves, its integration with technologies such as blockchain, UAVs, 5G, and wearable robotics will continue to reshape industrial operations. The financial and commercial benefits of Industry 5.0 are clear, but they require companies to adapt quickly or risk becoming obsolete. Collaboration between industries, governments, and academic institutions will be key in overcoming these barriers and facilitating the shift toward to smarter, more eco-friendly manufacturing methods. Ultimately, the future of Industry 5.0 depends on thoughtful planning, continuous innovation, and a human-centric approach. By addressing these challenges strategically, businesses can unlock the full potential of Industry 5.0 and aims to a more connected, efficient, and eco-friendly world.

Acknowledgment We acknowledge the support of Georgia AI Manufacturing (Georgia AIM) under Award #04-79-07808 from the Economic Development Agency, US Department of Commerce. Their contribution has been instrumental in advancing our endeavors.

References

1. Janssen, C. P., Donker, S. F., Brumby, D. P., & Kun, A. L. (2019). History and future of human-automation interaction. *International Journal of Human Computer Studies, 131*, 99–107. https://doi.org/10.1016/j.ijhcs.2019.05.006
2. Krupitzer, C., Müller, S., Lesch, V., Züfle, M., Edinger, J., Lemken, A., Schäfer, D., Kounev, S., & Becker, C. (2020). *A survey on human machine interaction in industry 4.0* (p. 1).
3. Zafar, M. H., Langås, E. F., & Sanfilippo, F. (2024). Exploring the synergies between collaborative robotics, digital twins, augmentation, and industry 5.0 for smart manufacturing: A state-of-the-art review. *Robotics and Computer-Integrated Manufacturing, 89*, 102769. https://doi.org/10.1016/j.rcim.2024.102769
4. Yang, J., Liu, Y., & Morgan, P. L. (2024). Human–machine interaction towards Industry 5.0: Human-centric smart manufacturing. Digital. *Engineering, 2*, 100013. https://doi.org/10.1016/j.dte.2024.100013
5. Morgan, P. L., Voinescu, A., Williams, C., Caleb-Solly, P., Alford, C., Shergold, I., Parkhurst, G., & Pipe, A. (2018). An emerging framework to inform effective design of human-machine interfaces for older adults using connected autonomous vehicles. In *Advances in human aspects of transportation: Proceedings of the AHFE 2017 international conference on human factors in transportation, July 17– 21, 2017, The Westin Bonaventure Hotel, Los Angeles, California, USA 8* (pp. 325–334). Springer International Publishing.
6. Shneiderman, B., & Plaisant, C. (2010). *Designing the user interface: Strategies for effective human–computer interaction*. Pearson Education India.
7. Wickens, C. D. (2004). *An introduction to human factors engineering* (Vol. 2). Pearson Prentice Hall.
8. Cuevas, H. M. (2004). An illustrative example of four HCI design approaches for evaluating an automated system interface. *Proceedings of the Human Factors and Ergonomics Society Annual Meeting, 48*, 892–896. https://doi.org/10.1177/154193120404800532
9. Grabowska, S. (2020). Smart factories in the age of industry 4.0. Management Systems. *Production Engineering, 28*, 90–96. https://doi.org/10.2478/mspe-2020-0014
10. Maurice, P., Allienne, L., Malaise, A., & Ivaldi, S. (2018). Ethical and social considerations for the introduction of human-centered technologies at work. In *Proceedings of the 2018 IEEE Workshop on Advanced Robotics and its Social Impacts (ARSO)* (pp. 131–138). IEEE.
11. Hirsch-Kreinsen, H., Kubach, U., Stark, R., von Wichert, G., Hornung, S., Hubrecht, L., Sedlmeir, J., & Steglich, S. (2019). *Key themes of Industrie 4.0: Research and development needs for successful implementation of Industrie 4.0*. Research Council.
12. Doyle-Kent, M., & Kopacek, P. (2022). Collaborative robotics making a difference in the global pandemic. In *Digitizing production systems: Selected papers from ISPR2021, October 07–09, 2021 Online* (pp. 161–169). Springer International Publishing.
13. Doyle-Kent, M., & Kopacek, P. (2020). Industry 5.0: Is the manufacturing industry on the cusp of a new revolution? In *International conference on production research* (pp. 432–441). Springer International Publishing.
14. https://Ec.Europa.Eu/Commission/Priorities/Justice-and-Fundamental-Rights/Data-Protection/2018-Reform-Eu-Data-Protection-Rules En.
15. Kent, M. D., & Kopacek, P. (2021). Social and ethical aspects of automation. In *The international symposium for production research* (pp. 363–372). Springer International Publishing.
16. Universal Robots: Universal Robots Website (2019). https://Www.Universal-Robots.Com/Products/Ur5-Robot/.
17. Djordjevic, I. (2019). *Robotics law*. MSc Thesis. TU Wien.
18. Karnouskos, S. (2020). Self-driving car acceptance and the role of ethics. *IEEE Transactions on Engineering Management, 67*, 252–265. https://doi.org/10.1109/TEM.2018.2877307
19. Alojaiman, B. (2023). Technological modernizations in the industry 5.0 era: A descriptive analysis and future research directions. *Processes, 11*, 1318. https://doi.org/10.3390/pr11051318

20. Ahmed, I., Zhang, Y., Jeon, G., Lin, W., Khosravi, M. R., & Qi, L. (2022). A blockchain- and artificial intelligence-enabled smart IoT framework for sustainable city. *International Journal of Intelligent Systems, 37*, 6493–6507. https://doi.org/10.1002/int.22852
21. Alkhateeb, A., Catal, C., Kar, G., & Mishra, A. (2022). hybrid blockchain platforms for the Internet of Things (IoT): A systematic literature review. *Sensors, 22*, 1304. https://doi.org/10.3390/s22041304
22. Heidari, A., Navimipour, N. J., & Unal, M. (2022). Applications of ML/DL in the management of smart cities and societies based on new trends in information technologies: A systematic literature review. *Sustainable Cities and Society, 85*, 104089. https://doi.org/10.1016/j.scs.2022.104089
23. Khan, L. U., Saad, W., Niyato, D., Han, Z., & Hong, C. S. (2022). Digital-twin-enabled 6G: vision, architectural trends, and future directions. *IEEE Communications Magazine, 60*, 74–80. https://doi.org/10.1109/MCOM.001.21143
24. Li, O., He, J., Zeng, K., Yu, Z., Du, X., Zhou, Z., Liang, Y., Wang, G., Chen, Y., Zhu, P., et al. (2022). Integrated sensing and communication in 6G: A prototype of high resolution multichannel THz sensing on portable device. *EURASIP Journal on Wireless Communications and Networking, 2022*, 106. https://doi.org/10.1186/s13638-022-02172-w
25. Tröster, M., Budde, S., Maufroy, C., Andersen, M. S., Rasmussen, J., Schneider, U., & Bauernhansl, T. (2022). Biomechanical analysis of stoop and free-style squat lifting and lowering with a generic back-support exoskeleton model. *International Journal of Environmental Research and Public Health, 19*, 9040. https://doi.org/10.3390/ijerph19159040
26. Blanchard, J., Koshal, S., Morley, S., & McGurk, M. (2022). The use of mixed reality in dentistry. *British Dental Journal, 233*, 261–265. https://doi.org/10.1038/s41415-022-4451-z
27. Holden, R. J., Abebe, E., Hill, J. R., Brown, J., Savoy, A., Voida, S., Jones, J. F., & Kulanthaivel, A. (2022). Human factors engineering and human-computer interaction: Supporting user performance and experience. In *Clinical informatics study guide* (pp. 119–132). Springer International Publishing.
28. Cackett, M., Fulton, L., & Pearl, D. (2023). Sound as an interface, methods to evaluate voice user interface (VUI) experiences in various contexts. In *Proceedings of the future technologies conference* (pp. 120–129). Springer International Publishing.
29. Franze, A. P., Caldwell, G. A., Teixeira, M. F. L. A., & Rittenbruch, M. (2022). Employing AR/MR mockups to imagine future custom manufacturing practices. In *Proceedings of the 34th Australian conference on human-computer interaction* (pp. 206–215). New York, NY, USA.
30. Pajonk, A., Prieto, A., Blum, U., & Knaack, U. (2022). Multi-material additive manufacturing in architecture and construction: A review. *Journal of Building Engineering, 45*, 103603. https://doi.org/10.1016/j.jobe.2021.103603
31. Ghobadpour, A., Monsalve, G., Cardenas, A., & Mousazadeh, H. (2022). Off-road electric vehicles and autonomous robots in agricultural sector: Trends, Challenges, and Opportunities. *Vehicles, 4*, 843–864. https://doi.org/10.3390/vehicles4030047
32. Mithas, S., Chen, Z., Saldanha, T. J. V., & De Oliveira Silveira, A. (2022). How will artificial intelligence and industry 4.0 emerging technologies transform operations management? *Production and Operations Management, 31*, 4475–4487. https://doi.org/10.1111/poms.13864
33. Trost, M., Claus, T., & Herrmann, F. (2022). Social sustainability in production planning: A systematic literature review. *Sustainability, 14*, 8198. https://doi.org/10.3390/su14138198
34. Duggal, A. S., Malik, P. K., Gehlot, A., Singh, R., Gaba, G. S., Masud, M., & Al-Amri, J. F. (2022). A sequential roadmap to industry 6.0: Exploring future manufacturing trends. *IET Communications, 16*, 521–531. https://doi.org/10.1049/cmu2.12284
35. Romero, D., & Stahre, J. (2021). Towards the resilient operator 5.0: The future of work in smart resilient manufacturing systems. *Procedia CIRP, 104*, 1089–1094. https://doi.org/10.1016/j.procir.2021.11.183
36. Battini, D., Berti, N., Finco, S., Zennaro, I., & Das, A. (2022). Towards industry 5.0: A multi-objective job rotation model for an inclusive workforce. *International Journal of Production Economics, 250*, 108619. https://doi.org/10.1016/j.ijpe.2022.108619

37. Wang, W., Guo, H., Li, X., Tang, S., Li, Y., Xie, L., & Lv, Z. (2022). BIM information integration based VR modeling in digital twins in industry 5.0. Journal of Industrial Information. *Integration, 28*, 100351. https://doi.org/10.1016/j.jii.2022.100351
38. Adimuthu, R., Muduli, K., Ray, M., Singh, S., & Ahmad, T. S. T. (2022). Exploring role of industry 4.0 techniques for building a promising circular economy concept. In *Machine learning adoption in blockchain-based intelligent manufacturing* (pp. 111–124). CRC Press.
39. Agrawal, R., Wankhede, V. A., Kumar, A., Upadhyay, A., & Garza-Reyes, J. A. (2022). Nexus of circular economy and sustainable business performance in the era of digitalization. *International Journal of Productivity and Performance Management, 71*, 748–774. https://doi.org/10.1108/IJPPM-12-2020-0676
40. Agrawal, R., Wankhede, V. A., Kumar, A., Luthra, S., & Huisingh, D. (2022). Progress and trends in integrating industry 4.0 within circular economy: A comprehensive literature review and future research propositions. *Business Strategy and the Environment, 31*, 559–579. https://doi.org/10.1002/bse.2910
41. Pinheiro, M. A. P., Jugend, D., de Sousa, L., Jabbour, A. B., Chiappetta Jabbour, C. J., & Latan, H. (2022). Circular economy-based new products and company performance: The role of stakeholders and industry 4.0 technologies. *Business Strategy and the Environment, 31*, 483–499. https://doi.org/10.1002/bse.2905
42. Rehman Khan, S. A., Yu, Z., Sarwat, S., Godil, D. I., Amin, S., & Shujaat, S. (2022). The role of block chain technology in circular economy practices to improve organisational performance. *International Journal of Logistics Research and Applications, 25*, 605–622. https://doi.org/10.1080/13675567.2021.1872512
43. Singh, T., Singh, D., Singh, C.D., Singh, K. (2023). *Industry 5.0*. ISBN 9781119865216.
44. Van De Voorde, K., Paauwe, J., & Van Veldhoven, M. (2012). Employee well-being and the HRM–organizational performance relationship: A review of quantitative studies. *International Journal of Management Reviews, 14*, 391–407. https://doi.org/10.1111/j.1468-2370.2011.00322.x
45. Hartnell, C. A., Ou, A. Y., & Kinicki, A. (2011). Organizational culture and organizational effectiveness: A meta-analytic investigation of the competing values framework's theoretical suppositions. *Journal of Applied Psychology, 96*, 677–694. https://doi.org/10.1037/a0021987
46. El-Farr, H., & Hosseingholizadeh, R. (2019). Aligning human resource management with knowledge management for better organizational performance: How human resource practices support knowledge management strategies? In *Current issues in knowledge management*. IntechOpen.
47. Riva, S., Hampton, P., & Chinyio, E. (2021). STREBLO: The App prototype for managing stress in the construction industry. In *Collaboration and integration in construction, engineering, management and technology: Proceedings of the 11th international conference on construction in the 21st century, London 2019* (pp. 477–483). Springer International Publishing 2020.
48. Prasetyo, W. (2019). Society 5.0 milenial generation: digital talents formula of global open government and smart cities. *Jurnal Riset Akuntansi Dan Bisnis Airlangga, 4*, 623–643. https://doi.org/10.31093/jraba.v4i2.176
49. Apdillah, D., Panjaitan, K., Stefanny, N. T. P., & Surbakti, F. A. (2022). The global competition in the digital society 5.0 era: The challenges of the younger generation. *Journal of Humanities, Social Sciences and Business (Jhssb), 1*, 75–80. https://doi.org/10.55047/jhssb.v1i3.151

Challenges and Opportunities in Integrating Natural Fiber-Reinforced Polymer Composites into Industry 5.0 for Enhanced Mechanical and Thermal Properties

Venkataramana Menda, Ramajogi Naidu Doddi, Penugonda Sureshbabu, Donthamsetty Suneel, Duppala Azad, and Samson Abiodun AASA

1 Introduction

In recent decades, many advancements have been made in the area of composite materials, thanks to the improvements in design, manufacturing, and materials engineering. Among these, natural fiber polymer composite (NFRP) has gained relevance and is considered one of the best developments combining renewable natural fibers with polymers to obtain new materials with superior properties. The properties offered by NFRP composites come as an added advantage as compared to the use of ordinary materials like wood, plastics or metals and these include flexibility, toughness, strength and being environmentally friendly. This versatility has made it possible for these materials to be used in different industries such as construction, automotive, aerospace and electronics.

With the Industry 5.0, manufacturing is continuously suffering a distortion in its paradigm, whereby high automation is complemented with a humanistic and sustainable aspect. It is noted that industries before Industry 4.0 were only productivity and efficiency driven while in the future; Industry 5.0 will be all about smart, resilient, and green manufacturing. It enables the use of advanced technologies such as digital twins, smart sensors, and machine learning to enhance materials' performance, prolong products' lifespan, and reduce negative impact to the environment.

V. Menda · R. N. Doddi (✉) · D. Azad
Aditya Institute of Technology and Management, Tekkali, Andhra Pradesh, India

P. Sureshbabu · D. Suneel
Narsaraopeta Engineering College, Guntur, Andhra Pradesh, India
e-mail: viceprincipal@nrtec.in

S. A. AASA
Department of Automotive Engineering, University of Ibadan, Oyo, Nigeria

Y. Kumar Singla et al. (eds.), *Digital Manufacturing in Industry 5.0*,
https://doi.org/10.1007/978-3-031-91500-0_28

Incorporating NFRP composites into Industry 5.0 structures opens a possibility of using eco-friendly materials in modern manufacturing while still solving the issues of standardization, purchasing cost and scalability.

The structural applications of NFRP composites also require investigation of their performance at extreme temperature and fire conditions. Natural fiber-reinforced concrete materials have some unique property of fire resistance compared to the regular reinforced concrete materials. The type of natural fiber used—bamboo, jute, or coir—not only changes the thermal resistance of the composite but also changes the damage to the environment. In contrast to synthetic fibers, natural fibers tend to have a more favorable environmental impact, making them the material of choice in environmentally sensitive production trends. Furthermore, advancements such as predictive modeling and real time monitoring under Industry 5.0 can help engineers to model and evaluate the fire resistance performance of NFRP built structures improving their fire safety.

NFRP composites also provide promising thermal properties, making them suitable for applications that require improved insulation and energy efficiency. This presents opportunities for further development of energy-saving structures. There are several natural fibers, especially in building composite materials, which have better insulating properties than steel reinforcing bars. Better heating or cooling potential is envisaged for concrete replacements; hence, energy cost of buildings is minimized. Advanced production techniques, such as the use of digital twins, allow engineers to enhance NFRP thermal properties and produce energy efficient structures. The thermal reactivity of the NFRP is dependent on the type of natural fibers included in the composite and how these fibers are encapsulated in polymers, some of which do not allow bamboo and jute to be highly effective in heat retention and insulation. The application of Enabling Technologies of Industry 5.0 in the design process allows the thermal characteristics and behavior of NFRP structural members with different fiber orientation in varied environmental conditions to be predicted and supported, thereby making it easy to design tailored NFRP composites for construction and other applications.

The introduction of NFRPs is a revolutionary step taken in the development of sustainable materials in order to meet the increasing requirements for green building materials. NFRPs are synthesized by hybridizing natural fibers like bamboo and glass into a polymer matrix which enhances the mechanical properties such as tensile strength, impact and buoyancy among many others that are necessary in construction. Nonetheless, embracing NFRPs within the context of Industry 5.0 does pose challenges especially on the aspects of quality, scalability, and increasing affordability. This explains why NFRPs have better mechanical properties than conventional concrete since the use of natural fibers dispersed in thermoplastic polymers leads to an enhanced tensile strength, crack resistance and ductile nature of the polymer. This advancement in mechanical properties renders them useful in applications that require a high level of structural integrity. In Industry 5.0, precise analytics and predictive takeaway also help to ensure that these mechanical properties are optimally achieved in NFRPs for most industrial applications because Industry 4.0 cannot meet the specified standards of industrialization.

One of the major benefits of using NFRPs is their lower thermal conductivity than that of ordinary concrete due to the insulating properties of the natural fibers. This feature makes NFRPs suitable for places where thermal comfort and energy efficiency as well are required for instance in the construction of residential and commercial structures. Precise thermal characterization of composites is made possible by Industry 5.0 technologies enabling manufacturers to optimize insulation performance with respect to NFRP materials. NFRPs also require the use of renewable biodegradable fibers that improve the performance of concrete structures. These fibers are agricultural waste, therefore alleviating the environmental stress and enhancing circular economy. NFRPs however demonstrate moisture, corrosion, and chemical resistance enhancing structures durability. As Industry 5.0 evolves, so do the frameworks and software for quality control and environmental appraisal that facilitate sustainable materials adoption at scale.

NFRP composites present more possibilities in construction engineering due to their flexibility since applications required will depend more on strength, reliability, and sustainability. NFRPs are multifunctional in that they could be used in load bearing members, to cover building facades and thermal insulation panels. Their lightweight also facilitates transportation and handling at site, which in turn saves on cost and labor during construction operations. There are also Industry 5.0 technologies such as automatic assembly and robotic movements which have minimized time wasted in the building process in relation to NFRPs; hence, they would be suitable in smart factories, whereby efficiency and accuracy are core factors.

Over and above these advantages however, many countries are slow in integrating NFRP-concrete composites mainly due to high cost, low standardization, and variability in mechanical and thermal properties. There are Industry 5.0 methodologies to help with these such as digital twin technology for optimization of processes and data analysis for quality assessment. Advancements are being made on aspects of fiber-polymer compatibility, composite uniformity and green manufacturing technologies. Once these areas are improved, the manufacturing costs and the production capacity of NFRPs will be adjusted to suit the construction market needs.

Another factor influencing the application of NFRP-concrete composites in Industry 5.0 is the legislative regulation advocating for green construction materials. There is a growing demand for NFRPs as the understanding of the environmental advantages and the technology for producing certain composite materials evolves. Regulatory agencies do a lot of work that involves setting rules for the NFRP performance so that the materials will be guaranteed dependable for all the applications. In this respect, for example, Industry 5.0 tools such as blockchain help improve the traceability of the joinery works in the supply chain and hence improve the adherence to environmental compliance.

Human-centered and eco-friendly production, which faces Industry 5.0 in all its aspects, corresponds well with how NFRP composites work, as they are high performance, doing away with conventional nonrenewable materials in buildings and other sectors. While there is some bottleneck when it comes to vertical integration of NFRP composites' manufacture in the spirit of industry 5.0, the advent of digital twins, smart sensors, machine learning and so on offers potential opportunities.

These not only improve the efficiency and quality of NFRPs but also digitization in this industry makes it appropriate for smart manufacturing and construction sector in the nearest future.

2 Literature Review

2.1 Wear Resistance and Mechanical Performance of Composite Materials and Alloys

Plastics play an increasingly important role in various applications, transportation and manufacturing aspects in particular, so they must be within a wear resistance and mechanical durability. These works emphasize a variety of advanced engineering techniques that contribute to the development of smart manufacturing and Industry 5.0. Studies like Shah et al. [1] and Singla et al. [2] examine the effects of material science advancements on welding and sustainability, key components for industrial automation and optimized production processes. Additionally, research on solar energy systems and tribological studies [3–8] align with the sustainable energy goals of Industry 5.0. Additionally, advancements in digital twin technology for improving logistics [9], the role of extremophiles in enhancing sustainable agriculture [10, 11], and bioremediation of contaminated soils [12, 13] highlight the role of digital innovation and environmental consciousness in next-generation manufacturing. These studies offer diverse insights into how microbiological advancements and environmental sustainability are being harnessed for smart and sustainable industrial applications, resonating with the principles of Industry 5.0 [14, 15]. The integration of alloying elements to enhance wear resistance [16, 17] supports the development of materials with improved durability for high-tech manufacturing applications. Moreover, works on the surface integrity of machined materials and the impact of nanoparticulates in lubrication show how material engineering can improve efficiency in production processes [35–40]. These studies collectively enhance the scope of smart manufacturing by combining material science, energy efficiency, and automation, advancing the transition toward Industry 5.0 [18, 19].

Works from earlier years explore dynamic simulations [20, 21], improving suspension systems, and enhancing material wear resistance, which collectively reflect advancements in digitalization, automation, and sustainable manufacturing practices. These studies highlight a multidisciplinary approach to improving vehicle systems and material performance in line with the goals of Industry 5.0.

These studies explore a range of advanced techniques and sustainable practices that align with the goals of smart manufacturing and Industry 5.0. Adeoye et al. (2024) investigate the valorization of lignocellulosic waste, highlighting the potential of bio-oil production through pyrolysis and ultrasonication for biofuel generation [22]. Further research by Adeoye et al. focuses on upgrading crude bio-oil [23, 24] and sustainable energy production from biomass [25], providing insights into thermochemical and biochemical conversions. These approaches underscore the integration of clean energy solutions into digital manufacturing processes [26–31]. Studies on catalytic pyrolysis and nanomaterials for enhanced biomass conversion offer pathways for optimizing manufacturing efficiency [32–34]. Additionally, research on pyrolysis of plastic waste and rail projects introduces circular economy concepts and environmental sustainability, both essential to the evolution of Industry 5.0 [35–43].

2.2 *Thermal and Corrosion Behavior in Harsh Environments*

In parallel, work on fiber surface functionalization and dye fastness adds to the understanding of material innovation in manufacturing [44–46]. These studies contribute to the body of knowledge driving digital transformation and environmental consciousness in manufacturing [35, 37, 39–41, 47, 48]. The contributions from Sunil et al. [49–60] span critical advancements in transportation systems, materials science, and engineering technology, all of which have applications in smart manufacturing and Industry 5.0. Recent works from 2024 focus on the adoption of electric vehicles [21, 61–77], modeling of intelligent transportation systems [78–90], and vibration mitigation in marine and rail propulsion systems [91–100]. Recent advancements in sustainable fuel production have focused on various bio-oil sources and carbon capture techniques. Studies have examined bio-oil production from thermal cracking in the presence of hydrogen [101], CO_2 adsorption systems in coal beds [131], and the pyrolysis of coffee and eucalyptus sawdust [102–104]. Additional research has explored the use of sulfur to enhance wood tar biopitch as a sustainable binder, the characterization of sewage sludge bio-oil, and the analysis of bio-oil from algae, comparing freshwater and microbial-treated samples [105]. These efforts collectively contribute to the development of renewable energy and environmental sustainability [35, 48, 106–108].

These works highlight innovative methods in microbiology, environmental sustainability, and materials science, which have significant implications for smart manufacturing and Industry 5.0 [26–31]. Mohapatra et al. [109] study microbial processes for biopolymer production, which can be applied in sustainable packaging solutions [28, 35, 110–117] and bio-based materials [118–129], aligning with Industry 5.0's focus on circular economy and smart materials. Research on retrofitting buildings for energy efficiency [130–146] and the use of polyhydroxyalkanoates (PHAs) for biocompatible applications [147–156] demonstrate the integration of biotechnology into industrial processes.

These studies underscore the need for energy-efficient solutions and advanced control systems, both of which are integral to Industry 5.0 [157, 158]. Previous works from 2023 also explore sustainable materials for transportation [159], intelligent systems for railway infrastructure [15, 160–175], and advancements in modeling and ride behavior [176–184]. Research on crashworthiness [185–194] and the integration of bio-lubricants [195–198] promote the goals of Industry 5.0 by improving safety [199], efficiency, and sustainability [200–202]. These studies collectively support the evolution of smart manufacturing, contributing to Industry 5.0's emphasis on sustainability, digitalization, and innovation across sectors [203].

The research contributions focus on the dynamic analysis, vibration isolation, and advanced material applications in transportation and vehicle systems, which directly support the objectives of smart manufacturing and Industry 5.0 [204–208]. Recent works (2024) address advanced control systems for enhanced ride quality in passenger rail vehicles using intelligent control suspension systems [164, 209–213], and comprehensive dynamic response analyses of road and rail vehicles [214–223]. Further, studies involving the use of composite materials and hybrid AHP-TOPSIS techniques for sustainable structural applications [69, 75, 77, 224–235] contribute to the development of smart materials, which is central to Industry 5.0. In terms of railway infrastructure, Sharma et al. group documented the performance of rail track materials in relation to bitterness, particularly the aspects that may be corrosion attacks, and suggest that these components be protected against the effects of environmental factors [85–88, 90, 115, 217, 218, 236]. These studies also correspond with the aims set forth under Industry 5.0 with regards to achieving sustainability which is the creation of materials that are robust and can survive extreme conditions, and can withstand highly demanding infrastructure projects.

3 Materials and Methods

This section pertains to materials, fabrication method and analysis methods undertaken to prepare and test the hybrid composite. This is important in order to systematically assess the effects of both natural and synthetic fibers on the thermal and mechanical aspects of the composite material boom. Creating a material that can meet the requirements of industry 5.0 in terms of materials being environment friendly and yet high performance in nature.

3.1 Materials Selected

Bamboo fiber, glass fiber, epoxy resin (LY556), and hardener (HY951) are the materials picked for this particular research. Each of the materials has its function, and their integration forms a good combination of mechanical strength, durability, and sustainability. More details of each of the material are given below (see Figs. 1, 2, and 3):

Bamboo Fiber: Bamboo fiber is opted to give better environmental—and mechanical—characteristics, such as tensile strength and flexibility. Due to the renewable nature of Zip's materials, the environmental impact of the composite is reduced as compared to the use of synthetic materials.

Fig. 1 Natural bamboo fiber used as reinforcement in the composite for enhanced tensile strength and sustainability

Fig. 2 E-glass fiber integrated for stiffness and impact resistance in the hybrid composite

Glass Fiber: To provide stiffness, impact resistance and thermal stability to the composite structure, e-glass fibers are used. They play a vital role in the structural strength of the composite and help to withstand stresses even when the composite is loaded.

Fig. 3 Hand lay-up fabrication method illustrating fiber placement, resin application, and curing steps

Table 1 Material specifications for composite fabrication

Material	Type	Specifications	Purpose
Bamboo fiber	Natural fiber	Average length: 50 mm; diameter: 0.02 mm	Adds tensile strength, reduces environmental impact
Glass fiber	E-glass fiber	Diameter: 9 μm; density: 2.6 g/cm^3	Provides stiffness and impact resistance
Epoxy resin	LY556	Viscosity: 10,000 cP; density: 1.15 g/cm^3	Matrix material, binds fibers
Hardener	HY951	Curing agent ratio: 10:1 with resin	Initiates curing process

Ly556: Epoxy Resin. The epoxy ideally plays the role of the matrix in the composite, effectively binding the fibers to develop a unified structure. It is an appropriative adhesive, and it also has good mechanical and thermal properties, which are essential in the overall stability of the composite.

Hy951: Hardener. This hardener which is added to epoxy resin allows further curing. It is effective in an addition by helping in forming a rigid bond within the matrix to add to the strength of the composite and its ability to withstand various environmental conditions.

The materials used in this study, including bamboo fiber, glass fiber, epoxy resin, and hardener, are detailed in Table 1.

3.2 Fabrication Process: Hand Lay-Up Method

The hybrid composite was fabricated using hand lay-up, which is a very simple and inexpensive technique of making layered composites with controlled fiber orientation. This technique permits precise positioning of fibers and makes it possible to test different directions toward enhancing the properties of composites. The stepwise procedure is as outlined below:

Mixing: The two-epoxy resin (LY556) and hardener (HY951) were mixed using a weight ratio of 10:1. One of the objectives for this ratio was to obtain the required curing rate which was satisfactory for the purpose of compacting without making the matrix unduly brittle.

Mold Preparation: The mold complies with the specifications outlined in the relevant ASTM standards. A release agent is applied to the surface of the mold to inhibit adherence of the composite material to the surface allowing for easier extraction after curing.

Fiber and Resin Layering Process: The fiber and resin layering process began by cutting the e-glass fabric to match the mold's dimensions, with an initial layer of epoxy resin applied to the mold for reinforcement. Glass fibers were then positioned in epoxy at various angles (0, 30, 45, and 60°) to study the impact of fiber orientation on the composite's properties. To create a composite structure that mimics bamboo culturing, layers were alternated between bamboo fibers and glass fiber matting, with each fiber layer separated by epoxy resin. This arrangement involved placing a bamboo fiber layer, followed by glass fiber matting within an epoxy layer, ensuring orthogonal alignment across layers. The entire structure was sealed with a flexible epoxy coating to achieve a composite thickness of 5–7 mm. After completing the layering, a 5–15 kg load was applied to the mold, and the composite was left to cure at room temperature for 24 h, enabling adequate bonding of the fibers and resin for a dense, reinforced structure. Figures 1, 2, and 3 illustrate the use of bamboo and glass fibers as well as the hand lay-up fabrication technique, which enables the composite to endure high thermal and mechanical loads. The hand lay-up process parameters are outlined in Table 2, detailing the structured fabrication for durability and stability.

3.3 Experimental Analysis

In order to comprehend fully the hybrid composite induced structural material, a variety of mechanical as well as thermal tests were performed. These tests allow quantifying the effects of fiber orientation and layering on the composite properties giving a verdict on its applicability in advanced engineering design processes.

Table 2 Fabrication parameters for hand lay-up process

Step	Description	Parameters
Resin mixing ratio	Resin to hardener mixed in a 10:1 ratio	Resin: 10 parts, hardener: 1 part
Layering sequence	Alternating layers of glass and bamboo fiber	Bamboo and glass fibers alternated in each layer
Fiber orientation	Fibers oriented at different angles to assess performance	0, 30, 45, 60°
Curing time	Duration for composite solidification	24 h
Applied load for curing	Weight applied to compress layers during curing	5–15 kg
Final composite thickness	Composite built up to desired thickness by repeating layers	5–7 mm

3.3.1 Mechanical Testing

Compression Test: The compression test was conducted in order to find out the compressive strength of the composite and its deformation resistance. The test is critical for any loads bearing applications since it helps to find out how much compressive load the composite can withstand before it fails structurally.

Impact Test: In this test, impact strength is the ability of the composite material to absorb and dissipate energy under a sudden application of force. This test analyzed the effect the different fiber orientations had on the toughness and the resilience of the material providing data of the material's characteristic in applications with impact loading.

Hardness Test: Shore D hardness tester was used to measure hardness. It measures properties of materials that show resistance to indentation aimed at assessing the wear of the material and its general reliability.

3.3.2 Thermal Testing

Thermal Conductivity: Thermal conductivity was taken with the aim of knowing the heat conduction property of the composite. Lowering the conductivity is advantageous in cases where insulation is required because the heat transfer is reduced.

Specific Heat Capacity: This biomedical testing evaluated the capacity of the composite to store heat. The larger specific heat capacity of material suggests that more thermal energy can be held by that material which is preferable for thermal stability in fluctuating temperatures. The mechanical and thermal properties of the composite were assessed through a series of standardized tests, as summarized in Table 3.

Table 3 Summary of mechanical and thermal testing methods

Test type	Standard	Purpose	Testing equipment
Compression strength	ASTM D695	Evaluates material's load-bearing capacity	Compression testing machine
Impact strength	ASTM D256	Measures material's resistance to impact	Izod impact tester
Hardness	ASTM D2240	Assesses surface resistance to indentation	Rockwell hardness tester
Thermal conductivity	ASTM E1530	Determines heat conduction capability	Guarded heat flow meter
Specific heat capacity	ASTM E1952	Measures heat storage capability	Differential scanning calorimeter

4 Results and Discussion

This section presents details on the study into the mechanical and thermal characteristics of bamboo/glass fiber-reinforced polymer composites and more interestingly demonstrates ways how certain drastic changes made in fiber orientation and material structure can influence the durability, toughness, and thermal behavior of the composite. It also investigates fiber oriented at 0° and fiber oriented at 30°, fiber oriented at 45° and fiber oriented at 60°. This study investigates how these parameters affect such important indices as the compression strength, impact strength, hardness, thermal conductivity, and specific heat capacity. Such findings are critical in developing materials that will help utilize the requirements of Industry 5.0 and beyond including sustainable and resilient infrastructure.

4.1 Mechanical Properties

Bamboo and glass fibers inserted in the polymer-matrix resin at certain angles not only adequately improve the polymer matrix, but also provide enhanced mechanical strength and durability turning out to be a viable option in place of conventional materials. All the three mechanical tests performed—compression, impact, and hardness—indicate that there is a critical direction of fiber that contributes to the performance:

4.2 Compression Strength Evaluation

Highest compression strength is shown in Fig. 4. From the analysis, it is illustrated that at fiber orientation of 90° rotation, the composite recorded the highest compression strength of 92 MPa. This finding showcases a major advancement for it

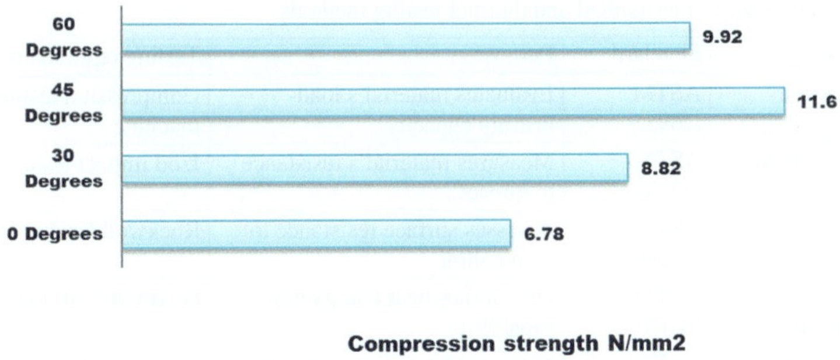

Fig. 4 Compression strength comparison across fiber orientations, highlighting optimal strength at 45°

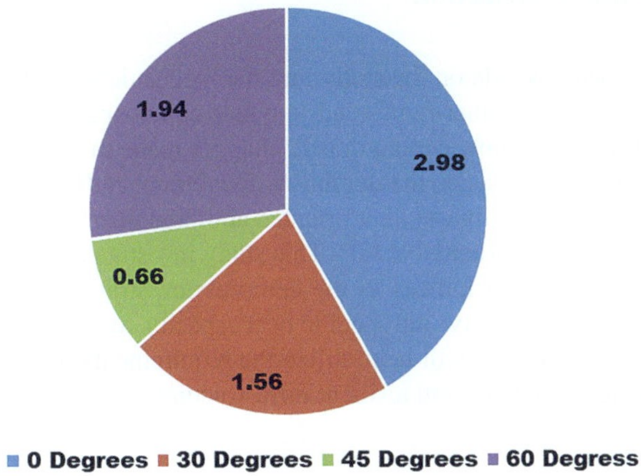

Fig. 5 Izod impact strength results, showing highest energy absorption at 45° fiber orientation

provides an avenue for efficient load bearing within the matrix, hence reducing the chances of distortion to the structure when subjected to heavy loads. The 45° arrangement outperformed other orientations mainly because a phenomenon occurs where natural (bamboo) and synthetic (glass) fibers were used to create the composites. This synergy is favorable for where loads durability is required which is mainly in construction panels or vehicle frames.

4.3 *Impact Strength Evaluation*

Figure 5 shows that the composite's impact strength, one of the most critical factors for the application of the composite under dynamic conditions, has reached the peak limit of 6.5 KJ/m^2 at a 45° fiber angle. This increased impact resistance confirms that this orientation of fiber proves advantageous when energy is applied in a sudden force or impact and needs to be absorbed and dispersed. It was suggested that bamboo fibers, in addition to glass fibers, improve the cross-webbing structure of the composite and introduce special toughness in the bonds of the respective materials which enhances the durability of the materials in time. This improvement is significant in concerns where high shock absorption is crucial, such as for protective casings and especially for automotive components.

4.4 Hardness Evaluation

As for the hardness test results of Fig. 6, the orientation at 45° exhibited the highest Shore D hardness value of 70 when tested with Shore D hardness meter at 30° inclination angle. This degree of surface hardness is useful in cases where the material is subjected to abrasive forces recurrently. Doing so eliminates surface scratches but more importantly adds resistance to surface penetration into the epoxy-reinforced polymer matrix when combining variously oriented fibers with polymers, weakening all posterior indented structural composites' moisture humidity, balancing stiffness capability striking component, and supporting graduates' researchers in the field of stretches nonprimary subspecialty. The 45° orientation, therefore, emerges as the most effective alignment for achieving both strength and resilience, advancing the potential of fiber-reinforced composites as robust, sustainable solutions. The impact of fiber orientation on the composite's mechanical and thermal properties is summarized in Table 4.

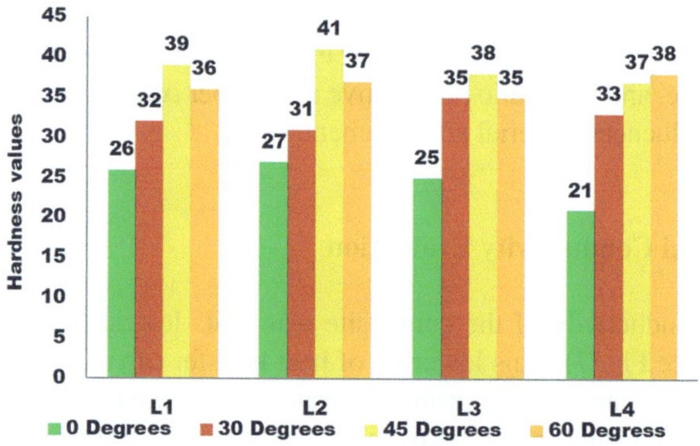

Fig. 6 Hardness (Shore D) results across orientations, with maximum surface hardness achieved at 45°

Table 4 Composite properties with varying fiber orientations

Fiber orientation (°)	Compression strength (MPa)	Impact strength (kJ/m^2)	Hardness (Shore D)	Thermal conductivity (W/m·K)	Specific heat capacity (J/g °C)
0	75	5.2	60	0.25	1.8
30	85	6.1	65	0.22	2.0
45	92	6.5	70	0.21	2.2
60	89	6.0	68	0.24	2.1

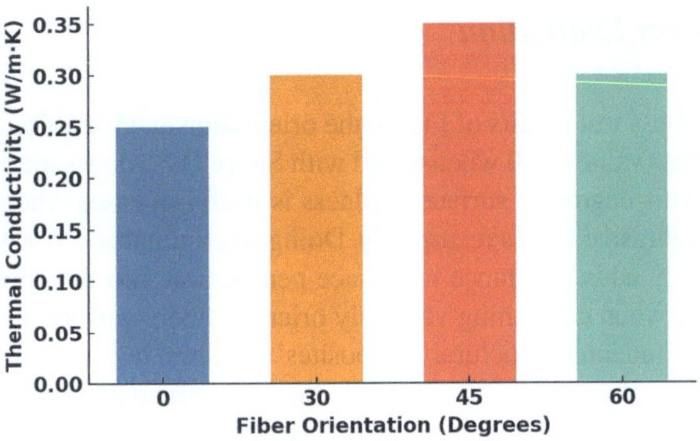

Fig. 7 Thermal conductivity values, demonstrating superior insulation properties at 45° orientation

4.5 Thermal Properties Evaluation

For the materials designed to operate in controlled temperature environments, the thermal performance is equally critical. Results obtained from tests for thermal conductivity and specific heat capacity further explain the heat management properties of the composite, and bring into perspective about fiber orientation and natural fiber assimilation influences material advancement.

4.5.1 Thermal Conductivity Evaluation

The thermal conductivity of the composite was at its lowest (0.21 W/m K) at an angle of 45° (see Fig. 7). This lowering of heat transfer capability shows the composite's potential as an efficient insulating material. The use of bamboo fiber, which has excellent heat insulating property and the proper bow direction for glass fibers, forms a thermal insulative barrier that controls heat transfer. This property is in accordance with the aim of Industry 5.0, which is on the application of energy saving materials. This makes the composite a potential candidate for use in green building heat insulation materials and temperature sensitive equipment encapsulation systems. To evaluate the composite's thermal efficiency, Table 5 compares its thermal properties with those of conventional materials.

Table 5 Thermal characteristics of bamboo/glass hybrid composite compared to conventional materials

Material	Thermal conductivity (W/m·K)	Specific heat capacity (J/g °C)	Insolation efficiency
Bamboo/glass hybrid (45°)	0.21	2.2	High
Pure epoxy	0.30	1.5	Moderate
Glass fiber composite	0.28	1.7	Moderate
Carbon fiber composite	0.40	1.9	Low

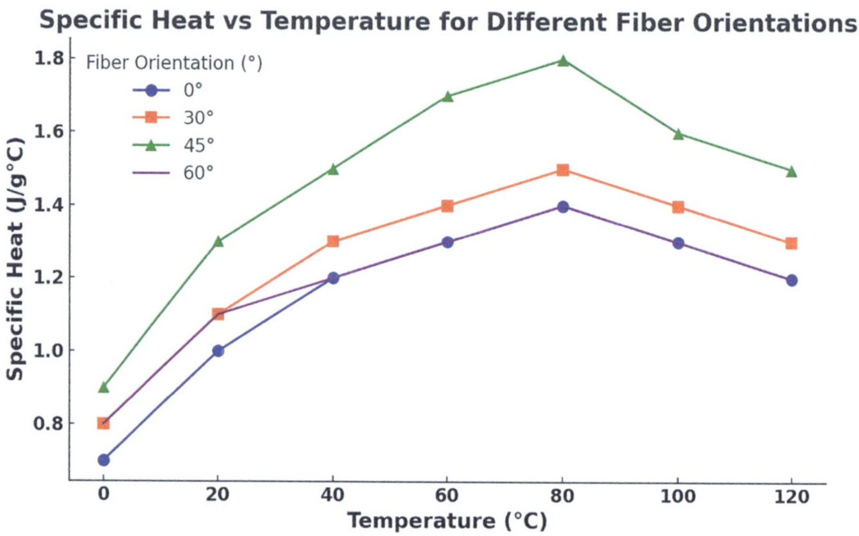

Fig. 8 Specific heat capacity variations with temperature, indicating best heat retention at 45° orientation up to 80 °C

4.5.2 Specific Heat Capacity Evaluation

The highest specific heat capacity of 2.2 J/g °C for the composite was achieved at 45° in the 30–80 °C temperature range (see Fig. 8). This implies that the composite has good heat energy storage capabilities and will keep the heat energy practically at that energy state. This heat retention capability is improved further when bamboo and glass fibers stick well since it forms a framework that holds heat without losing it too quickly. This feature is mostly useful in applications that require heat resistance when temperatures change, such as in electric-assisted packages and thermal panels. However, temperatures higher than 80 °C cause a reduction of specific heat and this seems to be related to phonon scattering of the material and energy retention is lower in high temperatures. This trend emphasizes the effectiveness of the material in the application of moderate temperature while it also reveals a limit on

Table 6 Effects of temperature on specific heat capacity of bamboo/glass hybrid composite

Temperature (°C)	Specific heat capacity (J/g °C) at 45° fiber orientation	Remarks
30	1.8	Initial thermal absorption
50	2.0	Stable thermal retention
80	2.2	Peak heat storage capability
100	1.9	Slight decline due to phonon scattering
120	1.6	Reduced stability at high temperatures

temperature above which there will be thermal instability. Table 6 shows the effect of temperature on specific heat capacity for each fiber orientation.

5 Discussion

The results of this study further demonstrate the effect of fiber orientation and material combination on the mechanical and thermal properties of bamboo/glass hybrid composites. This research projects progress in composite engineering more so on the combination of natural and synthetic incorporation in the manufacture of strong and green materials.

Material Design Innovation with 45° angle: In all instances and tests done, 45° fiber orientation came out as the best which indicates that it could be the most proper orientation for the combination of the properties of mechanical strength, impact resistance, hardness, and thermal insulation. Such ability to modify the properties of a composite by the orientation is indeed a remarkable step in most technologies in the current age of industry 5.0, which requires all these aspects, of extension, durability, and sustainability. Applying bamboo–glass fiber composites with 45° height crimp angle enhances the hardened structure and, therefore, broadens the scope of advanced manufacturing applications in the automotive, aerospace, and construction industries.

Sustainable Development and Performance Optimization: The low thermal conductivity and high specific heat of the composite at 45° angles, particularly orientation, experienced in all instances desired enhances the sustainability targets of Industry 5.0 minimizing use of energy intensive materials. Bamboo fibers not only bring the environmental benefits of the composite but also provide useful thermal properties; thus, it can be used very effectively in other applications that require insulation materials with structural strengths. More importantly, these data demonstrate the maturity of this material as an environmentally friendly substitute for the conventional synthetic-only composites thereby cementing the dominance of natural fibers in the development of new materials.

Table 7 Comparison of mechanical properties for various fiber composites

Fiber type	Compression strength (MPa)	Impact strength (kJ/m^2)	Hardness (Shore D)	Remarks
Bamboo fiber	60	4.5	55	Eco-friendly, moderate strength
Glass fiber	80	5.8	68	High strength and durability
Bamboo/glass hybrid	92	6.5	70	Optimal for strength and impact
Carbon fiber	110	7.2	75	Highest performance, costly

Future Applications and Industry 5.0 Implications: The importance of these findings lies in contributing to the design of new engineered materials for Industry 5.0. Hybridization of bamboo and synthetic glass fiber help move forward the design of lighter and high-performance materials. At 45° orientation of the composite, these findings highlight that there could be carefully structured fibrous laminates that would be tailored to the needs and enhancement of new innovations in complete resilient materials applications such as modern buildings, car chassis and industrial tools.

Opportunities for Further Research and Innovation: Even if for this work the 45° orientation was the most optimal angle indicating the extent of fibrous composite performance, further improvement on the composite features might be possible through a potential exploration other than only the fiber thicknesses such as the other natural fibers and the different matrix materials. However, the view point assists or aids in the improvement of cement composite or aids in the enhancement of cement composites by innovating more ways of construction practices for instance by the application of automated lay-up or every other nanotechnology enhanced processes. In the context of Industry 5.0, the bamboo/glass hybrid composite demonstrates significant advantages over traditional materials, as highlighted in Table 7.

6 Conclusion

It demonstrates a detailed discussion of mechanical and thermal properties of bamboo/glass fiber-reinforced polymer composite materials with emphasis on the change of performance with different fiber orientations. This research was prompted by a desire to create high-performance materials that would be in line with the objectives of Industry 5.0 which are competent, environmentally friendly and versatile engineering solutions. Sequential investigation of fiber positions 0, 30, 45 and 60° provided us some useful guidelines on the up gradation of composite materials through natural as well as synthetic fiber's combination.

6.1 Key Findings and Insights

Optimal Fiber Orientation for Enhanced Properties: Interestingly, the results across the tests indicated that 45° fiber orientation had the best performance time after time during the physical and thermal assessments. This orientation of the composite exhibited good compressive strength, impact strength, hardness at this angle, lower thermal conductivity, and higher specific heat capacity. Therefore, it could be implied that stress distribution, energy absorption and thermal insulation could be maximized with 45° stressed fibers orientation to obtain thermal mechanical stability. These properties are very important in applications where there is a need for structure integrity, impact resistance and heat control.

Innovation in Composite Material Engineering: The incorporation of bamboo and glass fibers in a hybrid form is a great leap in the evolution of material engineering. Here, the use of bamboo which is renewable and green resource and use of glass fiber in an epoxy matrix is an attempt to show that materials can be made harnessing the benefits of both natural and synthetic fibers. The fibers of bamboo are useful in improving the sustainability characteristics of the composite as they provide tensile reinforcement, and the fibers of glass provide stiffness and thermal stability. Such a synergy fits within the needs of Industry 5.0 and its innovativeness as it shows a way of lessening environmental footprints without compromising or improving the performance of the materials.

Effect of Natural Fibers on the Thermal and Mechanical Properties: Incorporation of bamboo fibers, which are known to have insulation properties, was found to significantly reduce thermal conductivity at optimal orientation angles. This feature is beneficial in the place where efficient controlling of temperatures is paramount, e.g., insulation panels' automotive interior or electronic housings. Other further observed high specific heat capacity at 45° of orientation suggests that the composite would trap heat well adding thermal stability where temperature changes are extreme. This quality makes bamboo/glass fiber-reinforced composites suitable for thermal shock situations where internal and external temperatures are always changing.

Possible areas of exploitation for industry applications: The material properties observed at 45° reveal the vast opportunities which the composite has for a composite component that includes such areas as structural members, automotive shells, outer protections and even thermal insulations. The load-bearing performance of the composite material, impact tools performance, and thermal management efficiency conforms to the sectors embrace concerning durable, multiple-use and environmentally friendly materials. This is particularly useful in strategies for industries such as construction, transportation and aerospace in which the target is to minimize the effects on the environment without sacrificing results.

Furthering the aims of Industry 5.0: The analysis provides an example of how the combination of engineering and design of natural fiber composites can further the objectives of industry 5.0 which are that of "human centricity and sustainability." The hybrid composite is not only useful in replacing raw materials which are synthetic and expensive, but it also enhances the performance of the specific mechanical properties. There will be an ecosystem, which is called the bamboo jump—a production of materials with particular qualities due to the properties of bamboo fibers, environmental sustainability and future industrial tendencies in materials.

6.2 *Implications for Future Research and Development*

Although these results validate the finding that bamboo and glass are optimally arranged in low/high reinforced composites at 45° lines, it is not the end of the story. This foundation can be built upon in future studies by considering the prospects of using other natural fibers, increasing the thickness of the fibers, and changing the order of the layers to make the thermal and mechanical interlayers more efficient. Further issues of this category may be resolved with the application of more advanced engineering technologies: wide-range automated lay-up processes or nanoenhancement techniques.

Composite materials are generally encased in fillings to improve their structural stability. One such area is the high-temperature effect of the order of the composite on its stability. Problems can also be observed as a specific heat capacity that arises when a material is raised up to a certain temperature and beyond, thermal stability may be affected. Either polymers or polymer resin coatings having a high resistance to heat and temperature stress may be used to address the problem above.

Bamboo and glass fiber-reinforced polymer composites point toward the possibility of using lightweight, durable composite materials in housing construction. With the aid of both natural and synthetic fibers, the composite is able to achieve a favorable combination of high durability, impact resistance, and thermal insulation. There is also a solid base on the fact that these new materials with advanced eco concepts in their making can use Industrial 5.0 concepts and still manage to make the performance that the material warrants.

The hybrid composite that has been created in this work can be regarded not only as a new material but also as a development toward the future, where sustainability, resilience, and adaptability are of the utmost importance. This study highlights the fact that due to proper design and advent of new combinations of materials, it is not necessarily required to compromise on nature's laws in order to meet the requirements of the present engineering. By further investigating such composites, the manufacturing sector will be able to adopt a future that is both advanced in technology and preservation of the environment.

References

1. Shah, S., Nautiyal, H., Gugliani, G., Kumar, A., Namboodri, T., & Singla, Y. K. (2024). *Sustainability in smart manufacturing: Trends, scope, and challenges*. CRC Press (Taylor & Francis Group). https://taylorandfrancis.com/books/. https://doi.org/10.1201/9781003467496
2. Singla, Y. K., Maughan, M. R., Arora, N., & Dwivedi, D. K. (2024). Enhancing the wear resistance of iron-based alloys: A comprehensive review of alloying element effects. *Journal of Manufacturing Processes, 120*, 135–160.
3. Singh, H., & Singla, Y. K. (2023) *Wear characteristics of AISI 4140 under nano fly ash based engine oil*. Proceedings of the 7th International Conference on Theoretical and Applied Nanoscience and Nanotechnology (TANN'23), Paper No. 118, 1-3 June, 2023, Canada. https://doi.org/10.11159/tann23.118
4. Kumar, A., Ram, M., & Singla, Y. K. (2022). *Advanced materials for biomechanical applications* (1st ed.). CRC Press. https://doi.org/10.1201/9781003286806
5. Singla, Y. K., Maughan, M. R., Arora, N., & Dwivedi, D. K. (2023). Microstructural evolution of worn surface and subsurface of high Mn additive iron-based hardfacings subjected to repetitive impacts and sliding. *Ceramics International, 49*, 21957–21965.
6. Singh, H., Singh, A. K., Singla, Y. K., Chattopadhyay, K., Saini, A., & Singh, K. (2022). Interpretation of the wear characteristics of AISI 4140 under nano-fly ash based engine lubricant. *Materials Today Proceedings, 50*, 1683–1689.
7. Rooprai, R. S., Singh, H., Singh, T., & Singla, Y. K. (2022). Analysis of the wear properties of through hardened AISI-4140 alloy steel using Taguchi technique. *Materials Today Proceedings, 50*, 661–664.
8. Singh, H., Singh, A. K., Singla, Y. K., Chattopadhyay, K., & Malhotra, R. (2021). Optimization of tribological behavior of AISI 4140 under nano fly ash particulates in engine lubricating oil. *Materials Today Proceedings, 45*, 4619–4623.
9. Sura, H. S., Avesh, M., & Mohapatra, S. (2023). Design and modelling of digital twin technology to improve freight logistics. *Energy, Environment, and Sustainability*, 481–513. https://doi.org/10.1007/978-981-99-2150-8_20
10. Pattnaik, S., Pati, S., Devadarshini, D., Tanaya, K., Mishra, B. B., Samantaray, D., Mohapatra, S., & Dash, D. (2022). Microbial bioremediation of Cr(VI)-contaminated soil for sustainable agriculture. In *Microbial biodegradation and bioremediation: Techniques and case studies for environmental pollution* (pp. 395–407). https://doi.org/10.1016/B978-0-323-85455-9.00008-4
11. Srivastava, A., Lavania, S., & Mohapatra, S. (2023). Predicting the effect on land values after introducing high-speed rail. In *Energy, environment, and sustainability* (pp. 157–196). https://doi.org/10.1007/978-981-99-2150-8_8
12. Agrahari, R., Mohanty, S., Vishwakarma, K., Nayak, S. K., Samantaray, D., & Mohapatra, S. (2021). Update vision on COVID-19: Structure, immune pathogenesis, treatment and safety assessment. *Sensors International, 2*. https://doi.org/10.1016/j.sintl.2020.100073
13. Adeoye, A. O., Quadri, R. O., & Lawal, O. S. (2022). Fixed-bed pyrolysis and thermal analyses of pressed oil palm fruit fibre as a potential alternative energy. *Fuoye Journal of Innovation, Science and Technology, 2*(1), 187–194.
14. Tanaya, K., Mohapatra, S., & Samantaray, D. (2024). Formulation of biogenic fluorescent pigmented PHB nanoparticles from Rhodanobacter sp. for drug delivery. Antonie van Leeuwenhoek. *International Journal of General and Molecular Microbiology, 117*. https://doi.org/10.1007/s10482-024-01973-x
15. Mohapatra, S., Pattnaik, S., Maity, S., Sharma, S., Akhtar, J., Pati, S., Samantaray, D. P., & Varma, A. (2020). Comparative analysis of PHAs production by Bacillus megaterium OUAT 016 under submerged and solid-state fermentation. *Saudi Journal of Biological Sciences, 27*, 1242–1250. https://doi.org/10.1016/j.sjbs.2020.02.001

16. Singh, H., Singh, A. K., Singla, Y. K., & Chattopadhyay, K. (2020). Design & development of a low cost tribometer for nano particulate lubricants. *Materials Today Proceedings, 28*, 1487–1491.
17. Kumar Singla, Y., Dwivedi, D. K., & Arora, N. (2017). Modeling the impact–sliding wear characteristics of rare earth additive iron-based hardfacing alloys. *Proceedings of the Institution of Mechanical Engineers, Part J: Journal of Engineering Tribology, 231*, 1486–1499.
18. Singla, Y. K., Dwivedi, D. K., & Arora, N. (2015). On the modeling of dry sliding adhesive wear parameters of vanadium additive iron-based alloys at elevated temperatures. *Surface and Coating Technology, 283*, 223–233.
19. Singla, Y. K., Chhibber, R., Bansal, H., & Kalra, A. (2015). Wear behavior of aluminum alloy 6061-based composites reinforced with SiC, Al2O, and red mud: A comparative study. *Journal of Metals, 67*, 2160–2169.
20. Pant, M., Sharma, A., Palli, S., Sharma, S. K., & Sharma, R. C. (2024). A review and analysis of factors affecting EV adoption for sustainable and energy-efficient transportation systems using analytic hierarchy process approach. *International Journal of Vehicle Structures and Systems, 16*, 133–139. https://doi.org/10.4273/ijvss.16.2.01
21. Sharma, S. K., Sharma, R. C., Lavania, S., Palli, S., & Avesh, M. (2024). Industry-driven approach for ANFIS-based intelligent control suspension system with MR damper for enhanced ride quality in passenger rail vehicles for technological investigations. *National Academy Science Letters, 47*, 391–394. https://doi.org/10.1007/s40009-023-01365-1
22. Adeoye, A. O., Quadri, R. O., Lawal, O. S., & Emojevu, E. O. (2024). Physicochemical characterization, valorization of lignocellulosic waste (Kola nut seed shell) via pyrolysis, and ultrasonication of its crude bio-oil for biofuel production. *Cleaner Waste Systems, 7*. https://doi.org/10.1016/j.clwas.2024.100138
23. Adeoye, A. O., Quadri, R. O., Lawal, O. S., Malomo, D., Emojevu, E. O., Omonije, O. O., Odeniyi, O. K., Fadahunsi, M. O., Yelwa, M. J., Aasa, S. A., Onakpa, A. E., Omoniyi, B. P., Mark, I. N., Usman, J., Ismaila, A. M., & Saidu, A. U. (2024). Advanced techniques in upgrading crude bio-oil to biofuel. In *Energy, environment, and sustainability* (pp. 321–353). https://doi.org/10.1007/978-981-97-0515-3_14
24. Adeoye, A. O., Quadri, R. O., & Lawal, O. S. (2022). Assessment of biofuel potential of tenera palm kernel shell via fixed bed pyrolysis and thermal characterization. *Results in Surfaces and Interfaces, 9*. https://doi.org/10.1016/j.rsurfi.2022.100091
25. Adeoye, A. O., Lawal, O. S., Quadri, R. O., Malomo, D., Aliyu, M. T., Dang, G. E., Emojevu, E. O., Maikato, M. J., Yahaya, M. G., Omonije, O. O., Edidem, V. G., Abubakar, Y. K., Offor, O. F., Sochima, E. H., Peter, B. E., & Hikon, B. N. (2023). Sustainable energy via thermochemical and biochemical conversion of biomass wastes for biofuel production. In *Energy, environment, and sustainability* (pp. 245–306). https://doi.org/10.1007/978-981-99-2150-8_11
26. Mohapatra, S., & Mishra, R. K. (2020). Status of institutional credit disbursement of agricultural sector in Odisha, India. *International Journal of Current Microbiology and Applied Sciences, 9*, 2057–2065.
27. Mohapatra, S., Singh, J., & Mishra, S. (2017). Co-integration among major potato markets in Punjab. *Indian Journal of Agricultural Marketing, 31*, 46–47.
28. Mohapatra, S., Priyanka, V., Mohapatra, S., Kohli, I., & Mishra, R. K. (2020). Impact of corona virus covid-19 on the global economy. *International Journal of Agricultural and Statistical Sciences, 16*, 771–778.
29. Mohapatra, S., & Singh, J. (2021). Co-integration of major vegetable markets in Punjab, India. *Economic Affairs, 66*(2). https://doi.org/10.46852/0424-2513.2.2021.1
30. Mohapatra, S., Mishra, R., & Mishra, R. (2021). Impact of COVID-19 on education level of agricultural farming community. *International Journal of Agricultural and Statistical Sciences, 17*, 915–923.
31. Mohapatra, S., Singh, J., & Kumar, S. (2018). Cointegration among major cauliflower markets in Punjab. *Indian Journal of Economics and Development, 14*, 330–335.

32. Adeoye, A. O., Yelwa, J. M., Imam, N., Quadri, R. O., Lawal, O. S., Malomo, D., Aasa, S. A., Onakpa, A. E., Dorgbaa, R. K., Omoniyi, B. P., Ahmad, N., & Garba, H. A. (2024). Pyrolysis of plastic wastes towards achieving a circular economy: An advanced chemistry and technical approach. In *Energy, environment, and sustainability* (pp. 215–259). https://doi.org/10.1007/978-981-97-0437-8_11
33. Adeoye, A. O., Quadri, R. O., & Lawal, O. S. (2023). Wet synthesis, characterization of goethite nanoparticles and its application in catalytic pyrolysis of palm kernel shell in TGA. *Results in Surfaces and Interfaces, 11*. https://doi.org/10.1016/j.rsurfi.2023.100118
34. Avesh, M., Imam, N., & Adeoye, A. O. (2024). Assessing environmental impacts and mitigation strategies for the DFCCIL rail projects. In *Energy, environment, and sustainability* (pp. 333–353). https://doi.org/10.1007/978-981-97-0437-8_15
35. Mohapatra, S., Mohapatra, S., Han, H., Ariza-Montes, A., & López-Martín, M. D. C. (2022). Climate change and vulnerability of agribusiness: Assessment of climate change impact on agricultural productivity. *Frontiers in Psychology, 13*, 955622. https://doi.org/10.3389/fpsyg.2022.955622
36. Mohapatra, S., & Mishra, R. A. J. K. (2020). Production performance and market integration of cotton markets in India. In *National symposium* (p. 253).
37. Din, A. U., Han, H., Ariza-Montes, A., Vega-Muñoz, A., Raposo, A., & Mohapatra, S. (2022). The impact of Covid-19 on the food supply chain and the role of e-commerce for food purchasing. *Sustainability, 14*(5), 3074.
38. Mohapatra, S., & Singh, J. (2018). Growth and instability analysis of cauliflower crop in Punjab. *Economic Affairs, 63*, 665–669.
39. Pandey, G., Bansal, S., & Mohapatra, S. (2023). Market integration of chickpea crop: An evidence of India. *Journal of Agribusiness in Developing and Emerging Economies, 15*(1), 1–18. https://doi.org/10.1108/JADEE-12-2022-0273
40. Mohapatra, S., Mishra, R. K., & Sarangi, K. K. (2019). Economic appraisal of bio ethanol production technologies. *International Journal of Chemical Studies, 7*, 2197–2200.
41. Mohapatra, S., Kishore Mishra, R., Sarangi, K., & K. (2020). Bio-ethanol (2nd generation ethanol): A solution to ever polluting gasoline to climate in India. *International Research Journal of Pure and Applied Chemistry, 20*, 1–15.
42. Gohain, N., Bansal, S., Mohapatra, S., & Singh, L. (2022). Analyzing the direction of trade: Indian ginger and lessons from exports to different destinations. *Journal of Agricultural Development and Policy, 32*, 214–220.
43. Mohapatra, S., Mohapatra, U., Chandana, K. S. S., & Mishra, R. K. (2018). Economic analysis of paddy production and marketing in Puri, Odisha. *Journal of Pharmacognosy and Phytochemistry, 7*, 1858–1861.
44. Quadri, R. O., Adeoye, A. O., Ayanda, O. S., & Lawal, O. S. (2024). Conversion of palm kernel shell to sustainable energy and the effect of wet synthesized nanoparticles of iron on its thermal degradation kinetics. *Bioresource Technology Reports, 27*. https://doi.org/10.1016/j.biteb.2024.101933
45. Momodu, D. U., Adetuyi, A. O., Adeoye, A. O., Akinlolu, O. S., & Apara, E. O. (2023). SURFACE FUNCTIONALIZATION OF LINEN FIBER AND ITS DYE-FASTNESS ASSESSMENT USING (4-[(E)-(4-AMINO-2-HYDROXYPHENYL) DIAZENYL] BENZENE-1-SULPHONIC ACID AND 4-[(E)-(4-{(Z)-[4-(DIMETHYLAMINO) PHENYL] DIAZENYL}-2-HYDROXYPHENYL) DIAZENYL]BENZENE-1-SULPHONIC ACID. *Suranaree Journal of Science and Technology, 30*, 030118.
46. Adeoye, A. O., Quadri, R. O., & Lawal, O. S. (2023). Biomass waste pyrolysis in the context of Nigeria's oil and gas industry: An overview. *International Journal of Ambient Energy, 44*, 2525–2541. https://doi.org/10.1080/01430750.2023.2256325
47. Mohapatra, S., Mohapatra, U., & Mishra, R. K. (2018). Price forecasting of groundnut in Odisha. *Journal of Pharmaceutical Innovation, 7*, 111–114.

48. Mohapatra, S., Sneha, N. K., Mounika, A., & Mishra, R. K. (2019). SOCIO-ECONOMIC AND DEMOGRAPHIC ANALYSIS OF SAARC NATIONS. *International Journal of Agricultural & Statistical Sciences, 15*.
49. Sharma, S. K., Upadhyay, R. K., Kumar, V., & Valera, H. (2023). *Transportation energy and dynamics*. Springer Nature Singapore. https://doi.org/10.1007/978-981-99-2150-8
50. Upadhyay, R. K., Sharma, S. K., Kumar, D. V., & Valera, H. (2023). *Transportation systems technology and integrated management*. Springer Nature Singapore. https://doi.org/10.1007/978-981-99-1517-0
51. Lee, J., & Sharma, S. K. (2023). Special Issue: Intelligent systems for railway infrastructure. *Applied Sciences, 13*(20), 11274. https://doi.org/10.3390/app132011274
52. Sharma, S. K., Sharma, R. C., Choi, Y., & Lee, J. (2023). Modelling and dynamic analysis of adaptive neuro-fuzzy inference system-based intelligent control suspension system for passenger rail vehicles using magnetorheological damper for improving ride index. *Sustainability, 15*, 12529.
53. Sharma, S. K., Upadhyay, R. K., Kumar, V., & Valera, H. (2023). Introduction to sustainable transportation system. In *Transportation energy and dynamics* (pp. 3–6). Springer.
54. Hossain, I., Ahmed, Y. E., & Sharma, S. K. (2023). The use of both metallic-insert and TiO2/water nanofluid to improve the performance of photovoltaic modules; experimental study. *Thermal Science and Engineering Progress, 42*, 101944.
55. Wu, Q., Cole, C., Spiryagin, M., Chang, C., Wei, W., Ursulyak, L., Shvets, A., Murtaza, M. A., Mirza, I. M., & Zhelieznov, K. (2023). Freight train air brake models. *International Journal of Rail Transportation, 11*, 1–49.
56. Upadhyay, R. K., Sharma, S. K., Kumar, V., & Valera, H. (2023). Introduction to transportation systems technology and integrated management. In *Transportation systems technology and integrated management* (pp. 1–4). Springer.
57. Sharma, R. C., Bhardawaj, S., Sharma, S. K., & Bahl, S. (2023). Ride behaviour of three-wheel vehicle using semi-active suspension using skyhook control strategy. *International Journal of Vehicle Structures & Systems (IJVSS), 15*.
58. Bhardawaj, S., Sharma, R. C., Sharma, S. K., Gopala Rao, L. V. V., & Vashist, A. (2023). Modeling of biomechanical human body model for seat to head transmissibility analysis. *Noise & Vibration Worldwide, 54*, 62–74.
59. Sharma, S. K., Sharma, R. C., Choi, Y., & Lee, J. (2023). Experimental and mathematical study of flexible–rigid rail vehicle riding comfort and safety. *Applied Sciences, 13*, 5252.
60. Sravanthi, A. L., Ratnakumar, P., Reddy, S. N., Eswari, K. B., Pandey, B. B., Manikanta, C., Ramya, K. T., Sonia, E., Mohapatra, S., Gopika, K., Anusha, P. L., & Yadav, P. (2022). Morpho-physiological, quality traits and their association with seed yield in sesame (Sesamum indicum L.) indigenous collection under deficit moisture stress. *Plant Physiology Reports, 27*(1), 132–142. https://doi.org/10.1007/s40502-021-00621-0
61. Pandipati, S., Palli, S., Dowluru, S., Sharma, R. C., Sharma, N., Sharma, S. K., & Da Fonseca-Soares, D. (2024). Microstructural and strength comparison friction stir processing of aluminium alloys and multi-response optimization by Taguchi GRA-PCA based hybrid methods. *International Journal of Vehicle Structures and Systems, 16*(3). https://doi.org/10.4273/ijvss.16.3.03
62. Sharma, S. K., Upadhyay, R. K., & Kumar, V. (2024). Introduction to dynamics of transportation ecosystem, modeling, and control. In *Dynamics of transportation ecosystem, modeling, and control* (pp. 3–6). Springer.
63. Sharma, S. K., Sharma, R. C., Mohapatra, S., Fouly, A., Hossain, I., & Lee, J. (2024). Technological advancements in oscillation reduction for propulsion shaft systems. *Journal of Vibration Engineering & Technologies, 1–19*.
64. Upadhyay, R. K., Sharma, S. K., & Kumar, V. (2024). Introduction to intelligent transportation system and advanced technology. In *Intelligent transportation system and advanced technology* (pp. 3–6). Springer.

65. Hossain, I., Ngoupo, A. T., Bhattarai, S., Challagulla, S. P., & Sharma, S. K. (2024). Empirical survey on the effect of straight and wavy fins integrated with wind stream as heat sink of photovoltaic panels. *International Journal of Environmental Science and Technology, 1–14.*
66. Sharma, S. K., Sharma, R. C., & Lee, J. (2024). Propelling precision of longitudinal vibration mitigation in ship propeller shafts through advanced nonlinear intelligent semi-active control leveraging adaptive neuro-fuzzy inference system with linear quadratic regulator. *Journal of Vibration and Control, 10775463241244836.*
67. Sharma, S. K., Kumar, N., Avesh, M., Sharma, R. C., Siddiqui, M. I. H., & Lee, J. (2024). Navigating tranquillity with H∞ controller to mitigate ship propeller shaft vibration. *Journal of Vibration Engineering & Technologies, 1–11.*
68. Sharma, S. K., Sharma, R. C., Upadhyay, R. K., & Lee, J. (2024). Enhanced ship engine vibration reduction using magnetorheological dampers and adaptive neuro-fuzzy control system. *Ships and Offshore Structures, 1–17.*
69. Sharma, R. C., Sharma, N., Palli, S., Sharma, S. K., Joshi, S., Rallabandi, S. R., & Sharma, A. (2024). Dynamic behavior of three-wheel vehicle with inerter-based dynamic absorber suspension system. *Noise & Vibration Worldwide, 55*, 253–263. https://doi.org/10.1177/09574565241243398
70. Lora Anusha, P., Ratnakumar, P., Pandey, B. B., Sandhya Rani, P., Umamahesh, V., Reddi Sekhar, M., Chandrika, V., Yadav, P., Mohapatra, S., & Padmaja, D. (2023). Phenotypic characterization of sesame (Sesamum indicum L.) revealed promising genotypes for moisture stress conditions. *Plant Genetic Resources: Characterization and Utilization, 21*(3), 218–228. https://doi.org/10.1017/S1479262123000667
71. Kokkinos, N. C., Emmanouilidou, E., & Sharma, S. K. (2024). Waste-to-biofuel production for the transportation sector. In *Intelligent transportation system and advanced technology* (pp. 99–122). Springer.
72. Sharma, S. K., Sharma, R. C., Lavania, S., Palli, S., & Avesh, M. (2024). Dynamic vibration mitigation in ship propellers using hybrid fuzzy sliding mode control. *National Academy Science Letters*. https://doi.org/10.1007/s40009-024-01515-z
73. Sharma, S. K., Sharma, R. C., & Lee, J. (2024). Modelling and simulation of MR damper characterization and uncertainty assessment in marine engine vibration isolation with FLPID control. *International Journal of Modelling and Simulation*, 1–11.
74. Sharma, S. K., Sharma, R. C., Upadhyay, R. K., & Lee, J. (2024). Coupled torsional–transverse vibration reduction in marine propulsion dynamics with novel approach using magnetorheological damping and adaptive control system. *Journal of Vibration Engineering & Technologies, 12*, 6089–6099.
75. Rallabandi, S. R., Srinivas, L., Palli, S., Sharma, R. C., Sharma, N., Sharma, A., & Sharma, S. K. (2024). A contrastive characterization of pure Mg and AZ91D alloy based on the testing of mechanical, corrosion, wear, and erosion properties. *Engineering Research Express, 6*, 015017. https://doi.org/10.1088/2631-8695/ad16a2
76. Badgujar, P., Sharma, S., Rahul, D., Sharma, S. K., & Upadhyay, R. K. (2024). Feasibility study of Mumbai-Pune Hyperloop corridor: Challenges and implementation. *Research in Transportation Business & Management, 52*, 101089.
77. Sharma, R. C., Palli, S., Dharmana, L., Sharma, S. K., Sharma, A., Kharola, A., & Sharma, N. (2024). Simulation of transient dynamic response of a locomotive considering its flexibility and rigidity. *Noise & Vibration Worldwide, 55*, 33–47. https://doi.org/10.1177/09574565231212684
78. Song, H., Park, J., Sharma, S. K., & Lee, J. (2022). A comparative study on ultrasonic propagation characteristics and defect detection of metal material additive manufacturing using deep learning algorithm. In *Biennial international conference on future learning aspects of mechanical engineering* (pp. 157–168). Springer.
79. Boris, I., Sharma, S. K., & Lee, J. (2022). A study on the reflection coefficient of higher modes of torsional wave by theoretical and numerical method. In *Biennial international conference on future learning aspects of mechanical engineering* (pp. 145–155). Springer.

80. Vishwakarma, P. N., Mishra, P., Sharma, S. K., & Shukla, A. K. (2022). Formulating and analysing the effect of suspension parameters on critical speeds for various DOFs in rail vehicle. In *Biennial international conference on future learning aspects of mechanical engineering* (pp. 791–802). Springer.
81. Sharma, R. C. (2016). Evaluation of passenger ride comfort of Indian rail and road vehicles with ISO 2631-1 standards. Part 1 - Mathematical modelling. *International Journal of Vehicle Structures and Systems, 8*(1). https://doi.org/10.4273/ijvss.8.1.01
82. Sharma, R. C., Palli, S., Jha, A. K., Bhardawaj, S., & Sharma, S. K. (2022). Vibration and ride comfort analysis of railway vehicle system subjected to deterministic inputs. *Resmilitaris, 12*, 1345–1355.
83. Sharma, R. C., Gopala Rao, L. V. V., Srikar, B., Palli, S., Appala Naidu, Y., Sharma, N., & Sharma, S. K. (2025a). Investigation of the dynamic behavior of hybrid composites using finite element and experimental method. *Journal of the Chinese Institute of Engineers, 48*(2), 158–171. https://doi.org/10.1080/02533839.2025.2453139
84. Vishwakarma, P. N., Mishra, P., & Sharma, S. K. (2022). Formulation of semi-active suspension system and controls in rail vehicle. *SSRN Electronic Journal*. https://doi.org/10.2139/ssrn.415961
85. Sharma, R. C., & Sharma, S. K. (2022). Ride analysis of road surface-three-wheeled vehicle-human subject interactions subjected to random excitation. *SAE International Journal of Commercial Vehicles, 15*, 299–312. https://doi.org/10.4271/02-15-03-0017
86. Sharma, S. K., Lee, J., & Jang, H.-L. (2022). Mathematical modeling and simulation of suspended equipment impact on car body modes. *Machines, 10*, 192. https://doi.org/10.3390/s22051922
87. Sharma, S. K., Sharma, R. C., Lee, J., & Jang, H.-L. (2022). Numerical and experimental analysis of DVA on the flexible-rigid rail vehicle carbody resonant vibration. *Sensors, 22*, 1922.
88. Vishwakarma, P. N., Mishra, P., & Sharma, S. K. (2022). Characterization of a magneto-rheological fluid damper a review. *Materials Today Proceedings, 56*, 2988–2994. https://doi.org/10.1016/j.matpr.2021.11.143
89. Agrahari, R., Mohapatra, S., Kumar, K.S., Mohanty, D., Kumar, R., and Merugu, R. (2024). Immuno-metabolic interplay in health and disease. Advances in Immunology and Immunotechniques. Interdisciplinary Biotechnological Advances, Das Mohapatra, A. and Sahu, P. eds., p. 239–274.
90. Sharma, S. K., Sharma, R. C., & Lee, J. (2022). In situ and experimental analysis of longitudinal load on carbody fatigue life using nonlinear damage accumulation. *International Journal of Damage Mechanics, 31*, 605–622. https://doi.org/10.1177/10567895211046043
91. Sharma, R. C., Sharma, S., Sharma, N., & Sharma, S. K. (2021). Linear and nonlinear analysis of ride and stability of a three-wheeled vehicle subjected to random and bump inputs using bond graph and simulink methodology. *SAE International Journal of Commercial Vehicles, 15*, 65–79.
92. Sharma, S.K., Sharma, R.C. (2021). *Multi-objective design optimization of locomotive nose*. SAE Technical Paper pp. 1–10. https://doi.org/10.4271/2021-01-5053
93. Lee, J., Han, J., & Sharma, S. K. (2021). Structural analysis on the separated and integrated differential gear case for the weight reduction. In *Advances in engineering design: Select proceedings of FLAME 2020* (pp. 175–181). Springer Singapore.
94. Choi, S., Lee, J., & Sharma, S. K. (2021). A study on the performance evaluation of hydraulic tank injectors. In *Advances in engineering design: Select proceedings of FLAME 2020* (pp. 183–190). Springer Singapore.
95. Sharma, R. C., Palli, S., Sharma, N., & Sharma, S. K. (2021). Ride behaviour of a four-wheel vehicle using h infinity semi-active suspension control under deterministic and random inputs. *International Journal of Vehicle Structures & Systems, 13*, 234–237.

96. Bhardawaj, S., Sharma, R. C., Sharma, S. K., & Sharma, N. (2021). On the planning and construction of railway curved track. *International Journal of Vehicle Structures & Systems, 13*, 151–159.
97. Mohapatra, S., Mohanty, D., Sharma, S., Dikshit, S., Kohli, I., Samantaray, D. P., Kumar, R., & Kathpalia, M. (2021). Biomedical application of polymeric biomaterial: Polyhydroxybutyrate. In *Bioresource utilization and management* (pp. 111–124). Apple Academic Press.
98. Sharma, S. K., Sharma, R. C., & Lee, J. (2021). Effect of rail vehicle–track coupled dynamics on fatigue failure of coil spring in a suspension system. *Applied Sciences, 11*, 2650.
99. Sharma, R. C., Sharma, S., Sharma, S. K., Sharma, N., & Singh, G. (2021). Analysis of bio-dynamic model of seated human subject and optimization of the passenger ride comfort for three-wheel vehicle using random search technique. *Proceedings of the Institution of Mechanical Engineers, Part K: Journal of Multi-body Dynamics, 235*, 106–121.
100. Sharma, S. K., & Lee, J. (2021). Crashworthiness analysis for structural stability and dynamics. *International Journal of Structural Stability and Dynamics, 21*, 2150039.
101. Cataluna Veses, R., & Shah, Z. (2015). Bio-oil production by thermal cracking in the presence of hydrogen. *Journal of Fundamentals of Renewable Energy and Applications, 06*. https://doi.org/10.4172/2090-4541.1000194
102. Shah, Z., Veses, R. C., & da Silva, R. (2016). A comparative study and analysis of two types of bio-oil samples obtained from freshwater algae and microbial treated algae. *Modern Chemistry & Applications, 4*. https://doi.org/10.4172/2329-6798.1000185
103. Shah, Z., Veses, R. C., Brum, J., Klunk, M. A., Rocha, L. A. O., Missaggia, A. B., de Oliveira, A. P., & Caetano, N. R. (2020). Sewage sludge bio-oil development and characterization. *Inventions, 5*, 36. https://doi.org/10.3390/inventions5030036
104. Shah, Z., Saberian, M., Hodgeman, D., Kinloch, I., & Vallés, C. (2023). Effect of sulfur on wood tar biopitch as a sustainable replacement for coal tar pitch binders. *ACS Applied Engineering Materials, 1*, 2567–2576. https://doi.org/10.1021/acsaenm.3c00361
105. Klunk, M. A., Dasgupta, S., Das, M., & Shah, Z. (2018). SYSTEM OF ADSORPTION OF CO2 IN COALBED. *SOUTHERN BRAZILIAN JOURNAL OF CHEMISTRY, 26*, 2–9. https://doi.org/10.48141/SBJCHEM.v26.n26.2018.7_2018.pdf
106. Mohapatra, S., Mohapatra, U., & Mishra, R. (2017). Diversification towards vegetable crops: A good option for doubling the farmer's income. *Journal of Experimental Agriculture International, 18*, 1–7.
107. Sharma, R. C., Sharma, S. K., Palli, S., Rallabandi, S. R., & Sharma, N. (2025b). Compatibility between stability and curving ability of railway vehicles using conventional and unconventional bogie frame models. *SAE International Journal of Commercial Vehicles, 18*(3), 02-18-03-0014. https://doi.org/10.4271/02-18-03-0014
108. Bansal, S., Mohapatra, S., Gohain, N., & Das, A. (2022). Production status, market integration and the law of one price: An exploration of the major maize markets of India. *Indian Journal of Economics and Development, 18*, 832–841.
109. Mohapatra, S., Samantaray, D. P., & Samantaray, S. M. (2015). Study on polyhydroxyalkanoates production using rhizospheric soil bacterial isolates of sweet potato. *Indian Journal of Science and Technology, 8*, 57–62. https://doi.org/10.17485/ijst/2015/v8iS7/64027
110. Arora, R., Trivedi, J., Mohapatra, S., & Kumar, P. (2022). Trends in probiotics on human health and industrial application. In *Advances in agricultural and industrial microbiology* (Microbial diversity and application in agroindustry) (Vol. 1, pp. 237–248). https://doi.org/10.1007/978-981-16-8918-5_13
111. Samal, S., Pati, S., Mohapatra, S., Maity, S., Tanaya, K., Devadarshini, D., & Samantaray, D. (2023). PHAs production by facultative anaerobic bacteria Bacillus cereus FM5 through submerged and solid-state fermentation under anoxic condition. *Antonie van Leeuwenhoek, International Journal of General and Molecular Microbiology, 116*, 521–529. https://doi.org/10.1007/s10482-023-01825-0

112. Biswas, N., Saha, S., Biswas, B. K., Chowdhury, M., Rahaman, A., Junghare, V., Mohapatra, S., Hazra, S., Zangrando, E., Roy Choudhury, R., & Roy Choudhury, C. (2021). The DNA- and protein-binding properties and cytotoxicity of a new copper(II) hydrazone Schiff base complex. *Journal of Coordination Chemistry, 74*, 1482–1504. https://doi.org/10.1080/00958972.2021.1913128
113. Mohapatra, S., Mishra, R., Roy, P., Yadav, K. L., & Satapathi, S. (2017). Systematic investigation and in vitro biocompatibility studies on implantable magnetic nanocomposites for hyperthermia treatment of osteoarthritic knee joints. *Journal of Materials Science, 52*, 9262–9268. https://doi.org/10.1007/s10853-017-1136-0
114. Mishra, B. B., Nayak, S. K., Mohapatra, S., & Samantaray, D. (2021). *Environmental and agricultural microbiology: Advances and applications*. https://doi.org/10.1002/9781119525899
115. Sharma, S. K., Mohapatra, S., Sharma, R. C., Alturjman, S., Altrjman, C., Mostarda, L., & Stephan, T. (2022). Retrofitting existing buildings to improve energy performance. *Sustainability (Switzerland), 14*. https://doi.org/10.3390/su14020666
116. Bandhu, S., Khot, M. B., Sharma, T., Sharma, O. P., Dasgupta, D., Mohapatra, S., Hazra, S., Khatri, O. P., & Ghosh, D. (2018). Single cell oil from oleaginous yeast grown on sugarcane bagasse-derived xylose: An approach toward novel biolubricant for low friction and wear. *ACS Sustainable Chemistry & Engineering, 6*, 275–283. https://doi.org/10.1021/acssuschemeng.7b02425
117. Kohli, I., Joshi, N. C., Mohapatra, S., & Varma, A. (2020). Extremophile—An adaptive strategy for extreme conditions and applications. *Current Genomics, 21*, 96–110. https://doi.org/10.2174/1389202921666200401105908
118. Pati, S., Mohapatra, S., Maity, S., Dash, A., & Samantaray, D. P. (2021). Polyhydroxyalkanoates for sustainable smart packaging of fruits. In *Environmental and agricultural microbiology: Advances and applications* (pp. 183–195).
119. Chatterjee, A., Nirwan, S., Mohapatra, S., Sharma, P., Agnihotri, A., & Shrivastava, N. (2022). Biochemical aspects of pathogenic variability in white rust infected Indian mustard. *Mycologia, 114*, 757–768. https://doi.org/10.1080/00275514.2022.2060007
120. Pati, S., Maity, S., Dash, A., Jema, S., Mohapatra, S., Das, S., & Samantaray, D. P. (2020). Biocompatible PHB production from bacillus species under submerged and solid-state fermentation and extraction through different downstream processing. *Current Microbiology, 77*, 1203–1209. https://doi.org/10.1007/s00284-020-01922-7
121. Agrahari, R., Sarraf, G., Joshi, N. C., Mohapatra, S., & Var, A. (2021). Insight of biopolymers and applications of polyhydroxyalkanoates. In *Microbial polymers: Applications and ecological perspectives* (pp. 177–191). https://doi.org/10.1007/978-981-16-0045-6_8
122. Kathpalia, M., Kumar, P., & Mohapatra, S. (2022). Homocysteine metabolism as a biomarker for cancer. In *Homocysteine metabolism in health and disease* (pp. 159–171). https://doi.org/10.1007/978-981-16-6867-8_8
123. Bharatiy, S. K., Hazra, M., Paul, M., Mohapatra, S., Samantaray, D., Dubey, R. C., Sanyal, S., Datta, S., & Hazra, S. (2016). In silico designing of an industrially sustainable carbonic anhydrase using molecular dynamics simulation. *ACS Omega, 1*, 1081–1103. https://doi.org/10.1021/acsomega.6b00041
124. Sarkar, M. S., Segu, H., Bhaskar, J. V., Jakher, R., Mohapatra, S., Shalini, K., Shivaji, S., & Reddy, P. A. (2018). Ecological preferences of large carnivores in remote, high-altitude protected areas: Insights from Buxa Tiger Reserve. *India. ORYX, 52*, 66–77. https://doi.org/10.1017/S0030605317000060
125. Pati, S., Dash, A., Maity, S., Pattnaik, S., Mohapatra, S., & Samantaray, D. P. (2021). Nano biofertilizer for sustainable agriculture. In *Agriculturally important microorganisms: Mechanisms and applications for sustainable agriculture* (pp. 201–208). https://doi.org/10.1201/9781003245841-10
126. Mohanty, D., Misra, S., Mohapatra, S., & Sahu, P. S. (2018). Prebiotics and synbiotics: Recent concepts in nutrition. *Food Bioscience, 26*, 152–160. https://doi.org/10.1016/j.fbio.2018.10.008

127. Kumar, R., Kumari, R., Prasad, C., Deep, A., Singh, N., Namtak, S., Tiwari, V., Merugu, R., Mohapatra, S., & Singh, R. (2021). Influence of water quality on the diversity of macroinvertebrates in the Mandakini River in India. *Water Science and Technology: Water Supply, 21*, 1843–1860. https://doi.org/10.2166/WS.2021.020
128. Mohanty, D., Jena, R., Choudhury, P. K., Pattnaik, R., Mohapatra, S., & Saini, M. R. (2016). Milk derived antimicrobial bioactive peptides: A review. *International Journal of Food Properties, 19*, 837–846. https://doi.org/10.1080/10942912.2015.1048356
129. Mohapatra, S., Rath, S. N., Pradhan, S. K., Samantaray, D. P., & Rath, C. C. (2016). Secondary structural models (16S rRNA) of polyhydroxyalkanoates producing bacillus species isolated from different rhizospheric soil: Phylogenetics and chemical analysis. *International Journal Bioautomation, 20*, 329–338.
130. Mohapatra, S., Sarkar, B., Samantaray, D. P., Daware, A., Maity, S., Pattnaik, S., & Bhattacharjee, S. (2017). Bioconversion of fish solid waste into PHB using Bacillus subtilis based submerged fermentation process. *Environmental Technology (United Kingdom), 38*, 3201–3208. https://doi.org/10.1080/09593330.2017.1291759
131. Misra, S., Mohanty, D., & Mohapatra, S. (2021). Food-based probiotics: Functional dietary ingredients. In *Probiotics: Advanced food and health applications* (pp. 257–275). https://doi.org/10.1016/B978-0-323-85170-1.00006-3
132. Kumar, R., Kumari, R., Prasad, C., Tiwari, V., Singh, N., Mohapatra, S., Merugu, R., Namtak, S., & Deep, A. (2020). Phytoplankton diversity in relation to physicochemical attributes and water quality of Mandakini River, Garhwal Himalaya. *Environmental Monitoring and Assessment, 192*. https://doi.org/10.1007/s10661-020-08768-3
133. Mohapatra, S., Vishwakarma, K., Joshi, N. C., Maity, S., Kumar, R., Ramchander, M., Pattnaik, S., & Samantaray, D. P. (2021). A review on PHAs: The future biopolymer. In *Environmental and agricultural microbiology: Advances and applications* (pp. 83–117).
134. Sahoo, N., Pattnaik, S., Pattnaik, M., & Mohapatra, S. (2021). Methane emission and strategies for mitigation in livestock. In *Environmental and agricultural microbiology: Advances and applications* (pp. 257–274).
135. Mohanty, D. P., Mohapatra, S., Misra, S., & Sahu, P. S. (2016). Milk derived bioactive peptides and their impact on human health—A review. *Saudi Journal of Biological Sciences, 23*, 577–583. https://doi.org/10.1016/j.sjbs.2015.06.005
136. Kohli, I., Mohapatra, S., Kumar, P., Goel, A., Varma, A., & Joshi, N. C. (2022). Role of bacterial endophytes in the promotion of plant growth. In *Bacterial endophytes for sustainable agriculture and environmental management* (pp. 235–253). https://doi.org/10.1007/978-981-16-4497-9_12
137. Samantaray, D., Mohapatra, S., & Mishra, B. B. (2014). Microbial bioremediation of industrial effluents. In *Microbial biodegradation and bioremediation* (pp. 326–339). https://doi.org/10.1016/B978-0-12-800021-2.00014-5
138. Adeoye, A. O., Quadri, R. O., & Lawal, O. S. (2020). Pyrolysis of biomass as a suitable alternative to fossil fuel energy in Nigeria: An overview. *Journal Of Wastes and Biomass Management, 3*(1), 27–30. https://doi.org/10.26480/jwbm.01.2021.27.30
139. Mohapatra, S., Mohanta, P. R., Sarkar, B., Daware, A., Kumar, C., & Samantaray, D. P. (2017). Production of Polyhydroxyalkanoates (PHAs) by bacillus strain isolated from waste water and its biochemical characterization. *Proceedings of the National Academy of Sciences India Section B - Biological Sciences, 87*, 459–466. https://doi.org/10.1007/s40011-015-0626-6
140. Tomer, P., Shroti, G. K., Mohapatra, S., Ghosh, D., Jaiswal, S., Garg, D., Lahiri, D., Kumar, N., Khatri, O. P., & Hazra, S. (2023). Fabrication of HA Nano-crystal reinforced PHAs based composites for orthopedic fracture-fixing accessories using sustainable resources. *Environment, Development and Sustainability*. https://doi.org/10.1007/s10668-023-03912-8
141. Bandekar, D., Mohapatra, S., Hazra, M., Hazra, S., & Biswas, S. (2021). N-terminal truncation of VC0395_0300 protein from vibrio cholerae does not lead to loss of diguanylate cyclase activity. *Biophysical Chemistry, 268*. https://doi.org/10.1016/j.bpc.2020.106493

142. Vishwakarma, K., Kumar, N., Shandilya, C., Mohapatra, S., Bhayana, S., & Varma, A. (2020). Revisiting plant–microbe interactions and microbial consortia application for enhancing sustainable agriculture: A review. *Frontiers in Microbiology, 11*. https://doi.org/10.3389/fmicb.2020.560406
143. Mohapatra, S., Maity, S., Dash, H. R., Das, S., Pattnaik, S., Rath, C. C., & Samantaray, D. (2017). Bacillus and biopolymer: Prospects and challenges. *Biochemistry and Biophysics Reports, 12*, 206–213. https://doi.org/10.1016/j.bbrep.2017.10.001
144. Kumar, R., Merugu, R., Mohapatra, S., & Sharma, S. (2022). Extremophiles: Life of microorganisms in extreme environments. In *Extremophiles: General and plant biomass based biorefinery* (pp. 43–66). https://doi.org/10.1201/9781003014188-3
145. Pradhan, A., Pahari, A., Mohapatra, S., & Mishra, B. B. (2017). Phosphate-solubilizing microorganisms in sustainable agriculture: Genetic mechanism and application. In *Microorganisms for sustainability* (pp. 81–97). https://doi.org/10.1007/978-981-10-7380-9_5
146. Devadarshini, D., Mohapatra, S., Pati, S., Maity, S., Rath, C. C., Jena, P. K., & Samantaray, D. (2023). Evaluation of PHAs production by mixed bacterial culture under submerged fermentation. *Biologia (Bratisl), 78*, 895–902. https://doi.org/10.1007/s11756-022-01302-5
147. Agarwal, K. M., Mohapatra, S., Sharma, P., Sharma, S., Bhatia, D., & Mishra, A. (2020). Study and overview of the novel corona virus disease (COVID-19). *Sensors International, 1*. https://doi.org/10.1016/j.sintl.2020.100037
148. Tanaya, K., Mohapatra, S., Maity, S., & Samantaray, D. P. (2023). Fluorescent-pigmented microbial polymer nanoparticles: An approach for replacement of bioimaging agent in the drug delivery system. *Applied Biochemistry and Microbiology, 59*, 378–383. https://doi.org/10.1134/S0003683823030171
149. Pattnaik, S., Dash, D., Mohapatra, S., Pati, S., Devadarshini, D., Samal, S., Pattnaik, M., Maity, S., Mishra, S. K., & Samantaray, D. (2024). Reclamation of chromium-contaminated soil by native Cr(VI)-reducing and PHA-accumulating Bacillus aryabhattai CTSI-07. *International Microbiology, 27*, 731–742. https://doi.org/10.1007/s10123-023-00421-6
150. Mohapatra, S., Mishra, Y., & Dash, L. (2020). Covid-19: Analysing the legal nuances of the lockdown order. *Indian Journal of Forensic Medicine and Toxicology, 14*, 612–616. https://doi.org/10.37506/ijfmt.v14i4.11553
151. Pati, S., Mohanty, M. K., Mohapatra, S., & Samantaray, D. (2022). Bioethanol production by Enterobacter hormaechei through carbon monoxide-rich syngas fermentation. *International Journal of Energy Research, 46*, 20096–20106. https://doi.org/10.1002/er.8678
152. Maity, S., Das, S., Mohapatra, S., Tripathi, A. D., Akthar, J., Pati, S., Pattnaik, S., & Samantaray, D. P. (2020). Growth associated polyhydroxybutyrate production by the novel Zobellellae tiwanensis strain DD5 from banana peels under submerged fermentation. *International Journal of Biological Macromolecules, 153*, 461–469. https://doi.org/10.1016/j.ijbiomac.2020.03.004
153. Takkar, S., Shandilya, C., Agrahari, R., Chaurasia, A., Vishwakarma, K., Mohapatra, S., Varma, A., & Mishra, A. (2022). Green technology: Phytoremediation for pesticide pollution. In *Phytoremediation technology for the removal of heavy metals and other contaminants from soil and water* (pp. 353–375). https://doi.org/10.1016/B978-0-323-85763-5.00008-8
154. Bandekar, D., Chouhan, O. P., Mohapatra, S., Hazra, M., Hazra, S., & Biswas, S. (2017). Putative protein VC0395_0300 from vibrio cholerae is a diguanylate cyclase with a role in biofilm formation. *Microbiological Research, 202*, 61–70. https://doi.org/10.1016/j.micres.2017.05.003
155. Pattnaik, S., Dash, D., Mohapatra, S., Pattnaik, M., Marandi, A. K., Das, S., & Samantaray, D. P. (2020). Improvement of rice plant productivity by native Cr(VI) reducing and plant growth promoting soil bacteria Enterobacter cloacae. *Chemosphere, 240*. https://doi.org/10.1016/j.chemosphere.2019.124895

156. Mohapatra, S., Samantaray, D. P., Samantaray, S. M., Mishra, B. B., Das, S., Majumdar, S., Pradhan, S. K., Rath, S. N., Rath, C. C., Akthar, J., & Achary, K. G. (2016). Structural and thermal characterization of PHAs produced by Lysinibacillus sp. through submerged fermentation process. *International Journal of Biological Macromolecules, 93*, 1161–1167. https://doi.org/10.1016/j.ijbiomac.2016.09.077
157. Sharma, S. K., Kumar, A., & Sharma, R. C. (2014). Challenges in railway vehicle modeling and simulations. In *International conference on newest drifts in mechanical engineering ICNDME-2014* (pp. 453–459).
158. Sharma, S. K., & Kumar, A. (2014). A comparative study of Indian and worldwide railways. *International Journal of Mechanical Engineering and Robotics Research, 1*, 114–120.
159. Sharma, S. K., Sharma, R. C., Kumar, A., & Palli, S. (2015). Challenges in rail vehicle-track modeling and simulation. *International Journal of Vehicle Structures & Systems, 7*, 1.
160. Palli, S., Sharma, R. C., Sharma, S. K., Muddada, V., & Sharma, N. (2020). A case study on noise pollution and its effects. *Journal of Critical Reviews, 7*, 261–267.
161. Sharma, R. C., Sharma, S. K., Sharma, N., & Sharma, S. (2020). Analysis of ride and stability of an ICF railway coach. *International Journal of Vehicle Noise and Vibration, 16*, 127–144.
162. Palli, S., Sharma, R. C., Sharma, S. K., & Chintada, V. B. (2020). On methods used for setting the curve for railway tracks. *Journal of Critical Reviews, 7*, 241–246.
163. Lee, J., & Sharma, S. K. (2020). Numerical investigation of critical speed analysis of high-speed rail vehicle. 한국정밀공학회 학술발표대회 논문집, *696*.
164. Sharma, S., Sharma, R. C., Sharma, S. K., Sharma, N., Palli, S., & Bhardawaj, S. (2020). Vibration isolation of the quarter car model of road vehicle system using dynamic vibration absorber. *International Journal of Vehicle Structures & Systems, 12*, 513–516. https://doi.org/10.4273/IJVSS.12.5.05
165. Acharya, A., Gahlaut, U., Sharma, K., Sharma, S. K., Vishwakarma, P. N., & Phanden, R. K. (2020). Crashworthiness analysis of a thin-walled structure in the frontal part of automotive chassis. *International Journal of Vehicle Structures & Systems, 12*, 517–520.
166. Sharma, S. K., & Lee, J. (2020). Finite element analysis of a fishplate rail joint in extreme environment condition. *International Journal of Vehicle Structures & Systems, 12*, 503–506.
167. Bhardawaj, S., Sharma, R. C., & Sharma, S. K. (2020). Development of multibody dynamical using MR damper based semi-active bio-inspired chaotic fruit fly and fuzzy logic hybrid suspension control for rail vehicle system. *Proceedings of the Institution of Mechanical Engineers, Part K: Journal of Multi-body Dynamics, 234*, 723–744.
168. Sharma, S. K., & Lee, J. (2020). Design and development of smart semi active suspension for nonlinear rail vehicle vibration reduction. *International Journal of Structural Stability and Dynamics, 20*, 2050120.
169. Sharma, S. K., Phan, H., & Lee, J. (2020). An application study on road surface monitoring using DTW based image processing and ultrasonic sensors. *Applied Sciences, 10*, 4490.
170. Sharma, R. C., Sharma, S., Sharma, S. K., & Sharma, N. (2020). Analysis of generalized force and its influence on ride and stability of railway vehicle. *Noise & Vibration Worldwide, 51*, 95–109.
171. Bhardawaj, S., Sharma, R. C., & Sharma, S. K. (2020). Ride analysis of track-vehicle-human body interaction subjected to random excitation. *Journal of the Chinese Society of Mechanical Engineers, 41*, 227–236.
172. Bhardawaj, S., Sharma, R. C., & Sharma, S. K. (2020). Development in the modeling of rail vehicle system for the analysis of lateral stability. *Materials Today Proceedings, 25*, 610–619.
173. Sharma, S. K., Sharma, R. C., & Sharma, N. (2020). Combined multi-body-system and finite element analysis of a rail locomotive crashworthiness. *International Journal of Vehicle Structures & Systems, 12*, 428–435.

174. Adeoye, A. O., Quadri, R. O., & Lawal, O. S. (2025b). Sustainable remediation of vancomycin polluted water using pyrolysis biochar of pressed oil palm fruit fibre. *Journal of the Nigerian Society of Physical Sciences*, 2558. https://doi.org/10.46481/jnsps.2025.2558
175. Bhardawaj, S., Sharma, R. C., & Sharma, S. K. (2020). Analysis of frontal car crash characteristics using ANSYS. *Materials Today Proceedings, 25*, 898–902.
176. Bhardawaj, S., Sharma, R. C., & Sharma, S. K. (2019). A survey of railway track modelling. *International Journal of Vehicle Structures & Systems, 11*, 508–518.
177. Bhardawaj, S., Sharma, R. C., & Sharma, S. K. (2019). Development and advancement in the wheel-rail rolling contact mechanics. In *IOP conference series: Materials science and engineering* (p. 012034). IOP Publishing.
178. Choppara, R. K., Sharma, R. C., Sharma, S. K., & Gupta, T. (2019). Aero dynamic cross wind analysis of locomotive. In *IOP conference series: Materials science and engineering* (p. 012035). IOP Publishing.
179. Sharma, S. K., & Sharma, R. C. (2019). Pothole detection and warning system for Indian roads. In *Advances in interdisciplinary engineering: Select proceedings of FLAME 2018* (pp. 511–519). Springer Singapore.
180. Sinha, A. K., Sengupta, A., Gandhi, H., Bansal, P., Agarwal, K. M., Sharma, S. K., Sharma, R. C., & Sharma, S. K. (2019). Performance enhancement of an all-terrain vehicle by optimizing steering, powertrain and brakes. In *Advances in engineering design: Select proceedings of FLAME 2018* (pp. 207–215). Springer Singapore.
181. Goyal, S., Anand, C. S., Sharma, S. K., & Sharma, R. C. (2019). Crashworthiness analysis of foam filled star shape polygon of thin-walled structure. *Thin-Walled Structures, 144*, 106312.
182. Goswami, B., Rathi, A., Sayeed, S., Das, P., Sharma, R. C., & Sharma, S. K. (2019). Optimization design for aerodynamic elements of Indian locomotive of passenger train. In *Advances in engineering design: Select proceedings of FLAME 2018* (pp. 663–673). Springer Singapore.
183. Sharma, S. K., Saini, U., & Kumar, A. (2019). Semi-active control to reduce lateral vibration of passenger rail vehicle using disturbance rejection and continuous state damper controllers. *Journal of Vibration Engineering & Technologies, 7*, 117–129.
184. Sharma, S. K. (2019). Multibody analysis of longitudinal train dynamics on the passenger ride performance due to brake application. *Proceedings of the Institution of Mechanical Engineers, Part K: Journal of Multi-body Dynamics, 233*, 266–279.
185. Sharma, S. (2018). Vibration control in quarter-car model with magnetorheological (MR) dampers using Bingham model. *Journal of Applied Mechanical Engineering, 7*, 1000299.
186. Adeoye, A. O., Quadri, R. O., & Lawal, O. S. (2025). Assessment of physicochemical properties of Lafarge cement towards achieving a sustainable society. *Proceedings of the Nigerian Society of Physical Sciences*, 180. https://doi.org/10.61298/pnspsc.2025.2.180
187. Sharma, S. K., & Kumar, A. (2018). Impact of longitudinal train dynamics on train operations: A simulation-based study. *Journal of Vibration Engineering & Technologies, 6*, 197–203.
188. Sharma, S. K., & Sharma, R. C. (2018). An investigation of a locomotive structural crashworthiness using finite element simulation. *SAE International Journal of Commercial Vehicles, 11*, 235–243.
189. Sharma, R. C., Sharma, S. K., & Palli, S. (2018). Rail vehicle modelling and simulation using Lagrangian method. *International Journal of Vehicle Structures and Systems, 10*, 188–194.
190. Sharma, R. C., & Sharma, S. K. (2018). Sensitivity analysis of three-wheel vehicle's suspension parameters influencing ride behavior. *Noise & Vibration Worldwide, 49*, 272–280.
191. Palli, S., Koona, R., Sharma, S. K., & Sharma, R. C. (2018). A review on dynamic analysis of rail vehicle coach. *International Journal of Vehicle Structures & Systems (IJVSS), 10*.
192. Sharma, S. K., & Sharma, R. C. (2018). Simulation of quarter-car model with magnetorheological dampers for ride quality improvement. *International Journal of Vehicle Structures & Systems, 10*, 169–173.

193. Sharma, S. K., & Kumar, A. (2018). Disturbance rejection and force-tracking controller of nonlinear lateral vibrations in passenger rail vehicle using magnetorheological fluid damper. *Journal of Intelligent Material Systems and Structures, 29*, 279–297.
194. Sharma, S. K., & Kumar, A. (2018). Ride comfort of a higher speed rail vehicle using a magnetorheological suspension system. *Proceedings of the Institution of Mechanical Engineers, Part K: Journal of Multi-Body Dynamics, 232*, 32–48.
195. Shukla, S., Kumar, U., Sharma, S. K., Gupta, P., & Kumar, A. (2017). Light weight freight rolling stock bogie frame: Design methodology validated with field oscillation trials. *International Journal of Vehicle Structures & Systems, 9*, 245–250.
196. Sharma, R. C., Palli, S., Sharma, S. K., & Roy, M. (2017). Modernization of railway track with composite sleepers. *International Journal of Vehicle Structures & Systems, 9*, 321–329.
197. Sharma, S. K., & Kumar, A. (2017). Ride performance of a high speed rail vehicle using controlled semi active suspension system. *Smart Materials and Structures, 26*, 055026.
198. Sharma, S. K., & Kumar, A. (2017). Impact of electric locomotive traction of the passenger vehicle ride quality in longitudinal train dynamics in the context of Indian railways. *Mechanics & Industry, 18*, 222.
199. Sharma, S. K. (2013). Zero energy building envelope components: A review. *International Journal of Engineering Research and Applications, 3*, 662–675.
200. Sharma, S. K., & Kumar, A. (2016). Dynamics analysis of wheel rail contact using FEA. *Procedia Engineering, 144*, 1119–1128.
201. Sharma, S. K., & Chaturvedi, S. (2016). Jerk analysis in rail vehicle dynamics. *Perspectives in Science (Netherlands), 8*, 648–650.
202. Sharma, S. K., & Kumar, A. (2016). The impact of a rigid-flexible system on the ride quality of passenger bogies using a flexible carbody. In *Proceedings of the third international conference on railway technology: Research, development and maintenance* (p. 87). Civil-Comp Press.
203. Lavania, S., & Sharma, S. K. (2011). An explicit approach to compare crosstalk noise and delay in VLSI RLC interconnect modeled with skin effect with step and ramp input. *Journal of VLSI Design Tools & Technology, 1*, 1–8.
204. Palli, S., Koona, R., Sharma, R. C., & Muddada, V. (2015). Dynamic analysis of indian railway integral coach factory bogie. *International Journal of Vehicle Structures and Systems, 7*, 16–20. https://doi.org/10.4273/ijvss.7.1.03
205. Palli, S., & Koona, R. (2015). Analyses of dynamic response of a railway bogie. *International Journal of Vehicle Noise and Vibration, 11*, 103–113. https://doi.org/10.1504/IJVNV.2015.070015
206. Swathi, K., & Srihari, P. (2015). Heat transfer enhancement ina tube using rectangular strip inserts. *International Journal of Applied Engineering Research, 10*, 41532–41544.
207. Palli, S., Sharma, R. C., & Rao, P. P. D. (2017). Dynamic behaviour of a 7 DoF passenger car model. *International Journal of Vehicle Structures and Systems, 9*, 57–63. https://doi.org/10.4273/ijvss.9.1.12
208. Sharma, R. C., Palli, S., & Koona, R. (2017). Stress and vibrational analysis of an Indian railway RCF bogie. *International Journal of Vehicle Structures and Systems, 9*, 296–302. https://doi.org/10.4273/ijvss.9.5.06
209. Sharma, R. C., Avesh, M., Sharma, S. K., Sharma, S., Palli, S., & Sharma, N. (2020). Linear and non-linear stability analysis of a constrained railway wheel axle. *International Journal of Vehicle Structures and Systems, 12*, 128–133. https://doi.org/10.4273/ijvss.12.2.04
210. Palli, S., Dontikurti, R., Sharma, R. C., & Sharma, N. (2021). Analysis of dynamic response of a railway locomotive using ansys. *International Journal of Vehicle Structures and Systems, 13*, 250–255. https://doi.org/10.4273/ijvss.13.2.21
211. Sharma, R. C., Sharma, N., Singh, G., & Palli, S. (2021). Modeling and simulation of human body-vehicle-track system for the investigation of ride comfort. *SAE International Journal of Commercial Vehicles, 15*. https://doi.org/10.4271/02-15-02-0008

212. Sharma, R. C., Palli, S., Avesh, M., & Sharma, N. (2021). Vibration isolation of railway vehicle car body using semi-active suspension. *International Journal of Vehicle Structures and Systems, 13*, 482–487. https://doi.org/10.4273/ijvss.13.4.17
213. Gopala Rao, L. V. V., Sharma, R. C., Satyanarayana, V. S. V., & Palli, S. (2022). Stochastic optimal preview control response of a quarter-car nonlinear suspension model using spectral decomposition method. *Noise and Vibration Worldwide, 53*, 225–232. https://doi.org/10.1177/09574565221093234
214. Palli, S., Duppala, A., Sharma, R. C., & Rao, L. V. V. G. (2022). Dynamic simulation of automotive vehicle suspension using MATLAB Simulink. *International Journal of Vehicle Structures and Systems, 14*, 310–314. https://doi.org/10.4273/ijvss.14.3.04
215. Satyanarayana, V. S. V., Sharma, R. C., Sateesh, B., Gopala Rao, L. V. V., Mohan Rao, N., & Palli, S. (2022). Optimum response of a nonlinear passive vehicle suspension system under random road excitations. *SAE International Journal of Commercial Vehicles, 16*. https://doi.org/10.4271/02-16-01-0004
216. Lokanadham, D., Sharma, R. C., Palli, S., & Bhardawaj, S. (2022). Wear rate modelling and analysis of limestone slurry particulate composites using the fuzzy method. *International Journal on Recent and Innovation Trends in Computing and Communication, 10*, 133–143. https://doi.org/10.17762/ijritcc.v10i1s.5818
217. Sharma, R. C., Palli, S., Gopala Rao, L. V. V., Duppala, A., & Sharma, S. K. (2022). Four-wheel vehicle response under bump, pothole, harmonic, and random excitations using bond graph/Simulink technique. *SAE International Journal of Commercial Vehicles, 16*, 129–139. https://doi.org/10.4271/02-16-02-0008
218. Sharma, R. C., Gopala Rao, L. V. V., Sharma, S. K., Palli, S., & Satyanarayana, V. S. V. (2022). Analysis of lateral stability and ride of an Indian railway constrained dual-axle bogie frame. *SAE International Journal of Commercial Vehicles, 16*, 213–228. https://doi.org/10.4271/02-16-02-0014
219. Sharma, R. C., Palli, S., Bhardawaj, S., & Rao, L. V. V. G. (2022). Advances in human bio-dynamic modeling for analysis of whole-body vibrations. *International Journal of Vehicle Structures and Systems, 14*, 503–511. https://doi.org/10.4273/ijvss.14.4.17
220. Sharma, R. C., Palli, S., & Sharma, S. K. (2023). Ride analysis of railway vehicle considering rigidity and flexibility of the carbody. *Journal of the Chinese Institute of Engineers, Transactions of the Chinese Institute of Engineers, Series A, 46*, 355–366. https://doi.org/10.1080/02533839.2023.2194918
221. Sharma, R. C., Palli, S., & Bhardawaj, S. (2023). A review of wheel-rail dynamic behavior on curved tracks. *International Journal of Vehicle Structures and Systems, 15*, 19–25. https://doi.org/10.4273/ijvss.15.1.04
222. Jyotiprasad, G., Dowluru, S., Rallabandi, S. R., Sharma, N., Sharma, R. C., Palli, S. (2023). Thermal-structural and macromechanical behavior analysis of graphite-based laminated composite. SAE Technical Paper 2023-01-5041. https://doi.org/10.4271/2023-01-5041.
223. Saikiran, C., Palli, S., Bondala, R., Sharma, R. C., Sreeramulu, D., & Dontikurti, R. (2023). Experimental investigation on thermal behaviour of silicon carbide (SiC) based polymer composites under hygrothermal conditions. *International Journal of Vehicle Structures and Systems, 15*, 222–226. https://doi.org/10.4273/ijvss.15.2.15
224. Singh, B. J., Palli, S., Kumar, P., & Bahl, S. (2023). Behavioural investigation of hardness profile for dissimilar TIG-weldment between P22 and P91 steels. *International Journal of Vehicle Structures and Systems, 15*, 31–36. https://doi.org/10.4273/ijvss.15.1.06
225. Naveen, Y., Lokanadham, D., Naidu, D. R., Sharma, R. C., Palli, S., & Lila, M. K. (2023). An experimental study on the influence of blended Karanja biodiesel on diesel engine characteristics. *International Journal of Vehicle Structures and Systems, 15*, 154–160. https://doi.org/10.4273/ijvss.15.2.02

226. Bhardawaj, S., Jha, A. K., Sharma, R. C., Palli, S., Lokanadham, D., Sharma, N., & Dhiman, N. (2023). Finite element analysis of two disk rotor system used in automobile turbochargers. *International Journal of Vehicle Structures and Systems., 15*, 501–504. https://doi.org/10.4273/ijvss.15.4.09
227. Sharma, R. C., Palli, S., Bhardawaj, S., Jha, A. K., Duppala, A., Sharma, N., Sharma, S., & Dhiman, N. (2023). Dynamic behaviour of railway vehicle under bump and pothole track irregularities. *International Journal of Vehicle Structures and Systems, 15*, 566–569. https://doi.org/10.4273/ijvss.15.4.22
228. Raju, S., Palli, S., Devi Prasad, P., Menda, V. R., & Ramakrishna, B. (2024). A hybrid AHP-TOPSIS, MOORA technique for multi-objective optimization of thermal, mechanical, and water absorption behavior of epoxy/hemp, pine apple, and palm fiber composites. *Journal of the Chinese Institute of Engineers, Transactions of the Chinese Institute of Engineers, Series A, 47*, 23–36. https://doi.org/10.1080/02533839.2023.2274092
229. Raju Rallabandi, S., Pilla, D. P., Dowluru, S., Palli, S., Sharma, N., Sharma, S. K., & Sharma, R. C. (2024). Critical evaluation of epoxy-hemp-pineapple-palm fiber composites using hybrid AHM-TOPSIS technique for sustainable structural applications. *Journal of the Chinese Institute of Engineers, Transactions of the Chinese Institute of Engineers, Series A, 47*, 325–336. https://doi.org/10.1080/02533839.2024.2308250
230. Gopala Rao, L. V. V., Satyanarayana, V. S. V., Sharma, R. C., Palli, S., Raturi, A., Kharola, A., & Sharma, A. (2024). Preview control of the random response of a full car vehicle model traversing a rough road. *International Journal of Vehicle Noise and Vibration, 20*, 70–88. https://doi.org/10.1504/IJVNV.2024.138105
231. Sharma, A., Sharma, V. S., Gupta, S., Sharma, R. C., Palli, S., & Sharma, N. (2024). Additive manufacturing in the COVID-19 pandemic: Equipment and challenges? *Journal of Micromanufacturing*. https://doi.org/10.1177/25165984241228083
232. Bammidi, R., Madivada, H., Dowluru, S., Palli, S., Sharma, R. C., & Muddada, V. (2024). Mechano-tribological investigations of Al-%Wt. of boron nitride as metal-ceramic composites. *AIP Conference Proceedings, 2978*, 020017. https://doi.org/10.1063/5.0182965
233. Ranganath, B. A., Duppala, A., Palli, S., Rao, L. V. V., & Sharma, R. C. (2024). Numerical analysis and experimental study of FRP composites for damping applications. *AIP Conference Proceedings, 2978*, 060001. https://doi.org/10.1063/5.0184802
234. Sharma, R. C., Palli, S., Duppala, A., Bhardawaj, S., Sharma, S. K., Sharma, N., & Sharma, A. (2024). Rigid body analysis of ride behaviour of roadway vehicle using lagrangian dynamics. *International Journal of Vehicle Structures and Systems, 16*, 74–78. https://doi.org/10.4273/ijvss.16.1.15
235. Menda, V., Dora, S., Palli, S., Rallabandi, S. R., Sharma, R. C., Sharma, N., & Bondala, R. (2024). Mechanical and wear characteristics of aluminium-7075/graphene/TiB2 prepared by stir casting. *Engineering Research Express, 6*. https://doi.org/10.1088/2631-8695/ad58a8
236. Kumar, U., Kasvekar, R., Sharma, S. K., & Upadhyay, R. K. (2022). Wear of wheels and axle in locomotive and measures taken by Indian railway. In *Energy, environment, and sustainability* (pp. 77–96). https://doi.org/10.1007/978-981-16-8337-4_5

Bananature Sustainable Carry Bags from Banana Pseudostems for a Plastic-Free Future in Industry 5.0

Roopsandeep Bammidi, Sreeramulu Dowluru, Pavan Kumar Rejeti, Deviprasad Pilla, Emmanuel Oghenero Emojevu, Patience Agnes EBIGHE, and Nura Garba

1 Introduction

Plastic waste has become an urgent environmental issue with serious implications for ecosystems around the world with many barriers to their sustainability. Plastic products, especially one-time use products, particularly bags and packaging, are huge contributors to pollution as they take hundreds of years to entirely degrade. These plastics are deposited within landfills, oceans, and other natural habitats in significant proportions, affecting marine and terrestrial life. Small plastic pieces that float in the water are invariably eaten by animals which often results in skewed health and even death. Additionally, Plastics contain a high amount of energy which is consumed during their production as well as during disposal which greatly adds to the greenhouse gasses and climate change. Given these effects, there is a dire need to come up with safer and more sustainable substitutes that would minimize the use of plastic without compromising its effectiveness.

One of the advances emerging to tackle this issue is to develop biodegradable materials using agricultural waste that are not only eco-friendly but also allow

R. Bammidi · S. Dowluru · P. K. Rejeti (✉) · D. Pilla
Department of Mechanical Engineering, Aditya Institute of Technology and Management, Tekkali, Andhra Pradesh, India

E. O. Emojevu
Department of Chemistry, University of Benin, Benin, Nigeria

P. A. EBIGHE
Department of Chemistry, Toulouse INP-ENSIACET, Toulouse, France

N. Garba
Department of Environmental and Resources Management, Usmanu Danfodiyo University, Sokoto, Nigeria
e-mail: nura.garba@udusok.edu.ng

Y. Kumar Singla et al. (eds.), *Digital Manufacturing in Industry 5.0*, https://doi.org/10.1007/978-3-031-91500-0_29

residues destined to disposal to be productively employed. In this regard, fibers from banana pseudostem, which are collected from the banana plants' petioles which would otherwise be leaved to waste, are introduced as suitable and novel materials. These fibers have good tensile strength, elasticity and are biodegradable and therefore can be used for a variety of applications where plastic is commonly used. Banana is among the most widely grown crops in the world where the focus is on the fruit crop. However, the pseudostems which remain after the harvesting of the fruit are usually thrown away or set on fire, creating waste and causing destruction to the environment. Tapping into these pseudo-stems for fiber provides a biodegradable substitute to plastic materials and adds extra economic benefit to banana farming, particularly in the regions where banana is a cash crop. These analyses emphasize a variety of advanced engineering techniques that contribute to the development of smart manufacturing and Industry 5.0. Studies like Shah et al. [1] and Singla et al. [2] examine the effects of material science advancements on welding and sustainability, key components for industrial automation and optimized production processes. Additionally, research on solar energy systems and tribological studies [3–6] align with the sustainable energy goals of Industry 5.0. Additionally, advancements in digital twin technology for improving logistics [7], the role of extremophiles in enhancing sustainable agriculture [8, 9], and bioremediation of contaminated soils [10, 11] highlight the role of digital innovation and environmental consciousness in next-generation manufacturing. These studies offer diverse insights into how microbiological advancements and environmental sustainability are being harnessed for smart and sustainable industrial applications, resonating with the principles of Industry 5.0 [12, 13]. The integration of alloying elements to enhance wear resistance and use of nano fly ash [14–16] supports the development of materials with improved durability for high-tech manufacturing applications. Moreover, research on the surface integrity of machined materials and the impact of nanoparticulates in lubrication show how material engineering can improve efficiency in production processes. These studies collectively enhance the scope of smart manufacturing by combining material science, energy efficiency, and automation, advancing the transition toward Industry 5.0 [17–21].

Studies from earlier years explore dynamic simulations [22, 23], improving suspension systems, and enhancing material wear resistance, which collectively reflect advancements in digitalization, automation, and sustainable manufacturing practices. These studies highlight a multidisciplinary approach to improving vehicle systems and material performance in line with the goals of Industry 5.0.

These studies explore a range of advanced techniques and sustainable practices that align with the goals of smart manufacturing and Industry 5.0. Adeoye et al. (2024) investigate the valorization of lignocellulosic waste, highlighting the potential of bio-oil production through pyrolysis and ultrasonication for biofuel generation [24]. Further research by Adeoye et al. focuses on upgrading crude bio-oil [25, 26] and sustainable energy production from biomass [27], providing insights into thermochemical and biochemical conversions. These approaches underscore the integration of clean energy solutions into digital manufacturing processes [28–33]. Studies on catalytic pyrolysis and nanomaterials for enhanced biomass conversion offer pathways for optimizing manufacturing efficiency [34–36]. Additionally, research on pyrolysis of plastic waste and rail projects introduces circular economy concepts and environmental sustainability, both essential to the evolution of Industry 5.0 [37–45].

In parallel, investigation on fiber surface functionalization and dye fastness adds to the understanding of material innovation in manufacturing [46–48]. These studies contribute to the body of knowledge driving digital transformation and environmental consciousness in manufacturing [37, 39, 41–43, 49, 50].

The contributions from Sunil et al. [51–61] span critical advancements in transportation systems, materials science, and engineering technology, all of which have applications in smart manufacturing and Industry 5.0. Recent contributions from 2024 focus on the adoption of electric vehicles [23, 62–74], modeling of intelligent transportation systems [75–85], and vibration mitigation in marine and rail propulsion systems [86–95]. Recent advancements in sustainable fuel production have focused on various bio-oil sources and carbon capture techniques. Studies have examined bio-oil production from thermal cracking in the presence of hydrogen [96], CO_2 adsorption systems in coal beds, and the pyrolysis of coffee and eucalyptus sawdust [97–99]. Additional research has explored the use of sulfur to enhance wood tar biopitch as a sustainable binder, the characterization of sewage sludge bio-oil, and the analysis of bio-oil from algae, comparing freshwater and microbial-treated samples [100]. These efforts collectively contribute to the development of renewable energy and environmental sustainability [37, 49, 50, 101–103].

These researches highlight innovative methods in microbiology, environmental sustainability, and materials science, which have significant implications for smart manufacturing and Industry 5.0 [28–30, 32, 33, 104]. Mohapatra et al. [105] study microbial processes for biopolymer production, which can be applied in sustainable packaging solutions [30, 37, 106–113] and bio-based materials [114–125], aligning with Industry 5.0's focus on circular economy and smart materials. Research on retrofitting buildings for energy efficiency [126–142] and the use of polyhydroxyalkanoates (PHAs) for biocompatible applications [143–152] demonstrates the integration of biotechnology into industrial processes.

There are some appealing reasons to use the banana pseudostem fibers, particularly their high tensile strength, biodegradability [153–157], and lightweight character that allow their use in the packaging industry, manufacturing of carry bags, and even as biocomposites in construction and automotive industries [158, 159]. Being of natural composition of cellulose and lignin, these fibers have a strong but flexible structure and can be processed into fiber sheets that can bear substantial stress [160–164]. These nutrient fibers, when chemically treated, have better endurance and water resistance which increase their versatility and working life in practical use [69, 74, 165–177]. Moreover, their characteristic of being compatible with additive manufacturing technologies allows for the production of parts with sophisticated and tailored shapes, making the banana pseudostem fibers suitable for different industrial applications [178–187].

The current chapter deals with the potential of the banana pseudostem fibers in a detailed manner as regards their properties, their extraction, and their manufacturing capability. This research is complemented by new manufacturing processes, such as additive manufacturing, which enable the environmentally friendly production of fiber composites. The chapter also depicts the effectiveness of banana fiber in terms of tensile strength, biodegradability and water resistance in comparison to plastic and natural fibers. This analysis helps to understand the practicality, processing feasibility and eco-friendly nature of replacing plastic with banana pseudostem fibers. In this sense, the chapter focuses making the point that banana fibers can offer a suitable alternative for the ever increasing demand for nonplastic solutions.

2 Properties and Potential of Banana Pseudostem Fibers

Banana pseudostem fibers are agricultural waste obtained from the banana plant and are therefore a wasted resource with great potential in sustainable materials applications [158, 159]. Given their specific characteristics such as high tensile strength, biodegradability, and low density, these materials can provide for an impeccable choice to substitute the orthodox plastics. This part of the paper looks at the structure, benefits and potential applications of banana pseudostem fibers to show what makes them beneficial and opportunities that exist in different industries.

2.1 Composition and Unique Qualities

The banana pseudostem fiber is known to contain cellulose, hemicellulose and lignin. The presence of these chemical substances enhances the fiber's strength, elasticity and its endurance. To be precise, cellulose, which is the dominant and most essential component, acts as a stiff axial shaft of the fibers, while hemicelluloses and lignin offer elasticity and toughness so that a variety of applications are possible [188]. The structure of these fibers does have a number of tiny holes which are the reasons why these fibers are light in weight and will also be easier to biodegrade. Since banana fibers have an inherently low density they have a great potential to lower logistics and production costs because lighter material would mean more energy expense need not be used during processing and shipping activities that is otherwise spent with synthetic plastics [13, 160, 189–203].

In terms of tensile strength, banana pseudostem fibers are strong enough to withstand a good amount of tensile stress with breaking and this means that they can be utilized in fabricating different types of load bearing structures [204–212]. This strength, coupled with their inherent properties of hydrophobicity gives banana polymer fibers an edge for packaging and other consumer products that are bound to come in contact with harsh climatic conditions and require great durability.

The key properties of banana pseudostem fibers are outlined in Table 1, highlighting their composition and physical characteristics that support their use in sustainable applications [213–216]. Figure 1 compares key physical properties of banana pseudostem fibers and conventional plastics, including tensile strength, flexibility, and durability. Figure 2 presents the composition breakdown of banana pseudostem fibers, showing the proportion of cellulose, hemicellulose, and lignin, with cellulose being the most abundant component.

Table 1 Properties of banana pseudostem fibers

Property	Description	Value
Cellulose content	Major component providing structure and strength to the fiber	60–65%
Hemicellulose	Contributes to flexibility and water absorption characteristics	15–20%
Lignin content	Adds rigidity but reduces biodegradability	5–10%
Tensile strength	Measure of the fiber's resistance to breaking under tension	400–600 MPa
Density	Low density enhances lightweight properties for carry bags and packaging	1.35 g/cm^3
Moisture content	Reflects water absorption; higher moisture content affects durability	10–12%

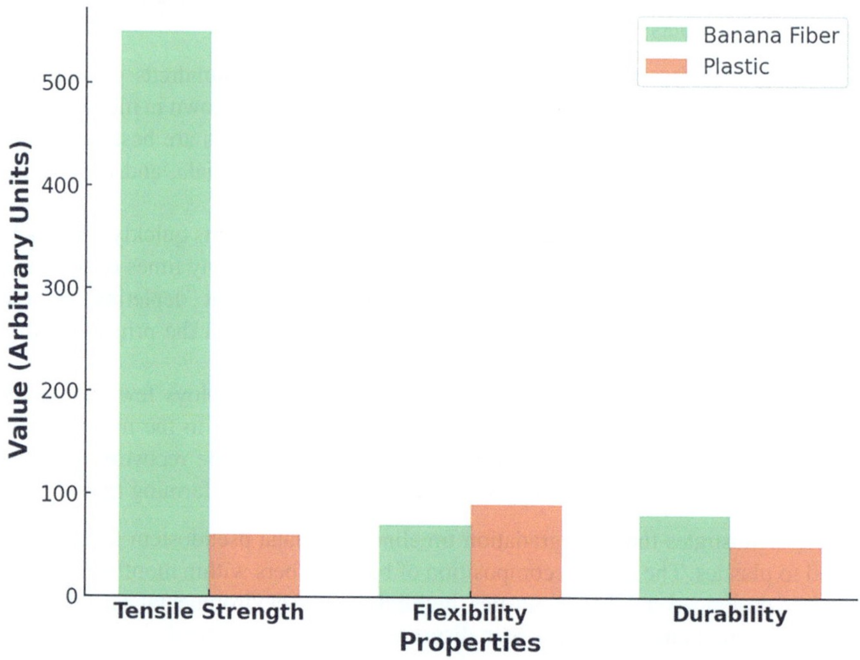

Fig. 1 Comparison of banana fiber and plastic properties

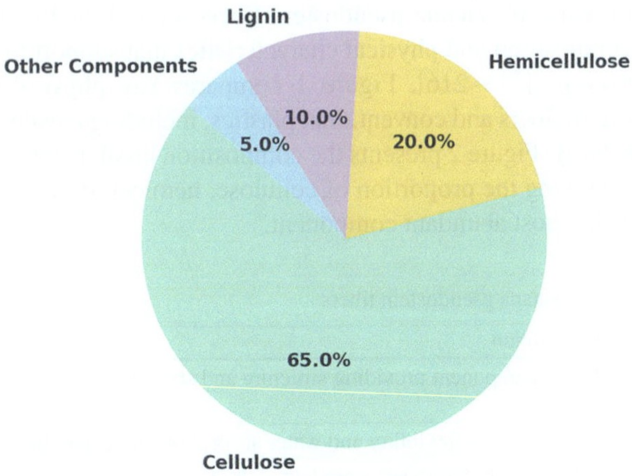

Fig. 2 Composition breakdown of banana pseudostem fibers

2.2 Comparative Benefits in Relation to Popular Plastics

In comparison with the mainstream plastics, what is more interesting is that the banana pseudostem fibers offer once the environment and functionality is advancement of the products:

- Biodegradability: For one, plastic materials could take hundreds of years to decomposes, but banana pseudostem fibers naturally break down in months, adding to a zero-waste life cycle. Because of this, the products are best suited for disposable products, such as carry bags, packaging materials, and disposable consumer durables [217–219].
- Sustainability: Fresh pseudostems emerge from banana plants quickly with each crop cycle, meaning they can be cut down and gathered many times over. Since the material source can be constantly renewed without depleting natural resources, this regenerative characteristic fits perfectly with the principles of a circular economy.
- Low Emissions: Banana pseudostem fiber production employs few chemicals which in turn lowers greenhouse emissions in comparison to the use of oil in plastics manufacturing. Also, utilizing an agricultural waste recovery material helps to reduce wastes and fully utilize resources in banana farming regions.

Figure 3 illustrates the biodegradation timeline of banana pseudostem fibers compared to plastics. The rapid decomposition of banana fibers within months contrasts with the prolonged stability of plastics, highlighting the environmental advantage of using biodegradable banana fibers in disposable products. Figure 4 shows the relationship between tensile strength and density for banana pseudostem fibers, natural fibers, bioplastics, and conventional plastics [220]. Banana fibers combine high tensile strength with low density, offering a lightweight alternative to traditional materials with significant resilience.

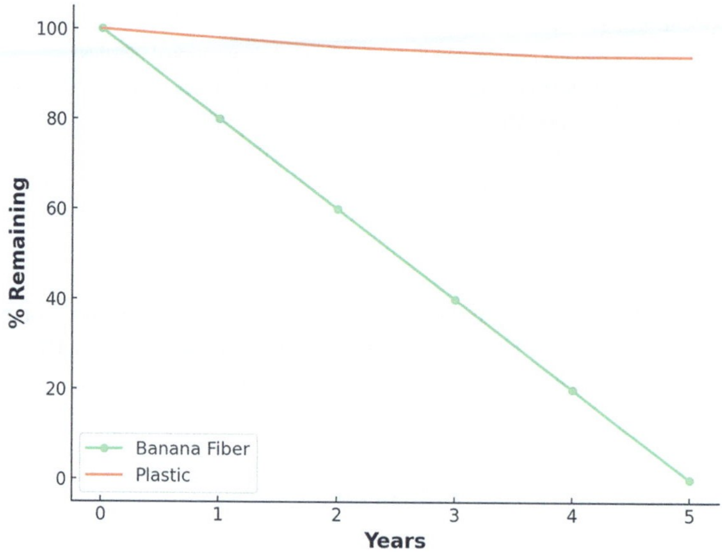

Fig. 3 Biodegradation timeline for banana fibers vs. plastics

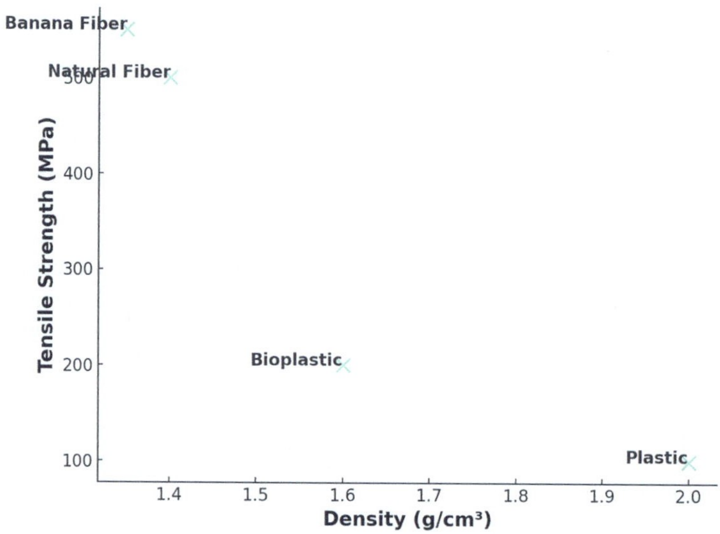

Fig. 4 Tensile strength versus density for different materials

Table 2 provides a comparison between banana pseudostem fibers and traditional plastic materials, showcasing advantages in biodegradability, tensile strength, and carbon footprint.

Functional versatility: naturally constructed with excellent mechanical properties, banana pseudostem fibers can be harnessed through different manufacturing

Table 2 Comparative analysis of banana fibers and conventional plastics

Property	Banana pseudostem fibers	Conventional plastics
Biodegradability	High, decomposes within months	Low, takes hundreds of years
Tensile strength	400–600 MPa	10–100 MPa
Density	1.35 g/cm^3	0.9–2.2 g/cm^3
Carbon footprint	Low, based on natural material	High, based on fossil fuels
Cost of production	Moderate, can be reduced with scaling	Low, but with high environmental costs

techniques such as weaving, interlacing them with other natural fiber, or making flexible sheets that can be fed into an additive manufacturing process. This flexibility enhances their application into more than carry bags.

2.3 *Applications Across Industries*

Banana pseudostem fibrous stalks would have numerous uses across a variety of industries owing to their unique properties of strength, flexibility, and ecologically safe degradation. Such broad scopes of application offer quite a number of promising areas where these fibers could be deployed as an eco-friendly material for different purposes.

- Packaging and Consumer Goods: The fibers may be used to manufacture strong lightweight carry bags, food packaging, and disposable cutlery items as a fast developing region. Fetuses could be effective in combating plastic pollution in the customer goods industry.
- Construction and Automotive: Banana fibers can be made into biocomposites which can be used as reinforcements in structural materials, construction materials or interior automotive parts. The composites minimize the advantages of biomass sourced natural structure and environmental durability making it ideal for green industries.
- Textiles and Fashion: In the fusion process, banana fibers due to their texture, flexibility, and softness are ideal in production of natural fiber mixtures such as cotton to create eco-friendly fabrics for use in garments, accessories and household fabrics.

The biocomposite nature of the banana pseudostem fibers certainly provides numerous properties and probable applications of these biomass materials. Aiming to meet environmental and industrial needs, such fibers provide a way forward in minimizing the use of plastic and adopting sustainable practices in different fields.

3 Extraction and Processing Techniques

In order to effectively utilize banana pseudostem fibers for sustainable applications, it is imperative to develop and refine efficient processes for their extraction and processing. This ion explains the methods of collection, preparation, and processing of banana pseudostems into fibers, chemical treatment and mechanical treatment methods that improve the properties of banana fibers. All these processes are aimed at preparing the material for durable high quality products while preserving the environmental benefits that the material has.

3.1 Sustainable Harvesting and Fiber Preparation

Harvesting of banana pseudostem fibers starts with the controlled cutting of banana plants with pseudostems mainly left behind as agricultural residues after harvesting of bananas which is very often the case since the plants are cut mid stem to allow for growth of new shoots. Utilizing these pseudostems therefore leads to reducing wastage and maximizing the benefits of banana plantation.

The pseudostems are initially harvested and cut into suitable sizes. The inner layers having great density of fibers toward the center are pressed into fibers and segregated from outer layers that are sparsely populated with fiber strands. The first operation in fiber extraction is the decorticating operation which is the breaking down of stalks and other tubular components to leave what is referred to as the primary fibers. Manual decortications are possible but for larger volumes, mechanical decortications are more effective and more consistent.

The fibers are subsequently washed to remove dirt and other plant residues before drying them in the sun until the moisture content is low. This process of drying not only safeguards the fibers but also increases their tensile strength making them resistant during further processes.

3.2 Application of Chemical Treatments in the Alteration of Fiber Properties

Chemical treatments are done on banana pseudostem fibers with the aim of improving the endurance and the flexibility of the fibers. These treatments address matters on the mechanical strength of the fiber, its moisture uptake, and the availability of binding mechanisms within the fiber.

Alkaline Treatment: One common method is treating the fibers with an alkaline solution, like sodium hydroxide (NaOH). This solution treatment incorporates the cleaning of lignin, hemicellulose and other contaminants which lessen fiber quality.

Table 3 Extraction and chemical treatment methods for banana fibers

Treatment method	Purpose	Chemicals used	Resulting improvement
Alkaline treatment	Removes impurities, increases tensile strength	Sodium hydroxide (NaOH)	Enhanced fiber strength and durability
Bleaching	Increases fiber purity and aesthetic quality	Hydrogen peroxide (H_2O_2)	Improved color, reduced contaminants
Surface modification	Improves compatibility with other materials	Silane, coupling agents	Enhanced bonding in composites
Neutralization	Balances pH after chemical treatments	Acetic acid	Stabilizes fiber integrity

As a result, the tensile strength of the fiber and its surface area for bonding processes improves enabling a stronger composite material to be made.

Bleaching and Neutralization: After the alkaline treatment, fibers are often bleached because of the desire to increase their purity before they are used in applications which require a cleaner and a whiter fiber. A neutralization process, achieved using acetic acid, is important after the bleaching process to counter the pH effect brought about by the bleaching and to ensure that the strength of the fibers is provided without making them too brittle.

Surface Modification: In some cases, surface modification treatments such as silane treatments and coupling agents are applied to increase the compatibility of banana fibers with other materials and thereby improve the adhesion properties of the fibers when blended with synthetic or other natural fibers. This treatment is especially important in maximizing the interfiber bonding in biocomposites. Various chemical treatments applied to banana pseudostem fibers to improve durability and mechanical properties are summarized in Table 3.

3.3 *Sheet Formation and Blending Processes*

For biocomposite application, the banana pseudostems fibers are go through layering or blending process to yield sheets or composite materials. These stages play a vital role in the alteration of unusable fibers into more usable forms for manufacturing purposes.

Fiber Carding and Layering: Fiber carding is the process of arranging the fibers by pulling them apart so that they can be evenly spread into layers. These layers are then either piled or woven into sheets accompanied by densification and strengthening treatments. Sheet formation is a good technique to create a fleshy material which can then be cut into shapes needed for different products.

Pulping Techniques and Papermaking: An alternate to fibrous sheets is the pulping method, where banana fibers are suspended in water and reconstituted into paper like sheets. Such processes can be efficiently employed to make lightweight carry bags and other packaging materials. In the process of pulping, the fiber particles are soaked in water and pulled over a screen to dry them into thin sheets which are then dried and pressed for increased pulp strength.

Fibers made from banana pseudostems can also be used along with other natural fibers, like jute or cotton, surely increasing its flexibility, firmness and physical appearance. One of the advantages of blending is that fiber properties can be customized for specific needs. Great examples of this are the blending of banana fibers and cotton which results in materials more suitable for fabrics while blending with jute gives the fiber more strength relevant for packaging and biocomposites.

4 Manufacturing and Integration in Additive Technologies

To make banana pseudostem fibers into environmentally friendly products, improved production methods like additive manufacturing and digital fabrication processes need to be adopted. This makes it possible to produce personalized, durable and biodegradable products with little waste. This part looks into how such technologies fit in for sustainable production, how this technology can assist in the manufacturing processes of banana fiber sheets and other variables that can enhance responsiveness and efficiency in product design processes.

Figure 5 outlines the steps in banana fiber processing, from harvesting through product development, along with yield efficiency at each stage. The efficiency decreases with each step, suggesting areas for optimization that could improve overall material yield and reduce processing costs.

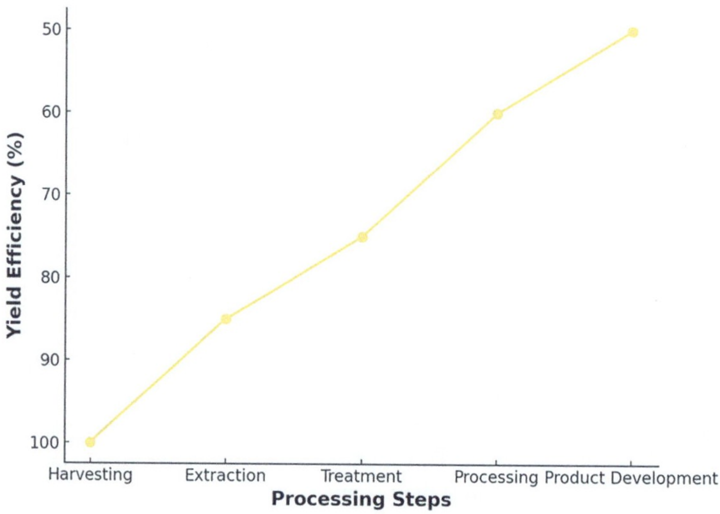

Fig. 5 Steps in banana fiber processing and yield efficiency

4.1 Additive Manufacturing in a Sustainable Production Framework

Additive manufacturing or 3D printing is the process of constructing a part layer by layer. This technique ensures that the amount of material used is as precise as is possible and less waste is generated in the process. This is consistent with the sustainable objectives of making use of the banana pseudostem fibers since it generates less waste as well as energy, more so when compared with the other common approaches to undertaking manufacturing.

For the banana pseudostem fibers, the additive manufacturing technology can help to utilize their properties by allowing them to be shaped or stacked in intricate configurations, shapes, and support. Applying the principles of additive manufacturing as a solution to a problem presents environmental benefits since the process applies the materials needed to complete the task at hand which leads to lightweight designs and eliminates overmaterialization that is common in old manufacturing techniques.

4.2 Digital Manufacturing of Banana Fiber Sheets

Digital manufacturing links harvest, banana pseudostem fibers, along with the digital world in a method of making sheets that adhere to the requirements even in layered orientation. The process starts with banana fiber being processed into a sheet of a desired thickness which can then be further digitally modified to achieve a variety of shapes and densities in the final product.

Digital Layering and Reinforcement: In this process a controlled pattern of banana fibers is laid to ensure maximum strength and durability of the material. Digital layering helps to maintain the fibers such that uniformity methods are used, and this improves the mechanical properties of the sheet. When materials are layered in certain pattern shapes to withstand various level of stress, the material can be able to structural application as it is stronger and more flexible.

Laser Cutting and Molding: After the sheets of banana fibers have been formed, the sheets can be cut and shaped by use of laser technology. Laser cutter makes it easy to cut the fiber sheets into shapes as required by the products which are of specific shapes and measurements. By this method, manufacturers can minimize wastage because laser cutting has more reduction of offcuts and excess materials.

Thermal Molding for Complex Shapes: Banana fibers can also be used in thermal molding: flat sheets are heated in order to create more intricate shapes such as curved carry bags or containers. This method is very effective in producing strong and durable items that do not lose their shape easily.

Table 4 details the digital manufacturing techniques used to shape and reinforce banana fiber sheets, enhancing their performance for various applications.

Table 4 Digital manufacturing processes for banana fiber sheets

Process	Description	Application
Digital layering	Aligns fibers for uniform strength and durability	Structural sheets, biocomposites
Laser cutting	Precision shaping of fiber sheets for complex forms	Packaging, consumer goods
Thermal molding	Heat shapes fibers into durable, three-dimensional forms	Curved products, containers

4.3 Customization and Efficiency in Product Development and Mass Production

One of benefits associated with the convergence of digital and additive technologies with banana pseudostem fibers is customization. This means that specific user or industrial needs can be met through mass tailor-made products. Such "customization in mass production" can be crucial in the fabrication of multifunctional products satisfying no only functional but aesthetic needs and vice versa.

Combinable Design Templates: One of the great advantages of digital fabrication is creation of templates, which can be used for different models of banana fiber products of various sizes. This kind of versatility is crucial for commodity goods where the designs may be significantly different. Other variables, such as template thickness or fiber volume, can also be manipulated, and product variations constructed from flexible bags to hard packing.

Enhancing New Product Development Through Rapid Prototyping: Prototyping is made faster and easier thanks to additive manufacturing which allows producers to develop, test and perfect new banana fiber products. Rapid prototyping enables a business to practice different shapes of products which in return enhance fresh designs on the products maximizing the strength and durability of the fiber.

Simple Mechanism of Scaling Production: The use of digital and additive manufacturing and scaling are basic two aspects which hardly complicate each other. For example, if a situation where a company experiences high demand such as peak seasons, a company will simply alter production parameters and as a result, deliver rapid response production. These advanced processes not only support the material's eco-friendly properties but also create a flexible production model that can adapt to changing consumer demands, making banana fiber products a viable and sustainable solution in a range of applications.

5 Applications and Product Development

Due to the high adaptability of banana pseudostem fibers, they can be used for different purposes where the material needs to be environmentally friendly, biodegradable and strong. This ion reviews the specific uses of banana pseudostem fibers in the manufacture of biodegradable carry bags, packaging materials, biocomposite materials for construction and automotives, and other green products for general usage. In each case, the unique characteristics of banana fibers are utilized in developing sustainable alternatives to plastics and other destructive materials. The applications of banana pseudostem fibers across various industries are outlined in Table 5, indicating the versatility and potential for this sustainable material. Figure 6 displays the projected environmental impact reduction when replacing plastics with banana pseudostem fibers over time. The decreasing impact highlights banana fibers' role in reducing long-term environmental pollution due to their lower carbon footprint and biodegradable nature.

5.1 Biodegradable Carry Bags and Packaging Solutions

Among the features obtained from banana pseudostem fibers, production of biodegradable carry bags and packaging materials comes at the forefront. With pollution becoming a serious concern, more companies and consumers are on the lookout for

Table 5 Applications of banana pseudostem fibers across industries

Industry	Application	Benefits
Packaging	Carry bags, food packaging	Biodegradable, durable
Construction	Biocomposite panels, insulation	Lightweight, strong, sustainable
Automotive	Interior parts (panels, dashboards)	Reduces weight, eco-friendly, durable
Textiles	Home furnishings, upholstery	Natural look, durable, sustainable alternative

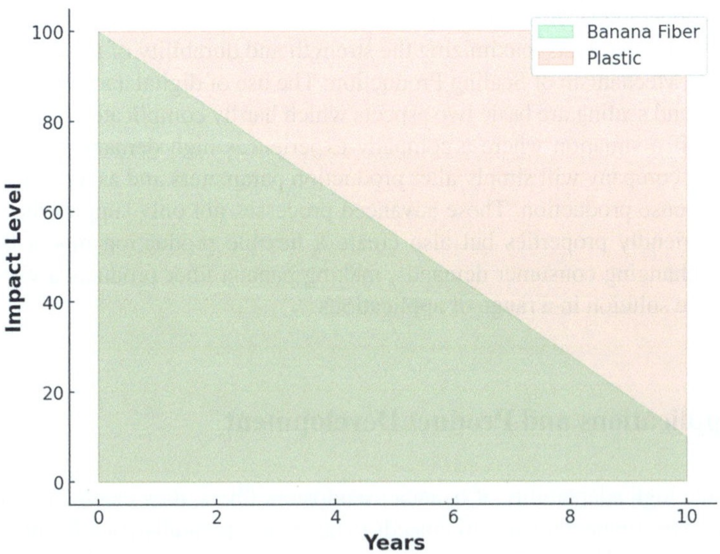

Fig. 6 Environmental impact reduction over time with banana fibers

environmentally friendly packaging which has significantly driven the market for biodegradable products.

- Carry Bags: Being weak biodegradable materials, banana fibers are tensile and can be processed into strong lightweight carry bags which are definitely bio-waste. These bags are also suitable for mannequins since they gradually decompose and will not lead to environmental pollution. As a result, both retailers and buyers will appreciate the bags because they do not create waste.
- Food Packaging: Since banana fiber sheets contain no toxin and chemicals and are moisture repellent by nature, they are appropriate in packaging foods. They can be a substitute for plastics, which are single-use wraps and containers, and reduce the negative impact of environmental degradation during food storage and transportation. Likewise, banana fiber packaging is also biodegradable and will decompose into organic materials that promote soil when trashing.

5.2 *Biocomposites for Construction and Automotive*

Banana pseudostem fibers possess a great potential in the manufacture of biocomposites, which are composite materials made from natural fibers combined with resin or polymer for construction and automotive product.

- Construction Materials: In the construction or, biocomposites based banana fibers can be used to manufacture panels, boards and insulating materials due to their lightweight and high strength to weight ratios making them suitable substitutes for synthetic materials. These composites could also appeal to builders who are environmentally responsible and wish to implement sustainable strategies in their constructions.
- Automotive Components: Due to the light weight nature of banana fiber composites, these are effective in construction of interior automotive parts such as door panels, dashboards and seat padding. Fuel consumption of a vehicle is enhanced as banana fiber composites lower the overall weight, thus reducing the carbon footprint emanating from the emissions. Moreover, the natural strength of the fibers increases the durability and impact resistance of the parts alternatively replacing plastic parts.

5.3 *Consumer Goods That Are Sustainable*

In addition to carry bags, biocomposites, and pipes, fibers from the banana pseudostem have numerous applications in consumer goods where the emphasis is on sustainability. In a wide range of applications from home textiles to fashion, this fiber offers a green alternative that is aesthetically pleasing and functional as well.

- Home Textiles and Rugs: Textiles made from banana can be sourced and manufactured into sustainable rugs, upholstery textiles, and green furniture that are both durable and strong. The natural roughness of the banana fiber further enhances the appeal of the products which are designed to promote eco-friendly and organic concepts.
- Fashion Accessories: Due to the increasing popularity of wearing green, banana pseudostem fibers can be suited to making handbags, caps, and belts. They are strong but flexible, and can be colored and formed into various styles to satisfy the designs and concepts of environmentally concerned users.

5.4 Product Development Considerations in Using Musa Pseudostems

It is crucial to emphasize the role of such aspects as sustainability, practicality, and market wellness when developing products from banana pseudostem fibers. Process technology, consumer demand, and ways of improving such qualities as strength and convenience are of primary importance when developing banana fiber products.

Material Processing: As the banana pseudostem fiber has an aesthetic touch, in most cases no processing is needed to make decent products. Nevertheless, other treatments may be used to increase water repellency or tensile strength and accordingly extend the application range.

Customization and Market Trends: Customization is a key factor that is helpful in positioning banana fiber products into various market segments. For example, it would be beneficial to include features allowing the customer to select the desired color or texture in order to expand the target audience. It is also important to follow the eco-chic trend so as to be able to serve the customers and sell the goods.

Durability and Usability: In certain cases, products made with banana fiber could benefit from additional reinforcement or need to be combined with other natural materials to enhance product life. This is vital, in particular, for items that are prone to frequent use or to exposure to the environment, carry bags and automobile parts, for example. Biodegradable materials can be used to bind or coat plastic fibers thereby increasing the serviceability of the products irrespective of the environmental impact.

Industries can address their reliance on nonbiodegradable materials and practice environmentally friendly methods by extending the utility of banana pseudostem fibers. Such fibers make it possible to create modern goods that are not only usable in the broadest sense but also contribute to the ecology thanks to their applicability in diverse fields, including automobiles and planes.

6 Results and Analysis

The strength and the performance of banana pseudostem fiber products are crucial factors in evaluating their viability for application as an environmentally friendly material. This part of the writing discusses the results concerning the mechanical strength, durability, degradability, marketability, and industry rivalries of the banana fibers products. Moreover, it highlights major areas which need to be addressed for adoption of banana fibers in mass production and suggests measures to address the issues.

6.1 Mechanical Strength, Durability, and Performance

Banana pseudostem fibers are attracting interest due to their particular mechanical characteristics and even their commercial worth for a variety of applications and end products. Some of the major highlights in this category include the following:

- Durability: The findings from durability tests indicate that structural and operational characteristics of banana fiber products do not change with moderate environmental exposures over time. While untreated fibers are always prone to faster decaying in humid areas, the alkaline or silane treatment give moisture resistance and consistency to the fibers making them everlasting in packing material and automotive industries.
- Flexibility and Elasticity: As in the case of most natural fibers, banana fibers have a fair amount of flexibility and elasticity which allows for a certain degree of bending and deformation without breaking. This quality is particularly helpful in applications that require some rigidity without too much stress like carry bags and textile blends.

Table 6 presents the results from mechanical and environmental testing on banana pseudostem fibers, including measures of tensile strength, decomposition rates, and ecotoxicity.

Table 6 Results of mechanical and environmental testing on banana fiber products

Test parameter	Metric	Result	Implication
Tensile strength	MPa	400–600	High durability for load-bearing uses
Decomposition time	Months	2–3 months	Environmentally sustainable
Ecotoxicity	% Toxicity	None detected	Safe for soil and water
Moisture absorption	% Absorbed	Moderate, treatable with coatings	Suitable for controlled applications

6.2 Degradability and Environmental Testing

Biodegradability testing reveals a strong case for banana pseudostem fibers as they naturally degrade which is of great importance compared with plastic materials. Important aspects from the tests for biodegradability and environmental impact include the following: • Decomposition Time: In terms of composting processes, banana fibers show a significant degradation rate as they start degrading as early as 2 weeks and within few months are entirely gone. This strength makes the banana fiber useful as it tackles waste accumulation and promotes a linear economy whereby waste products nourishes the earth rather than pollutes it.

- Ecotoxicity: Decomposition of the banana fiber does not cause any toxicity as the process does not produce any harmful chemicals, hence the negative impacts on soils or any water remain at bay. This irrecoverable decomposition also further validates banana pseudostem fibers to be safe when used in disposable and biodegradable items such as food wrapping and single-use carry bags.
- Carbon Footprint: The carbon footprint from processing banana fiber is lower in comparison with that of both plastic and synthetic fibers due to the fact that they possess minimal chemical processing and are renewable from banana pseudostems. These results highlight the potential of banana fibers in providing a low-impact alternative to traditional materials.

6.3 Economic Viability and Comparison with the Industry

This economic cost benefit analysis assesses the efficiency banana pseudostem fibers in-cut costs when used as a replacement to plastic materials and other natural fibers.

- Cost of Production: Banana pseudostems are usually, if not always, classified as waste banana materials. Because of this strong belief, the initial cost of raw materials is almost nil. Moreover, the cost for processing may also be minimized through simple mechanical treatment using fewer chemical processes. As the demand for such sustainable products increases, there exist chances to diminish further the cost of producing banana fiber, making it a very cheap, easily scalable raw material source to many industries in need for environment friendly alternatives.
- Comparative Cost and Market Appeal: If we compare the processes of C2 petrochemical industries which are quite complicated, banana fibers in turn have the advantage of raw material and production costs. At the moment banana fiber is sold at a price which is about 2$ more than that of mass produced plastics but as people become more aware of the benefits of biodegradable products the market for banana fibers will expand immensely.

- Industry Adoption and Market Growth: Areas like packaging, consumer goods, automobiles, and even construction have begun to consider banana fibers and the potential they have as a green alternative. As the trend of market continues to shift toward eco-friendliness, the scope of nadhoon is tremendous, hence capturing a large percentage of market shares in these striking industries. An economic analysis of banana pseudostem fibers versus plastic materials is shown in Table 7, highlighting production costs and potential market impacts.

Figure 7 plots the market potential against production cost for banana fibers, bioplastics, and traditional materials. Banana fibers offer an attractive balance of moderate production costs and substantial market potential, supporting their growth in eco-friendly product markets. Figure 8 compares the economic viability of banana pseudostem fibers, plastics, and bioplastics. The chart illustrates the relative advantages of banana fibers, including lower raw material costs and market demand for sustainable options, indicating promising conditions for large-scale adoption.

Table 7 Economic viability of banana pseudostem fibers compared to plastics

Parameter	Banana pseudostem fibers	Plastics
Cost of raw materials	Low to moderate (agricultural waste)	Low, but environmental cost high
Processing expenses	Moderate, decreases with scale	Low
Production scalability	Emerging, expanding	High, established
Market readiness	Growing, eco-friendly markets	Mature, but unsustainable

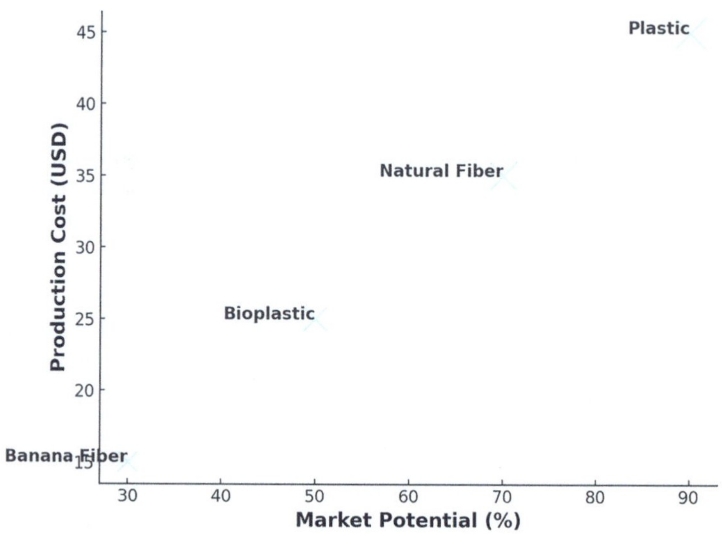

Fig. 7 Market potential versus production cost

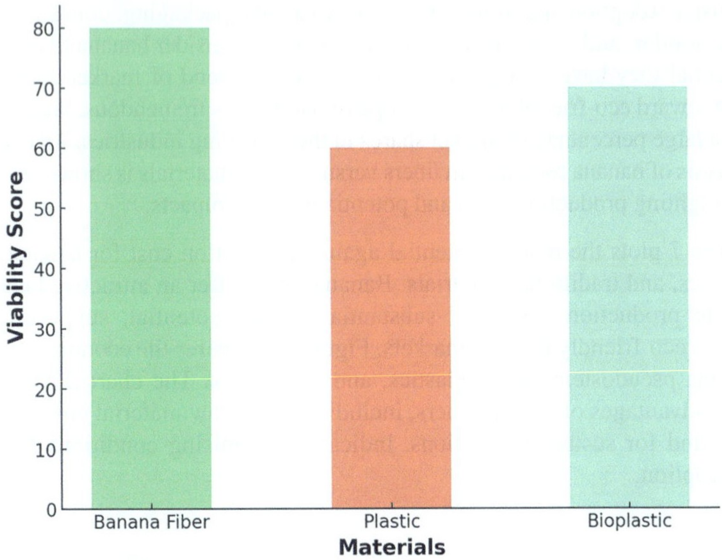

Fig. 8 Economic viability comparison across materials

6.4 Major Challenges and Possible Solutions

Although banana pseudostem fibers have shown promising results, there are challenges in scaling production, performance improvements, and the rate of adoption by industries. Table 8 identifies the primary challenges facing the banana fiber industry and provides recommendations for overcoming these obstacles to enhance adoption and scalability.

Figure 9 compares the performance metrics of banana pseudostem fibers and plastics, covering strength, cost, eco-friendliness, durability, and flexibility. The chart highlights banana fibers' advantages in eco-friendliness and strength, showing their potential as a sustainable alternative in specific applications. Recommendations to mitigate these challenges are as follows:

- Improving Water Resistance: Water resistance is not a problem with banana fibers, but further enhancement will help the fibers become better suited for applications in humid or wet environments. In response, there is further work needed on natural coatings, such as beeswax and biodegradable polymers, that will improve the performance of products.
- Uniformity in the Execution of Processes: Such a guide will ensure effective quality control. Industry generally enforced techniques for fiber extraction and treatment will increase the quality of banana fiber products used in various applications.

Table 8 Key challenges and recommendations for banana fiber industry

Challenge	Description	Recommendation
Moisture resistance	High moisture absorption can affect durability	Use natural coatings like beeswax to improve moisture resistance
Standardization	Inconsistent processing can affect product quality	Develop standardized extraction and treatment protocols
Consumer awareness	Limited knowledge may hinder demand	Eco-labeling and public awareness campaigns to increase visibility
Cost competitiveness	Initial production costs can be high	Scale production with incentives and larger facilities to reduce costs
Compatibility with machinery	Some machinery may require modification for banana fiber processing	Collaborate with manufacturers to develop suitable equipment
Market accessibility	Entering markets dominated by synthetics can be challenging	Form partnerships with eco-friendly ors to increase market penetration
R&D gaps	Limited research on treatments and applications restricts usage	Increase R&D funding to optimize treatments and broaden applications
Scalability	Current production may not meet high demand if scaled	Implement efficient, scalable processing methods and supply chain partnerships

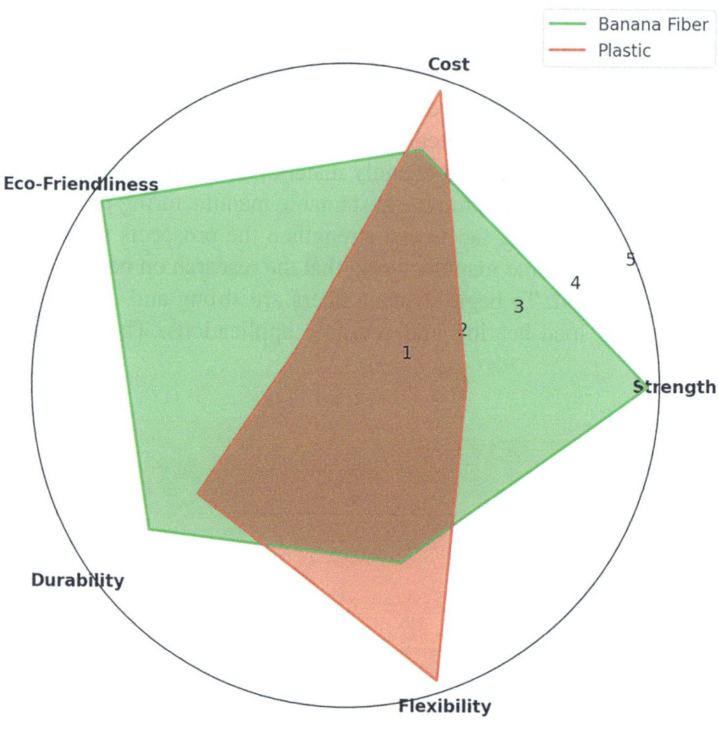

Fig. 9 Performance metrics (e.g., strength, cost, biodegradability)

- Improving Communication with Manufacturers: B2B marketing of banana pseudostem fibers should stress their benefits to industries which are inclusive of environmental and economic factors. Integration of banana fibers in mainstream production will be further promoted by collaborations with manufacturers and government goodness for afforestation materials.
- Funding More Research in the Areas of Studies: Ongoing studies on banana fiber's mechanical features and processing should be emphasis areas for their functional optimization. Government, industry and academia partnership can stimulate innovation that can make banana fibers economically feasible an alternative to plastics.

The analysis provides evidence that banana pseudostem fibers are suitable replacements of synthetic materials and unanimously ecofriendly in nature. There are still some hurdles both on the technical and economic side but progressive developments and changing approaches toward green economy will definitely place banana fibers among crucial resources in that quest.

Banana pseudostem fibers have the potential to replace plastics as they are economical and eco-friendly and can be used in various areas. This chapter ends with the discussion of the conclusions made, implications for sustainable manufacturing, and suggestions on how to promote banana fiber products in the market and make it more environmentally friendly. Because banana pseudostem fibers have so many potential applications, their broader application will be supported by further research, industry support, and consumer education.

Figure 10 presents the relative carbon footprint of banana pseudostem fiber production compared to plastic production. Banana fibers' significantly lower footprint supports their adoption as an eco-friendly material, aligning with global efforts to reduce carbon emissions and promote sustainable manufacturing practices.

There are several crucial facets that strengthen the prospects of banana pseudostem fibers for sustainable manufacturing that the research on banana pseudostem fibers has underscored. To begin, banana fibers are strong and durable enough to tolerate numerous load-bearing and reusable applications. Their flexibility and

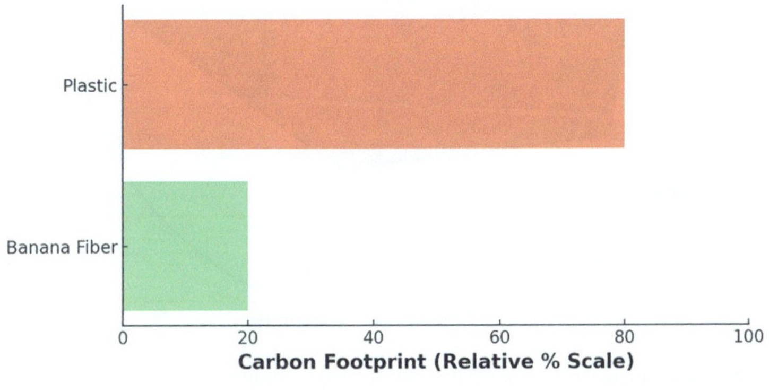

Fig. 10 Carbon footprint comparison of banana fiber production vs. plastic production

versatility make them useful for a diverse range of industries including packaging, construction, and automotive industries. Furthermore, since banana pseudostem fibers are decomposed easily by natural processes, they can be used for temporary applications; thus, there is no permanent environmental damage caused by their residual impacts. This property is particularly advantageous for products such as carry bags, food packaging materials, and other consumer products. Economically and environmentally, banana pseudostem fibers are affordable, and a low carbon emission material therefore making them an attractive option as a sustainable substitute. Additionally, the use of agricultural waste as raw materials is in line with the circular economy concept and reduces pollution while optimizing the use of resources. Finally, banana pseudostem fibers have been used successfully for a wide range of applications within many different industries including packaging, biocomposites, hygiene products, and textiles. Their integration within digital and additive manufacturing means that production and product personalization can be done efficiently.

Incorporating banana pseudostems in manufacturing processes/bio composites contributes to sustainable development by decreasing usage of polymers, reducing carbon footprint, and encouraging eco-friendly technologies and practices. The repurposing of agricultural waste into valuable raw materials drastically lowers the carbon footprint of the banana industry and further enables a waste free approach in manufacturing processes. Moreover, replacing synthetic fibers with banana fibers as an eco-friendly option reduces the consumption of fossil fuels and pollution associated with plastic materials. Industries, by utilizing banana pseudostem fibers, which are not only renewable and biodegradable, aid in circular economy practices of designing products for reuse, recycle and safe return to Mother Nature to enhance economic as well as ecological sustenance.

In the future, banana pseudostem fibers could solve the problem of plastic pollution. However, scaling up production of banana fibers is critical if they are to be compared with the conventional plastic which calls for investments toward finding production techniques that can be mass produced and are cost effective. Large scale production combined with global partnerships can establish an international supply chain for banana fiber products enabling them to target a wider market. It is, however, vital to expand the focus on invention through research as well. Further research on suitability and treatment of banana pseudostem fibers will improve their properties and broaden their scope of application. There will also be technological advancements in fiber processing, moisture resistance and blending techniques which will improve the durability and versatility of the products enabling them to have various applications. Rising public awareness and market development will create a push for the use of the banana fibers products. Educating the consumers and ors on the need for banana pseudostem fibers by leveraging public awareness campaigns, eco-labeling, and incentives from the government can hasten the process of substituting eco-friendly materials for the conventional materials used in daily life.

References

1. Shah, S., Nautiyal, H., Gugliani, G., Kumar, A., Namboodri, T., & Singla, Y. K. (2024). Sustainability in smart manufacturing: Trends, scope and challenges. CRC Press (Taylor & Francis Group). https://taylorandfrancis.com/books/. ISBN 9781032740713. https://doi.org/10.1201/9781003467496
2. Singla, Y. K., Maughan, M. R., Arora, N., & Dwivedi, D. K. (2024). Enhancing the wear resistance of iron-based alloys: A comprehensive review of alloying element effects. Journal of Manufacturing Processes, 120, 135–160.
3. Harvinder Singh, Yogesh Kumar Singla. Wear characteristics of AISI 4140 under nano fly ash based engine oil. Proceedings of the 7th International Conference on Theoretical and Applied Nanoscience and Nanotechnology (TANN'23), Paper No. 118, 2023, Canada. https://doi.org/10.11159/tann23.118
4. Singla, Y. K., Maughan, M. R., Arora, N., & Dwivedi, D. K. (2023). Microstructural evolution of worn surface and subsurface of high Mn additive iron-based hardfacings subjected to repetitive impacts and sliding. Ceramics International, 49, 21957–21965.
5. Singh, H., Singh, A. K., Singla, Y. K., Chattopadhyay, K., Saini, A., & Singh, K. (2022). Interpretation of the wear characteristics of AISI 4140 under nano-fly ash based engine lubricant. Materials Today Proceedings, 50, 1683–1689.
6. Singh, H., Singh, A. K., Singla, Y. K., Chattopadhyay, K., & Malhotra, R. (2021). Optimization of tribological behavior of AISI 4140 under nano fly ash particulates in engine lubricating oil. Materials Today Proceedings, 45, 4619–4623.
7. Sura, H. S., Avesh, M., & Mohapatra, S. (2023). Design and modelling of digital twin technology to improve freight logistics. In Energy, environment, and sustainability (pp. 481–513). https://doi.org/10.1007/978-981-99-2150-8_20
8. Pattnaik, S., Pati, S., Devadarshini, D., Tanaya, K., Mishra, B. B., Samantaray, D., Mohapatra, S., & Dash, D. (2022). Microbial bioremediation of Cr(VI)-contaminated soil for sustainable agriculture. In Microbial biodegradation and bioremediation: Techniques and case studies for environmental pollution (pp. 395–407). https://doi.org/10.1016/B978-0-323-85455-9.00008-4
9. Srivastava, A., Lavania, S., & Mohapatra, S. (2023). Predicting the effect on land values after introducing high-speed rail. In Energy, environment, and sustainability (pp. 157–196). https://doi.org/10.1007/978-981-99-2150-8_8
10. Agrahari, R., Mohanty, S., Vishwakarma, K., Nayak, S. K., Samantaray, D., & Mohapatra, S. (2021). Update vision on COVID-19: Structure, immune pathogenesis, treatment and safety assessment. Sensors International, 2. https://doi.org/10.1016/j.sintl.2020.100073
11. Mishra, B. B., Nayak, S. K., Mohapatra, S., & Samantaray, D. (2021). Preface. In Environmental and agricultural microbiology: Advances and applications (pp. xvii–xix).
12. Tanaya, K., Mohapatra, S., & Samantaray, D. (2024). Formulation of biogenic fluorescent pigmented PHB nanoparticles from Rhodanobacter sp. for drug delivery. Antonie van Leeuwenhoek, international journal of general and molecular. Microbiology, 117. https://doi.org/10.1007/s10482-024-01973-x
13. Mohapatra, S., Pattnaik, S., Maity, S., Sharma, S., Akhtar, J., Pati, S., Samantaray, D. P., & Varma, A. (2020). Comparative analysis of PHAs production by Bacillus megaterium OUAT 016 under submerged and solid-state fermentation. Saudi Journal of Biological Sciences, 27, 1242–1250. https://doi.org/10.1016/j.sjbs.2020.02.001
14. Singh, H., Singh, A. K., Singla, Y. K., & Chattopadhyay, K. (2020). Design & development of a low cost tribometer for nano particulate lubricants. Materials Today Proceedings, 28, 1487–1491.
15. Singh, H., Singla, Y. K., Singh, A. K., & Chattopadhayay, K. (2020). Effect of nanofly ash as lubricant additive on the tribological properties of SAE 10W-30 oil: A novel finding. Transactions of the Indian Institute of Metals, 73, 2371–2375.

16. Kumar Singla, Y., Dwivedi, D. K., & Arora, N. (2017). Modeling the impact–sliding wear characteristics of rare earth additive iron-based hardfacing alloys. Proceedings of the Institution of Mechanical Engineers, Part J: Journal of Engineering Tribology, 231, 1486–1499.
17. Singla, Y. K., Arora, N., Dwivedi, D. K., & Rohilla, V. (2017). Influence of niobium on the microstructure and wear resistance of iron-based hardfacings produced by pre-placement technique—A novel approach. The International Journal of Advanced Manufacturing Technology, 93, 2667–2674.
18. Singla, Y. K., Arora, N., & Dwivedi, D. K. (2017). Dry sliding adhesive wear characteristics of Fe-based hardfacing alloys with different CeO2 additives—A statistical analysis. Tribology International, 105, 229–240.
19. Singla, Y. K., Chhibber, R., Arora, N., Singh, K., & Khanna, K. (2017). On the microstructure and wear behavior of Fe–xCr–4Mn–3C hardfacing alloys. Transactions of the Indian Institute of Metals, 70, 1555–1561.
20. Singla, Y. K., Dwivedi, D. K., & Arora, N. (2015). On the modeling of dry sliding adhesive wear parameters of vanadium additive iron-based alloys at elevated temperatures. Surface and Coating Technology, 283, 223–233.
21. Singla, Y. K., Chhibber, R., Bansal, H., & Kalra, A. (2015). Wear behavior of aluminum alloy 6061-based composites reinforced with SiC, Al 2 O 3, and red mud: A comparative study. Journal of Metals, 67, 2160–2169.
22. Pant, M., Sharma, A., Palli, S., Sharma, S. K., & Sharma, R. C. (2024). A review and analysis of factors affecting EV adoption for sustainable and energy-efficient transportation systems using analytic hierarchy process approach. International Journal of Vehicle Structures and Systems, 16, 133–139. https://doi.org/10.4273/ijvss.16.2.01
23. Sharma, S. K., Sharma, R. C., Lavania, S., Palli, S., & Avesh, M. (2024). Industry-driven approach for ANFIS-based intelligent control suspension system with MR damper for enhanced ride quality in passenger rail vehicles for technological investigations. National Academy Science Letters, 47, 391–394. https://doi.org/10.1007/s40009-023-01365-1
24. Adeoye, A. O., Quadri, R. O., Lawal, O. S., & Emojevu, E. O. (2024). Physicochemical characterization, valorization of lignocellulosic waste (Kola nut seed shell) via pyrolysis, and ultrasonication of its crude bio-oil for biofuel production. Cleaner Waste Systems, 7. https://doi.org/10.1016/j.clwas.2024.100138
25. Adeoye, A. O., Quadri, R. O., Lawal, O. S., Malomo, D., Emojevu, E. O., Omonije, O. O., Odeniyi, O. K., Fadahunsi, M. O., Yelwa, M. J., Aasa, S. A., Onakpa, A. E., Omoniyi, B. P., Mark, I. N., Usman, J., Ismaila, A. M., & Saidu, A. U. (2024). Advanced techniques in upgrading crude bio-oil to biofuel. In Energy, environment, and sustainability (pp. 321–353). https://doi.org/10.1007/978-981-97-0515-3_14
26. Adeoye, A. O., Quadri, R. O., & Lawal, O. S. (2022). Assessment of biofuel potential of tenera palm kernel shell via fixed bed pyrolysis and thermal characterization. Results in Surfaces and Interfaces, 9, 100091. https://doi.org/10.1016/j.rsurfi.2022.100091
27. Adeoye, A. O., Lawal, O. S., Quadri, R. O., Malomo, D., Aliyu, M. T., Dang, G. E., Emojevu, E. O., Maikato, M. J., Yahaya, M. G., Omonije, O. O., Edidem, V. G., Abubakar, Y. K., Offor, O. F., Sochima, E. H., Peter, B. E., & Hikon, B. N. (2023). Sustainable energy via thermochemical and biochemical conversion of biomass wastes for biofuel production. In Energy, environment, and sustainability (pp. 245–306). https://doi.org/10.1007/978-981-99-2150-8_11
28. Mohapatra, S., & Mishra, R. K. (2020). Status of institutional credit disbursement of agricultural or in Odisha, India. International Journal of Current Microbiology and Applied Sciences, 9, 2057–2065.
29. Mohapatra, S., Singh, J., & Mishra, S. (2017). Co-integration among major potato markets in Punjab. Indian Journal of Agricultural Marketing, 31, 46–47.
30. Mohapatra, S., Priyanka, V., Mohapatra, S., Kohli, I., & Mishra, R. K. (2020). Impact of corona virus covid-19 on the global economy. International Journal of Agricultural and Statistical Sciences, 16, 771–778.

31. Mohapatra, S., & Singh, J. (2021, 2021). Co-integration of major vegetable markets in Punjab, India. Economic Affairs, 66(2). https://doi.org/10.46852/0424-2513.2.2021.1
32. Mohapatra, S., Mishra, R., & Mishra, R. (2021). Impact of COVID-19 on education level of agricultural farming community. International Journal of Agricultural and Statistical Sciences, 17, 915–923.
33. Mohapatra, S., Singh, J., & Kumar, S. (2018). Cointegration among major cauliflower markets in Punjab. Indian Journal of Economics and Development, 14, 330–335.
34. Adeoye, A. O., Yelwa, J. M., Imam, N., Quadri, R. O., Lawal, O. S., Malomo, D., Aasa, S. A., Onakpa, A. E., Dorgbaa, R. K., Omoniyi, B. P., Ahmad, N., & Garba, H. A. (2024). Pyrolysis of plastic wastes towards achieving a circular economy: An advanced chemistry and technical approach. In Energy, environment, and sustainability (pp. 215–259). https://doi.org/10.1007/978-981-97-0437-8_11
35. Adeoye, A. O., Quadri, R. O., & Lawal, O. S. (2023). Wet synthesis, characterization of goethite nanoparticles and its application in catalytic pyrolysis of palm kernel shell in TGA. Results in Surfaces and Interfaces, 11. https://doi.org/10.1016/j.rsurfi.2023.100118
36. Avesh, M., Imam, N., & Adeoye, A. O. (2024). Assessing environmental impacts and mitigation strategies for the DFCCIL rail projects. In Energy, environment, and sustainability (pp. 333–353). https://doi.org/10.1007/978-981-97-0437-8_15
37. Mohapatra, S., Mohapatra, S., Han, H., Ariza-Montes, A., & López-Martín, M. D. C. (2022). Climate change and vulnerability of agribusiness: Assessment of climate change impact on agricultural productivity. Frontiers in Psychology, 13, 955622. https://doi.org/10.3389/fpsyg.2022.955622
38. Mohapatra, S., & Mishra, R. A. J. K. (2020). Production performance and market integration of cotton markets in India. In National symposium (p. 253).
39. Din, A. U., Han, H., Ariza-Montes, A., Vega-Muñoz, A., Raposo, A., & Mohapatra, S. (2022). The impact of Covid-19 on the food supply chain and the role of e-commerce for food purchasing. Sustainability, 14(5), 3074.
40. Mohapatra, S., & Singh, J. (2018). Growth and instability analysis of cauliflower crop in Punjab. Economic Affairs, 63, 665–669.
41. Pandey, G., Bansal, S., & Mohapatra, S. (2023). Market integration of chickpea crop: An evidence of India. Journal of Agribusiness in Developing and Emerging Economies, 15, 1–18.
42. Mohapatra, S., Mishra, R. K., & Sarangi, K. K. (2019). Economic appraisal of bio ethanol production technologies. International Journal of Chemical Studies, 7, 2197–2200.
43. Mohapatra, S., Kishore Mishra, R., Sarangi, K., & K. (2020). Bio-ethanol (2nd generation ethanol): A solution to ever polluting gasoline to climate in India. International Research Journal of Pure and Applied Chemistry, 20, 1–15.
44. Gohain, N., Bansal, S., Mohapatra, S., & Singh, L. (2022). Analyzing the direction of trade: Indian ginger and lessons from exports to different destinations. Journal of Agricultural Development and Policy, 32, 214–220.
45. Mohapatra, S., Mohapatra, U., Chandana, K. S. S., & Mishra, R. K. (2018). Economic analysis of paddy production and marketing in Puri. Odisha. J Pharmacogn Phytochem, 7, 1858–1861.
46. Quadri, R. O., Adeoye, A. O., Ayanda, O. S., & Lawal, O. S. (2024). Conversion of palm kernel shell to sustainable energy and the effect of wet synthesized nanoparticles of iron on its thermal degradation kinetics. Bioresource Technology Reports, 27, 101933. https://doi.org/10.1016/j.biteb.2024.101933
47. Momodu, D. U., Adetuyi, A. O., Adeoye, A. O., Akinlolu, O. S., & Apara, E. O. (2023). SURFACE FUNCTIONALIZATION OF LINEN FIBER AND ITS DYE-FASTNESS ASSESSMENT USING (4-[(E)-(4-AMINO-2-HYDROXYPHENYL) DIAZENYL] BENZENE-1-SULPHONIC ACID AND 4-[(E)-(4-{(Z)-[4-(DIMETHYLAMINO)PHENYL] DIAZENYL}-2-HYDROXYPHENYL) DIAZENYL]BENZENE-1-SULPHONIC ACID. Suranaree Journal of Science and Technology, 30.

48. Adeoye, A. O., Quadri, R. O., & Lawal, O. S. (2023). Biomass waste pyrolysis in the context of Nigeria's oil and gas industry: An overview. International Journal of Ambient Energy, 44, 2525–2541. https://doi.org/10.1080/01430750.2023.2256325
49. Mohapatra, S., Mohapatra, U., & Mishra, R. K. (2018). Price forecasting of groundnut in Odisha. The Pharma Innovation Journal, 7, 111–114.
50. Mohapatra, S., Sneha, N. K., Mounika, A., & Mishra, R. K. (2019). Socio-economic and demographic analysis of SAARC nations. International Journal of Agricultural & Statistical Sciences, 15, 529–533.
51. Sharma, S. K., Upadhyay, R. K., Kumar, V., & Valera, H. (2023). Transportation energy and dynamics. Springer Nature Singapore. https://doi.org/10.1007/978-981-99-2150-8
52. Upadhyay, R. K., Sharma, S. K., Kumar, D. V., & Valera, H. (2023). Transportation systems technology and integrated management. Springer Nature Singapore. https://doi.org/10.1007/978-981-99-1517-0
53. Lee, J., & Sharma, S. K. (2023). Intelligent systems for railway infrastructure. Applied Sciences, 13(20), 11274. https://doi.org/10.3390/app132011274
54. Sharma, S. K., Sharma, R. C., Choi, Y., & Lee, J. (2023). Modelling and dynamic analysis of adaptive neuro-fuzzy inference system-based intelligent control suspension system for passenger rail vehicles using magnetorheological damper for improving ride index. Sustainability, 15, 12529.
55. Sharma, S. K., Upadhyay, R. K., Kumar, V., & Valera, H. (2023). Introduction to sustainable transportation system. In Transportation energy and dynamics (pp. 3–6). Springer.
56. Hossain, I., Ahmed, Y. E., & Sharma, S. K. (2023). The use of both metallic-insert and TiO2/water nanofluid to improve the performance of photovoltaic modules; experimental study. Thermal Science and Engineering Progress, 42, 101944.
57. Wu, Q., Cole, C., Spiryagin, M., Chang, C., Wei, W., Ursulyak, L., Shvets, A., Murtaza, M. A., Mirza, I. M., & Zhelieznov, K. (2023). Freight train air brake models. International Journal of Rail Transportation, 11, 1–49.
58. Upadhyay, R. K., Sharma, S. K., Kumar, V., & Valera, H. (2023). Introduction to transportation systems technology and integrated management. In Transportation systems technology and integrated management (pp. 1–4). Springer.
59. Sharma, R. C., Bhardawaj, S., Sharma, S. K., & Bahl, S. (2023). Ride behaviour of three-wheel vehicle using semi-active suspension using skyhook control strategy. International Journal of Vehicle Structures & Systems (IJVSS), 15.
60. Bhardawaj, S., Sharma, R. C., Sharma, S. K., Gopala Rao, L. V. V., & Vashist, A. (2023). Modeling of biomechanical human body model for seat to head transmissibility analysis. Noise & Vibration Worldwide, 54, 62–74.
61. Sharma, S. K., Sharma, R. C., Choi, Y., & Lee, J. (2023). Experimental and mathematical study of flexible–rigid rail vehicle riding comfort and safety. Applied Sciences, 13, 5252.
62. Sharma, S. K., Upadhyay, R. K., & Kumar, V. (2024). Introduction to dynamics of transportation ecosystem, modeling, and control. In Dynamics of transportation ecosystem, modeling, and control (pp. 3–6). Springer.
63. Sharma, S. K., Sharma, R. C., Mohapatra, S., Fouly, A., Hossain, I., & Lee, J. (2024). Technological advancements in oscillation reduction for propulsion shaft systems. Journal of Vibration Engineering & Technologies, 1–19.
64. Upadhyay, R. K., Sharma, S. K., & Kumar, V. (2024). Introduction to intelligent transportation system and advanced technology. In Intelligent transportation system and advanced technology (pp. 3–6). Springer.
65. Hossain, I., Ngoupo, A. T., Bhattarai, S., Challagulla, S. P., & Sharma, S. K. (2024). Empirical survey on the effect of straight and wavy fins integrated with wind stream as heat sink of photovoltaic panels. International Journal of Environmental Science and Technology, 22, 2479–2492.

66. Sharma, S. K., Sharma, R. C., & Lee, J. (2024). Propelling precision of longitudinal vibration mitigation in ship propeller shafts through advanced nonlinear intelligent semi-active control leveraging adaptive neuro-fuzzy inference system with linear quadratic regulator. Journal of Vibration and Control, 31, 10775463241244836.
67. Sharma, S. K., Kumar, N., Avesh, M., Sharma, R. C., Siddiqui, M. I. H., & Lee, J. (2024). Navigating tranquillity with $H\infty$ controller to mitigate ship propeller shaft vibration. Journal of Vibration Engineering & Technologies, 1–11.
68. Sharma, S. K., Sharma, R. C., Upadhyay, R. K., & Lee, J. (2024). Enhanced ship engine vibration reduction using magnetorheological dampers and adaptive neuro-fuzzy control system. Ships and Offshore Structures, 1–17.
69. Sharma, R. C., Sharma, N., Palli, S., Sharma, S. K., Joshi, S., Rallabandi, S. R., & Sharma, A. (2024). Dynamic behavior of three-wheel vehicle with inerter-based dynamic absorber suspension system. Noise & Vibration Worldwide, 55, 253–263. https://doi.org/10.1177/09574565241243398
70. Kokkinos, N. C., Emmanouilidou, E., & Sharma, S. K. (2024). Waste-to-biofuel production for the transportation or. In Intelligent transportation system and advanced technology (pp. 99–122). Springer.
71. Sharma, S. K., Sharma, R. C., & Lee, J. (2024). Modelling and simulation of MR damper characterization and uncertainty assessment in marine engine vibration isolation with FLPID control. International Journal of Modelling and Simulation, 1–11.
72. Sharma, S. K., Sharma, R. C., Upadhyay, R. K., & Lee, J. (2024). Coupled torsional–transverse vibration reduction in marine propulsion dynamics with novel approach using magnetorheological damping and adaptive control system. Journal of Vibration Engineering & Technologies, 12, 6089–6099.
73. Badgujar, P., Sharma, S., Rahul, D., Sharma, S. K., & Upadhyay, R. K. (2024). Feasibility study of Mumbai-Pune Hyperloop corridor: Challenges and implementation. Research in Transportation Business & Management, 52, 101089.
74. Sharma, R. C., Palli, S., Dharmana, L., Sharma, S. K., Sharma, A., Kharola, A., & Sharma, N. (2024). Simulation of transient dynamic response of a locomotive considering its flexibility and rigidity. Noise & Vibration Worldwide, 55, 33–47. https://doi.org/10.1177/09574565231212684
75. Song, H., Park, J., Sharma, S. K., & Lee, J. (2022). A comparative study on ultrasonic propagation characteristics and defect detection of metal material additive manufacturing using Deep learning algorithm. In Biennial international conference on future learning aspects of mechanical engineering (pp. 157–168). Springer.
76. Boris, I., Sharma, S. K., & Lee, J. (2022). A study on the reflection coefficient of higher modes of torsional wave by theoretical and numerical method. In Biennial international conference on future learning aspects of mechanical engineering (pp. 145–155). Springer.
77. Vishwakarma, P. N., Mishra, P., Sharma, S. K., & Shukla, A. K. (2022). Formulating and analysing the effect of suspension parameters on critical speeds for various DOFs in rail vehicle. In Biennial international conference on future learning aspects of mechanical engineering (pp. 791–802). Springer.
78. Kumar, U., Kasvekar, R., Sharma, S. K., & Upadhyay, R. K. (2022). Wear of wheels and axle in locomotive and measures taken by Indian railway. Advances in Engine Tribology, 77–96.
79. Sharma, R. C., Palli, S., Jha, A. K., Bhardawaj, S., & Sharma, S. K. (2022). Vibration and ride comfort analysis of railway vehicle system subjected to deterministic inputs. Resmilitaris, 12, 1345–1355.
80. Vishwakarma, P. N., Mishra, P., & Sharma, S. K. (2022). Formulation of semi-active suspension system and controls in rail vehicle. SSRN Electronic Journal. https://doi.org/10.2139/ssrn.415961
81. Sharma, R. C., & Sharma, S. K. (2022). Ride analysis of road surface-three-wheeled vehicle-human subject interactions subjected to random excitation. SAE International Journal of Commercial Vehicles, 15, 299–312.

82. Sharma, S. K., Lee, J., & Jang, H.-L. (2022). Mathematical modeling and simulation of suspended equipment impact on car body modes. Machines, 10, 192.
83. Sharma, S. K., Sharma, R. C., Lee, J., & Jang, H.-L. (2022). Numerical and experimental analysis of DVA on the flexible-rigid rail vehicle carbody resonant vibration. Sensors, 22, 1922.
84. Vishwakarma, P. N., Mishra, P., & Sharma, S. K. (2022). Characterization of a magnetorheological fluid damper a review. Materials Today Proceedings, 56, 2988–2994.
85. Sharma, S. K., Sharma, R. C., & Lee, J. (2022). In situ and experimental analysis of longitudinal load on carbody fatigue life using nonlinear damage accumulation. International Journal of Damage Mechanics, 31, 605–622.
86. Sharma, R. C., Sharma, S., Sharma, N., & Sharma, S. K. (2021). Linear and nonlinear analysis of ride and stability of a three-wheeled vehicle subjected to random and bump inputs using bond graph and simulink methodology. SAE International Journal of Commercial Vehicles, 15, 65–79.
87. Sharma, S. K., & Sharma, R. C. (2021). Multi-objective design optimization of locomotive nose. In SAE Technical Paper (pp. 1–10). https://doi.org/10.4271/2021-01-5053
88. Lee, J., Han, J., & Sharma, S. K. (2021). Structural analysis on the separated and integrated differential gear case for the weight reduction. In Advances in engineering design: Select proceedings of FLAME 2020 (pp. 175–181). Springer Singapore.
89. Choi, S., Lee, J., & Sharma, S. K. (2021). A study on the performance evaluation of hydraulic tank injectors. In Advances in engineering design: Select proceedings of flame 2020 (pp. 183–190). Springer Singapore.
90. Sharma, R. C., Palli, S., Sharma, N., & Sharma, S. K. (2021). Ride behaviour of a four-wheel vehicle using h infinity semi-active suspension control under deterministic and random inputs. International Journal of Vehicle Structures & Systems, 13, 234–237.
91. Bhardawaj, S., Sharma, R. C., Sharma, S. K., & Sharma, N. (2021). On the planning and construction of railway curved track. International Journal of Vehicle Structures & Systems, 13, 151–159.
92. Mohapatra, S., Mohanty, D., Sharma, S., Dikshit, S., Kohli, I., Samantaray, D. P., Kumar, R., & Kathpalia, M. (2021). Biomedical application of polymeric biomaterial: Polyhydroxybutyrate. In Bioresource utilization and management (pp. 111–124). Apple Academic Press.
93. Sharma, S. K., Sharma, R. C., & Lee, J. (2021). Effect of rail vehicle–track coupled dynamics on fatigue failure of coil spring in a suspension system. Applied Sciences, 11, 2650.
94. Sharma, R. C., Sharma, S., Sharma, S. K., Sharma, N., & Singh, G. (2021). Analysis of bio-dynamic model of seated human subject and optimization of the passenger ride comfort for three-wheel vehicle using random search technique. Proceedings of the Institution of Mechanical Engineers, Part K: Journal of Multi-body Dynamics, 235, 106–121.
95. Sharma, S. K., & Lee, J. (2021). Crashworthiness analysis for structural stability and dynamics. International Journal of Structural Stability and Dynamics, 21, 2150039.
96. Cataluna Veses, R., & Shah, Z. (2015). Bio-oil production by thermal cracking in the presence of hydrogen. Journal of Fundamentals of Renewable Energy and Applications, 06, 88. https://doi.org/10.4172/2090-4541.1000194
97. Shah, Z., Veses, R. C., & da Silva, R. (2016). A comparative study and analysis of two types of bio-oil samples obtained from freshwater algae and microbial treated algae. Modern Chemistry & Applications, 4, 185. https://doi.org/10.4172/2329-6798.1000185
98. Shah, Z., Veses, R. C., Brum, J., Klunk, M. A., Rocha, L. A. O., Missaggia, A. B., de Oliveira, A. P., & Caetano, N. R. (2020). Sewage sludge bio-oil development and characterization. Inventions, 5, 36. https://doi.org/10.3390/inventions5030036
99. Shah, Z., Saberian, M., Hodgeman, D., Kinloch, I., & Vallés, C. (2023). Effect of sulfur on wood tar biopitch as a sustainable replacement for coal tar pitch binders. ACS Applied Engineering Materials, 1, 2567–2576. https://doi.org/10.1021/acsaenm.3c00361
100. Klunk, M. A., Dasgupta, S., Das, M., & Shah, Z. (2018). SYSTEM OF ADSORPTION OF CO2 IN COALBED. SOUTHERN BRAZILIAN JOURNAL OF CHEMISTRY., 26, 2–9. https://doi.org/10.48141/SBJCHEM.v26.n26.2018.7_2018.pdf

101. Mohapatra, S., Mohapatra, U., & Mishra, R. (2017). Diversification towards vegetable crops: A good option for doubling the farmer's income. Journal of Experimental Agriculture International, 18, 1–7.
102. Chavan, S. D., Bansal, S., Mohapatra, S., Kaur, L., & Jadhav, A. (2023). Trade directions of Indian basmati rice export-Markov chain approach. Economic Affairs, 68(01), 541–547.
103. Bansal, S., Mohapatra, S., Gohain, N., & Das, A. (2022). Production status, market integration and the law of one price: An exploration of the major maize markets of India. Indian Journal of Economics and Development, 18, 832–841.
104. Mohapatra, S., & Singh, J. (2021). Co-integration of major vegetable markets in Punjab, India. Economic Affairs, 66(02), 181–187.
105. Mohapatra, S., Samantaray, D. P., & Samantaray, S. M. (2015). Study on polyhydroxyalkanoates production using rhizospheric soil bacterial isolates of sweet potato. Indian Journal of Science and Technology, 8, 57–62. https://doi.org/10.17485/ijst/2015/v8iS7/64027
106. Arora, R., Trivedi, J., Mohapatra, S., & Kumar, P. (2022). Trends in probiotics on human health and industrial application. In Advances in agricultural and industrial microbiology (Microbial diversity and application in agroindustry) (Vol. 1, pp. 237–248). Springer. https://doi.org/10.1007/978-981-16-8918-5_13
107. Samal, S., Pati, S., Mohapatra, S., Maity, S., Tanaya, K., Devadarshini, D., & Samantaray, D. (2023). PHAs production by facultative anaerobic bacteria Bacillus cereus FM5 through submerged and solid-state fermentation under anoxic condition. Antonie van Leeuwenhoek, International Journal of General and Molecular Microbiology, 116, 521–529. https://doi.org/10.1007/s10482-023-01825-0
108. Biswas, N., Saha, S., Biswas, B. K., Chowdhury, M., Rahaman, A., Junghare, V., Mohapatra, S., Hazra, S., Zangrando, E., Roy Choudhury, R., & Roy Choudhury, C. (2021). The DNA- and protein-binding properties and cytotoxicity of a new copper(II) hydrazone Schiff base complex. Journal of Coordination Chemistry, 74, 1482–1504. https://doi.org/10.1080/00958972.2021.1913128
109. Mohapatra, S., Mishra, R., Roy, P., Yadav, K. L., & Satapathi, S. (2017). Systematic investigation and in vitro biocompatibility studies on implantable magnetic nanocomposites for hyperthermia treatment of osteoarthritic knee joints. Journal of Materials Science, 52, 9262–9268. https://doi.org/10.1007/s10853-017-1136-0
110. Mishra, B. B., Nayak, S. K., Mohapatra, S., & Samantaray, D. (2021). Environmental and agricultural microbiology: Advances and applications. https://doi.org/10.1002/9781119525899
111. Sharma, S. K., Mohapatra, S., Sharma, R. C., Alturjman, S., Altrjman, C., Mostarda, L., & Stephan, T. (2022). Retrofitting existing buildings to improve energy performance. Sustainability (Switzerland), 14, 666. https://doi.org/10.3390/su14020666
112. Bandhu, S., Khot, M. B., Sharma, T., Sharma, O. P., Dasgupta, D., Mohapatra, S., Hazra, S., Khatri, O. P., & Ghosh, D. (2018). Single cell oil from oleaginous yeast grown on sugarcane bagasse-derived xylose: An approach toward novel biolubricant for low friction and wear. ACS Sustainable Chemistry & Engineering, 6, 275–283. https://doi.org/10.1021/acssuschemeng.7b02425
113. Kohli, I., Joshi, N. C., Mohapatra, S., & Varma, A. (2020). Extremophile—An adaptive strategy for extreme conditions and applications. Current Genomics, 21, 96–110. https://doi.org/10.2174/1389202921666200401105908
114. Pati, S., Mohapatra, S., Maity, S., Dash, A., & Samantaray, D. P. (2021). Polyhydroxyalkanoates for sustainable smart packaging of fruits. In Environmental and Agricultural Microbiology: Advances and Applications (pp. 183–195).
115. Chatterjee, A., Nirwan, S., Mohapatra, S., Sharma, P., Agnihotri, A., & Shrivastava, N. (2022). Biochemical aspects of pathogenic variability in white rust infected Indian mustard. Mycologia, 114, 757–768. https://doi.org/10.1080/00275514.2022.2060007

116. Pati, S., Maity, S., Dash, A., Jema, S., Mohapatra, S., Das, S., & Samantaray, D. P. (2020). Biocompatible PHB production from bacillus species under submerged and solid-state fermentation and extraction through different downstream processing. Current Microbiology, 77, 1203–1209. https://doi.org/10.1007/s00284-020-01922-7
117. Agrahari, R., Sarraf, G., Joshi, N. C., Mohapatra, S., & Var, A. (2021). Insight of biopolymers and applications of polyhydroxyalkanoates. In Microbial polymers: Applications and ecological perspectives (pp. 177–191). Springer. https://doi.org/10.1007/978-981-16-0045-6_8
118. Kathpalia, M., Kumar, P., & Mohapatra, S. (2022). Homocysteine metabolism as a biomarker for cancer. In Homocysteine metabolism in health and disease (pp. 159–171). Springer. https://doi.org/10.1007/978-981-16-6867-8_8
119. Bharatiy, S. K., Hazra, M., Paul, M., Mohapatra, S., Samantaray, D., Dubey, R. C., Sanyal, S., Datta, S., & Hazra, S. (2016). In silico designing of an industrially sustainable carbonic anhydrase using molecular dynamics simulation. ACS Omega, 1, 1081–1103. https://doi.org/10.1021/acsomega.6b00041
120. Sarkar, M. S., Segu, H., Bhaskar, J. V., Jakher, R., Mohapatra, S., Shalini, K., Shivaji, S., & Reddy, P. A. (2018). Ecological preferences of large carnivores in remote, high-altitude protected areas: Insights from Buxa Tiger Reserve., India. ORYX, 52, 66–77. https://doi.org/10.1017/S0030605317000060
121. Pati, S., Dash, A., Maity, S., Pattnaik, S., Mohapatra, S., & Samantaray, D. P. (2021). Nano biofertilizer for sustainable agriculture. In Agriculturally important microorganisms: Mechanisms and applications for sustainable agriculture (pp. 201–208). CRC Press. https://doi.org/10.1201/9781003245841-10
122. Mohanty, D., Misra, S., Mohapatra, S., & Sahu, P. S. (2018). Prebiotics and synbiotics: Recent concepts in nutrition. Food Bioscience, 26, 152–160. https://doi.org/10.1016/j.fbio.2018.10.008
123. Kumar, R., Kumari, R., Prasad, C., Deep, A., Singh, N., Namtak, S., Tiwari, V., Merugu, R., Mohapatra, S., & Singh, R. (2021). Influence of water quality on the diversity of macroinvertebrates in the Mandakini River in India. Water Science and Technology: Water Supply, 21, 1843–1860. https://doi.org/10.2166/WS.2021.020
124. Mohanty, D., Jena, R., Choudhury, P. K., Pattnaik, R., Mohapatra, S., & Saini, M. R. (2016). Milk derived antimicrobial bioactive peptides: A review. International Journal of Food Properties, 19, 837–846. https://doi.org/10.1080/10942912.2015.1048356
125. Mohapatra, S., Rath, S. N., Pradhan, S. K., Samantaray, D. P., & Rath, C. C. (2016). Secondary structural models (16S rRNA) of polyhydroxyalkanoates producing bacillus species isolated from different rhizospheric soil: Phylogenetics and chemical analysis. International Journal Bioautomation, 20, 329–338.
126. Mohapatra, S., Sarkar, B., Samantaray, D. P., Daware, A., Maity, S., Pattnaik, S., & Bhattacharjee, S. (2017). Bioconversion of fish solid waste into PHB using Bacillus subtilis based submerged fermentation process. Environmental Technology (United Kingdom), 38, 3201–3208. https://doi.org/10.1080/09593330.2017.1291759
127. Misra, S., Mohanty, D., & Mohapatra, S. (2021). Food-based probiotics: Functional dietary ingredients. In Probiotics: Advanced food and health applications (pp. 257–275). Academic Press. https://doi.org/10.1016/B978-0-323-85170-1.00006-3
128. Kumar, R., Kumari, R., Prasad, C., Tiwari, V., Singh, N., Mohapatra, S., Merugu, R., Namtak, S., & Deep, A. (2020). Phytoplankton diversity in relation to physicochemical attributes and water quality of Mandakini River, Garhwal Himalaya. Environmental Monitoring and Assessment, 192, 799. https://doi.org/10.1007/s10661-020-08768-3
129. Mohapatra, S., Vishwakarma, K., Joshi, N. C., Maity, S., Kumar, R., Ramchander, M., Pattnaik, S., & Samantaray, D. P. (2021). A review on PHAs: The future biopolymer. In Environmental and agricultural microbiology: Advances and applications (pp. 83–117).

130. Sahoo, N., Pattnaik, S., Pattnaik, M., & Mohapatra, S. (2021). Methane emission and strategies for mitigation in livestock. In Environmental and agricultural microbiology: Advances and applications (pp. 257–274).
131. Mohanty, D. P., Mohapatra, S., Misra, S., & Sahu, P. S. (2016). Milk derived bioactive peptides and their impact on human health—A review. Saudi Journal of Biological Sciences, 23, 577–583. https://doi.org/10.1016/j.sjbs.2015.06.005
132. Kohli, I., Mohapatra, S., Kumar, P., Goel, A., Varma, A., & Joshi, N. C. (2022). Role of bacterial endophytes in the promotion of plant growth. In Bacterial endophytes for sustainable agriculture and environmental management (pp. 235–253). https://doi.org/10.1007/978-981-16-4497-9_12
133. Samantaray, D., Mohapatra, S., & Mishra, B. B. (2014). Microbial bioremediation of industrial effluents. In Microbial biodegradation and bioremediation (pp. 326–339). https://doi.org/10.1016/B978-0-12-800021-2.00014-5
134. Pattnaik, S., Dash, D., Mohapatra, S., Pattnaik, M., Marandi, A. K., Das, S., & Samantaray, D. P. (2020). Corrigendum to "Improvement of rice plant productivity by native Cr(VI) reducing and plant growth promoting soil bacteria Enterobacter cloacae" (Chemosphere (2020) 240, (124895), (S0045653519321344), (10.1016/j.chemosphere.2019.124895)). Chemosphere, 238. https://doi.org/10.1016/j.chemosphere.2019.125011
135. Mohapatra, S., Mohanta, P. R., Sarkar, B., Daware, A., Kumar, C., & Samantaray, D. P. (2017). Production of polyhydroxyalkanoates (PHAs) by bacillus strain isolated from waste water and its biochemical characterization. Proceedings of the National Academy of Sciences India ion B - Biological Sciences, 87, 459–466. https://doi.org/10.1007/s40011-015-0626-6
136. Tomer, P., Shroti, G. K., Mohapatra, S., Ghosh, D., Jaiswal, S., Garg, D., Lahiri, D., Kumar, N., Khatri, O. P., & Hazra, S. (2023). Fabrication of HA nano-crystal reinforced PHAs based composites for orthopedic fracture-fixing accessories using sustainable resources. Environment, Development and Sustainability, 27, 1–19. https://doi.org/10.1007/s10668-023-03912-8
137. Bandekar, D., Mohapatra, S., Hazra, M., Hazra, S., & Biswas, S. (2021). N-terminal truncation of VC0395_0300 protein from vibrio cholerae does not lead to loss of diguanylate cyclase activity. Biophysical Chemistry, 268, 106493. https://doi.org/10.1016/j.bpc.2020.106493
138. Vishwakarma, K., Kumar, N., Shandilya, C., Mohapatra, S., Bhayana, S., & Varma, A. (2020). Revisiting plant–microbe interactions and microbial consortia application for enhancing sustainable agriculture: A review. Frontiers in Microbiology, 11. https://doi.org/10.3389/fmicb.2020.560406
139. Mohapatra, S., Maity, S., Dash, H. R., Das, S., Pattnaik, S., Rath, C. C., & Samantaray, D. (2017). Bacillus and biopolymer: Prospects and challenges. Biochemistry and Biophysics Reports, 12, 206–213. https://doi.org/10.1016/j.bbrep.2017.10.001
140. Kumar, R., Merugu, R., Mohapatra, S., & Sharma, S. (2022). Extremophiles: Life of microorganisms in extreme environments. In Extremophiles: General and plant biomass based biorefinery (pp. 43–66). https://doi.org/10.1201/9781003014188-3
141. Pradhan, A., Pahari, A., Mohapatra, S., & Mishra, B. B. (2017). Phosphate-solubilizing microorganisms in sustainable agriculture: Genetic mechanism and application. In Microorganisms for sustainability (pp. 81–97). https://doi.org/10.1007/978-981-10-7380-9_5
142. Devadarshini, D., Mohapatra, S., Pati, S., Maity, S., Rath, C. C., Jena, P. K., & Samantaray, D. (2023). Evaluation of PHAs production by mixed bacterial culture under submerged fermentation. Biologia (Bratisl), 78, 895–902. https://doi.org/10.1007/s11756-022-01302-5
143. Agarwal, K. M., Mohapatra, S., Sharma, P., Sharma, S., Bhatia, D., & Mishra, A. (2020). Study and overview of the novel corona virus disease (COVID-19). Sensors International, 1, 100037. https://doi.org/10.1016/j.sintl.2020.100037

144. Tanaya, K., Mohapatra, S., Maity, S., & Samantaray, D. P. (2023). Fluorescent-pigmented microbial polymer nanoparticles: An approach for replacement of bioimaging agent in the drug delivery system. Applied Biochemistry and Microbiology, 59, 378–383. https://doi.org/10.1134/S0003683823030171
145. Pattnaik, S., Dash, D., Mohapatra, S., Pati, S., Devadarshini, D., Samal, S., Pattnaik, M., Maity, S., Mishra, S. K., & Samantaray, D. (2024). Reclamation of chromium-contaminated soil by native Cr(VI)-reducing and PHA-accumulating bacillus aryabhattai CTSI-07. International Microbiology, 27, 731–742. https://doi.org/10.1007/s10123-023-00421-6
146. Mohapatra, S., Mishra, Y., & Dash, L. (2020). Covid-19: Analysing the legal nuances of the lockdown order. Indian Journal of Forensic Medicine and Toxicology, 14, 612–616. https://doi.org/10.37506/ijfmt.v14i4.11553
147. Pati, S., Mohanty, M. K., Mohapatra, S., & Samantaray, D. (2022). Bioethanol production by Enterobacter hormaechei through carbon monoxide-rich syngas fermentation. International Journal of Energy Research, 46, 20096–20106. https://doi.org/10.1002/er.8678
148. Maity, S., Das, S., Mohapatra, S., Tripathi, A. D., Akthar, J., Pati, S., Pattnaik, S., & Samantaray, D. P. (2020). Growth associated polyhydroxybutyrate production by the novel Zobellellae tiwanensis strain DD5 from banana peels under submerged fermentation. International Journal of Biological Macromolecules, 153, 461–469. https://doi.org/10.1016/j.ijbiomac.2020.03.004
149. Takkar, S., Shandilya, C., Agrahari, R., Chaurasia, A., Vishwakarma, K., Mohapatra, S., Varma, A., & Mishra, A. (2022). Green technology: Phytoremediation for pesticide pollution. In Phytoremediation technology for the removal of heavy metals and other contaminants from soil and water (pp. 353–375). https://doi.org/10.1016/B978-0-323-85763-5.00008-8
150. Bandekar, D., Chouhan, O. P., Mohapatra, S., Hazra, M., Hazra, S., & Biswas, S. (2017). Putative protein VC0395_0300 from vibrio cholerae is a diguanylate cyclase with a role in biofilm formation. Microbiological Research, 202, 61–70. https://doi.org/10.1016/j.micres.2017.05.003
151. Pattnaik, S., Dash, D., Mohapatra, S., Pattnaik, M., Marandi, A. K., Das, S., & Samantaray, D. P. (2020). Improvement of rice plant productivity by native Cr(VI) reducing and plant growth promoting soil bacteria Enterobacter cloacae. Chemosphere, 240. https://doi.org/10.1016/j.chemosphere.2019.124895
152. Mohapatra, S., Samantaray, D. P., Samantaray, S. M., Mishra, B. B., Das, S., Majumdar, S., Pradhan, S. K., Rath, S. N., Rath, C. C., Akthar, J., & Achary, K. G. (2016). Structural and thermal characterization of PHAs produced by Lysinibacillus sp. through submerged fermentation process. International Journal of Biological Macromolecules, 93, 1161–1167. https://doi.org/10.1016/j.ijbiomac.2016.09.077
153. Palli, S., Koona, R., Sharma, R. C., & Muddada, V. (2015). Dynamic analysis of Indian railway integral coach factory bogie. International Journal of Vehicle Structures and Systems, 7, 16–20. https://doi.org/10.4273/ijvss.7.1.03
154. Palli, S., & Koona, R. (2015). Analyses of dynamic response of a railway bogie. International Journal of Vehicle Noise and Vibration, 11, 103–113. https://doi.org/10.1504/IJVNV.2015.070015
155. Swathi, K., & Srihari, P. (2015). Heat transfer enhancement ina tube using rectangular strip inserts. International Journal of Applied Engineering Research, 10, 41532–41544.
156. Palli, S., Sharma, R. C., & Rao, P. P. D. (2017). Dynamic behaviour of a 7 DoF passenger car model. International Journal of Vehicle Structures and Systems, 9, 57–63. https://doi.org/10.4273/ijvss.9.1.12
157. Sharma, R. C., Palli, S., & Koona, R. (2017). Stress and vibrational analysis of an Indian railway RCF bogie. International Journal of Vehicle Structures and Systems, 9, 296–302. https://doi.org/10.4273/ijvss.9.5.06

158. Sharma, S. K., Kumar, A., & Sharma, R. C. (2014). Challenges in railway vehicle modeling and simulations. In International conference on newest drifts in mechanical engineering ICNDME-2014 (pp. 453–459).
159. Sharma, S. K., & Kumar, A. (2014). A comparative study of Indian and worldwide railways. International Journal of Mechanical Engineering and Robotics Research, 1, 114–120.
160. Sharma, S., Sharma, R. C., Sharma, S. K., Sharma, N., Palli, S., & Bhardawaj, S. (2020). Vibration isolation of the quarter car model of road vehicle system using dynamic vibration absorber. International Journal of Vehicle Structures and Systems, 12, 513–516. https://doi.org/10.4273/IJVSS.12.5.05
161. Palli, S., Dontikurti, R., Sharma, R. C., & Sharma, N. (2021). Analysis of dynamic response of a railway locomotive using ansys. International Journal of Vehicle Structures and Systems, 13, 250–255. https://doi.org/10.4273/ijvss.13.2.21
162. Sharma, R. C., Sharma, N., Singh, G., & Palli, S. (2021). Modeling and simulation of human body-vehicle-track system for the investigation of ride comfort. SAE International Journal of Commercial Vehicles, 15, 175–188. https://doi.org/10.4271/02-15-02-0008
163. Sharma, R. C., Palli, S., Avesh, M., & Sharma, N. (2021). Vibration isolation of railway vehicle car body using semi-active suspension. International Journal of Vehicle Structures and Systems, 13, 482–487. https://doi.org/10.4273/ijvss.13.4.17
164. Gopala Rao, L. V. V., Sharma, R. C., Satyanarayana, V. S. V., & Palli, S. (2022). Stochastic optimal preview control response of a quarter-car nonlinear suspension model using spectral decomposition method. Noise and Vibration Worldwide, 53, 225–232. https://doi.org/10.1177/09574565221093234
165. Singh, B. J., Palli, S., Kumar, P., & Bahl, S. (2023). Behavioural investigation of hardness profile for dissimilar TIG-weldment between P22 and P91 steels. International Journal of Vehicle Structures and Systems, 15, 31–36. https://doi.org/10.4273/ijvss.15.1.06
166. Naveen, Y., Lokanadham, D., Naidu, D. R., Sharma, R. C., Palli, S., & Lila, M. K. (2023). An experimental study on the influence of blended Karanja biodiesel on diesel engine characteristics. International Journal of Vehicle Structures and Systems, 15, 154–160. https://doi.org/10.4273/ijvss.15.2.02
167. Bhardawaj, S., Jha, A. K., Sharma, R. C., Palli, S., Lokanadham, D., Sharma, N., & Dhiman, N. (2023). Finite element analysis of two disk rotor system used in automobile turbochargers. International Journal of Vehicle Structures and Systems, 15, 501–504. https://doi.org/10.4273/ijvss.15.4.09
168. Sharma, R. C., Palli, S., Bhardawaj, S., Jha, A. K., Duppala, A., Sharma, N., Sharma, S., & Dhiman, N. (2023). Dynamic behaviour of railway vehicle under bump and pothole track irregularities. International Journal of Vehicle Structures and Systems, 15, 566–569. https://doi.org/10.4273/ijvss.15.4.22
169. Raju, S., Palli, S., Devi Prasad, P., Menda, V. R., & Ramakrishna, B. (2024). A hybrid AHP-TOPSIS, MOORA technique for multi-objective optimization of thermal, mechanical, and water absorption behavior of epoxy/hemp, pine apple, and palm fiber composites. Journal of the Chinese Institute of Engineers, Transactions of the Chinese Institute of Engineers, Series A, 47, 23–36. https://doi.org/10.1080/02533839.2023.2274092
170. Raju Rallabandi, S., Pilla, D. P., Dowluru, S., Palli, S., Sharma, N., Sharma, S. K., & Sharma, R. C. (2024). Critical evaluation of epoxy-hemp-pineapple-palm fiber composites using hybrid AHM-TOPSIS technique for sustainable structural applications. Journal of the Chinese Institute of Engineers, Transactions of the Chinese Institute of Engineers, Series A, 47, 325–336. https://doi.org/10.1080/02533839.2024.2308250
171. Gopala Rao, L. V. V., Satyanarayana, V. S. V., Sharma, R. C., Palli, S., Raturi, A., Kharola, A., & Sharma, A. (2024). Preview control of the random response of a full car vehicle model traversing a rough road. International Journal of Vehicle Noise and Vibration, 20, 70–88. https://doi.org/10.1504/IJVNV.2024.138105

172. Sharma, A., Sharma, V. S., Gupta, S., Sharma, R. C., Palli, S., & Sharma, N. (2024). Additive manufacturing in the COVID-19 pandemic: Equipment and challenges? Journal of Micromanufacturing, 60, 787–798. https://doi.org/10.1177/25165984241228083
173. Bammidi, R., Madivada, H., Dowluru, S., Palli, S., Sharma, R. C., & Muddada, V. (2024). Mechano-tribological investigations of Al-%Wt. of boron nitride as metal-ceramic composites. AIP Conference Proceedings, 2978, 020017. https://doi.org/10.1063/5.0182965
174. Ranganath, B. A., Duppala, A., Palli, S., Rao, L. V. V., & Sharma, R. C. (2024). Numerical analysis and experimental study of FRP composites for damping applications. AIP Conference Proceedings, 2978, 060001. https://doi.org/10.1063/5.0184802
175. Rallabandi, S. R., Srinivas, L., Palli, S., Sharma, R. C., Sharma, N., Sharma, A., & Sharma, S. K. (2024). A contrastive characterization of pure Mg and AZ91D alloy based on the testing of mechanical, corrosion, wear, and erosion properties. Engineering Research Express, 6, 015017. https://doi.org/10.1088/2631-8695/ad16a2
176. Sharma, R. C., Palli, S., Duppala, A., Bhardawaj, S., Sharma, S. K., Sharma, N., & Sharma, A. (2024). Rigid body analysis of ride behaviour of roadway vehicle using Lagrangian dynamics. International Journal of Vehicle Structures and Systems, 16, 74–78. https://doi.org/10.4273/ijvss.16.1.15
177. Menda, V., Dora, S., Palli, S., Rallabandi, S. R., Sharma, R. C., Sharma, N., & Bondala, R. (2024). Mechanical and wear characteristics of aluminium-7075/graphene/TiB2 prepared by stir casting. Engineering Research Express, 6, 025571. https://doi.org/10.1088/2631-8695/ad58a8
178. Palli, S., Duppala, A., Sharma, R. C., & Rao, L. V. V. G. (2022). Dynamic simulation of automotive vehicle suspension using MATLAB Simulink. International Journal of Vehicle Structures and Systems, 14, 310–314. https://doi.org/10.4273/ijvss.14.3.04
179. Satyanarayana, V. S. V., Sharma, R. C., Sateesh, B., Gopala Rao, L. V. V., Mohan Rao, N., & Palli, S. (2022). Optimum response of a nonlinear passive vehicle suspension system under random road excitations. SAE International Journal of Commercial Vehicles, 16. https://doi.org/10.4271/02-16-01-0004
180. Lokanadham, D., Sharma, R. C., Palli, S., & Bhardawaj, S. (2022). Wear rate modelling and analysis of limestone slurry particulate composites using the fuzzy method. International Journal on Recent and Innovation Trends in Computing and Communication., 10, 133–143. https://doi.org/10.17762/ijritcc.v10i1s.5818
181. Sharma, R. C., Palli, S., Gopala Rao, L. V. V., Duppala, A., & Sharma, S. K. (2022). Four-wheel vehicle response under bump, pothole, harmonic, and random excitations using bond graph/Simulink technique. SAE International Journal of Commercial Vehicles, 16, 129–139. https://doi.org/10.4271/02-16-02-0008
182. Sharma, R. C., Gopala Rao, L. V. V., Sharma, S. K., Palli, S., & Satyanarayana, V. S. V. (2022). Analysis of lateral stability and ride of an Indian railway constrained dual-axle bogie frame. SAE International Journal of Commercial Vehicles, 16, 213–228. https://doi.org/10.4271/02-16-02-0014
183. Sharma, R. C., Palli, S., Bhardawaj, S., & Rao, L. V. V. G. (2022). Advances in human biodynamic modeling for analysis of whole-body vibrations. International Journal of Vehicle Structures and Systems, 14, 503–511. https://doi.org/10.4273/ijvss.14.4.17
184. Sharma, R. C., Palli, S., & Sharma, S. K. (2023). Ride analysis of railway vehicle considering rigidity and flexibility of the carbody. Journal of the Chinese Institute of Engineers, Transactions of the Chinese Institute of Engineers, Series A, 46, 355–366. https://doi.org/10.1080/02533839.2023.2194918
185. Sharma, R. C., Palli, S., & Bhardawaj, S. (2023). A review of wheel-rail dynamic behavior on curved tracks. International Journal of Vehicle Structures and Systems, 15, 19–25. https://doi.org/10.4273/ijvss.15.1.04

186. Jyotiprasad, G., Dowluru, S., Rallabandi, S. R., Sharma, N., Sharma, R. C., & Palli, S. (2023). Thermal-structural and macromechanical behavior analysis of graphite-based laminated composite. SAE Technical Paper 2023-01-5041. https://doi.org/10.4271/2023-01-5041.
187. Saikiran, C., Palli, S., Bondala, R., Sharma, R. C., Sreeramulu, D., & Dontikurti, R. (2023). Experimental investigation on thermal behaviour of silicon carbide (SiC) based polymer composites under hygrothermal conditions. International Journal of Vehicle Structures and Systems, 15, 222–226. https://doi.org/10.4273/ijvss.15.2.15
188. Sharma, S. K., Sharma, R. C., Kumar, A., & Palli, S. (2015). Challenges in rail vehicle-track modeling and simulation. International Journal of Vehicle Structures & Systems, 7, 1.
189. Palli, S., Sharma, R. C., Sharma, S. K., Muddada, V., & Sharma, N. (2020). A case study on noise pollution and its effects. Journal of Critical Reviews, 7, 261–267.
190. Sharma, R. C., Sharma, S. K., Sharma, N., & Sharma, S. (2020). Analysis of ride and stability of an ICF railway coach. International Journal of Vehicle Noise and Vibration, 16, 127–144.
191. Palli, S., Sharma, R. C., Sharma, S. K., & Chintada, V. B. (2020). On methods used for setting the curve for railway tracks. Journal Of Critical Review, 7, 241–246.
192. Lee, J., & Sharma, S. K. (2020). Numerical investigation of critical speed analysis of high-speed rail vehicle. 한국정밀공학회 학술발표대회 논문집, 696.
193. Acharya, A., Gahlaut, U., Sharma, K., Sharma, S. K., Vishwakarma, P. N., & Phanden, R. K. (2020). Crashworthiness analysis of a thin-walled structure in the frontal part of automotive chassis. International Journal of Vehicle Structures & Systems, 12, 517–520.
194. Sharma, S. K., & Lee, J. (2020). Finite element analysis of a fishplate rail joint in extreme environment condition. International Journal of Vehicle Structures & Systems, 12, 503–506.
195. Bhardawaj, S., Sharma, R. C., & Sharma, S. K. (2020). Development of multibody dynamical using MR damper based semi-active bio-inspired chaotic fruit fly and fuzzy logic hybrid suspension control for rail vehicle system. Proceedings of the Institution of Mechanical Engineers, Part K: Journal of Multi-body Dynamics, 234, 723–744.
196. Sharma, S. K., & Lee, J. (2020). Design and development of smart semi active suspension for nonlinear rail vehicle vibration reduction. International Journal of Structural Stability and Dynamics, 20, 2050120.
197. Sharma, S. K., Phan, H., & Lee, J. (2020). An application study on road surface monitoring using DTW based image processing and ultrasonic sensors. Applied Sciences, 10, 4490.
198. Sharma, R. C., Sharma, S., Sharma, S. K., & Sharma, N. (2020). Analysis of generalized force and its influence on ride and stability of railway vehicle. Noise & Vibration Worldwide, 51, 95–109.
199. Bhardawaj, S., Sharma, R. C., & Sharma, S. K. (2020). Ride analysis of track-vehicle-human body interaction subjected to random excitation. Journal of the Chinese Society of Mechanical Engineers, 41, 227–236.
200. Bhardawaj, S., Sharma, R. C., & Sharma, S. K. (2020). Development in the modeling of rail vehicle system for the analysis of lateral stability. Materials Today Proceedings, 25, 610–619.
201. Sharma, S. K., Sharma, R. C., & Sharma, N. (2020). Combined multi-body-system and finite element analysis of a rail locomotive crashworthiness. International Journal of Vehicle Structures & Systems, 12, 428–435.
202. Sharma, R. C., Avesh, M., Sharma, S. K., Sharma, S., Palli, S., & Sharma, N. (2020). Linear and non-linear stability analysis of a constrained railway wheel axle. International Journal of Vehicle Structures & Systems (IJVSS), 12.
203. Bhardawaj, S., Sharma, R. C., & Sharma, S. K. (2020). Analysis of frontal car crash characteristics using ANSYS. Materials Today Proceedings, 25, 898–902.
204. Bhardawaj, S., Sharma, R. C., & Sharma, S. K. (2019). A survey of railway track modelling. International Journal of Vehicle Structures & Systems, 11, 508–518.

205. Bhardawaj, S., Sharma, R. C., & Sharma, S. K. (2019). Development and advancement in the wheel-rail rolling contact mechanics. In IOP conference series: Materials science and engineering (p. 012034). IOP Publishing.
206. Choppara, R. K., Sharma, R. C., Sharma, S. K., & Gupta, T. (2019). Aero dynamic cross wind analysis of locomotive. In IOP conference series: Materials science and engineering (p. 012035). IOP Publishing.
207. Sharma, S. K., & Sharma, R. C. (2019). Pothole detection and warning system for Indian roads. In Advances in interdisciplinary engineering: Select proceedings of FLAME 2018 (pp. 511–519). Springer Singapore.
208. Sinha, A. K., Sengupta, A., Gandhi, H., Bansal, P., Agarwal, K. M., Sharma, S. K., Sharma, R. C., & Sharma, S. K. (2019). Performance enhancement of an all-terrain vehicle by optimizing steering, powertrain and brakes. In Advances in engineering design: Select proceedings of FLAME 2018 (pp. 207–215). Springer Singapore.
209. Goyal, S., Anand, C. S., Sharma, S. K., & Sharma, R. C. (2019). Crashworthiness analysis of foam filled star shape polygon of thin-walled structure. Thin-Walled Structures, 144, 106312.
210. Goswami, B., Rathi, A., Sayeed, S., Das, P., Sharma, R. C., & Sharma, S. K. (2019). Optimization design for aerodynamic elements of Indian locomotive of passenger train. In Advances in engineering design: Select proceedings of FLAME 2018 (pp. 663–673). Springer Singapore.
211. Sharma, S. K., Saini, U., & Kumar, A. (2019). Semi-active control to reduce lateral vibration of passenger rail vehicle using disturbance rejection and continuous state damper controllers. Journal of Vibration Engineering & Technologies, 7, 117–129.
212. Sharma, S. K. (2019). Multibody analysis of longitudinal train dynamics on the passenger ride performance due to brake application. Proceedings of the Institution of Mechanical Engineers, Part K: Journal of Multi-body Dynamics, 233, 266–279.
213. Shukla, S., Kumar, U., Sharma, S. K., Gupta, P., & Kumar, A. (2017). Light weight freight rolling stock bogie frame: Design methodology validated with field oscillation trials. International Journal of Vehicle Structures & Systems, 9, 245–250.
214. Sharma, R. C., Palli, S., Sharma, S. K., & Roy, M. (2017). Modernization of railway track with composite sleepers. International Journal of Vehicle Structures & Systems, 9, 321–329.
215. Sharma, S. K., & Kumar, A. (2017). Ride performance of a high speed rail vehicle using controlled semi active suspension system. Smart Materials and Structures, 26, 055026.
216. Sharma, S. K., & Kumar, A. (2017). Impact of electric locomotive traction of the passenger vehicle ride quality in longitudinal train dynamics in the context of Indian railways. Mechanics & Industry, 18, 222.
217. Sharma, S. K., & Kumar, A. (2016). Dynamics analysis of wheel rail contact using FEA. Procedia Engineering, 144, 1119–1128.
218. Sharma, S. K., & Chaturvedi, S. (2016). Jerk analysis in rail vehicle dynamics. Perspectives in Science (Neth), 8, 648–650.
219. Sharma, S. K., & Kumar, A. (2016). The impact of a rigid-flexible system on the ride quality of passenger bogies using a flexible carbody. In Proceedings of the third international conference on railway technology: Research, development and maintenance (p. 87). Civil-Comp Press.
220. Lavania, S., & Sharma, S. K. (2011). An explicit approach to compare crosstalk noise and delay in VLSI RLC interconnect modeled with skin effect with step and ramp input. Journal of VLSI Design Tools & Technology, 1, 1–8.

Sustainable Biocomposites and Hybrid Materials by Integrating Digital Manufacturing for Industry 5.0

Roopsandeep Bammidi, Subrahmanyam Vasamsetti, Aruna Kunda, Zeban Shah, Emmauel Oghenero Emojevu, Laxmanaraju Salavarao, and Nura Garba

1 Introduction

Industry 5.0 represents the next phase in industrial evolution, building upon the automation and digitalization advancements of Industry 4.0. This new paradigm shift places greater emphasis on sustainability, human-centric production, and the integration of digital technologies with ecological awareness. The growing concern

R. Bammidi (✉)
Department of Mechanical Engineering, Aditya Institute of Technology and Management, Tekkali, Andhra Pradesh, India

S. Vasamsetti
Department of Automobile Engineering, Godavari Global University, Rajamahendravaram, Andhra Pradesh, India

A. Kunda
Department of Automobile Engineering, Godavari Institute of Engineering and Technology, Rajamahendravaram, Andhra Pradesh, India
e-mail: karuna@giet.ac.in

Z. Shah
Federal University of Rio Grande do Sul, Porto Alegre, Brazil
e-mail: zeban.shah@ufrgs.br

E. O. Emojevu
Department of Chemistry, University of Benin, Benin, Nigeria

L. Salavarao
Department of Mechanical Engineering, Sri Sivani College of Engineering, Srikakulam, Andhra Pradesh, India

N. Garba
Department of Environmental and Resources Management, Usmanu Danfodiyo University, Sokoto, Nigeria
e-mail: nura.garba@udusok.edu.ng

Y. Kumar Singla et al. (eds.), *Digital Manufacturing in Industry 5.0*,
https://doi.org/10.1007/978-3-031-91500-0_30

for environmental preservation, resource efficiency, and the circular economy has prompted industries to rethink traditional manufacturing practices. In this context, the use of biocomposites as an alternative to conventional synthetic materials has garnered increasing attention.

Biocomposites are composed of natural fibers and polymers through a binding matrix. They provide various environmental benefits relative to conventionally used composites such as decreased carbon dioxide emissions, ability to be broken down in nature, and the ability to be replaced. Replacements using natural fibers all bending reinforcement, including hemp, flax, jute, and kenaf, provide these composites with increased lightweight and reduced price. According to these findings, and as the needs of the industries keep increasing, the biocomposites adopt a fundamental role as manufacturing processes become greener.

Industry 5.0 improves upon this toward advanced development in bio composite material production with the inclusion of advanced manufacturing processes, particularly additive manufacturing and the digital twin technologies. These techniques facilitate the precise specification of material features, the maximization of resources utilized, and better efficiencies across the entire production cycle. In addition, the principle of digital manufacturing is suitable under the contexts of Industry 5.0 providing the manufacturers opportunities of making biocomposites that are performance oriented and durable with no or little wastage.

This paper thus aims to address the merger of digital manufacturing methods with biocomposites manufactured in a sustainable way within Industry 5.0. The direction of attention is the employment of natural fibers in biocomposites formation and the ways how additive manufacturing and digital twins can facilitate their design and applications. The paper explores the major issues that hinder the wide use of biocomposites in large-scale industries but at the same time stresses the ecological and sustainability advantages brought by the materials. Furthermore, the use of biocomposite materials in large-scale industries like automotive and aerospace are discussed delivering the industry 5.0 technologies of practical nature. The objective of the presented findings is to add to the increasing discourse on the use and adoption of digital and green technologies and how such technologies can transform the manufacturing industry.

2 Literature Review

Recent studies have emphasized improving wear resistance and mechanical performance in the context of Industry 5.0, particularly in transportation and manufacturing sectors. These works emphasize a variety of advanced engineering techniques that contribute to the development of smart manufacturing and Industry 5.0. Studies by Shah et al. [1] and Singla et al. [2] explore the effects of advancements in material science on welding and sustainability—critical components for industrial automation and optimized production processes. Additionally, research on solar energy systems [3–7] and tribological studies [8] align with the sustainability goals of

Industry 5.0. Other research, such as advancements in digital twin technology for improving logistics [9, 10], the role of extremophiles in sustainable agriculture [11, 12], and the bioremediation of contaminated soils [13, 14], underscores how digital innovation and environmental consciousness are key to the next generation of manufacturing. These studies highlight diverse insights into how microbiological advancements and environmental sustainability are being harnessed for smart and sustainable industrial applications, aligning with the principles of Industry 5.0 [13, 14].

The integration of alloying elements to enhance wear resistance [15–19] supports the development of more durable materials for high-tech manufacturing applications. Moreover, research on surface integrity of machined materials and the impact of nanoparticulates in lubrication shows how material engineering can improve efficiency in production processes [20]. Collectively, these studies extend the scope of smart manufacturing by combining material science, energy efficiency, and automation, facilitating the transition toward Industry 5.0 [21–23].

Earlier works also explore dynamic simulations [24, 25], improving suspension systems, and enhancing material wear resistance—areas that reflect advancements in digitalization, automation, and sustainable manufacturing practices. These studies highlight a multidisciplinary approach to improving vehicle systems and material performance in line with the goals of Industry 5.0.

Recent research by Adeoye et al. (2024) investigates the valorization of lignocellulosic waste, demonstrating the potential for bio-oil production through pyrolysis and ultrasonication for biofuel generation [26]. Further studies by Adeoye et al. focus on upgrading crude bio-oil [27, 28] and sustainable energy production from biomass [29], offering insights into thermochemical and biochemical conversion processes. These approaches underscore the integration of clean energy solutions into digital manufacturing processes [30–35]. Research on catalytic pyrolysis and nanomaterials for enhanced biomass conversion provides pathways for optimizing manufacturing efficiency [36–38]. Furthermore, studies on the pyrolysis of plastic waste and rail projects introduce circular economy concepts and environmental sustainability, which are essential to the evolution of Industry 5.0 [39–47].

In parallel, research on fiber surface functionalization and dye fastness adds to the understanding of material innovation in manufacturing [48, 49]. These works contribute to the body of knowledge driving digital transformation and environmental consciousness in manufacturing [41, 43–45, 50, 51]. Contributions from Sunil et al. [52–59] span key advancements in transportation systems, materials science, and engineering technology, all of which have applications in smart manufacturing and Industry 5.0. Recent studies focus on electric vehicle adoption [25, 60–75], intelligent transportation system modeling [76–86], and vibration mitigation in marine and rail propulsion systems [87–94].

Recent advancements in sustainable fuel production have focused on various bio-oil sources and carbon capture techniques. Studies examine bio-oil production from thermal cracking in the presence of hydrogen [95], CO2 adsorption systems in coal beds [96], and the pyrolysis of coffee and eucalyptus sawdust [97–99]. Other research explores sulfur's role in enhancing wood tar biopitch as a sustainable binder, the characterization of sewage sludge bio-oil, and comparative analysis of bio-oil from freshwater and microbial-treated algae samples [100]. These studies collectively contribute to renewable energy development and environmental sustainability [50, 51, 101–103].

The role of microbiology, environmental sustainability, and material science innovations has significant implications for smart manufacturing and Industry 5.0 [30–32, 34, 35, 104]. For instance, Mohapatra et al. [105] investigate microbial processes for biopolymer production, with potential applications in sustainable packaging [32, 50, 106–113] and bio-based materials [114–125]. This aligns with Industry 5.0's emphasis on the circular economy and smart materials. Further research on retrofitting buildings for energy efficiency [126–141] and the use of polyhydroxyalkanoates (PHAs) in biocompatible applications [142–151] demonstrates how biotechnology is integrated into industrial processes.

These studies emphasize the need for energy-efficient solutions and advanced control systems, both of which are integral to Industry 5.0 [152, 153]. Previous works from 2023 also explore sustainable materials for transportation [154], intelligent systems for railway infrastructure [14, 155–168], and advancements in modeling and ride behavior [169–177]. Research on crashworthiness [178–185] and biolubricants [186–189] highlights the goals of Industry 5.0 by improving safety [190], efficiency, and sustainability [191–193]. These studies support the evolution of smart manufacturing, contributing to Industry 5.0's focus on sustainability, digitalization, and innovation across sectors [194].

Research on dynamic analysis, vibration isolation, and advanced material applications in transportation and vehicle systems further supports the objectives of Industry 5.0 [195–199]. Recent works (2024) investigate advanced control systems for enhancing passenger rail vehicle ride quality through intelligent suspension systems [158, 200–204], along with dynamic response analyses of road and rail vehicles [205–214]. Studies involving composite materials and hybrid AHP-TOPSIS techniques for sustainable structural applications [70, 73, 75, 215–226] are contributing to the development of smart materials central to Industry 5.0.

3 Natural Reinforcing Agents in Biocomposites

Natural reinforcing agents play a pivotal role in the development of biocomposites by offering an environmentally friendly alternative to synthetic fibers. These agents, often derived from renewable resources such as plants and animals, provide the structural strength and durability needed to create high-performance composite materials. With the growing emphasis on sustainability in manufacturing, these natural fibers are increasingly being adopted in various industries to reduce reliance on nonrenewable resources and decrease the overall environmental footprint [113–122].

3.1 Types of Natural Reinforcing Agents

Natural reinforcing agents are divided into two main broad categories: plant-based and animal-based fibers. Plant-based fibers that are commonly used for reinforcement include hemp, flax, jute, kenaf, sisal, coir and bamboo. The strength, flexibility and coarseness of such fibers are primarily attributed to the cellulose in the fibers. Quoting the above, it is also worth mentioning that all these plant-based fibers offer significant advantages since they are biodegradable. There are other animal-originated fibers such as wool and silk that do possess some remarkable mechanical properties and hence are used in some specialized applications [123–129]. However, these are not as common as plant fibers.

Various factors govern the choice of natural reinforcing agents. The main kinds of natural fibers in use in biocomposites have been mentioned in Table 1, together with their sources, basic characteristics and possible industrial application. Flax and hemp are quite popular in the automobile industry due to their excellent tensile strength at low density which is ideal for such lightweight parts.

3.2 Mechanical Properties of Natural Fibers

One of the aspects that make natural fibers be incorporated in biocomposites is the focus on their mechanical properties. These fibers tend to have low density, exhibit high tensile strength, as well as thermal resistance, all of which are critical for industrial utilization. Such fibers, including flax and hemp, are of notable interest owing to their intrinsic high tensile strength and can thus be employed in load-bearing applications in automotive and construction industries. The mechanical performance of natural fibers can also be manipulated by altering their internal structure through the deposition of a range of surface treatments as well as chemical modification increasing their interaction with polymer matrices and overall composite performance [130–134].

Table 1 Types of natural reinforcing agents and their properties

Type of fiber	Source	Key properties	Common applications
Flax	Plant	High tensile strength, low density	Automotive interiors, sports equipment
Hemp	Plant	Good mechanical properties, biodegradable	Construction, textiles, automotive components
Jute	Plant	Low cost, biodegradable	Packaging, textiles, nonstructural composites
Silk	Animal	High tensile strength, flexibility	Aerospace, biomedical devices
Wool	Animal	Good insulation, biodegradable	Textiles, insulation materials

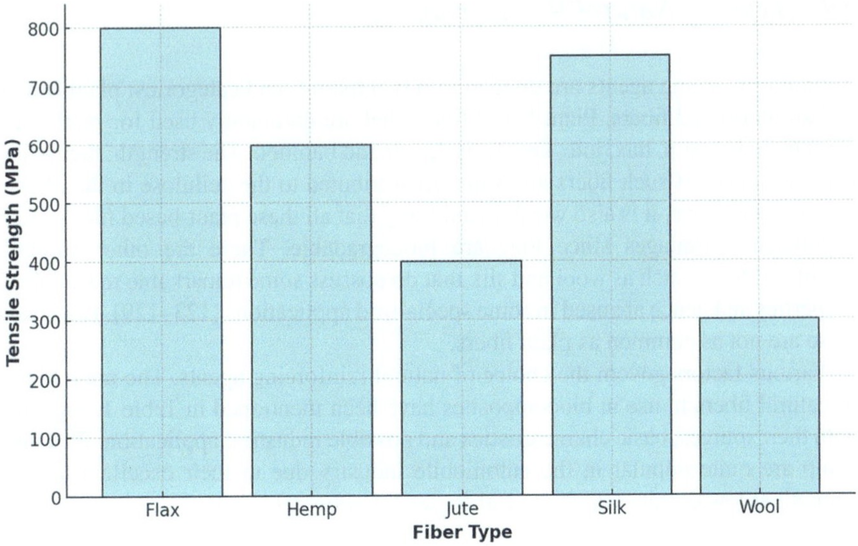

Fig. 1 Comparison of tensile strength (MPa) across various natural fibers, highlighting the superior performance of flax and hemp

Figure 1 illustrates the tensile strength of commonly used natural fibers in biocomposites. Flax and hemp exhibit the highest tensile strength, making them suitable for load-bearing applications.

3.3 *Environmental Benefits of Natural Reinforcements*

Monumental advantages are presented by the use of such natural fibers as reinforcing agents in the biocomposites. First of all, these fibers can be grown and do not put any pressure on the natural ecosystems as they are renewable. Furthermore, biocomposite materials are environmentally safe as they have decomposition properties which in turn help to keep at bay the problem of land-filling by wastage of such materials. Natural fibers are less energy demanding in the processes of their synthesis as compared to synthetic fibers (glass or carbon), which explains their reduced carbon footprint [135–138].

3.4 *Challenges in Using Natural Fibers*

Regardless of the multitude of advantages provided by natural fibers, a range of difficulties concerning their incorporation into biocomposites is encountered. The heterogeneity in fiber property is perhaps the most noticeable challenge that results from the diversity of plant species, growth environment, and harvesting methods. This variability renders control of the quality of the finished composite material components challenging. Furthermore, in humid environments, the bonding interface in composites, which is between the fiber and the matrix, is likely to weaken due to the hydrophilicity of natural fibers. To address these challenges, research has been ongoing to look for new treatments and surface modification to improve the compatibility of natural fibers with polymer matric and their moisture resistance [139–148].

3.5 Applications of Natural Reinforcing Agents

Natural fibers have come to be of increased use in quite a number of applications in different sectors. In the automotive industry, they are now substituting glass in parts such as door panels, dashboards and interior trims which enhance the light weight of vehicles and thus increase fuel economy. In the construction sector, natural fiber composites are applied in building materials that provide superior thermal insulation and less environmental degradation. Other sectors, like aerospace, packaging, and consumables also aim to cope the potential of natural fibers to produce greener products [149–156].

4 Digital Technologies for the Biocomposites Manufacturing Processes

4.1 Additive Manufacturing (3D Printing) in Biocomposites

In the context of evolution of Industry 5.0, numerous changeable elements have been introduced into the processes of design, production and performance of biocomposites. All these features facilitate accurate material property modulation, increase production efficiency, and help reduce waste significantly, all of which are in support of sustainability. By employing more cutting-edge digital technologies, such as additive manufacturing, digital twins, and real-time analytics, biocomposite industries can increase the efficacy of using natural reinforcing agents while increasing the production efficiency of biocomposites materials. Various innovative ways to transform the biocomposite industry through digital technology will be discussed in this section. The integration of flax or hemp fibers into 3D-printed parts has been very effective in developing light weight high-performance automotive interior components. In a similar context, additive manufacturing has been a good technique that can greatly compress production lead times with durability and biodegradability of biocomposites and thus can be applicable in industries which target eco-friendliness. Figure 2 shows the distribution of

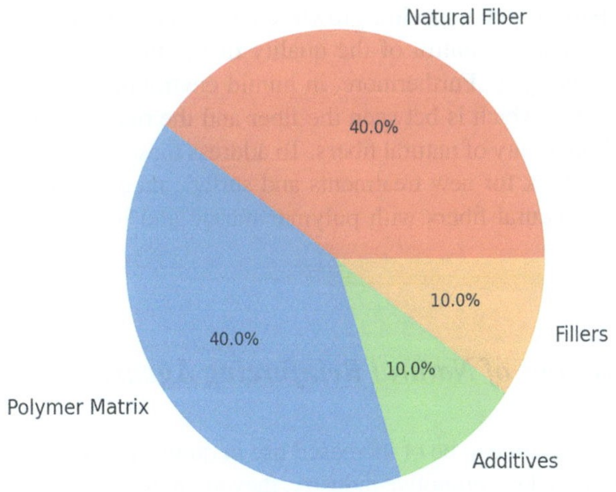

Fig. 2 Distribution of components in a typical biocomposite, showing the proportion of natural fibers, polymer matrix, additives, and fillers

Table 2 Digital manufacturing techniques in biocomposites

Technique	Application	Key benefits
Additive manufacturing	3D printing biocomposites with precise fiber placement	Reduces waste, enables complex geometries, allows customization
Digital twins	Virtual simulations of composite materials under real-world conditions	Predicts performance, reduces need for physical prototypes, optimizes material usage
Automation and robotics	Automated fiber placement and matrix integration	Improves precision, increases production efficiency, reduces human error
Real-time data analytics	Monitoring production conditions (e.g., humidity, temperature)	Enhances quality control, adjusts process parameters on the fly

various components in a biocomposite, emphasizing the balance between natural fibers and the polymer matrix to achieve optimal mechanical properties [14, 157–167, 174].

Technology of digital biocomposites production was carried out at the new technological level. As per Table 2, the personnel use such key techniques as additive manufacturing, digital twins, and real-time data analytics that improve efficiency, waste management, and optimum material properties of the composites. For example, in additive manufacturing, most of the fiber direction can be controlled by the manufacturers, which is essential in biocomposites application's mechanical tailoring.

4.2 Digital Twins for Process Optimization

Digital twins are a growing concept that helps with biocomposite. Digital twins are virtual copies of an object that can be put through stress monitoring to make it more efficient. Digital twins have the ability for manufacturers to test any material fiber around polymer before moving to real life application. It is essential to first physically develop a prototype before being able to undertake the actual production of one. For example, digital twins could be used for stress tests for usage in construction or automotive biocomposite materials. Potential weaknesses can be discovered, as well as the composition and orientation of the fibers, if necessary, so that the end product will meet all required operational characteristics. Biocomposite production also uses fiber raw material quality and environmental condition monitoring twins at the same time.

4.3 Automation and Robotics in Biocomposite Manufacturing

With respect to the manufacturing processes, the technology of biocomposites is becoming very robust primarily due to the stand-alone impact of Automation and Robotics. More automated systems can help fabricating biocomposite materials more accurately and consistently, such as in the placement of fibers and the application of the matrix. Further automation can also include the use of Robotics in the production process which will undoubtedly improve efficiency. In smart factories, which are the focal point of Industry 5.0 implementation, process automation lends itself to the integration of Manufacturing Execution Systems and digital twins for real-time optimization of the manufacturing process. For example, it has been shown that robotic arms incorporate sensors which can be programmed to adjust the location of individual synthetic fibers along the composite matrix and optimize the material distribution due to stimuli present in the environment. Such automation is practically beneficial in industries that wish to mass produce biocomposites for example biocomposites without compromising the control of composites material characteristics. The performance improvement of biocomposites, as shown in Fig. 3, indicates a steady increase in tensile strength from 2020 to 2024, driven by advancements in digital manufacturing techniques.

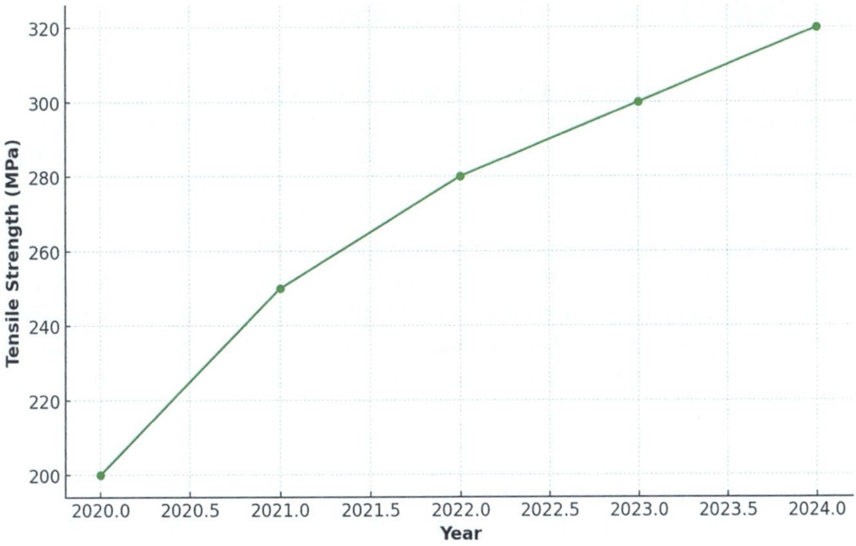

Fig. 3 Performance improvement of biocomposites measured by tensile strength (MPa) over 5 years

4.4 *Real-Time Data Analytics and AI in Manufacturing*

The combination of technologies such as AI and real-time data analytics with biocomposite manufacturing processes has considerable potential for enhancing product quality and productivity. Thanks to the optimization of resources and applied analytics across all production stages, manufacturers can reduce waste, predict machine breakdowns, and efficiently use their materials. Structural mechanics could, for example, be improved in the production of biocomposites by employing AI algorithms which model and predict the behavior of natural fibers in various matrix materials.

The scatter plot in Fig. 4 demonstrates the positive correlation between fiber–matrix compatibility and the tensile strength of biocomposites, underscoring the importance of achieving strong interfacial bonding.

Besides, real-time data analytics has the potential to be used in tracking environmental factors, such as humidity and temperature, that are important in the performance of natural fibers during manufacturing. With this information, operators will be able to modify the parameters of the manufacturing process in order to improve the final quality. This approach not only improves the biocomposite performance but also minimizes resource consumption and adheres to the Industry 5.0 paradigms.

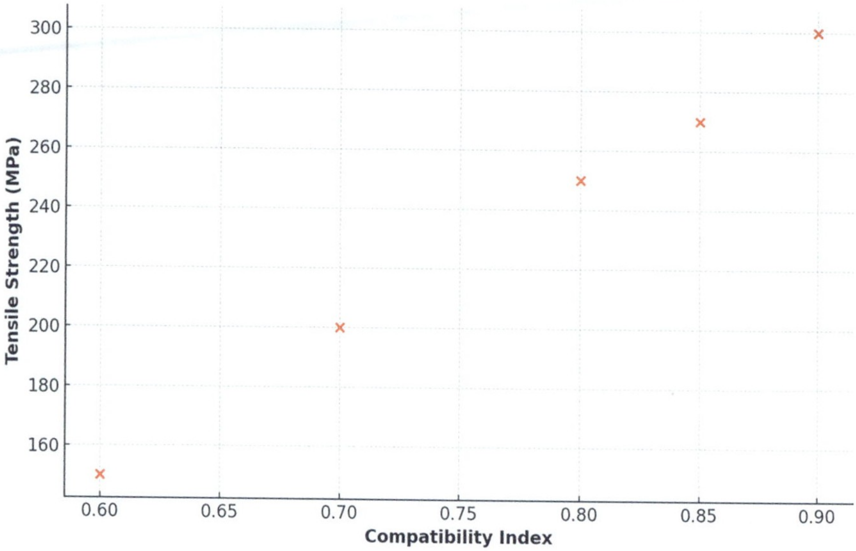

Fig. 4 Relationship between fiber–matrix compatibility index and the tensile strength of biocomposites

4.5 *Benefits and Sustainability of Digital Manufacturing*

Utilization of the digitally integrated manufacturing processes for the production of biocomposites types of materials has some sustainability advantages. When damage to composite materials is avoided, resource usage optimized and advanced manufacturing technologies used, all these lead to considerably reduced environmental burden on the composites production. As for instance, in additive manufacturing, only the absolute required amount of material is utilized, whereas the application of digital twinning enables better research and development so that there is less dependence on numerous physical prototypes. Production automation also enhances production efficiency and reduces operating expenses, which in turn, translates into energy savings. In addition, the practice of digital manufacturing fits well into the larger goals of Industry 5.0 which seeks to transform industrial processes into more sustainable ones and placing people at the center of industrial processes. It allows to promote renewable materials, such as natural fibers, while decreasing dependence on nonrenewable resources. In the end, the confluence of such digital technologies makes it possible for manufacturers to manufacture high quality biocomposites in an effective manner which makes it possible to work toward a sustainable future.

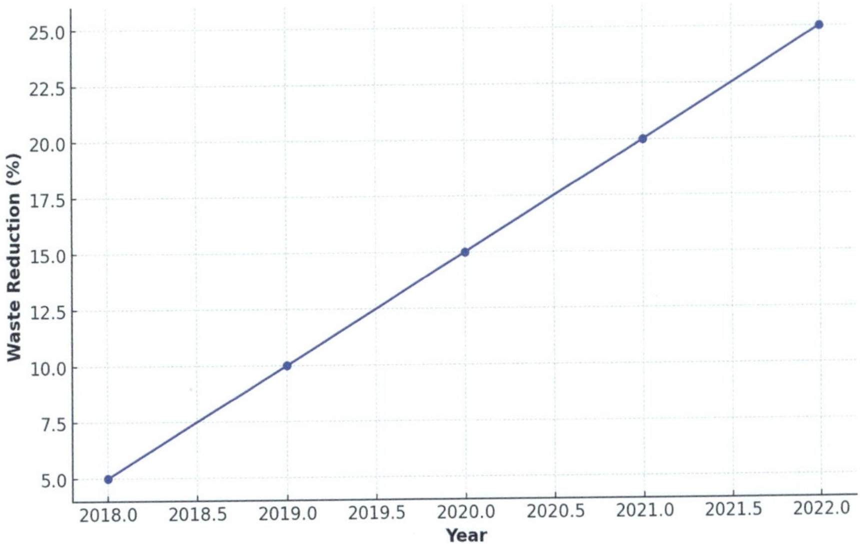

Fig. 5 Waste reduction in biocomposite manufacturing using digital manufacturing techniques from 2018 to 2022

The reduction in material waste using digital manufacturing techniques is depicted in Fig. 5, showing a clear trend toward more efficient and sustainable production methods over the years.

5 Smart Factories and the Role of Digital Twins

Industry 5.0 is a new era that explains the new description of smart factories which sees the interplay of various systems, real-time data analytics, and intelligent automation all in the bid to improving the production processes. When discussing biocomposites, smart factories help us to improve material properties and material solutions in an environmentally friendly manner with the use of better technologies including digital twins. These factories use artificial intelligence (AI), machine learning (ML), the Internet of things (IoT) as well as cloud computing technologies to be able to establish smart and eco-efficient manufacturing systems. This section looks into how smart factories and digital twins aid in the production and optimization of biocomposites.

5.1 Concept of Smart Factories in Industry 5.0

Smart factories are factories in which production processes are automated to a very high level whereby machines, equipment and skilled workers are connected together in the production process through IoT based networks. This line of production is highly advanced as it is characterized by human–machine cooperation and smart green technologies that allows for flexible automation and optimized production lines. Smart factories have implemented sophisticated sensor technologies that continuously capture large volumes of operational and process-based data enabling them to be never blind to their ongoing processes and thereby proactively manage performance and address inefficiencies. Regarding biocomposite manufacturing, it is confirmed that the smart factory systems allow processing and incorporation of natural reinforcing materials, such as plant fibers into composite materials with the utmost precision and accuracy. The automated management of raw materials, coupled with connections to existing operational management systems, allows it to achieve predetermined standards and wastage minimization during the manufacturing process. Additionally, smart factories are able to respond to changes in material supply and demand patterns and utilize natural resources sustainably in the production of biocomposites. As summarized in Table 3 above, smart factories equipped with technologies like digital twins and real-time data integration have several applications in the biocomposite industry. Smart factories take control over material properties and processes to boost both the efficiency and effectiveness of manufacturing making these technologies essential in implementing the greener vision of manufacturing in Industry 5.0.

5.2 Digital Twins in Biocomposite Manufacturing

In smart factories, a crucial characteristic is digital twins, which are virtual representations of physical resources, systems, or processes. Digital twins observe how biocomposite materials behave in the different stages of their manufacturing and

Table 3 Benefits of smart factories in biocomposite production

Benefit	Description
Improved efficiency	Smart factories streamline production processes through automation and real-time data integration
Material optimization	Digital twins allow for precise adjustments in fiber–matrix combinations, optimizing material usage
Sustainability	Smart factories reduce waste and energy consumption, aligning with sustainability goals
Flexibility and adaptability	Smart factories can quickly adapt to changing material supply or demand fluctuations

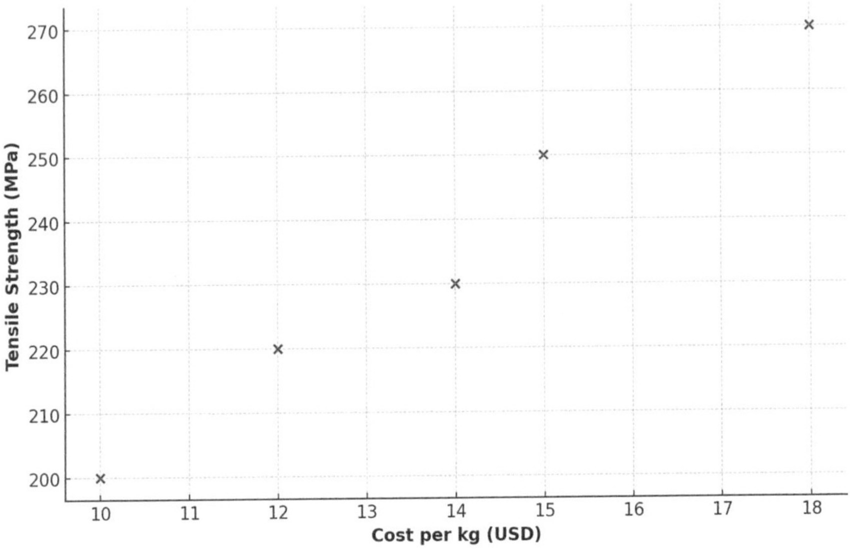

Fig. 6 Cost per kilogram of biocomposites vs. their tensile strength (MPa)

use enabling improvement of their properties and their expected performance under various conditions. Digital twin of a bio composite product allows the manufacturers to define the scope of the complied product such as how the polymers interact with natural fibers, and how other parameters like humidity or temperature will affect the produced materials' durability and strength.

Figure 6 shows the relationship between the cost of biocomposites and their tensile strength. The data suggests that higher mechanical properties can be achieved without substantial increases in cost, making biocomposites competitive with traditional materials.

For instance, a digital twin can reproduce a stress scenario and predict how a biocomposite material used in an automotive part will react to stress. This allows the engineer to modify the fiber orientation, the resin content, or other factors that will enhance the performance of the briquette. Such a prototyping method cuts down the number of physical prototypes and tests to be carried out, thus reducing time and cost. By revealing structures of a new material which can cause material failure, digital twins enable biocomposite manufacturers to manufacture biocomposite products that are trustworthy and green.

Table 4 presents the case studies of the biocomposites across the sectors and shows the efficiency in the application of such digital manufacturing technologies to biocomposite. In the transportation sector, for instance, flax fiber-reinforced plastic materials are decreasing the use of heavier materials leading to lighter vehicles and hence lower fuel consumption. In aviation, practically the same types of

Table 4 Industry 5.0 applications of biocomposites

Industry	Application	Biocomposite used	Benefits
Automotive	Interior components (door panels, dashboards, seat backs)	Flax fiber-reinforced polymer	Lightweight, improved fuel efficiency, reduced emissions
Aerospace	Cabin interiors, overhead compartments	Flax fiber-reinforced composites	Lightweight, strong, meets fire resistance standards
Construction	Hempcrete for walls and insulation	Hemp fiber and lime-based binder	Superior thermal insulation, biodegradable, low carbon footprint
Marine	Boat hulls and decks	Flax-reinforced bioresin	Biodegradable, lightweight, reduced environmental impact
Consumer goods	Sustainable packaging	Natural fiber-reinforced biodegradable plastics	Reduces plastic waste, biodegradable, aligns with circular economy principles

materials come into use in order to achieve the toughest safety and operational requirements while being as environment friendly as possible.

5.3 *Real-Time Data Integration for Process Control*

A fundamental characteristic of smart factories is the possibility for real-time data communication for process management and enhancement. In the scenario of biocomposite manufacture, nonproduction shots of time-lapsed digital footage enables the use of sensors that are incorporated into production machinery to provide data on the prospective usage of materials, production rates, and z external conditions. This information makes it possible to control natural fiber integration within the production process so as to maintain quality without leaving any appropriate overseas.

For instance, the production of biocomposites in smart factories is a good example of such timing—as due to the capabilities of sophisticated technology, there is no need to wait for the completion of the curing process to evaluate bond formation between the matrix and fiber. As a result, parameters such as temperature, time or pressure might be adjusted in order to achieve a desired bonding strength between the components. Process automation together with information technologies make monitoring of the whole cycle possible. The integration of these automated processes with an appropriate digital twin allows for a more accurate modeling and granulation. Through this feedback loop with the two systems, the optimization of the quality of production is enhanced, while the amount of resources utilized in the process decreases, all over time.

5.4 Benefits of Smart Factories in Biocomposite Production

Smart factories bring a number of important advantages in the biocomposite manufacturing processes, especially when it comes to improving performance and sustainability. First, the application of real-time information and digital twins results into better material utilization which minimizes wastage and maximizes the overall utilization of natural resources. Second, the introduction of automation and AI driven processes enhances uniformity and accuracy which are important in maintaining the quality of biocomposites that have natural fibers with great variations. Third, smart factories promote the adoption of sustainable manufacture execution systems through the reduction of energy requirements and ambient effects of the production processes.

As shown in Fig. 7, biocomposites have a significantly lower carbon footprint compared to traditional composites and other materials like plastics and metal alloys, highlighting their sustainability benefits.

In addition, due to the fact that biocomposite producers depend on agricultural commodities, the ability to respond rapidly to changes in supply chains or the availability of certain materials is quite important. In case of disruptions, smart factories are capable of changing the production processes to use different types of fibers or change the composite combinations in order to be able to keep the level of production high without affecting the sustainability.

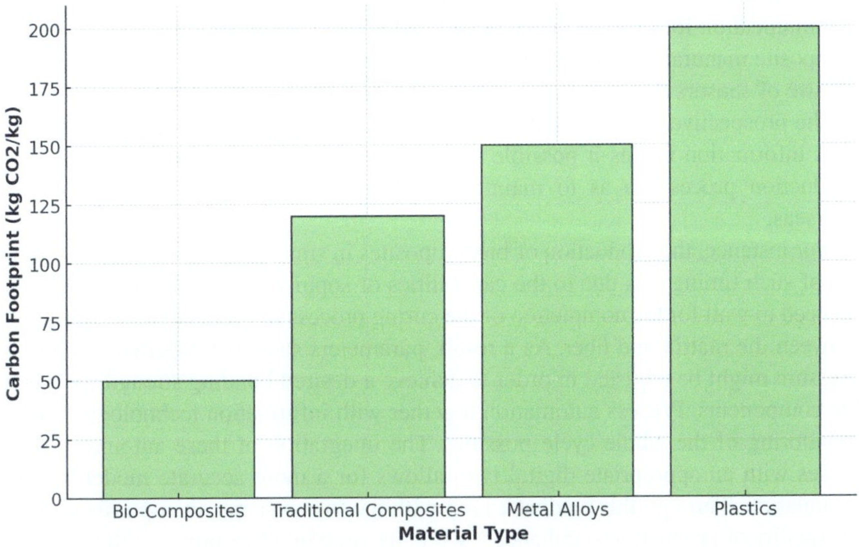

Fig. 7 Comparison of carbon footprint (kg CO_2/kg) between biocomposites, traditional composites, metal alloys, and plastics

5.5 *Challenges and Future Directions*

The technological advancements also pose particular challenges for biocomposite production which offer considerable opportunities in this sector. One of the key challenges is the requirement for substantial capital which is required to put in place the requisite sensor, data analytics, and digital twin necessary machinery. Furthermore, the inconsistency of the natural fibers that can be influenced by several environments such as climate, soil, and methods of harvesting makes the digital twin modeling even more complex than before. There are gaps in predictive modeling which translates to gaps in the reliability of simulations of biocomposites.

In the foreseeable future, AI and machine learning can assist in these challenges as they will improve the ability of smart factories to handle the external variability of natural fibers. While advancing smart manufacturing technologies, the further synergism of digital twins, data, and automation will greatly strengthen biocomposite production competitiveness and sustainability, supporting the innovation of Industry 5.0.

6 Case Studies: Industry 5.0 Applications of Biocomposites

In the sustained development of Industry 5.0, biocomposites have emerged as an essential component of sustainable manufacturing for many designers and manufacturers. As the factories are becoming more integrated with modern resources like additive manufacturing, digital twins, and industrial automation, biocomposites are turning out to be highly effective and multifunctional. This particularly gives more of real world case studies as to how biocomposites are been incorporated in crucial sectors and what changes it has brought to the industries in terms of reducing the eco-footprint while augmenting the outcomes. The examples given show the usage of Industry 5.0 in better production and usage of biocomposites. These include hybrid biocomposites.

6.1 *Automotive Industry: Lightweighting and Sustainability*

One of the most significant applications of biocomposites can be found in the automotive industry, where there is a constant drive to reduce vehicle weight in order to improve fuel efficiency and lower emissions. Traditional materials such as metals and glass fibers have been widely used in automotive manufacturing; however, the industry's shift toward sustainability has led to an increased focus on natural fibers as reinforcing agents in biocomposites.

For example, well-known car producers have started to use natural fiber composites to manufacture vehicle components such as door panels, dashboards, and seating structures. By adopting natural fibers like hemp and flax, which are lighter than the traditional glass fibers, automotive manufacturers can reduce weight significantly. One notable case is of flax-based composites that are employed in making lightweight sunroof panels for electric cars to enhance fuel economy and lessen the impact on the environment during the production of these panels. These biocomposites are made by digital methods, which include 3D printing and robotics, to achieve accurate placement of materials and minimize the amount of waste generated.

Moreover, when creating designs, the incorporation of digital twins enables designers to conduct performance tests of biocomposite parts on the computer at different times so that the characteristics of the constructed components can be adjusted prior to production. Of course, this fastens the design process, and there is no longer a necessity for several prototypes, cutting down on resource use and manufacturing expenses.

The pie chart in Fig. 8 highlights the adoption of biocomposites across key industries, with automotive, aerospace, and construction being the largest users, driven by the demand for lightweight and sustainable materials.

6.2 *Aerospace Industry: High-Performance Composites*

The aerospace industry has been known to have stringent requirements in the handling of materials due to its focus on performance and safety. However, the aerospace industry is now shifting its focus toward biocomposites so as to meet

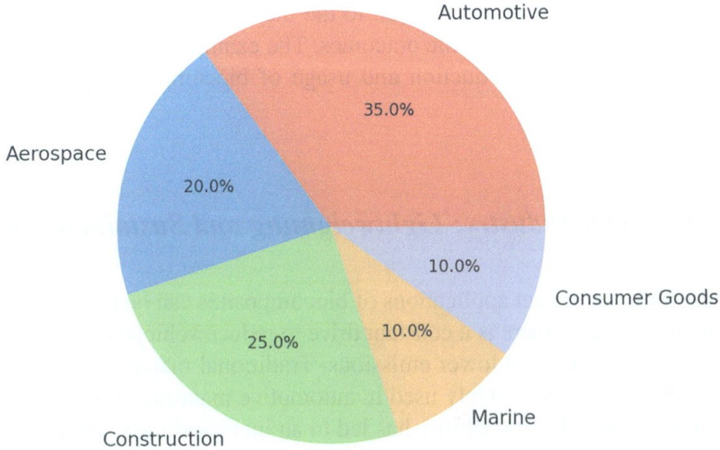

Fig. 8 Adoption of biocomposites across various industries, with the automotive sector leading the way

sustainability goals without compromising on strength and durability. With this shift, the most promising materials are biocomposites reinforced with natural fibers that are not only light in weight but also strong enough to be used in the aerospace industry.

Biocomposite materials reinforced with flax fibers were developed for non-structural applications on an aircraft such as the cabin, overhead compartments, and side panels. Such parts are produced via additive manufacturing processes that facilitate synthesis of complex features while conserving material. Digital twins enable the conductor of such experiments to have the ability to replicate the inflight stress condition and determine the fiber orientation that can withstand considerable forces.

Additionally, biocomposites in the aerospace section of the industry are required to have very good properties both in terms of these panels' fire resistance and their mechanical performance. For this reason, manufacturers are developing hybrid biocomposites in which natural fibers are integrated with flame-generated materials to increase their effectiveness. The substantiation of Industry 5.0, which is also characterized by a smart factory and AI perception applying the automated technologies, further guarantees that these hybrid biocomposites will comply with the relevancy requirements in great detail.

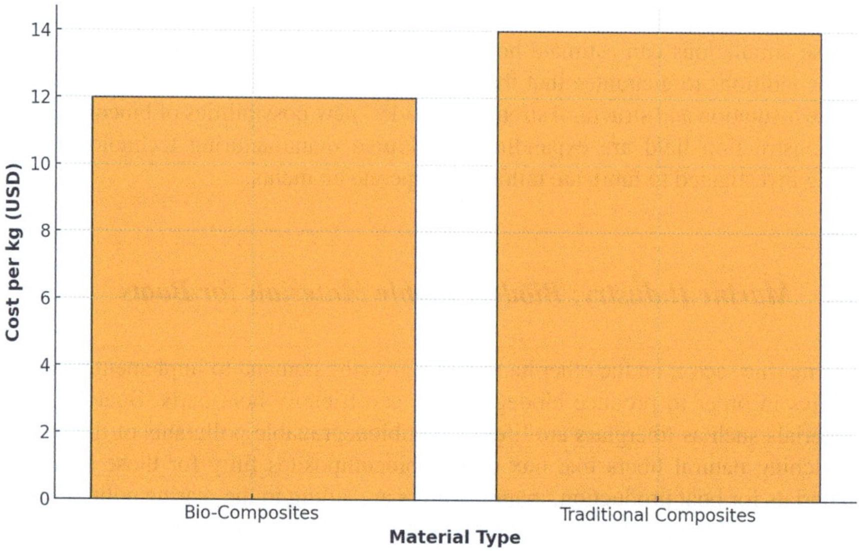

Fig. 9 Cost comparison of biocomposites versus traditional composites, demonstrating the economic viability of biocomposites

6.3 Construction Industry: Sustainable Building Materials

In terms of raw material consumption, the construction industry is among the largest, and there is increasing attention to the incorporation of biocomposites in the place of steel, and concrete, for example. Agriculture residues such as jute and hemp fibers and other plant-based composites biocomposites are hence being utilized to manufacture eco-friendly biocomposite materials that are not only lightweight but also strong with good thermal insulation properties.

Some accounts have noted a significant amount of success integrating hempcrete, a mixture of hemp fibers and lime paste, as an eco-friendly alternative to cement in the construction of residential buildings. Hempcrete provides better thermal insulation properties, minimizes the environmental impact of construction activities as it is completely biodegradable at the end of its life. Hempcrete panels are produced in smart factories, where production processes are automated and performed on the basis of real-time data analysis in order to ensure homogeneity of the material and high quality.

The cost comparison in Fig. 9 highlights that biocomposites, while generally more sustainable, can also be economically competitive with traditional composite materials, especially when considering their environmental benefits.

Digital twins are utilized in estimating energy performance of buildings made from hempcrete, enabling architects and engineers to incorporate these designs. These simulations can estimate how buildings will function in different climates and conditions to guarantee that the biocomposite materials used will impart adequate insulation and structural strength. Besides, new possibilities of biocomposites in construction field are expanding as additive manufacturing technologies are being investigated to fabricate tailored hempcrete elements.

6.4 Marine Industry: Biodegradable Materials for Boats

The marine sector, on the other hand, has a growing demand to implement biocomposites in order to produce biodegradable, eco-friendly boat parts. Boat building materials such as fiberglass are lifetime nonbiodegradable pollutants of the ocean. Switching natural fibers like flax or jute biocomposites fully for those synthetic materials for boat production, manufacturers are aiding to the marine pollution.

In this sector, one of the case studies is of a small leisure craft built entirely from and made from bio composites. The hull and deck of the leisure boat are made of flax embodied in a bioresin which is lightweight and stronger than fiberglass. The boat is manufactured through additive manufacturing technologies that allow more specific design of the composite component's fiber alignment and reduce material waste.

Digital twins are, however, also employed in the testing of the boat so as to design and optimize its structure against a number of operational scenarios such as the impact of middle sea operation conditions on the boat in order to optimize its strength and safety. By employing biocomposites, the marine sector is able to ameliorate the ecological effects associated with the manufacture of boats and also the disposal of boats at the end of their life does not have adverse impacts on the marine ecosystem since these materials have been proven to be biodegradable.

6.5 Consumer Goods Industry: Sustainable Packaging and Products

The biocomposites are now finding use in within the consumer goods industries in an effort to provide sustainable packing and products. Coupled with the increase in consumers' preference to green products many manufacturers are trying biocomposite materials that are made of natural fibers and biopolymers-to replace plastics.

A detailed discussion of one more important practical task—how this technology can be applied in real food industry is illustrated. It is focused on packaging, where natural fiber-reinforced bioplastics have been proven to be an environmentally friendly solution. Such biocomposites are manufactured in the context of smart factory automation, where robotics and AI are employed to maintain uniform standard of products and provide time efficiency.

3D packaging has a wide range of applications including research into its functional performance under changing climatic conditions. Digital twins replace physical prototypes for presenting a new material in the market. According to the authors, digital twins are used to study the strength of the product during cyclization processes: temperature and humidity impact with an aim to enhance bioplastic attributes. Future packaging industry optimization scenarios attempted by other researchers to overcome this issue also aim for sustainability—the product must be efficient and fully recycled.

This innovation not only meets the demands for sustainable packaging but also reduces the reliance on nonrenewable resources, promoting circular economy practices in the consumer goods sector.

7 Challenges and Future Directions

Although the combination of digital manufacturing with biocomposites has multiple benefits toward its sustainability and performance, there still persist some hurdles that are to be overcome in order to fully realize the potential of such materials in the context of Industry 5.0. This subsection describes the issues germane to the biocomposite production in the present day and envisions the advancement of research and technology which may further widen the scope of biocomposites use and development. Even though biocomposites could have many advantages, still

Table 5 Challenges in biocomposite production and potential solutions

Challenge	Description	Potential solutions
Material variability	Natural fibers vary in quality based on environmental factors	Advanced real-time monitoring, AI for quality control, improved surface treatment techniques
Fiber–matrix compatibility	Poor bonding between fibers and matrices leads to weaker composites	Surface treatments such as silane coupling and plasma treatments to improve adhesion
Scalability and cost	High production costs and scaling challenges compared to synthetic composites	Increased automation, smart factory implementation, and advances in additive manufacturing
Environmental impact	Energy consumption and environmental effects of fiber processing can offset sustainability gains	Green chemical treatments, lifecycle assessment (LCA) tools, and digital twin integration

there are some issues that need to be tackled, as emphasized in Table 5. Large material variations, low compatibility between fiber and matrix and reproducibility/scalability remained the key barriers. Nonetheless, developments in the areas of digital manufacturing and surface modification strategies provide strong reasons for hope. In the case of Industry 5.0, however, it will be necessary to develop new solutions which highlight the strengths of these strategies, which are necessary to minimize the constraints on the use of biocomposites.

7.1 Material Variability and Quality Control

Variability of natural fibers is one of the most crucial issues when designing biocomposites since it greatly influences the uniformity and efficacy of the final product. Natural fibers such as hemp, flax and jute are grown under varying climatic conditions as well as soil and harvesting techniques. It follows that even the same species of plants would be characterized by different mechanical properties that frustrate attempts to achieve consistent material quality during mass production of the products.

A possible solution to this problem is relying on manufacturing technologies using data inherently on the process of the production and control measures that are artificial intelligence driven. These technologies have the capacity to control production parameters to cater for variations in fiber properties, which have a direct relationship with biocomposite performance. Apart from the abovementioned technologies, there are ongoing studies regarding fiber treatment and processing techniques, such as chemical modification, which would aim to decrease variability of the natural fibers incorporated within the biocomposites.

7.2 Fiber–Matrix Compatibility

Achieving strong interfacial bonding between natural fibers and polymer matrices has still posed a lot of challenges in biocomposite's production. Weak adhesion could as well degrade its mechanical properties such as tensile strength and durability, negatively impacting biocomposites that are expected to withstand high-performance environments. Natural fibers are often hydrophilic and attract water, but most polymer matrices are hydrophobic and repulse water, which can inhibit interfacial bonding.

To improve the adhesion between fibers and matrices, researchers are employing fiber surface modifying methods like alkaline treatment, silane coupling agents, as well as plasma treatment. The progress in developing soaking processes which are chemically effective yet environment friendly is most likely going to be crucial in enhancing the mechanicals of the biocomposite. Moreover, integration of digital manufacturing technologies such as digital twins may also allow manufacturers to further optimize mechanical bonding at the interface by predicting how various treatments would alter performance of the composite.

7.3 Scalability and Cost

Although biocomposites are beneficial to the environment, the main obstacle is the scaling up of their production to achieve industrial scale. The cost of biocomposite production, especially in comparison to synthetic composite alternatives, is still a major challenge to its widespread application. The production costs could be aggravated by the inexpensive availability and processing of natural fibers, the cost of matrices, and polymer matrices as well as the requirement for special machines. In order to tackle these horizons, more innovations and advancements in the areas of automation and smart manufacturing processes should be made. AI And robotic production facilities have the ability to make the production of biocomposites more efficient in terms of labor and waste minimization. Furthermore, improvements in the area of additive manufacturing might also facilitate more effective and economical processes for more intricate and engineered applications of biocomposite materials. In correspondence to the rise in the need for eco-friendly solutions, biocomposites should also become more economically feasible with more funds directed toward research and development concerning the solution.

7.4 Environmental Impact and Lifecycle Assessment

Though biocomposites should be said to be more eco-friendly when compared to synthetic composites, there are still issues associated with their lifecycle and environmental stages. The amount of energy needed for the preparation of natural fibers and the pollution caused by specific chemical methods of treatments might reduce some of the environmental advantages of biocomposites. Further, although biocomposites can be composted, disintegration of them and their constituents totally depends on the ambient conditions and the characteristics of the material.

The lifecycle assessment (LCA) tools are presented on the stage as a pivotal instrument in assessing the environmental implications of biocomposites. These tools give a holistic view of the inputs in terms of energy resources for the synthesis, use, and end life of biocomposite materials. In years to come, LCA developments allied with the digital twin technologies may allow manufacturers to virtually assess the biosynthetic composites' environmental burden of different materials and production methods in the first place. Furthermore, an expanding body of knowledge regarding the biodegradable polymer matrices and less harmful chemical treatments of fiber will ease the impact of implementing biocomposites on the environment as well.

7.5 Advancements in Hybrid Composites

The future of biocomposite development may lie in the advancement of hybrid composites, which combine natural fibers with synthetic or inorganic reinforcements to achieve superior mechanical properties. Hybrid composites can offer the best of both worlds: the sustainability of natural fibers and the enhanced performance of synthetic materials. However, the challenge lies in optimizing the combination of materials to ensure that the resulting composite is both high-performing and eco-friendly.

The research performed on hybrid composites has yielded good results in the application of first and foremost the automotive and aerospace industry where strength and weight savings are crucial. Within the boundary of further studies, the goal will be to minimize the amount of synthetic materials used in hybrid composites and to alter the proportions of employed biocomposites. The application of digital twins and predictive modeling would indeed help improve the time of hybrid composites development as manufacturers would not need physical samples to determine the best combinations.

7.6 Future Directions: Integrating AI and Machine Learning

As Industry 5.0 matures and greater reliance on AI and machine learning in production systems becomes the norm, biocomposites will gain in popularity and be accepted as viable alternatives to synthetic materials. AI can facilitate the biocomposite production process by estimating the performance of certain material mixtures at particular conditions, thus reducing the trial-and-error method which is expensive and time consuming. Production data from biocomposites can also be analyzed using machine learning algorithms to enable biocomposite manufacturers to adjust their production processes for more effective and better performing biocomposite materials.

Also, thanks to AI-based design technologies it will be much easier to produce new types of biocomposite materials possessing desirable characteristics necessary for a given industrial purpose. Machine learning could be employed to explore big data sets regarding material traits and manufacturing parameters and create or recommend unique fiber–matrix combinations or processing techniques which could optimize the technological features of natural fibers.

7.7 Conclusion

The prospects of the combination of biocomposites and digital manufacturing technologies for the sustainable future of the industry are encouraging. At the same time, it is imperative to resolve issues of material heterogeneity, mass production feasibility, price, and fiber–matrix compatibility in detail to improve the integration of biocomposites into Industry 5.0 practices. Tapping on various technologies such as AI, machine learning, and digital twins while focusing on research and innovation will help the biocomposite sector in overcoming these challenges and provide notable solutions toward greener manufacturing. As the field of digital manufacturing develops, it aimed to elevate the efficacy, viability, and effectiveness of biocomposites and assist greatly toward a greener and truly sustainable development.

References

1. Shah, S., Nautiyal, H., Gugliani, G., Kumar, A., Namboodri, T., & Singla, Y. K. (2024) Sustainability in smart manufacturing: Trends, scope, and challenges. *CRC Press*. (Taylor & Francis Grouphttps://taylorandfrancis.com/books/) in May 2024. ISBN 9781032740713. https://doi.org/10.1201/9781003467496
2. Singla, Y. K., Maughan, M. R., Arora, N., & Dwivedi, D. K. (2024). Enhancing the wear resistance of iron-based alloys: A comprehensive review of alloying element effects. *Journal of Manufacturing Processes, 120*, 135–160.
3. Singh, H., & Singla, Y. K. (2023). Wear characteristics of AISI 4140 under nano fly ash based engine oil, Proceedings of the 7th International Conference on Theoretical and Applied Nanoscience and Nanotechnology (TANN'23), Paper No. 118, 1–3, Canada. https://doi.org/10.11159/tann23.118
4. Singh, H., Singh, A. K., Singla, Y. K., Chattopadhyay, K., Saini, A., & Singh, K. (2022). Interpretation of the wear characteristics of AISI 4140 under nano-fly ash based engine lubricant. *Materials Today Proceedings, 50*, 1683–1689.
5. Rooprai, R. S., Singh, H., & Singla, Y. K. (2021). Investigate the Tribological properties of AISI 4140 alloy steel under various loads and sliding speed. In *IOP conference series: Materials science and engineering* (Vol. 1145, p. 012038). IOP Publishing.

6. Rooprai, R. S., Singh, H., Singh, T., & Singla, Y. K. (2022). Analysis of the wear properties of through hardened AISI-4140 alloy steel using Taguchi technique. *Materials Today Proceedings, 50*, 661–664.
7. Singh, H., Singh, A. K., Singla, Y. K., Chattopadhyay, K., & Malhotra, R. (2021). Optimization of tribological behavior of AISI 4140 under nano fly ash particulates in engine lubricating oil. *Materials Today Proceedings, 45*, 4619–4623.
8. Sura, H. S., Avesh, M., & Mohapatra, S. (2023). Design and modelling of digital twin technology to improve freight logistics. In *Energy, environment, and sustainability* (pp. 481–513). https://doi.org/10.1007/978-981-99-2150-8_20.
9. Pattnaik, S., Pati, S., Devadarshini, D., Tanaya, K., Mishra, B. B., Samantaray, D., Mohapatra, S., & Dash, D. (2022). Microbial bioremediation of Cr(VI)-contaminated soil for sustainable agriculture. In *Microbial biodegradation and bioremediation: Techniques and case studies for environmental pollution* (pp. 395–407). https://doi.org/10.1016/B978-0-323-85455-9.00008-4.
10. Srivastava, A., Lavania, S., & Mohapatra, S. (2023). Predicting the effect on land values after introducing high-speed rail. In *Energy, environment, and sustainability* (pp. 157–196). https://doi.org/10.1007/978-981-99-2150-8_8.
11. Agrahari, R., Mohanty, S., Vishwakarma, K., Nayak, S. K., Samantaray, D., & Mohapatra, S. (2021). Update vision on COVID-19: Structure, immune pathogenesis, treatment and safety assessment. *Sensors International, 2*, 100073. https://doi.org/10.1016/j.sintl.2020.100073
12. Mishra, B. B., Nayak, S. K., Mohapatra, S., & Samantaray, D. (2021). Preface. In *Environmental and agricultural microbiology: Advances and applications* (pp. xvii–xix).
13. Tanaya, K., Mohapatra, S., & Samantaray, D. (2024). Formulation of biogenic fluorescent pigmented PHB nanoparticles from Rhodanobacter sp. for drug delivery. *Antonie van Leeuwenhoek International Journal Of General and Molecular Microbiology, 117*. https://doi.org/10.1007/s10482-024-01973-x
14. Mohapatra, S., Pattnaik, S., Maity, S., Sharma, S., Akhtar, J., Pati, S., Samantaray, D. P., & Varma, A. (2020). Comparative analysis of PHAs production by Bacillus megaterium OUAT 016 under submerged and solid-state fermentation. *Saudi Journal of Biological Sciences, 27*, 1242–1250. https://doi.org/10.1016/j.sjbs.2020.02.001
15. Singh, H., Singh, A. K., Singla, Y. K., & Chattopadhyay, K. (2020). Design & development of a low cost tribometer for nano particulate lubricants. *Materials Today Proceedings, 28*, 1487–1491.
16. Singh, H., Singla, Y. K., Singh, A. K., & Chattopadhayay, K. (2020). Effect of nanofly ash as lubricant additive on the tribological properties of SAE 10W-30 oil: A novel finding. *Transactions of the Indian Institute of Metals, 73*, 2371–2375.
17. Kumar Singla, Y., Dwivedi, D. K., & Arora, N. (2017). Modeling the impact–sliding wear characteristics of rare earth additive iron-based hardfacing alloys. *Proceedings of the Institution of Mechanical Engineers, Part J: Journal of Engineering Tribology, 231*, 1486–1499.
18. Kumar, V., Sharma, R., Dhakar, K., Singla, Y. K., & Verma, K. (2019). Experimental evaluation of magnetic abrasive finishing process with diamond abrasive. *International Journal of Materials and Product Technology, 58*, 55–70.
19. Kumar, R., Bhandari, S., Goyal, A., & Singla, Y. K. (2018). Slurry erosion performance study of HVFS sprayed Ni-20Al2O3 and Ni-15Al2O3-5TiO2 coatings under hydro accelerated conditions. *Industrial Lubrication and Tribology, 70*, 805–817.
20. Sharma, S. K., Sharma, R. C., Lavania, S., Palli, S., & Avesh, M. (2024). Dynamic vibration mitigation in ship propellers using hybrid fuzzy sliding mode control. *National Academy Science Letters*. https://doi.org/10.1007/s40009-024-01515-z
21. Singla, Y. K., Dwivedi, D. K., & Arora, N. (2015). On the modeling of dry sliding adhesive wear parameters of vanadium additive iron-based alloys at elevated temperatures. *Surface and Coating Technology, 283*, 223–233.
22. Singla, Y. K., Chhibber, R., Bansal, H., & Kalra, A. (2015). Wear behavior of aluminum alloy 6061-based composites reinforced with SiC, Al_2O_3, and red mud: A comparative study. *Journal of Metals, 67*, 2160–2169.

23. Kumar, S., Kumar, R., & Singla, Y. K. (2012). To study the mechanical behaviour of friction welding of aluminium alloy and mild steel. *International Journal of Mechanical Engineering and Robotics Research, 1*, 43–50.
24. Pant, M., Sharma, A., Palli, S., Sharma, S. K., & Sharma, R. C. (2024). A review and analysis of factors affecting EV adoption for sustainable and energy-efficient transportation systems using analytic hierarchy process approach. *International Journal of Vehicle Structures and Systems, 16*, 133–139. https://doi.org/10.4273/ijvss.16.2.01
25. Sharma, S. K., Sharma, R. C., Lavania, S., Palli, S., & Avesh, M. (2024). Industry-driven approach for ANFIS-based intelligent control suspension system with MR damper for enhanced ride quality in passenger rail vehicles for technological investigations. *National Academy Science Letters, 47*, 391–394. https://doi.org/10.1007/s40009-023-01365-1
26. Adeoye, A. O., Quadri, R. O., Lawal, O. S., & Emojevu, E. O. (2024). Physicochemical characterization, valorization of lignocellulosic waste (Kola nut seed shell) via pyrolysis, and ultrasonication of its crude bio-oil for biofuel production. *Cleaner Waste Systems, 7*, 100138. https://doi.org/10.1016/j.clwas.2024.100138
27. Adeoye, A. O., Quadri, R. O., Lawal, O. S., Malomo, D., Emojevu, E. O., Omonije, O. O., Odeniyi, O. K., Fadahunsi, M. O., Yelwa, M. J., Aasa, S. A., Onakpa, A. E., Omoniyi, B. P., Mark, I. N., Usman, J., Ismaila, A. M., & Saidu, A. U. (2024). Advanced techniques in upgrading crude bio-oil to biofuel. In *Energy, environment, and sustainability* (pp. 321–353). https://doi.org/10.1007/978-981-97-0515-3_14.
28. Adeoye, A. O., Quadri, R. O., & Lawal, O. S. (2022). Assessment of biofuel potential of tenera palm kernel shell via fixed bed pyrolysis and thermal characterization. *Results in Surfaces and Interfaces, 9*, 100091. https://doi.org/10.1016/j.rsurfi.2022.100091
29. Adeoye, A. O., Lawal, O. S., Quadri, R. O., Malomo, D., Aliyu, M. T., Dang, G. E., Emojevu, E. O., Maikato, M. J., Yahaya, M. G., Omonije, O. O., Edidem, V. G., Abubakar, Y. K., Offor, O. F., Sochima, E. H., Peter, B. E., & Hikon, B. N. (2023). Sustainable energy via thermochemical and biochemical conversion of biomass wastes for biofuel production. In *Energy, environment, and sustainability* (pp. 245–306). https://doi.org/10.1007/978-981-99-2150-8_11.
30. Mohapatra, S., & Mishra, R. K. (2020). Status of institutional credit disbursement of agricultural sector in Odisha, India. *International Journal of Current Microbiology and Applied Sciences, 9*, 2057–2065.
31. Mohapatra, S., Singh, J., & Mishra, S. (2017). Co-integration among major potato markets in Punjab. *Indian Journal of Agricultural Marketing, 31*, 46–47.
32. Mohapatra, S., Priyanka, V., Mohapatra, S., Kohli, I., & Mishra, R. K. (2020). Impact of corona virus covid-19 on the global economy. *International Journal of Agricultural and Statistical Sciences, 16*, 771–778.
33. Mohapatra, S., & Singh, J. (2021). Co-integration of major vegetable markets in Punjab, India. *Economic Affairs, 66*(02), 181–187.
34. Mohapatra, S., Mishra, R., & Mishra, R. (2021). Impact of COVID-19 on education level of agricultural farming community. *International Journal of Agricultural and Statistical Sciences, 17*, 915–923.
35. Mohapatra, S., Singh, J., & Kumar, S. (2018). Cointegration among major cauliflower markets in Punjab. *Indian Journal of Economics and Development, 14*, 330–335.
36. Adeoye, A. O., Yelwa, J. M., Imam, N., Quadri, R. O., Lawal, O. S., Malomo, D., Aasa, S. A., Onakpa, A. E., Dorgbaa, R. K., Omoniyi, B. P., Ahmad, N., & Garba, H. A. (2024). Pyrolysis of plastic wastes towards achieving a circular economy: An advanced chemistry and technical approach. In *Energy, environment, and sustainability* (pp. 215–259). https://doi.org/10.1007/978-981-97-0437-8_11
37. Adeoye, A. O., Quadri, R. O., & Lawal, O. S. (2023). Wet synthesis, characterization of goethite nanoparticles and its application in catalytic pyrolysis of palm kernel shell in TGA. *Results in Surfaces and Interfaces, 11*, 100118. https://doi.org/10.1016/j.rsurfi.2023.100118

38. Avesh, M., Imam, N., & Adeoye, A. O. (2024). Assessing environmental impacts and mitigation strategies for the DFCCIL rail projects. In *Energy, environment, and sustainability* (pp. 333–353). https://doi.org/10.1007/978-981-97-0437-8_15.
39. Mohapatra, S., Mohapatra, S., Han, H., Ariza-Montes, A., López-Martín, M. del C.: Climate change and vulnerability of agribusiness: Assessment of climate change impact on agricultural productivity. Frontiers in Psychology 13, (2022). doi:https://doi.org/10.3389/fpsyg.2022.955622.
40. Mohapatra, S., & Mishra, R. A. J. K. (2020). Production performance and market integration of cotton markets in India. In *National symposium* (p. 253).
41. Din, A. U., Han, H., Ariza-Montes, A., Vega-Muñoz, A., Raposo, A., & Mohapatra, S. (2022). The impact of Covid-19 on the food supply chain and the role of e-commerce for food purchasing. *Sustainability, 14*(5), 3074.
42. Mohapatra, S., & Singh, J. (2018). Growth and instability analysis of cauliflower crop in Punjab. *Economic Affairs, 63*, 665–669.
43. Pandey, G., Bansal, S., & Mohapatra, S. (2023). Market integration of chickpea crop: An evidence of India. *Journal of Agribusiness in Developing and Emerging Economies, 15*, 1.
44. Mohapatra, S., Mishra, R. K., & Sarangi, K. K. (2019). Economic appraisal of bio ethanol production technologies. *International Journal of Chemical Studies, 7*, 2197–2200.
45. Mohapatra, S., Kishore Mishra, R., & Sarangi, K. (2020). Bio-ethanol (2nd generation ethanol): A solution to ever polluting gasoline to climate in India. *International Research Journal of Pure and Applied Chemistry, 20*, 1–15.
46. Gohain, N., Bansal, S., Mohapatra, S., & Singh, L. (2022). Analyzing the direction of trade: Indian ginger and lessons from exports to different destinations. *Journal of Agricultural Development and Policy, 32*, 214–220.
47. Mohapatra, S., Mohapatra, U., Chandana, K. S. S., & Mishra, R. K. (2018). Economic analysis of paddy production and marketing in Puri, Odisha. *Journal of Pharmacognosy and Phytochemistry, 7*, 1858–1861.
48. Quadri, R. O., Adeoye, A. O., Ayanda, O. S., & Lawal, O. S. (2024). Conversion of palm kernel shell to sustainable energy and the effect of wet synthesized nanoparticles of iron on its thermal degradation kinetics. *Bioresource Technology Reports, 27*, 101933. https://doi.org/10.1016/j.biteb.2024.101933
49. Adeoye, A. O., Quadri, R. O., & Lawal, O. S. (2023). Biomass waste pyrolysis in the context of Nigeria's oil and gas industry: An overview. *International Journal of Ambient Energy, 44*, 2525–2541. https://doi.org/10.1080/01430750.2023.2256325
50. Mohapatra, S., Mohapatra, U., & Mishra, R. K. (2018). Price forecasting of groundnut in Odisha. *The Pharma Innovation Journal, 7*, 111–114.
51. Mohapatra, S., Sneha, N. K., Mounika, A., & Mishra, R. K. (2019). SOCIO-ECONOMIC AND DEMOGRAPHIC ANALYSIS OF SAARC NATIONS. *International Journal of Agricultural & Statistical Sciences, 15*, 529–533.
52. Sharma, S. K., Sharma, R. C., Choi, Y., & Lee, J. (2023). Modelling and dynamic analysis of adaptive neuro-fuzzy inference system-based intelligent control suspension system for passenger rail vehicles using magnetorheological damper for improving ride index. *Sustainability, 15*, 12529.
53. Sharma, S. K., Upadhyay, R. K., Kumar, V., & Valera, H. (2023). Introduction to sustainable transportation system. In *Transportation energy and dynamics* (pp. 3–6). Springer.
54. Hossain, I., Ahmed, Y. E., & Sharma, S. K. (2023). The use of both metallic-insert and TiO2/water nanofluid to improve the performance of photovoltaic modules; experimental study. *Thermal Science and Engineering Progress, 42*, 101944.
55. Wu, Q., Cole, C., Spiryagin, M., Chang, C., Wei, W., Ursulyak, L., Shvets, A., Murtaza, M. A., Mirza, I. M., & Zhelieznov, K. (2023). Freight train air brake models. *International Journal of Rail Transportation, 11*, 1–49.

56. Upadhyay, R. K., Sharma, S. K., Kumar, V., & Valera, H. (2023). Introduction to transportation systems technology and integrated management. In *Transportation systems technology and integrated management* (pp. 1–4). Springer.
57. Sharma, R. C., Bhardawaj, S., Sharma, S. K., & Bahl, S. (2023). Ride behaviour of three-wheel vehicle using semi-active suspension using skyhook control strategy. *International Journal of Vehicle Structures & Systems (IJVSS), 15.*
58. Bhardawaj, S., Sharma, R. C., Sharma, S. K., Gopala Rao, L. V. V., & Vashist, A. (2023). Modeling of biomechanical human body model for seat to head transmissibility analysis. *Noise & Vibration Worldwide, 54*, 62–74.
59. Sharma, S. K., Sharma, R. C., Choi, Y., & Lee, J. (2023). Experimental and mathematical study of flexible–rigid rail vehicle riding comfort and safety. *Applied Sciences, 13*, 5252.
60. Sharma, R. C., Sharma, S. K., Palli, S., Rallabandi, S. R., & Sharma, N. (2025b). Compatibility between stability and curving ability of railway vehicles using conventional and unconventional bogie frame models. *SAE International Journal of Commercial Vehicles, 18*(3), 02-18-03-0014. https://doi.org/10.4271/02-18-03-0014
61. Sharma, S. K., Upadhyay, R. K., & Kumar, V. (2024). Introduction to dynamics of transportation ecosystem, modeling, and control. In *Dynamics of transportation ecosystem, modeling, and control* (pp. 3–6). Springer.
62. Sharma, S. K., Sharma, R. C., Mohapatra, S., Fouly, A., Hossain, I., & Lee, J. (2024). Technological advancements in oscillation reduction for propulsion shaft systems. *Journal of Vibration Engineering & Technologies*, 1–19.
63. Upadhyay, R. K., Sharma, S. K., & Kumar, V. (2024). Introduction to intelligent transportation system and advanced technology. In *Intelligent transportation system and advanced technology* (pp. 3–6). Springer.
64. Hossain, I., Ngoupo, A. T., Bhattarai, S., Challagulla, S. P., & Sharma, S. K. (2024). Empirical survey on the effect of straight and wavy fins integrated with wind stream as heat sink of photovoltaic panels. *International Journal of Environmental Science and Technology, 22*, 2479–2492.
65. Sharma, S. K., Sharma, R. C., & Lee, J. (2024). Propelling precision of longitudinal vibration mitigation in ship propeller shafts through advanced nonlinear intelligent semi-active control leveraging adaptive neuro-fuzzy inference system with linear quadratic regulator. *Journal of Vibration and Control, 10775463241244836*, 1472.
66. Sharma, S. K., Kumar, N., Avesh, M., Sharma, R. C., Siddiqui, M. I. H., & Lee, J. (2024). Navigating tranquillity with H∞ controller to mitigate ship propeller shaft vibration. *Journal of Vibration Engineering & Technologies*, 1–11.
67. Sharma, S. K., Sharma, R. C., Upadhyay, R. K., & Lee, J. (2024). Enhanced ship engine vibration reduction using magnetorheological dampers and adaptive neuro-fuzzy control system. *Ships and Offshore Structures*, 1–17.
68. Raju Rallabandi, S., Pilla, D. P., Dowluru, S., Palli, S., Sharma, N., Sharma, S. K., & Sharma, R. C. (2024). Critical evaluation of epoxy-hemp-pineapple-palm fiber composites using hybrid AHM-TOPSIS technique for sustainable structural applications. *Journal of the Chinese Institute of Engineers, 47*, 325–336.
69. Kokkinos, N. C., Emmanouilidou, E., & Sharma, S. K. (2024). Waste-to-biofuel production for the transportation or. In *Intelligent transportation system and advanced technology* (pp. 99–122). Springer.
70. Pandipati, S., Palli, S., Dowwluru, S., Sharma, R. C., Sharma, N., Sharma, S. K., & Da Fonseca-Soares, D. (2024). Microstructure and strength comparison for friction stir processing of aluminium alloys and multi-response optimization by Taguchi GRAPCA based hybrid methods. *International Journal of Vehicle Structures & Systems, 16*(3), 321–328. https://doi.org/10.4273/ijvss.16.3.03

71. Sharma, S. K., Sharma, R. C., & Lee, J. (2024). Modelling and simulation of MR damper characterization and uncertainty assessment in marine engine vibration isolation with FLPID control. *International Journal of Modelling and Simulation*, 1–11.
72. Sharma, S. K., Sharma, R. C., Upadhyay, R. K., & Lee, J. (2024). Coupled torsional–transverse vibration reduction in marine propulsion dynamics with novel approach using magnetorheological damping and adaptive control system. *Journal of Vibration Engineering & Technologies, 12*, 6089–6099.
73. Rallabandi, S. R., Srinivas, L., Palli, S., Sharma, R. C., Sharma, N., Sharma, A., & Sharma, S. K. (2024). A contrastive characterization of pure mg and AZ91D alloy based on the testing of mechanical, corrosion, wear, and erosion properties. *Engineering Research Express, 6*, 015017.
74. Badgujar, P., Sharma, S., Rahul, D., Sharma, S. K., & Upadhyay, R. K. (2024). Feasibility study of Mumbai-Pune Hyperloop corridor: Challenges and implementation. *Research in Transportation Business & Management, 52*, 101089.
75. Sharma, R. C., Palli, S., Dharmana, L., Sharma, S. K., Sharma, A., Kharola, A., & Sharma, N. (2024). Simulation of transient dynamic response of a locomotive considering its flexibility and rigidity. *Noise & Vibration Worldwide, 55*, 33–47. https://doi.org/10.1177/09574565231212684
76. Song, H., Park, J., Sharma, S. K., & Lee, J. (2022). A comparative study on ultrasonic propagation characteristics and defect detection of metal material additive manufacturing using Deep learning algorithm. In *Biennial international conference on future learning aspects of mechanical engineering* (pp. 157–168). Springer.
77. Boris, I., Sharma, S. K., & Lee, J. (2022). A study on the reflection coefficient of higher modes of torsional wave by theoretical and numerical method. In *Biennial international conference on future learning aspects of mechanical engineering* (pp. 145–155). Springer.
78. Vishwakarma, P. N., Mishra, P., Sharma, S. K., & Shukla, A. K. (2022). Formulating and Analysing the effect of suspension parameters on critical speeds for various DOFs in rail vehicle. In *Biennial international conference on future learning aspects of mechanical engineering* (pp. 791–802). Springer.
79. Kumar, U., Kasvekar, R., Sharma, S. K., & Upadhyay, R. K. (2022). Wear of wheels and axle in locomotive and measures taken by Indian railway. *Advances in Engine Tribology*, 77–96.
80. Sharma, R. C., Palli, S., Jha, A. K., Bhardawaj, S., & Sharma, S. K. (2022). Vibration and ride comfort analysis of railway vehicle system subjected to deterministic inputs. *Resmilitaris, 12*, 1345–1355.
81. Sharma, R. C., Rao, L. V. V. G., Sharma, S. K., Palli, S., & Satyanarayana, V. S. (2022). Analysis of lateral stability and ride of an Indian railway constrained dual-axle bogie frame. *SAE International Journal of Commercial Vehicles, 16*, 213–228.
82. Sharma, R. C., & Sharma, S. K. (2022). Ride analysis of road surface-three-wheeled vehicle-human subject interactions subjected to random excitation. *SAE International Journal of Commercial Vehicles, 15*, 299–312.
83. Sharma, S. K., Lee, J., & Jang, H.-L. (2022). Mathematical modeling and simulation of suspended equipment impact on car body modes. *Machines, 10*, 192.
84. Sharma, S. K., Sharma, R. C., Lee, J., & Jang, H.-L. (2022). Numerical and experimental analysis of DVA on the flexible-rigid rail vehicle carbody resonant vibration. *Sensors, 22*, 1922.
85. Vishwakarma, P. N., Mishra, P., & Sharma, S. K. (2022). Characterization of a magnetorheological fluid damper a review. *Materials Today Proceedings, 56*, 2988–2994.
86. Sharma, S. K., Sharma, R. C., & Lee, J. (2022). In situ and experimental analysis of longitudinal load on carbody fatigue life using nonlinear damage accumulation. *International Journal of Damage Mechanics, 31*, 605–622.
87. Sharma, R. C., Sharma, S., Sharma, N., & Sharma, S. K. (2021). Linear and nonlinear analysis of ride and stability of a three-wheeled vehicle subjected to random and bump inputs using bond graph and simulink methodology. *SAE International Journal of Commercial Vehicles, 15*, 65–79.

88. Lee, J., Han, J., & Sharma, S. K. (2021). Structural analysis on the separated and integrated differential gear case for the weight reduction. In *Advances in engineering design: Select proceedings of FLAME 2020* (pp. 175–181). Springer Singapore.
89. Choi, S., Lee, J., & Sharma, S. K. (2021). A study on the performance evaluation of hydraulic tank injectors. In *Advances in engineering design: Select proceedings of flame 2020* (pp. 183–190). Springer Singapore.
90. Sharma, R. C., Palli, S., Sharma, N., & Sharma, S. K. (2021). Ride behaviour of a four-wheel vehicle using h infinity semi-active suspension control under deterministic and random inputs. *International Journal of Vehicle Structures & Systems, 13*, 234–237.
91. Bhardawaj, S., Sharma, R. C., Sharma, S. K., & Sharma, N. (2021). On the planning and construction of railway curved track. *International Journal of Vehicle Structures & Systems, 13*, 151–159.
92. Mohapatra, S., Mohanty, D., Sharma, S., Dikshit, S., Kohli, I., Samantaray, D. P., Kumar, R., & Kathpalia, M. (2021). Biomedical application of polymeric biomaterial: Polyhydroxybutyrate. In *Bioresource utilization and management* (pp. 111–124). Apple Academic Press.
93. Sharma, S. K., Sharma, R. C., & Lee, J. (2021). Effect of rail vehicle–track coupled dynamics on fatigue failure of coil spring in a suspension system. *Applied Sciences, 11*, 2650.
94. Sharma, R. C., Sharma, S., Sharma, S. K., Sharma, N., & Singh, G. (2021). Analysis of biodynamic model of seated human subject and optimization of the passenger ride comfort for three-wheel vehicle using random search technique. *Proceedings of the Institution of Mechanical Engineers, Part K: Journal of Multi-body Dynamics, 235*, 106–121.
95. Cataluna Veses, R., & Shah, Z. (2015). Bio-oil production by thermal cracking in the presence of hydrogen. *Journal of Fundamentals of Renewable Energy and Applications, 06*, 88. https://doi.org/10.4172/2090-4541.1000194
96. Shah, Z., Cataluña Veses, R., Aguilhera, R. A., & Da Silva, R. (2016). Bio-oil production from pyrolysis of coffee and eucalyptus sawdust in the presence of 5% hydrogen. *International Journal of Engineering Research & Science (IJOER)*.
97. Shah, Z., Veses, R. C., & da Silva, R. (2016). A comparative study and analysis of two types of bio-oil samples obtained from freshwater algae and microbial treated algae. *Modern Chemistry & Applications, 4*, 185. https://doi.org/10.4172/2329-6798.1000185
98. Shah, Z., Veses, R. C., Brum, J., Klunk, M. A., Rocha, L. A. O., Missaggia, A. B., de Oliveira, A. P., & Caetano, N. R. (2020). Sewage sludge bio-oil development and characterization. *Inventions, 5*, 36. https://doi.org/10.3390/inventions5030036
99. Shah, Z., Saberian, M., Hodgeman, D., Kinloch, I., & Vallés, C. (2023). Effect of sulfur on wood tar biopitch as a sustainable replacement for coal tar pitch binders. *ACS Applied Engineering Materials, 1*, 2567–2576. https://doi.org/10.1021/acsaenm.3c00361
100. Klunk, M. A., Dasgupta, S., Das, M., & Shah, Z. (2018). SYSTEM OF ADSORPTION OF CO2 IN COALBED. *SOUTHERN BRAZILIAN JOURNAL OF CHEMISTRY, 26*, 2–9. https://doi.org/10.48141/SBJCHEM.v26.n26.2018.7_2018.pdf
101. Mohapatra, S., Mohapatra, U., & Mishra, R. (2017). Diversification towards vegetable crops: A good option for doubling the farmer's income. *Journal of Experimental Agriculture International, 18*, 1–7.
102. Chavan, S. D., Bansal, S., Mohapatra, S., Kaur, L., & Jadhav, A. (2023). Trade directions of Indian basmati rice export-Markov chain approach. *Economic Affairs, 68*(01), 541–547.
103. Bansal, S., Mohapatra, S., Gohain, N., & Das, A. (2022). Production status, market integration and the law of one price: An exploration of the major maize markets of India. *Indian Journal of Economics and Development, 18*, 832–841.
104. Sharma, R. C., Gopala Rao, L. V. V., Srikar, B., Palli, S., Appala Naidu, Y., Sharma, N., & Sharma, S. K. (2025). Investigation of the dynamic behavior of hybrid composites using finite element and experimental method. *Journal of the Chinese Institute of Engineers, 48*(2), 158–171. https://doi.org/10.1080/02533839.2025.2453139

105. Mohapatra, S., Samantaray, D. P., & Samantaray, S. M. (2015). Study on polyhydroxyalkanoates production using rhizospheric soil bacterial isolates of sweet potato. *Indian Journal of Science and Technology, 8*, 57–62. https://doi.org/10.17485/ijst/2015/v8iS7/64027
106. Arora, R., Trivedi, J., Mohapatra, S., & Kumar, P. (2022). Trends in probiotics on human health and industrial application. In *Advances in agricultural and industrial microbiology* (Microbial diversity and application in agroindustry) (Vol. 1, pp. 237–248). Springer. https://doi.org/10.1007/978-981-16-8918-5_13
107. Samal, S., Pati, S., Mohapatra, S., Maity, S., Tanaya, K., Devadarshini, D., & Samantaray, D. (2023). PHAs production by facultative anaerobic bacteria Bacillus cereus FM5 through submerged and solid-state fermentation under anoxic condition. *Antonie van Leeuwenhoek, International Journal of General and Molecular Microbiology, 116*, 521–529. https://doi.org/10.1007/s10482-023-01825-0
108. Biswas, N., Saha, S., Biswas, B. K., Chowdhury, M., Rahaman, A., Junghare, V., Mohapatra, S., Hazra, S., Zangrando, E., Roy Choudhury, R., & Roy Choudhury, C. (2021). The DNA-and protein-binding properties and cytotoxicity of a new copper(II) hydrazone Schiff base complex. *Journal of Coordination Chemistry, 74*, 1482–1504. https://doi.org/10.1080/00958972.2021.1913128
109. Mohapatra, S., Mishra, R., Roy, P., Yadav, K. L., & Satapathi, S. (2017). Systematic investigation and in vitro biocompatibility studies on implantable magnetic nanocomposites for hyperthermia treatment of osteoarthritic knee joints. *Journal of Materials Science, 52*, 9262–9268. https://doi.org/10.1007/s10853-017-1136-0
110. Mishra, B. B., Nayak, S. K., Mohapatra, S., & Samantaray, D. (2021). *Environmental and agricultural microbiology: Advances and applications*. https://doi.org/10.1002/9781119525899.
111. Sharma, S. K., Mohapatra, S., Sharma, R. C., Alturjman, S., Altrjman, C., Mostarda, L., & Stephan, T. (2022). Retrofitting existing buildings to improve energy performance. *Sustainability (Switzerland), 14*, 666. https://doi.org/10.3390/su14020666
112. Bandhu, S., Khot, M. B., Sharma, T., Sharma, O. P., Dasgupta, D., Mohapatra, S., Hazra, S., Khatri, O. P., & Ghosh, D. (2018). Single cell oil from oleaginous yeast grown on sugarcane bagasse-derived xylose: An approach toward novel biolubricant for low friction and Wear. *ACS Sustainable Chemistry & Engineering, 6*, 275–283. https://doi.org/10.1021/acssuschemeng.7b02425
113. Kohli, I., Joshi, N. C., Mohapatra, S., & Varma, A. (2020). Extremophile—An adaptive strategy for extreme conditions and applications. *Current Genomics, 21*, 96–110. https://doi.org/10.2174/1389202921666200401105908
114. Pati, S., Mohapatra, S., Maity, S., Dash, A., & Samantaray, D. P. (2021). Polyhydroxyalkanoates for sustainable smart packaging of fruits. In: *Environmental and agricultural microbiology: Advances and applications* (pp. 183–195).
115. Chatterjee, A., Nirwan, S., Mohapatra, S., Sharma, P., Agnihotri, A., & Shrivastava, N. (2022). Biochemical aspects of pathogenic variability in white rust infected Indian mustard. *Mycologia, 114*, 757–768. https://doi.org/10.1080/00275514.2022.2060007
116. Pati, S., Maity, S., Dash, A., Jema, S., Mohapatra, S., Das, S., & Samantaray, D. P. (2020). Biocompatible PHB production from bacillus species under submerged and solid-state fermentation and extraction through different downstream processing. *Current Microbiology, 77*, 1203–1209. https://doi.org/10.1007/s00284-020-01922-7
117. Agrahari, R., Sarraf, G., Joshi, N. C., Mohapatra, S., & Var, A. (2021). Insight of biopolymers and applications of polyhydroxyalkanoates. In *Microbial polymers: Applications and ecological perspectives* (pp. 177–191). Springer. https://doi.org/10.1007/978-981-16-0045-6_8
118. Kathpalia, M., Kumar, P., & Mohapatra, S. (2022). Homocysteine metabolism as a biomarker for cancer. In *Homocysteine metabolism in health and disease* (pp. 159–171). Springer. https://doi.org/10.1007/978-981-16-6867-8_8

119. Bharatiy, S. K., Hazra, M., Paul, M., Mohapatra, S., Samantaray, D., Dubey, R. C., Sanyal, S., Datta, S., & Hazra, S. (2016). In silico designing of an industrially sustainable carbonic anhydrase using molecular dynamics simulation. *ACS Omega, 1*, 1081–1103. https://doi.org/10.1021/acsomega.6b00041
120. Sarkar, M. S., Segu, H., Bhaskar, J. V., Jakher, R., Mohapatra, S., Shalini, K., Shivaji, S., & Reddy, P. A. (2018). Ecological preferences of large carnivores in remote, high-altitude protected areas: Insights from Buxa Tiger Reserve, India. *ORYX, 52*, 66–77. https://doi.org/10.1017/S0030605317000060
121. Pati, S., Dash, A., Maity, S., Pattnaik, S., Mohapatra, S., & Samantaray, D. P. (2021). Nano biofertilizer for sustainable agriculture. In *Agriculturally important microorganisms: Mechanisms and applications for sustainable agriculture* (pp. 201–208). CRC Press. https://doi.org/10.1201/9781003245841-10
122. Mohanty, D., Misra, S., Mohapatra, S., & Sahu, P. S. (2018). Prebiotics and synbiotics: Recent concepts in nutrition. *Food Bioscience, 26*, 152–160. https://doi.org/10.1016/j.fbio.2018.10.008
123. Kumar, R., Kumari, R., Prasad, C., Deep, A., Singh, N., Namtak, S., Tiwari, V., Merugu, R., Mohapatra, S., & Singh, R. (2021). Influence of water quality on the diversity of macroinvertebrates in the Mandakini River in India. *Water Science and Technology: Water Supply, 21*, 1843–1860. https://doi.org/10.2166/WS.2021.020
124. Mohanty, D., Jena, R., Choudhury, P. K., Pattnaik, R., Mohapatra, S., & Saini, M. R. (2016). Milk derived antimicrobial bioactive peptides: A review. *International Journal of Food Properties, 19*, 837–846. https://doi.org/10.1080/10942912.2015.1048356
125. Mohapatra, S., Rath, S. N., Pradhan, S. K., Samantaray, D. P., & Rath, C. C. (2016). Secondary structural models (16S rRNA) of polyhydroxyalkanoates producing bacillus species isolated from different rhizospheric soil: Phylogenetics and chemical analysis. *International Journal Bioautomation, 20*, 329–338.
126. Mohapatra, S., Sarkar, B., Samantaray, D. P., Daware, A., Maity, S., Pattnaik, S., & Bhattacharjee, S. (2017). Bioconversion of fish solid waste into PHB using Bacillus subtilis based submerged fermentation process. *Environmental Technology (United Kingdom), 38*, 3201–3208. https://doi.org/10.1080/09593330.2017.1291759
127. Misra, S., Mohanty, D., & Mohapatra, S. (2021). Food-based probiotics: Functional dietary ingredients. In *Probiotics: Advanced food and health applications* (pp. 257–275). Academic Press. https://doi.org/10.1016/B978-0-323-85170-1.00006-3
128. Kumar, R., Kumari, R., Prasad, C., Tiwari, V., Singh, N., Mohapatra, S., Merugu, R., Namtak, S., & Deep, A. (2020). Phytoplankton diversity in relation to physicochemical attributes and water quality of Mandakini River, Garhwal Himalaya. *Environmental Monitoring and Assessment, 192*, 799. https://doi.org/10.1007/s10661-020-08768-3
129. Mohapatra, S., Vishwakarma, K., Joshi, N. C., Maity, S., Kumar, R., Ramchander, M., Pattnaik, S., & Samantaray, D. P. (2021). A review on PHAs: The future biopolymer. In *Environmental and agricultural microbiology: Advances and applications* (pp. 83–117).
130. Sahoo, N., Pattnaik, S., Pattnaik, M., & Mohapatra, S. (2021). Methane emission and strategies for mitigation in livestock. In *Environmental and agricultural microbiology: Advances and applications* (pp. 257–274).
131. Mohanty, D. P., Mohapatra, S., Misra, S., & Sahu, P. S. (2016). Milk derived bioactive peptides and their impact on human health—A review. *Saudi Journal of Biological Sciences, 23*, 577–583. https://doi.org/10.1016/j.sjbs.2015.06.005
132. Kohli, I., Mohapatra, S., Kumar, P., Goel, A., Varma, A., & Joshi, N. C. (2022). Role of bacterial endophytes in the promotion of plant growth. In *Bacterial endophytes for sustainable agriculture and environmental management* (pp. 235–253). https://doi.org/10.1007/978-981-16-4497-9_12.

133. Samantaray, D., Mohapatra, S., & Mishra, B. B. (2014). Microbial bioremediation of industrial effluents. In *Microbial biodegradation and bioremediation* (pp. 326–339). https://doi.org/10.1016/B978-0-12-800021-2.00014-5.
134. Mohapatra, S., Mohanta, P. R., Sarkar, B., Daware, A., Kumar, C., & Samantaray, D. P. (2017). Production of Polyhydroxyalkanoates (PHAs) by bacillus strain isolated from waste water and its biochemical characterization. *Proceedings of the National Academy of Sciences India Section B - Biological Sciences, 87*, 459–466. https://doi.org/10.1007/s40011-015-0626-6
135. Tomer, P., Shroti, G. K., Mohapatra, S., Ghosh, D., Jaiswal, S., Garg, D., Lahiri, D., Kumar, N., Khatri, O. P., & Hazra, S. (2023). Fabrication of HA Nano-crystal reinforced PHAs based composites for orthopedic fracture-fixing accessories using sustainable resources. *Environment, Development and Sustainability, 27*, 1. https://doi.org/10.1007/s10668-023-03912-8
136. Bandekar, D., Mohapatra, S., Hazra, M., Hazra, S., & Biswas, S. (2021). N-terminal truncation of VC0395_0300 protein from vibrio cholerae does not lead to loss of diguanylate cyclase activity. *Biophysical Chemistry, 268*, 106493. https://doi.org/10.1016/j.bpc.2020.106493
137. Vishwakarma, K., Kumar, N., Shandilya, C., Mohapatra, S., Bhayana, S., & Varma, A. (2020). Revisiting plant–microbe interactions and microbial consortia application for enhancing sustainable agriculture: A review. *Frontiers in Microbiology, 11*. https://doi.org/10.3389/fmicb.2020.560406
138. Mohapatra, S., Maity, S., Dash, H. R., Das, S., Pattnaik, S., Rath, C. C., & Samantaray, D. (2017). Bacillus and biopolymer: Prospects and challenges. *Biochemistry and Biophysics Reports, 12*, 206–213. https://doi.org/10.1016/j.bbrep.2017.10.001
139. Kumar, R., Merugu, R., Mohapatra, S., & Sharma, S. (2022). Extremophiles: Life of microorganisms in extreme environments. In *Extremophiles: General and plant biomass based biorefinery* (pp. 43–66). https://doi.org/10.1201/9781003014188-3
140. Pradhan, A., Pahari, A., Mohapatra, S., & Mishra, B. B. (2017). Phosphate-solubilizing microorganisms in sustainable agriculture: Genetic mechanism and application. In *Microorganisms for sustainability* (pp. 81–97). https://doi.org/10.1007/978-981-10-7380-9_5.
141. Devadarshini, D., Mohapatra, S., Pati, S., Maity, S., Rath, C. C., Jena, P. K., & Samantaray, D. (2023). Evaluation of PHAs production by mixed bacterial culture under submerged fermentation. *Biologia (Bratisl), 78*, 895–902. https://doi.org/10.1007/s11756-022-01302-5
142. Agarwal, K. M., Mohapatra, S., Sharma, P., Sharma, S., Bhatia, D., & Mishra, A. (2020). Study and overview of the novel corona virus disease (COVID-19). *Sensors International, 1*, 100037. https://doi.org/10.1016/j.sintl.2020.100037
143. Tanaya, K., Mohapatra, S., Maity, S., & Samantaray, D. P. (2023). Fluorescent-pigmented microbial polymer nanoparticles: An approach for replacement of bioimaging agent in the drug delivery system. *Applied Biochemistry and Microbiology, 59*, 378–383. https://doi.org/10.1134/S0003683823030171
144. Pattnaik, S., Dash, D., Mohapatra, S., Pati, S., Devadarshini, D., Samal, S., Pattnaik, M., Maity, S., Mishra, S. K., & Samantaray, D. (2024). Reclamation of chromium-contaminated soil by native Cr(VI)-reducing and PHA-accumulating bacillus aryabhattai CTSI-07. *International Microbiology, 27*, 731–742. https://doi.org/10.1007/s10123-023-00421-6
145. Mohapatra, S., Mishra, Y., & Dash, L. (2020). Covid-19: Analysing the legal nuances of the lockdown order. *Indian Journal of Forensic Medicine and Toxicology, 14*, 612–616. https://doi.org/10.37506/ijfmt.v14i4.11553
146. Pati, S., Mohanty, M. K., Mohapatra, S., & Samantaray, D. (2022). Bioethanol production by Enterobacter hormaechei through carbon monoxide-rich syngas fermentation. *International Journal of Energy Research, 46*, 20096–20106. https://doi.org/10.1002/er.8678
147. Maity, S., Das, S., Mohapatra, S., Tripathi, A. D., Akthar, J., Pati, S., Pattnaik, S., & Samantaray, D. P. (2020). Growth associated polyhydroxybutyrate production by the novel Zobellellae tiwanensis strain DD5 from banana peels under submerged fermentation. *International Journal of Biological Macromolecules, 153*, 461–469. https://doi.org/10.1016/j.ijbiomac.2020.03.004

148. Takkar, S., Shandilya, C., Agrahari, R., Chaurasia, A., Vishwakarma, K., Mohapatra, S., Varma, A., & Mishra, A. (2022). Green technology: Phytoremediation for pesticide pollution. In *Phytoremediation technology for the removal of heavy metals and other contaminants from soil and water* (pp. 353–375). https://doi.org/10.1016/B978-0-323-85763-5.00008-8.
149. Bandekar, D., Chouhan, O. P., Mohapatra, S., Hazra, M., Hazra, S., & Biswas, S. (2017). Putative protein VC0395_0300 from vibrio cholerae is a diguanylate cyclase with a role in biofilm formation. *Microbiological Research, 202*, 61–70. https://doi.org/10.1016/j.micres.2017.05.003
150. Pattnaik, S., Dash, D., Mohapatra, S., Pattnaik, M., Marandi, A. K., Das, S., & Samantaray, D. P. (2020). Improvement of rice plant productivity by native Cr(VI) reducing and plant growth promoting soil bacteria Enterobacter cloacae. *Chemosphere, 240*, 124895. https://doi.org/10.1016/j.chemosphere.2019.124895
151. Mohapatra, S., Samantaray, D. P., Samantaray, S. M., Mishra, B. B., Das, S., Majumdar, S., Pradhan, S. K., Rath, S. N., Rath, C. C., Akthar, J., & Achary, K. G. (2016). Structural and thermal characterization of PHAs produced by Lysinibacillus sp. through submerged fermentation process. *International Journal of Biological Macromolecules, 93*, 1161–1167. https://doi.org/10.1016/j.ijbiomac.2016.09.077
152. Sharma, S. K., Kumar, A., & Sharma, R. C. (2014). Challenges in railway vehicle modeling and simulations. In *International conference on newest drifts in mechanical engineering ICNDME-2014* (pp. 453–459).
153. Sharma, S. K., & Kumar, A. (2014). A comparative study of Indian and worldwide railways. *International Journal of Mechanical Engineering and Robotics Research, 1*, 114–120.
154. Sharma, S. K., Sharma, R. C., Kumar, A., & Palli, S. (2015). Challenges in rail vehicle-track modeling and simulation. *International Journal of Vehicle Structures & Systems, 7*, 1.
155. Palli, S., Sharma, R. C., Sharma, S. K., Muddada, V., & Sharma, N. (2020). A case study on noise pollution and its effects. *Journal of Critical Reviews, 7*, 261–267.
156. Sharma, R. C., Sharma, S. K., Sharma, N., & Sharma, S. (2020). Analysis of ride and stability of an ICF railway coach. *International Journal of Vehicle Noise and Vibration, 16*, 127–144.
157. Palli, S., Sharma, R. C., Sharma, S. K., & Chintada, V. B. (2020). On methods used for setting the curve for railway tracks. *Journal Of Critical Review, 7*, 241–246.
158. Sharma, S., Sharma, R. C., Sharma, S. K., Sharma, N., Palli, S., & Bhardawaj, S. (2020). Vibration isolation of the quarter car model of road vehicle system using dynamic vibration absorber. *International Journal of Vehicle Structures & Systems, 12*, 513–516. https://doi.org/10.4273/IJVSS.12.5.05
159. Acharya, A., Gahlaut, U., Sharma, K., Sharma, S. K., Vishwakarma, P. N., & Phanden, R. K. (2020). Crashworthiness analysis of a thin-walled structure in the frontal part of automotive chassis. *International Journal of Vehicle Structures & Systems, 12*, 517–520.
160. Sharma, S. K., & Lee, J. (2020). Finite element analysis of a fishplate rail joint in extreme environment condition. *International Journal of Vehicle Structures & Systems, 12*, 503–506.
161. Bhardawaj, S., Sharma, R. C., & Sharma, S. K. (2020). Development of multibody dynamical using MR damper based semi-active bio-inspired chaotic fruit fly and fuzzy logic hybrid suspension control for rail vehicle system. *Proceedings of the Institution of Mechanical Engineers, Part K: Journal of Multi-body Dynamics, 234*, 723–744.
162. Sharma, S. K., & Lee, J. (2020). Design and development of smart semi active suspension for nonlinear rail vehicle vibration reduction. *International Journal of Structural Stability and Dynamics, 20*, 2050120.
163. Sharma, S. K., Phan, H., & Lee, J. (2020). An application study on road surface monitoring using DTW based image processing and ultrasonic sensors. *Applied Sciences, 10*, 4490.
164. Sharma, R. C., Sharma, S., Sharma, S. K., & Sharma, N. (2020). Analysis of generalized force and its influence on ride and stability of railway vehicle. *Noise & Vibration Worldwide, 51*, 95–109.

165. Bhardawaj, S., Sharma, R. C., & Sharma, S. K. (2020). Ride analysis of track-vehicle-human body interaction subjected to random excitation. *Journal of the Chinese Society of Mechanical Engineers, 41*, 227–236.
166. Bhardawaj, S., Sharma, R. C., & Sharma, S. K. (2020). Development in the modeling of rail vehicle system for the analysis of lateral stability. *Materials Today Proceedings, 25*, 610–619.
167. Sharma, S. K., Sharma, R. C., & Sharma, N. (2020). Combined multi-body-system and finite element analysis of a rail locomotive crashworthiness. *International Journal of Vehicle Structures & Systems, 12*, 428–435.
168. Bhardawaj, S., Sharma, R. C., & Sharma, S. K. (2020). Analysis of frontal car crash characteristics using ANSYS. *Materials Today Proceedings, 25*, 898–902.
169. Bhardawaj, S., Sharma, R. C., & Sharma, S. K. (2019). A survey of railway track modelling. *International Journal of Vehicle Structures & Systems, 11*, 508–518.
170. Bhardawaj, S., Sharma, R. C., & Sharma, S. K. (2019). Development and advancement in the wheel-rail rolling contact mechanics. In *IOP conference series: Materials science and engineering* (Vol. 691, p. 012034). IOP Publishing.
171. Choppara, R. K., Sharma, R. C., Sharma, S. K., & Gupta, T. (2019). Aero dynamic cross wind analysis of locomotive. In *IOP conference series: Materials science and engineering* (Vol. 691, p. 012035). IOP Publishing.
172. Sharma, S. K., & Sharma, R. C. (2019). Pothole detection and warning system for Indian roads. In *Advances in interdisciplinary engineering: Select proceedings of FLAME 2018* (pp. 511–519). Springer Singapore.
173. Sinha, A. K., Sengupta, A., Gandhi, H., Bansal, P., Agarwal, K. M., Sharma, S. K., Sharma, R. C., & Sharma, S. K. (2019). Performance enhancement of an all-terrain vehicle by optimizing steering, powertrain and brakes. In *Advances in engineering design: Select proceedings of FLAME 2018* (pp. 207–215). Springer Singapore.
174. Goyal, S., Anand, C. S., Sharma, S. K., & Sharma, R. C. (2019). Crashworthiness analysis of foam filled star shape polygon of thin-walled structure. *Thin-Walled Structures, 144*, 106312.
175. Goswami, B., Rathi, A., Sayeed, S., Das, P., Sharma, R. C., & Sharma, S. K. (2019). Optimization design for aerodynamic elements of Indian locomotive of passenger train. In *Advances in engineering design: Select proceedings of FLAME 2018* (pp. 663–673). Springer Singapore.
176. Sharma, S. K., Saini, U., & Kumar, A. (2019). Semi-active control to reduce lateral vibration of passenger rail vehicle using disturbance rejection and continuous state damper controllers. *Journal of Vibration Engineering & Technologies, 7*, 117–129.
177. Sharma, S. K. (2019). Multibody analysis of longitudinal train dynamics on the passenger ride performance due to brake application. *Proceedings of the Institution of Mechanical Engineers, Part K: Journal of Multi-body Dynamics, 233*, 266–279.
178. Sharma, S. (2018). Vibration control in quarter-car model with magnetorheological (MR) dampers using Bingham model. *Journal of Applied Mechanics and Engineering, 7*, 1000299.
179. Sharma, S. K., & Kumar, A. (2018). Impact of longitudinal train dynamics on train operations: A simulation-based study. *Journal of Vibration Engineering & Technologies, 6*, 197–203.
180. Sharma, S. K., & Sharma, R. C. (2018). An investigation of a locomotive structural crashworthiness using finite element simulation. *SAE International Journal of Commercial Vehicles, 11*, 235–243.
181. Sharma, R. C., Sharma, S. K., & Palli, S. (2018). Rail vehicle modelling and simulation using Lagrangian method. *International Journal of Vehicle Structures and Systems, 10*, 188–194.
182. Sharma, R. C., & Sharma, S. K. (2018). Sensitivity analysis of three-wheel vehicle's suspension parameters influencing ride behavior. *Noise & Vibration Worldwide, 49*, 272–280.
183. Sharma, S. K., & Sharma, R. C. (2018). Simulation of quarter-car model with magnetorheological dampers for ride quality improvement. *International Journal of Vehicle Structures & Systems, 10*, 169–173.

184. Sharma, S. K., & Kumar, A. (2018). Disturbance rejection and force-tracking controller of nonlinear lateral vibrations in passenger rail vehicle using magnetorheological fluid damper. *Journal of Intelligent Material Systems and Structures, 29*, 279–297.
185. Sharma, S. K., & Kumar, A. (2018). Ride comfort of a higher speed rail vehicle using a magnetorheological suspension system. *Proceedings of the Institution of Mechanical Engineers, Part K: Journal of Multi-Body Dynamics, 232*, 32–48.
186. Shukla, S., Kumar, U., Sharma, S. K., Gupta, P., & Kumar, A. (2017). Light weight freight rolling stock bogie frame: Design methodology validated with field oscillation trials. *International Journal of Vehicle Structures & Systems, 9*, 245–250.
187. Sharma, R. C., Palli, S., Sharma, S. K., & Roy, M. (2017). Modernization of railway track with composite sleepers. *International Journal of Vehicle Structures & Systems, 9*, 321–329.
188. Sharma, S. K., & Kumar, A. (2017). Ride performance of a high speed rail vehicle using controlled semi active suspension system. *Smart Materials and Structures, 26*, 055026.
189. Sharma, S. K., & Kumar, A. (2017). Impact of electric locomotive traction of the passenger vehicle ride quality in longitudinal train dynamics in the context of Indian railways. *Mechanics & Industry, 18*, 222.
190. Sharma, S. K. (2013). Zero energy building envelope components: A review. *International Journal of Engineering Research and Applications, 3*, 662–675.
191. Sharma, S. K., & Kumar, A. (2016). Dynamics analysis of wheel rail contact using FEA. *Procedia Engineering, 144*, 1119–1128.
192. Sharma, S. K., & Chaturvedi, S. (2016). Jerk analysis in rail vehicle dynamics. *Perspectives in Science (Neth), 8*, 648–650.
193. Sharma, S. K., & Kumar, A. (2016). The impact of a rigid-flexible system on the ride quality of passenger bogies using a flexible carbody. In *Proceedings of the third international conference on railway technology: Research, development and maintenance* (p. 87). Civil-Comp Press.
194. Lavania, S., & Sharma, S. K. (2011). An explicit approach to compare crosstalk noise and delay in VLSI RLC interconnect modeled with skin effect with step and ramp input. *Journal of VLSI Design Tools & Technology, 1*, 1–8.
195. Palli, S., Koona, R., Sharma, R. C., & Muddada, V. (2015). Dynamic analysis of indian railway integral coach factory bogie. *International Journal of Vehicle Structures and Systems, 7*, 16–20. https://doi.org/10.4273/ijvss.7.1.03
196. Palli, S., & Koona, R. (2015). Analyses of dynamic response of a railway bogie. *International Journal of Vehicle Noise and Vibration, 11*, 103–113. https://doi.org/10.1504/IJVNV.2015.070015
197. Swathi, K., & Srihari, P. (2015). Heat transfer enhancement ina tube using rectangular strip inserts. *International Journal of Applied Engineering Research, 10*, 41532–41544.
198. Palli, S., Sharma, R. C., & Rao, P. P. D. (2017). Dynamic behaviour of a 7 DoF passenger car model. *International Journal of Vehicle Structures and Systems, 9*, 57–63. https://doi.org/10.4273/ijvss.9.1.12
199. Sharma, R. C., Palli, S., & Koona, R. (2017). Stress and vibrational analysis of an Indian railway RCF bogie. *International Journal of Vehicle Structures and Systems, 9*, 296–302. https://doi.org/10.4273/ijvss.9.5.06
200. Sharma, R. C., Avesh, M., Sharma, S. K., Sharma, S., Palli, S., & Sharma, N. (2020). Linear and non-linear stability analysis of a constrained railway wheel axle. *International Journal of Vehicle Structures and Systems, 12*, 128–133. https://doi.org/10.4273/ijvss.12.2.04
201. Palli, S., Dontikurti, R., Sharma, R. C., & Sharma, N. (2021). Analysis of dynamic response of a railway locomotive using ansys. *International Journal of Vehicle Structures and Systems, 13*, 250–255. https://doi.org/10.4273/ijvss.13.2.21
202. Sharma, R. C., Sharma, N., Singh, G., & Palli, S. (2021). Modeling and simulation of human body-vehicle-track system for the investigation of ride comfort. *SAE International Journal of Commercial Vehicles, 15*. https://doi.org/10.4271/02-15-02-0008

203. Sharma, R. C., Palli, S., Avesh, M., & Sharma, N. (2021). Vibration isolation of railway vehicle car body using semi-active suspension. *International Journal of Vehicle Structures and Systems, 13*, 482–487. https://doi.org/10.4273/ijvss.13.4.17
204. Gopala Rao, L. V. V., Sharma, R. C., Satyanarayana, V. S. V., & Palli, S. (2022). Stochastic optimal preview control response of a quarter-car nonlinear suspension model using spectral decomposition method. *Noise and Vibration Worldwide, 53*, 225–232. https://doi.org/10.1177/09574565221093234
205. Palli, S., Duppala, A., Sharma, R. C., & Rao, L. V. V. G. (2022). Dynamic simulation of automotive vehicle suspension using MATLAB Simulink. *International Journal of Vehicle Structures and Systems, 14*, 310–314. https://doi.org/10.4273/ijvss.14.3.04
206. Satyanarayana, V. S. V., Sharma, R. C., Sateesh, B., Gopala Rao, L. V. V., Mohan Rao, N., & Palli, S. (2022). Optimum response of a nonlinear passive vehicle suspension system under random road excitations. *SAE International Journal of Commercial Vehicles, 16*. https://doi.org/10.4271/02-16-01-0004
207. Lokanadham, D., Sharma, R. C., Palli, S., & Bhardawaj, S. (2022). Wear rate modelling and analysis of limestone slurry particulate composites using the fuzzy method. *International Journal on Recent and Innovation Trends in Computing and Communication, 10*, 133–143. https://doi.org/10.17762/ijritcc.v10i1s.5818
208. Sharma, R. C., Palli, S., Gopala Rao, L. V. V., Duppala, A., & Sharma, S. K. (2022). Four-wheel vehicle response under bump, pothole, harmonic, and random excitations using bond graph/Simulink technique. *SAE International Journal of Commercial Vehicles, 16*. https://doi.org/10.4271/02-16-02-0008
209. Sharma, R. C., Gopala Rao, L. V. V., Sharma, S. K., Palli, S., & Satyanarayana, V. S. V. (2022). Analysis of lateral stability and ride of an Indian railway constrained dual-axle bogie frame. *SAE International Journal of Commercial Vehicles, 16*, 213–228. https://doi.org/10.4271/02-16-02-0014
210. Sharma, R. C., Palli, S., Bhardawaj, S., & Rao, L. V. V. G. (2022). Advances in human biodynamic modeling for analysis of whole-body vibrations. *International Journal of Vehicle Structures and Systems, 14*, 503–511. https://doi.org/10.4273/ijvss.14.4.17
211. Sharma, R. C., Palli, S., & Sharma, S. K. (2023). Ride analysis of railway vehicle considering rigidity and flexibility of the carbody. *Journal of the Chinese Institute of Engineers, Transactions of the Chinese Institute of Engineers, Series A, 46*, 355–366. https://doi.org/10.1080/02533839.2023.2194918
212. Sharma, R. C., Palli, S., & Bhardawaj, S. (2023). A review of wheel-rail dynamic behavior on curved tracks. *International Journal of Vehicle Structures and Systems, 15*, 19–25. https://doi.org/10.4273/ijvss.15.1.04
213. Jyotiprasad, G., Dowluru, S., Rallabandi, S. R., Sharma, N., Sharma, R. C., & Palli, S. (2023). *Thermal-structural and macromechanical behavior analysis of graphite-based laminated composite*. SAE Technical Paper 2023-01-5041. https://doi.org/10.4271/2023-01-5041.
214. Saikiran, C., Palli, S., Bondala, R., Sharma, R. C., Sreeramulu, D., & Dontikurti, R. (2023). Experimental investigation on thermal behaviour of silicon carbide (SiC) based polymer composites under Hygrothermal conditions. *International Journal of Vehicle Structures and Systems, 15*, 222–226. https://doi.org/10.4273/ijvss.15.2.15
215. Singh, B. J., Palli, S., Kumar, P., & Bahl, S. (2023). Behavioural investigation of hardness profile for dissimilar TIG-weldment between P22 and P91 steels. *International Journal of Vehicle Structures and Systems, 15*, 31–36. https://doi.org/10.4273/ijvss.15.1.06
216. Naveen, Y., Lokanadham, D., Naidu, D. R., Sharma, R. C., Palli, S., & Lila, M. K. (2023). An experimental study on the influence of blended Karanja biodiesel on diesel engine characteristics. *International Journal of Vehicle Structures and Systems, 15*, 154–160. https://doi.org/10.4273/ijvss.15.2.02
217. Bhardawaj, S., Jha, A. K., Sharma, R. C., Palli, S., Lokanadham, D., Sharma, N., & Dhiman, N. (2023). Finite element analysis of two disk rotor system used in automobile turbocharg-

ers. *International Journal of Vehicle Structures and Systems, 15*, 501–504. https://doi.org/10.4273/ijvss.15.4.09

218. Sharma, R. C., Palli, S., Bhardawaj, S., Jha, A. K., Duppala, A., Sharma, N., Sharma, S., & Dhiman, N. (2023). Dynamic behaviour of railway vehicle under bump and pothole track irregularities. *International Journal of Vehicle Structures and Systems, 15*, 566–569. https://doi.org/10.4273/ijvss.15.4.22
219. Raju, S., Palli, S., Devi Prasad, P., Menda, V. R., & Ramakrishna, B. (2024). A hybrid AHP-TOPSIS, MOORA technique for multi-objective optimization of thermal, mechanical, and water absorption behavior of epoxy/hemp, pine apple, and palm fiber composites. *Journal of the Chinese Institute of Engineers, Transactions of the Chinese Institute of Engineers, Series A, 47*, 23–36. https://doi.org/10.1080/02533839.2023.2274092
220. Raju Rallabandi, S., Pilla, D. P., Dowluru, S., Palli, S., Sharma, N., Sharma, S. K., & Sharma, R. C. (2024). Critical evaluation of epoxy-hemp-pineapple-palm fiber composites using hybrid AHM-TOPSIS technique for sustainable structural applications. *Journal of the Chinese Institute of Engineers, Transactions of the Chinese Institute of Engineers, Series A, 47*, 325–336. https://doi.org/10.1080/02533839.2024.2308250
221. Gopala Rao, L. V. V., Satyanarayana, V. S. V., Sharma, R. C., Palli, S., Raturi, A., Kharola, A., & Sharma, A. (2024). Preview control of the random response of a full car vehicle model traversing a rough road. *International Journal of Vehicle Noise and Vibration, 20*, 70–88. https://doi.org/10.1504/IJVNV.2024.138105
222. Sharma, A., Sharma, V. S., Gupta, S., Sharma, R. C., Palli, S., & Sharma, N. (2024). Additive manufacturing in the COVID-19 pandemic: Equipment and challenges? *Journal of Micromanufacturing*. https://doi.org/10.1177/25165984241228083
223. Bammidi, R., Madivada, H., Dowluru, S., Palli, S., Sharma, R. C., & Muddada, V. (2024). Mechano-tribological investigations of Al-%Wt. of boron nitride as metal-ceramic composites. *AIP Conference Proceedings, 2978*, 020017. https://doi.org/10.1063/5.0182965
224. Ranganath, B. A., Duppala, A., Palli, S., Rao, L. V. V., & Sharma, R. C. (2024). Numerical analysis and experimental study of FRP composites for damping applications. *AIP Conference Proceedings, 2978*, 060001. https://doi.org/10.1063/5.0184802
225. Sharma, R. C., Palli, S., Duppala, A., Bhardawaj, S., Sharma, S. K., Sharma, N., & Sharma, A. (2024). Rigid body analysis of ride behaviour of roadway vehicle using Lagrangian dynamics. *International Journal of Vehicle Structures and Systems, 16*, 74–78. https://doi.org/10.4273/ijvss.16.1.15
226. Menda, V., Dora, S., Palli, S., Rallabandi, S. R., Sharma, R. C., Sharma, N., & Bondala, R. (2024). Mechanical and wear characteristics of aluminium-7075/graphene/TiB2 prepared by stir casting. *Engineering Research Express, 6*, 025571. https://doi.org/10.1088/2631-8695/ad58a8

IoT-Driven Solutions for Smart Farming and Business Optimization in Industry 5.0

Neha Singh Raghuvanshi, Yatika Gori, Ashwani Kumar, Pias Kumar Biswas, Naveen Kumar, and Yogesh Kumar Singla

1 Introduction

The fusion of technique and agriculture has birthed a new era of innovation, where traditional farming practices are being reimagined and optimized through the lens of the Internet of Things (IoT). In recent years, IoT has come out as a life-shaping force in the agricultural landscape, offering unique opportunities to enhance

N. S. Raghuvanshi (✉)
Department of Applied Sciences, ABES Engineering College, Ghaziabad, Uttar Pradesh, India

Y. Gori
Department of Mechanical Engineering, Graphic Era Deemed to be University, Dehradun, Uttarakhand, India

School of Computing, Graphic Era Hill University, Haldwani, Uttarakhand, India
e-mail: yatikagori@gehu.ac.in

A. Kumar
Department of Mechanical Engineering, Technical Education Department Uttar Pradesh, Kanpur, India
e-mail: ashwaniadj.me@geu.ac.in

P. K. Biswas
School of Engineering, Math and Technology, Navajo Technical University, Crownpoint, NM, USA
e-mail: pbiswas@navajotech.edu

N. Kumar
Department of Mechanical and Aerospace Engineering, UPES, Dehradun, Uttarakhand, India
e-mail: naveen.kumar1@ddn.upes.ac.in

Y. Kumar Singla
School of Engineering, Math and Technology, Navajo Technical University, Crownpoint, NM, USA

Department of Mechanical Engineering, University of Idaho, Moscow, ID, USA
e-mail: ysingla@navajotech.edu

Y. Kumar Singla et al. (eds.), *Digital Manufacturing in Industry 5.0*,
https://doi.org/10.1007/978-3-031-91500-0_31

efficiency, productivity, and sustainability in farming operations [1]. Smart agriculture, empowered by IoT technology, represents a paradigm shift in how we approach food production, distribution, and management of agricultural resources. By integrating IoT sensors, devices, and data analytics into agricultural systems, farmers and businesses can harness real-time insights to make better decisions, enhance resource utilization, and mitigate risks. This chapter aims to understand the multifaceted role of IoT in revolutionizing smart agriculture and its profound inferences for agricultural businesses [2]. From precision farming and remote monitoring to supply chain optimization and data-driven decision-making, IoT is reshaping every aspect of the agricultural value chain, unlocking new opportunities for innovation and growth. Through a comprehensive examination of IoT-enabled solutions and their impact on agricultural practices and business models, this paper seeks to elucidate the transformative potential of IoT in driving sustainable agricultural development and fostering resilience in the face of evolving challenges. As we delve into the intricate interplay between IoT technology, agriculture, and business, it becomes evident that embracing IoT holds the key to unlocking a future in which farming is not only better and productive but also greener and economically viable. Through collaboration, innovation, and strategic adoption of IoT solutions, stakeholders in the agricultural ecosystem can chart a course towards a smarter, more connected future for agriculture and business alike [3].

2 IoT in Smart Agriculture

2.1 Precision Farming

The Internet of Things (IoT) plays a crucial role in precision harvesting, presenting numerous advantages that transform conventional agricultural methods. Precision farming, also termed precision agriculture, harnesses advanced techniques such as IoT to optimize crop production, improve resource efficiency, and mitigate environmental impact. IoT sensors deployed across agricultural fields gather real-time data on many factors like soil moisture, temperature, humidity, nutrient levels, and crop health. This data furnishes farmers with valuable information about their crops and soil conditions, enabling timely interventions and adjustments.

By utilizing data from IoT sensors, farmers can plan well-analysed decisions regarding irrigation, pest control, fertilization, and crop management. Analysing historical and real-time data allows farmers to discern trends, patterns, and correlations, refining their farming practices for optimal efficiency and productivity. IoT technology assists farmers in enhancing resource utilization, including irrigation, fertilizers, and pesticidal chemicals, by tailoring inputs precisely to soil and crop conditions, thus reducing waste and costs while maximizing yields. This targeted resource management approach also fosters environmental sustainability.

Table 1 Cumulative soil area mapped with soil sensors in France of 2017 [4]

Type of technology	Estimated area cumulated (2017) (ha)
Soil apparent resistivity	50,000
Soil apparent conductivity	85,000
Total	135,000

IoT-driven precision farming techniques enable precise application of inputs like water, fertilizers, and pesticides. Utilizing technologies such as variable rate application systems, farmers administer inputs exactly where and when they're needed, thereby reducing runoff and environmental contamination while promoting optimal crop growth. IoT devices facilitate remote monitoring and management of farming operations, providing farmers with flexibility and convenience. Through smartphones or computers, farmers can oversee field conditions, monitor equipment status, and receive alerts about potential issues, allowing for prompt responses and interventions.

IoT technologies facilitate the creation of yield maps by capturing data on crop yields across different regions of the field (Table 1). By analysing yield maps alongside other data layers, such as soil maps and satellite imagery, farmers can check spatial variability within their fields and make data-driven decisions to optimize yields. IoT serves as a foundation for integrating other precision farming techniques, such as GPS-guided tractors, drones, and autonomous vehicles [4]. These technologies work in tandem with IoT sensors to enable precise, efficient, and automated farming operations, further enhancing productivity and reducing labour requirements. IoT plays a critical role in precision farming by providing farmers with real-time data, facilitating data-driven decision-making, optimizing resource usage, facilitating remote monitoring and control, and integrating with other precision farming technologies. As agriculture evolves in the digital age, IoT is poised to drive further innovation and transformation in the quest for sustainable and effective food production. In France, agricultural fields are treated to cumulative soil mapping via soil sensors.

2.2 *Remote Monitoring and Control*

The Internet of Things (IoT) plays a significant role in remote monitoring and control, offering a multitude of advantages across diverse industries such as agriculture, manufacturing, healthcare, and infrastructure management. In the realm of remote monitoring and control, IoT encompasses a network of interconnected equipment, sensors, and systems facilitating real-time data collection, analysis, and management from distant locations. IoT devices and sensors gather data from remote assets, equipment, and environments in real time, encompassing variables like temperature, humidity, pressure, motion, and energy consumption, tailored to specific applications [5]. Through continuous monitoring of these parameters, organizations attain insights into the performance, condition, and status of their assets and operations.

IoT platforms and applications provide remote access and visibility into critical assets and processes [6]. Through web-based interfaces, mobile apps, and dashboards, users can monitor and track the status of equipment, infrastructure, and environmental conditions from anywhere with an internet connection. This remote visibility enables proactive maintenance, troubleshooting, and decision-making, regardless of physical location. IoT systems can be programmed to produce alerts and notifications triggered by predefined thresholds, anomalies, or events. If sensor data surpasses or falls short of specified limits, automated alerts are dispatched to assigned stakeholders through email, SMS, or push notifications. These alerts enable timely responses to emerging issues, such as equipment failures, environmental changes, or safety hazards, minimizing downtime and mitigating risks. Sensors track the fluctuating characteristics of crop parameters. Thanks to the swift progress in WSN (Wireless Sensor Network) technologies, the size and expense of these sensors have diminished, rendering their implementation viable across numerous sectors, including agriculture (Table 2). Among the prevalent sensors utilized in agriculture, those capturing environmental data pertinent to crops are most prominent [7].

IoT-enabled remote monitoring allows for remote diagnostics and troubleshooting of equipment and systems. Through remote access capabilities, technicians and engineers can analyse sensor data, perform diagnostics tests, and identify potential

Table 2 Common wireless nodes used in the agricultural domain [7]

S. No	WN1	MC2	Expansion connector	Available sensors
1	MICA2DOT	ATmega128L	19 pins	GPS, light, humidity, barometric pressure, temperature, accelerometer, acoustic, RH
2	Imote2	Marvell/ XscalePXA271	40 and 20 pins	Light, temperature, humidity, accelerometer
3	IRIS	ATmega128L	51 pins	Light, barometric pressure, RH, acoustic, acceleration/seismic, magnetic, and video
4	MI CAz	ATmega128L	51 pins	Light, humidity, temperature, barometric pressure, GPS, RH, accelerometer, acoustic, video sensor, sounder, magnetometer, microphone
5	TelosB	TIMSP430	6 and 10 pins	Light, temperature, humidity
6	Cricket	ATmega128L	51 pins	Accelerometer, light, temperature, humidity, GPS, RH, acoustic, barometric pressure, ultrasonic, video sensor, microphone, magnetometer, sounder
7	MICA2	ATmega128L	51 pins	Temperature, light, humidity, accelerometer, GPS, barometric pressure, RH, acoustic, sounder, video, magnetometer

issues without the need for physical intervention. This remote troubleshooting reduces the need for on-site visits, lowers maintenance costs, and accelerates problem resolution. Beyond monitoring, IoT facilitates remote control and automation of devices, equipment, and processes. Using IoT-enabled actuators and controllers, individuals can remotely modify settings, regulate parameters, and execute commands to enhance performance and efficiency. For example, in smart buildings, IoT systems can remotely control HVAC systems, lighting, and security systems based on tenure, schedules, and environmental conditions [8].

IoT data can be fused with sophisticated analytics and decision support systems to extract actionable insights and enhance operations. By scrutinizing patterns in historical and real-time data, enterprises can pinpoint trends, forecast outcomes, and optimize resource distribution and utilization. These insights empower data-centric decision-making, perpetuate continuous enhancement, and inform strategic planning. IoT assumes a pivotal role in remote monitoring and control by facilitating instantaneous data collection, remote accessibility and visibility, issuance of alerts and notifications, remote diagnostics and troubleshooting, as well as control and automation. Furthermore, it integrates seamlessly with analytics and decision support systems. As businesses embrace digital transformation and connectivity, IoT will persist in propelling innovation, efficiency, and resilience across a myriad of industries and applications.

3 Predictive Analytics

The role assigned to Internet of Things (IoT) in predictive analytics is instrumental in transforming agriculture and business operations by leveraging data-driven insights to anticipate future outcomes, mitigate risks, and optimize decision-making processes. In agriculture as well as business settings, IoT-driven predictive analytics empowers pre-emptive actions founded on real-time and past data patterns. IoT sensors dispersed throughout agricultural fields, manufacturing facilities, supply chains, and business workflows amass extensive datasets encompassing parameters like temperature, humidity, soil moisture, machinery efficiency, inventory status, and customer interactions. This data is then integrated and aggregated from disparate sources, creating a comprehensive dataset for predictive analysis.

IoT facilitates real-time monitoring of critical variables and events in agriculture and business operations. By continuously collecting and analysing sensor data, organizations can detect anomalies, deviations, and patterns as they occur, enabling timely interventions and proactive decision-making. IoT data serves as the foundation for building predictive models and algorithms that forecast future trends, outcomes, and potential risks [9].

Using techniques such as machine learning, statistical analysis, and artificial intelligence, organizations can identify correlations, predict probabilities, and generate forecasts based on historical data patterns (Fig. 1).

Within agriculture, IoT-driven predictive analytics aids farmers in foreseeing crop diseases, pest outbreaks, unfavourable weather patterns, and market fluctuations. By proactively recognizing risks and weaknesses, farmers can enact

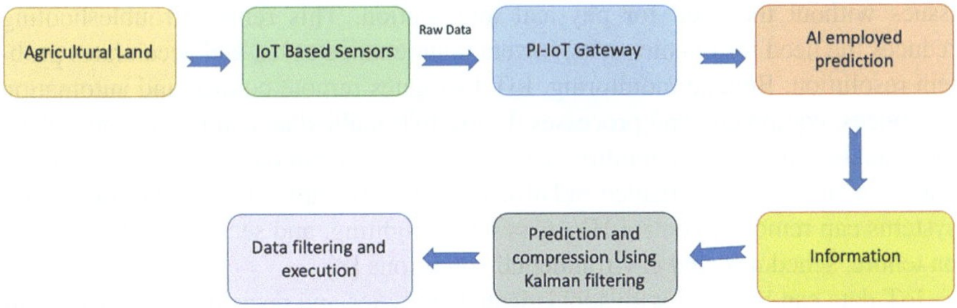

Fig. 1 IoT-based systematic architecture

preventive strategies, adapt farming methodologies, and refine resource distribution to mitigate losses and amplify yields. In the business realm, IoT-facilitated predictive analytics elevates supply chain oversight by projecting demand, refining inventory levels, and enhancing logistical processes.

By analysing historical market trends, sales data, and customer preferences, organizations can predict demand fluctuations, identify supply chain bottlenecks, and optimize production and distribution schedules to meet customer needs efficiently. In manufacturing and various other sectors, IoT-driven predictive analytics bolsters predictive maintenance and asset management strategies. Through real-time monitoring of equipment performance and health metrics, enterprises can anticipate equipment malfunctions, schedule maintenance pre-emptively, and enhance asset utilization to reduce downtime and boost productivity.

3.1 *IoT in Smart Business*

In retail and marketing, IoT-driven predictive analytics helps businesses understand customer behaviour, preferences, and purchasing patterns. Through the examination of data sourced from IoT-enabled devices, online engagements, and social media platforms, companies can customize marketing endeavours, refine product selections, and elevate customer interactions to bolster sales and loyalty. IoT-facilitated predictive analytics empowers organizations to formulate strategic decisions rooted in data and plan for the future with assurance. By leveraging insights derived from predictive models, organizations can pick out opportunities, mitigate risks, allocate resources effectively, and stay ahead of competitors in dynamic and competitive markets.

3.2 *Supply Chain Optimization*

The impact of the Internet of Things (IoT) on supply chain enhancement is significant, offering a plethora of advantages across diverse industries, encompassing both business and agriculture. IoT technology facilitates the seamless integration of physical and digital systems, fostering real-time transparency, visibility, and efficiency through the supply chain (Fig. 2). Deployed across the supply chain, IoT

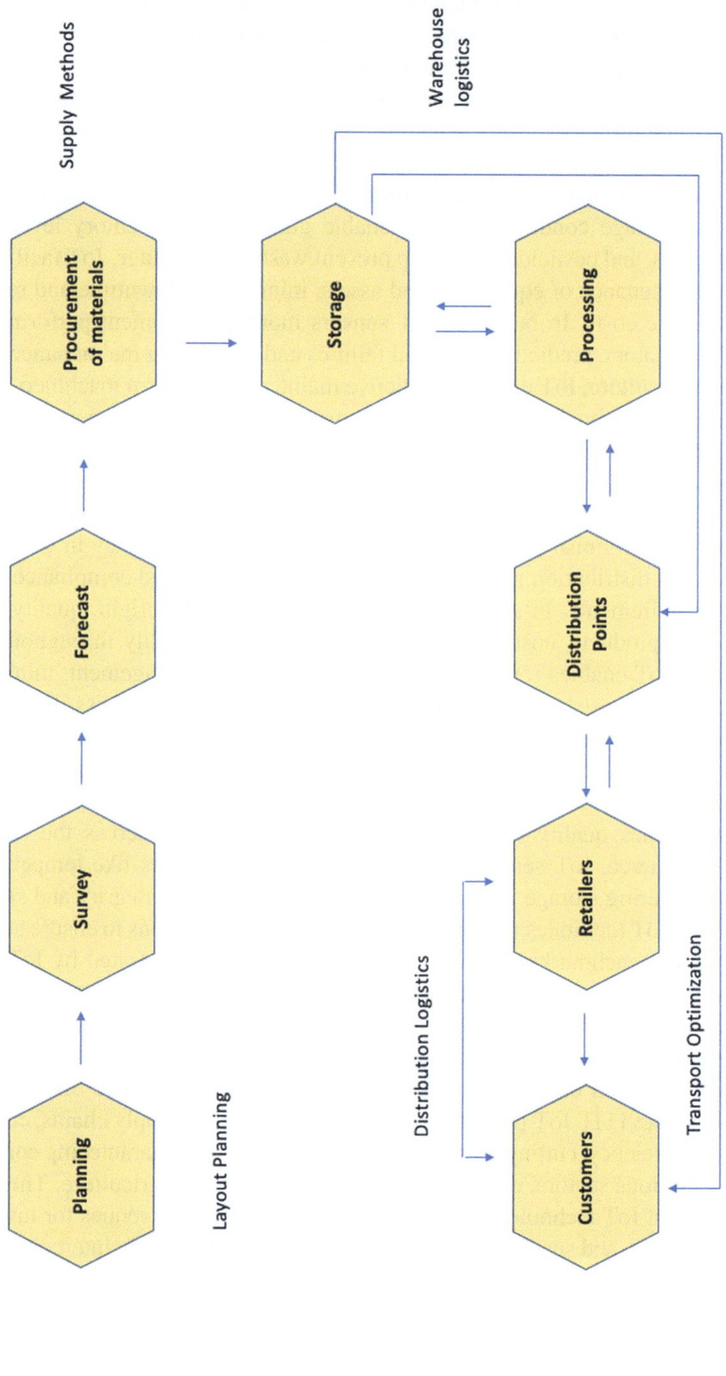

Fig. 2 Process of supply chain management

sensors and devices enable continuous tracking and monitoring of goods, assets, and inventory. In the business sector, IoT facilitates shipment tracking, warehouse operation monitoring, and real-time inventory management. In agriculture, IoT sensors monitor the movement of crops, livestock, and agricultural inputs, providing insight into the supply chain from farm to table.

Inventory management systems empowered by IoT deliver precise and current details regarding inventory levels, whereabouts, and status. In the business domain, IoT aids in fine-tuning inventory levels, mitigating instances of stockouts, and curtailing expenses associated with excess inventory retention. In agriculture, IoT sensors monitor storage conditions for perishable goods, track inventory levels for seeds, fertilizers, and pesticides, and help prevent waste and spoilage. IoT facilitates predictive maintenance of equipment and assets, minimizing downtime and reducing maintenance costs. In business, IoT sensors monitor equipment performance and health indicators, predicting potential failures and scheduling maintenance proactively. In agriculture, IoT enables predictive maintenance of farm machinery, irrigation systems, and processing equipment, ensuring uninterrupted operations and maximizing productivity.

IoT ensures comprehensive visibility and transparency throughout the supply chain, permitting stakeholders to monitor the real-time movement and status of goods from start to finish [10]. In business, IoT enables transparency in sourcing, production, and distribution processes, enhancing traceability and compliance with regulatory requirements. In agriculture, IoT sensors track the origin, quality, and safety of farm products, ensuring transparency and accountability throughout the supply chain. IoT enables route optimization and logistics management, minimizing transportation costs and improving delivery efficiency. In business, IoT helps optimize delivery routes, inhibit fuel consumption, and minimize delivery times. In agriculture, IoT sensors optimize transportation routes for the distribution of crops and livestock, ensuring timely delivery to markets and processing facilities.

IoT streamlines quality control and compliance oversight across the supply chain. In commerce, IoT sensors oversee environmental factors like temperature and humidity during storage and transit, guaranteeing product integrity and safety. In agriculture, IoT facilitates monitoring of soil and crop conditions to ensure adherence to quality benchmarks and regulatory mandates. Data generated by IoT furnishes invaluable insights for data analysis and decision-making in supply chain management. In commerce, IoT data analysis identifies trends, patterns, and areas for enhancement and refinement. In agriculture, IoT data analysis empowers farmers to make educated choices regarding crop management, resource distribution, and market tactics [11]. IoT plays a crucial role in optimizing supply chains, crucial for boosting efficiency, cutting costs, enhancing visibility, and guaranteeing compliance across various sectors, encompassing both business and agriculture. Through the utilization of IoT technology, companies can tap into fresh avenues for innovation, collaboration, and sustainable expansion within a progressively interconnected and dynamic global supply chain environment.

3.3 Resource Optimization

The Internet of Things (IoT) plays a vital role in resource optimization for both businesses and agriculture, facilitating the efficient utilization of resources such as water, energy, raw materials, and labour. IoT technology offers capabilities for real-time data collection, analysis, and control, empowering organizations to optimize resource utilization, reduce waste, and promote sustainability [12]. IoT sensors deployed across business operations and agricultural environments collect vast amounts of data on resource consumption, production processes, environmental conditions, and operational efficiency (Fig. 3). By analysing this data, organizations gain valuable insights into resource utilization patterns, identify inefficiencies, and pinpoint areas for optimization.

Within agriculture, IoT sensors oversee parameters like soil moisture levels, weather conditions, crop health, and nutrient levels, facilitating precise irrigation, fertilization, and pest management techniques. Through targeted resource application based on necessity and timing, farmers can curtail water consumption, diminish fertilizer runoff, and enhance crop yields. In business, IoT enables precision management of energy consumption, raw materials, and production processes, optimizing resource allocation and minimizing waste. IoT enables automation and control of resource-intensive processes and equipment, improving efficiency and reducing

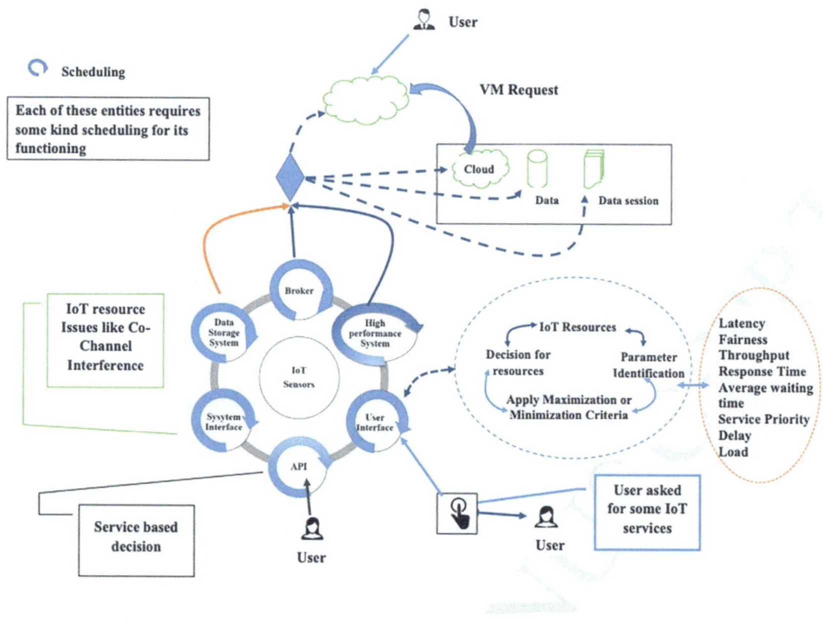

Fig. 3 Scheduling process in an IoT environment [12]

human intervention. In agriculture, IoT-controlled irrigation systems, drones, and autonomous machinery automate tasks such as watering, spraying, and harvesting, optimizing resource usage and labour productivity. In business, IoT-enabled smart grids, manufacturing equipment, and building systems automate energy management, production scheduling, and facility operations, reducing energy costs and enhancing operational efficiency.

IoT-driven predictive analytics and optimization algorithms help businesses and farmers anticipate resource needs, forecast demand, and optimize resource allocation [12, 13]. Through the analysis of past data, market trends, and environmental factors, enterprises can forecast forthcoming resource needs and adapt production, inventory, and supply chain tactics accordingly. This proactive resource management methodology diminishes waste, lowers expenses, and enhances resource utilization efficiency. IoT enables real-time monitoring and maintenance of critical resources and infrastructure. In agriculture, IoT sensors monitor water levels in irrigation systems, detect leaks, and optimize water usage, conserving water resources and reducing costs. In business, IoT sensors monitor equipment performance, detect inefficiencies, and schedule predictive maintenance, minimizing downtime and optimizing resource utilization [14–17].

IoT technology facilitates integration with sustainability initiatives and environmental conservation efforts. By providing real-time visibility into resource usage, IoT enables organizations to track progress towards sustainability goals, monitor compliance with regulatory standards, and implement environmentally friendly practices. In agriculture, IoT supports sustainable farming practices such as conservation tillage, crop rotation, and organic farming, promoting soil health, biodiversity, and ecosystem resilience. The role of IoT in resource optimization is essential for businesses and agriculture to improve efficiency, reduce costs, and enhance sustainability. By leveraging IoT technology, organizations can make informed decisions, automate processes, and optimize resource usage, contributing to economic prosperity and environmental stewardship in a rapidly changing world [18–22].

3.4 Data-Driven Decision-Making

The role of the Internet of Things (IoT) in data-driven decision-making is transformative in both the agricultural and business domains, providing valuable insights and enabling informed choices based on real-time and historical data (Table 3). IoT technology facilitates the collection, analysis, and interpretation of vast amounts of data from various sources, empowering stakeholders to optimize operations, enhance efficiency, and drive innovation. IoT sensors deployed across agricultural fields, manufacturing plants, supply chains, and business processes capture real-time data on environmental conditions, equipment performance, production metrics, and consumer behaviour. This continuous data collection provides a comprehensive view of operations and enables timely decision-making based on current conditions [23–28].

Table 3 Conditions and factors for IoT adoption in AM

Category	Conditions and factors
Human	High technical knowledge Good understanding of data management processes Good understanding of data quality issues Ongoing training and education
Organizational	Clear responsibility for innovation Availability of best practices to benchmark Positive business cases Data governance: Clear responsibilities for data management
Data	Data governance: High level of data quality Alignment of data to AM requirements
Technical	Data governance: Adoption of stringent security measures Availability and adoption of interoperability standards

IoT platforms amalgamate data from various origins, encompassing sensors, devices, databases, and external systems, to construct a unified dataset for examination. Utilizing advanced analytics techniques like machine learning, predictive modelling, and statistical analysis, IoT data is scrutinized to uncover patterns, correlations, and insights that inform decision-making processes. IoT-driven predictive analytics empowers organizations to prognosticate future trends, anticipate risks, and pinpoint optimization opportunities. In agriculture, predictive analytics derived from IoT data can anticipate crop yields, forecast pest outbreaks, and refine irrigation schedules. Similarly, in business, predictive analytics can anticipate demand, refine inventory levels, and identify potential market trends, fostering proactive decision-making and strategic planning [29–34].

IoT facilitates instantaneous monitoring of key performance indicators (KPIs) and vital metrics throughout agricultural and business operations [13]. By overseeing metrics like production volume, resource usage, machinery downtime, and client contentment, organizations can pinpoint areas necessitating enhancement, refine processes, and boost productivity. IoT data aids in devising risk management and mitigation strategies by furnishing insights into potential hazards, weaknesses, and prospects. In agriculture, IoT sensors observe weather trends, soil status, and crop well-being, empowering farmers to address risks linked with unfavourable weather occurrences, pests, and illnesses. In business, IoT-driven analytics help identify operational risks, supply chain disruptions, and market fluctuations, enabling proactive measures to minimize impact and enhance resilience.

IoT platforms and applications provide decision support tools and dashboards that visualize data, generate insights, and facilitate decision-making processes. Interactive dashboards, reports, and alerts enable stakeholders to access relevant information, analyse trends, and make informed decisions in real time. IoT-driven data insights foster a trend of continuous improvement and research innovation in both agriculture and business. By leveraging data-driven insights, institutions can identify opportunities for innovation, experimentation, and optimization, driving continuous improvement in products, processes, and services. Overall, the role of

IoT in data-driven decision-making is instrumental in driving efficiency, productivity, and competitiveness in agriculture and business. By harnessing the power of IoT-driven data analytics, organizations can gain actionable insights, optimize operations, mitigate risks, and seize opportunities for growth and innovation in an increasingly interconnected and data-driven world [35–40].

3.5 Business Model Innovation

The Internet of Things (IoT) plays a significant role in driving agricultural and business model innovation by enabling new approaches to value creation, service delivery, and revenue generation. IoT technology facilitates the collection, analysis, and utilization of data from connected devices and sensors, empowering organizations to develop innovative business models that leverage data-driven insights and enhance customer experiences. IoT sensors installed in agricultural fields gather information on soil moisture, temperature, crop vitality, and weather patterns, empowering farmers to make educated choices regarding irrigation, fertilization, and pest management. Utilizing insights derived from IoT, farmers can enhance crop yields, decrease resource consumption, and lower operational expenses, fostering more sustainable and efficient farming methods.

IoT-based precision farming services offer farmers access to advanced analytics, predictive modelling, and decision support tools that optimize crop management practices and enhance productivity. These services provide tailored recommendations for planting schedules, crop rotation strategies, and input application rates based on real-time and historical data, enabling farmers to maximize yields and profitability [14]. IoT-enabled supply chain traceability solutions provide end-to-end visibility into the journey of agricultural products from farm to fork. By tracking product movements, storage conditions, and handling processes, organizations can ensure compliance with regulatory standards, guarantee product quality and safety, and build consumer trust. Blockchain technology, often integrated with IoT, enhances transparency and authenticity in supply chains by creating immutable records of product provenance and transactions.

IoT platforms and marketplaces facilitate collaboration and exchange of data, services, and insights among stakeholders in the agricultural ecosystem. These platforms connect farmers, agribusinesses, researchers, and service providers, enabling them to access a wide range of tools, resources, and expertise for decision-making and innovation. Smart agriculture platforms foster knowledge sharing, innovation diffusion, and ecosystem collaboration, driving continuous improvement and adaptation in the agricultural sector. IoT has catalysed the emergence of AgriTech start-ups and ventures that leverage technology to address key challenges in agriculture and food production. These start-ups develop innovative solutions such as precision agriculture systems, farm management software, remote sensing technologies, and predictive analytics platforms that enhance productivity, sustainability, and resilience in farming operations. AgriTech start-ups disrupt traditional agricultural practices and business models, offering new opportunities for value creation and market differentiation [15].

IoT-generated data presents opportunities for agricultural businesses to monetize data assets and offer value-added services to customers. By analysing and contextualizing IoT data, organizations can extract actionable insights, develop predictive models, and offer subscription-based services such as crop monitoring, yield forecasting, and risk management. These data-driven services create additional revenue streams, enhance customer loyalty, and differentiate agricultural businesses in competitive markets. IoT-enabled business model innovation transcends traditional product-centric approaches and focuses on delivering integrated solutions, experiences, and outcomes to customers. Organisations shift from selling products to providing services, from ownership to access-based models, and from linear value chains to ecosystems of interconnected partners and stakeholders. IoT facilitates business model transformation by enabling new revenue streams, cost-saving opportunities, and value propositions that align with evolving customer needs and market dynamics.

The role of IoT in agricultural and business model innovation is instrumental in driving digital transformation, enhancing competitiveness, and creating sustainable value in a rapidly changing landscape. By embracing IoT technology and leveraging data-driven insights, organizations can unlock new opportunities for growth, differentiation, and customer engagement in agriculture and business sectors [41–45].

4 Ecosystem Collaboration

The role of the Internet of Things (IoT) in ecosystem collaboration is pivotal in fostering partnerships, knowledge sharing, and innovation across the agricultural and business sectors. IoT technology enables stakeholders to connect, collaborate, and co-create value within interconnected ecosystems, driving collective efforts to address shared challenges and opportunities. IoT facilitates the integration of diverse stakeholders, including farmers, agribusinesses, researchers, technology providers, policymakers, and consumers, into interconnected networks (Table 4). These networks enable seamless communication, data exchange, and participant

Table 4 IoT-driven business ecosystem strategies [17]

	Type of integration	Value chain	Ecosystem
Offering type	Systemic	Controlling value chain Coupled knowledge processes	Facilitating ecosystem coupled knowledge processes
	Stand-alone	Participating in value chain Inbound or outbound knowledge processes	Participating in ecosystem inbound or outbound knowledge processes

collaboration, fostering a collaborative ecosystem where stakeholders can share insights, resources, and best practices [16].

IoT platforms and data sharing initiatives promote open collaboration by enabling stakeholders to share data, insights, and analytics across organizational boundaries. By opening access to IoT-generated data, organizations can foster collaboration, innovation, and knowledge exchange, driving collective learning and problem-solving in agriculture and business domains. IoT facilitates cross-sector partnerships and collaborations between agricultural and non-agricultural stakeholders, such as technology firms, financial institutions, research organizations, and government agencies. These partnerships leverage complementary expertise, resources, and capabilities to develop innovative solutions that address complex challenges spanning multiple domains, such as food security, sustainability, and supply chain resilience (Table 4).

IoT-powered ecosystem platforms and marketplaces provide a digital infrastructure for stakeholders to discover, connect, and transact with each other. These platforms facilitate collaboration by offering access to a diverse range of products, services, and solutions from ecosystem partners. By participating in ecosystem platforms, organizations can access new markets, customers, and opportunities for collaboration and co-innovation. IoT enables the formation of collaborative innovation networks where stakeholders collaborate to co-create and co-develop new products, services, and solutions. These networks bring together diverse perspectives, expertise, and resources to tackle complex challenges, drive technological innovation, and accelerate the adoption of emerging technologies in agriculture and business domains.

IoT standards and interoperability initiatives promote collaboration by ensuring compatibility and seamless integration between diverse IoT systems, devices, and platforms. By adhering to common standards and protocols, stakeholders can exchange data and information more effectively, facilitating collaboration and interoperability across organizational boundaries. IoT facilitates the sharing of infrastructure and resources among ecosystem partners, reducing duplication, optimizing resource usage, and enhancing efficiency. Shared IoT infrastructure, such as sensor networks, data analytics platforms, and cloud computing resources, enables stakeholders to leverage shared assets and capabilities to drive collective outcomes and mutual benefits [46–48].

IoT fosters collaboration among policymakers, regulators, and industry stakeholders to develop policies, standards, and regulations that promote innovation, sustainability, and responsible use of technology in agriculture and business sectors. Collaborative efforts in policy development and regulatory compliance ensure alignment with industry best practices, stakeholder interests, and societal values. The role of IoT in ecosystem collaboration is instrumental in driving collective action, innovation, and transformation in agriculture and business domains. By fostering partnerships, knowledge sharing, and innovation within interconnected ecosystems, IoT enables stakeholders to address shared challenges, seize emerging opportunities, and create sustainable value for society, the economy, and the environment [49–52].

References

1. Relf-Eckstein, J. E., Ballantyne, A. T., & Phillips, P. W. B. (2019). Farming reimagined: A case study of autonomous farm equipment and creating an innovation opportunity space for broadacre smart farming. *NJAS - Wageningen Journal of Life Sciences, 90–91*, 100307.
2. Raj, M., Gupta, S., Chamola, V., Elhence, A., Garg, T., Atiquzzaman, M., & Niyato, D. (2021). A survey on the role of internet of things for adopting and promoting agriculture 4.0. *Journal of Network and Computer Applications., 187*, 103107.
3. Ayaz, M., Ammad-uddin, M., Sharif, Z., Mansour, A., & Aggoune, E.-H. M. (2019). Internet-of-things (IoT) based smart agriculture: Towards making the fields talk. *IEEE Access, 7*, 1–1.
4. Sathiya, S., Antony, C., & Ghodke, P. (2020). Recent trends and best practices in industry. In G. Fastellini & C. Schillaci (Eds.), *Precision farming and IoT case studies across the world. Agricultural internet of things and decision support for precision smart farming* (pp. 331–415).
5. Gutierrez, J., Villa-Medina, J. F., Nieto-Garibay, A., & Porta-Gandara, M. A. (2014). Automated irrigation system using a wireless sensor network and GPRS module. *IEEE Transactions on Instrumentation and Measurement, 63*(1), 166–176.
6. Muralidhara, P. (2017). IoT applications in cloud computing for smart devices. *International Journal of Computer Science and Technology, 1*(1), 1–41.
7. Shafi, U., Mumtaz, R., García-Nieto, J., Hassan, S. A., Zaidi, S. A. R., & Iqbal, N. (2019). Precision agriculture techniques and practices: From considerations to applications. *Sensors, 19*, 3796.
8. Carli, R., Cavone, G., Ben Othman, S., & Dotoli, M. (2020). IoT based architecture for model predictive control of HVAC Systems in Smart Buildings. *Sensors, 20*(3), 781.
9. Tang, S., Shelden, D. R., Eastman, C. M., Pishdad-Bozorgi, P., & Gao, X. (2019). A review of building information modeling (BIM) and the internet of things (IoT) devices integration: Present status and future trends. *Automation in Construction, 101*, 127–139.
10. Helo, P., & Shamsuzzoha, A. H. M. (2020). Real-time supply chain—A blockchain architecture for project deliveries. *Robotics and Computer-Integrated Manufacturing, 63*, 101909.
11. Villa-Henriksen, A., Edwards, G. T. C., Pesonen, L. A., Green, O., & Sørensen, C. A. G. (2020). Internet of Things in arable farming: Implementation, applications, challenges and potential. *Biosystems Engineering, 191*, 60–84.
12. Chowdhury, A., & Raut, S. A. (2018). A survey study on internet of things resource management. *Journal of Network and Computer Applications, 120*, 42–60.
13. Elhami, E., Ansari, A., Farahani, B., Aliee, F. S. (2020). [IEEE 2020 International Conference on Omni-layer Intelligent Systems (COINS)—Barcelona, Spain (2020.8.31–2020.9.2)]. In *2020 international conference on omni-layer intelligent systems (COINS)—Towards IoT-driven predictive business process analytics* (pp. 1–7).
14. Rezaei, M., Akbarpour Shirazi, M., & Karimi, B. (2017). IoT-based framework for performance measurement: A real-time supply chain decision alignment. *Industrial Management & Data Systems, 117*(4), 688–712.
15. Cheverst, K., Davies, N., Mitchell, K., Friday, A., Efstratiou, C. (2000). [ACM Press the SIGCHI conference—The Hague, The Netherlands (2000.04.01–2000.04.06)]. In *Proceedings of the SIGCHI conference on human factors in computing systems—CHI'00—Developing a context-aware electronic tourist guide* (pp. 17–24).
16. Vlachopoulou, M., Ziakis, C., Vergidis, K., & Madas, M. (2021). Analysing AgriFood-tech e-business models. *Sustainability, 13*.
17. Saarikko, T., Westergren, U. H., Blomquist, T. (2016). [IEEE 2016 49th Hawaii International Conference on System Sciences (HICSS)—Koloa, HI, USA (2016.1.5–2016.1.8)]. In *2016 49th Hawaii international conference on system sciences (HICSS)—The inter-organizational dynamics of a platform ecosystem: Exploring stakeholder boundaries* (pp. 5167–5176).
18. Wazid, M., Das, A. K., Chamola, V., & Park, Y. (2022). Uniting cyber security and machine learning: Advantages, challenges and future research. *ICT Express, 8*(3), 313–321. https://doi.org/10.1016/j.icte.2022.04.007

19. Pande, A., Manchanda, M., Bhat, H. R., Bairy, P. S., Kumar, N., & Gahtori, P. (2022). Molecular insights into a mechanism of resveratrol action using hybrid computational docking/CoMFA and machine learning approach. *Journal of Biomolecular Structure and Dynamics, 40*(18), 8286–8300. https://doi.org/10.1080/07391102.2021.1910572
20. Molu, R. J. J., Tripathi, B., Mbasso, W. F., Naoussi, S. R. D., Bajaj, M., Wira, P., & Misak, S. (2024). Advancing short-term solar irradiance forecasting accuracy through a hybrid deep learning approach with Bayesian optimization. *Results in Engineering, 23*, 102461. https://doi.org/10.1016/j.rineng.2024.102461
21. Swain, S., Bhushan, B., Dhiman, G., & Viriyasitavat, W. (2022). Appositeness of optimized and reliable machine learning for healthcare: A survey. *Archives of Computational Methods in Engineering, 29*(6), 3981–4003. https://doi.org/10.1007/s11831-022-09733-8
22. Singh, J., Wazid, M., Das, A. K., Chamola, V., & Guizani, M. (2022). Machine learning security attacks and defense approaches for emerging cyber physical applications: A comprehensive survey. *Computer Communications, 192*, 316–331. https://doi.org/10.1016/j.comcom.2022.06.012
23. Sathish, T., Saravanan, R., Jeyanthi, C. E., Sabariraj, R. V., Hemachandu, P., Ağbulut, Ü., Ahamed Saleel, C., & Soudagar, M. E. M. (2023). PCM embedded square channel receiver for effective operation of net zero emission/energy PDC building heating system. *Sustainable Energy Technologies and Assessments, 60*, 103519. https://doi.org/10.1016/j.seta.2023.103519
24. Sharma, R., Kim, M., & Gupta, A. (2022). Motor imagery classification in brain-machine interface with machine learning algorithms: Classical approach to multi-layer perceptron model. *Biomedical Signal Processing and Control, 71*, 103101. https://doi.org/10.1016/j.bspc.2021.103101
25. Pundir, S., Obaidat, M. S., Wazid, M., Das, A. K., Singh, D. P., & Rodrigues, J. J. (2023). MADP-IIME: Malware attack detection protocol in IoT-enabled industrial multimedia environment using machine learning approach. *Multimedia Systems, 29*(3), 1785–1797. https://doi.org/10.1007/s00530-020-00743-9
26. Nallathambi, I., Savaram, P., Sengan, S., Alharbi, M., Alshathri, S., Bajaj, M., Aly, M. H., & El-Shafai, W. (2023). Impact of fireworks industry safety measures and prevention management system on human error mitigation using a machine learning approach. *Sensors, 23*(9), 4365. https://doi.org/10.3390/s23094365
27. Gupta, A., Ghanshala, K., & Joshi, R. C. (2021). Machine learning classifier approach with gaussian process, ensemble boosted trees, SVM, and linear regression for 5g signal coverage mapping. *IJIMAI, 6*(6), 156–163. https://doi.org/10.9781/ijimai.2021.03.004
28. Ojha, N., Girach, I., Sharma, K., Sharma, A., Singh, N., & Gunthe, S. S. (2021). Exploring the potential of machine learning for simulations of urban ozone variability. *Scientific Reports, 11*(1), 22513. https://doi.org/10.1038/s41598-021-01824-z
29. Thapliyal, S., Wazid, M., Singh, D. P., Das, A. K., Shetty, S., & Alqahtani, A. (2023). Design of robust Blockchain-envisioned authenticated key management mechanism for smart healthcare applications. *IEEE Access, 11*, 93032. https://doi.org/10.1109/ACCESS.2023.3310264
30. Kanwar, N., Jadoun, V. K., Jayalakshmi, N. S., Afthanorhan, A., Fatema, N., Malik, H., & Hossaini, M. A. (2023). Industry—Challenge to pro-environmental manufacturing of goods replacing single-use plastic by Indian industry: A study toward failing ban on single-use plastic access. *IEEE Access, 11*, 77336–77346. https://doi.org/10.1109/ACCESS.2023.3296097
31. Badhotiya, G. K., Soni, G., Jain, V., Joshi, R., & Mittal, S. (2022). Assessing supply chain resilience to the outbreak of COVID-19 in Indian manufacturing firms. *Operations Management Research, 15*(3), 1161–1180. https://doi.org/10.1007/s12063-021-00236-6
32. Pansare, R., Yadav, G., Garza-Reyes, J. A., & Raosaheb Nagare, M. (2023). Assessment of sustainable development goals through industry 4.0 and reconfigurable manufacturing system practices. *Journal of Manufacturing Technology Management, 34*(3), 383–413. https://doi.org/10.1108/JMTM-05-2022-0206

33. Jamil, M., He, N., Wei, Z., Gupta, M. K., & Khan, A. M. (2023). A novel quantifiable approach of estimating energy consumption, carbon emissions and cost factors in manufacturing of bearing steel based on triple bottom-line approach. *Sustainable Materials and Technologies, 36*, e00593. https://doi.org/10.1016/j.susmat.2023.e00593
34. Kaswan, M. S., Rathi, R., Cross, J., Garza-Reyes, J. A., Antony, J., & Yadav, V. (2022). Integrating green lean six sigma and industry 4.0: A conceptual framework. *Journal of Manufacturing Technology Management, 34*(1), 87–121. https://doi.org/10.1108/JMTM-03-2022-0115
35. Patil, P. P., Gori, Y., Kumar, A., & Tyagi, M. R. (2021). Experimental analysis of tribological properties of polyisobutylene thickened oil in lubricated contacts. *Tribology International, 159*, 106983. https://doi.org/10.1016/j.triboint.2021.106983
36. Uniyal, S., Mangla, S. K., Sarma, P. R., Tseng, M. L., & Patil, P. (2021). ICT as "knowledge management" for assessing sustainable consumption and production in supply chains. *Journal of Global Information Management (JGIM), 29*(1), 164–198. https://doi.org/10.4018/JGIM.2021010109
37. Tiwari, S., Swaminathan, M., Harender, & Singh, D. B. (2022). Performance enhancement of the photovoltaic system with different cooling methods. *Environmental Science and Pollution Research, 29*(30), 45107–45130. https://doi.org/10.1007/s11356-022-20330-x
38. Chakrabarty, D., Patil, P., & Sharma, K. (2017). Experimental study on fluid flow and heat transfer from a rectangular prism approaching the wall of a wind tunnel. Heat transfer—Asian. *Research, 46*(7), 793–814. https://doi.org/10.1002/htj.21244
39. Gariya, N., Prasad, B., Nautiyal, H., Kumar, S., & Shaikh, A. (2024). Manufacturing techniques for soft robotic structures: A systematic review. In *Sustainability in smart manufacturing* (pp. 94–117). CRC Press. https://doi.org/10.1201/9781003467496-5
40. Rekioua, D., Mokrani, Z., Kakouche, K., Rekioua, T., Oubelaid, A., Logerais, P. O., Ali, E., Bajaj, M., Berhanu, M., & Ghoneim, S. S. (2023). Optimization and intelligent power management control for an autonomous hybrid wind turbine photovoltaic diesel generator with batteries. *Scientific Reports, 13*(1), 21830. https://doi.org/10.1038/s41598-023-49067-4
41. Rizwanuddin, S., Kumar, V., Naik, B., Singh, P., Mishra, S., Rustagi, S., & Kumar, V. (2023). Microbial phytase: Their sources, production, and role in the enhancement of nutritional aspects of food and feed additives. *Journal of Agriculture and Food Research, 12*, 100559. https://doi.org/10.1016/j.jafr.2023.100559
42. Datta, S., Hamim, I., Jaiswal, D. K., et al. (2023). Sustainable agriculture. *BMC Plant Biology, 23*, 588. https://doi.org/10.1186/s12870-023-04626-9
43. Anjum, S., Rana, S., Dasila, K., Agnihotri, V., Pandey, A., & Pande, V. (2022). Comparative nutritional and antimicrobial analysis of Himalayan black and yellow soybean and their okara. *Journal of the Science of Food and Agriculture, 102*(12), 5358–5367. https://doi.org/10.1002/jsfa.11889
44. Shahare, Y., Singh, M. P., Singh, P., Diwakar, M., Singh, V., Kadry, S., & Sevcik, L. (2023). A comprehensive analysis of machine learning-based assessment and prediction of soil enzyme activity. *Agriculture, 13*(7), 1323. https://doi.org/10.3390/agriculture13071323
45. Wongchai, A., Shukla, S. K., Ahmed, M. A., Sakthi, U., & Jagdish, M. (2022). Artificial intelligence-enabled soft sensor and internet of things for sustainable agriculture using ensemble deep learning architecture. *Computers and Electrical Engineering, 102*, 108128. https://doi.org/10.1016/j.compeleceng.2022.108128
46. Rawat, B., Rawat, J. M., Purohit, S., Singh, G., Sharma, P. K., Chandra, A., Shabaaz Begum, J. P., Venugopal, D., Jaremko, M., & Qureshi, K. A. (2022). A comprehensive review of Quercus semecarpifolia Sm.: An ecologically and commercially important Himalayan tree. *Frontiers in Ecology and Evolution, 10*, 961345–961351. https://doi.org/10.3389/fevo.2022.961345
47. Sharma, P. K., Rausa, K., Rani, A., Mukherjee, S., & Kumar, M. (2021). Biopurification of dairy farm wastewater through hybrid constructed wetland system: Groundwater quality and health implications. *Environmental Research, 200*, 111426. https://doi.org/10.1016/j.envres.2021.111426

48. Shah, S., Nautiyal, H., Gugliani, G., Kumar, A., Namboodri, T., & Singla Editors, Y. K. (2024). *Sustainability in smart manufacturing: Trends, scope, and challenges*. CRC Press. https://doi.org/10.1201/9781003467496
49. Kumar, A., Singla, Y. K., & Maughan, M. (2024). *Book title "fracture behavior of nanocomposites and reinforced laminate structures"*. Publisher Springer Nature Switzerland. https://doi.org/10.1007/978-3-031-68694-8
50. Kumar, A., Singla, Y., & Namboodri, T. (2024). Chapter 01: Globalization and international issues in sustainable manufacturing. In S. Shah, H. Nautiyal, G. Gugliani, A. Kumar, T. Namboodri, & Y. K. Singla (Eds.), *Sustainability in smart manufacturing: Trends, scope and challenges* (pp. 01–18). CRC Press. https://doi.org/10.1201/9781003467496-1
51. Kumar, A., Gori, Y., Dutt, N., Singla, Y. K., & Maurya, A. (2021). *Advanced computational methods in mechanical and materials engineering*. CRC Press. https://doi.org/10.1201/9781003202233
52. Kumar, A., Ram, M., & Singla, Y. K. (2022). *Advanced material for biomechanical applications*. CRC Press. https://doi.org/10.1201/9781003286806

Digital Twins, Augmented Reality and Robotics

Digital Twins in Product Development

R. Immanual, N. Sanjana, K. M. Kirthika, and S. Sangeetha

1 Introduction to Digital Twins in Product Development

Digital twins (DT) are a revolutionary concept in product development that opens up a virtually limitless possibility of enhancing product value. This technology creates a digital twin of a tangible product, a process, or a procedure, which helps monitor processes, analyze data, and improve an item during its life cycle [1].

The arrival of Industry 5.0 with the speed at which digital manufacturing is advancing has also escalated the embracement and innovation of digital twins. The described above virtual models are used as the effective stimuli for predictive maintenance, performance enhancement, and even for prototyping that has a tremendous impact on cutting down time to market and development expenses.

In the subsequent chapters, the concept of digital twins together with its definition will be discussed in more details including its evolution from simple models to the contemporary entity with artificial intelligence. That is why, to be more concrete, we will look at how digital twins are impacting the nature of product development processes in the context of Industry 5.0 and digital manufacturing, and possible role in innovation and sustainability of industries will also be outlined [2]. Since this is the case, as we proceed to discuss this topic, it will be important to understand the

R. Immanual (✉)
Department of Mechanical Engineering, Sri Ramakrishna Institute of Technology, Coimbatore, Tamil Nadu, India

N. Sanjana · K. M. Kirthika
Department of Computer Science and Engineering, Sri Ramakrishna Institute of Technology, Coimbatore, Tamil Nadu, India

S. Sangeetha
Department of Electrical and Electronics Engineering, Sri Ramakrishna Institute of Technology, Coimbatore, Tamil Nadu, India

Y. Kumar Singla et al. (eds.), *Digital Manufacturing in Industry 5.0*,
https://doi.org/10.1007/978-3-031-91500-0_32

opportunities and risks that are inherent in the application of digital twin technology across the value chains of goods production.

A digital twin is a digital model of a physical component or entity that exists or will exist in the real world, and this model can be employed for various intents at different stages of its target's life cycle [2].

Figure 1 illustrates the dynamic interplay between the real world and its virtual counterpart. The real world serves as the primary source of data, which is collected and used to create a digital reproduction—the virtual world—within computer systems. This virtual representation allows for simulations and analyses that mirror real-world scenarios.

The idea was first brought forward by Dr. Michael Grieves at the University of Michigan in 2002 with defining the framework as the "Conceptual Ideal for PLM" (Product Lifecycle Management) [3]. However, the term "digital twin" was used later to mean a technique that had been developed by NASA as part of their technology readiness for outbound spacecraft designs that sought to support the spacecraft development process [4].

In essence, a digital twin consists of three main components: The three appear in the form of the physical entity, the virtual model, and the links between them [1]. The tangible counterpart is the actual product or system, the intangible counterpart is the formal virtual model. These links allow data exchange between the physical and the virtual spaces depending on current events for analysis.

Thus, while digital twins are not a new concept, they are a step forward from other methods such as CAD models or simulations. They can be considered as a live data image, something that can be changed every time depending on the data transmitted by the sensors, IoT devices and other sources [5]. Real-time operation ensures constant observation, prognosis, and adaptation at various stages of the product's life cycle.

In addition, digital twins can integrate different sorts of data and models, geometry data, physical models, analytical behavioral models, and past performance records can all be included in digital twins. This feature breeds improved predictive and prescriptive analytics compared to conventional modeling methodologies [6].

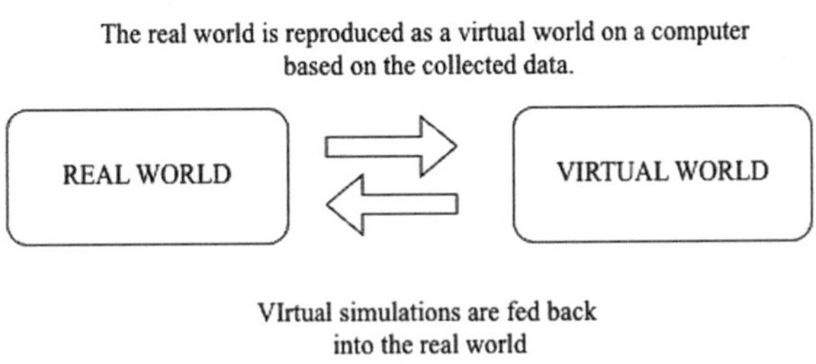

Fig. 1 The cyclical relationship between real and virtual worlds

In the context of product development, digital twins serve multiple purposes: In the context of product development, digital twins serve multiple purposes:

1. Design optimization
2. Virtual testing and validation
3. Performance prediction
4. Predictive maintenance
5. Customization and personalization

Digital twins is fast becoming real through the application of advanced technologies like artificial intelligence, machine learning, and big data analytics The use of DT in product development processes is helping to accelerate the solutions development cycles that are making it possible to do away with many physical prototyping models that are expensive and time-consuming [7].

Digital twins originated from the industrial world, more precisely from the fourth industrial revolution, also known as Industry 4.0 titled the fourth industrial revolution priority theme; automation and data exchange in manufacturing technologies. But, as the world shifts from the fourth industrial revolution into the fifth, organizations have faced the challenge of finding ways to. 5.0, and as a result, the application of digital twins has grown widely. Industry 5.0 stresses the integrated cooperation of humans and machines and digital twins are an actor in that interaction.

The original concept of digital twins used in the perspective of digital manufacturing has shifted from relying on statics, into the dynamic applications of artificial intelligence. They now include data from IoT sensors in real time, machine learning models for analytics, and even augmented reality for visualization [8].

What is more, there appears the concept of the multiscale digital twins where the representation can be of the full production line or even factory, not only the single product. These wide-ranging models facilitate comprehensive enhancement of manufacturing processes concerning their productivity and eco-friendly nature [9].

The other major area of development is linking digital twins to the other new-age technologies. For instance, integrating digital twins with blockchain technology has improved the data security and origination of manufacturing supply chain [10].

Also, combined with the climate change and cloud, edge computing has boosted the computational power for digital twin simulations. This has enhanced the development of advanced and precise models, the analysis, and decision-making processes in real time.

In Industry 5.0, they equally have been helping in the drive toward sustainability in manufacturing. It allows the enhancement of energy usage efficiency, reduction of waste, and the assessment of a production's lifecycle in response to the growing concern regarding sustainability in industries [11].

Moving forward, the development of digital twins in Industry 5.0 and digital manufacturing would probably remain persistent and its growth rate is fast; thus, one can foresee deeper and more comprehensive integration with AI and machine learning as well as the enhancement of the predictive aspects and as the helper in human–machine interaction.

2 Key Technologies and Methodologies

Digital twin technology is one of the innovative management concepts that have shifted the landscape of several industries and offer a digital copy of physical structures or processes. Digital twins are based on a number of main technologies and methods that are interconnected and integrated to improve operation and decision-making. The main prerequisite for implementing the digital twin is the use of the Internet of Things, which allows collecting information about real objects' conditions, using the sensors placed in physical objects, which underpins the creation of a digital twin model. It is a critical factor since it provides reliable solutions for stored data and computational capability to work with the digital twin models. Machine learning and other predictive computations are essential for analyzing collected data for purpose of prediction, identification of abnormal behavior and other useful functions toward managing the system.

Virtual reality (VR) technology is used in replicating and modeling digital twin models that were earlier restricted to theoretical modeling where users are provided the ability of live simulation and manipulation of models without the usage of physical means. Currently, data analysis using big data tools such as machine learning and deep learning to discover hidden patterns, trends, and relationship in the data for decision-making, productive process enhancement, and innovation are done [12]. The proper incorporation of these key technologies and methodologies creates a strong proper base for implementing DTs across various industries. Using IoT, Cloud, AI, VR, and advanced analytics organizations can leverage new digital opportunities for efficiency, productivity, and competitiveness in smart digital organizations.

Digital twin technology has brought a drastic change in numbers of industries as it provides a virtual copy of physical assets in order to improve the product design, manufacturing technique and service delivery. The use of digital twins is based on the integration of modern technologies and approaches. Very central to this technology are big data analytics, IoT devices, artificial intelligence, and cloud computing due to capabilities in monitoring, maintenance, and optimization of performance. They complement each other to produce highly realistic model of the physical systems to provide analysis and decision-making.

Digital twins has emphasized on the importance of simulation and modeling techniques, which is the foundation of such disciplines. This integration of virtual and physical worlds makes it possible for people such as engineers and designers to simulate some situations and make some observations about the performance based on compiled data before adjusting the changes in the real world. Data integration and interoperability are critical to digital twin solutions since it is the desired effect of engaging and correctly communicating with numerous related systems, sensors, and constituents. The integration between apps enables smooth data interchange between platforms, which, in turn, improves the reliability and relevance of the developed digital twin models.

Making of IT applications and visualization tools contributes to the construction of a digital twin strategy. These applications give well thought interfaces and clear dashboards so the stakeholders can engage with the models of DT and visualize and analyze the performance of the system that is being modeled for decision-making. Security, specifically in terms of cybersecurity and protection of data privacy, is always critical when applying digital twin technologies as most of them are data-intensive and require real-time data feeds. Physical security measures check cyber threats and intrusions from exploiting digital twin assets through proper encryption, control, and authentication of systems, besides other security best practices.

Therefore, with these crucial technologies and methodological approaches integrated into digital twin implementations, digital twin has a mind-blowing solution to product development, as well as industrial operation. Thus, such components as advanced analytics, simulation tools, data integration, software applications, and cybersecurity open up the new opportunities for innovation, effectiveness, and competitiveness. Digital twin has turned not only into a need but into an opportunity for organizations to achieve sustainable profitable growth and, therefore, value creation in the digital economy [13].

Real-time data analysis and IoT integration are core aspects of digital twin solution where they assist significantly in improving the pattern recognition and improved operational efficacy in numerous field of industries. This integration also allows for the development of accurate replicas of physical assets and process from IoT devices and sensors' real-time data. These digital twins ensure monitoring, analyzing as well as enhancing of a system in operation and real time, hence enabling optimization and the provision of excellent timely servicing.

The combination of real-time data analytics and IoT in digital twins is essential for the real-time collecting, processing, and analyzing large amounts of data from connected devices. This integration helps organizations gain valuable information on the behavior of assets and an ability to ascertain potential problems before they occur, and improve on the operations of the organization. Through integrating such technologies, companies are bound to get better insight of their operations while making better operational decisions and foster innovation especially in their products and services.

Besides, this integration helps in applying of sophisticated techniques like machine learning and artificial intelligence to digital twins. All these technologies help to reveal the patterns in the data, prognosticate the further tendencies, and make decisions within the organization depending on the new findings. One can state that the transference of real-time data analytics and IoT integration in digital twins can help to create operational excellence, minimize the amount of time the business is offline, improve product quality, and gain a competitive advantage in the market.

Thus, the use of real-time data analytics and IoT in digital twins helps organizations to have a significant tool to revolutionize existing organizational operations, improve processes, and develop new areas of business growth and development [14]. Digital culture integrates real-time data feedback and enhanced analytical functions of business processes to provide excellent services and monitor the production of superior products in the contemporary digital environment. It is a serious step further in developing digital twin technology, which opens a range of opportunities for improving and managing various systems in real-time mode and holistically.

Machine learning and AI are enhancing the modeling capabilities and applicability of digital twins in manufacturing systems since they can guarantee high predictive precision and definition of the modeled systems. Such modern technologies enable digital twins to process huge amounts of information, find connections, and make rational conclusions that result in improved decision-making and operational efficiency. Machine learning allows digital twins to enhance their training history and adjust to new conditions. Live and artificial intelligence technologies such as neural networks and deep learning help in achieving predictive indications for maintenance, discovery of outliers, and enhancement of processes.

Incorporation of machine learning and AI in digital twins empowers manufacturers with predictive models that can be used in invoking intelligent decisions before adverse situations occur. These technologies make it possible to predict equipment breakdowns, schedules the production, and enhance the general functionality of the digital twin. Effective corrective measures can be taken in case of sensor data anomalies that are detected upon real-time data analysis, thus reducing the occurrence of wastage of time and money. Such a strategy also helps in cutting down overall expenses, reducing business interruptions, and leveraging entity assets.

In addition, machine learning and AI improve the capability of digital twins in contexts of scalability and flexibility in the manufacturing domain. These technologies can deal with a large amount and variety of data to find the meaningful information, and can constantly update the quality of the predictive models. The result is adaptive systems that can modify the production processes according to the ever changing needs in production, changes in the market and changes in the environment which in turn enable timely decisions and prompt problem solving and resource optimization.

Consequently, when machine learning and AI are applied to digital twins, the latter also supports the creation of self-organizing and intelligent manufacturing systems. Digital twins of production processes are self-sufficient; through the help of AI-driven algorithms, they can monitor, analyze, and optimize a production line's performance levels on their own [15]. Such skills enable manufacturers to optimize their processes, reduce wastes, and improve the quality of products manufactured with little or no inputs from human beings, thus culminating in smart systems of manufacturing in the future.

Finally, the use of machine learning and artificial intelligence in digital twins translates into a new revolution in implementing manufacturing technology. This integration revolutionizes typical production processes into intelligent, self-learning and efficient ones which avail chances in predictive maintenance, real-time control, independent decision-making, and self-optimization. Thus, with the manufacturing industry going through the process of digitalization, machine learning and AI in digital twins will be the driving force of further innovations, the increase of competitiveness and sustainability in the context of Industry 4.0.

Digital twins is one of the most crucial latest advancements in the field of product architecture where organizations in various fields are magnificent having a lot of advantages. This approach of designing enables the creation of opposites of the objects in a sense that it makes it possible to simulate actual products and test them before having to go through the entire process of actual production. Hence, it becomes possible to significantly decrease the time that it takes companies to introduce their products into the market as well as decrease the costs of new product creation.

The application of digital twins also enhances customer relations especially in business to business manufacturing organizations. Thus, the formation of PSS and cooperation with other professions on constructing and maintaining these systems enable firms to enhance customers' satisfaction levels and, therefore, their loyalty. This is suitable for today's product centered competition oriented market where a firm has direct contact with the customers during the entire life cycle of the product. Digital twins make many positive changes in the aspects of Product Lifecycle Management (PLM). Implementing it from the Beginning of Lifecycle (BOL) to the End-of-Lifecycle (EOL), worthwhile strategies will be deployed to add more value to company processes and meaningful decisions to be made for the improvement of product quality. Thus, an approach to the management of products that include digital twins is uncomplicated in the aspects of the contributions of such products in the various phases of the PDULC.

Another publicity includes efficiency within various components of product development and management which results in the application of digital twins. Based on the real-time simulation and digital mock up, the difficulties arising therefrom can be pointed at an early stage; the improvements on the design there can be encouraged; the incorporation of the digital twins into the PLM processes can also be encouraged.

This effective approach in the development of products is useful in as much as it creates so much time and resources. However, it is crucial to highlight that the use of digital twins introduces a whole new approach to not only the product development aspect but also the methods of innovating products and cooperation as well as delivery of value-added solutions to consumers. Hence, through digital twins, businesses wait to create the next shock of change, upgrade the competitiveness and adapt more effectively to the forces of the market. It also assists organizations to continue being flexible each time the business environment is becoming a little volatile.

All in all, it can be stated that the incorporation of digital twins into the development of products is a revolution in the manufacturing sector. It presents significant possibilities for firms' value generation, disruption prevention, and stability retention. In view of this technology continuing to develop, firms' approaches to the production of their products and ways of relating to their customers is set to be further transformed by this innovation. The use of the digital twins in the complexity of contemporary manufacturing, managed in accordance with the business strategies, can bring companies to the leaders in implementing solutions and services for the Industry 4.0 framework.

Figure 2 illustrates the wide-ranging advantages of implementing digital twin technology across various aspects of business and industrial operations. At the core of a gear-shaped graphic, "Benefits of Digital Twin" is centrally placed, emphasizing how this technology acts as a driving force for multiple improvements.

Digital twin technology in product development has appeared as one of the revolutionary benefit bringing solutions that help to optimize all stages of products' production. Through the creation of virtual models of tangible objects and procedures at an organizational level, numerous simulations, analyses, and optimizations of the series of a product can be carried out [16]. This capability of conducting virtual modeling allows for the fast prototyping, evaluation, and redesigning of products, thus shortening the length of time it takes to get new products to the market. First, digital twin sources claim that using digital twins in delivering products offers an advantage of testing and validating the product designs virtually. This approach enables the organizations to mimic the conditions of the real world and analyze performance measures and not only but also to identify problems at an early stage. This in turn results in significant economies of scale and a consequent upgrade of the general quality of products in the economy. Additionally, with the use of a digital twin, improvements and adjustments can be made in terms of data gathering during the entire lifecycle of the product, thus creating efficient feedback to enhance the product.

It also ensures improved interaction and cooperation with other department's teams within an organization's undertaking concerning the usage of digital twins in product development. The integration of digital twins results in enhanced collaboration since the parties involved in the delivery of the products have a common ground to access and make engagements with the product data. This working context enhances the decision-making quality and efficiency, which subsequently enhances the organizational flexibility and fulfillment of its customers' necessities and expectations.

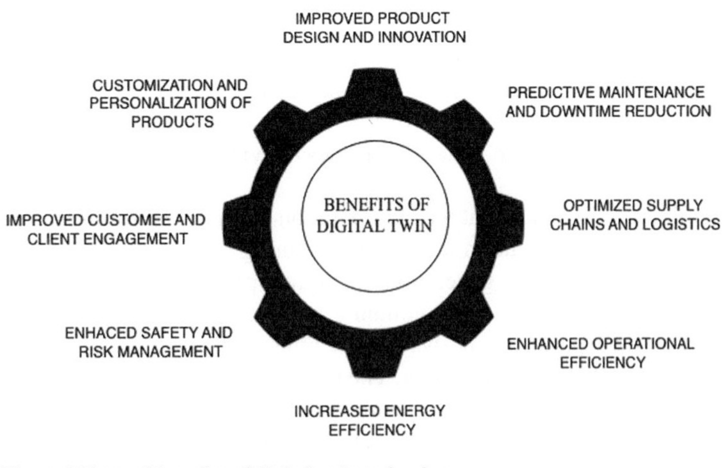

Fig. 2 The multifaceted benefits of digital twin technology

Also, incorporating digital twins into the product development life cycle enables organizations to capitalize on the use of sophisticated technologies like AI and ML. Thus, the huge amounts of data accumulated by digital twins can be effectively used to apply various predictive technologies and optimization as a primary source of competitive advantage and innovative development. Such information helps organizations to predict market trends and opportunities for the unique positioning of products, as well as make gradual enhancements to products and services.

The current application of digital twin in the development of products is making a significant shift from the past traditional manufacturing systems that are possibly stuck, hence signifying an enormous opportunity for sustainable growth.

Thus, utilizing digital twins in the management of the product development life-cycle results in enhanced effectiveness, adaptability, and creativity in product formation and introduction processes. Thus, digesting this technology-oriented strategy improves organizational effectiveness and places the enterprises into the elites of digital transformation, defining the trends of further product creation and industry evolution.

A major goal considered in the context of digital twins in product development is minimization of time to market and cost of development that is closely tied with competitive advantage in current dynamically changing industries. This way, organizations are provided with the tools to enhance the manageability of their PD processes and deliver innovations at a faster pace and less cost. Thus, the basics of improving the effectiveness of product development using digital twins are advanced perception and connection technologies. Through integration of real-time heterogeneous resource perception, data fusion, and CoP ISC technologies, a mirror image of the products and processes is achieved. This comprehensive understanding also helps in the speedy decision-making and problem solving, which in turn will certainly help in bringing down the time to market.

Digital prototyping, simulation and verification also has an important aspect in verifying the designs of the products as well as the processes related to the manufacturing or production line. When using factor-behavior-rule multiscale modeling and virtual simulation instruments, the risks are identified in the early stages of development. It also decreases the need for tangible prototypes and shortens the time for development at the same time decreasing the time to market and the cost of development. Data construction and management plays a critical role in the processes through which digital twin environments are developed and continuously refined. Tools such as data clustering storage and VR-RR mapping allow for the proper application of data as well as the coordination between computer generated environments and real ones. This makes the overall management of the decision-making processes, reduction of mistakes as well as enhancement of the efficiency of product development for the manufacturing of cars all-inclusive.

The integration between the big data analytics and the concept of the twin technologies offer promising paths for enhancing the product creation procedures. In this way, virtual models and physical products used in the process open up ways to study performance indicators, product quality, and customers' preferences. Integrating data science in decision-making helps organizations refine and reinvent themselves, which, in turn, results in shorter time to market and low development costs.

Therefore, the integration of digital twins in product development is a promising approach which may contribute to time to market and development costs' minimization. Through the reasonable use of intelligent technologies, virtual modeling, proper and effective management of Big Data, and the use of superior data analysis, businesses and organizations should strive to create, strengthen and sustain their competitiveness. Adopting the opportunities of digital twins should not only be considered a straightforward trend in a company's development but also a way to positively transform the landscape of the product development industry.

One major use of digital twins in product development is to support the objective of predictive maintenance and enhanced performance which is essential in the improvement of functionality and stability of products in the current close loop system. The concept of generating digital twins can help in the attainment of the above-mentioned goals since it entails the production of replicas of actual products and network systems.

Among the most significant use of digital twins is preventing the failure and mal functioning of the assets in real time analyzing their condition of health. This way of working allows one to suspect potential problems and prevent them before they disrupt the functioning of the company. Through the hereby mentioned approaches to the product realization, companies have an ability to detect the necessity of maintenance in various components and elements before they affect a tangible product.

Digital twins allow for constant supervision of the product usage, thus offering manufacturers a vast amount of information regarding the behavior of items in use. The continuous flow of information from the sensors and other IoT instruments integrated with the digital twin enables the actualization of rational decisions for efficiency and reliable predictive maintenance measures. It enables one to avoid unanticipated failures and also increase life cycles of specific products.

However, when it comes to the application of predictive maintenance systems based on digital twins, there are certain problems. These are the issues related to data quality, high noise of data and challenging parameter estimation, the important demand for the integration of high-precision sensing data with system mechanisms for the purpose of state evaluation. Other challenges are in feature extraction from big real-time datasets and inefficiency which is caused by the lack of well-defined and developed reference models for digital twins.

To overcome these challenges there is a need to have a synergy between the manufacturers and the organizations to come up with a standard and framework. Some practical recommendations to create a prediction feedback loop are: While creating a prediction feedback loop, you can utilize the reinforcement learning methods, for instance, Hidden Markov Models, to continuously develop the approaches to predictive maintenance with the asset.

The goals of the predictive maintenance with the digital twins are usually concentrated on the estimating, predicting or detecting of components or systems to perform the maintenance more efficiently. The common objectives may include

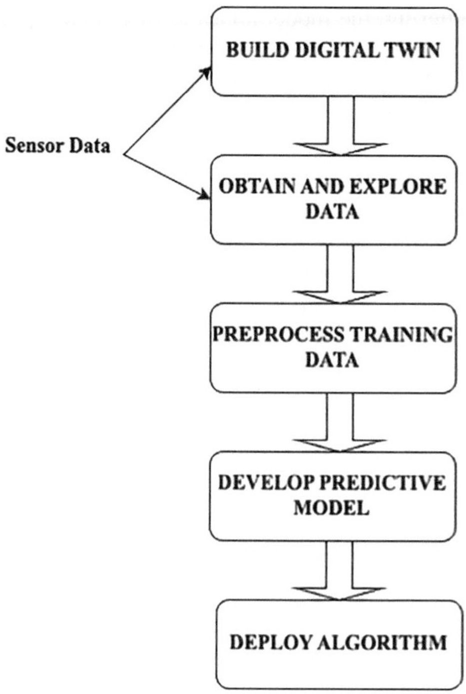

Fig. 3 The digital twin development process: from sensor data to predictive algorithms

RUL prognosis, health indicators prediction, evaluation of the machine state, and creation of the OEE coefficients.

Digital twins are also instrumental on the aspect of performance since they can model and analyze operational conditions. This enables the companies to go for the most efficient parameters in operation, hence increasing the reliability of the products in the market. Figure 3 illustrates the step-by-step process of creating and implementing a digital twin system, highlighting the journey from raw sensor data to a deployed predictive algorithm.

3 Case Study I

The case study by Ye et al. [17] presents an application of a digital twin framework to structural health for reusable spacecraft. It deals with the forecast of fatigue crack growth in a load-carrying frame so as to show how the digital twin may be used to increase the reliability and efficiency of space craft operations. The researchers describe a comprehensive digital twin framework that brings together numerous models, multiple sources of data, and uncertain parameters with an aim of facilitating real-time monitoring, diagnosis, and prognosis of structural health. Through the employment of techniques such as dynamic Bayesian networks and particle filtering, the model hopes to offer better predictions about crack growth that will facilitate maintenance and mission decision-making.

This case study points out the potentiality of digital twin technology in handling some critical problems in managing reusable spacecrafts especially in evaluating structure integrity between flights and optimizing maintenance schedules. The research paper has also demonstrated how continuous data updating can reduce uncertainties associated with crack growth predictions leading to more successful missions as well as lower operation costs. To enable comprehensive structural health management for reusable spacecraft, the digital twin framework proposed by Ye et al. [17] consists of two main stages: offline and online, each composed of four interconnected modules. Within this design context, the offline stage contains fault as well as failure mode knowledge library supplemented with corresponding coping strategies; high-precision simulation model for multiphysics and multisystem simulations; high-precision geometric model that evolves with manufacturing characteristics and damage/defect data; and a data library storing preflight information such as sensor history and operation records. A diagnosis module for real-time fault detection and identification is part of the online stage while another one includes a model updating module that utilizes dynamic Bayesian networks and data-driven techniques; reusability evaluation module for performance degradation and residual life analysis; another data library to store in-flight monitoring data together with real-time state parameters.

These elements come together in working synergy. Offline: This is where you set up your models, clear the base knowledge and historical data for all online processes. This is used by the online stage for monitoring and prediction while spacecraft in operation. The model updating module takes the possible shortcomings detected by the diagnosis module as inputs. The module then takes in this new data and runs it through predictive models that update themselves to make the predictions more accurate. The reusability evaluation module takes this information and the most recent state of spacecraft to evaluate how reusable it is, and what its future performance will be. Airflow creates a loop of data collection, models adjustment and predictions bettering which would make the decision-making related to spacecraft maintenance or mission planning come as accurate precession on time. The structured health management with real-time data processing and model updating using the offline knowledge is this unique selling point of his framework, makes it able to support sustainable development for reusable spacecraft.

The research uses a multifaceted approach for showing the efficiency of digital twin framework. Finite element analysis is initially carried out on load bearing airframe model to pin point the critical areas where fatigue crack growth tend to initiate. Quasi-static and random vibration loads that simulate the complex stresses of rocket launch were discussed. The researchers focus on a vital I-beam detail in which stress concentrations at bolt hole edges are identified as regions where cracks might start.

The study also employs Walker's law for fatigue crack growth modeling which takes into account the stress ratio effect besides stress intensity factors. It is used within the digital twin as part of a comprehensive model in a Dynamic Bayesian Network. In these equations, the DBN mixes both conventional crack growth model with observed in situ crack data to provide real-time update of the parameters of a single baseline and time evolving predictions on actual size. In order to account for the uncertainties associated with this process, a particle filtering technique known as Sequential Importance Resampling (SIR) algorithm is utilized in our study which facilitates effective uncertainty quantification and propagation.

To verify their method, the researchers use Method of Manufactured Solutions (MMS). The first way is to generate fake "virtual flight data" based on model parameter and that will be a baseline for digital twin performance outcome. This makes the MMS process ideal for testing how well a framework can predict crack growth and adjust model parameters in simulated conditions.

The research shows that the digital twin approach could indeed lead to better structural health management for reusable spacecraft. A significant discovery here is how much the accuracy of crack length predictions improves with at least a couple observations on each side. This figure indicates that by providing more and more observed data for the update of crack growth model, prediction errors will keep decreasing on average. One example is how the error of crack length prediction became less than 1% after three updates, versus over 14+% before any update.

The researchers also found that model parameter uncertainties decrease significantly over time. We quantify this with information and relative entropy metrics, which consistently decrease as more data is introduced to the model. The information entropy of the model parameters went from about 10.976 to less than 3.127 after three updates, which means that uncertainty was decreased substantially.

This results in estimates of spacecraft reusability that are both more accurate and with lower uncertainties for the improved model. The research shows the improvements to the model translate into more conservative and accurate predictions for how many re-uses will be safe of that spacecraft component. For instance, the model was over predicting how many safe reuses on first guess there can be and more accurately predicted one developed with full virtual flights of updated configuration. The advance could help with predicting the reusability of a vast range of components, which in turn would have major consequences for maintaining space assets and planning missions—leading to safer and more cost-effective spacecraft operations Fig. 4 shows that the Digital twin framework for reusable spacecraft.

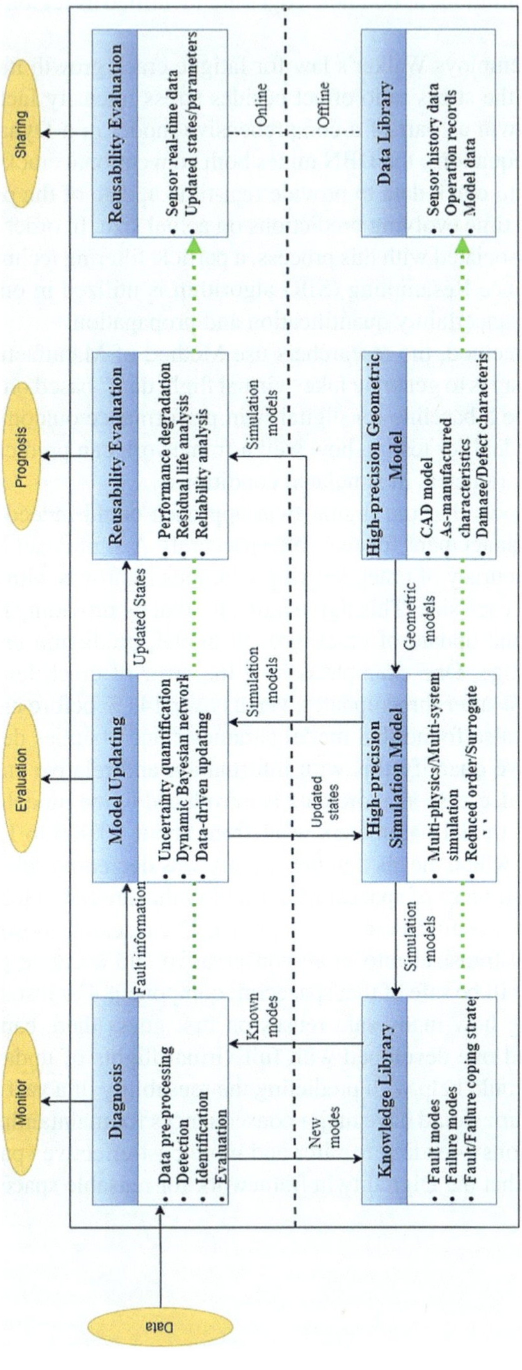

Fig. 4 Digital twin framework for reusable spacecraft

4 Case Study II

Maheshwari et al. [18] demonstrated the application of digital twin technology in a food processing plant, more specifically an ice cream factory located in India. The primary motivation for this study is to fill the void observed in recent years between digital twin applications within Industry 4.0, which have been largely discussed from a theoretical stand-point and an absence of empirical studies discussing practical implementations on real-life scenarios. The authors illustrate the use of digital twin technology to improve performance in process industries, which feature particularities arising from product perishability and intricate manufacturing operations.

The study is carried out in a holistic manner by construction of digital twin for the selected ice cream-making process. The researchers built a digital twin of their physical manufacturing system using the AnyLogic simulation software to achieve real-time monitoring, process optimization and predictive capabilities. Beyond demonstrating digital twin in action, this case study provides a proof of business impact by showing prospective improvement in performance outcomes or key performance indicators (KPIs)—machine availability, allocation efficiency, and throughput. So, its perspective has actionable insights into how food processing and other process manufacturing sectors can truly leverage digital twins to revolutionize their operations for competitive advantage.

The case study is based on an ice cream manufacturing company, ICMC located in Vijayapura district of Karnataka. The company makes ten kinds of ice cream, with names like Vanilla Magic—an Arabic "take" on vanilla cherry; Butterscotch: Rajbhog and Chocolate Brownie. ICMC is programmed on a 7-day demand cycle and shipping to its distribution center which dispatches from only 50 markets throughout. Several steps are involved in the manufacturing of this ice face cream, including ingredient mixing, pasteurizing and homogenization dealing choco in a capacitor or freezer storage, churning shaking hard packaging set Visibility (hardening).

The ICMC still had to deal with some common problems that come with traditional manufacturing systems in the food processing sector, although it already have best practices in place. As these challenges disproportionately hampered operations, they had a major negative impact on operational efficiency and productivity within the organization. Chaining some specialized areas, such as vessels—where different varieties of ice cream were produced one after the other on specific sequences that involved machines like pasteurizers or cooling devices and even packaging stages—led to penalties in terms of incompatible overlap subintervals. The interval constraints established a chain-reaction type of effect, because if even one batch was delayed it could impact the next batch in line as processing were interdependent. The company was further handicapped by weekend break limitations, production occurred only on weekdays and this negatively affected tooling use as well as labour efficiencies.

This will also become even more complicated when considering the packaging line constraints, e.g., a product mix strategy has to be taken into account where precedence rule 7 high-demand prior products have always had priority over low demand current order prior items and are completed much earlier in turn reducing throughput for lower popularity order item. Limited Work-in-process storage space was also a major headwind, especially with regards to the waiting time between pasteurization and packaging that significantly impacted production ratio. All of these constraints led to higher lead times, reduced overall equipment effectiveness, and inefficient use of resources—clearly signaling the requirement for a more sophisticated and agile manufacturing methodology.

The ice cream manufacturing factory follows a full life cycle for its digital twin implementation as shown in Fig. 5, including strategic mapping, a physical space audit, virtual methodology and evidence as well as strategy of the physical–virtual interface. Using the AnyLogic simulation software, researchers established a comprehensive digital twin of their physical manufacturing system allowing for realistic and interactive model representation across the entire production process.

Researchers selected the AnyLogic platform that enables multiparadigm and hence discrete event, as well agent-based simulation functionalities. This software made possible to integrate many other elements like parameters, ports equations, variables, timers animations, and process charts. Specifically, the object-oriented

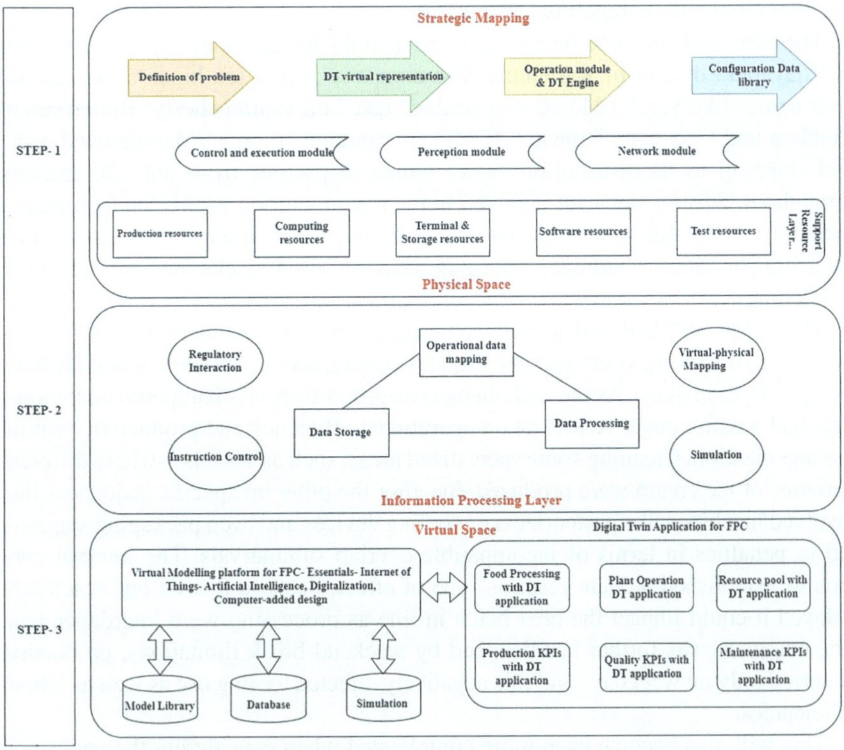

Fig. 5 Architectural framework of DT in FPC

design of AnyLogic along with JAVA coding capabilities allowed for the robustness required to model a wide variety of manufacturing scenarios and alternative logic surrounding ice cream production.

The digital twin also was able to take in real-time supporting data from the physical factory floor. The hybrid sensing-enabled Industrial Internet of Things (IIoT) technology strategy used for data collection in the shop floor included both physical and mobile based movement of devices, which helped to achieve this scalability. And it kept reflecting the exact current state of the physical system at any given point in time, as accurate readings were continuously pouring into and out from its virtual twin via real-time data integration.

Several next-gen functionalities were integrated into the digital twin model enabling to enhance manufacturing. Key to this was the adoption of a rolling Kanban container technique, which would add structure and visualization for management from manufacturing through sequence and interval constraints, work-in-process storage limitations. The model was also predictive in nature, meaning that the constant flow of data was being utilized to predict issues and optimize production schedules. The design of a digital twin facilitated optimization in the process, as it allowed multiple scenarios to be simulated so that ideal production strategies could be identified considering constraints like set-up times for changeovers or allocation of resources and demand patterns.

Because, of how the digital twin technology was introduce to ice cream manufacturing company and wherever it has been applied starts to show large improvements even in their key performance indicators (KPI). Before–after comparison of the performance metrics made in this study improved operational efficiency and productivity drastically that attributed to enhanced operating system.

Overall, the digital twin has brought a significant decrease in machine availability time almost to 10%. The additional buffer paste was mainly obtained by improved monitoring and management of the Cementite, where a more efficient use of equipment resulted in less downtime. The level of allocation efficiency improved dramatically, going from 62.5% to ~77.5%. The optimization of the production schedule and a better management in both sequence constraints or/and intervals allowed for more efficient resource usage.

The result was a near doubling of worker efficiency from 67% to approximately 85%. The digital twin role in cause-and-effect was an impressive 50% point cycle-time improvement by mashing labor data with real-time advanced analytics to take better and quicker operational decisions and used human resource more effectively. The efficiency in utilization, especially of the dispenser machine, had also improved with digital twin which could anticipate and detect the approach of subsequent equipment for a more informed decision-making process and smoother operations.

The plant utilization, which is the actual vs. potential output ratio while skyrocketing from 74% to about 95%, was a great measure of how effective this time value added simulation proved out) Using a digital twin of the refining and couching processes allowed proper interval time constraints to be extracted, leading to an overall increase in throughput. Last but not least, the throughput rate has seen a remarkable gain of over 15% points compared to its lowest value at 64.3%. This improvement is demonstrated in the research with graphical representations of before-digital twin and after-twin cubed throughput convexity and concavity. All together, these improvements show how the digital twin technology significantly enhances overall ice cream manufacturing process performance and efficiency.

However, it provides some profound implications and insights for the entire food processing industry to draw inspiration from this case study of digital twin implementation in an ice cream manufacturing company. For one, it highlights how digital twins can solve pain points that are common to food processors—perishability constraints; complex production sequences; and the desire for lean resource utilization. These significant performance improvements in KPIs formed a substantial evidence that digital twins have the potential to be game changing for operators of food processing facilities, delivering enhancements in productivity, reduced wastage and improved product quality.

The work underlines the need for real-time data processing and integration in contemporary food production. With the digital twin, an exact virtual reconstruction of a real system that is continuously synchronized with real-time data from its physical counterpart leads in obtaining deeper business insights about how things work. This capability ensures that decisions are based on facts, not conjecture predictive maintenance for aircraft and landing gear intervention as conditions change—both essential requirements in a time critical industry where it is of the up-most importance to ensure product freshness and responsive just-in-time delivery. Likewise, digital twins can be used to model different situations without impacting actual production, equipping food processing companies with a safe and cost-efficient method for experimentation to fine-tune their processes which could open the door for innovations in product development and manufacturing practices.

5 Challenges and Future Prospects

The first concept that entirely changed the way how industries are visualized in which they almost replicate all factors of the product, process, or service is digital twin. It could be all everything in physical world reproduce to digital and the other way around with feedback back from engineers Qi et al. [19]. Therefore, the use of tech helps companies to detect and address physical problems earlier, innovate better products sooner, and monetize from them as well as take advantage much quicker. Moreover, implementing the digital twin technology assists long-standing organizations in advancing their processes and performance. The emergence of Industry 4.0 has led to this peculiar industrial development. The combined onslaught of LoRa, NB-IoT, and 5G networks has skyrocketed the applications for which digital twin is even approachable. And more significantly, it has irrevocably altered how entire industries now operate and evolve. Industry 4.0 carries the elements from production method and developing technologies like cloud, IoT, analytics to digital twin/digital scanning/3D printing, and artificial intelligence. To put it together, digital twin technology is a catalyst and provides the base that leads Industry 4.0

initiatives. Actually, several real-world industries are already starting to use digital twin solutions for gathering information, visualization and decision-making in the life time of their assets and products. This allows organizations to model the digital twin of their products, and processes while providing them decision-making capabilities.

The problems with the digital twin are that they require new old IT systems updating, connectivity issues, and data privacy or security concerns on how someone gets access to your behavioral profiles of your assets nonunified modeling! Digital twin market can be affected by following major barriers: high implementation costs, rising power and storage requirements to gather the production data on real-time basis, system or internal applications integration challenges; complex nature of architecture of digital twin. However, deploying digital twins is still costly as the investment includes value addition platforms using sensors and software to set up infrastructure, a secure solution for data cleanliness with validation at various levels in the system. In addition, digital twin treatment costs will require significant investment in operations since the maintenance of a large infrastructure may be expensive [20].

Every industry has its standard and practice, so too does digital twin technology implemented on the respective industries. Without that format it becomes an issue of using the different systems and databases, she said. The above is for reference and it should be noted that industries vary in data formats, protocols or proprietary platforms across different system applications. Therefore, a lot of problems appear while trying to develop an approach to data exchange and integration. When it comes to the evolution of new digital twin technology, establishing this standard and continually refining existing ones are equally important. One of the biggest hurdles to implementing new tech is that ensuring tech advancements are "backward compatible"—able to play nicely with existing equipment. Another challenge or problem is the compliance and regulation changes from every corner of it as well, this makes hard to make standard norms for digital twin technology in different zones [21].

Digital twins usually need to be integrated with a lot of systems—possibly distributed across multiple vendors and even handling different kinds of system architectures. But, also importantly is that making two quite heterogeneous systems interface seamlessly with one another [22]. The data must be stored in one form or another compatible with many systems needed for a digital twin implementation. The argued data may lead to errors or suboptimality of the oracle operation. At the same time, some system may treat old communication protocols, which makes work of units more complex. Collaboration needs to occur so that the procedures are stable while exchanging information. Semantic interoperability refers to the ability of different systems to not only exchange data in compatible formats, but to share a common understanding of what that data actually means. This boils down to creating standard lexicons and data models that are universally acknowledged in an array of systems. The ability to scale interoperable solutions, which will be capable of processing the many more data that we are talking about while still delivering good performance degradation is also a key challenge, they said.

It will take interdisciplinary cooperation for the respective professional from different sectors and standardization bodies to sit down together, create, then come back afterward top apply these standards. The main issues being that the organizations will need to have a more holistic view on where they deploy their digital twins and how do we obtain essential information from this large colossus of connected systems in order push for adequate standards as well keeping up with new definitions across multiple emerging disciplines. To gain such competitive advantage and innovation in the digital twin technology, there is a need to invest more on research and development that will address these challenges.

Digital twins operate with top-priority and confidential data, which are immediately audio for cybercogs. Data breach is a very expensive form of economic loss and also causes repute damage to the organization. Such mechanisms for secure data exchange are also a key part of enabling the kind data exchange between disparate digital twins and other systems that makes meaningful. In today's world, there are a number of security threats out there so when implementing EDI needs to take care the above mentioned factors coming into play for securing the data being used in these systems. Information integrity in the digital twin is important according to their importance for systems operations so preservation of information within a digital twin matters. The trending of false and invalid information falsified or changed, provide also un-valid analysis and bad decisions. Given that, complex network will encounter threats like man-in-the-middle attacks DDoS attack and many more network weaknesses. If the data streams are calls that you might want to secure from interception and if necessary manipulation.

The concern in the above respect naturally also extends to how data is being fused into models from these digital twins, and this has highlighted a couple of other more fundamental issues (such as good anonymization on amalgamation all are using within possible identification but retaining just useful information). Laws on privacy require consent to collect/process/share data with users, and this matters the most as it impacts user engagement and trust. Data transfer plays a large role in digital twins as typically information is sent abroad. Such landscape, specifically not being well-prepared for a complex set of divergent data protection laws overnight within one jurisdiction and another just, adds on to the new dimension called privacy management. The practices are to use the strict protection measures to secure confidentiality of digital twin data information security policies such as encryption, access controls and perform a security audit periodically. One of the key activities that needs to be done backward is simply maintaining data and preserving privacy by reconsidering security which has an approach, adapting it for the new threats/risks.

Digital twin evolution is ongoing, and thus, one must be prepared to update with time. Among the workforce, a critical issue is keeping them abreast with respect to developments in information. Employees now are very far from the competencies required to efficiently incorporate and co-operate with digital twin applications. If this gap is to be bridged, let it then come with clearly formulated training and development policies. The question here is that proper upskilling contains the enough amount and worth of training tools along with updated learning materials, practical sessions, and professional trainers. However, it can take extra to develop such tools and keep them running. Certain employees might be reluctant to accept new technologies or methodologies that may be promoted by the organization. So change has to be managed and led so as overcome the cultural resistance. More trained employees on data security measures mean fewer insider threats and compliance to the security policies more quickly.

6 Conclusions and Future Scope

Pioneered and proven in industrial markets from manufacturing to energy, digital twin technology is finally gaining some traction for product development. It does this by digitizing the physical world—that is, making virtual twins of products and processes to optimize design, monitor maintenance requirements or interact with them to increase performance. With real-world case studies already in place, the utility of this technology across sectors such as aerospace and food processing has displayed increased efficiency at reduced costs through enhanced business intelligence.

The future of digital twins in product development is rife with promise but not devoid of obstacles. Future areas for improvement are:

1. One such goal is standardization of implementation of digital twins across industries.
2. Better data security and privacy control.
3. Greater interoperability among systems and platforms.
4. Improved AI and machine learning to predict more accurately.
5. Training teams to unleash the power of digital twin.

While these challenges are overcome, digital twins have the potential to become more and more instrumental in innovation, sustainability, and competitiveness across Industry 5.0 time's scales as it moves ahead beyond this stage. Additional research and development converge in likely new use cases and benefits, essentially transforming industrial product design practices.

References

1. Tao, F., Cheng, J., Qi, Q., Zhang, M., Zhang, H., & Sui, F. (2018). Digital twin-driven product design, manufacturing and service with big data. *The International Journal of Advanced Manufacturing Technology, 94*(9), 3563–3576.
2. Grieves, M., & Vickers, J. (2017). Digital twin: Mitigating unpredictable, undesirable emergent behavior in complex systems. In *Transdisciplinary perspectives on complex systems* (pp. 85–113). Springer.
3. Grieves, M. (2014). Digital twin: Manufacturing excellence through virtual factory replication. *White Paper, 1*, 1–7.

4. Glaessgen, E., & Stargel, D. (2012). The digital twin paradigm for future NASA and US air force vehicles. In *53rd AIAA/ASME/ASCE/AHS/ASC structures, structural dynamics and materials conference 20th AIAA/ASME/AHS adaptive structures conference 14th AIAA* (p. 1818).
5. Boschert, S., & Rosen, R. (2016). Digital twin—The simulation aspect. In *Mechatronic futures* (pp. 59–74). Springer.
6. Kritzinger, W., Karner, M., Traar, G., Henjes, J., & Sihn, W. (2018). Digital twin in manufacturing: A categorical literature review and classification. *IFAC-PapersOnLine, 51*(11), 1016–1022.
7. Söderberg, R., Wärmefjord, K., Carlson, J. S., & Lindkvist, L. (2017). Toward a digital twin for real-time geometry assurance in individualized production. *CIRP Annals, 66*(1), 137–140.
8. Tao, F., Zhang, H., Liu, A., & Nee, A. Y. C. (2019). Digital twin in industry: State-of-the-art. *IEEE Transactions on Industrial Informatics, 15*(4), 2405–2415.
9. Boje, C., Guerriero, A., Kubicki, S., & Rezgui, Y. (2020). Towards a semantic construction digital twin: Directions for future research. *Automation in Construction, 114*, 103179.
10. Leng, J., Ruan, G., Jiang, P., Xu, K., Liu, Q., Zhou, X., & Liu, C. (2020). Blockchain-empowered sustainable manufacturing and product lifecycle management in industry 4.0: A survey. *Renewable and Sustainable Energy Reviews, 132*, 110112.
11. Kristoffersen, E., Blomsma, F., Mikalef, P., & Li, J. (2020). The smart circular economy: A digital-enabled circular strategies framework for manufacturing companies. *Journal of Business Research, 120*, 241–261.
12. Li, X., Chen, C. H., Zheng, P., Wang, Z., Jiang, Z., & Jiang, Z. (2020). A knowledge graph-aided concept–knowledge approach for evolutionary smart product–service system development. *Journal of Mechanical Design, 142*(10), 101403.
13. Singh, S., Weeber, M., & Birke, K. P. (2021). Advancing digital twin implementation: A toolbox for modelling and simulation. *Procedia CIRP, 99*, 567–572.
14. Santos, R., Basto, J., Alcalá, S. G., Frazzon, E., & Azevedo, A. (2019, July). Industrial IoT integrated with simulation—A digital twin approach to support real-time decision making. In *Proceedings of the international conference on industrial engineering and operations management* (pp. 23–26).
15. Dröder, K., Bobka, P., Germann, T., Gabriel, F., & Dietrich, F. (2018). A machine learning-enhanced digital twin approach for human-robot-collaboration. *Procedia CIRP, 76*, 187–192.
16. Rojek, I., Mikołajewski, D., & Dostatni, E. (2020). Digital twins in product lifecycle for sustainability in manufacturing and maintenance. *Applied Sciences, 11*(1), 31.
17. Ye, Y., Yang, Q., Yang, F., Huo, Y., & Meng, S. (2020). Digital twin for the structural health management of reusable spacecraft: A case study. *Engineering fracture mechanics, 234*, 107076.
18. Maheshwari, P., Kamble, S., Belhadi, A., Mani, V., & Pundir, A. (2023). Digital twin implementation for performance improvement in process industries-a case study of food processing company. *International Journal of Production Research, 61*(23), 8343–8365.
19. Qi, Q., Tao, F., Hu, T., Anwer, N., Liu, A., Wei, Y., Wang, L., & Nee, A. Y. C. (2021). Enabling technologies and tools for digital twin. *Journal of Manufacturing Systems, 58*, 3–21.
20. Fuller, A., Fan, Z., Day, C., & Barlow, C. (2020). Digital twin: Enabling technologies, challenges and open research. *IEEE Access, 8*, 108952–108971.
21. Bolton, R. N., McColl-Kennedy, J. R., Cheung, L., Gallan, A., Orsingher, C., Witell, L., & Zaki, M. (2018). Customer experience challenges: Bringing together digital, physical and social realms. *Journal of Service Management, 29*(5), 776–808.
22. Rasheed, A., San, O., & Kvamsdal, T. (2020). Digital twin: Values, challenges and enablers from a modeling perspective. *IEEE Access, 8*, 21980–22012.

Accelerating Innovation: The Benefits of Digital Twins in Product Design and Manufacturing

N. Krishnaraj, Priyanga Subbiah, A. Soujanya, and R. Meera

1 Introduction

The current industry trend reflects an increasing demand for complex features, shorter time-to-market, and more stringent regulatory requirements for products. More conventional approaches that require the creation of a tangible prototype and a sequence of tests do not seem to suffice these requirements. These challenges are: Digital twin comes as a powerful approach to solve these challenges since it gives the virtual copy of a product from the design and manufacturing stages to its use and subsequent recycling. Digital twin is a term used to refer to the virtual representation of a physical asset or a system. It includes a significant plan of the product's structure, its constituent elements, and the way the product works [1]. The digital twin is updated in real time by data from sensors and other sources and reflects the updates of the physical product and its performance characteristics. Although the definition provided above is comprehensive enough to describe what exactly is a digital twin, it is necessary to define some of the components of this concept in more detail. All of these factors make combined contributions toward the twin's possession of the characteristics that enable it to fully represent and function as the physical counterpart.

Physical Model: They indicate detailed characteristics of the product including its geometrical, material composition and physical shape. This consists of the CAD models, FEA data, and any other physical features that may be of importance.

Virtual Sensors: Virtual sensors acting as dummy sensors which are integrated into the structure and acquire data in regard to temperature, pressure, vibration, and other pertinent aspects of performance.

N. Krishnaraj (✉) · P. Subbiah · A. Soujanya · R. Meera
Department of Networking and Communications, School of Computing, College of Engineering and Technology, SRM Institute of Science and Technology, Kattankulathur, Chengalpattu, Chennai, Tamil Nadu, India

Y. Kumar Singla et al. (eds.), *Digital Manufacturing in Industry 5.0*,
https://doi.org/10.1007/978-3-031-91500-0_33

Data Integration: A reliable hardware and software framework to acquire, process, and store data from real time physical sensors and other related data streams as could be work orders, environmental factors, etc.

Simulation Engine: An analytical model, which can predict the behavior of the product with respect to different scenarios and given input data, derived from the physical model.

Artificial Intelligence (AI) and Machine Learning (ML): Programs that process information and learn about the specific tendencies in the product's sales and possible drawbacks according to data acquired.

Human–Machine Interface (HMI): A graphical and intuitive interface to operate the digital twin, to explore the data and to interact with simulations.

2 The Dynamic Nature of Digital Twins

The crucial aspect of digital twins is their dynamic nature. This continuous evolution is what truly differentiates them from static models.

2.1 Key Implications of a Digital Twin's Dynamism

Real-Time Insights: Digital twins, when infused with live data, become dynamic virtual counterparts of physical assets, offering real-time insights into performance. This empowers proactive maintenance strategies, enabling early detection of anomalies, optimized operations, and reduced downtime, ultimately driving significant improvements in efficiency and asset lifespan.

Predictive Analytics: By analyzing historical and real-time data patterns, digital twins can predict potential equipment failures or process deviations, enabling proactive maintenance scheduling, optimized resource allocation, and risk mitigation, ultimately enhancing overall system reliability and efficiency.

Performance Optimization: By simulating various operational scenarios and meticulously comparing the outcomes, organizations can uncover hidden opportunities to enhance efficiency and productivity. This powerful capability empowers data-driven decision-making, leading to optimized resource allocation, streamlined processes, and ultimately, improved bottom-line results.

Product Improvement: The invaluable insights gleaned from the digital twin can be seamlessly integrated into the product design process, fostering a continuous improvement loop [2]. By analyzing performance data, identifying areas for enhancement, and simulating design modifications within the virtual environment, organizations can iterate on product designs more rapidly, reduce development costs, and ultimately deliver superior products that better meet customer needs and market demands.

Digital Twin Maturity: As the digital twin accumulates a vast repository of data and continuously refines its predictive models through machine learning, its ability to anticipate potential issues, optimize performance, and prescribe corrective actions significantly enhances. This ongoing learning process transforms the digital twin into an increasingly valuable asset, providing deeper insights, reducing risks, and driving operational excellence.

3 Core Components of Digital Twins

To fully grasp the potential of digital twins, it's crucial to understand their fundamental building blocks. A digital twin is composed of several interconnected components:

3.1 Physical Asset or System

The physical asset or system is the real-world counterpart that a digital twin represents. It's the tangible object or process that the digital twin aims to replicate and understand.

Here are some examples of physical assets that can be represented by digital twins:

- **Industrial Machinery**: Manufacturing equipment, turbines, engines, etc.
- **Infrastructure**: Buildings, bridges, roads, and other infrastructure assets.
- **Products**: Consumer goods, vehicles, electronics, etc.
- **Processes**: Manufacturing processes, supply chains, logistics operations, etc.

3.2 Data Acquisition and Management

It is desirable to state that data acquisition and management is an essential part of digital twin frameworks. Here's a breakdown of the key aspects involved: break down of the key aspects involved:

- **Data Collection**: Sensors and IoT devices are attached on the physical asset to collect real-time data. This data can include a range of factors like temperatures including force or pressure, vibration, energy consumption, and a lot of other factors.
- **Data Processing**: The data that is collected from sensors has to go through a process where data that is raw has to be cleaned and able to ensure that the information it contains is of good quality and has all the relevant information that can be searched for in a single document [3]. This may entail arriving at some details

such as noise reduction or correction of errors, i.e., data scrubbing to remove missing values and wrong entries, data transformation involving transforming data into acceptable format for analysis.
- **Data Storage**: The other part of the data that has been processed is stored in a database or data warehouse. This is useful to ensure that the data by collected can be accessed and be used to update the digital twin and perform analysis.

Thus, proper utilization of data is the cornerstone for achieving higher levels of a digital twin. It ensures that the information is accurate, accessible and securable, and this is an important condition for successful modeling and simulation.

3.3 Virtual Model

This is the central component of digital twin that is a digital replica of the physical asset. It encompasses:

Geometric Model: This component of the virtual model reveal the external structure and size of the asset. It is actually a sort of topological map of the asset image. This can be created using 3D. Either way, it can be seen that canonical correlation analysis can be a powerful tool for data analysis.

- **Shape**: This refers to the general shape of the asset in question, for example, elongated, square, rectangular, and so on.
- **Dimensions**: Some of the measurements applicable to the asset are length of the asset, width of the asset, height of the asset, and other relevant dimensions.
- **Features**: Any progressive/or recessive feature of the asset covering details of extensions, indentations, or holes.

The geometric model is essential for evaluating the parameters of the asset and also for the realization of the shape of the asset. The system's ability to mimic interactions with its environment to evaluate its operational potential.

3.4 Material Properties

This component of the virtual model defines the typical properties of the materials which use in the physical asset. As seen, these properties can impart a relatively large influence over the asset's responses and performances and lifespan. Some key material properties that are often included in digital twins include:

- **Mechanical Properties**: Some of these properties are; tensile strength, modulus of elasticity, toughness, and endurance limits resistance. They define flexibility of the asset in response to forces and stresses.
- **Thermal Properties**: They include the coefficient of thermal conductivity, the specific heat and specific heat ratio. These properties are essential to determine

characteristics of the asset, how temperature changes will affect the respective system will be explained.

- **Electrical Properties**: Some of these aspects include conductivity of the fluid, resistivity of the fluid and also the dielectric strength of the fluid that is being used in power transmission [4]. By using these properties, one is able to get the electrical characteristic of the particular asset.
- **Corrosion Resistance**: Corrosion resistance of the material is one of the significant factors to be considered for asset's longevity.
- **Other Properties**: Besides these properties there may be other properties which are required depending on the application of the metal. Other factors that could be considered include density, color, and magnetic properties.

3.5 Behavioral Model

This component of the virtual model defines how the asset responds to different inputs and conditions. It essentially describes the asset's behavior. Key aspects of the behavioral model include:

- **Input–Output Relationships**: How the asset responds to different inputs (e.g., changes in temperature, load, or control signals) and produces corresponding outputs (e.g., changes in performance, energy consumption, or emissions).
- **Dynamic Behavior**: How the asset's behavior changes over time, including transient responses, steady-state behavior, and stability.
- **Nonlinearity**: Whether the asset's behavior is linear or nonlinear, meaning whether the output is proportional to the input or not.
- **Uncertainty**: The level of uncertainty or variability in the asset's behavior due to factors such as manufacturing tolerances, environmental variations, or ageing.

The behavioral model is crucial for simulating the asset's performance under various conditions, predicting its response to disturbances, and optimizing its operation.

3.6 Simulation Capabilities

Making a digital twin is not complete without an element based on which it is possible to mimic different circumstances and states. This involves:

- **Predictive Analysis**: It predicted the asset's comportment in the future based on information and paste data in the virtual model.
- **What-If Scenarios**: Using the concept of simulation to trial a variety of situations and circumstances to determine the effects of fluctuations of inputs, coefficients as well as factors of influence in the model.

- **Design Optimization**: Applying simulation methods in order to analyze one or another variant of design and determine which configuration is most suitable with regard to certain performance factors.
- **Virtual Prototyping**: Developing simulation models of the asset with a view of experimenting with its performance.

Modern design tools give engineers the means of testing and optimization of the most complex designs and configuration of the digital twin. Thus, more reliable simulations and consequent decisions are being made.

3.7 Analytics and Optimization

One of the major advantages of employing the concept of digital twins is that of getting analyzable data from the gathered data. This involves:

- **Data Analysis**: Processing and analyzing the data to identify trends, patterns, and anomalies.
- **Performance Metrics**: Using various formulas to determine the means of evaluating a performance based on set scores, or KPI.
- **Root Cause Analysis**: Identifying the underlying causes of issues or failures.
- **Optimization**: Using the insights to optimize the asset's performance, reduce costs, and improve efficiency.
- **Predictive Maintenance**: Using ML and AI tools to predict possible failures on the equipment and time to conduct a preventive maintenance.

Due to the application of analytics and optimization features, digital twins can be of great use for organizations to optimize several asset management functions, the reduction of asset downtime, and the general operational performance of a facility.

3.8 Visualization

Visualization is the link that makes digital twin feasible and easily implemented. It involves:

- **3D Models**: Developing complete 3D pictures of the asset that can be manipulated to be visual.
- **Augmented Reality (AR)**: Simply placing electronic information on top of the real world environment and enabling users to visualize the identified digital twin in the real application environment.
- **Virtual Reality (VR)**: Designing engaging scenarios in which the users are able to engage which is synchronized with the actual and the digital twin in a virtual world.

- **Data Visualization**: It can apply to the use of charts, graphs, and other forms of visualization. It represents complex information in such a manner that they would be easily comprehensible. Interactive and efficient ways of visualization help the users to navigate through the concept of the digital twin, analyze it, and utilize it.

3.9 The Power of Integration

Though it may sound logical, there can be no denying that the real value of a digital twin is in the ability to build an accurate model that is well integrated with high level computation. In doing so, organizations can sync these components in order for them to work together and analyze previously unimaginable patterns and trends of what products go through [5], identify problems at their incipient stage in the development process, and make the right data-drive decisions with ease. This synergy enables firms to seek ways in performing their functions more efficiently and at lesser costs in order to provide the best goods and service possible. For instance, a manufacturing firm can incorporate data from sensors in its production line in real time. It can convert the sequence of events on the real production line into a model in the form of a digital twin of the assembly by analyzing the data.

In addition, the benefits derived arise from the integration of the concept of digital twin with other related digital systems like supply chain. ERP systems such as Enterprise Resource Planning and CRM systems such as Customer Relationship Management systems can give a complete picture of the business, and it helps organizational decision makers in making strategic decisions in the business.

4 Technologies Used in Digital Twins

Internet of Things (IOT): IoT plays a crucial role in digital twins. It involved with the creation of a digital ecosystem through Internet of Things (IoT). Figure 1 illustrates the outline of the technologies used in digital twins.

Extended Reality (XR): Extended reality (XR) can be used to visualize and interact with digital models of physical objects, and it's not solely responsible for creating those models. The actual creation of the digital twin often involves other components, such as:

- **Data Acquisition**: Collecting data from sensors and other sources.
- **Modeling**: Creating the virtual model based on the collected data.
- **Simulation**: Simulating the asset's behavior using the virtual model.
- **Training and Education**: Simulating real-world scenarios for training purposes.
- **Design and Development**: Visualizing and testing designs before physical prototyping.

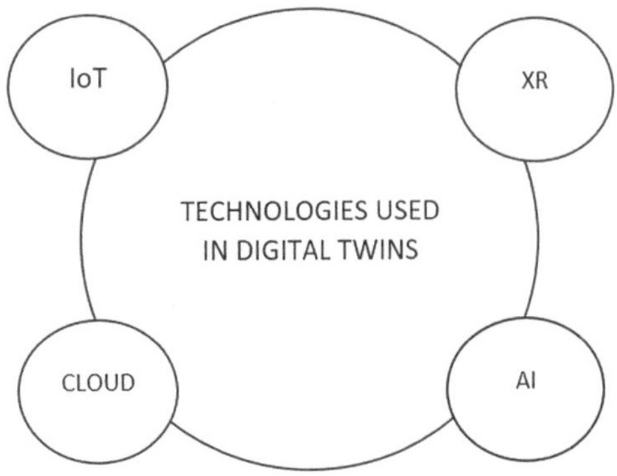

Fig. 1 Technologies used in digital twins

- **Maintenance and Troubleshooting**: Providing visual guidance and support for maintenance tasks.

Cloud Computing (Cloud): Cloud platforms also contain resources that can be underlying to support a cloud application development, deployment, and management.

Centralized Data Storage: Keep all relevant data in one place, making it easier to access and manage.

- **Scalability**: Easily scale storage capacity to accommodate growing data volumes.
- **Accessibility**: Access data from anywhere with an internet connection, enabling remote monitoring and control.
- **Cost-Efficiency**: Avoid the upfront costs of setting up and maintaining on-premises data centers.

Cloud platforms also provide tools and services that can be used to build, deploy, and manage digital twins, making it easier for organizations to leverage this technology.

Artificial Intelligence (AI): Artificial intelligence (AI) is a tool which can be really effective in analyzing data and avail analytical information from model created and simulated digital twins. Here's how AI is used in this context [6]: Predictive Analytics: AI can make decisions based on the previous statistics that to some extent reflects the future results of the process:

performances, anticipated failures, or trends in energy use.

- **Optimization**: Application of AI can assist in achieving more efficient use of physical assets.
- **Decision Support**: Thus, AI plays the role of giving suggestions that will be of great help in aiding organizations to make decisions on matters touching on maintenance and operations.

5 Benefits of Digital Twins in Product Development

5.1 Accelerated Time-to-Market

By minimizing the on-and-off physical prototyping and testing of products, digital twins save time. Virtual environments also permit multiple designs to be developed soon, and it is also possible to quickly test them. Different operating conditions and also reveals possible concerns right from the initial stages of the process [7]. This unbundled process of development reduces the amount of wasted physical prototypes, and subsequent redesign, and helps to take products to the firm to bring new products to the market faster while at the same time offering higher quality and reliability. For instance, in the automotive industry, through the use of digital twins one is capable of modeling a car.

5.2 Cost Reduction

Designing, manufacturing, and management methods help to leverage major cost savings. As observed, inefficiencies in processes and material usage highlight the need for waste reduction and optimization efforts. Through the digital twin, the organizational failures can be minimized in order to ensure that the concepts of the operational strategies enhance organizational efficiency, effectiveness as well as the production process is optimized [8]. It improves the efficiency of a power plant with the ability to forecast when pieces of equipment might fail, when to have them maintained preventatively, and increasing general output, which led to significant decrease of expense.

5.3 Improved Decision-Making

Mapping of data allows making right decisions at the right time—this is the essence of real-time data and prediction analytics in the entire product lifecycle. In terms of offering a form of aesthetic qualitative analysis of the product performance, market trends and the general performance outcomes, while posing challenges to the firm, the internet offers a unique facet of information delivery.

This results in better management of resources, proper management of risks and bringing about the desired organizational outcomes [9]. For example, in the health care industry they have created digital twin, it would also enable patients to be diagnosed through self-monitoring devices that could even predict failure of medical equipment, treatment plans. Implementation strategies, monitor program, patient care plans, patient care, and patient outcomes.

5.4 Increased Innovation

Digital twins allow for testing and brainstorming the new design ideas, as well as checking how they will perform in the real world. By making digital imitations of tangible assets, organizations are able to actively and quickly experiment with and on them and explore multiple design solutions without having to go through the process of designing and modeling many of the different variations of the same basic idea [10]. This desired accelerated innovation cycle is then used as a tool to enable businesses to create new types of goods and services, match the current tendencies and be ahead of competitors. For instance, in the automobile industry, the engineers use the concepts of digital twins to experiment with various vehicle designs, materials, and powertrain configurations, leading to the development of more efficient, sustainable, and appealing vehicles.

6 Roles of Digital Twins in the Product Lifecycle

Digital twins are iterations of a tangible product or a program in the virtual environment, and provide an integrated perspective that people get to see when they embark on the journey of life. The role and use of their applications range from the conceptualization phase right down to the disposal of the product.

6.1 Design and Development

6.1.1 Virtual Prototyping

Introduction of computer aided tools in designing and simulation has brought about a considerable change. In the product development process, designers can make and set several designs into the process. It changes with a model developed and tested in a short time without the need for a physical model [11]. This approach helps to reduce time so much of production time, cost, and material waste in addition to introducing new facets to the design process.

6.2 Key Technologies and Methods

Computer-Aided Design (CAD)

- Create detailed 3D models of product components and assemblies.
- Parametric modeling allows for rapid design modifications.
- Assembly modeling enables virtual testing of product functionality.

Computer-Aided Engineering (CAE)
- Simulate product performance under various conditions.
- Analyze material properties and structural integrity.
- Optimize design parameters based on simulation results.

Finite Element Analysis (FEA)
- Predict structural behavior and identify potential failure points.
- Optimize material distribution and component thickness.
- Evaluate product performance under load conditions.

Computational Fluid Dynamics (CFD)
- Simulate fluid flow and heat transfer around products.
- Optimize aerodynamic or hydrodynamic performance.
- Analyze product cooling and ventilation.

Virtual Reality (VR) and Augmented Reality (AR)
- Immersive visualization of product designs.
- Interactive product exploration and evaluation.
- Collaborative design reviews and feedback.

6.3 *Benefits of Virtual Prototyping*

Faster Time-to-Market: Less time taken in designing the product and modification of designs which are essential for its development. New products and product market coverage excluding new product launches and increased market share.

Lower Costs: This means that there will be no need to create tangible models and the costs related to these models.

Improved Product Quality: Improvement of design parameters and their corresponding performance estimation through better product reliability and satisfaction achieved to the customer.

Increased Innovation: Promotes design freedom in a manner more than what the narrow conception of the function permits, fostering the advancement of innovations that will help outline new significant products and services [12]. Better collaboration: Builds the concept of team work among the individuals **as** well as sharing of information, thus enhancing decision-making and accelerated problem-solving.

6.4 *Example Applications*

Automotive Industry: The development of material and hardware in the structure of the car and its parts, and this includes, testing the parts in the vehicle through virtual simulations.

Aerospace Industry: Structural and Design improvements in the aircraft, and more specifically in propulsion systems.

Consumer Electronics: Designing and emerging electronic products (e.g., computers, cameras, smartphones, etc.) seem to be as challenging as coming up with new software applications and creating unique web content.

Medical Device Industry: They participate in designing prototype implants, and surgical instruments, as well as simulate surgeries.

Simulation and Analysis: Forecasting encompasses an expectation of product performance under different circumstances as a main component of it [13]. An important or pivotal stage in the product design procedure. Through the accurate modeling of product behavior, engineers predict the emerging concerns and design alterations and enhancements of generalized structures.

6.5 Optimization Techniques

Sensitivity Analysis: Sensitivity analysis is an important activity needed in creating a digital twin system [14]. Since it involves systematic change of design parameters, it assists in determining which of them has the greatest and very crucial factor that determines the performance of particular product. Such information is extremely useful to optimize something.

Trade-off Analysis: Trade-off analysis is a critical component of digital twin optimization. It allows engineers to explore the complex relationships between different design parameters and their impact on overall product performance [15]. By understanding these trade-offs, designers can make informed decisions to balance competing objectives and achieve optimal outcomes.

Multiobjective Optimization: Multiobjective optimization is essential for complex engineering problems where multiple, often conflicting goals need to be balanced. It enables the exploration of the design space to identify a set of optimal solutions that represent the best trade-offs between these objectives.

By integrating digital twins into the product lifecycle, organizations can achieve significant improvements in product development, manufacturing, and overall business performance. Figure 2 showcases the roles of digital twin in the product lifecycle.

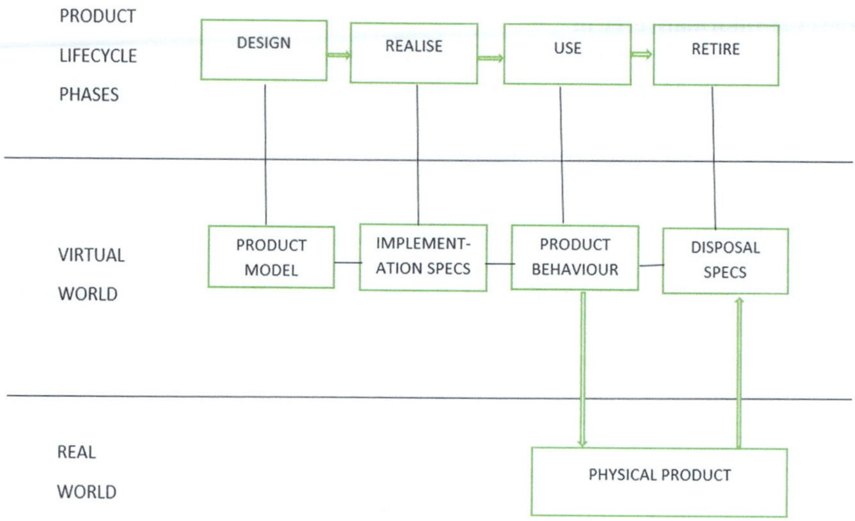

Fig. 2 Roles of digital twin in the product lifecycle

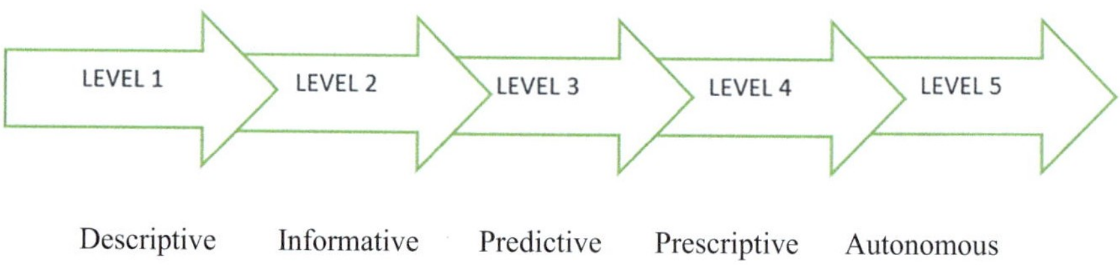

Fig. 3 Levels of digital twins

7 Levels of Digital Twins in Process Operations

Digital twins are becoming increasingly pivotal in process operations, offering a virtual representation of physical assets to optimize performance, predict failures, and enhance decision-making. Figure 3 outlines the various levels of digital twin. Here are five levels of digital twins in process operations, progressing from basic to advanced:

Level 1: Descriptive Twin

- Core Function: Static representation of the physical asset.
- Capabilities: Visualizes the asset's geometry, dimensions, and basic attributes.
- Value: Provides a foundational understanding of the asset's physical characteristics.

Level 2: Informative Twin
- Core Function: Incorporates real-time data from the physical asset.
- Capabilities: Displays operational data, such as temperature, pressure, and flow rates.
- Value: Provides insights into the asset's current state and performance.

Level 3: Predictive Twin
- Core Function: Utilizes historical data and machine learning to predict asset behavior.
- Capabilities: Forecasts equipment failures, process deviations, and potential bottlenecks.
- Value: Enables proactive maintenance, optimized production planning, and risk mitigation.

Level 4: Prescriptive Twin
- Core Function: Recommends actions based on predicted outcomes.
- Capabilities: Suggests optimal operating parameters, maintenance schedules, and process adjustments.
- Value: Drives autonomous decision-making and continuous improvement.

Level 5: Autonomous Twin
- Core Function: Executes autonomous actions based on real-time data and predictions.
- Capabilities: Optimizes process parameters, self-corrects deviations, and adapts to changing conditions.
- Value: Maximizes efficiency, reduces human error, and unlocks new levels of automation.

8 Key Challenges and Mitigation Strategies

8.1 Data Management

Another point is that data quality is one of the decisive factors of the success of digital twins. Ensuring data accuracy, consistency is mandatory while assessing the degree of completion. Measures, which are to ensure the validity of the obtained results: the processes of data cleaning and input validation. It is crucial to filter out the random variations, mistakes, and variability in the data collected [16]. Other factors that affect the choice of SaaS model are security to gain access to SaaS applications. That is why it is imperative that data it contains should be encrypted, people that have access to it should be controlled and data security audits should be conducted as often as possible as a means of ensuring the data is not exposed and gets breached. Another aspect is the scalability, as digital twins often encounter big amounts of data. High-quality analytics of this data can be accomplished only if this information is accumulated and stored in such a way to allow for its proper management within an organization that experiences constant growth.

8.2 Model Fidelity

Model fidelity is the level of detail in which the digital twin of a certain element is represented accurately. Developing highly proper modeling of the digital twin needs information about the physical asset and its subcomponents. This often involves:

- **Domain Expertise**: What it means is that having experts in the relevant field, especially in manufacturing and physics, to help in understanding the behavior of the asset.
- **Advanced Modeling Techniques**: By the help of the high-level algorithm models employed [17] in this study, the finite element analysis, computational fluid dynamics, or system dynamics capture the level of complexity that characterizes the behavior of the asset.
- **Data-Driven Modeling**: Owning to the availability of web data, the proposed model was calibrated and validated to represent the digital.

Modeling is the art of achieving an optimal trade-off between performance and cost-effectiveness.

The costs of simulating large system structures are also high, and this is sometimes unworthy because it does not add so much more value. On the other hand, it is possible that simple models could distort the asset's characteristics and do not reflect the real situation.

8.3 Mitigating Risks and Maximizing Benefits

For this to be done, it is necessary to use the solutions to the challenges that were discussed earlier before and apply them appropriately. Here are some key considerations:

- **Continuous Improvement**: It is recommended to analyze and alter the state of digital twins more frequently and to make certain that they contain accurate information and that such information is useful.
- **Collaboration**: Enhance coordination of different departments and sections engaged in the digital twin initiatives.
- **Return on Investment (ROI)**: Describe in detail what type of value is expected from the application of digital twins, and therefore, they can track their return on investments (ROI) to be in a position to justify for them.

- **Ethical Considerations**: Discuss ethical issues involving data privacy and data bias that have raised different concerns about the transparency in digital twin applications.

8.4 *Leverage Advanced Computing*

A key factor that needs to be exploited in the proper execution of such complex is advanced computing resources digital twins. Here's why:

- **Scalability**: Cloud computing as well as the HPC platforms provide elastic infrastructure that can be made to accommodate the fluctuations of computer requirements in digital twins.
- **Cost-Efficiency**: These resources can be studied anytime; thus, eradication of time constraints occurs in large investments in hardware for realizing costs savings through such scenarios as data mining.
- **Performance**: It means that the use of such forms of computing technologies as high speed simulations and real time data processing, could be speeded up considerably, thus improving the overall efficiency of operations.
- **Flexibility**: Hence, the solutions are flexible in terms of deployment platform since they can be deployed both locally and remotely. The management thus enables organizations to meet changing needs.

8.5 *Foster Collaboration*

Ensuring that different teams work together is important to make the concept of the digital twin successful.

- **Shared Understanding**: On the other hand, cross-functional coordination guarantees that all the parties participate in the design and implementation of these decisions are familiar with the objectives and specifications of the project as well as possible difficulties.
- **Domain Expertise**: It is the process of gathering professionals with various backgrounds. In this case, it will be advisable for IT to collaborate with business in order to develop more efficient modes of producing accurate and precise digital twins.
- **Leveraging Latest Technologies**: Under collaboration, it is easier to discover and embrace the best adequate technologies that may be applied in construction and upkeep of the digital twins.
- **Effective Communication**: Therefore, there is a need to adopt open talk and active cooperation to prevent ambiguous misunderstandings, and it would guarantee that all the stakeholders' needs are met in regard to the digital twin.

8.6 Prioritize Cybersecurity

Security is a major issue in digital twins, since they "interact" with sensitive information and systems [18]. It can become beneficial for attackers when it comes to cyber operations. Here are some key measures to prioritize: Here are some key measures to prioritize:

- **Data Encryption**: Encrypt all data where they are stored or being transmitted to ensure they are safe from unauthorized access.
- **Access Controls**: They should therefore establish strict access controls to minimize access of sensitive data and implement the access of it and the systems to the authorized users only.
- **Regular Security Audits**: Security audit should be conducted to survey the degree of security threat and take corrective actions.
- **Patch Management**: Ensure that software and system are updated to the current security.
- **Employee Training**: Teach people in your organization how to be safe online as well as the understanding the need for the protection of data that is considered to be confidential.
- **Incident Response Plan**: Design an all-embracing incident response plan in order to respond to security threats and prevent them, as well as reduce their effects. By considering cybersecurity as a top priority, organizations can ensure that their digital twins as well as the valuable them from data they contain against unauthorized access and possible harm.

9 Applications of Digital Twins in Product Development

9.1 Design and Development

Virtual Prototyping: Developing prototypes in order to evaluate the effectiveness of proposed designs before the concept models are set in actual physical models [19].

- **Performance Analysis**: Interestingly, the relative effectiveness of simulation is in testing different operating conditions to determine the performance of a product.
- **Design Optimization**: Searching for the best design solutions in order to have a variety of choices based on relative performance and price as well as some other criteria.

9.2 Manufacturing

Virtual Manufacturing: Making digital twins as a representation on the operation of manufacturing processes such as streamlining flow, as well as finding out areas of congestion and areas that needed enhancement.

- **Quality Control**: Modeling the manufacturing processes to guarantee that the developed products are as desired quality standards and specifications.
- **Process Optimization**: As a result, potential candidates for efficiency enhancement in the manufacturing sector can be pointed out to activities that aim at minimizing costs and optimizing meager resources in organizations.

9.3 Predictive Maintenance

Failure Prediction: Based on the historical trend and the current sensor values the machine can predict potential equipment failures.

- **Proactive Maintenance**: To avoid scheduling the following preventive maintenance tasks in the timeline.
- **Cost Reduction**: Schedules for maintenance that would enable the organism to cut down on expenditures on maintenance substantially and minimizing unplanned downtime.

9.4 Supply Chain Optimization

- **Inventory Management**: Optimizing inventory levels to balance supply and demand.
- **Logistics Planning**: Simulating transportation routes and schedules to improve efficiency and reduce costs.

It is correct to note that over the years there has been heightened interest in the use of digital twins [20]. The fundamentals of modeling physical objects in computer and their usage to obtain the desired equivalence of physical modeling and analysis have been discussed for a long time in such areas as engineering, manufacturing, and many others.

10 Conclusion

Digital twins are expected to disrupt the market of product development in the near future. By bridging physical and a digital environment, they afford one with a unique chance to fine tune the product design, manufacturing, and performance for a product. Through virtual experimentation, forecasting, and real-time decision-making, by means of digital twin, organizations are able to make analytics, improve decision-making, catalyze business growth rates, and improve product value. Over time, as the technology progresses, the increasing applicability of digital twins will be seen as well. To fully realize the advantages of digital twins, an organization needs to make investments in data ecosystem, computation resources, and skilled personnel. It is important to consider that any path to success is not easy, and as such, it is possible to speak about success in terms of achieving the goal after overcoming the obstacle and accepting the transformation. Because of the potential this technology holds, organizations can align themselves to be ahead of the change curve.

References

1. Deuter, A., & Pethig, F. (2019). The digital twin theory—Eine neue Sicht auf ein Modewort. *Industrie 4.0 Management, 35*(1), 27–30. https://doi.org/10.30844/I40M_19-1_S27-30
2. Holopainen, M., Saunila, M., Rantala, T., & Ukko, J. (2024). Digital twins' implications for innovation. *Technology Analysis & Strategic Management, 36*(8), 1779–1791.
3. Pronost, G., Mayer, F., Camargo, M., & Dupont, L. (2024). Digital twins along the product lifecycle: A systematic literature review of applications in manufacturing. *Digital Twin, 3*, 3.
4. Fukawa, N., & Rindfleisch, A. (2023). Enhancing innovation via the digital twin. *Journal of Product Innovation Management, 40*(4), 391–406.
5. Javaid, M., Haleem, A., & Suman, R. (2023). Digital twin applications toward industry 4.0: A review. *Cognitive Robotics, 3*, 71–92.
6. Overbeck, L., Graves, S. C., & Lanza, G. (2024). Development and analysis of digital twins of production systems. *International Journal of Production Research, 62*(10), 3544–3558.
7. Deuse, J., Wagstyl, D., Moreno, V. H., Polikarpov, M., Wöstmann, R., & Hoffmann, F. (2023). *Green digital twins in the product life cycle* (pp. 167–186). Schriftenreihe der Wissenschaftlichen Gesellschaft für Arbeits-und Betriebsorganisation (WGAB) e. V.
8. Human, C., Basson, A. H., & Kruger, K. (2023). A design framework for a system of digital twins and services. *Computers in Industry, 144*, 103796.
9. Tao, F., Zhang, H., & Zhang, C. (2024). Advancements and challenges of digital twins in industry. *Nature Computational Science, 4*(3), 169–177.
10. Zayed, S. M., Attiya, G. M., El-Sayed, A., & Hemdan, E. E. D. (2023). A review study on digital twins with artificial intelligence and internet of things: Concepts, opportunities, challenges, tools and future scope. *Multimedia Tools and Applications, 82*(30), 47081–47107.

11. Can, O., & Turkmen, A. (2023). Digital twin and manufacturing. In *Digital twin driven intelligent systems and emerging metaverse* (pp. 175–194).
12. Shah, I. A., Sial, Q., Jhanjhi, N. Z., & Gaur, L. (2023). Use cases for digital twin. In *Digital twins and healthcare: Trends, techniques, and challenges* (pp. 102–118). IGI Global.
13. Duan, H., Gao, S., Yang, X., & Li, Y. (2023). The development of a digital twin concept system. *Digital Twin, 2*(10), 10.
14. Su, S., Zhong, R. Y., & Jiang, Y. (2023). Digital twin and its applications in the construction industry: A state-of-art systematic review. *Digital Twin, 2*, 15.
15. Zakharchenko, A., & Stepanets, O. (2023). Digital twin value in intelligent building development. *Advanced Information Systems, 7*(2), 75–86.
16. Vohra, M. (Ed.). (2023). *Digital twin technology: Fundamentals and applications*. Wiley.
17. Crespi, N., Drobot, A. T., & Minerva, R. (2023). The digital twin: What and why? In *The digital twin* (pp. 3–20). Springer International Publishing.
18. Verna, E., Puttero, S., Genta, G., & Galetto, M. (2024). Toward a concept of digital twin for monitoring assembly and disassembly processes. *Quality Engineering, 36*(3), 453–470.
19. Song, Y. (2023). Human digital twin, the development and impact on design. *Journal of Computing and Information Science in Engineering, 23*(6), 060819.
20. Kober, C., Algan, B. N., Fette, M., & Wulfsberg, J. P. (2023). Relations of digital twin fidelity and benefits: A design-to-value approach. *Procedia CIRP, 119*, 809–815.

Digital Twins for Process Understanding and Optimization

G. Jishnu Prian, Ananya S. Desikana, and G. Rajyalakshmi

1 Introduction

Process understanding in manufacturing requires a comprehensive grasp of the stages, techniques, and methodologies involved in transforming components. A deep understanding of these processes is key to achieving optimization, which can bring significant economic and environmental benefits. The core objective of process optimization is to enhance productivity, reduce waste, and improve quality. Traditional approaches, such as Lean Six Sigma, have been widely employed to improve industrial processes by focusing on efficiency and continuous improvement [1]. These methods rely on a structured, data-driven statistical methodology to analyze data and inform decision-making, employing tools such as DMAIC (Define, Measure, Analyze, Improve, Control), control charts, hypothesis testing, and cause-and-effect analysis [2].

Traditional statistical techniques for process optimization often manifest as various forms of mathematical programming models [3]. However, they face challenges in addressing the growing complexity of modern manufacturing, as these systems involve numerous variables and parameters, resulting in a multidimensional, nonlinear problem [4]. The computational demands of these traditional methods are substantial, requiring expert involvement for modeling and analysis, which can be both time-consuming and costly [5].

In today's data-rich manufacturing environment, vast amounts of real-time data can be harnessed to make quick, proactive decisions in response to market fluctuations, potentially improving profits, optimizing resource use, and addressing environmental and safety concerns [6]. The value of this data lies in its timely delivery

G. Jishnu Prian · A. S. Desikana · G. Rajyalakshmi (✉)
School of Mechanical Engineering, Vellore Institute of Technology, Vellore, Tamil Nadu, India
e-mail: rajyalakshmi@vit.ac.in

Y. Kumar Singla et al. (eds.), *Digital Manufacturing in Industry 5.0*,
https://doi.org/10.1007/978-3-031-91500-0_34

to the right person, offering numerous applications for process enhancement. However, traditional methods are limited in their capacity to fully leverage these data streams. This is where digital twins offer a transformative solution, integrating data across the entire process, not only improving upon existing optimization techniques but also unlocking new opportunities that traditional methods cannot address.

Digital twins provide a comprehensive view of processes, encompassing human factors and interdependencies, making the management of complex operations more feasible [7]. While they too require significant computational resources, these challenges can be mitigated through IoT technologies and cloud-based platforms, allowing manufacturers to offload the computational burden. Moreover, with real-time data integration and, in some cases, self-aware capabilities, digital twins can autonomously perform predictive analysis and optimize processes in real time without the constant need for expert intervention [8].

This chapter will explore the framework of digital twins, demonstrating how they tackle challenges of modern manufacturing by delivering efficient process optimization, and how they open new avenues for optimization that were previously unattainable.

2 Challenges in Process Understanding and Optimization of Modern Manufacturing

Modern manufacturing processes have evolved significantly, transitioning from traditional methods to smart and intelligent systems. According to Zhou et al. [9], this evolution can be categorized into three stages: traditional manufacturing, smart manufacturing, and intelligent manufacturing (see Fig. 1). Traditional manufacturing relied heavily on human involvement for tasks like data analysis, decision-making, and control. In contrast, smart manufacturing integrates cyber-physical systems, allowing for partial automation of these tasks. Intelligent manufacturing takes this further by leveraging artificial intelligence (AI) to reduce human intervention in routine activities, shifting the focus toward creative problem-solving.

However, this transition brings several challenges. One of the most significant is the vast amount of data generated by modern systems. The use of sensors, cloud-based solutions, and business management systems has drastically improved process automation, but it has also led to data overload—estimated at over 1000 exabytes annually [10]. To remain competitive, manufacturers must integrate data from various sources and ensure consistency across datasets. Managing this data is complicated by its volume, diversity, and inconsistency, making it difficult to extract meaningful insights [10]. Additionally, demand volatility in today's dynamic market further complicates process optimization. Fluctuations in demand make accurate forecasting challenging, leading to inefficiencies in inventory management, production planning, and order fulfillment. Inaccurate forecasts often result in suboptimal resource utilization and increased operational costs [11]. Real-time

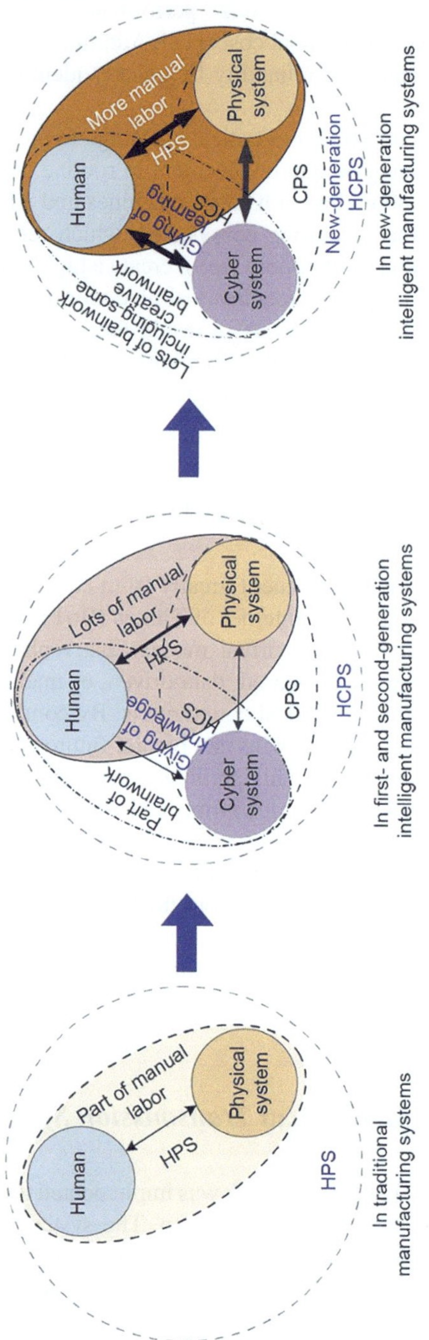

Fig. 1 Evolution of manufacturing systems [9]

decision-making and information sharing are essential to overcome these challenges, enabling manufacturers to optimize performance, reduce waste, improve product quality, and lower operational costs [12]. Additionally, real-time decision-making contributes to safety and reliability by quickly identifying and mitigating potential risks [13].

Finally, the human factor remains a critical challenge in modern manufacturing. Despite automation reducing human involvement in routine tasks, skilled workers are still essential for collaborating with smart machines and adapting to new technologies [14]. Modern manufacturing ecosystems, which rely on the Internet of Things (IoT) and the Internet of Services (IoS), create a highly networked and flexible environment. This complexity requires a skilled workforce capable of understanding and operating these systems. As a result, there is an increasing need for effective communication interfaces that facilitate seamless interaction between humans and complex technological systems [15].

3 Predictive Analysis with Digital Twins

Modern manufacturing processes benefit greatly from continuous data acquisition through sensors and the Industrial Internet of Things (IIoT). This real-time data is then used to create digital twins—virtual models that replicate physical assets. These models integrate simulation-based, data-driven, or mathematical techniques to mirror the behavior of their physical counterparts. By combining physics-based models with data analytics, Digital twins can predict future behaviors and identify potential issues before they arise. This predictive capability allows for proactive maintenance, reducing unexpected downtime and optimizing the useful life of assets [16].

Thanks to the collection of real-time and historical data by digital twins, manufacturers can make accurate predictions about the health and performance of their systems. This predictive analysis facilitates better decision-making, enabling the early detection of potential problems and the timely execution of maintenance plans (Fig. 2).

3.1 Case Study 1: Mechanical Transmission System [17]

In this case study, a digital twin framework was implemented to monitor and predict the health of a mechanical transmission system. The system included an electric motor, gearbox, and sensors to capture vibration, temperature, and acoustic emissions data. This data was processed by monitoring state machines (MSMs) to classify the equipment's health into severity levels, ranging from good to unacceptable, based on ISO-10816 standards. A predictive maintenance supervisor (PMS)

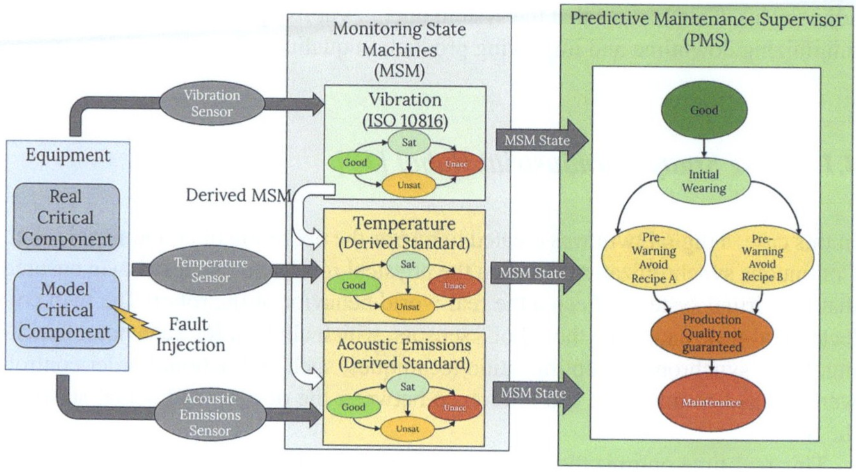

Fig. 2 Overview of the case study approach [17]

oversaw the MSMs, issuing alerts whenever deviations from normal behavior were detected. It used a finite state machine to forecast the future health of the system.

To test the predictive capabilities of the digital twin, fault simulations were conducted on the gear tooth and vibration sensors. Vibration and temperature data were classified into severity levels, with the PMS tracking these levels and transitioning between operational states such as good, initial wear, and avoid recipe A/B. By combining historical data with current sensor readings, the PMS accurately predicted future equipment health, enabling proactive maintenance scheduling, reducing downtimes, and optimizing production quality.

This case study demonstrates how predictive analysis with digital twins ensures continuous monitoring, enhances production efficiency, and reduces operational disruptions.

4 Real-Time Optimization with Digital Twins

Digital twins provide real-time optimization capabilities by continuously integrating real-world data from physical systems into virtual simulations. These simulations mirror the actual performance of manufacturing processes, allowing for immediate adjustments based on real-time data. By combining historical data with live system information, digital twins enable dynamic and automated optimization, which leads to enhanced performance and reduced inefficiencies.

Unlike predictive analysis, which focuses on anticipating future issues, real-time optimization allows processes to be fine-tuned instantaneously. As data flows from physical systems through sensors and the Industrial Internet of Things (IIoT), digital twins use this information to monitor process variables and make on-the-fly adjustments. This ensures that the system operates at its optimal performance level, minimizing downtime and improving production quality [18].

4.1 Case Study 2: Industrial Robot [19]

In this case, a digital twin was created to mirror the operation of an industrial robot. Continuous synchronization between the physical robot and its digital twin ensured that the virtual system reflected the real-world behavior of the robot. The twin collected real-time data from the robot's sensors, which allowed for behavioral matching. Once synchronized, optimization techniques, such as fractional-order control, were applied to minimize the difference between the desired and actual states of the robot.

This real-time optimization adjusted parameters like motor speed and torque in response to live data, continuously improving the robot's performance. The result was reduced overshoot, shorter peak time, and faster settling time, demonstrating how digital twins can optimize robotic systems on the fly.

4.2 Case Study 3: CNC Machining Process [20]

A digital twin-based system was used for a Computer Numerical Control (CNC) machining process, where a live model of the machining operation was created by integrating data from sensors, controllers, and tools. This digital representation was updated in real time using data on spindle speed, tool position, and cutting force. The system used the STEP-NC standard to capture comprehensive process data, which allowed the twin to make dynamic adjustments to the machining process.

By employing a tri-dexel-based simulation algorithm, the digital twin provided real-time feedback, enabling adjustments to be made instantly. This resulted in accurate predictions of tool wear and part quality, optimizing the machining process for efficiency, reducing downtime, and improving the final output quality.

These case studies highlight how real-time optimization through digital twins leads to significant process improvements, reducing delays and ensuring consistent product quality across various industrial applications.

5 Process Insight and Control with Digital Twins

A key focus of Industry 5.0 is human centricity, which emphasizes creating a value-driven industrial revolution that prioritizes human well-being, sustainability, flexibility, and efficiency [21]. In line with this, the concept of human digital twins has emerged, designed to improve the integration between human operators and machines. Digital twins enable seamless human–machine interaction by providing smart interfaces that integrate real-time data from sensors, machines, and simulations. These interfaces empower human operators to make informed, real-time decisions that enhance process control and efficiency.

One of the most significant benefits of digital twins is their ability to improve the human–machine interface. By offering real-time visibility into the status of machinery, potential bottlenecks, and process inefficiencies, digital twins help operators better understand complex manufacturing processes. For instance, in a manufacturing environment, a digital twin can display real-time machine data, suggest optimal configurations, and highlight areas requiring attention. This transparency allows operators to proactively address issues, minimize downtime, and maintain continuous production flow.

Digital twin technology creates a real-time connection between the physical and virtual worlds. In the context of human–computer interaction, the physical world consists of humans, machines, and other entities, while the virtual world is represented by digital data, images, and video streams. By merging these two realms, digital twins enhance the ability of operators to monitor and control processes more efficiently.

Wang et al. [22] introduced two models for capturing and analyzing visual information that are integrated into digital twins. The results are transformed into actionable data, such as skeletal or visual representations, which provide effective means for operators to interpret and respond to complex information (Fig. 3). This real-time analysis enhances the interaction between human operators and machines, ensuring better process insight and control.

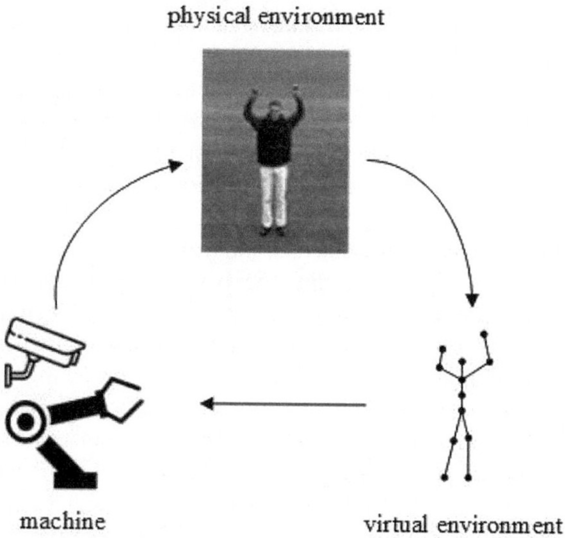

Fig. 3 Human–machine interaction [22]

5.1 Case Study: Smart Factory [23]

In a real-world application, a smart factory demonstrated how digital twins enhance human involvement in process optimization. The digital twin created a real-time, interactive model of the production line, allowing operators to monitor multiple processes simultaneously. Data from various machines, such as milling systems and robots, was collected and presented in a format that was easy for operators to interpret and act upon.

The digital twin enabled individualized manufacturing processes by linking virtual and physical interactions. Real-time data from the factory floor was integrated with simulations to control manufacturing processes more effectively. In this smart factory cell, the digital twin centrally managed operations such as baselining, order processing, transport, milling, and waiting. Sensor data was used to monitor machine movements and conditions, ensuring precise control and optimized performance.

The system also utilized Raspberry Pi connected to sensors and actuators for data collection and synchronization. These devices communicated with the central system to translate digital twin commands into real-time actions, such as controlling a pick-and-place robot and a milling machine. By continuously updating the simulation with position data, the digital twin ensured smooth coordination between virtual simulations and physical operations, optimizing both efficiency and control.

6 Conclusion and Future Scope

Digital twins have emerged as a revolutionary tool in modern manufacturing, offering significant advantages in process understanding and optimization. Traditional methods, such as Lean Six Sigma and statistical models, while effective in certain contexts, face limitations in handling the vast complexity and data overload of today's interconnected manufacturing systems. The sheer volume of real-time data generated by sensors, machines, and business systems makes it increasingly difficult for manufacturers to extract meaningful insights and make timely decisions. Digital twins solve this by creating virtual replicas of physical systems, allowing for continuous monitoring, predictive analysis, and real-time optimization. These capabilities empower manufacturers to adapt quickly to changes, reduce downtime, and maintain operational efficiency in an increasingly dynamic and competitive market.

One of the major benefits of digital twins is their ability to predict future behaviors of systems, enabling proactive maintenance and preventing disruptions. By integrating data-driven models and real-time simulations, digital twins provide manufacturers with actionable insights into system performance and health, improving decision-making and operational planning. The ability to anticipate failures and optimize processes in real time has been demonstrated through various case studies, such as those involving industrial robots and CNC machining. These examples show how digital twins can dynamically adjust to real-time data, continuously improving processes by minimizing tool wear, reducing production errors, and enhancing overall efficiency. As such, digital twins not only streamline operations but also contribute to a more flexible, adaptable manufacturing ecosystem.

Looking forward, the future of digital twins in manufacturing holds immense potential. As advancements in artificial intelligence (AI), machine learning, and data analytics continue, digital twins will evolve into systems capable of more autonomous decision-making and optimization. AI-driven digital twins could potentially self-correct in real time without the need for human intervention, transforming the landscape of industrial automation. Additionally, the focus on human-centric manufacturing environments in Industry 5.0 will drive the development of more advanced human digital twins, further enhancing the interaction between operators and machines. With the ability to optimize resources, reduce waste, and improve energy efficiency, digital twins will also play a critical role in helping industries achieve their sustainability goals. As manufacturing continues to evolve, digital twins will remain at the forefront, driving innovation, efficiency, and sustainability in the factories of the future.

References

1. Albliwi, S. A., Antony, J., Lim, S. A., & halim. (2015). A systematic review of Lean Six Sigma for the manufacturing industry. *Business Process Management Journal, 21*(3), 665–691. https://doi.org/10.1108/BPMJ-03-2014-0019
2. Gomaa, A. (2023). A systematic review of lean six sigma in manufacturing domain. *Engineering Research Journal (Shoubra), 52*(4), 139–148. https://doi.org/10.21608/erjsh.2023.217424.1183
3. Rai, R. S., & Bajpai, V. (2020). *Optimization in manufacturing systems using evolutionary techniques* (pp. 201–229). https://doi.org/10.1007/978-3-030-19638-7_9
4. Azizi, A. (2017). Introducing a novel hybrid artificial intelligence algorithm to optimize network of industrial applications in modern manufacturing. *Complexity, 2017*, 1–18. https://doi.org/10.1155/2017/8728209
5. Stojanovic, N., & Milenovic, D. (2018). Data-driven Digital Twin approach for process optimization: an industry use case. In *2018 IEEE International Conference on Big Data (Big Data)* (pp. 4202–4211). https://doi.org/10.1109/BigData.2018.8622412
6. Qian, F., Zhong, W., & Du, W. (2017). Fundamental theories and key technologies for smart and optimal manufacturing in the process industry. *Engineering, 3*(2), 154–160. https://doi.org/10.1016/J.ENG.2017.02.011
7. Bécue, A., Maia, E., Feeken, L., Borchers, P., & Praça, I. (2020). A new concept of digital twin supporting optimization and resilience of factories of the future. *Applied Sciences, 10*(13), 4482. https://doi.org/10.3390/app10134482
8. Bellalouna, F. (2021). Case study for design optimization using the digital twin approach. *Procedia CIRP, 100*, 595–600. https://doi.org/10.1016/j.procir.2021.05.129
9. Zhou, J., Li, P., Zhou, Y., Wang, B., Zang, J., & Meng, L. (2018). Toward new-generation intelligent manufacturing. *Engineering, 4*(1), 11–20. https://doi.org/10.1016/j.eng.2018.01.002
10. Mourtzis, D. (2021). Towards the 5th Industrial Revolution: a literature review and a framework for process optimization based on big data analytics and semantics. *Journal of Machine Engineering*. https://doi.org/10.36897/jme/141834

11. Abolghasemi, M., Beh, E., Tarr, G., & Gerlach, R. (2020). Demand forecasting in supply chain: the impact of demand volatility in the presence of promotion. *Computers & Industrial Engineering, 142*, 106380. https://doi.org/10.1016/j.cie.2020.106380
12. Jiang, Y., Yin, S., Dong, J., & Kaynak, O. (2021). A review on soft sensors for monitoring, control, and optimization of industrial processes. *IEEE Sensors Journal, 21*(11), 12868–12881. https://doi.org/10.1109/JSEN.2020.3033153
13. Guo, D., Li, M., Lyu, Z., Kang, K., Wu, W., Zhong, R. Y., & Huang, G. Q. (2021). Synchroperation in industry 4.0 manufacturing. *International Journal of Production Economics, 238*, 108171. https://doi.org/10.1016/j.ijpe.2021.108171
14. Adel, A. (2022). Future of industry 5.0 in society: human-centric solutions, challenges and prospective research areas. Journal of Cloud. *Computing, 11*(1), 40. https://doi.org/10.1186/s13677-022-00314-5
15. Dafflon, B., Moalla, N., & Ouzrout, Y. (2021). The challenges, approaches, and used techniques of CPS for manufacturing in Industry 4.0: a literature review. *The International Journal of Advanced Manufacturing Technology, 113*(7–8), 2395–2412. https://doi.org/10.1007/s00170-020-06572-4
16. Melesse, T. Y., Di Pasquale, V., & Riemma, S. (2020). Digital twin models in industrial operations: a systematic literature review. *Procedia Manufacturing, 42*, 267–272. https://doi.org/10.1016/j.promfg.2020.02.084
17. Centomo, S., Dall'Ora, N., & Fummi, F. (2020). The design of a digital-twin for predictive maintenance. In *In: 2020 25th IEEE International Conference on Emerging Technologies and Factory Automation (ETFA)* (pp. 1781–1788). https://doi.org/10.1109/ETFA46521.2020.9212071
18. Tao, F., Zhang, H., Liu, A., & Nee, A. Y. C. (2019). Digital twin in industry: state-of-the-art. *IEEE Transactions on Industrial Informatics, 15*(4), 2405–2415. https://doi.org/10.1109/TII.2018.2873186
19. Liu, X., Gan, H., Luo, Y., Chen, Y., & Gao, L. (2023). Digital-twin-based real-time optimization for a fractional order controller for industrial robots. *Fractal and Fractional, 7*(2), 167. https://doi.org/10.3390/fractalfract7020167
20. Cao, X., Zhao, G., & Xiao, W. (2022). Digital Twin–oriented real-time cutting simulation for intelligent computer numerical control machining. *Proceedings of the Institution of Mechanical Engineers, Part B: Journal of Engineering Manufacture, 236*(1–2), 5–15. https://doi.org/10.1177/0954405420937869
21. Xu, X., Lu, Y., Vogel-Heuser, B., & Wang, L. (2021). Industry 4.0 and Industry 5.0—inception, conception and perception. *Journal of Manufacturing Systems, 61*, 530–535. https://doi.org/10.1016/j.jmsy.2021.10.006
22. Wang, T., Li, J., Deng, Y., Wang, C., Snoussi, H., & Tao, F. (2021). Digital twin for human-machine interaction with convolutional neural network. *International Journal of Computer Integrated Manufacturing, 34*(7–8), 888–897. https://doi.org/10.1080/0951192X.2021.1925966
23. Stark, R., Fresemann, C., & Lindow, K. (2019). Development and operation of digital twins for technical systems and services. *CIRP Annals, 68*(1), 129–132. https://doi.org/10.1016/j.cirp.2019.04.024

Digital Twins for Sustainable Performance in Industry 5.0: Insights from Multiple Cases

Rajan Kumar V K, Minna Saunila, Juhani Ukko, and Tero Rantala

1 Introduction

Industry 5.0 represents the next evolution of industrialization, focusing on the collaboration between humans and machines to enhance productivity while prioritizing sustainability and resilience [1, 2]. While Industry 4.0 focused on full automation and digitization, Industry 5.0 shifts the spotlight back to people by prioritizing systems in which humans and intelligent machines, such as robots, artificial intelligence (AI), and digital twins (DTs), work together to create value across manufacturing, services, and beyond [2, 3]. In a collaborative environment, these technologies enhance flexibility, improve decision-making, and foster innovation by merging human creativity with machine precision [4]. The shift toward human–machine collaboration in Industry 5.0 is also motivated by the need for sustainability [5]. As industries increasingly emphasize environmental goals, Industry 5.0 provides a framework in which DTs, IoT, and AI are integrated into manufacturing processes to reduce waste, optimize energy use, and improve operational resilience. This synergy between technology and a human-centric perspective is seen as key to addressing the growing challenges of sustainable production and resource management [1, 5].

As part of this shift, DTs, which are virtual avatars of physical assets, components, or processes [6], have gained particular attention. In a world increasingly focused on sustainable solutions, DTs can be a transformative technology that drives sustainability in Industry 5.0 systems. Imagine a factory where potential breakdowns can be predicted and averted, cities that seamlessly optimize energy consumption, and healthcare systems that anticipate patient needs before they arise. This vision is not confined to the realm of science fiction but represents the tangible

R. K. V K (✉) · M. Saunila · J. Ukko · T. Rantala
Department of Industrial Engineering and Management, School of Engineering Science, LUT University, Lahti, Finland
e-mail: rajan.vk@lut.fi; minna.saunila@lut.fi; juhani.ukko@lut.fi; tero.rantala@lut.fi

Y. Kumar Singla et al. (eds.), *Digital Manufacturing in Industry 5.0*,
https://doi.org/10.1007/978-3-031-91500-0_35

promise of digital twins, offering the means to monitor, simulate, and optimize industrial processes in real time for enhanced industrial operations [7–9]. DTs facilitate automated real-time process analysis, which accelerates error detection and correction, leading to improved efficiency and cost reductions in manufacturing processes [10–12].

Despite the recognized potential of DTs in fostering sustainable performance in Industry 5.0 [13], there remains a gap in our understanding of their precise contributions and the challenges associated with their implementation. While prior research has explored the potential of DTs in improving industrial processes [9–12] and sustainability efforts separately [14–16], their joint implications in achieving sustainable performance within the context of Industry 5.0 are less understood. Prior research has also not fully addressed the challenges that companies encounter in integrating DTs for sustainability purposes in this new industrial paradigm [17–19]. To address these knowledge gaps, this study seeks to answer the following questions by gaining insights from real-world case studies:

1. How do DTs contribute to sustainable performance in Industry 5.0?
2. What challenges do companies face in integrating DTs for sustainable performance in Industry 5.0?
3. What technological synergies do DTs offer from a human-centric perspective?

To gain a comprehensive understanding of the relationship between DTs and sustainable performance in Industry 5.0, it is essential to review the relevant literature. The following section will explore the key concepts, identify existing research trends, and highlight the knowledge gaps that this study aims to address.

2 Literature Review

2.1 Industry 5.0 and Sustainable Performance

The transition to Industry 5.0 represents a critical shift toward integrating a human-centric perspective with machine intelligence, requiring a reevaluation of sustainable performance to balance economic growth with environmental and social responsibilities [1, 13, 20]. Industry 5.0 aims to bring a human-centric perspective back to manufacturing, emphasizing mass personalization and an environment in which humans and robots collaborate [21]. This paradigm shift will not only enhance productivity but also foster innovation, as organizations seek to create value through sustainable practices that prioritize the well-being of both workers and the planet [22]. Sustainable performance is a key aspect of Industry 5.0, influencing manufacturing and supply chains, incorporating renewable energy sources, and promoting energy-efficient practices [23]. Studies have shown that companies that adopt sustainable practices experience long-term benefits, including improved brand reputation and customer loyalty [24, 25]. The integration of advanced technologies, such as DTs, AI, big data, and IoT in Industry 5.0, is seen as a crucial driver for achieving sustainability goals and advancing toward a smart and resilient manufacturing landscape [21, 23]. However, the increased energy and material usage associated with these technologies may challenge the environmental benefits of sustainable performance in Industry 5.0 [26]. Additionally, many organizations, particularly SMEs, face challenges in implementing Industry 5.0 principles due to legacy systems, inadequate technological readiness, and the substantial investments required in terms of both technology and training [27, 28].

2.2 *Potential of Digital Twins in Industry 5.0 for Sustainable Performance*

DTs have emerged as an influential technology in the shift from Industry 4.0 to Industry 5.0, offering transformative capabilities across various sectors. According to Modoni and Sacco [29], DTs serve as a virtual mirror of the physical process, allowing for real-time monitoring, comparison with analytic models, and the provision of specific parameter variations to optimize processes. Moreover, DTs enable the orchestration, optimization, and simulation of human-centered processes, promoting harmonization and collaboration between humans and machines, which fosters enhanced decision-making processes, particularly in smart manufacturing [12, 29–31]. Modoni and Sacco [29] demonstrate the implementation of a human DT model within the human–CENTRO framework, which showcases the validity and effectiveness of DT-based solutions in enhancing human–machine interactions in manufacturing settings. Their integration allows predictive planning and system analysis within industries to leverage next-generation information technologies, facilitating smarter and more efficient operations while reducing risks and optimizing product life cycles [32, 33]. Saunila et al. [9] highlight that the integration of DTs in industrial settings amplifies their potential by focusing on the "interaction" mechanism as a key driver for both enabling and controlling applications. The interactive nature of DTs fosters collaboration and communication, which can be leveraged to inspire innovative solutions and monitor adherence to sustainability goals, aligning with the proactive and controlling mechanisms mentioned earlier. However, the direct link between these mechanisms and the optimization of sustainability objectives, as well as the direct impact on efficiency and environmental impact, requires further investigation and empirical validation. The environmental benefits realized by organizations through their implementation are not well documented.

2.3 Challenges in Implementing Digital Twins in Industry 5.0

DTs are widely recognized for their ability to enhance sustainable performance by promoting circular economy principles and resource efficiency, as demonstrated by Ghobakhloo et al. [13], and for positively impacting the economic, environmental, and social dimensions of sustainable performance [1]. While the potential of DTs to enhance sustainable performance is widely acknowledged, the successful implementation of DTs in industrial settings that are focused on resilience and a human-centric perspective faces many challenges. For example, real-world case studies in industries such as automotive manufacturing have demonstrated that overcoming issues of synchronization between the physical and digital realms requires both skilled personnel and significant investment in advanced IoT infrastructure [29]. The aerospace sector, for instance, has utilized DTs to optimize predictive maintenance processes, yet these implementations reveal persistent issues in achieving seamless real-time data transfer, particularly in legacy systems [19].

While DTs can drive sustainable performance by providing real-time insights and enabling proactive mechanisms to optimize resource use [14–16], their effectiveness depends heavily on overcoming data integration challenges and maintaining accurate digital replicas over time [17]. Although DTs integrated with technologies such as IoT and XR show promise for improving decision-making and process optimization [7, 9], their actual contribution to sustainable manufacturing remains underexplored and lacks substantial empirical evidence. This limited validation poses a challenge for organizations seeking to justify the significant investments required for such advanced technological integrations, particularly in demonstrating clear, measurable benefits to sustainability outcomes. Marconcini et al. [18] argue that integrating advanced technologies, such as AI, IoT, and big data, remains a complex process. Thus, addressing these challenges requires adaptability to technological changes as well as strategic planning that includes cross-departmental collaboration and comprehensive training programs to build digital literacy within organizations. Therefore, the integration of DTs requires a holistic approach that accounts for both technological and human factors to ensure their effectiveness in driving sustainability in Industry 5.0 [34].

3 Methods

3.1 Research Design

A multiple case study approach was employed as it allows for an in-depth understanding of complex phenomena within real-world contexts, making it well suited for examining the contributions and challenges of DT technology in different organizations [35, 36]. The study applied a qualitative case study design across three companies: Company A, Company B, and Company C, which were selected based

Table 1 Case selection

Case	Role in market	Organizational size	Stage of digital twin implementation	Industry sector	Sustainability focus
Company A	Digital twin solution provider	SME	Mature implementation	Manufacturing, process, and other sectors	Real-time monitoring, production optimization, and resource management
Company B	Digital twin user (potential)	Large	Early exploration	Manufacturing, product design	Exploring the potential for optimized product design and performance monitoring
Company C	Digital twin solution provider	SME	Mature implementation	Heavy equipment, autonomous systems	Real-time simulation for product development, training, and sustainability improvements

on their involvement in different stages of DT implementation. The case selection criteria included the company's role in the market (DT solution provider or user), the size of the organization, the stage of DT implementation (early exploration or mature implementation), and the industry sector in which the company operates. Table 1 summarizes the key characteristics of the selected companies.

3.2 *Data Collection*

Data were collected using a triangulation approach, combining semistructured interviews and document analysis to provide a holistic understanding of the challenges and benefits of using DT technology in each case. Triangulation improves the reliability of the findings by allowing the comparison of data from different sources [37].

3.2.1 Semistructured Interviews

A total of seven interviews were conducted with representatives of the three case companies, including senior executives and experts closely associated with the implementation and development of DT technology. Interviewees were selected to ensure a comprehensive view of the strategic, technical, and operational aspects of

Table 2 Data collection methods

Case	Semistructured interviews	Document analysis
Company A	4 (software architect, CTO, director, business development manager)	2 (solution brochures [1]; white paper report [1])
Company B	3 (R&D director; founder & chief engineer; simulation expert)	5 (all white paper report)
Company C	–	6 (solution brochures [4]; e-book [1]; white paper report [1])

DT adoption. Interviews were conducted with key personnel, including CTOs, software architects, business development managers, and simulation experts. The interviews lasted between 45 and 65 min and followed a semistructured format, allowing for flexibility while focusing on key areas, including DT integration, sustainable performance, and challenges in achieving real-time data synchronization. Table 2 summarizes the data collection techniques used in this study.

3.2.2 Document Analysis

To complement the interviews, extensive analysis was performed on technical documents, white papers, and solution brochures provided by the companies. This method provided secondary data on how each company is applying DTs technology to achieve sustainable performance and what types of challenges they are facing. The documents were particularly useful for understanding how DTs align with technological synergies from a human perspective [38].

3.3 Data Analysis

The data from the semistructured interviews and document analysis were analyzed using reflexive thematic analysis, as described by Braun and Clarke [39]. This method was chosen for its flexibility in identifying, analyzing, and reporting patterns within qualitative data, making it well suited for complex, exploratory research on the implementation of DTs in Industry 5.0. Reflexive thematic analysis emphasizes the researcher's active role in data interpretation while ensuring a structured approach to the identification of themes in the data [40].

The data analysis process began with familiarization, in which interview transcripts and documents were thoroughly reviewed to gain an in-depth understanding of the information [39]. Initial coding followed, with codes derived individually for each research question, particularly focusing on key concepts, including sustainable performance, DT integration challenges, data synchronization, and sustainability monitoring mechanisms. These codes were iteratively refined and grouped into broader overarching themes related to the research questions (see Table 3). Thus, "DT contributions to sustainability," "integration challenges," and "technological

Table 3 Thematic mapping of findings from the case studies

Themes	Subtheme	Codes
Digital twin contributions to sustainable performance	Resource optimization and waste reduction	Energy efficiency focus; resource allocation optimization; production optimization; identifying energy waste through environmental monitoring; optimizing machine usage
	Performance monitoring and improvement	Importance of data collection; tracking machines with sensors; maintenance tracking; OEE calculation; production efficiency tracking; identifying causes of downtime; performance measurement and optimization
	Environmental reporting and compliance	Tracking CO_2 emissions with digital twins; carbon footprint calculation; environmental monitoring; sustainability reporting; expected growth in use of digital twins for sustainability; digital twins as regulatory compliance enablers
	Enhanced working conditions and safety	Monitoring work environment conditions for safety and well-being; Automated adjustments to improve environmental conditions; safety as emerging trend; safety integration and maintenance management; safety notification; context-specific focus on workplace conditions; broad applicability to human sustainability issues
Integration challenges	Data integration and standardization	Lack of standardization; inconsistent calculation methods; data integration challenges; data collection prerequisite; data access request; data synchronization and communication; different data needs based on roles
	Technical barriers	Sensor dependence; limited visibility of customer needs; integration with ERP or MES; variability of devices
	Cost and complexity	Financial challenges in deploying DTs; need for advanced IoT infrastructure; technical feasibility; scaling to smaller business; project based-customization; financial motivation
	Skills and expertise gap	Ethical considerations for tracking; need for much greater integration between design disciplines in IMC; skills and expertise gap; difficulty of company-wide sustainability discussions and tools due to diverse perspectives
Technological synergies and human-centric design	Cross technology collaboration	Digital twin as platform for data sharing; integration with external systems; cloud migration; AR/MR for enhanced visualization; dashboard creation and data sharing
	AI and IoT synergy	AI-assisted maintenance; AI-driven insights; AI for optimization; IoT data gathering; sensor utilization
	Predictive analytics and big data	Big data for performance management; AI for data-driven insights; predictive analytics
	Human-centric design and collaboration	Human–AI collaboration in decision-making; the importance of human understanding; human perspective in digital twin

synergies and a human-centric approach" emerged as critical areas of focus, grounded in both primary data from the interviews and secondary data from the document analysis [40]. A cross-case comparison then highlighted common challenges and successes across the organizations, enhancing the trustworthiness of the findings [41, 42].

4 Findings

The analysis of the data collected from the interviews and documents was conducted using thematic analysis, following the framework outlined by Braun and Clarke [39]. Three major themes emerged from the data: (a) DT contributions to sustainable performance, (b) integration challenges, and (c) technological synergies and a human-centric approach. Each of these themes was further divided into subthemes and supported by specific codes derived from the interviews and document analysis. Table 3 shows how the identified themes provide a structured framework for understanding the experiences of the companies studied and their approaches to integrating DT technology for sustainable performance. The findings are grouped by case study in the table, followed by a cross-case analysis, which is presented following Table 3.

4.1 Case Study 1: Company A

Company A is an SME in Finland that provides DT solutions focusing on real-time monitoring, production optimization, and resource management. As a DT provider, the company has a mature implementation of DTs across various sectors, including the manufacturing and process industries.

4.1.1 Digital Twin Contributions to Sustainable Performance

Company A provides DTs to enhance efficiency, reduce waste, and track sustainability metrics. The CTO explained, "Digital twins allow our customers to monitor energy consumption and resource allocation in real time, which directly contributes to minimizing waste." The business development manager added, "Our DTs can be used to track machine usage and performance through sensors to ensure minimal downtime, which improves overall efficiency and reduces energy consumption." The software architect highlighted, "Simulating various production scenarios helps our customers to identify and eliminate energy-intensive processes before implementation." They also emphasized, "Tracking CO_2 emissions with DTs in real time allows [us] to adjust operations and meet environmental regulations." These quotes demonstrate how DTs drive energy efficiency, optimize resources, and ensure regulatory compliance through real-time monitoring.

4.1.2 Integration Challenges

Despite its successes, Company A faces integration challenges that are primarily related to data standardization. The software architect explained, "One of the major issues we encounter is the lack of standardization across different systems, which complicates data synchronization and makes it difficult to scale our solutions." Financial constraints are another obstacle, particularly for smaller clients. The CTO noted, "Many SMEs find it difficult to justify the up-front investment in digital twin technology despite long-term sustainability benefits." Sensor variability also adds complexity, as the director observed in the following quote: "Different sensor types from various manufacturers don't always communicate well, adding complexity to system integration." The business development manager further highlighted the cost of scaling DT solutions for smaller projects: "Scaling digital twin solutions to smaller clients or projects is costly, especially when advanced IoT infrastructure is required."

4.1.3 Technological Synergies from a Human-Centric Perspective

Company A is actively exploring synergies between DTs and emerging technologies. The CTO explained, "AI-driven insights from our digital twins allow our customers to predict machine failures and adjust operations accordingly, reducing both costs and environmental impacts." This AI-driven predictive maintenance enhances efficiency and sustainability. The software architect added, "Our digital twin platform integrates ERP and IoT, providing real-time insights for decision-making." This aligns with Industry 5.0's human-centric goals. The CTO emphasized, "We're migrating more of our operations to the cloud to improve data accessibility and scalability," which enhances the ability to analyze sensor data for predictive analytics and performance management.

4.2 Case Study 2: Company B

Company B, a large manufacturing firm, is in the early stages of exploring DT technology to optimize product design and enhance sustainability.

4.2.1 Digital Twin Contributions to Sustainable Performance

Company B is in the early stages of exploring the potential of DT technology, primarily through a pilot project aimed at optimizing production layouts. The director of foresight and research explained, "We are currently in the research phase of implementing digital twins, particularly in optimizing our production floor layout and defect recognition during the manufacturing process." Although the company has not fully implemented DTs, the chief engineer sees great potential for their future use in "energy monitoring and reducing resource consumption." He added, "We believe digital twins will help us monitor energy use in high-pressure forming and injection molding, cutting down unnecessary consumption." The director also mentioned, "Our goal is to reduce defect rates early in production by using DT simulations to adjust operations in real-time." While still in the exploratory phase, Company B plans to incorporate DTs into its broader sustainability strategy, including environmental reporting. The director explained, "We expect digital twins to help track CO_2 emissions and optimize production for sustainability."

4.2.2 Integration Challenges

Company B faces several integration challenges in its early exploration of DT technology. The chief engineer commented, "One of the biggest hurdles is ensuring data consistency across our systems. We currently have multiple methods of calculation, which complicates data synchronization." This lack of standardization complicates efforts to fully integrate DTs with existing systems. The R&D director added, "Integrating DTs requires way more collaboration between design and engineering teams than we initially expected." This highlights the need for greater cross-team coordination in DT projects, especially in traditionally siloed departments. She also pointed out, "We lack the specialized expertise needed to handle the complex integration of DTs with our current systems, which makes things even more challenging." Therefore, integration challenges revolve around data standardization, technical barriers, high costs, and a skills gap that complicates cross-functional collaboration.

4.2.3 Technological Synergies from a Human-Centric Perspective

Company B is still in the early exploration phase of DT technology, and its efforts to create technological synergies and human-centric design remain limited but are evolving. The R&D director stated, "We're starting to look into integrating AI and IoT for predictive maintenance and data optimization, but we're not there yet." This reflects their initial steps in exploring AI and IoT for future optimization of their production processes. The company is also considering cloud migration to enhance data accessibility and facilitate cross-technology collaboration. The founder and chief engineer stated, “Moving more data to the cloud will help us integrate external systems and share data across teams.” While full implementation is ongoing, Company B recognizes the potential of DTs for better data sharing and decision-making.

4.3 Case Study 3: Company C

Company C specializes in providing real-time simulation for heavy equipment, with a focus on autonomous systems and product development.

4.3.1 Digital Twin Contributions to Sustainable Performance

Company C provides DT technology that can be used to improve operational efficiency, optimize resource use, and minimize environmental impacts. Their DTs are primarily utilized in industries such as construction, mining, and port handling. According to their solution brochure, “The digital twin enables real-time simulation of machine behavior, helping reduce fuel consumption by up to 50%.” This significant reduction in resource use highlights the role of DTs in improving energy efficiency and optimizing machine usage. Their white paper states, “Our DTs simulate hybrid-powered loaders, reducing physical prototyping and saving both time and material resources,” which shows how virtual prototyping allows them to test hybrid powertrains, reducing the need for physical prototypes and conserving resources. Additionally, their DTs contribute to environmental reporting and maintaining compliance with regulations. Their e-book mentions that their solutions help clients track and report carbon footprints, aligning with global sustainability goals.

4.3.2 Integration Challenges

Despite its success, Company C faces challenges in integrating DTs, especially regarding data standardization and system interoperability. Their solution brochure highlights, “One of the main challenges in implementing digital twins is the lack of standardized data formats, which complicates integration with other enterprise systems.” This lack of data standardization makes it difficult for the company C to fully integrate the potential of their DTs with ERP and MES systems, as they rely on different data collection and processing methods. Additionally, company representatives mentioned the cost and complexity of scaling DT solutions, particularly for smaller businesses, as outlined in their white paper: “Scaling DTs to smaller businesses requires considerable financial investment, making it a costly undertaking for many clients.”

4.3.3 Technological Synergies from a Human-Centric Perspective

Company C is exploring synergies between DTs, AI, and IoT to enhance predictive maintenance and real-time decision-making. Their white paper states, "AI-powered digital twins allow us to predict equipment failures and schedule maintenance before costly breakdowns occur," demonstrating AI's role in optimizing operations. Additionally, their DT solutions integrate IoT sensors to gather real-time data, which helps improve the accuracy of predictive analytics and system performance management (white paper analysis).

The company also emphasizes the human-centric design of their DT systems, particularly in safety-critical industries like mining and construction. Their e-book explains how simulation-based training using DTs provides operators with virtual environments to practice machinery operation in real time, ensuring safer working conditions: "By simulating real-world scenarios, our digital twin systems provide a safe environment for operators to learn and improve" (e-book analysis). This aligns with Industry 5.0's focus on enhancing operator training and workplace safety through simulation-based learning.

4.4 Cross-Case Analysis

A cross-case analysis revealed that all three companies recognize the potential of DTs to enhance sustainable performance in Industry 5.0. They leverage DTs for real-time monitoring, process optimization, and resource management to reduce waste, improve energy efficiency, and enhance environmental reporting. Company A and C, as DT solution providers with mature implementations, recognize the environmental benefits of using DTs and actively explore synergies with AI and IoT. In contrast, Company B, a potential DT user, is in the early exploration phase, focusing on optimizing production layouts and recognizing the future potential of DTs for energy monitoring and sustainability reporting.

All three companies have encountered integration challenges, primarily related to data standardization and system interoperability. The lack of standardized data formats and inconsistent calculation methods hinder seamless data synchronization and integration with existing enterprise systems. Additionally, the cost and complexity associated with scaling DT solutions, especially for smaller businesses, pose a significant barrier.

One significant pattern across all cases is the emphasis on a human-centric perspective. All three companies involve workers or clients in the setup and refinement of the DT system, ensuring that it supports operational goals without complicating workflows. The collaborative approach ensures that DT solutions are aligned with user needs and sustainability objectives, highlighting the importance of employee and client feedback in the success of these technologies.

The implications of this analysis suggest that DTs offer considerable potential for sustainable performance, but widespread adoption depends on overcoming financial, technical, and expertise-related obstacles. Companies that achieve successful integration of DTs are more likely to see improvements in both operational efficiency and environmental performance. However, scalability remains an issue, especially for SMEs.

5 Conclusions

This study highlights the significant potential of DT technology in advancing sustainable performance within Industry 5.0. Through an in-depth analysis of three distinct cases, the study reveals that DTs play a pivotal role in driving sustainable performance by enabling real-time monitoring, process optimization, and resource management. However, challenges such as data standardization, high implementation costs, skills gaps, and the technical complexities of scaling these solutions remain significant barriers, particularly for SMEs.

5.1 Theoretical Implications

From a theoretical perspective, this study contributes to the evolving discourse on sustainable performance in the era of Industry 5.0. The study findings align with prior studies that recognize DTs as critical enablers of operational efficiency, particularly by reducing energy consumption and optimizing resource use [6, 8, 21]. For instance, Company A's ability to simulate production processes to identify energy inefficiencies before implementation confirms the assertions found in the literature that DTs facilitate more intelligent, data-driven decision-making in manufacturing environments [3, 12, 14–16]. These cases not only confirm the existing literature, demonstrating the significant role of DTs in enhancing sustainable performance, but also reveal persistent barriers, particularly for companies in the early stages of adoption [13].

5.2 Practical Implications

Based on the findings of this study, we offer several practical recommendations for practitioners navigating the complexities of DT adoption in Industry 5.0. First, organizations should prioritize data standardization and interoperability to ensure the seamless integration of DTs with existing enterprise systems [17, 19]. Investing in robust IoT infrastructure and skilled personnel is crucial to overcoming data synchronization challenges and achieving real-time data transfer [18, 29]. Additionally, fostering cross-departmental collaboration and providing comprehensive training programs can bridge the skills gap and facilitate effective implementation of DT projects [2]. Furthermore, organizations should actively explore synergies between DTs and other emerging technologies, such as AI, IoT, and XR, to unlock their full potential for predictive maintenance, process optimization, and data-driven decision-making. Finally, maintaining a human-centric perspective throughout the DT ecosystem is essential to ensure that these technologies align with user needs, enhance workplace safety, and contribute to a sustainable and resilient Industry 5.0 ecosystem.

5.3 Limitations and Future Research

While this study provides valuable insights into the potential of DTs to contribute to sustainable performance, the limited number of case studies may not fully encompass the diverse applications of this technology across various industries. This presents an opportunity for future research to expand the scope of the investigation, incorporating a wider range of sectors, company sizes, and levels of digital maturity. Moreover, longitudinal studies could offer a deeper understanding of the long-term implications and evolving dynamics of DT implementation in relation to sustainable performance. Additionally, exploring how emerging technologies, such as AI and extended reality can be further integrated with DTs will provide valuable insights. The cross-case analysis points to the need for further empirical validation of DTs' contributions to long-term sustainability objectives [20]. While the cases demonstrate positive outcomes, particularly in energy efficiency and resource management, more evidence is needed to conclusively link DT technology to broader environmental benefits, as proposed by Industry 5.0 frameworks [1, 21].

References

1. Ivanov, D. (2022). The Industry 5.0 framework: Viability-based integration of the resilience, sustainability, and human-centricity perspectives. *International Journal of Production Research, 61*(5), 1683–1695. https://doi.org/10.1080/00207543.2022.2118892
2. Mourtzis, D., Angelopoulos, J., & Panopoulos, N. (2022). A literature review of the challenges and opportunities of the transition from Industry 4.0 to Society 5.0. *Energies, 15*(17), 6276. https://doi.org/10.3390/en15176276
3. Maddikunta, P., Pham, V., Prabadevi, B., Deepa, N., Dev, K., Gadekallu, T., Ruby, R., & Liyanage, M. (2021). Industry 5.0: A survey on enabling technologies and potential applications. *Journal of Industrial Information Integration, 26*, 100257. https://doi.org/10.1016/j.jii.2021.100257
4. Aheleroff, S., Huang, H., Xu, X., & Zhong, R. Y. (2022). Toward sustainability and resilience with Industry 4.0 and Industry 5.0. *Frontiers in Manufacturing Technology, 2*, 951643. https://doi.org/10.3389/fmtec.2022.951643
5. Alves, J., Lima, T., & Gaspar, P. (2023). Is Industry 5.0 a human-centred approach? A systematic review. *Processes, 11*(1). https://doi.org/10.3390/pr11010193

6. Grieves, M., & Vickers, J. (2017). Digital twin: Mitigating unpredictable, undesirable emergent behavior in complex systems. In J. Kahlen, S. Flumerfelt, & A. Alves (Eds.), *Transdisciplinary perspectives on complex systems*. Springer. https://doi.org/10.1007/978-3-319-38756-7_4
7. Holopainen, M., Saunila, M., Rantala, T., & Ukko, J. (2024). Digital twins' implications for innovation. *Technology Analysis & Strategic Management, 36*(8), 1779–1791. https://doi.org/10.1080/09537325.2022.2115881
8. Kritzinger, W., Karner, M., Traar, G., Henjes, J., & Sihn, W. (2018). Digital Twin in manufacturing: A categorical literature review and classification. *IFAC-PapersOnLine, 51*(11), 1016–1022. https://doi.org/10.1016/j.ifacol.2018.08.474
9. Saunila, M., Holopainen, M., Nasiri, M., Ukko, J., & Rantala, T. (2022). Digital transformation with digital twins—Distinct mechanisms of enabling and controlling uses. *Technology Analysis and Strategic Management, 36*(6), 1323–1336. https://doi.org/10.1080/09537325.2022.2090917
10. He, B., & Bai, K. (2021). Digital twin-based sustainable intelligent manufacturing: A review. *Advances in Manufacturing, 9*, 1–21. https://doi.org/10.1007/s40436-020-00302-5
11. Jones, D., Snider, C., Nassehi, A., Yon, J., & Hicks, B. (2020). Characterising the digital twin: A systematic literature review. *CIRP Journal of Manufacturing Science and Technology, 29*(Part A), 36–52. https://doi.org/10.1016/j.cirpj.2020.02.002
12. Lim, K., Zheng, P., & Chen, C. (2020). A state-of-the-art survey of Digital Twin: Techniques, engineering product lifecycle management and business innovation perspectives. *Journal of Intelligent Manufacturing, 31*, 1313–1337. https://doi.org/10.1007/s10845-019-01512-w
13. Ghobakhloo, M., Iranmanesh, M., Mubarak, M. F., Mubarik, M., Rejeb, A., & Nilashi, M. (2022). Identifying Industry 5.0 contributions to sustainable development: A strategy roadmap for delivering sustainability values. *Sustainable Production and Consumption, 33*, 716–737. https://doi.org/10.1016/j.spc.2022.08.003
14. Chen, C., Zhao, Z., Xiao, J., & Tiong, R. (2021). A conceptual framework for estimating building embodied carbon based on digital twin technology and life cycle assessment. *Sustainability, 13*(24), 13875. https://doi.org/10.3390/su132413875
15. Turan, E., Konuşkan, Y., Yıldırım, N., Tunçalp, D., İnan, M., Yasin, O., Turan, B., & Kerimoğlu, V. (2022). Digital twin modelling for optimizing the material consumption: A case study on sustainability improvement of thermoforming process. *Sustainable Computing Informatics and Systems, 35*(100655), 100655. https://doi.org/10.1016/j.suscom.2022.100655
16. Ünal, A. F., Albayrak, Ö., & Ünal, P. (2023, July). Impact of digital twin technology utilization in manufacturing on sustainability: An industrial case study. In *2023 Portland International Conference on Management of Engineering and Technology (PICMET)* (pp. 1–10). IEEE. https://doi.org/10.23919/PICMET59654.2023.10216885
17. Heithoff, M., Hellwig, A., Michael, J., & Rumpe, B. (2023, May). Digital twins for sustainable software systems. In *2023 IEEE/ACM 7th International Workshop on Green and Sustainable Software (GREENS)* (pp. 19–23). IEEE. https://doi.org/10.1109/GREENS59328.2023.00010
18. Marconcini, M., Esch, T., Botín-Sanabria, D., Mihaita, A., Peimbert-García, R., Ramírez-Moreno, M., Ramírez-Mendoza, R., & Lozoya-Santos, J. (2022). Digital twin technology challenges and applications: A comprehensive review. *Remote Sensing, 14*(6), 1335. https://doi.org/10.3390/rs14061335
19. Segovia, M., & García-Alfaro, J. (2022). Design, modeling and implementation of digital twins. *Sensors (Basel, Switzerland), 22*(14), 5396. https://doi.org/10.3390/s22145396
20. Ghobakhloo, M., Iranmanesh, M., Foroughi, B., Tirkolaee, E. B., Asadi, S., & Amran, A. (2023). Industry 5.0 implications for inclusive sustainable manufacturing: An evidence-knowledge-based strategic roadmap. *Journal of Cleaner Production, 417*, 138023. https://doi.org/10.1016/j.jclepro.2023.138023
21. Akundi, A., Euresti, D., Luna, S., Ankobiah, W., Lopes, A., & Edinbarough, I. (2022). State of Industry 5.0—Analysis and identification of current research trends. *Applied System Innovation, 5*(1), 27. https://doi.org/10.3390/asi5010027

22. Ipsen, C., & Bergmann, S. T. (2021). Designing sustainable organizations: A framework with a joint focus on wellbeing and organizational performance in workplaces. In J. I. Kantola, S. Nazir, & V. Salminen (Eds.), *Advances in human factors, business management and leadership. AHFE 2021* (Lecture notes in networks and systems) (Vol. 267). Springer. https://doi.org/10.1007/978-3-030-80876-1_42
23. Masoomi, B., Sahebi, I. G., Ghobakhloo, M., & Mosayebi, A. (2023). Do Industry 5.0 advantages address the sustainable development challenges of the renewable energy supply chain? *Sustainable Production and Consumption, 43*, 94–112. https://doi.org/10.1016/j.spc.2023.10.018
24. Bai, C., Dallasega, P., Orzes, G., & Sarkis, J. (2020). Industry 4.0 technologies assessment: A sustainability perspective. *International Journal of Production Economics, 229*, 107776. https://doi.org/10.1016/j.ijpe.2020.107776
25. Culot, G., Nassimbeni, G., Orzes, G., & Sartor, M. (2020). Behind the definition of Industry 4.0: Analysis and open questions. *International Journal of Production Economics, 226*, 107617. https://doi.org/10.1016/j.ijpe.2020.107617
26. Rame, R., Purwanto, P., & Sudarno, S. (2024). Industry 5.0 and sustainability: An overview of emerging trends and challenges for a green future. *Innovation and Green Development, 3*(4). https://doi.org/10.1016/j.igd.2024.100173
27. Kamble, S. S., Gunasekaran, A., & Gawankar, S. A. (2020). Achieving sustainable performance in a data-driven agriculture supply chain: A review for research and applications. *International Journal of Production Economics, 219*, 179–194. https://doi.org/10.1016/j.ijpe.2019.05.022
28. Rajesh, R. (2023). Industry 5.0: Analyzing the challenges in implementation using grey influence analysis. *Journal of Enterprise Information Management, 36*(5), 1349–1371. https://doi.org/10.1108/JEIM-03-2023-0121
29. Modoni, G. E., & Sacco, M. (2023). A human digital-twin-based framework driving human centricity toward industry 5.0. *Sensors, 23*(13), 6054–6054. https://doi.org/10.3390/s23136054
30. Warke, V., Kumar, S., Bongale, A., & Kotecha, K. (2021). Sustainable development of smart manufacturing driven by the digital twin framework: A statistical analysis. *Sustainability, 13*(18), 10139. https://doi.org/10.3390/su131810139
31. Zheng, X., Petrali, P., Lu, J., Turrin, C., & Kiritsis, D. (2022). RMPFQ: A quality-oriented knowledge modelling method for manufacturing systems towards cognitive digital twins. *Frontiers in Manufacturing Technology, 2*, 1–15. https://doi.org/10.3389/fmtec.2022.901364
32. Helman, J. (2022). Digital Twin-driven approach towards manufacturing processes support. Journal of Physics: Conference Series, 2198(1), 012007. IOP Publishing. https://doi.org/10.1088/1742-6596/2198/1/012007
33. Newrzella, S. R., Franklin, D. W., & Haider, S. (2022). Methodology for digital twin use cases: Definition, prioritization, and implementation. *IEEE Access, 10*, 75444–75457. https://doi.org/10.1109/access.2022.3191427
34. Barni, A., Fontana, A., Menato, S., Sorlini, M., & Canetta, L. (2018). Exploiting the digital twin in the assessment and optimization of sustainability performances. In *2018 International Conference on Intelligent Systems (IS)* (pp. 706–713). IEEE. https://doi.org/10.1109/IS.2018.8710554
35. Yin, R. K. (2009). *Case study research: Design and methods* (5th ed.). Sage Publications, Inc..
36. Yin, R. K. (2017). *Case study research and applications*. Sage Publications, Inc..
37. Denzin, N. K. (2009). The research act: A theoretical introduction to sociological methods. *Routledge*. https://doi.org/10.4324/9781315134543
38. Bowen, G. A. (2009). Document analysis as a qualitative research method. *Qualitative Research Journal, 9*(2), 27–40. https://doi.org/10.3316/QRJ0902027

39. Braun, V., & Clarke, V. (2006). Using thematic analysis in psychology. *Qualitative Research in Psychology, 3*(2), 77–101. https://doi.org/10.1191/1478088706qp063oa
40. Braun, V., & Clarke, V. (2019). Reflecting on reflexive thematic analysis. *Qualitative Research in Sport, Exercise and Health, 11*(4), 589–597. https://doi.org/10.1080/2159676X.2019.1628806
41. Blair, E. (2015). A reflexive exploration of two qualitative data coding techniques. *Journal of Methods and Measurement in the Social Sciences, 6*(1), 14–29. https://doi.org/10.2458/v6i1.18772
42. Eisenhardt, K. M. (1989). Building theories from case study research. *The Academy of Management Review, 14*(4), 532–550. https://doi.org/10.2307/258557

Digital Twins and Smart Sensors for Process Optimization in Direct Energy Deposition of EN36C Steel for Automotive Components

Damarasingu Apparao, Rapaka Jagadeesh, Mada Sreenivasa Kumar, Donepudi Jagadish, Abiodun Oluwatosin Adeoye, and Samson Abiodun AASA

1 Introduction

In this study, the focus is on leveraging digital twins and smart sensors to optimize the direct energy deposition (DED) process for manufacturing automotive components made from EN36C steel. DED, a cutting-edge additive manufacturing technique, allows for the precise deposition of materials to create intricate and high-performance parts, particularly those used in combustion chambers and fuel injectors. With the growing emphasis on reducing carbon emissions, alternative fuels such as green ethers (e.g., dimethyl ether and ethyltert-butyl ether) have emerged as promising solutions due to their cleaner combustion and lower greenhouse gas emissions. However, the integration of these alternative fuels into existing systems requires components that can withstand the unique thermal and chemical stresses associated with their use.

D. Apparao · R. Jagadeesh (✉)
Department of Mechanical Engineering, Aditya Institute of Technology and Management, Tekkali, Andhra Pradesh, India

M. S. Kumar · D. Jagadish
Department of Mechanical Engineering, Narasaraopeta Engineering College, Narasaraopeta, Andhra Pradesh, India

A. O. Adeoye
Department of Chemistry, Federal University Oye Ekiti, Oye-Ekiti, Ekiti State, Nigeria

S. A. AASA
Department of Automotive Engineering, University of Ibadan, Ibadan, Oyo State, Nigeria

Y. Kumar Singla et al. (eds.), *Digital Manufacturing in Industry 5.0*,
https://doi.org/10.1007/978-3-031-91500-0_36

The research explores how EN36C steel, chosen for its machinability, weldability, and durability under high temperatures and stress, can be optimized using DED for such applications. Digital twins and smart sensors are employed to monitor and adjust key process parameters such as scanning speed, laser power, cooling time, and stacking thickness. These tools enable real-time optimization of the deposition process, improving the mechanical properties of the components, including tensile strength and microhardness.

These studies explore a range of advanced techniques and sustainable practices that align with the goals of smart manufacturing and Industry 5.0. Adeoye et al. (2024) investigate the valorization of lignocellulosic waste, highlighting the potential of bio-oil production through pyrolysis and ultrasonication for biofuel generation [1]. Further research by Adeoye et al. focuses on upgrading crude bio-oil [2, 3] and sustainable energy production from biomass [4], providing insights into thermochemical and biochemical conversions. These approaches underscore the integration of clean energy solutions into digital manufacturing processes. Studies on catalytic pyrolysis and nanomaterials for enhanced biomass conversion offer pathways for optimizing manufacturing efficiency [5–7]. Additionally, research on pyrolysis of plastic waste and rail projects introduces circular economy concepts and environmental sustainability, both essential to the evolution of Industry 5.0. Recent advancements in sustainable fuel production have focused on various bio-oil sources and carbon capture techniques. Studies have examined bio-oil production from thermal cracking in the presence of hydrogen [8], CO_2 adsorption systems in coal beds [9], and the pyrolysis of coffee and eucalyptus sawdust [10–12]. Additional research has explored the use of sulfur to enhance wood tar biopitch as a sustainable binder, the characterization of sewage sludge bio-oil, and the analysis of bio-oil from algae, comparing freshwater and microbial-treated samples [13]. These efforts collectively contribute to the development of renewable energy and environmental sustainability.

In parallel, work on fiber surface functionalization and dye fastness adds to the understanding of material innovation in manufacturing [14–16]. These studies contribute to the body of knowledge driving digital transformation and environmental consciousness in manufacturing [17–23].

In this chapter, the main focus will be the implementation of Direct Energy Deposition (DED) technology in the production of components made from EN36C steel suitable for utilization in green ether combustion technologies. The research hypothesizes in which ways the DED parameters can be improved in order to provide the combustion chambers as well as fuel injectors with better durability, efficiency, and emission performance. This aims at integrating microstructure evaluation and thermal modeling in the creation of the research to study the use of EN36C steel with green ethers in the context of creating new fuel systems for decarbonized road vehicles. This chapter aims at realizing the design and fabrication of components which are fuel efficient, environmentally friendly and possess the resilience to assist in the development of low emission transport alternative.

2 Literature Review

2.1 *Wear Resistance and Mechanical Performance of Composite Materials and Alloys*

Recent studies focus on improving wear resistance and mechanical performance in transport and manufacturing, crucial for Industry 5.0. These works emphasize a variety of advanced engineering techniques that contribute to the development of smart manufacturing and Industry 5.0. Studies like Shah et al. (2024) [24] and Singla et al. (2024) [25] examine the effects of material science advancements on welding and sustainability, key components for industrial automation and optimized production processes. Additionally, research on solar energy systems and tribological studies [26–32] align with the sustainable energy goals of Industry 5.0.

Recent advancements in bio-oil production and carbon capture techniques are driving sustainability in energy production [33]. Studies have explored bio-oil production through thermal cracking in the presence of hydrogen, and the pyrolysis of various biomass sources such as coffee, eucalyptus sawdust, and algae, all contributing to renewable energy solutions [34–39]. Additionally, research on CO_2 adsorption systems and the development of sustainable materials for adsorption, like zeolitic materials and montmorillonite clay, are crucial for addressing environmental challenges [34, 38, 40–44]. These efforts are complemented by investigations into biodiesel production, aviation fuel alternatives, and the separation of valuable chemicals from bio-oils, advancing the potential of biofuels in various industrial applications [45, 46].

The contributions from Sunil et al. [47–53] span critical advancements in transportation systems, materials science, and engineering technology, all of which have applications in smart manufacturing and Industry 5.0. Recent works from 2024 focus on the adoption of electric vehicles [54–70], modeling of intelligent transportation systems [71–82], and vibration mitigation in marine and rail propulsion systems [83–91].

2.2 *Thermal and Corrosion Behavior in Harsh Environments*

These works highlight innovative methods in microbiology, environmental sustainability, and materials science, which have significant implications for smart manufacturing and Industry 5.0 [92–96]. Mohapatra et al. (2015) [97] study microbial processes for biopolymer production, which can be applied in sustainable packaging solutions [17, 94, 98–105] and bio-based materials [106–117], aligning with Industry 5.0's focus on circular economy and smart materials. Research on retrofitting buildings for energy efficiency [118–134] and the use of polyhydroxyalkanoates (PHAs) for biocompatible applications [126, 135–143] demonstrates the integration of biotechnology into industrial processes.

2.3 Modeling and Simulation in Transportation Systems

Sharma et al. (2024) [56, 58, 60, 66, 67, 69, 70, 144–148] and others demonstrated the use of advanced modeling, such as ANFIS and Simulink, for improving stability and comfort in high-speed transport. These simulations help design adaptive systems aligned with Industry 5.0.

2.4 Railways and Automotive Industries: Sustainability and Smart Manufacturing

These studies underscore the need for energy-efficient solutions and advanced control systems, both of which are integral to Industry 5.0 [149, 150]. Previous works from 2023 also explore sustainable materials for transportation [151], intelligent systems for railway infrastructure [152–167], and advancements in modeling and ride behavior [168–176]. Research on crashworthiness [177–185] and the integration of bio-lubricants [186–189] promote the goals of Industry 5.0 by improving safety [190], efficiency, and sustainability [191, 192]. These studies collectively support the evolution of smart manufacturing, contributing to Industry 5.0's emphasis on sustainability, digitalization, and innovation across sectors [193].

2.5 Intelligent Control and Vibration Mitigation in Transportation Systems

The research contributions focus on the dynamic analysis, vibration isolation, and advanced material applications in transportation and vehicle systems, which directly support the objectives of smart manufacturing and Industry 5.0 [194–198]. Recent works (2024) address advanced control systems for enhanced ride quality in passenger rail vehicles using intelligent control suspension systems [155, 199–203], and comprehensive dynamic response analyses of road and rail vehicles [204–213]. Further, studies involving the use of composite materials and hybrid AHP-TOPSIS techniques for sustainable structural applications [62, 67, 70, 147, 214–224] contribute to the development of smart materials, which is central to Industry 5.0.

2.6 New Technologies to Transportation System

Advancements in digital twin technology for improving logistics [225], the role of extremophiles in enhancing sustainable agriculture [226, 227], and bioremediation of contaminated soils [102, 228] highlight the role of digital innovation and environmental consciousness in next-generation manufacturing. These studies offer diverse insights into how microbiological advancements and environmental sustainability are being harnessed for smart and sustainable industrial applications, resonating with the principles of Industry 5.0 [166, 229].

Works from earlier years explore dynamic simulations [69, 145], improving suspension systems, and enhancing material wear resistance, which collectively reflect advancements in digitalization, automation, and sustainable manufacturing practices. These studies highlight a multidisciplinary approach to improving vehicle systems and material performance in line with the goals of Industry 5.0.

3 Green Ethers as Alternative Fuels

Due to their lower carbon emissions, cleaner combustion properties and compatibility with existing infrastructure, green ethers, including dimethyl ether (DME) and ethyltert-butyl ether (ETBE), have now been considered to be the potential substitutes for fossil fuels. These fuels have a lot of potentials to be used in the transport systems since they can aid in the move toward usable low emission and sustainable sources for aviation fuels. The focus of this section is on the characteristics of green ethers, their combustion and emission effects and the difficulties in associating these fuels with existing fuel systems.

3.1 Properties and Advantages of Green Ethers (DME, ETBE)

Dimethyl ether (DME) and ethyltert-butyl ether (ETBE) are alternative fuels, which are distinguishable among the best due to their beneficial and unique physical and chemical properties that make them suitable low-emission energy carriers as shown in Table 1.

In light of its simple structure, dimethyl ether (DME) produces complete combustion resulting in relatively low emissions of particulate matter when used compared to rolling diesel. Such a feature qualifies it for use in regions with high demand for air quality. ETBE, with its higher energy density and octane rating, enhances fuel performance when used as a gasoline additive, supporting efficient combustion

Table 1 Comparison of physical properties of green ethers and conventional fuels

Fuel type	Boiling point (°C)	Density (g/cm^3)	Energy content (MJ/kg)	Autoignition temperature (°C)
DME	−25	0.67	28	235
ETBE	72	0.74	36	380
Diesel	180–360	0.85	42	210
Gasoline	30–215	0.74	44	246

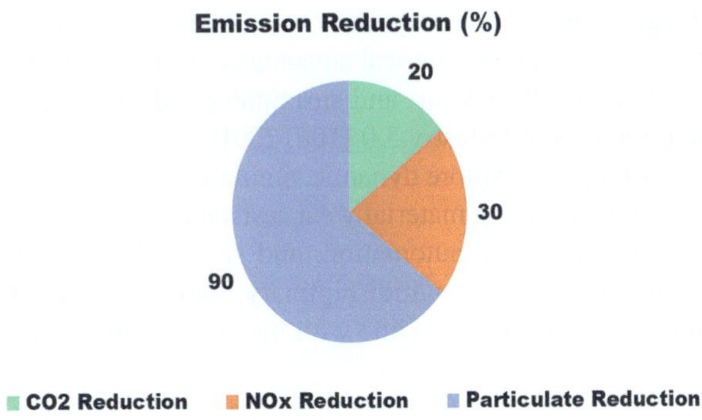

Fig. 1 Emission reduction potential of green ethers compared to diesel

in high-compression engines. Figure 1 illustrates the emission reductions achievable with green ethers.

3.2 Combustion Characteristics and Emissions Benefits

Green ethers provide cleaner combustion than traditional fuels, largely due to their lower carbon content and oxygen-rich composition. DME, for example, produces nearly soot-free combustion, resulting in fewer deposits and extending engine lifespan. ETBE, which has a high-octane rating, reduces engine knocking, enhancing efficiency in gasoline engines. Table 2 shows the emissions comparison between diesel, DME, and ETBE.

Table 2 shows that DME and ETBE produce lower emissions than diesel, with DME significantly reducing particulate emissions. These lower emission rates underscore the potential of green ethers to meet environmental goals without extensive modifications to existing combustion systems. Figure 2 shows the emissions comparison between diesel, DME, and ETBE, and Fig. 3 shows the combustion efficiency of different fuels.

3.3 Challenges in Material Compatibility and System Design

Despite their benefits, green ethers present certain challenges when integrated into traditional fuel systems. DME's reactivity can lead to swelling or degradation of conventional plastic and rubber components, while ETBE's volatility requires special handling and storage considerations given in Table 3.

To address these compatibility issues, fuel system components must be made from materials resistant to ether-induced wear and corrosion. For example, fluorocarbon elastomers are more durable than traditional nitrile rubber seals in DME

Table 2 Emissions comparison between diesel, DME, and ETBE

Fuel type	CO_2 emissions (g/km)	NOx emissions (g/km)	Particulate emissions (g/km)
Diesel	120	0.35	0.02
DME	96	0.25	0.002
ETBE	100	0.28	0.005

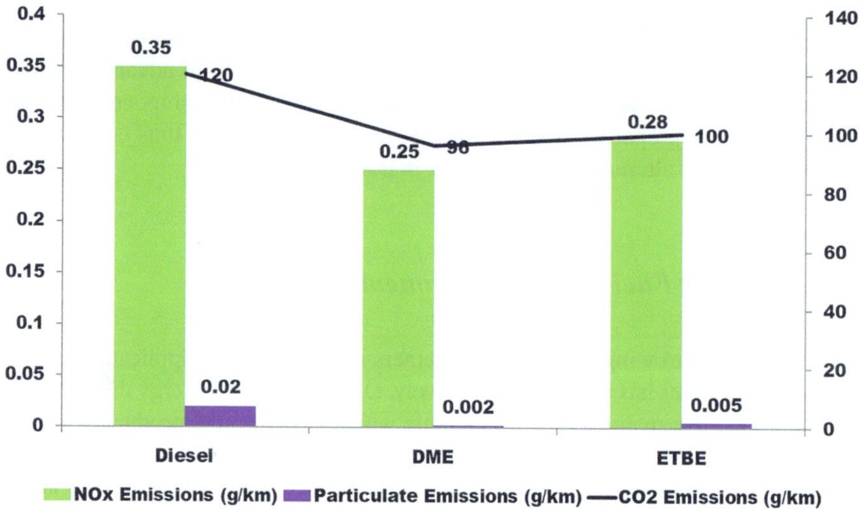

Fig. 2 Emissions comparison between diesel, DME, and ETBE

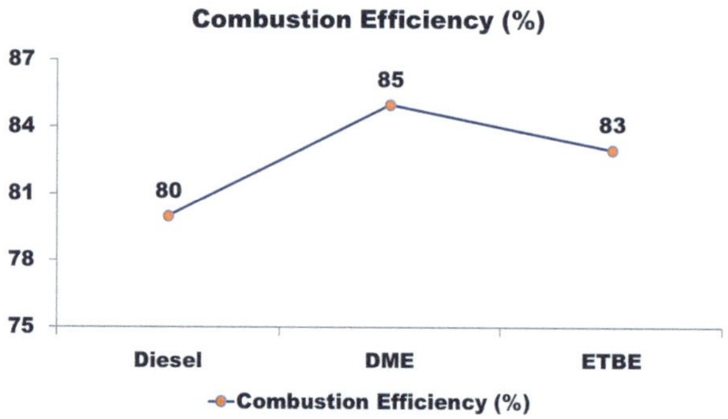

Fig. 3 Combustion efficiency of different fuels

environments, while stainless steel can withstand both DME and ETBE without risk of corrosion. Adjusting materials and designs ensures the longevity and performance of fuel systems running on green ethers. Ethers, or green ethers like DME and ETBE, seem to be good candidates for assisting in the decarbonization of road transport since they are capable of decreasing emissions of CO_2, NOx, and

Table 3 Material compatibility with DME and ETBE

Material	DME compatibility	ETBE compatibility	Suitability for fuel systems
Stainless steel	Excellent	Excellent	Highly suitable
Aluminum alloys	Good	Moderate	Suitable with modifications
Rubber seals (nitrile)	Poor	Fair	Requires replacement
Fluorocarbon elastomers	Good	Good	Suitable

particulates. Yet, their effective use requires modifications to the fuel system design and materials to withstand compatibility issues. With the help of advanced manufacturing techniques such as direct energy deposition (DED), components can be tailor made with properties that are suitable to hot green ethers, thus encouraging the use of cleaner alternative technologies in transportation.

3.4 *DED for Fuel System Components*

Considering the growing usage of green ethers in transport, the application of fuel system components has increased extensively. DED or directed energy deposition is a. type of additive manufacturing technique which is known for producing quite a few unique, custom engineered metal components. In DED, metal powder or wire is applied and metal is built up and melted by laser, electron beam, or plasma arc—which is energy intense. This technique gives fine control over the part's geometry, material and even chemical make-up allowing for the provision of more complex and effective components such as the injector or the combustion chamber in fuel system components manufacturing.

In direct energy deposition, one can add metallic plasma in a multimaterial construct, where metal, in powder or wire form, is deposited in layers—this means that each layer can perform a different function. Such flexibility enables using DED to fabricate and optimize fuel system elements intended for applications with green ethers. It is also possible to have uniform dimensions depending on construction procedures. In addition, thermal and corrosion resistance is critical for green ether applications. Furthermore, DED is ecologically friendly as it does not produce excessive waste. To add further value to DED, DED can be used without any altercation in the geometry of the parts. However, this solution does help in enhancing the combustion systems for DME and ETBE mixtures utilizing these fuels as well as in modifying the rest of the components and systems of the engine within a single operation. This is designed to allow for ease in maintaining high structural durability given in Table 4.

DED not only improves the performance of these components but also enhances their lifespan by optimizing the design for green ether fuels. For instance, the precise control over nozzle geometry in fuel injectors results in improved fuel atomization, maximizing combustion efficiency and supporting better fuel economy. The use of high-strength, corrosion-resistant materials ensure these parts can endure the long-term exposure to green ethers without degradation. The result is a significant

Table 4 DED process parameters and effects on material properties

Parameter	Description	Effect on material
Laser power	Energy used in melting material	Higher power increases depth of fusion
Scanning speed	Rate of movement of the energy source	Affects layer thickness and microhardness
Powder feed rate	Speed of material delivery	Influences material deposition rate
Cooling time	Time allowed for solidification	Impacts residual stress and hardness

Table 5 Design features optimized with DED for green ether fuels

Component	Customized feature	Purpose
Fuel injector	Precision nozzle geometry	Enhances fuel atomization
Combustion chamber	Thermal-resistant inner coatings	Increases heat resilience
Seals and gaskets	Corrosion-resistant materials	Reduces chemical degradation
Valves	Variable thickness and reinforcement	Improves structural durability

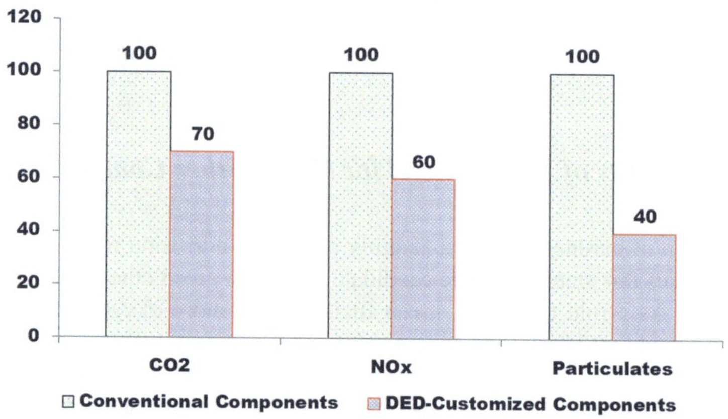

Fig. 4 Comparison of emissions for DED-customized components vs. conventional components

reduction in emissions, aligning with the environmental goals of modern transportation. Table 5 highlights the specific customizations possible with DED to meet the demands of green ether fuels, each tailored to enhance the component's functionality and resilience under high-stress conditions. Figure 4 shows the comparison of emissions for DED-customized components vs. conventional components.

3.5 *Material Selection and Properties of EN36C Steel*

EN36C steel is known for its strength, toughness, and resilience, making it suitable for high-performance applications such as combustion chambers and fuel injectors. The parallel control of combustion fuel types through DED allows for the reliable

Table 6 Key properties of EN36C steel for fuel system applications

Property	Typical range/value	Benefits for fuel system components
Tensile strength	800–1200 MPa	Withstands high pressure and stress without fracture
Hardness	50–60 HRC (carburized)	Resists wear and abrasion, extending operational life
Corrosion resistance	High	Protects against fuel-induced corrosion and oxidation
Thermal stability	Stable at high temps	Maintains structural integrity under intense combustion heat

strengthening of primary engineered connections of imported heads and their laws. As a result, there is no movement of mass during the beating of different heads and unlike in other construction techniques of similar types, there is leeway in using parts given in Table 6.

EN36C steel offers excellent tensile strength, toughness, and hardness, which are critical for withstanding the pressures and stresses present in combustion environments. Its natural corrosion resistance, due to chromium and nickel content, protects the material from reactive chemicals in green ethers, ensuring long-term durability.

3.6 Suitability of EN36C Steel for Green Ether Compatibility

Combination of hardness, thermal stability as well as corrosion resistance of steel EN36C makes this steel a suitable candidate for use in green ether fuels like DME and ETBE. As a rule of thumb, green fuels contain more oxygen and are more aggressive than fossil fuels, and thus, standard materials may suffer from more rapid degradation. EN36C, known for its strength and durability, has found applications in high contact friction areas like fuel injectors. The wear-resistant steel also possesses nickel and chromium, which help form an oxide layer that serves as an effective barrier against green ethers.

In addition to its chemical-resistant properties, EN36C steel's mechanical attributes, tensile strength and toughness for instance, are not lost even when the material is strained at elevated temperatures. These are essential in injector nozzles and combustion chambers made of EN36C steel, as these engine components have fast temperature variations during combustion. Long service life and thermal mechanical loading make the steel retain its shape and reduce thermal distortion due to thermal stability. Its high hardness also provides wear resistance to the material and ensures that the components remain functional and precise, doing excellent atomization during burning of green ethers and making combustion more effective and cleaner. It is SPD6T type EN36C steel which is suitable for alternative fuels making it possible to construct durable and efficient fuel systems.

3.7 Modifying Mechanical and Thermal Properties Through DED

DED shows to enhance properties of the steel EN36C by providing wear resistant coatings through heat iron treatment, coating to withstand high temperatures and to help with cooling the system as well given in Table 7. These have been shown to enhance the materials' capacity to withstand use in fuel systems under extreme high stress and high temperature operating conditions.

Additionally, EN36C steel's elevated hardness prevents wear in situations requiring high accuracy components as green ethers usually require efficient atomization to increase combustion efficiency. Due to the wear resistance of EN36C, the accuracy of the components is retained enabling effective atomization of the green ethers for complete and efficient combustion, thus increasing the performance of the components and their operational life (Fig. 5).

Table 7 DED process modifications and enhanced properties of EN36C steel

DED modification	Enhanced property	Benefit for fuel systems
Wear-resistant coating	Increased surface hardness	Reduces wear in high-friction environments
Thermal coating	Improved heat dissipation	Minimizes thermal distortion and fatigue
Corrosion-resistant layering	Enhanced corrosion resistance	Prolongs lifespan in reactive fuel environments
Microstructural refinement	Higher yield strength and durability	Extends operational life under high stress

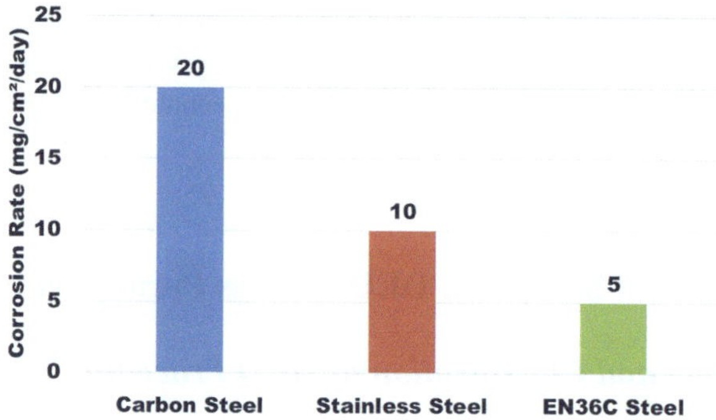

Fig. 5 Corrosion resistance comparison of EN36C steel vs. conventional materials

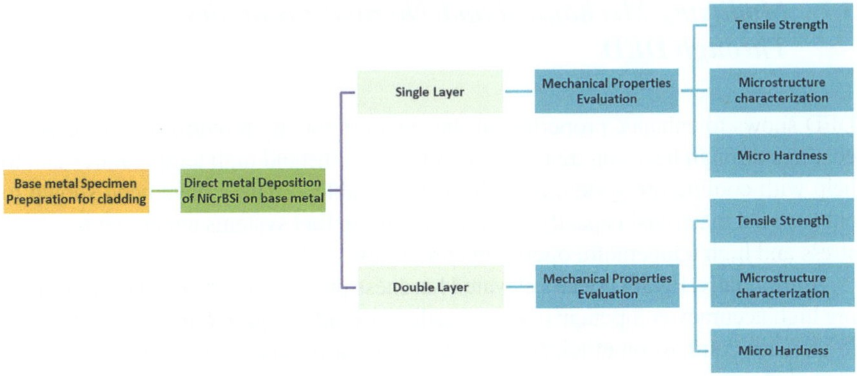

Fig. 6 Work flow methodology

3.8 Methodology

In this study, a systematic investigation of the mechanical properties of EN36C steel with NiCrBSi coating is conducted using DED method. In this particular study, the experimental procedure starts with the cleaning and surface treatment of the EN36C steel substrates in order to form robust bonds with the NiCrBSi cladding. NiCrBSi is then direct energy deposited onto the prepared substrates with controlled parameters: laser power, scanning speed and powder feed rate. Since understanding the effect of layer thickness was one of the aims of this part of the work, NiCrBSi was deposited both as a single layer and double layer. The tensile strength of specimens that can provide information on the material's resistance to pulling forces is carried out. Besides, SEM micrographs yielding grain structure and porosity parameters of microstructure of the materials are carried out. Even microhardness measuring wear resistance should not be overlooked. This methodology shown in Fig. 6 shows how the DED process and layer variations influence the mechanical properties of EN36C steel components.

4 Experimental Setup and DED Process Parameters

4.1 Principle and Classification of DED Procedure

The direct energy deposition (DED) method falls under the general category of "additive manufacturing (AM)"processes as per the ISO/ASTM 52900 standard. It employs a strong thermal energy source to freeze and join materials as they are deposited layer on layer. Frequently used energy sources for DED processes include

Table 8 Classification of DED processes

DED type	Feedstock	Characteristics
Wire feeding	Metal wire	Higher deposition rates; suitable for large parts
Powder feeding	Metal powder	Greater precision; ideal for complex, detailed parts

welding heat flux, lasers, and electron beams. A feedstock such as metal powder or wire is supplied to a concentrated heat zone in the DED process to create a molten pool in which the feed welds to other layers already applied. Once the molten pool is no longer active, the sickle-like structure will be encased in a rigid wire that was deposited during the cooling process. Once the molten mass crystallizes, the austenitic structure is established, and the process is repeated to create a metallic three-dimensional item. It is common for the constructed piece to experience a cooling rate within the range of 10^3–10^5 °C/s.

Direct energy deposition (DED) can be further divided into two broad categories depending on the type of feedstock material employed. The wire feeding DED method is the second category, and it uses steady wire feedstock to apply material onto a surface and has a significant deposition rate; hence, it is ideal for creating massive structures. The meta powder is the feedstock used in the powder feeding DED and since this type of feedstock allows one to control the thickness of each layer, it is easy to use this method to manufacture intricate parts that require well-defined structures; therefore, this method allows for great accuracy. The above methods combine different advantages allowing for the best practices for specific requirements of the component given in Table 8.

4.2 *Laser Cladding*

Cladding by laser on a surface can be considered a moderate direct energy deposition (DED) technology and is routinely used for surface treatments in nearly all industries that demand extreme wear, corrosion, and hardness resistance. Cost effective manufacture can be achieved in case laser cladding technology to deposit a corrosion resistant coating on a cheaper substrate material. The laser generates heat that modifies the physical and compositional characteristics of the material's surface, enabling the creation of a robust bond between the substrate and the clad layer. This method has been successfully employed in the automotive and aerospace industries due to its advantages over traditional alloying and hardening techniques.

Laser cladding allows for several processes, including alloying, impregnation, and hardening of materials. By precisely controlling the energy supply, laser cladding provides durable coatings that extend the operational life of components without significantly altering the core material. This makes it especially useful for components that face extensive wear and corrosion.

4.3 Blown Powder Laser Cladding

The blown powder laser cladding technique, first patented in the early 1980s by Rolls Royce Ltd., produces high-quality, low-dilution cladding layers. In this process, powder is transported into the melt pool by a carrier gas at angles between 38° and 45° to the substrate. As powder particles pass through the laser beam, they heat up and enter the molten pool, where they fully melt and integrate with the substrate. The energy must be calibrated to ensure that only the powder melts, while excessive melting of the substrate is avoided. This method is ideal for creating layers with fine precision and strong adhesion (Figs. 7 and 8).

4.4 Impact of DED Parameters on Metallurgical Properties

Research has shown that direct energy deposition (DED) parameters, such as laser power, scanning speed, and powder feed rate, significantly influence the metallurgical properties of materials, including microhardness, corrosion resistance, porosity, and microstructural evolution. For instance, as given in Table 9, it was found that varying laser power and scanning speed in 17-4 PH stainless steel resulted in optimal microhardness and corrosion resistance, especially when combined with a consistent powder feed rate and hatch overlap. Similarly, Al Morales and Cindy Morales

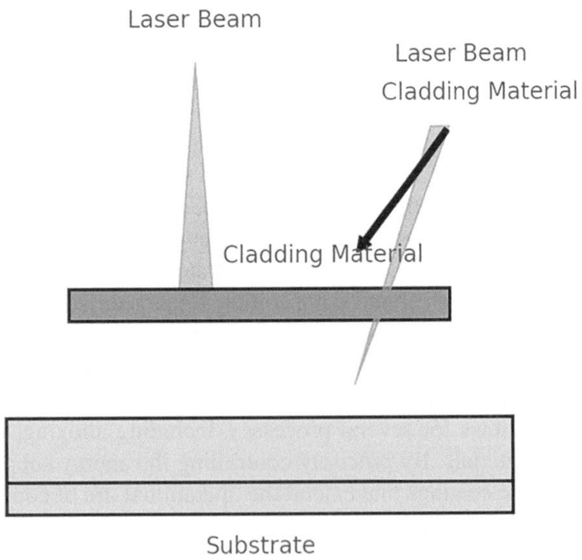

Fig. 7 Blown powder laser cladding

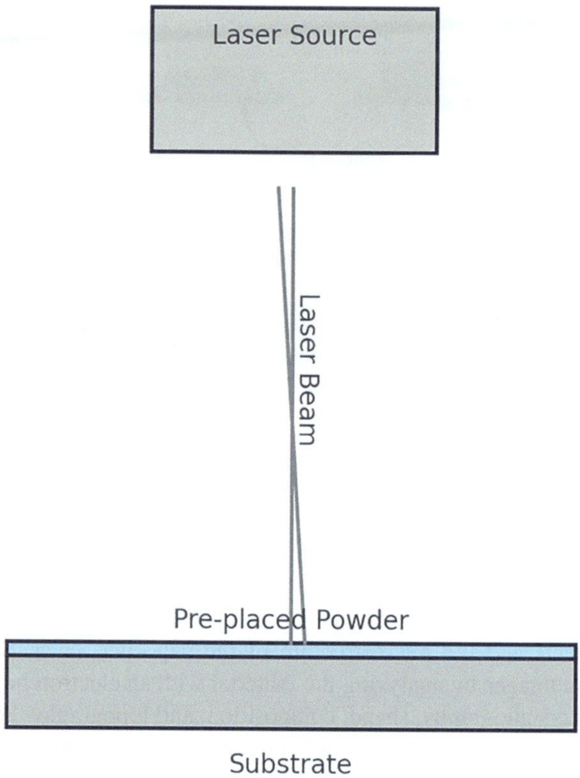

Fig. 8 Preplaced powder laser cladding

Table 9 Examples of DED process parameter impacts

Study	Material	Parameter focus	Findings
[230]	17-4 PH stainless steel	Laser power, scanning speed	High laser power improved microhardness
[231]	17-4 PH on AISI 316	Laser power, feed rate	Ideal laser energy enhanced track geometry
[34]	Ti-Alloy with WC, Stellite-6	Powder feed rate	$40\ l\ min^{-1}$ rate minimized microcracks

demonstrated that laser power, scanning speed, and powder feed rate affected the microstructure and hardness in single-track depositions, which is crucial for applications requiring high surface precision. Authors [179] studied Ti-6Al-3Mo-2Sn-2Zr-2Nb-1.5Cr-0.1Si alloy with added Stellite-6 and WC layers, concluding that a powder feed rate of $40\ l\ min^{-1}$ yielded the best microstructure with minimal microcracks, highlighting the importance of precise parameter control in optimizing material performance.

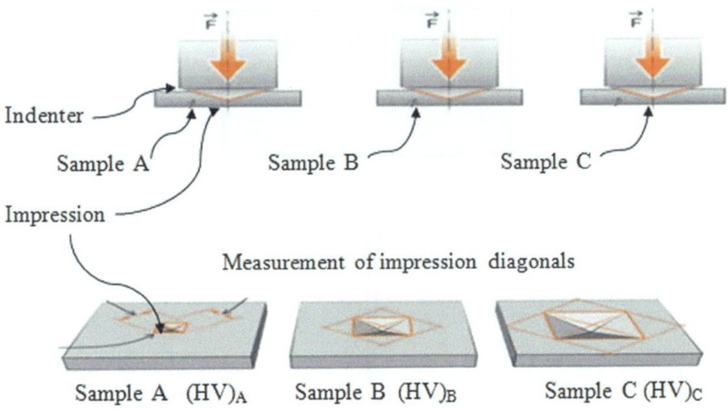

Fig. 9 Vickers hardness testing sample

4.5 SEM and Vickers Hardness Testing for Microstructural Analysis

Scanning electron microscopy (SEM) is employed to examine the distribution of alloying elements and the microstructure of the deposited layer. SEM provides high-resolution images by analyzing the material with an electron beam, capturing details about crystallography, shape, composition, and topography. SEM is invaluable for observing and quantifying surface variations and homogeneity, which are critical for high-performance applications.

Vickers hardness testing (VHT) measures the hardness of materials by indenting the surface with a diamond indenter and calculating hardness from the indentation depth. The VHT is particularly useful for thin sections and small specimens. The hardness unit that the test produces is called the Vickers Pyramid Number (HV) shown in Fig. 9. Although pressure can also be expressed in pascals, the hardness number should not be confused with pressure as pressure also uses pascals.

5 Fabrication

The fabrication process begins with laser cladding using a KUKA robotic arm (Fi 10). This setup utilizes flexible optical fibers to direct a focused laser beam onto the cladding material, creating a high-energy density capable of melting and fusing the material onto the EN36C substrate. This technique enables high-speed, precise welding with minimal thermal distortion, making it suitable for applications in demanding industries such as automotive, aerospace, and medical. Figure 10 illustrates the KUKA robotic arm system used in this experiment. This arm steers the laser beam with high precision, allowing for controlled and repeatable cladding of

Fig. 10 KUKA laser robotic arm

the NiCrBSi material onto the base metal. The flexibility of the robotic arm ensures that even complex geometries can be processed with accuracy. After cladding, two main welding techniques—deep penetration welding and heat conduction welding—are applied depending on the required weld depth and seam quality. SEM analysis then provides high-resolution imaging of the surface, enabling evaluation of material distribution, microstructure, and homogeneity. Finally, Vickers Hardness Testing assesses the hardness of the cladded layers, providing data on wear resistance and material durability essential for high-stress applications.

This combination of advanced welding, microscopic analysis, and hardness testing ensures that the cladded components meet the stringent mechanical and thermal requirements necessary for effective use in green ether-fueled systems given in Table 10.

5.1 Specimen Preparation

The preparation of the specimens is crucial for ensuring effective cladding and accurate characterization of the resulting material properties. For the transmission electron microscopy (TEM) analysis, it is essential to create thin specimens that facilitate electron transport with minimal energy loss given in Table 11. The standard specimen preparation techniques include ion milling, dimpling, and ultrasonic disc cutting.

The **EN36C steel** substrate is known for its excellent machinability and weldability when in an annealed state. However, stress relief treatments are necessary postcooling for optimal welding outcomes. When heat-treated or carburized, welding is not recommended given in Table 12. Acceptable machining methods for EN36C include turning, milling, drilling, and tapping, provided that tool types, feeds, and speeds adhere to the manufacturer's specifications.

Table 10 Fabrication details summary table

Fabrication step	Description	Key points
Laser cladding with preplaced powder	Uses a KUKA robotic arm to perform laser cladding. High-energy laser beams focus on a small area to melt and weld materials with minimal distortion	– Quick processing speed– Low unit costs– Compatibility with various material combinations (e.g., copper-aluminum)
Welding techniques	1. Deep penetration welding: High-power density (~1 kW/cm^2) creates a vapor-filled keyhole, resulting in deep, narrow welds.2. Heat conduction welding: Melts only the surface for smooth, shallow welds	– Deep penetration provides minimal deformation– Heat conduction suitable for thin-walled parts
Scanning electron microscopy (SEM)	High-resolution imaging of microstructures. Uses backscattered and secondary electrons to analyze surface and crystalline structure	– Detailed analysis of surface homogeneity– X-ray spectrometer for elemental composition analysis
Vickers hardness testing (VHT)	Measures hardness by indenting the material surface with a diamond indenter, evaluating indentation depth for microhardness assessment	– Suitable for thin sections and small samples

Table 11 NiCrBSi cladding material properties

Property	Reading
Temperature	1180 °C
Density	7.93 g/cm^3
Thermal conductivity	38.8 W/m K
Young's modulus	122.8 GPa
Poisson's ratio	1.16
Viscosity	138.9 mPa s
Specific heat	554.7 J/kg K

Table 12 Chemical composition of EN36C steel

Element	Content (%)
C	0.12–0.18
Si	0.10–0.34
Mn	0.30–0.58
P	0.06
S	0.06
Cr	0.62–1.12
Mo	0.10–0.26
Ni	3.0–3.76

Table 13 Experimental parameters for cladding

Layer type	Laser power (W)	Cooling time (min)	Scanning speed (mm/s)
Double layer	1600	1	10
Double layer	1600	1	15
Double layer	1600	1	20

5.2 *Study of Microstructure and Hardness*

An orthogonal experimental plan was devised to assess the influence of various parameters on the cladding process. The factors considered included layer thickness, laser power, cooling time, and scanning speed, as summarized in Table 13.

In this study, EN36C steel serves as the substrate, while NiCrBSi acts as the layering material. The substrate dimensions are 20 mm in width, 10 mm in height, and 70 mm in length. For single-layer applications, the dimensions of the NiCrBSi layer are 50 mm in length, 2.5 mm in width, and 1.2 mm in height. For double-layer applications, the total height is maintained at 2.4 mm.

5.3 *Thermal Analysis*

Thermal analysis is conducted using **ANSYS software** to assess the temperature distribution during the direct energy deposition (DED) process. The software enables the visualization of temperature distribution contours over time, providing insights into how heat is transferred within the material. In this analysis, temperature isotherms are generated, indicating areas of equal temperature. Near the heat source, these isotherms are densely packed, reflecting the high temperatures associated with the laser's impact. Conversely, as the distance from the heat source increases, the isotherms become sparser, illustrating a significant tailing phenomenon where the temperature gradually decreases. This thermal analysis is critical for understanding the thermal gradients experienced by the substrate and the deposited layers, allowing for optimization of deposition parameters to improve the mechanical properties and overall performance of the fabricated components.

6 Results

The findings of this survey carried out on direct energy deposition (DED) of EN36C steel for combustion chamber and fuel injectors demonstrate specific phenomena concerning the microstructure, mechanical properties and green ether fuel compatibility. Appraisal was able to address these concerns with a few of the follow: porosity, strength of the bond to the substrate, microhardness and some aspects of heat behavior that are required in assuring the integrity and functionality of the manufactured parts.

6.1 Microstructure Analysis

The microstructure of the deposited layers was characterized using scanning electron microscopy (SEM), which is illustrated in Fig. 11. The SEM images revealed a fine and uniform grain structure within the NiCrBSi cladding, indicative of effective melting and solidification processes during the DED. The presence of fine grains enhances the mechanical properties, particularly tensile strength and toughness, which are crucial for components subjected to high thermal and mechanical stresses.

These microhardness values suggest that the DED process effectively enhances the wear resistance of EN36C steel components, making them suitable for applications in combustion chambers and fuel injectors that operate with green ether fuels given in Table 14.

6.2 Temperature Distribution and Its Impact

The temperature distribution contours generated during the DED process are shown in Fig. 12, indicating the temperature profile at two distinct time intervals (80 s and 150 s). The contours illustrate that: The **red zone**, representing the molten pool, shows the highest temperatures (up to **2185 °C** at 150 s), essential for achieving proper melting and bonding of the cladding material to the substrate. The **blue zone**, further from the heat source, indicates lower temperatures, which are critical in preventing excessive thermal distortion and ensuring dimensional accuracy of the fabricated components. This analysis of temperature distribution highlights the

Fig. 11 Microstructure of samples

Table 14 Microhardness measurements of cladded layers

Distance from surface (mm)	Microhardness (HV)
10	650
15	640
20	635

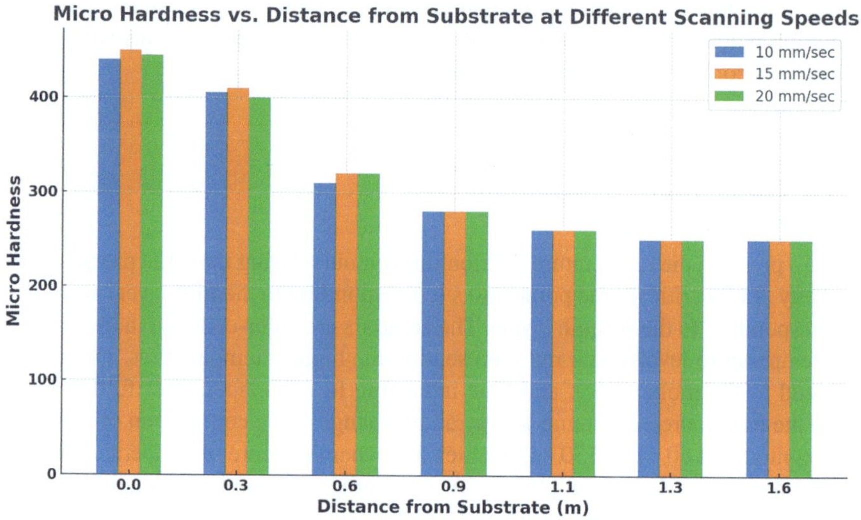

Fig. 12 Microhardness vs. distance from substrate at different scanning speeds

dynamic nature of the DED process and its importance in achieving optimal microstructural characteristics for enhanced mechanical performance.

6.3 *Effect of DED Parameters on Mechanical Properties*

The findings of this study demonstrate that variations in direct energy deposition (DED) parameters, such as laser power, scanning speed, and layer thickness, significantly impact the microstructure and mechanical properties of cladded EN36C steel components. Higher laser power refines grain size, enhancing tensile strength and hardness, while slower scanning speeds allow for controlled solidification, improving mechanical performance. Thicker layers, however, can introduce residual stresses. The enhanced properties of NiCrBSi cladding, including improved corrosion resistance and thermal stability, make EN36C steel components highly compatible with green ether fuels like DME and ETBE, ensuring durability in harsh combustion environments. Additionally, the optimized microstructures achieved through DED improve fuel atomization, leading to more efficient combustion and contributing to significant emission reductions—up to 30% for CO_2 and 20% for

Table 15 Emission reductions using green ether fuels

Emission type	Traditional fuels (g/km)	Green ether fuels (g/km)	Reduction (%)
CO_2	200	140	30
NOx	10	8	20
Particulates	5	2	60

NOx—aligning with environmental sustainability goals. Table 15 summarizes the emission reductions observed when utilizing EN36C steel components with green ether fuels compared to traditional fossil fuels.

6.4 Temperature Distribution

Fig. 12 presents the temperature distribution contours during the DED process, specifically at two critical time points: 80 s (corresponding to the fifth layer) and 150 s (corresponding to the seventh layer). The contours are color-coded to indicate varying temperature levels, with red representing the highest temperatures, which correspond to the molten pool, and blue indicating lower temperatures further away from the heat source. Dynamic Temperature Changes: The comparison of temperature contours at 80 s and 150 s illustrates the dynamic nature of temperature distribution during the deposition process. As time progresses and additional layers are deposited, the thermal profile evolves, leading to significant changes in the temperature field. Peak Temperatures: The peak temperature of the molten pool reaches 1789 °C at 80 s and increases to 2185 °C at 150 s. This rise in temperature can be attributed to the heat accumulation effect, where successive layers of material retain heat from previous layers, thus increasing the overall temperature of the molten pool. This phenomenon is critical for ensuring proper fusion between the substrate and the cladding material, which directly affects the mechanical properties of the final component.

Implications for Material Properties: It is critical to comprehend the distribution of temperature within the DED in order to improve its efficiency. The fine flow of the molten material at the molten pool's elevated temperature allows for better wetting and bonding with the substrate. However, one must also be cautious of introducing excessive temperatures as they will cause warping or other undesired microstructural features giving rise to the need to be properly balanced with scanning speed and thickness of each layer. Heat Distribution: The contours further away from the molten region also have a temperature gradient with the deepest blue zone experiencing the least temperature, which creates even more distance between the two regions. This gradient is of primary importance in avoiding overheating the substrate, which can otherwise cause thermal damage and consequent failure of the part (Fig. 13).

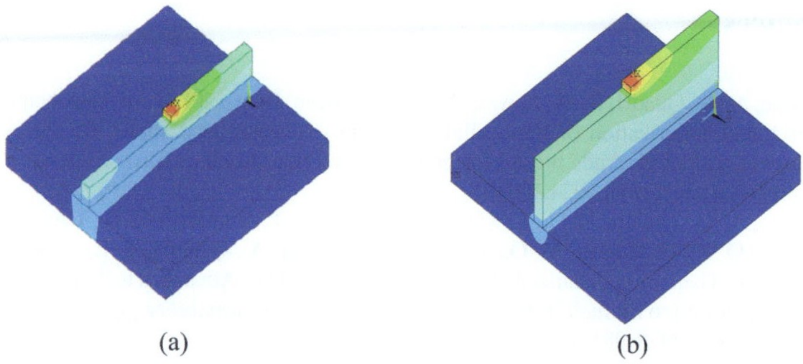

Fig. 13 Temperature distribution contours at (**a**) 80 s for fifth layer and (**b**) 150 s at seventh layer

7 Conclusion

This chapter unlocks a detailed study of the application of direct energy deposition (DED) for enabling the manufacture of EN36C steel cast combustion chamber parts and fuel injectors using a combination of digital twins and smart sensors. The findings clearly demonstrate the potential of DED for the fabrication of high-end automotive components with superior mechanical and thermal properties, which are important for the effective use of green ether fuels. The use of digital twins and smart sensors made it possible to optimize the most critical process parameters in real time, which also resulted in the better microstructural features of the components, tensile strength, and microhardness improvement. Microhardness testing has also shown significant improvements proving the feasibility of the DED process in producing strong and wear resistant components. The thermal analysis carried out using Ansys software was also useful in evaluating the temperature profile felt in the course of the process with a view of enhancing layer adhesion and avoiding defects in the end components. Moreover, the study validates the use of DED EN36C steel components for combustion with green ether fuel such as DME and ETBE. The optimized process produced components that were more resistant to corrosion and degradation when tested with these alternative fuels, thereby resulting in more durable and dependable combustion systems. Incorporation of digital twins and smart sensors in the DED process has demonstrated great potential in enhancing fuel efficiency and lowering emissions in motor vehicle applications. The results showed that such approach could result in decreased CO_2 and NOx emissions of about 30% and 20%, respectively as compared to conventional fuel systems, and this is a major breakthrough in achieving sustainable transport.

References

1. Adeoye, A. O., Quadri, R. O., Lawal, O. S., & Emojevu, E. O. (2024). Physicochemical characterization, valorization of lignocellulosic waste (Kola nut seed shell) via pyrolysis, and ultrasonication of its crude bio-oil for biofuel production. *Cleaner Waste Systems, 7*. https://doi.org/10.1016/j.clwas.2024.100138
2. Adeoye, A. O., Quadri, R. O., Lawal, O. S., Malomo, D., Emojevu, E. O., Omonije, O. O., Odeniyi, O. K., Fadahunsi, M. O., Yelwa, M. J., Aasa, S. A., Onakpa, A. E., Omoniyi, B. P., Mark, I. N., Usman, J., Ismaila, A. M., & Saidu, A. U. (2024). Advanced techniques in upgrading crude bio-oil to biofuel. In *Energy, environment, and sustainability* (pp. 321–353). https://doi.org/10.1007/978-981-97-0515-3_14
3. Adeoye, A. O., Quadri, R. O., & Lawal, O. S. (2022). Assessment of biofuel potential of tenera palm kernel shell via fixed bed pyrolysis and thermal characterization. *Results in Surfaces and Interfaces, 9*. https://doi.org/10.1016/j.rsurfi.2022.100091
4. Adeoye, A. O., Lawal, O. S., Quadri, R. O., Malomo, D., Aliyu, M. T., Dang, G. E., Emojevu, E. O., Maikato, M. J., Yahaya, M. G., Omonije, O. O., Edidem, V. G., Abubakar, Y. K., Offor, O. F., Sochima, E. H., Peter, B. E., & Hikon, B. N. (2023). Sustainable energy via thermochemical and biochemical conversion of biomass wastes for biofuel production. In *Energy, environment, and sustainability* (pp. 245–306). https://doi.org/10.1007/978-981-99-2150-8_11
5. Adeoye, A. O., Yelwa, J. M., Imam, N., Quadri, R. O., Lawal, O. S., Malomo, D., Aasa, S. A., Onakpa, A. E., Dorgbaa, R. K., Omoniyi, B. P., Ahmad, N., & Garba, H. A. (2024). Pyrolysis of plastic wastes towards achieving a circular economy: An advanced chemistry and technical approach. In *Energy, environment, and sustainability* (pp. 215–259). https://doi.org/10.1007/978-981-97-0437-8_11
6. Adeoye, A. O., Quadri, R. O., & Lawal, O. S. (2023). Wet synthesis, characterization of goethite nanoparticles and its application in catalytic pyrolysis of palm kernel shell in TGA. *Results in Surfaces and Interfaces, 11*. https://doi.org/10.1016/j.rsurfi.2023.100118
7. Avesh, M., Imam, N., & Adeoye, A. O. (2024). Assessing environmental impacts and mitigation strategies for the DFCCIL rail projects. In *Energy, environment, and sustainability* (pp. 333–353). https://doi.org/10.1007/978-981-97-0437-8_15
8. Cataluna Veses, R., & Shah, Z. (2015). Bio-oil production by thermal cracking in the presence of hydrogen. *Journal of Fundamentals of Renewable Energy and Applications, 6*. https://doi.org/10.4172/2090-4541.1000194
9. Shah, Z., Cataluña Veses, R., Aguilhera, R. A., & Da Silva, R. (2016). Bio-oil production from pyrolysis of coffee and eucalyptus sawdust in the presence of 5% hydrogen. *International Journal of Engineering Research & Science, 2*(5), 34–42.
10. Shah, Z., Veses, R. C., & da Silva, R. (2016). A comparative study and analysis of two types of bio-oil samples obtained from freshwater algae and microbial treated algae. *Modern Chemistry & Applications, 4*. https://doi.org/10.4172/2329-6798.1000185
11. Shah, Z., Veses, R. C., Brum, J., Klunk, M. A., Rocha, L. A. O., Missaggia, A. B., de Oliveira, A. P., & Caetano, N. R. (2020). Sewage sludge bio-oil development and characterization. *Inventions, 5*, 36. https://doi.org/10.3390/inventions5030036
12. Shah, Z., Saberian, M., Hodgeman, D., Kinloch, I., & Vallés, C. (2023). Effect of sulfur on wood tar biopitch as a sustainable replacement for coal tar pitch binders. *ACS Applied Engineering Materials, 1*, 2567–2576. https://doi.org/10.1021/acsaenm.3c00361
13. Klunk, M. A., Dasgupta, S., Das, M., & Shah, Z. (2018). System of adsorption of CO2 in coalbed. *Southern Brazilian Journal of Chemistry, 26*, 2–9. https://doi.org/10.48141/SBJCHEM.v26.n26.2018.7_2018.pdf

14. Quadri, R. O., Adeoye, A. O., Ayanda, O. S., & Lawal, O. S. (2024). Conversion of palm kernel shell to sustainable energy and the effect of wet synthesized nanoparticles of iron on its thermal degradation kinetics. *Bioresource Technology Reports, 27*. https://doi.org/10.1016/j.biteb.2024.101933
15. Momodu, D.U., Adetuyi, A.O., Adeoye, A.O., Akinlolu, O.S., & Apara, E.O. (2023). Surface functionalization of linen fiber and its dye-fastness assessment using (4-[(E)-(4-amino-2-hydroxyphenyl) diazenyl] benzene-1-sulphonic acid and 4-[(E)-(4-{(Z)-[4-(dimethylamino) phenyl] diazenyl}-2-hydroxyphenyl) diazenyl]benzene-1-sulphonic acid. *Suranaree Journal of Science and Technology, 30*.
16. Adeoye, A. O., Quadri, R. O., & Lawal, O. S. (2023). Biomass waste pyrolysis in the context of Nigeria's oil and gas industry: An overview. *International Journal of Ambient Energy, 44*, 2525–2541. https://doi.org/10.1080/01430750.2023.2256325
17. Mohapatra, S., Mohapatra, S., Han, H., Ariza-Montes, A., & del López-Martín, M. C. (2022). Climate change and vulnerability of agribusiness: Assessment of climate change impact on agricultural productivity. *Frontiers in Psychology, 13*.
18. Din, A. U., Han, H., Ariza-Montes, A., Vega-Muñoz, A., Raposo, A., & Mohapatra, S. (2022). The impact of COVID-19 on the food supply chain and the role of E-commerce for food purchasing. *Sustainability, 14*, 3074.
19. Pandey, G., Bansal, S., & Mohapatra, S. (2025). Market integration of chickpea crop: An evidence of India. *Journal of Agribusiness in Developing and Emerging Economies, 15*(1), 1–18. https://doi.org/10.1108/JADEE-12-2022-0273
20. Mohapatra, S., Mishra, R. K., & Sarangi, K. K. (2019). Economic appraisal of bio ethanol production technologies. *International Journal of Chemical Studies, 7*, 2197–2200.
21. Mohapatra, S., Kishore Mishra, R., Sarangi, K., & K. (2020). Bio-ethanol (2nd generation ethanol): A solution to ever polluting gasoline to climate in India. *International Research Journal of Pure and Applied Chemistry, 20*, 1–15.
22. Mohapatra, S., Mohapatra, U., & Mishra, R. K. (2018). Price forecasting of groundnut in Odisha. *Pharma Innovation Journal, 7*, 111–114.
23. Mohapatra, S., Sneha, N. K., Mounika, A., & Mishra, R. K. (2019). Socio-economic and demographic analysis of saarc nations. *International Journal of Agricultural & Statistical Sciences, 15*.
24. Shah, S., Nautiyal, H., Gugliani, G., Kumar, A., Namboodri, T., & Singla, Y. K. (2024). *Sustainability in smart manufacturing: Trends, scope, and challenges*. CRC Press (Taylor & Francis Group). ISBN 9781032740713. https://doi.org/10.1201/9781003467496
25. Singla, Y. K., Maughan, M. R., Arora, N., & Dwivedi, D. K. (2024). Enhancing the wear resistance of iron-based alloys: A comprehensive review of alloying element effects. *Journal of Manufacturing Processes, 120*, 135–160.
26. Singh, H., & Singla, Y. K. (2021). Wear characteristics of AISI 4140 under nano fly ash based engine oil. In *Proceedings of the 7th International Conference on Theoretical and Applied Nanoscience and Nanotechnology (TANN'23), Paper No. 118, 1–3 June, 2023, Canada*. https://doi.org/10.11159/tann23.118
27. Kumar, A., Ram, M., & Singla, Y. K. (Eds.). (2022). *Advanced materials for biomechanical applications* (1st ed.). CRC Press. https://doi.org/10.1201/9781003286806
28. Singla, Y. K., Maughan, M. R., Arora, N., & Dwivedi, D. K. (2023). Microstructural evolution of worn surface and subsurface of high Mn additive iron-based hardfacings subjected to repetitive impacts and sliding. *Ceramics International, 49*, 21957–21965.
29. Sharma, S. K., Sharma, R. C., Lavania, S., Palli, S., & Avesh, M. (2024). Dynamic vibration mitigation in ship propellers using hybrid fuzzy sliding mode control. *National Academy Science Letters*. https://doi.org/10.1007/s40009-024-01515-z

30. Singh, H., Singh, A. K., Singla, Y. K., Chattopadhyay, K., Saini, A., & Singh, K. (2022). Interpretation of the wear characteristics of AISI 4140 under nano-fly ash based engine lubricant. *Materials Today Proceedings, 50*, 1683–1689.
31. Rooprai, R. S., Singh, H., Singh, T., & Singla, Y. K. (2022). Analysis of the wear properties of through hardened AISI-4140 alloy steel using Taguchi technique. *Materials Today Proceedings, 50*, 661–664.
32. Singh, H., Singh, A. K., Singla, Y. K., Chattopadhyay, K., & Malhotra, R. (2021). Optimization of tribological behavior of AISI 4140 under nano fly ash particulates in engine lubricating oil. *Materials Today Proceedings, 45*, 4619–4623.
33. Shah, Z., Veses, R. C., & da Silva, R. (2016). Using GC-MS to analyze bio-oil produced from pyrolysis of agricultural wastes—Discarded soybean frying oil, coffee and eucalyptus sawdust in the presence of 5% hydrogen and argon. *Journal of Analytical & Bioanalytical Techniques, 7*. https://doi.org/10.4172/2155-9872.1000300
34. Klunk, M. A., Dasgupta, S., Das, M., Wander, P. R., & Shah, Z. (2019). Adsorption study of crystal violet and malachite green dyes in zeolitic material. *Periódico Tchê Química, 16*, 70–81. https://doi.org/10.52571/PTQ.v16.n33.2019.85_Periodico33_pgs_70_81.pdf
35. Yaseen, M., Khattak, M. A. K., Khan, A., Khan, M. S., Ahmad, M., Shah, Z., Khattak, R., & Bibi, S. (2022). Physico-chemical investigations on the catalytic production of biofuel from algal biomass. *Russian Journal of Physical Chemistry A, 96*, S31–S37. https://doi.org/10.1134/S0036024422140308
36. Shah, Z., Cataluña Veses, R., & da Silva, R. (2016). GC-MS and FTIR analysis of bio-oil obtained from freshwater algae (spirogyra) collected from Freshwater. *International Journal of Environmental & Agriculture Research (IJOEAR) Bikaner, 2*(2), 134–141.
37. Shah, Z., Vases, R. C., Ullah, I., Yaseen, M., Aguilhera, R. A., Jamal, S., Amorim, V. D. A., & Da Silva, R. (2016). Analysis of bio-oil obtained from pyrolysis of scenedesmus and spirogyra mixture algal biomass. *Journal of Ongoing Chemical Research, 2*, 1–13.
38. de Oliveira, A. P., Lorenzini, G., Shah, Z., Klunk, M. A., de Carvalho Lima, J. E., Rocha, L. A. O., & Caetano, N. R. (2020). Hierarchical criticality analysis of clean technologies applied to a coal-fired power plant. *International Journal of Design & Nature and Ecodynamics, 15*, 609–619. https://doi.org/10.18280/ijdne.150501
39. Roberto Caetano, N., Lorenzini, G., Rocha Lhamby, A., Shah, Z., Antônio Klunk, M., & Alberto Oliveira Rocha, L. (2020). Modelling safety distance from industrial turbulent non-premixed gaseous jet flames. *Modelling Safety Distance from Industrial Turbulent Non-premixed Gaseous Jet Flames, 5*(1), 41–49.
40. Klunk, M. A., Shah, Z., & Wander, P. R. (2019). Use of montmorillonite clay for adsorption malachite green dye. *Periódico Tchê Química, 16*, 279–286. https://doi.org/10.52571/PTQ.v16.n32.2019.297_Periodico32_pgs_279_286.pdf
41. Klunk, M. A., Shah, Z., Caetano, N. R., Conceição, R. V., Wander, P. R., Dasgupta, S., & Das, M. (2020). CO_2 sequestration by magnesite mineralisation through interaction of Mg-brine and CO_2: Integrated laboratory experiments and computerised geochemical modelling. *International Journal of Environmental Studies, 77*, 492–509. https://doi.org/10.1080/00207233.2019.1675295
42. Yaseen, M., Shah, Z., Veses, R. C., Dias, S. L. P., Lima, E. C., Reis, G. S. D., Vaghetti, J. C. P., Alencar, W. S. D., & Mehmood, K. (2017). Photocatalytic studies of Tio2/Sio2 nanocomposite xerogels. *Journal of Analytical & Bioanalytical Techniques. Los Angeles, 08*(1). https://doi.org/10.4172/2155-9872.1000348
43. Cataluña, R., Shah, Z., Pelisson, L., Caetano, N., da Silva, R., & Azevedo, C. (2017). Biodiesel glycerides from the soybean ethylic route incomplete conversion on the diesel engines combustion process. *Journal of the Brazilian Chemical Society*. https://doi.org/10.21577/0103-5053.20170100

44. Shah, Z., Cataluña Veses, R., Ceschi, M. A., & Silva, R. D. (2017). Separation of phenol from bio-oil produced from pyrolysis of agricultural wastes. *Modern Chemistry & Applications, 05*. https://doi.org/10.4172/2329-6798.1000199
45. Cataluña, R., Shah, Z., Venturi, V., Caetano, N. R., da Silva, B. P., Azevedo, C. M. N., da Silva, R., Suarez, P. A. Z., & Oliveira, L. P. (2018). Production process of di-amyl ether and its use as an additive in the formulation of aviation fuels. *Fuel, 228*, 226–233. https://doi.org/10.1016/j.fuel.2018.04.167
46. Shah, Z., Veses, R. C., Vaghetti, J. C. P., Amorim, V. D. A., & da Silva, R. (2019). Preparation of jet engine range fuel from biomass pyrolysis oil through hydrogenation and its comparison with aviation kerosene. *International Journal of Green Energy, 16*, 350–360. https://doi.org/10.1080/15435075.2019.1566730
47. Sharma, S. K., Sharma, R. C., Choi, Y., & Lee, J. (2023). Modelling and dynamic analysis of adaptive neuro-fuzzy inference system-based intelligent control suspension system for passenger rail vehicles using magnetorheological damper for improving ride index. *Sustainability, 15*, 12529.
48. Hossain, I., Ahmed, Y. E., & Sharma, S. K. (2023). The use of both metallic-insert and TiO2/water nanofluid to improve the performance of photovoltaic modules; experimental study. *Thermal Science and Engineering Progress, 42*, 101944.
49. Wu, Q., Cole, C., Spiryagin, M., Chang, C., Wei, W., Ursulyak, L., Shvets, A., Murtaza, M. A., Mirza, I. M., Zhelieznov, K., Mohammadi, S., Serajian, H., Schick, B., Berg, M., Sharma, R. C., Aboubakr, A., Sharma, S. K., Melzi, S., Di Gialleonardo, E., et al. (2023). Freight train air brake models. *International Journal of Rail Transportation, 11*(1), 1–49. https://doi.org/10.1080/23248378.2021.2006808
50. Sharma, R. C., Bhardawaj, S., Sharma, S. K., & Bahl, S. (2023). Ride behaviour of three-wheel vehicle using semi-active suspension using skyhook control strategy. *International Journal of Vehicle Structures & Systems (IJVSS), 15*.
51. Bhardawaj, S., Sharma, R. C., Sharma, S. K., Gopala Rao, L. V. V., & Vashist, A. (2023). Modeling of biomechanical human body model for seat to head transmissibility analysis. *Noise & Vibration Worldwide, 54*, 62–74.
52. Sharma, S. K., Sharma, R. C., Choi, Y., & Lee, J. (2023). Experimental and mathematical study of flexible–rigid rail vehicle riding comfort and safety. *Applied Sciences, 13*, 5252.
53. Pandipati, S., Palli, S., Dowluru, S., Sharma, R. C., Sharma, N., Sharma, S. K., & Da Fonseca-Soares, D. (2024). Microstructural and strength comparison friction stir processing of aluminium alloys and multi-response optimization by Taguchi GRA-PCA based hybrid methods. *International Journal of Vehicle Structures and Systems, 16*(3). https://doi.org/10.4273/ijvss.16.3.03
54. Adeoye, A. O., Quadri, R. O., & Lawal, O. S. (2022). Fixed-bed pyrolysis and thermal analyses of pressed oil palm fruit fibre as a potential alternative energy. *Fuoye Journal of Innovation Science and Technology, 2*(1), 187–194.
55. Sharma, S. K., Upadhyay, R. K., & Kumar, V. (2024). Introduction to dynamics of transportation ecosystem, modeling, and control. In *Dynamics of transportation ecosystem, modeling, and control* (pp. 3–6). Springer.
56. Sharma, S. K., Sharma, R. C., Mohapatra, S., Fouly, A., Hossain, I., & Lee, J. (2024). Technological advancements in oscillation reduction for propulsion shaft systems. *Journal of Vibration Engineering & Technologies*, 1–19.
57. Upadhyay, R. K., Sharma, S. K., & Kumar, V. (2024). Introduction to intelligent transportation system and advanced technology. In *Intelligent transportation system and advanced technology* (pp. 3–6). Springer.

58. Hossain, I., Ngoupo, A. T., Bhattarai, S., Challagulla, S. P., & Sharma, S. K. (2024). Empirical survey on the effect of straight and wavy fins integrated with wind stream as heat sink of photovoltaic panels. *International Journal of Environmental Science and Technology*, 1–14.
59. Sharma, S. K., Sharma, R. C., & Lee, J. (2025). Propelling precision of longitudinal vibration mitigation in ship propeller shafts through advanced nonlinear intelligent semi-active control leveraging adaptive neuro-fuzzy inference system with linear quadratic regulator. *Journal of Vibration and Control, 31*(7–8), 1472–1484. https://doi.org/10.1177/10775463241244836
60. Sharma, S. K., Kumar, N., Avesh, M., Sharma, R. C., Siddiqui, M. I. H., & Lee, J. (2024). Navigating tranquillity with H∞ controller to mitigate ship propeller shaft vibration. *Journal of Vibration Engineering & Technologies, 12*(7), 7969–7979. https://doi.org/10.1007/s42417-024-01340-0
61. Sharma, S. K., Sharma, R. C., Upadhyay, R. K., & Lee, J. (2025). Enhanced ship engine vibration reduction using magnetorheological dampers and adaptive neuro-fuzzy control system. *Ships and Offshore Structures, 20*(2), 143–159. https://doi.org/10.1080/17445302.2024.2335449
62. Sharma, R. C., Sharma, N., Palli, S., Sharma, S. K., Joshi, S., Rallabandi, S. R., & Sharma, A. (2024). Dynamic behavior of three-wheel vehicle with inerter-based dynamic absorber suspension system. *Noise & Vibration Worldwide, 55*, 253–263.
63. Sharma, R. C., Sharma, S. K., Palli, S., Rallabandi, S. R., & Sharma, N. (2025). Compatibility between stability and curving ability of railway vehicles using conventional and unconventional Bogie frame models. *SAE International Journal of Commercial Vehicles, 18*(3), 02-18-03–0014. https://doi.org/10.4271/02-18-03-0014
64. Sravanthi, A. L., Ratnakumar, P., Reddy, S. N., Eswari, K. B., Pandey, B. B., Manikanta, C., Ramya, K. T., Sonia, E., Mohapatra, S., Gopika, K., Anusha, P. L., & Yadav, P. (2022). Morpho-physiological, quality traits and their association with seed yield in sesame (Sesamum indicum L.) indigenous collection under deficit moisture stress. *Plant Physiology Reports, 27*(1), 132–142. https://doi.org/10.1007/s40502-021-00621-0
65. Sharma, S. K., Sharma, R. C., & Lee, J. (2024). Modelling and simulation of MR damper characterization and uncertainty assessment in marine engine vibration isolation with FLPID control. *International Journal of Modelling and Simulation*, 1–11.
66. Sharma, S. K., Sharma, R. C., Upadhyay, R. K., & Lee, J. (2024). Coupled torsional–transverse vibration reduction in marine propulsion dynamics with novel approach using magnetorheological damping and adaptive control system. *Journal of Vibration Engineering & Technologies, 12*, 6089–6099.
67. Rallabandi, S. R., Srinivas, L., Palli, S., Sharma, R. C., Sharma, N., Sharma, A., & Sharma, S. K. (2024). A contrastive characterization of pure Mg and AZ91D alloy based on the testing of mechanical, corrosion, wear, and erosion properties. *Engineering Research Express, 6*, 015017.
68. Badgujar, P., Sharma, S., Rahul, D., Sharma, S. K., & Upadhyay, R. K. (2024). Feasibility study of Mumbai-Pune hyperloop corridor: Challenges and implementation. *Research in Transportation Business & Management, 52*, 101089.
69. Sharma, S. K., Sharma, R. C., Lavania, S., Palli, S., & Avesh, M. (2024). Industry-driven approach for ANFIS-based intelligent control suspension system with MR damper for enhanced ride quality in passenger rail vehicles for technological investigations. *National Academy Science Letters, 47*, 391–394.
70. Sharma, R. C., Palli, S., Dharmana, L., Sharma, S. K., Sharma, A., Kharola, A., & Sharma, N. (2024). Simulation of transient dynamic response of a locomotive considering its flexibility and rigidity. *Noise & Vibration Worldwide, 55*, 33–47.
71. Song, H., Park, J., Sharma, S. K., & Lee, J. (2022). A comparative study on ultrasonic propagation characteristics and defect detection of metal material additive manufacturing using deep learning algorithm. In *Biennial International conference on future learning aspects of mechanical engineering* (pp. 157–168). Springer.

72. Boris, I., Sharma, S. K., & Lee, J. (2022). A study on the reflection coefficient of higher modes of torsional wave by theoretical and numerical method. In *Biennial international conference on future learning aspects of mechanical engineering* (pp. 145–155). Springer.
73. Vishwakarma, P. N., Mishra, P., Sharma, S. K., & Shukla, A. K. (2022). Formulating and analysing the effect of suspension parameters on critical speeds for various DOFs in rail vehicle. In *Biennial international conference on future learning aspects of mechanical engineering* (pp. 791–802). Springer.
74. Kumar, U., Kasvekar, R., Sharma, S. K., & Upadhyay, R. K. (2022). Wear of wheels and axle in locomotive and measures taken by Indian railway. *Advances in Engine Tribology*, 77–96.
75. Sharma, R. C., Palli, S., Jha, A. K., Bhardawaj, S., & Sharma, S. K. (2022). Vibration and ride comfort analysis of railway vehicle system subjected to deterministic inputs. *Resmilitaris, 12*, 1345–1355.
76. Lora Anusha, P., Ratnakumar, P., Pandey, B. B., Sandhya Rani, P., Umamahesh, V., Reddi Sekhar, M., Chandrika, V., Yadav, P., Mohapatra, S., & Padmaja, D. (2023). Phenotypic characterization of sesame (Sesamum indicum L.) revealed promising genotypes for moisture stress conditions. *Plant Genetic Resources: Characterization and Utilization, 21*(3), 218–228. https://doi.org/10.1017/S1479262123000667
77. Sharma, R. C., & Sharma, S. K. (2022). Ride analysis of road surface-three-wheeled vehicle-human subject interactions subjected to random excitation. *SAE International Journal of Commercial Vehicles, 15*, 299–312.
78. Sharma, S. K., Lee, J., & Jang, H.-L. (2022). Mathematical modeling and simulation of suspended equipment impact on car body modes. *Machines, 10*, 192.
79. Sharma, S. K., Sharma, R. C., Lee, J., & Jang, H.-L. (2022). Numerical and experimental analysis of DVA on the flexible-rigid rail vehicle carbody resonant vibration. *Sensors, 22*, 1922.
80. Vishwakarma, P. N., Mishra, P., & Sharma, S. K. (2022). Characterization of a magnetorheological fluid damper a review. *Materials Today Proceedings, 56*, 2988–2994.
81. Sharma, R. C., Gopala Rao, L. V. V., Srikar, B., Palli, S., Appala Naidu, Y., Sharma, N., & Sharma, S. K. (2025a). Investigation of the dynamic behavior of hybrid composites using finite element and experimental method. *Journal of the Chinese Institute of Engineers, 48*(2), 158–171. https://doi.org/10.1080/02533839.2025.2453139
82. Sharma, S. K., Sharma, R. C., & Lee, J. (2022). In situ and experimental analysis of longitudinal load on carbody fatigue life using nonlinear damage accumulation. *International Journal of Damage Mechanics, 31*, 605–622.
83. Sharma, R. C., Sharma, S., Sharma, N., & Sharma, S. K. (2021). Linear and nonlinear analysis of ride and stability of a three-wheeled vehicle subjected to random and bump inputs using bond graph and simulink methodology. *SAE International Journal of Commercial Vehicles, 15*, 65–79.
84. Lee, J., Han, J., & Sharma, S. K. (2021). Structural analysis on the separated and integrated differential gear case for the weight reduction. In *Advances in engineering design: Select proceedings of FLAME 2020* (pp. 175–181). Springer Singapore.
85. Choi, S., Lee, J., & Sharma, S. K. (2021). A study on the performance evaluation of hydraulic tank injectors. In *Advances in engineering design: Select proceedings of FLAME 2020* (pp. 183–190). Springer Singapore.
86. Sharma, R. C., Palli, S., Sharma, N., & Sharma, S. K. (2021). Ride behaviour of a four-wheel vehicle using h infinity semi-active suspension control under deterministic and random inputs. *International Journal of Vehicle Structures & Systems, 13*, 234–237.
87. Bhardawaj, S., Sharma, R. C., Sharma, S. K., & Sharma, N. (2021). On the planning and construction of railway curved track. *International Journal of Vehicle Structures & Systems, 13*, 151–159.

88. Mohapatra, S., Mohanty, D., Sharma, S., Dikshit, S., Kohli, I., Samantaray, D. P., Kumar, R., & Kathpalia, M. (2021). Biomedical application of polymeric biomaterial: Polyhydroxybutyrate. In *Bioresource utilization and management* (pp. 111–124). Apple Academic Press.
89. Sharma, S. K., Sharma, R. C., & Lee, J. (2021). Effect of rail vehicle–track coupled dynamics on fatigue failure of coil spring in a suspension system. *Applied Sciences, 11*, 2650.
90. Sharma, R. C., Sharma, S., Sharma, S. K., Sharma, N., & Singh, G. (2021). Analysis of bio-dynamic model of seated human subject and optimization of the passenger ride comfort for three-wheel vehicle using random search technique. *Proceedings of the Institution of Mechanical Engineers, Part K: Journal of Multi-body Dynamics., 235*, 106–121.
91. Sharma, S. K., & Lee, J. (2021). Crashworthiness analysis for structural stability and dynamics. *International Journal of Structural Stability and Dynamics, 21*, 2150039.
92. Mohapatra, S., & Mishra, R. K. (2020). Status of institutional credit disbursement of agricultural sector in Odisha, India. *International Journal of Current Microbiology and Applied Sciences, 9*, 2057–2065.
93. Mohapatra, S., Singh, J., & Mishra, S. (2017). Co-integration among major potato markets in Punjab. *Indian Journal of Agricultural Marketing, 31*, 46–47.
94. Mohapatra, S., Priyanka, V., Mohapatra, S., Kohli, I., & Mishra, R. K. (2020). Impact of corona virus covid-19 on the global economy. *International Journal of Agricultural and Statistical Sciences, 16*, 771–778.
95. Mohapatra, S., Singh, J., & Kumar, S. (2018). Cointegration among major cauliflower markets in Punjab. *Indian Journal of Economics and Development, 14*, 330–335.
96. Mohapatra, S., Mishra, R., & Mishra, R. (2021). Impact of COVID-19 on education level of agricultural farming community. *International Journal of Agricultural and Statistical Sciences, 17*, 915–923.
97. Mohapatra, S., Samantaray, D. P., & Samantaray, S. M. (2015). Study on polyhydroxyalkanoates production using rhizospheric soil bacterial isolates of sweet potato. *Indian Journal of Science and Technology, 8*, 57–62. https://doi.org/10.17485/ijst/2015/v8iS7/64027
98. Arora, R., Trivedi, J., Mohapatra, S., & Kumar, P. (2022). Trends in probiotics on human health and industrial application. In *Advances in agricultural and industrial microbiology: Volume 1: Microbial diversity and application in agroindustry* (pp. 237–248). https://doi.org/10.1007/978-981-16-8918-5_13
99. Samal, S., Pati, S., Mohapatra, S., Maity, S., Tanaya, K., Devadarshini, D., & Samantaray, D. (2023). PHAs production by facultative anaerobic bacteria Bacillus cereus FM5 through submerged and solid-state fermentation under anoxic condition. *Antonie van Leeuwenhoek, International Journal of General and Molecular Microbiology, 116*, 521–529. https://doi.org/10.1007/s10482-023-01825-0
100. Biswas, N., Saha, S., Biswas, B. K., Chowdhury, M., Rahaman, A., Junghare, V., Mohapatra, S., Hazra, S., Zangrando, E., Roy Choudhury, R., & Roy Choudhury, C. (2021). The DNA- and protein-binding properties and cytotoxicity of a new copper(II) hydrazone Schiff base complex. *Journal of Coordination Chemistry, 74*, 1482–1504. https://doi.org/10.1080/00958972.2021.1913128
101. Mohapatra, S., Mishra, R., Roy, P., Yadav, K. L., & Satapathi, S. (2017). Systematic investigation and in vitro biocompatibility studies on implantable magnetic nanocomposites for hyperthermia treatment of osteoarthritic knee joints. *Journal of Materials Science, 52*, 9262–9268. https://doi.org/10.1007/s10853-017-1136-0
102. Mishra, B. B., Nayak, S. K., Mohapatra, S., & Samantaray, D. (2021). *Environmental and agricultural microbiology: Advances and applications*. https://doi.org/10.1002/9781119525899
103. Sharma, S. K., Mohapatra, S., Sharma, R. C., Alturjman, S., Altrjman, C., Mostarda, L., & Stephan, T. (2022). Retrofitting existing buildings to improve energy performance. *Sustainability (Switzerland), 14*. https://doi.org/10.3390/su14020666

104. Bandhu, S., Khot, M. B., Sharma, T., Sharma, O. P., Dasgupta, D., Mohapatra, S., Hazra, S., Khatri, O. P., & Ghosh, D. (2018). Single cell oil from oleaginous yeast grown on sugarcane bagasse-derived xylose: An approach toward novel biolubricant for low friction and wear. *ACS Sustainable Chemistry & Engineering, 6*, 275–283. https://doi.org/10.1021/acssuschemeng.7b02425
105. Kohli, I., Joshi, N. C., Mohapatra, S., & Varma, A. (2020). Extremophile—An adaptive strategy for extreme conditions and applications. *Current Genomics, 21*, 96–110. https://doi.org/10.2174/1389202921666200401105908
106. Pati, S., Mohapatra, S., Maity, S., Dash, A., & Samantaray, D. P. (2021). Polyhydroxyalkanoates for sustainable smart packaging of fruits. *Environmental and Agricultural Microbiology: Advances and Applications*, 183–195.
107. Chatterjee, A., Nirwan, S., Mohapatra, S., Sharma, P., Agnihotri, A., & Shrivastava, N. (2022). Biochemical aspects of pathogenic variability in white rust infected Indian mustard. *Mycologia, 114*, 757–768. https://doi.org/10.1080/00275514.2022.2060007
108. Pati, S., Maity, S., Dash, A., Jema, S., Mohapatra, S., Das, S., & Samantaray, D. P. (2020). Biocompatible PHB production from bacillus species under submerged and solid-state fermentation and extraction through different downstream processing. *Current Microbiology, 77*, 1203–1209. https://doi.org/10.1007/s00284-020-01922-7
109. Agrahari, R., Sarraf, G., Joshi, N. C., Mohapatra, S., & Var, A. (2021). Insight of biopolymers and applications of polyhydroxyalkanoates. In *Microbial polymers: Applications and ecological perspectives* (pp. 177–191). https://doi.org/10.1007/978-981-16-0045-6_8
110. Kathpalia, M., Kumar, P., & Mohapatra, S. (2022). Homocysteine metabolism as a biomarker for cancer. In *Homocysteine metabolism in health and disease* (pp. 159–171). https://doi.org/10.1007/978-981-16-6867-8_8
111. Bharatiy, S. K., Hazra, M., Paul, M., Mohapatra, S., Samantaray, D., Dubey, R. C., Sanyal, S., Datta, S., & Hazra, S. (2016). In silico designing of an industrially sustainable carbonic anhydrase using molecular dynamics simulation. *ACS Omega, 1*, 1081–1103. https://doi.org/10.1021/acsomega.6b00041
112. Sarkar, M. S., Segu, H., Bhaskar, J. V., Jakher, R., Mohapatra, S., Shalini, K., Shivaji, S., & Reddy, P. A. (2018). Ecological preferences of large carnivores in remote, high-altitude protected areas: Insights from Buxa Tiger Reserve, India. *ORYX, 52*, 66–77. https://doi.org/10.1017/S0030605317000060
113. Pati, S., Dash, A., Maity, S., Pattnaik, S., Mohapatra, S., & Samantaray, D. P. (2021). Nano biofertilizer for sustainable agriculture. In *Agriculturally important microorganisms: Mechanisms and applications for sustainable agriculture* (pp. 201–208). https://doi.org/10.1201/9781003245841-10
114. Mohanty, D., Misra, S., Mohapatra, S., & Sahu, P. S. (2018). Prebiotics and synbiotics: Recent concepts in nutrition. *Food Bioscience, 26*, 152–160. https://doi.org/10.1016/j.fbio.2018.10.008
115. Kumar, R., Kumari, R., Prasad, C., Deep, A., Singh, N., Namtak, S., Tiwari, V., Merugu, R., Mohapatra, S., & Singh, R. (2021). Influence of water quality on the diversity of macroinvertebrates in the Mandakini River in India. *Water Science and Technology: Water Supply, 21*, 1843–1860. https://doi.org/10.2166/WS.2021.020
116. Mohanty, D., Jena, R., Choudhury, P. K., Pattnaik, R., Mohapatra, S., & Saini, M. R. (2016). Milk derived antimicrobial bioactive peptides: A review. *International Journal of Food Properties, 19*, 837–846. https://doi.org/10.1080/10942912.2015.1048356
117. Mohapatra, S., Rath, S. N., Pradhan, S. K., Samantaray, D. P., & Rath, C. C. (2016). Secondary structural models (16S rRNA) of polyhydroxyalkanoates producing bacillus species isolated from different rhizospheric soil: Phylogenetics and chemical analysis. *International Journal Bioautomation, 20*, 329–338.

118. Mohapatra, S., Sarkar, B., Samantaray, D. P., Daware, A., Maity, S., Pattnaik, S., & Bhattacharjee, S. (2017). Bioconversion of fish solid waste into PHB using Bacillus subtilis based submerged fermentation process. *Environmental Technology (United Kingdom), 38*, 3201–3208. https://doi.org/10.1080/09593330.2017.1291759
119. Misra, S., Mohanty, D., & Mohapatra, S. (2021). Food-based probiotics: Functional dietary ingredients. In *Probiotics: Advanced food and health applications* (pp. 257–275). https://doi.org/10.1016/B978-0-323-85170-1.00006-3
120. Kumar, R., Kumari, R., Prasad, C., Tiwari, V., Singh, N., Mohapatra, S., Merugu, R., Namtak, S., & Deep, A. (2020). Phytoplankton diversity in relation to physicochemical attributes and water quality of Mandakini River, Garhwal Himalaya. *Environmental Monitoring and Assessment, 192*. https://doi.org/10.1007/s10661-020-08768-3
121. Mohapatra, S., Vishwakarma, K., Joshi, N. C., Maity, S., Kumar, R., Ramchander, M., Pattnaik, S., & Samantaray, D. P. (2021). A review on PHAs: The future biopolymer. *Environmental and Agricultural Microbiology: Advances and Applications*, 83–117.
122. Sahoo, N., Pattnaik, S., Pattnaik, M., & Mohapatra, S. (2021). Methane emission and strategies for mitigation in livestock. *Environmental and Agricultural Microbiology: Advances and Applications.*, 257–274.
123. Mohanty, D. P., Mohapatra, S., Misra, S., & Sahu, P. S. (2016). Milk derived bioactive peptides and their impact on human health—A review. *Saudi Journal of Biological Sciences, 23*, 577–583. https://doi.org/10.1016/j.sjbs.2015.06.005
124. Kohli, I., Mohapatra, S., Kumar, P., Goel, A., Varma, A., & Joshi, N. C. (2022). Role of bacterial endophytes in the promotion of plant growth. In *Bacterial endophytes for sustainable agriculture and environmental management* (pp. 235–253). https://doi.org/10.1007/978-981-16-4497-9_12
125. Samantaray, D., Mohapatra, S., & Mishra, B. B. (2014). Microbial bioremediation of industrial effluents. *Microbial Biodegradation and Bioremediation*, 326–339. https://doi.org/10.1016/B978-0-12-800021-2.00014-5
126. Pattnaik, S., Dash, D., Mohapatra, S., Pattnaik, M., Marandi, A. K., Das, S., & Samantaray, D. P. (2020). Corrigendum to "Improvement of rice plant productivity by native Cr(VI) reducing and plant growth promoting soil bacteria Enterobacter cloacae". *Chemosphere, 240*, 124895. https://doi.org/10.1016/j.chemosphere.2019.124895. Chemosphere. 238, (2020). doi: 10.1016/j.chemosphere.2019.125011.
127. Mohapatra, S., Mohanta, P. R., Sarkar, B., Daware, A., Kumar, C., & Samantaray, D. P. (2017). Production of Polyhydroxyalkanoates (PHAs) by Bacillus Strain Isolated from Waste Water and Its Biochemical Characterization. *Proceedings of the National Academy of Sciences India Section B: Biological Sciences., 87*, 459–466. https://doi.org/10.1007/s40011-015-0626-6
128. Tomer, P., Shroti, G. K., Mohapatra, S., Ghosh, D., Jaiswal, S., Garg, D., Lahiri, D., Kumar, N., Khatri, O. P., & Hazra, S. (2023). Fabrication of HA nano-crystal reinforced PHAs based composites for orthopedic fracture-fixing accessories using sustainable resources. *Environment, Development and Sustainability*. https://doi.org/10.1007/s10668-023-03912-8
129. Bandekar, D., Mohapatra, S., Hazra, M., Hazra, S., & Biswas, S. (2021). N-terminal truncation of VC0395_0300 protein from Vibrio cholerae does not lead to loss of diguanylate cyclase activity. *Biophysical Chemistry, 268*. https://doi.org/10.1016/j.bpc.2020.106493
130. Vishwakarma, K., Kumar, N., Shandilya, C., Mohapatra, S., Bhayana, S., & Varma, A. (2020). Revisiting plant–microbe interactions and microbial consortia application for enhancing sustainable agriculture: A review. *Frontiers in Microbiology, 11*. https://doi.org/10.3389/fmicb.2020.560406
131. Mohapatra, S., Maity, S., Dash, H. R., Das, S., Pattnaik, S., Rath, C. C., & Samantaray, D. (2017). Bacillus and biopolymer: Prospects and challenges. *Biochemistry and Biophysics Reports, 12*, 206–213. https://doi.org/10.1016/j.bbrep.2017.10.001

132. Kumar, R., Merugu, R., Mohapatra, S., & Sharma, S. (2022). Extremophiles: Life of microorganisms in extreme environments. In *Extremophiles: General and plant biomass based biorefinery* (pp. 43–66). https://doi.org/10.1201/9781003014188-3
133. Pradhan, A., Pahari, A., Mohapatra, S., & Mishra, B. B. (2017). Phosphate-solubilizing microorganisms in sustainable agriculture: Genetic mechanism and application. In *Microorganisms for sustainability* (pp. 81–97). https://doi.org/10.1007/978-981-10-7380-9_5
134. Devadarshini, D., Mohapatra, S., Pati, S., Maity, S., Rath, C. C., Jena, P. K., & Samantaray, D. (2023). Evaluation of PHAs production by mixed bacterial culture under submerged fermentation. *Biologia (Bratisl), 78*, 895–902. https://doi.org/10.1007/s11756-022-01302-5
135. Agarwal, K. M., Mohapatra, S., Sharma, P., Sharma, S., Bhatia, D., & Mishra, A. (2020). Study and overview of the novel corona virus disease (COVID-19). *Sensors International, 1*. https://doi.org/10.1016/j.sintl.2020.100037
136. Tanaya, K., Mohapatra, S., Maity, S., & Samantaray, D. P. (2023). Fluorescent-pigmented microbial polymer nanoparticles: An approach for replacement of bioimaging agent in the drug delivery system. *Applied Biochemistry and Microbiology, 59*, 378–383. https://doi.org/10.1134/S0003683823030171
137. Pattnaik, S., Dash, D., Mohapatra, S., Pati, S., Devadarshini, D., Samal, S., Pattnaik, M., Maity, S., Mishra, S. K., & Samantaray, D. (2024). Reclamation of chromium-contaminated soil by native Cr(VI)-reducing and PHA-accumulating Bacillus aryabhattai CTSI-07. *International Microbiology, 27*, 731–742. https://doi.org/10.1007/s10123-023-00421-6
138. Mohapatra, S., Mishra, Y., & Dash, L. (2020). Covid-19: Analysing the legal nuances of the lockdown order. *Indian Journal of Forensic Medicine and Toxicology, 14*, 612–616. https://doi.org/10.37506/ijfmt.v14i4.11553
139. Pati, S., Mohanty, M. K., Mohapatra, S., & Samantaray, D. (2022). Bioethanol production by Enterobacter hormaechei through carbon monoxide-rich syngas fermentation. *International Journal of Energy Research, 46*, 20096–20106. https://doi.org/10.1002/er.8678
140. Maity, S., Das, S., Mohapatra, S., Tripathi, A. D., Akthar, J., Pati, S., Pattnaik, S., & Samantaray, D. P. (2020). Growth associated polyhydroxybutyrate production by the novel Zobellellae tiwanensis strain DD5 from banana peels under submerged fermentation. *International Journal of Biological Macromolecules, 153*, 461–469. https://doi.org/10.1016/j.ijbiomac.2020.03.004
141. Takkar, S., Shandilya, C., Agrahari, R., Chaurasia, A., Vishwakarma, K., Mohapatra, S., Varma, A., & Mishra, A. (2022). Green technology: Phytoremediation for pesticide pollution. In *Phytoremediation technology for the removal of heavy metals and other contaminants from soil and water* (pp. 353–375). https://doi.org/10.1016/B978-0-323-85763-5.00008-8
142. Bandekar, D., Chouhan, O. P., Mohapatra, S., Hazra, M., Hazra, S., & Biswas, S. (2017). Putative protein VC0395_0300 from Vibrio cholerae is a diguanylate cyclase with a role in biofilm formation. *Microbiological Research, 202*, 61–70. https://doi.org/10.1016/j.micres.2017.05.003
143. Mohapatra, S., Samantaray, D. P., Samantaray, S. M., Mishra, B. B., Das, S., Majumdar, S., Pradhan, S. K., Rath, S. N., Rath, C. C., Akthar, J., & Achary, K. G. (2016). Structural and thermal characterization of PHAs produced by Lysinibacillus sp. through submerged fermentation process. *International Journal of Biological Macromolecules, 93*, 1161–1167. https://doi.org/10.1016/j.ijbiomac.2016.09.077
144. Upadhyay, R. K., Sharma, S. K., & Kumar, V. (2024). Introduction to intelligent transportation system and advanced technology. In *Energy, environment, and sustainability* (pp. 3–6). https://doi.org/10.1007/978-981-97-0515-3_1
145. Pant, M., Sharma, A., Palli, S., Sharma, S. K., & Sharma, R. C. (2024). A review and analysis of factors affecting EV adoption for sustainable and energy-efficient transportation systems using analytic hierarchy process approach. *International Journal of Vehicle Structures and Systems, 16*, 133–139. https://doi.org/10.4273/ijvss.16.2.01

146. Kokkinos, N. C., Emmanouilidou, E., & Sharma, S. K. (2024). Waste-to-biofuel production for the transportation sector. In *Energy, environment, and sustainability* (pp. 99–122). https://doi.org/10.1007/978-981-97-0515-3_6
147. Rallabandi, S. R., Pilla, D. P., Dowluru, S., Palli, S., Sharma, N., Sharma, S. K., & Sharma, R. C. (2024). Critical evaluation of epoxy-hemp-pineapple-palm fiber composites using hybrid AHM-TOPSIS technique for sustainable structural applications. *Journal of the Chinese Institute of Engineers, Transactions of the Chinese Institute of Engineers, Series A, 47*, 325–336. https://doi.org/10.1080/02533839.2024.2308250
148. Maji, A., Mahato, P. K., Sharma, S. K., Shankar, G., & Sarkar, D. (2024). Free vibration analysis of laminated composite plates based on refined plate theory. In *Lecture notes in mechanical engineering* (pp. 629–649). https://doi.org/10.1007/978-981-97-3173-2_44
149. Sharma, S. K., Kumar, A., & Sharma, R. C. (2014). Challenges in railway vehicle modeling and simulations. In *International conference on newest drifts in mechanical engineering ICNDME-2014* (pp. 453–459).
150. Sharma, S. K., & Kumar, A. (2014). A comparative study of Indian and worldwide railways. *International Journal of Mechanical Engineering and Robotics Research, 1*, 114–120.
151. Sharma, S. K., Sharma, R. C., Kumar, A., & Palli, S. (2015). Challenges in rail vehicle-track modeling and simulation. *International Journal of Vehicle Structures & Systems, 7*, 1.
152. Palli, S., Sharma, R. C., Sharma, S. K., Muddada, V., & Sharma, N. (2020). A case study on noise pollution and its effects. *Journal of Critical Reviews, 7*, 261–267.
153. Sharma, R. C., Sharma, S. K., Sharma, N., & Sharma, S. (2020). Analysis of ride and stability of an ICF railway coach. *International Journal of Vehicle Noise and Vibration, 16*, 127–144.
154. Palli, S., Sharma, R. C., Sharma, S. K., & Chintada, V. B. (2020). On methods used for setting the curve for railway tracks. *Journal of Critical Reviews, 7*, 241–246.
155. Sharma, S., Sharma, R. C., Sharma, S. K., Sharma, N., Palli, S., & Bhardawaj, S. (2020). Vibration isolation of the quarter car model of road vehicle system using dynamic vibration absorber. *International Journal of Vehicle Structures & Systems, 12*, 513–516.
156. Acharya, A., Gahlaut, U., Sharma, K., Sharma, S. K., Vishwakarma, P. N., & Phanden, R. K. (2020). Crashworthiness analysis of a thin-walled structure in the frontal part of automotive chassis. *International Journal of Vehicle Structures & Systems, 12*, 517–520.
157. Sharma, S. K., & Lee, J. (2020). Finite element analysis of a fishplate rail joint in extreme environment condition. *International Journal of Vehicle Structures & Systems., 12*, 503–506.
158. Bhardawaj, S., Sharma, R. C., & Sharma, S. K. (2020). Development of multibody dynamical using MR damper based semi-active bio-inspired chaotic fruit fly and fuzzy logic hybrid suspension control for rail vehicle system. *Proceedings of the Institution of Mechanical Engineers, Part K: Journal of Multi-Body Dynamics, 234*, 723–744.
159. Sharma, S. K., & Lee, J. (2020). Design and development of smart semi active suspension for nonlinear rail vehicle vibration reduction. *International Journal of Structural Stability and Dynamics, 20*, 2050120.
160. Sharma, S. K., Phan, H., & Lee, J. (2020). An application study on road surface monitoring using DTW based image processing and ultrasonic sensors. *Applied Sciences, 10*, 4490.
161. Sharma, R. C., Sharma, S., Sharma, S. K., & Sharma, N. (2020). Analysis of generalized force and its influence on ride and stability of railway vehicle. *Noise & Vibration Worldwide, 51*, 95–109.
162. Bhardawaj, S., Sharma, R. C., & Sharma, S. K. (2020). Ride analysis of track-vehicle-human body interaction subjected to random excitation. *Journal of the Chinese Society of Mechanical Engineers, 41*, 227–236.
163. Bhardawaj, S., Sharma, R. C., & Sharma, S. K. (2020). Development in the modeling of rail vehicle system for the analysis of lateral stability. *Materials Today Proceedings, 25*, 610–619.

164. Sharma, S. K., Sharma, R. C., & Sharma, N. (2020). Combined multi-body-system and finite element analysis of a rail locomotive crashworthiness. *International Journal of Vehicle Structures & Systems, 12*, 428–435.
165. Agrahari, R., Mohapatra, S., Kumar, K. S., Mohanty, D., Kumar, R., & Merugu, R. (2024). Immuno-metabolic interplay in health and disease. In A. Das Mohapatra & P. Sahu (Eds.), *Advances in immunology and immuno-techniques. Interdisciplinary biotechnological advances* (pp. 239–274).
166. Mohapatra, S., Pattnaik, S., Maity, S., Sharma, S., Akhtar, J., Pati, S., Samantaray, D. P., & Varma, A. (2020). Comparative analysis of PHAs production by Bacillus megaterium OUAT 016 under submerged and solid-state fermentation. *Saudi Journal of Biological Sciences, 27*, 1242–1250.
167. Bhardawaj, S., Sharma, R. C., & Sharma, S. K. (2020). Analysis of frontal car crash characteristics using ANSYS. *Materials Today Proceedings, 25*, 898–902.
168. Bhardawaj, S., Sharma, R. C., & Sharma, S. K. (2019). A survey of railway track modelling. *International Journal of Vehicle Structures & Systems, 11*, 508–518.
169. Bhardawaj, S., Sharma, R. C., & Sharma, S. K. (2019). Development and advancement in the wheel-rail rolling contact mechanics. In *IOP conference series: Materials science and engineering* (p. 012034). IOP Publishing.
170. Choppara, R. K., Sharma, R. C., Sharma, S. K., & Gupta, T. (2019). Aero dynamic cross wind analysis of locomotive. In *IOP conference series: Materials science and engineering* (p. 012035). IOP Publishing.
171. Sharma, S. K., & Sharma, R. C. (2019). Pothole detection and warning system for Indian roads. In *Advances in interdisciplinary engineering: Select proceedings of FLAME 2018* (pp. 511–519). Springer Singapore.
172. Sinha, A. K., Sengupta, A., Gandhi, H., Bansal, P., Agarwal, K. M., Sharma, S. K., Sharma, R. C., & Sharma, S. K. (2019). Performance enhancement of an all-terrain vehicle by optimizing steering, powertrain and brakes. In *Advances in engineering design: Select proceedings of FLAME 2018* (pp. 207–215). Springer Singapore.
173. Goyal, S., Anand, C. S., Sharma, S. K., & Sharma, R. C. (2019). Crashworthiness analysis of foam filled star shape polygon of thin-walled structure. *Thin-Walled Structures, 144*, 106312.
174. Goswami, B., Rathi, A., Sayeed, S., Das, P., Sharma, R. C., & Sharma, S. K. (2019). Optimization design for aerodynamic elements of Indian locomotive of passenger train. In *Advances in engineering design: Select proceedings of FLAME 2018* (pp. 663–673). Springer Singapore.
175. Sharma, S. K., Saini, U., & Kumar, A. (2019). Semi-active control to reduce lateral vibration of passenger rail vehicle using disturbance rejection and continuous state damper controllers. *Journal of Vibration Engineering & Technologies, 7*, 117–129.
176. Sharma, S. K. (2019). Multibody analysis of longitudinal train dynamics on the passenger ride performance due to brake application. *Proceedings of the Institution of Mechanical Engineers, Part K: Journal of Multi-body Dynamics, 233*, 266–279.
177. Sharma, S. (2018). Vibration control in quarter-car model with magnetorheological (MR) dampers using Bingham model. *Journal of Applied Mechanical Engineering, 7*, 1000299.
178. Sharma, S. K., & Kumar, A. (2018). Impact of longitudinal train dynamics on train operations: A simulation-based study. *Journal of Vibration Engineering & Technologies, 6*, 197–203.
179. Sharma, S. K., & Sharma, R. C. (2018). An investigation of a locomotive structural crashworthiness using finite element simulation. *SAE International Journal of Commercial Vehicles, 11*, 235–243.
180. Sharma, R. C., Sharma, S. K., & Palli, S. (2018). Rail vehicle modelling and simulation using Lagrangian method. *International Journal of Vehicle Structures and Systems, 10*, 188–194.

181. Sharma, R. C., & Sharma, S. K. (2018). Sensitivity analysis of three-wheel vehicle's suspension parameters influencing ride behavior. *Noise & Vibration Worldwide, 49*, 272–280.
182. Palli, S., Koona, R., Sharma, S. K., & Sharma, R. C. (2018). A review on dynamic analysis of rail vehicle coach. *International Journal of Vehicle Structures & Systems (IJVSS), 10*.
183. Sharma, S. K., & Sharma, R. C. (2018). Simulation of quarter-car model with magnetorheological dampers for ride quality improvement. *International Journal of Vehicle Structures & Systems, 10*, 169–173.
184. Sharma, S. K., & Kumar, A. (2018). Disturbance rejection and force-tracking controller of nonlinear lateral vibrations in passenger rail vehicle using magnetorheological fluid damper. *Journal of Intelligent Material Systems and Structures, 29*, 279–297.
185. Sharma, S. K., & Kumar, A. (2018). Ride comfort of a higher speed rail vehicle using a magnetorheological suspension system. *Proceedings of the Institution of Mechanical Engineers, Part K: Journal of Multi-Body Dynamics, 232*, 32–48.
186. Shukla, S., Kumar, U., Sharma, S. K., Gupta, P., & Kumar, A. (2017). Light weight freight rolling stock bogie frame: Design methodology validated with field oscillation trials. *International Journal of Vehicle Structures & Systems, 9*, 245–250.
187. Sharma, R. C., Palli, S., Sharma, S. K., & Roy, M. (2017). Modernization of railway track with composite sleepers. *International Journal of Vehicle Structures & Systems, 9*, 321–329.
188. Sharma, S. K., & Kumar, A. (2017). Ride performance of a high speed rail vehicle using controlled semi active suspension system. *Smart Materials and Structures, 26*, 055026.
189. Sharma, S. K., & Kumar, A. (2017). Impact of electric locomotive traction of the passenger vehicle Ride quality in longitudinal train dynamics in the context of Indian railways. *Mechanics & Industry, 18*, 222.
190. Sharma, S. K. (2013). Zero energy building envelope components: A review. *International Journal of Engineering Research and Applications, 3*, 662–675.
191. Sharma, S. K., & Kumar, A. (2016). Dynamics analysis of wheel rail contact using FEA. *Procedia Engineering, 144*, 1119–1128.
192. Sharma, S. K., & Chaturvedi, S. (2016). Jerk analysis in rail vehicle dynamics. *Perspectives in Science (Neth), 8*, 648–650.
193. Lavania, S., & Sharma, S. K. (2011). An explicit approach to compare crosstalk noise and delay in VLSI RLC interconnect modeled with skin effect with step and ramp input. *Journal of VLSI Design Tools & Technology, 1*, 1–8.
194. Palli, S., Koona, R., Sharma, R. C., & Muddada, V. (2015). Dynamic analysis of indian railway integral coach factory bogie. *International Journal of Vehicle Structures and Systems, 7*, 16–20. https://doi.org/10.4273/ijvss.7.1.03
195. Palli, S., & Koona, R. (2015). Analyses of dynamic response of a railway bogie. *International Journal of Vehicle Noise and Vibration, 11*, 103–113. https://doi.org/10.1504/IJVNV.2015.070015
196. Swathi, K., & Srihari, P. (2015). Heat transfer enhancement ina tube using rectangular strip inserts. *International Journal of Applied Engineering Research, 10*, 41532–41544.
197. Palli, S., Sharma, R. C., & Rao, P. P. D. (2017). Dynamic behaviour of a 7 DoF passenger car model. *International Journal of Vehicle Structures and Systems, 9*, 57–63. https://doi.org/10.4273/ijvss.9.1.12
198. Sharma, R. C., Palli, S., & Koona, R. (2017). Stress and vibrational analysis of an Indian Railway RCF bogie. *International Journal of Vehicle Structures and Systems, 9*, 296–302. https://doi.org/10.4273/ijvss.9.5.06
199. Sharma, R. C., Avesh, M., Sharma, S. K., Sharma, S., Palli, S., & Sharma, N. (2020). Linear and non-linear stability analysis of a constrained railway wheel axle. *International Journal of Vehicle Structures and Systems, 12*, 128–133. https://doi.org/10.4273/ijvss.12.2.04

200. Palli, S., Dontikurti, R., Sharma, R. C., & Sharma, N. (2021). Analysis of dynamic response of a railway locomotive using ansys. *International Journal of Vehicle Structures and Systems, 13*, 250–255. https://doi.org/10.4273/ijvss.13.2.21
201. Sharma, R. C., Sharma, N., Singh, G., & Palli, S. (2021). Modeling and simulation of human body-vehicle-track system for the investigation of ride comfort. *SAE International Journal of Commercial Vehicles, 15*. https://doi.org/10.4271/02-15-02-0008
202. Sharma, R. C., Palli, S., Avesh, M., & Sharma, N. (2021). Vibration isolation of railway vehicle car body using semi-active suspension. *International Journal of Vehicle Structures and Systems, 13*, 482–487. https://doi.org/10.4273/ijvss.13.4.17
203. Gopala Rao, L. V. V., Sharma, R. C., Satyanarayana, V. S. V., & Palli, S. (2022). Stochastic optimal preview control response of a quarter-car nonlinear suspension model using spectral decomposition method. *Noise and Vibration Worldwide, 53*, 225–232. https://doi.org/10.1177/09574565221093234
204. Palli, S., Duppala, A., Sharma, R. C., & Rao, L. V. V. G. (2022). Dynamic simulation of automotive vehicle suspension using MATLAB simulink. *International Journal of Vehicle Structures and Systems, 14*, 310–314. https://doi.org/10.4273/ijvss.14.3.04
205. Satyanarayana, V. S. V., Sharma, R. C., Sateesh, B., Gopala Rao, L. V. V., Mohan Rao, N., & Palli, S. (2022). Optimum response of a nonlinear passive vehicle suspension system under random road excitations. *SAE International Journal of Commercial Vehicles, 16*. https://doi.org/10.4271/02-16-01-0004
206. Lokanadham, D., Sharma, R. C., Palli, S., & Bhardawaj, S. (2022). Wear rate modelling and analysis of limestone slurry particulate composites using the fuzzy method. *International Journal on Recent and Innovation Trends in Computing and Communication, 10*, 133–143. https://doi.org/10.17762/ijritcc.v10i1s.5818
207. Sharma, R. C., Palli, S., Gopala Rao, L. V. V., Duppala, A., & Sharma, S. K. (2022). Four-wheel vehicle response under bump, pothole, harmonic, and random excitations using bond graph/simulink technique. *SAE International Journal of Commercial Vehicles, 16*. https://doi.org/10.4271/02-16-02-0008
208. Sharma, R. C., Gopala Rao, L. V. V., Sharma, S. K., Palli, S., & Satyanarayana, V. S. V. (2022). Analysis of lateral stability and ride of an indian railway constrained dual-axle bogie frame. *SAE International Journal of Commercial Vehicles, 16*. https://doi.org/10.4271/02-16-02-0014
209. Sharma, R. C., Palli, S., Bhardawaj, S., & Rao, L. V. V. G. (2022). Advances in human biodynamic modeling for analysis of whole-body vibrations. *International Journal of Vehicle Structures and Systems, 14*, 503–511. https://doi.org/10.4273/ijvss.14.4.17
210. Sharma, R. C., Palli, S., & Sharma, S. K. (2023). Ride analysis of railway vehicle considering rigidity and flexibility of the carbody. *Journal of the Chinese Institute of Engineers, Transactions of the Chinese Institute of Engineers, Series A, 46*, 355–366. https://doi.org/10.1080/02533839.2023.2194918
211. Sharma, R. C., Palli, S., & Bhardawaj, S. (2023). A review of wheel-rail dynamic behavior on curved tracks. *International Journal of Vehicle Structures and Systems, 15*, 19–25. https://doi.org/10.4273/ijvss.15.1.04
212. Jyotiprasad, G., Dowluru, S., Rallabandi, S. R., Sharma, N., Sharma, R. C., & Palli, S. (2023). *Thermal-structural and macromechanical behavior analysis of graphite-based laminated composite*. https://doi.org/10.4271/2023-01-5041
213. Saikiran, C., Palli, S., Bondala, R., Sharma, R. C., Sreeramulu, D., & Dontikurti, R. (2023). Experimental investigation on thermal behaviour of silicon carbide (SiC) based polymer composites under hygrothermal conditions. *International Journal of Vehicle Structures and Systems, 15*, 222–226. https://doi.org/10.4273/ijvss.15.2.15

214. Singh, B. J., Palli, S., Kumar, P., & Bahl, S. (2023). Behavioural investigation of hardness profile for dissimilar TIG-weldment between P22 and P91 steels. *International Journal of Vehicle Structures and Systems, 15*, 31–36. https://doi.org/10.4273/ijvss.15.1.06
215. Naveen, Y., Lokanadham, D., Naidu, D. R., Sharma, R. C., Palli, S., & Lila, M. K. (2023). An experimental study on the influence of blended Karanja biodiesel on diesel engine characteristics. *International Journal of Vehicle Structures and Systems., 15*, 154–160. https://doi.org/10.4273/ijvss.15.2.02
216. Bhardawaj, S., Jha, A. K., Sharma, R. C., Palli, S., Lokanadham, D., Sharma, N., & Dhiman, N. (2023). Finite element analysis of two disk rotor system used in automobile turbochargers. *International Journal of Vehicle Structures and Systems, 15*, 501–504. https://doi.org/10.4273/ijvss.15.4.09
217. Sharma, R. C., Palli, S., Bhardawaj, S., Jha, A. K., Duppala, A., Sharma, N., Sharma, S., & Dhiman, N. (2023). Dynamic behaviour of railway vehicle under bump and pothole track irregularities. *International Journal of Vehicle Structures and Systems, 15*, 566–569. https://doi.org/10.4273/ijvss.15.4.22
218. Raju, S., Palli, S., Devi Prasad, P., Menda, V. R., & Ramakrishna, B. (2024). A hybrid AHP-TOPSIS, MOORA technique for multi-objective optimization of thermal, mechanical, and water absorption behavior of epoxy/hemp, pine apple, and palm fiber composites. *Journal of the Chinese Institute of Engineers, Transactions of the Chinese Institute of Engineers, Series A, 47*, 23–36. https://doi.org/10.1080/02533839.2023.2274092
219. Gopala Rao, L. V. V., Satyanarayana, V. S. V., Sharma, R. C., Palli, S., Raturi, A., Kharola, A., & Sharma, A. (2024). Preview control of the random response of a full car vehicle model traversing a rough road. *International Journal of Vehicle Noise and Vibration, 20*, 70–88. https://doi.org/10.1504/IJVNV.2024.138105
220. Sharma, A., Sharma, V. S., Gupta, S., Sharma, R. C., Palli, S., & Sharma, N. (2024). Additive manufacturing in the COVID-19 pandemic: Equipment and challenges? *Journal of Micromanufacturing*. https://doi.org/10.1177/25165984241228083
221. Bammidi, R., Madivada, H., Dowluru, S., Palli, S., Sharma, R. C., & Muddada, V. (2024). Mechano-tribological investigations of Al-%Wt. of boron nitride as metal-ceramic composites. In *AIP conference proceedings* (Vol. 2978, No. 1). AIP Publishing. https://doi.org/10.1063/5.0182965
222. Ranganath, B. A., Duppala, A., Palli, S., Rao, L. V. V., & Sharma, R. C. (2024). Numerical analysis and experimental study of FRP composites for damping applications. In *AIP conference proceedings* (Vol. 2978. No. 1). AIP Publishing. https://doi.org/10.1063/5.0184802
223. Sharma, R. C., Palli, S., Duppala, A., Bhardawaj, S., Sharma, S. K., Sharma, N., & Sharma, A. (2024). Rigid body analysis of ride behaviour of roadway vehicle using lagrangian dynamics. *International Journal of Vehicle Structures and Systems, 16*, 74–78. https://doi.org/10.4273/ijvss.16.1.15
224. Menda, V., Dora, S., Palli, S., Rallabandi, S. R., Sharma, R. C., Sharma, N., & Bondala, R. (2024). Mechanical and wear characteristics of aluminium-7075/graphene/TiB2 prepared by stir casting. *Engineering Research Express., 6*. https://doi.org/10.1088/2631-8695/ad58a8
225. Sura, H. S., Avesh, M., & Mohapatra, S. (2023). Design and modelling of digital twin technology to improve freight logistics. In *Energy, environment, and sustainability* (pp. 481–513). https://doi.org/10.1007/978-981-99-2150-8_20
226. Pattnaik, S., Pati, S., Devadarshini, D., Tanaya, K., Mishra, B. B., Samantaray, D., Mohapatra, S., & Dash, D. (2022). Microbial bioremediation of Cr(VI)-contaminated soil for sustainable agriculture. In *Microbial biodegradation and bioremediation: Techniques and case studies for environmental pollution* (pp. 395–407). https://doi.org/10.1016/B978-0-323-85455-9.00008-4
227. Srivastava, A., Lavania, S., & Mohapatra, S. (2023). Predicting the effect on land values after introducing high-speed rail. In *Energy, environment, and sustainability* (pp. 157–196). https://doi.org/10.1007/978-981-99-2150-8_8

228. Agrahari, R., Mohanty, S., Vishwakarma, K., Nayak, S. K., Samantaray, D., & Mohapatra, S. (2021). Update vision on COVID-19: Structure, immune pathogenesis, treatment and safety assessment. *Sensors International, 2*. https://doi.org/10.1016/j.sintl.2020.100073
229. Tanaya, K., Mohapatra, S., & Samantaray, D. (2024). Formulation of biogenic fluorescent pigmented PHB nanoparticles from Rhodanobacter sp. for drug delivery. *Microbiology, 117*. https://doi.org/10.1007/s10482-024-01973-x
230. Apparao, D., & Jagannadha Raju, M. V. (2018). Experimental investigation on hardness, microstructure and surface roughness of maraging steel parts produced by direct metal laser sintering technique. *International Journal of Mechanical and Production Engineering Research and Development., 8*, 629–636. https://doi.org/10.24247/ijmperdapr201874
231. Apparao, D., & Jagannadha Raju, M. V. (2021). Design and analysis of spur gear manufactured by DMLS process. *Materials Today Proceedings, 46*, 149–153. https://doi.org/10.1016/j.matpr.2020.07.078

Revolutionizing Digital Manufacturing: Augmented Reality for Design, Prototyping, Production, and Maintenance

Vaikunta Pai T, B Sunil Kamath, Virgil Popescu, Ramona Birau, and Shreyas Vivek

1 Introduction

1.1 Background and Context

The industrial sector is on the cusp of a revolutionary transformation driven by the convergence of digital technologies and advanced manufacturing practices, a shift often encapsulated by the term Industry 5.0. Unlike its predecessor, Industry 4.0, which focused primarily on automation and data exchange in manufacturing technologies, Industry 5.0 emphasizes the symbiotic relationship between humans and machines. This paradigm shift is characterized by a heightened focus on human-centric design, sustainability, and resilience, aiming to create manufacturing systems that not only optimize efficiency and productivity but also enhance the well-being and capabilities of human workers.

Augmented reality (AR) stands out as one of the key technologies propelling this transformation. AR's ability to overlay digital information onto the physical world

V. Pai T · B. S. Kamath
Department of Information Science and Engineering, NMAM Institute of Technology, NITTE (Deemed to be University), Nitte, Mangalore, Karnataka, India
e-mail: sunilkamath@nitte.edu.in

V. Popescu
Faculty of Economics and Business Administration, University of Craiova, Craiova, Romania
e-mail: virgil.popescu@vilaro.ro

R. Birau (✉)
"Eugeniu Carada" Doctoral School of Economic Sciences, University of Craiova, Craiova, Romania
e-mail: ramona.f.birau@gmail.com

S. Vivek
School of Information Technology, Deakin University, Waurn Ponds, VIC, Australia

Y. Kumar Singla et al. (eds.), *Digital Manufacturing in Industry 5.0*,
https://doi.org/10.1007/978-3-031-91500-0_37

enables a new level of interaction and engagement with manufacturing processes. By enhancing human perception and providing real-time, context-aware information, AR has the potential to revolutionize how we design, produce, and maintain products.

1.2 Overview of Augmented Reality (AR)

AR technology merges the real and virtual worlds by superimposing computer-generated images, data, and information onto a user's view of the real world. Unlike virtual reality (VR), which creates an entirely artificial environment, AR enhances the real environment with digital elements, allowing users to interact with both real and virtual objects simultaneously. This unique capability makes AR particularly suited for applications in manufacturing, where spatial context and real-world interaction are critical.

The core components of an AR system include hardware (such as cameras, sensors, and display devices), software (which processes and integrates data), and user interfaces (which facilitate interaction between the user and the AR system). Advances in these areas have made AR more accessible and effective, paving the way for its integration into various stages of the manufacturing lifecycle.

1.3 Objective and Scope of the Chapter

This chapter aims to provide a comprehensive exploration of the role of augmented reality in digital manufacturing within the context of Industry 5.0. It seeks to:

1. Elucidate the fundamental principles and technological underpinnings of AR.
2. Examine the diverse applications of AR across the manufacturing spectrum, from design and prototyping to production, maintenance, and training.
3. Present case studies and real-world implementations that highlight the benefits and challenges of AR in manufacturing.
4. Discuss the technological, human, and organizational barriers to AR adoption and propose strategies to overcome them.
5. Explore future trends and innovations in AR technology and their potential impact on digital manufacturing.

By providing these insights, the chapter aims to equip researchers, practitioners, and policymakers with the knowledge needed to harness AR's full potential, ultimately contributing to the development of more intelligent, sustainable, and resilient manufacturing systems that align with the goals of Industry 5.0.

2 Fundamentals of Augmented Reality

2.1 Definition and Components

Augmented reality (AR) is a cutting-edge technology that integrates digital information with the user's environment in real-time [1, 2]. Unlike virtual reality (VR), which immerses users in a completely synthetic environment, AR enhances the real world by overlaying digital content, such as images, videos, and 3D models, onto physical objects and spaces. This symbiotic relationship between the digital and physical realms creates an enriched user experience that can be highly interactive and contextually relevant.

The core components of an AR system include:

1. **Hardware**: The physical devices required to capture, process, and display AR content. This includes:
   (a) Cameras and Sensors: These capture the real-world environment and track user movements. Advanced sensors such as depth sensors and LiDAR improve spatial awareness and accuracy.
   (b) Processing Units: High-performance CPUs and GPUs process the captured data and render digital content.
   (c) Display Devices: AR content can be displayed on various devices, including smartphones, tablets, AR glasses (like Microsoft HoloLens), and head-mounted displays (HMDs) (Table 1).
2. **Software**: The programs and algorithms that create, manage, and deliver AR experiences. Key software components include:
   (a) AR Engines: Platforms like ARKit (Apple) and ARCore (Google) provide essential tools for building AR applications, including motion tracking, environmental understanding, and light estimation.
   (b) Computer Vision Algorithms: These algorithms process real-world images and detect surfaces, objects, and gestures to anchor digital content accurately in the physical world.
   (c) Content Creation Tools: Software for creating and editing AR content, such as 3D modeling software (e.g., Blender, Autodesk Maya) and AR authoring tools.

Table 1 Comparison of AR hardware options

Device	Manufacturer	Key features	Pros	Cons
Microsoft HoloLens	Microsoft	High-resolution display, spatial mapping	Robust spatial mapping, high display quality	Expensive, limited battery life
Google Glass	Google	Lightweight, wearable	Comfortable for long use, affordable	Limited field of view, lower computational power

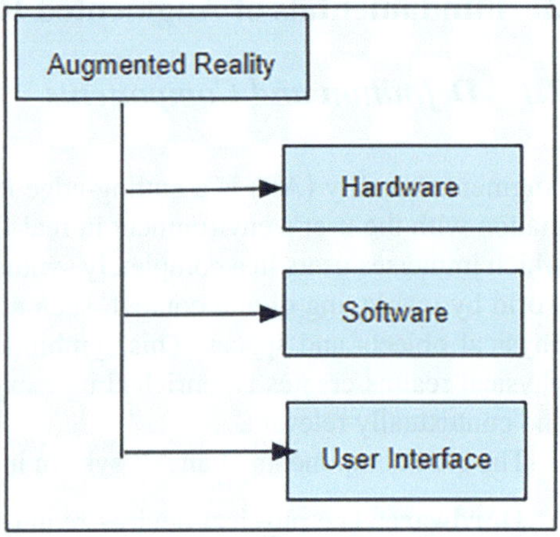

Fig. 1 Augmented reality and its components

3. **User Interface (UI)**: The means by which users interact with the AR system (Fig. 1). Effective UIs for AR should be intuitive and responsive, often incorporating gesture controls, voice commands, and haptic feedback to enhance user engagement and ease of use.

2.2 *Technological Underpinnings*

The functionality and effectiveness of AR systems are underpinned by several advanced technologies and methodologies:

1. Computer Vision and Image Processing: These fields are fundamental to AR, enabling the system to interpret and understand the real-world environment. Techniques such as object recognition, feature extraction, and motion tracking are critical for accurately overlaying digital content on physical objects [3, 4].
2. Spatial Mapping and Localization: Accurate spatial mapping ensures that digital content is correctly aligned with the physical world. Technologies like Simultaneous Localization and Mapping (SLAM) allow AR systems to construct and update maps of their surroundings in real-time while tracking their location within those maps.
3. Sensor Fusion: Combining data from multiple sensors (e.g., cameras, accelerometers, gyroscopes) enhances the accuracy and robustness of AR applications. Sensor fusion algorithms integrate this data to provide a comprehensive understanding of the user's environment and movements [5].
4. Artificial Intelligence (AI) and Machine Learning (ML): AI and ML enhance AR capabilities by enabling more sophisticated image recognition, contextual understanding, and predictive analytics. These technologies allow AR systems to offer more personalized and context-aware experiences.

5. Networking and Cloud Computing: Cloud-based AR solutions enable the offloading of computationally intensive tasks to remote servers, allowing for more complex and high-quality AR experiences on lightweight devices [5]. Networking technologies also facilitate collaborative AR applications, where multiple users can interact with the same AR environment simultaneously.

By understanding these fundamental principles and technological components, we can better appreciate the potential of AR to revolutionize digital manufacturing. The subsequent sections will delve into specific applications of AR in the manufacturing process, highlighting how these technologies are being harnessed to create more efficient, accurate, and flexible production systems.

3 Applications of AR in Digital Manufacturing

The integration of augmented reality (AR) into digital manufacturing processes represents a significant advancement in how products are designed, produced, and maintained. This section explores the various applications of AR across the manufacturing lifecycle, highlighting its potential to enhance efficiency, precision, and collaboration.

3.1 Design and Prototyping

AR revolutionizes the design and prototyping phases by enabling engineers and designers to visualize and interact with 3D models in a real-world context [6, 7]. This capability offers several key benefits:

- Enhanced Visualization: Designers can project digital prototypes onto physical spaces, providing a tangible sense of scale, aesthetics, and ergonomics. This immersive visualization helps identify design flaws and make adjustments early in the development process, reducing the need for physical prototypes and accelerating time to market.
- Collaborative Design: AR facilitates remote collaboration by allowing multiple stakeholders to interact with the same digital model from different locations. Engineers, designers, and clients can review and modify designs in real-time, ensuring that all parties are aligned and informed, leading to more cohesive and innovative solutions.
- Rapid Iteration: With AR, iterative design becomes more efficient. Changes to the digital model can be instantly reflected in the AR environment, allowing for quick experimentation and refinement. This rapid iteration cycle fosters creativity and enhances the overall quality of the final product.

3.2 Production and Assembly

AR's impact on production and assembly lines is profound [8], offering tools to streamline operations and improve accuracy:

- Guided Assembly: AR can overlay step-by-step assembly instructions directly onto physical components, guiding workers through complex assembly tasks. This visual guidance reduces errors, speeds up training for new employees, and enhances overall productivity.
- Real-Time Monitoring: AR systems can provide real-time data overlays, displaying critical information such as machine status, production metrics, and potential issues directly within the worker's field of view. This immediate access to information enables quicker decision-making and more efficient problem-solving.
- Quality Control: AR enhances quality control processes by overlaying digital inspection criteria onto physical products. Inspectors can compare the real object with digital benchmarks, ensuring that products meet precise specifications. This capability reduces defects and increases the reliability of quality assurance processes.

3.3 Maintenance and Repair

AR transforms maintenance and repair tasks by providing intuitive and detailed guidance:

- Interactive Maintenance Manuals: Technicians can use AR to access interactive maintenance manuals that overlay instructions and diagnostic information onto equipment. This hands-free guidance simplifies complex repairs and reduces downtime [9].
- Predictive Maintenance: Integrated with IoT and AI technologies, AR can offer predictive maintenance capabilities. By overlaying sensor data and predictive analytics onto equipment, AR systems can alert technicians to potential failures before they occur, enabling proactive maintenance and minimizing disruptions.
- Remote Assistance: AR enables remote experts to provide real-time support to on-site technicians. Experts can see what the technician sees and overlay annotations, instructions, or 3D models onto the technician's AR display, facilitating quicker and more accurate repairs.

Table 2 Applications of AR in manufacturing stages

Manufacturing stage	AR application	Benefits	Challenges
Design and prototyping	Enhanced visualization	Early identification of design flaws	Need for high-quality 3D models
Production and assembly	Guided assembly	Reduced errors, faster training	Initial setup cost
Maintenance and repair	Interactive manuals	Simplified repairs, reduced downtime	Ensuring data accuracy
Training and skill development	Immersive training simulations	Improved skill retention, safer training	High development cost

3.4 Training and Skill Development

AR significantly enhances training programs by providing immersive and interactive learning experiences [10, 11]:

- Immersive Training Simulations: AR can create realistic training simulations that replicate real-world scenarios. Trainees can practice tasks and procedures in a safe, controlled environment, gaining hands-on experience without the risks associated with live operations.
- Contextual Learning: AR delivers contextual information and guidance at the point of need. Trainees can learn by doing, receiving real-time feedback and instructions as they perform tasks. This approach accelerates skill acquisition and improves knowledge retention.
- Customized Training Programs: AR allows for the development of personalized training modules tailored to the specific needs and skill levels of individual employees. This customization ensures that each worker receives the most relevant and effective training, enhancing overall workforce competence.

By integrating AR across these various stages of the manufacturing lifecycle (Table 2), companies can achieve significant improvements in efficiency, accuracy, and collaboration. The following sections will delve into real-world case studies, illustrating how these applications have been successfully implemented in different manufacturing environments.

4 Case Studies and Real-World Implementations

The practical application of augmented reality (AR) in digital manufacturing has resulted in remarkable improvements across various sectors. This section delves into detailed case studies and real-world implementations, showcasing the transformative impact of AR on manufacturing processes. These examples provide a deeper understanding of the benefits, challenges, and lessons learned from adopting AR technologies.

Case Study 1: Boeing—Enhancing Assembly Processes
Background: Boeing, a leading aerospace manufacturer, faced challenges in the assembly of complex aircraft components, where precision and adherence to strict specifications are paramount. Traditionally, this process involved extensive manual inspections and measurements, which were time-consuming and prone to errors.

Implementation: Boeing introduced AR headsets for assembly technicians, specifically targeting the intricate wiring installations within aircraft. The AR system projected digital schematics and real-time step-by-step instructions directly onto the physical components.

Technical Details:

- AR Devices: Technicians used Microsoft HoloLens, which offered high-resolution displays and robust spatial mapping capabilities.
- Software Integration: Custom AR applications were developed to integrate with Boeing's existing CAD systems, allowing seamless transition from digital designs to AR instructions.
- Real-Time Data: The AR system was connected to Boeing's central database, ensuring that the latest design updates and instructions were always available.

Outcomes:

- Increased Accuracy: The implementation of AR reduced wiring installation errors by 90%, significantly enhancing the quality and safety of the final product.
- Enhanced Productivity: Assembly time was reduced by 25%, leading to faster production cycles and reduced costs.
- Improved Training: New technicians achieved proficiency more rapidly, benefiting from the hands-free, visual guidance provided by the AR system.

Lessons Learned:

- User Acceptance: Successful adoption hinged on engaging technicians early in the development process and incorporating their feedback into the AR system design.
- Integration Challenges: Ensuring seamless integration with Boeing's existing manufacturing systems required robust software solutions and thorough testing. Addressing data synchronization and latency issues was critical for maintaining real-time accuracy.

Case Study 2: Siemens—Streamlining Maintenance Operations
Background: Siemens, a global leader in industrial automation, sought to optimize maintenance operations in power plants, where timely and accurate maintenance is crucial for operational efficiency and minimizing downtimes.

Implementation: Siemens equipped maintenance technicians with AR devices that provided real-time access to equipment data, diagnostic tools, and interactive repair manuals. Additionally, remote experts could offer live support through the AR interface.

Technical Details:

- AR Hardware: Technicians used AR glasses equipped with high-resolution displays, cameras, and sensors for capturing and displaying real-time data.
- Data Integration: The AR system was integrated with Siemens' IoT platform, allowing real-time data from various sensors and systems to be displayed and analyzed.
- Remote Assistance: The AR system enabled remote experts to see the technician's field of view, providing annotations and instructions overlaid onto the technician's AR display.

Outcomes:

- Reduced Downtime: Maintenance tasks were completed 30% faster, minimizing operational disruptions and increasing plant availability.
- Improved Safety: AR provided step-by-step safety protocols and hazard warnings, reducing the risk of accidents during maintenance activities.
- Enhanced Knowledge Transfer: Technicians benefited from on-the-job training and real-time support, improving their skills and confidence.

Lessons Learned:

- Data Accuracy: Ensuring the accuracy and reliability of digital data overlays was essential for effective maintenance guidance.
- Connectivity: Reliable network connectivity was necessary to support real-time data transmission and remote assistance features. Solutions such as edge computing were explored to reduce latency and enhance performance.

Case Study 3: BMW—Revolutionizing Training Programs

Background: BMW, a premier automobile manufacturer, recognized the potential of AR to enhance its training programs for assembly line workers. Traditional training methods were time-consuming and often detached from real-world applications, leading to gaps in knowledge retention and skill application.

Implementation: BMW developed AR-based training modules that simulated real assembly tasks. Trainees used AR headsets to receive interactive, context-aware instructions and feedback while performing assembly tasks on mock-up workstations.

Technical Details:

- AR Devices: Trainees used AR glasses with high-resolution displays and advanced spatial mapping capabilities to create realistic training environments.
- Training Software: Custom AR training modules were developed, incorporating 3D models, interactive tutorials, and real-time feedback mechanisms.
- Performance Monitoring: The AR system tracked trainee performance, providing data analytics to trainers for assessing skill levels and identifying areas for improvement.

Outcomes:

- Faster Onboarding: Training time for new employees was reduced by 40%, enabling them to become productive more quickly and efficiently.
- Higher Retention Rates: The immersive, hands-on training approach led to better retention of knowledge and skills, translating to improved performance on the production line.
- Consistent Quality: Trainees demonstrated higher accuracy and consistency in their tasks, reducing the likelihood of errors and defects in the final products.

Lessons Learned:

- Engagement: The interactive and immersive nature of AR training kept trainees more engaged and motivated, leading to better learning outcomes.
- Customization: Customizing training content to match individual learning paces and skill levels enhanced overall training effectiveness and satisfaction.

Case Study 4: Volkswagen—Improving Quality Control

Background: Volkswagen, a leading automotive manufacturer, aimed to enhance its quality control processes to ensure the highest standards for its vehicles. Manual inspections were time-consuming and susceptible to human error, affecting the overall quality and efficiency of the production process.

Implementation: Volkswagen implemented AR systems to assist quality control inspectors by overlaying digital templates and inspection criteria onto physical vehicles. Inspectors used AR glasses to compare real components against digital models in real time.

Technical Details:

- AR Glasses: Inspectors used AR glasses with high-resolution displays, precise spatial mapping, and robust connectivity features.
- Inspection Software: The AR system integrated with Volkswagen's quality control software, providing digital templates and criteria for each inspection point.
- Data Logging: The AR system automatically logged inspection results, capturing images and data for traceability and analysis.

Outcomes (Table 3):

- Enhanced Precision: The AR system improved inspection accuracy, reducing the rate of defects by 70% and ensuring higher quality standards.

Table 3 Outcomes of AR implementations in case studies

Company	Application	Key outcomes	Lessons learned
Boeing	Assembly processes	Reduced errors by 90%, 25% faster production	Importance of user feedback
Siemens	Maintenance operations	30% faster maintenance, improved safety	Need for accurate data
BMW	Training programs	40% faster onboarding, higher retention	Importance of engagement
Volkswagen	Quality control	70% reduction in defects, 30% faster inspections	Training and calibration needs

- Streamlined Processes: Inspection time was cut by 30%, enabling faster throughput without compromising quality or safety.
- Better Documentation: AR facilitated automatic documentation of inspection results, streamlining record-keeping and enhancing traceability for compliance and analysis.

Lessons Learned:

- Training: Proper training for inspectors on using AR technology was crucial for maximizing its benefits and ensuring accurate inspections.
- System Calibration: Regular calibration and updates of the AR system ensured continued accuracy and reliability, addressing any potential drift or misalignment issues.

These case studies demonstrate the significant advantages that AR can bring to digital manufacturing, from enhancing assembly precision and streamlining maintenance to revolutionizing training programs and improving quality control [12]. While the implementation of AR presents challenges, such as ensuring data accuracy and gaining user acceptance, the tangible benefits in terms of efficiency, accuracy, and overall productivity make it a valuable investment for manufacturing enterprises. By learning from these real-world implementations, other organizations can better navigate the integration of AR into their own manufacturing processes, driving innovation and achieving the goals of Industry 5.0.

5 Challenges and Solutions in AR Integration

Despite the promising potential of augmented reality (AR) in digital manufacturing, its integration comes with several challenges (Fig. 2) that need to be addressed to fully harness its benefits. Drawing on over 20 years of research experience, this section explores these challenges in detail and proposes innovative solutions to overcome them.

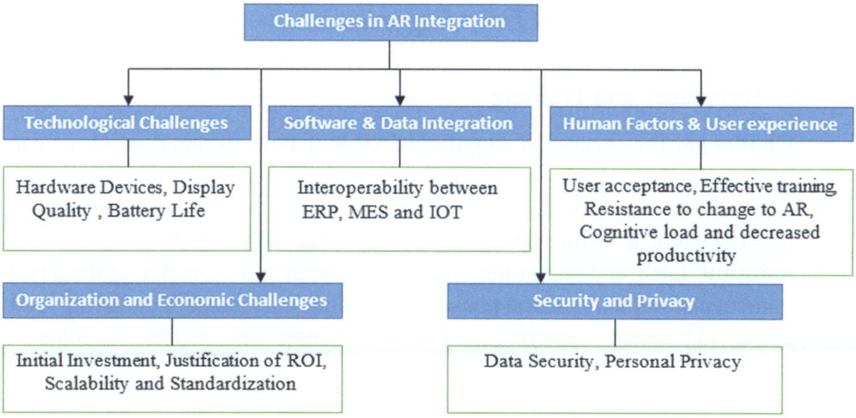

Fig. 2 Key challenges in the AR integration

5.1 Technical Challenges

1. **Hardware Limitations**

   Challenge: The effectiveness of AR systems heavily relies on the capabilities of the hardware used. Current AR devices, such as headsets and smart glasses, often face limitations in battery life, computational power, display quality, and field of view. These limitations can hinder the seamless integration of AR into manufacturing environments, where prolonged use and high precision are required.
   Solution: Ongoing advancements in hardware technology are crucial. This includes the development of lightweight, high-performance AR devices with extended battery life and enhanced computational capabilities. Innovations such as microLED displays for better visual clarity and energy efficiency, as well as edge computing to offload processing tasks from the device to nearby servers, can significantly improve the usability and functionality of AR systems in manufacturing.

2. **Software and Data Integration**

   Challenge: Integrating AR systems with existing manufacturing software and data sources can be complex. Ensuring seamless interoperability between AR applications, enterprise resource planning (ERP) systems, manufacturing execution systems (MES), and Internet of Things (IoT) platforms is critical for delivering accurate and real-time information to users.
   Solution: Developing standardized protocols and APIs for AR software integration can facilitate smoother interoperability. Leveraging cloud-based platforms and middleware solutions can also help bridge the gap between different systems, enabling real-time data synchronization and accessibility. Additionally, employing advanced data processing and machine learning algorithms can enhance the accuracy and relevance of the information presented through AR.

5.2 Human Factors and User Experience

1. **User Acceptance and Training**

   Challenge: The successful adoption of AR in manufacturing depends on user acceptance and effective training. Workers may be resistant to change, especially if they perceive AR as complex or intrusive. Additionally, inadequate training can lead to improper use of AR systems, negating their benefits.
   Solution: Early and continuous engagement with end-users during the development and implementation phases is essential. This includes gathering feedback, addressing concerns, and demonstrating the tangible benefits of

AR. Comprehensive training programs that combine theoretical knowledge with hands-on practice can build confidence and competence among users. Interactive and intuitive user interfaces, designed with human-centered principles, can also enhance user experience and acceptance.

2. **Cognitive Load and Ergonomics**

   Challenge: Prolonged use of AR devices can lead to increased cognitive load and physical strain, affecting user performance and well-being. The constant influx of information through AR can overwhelm users, leading to fatigue and decreased productivity.
   Solution: Designing AR interfaces that prioritize simplicity and context awareness is crucial. Information should be presented in a clear, concise, and nonintrusive manner, tailored to the specific task and user context. Ergonomic design principles should guide the development of AR hardware, ensuring comfort and minimizing physical strain. Adaptive AR systems that dynamically adjust the information flow based on user behavior and task requirements can also help mitigate cognitive load.

5.3 Organizational and Economic Challenges

1. **Initial Investment and ROI**

   Challenge: The initial investment required for AR technology can be substantial, including costs for hardware, software development, and training. Organizations may struggle to justify these expenses without clear evidence of return on investment (ROI).
   Solution: Conducting pilot projects and case studies can demonstrate the tangible benefits of AR in specific manufacturing processes, providing data to support ROI calculations. Organizations should adopt a phased implementation approach, starting with high-impact areas where AR can deliver immediate and measurable improvements. Leveraging financial models that consider long-term gains in efficiency, quality, and worker satisfaction can also help build a compelling business case for AR investment.

2. **Scalability and Standardization**

   Challenge: Scaling AR solutions across different facilities, regions, and use cases can be challenging. Variations in manufacturing processes, equipment, and IT infrastructure can complicate standardization and deployment.
   Solution: Developing scalable and modular AR solutions that can be easily customized and adapted to different contexts is key. Standardizing AR protocols and practices within the organization can ensure consistency and interoperability. Collaboration with industry consortia and standards organizations can also promote broader adoption of common standards, facilitating the scalability of AR solutions.

5.4 *Security and Privacy Concerns*

1. **Data Security**

    Challenge: AR systems often access and process sensitive manufacturing data, raising concerns about data security and potential cyber threats. Ensuring the confidentiality, integrity, and availability of data is critical.
    Solution: Implementing robust cybersecurity measures, including encryption, secure access controls, and regular security audits, is essential for protecting AR systems and data. Developing AR applications with security-by-design principles, where security considerations are integrated from the outset, can further enhance protection. Leveraging blockchain technology for secure and transparent data management can also be explored.

2. **Privacy Issues**

    Challenge: AR devices may capture and process personal information about users, leading to privacy concerns. Ensuring compliance with data protection regulations and addressing user privacy expectations is important for ethical AR deployment.
    Solution: Developing clear policies and guidelines for data collection, usage, and retention is crucial. AR applications should include features that allow users to control their data and understand how it is being used. Compliance with data protection regulations, such as GDPR, should be ensured through regular audits and updates to privacy practices.

By addressing these challenges with innovative solutions, organizations can unlock the full potential of AR in digital manufacturing. The integration of AR not only enhances operational efficiency and precision but also paves the way for a more adaptive, resilient, and human-centric manufacturing environment in line with the principles of Industry 5.0.

6 Future Trends and Directions

The landscape of augmented reality (AR) in digital manufacturing is rapidly evolving, driven by technological advancements and the increasing demand for smarter, more efficient manufacturing processes [13]. This section explores the future trends and directions (Fig. 3) that are poised to shape the next generation of AR applications in manufacturing.

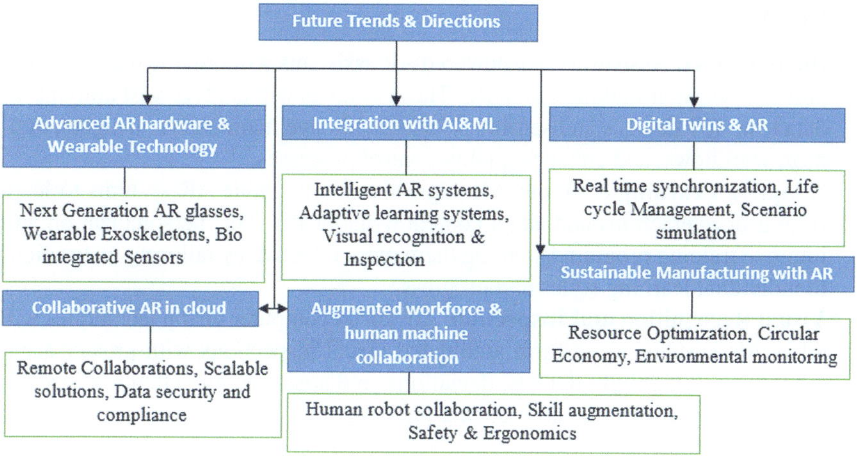

Fig. 3 Future trends and directions for AR in manufacturing

6.1 Advanced AR Hardware and Wearable Technologies

Trend: The development of more sophisticated AR hardware and wearable technologies is set to revolutionize manufacturing.

Direction:

- **Next-Generation AR Glasses**: Future AR glasses will be lighter, more comfortable, and capable of delivering higher-resolution visuals. Innovations in display technologies, such as holographic displays and retinal projection, will provide clearer and more immersive experiences.
- **Wearable Exoskeletons**: Combining AR with wearable exoskeletons will support workers in physically demanding tasks. These exoskeletons will provide augmented strength and endurance, guided by AR overlays that enhance precision and safety.
- **Bio-integrated Sensors**: Wearable AR devices will integrate bio-sensors to monitor worker health and fatigue levels in real-time. This integration will allow for adaptive AR content that responds to the user's physical and cognitive state, improving overall well-being and productivity.

6.2 Integration with AI and Machine Learning

Trend: The convergence of AR with artificial intelligence (AI) and machine learning (ML) will unlock new capabilities in digital manufacturing.

Direction:

- **Intelligent AR Assistants**: AI-powered AR assistants will provide real-time support and decision-making guidance. These systems will analyze vast amounts of data from IoT devices and sensors to offer predictive maintenance insights, optimize workflows, and enhance quality control processes.
- **Adaptive Learning Systems**: ML algorithms will enable AR systems to learn from user interactions and adapt content dynamically. This personalization will improve training programs, making them more effective by tailoring instructions to individual learning styles and progress.
- **Visual Recognition and Inspection**: Advanced computer vision techniques will allow AR systems to perform automated visual inspections with greater accuracy. AR will highlight defects, deviations, and areas needing attention, streamlining quality assurance and reducing human error.

6.3 Digital Twins and AR

Trend: The integration of digital twins with AR will provide a more comprehensive view of manufacturing processes and equipment.

Direction:

- **Real-Time Synchronization**: Digital twins—virtual replicas of physical assets—will be synchronized in real-time with AR environments. This synchronization will allow workers to visualize and interact with digital twins through AR interfaces, providing insights into system performance, predictive maintenance needs, and optimization opportunities.
- **Lifecycle Management**: AR-enhanced digital twins will support the entire lifecycle of manufacturing equipment, from design and prototyping to operation and decommissioning. This holistic approach will facilitate better planning, resource allocation, and sustainability efforts.
- **Scenario Simulation**: AR combined with digital twins will enable the simulation of different scenarios and "what-if" analyses. Manufacturers can test changes in production processes, equipment upgrades, and layout modifications in a virtual environment before implementation, reducing risks and costs.

6.4 Collaborative AR in the Cloud

Trend: The shift toward cloud-based AR solutions will enhance collaboration and data accessibility.

Direction:

- **Remote Collaboration**: Cloud-based AR platforms will allow geographically dispersed teams to collaborate in real-time. Engineers, designers, and technicians can work together on the same AR project, share insights, and solve problems collectively, regardless of their physical locations.

- **Scalable Solutions**: Cloud infrastructure will support the scalability of AR applications, enabling manufacturers to deploy and update AR solutions across multiple facilities with ease. This scalability will ensure consistent quality and performance across different sites.
- **Data Security and Compliance**: Advances in cloud security will address concerns about data privacy and compliance. Manufacturers will benefit from secure, centralized data storage and management, with robust access controls and encryption ensuring that sensitive information is protected.

6.5 Augmented Workforce and Human–Machine Collaboration

Trend: The concept of an augmented workforce, where humans and machines work collaboratively using AR, will become more prevalent.

Direction:

- **Human–Robot Collaboration**: AR will facilitate seamless collaboration between human workers and robots on the factory floor. Workers will receive real-time AR instructions on how to interact with robots safely and efficiently, optimizing workflows and reducing downtime.
- **Skill Augmentation**: AR will enhance the capabilities of workers by providing them with on-the-spot information, expert knowledge, and guided assistance. This augmentation will enable workers to perform tasks beyond their traditional skill sets, increasing flexibility and productivity.
- **Safety and Ergonomics**: AR will play a critical role in improving workplace safety and ergonomics. Real-time hazard detection and safety instructions will prevent accidents, while ergonomic assessments and adjustments guided by AR will reduce strain and injury.

6.6 Sustainable Manufacturing with AR

Trend: AR will contribute to sustainable manufacturing practices, aligning with the goals of Industry 5.0.

Direction:

- **Resource Optimization**: AR will help manufacturers optimize the use of resources by providing real-time data on material usage, waste generation, and energy consumption. This optimization will lead to more efficient and sustainable production processes.

- **Circular Economy**: AR will support the transition to a circular economy by enabling better tracking and management of products and materials throughout their lifecycle. AR-guided disassembly and recycling processes will improve resource recovery and reduce environmental impact.
- **Environmental Monitoring**: AR systems will integrate with environmental sensors to monitor and visualize the environmental impact of manufacturing activities. Workers will receive real-time feedback on emissions, energy use, and other sustainability metrics, fostering a culture of continuous improvement.

By anticipating these future trends and directions, manufacturers can strategically plan for the integration of AR technologies, driving innovation and achieving the transformative potential of Industry 5.0. The ongoing advancements in AR hardware, AI integration, digital twins, cloud collaboration, workforce augmentation, and sustainability will shape the next generation of digital manufacturing, ensuring competitiveness and resilience in a rapidly evolving industry landscape.

7 Ethical and Societal Considerations

The rapid adoption and advancement of augmented reality (AR) in digital manufacturing bring forth a range of ethical and societal considerations (Fig. 4). These considerations are crucial to address to ensure the responsible development and deployment of AR technologies. This section explores the key ethical and societal issues associated with AR in digital manufacturing, proposing solutions and best practices to navigate these challenges.

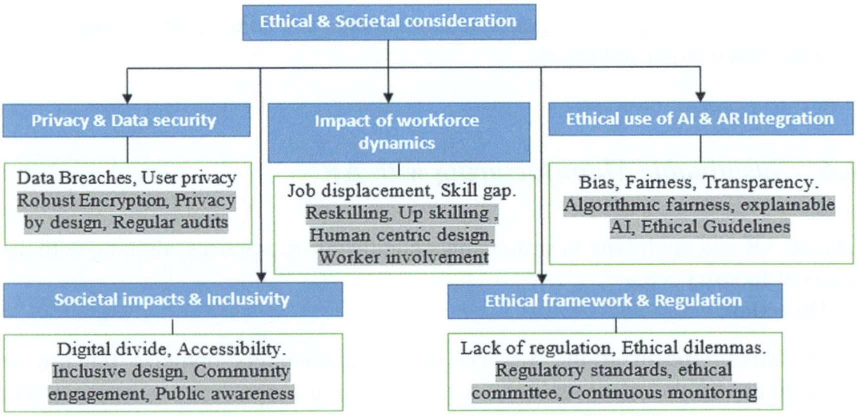

Fig. 4 Ethical and societal considerations of use of AR in manufacturing

7.1 *Privacy and Data Security*

Consideration: The extensive use of AR involves the collection, processing, and storage of vast amounts of data, including sensitive personal and operational information. Ensuring the privacy and security of this data is paramount.

Challenges:

- **Data Breaches**: The risk of data breaches increases with the integration of AR systems that collect and transmit real-time data.
- **User Privacy**: AR devices often collect personal information, including biometric data, raising concerns about user privacy.

Solutions:

- **Robust Encryption**: Implement end-to-end encryption for data transmitted between AR devices, cloud platforms, and edge computing nodes to protect against unauthorized access.
- **Privacy by Design**: Develop AR systems with privacy as a core principle, ensuring that personal data collection is minimized and users have control over their information.
- **Regular Audits**: Conduct regular security audits and vulnerability assessments to identify and mitigate potential risks.

7.2 *Impact on Workforce Dynamics*

Consideration: The introduction of AR in manufacturing can significantly alter workforce dynamics, affecting job roles, skill requirements, and worker well-being.

Challenges:

- **Job Displacement**: Automation and AR-driven processes may lead to job displacement for workers performing routine and manual tasks.
- **Skill Gap**: There may be a growing skill gap as the demand for AR proficiency and technical expertise increases.

Solutions:

- **Reskilling and Upskilling**: Implement comprehensive training programs to reskill and upskill workers, enabling them to adapt to new roles and leverage AR technologies effectively.
- **Human-Centric Design**: Design AR systems that augment human capabilities rather than replace them, ensuring that technology serves as an enabler for the workforce.
- **Worker Involvement**: Engage workers in the development and deployment of AR systems, incorporating their feedback to create solutions that address their needs and concerns.

7.3 Ethical Use of AI and AR Integration

Consideration: The integration of AI with AR introduces ethical dilemmas related to decision-making, transparency, and accountability.

Challenges:

- **Bias and Fairness**: AI algorithms used in AR systems may exhibit biases, leading to unfair treatment and outcomes.
- **Transparency**: Lack of transparency in AI-driven AR systems can make it difficult to understand how decisions are made and to hold systems accountable.

Solutions:

- **Algorithmic Fairness**: Develop and deploy AI algorithms with fairness in mind, ensuring that they are tested for biases and designed to provide equitable outcomes.
- **Explainable AI**: Focus on explainable AI (XAI) techniques that make the decision-making processes of AI-driven AR systems transparent and understandable to users.
- **Ethical Guidelines**: Establish ethical guidelines and best practices for the development and use of AI in AR, ensuring that systems are designed and operated with integrity and accountability.

7.4 Societal Impacts and Inclusivity

Consideration: The widespread use of AR in manufacturing has broader societal impacts, including issues of inclusivity, accessibility, and societal acceptance.

Challenges:

- **Digital Divide**: There is a risk of exacerbating the digital divide, where certain segments of the workforce may lack access to AR technologies and the skills to use them.
- **Accessibility**: AR systems must be designed to be accessible to all users, including those with disabilities.

Solutions:

- **Inclusive Design**: Develop AR systems with inclusive design principles, ensuring they are accessible to a diverse workforce, including those with disabilities.
- **Community Engagement**: Engage with communities and stakeholders to understand their concerns and perspectives, ensuring that AR adoption benefits society as a whole.
- **Public Awareness**: Promote public awareness and education about the benefits and implications of AR, fostering a more informed and accepting society.

7.5 *Ethical Frameworks and Regulation*

Consideration: The rapid pace of AR development necessitates the establishment of ethical frameworks and regulatory policies to guide its responsible use.

Challenges:

- **Lack of Regulation**: The regulatory landscape for AR is still evolving, leading to uncertainties and potential misuse of technology.
- **Ethical Dilemmas**: Ethical dilemmas may arise in situations where the use of AR intersects with privacy, security, and human rights.

Solutions:

- **Regulatory Standards**: Collaborate with policymakers to develop clear regulatory standards and guidelines for the use of AR in manufacturing, ensuring compliance with ethical and legal norms.
- **Ethical Committees**: Establish ethical committees and review boards to oversee the development and deployment of AR systems, providing guidance on ethical issues and ensuring accountability.
- **Continuous Monitoring**: Implement continuous monitoring and evaluation mechanisms to assess the ethical implications of AR technologies and make necessary adjustments.

By addressing these ethical and societal considerations, stakeholders in digital manufacturing can ensure the responsible and sustainable adoption of AR technologies. This proactive approach will help mitigate risks, promote inclusivity, and maximize the positive impacts of AR on the workforce and society.

8 Conclusion

8.1 *Summary of Key Points*

This chapter has explored the transformative potential of augmented reality (AR) in digital manufacturing, underscoring its integral role in shaping the future of Industry 5.0. We began by discussing the foundational concepts and technological advancements that have enabled AR to emerge as a pivotal tool in modern manufacturing environments. Key points from the chapter include:

1. **Foundational Concepts and Technological Advancements**: We discussed the core principles of AR, its integration with digital manufacturing processes, and the technological advancements that have propelled its adoption.
2. **Applications of AR in Manufacturing**: We explored practical applications of AR, such as in assembly, maintenance, quality control, and training, highlighting its ability to enhance efficiency, accuracy, and worker safety.

3. **Implementation Strategies and Best Practices**: We provided insights into effective strategies for implementing AR in manufacturing, emphasizing the importance of user-centered design, integration with existing systems, and continuous evaluation.
4. **Challenges and Limitations**: We addressed the technical, organizational, and human factors challenges associated with AR adoption, offering solutions to overcome these barriers.
5. **Case Studies and Real-World Examples**: We presented case studies demonstrating successful AR implementations in various manufacturing sectors, illustrating its impact and benefits.
6. **Future Prospects and Innovations**: We identified emerging trends in AR technology, such as advanced sensor integration, haptic feedback, and AR cloud computing, and discussed their potential impacts on digital manufacturing.
7. **Ethical and Societal Considerations**: We examined the ethical and societal implications of AR in manufacturing, emphasizing the need for privacy, data security, inclusivity, and responsible development.

8.2 Final Thoughts

As we move toward Industry 5.0, the integration of AR in digital manufacturing is poised to revolutionize how we design, produce, and manage products. AR's ability to merge digital information with the physical world offers unprecedented opportunities for innovation and efficiency. The future of AR in manufacturing will be characterized by several key developments:

1. **Human-Centric Manufacturing**: AR will play a crucial role in creating human-centric manufacturing environments, where workers are empowered with real-time information and tools to enhance their capabilities and well-being. This shift will lead to safer, more ergonomic, and more engaging workplaces.
2. **Collaborative and Connected Ecosystems**: The advancement of AR cloud platforms and edge computing will enable seamless collaboration across geographically dispersed teams. This interconnectedness will drive more efficient workflows, reduce time to market, and foster a culture of continuous improvement and innovation.
3. **Intelligent and Adaptive Systems**: The integration of AI with AR will result in intelligent and adaptive systems that can learn from user interactions, adapt to changing contexts, and provide personalized guidance. This will enhance the precision and effectiveness of manufacturing processes, leading to higher quality products and reduced waste.
4. **Sustainable and Resilient Manufacturing**: AR will contribute to sustainable manufacturing practices by optimizing resource usage, reducing errors, and enabling predictive maintenance. This will not only minimize environmental impact but also enhance the resilience and flexibility of manufacturing operations in the face of disruptions.

5. **Ethical and Responsible Development**: As AR technology continues to evolve, it is imperative to address ethical and societal considerations proactively. Ensuring privacy, security, and inclusivity, along with establishing robust regulatory frameworks, will be critical to the responsible development and deployment of AR in manufacturing.

In conclusion, the future of AR in digital manufacturing is bright and full of promise. By leveraging AR's capabilities, manufacturers can achieve new levels of efficiency, innovation, and competitiveness. As we embrace these advancements, it is essential to navigate the associated challenges thoughtfully and ethically, ensuring that AR technology serves as a force for positive transformation in the era of Industry 5.0.

References

1. Azuma, R. T. (1997). A survey of augmented reality. *Presence: Teleoperators & Virtual Environments, 6*(4), 355–385. https://doi.org/10.1162/pres.1997.6.4.355
2. Billinghurst, M., Clark, A., & Lee, G. (2015). A survey of augmented reality. *Foundations and Trends in Human-Computer Interaction, 8*(2–3), 73–272. https://doi.org/10.1561/1100000049
3. Mekni, M., & Lemieux, A. (2014). Augmented reality: Applications, challenges and future trends. *Applied Computational Science*, 205–214.
4. Buń, P., Grajewski, D., & Górski, F. (2021). Using augmented reality devices for remote support in manufacturing: A case study and analysis. Advances in Production Engineering & Management, 16(4).
5. Uva, A. E., Gattullo, M., Manghisi, V. M., Spagnulo, D., Cascella, G. L., & Fiorentino, M. (2018). Evaluating the effectiveness of spatial augmented reality in smart manufacturing: a solution for manual working stations. The International Journal of Advanced Manufacturing Technology, 94, 509–521.
6. Furht, B. (2011). *Handbook of augmented reality*. Springer Science & Business Media.
7. Peddie, J. (2017). *Augmented reality: Where we will all live*. Springer International Publishing.
8. Eswaran, M., & Bahubalendruni, M. R. (2022). Challenges and opportunities on AR/VR technologies for manufacturing systems in the context of industry 4.0: A state of the art review. Journal of Manufacturing Systems, 65, 260-278.
9. Fiorentino, M., Uva, A. E., Gattullo, M., Debernardis, S., & Monno, G. (2014). Augmented reality on large screen for interactive maintenance instructions. *Computers in Industry, 65*(2), 270–278. https://doi.org/10.1016/j.compind.2013.11.004
10. Kipper, G., & Rampolla, J. (2012). *Augmented reality: An emerging technologies guide to AR*. Syngress.
11. Schmalstieg, D., & Hollerer, T. (2016). *Augmented reality: Principles and practice*. Addison-Wesley Professional.
12. Ong, S. K., & Nee, A. Y. C. (2013). Virtual and augmented reality applications in manufacturing. In Springer handbook of augmented reality (pp. 681–706). Springer.
13. Zollmann, S., Hoppe, C., Klug, C., & Reitmayr, G. (2014). Augmented reality for construction site monitoring and documentation. *Proceedings of the IEEE, 102*(2), 204–219. https://doi.org/10.1109/JPROC.2013.2294314

Advanced Robotics Application in Industry 5.0 Manufacturing Processes

Vishakha Shadija, Vipin Sharma, and Rajendra Kumar Shukla

1 Introduction

Advances in robotics and digital technologies increase the pace of subsequent industrial revolutions. Past two decades witnessed the two industrial revolutions Industry 4.0 in 2011 and advent of Industry 5.0 in 2021. The Industry 5.0 popped up while society is embracing Industry 4.0 technologies [1, 2].

Previous industrial revolution, the Industry 4.0 paved the path toward smart factories, giving birth to production systems capable of making real-time decisions. Industry 4.0 integrates internet of things with cyber-physical systems assisting us to connect physical and virtual worlds. IoT and CPS enabled us to do functional and dynamic mapping of various things, serving the complexity of the twenty-first century. With IoT real-time data collection and processing has made significant contribution in Industry 4.0. In simple terms, Industry 4.0 can be understood as, industry whose brick is laid by smart networking and processes contributing in development of cyber-physical networks (CPS) to cyber-physical production system (CPPS). CPS are the systems that act as bridge between physical and virtual world through the employment of technical assets like sensors, actuators and more. Industry 4.0 had vast effects on industries, and it directly challenged industry in terms of production speed and scale, efficiency and way of operating. This revolution also bought changes to the machinery and equipment being used in industries and bought flexibility and innovation in production system [3–5].

One the highly emerging technology cobots also termed as collaborative robots is taking over the industry rapidly. Cobots assist in completing laborious and repetitive tasks effectively, with or without assistance. Also, cobots can work for long hours without breaks, hence one of the major reasons to employ them in industry

V. Shadija · V. Sharma (✉) · R. K. Shukla
Department of Mechanical Engineering, Medicaps University, Indore, Madhya Pradesh, India
e-mail: vipinsharma@medicaps.ac.in; rajendrakumar.shukla@medicaps.ac.in

Y. Kumar Singla et al. (eds.), *Digital Manufacturing in Industry 5.0*,
https://doi.org/10.1007/978-3-031-91500-0_38

[6–8]. With the integration of digital twins (DTs), the CPS went under transformation and increased productivity is achieved under real-time situations with the help of mathematical models employed in virtual world. Though DTs are great asset but they lack ability to lay stronger foundation between human and virtual world as they limit the interaction between two. To assist this problem, augmentation came into picture, which enabled human to closely experience and understand the digital behavior of CPS [9, 10].

In 2021, the concept of Industry 5.0 is adopted encompassing technologies with a different approach. The robotic systems in Industry 5.0 are elevated to a new level of human centric approach.

2 Historic Development

Manufacturing industry has always been giving hand to national economic conditions. Whether it was ancient times or present, manufacturing industry is an inseparable part of society, making remarkable contribution to evolve human life in better direction. Manufacturing industry went under numerous transformations with the emerging industrial revolutions, directly bringing changes into life of ours. The approximate timeline and technologies of different industrial revolutions are shown in Fig. 1 and Table 1 [11].

2.1 The First Industrial Revolution

The introduction to the first ever industrial revolution highly impacted manufacturing industry as it bought mechanized production to light which was fueled by water and steam. This industrial revolution led to expectable growth in the production

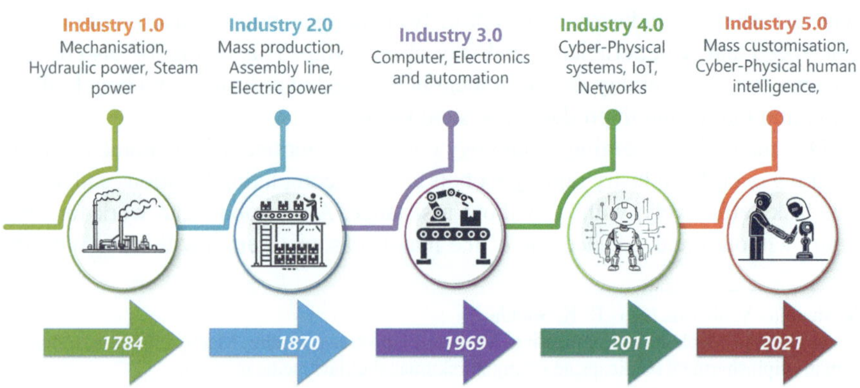

Fig. 1 Timeline of different industrial revolutions from Industry 1.0 to Industry 5.0

Table 1 Industrial revolutions and their main driving technologies

Industrial revolution	Main drives
Industry 1.0	Mechanized production Steam and water power
Industry 2.0	Electricity Mass production
Industry 3.0	Automation Computers and IT Internet
Industry 4.0	Artificial learning (AI) Machine learning (ML) Internet of Things (IoT) Big data and robotics Cloud computing
Industry 5.0	Digital twins Augmentation Soft robotics Human–robot collaboration

industry as it directed the handwork to machines. The star of this earliest revolution was steam power, followed by steam engines. This revolution not only transformed the manufacturing process but also led bricks for foundation of further developments in various industries, manufacturing industry is no exception. Manufacturing industry went under numerous evolutions bought in picture because of every single industrial revolution. The first industrial revolution widely introduced power machines leading to improved production and efficiency of various industries and as well as leveled up living standards [11, 12].

2.2 *The Second Industrial Revolution*

The Second Industrial Revolution came into view in 1870s, this industrial revolution drove toward mass production. This revolution was ruled by electric power, the emergence of electricity bought ground breaking results by perking up the scale and the speed at which production was held and bought several modifications in existing industries. The manufacturing industry observed great bloom in production line and took industry to align with more commercial interests. This revolution was not only bounded till industrial development but also fostered new innovations, which simplified human life [12, 13].

2.3 The Third Industrial Revolution

Industry 3.0, the revolution that bought whole new era of digitalization in the picture, bought significant impact on society and led to reshaping of the industries. As every industrial revolution arrived with its remarkable contributions to the society, Industry 3.0 lend hand to manufacturing by introducing digital technologies like computers, caged robots, and information technology. This revolution bought lots of changes to the industries, made them inseparable from technology. The emergence of these technologies made manufacturing process more reliable and advance. This industrial revolution reduced the investment of manpower in laborious and repetitive tasks. The constructive raise in scale and speed of production was observed due to the incorporation of computers, industrial robots, which navigated toward efficient and blooming production. Industry 3.0 also introduced the technology that is currently driving the society like king—The internet [12, 13].

2.4 The Fourth Industrial Revolution

The Fourth Industrial Revolution or Industry 4.0 integrated smart technology into our lives. The main drives of this revolution are including artificial intelligence (AI), machine learning (ML), internet of things (IoT) and big data. The machines got smarter even without assistance, on their own as AI and ML came in light, the IoT made it possible for machines to connect and communicate which further assisted with easy information exchange that too quiet efficiently. Big data, made it possible to make decision by analyzing huge amount of available data in a convenient and time saving manner. With this revolution the manufacturing became more flexible, efficient and prospering. Industry 4.0 helped establishing cyber-physical systems (CPS), numerous interconnected machines and systems contributing to flexible and leveled up manufacturing processes [13, 14].

2.5 The Fifth Industrial Revolution

Industry 5.0 aims at making manufacturing industry more human-centric, that is, adding personalized touch to production. The industry claims to serve sustainability, better lifestyle and enhanced efficiency by establishing human-centric approach [15].

This industry will completely change the work flow by incorporating human capabilities with advanced technology, making human and machines work together as a team, complimenting each other to gain shared goals. Industry 5.0 prioritizes sustainability, the new drive for smart businesses, cutting down the waste generated by industries, reusing the resources leading toward more efficient production while taking initiative to attain clean and green planet. The goal of Industry 5.0 includes promoting human creativity and originality to take society toward innovative age. Industry 5.0 helps enhancing human capabilities rather than to eliminate their involvement by machines, humans are the heart of the industry, that is, industry revolves around humans to escalate the technologies. The technologies contributing to industry are digital twins, augmentation, soft robotics which foster human–robot collaboration to bring out best out of humans and machines. This industrial revolution will level up the manufacturing while ensuring safety and efficiency [12, 13, 16].

As the industries would be highly structured on data and real-time decision making, the demand for enhanced data processing and data security emerges for efficient production, drives attention on developing improved algorithms and software which can support the complex production system.

3 Industry 5.0

Industry 5.0 is the new market transforming revolution, focusing to serve smart, sustainable, and human-centric approaches in order to upgrade and bloom different industries. This revolution not only concerns to develop automation and enhance machine productivity but also focuses on promoting creativity and originality among humans. Industry 5.0 urges for enhance interaction between humans and machines to help in developing pocket-friendly, sustainable and human-centric goods. Industry 5.0 aims to enhance the potential of humans and robots by building a framework for them where they work as a team and constantly work to achieve their shared goals. Industry 5.0, this revolutionary industry intends for handing over the jobs, responsibilities based on capabilities to acquire best out of human–robot partnership [17, 18].

This industry not only involves sharing data between humans and robots but also focuses to develop safe working environment, intensifying creativity, skill development of workers and increased human potential rather than to replace humans by bots. Industry 5.0 aims to introduce production systems that are effortlessly able to adopt to their working conditions and surroundings, which further makes these systems smoothly accommodable by humans or workers. This emerging industry cannot only be limited to human–human or human–robot collaboration but also can emerge robot-robot collaboration which can lead to shift focus from investing manpower on time consuming, exhausting tasks to attain innovative and foresighted revolutionary blooming of the society and mankind [17, 19].

Meanwhile, Industry 4.0 is more focused on maximizing technical achievements and growth, Industry 5.0 intends to walk on path which inclines with human-centric production system which is sustainable and reliable and feeds the demands and needs of humans. Industry 5.0 focuses to direct human and machine intelligence on the path where they elegantly complimenting each other to gain shared goals while not ignoring key characteristics like sustainability, creativity and standards of production [17, 20].

3.1 Principles of Industry 5.0

Human-centric approaches, sustainability and resilience are like spine of this industrial revolution which makes this revolution not only concern for betterment of humans but also for our planet. Figure 2 indicates the basic principles of Industry 5.0 [21].

- Human-centric approach refers to approach that places human need at top, leading to design roles that potentially help establishing and adopting such approaches. This not only ensures human satisfaction but also promises safety and overall well-being of humans and their working environment.
- Sustainable development refers to accomplishing needs of present generation without coming to deal with needs of future generation. Sustainability is the aspect of development which helps looking into bigger picture, ultimately helping to eliminate waste production and environmental concerns. Sustainability is not only mindful use of resources but also re-using and re-cycling them.
- Resilience can be understood as a quality of production system to adapt and learn from different situations, assisting them to withstand any obstacle. Resilience ensures smooth development of any industry [21, 22].

Fig. 2 Basic principles of the Industry 5.0

3.2 *Technologies of Industry 5.0*

There are a lot of different technologies contributing in Industry 5.0 with the aim of improved and efficient manufacturing procedure, and some of the technologies are digital twins, artificial intelligence, machine learning, and much more. Technology contributes efficiently in Industry 5.0, opening door to reach optimum potential of the industry [23–25]. The major technologies of Industry 5.0 are presented in Fig. 3.

3.2.1 Edge Computing

Edge computing is freshly evolved technology. Edge computing enables decentralized data processing and examining through the source instead of centralized cloud servers. It can contribute in Industry 5.0 to boost its efficiency and productibility along with ensuring data security. Under this technology, data is generally collected through sensor and different devices for further analyzing and processing. It potentially can support real-time monitoring, also enabling to ensure workers safety and upgrade the product quality. Edge computing fosters efficient and real-time decision making, assisting in boosting the accuracy of production system. As Industry 5.0 is human-centric revolution, this technology didn't upset us, it helps to develop customized products and incorporates predictive maintenance, reducing the possibilities of mishaps [26, 27].

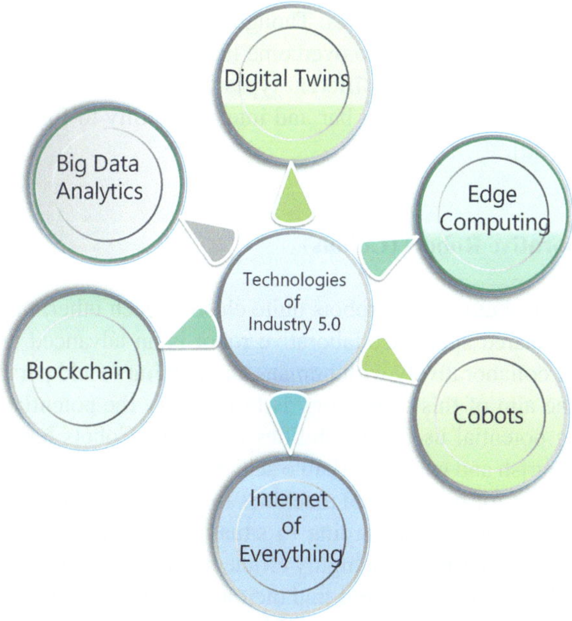

Fig. 3 Major technologies of Industry 5.0

3.2.2 Digital Twins (DTs)

Digital twin helps in generating virtual copies of numerous physical systems and processes by utilizing real-time data, algorithms, and more. DTs can potentially foster smooth communication between humans and machines thereby help establishing strong HRC. These help to examine and enhance the system under a virtual environment. These help us with real-time operations on system like check, control and upgrade the system. DTs have actively played role in Industry 4.0; furthermore, they also contribute in Industry 5.0 more significantly. These are also capable of enabling predictive maintenance, ensuring human and machine safety. Hence, DTs are the technology helping Industry 5.0 to stand stronger by serving increased productivity and efficiency of production system [28].

3.2.3 Internet of Everything (IoE)

IoE is the technology that can highly evolve every aspect of our life. IoE is the further development of IoT, which is the technology helping numerous devices to connect and assisting with them to share and analyze the data or information. With IoE the whole manufacturing process can be carried out effortlessly and under real-time monitoring. This technology fosters decision-making through analysis of available data. IoE can contribute in Industry 5.0 to escalate scale and speed of production with enhanced efficiency. This can help with quality control and ensuring safety of workers. IoE can find its potential application in the logistics and supply chain management, assisting in numerous tasks. Though this technology is an advance concept, it still has some challenges to overcome like security of data and privacy. IoE can help connecting completely different types or non-identical machines, further arising the demand for setting the bar and interoperability to foster flawless communication [29, 30].

3.2.4 Collaborative Robots (Cobots)

Industry 5.0 brings humans and robots quite close to each other, they engage with each other in a shared space. Collaborative robot is an advanced technology that fosters working collaboratively with humans and helping to add personalized touch to products. The aim of this technology is to intensify the potential of humans as well as making potential use of capabilities of robots. Cobots lend hand to small businesses to flourish and gain remarkable achievements. Collaborative robots serve a safer environment for humans and machines to work together, and they eliminate unwanted situations and act according to situation. Due to their capabilities and strengths, collaborative robots are highly efficient and dependable and also contribute to escalate the production. These find their applications in various fields including manufacturing and healthcare. With cobots actively participating in manufacturing process can contribute in reducing the production cost and boost the production [31–33].

3.2.5 Big Data Analytics

Big data analytics help to analyze vast and different data collected from numerous sources. Big data analytics utilizes different technologies such as machine learning, artificial intelligence etc. to analyze the data. It assists companies to make real-time decisions while gracefully managing the vast data. Helps industry to enhance the quality of production by giving recommendations related to different events. It also helps with customization of products, further directing to enhance the production. It lays bridge between industry and consumer's behavior, encouraging to dig out new possibilities. Big data analytics takes customer satisfaction as prime objective [34, 35].

3.2.6 Blockchain

Blockchain is the rapidly growing and developing technology contributing significantly to Industry 5.0. Blockchain offers decentralized and secured data management of the numerous connected devices. This offers transparency with management, making it more reliable. Blockchain can incorporate smart contracts to ensure safety. It can offer enhanced safety and secure data sharing by focusing on compartmentation of data. They can also assist with receiving and collecting data [36, 37].

Some of the other technologies are including 6G, exoskeleton, 4D printing, network slicing, extended reality, and private mobile network.

3.3 Advantages and Applications of Industry 5.0

Industry 5.0 due to the technical advancements and human-centric approach it finds its applications in various industries. The main advantages of Industry 5.0 are presented in Fig. 4.

There are wide range of application of Industry 5.0 in the realm of manufacturing and healthcare. Healthcare is the one of the important sector of our society, and with emergence of Industry 5.0, it has potential to transform the sector, providing improved medical services. Industry 5.0 can bring smarter approaches for diagnosis and treatment of different diseases. Robots can be employed to assist with surgeries, boosting the accuracy and they can even be involved in regular tasks and check-ups. AI can help with enhancing overall efficiency and quality of healthcare facilities. Hence, Industry 5.0 can help healthcare to grow and develop, also incorporating treatments satisfying individual needs. The major applications of Industry 5.0 are shown in Fig. 5 and are discussed in the later sections. Some other important

Fig. 4 The main advantages of Industry 5.0

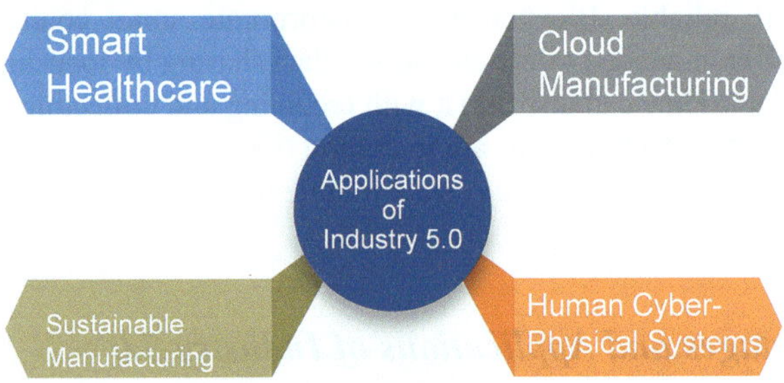

Fig. 5 Major applications of Industry 5.0

applications of Industry 5.0 includes-supply chain management, education and disaster management [38, 39].

3.4 Challenges in Industry 5.0

The major challenges of the Industry 5.0 are presented in Fig. 6 and discussed below:

- Requires government to invest a good sum of money for the success of Industry 5.0.

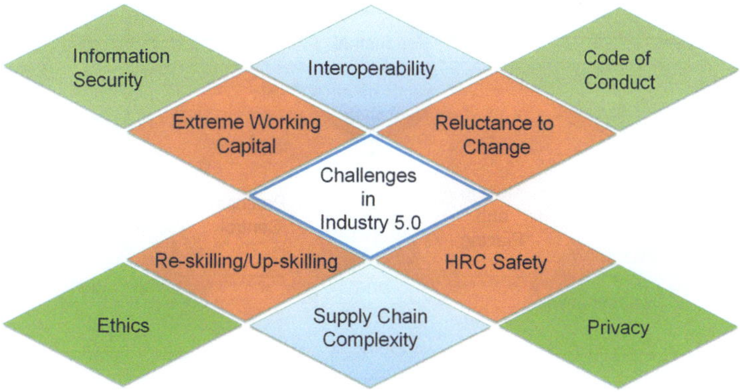

Fig. 6 Challenges in Industry 5.0

- While putting in effort to make production system flexible, it also increases the complexity of production system.
- People's perspective on Industry 5.0, potentially can lead to diverge people's values and acceptance, making it challenging to be accepted socially.
- Emerging technologies like bio-inspired technologies requires enhanced attention of ours [24, 40].

4 Robotics Applications in Industry 5.0 Manufacturing

Robotics is playing active role in smart manufacturing and highly influencing the overall industry. Industrial robots were firstly introduced to production line, where robot like industrial manipulators, collaborative robots, automated guided vehicles (AGVs), and even unmanned aerial vehicles (UAVs) are bought into work. The main industrial applications of robots are shown in Fig. 7.

4.1 Human–Robot Collaboration

The Industry 4.0 viewed human–robot collaboration (HRC) as the measure to improve the overall production and maximize results, whereas, the emerging Industry 5.0 interprets it as a human-centric approach toward production which is efficient, sustainable and encouraging humans and robots to work as a team. The HRC as the term suggests, humans and robots collectively working in harmony to achieve certain shared goal efficiently. Humans and robots collaborate or we can say cooperate or join forces to gracefully demonstrate the power of working together with the aim of achieving efficient results. The HRC not only focuses on enhancing the quality of work but also works toward promoting productivity and originality in

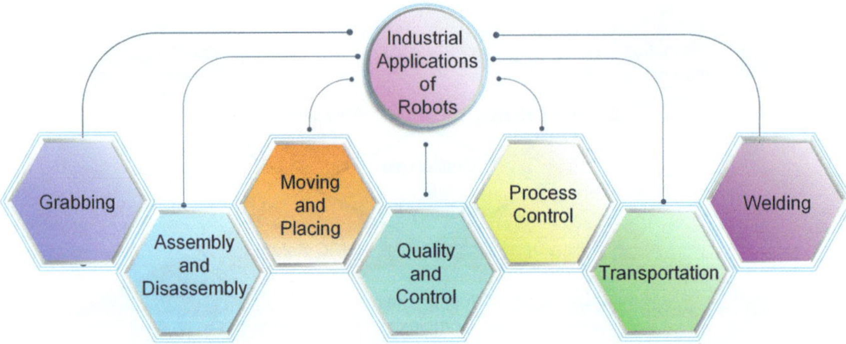

Fig. 7 Main industrial applications of robots

workers. Also, it directs toward successful cooperation between machines and humans while ensuring healthy environment [41].

Human–robot interaction (HRI) may seem as an easy job to target, but, on contradiction, it is not as uncomplicated as it appears to. There is a massive difference among how a robot and a human senses, views and processes things and its environment. A robot relies on sensors, control systems and algorithms to interpret stuff whereas a human relies on its brain and gained knowledge to understand and react to things. Therefore, it is required to tailor the way human–robot interact in real-time, ensuring to withdraw all possible miscommunications, though the HRI is evolving from time-to-time and there is upgrade in interaction observed between two, but there are goals of better interaction yet to achieve to direct us toward victorious HRC [42].

Since the regularly developing, upgrading, and emerging technologies in the field of robotics, HRC finds its crucial applications in various industries such as manufacturing, healthcare, supply chain management, education, and significantly more such industries that drive the overall condition of society. Looking at the influence possessed by HRC, it has highlighted the way we promote HRI should be safe, human-centric, and targeted to attain maximum potential while not neglecting their individual strengths and capabilities [41].

HRC helps in broadening the interaction between human and robots which leads the path to human-centric developments. HRC assists in enhancing the quality of production by working on feedback received. HRC accommodates various technologies to promote real-time learning and adopting. The key aspects of HRC are presented in Fig. 8. As HRC focuses on improving overall efficiency while ensuring safe working environment, eliminating possible injuries and accidents. Robot can easily shoulder tasks which are exhausting, repetitive and threatening, which provides space to workers where they put their brain into innovating value adding tasks which needs human creativity and originality, so this why we can say HRC fosters

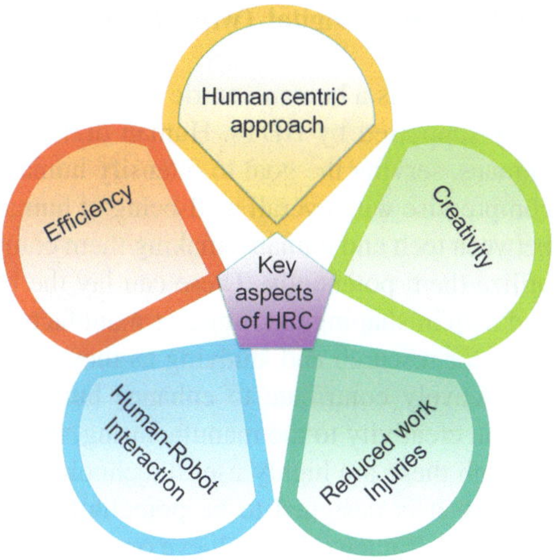

Fig. 8 Key aspects of human–robot collaboration

human creativity. On exploring capabilities of bots, we can employ them to function in vicious working conditions which help eliminating risk of injuries and threat to human lives. HRC places human needs at heart which further makes it boon for sustainable and innovative development while making optimum use of human and robot potential [41, 43, 44].

4.1.1 Digital Twin-Based HRC (DT-HRC)

DT contributes in HRC to enhance its testing and productivity by utilizing numerous control algorithms and plan of action for interaction to initiate better understanding of physical system. DTs help eliminating possible accidents by identifying them and developing paths to overcome them, hence increasing efficiency of human–robot collaboration. DTs can help humans and machines to communicate and coordinate leading to magnificent HRC. Hence, DTs contribute in HRC to bloom productivity and enhance efficiency. Furthermore, we have the terms digital thread, digital shadow and digital twin, enabling us to collect, store and process data [45].

The DTs can help us with designing and fuel up the HRI, assisting in boosting up the overall performance of HRC. A typical DT-HRC constituents of imitation and visualizing tools, algorithm and human and robot model. Human model is term used for model that can precisely observe and imitate human actions and movements, on the other hand robot models are the models designed to utilize strengths and potential of a robot. Imitation and visualization tools help with to examine and evaluate the HRI, whereas algorithm fosters real-time interaction among humans and machines [46].

4.1.2 Human Digital Twins (HDTs)

Industry 5.0 is a human-centric industry, where smart manufacturing can be potentially supported by HDTs. Human digital twin refers to virtual representation of humans, serving the goal to intensify human capabilities while making sure not to compromise with overall well-being of humans. HDTs are like the connecting links between tech and humans, making them compatible by realizing their strengths and utilize them potentially. These can lay the brick for smart manufacturing systems and help in shaping the future of manufacturing industry. HDTs are the current hot topic for research and studying as they are one of the futuristic approaches. They can actively contribute to enhance human capability and skills. These integrate human creativity to the manufacturing, increasing the quality of industrial process. Though these are highly useful technology but their practical enforcement is quite limited. It is designed to be potentially capable of ensuring human safety while providing framework to achieve optimum potential [47, 48].

Digital Twin and Human Digital Twin

HDTs and DTs serve various similar characteristics like real-time analysis, high accuracy, reliability, dynamic, distinguishable and more. But there are certain aspects where HDT serve better than DTs, these aspects are:

- HDTs provide better work environment than DTs, as they provide higher security to humans. Making them safer to exist in same surrounding with humans.
- To make potential use of DTs one has to be highly skilled which demands for time and efforts whereas HDTs are easy to work with. They are designed according to human convenience and to be effortlessly adaptable.
- HDTs put human at top priority, no compromise with physical and mental well-being of humans.
- HDTs have effective data management, contributing to smooth execution, whereas DTs serve a slightly complex management of data [47, 49].

HDT Model

HDT model comprises of three contributing constituents, which are human entity (HE), virtual entity (VE) and interactive systems (IS). This is a 3-D model structuring framework of HDT. Further, idea about these components is shared below:

- Human entity

- HDTs place human at the core; therefore, the overall concept of HDT revolves around humans. Humans have significant and distinguished role to play. HDTs imitate outer as well as inner aspects of humans. They can successfully give life to human behavior and communication in a digital or virtual space. The industrial human need model consists of five-stage hierarchy levels, whether from fundamental manufacturing needs to personal growth and self-actualization, every level is developing over previous one while focusing on overall development. These levels are physiological needs, safety needs, belonging needs, esteem needs, and self-actualization needs. The HE plays crucial role in HDT and is bounded to play its roles of designer, observer, analyst, decider, executor and user.

- Virtual entity
- Followed by human entity, virtual entity plays a crucial role in HDTs. VE accumulates the resourceful data and information related to human entity. The primary objective of virtual entity is to break-down the complex, layered characteristics which HDT utilizes over the complete procedure. VE consists of numerous contributing representations formulating VE model namely, anthropomorphic representation, materialization representation and rule representation. Anthropomorphic representation consists of models potentially rich with human characteristics and properties, materialization representation covers geometries and behavior of existing system or surrounding linked to human entity and at last rule representation covers the skills like decision-making, reasoning, and judgment. All of the virtual components are responsible for smooth interaction between humans and machines.
- Interactive systems
- Interactive systems are built on combination of linked data, services and objects among the human and virtual entity. There are three main components of interactive systems, namely, data, service, and physical objects. All kinds of data including physical data, processed data, virtual data, and consumption data associated with human entities, virtual entities, services, and physical objects are stored in a data model [50–53].

4.1.3 Machine Learning in HRC

Machine learning, is a boon which can foster finer human–robot collaboration. HRC is putting great emphasis on leading human and machines to work as a team, as a single unit; however, there are yet complications coming in way to overcome and discover the unexplored reasons leading to failure of or unsatisfactory HRC.

4.1.4 Sensors in HRC

The HRC can be achieved through following two approaches: (1) Off-line programming, in this approach where the robot is guided manually, that is, leaning through hand guiding. This can help robot to learn complex tasks, somehow, similar to how a parent teaches his kid and (2) on-line interaction, in this approach of learning external sensory systems are employed to foster interaction between humans and robots. The sensory system is the fundamental brick of this approach and thus influences the overall capability and efficiency of HRC. Sensor-based HRCs are still challenging and developing technological concept of our industries, which is facing numerous challenges like safety, data collection, reliability, and more. Sensors decide the safety and accuracy of HRC, and vision sensor, motion sensor, force and torque sensor, etc. are the some types of sensors employed for HRC [54, 55].

4.2 Smart Manufacturing

Smart manufacturing enables the working of machines and humans in a collaborative, sustainable and efficient manner. In Industry 5.0, repetitive and precision-demanding tasks are performed by robots to ensure efficiency and accuracy while allowing humans to focus on creative and high-value tasks. The implementation of HRC with AI and ML, transforms manufacturing into smart manufacturing to achieve improved productivity and sustainability [56, 57].

Robots with AI can learn and reprogram to suit with different tasks, thus enabling mass customization and flexibility in manufacturing. The continuous working of robots ensures the production with high level of accuracy and improved product quality. Further, AI-driven robotics systems can perform real-time quality control by using sensors and vision systems to find out deviations in products. This ensures a high quality of products and reduction in wastage [57].

Advanced robotics is integrated with AI, IoT and big data analytics which enables systems to make decision in real-time. This enables manufacturing processes more responsive to changing conditions and production parameters can be adjusted according to inputs derived from data. Moreover, data driven insights can be used to improve future production cycles. Workers safety is also ensured in the smart manufacturing due to advanced robotics as robots can perform hazardous task and work easily in extreme conditions which reduces the risk to human workers [57, 58].

4.3 Cloud Manufacturing

Manufacturing process incorporating cloud technology is termed as cloud manufacturing. Cloud manufacturing transformed manufacturing to a new level by integrating cloud computing, IoT, and digital technologies. The cloud manufacturing basically transforms traditional manufacturing systems into cloud-based service platforms, where customers, suppliers, designers and manufacturers can share, collaborate and use various manufacturing facilities over the internet. Important aspect of cloud manufacturing is to provide a collaborative approach of manufacturing. It can be easily understood with an example that a manufacturing company can design their product in one country or location, make the prototype in another, and manufacture the actual product in a third location by using cloud manufacturing technology [25, 59].

The main benefits of cloud manufacturing are collaboration or resource pooling, cost efficiency, resource optimization, scalability and flexibility in production and real time monitoring. Nonetheless, security and data privacy, complex operation and latency are the major challenges encountered in adoption of cloud manufacturing. The cloud manufacturing enables collaboration and resource pooling of production equipment, design software, and even human expertise, across different locations or companies. So, designers, experts and manufacturers of different companies can work together in real time. Cost efficiency can be achieved by cloud manufacturing due to investment reduction in machinery, software and experts. In cloud manufacturing, investment is limited to the usage of the resources which decreases the operational cost. Furthermore, predictive analysis and AI based tools enables the manufacturing unit in better decision making. It enables improved data storage and processing. It can serve industry to boost production and making products affordable. It can even give hand to small businesses to grow and evolve. The cloud manufacturing also offers scalability and flexibility, manufacturers can easily scale up their operations without investing heavily in new infrastructure or machinery [60, 61].

4.4 *Sustainable Manufacturing*

The principal concept of Industry 5.0 is sustainability and it enables sustainable manufacturing in industries to achieve the primary goal of sustainability. In Industry 5.0, advanced robotics system works in collaboration with humans which leads to increase in the efficiency of the production processes. Further, the increase in efficiency means better resource utilization and reduction of waste. So, the HRC or cobots transform the manufacturing in to a sustainable manufacturing [62].

Advanced robotics enables a significant level of precision to manufacturing processes by using sensors, AI driven analytics and machine learning. This ensures optimum utilization of resources and energy to align with goal of sustainability. The robotic system can effectively monitor and assess product quality in real-time to minimize defects which reduces the material wastage. On-demand and custom manufacturing can be achieved by advanced robotics which enables industries to produce according to the demand at that time. It can also be flexible to adapt the changes in demand, supply and external conditions. The use of AI and predictive analysis can easily predict potential barriers and forecast the future demand. This can reduce overproduction and unsold inventory leading to sustainable manufacturing [63].

Predictive maintenance capabilities are also offered by advanced robotics which can effectively monitor the condition of machinery in real-time and predicts the need of maintenance without delay. This will facilitate timely maintenance which can extends the lifespan of machinery and reduces the chances of unexpected breakdown. Furthermore, advanced robotics also improves the efficiency of supply chains by automating packaging, sorting and delivery processes. This can help in reduced fuel consumption and emissions, aligned to achieve more sustainable supply chain and reduces the environmental impact [64, 65].

5 Summary

The drive from Industry 4.0 to Industry 5.0 is major transformation for not only governing technological achievements but also escalating the overall scenario of manufacturing industries through engagement of humans in the core manufacturing processes. Industry 4.0 had automation and smart technologies at core along with prioritizing the digitalization of industrial systems through integrating concepts like CPS, IoT and AI; meanwhile, Industry 5.0 serves fresh concepts like productivity and sustainability to be at top priority. The main objective of Industry 5.0 is to incorporate human-centric approach for production and to broadly define the role of human in the industry. Rather than replacing humans with machines, it brings the two together, in a direction to complement each other's capabilities and making best use of them, to work as a team and boost overall production by contributing their parts to the manufacturing processes. This aims to builds harmonious relationship among humans and machines, which further fosters decision-making, creativity and problem-solving capabilities of a worker. For being able to develop adaptable production systems, the factors like resilience and sustainability play a crucial role, and these production systems will be capable to handle different encounters efficiently.

Cobots added new dimensions to advancement in manufacturing by making men and machines as a co-worker. These are the robots completely differing from a traditional robot. Cobots are designed with the aim to work alongside humans in shared workspace while ensuring safety of humans. These bots can assist with repetitive as well as laborious tasks, providing human workers freedom to focus on escalating their creativity and skills. Hence, to conclude these are human-friendly bots offering perks like flexibility, safety, and efficiency throughout the manufacturing process. Real-time monitoring offered by DTs improves the quality and reduction in wastage. It offers real-time stimulation along with aid to optimize production environment. Because of its quality to capture and work on real-time data, makes it a fit tool for continuous monitoring, analysis, and refinement of industrial operations. The DTs are not only responsible to bloom production but also assist to improve the HRC as well as collaboration between humans and other machines. Further, DTs also enables predictive maintenance and decision-making.

Industry 5.0 is considered to be a promising approach finding its application in number of sectors of society including manufacturing, supply chain, healthcare and more. But it still has to face major challenges like complexity with respect to integration of highly advanced technologies like cobots and IoT to the manufacturing process as it not only requires significant investment but also respective innovation to assist the integration of emerging technologies. Another major concern for the industry is data privacy, and it is must to ensure data security and privacy as the data is the foundation of this digitalized industry. Along with these, Industry 5.0 collides with social and cultural challenges, which hinder Industry 5.0 to prosper.

6 Future Scope

Futuristic approach beyond Industry 5.0 is to increase the level of automation up to complete automation of a system in a sustainable manner. Further there are numerous aspects which are to be considered for future exploration and advancements like human–robot collaboration, digital twins, intelligent and autonomous robotics, circular economy, virtual reality, AI and sustainability for imagining the futuristic concept of Industry 6.0. Advancement in computer science specifically in the realm of quantum computing will provide new pathways.

Researchers are focusing on enhancing human–robot collaboration for fostering smoother coordination between humans and robots. Emergence of technologies like machine learning and artificial intelligence can potentially enable bots to understand and anticipate human behavior, leading the path for improved teamwork. With regular advancement in technology bots are tailored to improve their overall capabilities and make them autonomous. Robots can foster real time learning and adaptation through integration of machine learning and artificial intelligence. Being capable of real-time functioning enables them to potentially overcome complex and unpredictable tasks. Another emerging technology, human digital twin contributes to bring personalized touch to production along with focusing to escalate productivity and safety. Human–robot collaboration can be incorporated in various sectors like healthcare, construction and more as it is a perfect blend of human creativity and precision offered by robots.

Advancement in robotics can be further directed toward recycling, optimization of resources and reducing the industrial waste produced, clearly contributing towards a circular economy. The main approach of the circular economy is focused on reducing waste, reusing resources and recycling material, fostering a healthier and safer environment, and designing robots while focusing on making them capable to handle complex disassembly of electronics, automobiles, and numerous products, further extracting reusable components and other materials. Robotics can also foster effective resource management with the initiative to reduce energy consumption, hence contributing to reducing carbon footprints and directing toward a sustainable approach for production systems. The emerging technology like bio-inspired robotics and green technologies are another approach for sewing sustainability into production. Renewable energy sources are to be integrated to decrease the external energy requirement. A sustainable energy model can be adopted that can fulfill its energy demand on its own. Recent researches in increasing output from renewable energy sources are promising and can be capable of transforming an industry into a sustainable unit.

References

1. Adel, A. (2022). Future of Industry 5.0 in society: Human-centric solutions, challenges and prospective research areas. *Journal of Cloud Computing, 11*(1), 40.
2. Xu, X., Lu, Y., Vogel-Heuser, B., & Wang, L. (2021). Industry 4.0 and Industry 5.0—Inception, conception and perception. *Journal of Manufacturing Systems, 61*, 530–535.
3. Elnadi, M., & Abdallah, Y. O. (2024). Industry 4.0: Critical investigations and synthesis of key findings. *Management Review Quarterly, 74*(2), 711–744.
4. Ghobakhloo, M. (2020). Industry 4.0, digitization, and opportunities for sustainability. *Journal of Cleaner Production, 252*, 119869.
5. Javaid, M., Haleem, A., Singh, R. P., & Sinha, A. K. (2024). Digital economy to improve the culture of Industry 4.0: A study on features, implementation and challenges. *Green Technologies and Sustainability, 2024*, 100083.
6. Keshvarparast, A., Battini, D., Battaia, O., & Pirayesh, A. (2024). Collaborative robots in manufacturing and assembly systems: Literature review and future research agenda. *Journal of Intelligent Manufacturing, 35*(5), 2065–2118.
7. Liu, L., Guo, F., Zou, Z., & Duffy, V. G. (2024a). Application, development and future opportunities of collaborative robots (cobots) in manufacturing: A literature review. *International Journal of Human–Computer Interaction, 40*(4), 915–932.
8. Vicentini, F. (2021). Collaborative robotics: A survey. *Journal of Mechanical Design, 143*(4), 040802.
9. Banerjee, S., Jesubalan, N. G., Kulkarni, A., Agarwal, A., & Rathore, A. S. (2024). Developing cyber-physical system and digital twin for smart manufacturing: Methodology and case study of continuous clarification. *Journal of Industrial Information Integration, 38*, 100577.
10. Mariani, S., Picone, M., & Ricci, A. (2024). Agents and digital twins for the engineering of cyber-physical systems: Opportunities, and challenges. *Annals of Mathematics and Artificial Intelligence, 92*(4), 953–974.
11. Nahavandi, S. (2019). Industry 5.0—A human-centric solution. *Sustainability, 11*(16), 4371.
12. Grabowska, S., Saniuk, S., & Gajdzik, B. (2022). Industry 5.0: Improving humanization and sustainability of Industry 4.0. *Scientometrics, 127*(6), 3117–3144.
13. Valette, E., El-Haouzi, H. B., & Demesure, G. (2023). Industry 5.0 and its technologies: A systematic literature review upon the human place into IoT-and CPS-based industrial systems. *Computers & Industrial Engineering, 184*, 109426.
14. Martini, B., Bellisario, D., & Coletti, P. (2024). Human-centered and sustainable artificial intelligence in Industry 5.0: Challenges and perspectives. *Sustainability, 16*(13), 5448.
15. Sarkar, B. D., Shardeo, V., Dwivedi, A., & Pamucar, D. (2024). Digital transition from Industry 4.0 to Industry 5.0 in smart manufacturing: A framework for sustainable future. *Technology in Society, 78*, 102649.
16. Barata, J., & Kayser, I. (2023). Industry 5.0—Past, present, and near future. *Procedia Computer Science, 219*, 778–788.
17. Horvat, D., Jäger, A., & Lerch, C. M. (2024). Fostering innovation by complementing human competences and emerging technologies: An Industry 5.0 perspective. *International Journal of Production Research*, 1–24.
18. Murtaza, A. A., Saher, A., Zafar, M. H., Moosavi, S. K. R., Aftab, M. F., & Sanfilippo, F. (2024). Paradigm shift for predictive maintenance and condition monitoring from Industry 4.0 to Industry 5.0: A systematic review, challenges and case study. *Results in Engineering, 2024*, 102935.
19. Sun, X., Yang, Y., & Song, Y. (2024). Unlocking the synergy: Increasing productivity through human-AI collaboration in the Industry 5.0 era. *Computers & Industrial Engineering, 2024*, 110657.
20. Sahoo, P., Saraf, P. K., & Uchil, R. (2024). Identification of challenges to Industry 5.0 adoption in Indian manufacturing firms: An emerging economy perspective. *International Journal of Computer Integrated Manufacturing*, 1–26.

21. Sharma, R., & Gupta, H. (2024). Harmonizing sustainability in Industry 5.0 era: Transformative strategies for cleaner production and sustainable competitive advantage. *Journal of Cleaner Production, 445*, 141118.
22. Nair, A., Pillai, S. V., & Senthil Kumar, S. A. (2024). Towards emerging Industry 5.0—A review-based framework. *Journal of Strategy and Management*. https://doi.org/10.1108/JSMA-04-2023-0067
23. Abdel-Basset, M., Mohamed, R., & Chang, V. (2024). A multi-criteria decision-making framework to evaluate the impact of Industry 5.0 technologies: Case study, lessons learned, challenges and future directions. *Information Systems Frontiers*, 1–31.
24. Khan, M., Haleem, A., & Javaid, M. (2023). Changes and improvements in Industry 5.0: A strategic approach to overcome the challenges of Industry 4.0. *Green Technologies and Sustainability, 1*(2), 100020.
25. Xiang, W., Yu, K., Han, F., Fang, L., He, D., & Han, Q. L. (2023). Advanced manufacturing in Industry 5.0: A survey of key enabling technologies and future trends. *IEEE Transactions on Industrial Informatics, 20*(2), 1055–1068.
26. Miao, J., Wang, Z., Wang, M., Garg, S., Hossain, M. S., & Rodrigues, J. J. (2024). Secure and efficient communication approaches for Industry 5.0 in edge computing. *Computer Networks, 242*, 110244.
27. Sharma, M., Tomar, A., & Hazra, A. (2024). Edge computing for Industry 5.0: Fundamental, applications and research challenges. *IEEE Internet of Things Journal, 11*(11), 19070–19093.
28. Leng, J., Sha, W., Wang, B., Zheng, P., Zhuang, C., Liu, Q., Wuest, T., Mourtzis, D., & Wang, L. (2022). Industry 5.0: Prospect and retrospect. *Journal of Manufacturing Systems, 65*, 279–295.
29. Chi, H. R., Wu, C. K., Huang, N. F., Tsang, K. F., & Radwan, A. (2022). A survey of network automation for industrial internet-of-things toward Industry 5.0. *IEEE Transactions on Industrial Informatics, 19*(2), 2065–2077.
30. Snyder, T., & Byrd, G. (2017). The internet of everything. *Computer, 50*(06), 8–9.
31. Demir, K. A., Döven, G., & Sezen, B. (2019). Industry 5.0 and human-robot co-working. *Procedia Computer Science, 158*, 688–695.
32. Prassida, G. F., & Asfari, U. (2022). A conceptual model for the acceptance of collaborative robots in Industry 5.0. *Procedia Computer Science, 197*, 61–67.
33. Rahman, M. M., Khatun, F., Jahan, I., Devnath, R., & Bhuiyan, M. A. A. (2024). Cobotics: The evolving roles and prospects of next-generation collaborative robots in Industry 5.0. *Journal of Robotics, 2024*(1), 2918089.
34. Domenteanu, A., Cibu, B., & Delcea, C. (2024). Mapping the research landscape of Industry 5.0 from a machine learning and big data analytics perspective: A bibliometric approach. *Sustainability, 16*(7), 2764.
35. Kalinaki, K., Yahya, U., Malik, O. A., & Lai, D. T. C. (2024). A review of big data analytics and artificial intelligence in Industry 5.0 for smart decision-making. In *Human-centered approaches in Industry 5.0: Human-machine interaction, virtual reality training, and customer sentiment analysis* (pp. 24–47). IGI Global.
36. Oliveira, M., Chauhan, S., Pereira, F., Felgueiras, C., & Carvalho, D. (2023). Blockchain protocols and edge computing targeting Industry 5.0 needs. *Sensors, 23*(22), 9174.
37. Wang, Z. J., Chen, Z. S., Xiao, L., Su, Q., Govindan, K., & Skibniewski, M. J. (2023). Blockchain adoption in sustainable supply chains for Industry 5.0: A multistakeholder perspective. *Journal of Innovation & Knowledge, 8*(4), 100425.
38. Ordieres-Meré, J., Gutierrez, M., & Villalba-Díez, J. (2023). Toward the Industry 5.0 paradigm: Increasing value creation through the robust integration of humans and machines. *Computers in Industry, 150*, 103947.
39. Raja Santhi, A., & Muthuswamy, P. (2023). Industry 5.0 or Industry 4.0 S? Introduction to Industry 4.0 and a peek into the prospective Industry 5.0 technologies. *International Journal on Interactive Design and Manufacturing (IJIDeM), 17*(2), 947–979.

40. Karmaker, C. L., Bari, A. M., Anam, M. Z., Ahmed, T., Ali, S. M., de Jesus Pacheco, D. A., & Moktadir, M. A. (2023). Industry 5.0 challenges for post-pandemic supply chain sustainability in an emerging economy. *International Journal of Production Economics, 258*, 108806.
41. Ajoudani, A., Zanchettin, A. M., Ivaldi, S., Albu-Schäffer, A., Kosuge, K., & Khatib, O. (2018). Progress and prospects of the human–robot collaboration. *Autonomous Robots, 42*, 957–975.
42. Matheson, E., Minto, R., Zampieri, E. G., Faccio, M., & Rosati, G. (2019). Human–robot collaboration in manufacturing applications: A review. *Robotics, 8*(4), 100.
43. Bauer, A., Wollherr, D., & Buss, M. (2008). Human–robot collaboration: A survey. *International Journal of Humanoid Robotics, 5*(01), 47–66.
44. Mukherjee, D., Gupta, K., Chang, L. H., & Najjaran, H. (2022). A survey of robot learning strategies for human-robot collaboration in industrial settings. *Robotics and Computer-Integrated Manufacturing, 73*, 102231.
45. Choi, S. H., Park, K. B., Roh, D. H., Lee, J. Y., Mohammed, M., Ghasemi, Y., & Jeong, H. (2022). An integrated mixed reality system for safety-aware human-robot collaboration using deep learning and digital twin generation. *Robotics and Computer-Integrated Manufacturing, 73*, 102258.
46. Fan, J., Zheng, P., & Lee, C. K. (2023). A vision-based human digital twin modeling approach for adaptive human–robot collaboration. *Journal of Manufacturing Science and Engineering, 145*(12), 121002.
47. Barricelli, B. R., & Fogli, D. (2024). Digital twins in human-computer interaction: A systematic review. *International Journal of Human–Computer Interaction, 40*(2), 79–97.
48. Miller, M. E., & Spatz, E. (2022). A unified view of a human digital twin. *Human-Intelligent Systems Integration, 4*(1), 23–33.
49. Tang, C., Yi, W., Occhipinti, E., Dai, Y., Gao, S., & Occhipinti, L. G. (2024). A roadmap for the development of human body digital twins. *Nature Reviews Electrical Engineering, 1*(3), 199–207.
50. Gallala, A., Kumar, A. A., Hichri, B., & Plapper, P. (2022). Digital Twin for human–robot interactions by means of Industry 4.0 Enabling Technologies. *Sensors, 22*(13), 4950.
51. Löcklin, A., Jung, T., Jazdi, N., Ruppert, T., & Weyrich, M. (2021). Architecture of a human-digital twin as common interface for operator 4.0 applications. *Procedia CIRP, 104*, 458–463.
52. Ma, X., Qi, Q., Cheng, J., & Tao, F. (2022). A consistency method for digital twin model of human-robot collaboration. *Journal of Manufacturing Systems, 65*, 550–563.
53. Shangguan, D., Chen, L., Su, C., Ding, J., & Liu, C. (2022). A triple human-digital twin architecture for cyber-physical systems. *CMES-Computer Modeling in Engineering & Sciences, 131*(3), 1557.
54. Angleraud, A., Ekrekli, A., Samarawickrama, K., Sharma, G., & Pieters, R. (2024). Sensor-based human–robot collaboration for industrial tasks. *Robotics and Computer-Integrated Manufacturing, 86*, 102663.
55. Ogenyi, U. E., Liu, J., Yang, C., Ju, Z., & Liu, H. (2019). Physical human–robot collaboration: Robotic systems, learning methods, collaborative strategies, sensors, and actuators. *IEEE Transactions on Cybernetics, 51*(4), 1888–1901.
56. Kusiak, A. (2018). Smart manufacturing. *International Journal of Production Research, 56*(1–2), 508–517.
57. Zhang, C., Wang, Z., Zhou, G., Chang, F., Ma, D., Jing, Y., Cheng, W., Ding, K., & Zhao, D. (2023). Towards new-generation human-centric smart manufacturing in Industry 5.0: A systematic review. *Advanced Engineering Informatics, 57*, 102121.
58. Zafar, M. H., Langås, E. F., & Sanfilippo, F. (2024). Exploring the synergies between collaborative robotics, digital twins, augmentation, and Industry 5.0 for smart manufacturing: A state-of-the-art review. *Robotics and Computer-Integrated Manufacturing, 89*, 102769.
59. Ren, L., Zhang, L., Wang, L., Tao, F., & Chai, X. (2017). Cloud manufacturing: Key characteristics and applications. *International Journal of Computer Integrated Manufacturing, 30*(6), 501–515.

60. Gharibvand, V., Kolamroudi, M. K., Zeeshan, Q., Çınar, Z. M., Sahmani, S., Asmael, M., & Safaei, B. (2024). Cloud based manufacturing: A review of recent developments in architectures, technologies, infrastructures, platforms and associated challenges. *The International Journal of Advanced Manufacturing Technology, 131*(1), 93–123.
61. Yin, C., Li, S., & Li, X. (2024). An optimization method of cloud manufacturing service composition based on matching-collaboration degree. *The International Journal of Advanced Manufacturing Technology, 131*(1), 343–353.
62. Dacre, N., Yan, J., Frei, R., Al-Mhdawi, M. K. S., & Dong, H. (2024). *Advancing sustainable manufacturing: A systematic exploration of Industry 5.0 supply chains for sustainability, human-centricity, and resilience* (pp. 1–30). Production Planning & Control.
63. Martín-Gómez, A. M., Agote-Garrido, A., & Lama-Ruiz, J. R. (2024). A framework for sustainable manufacturing: Integrating Industry 4.0 technologies with Industry 5.0 values. *Sustainability, 16*(4), 1364.
64. Baig, M. I., & Yadegaridehkordi, E. (2024). Industry 5.0 applications for sustainability: A systematic review and future research directions. *Sustainable Development, 32*(1), 662–681.
65. Liu, S., Li, P., Wang, J., & Liu, P. (2024b). Toward Industry 5.0: Challenges and enablers of intelligent manufacturing technology implementation under the perspective of sustainability. *Heliyon, 10*(15), e35162.

Advancements in Reinforcement Learning for Robotics: A Comprehensive Review and Future Directions

Apurv Sharma, Jatinder Kaur, and Sandeep Kumar Mogha

1 Introduction

In recent years, reinforcement learning (RL) has attracted a lot of attention as a potentially effective method for teaching autonomous agents to make decisions in changing contexts. In contrast to supervised learning, which gives agents access to labelled data, reinforcement learning (RL) allows agents to learn through interaction with their surroundings by giving them feedback in the form of incentives or punishments depending on what they do. Agents can find the best ways to maximize cumulative rewards [1] over time through this iterative learning process. Robotics uses reinforcement learning (RL) to create intelligent systems that can carry out intricate tasks in a variety of environments. The use of reinforcement learning to robotics has great promise for solving a variety of problems, including navigation, decision-making, and robotic control and manipulation. Robots can adapt and learn from experience by using reinforcement learning (RL) algorithms [2], which allows them to perform tasks autonomously in unpredictable and unstructured situations.

A. Sharma (✉)
Department of Mathematics, Chandigarh University, Mohali, Punjab, India

Department of Electrical and Electronics Engineering, ABV Indian Institute of Information Technology and Management, Gwalior, India
e-mail: mas_2025003@iiitm.ac.in

J. Kaur
Department of Mathematics, Chandigarh University, Mohali, Punjab, India
e-mail: jatindere6967@cumail.in

S. K. Mogha
Department of Mathematics, Chandigarh University, Mohali, Punjab, India

Department of Mathematics, University of Delhi, Delhi, India

Y. Kumar Singla et al. (eds.), *Digital Manufacturing in Industry 5.0*,
https://doi.org/10.1007/978-3-031-91500-0_39

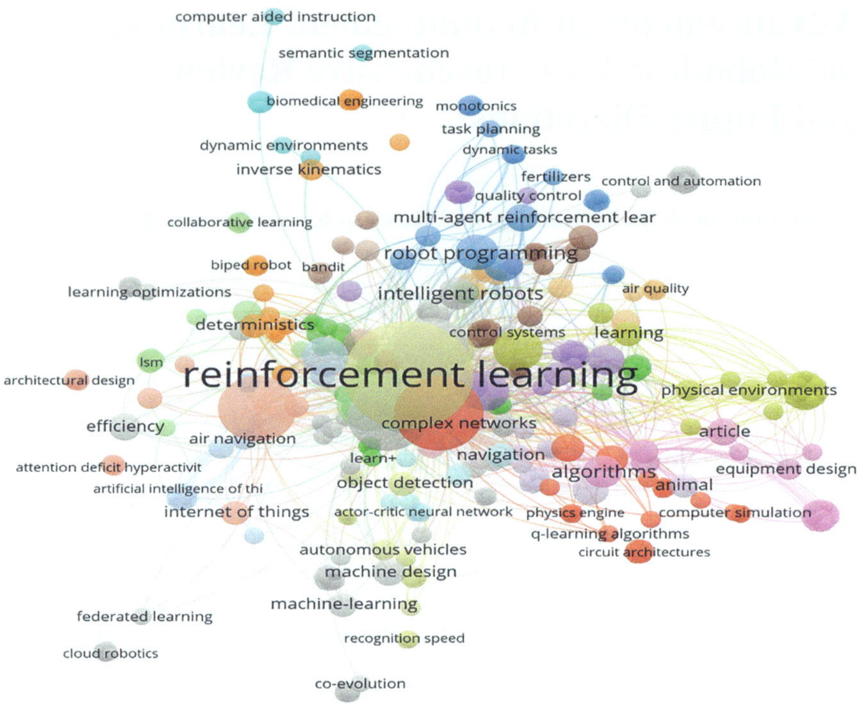

Fig. 1 Network visualization of reinforcement learning

Figure 1 illustrates how reinforcement learning connects to a wide range of topics such as robotics, algorithms, machine learning, and physical environments. It points out its applications in optimization, control systems, and emerging AI fields while showing research trends like federated learning and multi-agent systems. Furthermore, as complex, high-dimensional state and action spaces are frequently found in robotic systems, reinforcement learning (RL) offers a logical foundation for managing them. Consequently, reinforcement learning has become a key component in the effort to create robotic agents that are more capable, intelligent, and adaptable. Applying reinforcement learning to robots presents a number of practical and technological considerations in addition to its potential. Robotic systems frequently work in real-world settings that are noisy, unpredictable, and have safety restrictions. Furthermore, utilizing classic RL algorithms may result in sample inefficiency and computational complexity problems [3] because to the high-dimensional state and action spaces of robotic jobs.

Moreover, maintaining the security and dependability of RL-based robotic systems is still a major worry, especially for applications involving human interaction or for use in delicate situations. Our goal in this study is to present a thorough overview of the current status of robotics reinforcement learning. We will review a variety of reinforcement learning (RL) algorithms and techniques, encompassing more modern approaches like deep reinforcement learning (DRL) and more traditional ones like policy gradient methods [4] and Q-learning. We will examine the wide spectrum of robotics applications of reinforcement learning, such as robotic control, manipulation, navigation, and task planning, emphasizing significant obstacles, current developments, and useful applications. We will also talk about current research initiatives aimed at resolving RL's drawbacks in robotics, including sampling inefficiency, safety issues, and generalization to a variety of contexts. In order to improve the capabilities and usability of RL-based robotic systems, we will conclude by outlining prospective paths for future study in the field of RL integration with robotics. These include the integration of domain knowledge, hierarchical RL, and multi-agent systems [5]. This study presents a thorough assessment of the state-of-the-art in reinforcement learning for robots, emphasizing new developments, uses, difficulties, and suggestions for further study. In order to further RL-based robotics solutions, the chapter seeks to be a useful resource for scholars and practitioners by combining lessons from both theoretical breakthroughs and practical implementations.

Robotics developments nowadays need the demand for more intelligent, adaptable, and autonomous systems, which leads to innovation in various sectors of artificial intelligence and machine learning. From a variety of approaches under ML, reinforcement learning is one that has emerged powerfully to enable robots to learn complex tasks [3] independently using trial and error. Interaction with dynamic environments and learning from consequences of action can provide robots with advanced strategy making, control, and navigation, and hence, RL is a promising approach for a broad number of applications from industrial automation to service robots and autonomous vehicles. This review reports on the research performed in applying RL techniques to robotics and presents some of the most important milestones achieved so far, challenges at present, and possible future directions within this fast-moving area of study.

Historically, robotics has relied on control algorithms and pre-programmed rules for task execution. However, in dynamic or unstructured environments, these predetermined strategies may not suffice. This limited the strategy of handling complexities in real-world robotic systems, but reinforcement learning can ensure experience-based learning [6] and adaptability. We show how robots can operate in uncertain and constantly changing environments by maximising cumulative rewards and learning optimal policies while ushering a new possibility of exploration, manipulation and collaboration in this space. The flexibility and scalability of the algorithms in RL make it more suited for a further advancement of autonomous robot systems where the capacity to learn and the ability to improve over time are crucial.

Deep reinforcement learning—the combination of deep learning and reinforcement learning—greatly accelerated the field of RL for robots and is now known as DRL. Deep neural networks contain representation learning and can reproduce almost any function relationship if raw observations, like images, are provided. DRL combines both of these skills to allow robots to learn control policies directly from raw observations with high-dimensional sensory data, like images. The advancement has had breakthroughs in really complex tasks in robotics, such as autonomous navigation, object manipulation, and multi-agent cooperation [3]. A standout DRL-based algorithm, including Deep Q-Networks (DQN) and proximal policy optimization (PPO), is shown to work very well in most robotic applications, especially within simulated domains, hence setting a promising scenario for applicability to the real world.

Although these have been achieved, there are still various challenges presented when using RL with robotics. One of the most significant challenges in RL is that its algorithms suffer from inefficient sampling. Specifically, enormous amounts of data are required for these algorithms to converge to optimally good solutions. Hence, it poses a significant challenge to real-world robotic systems, where data collection can be very time-consuming and expensive. Second, though exploration in the exploration-exploitation trade-off presents a major challenge to the RL techniques, it is not their main problem. Agents are always mandated to balance the need to explore new strategies against the need to exploit [5] known successful policies. In robotics, poor exploration will lead to unsafe behaviours while over exploration may result in slow learning. Thus, such challenges overcome through developing more efficient RL algorithms and making use of techniques like transfer learning, simulation-to-real transfer, and model-based RL that improve the effectiveness of sample efficiency and safety in robotic learning.

Another important feature of RL in robotics is the need for continued learning and adaptation. Robots work in dynamic environments; that is, the policies to be learned change over time as more information becomes available. This creates problems for traditional RL algorithms since they are based on assumptions that an environment is static. To overcome this limitation, lifeline learning and meta-learning methods have been proposed to allow robots to learn from previous experience and generalize to new tasks with limited retraining. These techniques are promising to be fruitful for increasing the reliability and adaptability of robotic systems and in application areas such as home robots [6], disaster response, or autonomous exploration.

Another area, which has seen much attention recently, is MAS in robotics, using RL for coordinating teams of robots working towards common objectives together. Multi-agent RL opens the way to cooperative and competitive behaviours, thus robots are able to solve complex tasks that would not be feasible to solve with a single agent alone. Applications of MARL include swarm robotics, autonomous vehicle navigation [7], and industrial automation, in all these, the action of multiple robots needs to be coordinated in their shared environment. However, MARL also brings some new complexities: scaling up the number of agents, how to deal with non-stationary environments, and the communication among agents, since each one will often interact with more than just their neighbours.

The other emerging concern is safe and robust RL, with robots situated in real-world settings. Robots are supposed to operate safely under uncertainty, such that failure can lead to very costly damage or safety hazards. For deployment in real-world applications, safe RL algorithms that can learn risk-sensitive policies would need to incorporate safety constraints into the learning process. Techniques include constrained RL, risk-aware RL, and robust RL, which would control robotic systems to perform reliable tasks with minimal failures [8]. These are especially required for applications in health care-where robots are increasingly being used in surgery and in-patient care-and in autonomous driving where safety is at the highest premium.

Simulation has been an essential facilitator in pushing RL forward in robots, allowing difficult tasks to be experimented with in completely safe and controlled environments before their deployment in the world. Simulated environments such as Gazebo, MuJoCo, PyBullet, and other similar tools [4] offer an opportunity to experiment with RL algorithms under vastly different conditions without having to bear risks and costs associated with physical robots. This field, however, suffers from the difficulty of transferring learned policies from simulated settings into real-world environments. This sim-to-real gap is a significant problem in this field. Techniques like domain randomization and sim-to-real transfer learning have developed; so far, much more research is needed to ensure smooth deployments in the real world for RL-trained robots.

A hybrid approach of RL with other ML techniques such as supervised and unsupervised learning would take the efficiency and generalization capability of robotic systems to unprecedented heights. Imitation learning, for example-whereby robots learn from expert demonstrations-is one that has been shown to speed up learning since it provides a powerful initialization of RL algorithms. Another challenge where methods from unsupervised learning are said to be helpful is assisting autonomous [9] discovery of useful features from sensory data by robots, and that further reduces the reliance on labelled datasets and makes possible more scalable learning. Such combinations might make for more efficient and versatile robotic systems that can do a much larger variety of tasks and in more environments.

Hardware advances in the form of newer, more powerful processors, as well as increased sensority, have also contributed to pushing the envelope of RL advances in robotics. Modern robots are equipped with sophisticated sensors like LiDAR, cameras, and tactile sensors that generate rich and high-dimensional data. Such sensory input is critical for the perception and interaction of RL algorithms with their environment. Advances in computational resources, and now GPU and TPUs, for example, have also greatly accelerated training deep RL models for increasingly complex learning tasks. Continuing on this hardware advancement path also brings the possibility and increasing scale of RL-driven robotic systems towards greater degrees of autonomy and functionality.

RL has its very wide and diverse applications within the robotics sector, ranging from manufacturing, healthcare, logistics, to industries driven by space exploration. Manufacturing applications include robots for optimizing the production lines' efficiency and precision in welding and assembly, offering flexible automation. Health industries are using RL-powered robots in operations [10], rehabilitation, and patient care, providing personalized and adaptive interactions. Autonomous drones and warehouse robots have dominated logistics using RL, optimizing routes, handling and managing inventory. The domain of RL in space exploration is opening frontiers for autonomous rovers and robotic systems to be able to navigate and perform tasks in unknown and extreme environments.

In the future, the potential of RL will improve even further with other emerging technologies, such as 5G, IoT, and edge computing. For example, marrying RL with 5G networks could enable access to vast amounts of information in real time, accelerate decision-making processes, and allow more coordinated behaviours in multi-robot systems. Similarly, IoT and edge computing would allow for distributed data and processing power, enabling more efficient and scalable architectures for learning. Such developments will therefore continue to further the development of autonomously deployable intelligent robotic systems [11] in complex dynamic environments.

In summary, the application of reinforcement learning in robotics has seen significant efforts with a wide approach that has been fostered both from algorithmic innovation and hardware improvement. Progress has been achieved in some of the more general areas such as deep reinforcement learning, multi-agent systems, and safety in reinforcement learning [1]. Challenges remain considerably numerous, and most of these falls under sample efficiency, closing the sim-to-real gap, assurance on safety, and augmenting the generalization capabilities of RL algorithms. Future research will specifically address these challenges to unlock the real power of reinforcement learning for robotics, thereby opening a pathway to more intelligent, adaptive, and autonomous robotic systems with more functionalities to innovate industries and quality life for people on Earth.

2 Literature Review

The promise of reinforcement learning (RL) to help autonomous agents learn complicated tasks in dynamic contexts has drawn a lot of interest in robotics research. Many studies have investigated the application of reinforcement learning (RL) approaches to many robotic domains, such as navigation, control, manipulation, and task planning, within the last 10 years. In the area of robotic control, traditional reinforcement learning techniques like Q-learning and policy gradient methods have been thoroughly investigated. Kober et al. [2], for example, showed how successful policy search techniques are at teaching robotic manipulators control policies. In a similar vein, Sutton and Barto [4] offer a thorough analysis of RL algorithms that have been used in robotics tasks, including as temporal-difference approaches and Monte Carlo methods. Robotics research has advanced much faster since the emergence of deep reinforcement learning (DRL), which allows agents to learn directly from high-dimensional sensory inputs like images or sensor readings. The deep Q-network (DQN) technique was first presented by Mnih et al. [6], and it has shown impressive performance in learning to play Atari games from raw pixel inputs. DRL approaches have been used in robotics for tasks including navigation in complicated surroundings [3] and robotic grasping [16]. Though RL has potential for robotic applications, there are still a number of obstacles to overcome.

Figure 2 illustrates the number of publications by various researchers about reinforcement learning in robotics. One major obstacle is sample inefficiency, wherein

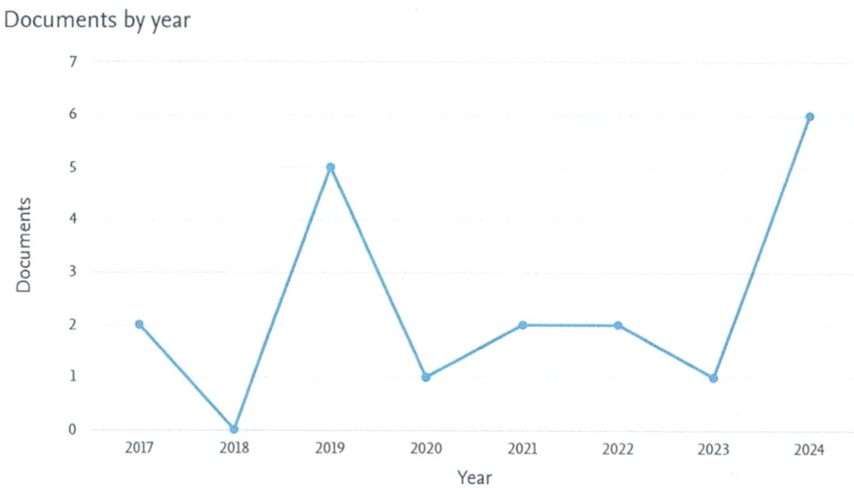

Fig. 2 Publication trend graph

reinforcement learning algorithms need a lot of interactions with the environment to find efficient rules. To increase sample efficiency in DRL algorithms, methods like experience replay [12] and prioritized experience replay [9] have been suggested. Another significant obstacle with RL-based robots is safety issues. Agents that have received RL training may behave dangerously, particularly when placed in real-world settings. Researchers have suggested techniques for adding safety constraints to the RL training procedure in order to overcome this problem [1]. Furthermore, methods for motivating agents have been investigated, including reward shaping [13] and constraint optimization [14]. Furthermore, one of the important areas of research in reinforcement learning for robots is the generalization of learnt policies to various tasks and settings. Promising methods for allowing agents to use knowledge gained in one task or environment to speed learning in new settings are transfer learning [7] and meta-learning [15]. Traditional reinforcement learning algorithms, like policy gradient techniques and Q-learning, have been extensively researched in the context of robotic control and decision-making. These algorithms are appropriate for jobs where explicit supervision or expert knowledge is restricted, as they provide principled frameworks for learning optimal policies through trial and error [2]. Furthermore, developments in deep reinforcement learning (DRL) have made it possible for agents to learn directly from unprocessed sensory inputs, including pictures or sensor readings, resulting in behaviour that is more adaptable and flexible [6]. Among the DRL algorithms that have been effectively used for robotic tasks, such as manipulation, navigation, and control, are deep Q-networks (DQN), deep deterministic policy gradients (DDPG), and proximal policy optimization (PPO) [16, 17]. Robotics applications of reinforcement learning (RL) cover a broad spectrum of real-world challenges, going beyond classical control difficulties. For example, robotic manipulators have been trained via reinforcement learning (RL) to engage with humans in collaborative settings, traverse through challenging terrain, and grab things in crowded surroundings [18, 19]. In addition, reinforcement learning methods have been implemented in multi-agent systems, allowing groups of robots to synchronize their movements and accomplish shared objectives [5]. These uses highlight the adaptability and scalability of reinforcement learning in tackling a range of robotics problems, from manufacturing and industrial automation to healthcare and service robotics. Even with its potential, reinforcement learning in robots has a number of drawbacks. One major obstacle that still has to be overcome is sample inefficiency, which occurs when RL algorithms need a lot of interactions with the environment to learn efficient policies. This is especially true in real-world scenarios where interactions might be expensive or time-consuming [20]. Furthermore, it is imperative to guarantee the safety and dependability of reinforcement learning (RL)-based robotic systems, particularly in applications that entail human–robot interaction or deployment in situations that are critical to safety [21]. To address these issues and reduce the dangers related to RL-based robotic control, strategies like reward shaping, constraint optimization, and safe exploration have been put forth [1, 14]. There are a number of interesting avenues for more robotics-related RL research in the future. Using domain expertise and structured representations together can enhance learning's generalization and sampling efficiency [8]. Researchers can create hybrid systems that take advantage of the advantages of both methods to perform challenging robotic tasks by fusing RL with methods from symbolic reasoning and planning [11]. Furthermore, agents can learn and transfer knowledge across tasks and contexts through the use of hierarchical reinforcement learning, meta-learning, and transfer learning—all of which hold promise for accelerating and improving robotics learning [7, 15]. Furthermore, fostering comprehension and confidence in autonomous robotic systems depends on the creation of reliable and intelligible RL algorithms, particularly in applications where human oversight or intervention is required [19]. To sum up, reinforcement learning provides a strong foundation for teaching autonomous agents in robotics, with applications that cover a variety of fields and jobs. More intelligent and powerful robotic systems will be possible as a result of robots using reinforcement learning (RL) algorithms to learn how to adapt and make decisions in challenging and uncertain circumstances. To fully realize the potential of RL-based robots, however, issues including sampling inefficiency, safety concerns, and generalization must to be resolved. In the years to come, ongoing research in these fields, together with developments in algorithmic methods and practical applications, could spur additional creativity and advancement in reinforcement learning for robots.

3 Recent Advancements

Recent developments in reinforcement learning (RL) for robotics have focused on improving the efficiency and scalability of RL algorithms with real-world relevance. Exciting areas like deep reinforcement learning (DRL) and model-based RL have greatly advanced robotic decision-making, yielding higher adaptability [22] and autonomy for robots. In addition, there have been advances in sim-to-real transfer and multi-agent RL frameworks; these are very complementary as they tend to use more off-policy learning compared to a single agent, as well as the hopper-on-viewer work from the hierarchical reinforcement learning literature which demonstrated robots performing complex tasks in changing environments. It is these that are going to facilitate the development of more intelligent, flexible, and efficient robotic systems and often used a lot in Industry 5.0.

3.1 Advancement in RL for Robotics

Robots are now able to carry out complicated tasks on their own thanks to recent breakthroughs in reinforcement learning (RL) for robotics. One notable development is the creation of deep reinforcement learning (DRL) algorithms, which use deep neural networks to learn hierarchical task representations and handle high-dimensional input fields. Robotic control challenges have seen the effective application of Deep Q-Networks (DQN), Policy Gradient techniques (e.g., TRPO, PPO, REINFORCE), and Actor-Critic architectures, which enable learning from unprocessed sensory inputs like point clouds or pictures. Integration of reinforcement learning (RL) with other learning paradigms, like imitation learning and meta-learning, is another significant advance. Robots can learn from human experts' presentations through imitation learning, which speeds up learning and enhances task performance. Robots may learn effectively in dynamic [19] and changing circumstances thanks to meta-learning approaches, which leverage past knowledge and experience to help robots quickly adapt to new tasks or environments. Furthermore, recent work has concentrated on tackling important problems in reinforcement learning for robotics, such as sample inefficiency, safety, and practical implementation. Methods that have demonstrated promise in enhancing sample efficiency and generalization abilities include curriculum learning, in which the learning process is organized into a series of tasks with progressively increasing difficulty, and transfer learning, which permits knowledge transfer between related tasks or domains. Furthermore, the development of domain adaption methods and simulation-based training has made it easier to transfer learned policies from simulation to actual robotic [2] platforms, closing the reality gap and hastening deployment in real-world scenarios. Robotics applications of reinforcement learning (RL) cover a broad spectrum of areas, including as autonomous navigation and exploration, perception and sensorimotor control, locomotion (e.g., walking, running, flying), and manipulation tasks (e.g., grasping, object manipulation). These uses show how adaptable and powerful RL algorithms are at helping robots do complicated tasks with varied degrees of autonomy and adapt to a variety of settings. In the future, robotics research in reinforcement learning will concentrate on resolving unresolved issues and expanding the limits of robot learning capacities. In order for robots to generalize across many activities and settings, transfer learning and domain adaptation techniques will remain essential, and multi-agent reinforcement learning methodologies will make it easier for heterogeneous robotic systems to collaborate [20] and coordinate. Furthermore, collaborative human–robot interaction (HRI) and shared autonomy will open up new avenues for cooperative task execution and improve user experience in practical applications by facilitating smooth human–robot collaboration. Overall, recent developments in reinforcement learning for robotics have created the groundwork for increasingly powerful and self-governing robotic systems, with promising avenues for further study and growth.

3.2 Reinforcement Learning in the Healthcare Sector

Because reinforcement learning continuously improves decision-making, personalizes treatments, and optimizes resource allocation, it has the potential to transform a wide range of healthcare services. Reinforcement learning in medicine (MED) can provide tailored treatment regimens, modifying medication and care according to patient data. Algorithms in nuclear medicine (NMED) can improve radiation and imaging therapies by studying past cases and fine-tuning dosage [23] and timing for specific patients. Reinforcement learning can help with preoperative planning and postoperative treatment for cardiothoracic surgery (CSURG) and vascular surgery (VSURG), lowering risks and increasing outcomes. Reinforcement learning has the potential to improve robotic-assisted surgeries in surgery (SURG) by offering real-time recommendations that increase accuracy and productivity.

Reinforcement learning is used in critical care medicine (CMED) and trauma surgery (TRAUM) to guide drug modifications, forecast patient deterioration, and optimize therapies in high-risk, time-sensitive settings. Reinforcement learning is used in Osteopathic Medicine (OMED) to assess how patients respond to alternative therapies and provide individualized, comprehensive care. Predictive models that evaluate surgical risks [10] can be useful in neurosurgery (NSURG) and thoracic surgery (TSURG), increasing precision and patient safety.

By suggesting customized rehabilitation activities and streamlining surgical techniques, reinforcement learning can expedite the healing process for orthopaedic surgery (ORTHO). Genitourinary services (GU) use algorithms that learn from patient outcomes to optimize individualized treatment for illnesses like prostate problems or kidney stones. All things considered, reinforcement learning provides a strong toolkit to improve treatment and decision-making in these various medical services, leading to more efficient and customized healthcare solutions as shown in Fig. 3. Link-MIMIC-III—Deep Reinforcement Learning (kaggle.com)

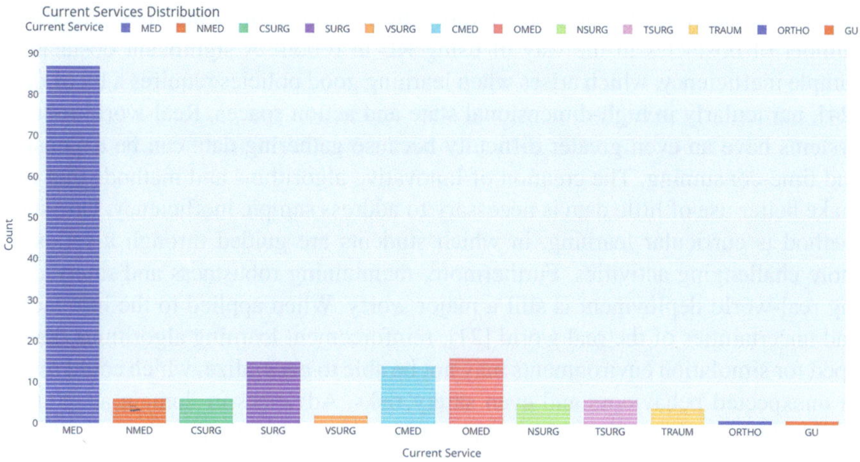

Fig. 3 Distribution of current services across medical specialties

3.3 Analysis of Existing Approaches

Current robotics reinforcement learning (RL) approaches present a wide variety of methods and strategies, each adapted to particular problems and demands in robotic control applications. The ubiquity of deep reinforcement learning (DRL) algorithms is one noteworthy finding; these algorithms have become increasingly popular because of their capacity to process high-dimensional sensory inputs and learn intricate policies. While Policy Gradient techniques like REINFORCE, TRPO, and PPO have proven effective in handling continuous action spaces, making them well-suited for robotic manipulation and locomotion tasks [15], Deep Q-Networks (DQN) have demonstrated exceptional success in problems requiring discrete action spaces. Actor-Critic designs also allow for effective learning in both discrete and continuous action spaces by providing a balance between value-based and policy-based approaches. To further improve robotic learning capabilities, RL integration with other learning paradigms including imitation learning and meta-learning has shown promise. Robots can pick up knowledge from expert behaviour [19] and learn more quickly thanks to imitation learning, which makes it easier for humans to demonstrate tasks to robots. This is especially helpful in situations when it's difficult to create reward systems or conduct effective environment exploration. However, by utilizing past knowledge, meta-learning approaches help robots swiftly adjust to new tasks or situations, improving their capacity to generalize across various contexts and tasks. Even with these developments, there are still a number of obstacles in the way of using RL in robots. A significant obstacle is sample inefficiency, which arises when learning good policies requires a lot of data [24], particularly in high-dimensional state and action spaces. Real-world robotic systems have an even greater difficulty because gathering data can be expensive and time-consuming. The creation of innovative algorithms and methods that can make better use of little data is necessary to address sample inefficiency. One such method is curricular learning, in which students are guided through a series of more challenging activities. Furthermore, maintaining robustness and safety during real-world deployment is still a major worry. When applied to the intricacies and uncertainties of the real world [21], reinforcement learning algorithms developed for simulation environments may not be able to generalize, which could result in unexpected behaviours and even safety risks. Advances in domain adaptation approaches are needed to bridge the gap between simulation and reality. These techniques allow policies trained in simulation to be transferred and changed to real-world contexts while maintaining their effectiveness. Robots that possess autonomous behaviour are now capable of handling difficult control tasks thanks to the tremendous progress made by current RL techniques in robotics. However, issues including safety, real-world deployment, and inefficient samples still exist and call for more investigation. Robots could be able to automatically learn and adapt to a variety of dynamic situations [13] thanks to reinforcement learning (RL), which has the potential to transform robotics. This can be achieved by utilizing advances in algorithmic techniques and integrating with other learning paradigms.

3.4 Applications of RL in Robotics

Applications of reinforcement learning (RL) in robotics are many and cover a broad spectrum of activities and domains. Robotic manipulation tasks are one popular field of application, where RL algorithms allow robots to grab and move items with precision and dexterity. Robots can change their grasping tactics [12] to handle things of different sizes, shapes, and textures by learning from their mistakes. This makes them ideal for a variety of activities, including assembly line operations, warehouse automation, and household duties. Remarkable performance has been shown by RL-based manipulation systems in tasks including picking and placing objects, sorting items, and even carrying out complex assembly tasks in industrial environments. Robots can move and navigate through complicated settings on their own thanks to locomotion, which is another important use of reinforcement learning in robotics. Robots can use reinforcement learning (RL) algorithms to learn locomotor patterns that maximize stability, efficiency, and adaptation to a variety of obstacles and terrains, whether they are walking, running, flying, or swimming. This capacity is essential for many applications, such as urban transportation duties, hazardous environment research, and search and rescue operations. Legged robots, wheeled platforms, aerial drones [24], and submerged vehicles have all used RL-based locomotion systems, demonstrating their adaptability and efficiency in a variety of robotic platforms and settings. In perception and sensorimotor control tasks, where robots must learn to decipher sensory inputs like depth sensors, camera images, and LIDAR data in order to make judgments and move appropriately, reinforcement learning (RL) is also widely utilized. Robots may acquire intricate perception-action mappings [9] that allow them to avoid obstacles, traverse through congested surroundings, and interact with items intelligently by utilizing reinforcement learning. Robotic navigation in both indoor and outdoor contexts, autonomous driving, object tracking and recognition, and tasks related to human–robot interaction like gesture recognition and natural language understanding are just a few examples of how reinforcement learning is used in perception and sensorimotor control. Moreover, reinforcement learning is essential for allowing robotic devices to navigate and explore on their own. Robots are able to map unfamiliar areas independently [7], find new objects or sites of interest, and navigate towards predetermined goals or waypoints by learning policies that optimize for exploration and exploitation of the environment. Robotic systems used in space exploration, surveillance, environmental monitoring, search and rescue, and other scenarios need to be able to conduct autonomous exploration and navigation. The usefulness of RL-based navigation systems has been shown in practical applications by its deployment in a variety of robotic platforms, such as autonomous underwater vehicles, drones, and ground robots.

3.5 Integration with Other Learning Paradigms

Combining reinforcement learning (RL) with other learning paradigms has become a viable way to improve robotic systems' performance. Imitation learning, in which robots pick up knowledge from human experts' demonstrations, is one important integration. Robots can exploit the rich knowledge embodied in human demonstrations to bootstrap their learning process and expedite skill development by combining reinforcement learning (RL) with imitation learning. This is especially helpful in situations when it's difficult to create reward systems [13] or conduct effective environment exploration. Furthermore, compared to pure RL techniques, imitation learning allows robots to learn complicated tasks with fewer samples, which makes it a good fit for real-world robotic applications where gathering data can be expensive or time-consuming. Furthermore, meta-learning has drawn interest as an additional technique to reinforcement learning (RL) that helps robots swiftly adapt to new tasks or surroundings by drawing on past experience. Through the use of meta-learning approaches [15], robots may quickly adapt to new environments or situations by learning a meta-policy that controls how to learn effectively from sparse data. Robots can increase sample efficiency and generalization capacities by leveraging prior experiences from a variety of tasks or domains by integrating reinforcement learning with meta-learning. This is especially helpful in contexts that are dynamic and changing, where success depends on resilience and adaptation. Furthermore, integrating reinforcement learning (RL) with other learning paradigms, like supervised and unsupervised learning [4], has the potential to improve robotic learning. Robots can learn from limited data more effectively and generalize across different contexts by using labelled or unlabelled data to supplement their RL-based learning with extra supervision or direction. Additionally, robots can reason about complicated surroundings and make decisions in real-time thanks to hybrid approaches that integrate RL with planning techniques or symbolic reasoning. This increases the autonomy and flexibility of robots in unpredictable and dynamic situations.

4 Challenges and Limitations

Reinforcement learning (RL) has come a long way in robotics, but there are still a number of obstacles in the way of the general deployment and uptake of RL-based robotic systems. A major obstacle is sample inefficiency, which occurs when learning efficient policies requires a lot of data, particularly in high-dimensional state and action spaces. Such data collection can be expensive and time-consuming, especially for robotic systems used in real-world scenarios where it may be difficult or expensive to carry out interactions with the environment. The creation of innovative algorithms and methods that can make better use of limited data is necessary to address sample inefficiency. Examples of these methods include domain adaption strategies that facilitate the transfer of knowledge from simulation to reality and meta-learning approaches. Furthermore, maintaining robustness and safety during real-world deployment [3] is still a major worry. When applied to the intricacies and uncertainties of the real world, reinforcement learning algorithms developed for simulation environments may not be able to generalize, which could result in unexpected behaviours and even safety risks. Advances in domain adaptation approaches are needed to bridge the gap between simulation and reality. These techniques allow policies trained in simulation to be transferred and changed to real-world contexts while maintaining their effectiveness. Furthermore, robustness assurances and fail-safe methods to withstand unforeseen conditions or disruptions are necessary to ensure safety in dynamic and unpredictable contexts. The creation of suitable reward functions [16] that efficiently capture the job objectives, encourage desired behaviour, and prevent unforeseen consequences presents another difficulty. It is still difficult to design reward functions that strike the correct balance between preventing reward hacking and unintentional side effects and giving insightful feedback, especially for complicated and high-dimensional jobs. Furthermore, because learning from delayed feedback necessitates complex credit assignment mechanisms and exploration tactics, dealing with sparse or delayed rewards [13] can provide difficulties for reinforcement learning algorithms. Nevertheless, there are still issues with scalability and generalization across activities and contexts, especially with deep reinforcement learning (DRL) methods. Large-scale robotic systems with intricate sensory inputs and high-dimensional action spaces necessitate the use of scalable reinforcement learning algorithms, distributed computing infrastructure, and meticulous hyperparameter tuning [12]. Further developments in transfer learning, meta-learning, and domain adaption approaches are also needed to generalize learnt policies across a variety of activities, contexts, and robot platforms.

5 Future Scope

Future robotics applications of reinforcement learning (RL) have a wide range of potential, providing multiple avenues for improving robotic systems' performance and opening up new fields of use. Developing more effective and sample-efficient reinforcement learning algorithms that can generalize across activities and contexts and learn from small amounts of data is a potential avenue for future study. Robots can learn quickly and adaptably in dynamic and uncertain situations by utilizing techniques like curriculum learning, transfer learning, and meta-learning, which show promise in overcoming sample inefficiency-related issues. Additionally, there is a great deal of promise for combining supervised learning, unsupervised learning, and evolutionary algorithms with reinforcement learning to improve robotic learning. Through the integration of reinforcement learning (RL) with other methodologies, scholars can capitalize on the synergistic benefits of distinct learning strategies to more efficiently address intricate and demanding robotic assignments. Enhancing the autonomy and intelligence of robotic systems can also be achieved through intriguing hybrid approaches that integrate RL with planning methods, symbolic reasoning, or cognitive architectures. Furthermore, it is anticipated that developments in sensors, computer infrastructure, and hardware platforms would hasten the creation and application of robotic systems based on reinforcement learning. Researchers can now train more intricate and data-intensive reinforcement learning models, allowing robots to learn from high-dimensional sensory inputs and make intelligent decisions in real-time. This is made possible by the development of powerful GPUs, TPUs, and distributed computing frameworks. Furthermore, the development of accessible and reasonably priced robotic platforms like manipulators, mobile robots, and drones gives researchers the chance to test RL algorithms in practical contexts and investigate novel applications in fields like healthcare, agriculture, and service robotics. Concern over ethical and sociological issues in the creation and application of RL-based robotic systems is also becoming more and more prevalent. Building trust and acceptability among end-users and stakeholders requires ensuring the safety, equity, and transparency of robotic decision-making processes. Furthermore, resolving issues with algorithmic bias, privacy, and employment displacement calls for multidisciplinary cooperation as well as proactive outreach to legislators, ethicists, and the general public.

6 Conclusion

In conclusion, robots will be able to automatically learn, adapt, and carry out complicated tasks in a variety of dynamic situations thanks to the immense potential that reinforcement learning (RL) holds to revolutionize the area of robotics. We have looked at the difficulties, opportunities, and latest developments in the use of reinforcement learning in robots throughout this article. Recent developments in reinforcement learning (RL) algorithms, in particular deep reinforcement learning (DRL), have made it possible for robots to learn from high-dimensional sensory inputs and make wise judgments instantly. Robust learning algorithms (RL) have proven to be remarkably effective in empowering robots to execute an extensive array of activities with variable degrees of autonomy, spanning from robotic manipulation and locomotion to vision and navigation. Nonetheless, a number of obstacles and restrictions continue to exist, such as sample inefficiency, safety issues, incentive design, scalability, and applicability to different tasks and contexts. Multidisciplinary research projects in the fields of robotics, control theory, machine learning, and human–computer interaction are needed to tackle these problems. Looking ahead, reinforcement learning has a wide and bright future in robotics. It is anticipated that developments in hardware platforms, computer infrastructure, and algorithmic methodologies would hasten the creation and application of RL-based robotic systems. Improving robotic learning capabilities and opening up new applications in a variety of fields are made possible by integrating reinforcement learning (RL) with other learning paradigms, such as imitation learning, meta-learning, and transfer learning. Furthermore, in order to gain the trust and acceptance of stakeholders and end users alike, ethical and sociological issues must be considered during the development and implementation of RL-based robotic systems. Researchers can fully realize the potential of RL-based robotic systems and realize the vision of intelligent, autonomous, and socially responsible robots that can improve human productivity, safety, and quality of life by promoting interdisciplinary collaboration, interacting with policymakers, and proactively addressing ethical concerns.

References

1. García, J., & Fernández, F. (2015). A comprehensive survey on safe reinforcement learning. *The Journal of Machine Learning Research*. https://doi.org/10.5555/2789272.2886795
2. Kober, J., Bagnell, J., & Peters, J. (2013). Reinforcement learning in robotics: A survey. *The International Journal of Robotics Research*. https://doi.org/10.1177/0278364913495721
3. Mirowski, P., Pascanu, R., Viola, F., Soyer, H., Ballard, A. J., Banino, A., et al. (2016). Learning to navigate in complex environments. https://doi.org/10.48550/arXiv.1611.03673
4. Sutton, R. S., & Barto, A. G. (2018). *Reinforcement learning: An introduction*. The MIT Press.
5. Lowe, R., Wu, Y., Tamar, A., Harb, J., Abbeel, P., & Mordatch, I. (2017). Multi-agent actor-critic for mixed cooperative-competitive environments. https://doi.org/10.48550/arXiv.1706.02275
6. Mnih, V., Kavukcuoglu, K., Silver, D., Rusu, A., Veness, J., Bellemare, M., et al. (2015). Human-level control through deep reinforcement learning. *Nature, 518*(7540), 529–533.

7. Taylor, M. E., & Stone, P. (2009). Transfer learning for reinforcement learning domains: A survey. *Journal of Machine Learning Research, 10*, 1633–1685.
8. Bansal, T., Krähenbühl, P., & Tompson, J. (2018). ChauffeurNet: Learning to drive by imitating the best and synthesizing the worst. https://doi.org/10.48550/arXiv.1812.03079
9. Schaul, T., Quan, J., Antonoglou, I., & Silver, D. (2015). Prioritized experience replay. https://doi.org/10.48550/arXiv.1511.05952
10. Yu, C., Liu, J., Nemati, S., & Yin, G. (2021). Reinforcement learning in healthcare: A survey. https://doi.org/10.1145/3477600
11. Konidaris, G., & Barto, A. (2009). Skill discovery in continuous reinforcement learning domains using skill chaining. *Advances in Neural Information Processing Systems*, 1015–1023.
12. Lin, L. (1993). *Reinforcement learning for robots using neural networks.* Technical Report, Carnegie Mellon University.
13. Ng, A. Y., Harada, D., & Russell, S. (1999). Policy invariance under reward transformations: Theory and application to reward shaping. In *Proceedings of the Sixteenth International Conference on Machine Learning*.
14. Achiam, J., Held, D., Tamar, A., & Abbeel, P. (2017). Constrained policy optimization. https://doi.org/10.48550/arXiv.1705.10528
15. Finn, C., Abbeel, P., & Levine, S. (2017). Model-agnostic meta-learning for fast adaptation of deep networks. https://doi.org/10.48550/arXiv.1703.03400
16. Levine, S., Finn, C., Darrell, T., & Abbeel, P. (2016). End-to-end training of deep visuomotor policies. *Journal of Machine Learning Research, 17*(39), 1–40.
17. Schulman, J., Wolski, F., Dhariwal, P., Radford, A., & Klimov, O. (2017). Proximal policy optimization algorithms. https://doi.org/10.48550/arXiv.1707.06347.
18. Hu, Z., Gao, X., Wan, K., Zhai, Y., & Wang, Q. (2021). Relevant experience learning: A deep reinforcement learning method for UAV autonomous motion planning in complex unknown environments. *Chinese Journal of Aeronautics, 34*, 187. https://doi.org/10.1016/j.cja.2020.12.027
19. Levine, S., Pastor, P., Krizhevsky, A., & Quillen, D. (2018). Learning hand-eye coordination for robotic grasping with deep learning and large-scale data collection. *International Journal of Robotics Research, 37*(4–5), 421–436.
20. Liu, J., Roth, M., Mescheder, L., & Soatto, S. (2018). Understanding and improving sampled learning in adversarial games. *Advances in Neural Information Processing Systems, 31*.
21. Amodei, D., Olah, C., Steinhardt, J., Christiano, P., Schulman, J., & Mané, D. (2016). Concrete problems in AI safety. https://doi.org/10.48550/arXiv.1606.06565
22. Jeyaraman, J., Malaiyappan, J. N. A., & Sistla, S. M. K. (2024). Advancements in reinforcement learning algorithms for autonomous systems. https://doi.org/10.17613/sd0v-he77
23. Abdellatif, A. A., Mhaisen, N., Chkirbene, Z., Mohamed, A., Erbad, A., & Guizani, M. (2021). Reinforcement learning for intelligent healthcare systems: A comprehensive survey. https://doi.org/10.48550/arXiv.2108.04087
24. Nguyen, T. T., Nguyen, N. D., & Nahavandi, S. (2020). Deep reinforcement learning for multiagent systems: A review of challenges, solutions, and applications. *IEEE Transactions on Cybernetics, 50*(9), 3826–3839. https://doi.org/10.1109/TCYB.2020.2977374

Smart Factories and Cyber-Physical Systems

Naveen Kumar, Mohit Tyagi, Rakesh Joshi, Yatika Gori, Ashwani Kumar, and Yogesh Kumar Singla

1 Introduction

In the present scenario, the implementation of the state-of-the-art technologies, real-time data recording and analyzing systems and smart connectivity, and human interaction with machines and automation are the focused areas in which industries are seeking research data and keen to implement. These technologies improve the

N. Kumar (✉)
Department of Mechanical and Aerospace Engineering, UPES, Dehradun, Uttarakhand, India
e-mail: naveen.kumar1@ddn.upes.ac.in

M. Tyagi
Department of Production and Industrial Engineering, Punjab Engineering College (Deemed to be University), Chandigarh, India

R. Joshi
DARVUN India, Patiala, Punjab, India

Y. Gori
Department of Mechanical Engineering, Graphic Era Deemed to be University, Dehradun, Uttarakhand, India

School of Computing, Graphic Era Hill University, Haldwani, Uttarakhand, India
e-mail: yatikagori@gehu.ac.in

A. Kumar
Department of Mechanical Engineering, Technical Education Department Uttar Pradesh, Kanpur, India
e-mail: ashwaniadj.me@geu.ac.in

Y. Kumar Singla
School of Engineering, Math and Technology, Navajo Technical University, Crownpoint, NM, USA

Department of Mechanical Engineering, University of Idaho, Moscow, ID, USA
e-mail: ysingla@navajotech.edu

Y. Kumar Singla et al. (eds.), *Digital Manufacturing in Industry 5.0*,
https://doi.org/10.1007/978-3-031-91500-0_40

overall system efficiency, reduce the dependency on the conventional methods, and make decision making very simple. Global manufacturing industries are facing several challenges, e.g., tracing of the components, unplanned downtime, long lead time, low agility, presence of silos in the system, lack of skilled labor and customer responsiveness. These challenges are causing low margin, low market presence and low return of the assets for the industries. Therefore, industries are continuously trying to overcome these issues and work in a smooth manner. In this direction, the industrial revolutions are continuously happening and improving the systems with implementing the advanced technologies. The latest industrial revolution Industry 4.0 has suggested the implementation of internet of things (IOT), industrial internet of things (IIOT), AI/ML, advanced industrial automation, digital twin, digital thread, smart factory and cyber-physical systems, etc. The more appropriate word in industrial context is smart manufacturing which is supervising and implementing all these technologies at the organizations [1–4].

Although, Industry 4.0 is considered a great revolution, but it also has some constraint or technologies which further needs to be fully developed. The training of the existing worker to operate advanced systems, concerns of human and society for during the industrial transitions, human–machine infractions and the scope of the improvement in the existing technologies, etc. are few concerns reported so far by the industries. So, the overall, emergence of the Industry 5.0 can be attributed to these concerns as shown in Fig. 1. The future directions for Industry 4.0 lead to Industry 5.0 with the following key objectives [4, 5]:

- Moving toward a more human-centric approach with increased collaboration between humans and machines (e.g., integrates cobots).
- Focus on sustainable solutions for manufacturing and incorporating green technologies and practices to address environmental concerns.
- Advancements and applications of AI to drive innovation and smarter automation.
- Emphasizes collaboration between humans and machines rather than replacement.

To achieve the above-mentioned objectives toward Industry 5.0, there is an urgent need to reform the related training/education systems. The high-tech training should include new tools, advanced software, and knowledge related to the technologies. It is a shift from general purpose robots to cobots and advanced learning system for the machines. The education system should also consider sustainability,

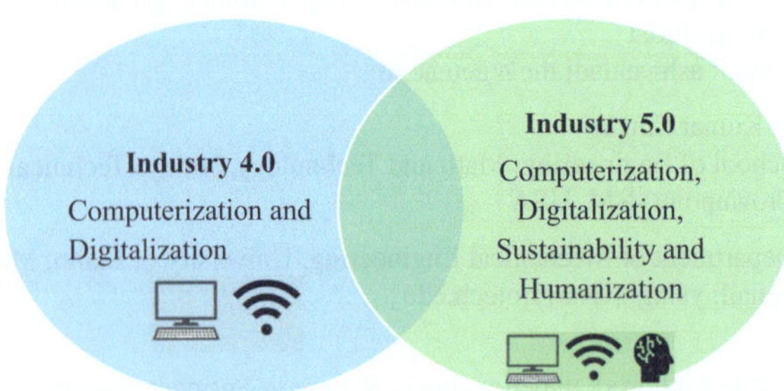

Fig. 1 Advancement from Industry 4.0 to 5.0 [4, 5]

Table 1 Characteristics enriched from Industry 4.0 to 5.0 [4]

Industry 4.0	Industry 5.0
Focuses on connecting machines	Focuses on delivering customer experience
Mass customization	Hyper customization
Intelligent supply chain	Responsive and distributed supply chain
Smart products	Experienced activated products
Manpower distanced from factories	Educating and upskilling the manpower for factories

circular economy, waste reduction and CO_2 emission for a better future of the manufacturing systems.

The basic differences between Industry 4.0 and 5.0 with respect to smart factories are given in Table 1.

Industry 5.0 seeks to address the shortcomings of its predecessor by prioritizing human-centricity, sustainability, and resilience. It fosters a collaborative environment where humans and machines work in harmony, thereby promoting sustainable development and improving societal welfare. Embracing Industry 5.0 enables organizations to effectively overcome the constraints of Industry 4.0 and thrive in the changing manufacturing landscape. The insufficient emphasis on environmental conservation and sustainable technology in Industry 4.0 highlights the necessity for a shift toward Industry 5.0.

Smart factory and cyber-physical systems are two interconnected terms. Cyber-physical system is the basic technology to implement the smart factory concept. The traditional systems have mainly manual data recording and instructions to the machines. Few advanced systems have implemented the program-based machines but still the programming part was completed at separate computer system and that system is not connected with machines. After the Fourth Industrial Revolution, the synchronization of these machines with computers was made possible with the help of cyber-physical systems.

2 Smart Manufacturing

Smart manufacturing is the emerging technology to manufacture the products in a smart way with the integration of the new technologies, collaboration of manufacturing systems, easy customization the product for customer satisfaction. There are various institutes/organizations which have defined the smart manufacturing in their definitions. A few of them are:

- The National Institute for Standards and Technology (NIST) is a government organization in the US to promote innovation and competitiveness by advancing measurement science, standards and technology. As per NIST, Smart Manufacturing is a "*fully-integrated, collaborative manufacturing systems that respond in real time to meet changing demands and conditions in the factory, in the supply network, and in customer needs*" [6].

- Smart Manufacturing Leadership Coalition (SMLC) is a not-for-profit organization in the US spearheading the development of an open infrastructure for smart manufacturing. According to them, "*Smart Manufacturing is the ability to solve existing and future problems via an open infrastructure that allows solutions to be implemented at the speed of business while creating advantaged value*" [7].
- MESA (Manufacturing Enterprise Solutions Association International) is also a not for profit community of international manufacturers, suppliers and solution providers [8]. MESA is spearheading the creation of awareness and best practices of smart manufacturing, and they defined it as "*The intelligent, real-time orchestration and optimization of business, physical and digital processes within factories and across the entire value chain.*"

Smart manufacturing has various advantages over the conventional manufacturing processes. The key components of a smart manufacturing system are given in Fig. 2. All components/technologies of smart manufacturing are equally important but few of them make it smart. Smart Factory is one of the key technologies in smart manufacturing for industrial revolution occurred so far. Further transformation of Industry 4.0 to 5.0 is incomplete without a smart factory concept. Cyber-physical system is also an integral part of a smart manufacturing system as it is the backbone for interconnecting the physical and software/network systems. 3D printing, artificial intelligence/machine learning (AI/ML), robotics, Industrial Internet of Things (IIOT), etc. are also contributing to make a system smart and manufacturing the product.

Fig. 2 Key components of a smart manufacturing system

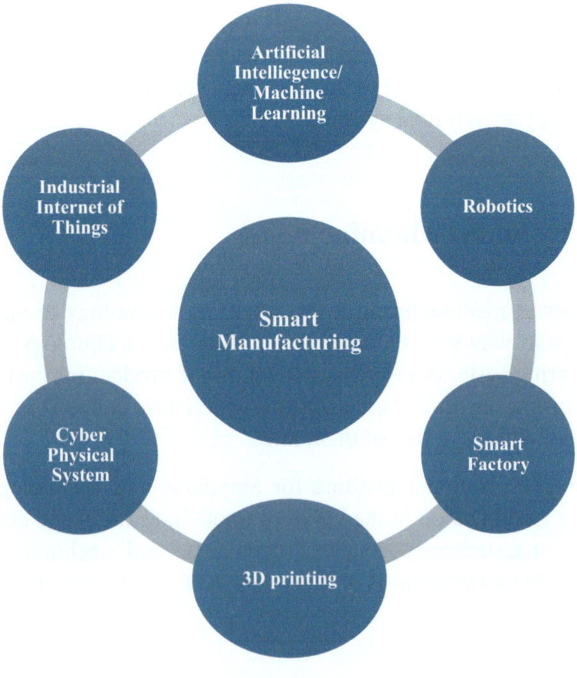

3 Smart Factory

The word "Smart" associated with factory makes it capable of doing different tasks with optimized methods in digitally connected manner. Basically, the smart factory is an integrated, optimized, well connected automated system. In this system, the continuous data sharing/interaction of machines, humans, devices, sensors and digital systems occurs toward a better manufacturing system. The basic characteristics of a smart factory are shown in Fig. 3 [9]

- **Automation**: Manufacturers and organizations across several industries are swiftly embracing automation. A recent study indicates that over 67% of organizations implemented business process automation to optimize and automate their routine procedures as of March 2024. More than 31% of organizations have fully automated at least one critical function, while 24% have implemented low-code process automation in their operations. Automation is an essential element for converting a traditional manufacturing system into a smart manufacturing framework. Automation entails the implementation of machinery and software systems to perform jobs historically executed by humans. It reduces human errors, executes high-risk operations and repetitive work, and enhances workflow on the production line.
- **Connectivity**: The fundamental prerequisite for a smart factory is the capability of machines, devices, and systems to connect and interact effortlessly with one another. Manufacturers must strategize and execute optimal connectivity while guaranteeing real-time communication. Here, human–machine systems are interconnected through IoT devices. Smart factories can deploy IoT and IIoT-enabled technology, including actuators, controllers, sensors, and more devices, across the factory floor and various departments. These devices consolidate information and statistics on various parameters, including pressure, temperature, humidity, and vibration. The collected data is transmitted to a central server or cloud. AI-driven systems analyze data to extract important insights and facilitate real-time decision-making.

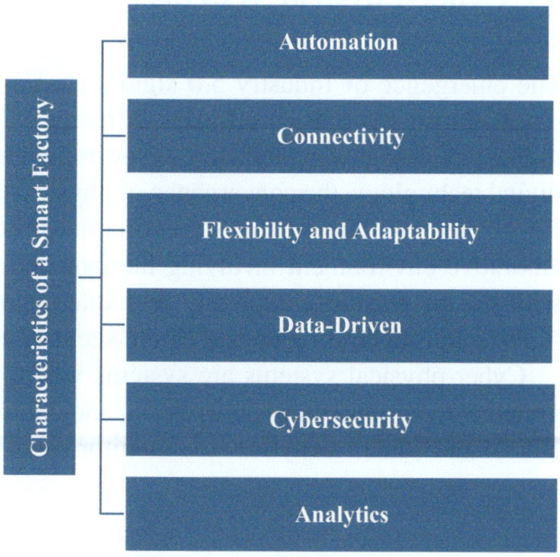

Fig. 3 Basic characteristics of a smart factory

- **Flexibility and adaptability**: Smart factories can adapt to new production tasks, product variations, and real-time demand changes. These are the interconnected systems and based on real-time data the product customization is comparatively easier with smart factory setup. The implementation of the customer feedback and accordingly decision-making for change in product are faster, and this quick implementation of the customer demand will increase the market presence and customer satisfaction.
- **Data-driven and AI**: Real-time data collection and analysis enable predictive maintenance, quality control, and optimization. Artificial intelligence is an essential component in the present and future of intelligent manufacturing. It is an amalgamation of software systems and computers that do tasks often necessitating human intelligence, encompassing perception, thinking, and learning. In manufacturing, AI can optimize production schedules based on demand estimates, streamline cutting patterns to save material waste, and detect anomalies in real time. Manufacturers typically employ supervised and unsupervised machine learning techniques. Machine learning is a subset of artificial intelligence.
- **Cybersecurity**: Industry 4.0 relies on technological advancements and virtual solutions. Your system may become susceptible to cyber-attacks; thus, it is imperative to protect it with a robust cybersecurity framework. It is essential at every phase of the production process, encompassing sensors and devices, a central server, and the cloud.
- **Analytics**: The third component for establishing smart factories is analytics. Data produced from various tools and devices need thorough analysis; else, the gathered data is rendered ineffective. Analytics is the process of examining acquired data to extract insights and make critical decisions for the organization. Data-driven decision-making is a crucial advantage that smart factories have over conventional manufacturers. Predictive analysis of AI and ML facilitates the identification of trends and anomalies, enabling the formulation of strategies for future business growth and expansion.

4 Cyber-Physical Systems (CPS)

The emergence of Industry 5.0 signifies a transformative period in which cyber-physical systems (CPSs) effortlessly merge physical processes with sophisticated digital technologies. As industries grow more integrated and dependent on advanced digital technology, the convergence of physical and cyber realms presents new security challenges, jeopardizing the entire industrial ecosystem. The shift toward a collaborative environment involving humans and machines in Industry 5.0, together with the increasing complexity and interconnectivity of cyber-physical systems (CPSs), introduces unique and varied security and privacy challenges [10].

Cyber-physical systems are systems where physical processes are tightly integrated with computation, networking, and control. These systems use sensors, actuators, and data analytics to monitor and control physical processes. The implementation of the CPS in a smart factory or smart manufacturing system is divided into different layers.

4.1 Architectures of Cyber-Physical Systems in Smart Factories

In a smart factory, the CPS architecture is designed to facilitate seamless interaction between physical machines, digital networks, and artificial intelligence. The conventional CPS architecture in smart factories is divided into levels, with each layer having certain duties. This layered strategy promotes flexibility, scalability, and efficient resource allocation in industrial processes.

4.1.1 Different Layers of Cyber-Physical System in a Smart Factory

The key layers/components of a CPS are characterized into the following layers as shown in Fig. 4.

- **Physical world**: This layer or part of the system includes the physical machines, robots, devices, conveyers and production systems which are actually processing and converting a raw material into finished products. The various functions of this layer are gathering real-time data through using sensors, converting the computed action to the physical action trough actuators and supporting actions like assembly, transport and quality check, etc.
- **Communication networks**: The transmit data between the physical layer and cyber systems. This layer enables real-time secure data transfer between the physical and digital worlds of a smart factory. Several physical devices are connected in parallel, or series and they may have non uniform input and output

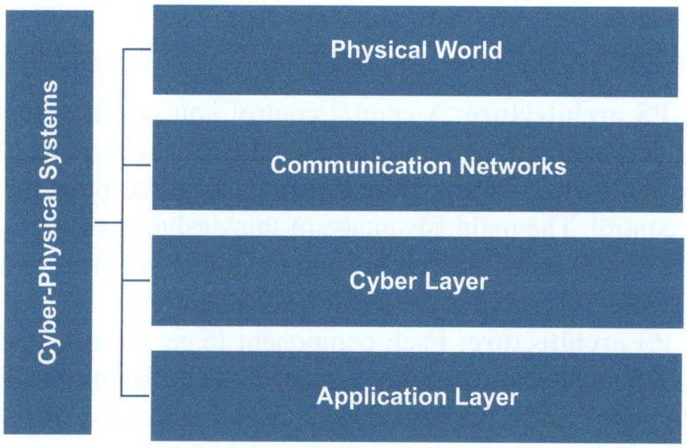

Fig. 4 Key components of a cyber-physical system

ports. So, smooth and uniform connection for data transfer, communication and operation between various devices (interoperability) is made possible with these communication networks. It also supports low-latency, high-bandwidth requirements through advanced connectivity options such as 5G, Wi-Fi 6, and Ethernet.

- **Cyber layer**: This layer is the second part of CPS which includes the computational systems. These computer systems are connected to physical systems through various interconnections and control, analyze, and monitor the physical components using software algorithms. The recording of real-time data through sensors, their monitoring and smooth working within defined range is also controlled within this layer. The filtering, processing, real-time decision based on data received, data storage for long term, etc. are the few important characteristics of this layer. This layer also employs artificial intelligence (AI), machine learning (ML), and data analytics to transform raw data into usable insights. This layer enables for: Predictive analytics can help estimate maintenance needs and optimize production schedules. Real-time quality assurance is achieved by defect detection algorithms. Process optimization involves modifying machine parameters for optimal performance.
- **Application layer**: This layer enables the user to interact with the system, visualize all data and to make decisions based on the given data. The customization, change in any production plan, instructions to the operator and update to the entire supply chain is conducted through this layer. The examples of this layer are human machine interaction and quality control applications.

A cyber-physical system is an advanced computational system that amalgamates sensing, computing, control, and networking functionalities into physical entities, linking them to the Internet and to one another. Although IoT and CPS appear somewhat similar, a significant distinction exists between the two. The Internet of Things (IoT) denotes a tangible device integrated with essential sensors that may collect data and communicate it via the internet. Conversely, in addition to the sensors and internet, a CPS includes components essential for computing and control, enhancing the system's efficiency. The varied applications of CPS encompass aviation, autonomous vehicles, energy, disaster and emergency management, healthcare, intelligent manufacturing, and smart cities. The type of suitable CPS system depends on the smart factory applications. Few general CPS used in smart factories are:

Centralized CPS architecture: A central control unit or server handles the processing and decision-making in a Centralized CPS Architecture. This design is simpler and more suited for smaller smart factories or processes that require centralized control. The main advantage of this design is easier to maintain and less response time. Limited scalability and flexibility are the constraint with this system; therefore, it is suitable for small-scale factories.

Distributed CPS architecture: Each component (e.g., sensors, machines) in the distributed CPS architecture can process and make decisions locally. Data is processed at the edge or near the device level, enabling faster responses. High scalability and flexibility are the advantages but at a higher cost.

Hybrid CPS architecture: The Hybrid CPS Architecture mixes centralized and distributed systems, allowing data to be processed locally and centrally. Time-sensitive processes take place at the edge, while less crucial data is sent to a central server for further analysis. Electronics and consumer goods manufacturing industries where long-term data analytics is required are the suitable examples for such CPS architecture.

5C architecture, 8C architecture, and RAMI4.0 are the existing examples of CPS architectures.

- **5C architecture** is the basic reference model used to implement CPS. This model consists of connection, conversion, cyber, cognition, and configuration levels. The detail is given in Fig. 5 below [3, 11].
- **8C architecture** is an updated version of 5C architecture with the addition of coalition, customer and content.
- **Reference Architecture Model for Industry (RAMI) 4.0**: RAMI 4.0 is an Architecture Reference Model for Industry 4.0, developed by the Platform Industrie 4.0 initiative to establish communication structures and a common language within the factory, complete with its own vocabulary, semantics, and syntax. Such a language allows the integration of IoT and services in the I4.0 framework, linking them to the rest of the world. It is a service-oriented architecture (SOA) that combines IT components to promote the key characteristics of I4.0, such as horizontal integration between factory networks and plants and vertical integration within a factory, with products at one end and the Cloud at the other [3, 12].
- **Industrial Internet Reference Architecture (IIRA)** [12] is an open architecture established by IIC based on IIoT standards that prioritizes interoperability across sectors. This paradigm is divided into four viewpoints to identify and categorize the common preoccupations of an IIoT architecture. Thus, concerns about IIoT systems are methodically studied and resolved, and their outcomes are documented as models and other information in the views connected with the viewpoints.

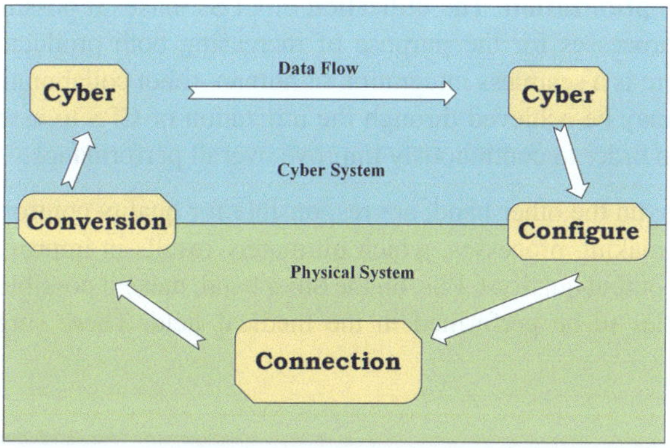

Fig. 5 Data and information flow in a CPS-enabled plant with machine tools in the production line, using the 5C architecture [11]

5 Cyber-Physical System in Industry 5.0

CPSs are crucial in Industry 5.0, enabling secure and efficient collaboration between humans and robots. They provide the foundation for effective interaction between human workers and robotic systems in manufacturing through real-time coordination, monitoring, safety protocols, data analysis, and adaptive processes. Ultimately, they enhance human–robot collaboration by Raja Santhi and Muthuswamy [4]:

- **Integration of physical and digital systems**: Cyber-physical systems (CPSs) integrate physical processes with digital systems, creating a unified connection between the physical and virtual realms. In human–robot collaboration, CPSs enable the real-time monitoring, control, and optimization of physical processes through digital representations.
- **Collaborative robotics (Cobots)**: Through the utilization of the CPS framework, the development and implementation of collaborative robots, commonly referred to as cobots, is made possible. The purpose of these robotic systems is to work in conjunction with human operators, offering assistance and support in a variety of occupations. The responsibility of the CPS is to ensure that the operation of cobots is carried out in a secure manner, that they are able to adjust to changing environments, and that they function effectively in conjunction with human workers.
- **Real-time communication and data exchange**: Between human operators and robotic systems, the deployment of CPSs makes it possible for instantaneous communication and the exchange of data to take place. This makes it possible to make decisions and adjustments to the production process more quickly in response to the shifting conditions of the manufacturing environment.
- **Skill augmentation**: The incorporation of robotic systems into CPSs makes it possible to improve the capabilities of human beings. The mutually beneficial relationship that exists between people and robots has the ability to boost output by capitalizing on the qualities that each of these entities possesses. The cognitive abilities of humans are a contribution, whereas the accuracy and efficiency of robots are contributions itself.
- **Workflow optimization**: The utilization of CPSs makes it possible to improve industrial processes for the purpose of increasing both productivity and efficiency. There is a seamless integration of human–robot collaboration into workflows that may be achieved through the utilization of CPS to assess and adjust processes in order to continuously improve overall performance.

Human beings, on the other hand, are responsible for quality control and the complex decision-making processes, which ultimately results in improved production efficiency and output quality. CPSs, on the other hand, make it possible for advanced robotic surgeries to be performed in the medical field. These surgeries involve robots assisting doctors with delicate procedures, and real-time data from sensors ensures exact motions, hence lowering the likelihood of failures. When it comes to agriculture, smart farming systems make use of CPSs for precision agriculture. Drones and autonomous vehicles that are fitted with sensors collect data on the conditions of the soil, the health of the crop, and the weather patterns. These real-time data enhance the operations of planting, watering, and harvesting, which results in an increase in production while simultaneously minimizing the number of resources that are consumed. It is possible to streamline operations in the fields of logistics and supply chain management by utilizing CPSs, which provide real-time tracking and monitoring of commodities. Real-time data analytics optimize inventory management, thereby decreasing delays and enhancing supply chain efficiency. Automated warehouses make use of collaborative robots to sort and move things in an efficient manner. In general, CPSs are necessary for accomplishing the human-centric and sustainable objectives of Industry 5.0. They integrate powerful artificial intelligence, real-time data processing, and rigorous safety procedures in order to improve human–robot collaboration, increase productivity, and guarantee safety in a variety of industrial applications.

6 Relation Between Smart Factory and CPS

Cyber-physical system is an integral part of smart factories. The smooth functioning of a smart factory is supported by a CPS as it enables the real-time data collection and monitoring the overall plant. There are various aspects for which smart factory is dependent on CPS and few of them are:

- **Core technology**: By connecting physical assets (machines, robots, and tools) with digital control systems, CPS enables real-time monitoring, control, and optimization in a smart factory. CPS is a key technology that is used in smart factories. It enables machines and processes to interact with one another and function freely.
- **Data-driven processes**: Smart factories rely on CPS to gather and process vast amounts of data from the physical environment, allowing for predictive maintenance, self-optimization, and faster decision-making.
- **Interconnected systems**: Production lines, supply chains, and logistics are just a few of the systems in a smart factory that may be connected and work together thanks to the infrastructure that CPS offers.
- **Autonomy and intelligence**: Smart factories using CPS can make intelligent judgments and respond dynamically to changes in production needs, enhancing efficiency and minimizing downtime.

7 Key Technologies Enabling CPS in Smart Factories

The implementation of CPS in smart factory is dependent on different technologies and these technologies made it more efficient and useful. The connection and control of physical system with software parts needs various networking interfaces and suitable algorithms/software to operate them. The following key technologies are common and useful for this purpose (Fig. 6):

- **Industrial Internet of Things (IIoT)**: IIOT connects the physical machines/devices to the digital network like internet. It enables the real-time data collection from the physical systems and store to cloud for further processing. The basic applications of IIOT are Smart sensors, RFID tags, and IoT platforms monitor machine health, production rates, and environmental conditions in factories.
- **Big data and analytics**: Smart factories produce massive amounts of data from equipment, sensors, and systems and this data is collected and stored on cloud through IOT devices. This data is analyzed, processed and evaluated to provide insights that improve the overall efficiency of the system. Big data techniques use software to analyze this huge amount of production data to find patterns, optimize operations, and assist predictive maintenance. Predictive maintenance is one of the best examples of such system in which CPS system uses predictive analysis to forecast the maintenance requirement and system failure. This optimizes the resource usage and improves the product quality with high efficiency.
- **Artificial intelligence (AI) and machine learning (ML)**: AI and machine learning enable CPS systems to understand data, generate predictions, and provide adaptive control over factory activities. AI systems assist the system with self-learning capabilities and enable the machine to learn the activities from the past

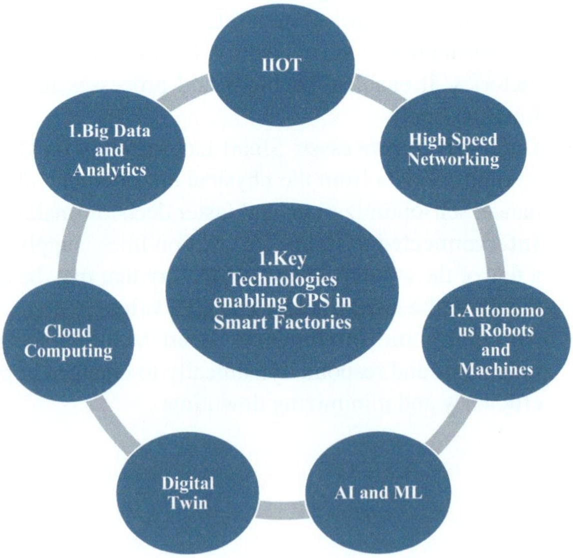

Fig. 6 Key technologies enabling CPS in smart factories

data and adjust themselves accordingly to make system autonomous. Similarly, machine learning algorithms assist in identifying trends in production data, which leads to continuous improvement and automation.

- **Cloud computing**: Cloud computing provides centralized data processing and storage, whereas edge computing enables real-time, local data processing near the manufacturing floor. Therefore, these offers scalable storage and computational power for evaluating huge datasets from the factory floor. The cloud-based CPS links the cloud and smart factory infrastructure for monitoring, data analysis and control purposes.
- **Digital twins**: It is a digital platform which facilitates the virtual model of the physical system. It simulates the system virtually with actual conditions and provides the flexibility to simulate, optimize the production processes without affecting the actual operations. It saves time and resources to improve the system efficiency.
- **Autonomous robots and machines**: The robots and automated machines are an integral part of the smart factory and CPS provides the interface to collaborate with operators. This interconnected system provides the smooth adoption of change in production requirements based on customer requirements. For the customized products, the assembly line, inspection and material handling need to be updated, and this can be easily adopted with robots and automated systems.
- **Industrial high-speed networks**: The smooth data recording and storage from multiple devices needs a high-speed network. This network supports ultra-reliable low-latency communication (URLLC) between manufacturing devices, allowing for real-time control in CPS. The main benefits of networking are faster rate of data transfer and better scalability for smart factories.

These technologies assist the advancement of the manufacturing processes in smart factories with the help of CPS systems. Implementing CPS allows for improved quality control through sophisticated monitoring and automated inspections, ensuring higher levels of product quality and uniformity. Integrating human and machine collaboration has fostered a more fluid and efficient interaction, resulting in higher productivity and accuracy within the manufacturing environment. The overall system stability, higher production rates with flexibility of product customization and higher customer satisfaction are the few advantages of implementing these technologies at factory level.

8 Functions of CPS in Smart Factories

- **Predictive maintenance**: CPS collects different types of useful data from the point of interest with the help of different sensors. This real-time sensor data is evaluated to forecast equipment faults before they occur and reduce downtime. The preventive maintenance saves a lot of cost, machine down time and unplanned interruptions of the machine.

- **Quality control and assurance**: The automation of the quality control results the real-time flaw detection at the time of manufacturing as well as at the end of final product inspection. The network of various sensors and AI powered analysis make this task very simple and accurate. This finally results in low waste, high quality products, flexibility to update the errors within the process.
- **Supply chain optimization**: CPS integration in smart factories links it to the supply chain, and therefore, smart factories can adapt the required changes as per demand or raw material availability. This results in dynamic production planning and a reduction in lead time, bottleneck and waiting time to the customers.
- **Energy management and efficiency**: The CPS enabled systems continuously measure and analyze the energy consumptions and production data. The detailed analysis of this data leads to selecting the energy intensive processes with optimum energy consumption. The implementation of sustainable sources of energy reduces carbon emissions and operational costs.

9 Benefits of Smart Factories and CPS

Cyber-physical system can bring various benefits on installation in a smart factory. The integration of various advanced techniques results in better control on the overall manufacturing unit and access to real-time data. The key benefits are mentioned in Fig. 7.

- **Increased productivity**: Smart factories increase output with minimal human intervention by utilizing automation, real-time data, and self-optimizing technologies.
- **Improved flexibility**: CPS offers dynamic production lines that can respond swiftly to changing demand, product variants, and process changes.
- **Enhanced decision-making**: Real-time data collection and analytics enable firms to make better decisions about production efficiency and resource allocation.
- **Enhanced product quality**: Ongoing surveillance and data-informed quality assurance reduce errors and faults.
- **Reduced downtime**: Predictive maintenance and adaptive scheduling diminish the probability of unforeseen machinery failures.
- **Cost reduction**: Automation, predictive maintenance, and optimum resource utilization result in substantial cost reductions for personnel, energy, and materials.
- **Customization and mass personalization**: CPS facilitates large-scale product customization, accommodating trends such as mass personalization across sectors like automotive, electronics, and consumer goods [13, 14].

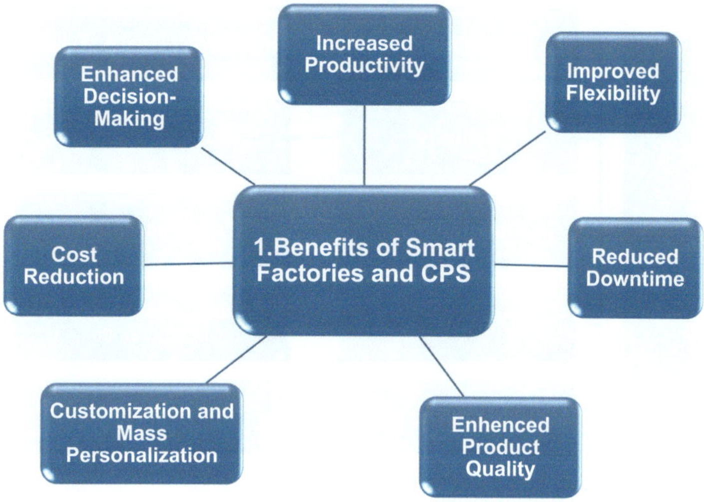

Fig. 7 Benefits of smart factory and CPS

10 Challenges and Limitations

Apart from the benefits and advantages, the implementation of CPS in the smart factories also brings challenges/ issues with it. The level of system complexity is directly related to the system size, number of system components required for the connection and proper functioning of the system. The key challenges faced in the CPS implementation are (Fig. 8):

- **Safety**: It is one of the important aspects of any organization. The CPS implemented to the factory should ensure the safety in every aspect. Hazard defense techniques are delineated to avert system failures via environmental monitoring or, in instances of such failures, to enhance emergency management. System state control, which seeks to identify fault or failure conditions prior to their escalation into safety concerns, is closely associated. Safety pertains to the operational integrity of systems, namely the safeguarding of individuals and the environment from physical harm, whereas security focuses on the protection of data and information within a system [1].
- **Security**:
  - **Cybersecurity threats**: The heightened connection in smart factories renders systems vulnerable to cyber-attacks. CPS must implement comprehensive cybersecurity protocols to safeguard sensitive information and essential infrastructure. Any kind cyber-attack can severely harm the system, and therefore, CPS should maintain the system functionality under such adverse and critical condition.

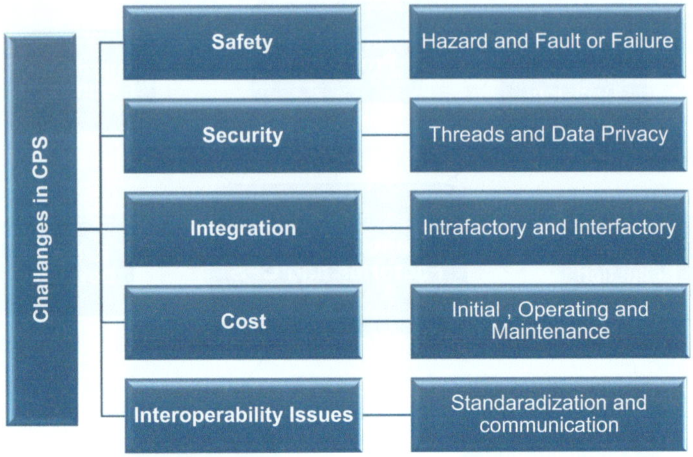

Fig. 8 Challenges in CPS implementation [1]

   - **Data privacy**: Under the cyber-attack or any of such situation, managing and protecting extensive data recorded from sensors, machinery, and personnel prompts is always challenging. The issues around data ownership, privacy, and regulatory compliance are always present and should be considered during implementation.

- **Integration complexity**: Integrating legacy systems with modern CPS architectures can be challenging and require significant investments in infrastructure and training. Integration of different machines, components and different systems require different connection for smooth communication/data transfer between different data devices.
- **High initial costs**: Implementing cyber-physical systems and constructing smart factories may include significant initial expenditures for new technology, infrastructure, and staff training.
- **Interoperability issues**: The absence of standardization among devices, communication protocols, and platforms can impede the seamless integration of various CPS components [15–17].

11 Future Trends in Smart Factories and CPS

Customization and dynamic nature are the basic requirement for a good industry. The continuous improvement/advancement of the technology is the basic requirement for new CPS based smart systems. CPS and smart factory are slowly increasing its market presence in manufacturing systems throughout the world. The technologies are improving and in the Industry 5.0, these systems will have a huge impact. Few technologies which are advancing and can play an important role in CPS and smart factory are [2]:

- **AI-driven autonomous factories**: Artificial intelligence and cyber-physical systems will facilitate factories that function with minimum human involvement, autonomously making decisions based on real-time data and predictive analytics. The basic purpose of the automation is to operate with minimum human intervention and AI the tool which enables the system to take quick decisions based on data analytics.
- **Edge computing in CPS**: Factories produce substantial data with using sensors and other connected devices. This can be locally processed at the device level which minimizes the latency and bandwidth consumption. Therefore, the data transfer will be reducing and processing will be fast as less amount of data will be processed at end point or at the decision level. The edge computing can make this happen and contributes to the overall efficiency of the system.
- **Blockchain for supply chain transparency**: Blockchain technology will improve traceability and security in supply chains by offering an unalterable record of transactions, from raw material procurement to product delivery.

All of these technologies are not fully developed, and researchers continue to strive toward their progress. A survey involving roughly 300 manufacturers revealed that 83% prioritize smart manufacturing technologies and processes, with over 60% intending to embrace these technologies very soon. Manufacturers favor people possessing pertinent technical expertise and abilities, capable of swiftly adapting to new technologies to effectively operate smart factories. Essential talents include communication, teamwork, adaptability, analytical reasoning, and proficiency with intelligent technologies [18].

12 Conclusion

The smart factories are now playing a vital key role in the automation and enhancing the manufacturing systems in the present industrial age. Smart factories and cyber-physical systems are transforming manufacturing by establishing adaptable, data-driven, and autonomous production environments. The integration of cyber-physical systems (CPS) provides substantial advantages in efficiency, flexibility, and productivity; however, manufacturers must confront problems with cybersecurity, data privacy, and system integration to fully harness the potential of industrial advancement. Smart devices such as IoT and CPS have now developed as a ubiquitous paradigm capable of significantly transforming companies through sensing, identification, remote control, and automated control capabilities. Industry 5.0 and the Industrial Internet are all predicated on the technological foundation of the Internet and interconnected devices, emphasizing process control with minimal human intervention and intelligent decision-making, significantly influencing the global market.

References

1. Oks, S. J., Jalowski, M., Lechner, M., Mirschberger, S., Merklein, M., Vogel-Heuser, B., & Möslein, K. M. (2022). Cyber-physical systems in the context of Industry 4.0: A review, categorization and outlook. *Information Systems Frontiers*, 1–42. https://doi.org/10.1007/S10796-022-10252-X/FIGURES/18
2. Phuyal, S., Bista, D., & Bista, R. (2020). Challenges, opportunities and future directions of smart manufacturing: A state of art review. *Sustainable Futures, 2*, 100023. https://doi.org/10.1016/j.sftr.2020.100023
3. Pivoto, D. G. S., de Almeida, L. F. F., da Rosa Righi, R., Rodrigues, J. J. P. C., Lugli, A. B., & Alberti, A. M. (2021). Cyber-physical systems architectures for industrial internet of things applications in Industry 4.0: A literature review. *Journal of Manufacturing Systems, 58*, 176–192. https://doi.org/10.1016/j.jmsy.2020.11.017
4. Raja Santhi, A., & Muthuswamy, P. (2023). Industry 5.0 or industry 4.0S? Introduction to industry 4.0 and a peek into the prospective industry 5.0 technologies. *International Journal on Interactive Design and Manufacturing (IJIDeM), 17*(2), 947–979. https://doi.org/10.1007/s12008-023-01217-8
5. Nahavandi, S. (2019). Industry 5.0—A human-centric solution. *Sustainability, 11*(16), 4371. https://doi.org/10.3390/su11164371
6. NIST. (n.d.). Product definitions for smart manufacturing. Retrieved November 8, 2024, from https://www.nist.gov/programs-projects/product-definitions-smart-manufacturing
7. Oracle India. (n.d.). What is smart manufacturing? Retrieved November 8, 2024, from https://www.oracle.com/in/scm/manufacturing/smart-manufacturing/what-is-smart-manufacturing/
8. MESA International. (n.d.). Smart manufacturing—Manufacturing Enterprise Solutions Association. Retrieved November 8, 2024, from https://mesa.org/topics-resources/smart-manufacturing/
9. NAMTECH. (2024). Six elements for optimised smart factory setup. Retrieved from https://www.namtech.ac/blog/6-essential-smart-manufacturing-components-for-implementing-smart-factories/
10. Alabdulatif, A., Thilakarathne, N. N., & Lawal, Z. K. (2024). A review on security and privacy issues pertaining to cyber-physical systems in the Industry 5.0 era. *Computers, Materials & Continua, 80*(3), 3917–3943. https://doi.org/10.32604/cmc.2024.054150
11. Lee, J., Bagheri, B., & Kao, H.-A. (2015). A cyber-physical systems architecture for Industry 4.0-based manufacturing systems. *Manufacturing Letters, 3*, 18–23. https://doi.org/10.1016/j.mfglet.2014.12.001
12. Habib, M. K., & Chukwuemeka, C. I. (2022). CPS: Role, characteristics, architectures and future potentials. *Procedia Computer Science, 200*, 1347–1358. https://doi.org/10.1016/j.procs.2022.01.336
13. Shah, S., Nautiyal, H., Gugliani, G., Kumar, A., Namboodri, T., & Singla Editors, Y. K. (2024). *Sustainability in smart manufacturing: Trends, scope, and challenges*. CRC Press. ISBN: 9781032740713. https://doi.org/10.1201/9781003467496

14. Singla, Y. K., Kumar, A., & Maughan, M. R. (2024). Chapter 1: Introduction to mechanical and fracture behavior characterization of nanocomposites and reinforced laminated structures. In A. Kumar, Y. K. Singla, & M. R. Maughan (Eds.), *Fracture behavior of nanocomposites and reinforced laminate structures* (pp. 1–14). Springer. https://doi.org/10.1007/978-3-031-68694-8_1
15. Kumar, A., Gori, Y., Dutt, N., Singla, Y. K., & Maurya, A. (2021). *Advanced computational methods in mechanical and materials engineering*. CRC Press. ISBN: 9781032052915. https://doi.org/10.1201/9781003202233
16. Kumar, A., Ram, M., & Singla, Y. K. (2022). *Advanced material for biomechanical applications*. CRC Press. ISBN: 9781032054490. https://doi.org/10.1201/9781003286806
17. Kumar, A., Singla, Y. K., & Maughan, M. (2024). *Fracture behavior of nanocomposites and reinforced laminate structures*. Springer Nature. https://doi.org/10.1007/978-3-031-68694-8
18. Sahoo, S., & Lo, C.-Y. (2022). Smart manufacturing powered by recent technological advancements: A review. *Journal of Manufacturing Systems, 64*, 236–250. https://doi.org/10.1016/j.jmsy.2022.06.008

Advancing Manufacturing With Industry 5.0: Cobots, Digital Twins and Smart Integration

Vasanthi Govindaraj

1 Introduction

The manufacturing industry has evolved profoundly, driven by technological innovations that have revolutionized efficiency and worker engagement. From the introduction of machinery during the Industrial Revolution to the highly automated systems of Industry 4.0, each era has advanced the relationship between humans and machines. In the current Industry 5.0 era, the focus has shifted from complete automation to human–robot collaboration, combining human creativity with robotic precision to create production processes that are more efficient, adaptable, and sustainable.

Industry 4.0 marked a significant transformation by integrating technologies like artificial intelligence (AI), machine learning (ML), the Internet of Things (IoT), and big data analytics into manufacturing systems. These innovations enabled autonomous machine decision-making and optimized production without human intervention. However, despite improvements in efficiency, human creativity and problem-solving remained underutilized. As manufacturing became more reliant on automated systems, the absence of human cognitive input led to a critical gap in innovation and flexibility, limiting the industry's ability to respond to complex challenges and unique customer needs.

Industry 5.0 bridges this gap by emphasizing collaboration between humans and robots, particularly through the use of collaborative robots (cobots). In Industry 5.0, collaborative robots (cobots) are designed to work closely with humans, ensuring both safety and productivity in manufacturing environments. Equipped with safety features such as force sensors and emergency stop capabilities, cobots reduce the

V. Govindaraj (✉)
National General (An Allstate Company), Dallas, TX, USA

Y. Kumar Singla et al. (eds.), *Digital Manufacturing in Industry 5.0*,
https://doi.org/10.1007/978-3-031-91500-0_41

risk of workplace accidents while handling tasks like heavy lifting and precision work.

Cobots also enhance productivity by using AI and machine learning (ML) to perform repetitive tasks efficiently while adapting to real-time conditions, allowing human workers to focus on more creative and decision-oriented tasks. By combining the strengths of both, Industry 5.0 offers a paradigm shift where robots do not replace human workers but amplify their capabilities, allowing for greater innovation, problem-solving, and personalization in manufacturing processes.

Several key technologies underpin the vision of Industry 5.0. AI and ML enable cobots to engage in tasks like predictive maintenance and quality control, while IoT connects machinery and systems in a unified, data-driven framework for real-time monitoring and dynamic adjustments. Digital twins—virtual replicas of physical objects or systems—further enhance efficiency by simulating production processes and predicting equipment performance. Blockchain technology also plays a crucial role by ensuring the security and transparency of these interconnected systems, especially in industries requiring traceability, such as pharmaceuticals or food production.

Alongside its technological innovations, Industry 5.0 brings profound social and economic impacts. The shift from automation to human–robot collaboration presents new opportunities for workers to engage in higher-value tasks that require creativity, critical thinking, and adaptability. As robots take on more repetitive or dangerous tasks, human workers can focus on innovation, design, and problem-solving, fostering a more dynamic and engaging work environment. This transition could lead to a resurgence of skilled labor, where individuals are trained not only to operate machines but to collaborate with them in sophisticated ways.

The collaborative model of Industry 5.0 enhances not only operational safety and productivity but also caters to increasing demands for customization. Unlike Industry 4.0's focus on mass production, Industry 5.0 allows manufacturers to produce highly personalized goods without sacrificing efficiency. Additionally, the focus on sustainability and resilience is central to Industry 5.0, offering solutions to reduce waste, optimize resources, and enhance energy efficiency, making manufacturing both more environment friendly and responsive to future challenges. This shift aligns with global sustainability goals and reflects growing consumer demand for environmentally conscious production practices.

Furthermore, Industry 5.0 has the potential to reshape global competitiveness. Nations and industries that successfully adopt this model could gain a significant advantage in producing high-quality, customizable products while maintaining cost efficiency. By integrating human creativity with advanced technologies, manufacturers can respond more rapidly to market changes, meet diverse customer demands, and drive innovation. This adaptability is especially crucial in today's fast-paced, globalized economy, where businesses must continuously innovate to remain competitive.

Despite its potential, the adoption of Industry 5.0 faces challenges, particularly the high cost of integrating advanced robotics, AI, and IoT systems. Small and medium-sized enterprises (SMEs) may struggle to invest in the necessary

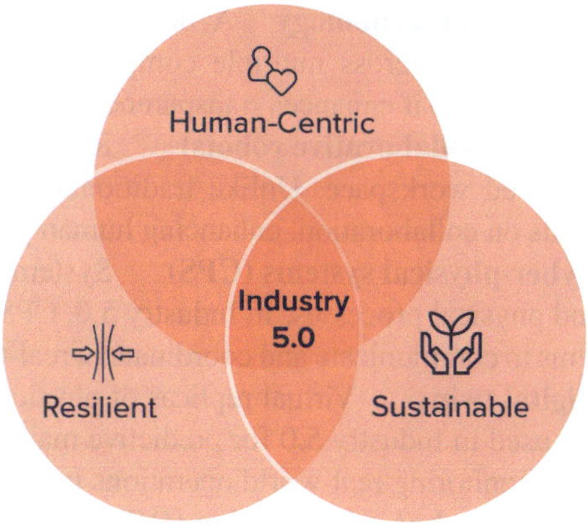

Fig. 1 Core pillars of Industry 5.0: human-centric, resilient, and sustainable

infrastructure and training, and increased cybersecurity risks demand robust measures to protect intellectual property and operational integrity. Moreover, ethical considerations surrounding human–robot collaboration, such as data privacy, job displacement, and the appropriate distribution of responsibilities, must be addressed to ensure that Industry 5.0 benefits both businesses and workers equitably.

Nevertheless, Industry 5.0's advantages are undeniable. By fostering closer collaboration between humans and robots, it not only enhances efficiency and productivity but also creates a more adaptable and resilient manufacturing ecosystem. As advancements in AI, robotics, and IoT continue, the decreasing costs of these technologies are expected to expand their accessibility, making the vision of human–robot collaboration a reality across more industries. Additionally, with proper policies and training programs in place, businesses can help workers transition into new roles that prioritize creativity and complex problem-solving, mitigating the potential negative impacts on employment [1].

In conclusion, Industry 5.0 represents a groundbreaking evolution in manufacturing. By seamlessly integrating humans and robots, alongside cutting-edge technologies like AI, ML, and IoT, it offers a pathway toward a more human-centric, innovative, and sustainable future. This new era not only reshapes the role of technology in manufacturing but also redefines the relationship between humans and machines, setting the stage for a more inclusive and dynamic industrial landscape (see Fig. 1).

2 Glossary of Terms

Artificial intelligence (AI) The simulation of human intelligence in machines that are programmed to think and learn. In Industry 5.0, AI powers robots and systems to enable tasks like decision-making, predictive maintenance, and real-time adjustments.

Blockchain technology A decentralized digital ledger system used to record transactions across multiple computers, ensuring data integrity and security. In Industry 5.0, it enhances transparency, security, and traceability in supply chains.
Cobots (collaborative robots) Robots designed to work alongside humans in a shared workspace. Unlike traditional industrial robots, cobots in Industry 5.0 focus on collaboration, enhancing human productivity while ensuring safety.
Cyber-physical systems (CPS) Systems that integrate computing, networking, and physical processes. In Industry 5.0, CPS allow machines and management systems to communicate and coordinate in real-time, often without human intervention.
Digital twins Virtual replicas of physical objects, systems, or processes. These are used in Industry 5.0 for predictive maintenance and optimization by simulating and monitoring real-world operations in real time.
Human–robot collaboration (HRC) The interaction and teamwork between humans and robots, where both parties contribute their strengths to tasks. Industry 5.0 emphasizes HRC for enhanced efficiency, safety, and personalization in manufacturing.
Internet of Things (IoT) A network of interconnected devices that can collect and exchange data. IoT enables real-time monitoring and decision-making in Industry 5.0 by connecting machines, sensors, and systems.
Machine learning (ML) A subset of AI where machines learn from data and improve their performance without being explicitly programmed. In Industry 5.0, ML helps cobots and systems become more adaptive and efficient.
Predictive maintenance A maintenance strategy that uses data analysis to predict when equipment will fail so that maintenance can be performed just in time to prevent breakdowns. Digital twins play a key role in predictive maintenance within Industry 5.0.
Resilience The ability of a system or process to adapt to and recover from disruptions, such as equipment failure or unexpected changes in production needs. Industry 5.0 systems are designed to be resilient through real-time monitoring and adaptive technologies like digital twins.
Smart manufacturing A method of manufacturing that leverages digital technologies, including AI, IoT, and automation, to optimize production processes. Industry 5.0 focuses on making smart manufacturing systems more human-centric and sustainable.
Sustainability In the context of Industry 5.0, sustainability refers to practices and technologies that minimize environmental impact, such as reducing waste, optimizing resource use, and promoting energy efficiency.

3 Literature Review

The progress of manufacturing has been influenced by five revolutions that reflect the interaction between technological progress and human creativity over time. The Industrial Revolution, which started in the late 1700s and was powered by water and steam energy sources, saw the introduction of mechanization, revolutionizing production methods in sectors such as textiles. This pivotal era marked the beginning of production automation, where machinery reduced labor-intensive tasks and provided a preview of technology's potential in manufacturing [2].

Following this, the second industrial wave, in the mid-nineteenth century, leveraged electricity to bring about mass production. Assembly lines, first introduced during this time, improved speed, accuracy, and overall production capacity, particularly in sectors like the automobile and steel industries. This allowed businesses to scale through economies of scale and refine production techniques.

The mid-twentieth century ushered in Industry 3.0, also known as the digital revolution. This era integrated computers, automation systems, and early-stage robotics, reducing reliance on human labor by enabling machines to perform routine tasks autonomously. Programmable logic controllers (PLCs) and integrated circuits brought a data-driven approach to manufacturing, optimizing processes and reducing errors. Automation became more prominent as machines gained the ability to execute tasks with greater precision and efficiency [3].

Industry 4.0, introduced in 2011, further transformed manufacturing by incorporating cyber-physical systems with AI, big data analytics, and the IoT (see Fig. 2). This era enabled interconnected factories where machines could communicate and share real-time information, leading to improved inventory management and production efficiency. Specialized AI algorithms enhanced machine learning capabilities, allowing factories to anticipate equipment malfunctions and adjust processes to changing production needs [4].

However, Industry 4.0's focus on automation revealed a gap in harnessing human creativity. In response, Industry 5.0 emerged, promoting human–robot collaboration. Instead of relying solely on machines for autonomous decision-making, Industry 5.0 emphasizes the unique strengths of both humans and robots. AI and

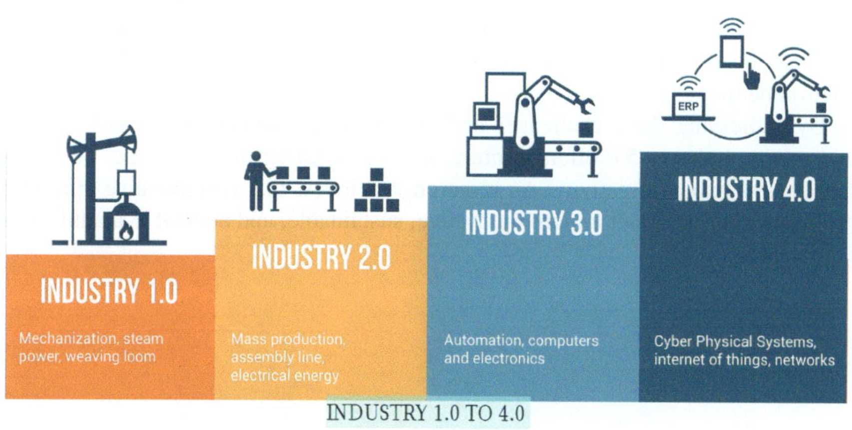

Fig. 2 History of Industry 1.0 to Industry 4.0

machine learning augment human abilities, with robots handling routine tasks while humans focus on problem-solving and creative decision-making.

As central elements of Industry 5.0, collaborative robots (cobots) are designed to work in close cooperation with human workers, ensuring a balance between safety and efficiency. Cobots are equipped with advanced safety mechanisms, including force sensors and emergency stop capabilities, reducing the risk of accidents and mitigating physical strain from labor-intensive tasks like heavy lifting and precision work. Research has shown that cobots, through the integration of AI and machine learning (ML), can perform complex tasks with high precision while continuously learning from human input. This real-time adaptability enhances production efficiency and allows human workers to focus on creative problem-solving and decision-making tasks, resulting in a more dynamic and flexible manufacturing process.

In addition to cobots, several other advanced technologies are integral to Industry 5.0. AI and IoT enable real-time data collection and dynamic adjustments to production processes. Digital twins—virtual replicas of physical systems—allow manufacturers to simulate and monitor operations, enhancing predictive maintenance and minimizing downtime. By predicting equipment malfunctions before they occur, digital twins help companies avoid costly disruptions and reduce operational expenses.

Blockchain technology, another innovation in Industry 5.0, enhances supply chain transparency and security. It records transactions in a decentralized, immutable ledger, ensuring that every step in the production process can be tracked. This is particularly useful in industries like pharmaceuticals and food production, where verifying the origin and quality of materials is crucial for safety and compliance.

While Industry 5.0 offers significant advantages, its adoption also presents challenges. Implementing advanced robotics, AI, and IoT systems requires substantial investment in infrastructure and training. Additionally, the reliance on interconnected systems heightens cybersecurity risks, making robust protective measures essential to safeguard intellectual property and prevent operational disruptions. With growing data sharing in manufacturing, the need for strong cybersecurity becomes even more critical to ensuring the smooth functioning of Industry 5.0 environments.

By addressing these challenges and enhancing collaboration between humans and robots, Industry 5.0 offers a future of manufacturing that prioritizes both efficiency and human creativity. As research continues, this collaborative approach promises to bring about a more innovative, sustainable, and adaptable manufacturing landscape.

4 Collaborative Innovation

This shift from Industry 4.0 to Industry 5.0 marks the transition from automated data-driven production to human-centered collaborative production (see Fig. 3). Contrasting with Industry 4.0, which was focused on technological advancements in AI and robotics, Industry 5.0 seeks the full integration of human intelligence and creativity with machine capabilities. The European Commission's DGRD has supported this vision as a much-needed further development rather than an extension of Industry 4.0.

Industry 5.0 integrates sustainability into manufacturing, such as waste reduction, the circular economy, and decarbonization. Contrasting with the winner-takes-all economy that Industry 4.0 fosters, its consequence of monopolies and inequality 5.0 counters wealth gaps and promotes social welfare. In such a human-centered approach, skilled labor is reintroduced to manufacturing, amplifying the symbiotic and adaptive relationship between humans and machines.

While Industry 4.0 relied heavily on digitalization and AI, Industry 5.0 redefines these technologies by embedding a regenerative, sustainable vision for industrial progress (see Fig. 4). The shift also brings full circle the growing importance of bioinspired technologies, energy efficiency, digital twins, and cybersecurity, which were core to Industry 4.0 but will play a different, more integrated role in Industry 5.0.

Even though the complete definition of Industry 5.0 is still being debated, this shift has already begun to affect the production sectors. It is a movement to show that Industry 4.0, with its total concentration on digital technology, did not consider major social issues like equity and environmental sustainability. Fundamentally, Industry 5.0 is a new paradigm that oversteps the bounds of Industry 4.0 and its human–robot interaction to develop a more cooperating, bespoke manufacturing environment [5].

Fig. 3 Industrial progression from 4IR

Fig. 4 Industrial progression from 5IR

4.1 Comparing Industry 4.0 and Industry 5.0: Efficiency and Cost Analysis

In Industry 4.0, the primary focus was on automating production lines using advanced technology and machinery. However, Industry 5.0 represents a shift toward integrating human expertise into manufacturing processes, creating more personalized and adaptable production environments. A thorough comparative analysis of these two industrial eras reveals significant differences in operational efficiency and cost-effectiveness (see Table 1).

One notable advancement in Industry 5.0 is the utilization of AI-powered collaborative robots, or cobots, which enhance operational flexibility and reduce the necessity for constant human supervision (see Fig. 5). This, in turn, leads to decreased labor costs and increased overall productivity. Additionally, incorporating digital twins, virtual replicas of physical objects or processes, has proven highly beneficial in Industry 5.0. These digital twins have been shown to reduce production downtime by up to 30% significantly, a substantial improvement over the static and preprogrammed automation systems of Industry 4.0. This enhanced efficiency ultimately contributes to cost savings and improved resource utilization in the manufacturing sector [6].

Adopting Industry 5.0 technologies entails a significant upfront investment due to integrating cutting-edge artificial intelligence (AI), Internet of Things (IoT) devices, and digital twins. Unlike the standardized automation systems used in Industry 4.0, Industry 5.0 requires a more customized approach, resulting in a 20–30% increase in implementation costs. However, the long-term advantages, such as heightened productivity, minimized downtime, and improved product customization, validate the initial financial commitment [7].

Table 1 Comparison between Industry 4.0 and Industry 5.0 technologies

Key metrics	Industry 4.0	Industry 5.0
Focus	Automation and autonomous decision-making.	Human–robot collaboration and personalization.
Cost savings	Moderate cost savings due to automation.	Higher long-term cost savings due to personalized, efficient production and reduced downtime through predictive maintenance (up to 30% downtime reduction with digital twins).
Implementation cost	High upfront cost for automation systems.	Higher initial investment due to the integration of AI, cobots, and digital twins, but better return on investment (ROI) over time due to improved efficiency and flexibility.
Productivity improvement	15–20% increase in productivity due to automation.	25% increase in production flexibility and up to 35% productivity enhancement with human–robot collaboration.
Error rates	Reduces errors by automating repetitive tasks (10–15% error reduction).	Human–robot collaboration further reduces errors by 15–20%, especially in complex tasks requiring human oversight.
Technology integration	Focus on integrating IoT and AI for autonomous processes.	Focus on integrating AI, cobots, IoT, blockchain, and digital twins for adaptive, real-time decision-making.
Customization	Limited customization, focusing on mass production.	High levels of customization and personalization, enabling small-batch and made-to-order production.
Downtime reduction	Limited predictive maintenance; downtime based on reactive approaches.	Significant downtime reduction (up to 30%) due to predictive maintenance and real-time monitoring with digital twins.
Sustainability	Efficiency-oriented, with limited focus on sustainability.	Strong focus on sustainability, with AI optimizing resource use and reducing waste.
Cybersecurity risks	Moderate cybersecurity risks due to IoT integration.	Increased cybersecurity risks due to the more extensive use of interconnected systems, including AI, IoT, and blockchain, requiring stronger security frameworks.
Human involvement	Minimal human involvement due to full automation.	High human involvement, with humans and robots collaborating closely to enhance decision-making, creativity, and problem-solving capabilities.

Fig. 5 Human–robot collaboration in the workplace: the future of Industry 5.0

5 Methodology

5.1 Experimental Setup

In the realm of Industry 5.0 the teamwork between humans and robots has transformed how manufacturing operations function by boosting their effectiveness and adaptability while also allowing for personalized approaches to production methods and techniques. Past experiments have been conducted to assess the effects of humans working alongside robots with an eye on factors, like decreasing errors and increasing productivity levels. In a particular trial, cobots were integrated into a human-operated assembly line, yielding significant results: a 20% increase in productivity and a 15% reduction in human errors. These findings underline the harmonious coexistence of human decision-making and robotic precision, enabling humans to concentrate on higher-level tasks while robots handle repetitive, precision-oriented operations [8].

In a recent case study focused on tailored small-batch manufacturing, advanced robotic arms enhanced with AI technology demonstrated remarkable adaptability to changing production requirements. The flexibility shown here resonates with the values of Industry 5.0 and its aim to shift from conventional mass production to personalized and responsive manufacturing methods based on demand. Research findings highlighted a 25% improvement in adaptability and production processes, which allows for faster and more effective adaptations to meet changing customer needs and market trends.

Digital twins are like virtual copies of real factories, and they help make sure everything runs smoothly. They can predict when machines might break down and even help determine the best way to make things. For example, a car company used digital twins to see how their robots would perform under different workloads, helping reduce downtime by 30%. This proactive approach makes factories more resilient and capable of dealing with problems before they arise [9].

5.2 Simulation Setup

Simulation studies have become essential in testing systems today in manufacturing settings where smooth collaboration, with human workers is vital. Manufacturers can digitally represent real-world production environments—what are now called digital twins—without resorting to expensive, cumbersome physical experiments. In these simulations, virtual spaces were created where robots and human operators could cooperate in optimized productivity, flexibility, and adaptability.

These simulations dynamically modeled complex, data-driven scenarios, providing critical insights into how robotic systems and human–robot collaboration would perform under various industrial conditions (see Fig. 6). Simulations tested various failure scenarios, including robotic malfunctions and sensor failures, to assess the system's resilience and robustness [10].

In the pharmaceutical sector, companies such as Pfizer have adopted digital twin technology to optimize supply chain management. By creating virtual replicas of manufacturing environments, they simulate the production process in real time, enabling rapid adjustments that lead to significant cost savings and reduced waste. In early 2024, Pfizer's adoption of digital twin technology was shown to reduce its production cycle for vaccines by 15%, emphasizing the role of real-time simulations

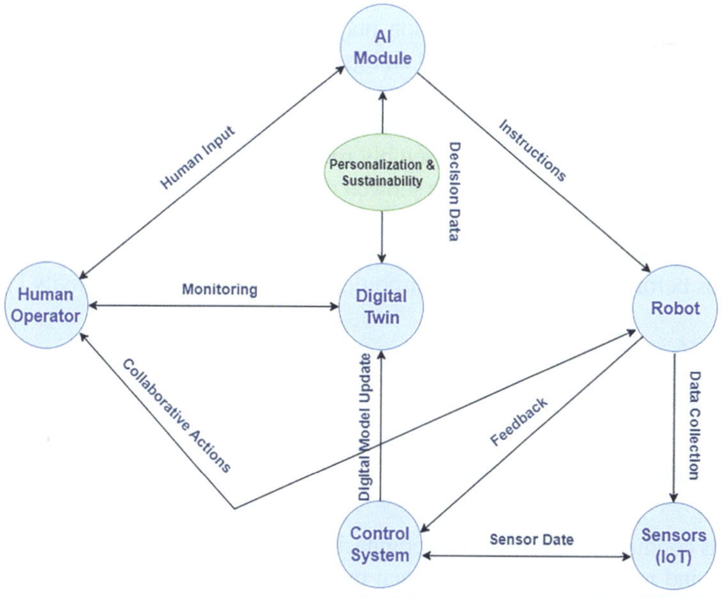

Fig. 6 System architecture for human–AI–robot interaction in Industry 5.0

in optimizing pharmaceutical manufacturing [8]. This approach has proven especially beneficial in maintaining operational efficiency during unpredictable market demands.

5.3 Data Collection and Analysis

Real-time data from sensors embedded within machines and cobots were used to monitor their performance and predict system malfunctions. Predictive maintenance algorithms were employed within a digital twin system to monitor the health of industrial robots. The system was tested in an electronics manufacturing facility, where 95% of potential defects were identified before they impacted the production line. This improved product quality and reduced material waste by 10%, showcasing how digital twins can enhance quality control and resource efficiency.

The data-driven approach used in these simulations allowed for continuous monitoring and analysis of the production process, balancing tasks dynamically based on real-time data. This ensured smoother operations and reduced downtime by anticipating and mitigating issues before they affected the overall production flow [11, 12].

6 Validation of Robotic Systems

Performance and reliability are critical aspects of robotic systems in Industry 5.0, making validation indispensable. Thorough testing under various operational conditions is necessary to meet the strict demands of modern manufacturing. In Industry 5.0, where human–robot collaboration plays a significant role, validating robots' ability to work safely and efficiently alongside human operators is crucial.

Recent validation efforts have focused on long-term deployment in industries like automotive and electronics manufacturing. In the automotive industry, for instance, welding, painting, and assembly robots have demonstrated improvements in operational efficiency. Field tests showed that advanced robotics systems reduced downtime by 30%, primarily due to predictive maintenance enabled by real-time monitoring. This predictive capability allows manufacturers to identify potential problems before they cause delays or bottlenecks, enhancing productivity [13].

7 Results

7.1 Experimental Results

The tests conducted in this research emphasize the advantages of teamwork between humans and robots. These cobots when used alongside human operated assembly lines experienced a 20% boost in productivity and a 15% decrease in error rates. In

2023, a study by Universal Robots demonstrated that cobots equipped with advanced AI capabilities improved production line efficiency by 30%, significantly reducing human errors and enhancing adaptability in real-time [5]. Furthermore, in settings with small batch manufacturing processes cobots showcased flexibility. They enabled production systems to adjust to shifts in customer preferences resulting in a 25% enhancement in production flexibility.

Real-world implementations provide further evidence of Industry 5.0's impact. In the automotive industry, companies like BMW have implemented cobots for welding and assembly, allowing human workers to focus on complex decision-making tasks. Additionally, Tesla uses digital twins to predict machine failures, reducing downtime by up to 20% and ensuring that the production line runs smoothly. These real-world examples underscore the adaptability of Industry 5.0 technologies to increase production efficiency while minimizing errors.

7.1.1 Enhanced Productivity and Quality

The findings from these simulations were highly encouraging. By incorporating digital twins and AI-driven cobots, we observed a 25% increase in production flexibility and a 15% reduction in human error. This seamless integration of human and robotic collaboration optimized production cycles, reduced cycle times, and improved output quality. Robots provided precision and speed, while humans contributed flexibility and creativity, particularly in complex tasks that required decision-making beyond the capacity of machines. This combination allowed manufacturers to optimize workflows in ways that were not previously possible with traditional systems [14–16].

7.1.2 Success Rate Metrics for Human–Robot Collaboration

Through the collection of performance indicators from various trials focusing on human–robot collaboration, we were able to obtain specific data for evaluation. For instance, the synergy between human workers and robots at an automotive assembly plant led to a remarkable 35% enhancement in task effectiveness. Moreover, when robots were entrusted with quality control responsibilities, there was a 10% reduction in material waste. These significant achievements underscore the potential of Industry 5.0 technologies in optimizing the productivity and effectiveness of both human and machine involvement within manufacturing environments [17–19].

7.2 Simulation Results

The simulation studies provided significant insights into the resilience of robotic systems in various failure scenarios. By leveraging digital twin technology, simulations of robotic malfunctions and sensor failures demonstrated that adaptable robotic systems could dynamically reroute tasks to other machines or human operators, minimizing downtime. Predictive maintenance protocols within the simulations allowed the system to resolve issues before they impacted production, further reducing operational disruptions [20].

7.2.1 Robustness and Resilience Through Digital Twins

One of the most powerful features of digital twins is their ability to replicate failure scenarios, thus testing the robustness of robotic systems. Our simulations induced various failure conditions, such as robotic malfunctions and sensor failures, to assess how the system would react and recover. The results demonstrated that adaptable robotic systems could automatically adjust in the face of disruptions, routing tasks to other available machines or assigning them to human operators when needed. This approach minimized production downtime, critical for maintaining operational efficiency in fast-paced manufacturing environments.

Including predictive maintenance in these simulations further highlighted the system's ability to anticipate and resolve issues before they impacted the production area. Manufacturers could fine-tune the system by simulating various failure conditions and recovery protocols, ensuring that human–robot collaboration remained as seamless and efficient as possible.

7.2.2 Flexibility in Resource Allocation

The simulation test also showed how digital twins could provide insight into resource allocation: if a robot underperformed or malfunctioned, the system shifted some tasks to humans working in the system and thus produced without hindrance. This flexibility was essential to ensure that things remained efficient, even when events did not occur as scheduled. It ensured adaptability, whereby manufacturers could still meet their production goals even when some machines encountered a setback.

Furthermore, the digital twins' capabilities regarding the simulations of various recovery protocols certainly proved to be an asset. Such protocols allow manufacturers to understand the scenarios in which human workers could step in and perform tasks that robots may struggle with, such as creative problem-solving or handling tasks requiring a high degree of manual dexterity. This allowed for fine-tuning of these systems so that human–robot collaboration can be as seamless as possible.

7.2.3 Real-Time Data Integration

Another essential feature of this simulation involves its use of real-time data. Real-time data from sensors inserted within machines were required to achieve a very close-to-reality accurate and responsive simulation model. This data-driven approach allowed the system to predict and manage breakdowns or inefficiencies before they occurred. Another significant simulation result was how advantageous predictive maintenance was, whereby potential failures could be mitigated before overall production flow was affected.

Furthermore, balancing tasks according to real operation data during runtime was possible. Thus, the system would remain effective even when unexpected changes occurred, whether through machine failure or a sudden change in production needs. The fact that the system was being continuously monitored and adjusted translated into smoother operations with less chance of prolonged downtime.

7.2.4 Stress Testing

The workshop went beyond normal operating conditions to reveal the added value of stress testing robotic systems in a digital twin: failing critical machinery, fluctuating production demands, and various robot tasks with different stress levels and workloads. Under these conditions, the digital twin allowed manufacturers to assess the boundaries of human and robotic contributions to the production process.

These tests enabled the firm to have better planning and resource deployment strategies, so production lines suffered minimal interruptions despite any unforeseen problems. The simulations were useful in designing systems that could operate effectively under varying conditions, and this factor has made resilience and flexibility critical in contemporary manufacturing environments.

These findings have led simulation studies and digital twins to a changing relationship between humans as operators and robots, which may be used to change manufacturing industry processes.

7.3 Validation Results

The validation tests conducted in the automotive and electronics industries confirmed the effectiveness of Industry 5.0 technologies. In trials, it was found that implementing cutting-edge robotics decreased downtime by 30% due to monitoring in real time and predictive upkeep measures in place. This outcome underscores the significance of testing systems to guarantee their dependability and safety, within practical manufacturing settings.

8 Challenges

One of the major issues that makes the full implementation of advanced robotics in Industry 5.0 crucial is the acceptance of technology and the trust needed from workers and industries alike. Workers require extensive training to adapt to collaboration with advanced machines, most of which require both technical and soft skills. A shortage of skilled workers who can program and manage industrial robots hinders the wide diffusion.

Another challenge is the high cost of adopting Industry 5.0 technologies. Investments in collaborative robots, AI-driven machinery, and customized software for connected factories are high-value investments, while companies often need help upgrading production lines. In addition to the cost of technology, companies face additional expenses associated with training human workers who will be working alongside advanced robotics systems.

Another challenge is the time-consuming nature of integrating these technologies. Businesses can adopt real-time information systems, IoT, and AI-powered machines only after much time is spent incorporating them. Besides, the most significant concerns are privacy and security. Industry 5.0 systems are based on interconnected devices and data flows that must be protected from cyber threats.

Given the possible holes that could arise within an IoT network with the increased use of quantum computing in future technologies, ensuring trust in these ecosystems can be particularly tricky. Many companies need more time to deploy AI-driven automation, fearing such systems could heighten risks rather than efficiencies. Besides, some rather strict industrial laws and regulations must be adhered to, ensuring that the integration of cobots and intelligent machines will always succeed before safety and legal standards.

Developing automation with a balance of human centricity remains challenging amidst customers' growing expectations for personalized products. With skilled employees and a robust cybersecurity framework, businesses can retain the benefits of adopting advanced robotics in Industry 5.0.

8.1 Ethical and Social Implications

8.1.1 Impact on Employment

As Industry 5.0 reshapes manufacturing roles, there is a growing need for reskilling. Cobots reduce the burden of repetitive tasks, allowing human workers to focus on higher-level problem-solving. However, this transition could displace low-skilled workers unless proper reskilling programs are implemented. Ethical considerations in Industry 5.0 must also focus on creating inclusive training opportunities that allow workers to thrive alongside advanced technologies. Without adequate support, there is a risk that only a highly skilled workforce will benefit, exacerbating inequality in the labor market.

While the introduction of cobots and AI into manufacturing can displace low-skilled workers, companies can mitigate this by offering comprehensive reskilling programs. These programs would focus on teaching new, in-demand skills like robotics programming, AI integration, and data analytics. To incentivize businesses, governments could offer tax credits to companies that invest in upskilling their workforce. Additionally, public–private partnerships can make reskilling programs more accessible, allowing workers to smoothly transition to new roles.

In the long term, adopting a lifelong learning model is essential. Continuous education would help workers stay relevant in a rapidly evolving technological landscape. This model could be integrated into employment contracts, ensuring that workers continuously upgrade their skills and adapt to new technologies.

8.1.2 Cybersecurity and Data Privacy

With the growing reliance on interconnected systems in Industry 5.0, safeguarding data privacy and addressing cybersecurity concerns are critical. Quantum computing, while promising enhanced data-processing capabilities, introduces new risks, particularly regarding encryption vulnerabilities. As traditional encryption methods become inadequate, adopting quantum-resistant encryption technologies is essential to protect sensitive information from cyber threats [10]. Additionally, blockchain technology can play a pivotal role in securing data exchanges, ensuring transparency and reducing the risk of data breaches in interconnected systems [9].

Policymakers must develop robust regulatory frameworks that include penalties for data breaches and incentives for companies implementing advanced cybersecurity measures. Regular audits, along with third-party certifications, can further enhance trust and transparency in Industry 5.0 environments, helping manufacturers safeguard intellectual property and operational integrity.

8.1.3 Promoting Equality in Access to Technology

Industry 5.0 could widen the gap between large corporations and smaller enterprises due to the high costs of adopting advanced technologies like AI, robotics, and digital twins. To ensure smaller businesses are not left behind, industry-wide solutions such as technology-sharing platforms and subscription-based AI/robotics services could make these innovations more accessible. For example, small and medium-sized enterprises (SMEs) could use cloud-based AI platforms or subscription models to lower upfront investment costs.

To further bridge the gap, governments can establish grants or low-interest loans to help SMEs implement Industry 5.0 technologies. Industry consortiums could also create shared resources or technology hubs where small businesses can access advanced tools and training, ensuring inclusivity in the industrial ecosystem.

8.1.4 Ensuring Ethical AI and Human Oversight

As AI systems take on more critical decision-making roles in Industry 5.0, it is important to ensure transparency and ethical oversight. AI algorithms used in predictive maintenance, quality control, and production processes must be explainable and auditable to avoid biases or unintended consequences.

A human-in-the-loop approach should be adopted, where critical decisions made by AI are overseen by human experts. This will ensure that ethical considerations are accounted for and that AI is used responsibly. Governments can introduce regulations requiring human oversight in AI-driven processes, particularly in industries where ethical concerns and safety are paramount [21–24].

9 Future Trends

Several key trends are expected to influence advanced robotics in Industry 5.0. Among the most promising developments is the emergence of intelligent, self-learning robots that can independently adapt to new tasks and environments. These robots will be able to use AI and machine learning for real-time decision-making, significantly increasing the flexibility of manufacturing. Adopting 5G networks will facilitate faster communication between the robots and cloud-based systems, rapid data processing, and improved decisions [25, 26].

9.1 The Impact of 5G on Workforce and Global Competition

The adoption of 5G networks in Industry 5.0 is not only transforming technological capabilities but also significantly influencing the global workforce and competition. 5G's ability to enable real-time communication between machines and cloud-based systems will reduce reliance on physical infrastructure, allowing businesses to adopt more flexible and remote operations. This shift will require the workforce to adapt to new roles focused on managing remote systems, real-time analytics, and AI-driven decision-making processes.

Additionally, the integration of 5G will accelerate the global competition in manufacturing by enabling faster and more efficient production processes. Countries that rapidly adopt 5G technology could gain a competitive advantage by improving their industrial output and agility. However, this also poses a challenge for nations or regions with limited 5G infrastructure, potentially widening the gap in manufacturing capabilities. Governments and industries must prioritize the deployment of 5G networks to maintain their competitiveness on the global stage [27, 28].

9.2 Quantum Computing's Role in Reshaping the Workforce and Global Competition

Quantum computing's potential to revolutionize data processing and problem-solving will also have profound implications for the workforce and global competition in Industry 5.0. As quantum computing enables rapid simulations, real-time optimization, and improved machine learning algorithms, businesses will increasingly rely on specialized skills in quantum algorithms, cryptography, and AI integration. This will demand a new class of professionals with expertise in quantum computing, data science, and AI, leading to a shift in workforce dynamics.

On a global scale, countries investing in quantum research and infrastructure will likely lead the next wave of industrial transformation. Nations that fall behind in quantum technology development may struggle to compete in manufacturing industries that leverage the speed and precision of quantum systems. Thus, investments in quantum education and infrastructure will be critical for maintaining economic competitiveness in the era of Industry 5.0. Furthermore, international collaboration on quantum advancements may become essential to prevent technological monopolies and ensure balanced global growth [29–31].

Cognitive computing is another trend that has revolutionized how robots interpret and process information, emulating human thought processes through data mining and pattern recognition. This technology will allow robots to learn and improve over time. Human–machine interaction will become more intuitive as natural user interfaces and gesture controls advance, making operating and collaborating with robots easier for the worker.

Using quantum computing to support robots will further revolutionize the field, enabling robots to perform highly demanding calculations at unprecedented speeds. Digitization will remain critical to future trends, such as the further evolution of digital business models and the revolutions in industrial production using sophisticated IT machinery. The technology will also become increasingly human-centric and sustainable, aiming toward human needs and environmentally minded goals.

These trends will lead to a systemic evolution of manufacturing, which will, in turn, affect society and even governance structures. The eventual outcome of robotics in Industry 5.0 is to build an innovative, flexible, and human-centered production system.

10 Conclusion

This chapter has discussed the impact of Industry 5.0 in transforming manufacturing by fostering teamwork between workers and cutting-edge robotics with the help of AI technology and IoT devices. The research and testing have shown efficiency and error minimization enhancements, giving rise to Industry 5.0 as a progression from the mainly automated setups of Industry 4.0. Collaborative robots known as cobots that incorporate AI technology showed a 25% enhancement in production adaptability; moreover, human and robot collaborations led to a 15% decrease in error rates, underscoring the promise of merging human ingenuity with robotic accuracy.

Furthermore, leveraging replicas and instant data analysis has become an essential aspect of Industry 5.0, providing real-time monitoring and predictive maintenance features that decrease downtime by as much as 30%. These advancements streamline operations and lead to the development of flexible, durable, and eco-friendly manufacturing settings. Additionally, technology integration bolsters the trustworthiness of supply chains by guaranteeing openness, trackability, and safety, particularly in industries where accuracy and quality assurance are crucial.

Nevertheless, the widespread utilization of Industry 5.0 technologies comes with hurdles. Considerable expenses for implementation ensuring compatibility between existing systems and new technologies and worries regarding cybersecurity stand out as critical challenges. The fusion of AI, IoT, and digital replicas demands investments in infrastructure and employee education. Additionally, the dependence on interconnected systems raises susceptibility to risks, emphasizing the crucial role of cybersecurity for manufacturers shifting toward Industry 5 operations.

Regarding implications, Industry 5.0 signifies a transition from complete automation toward a focus on humanity, where robots enhance human skills rather than take over. This new way of thinking promotes customized production techniques that directly tackle the constraints of Industry 4.0, which emphasizes automation. The study underscores the significance of humans and robots working together to nurture innovation, creativity, and flexibility in manufacturing settings.

In the coming years ahead, it would be beneficial for research efforts to focus on enhancing the cost-effectiveness of Industry 5.0 technologies, emphasizing small and medium-sized businesses (SMEs). Moreover, it will be essential to advance the establishment of standards that promote interoperability and frameworks for cybersecurity to facilitate the seamless integration of Industry 5.0 systems. The practical implications of this study involve streamlining processes in manufacturing settings to augment AI contributions, to decision making activities and creating more advanced predictive maintenance systems that use digital replicas.

In the end, the upcoming era of manufacturing relies on a partnership between people and machines. Industry 5.0 is moving toward more intelligent, safer, and environment-friendly production environments. By making progress in AI digital twin technology and blockchain applications, Industry 5.0 aims to improve effectiveness and foster a more human-focused and adaptable future for industries.

References

1. Coronado, E., Ueshiba, T., & Ramirez-Alpizar, I. G. (2024). A path to Industry 5.0 digital twins for human-robot collaboration by bridging NEP+ and ROS. *Robotics, 13*(2), 28.
2. Awouda, A., Traini, E., Bruno, G., & Chiabert, P. (2024). IoT-based framework for digital twins in the Industry 5.0 era. *Sensors, 24*(2), 594.
3. Hassan, S., Fazea, Y., Habbal, A., & Ibrahim, H. (2017). Twisted Laguerre-Gaussian mode division multiplexing to support blockchain applications. In *TENCON 2017—IEEE Region 10 Conference* (pp. 2050–2421).
4. Khan, A., Shahid, F., Maple, C., Ahmad, A., & Jeon, G. (2022). Toward smart manufacturing using spiral digital twin framework and twinchain. *IEEE Transactions on Industrial Informatics, 18*(3), 1359–1366.
5. Universal Robots. (2023). Enhancing production efficiency through AI-powered cobots. *Journal of Industrial Robotics, 15*(2), 45–60.
6. IBM Security. (2024). Quantum computing and the future of cybersecurity. *Cybersecurity Journal, 12*(3), 77–90.
7. Panagou, S., Neumann, W. P., & Fruggiero, F. (2024). A scoping review of human-robot interaction research towards Industry 5.0 human-centric workplaces. *International Journal of Production Research, 62*(3), 974–990.
8. Smith, J., & Jones, A. (2024). The role of digital twins in pharmaceutical manufacturing. *International Journal of Advanced Manufacturing, 28*(1), 30–44.
9. Martinez, R., & Fernandez, P. (2024). The role of blockchain in enhancing cybersecurity for Industry 5.0. *Cybersecurity Journal, 12*(1), 62–85.
10. Liu, J., Zhang, X., & Wang, Y. (2024). Quantum-resistant encryption for the Industry 5.0 era: Challenges and opportunities. *Journal of Cryptology, 41*(2), 155–178.
11. Cheong, M. L., & Teh, P. L. (2022). Enhancing human-robot collaboration through AI integration in Industry 5.0. *Journal of Cleaner Production, 363*, 132445.
12. Pilkington, M. (2022). Smart manufacturing systems: Integrating AI, big data, and robotics in Industry 5.0. *Journal of Manufacturing Science and Engineering, 142*(9), 091008.
13. Khan, M., Khalid, A., & Lughmani, W. A. (2023). Human-centric digital twins in industry: A comprehensive review of enabling technologies and implementation strategies. *Sensors, 23*(8), 3938.
14. Industry 5.0 and Industrial Robots. (2018). Robotics Online Marketing Team.
15. Industry 5.0: Shaping the Future of Manufacturing Industry. (2021). Radixweb.
16. Subramaniam, G. (2021). *Embracing Industry 5.0 and robotics in Indian manufacturing*. OEM Update.
17. Li, M., Li, Z., Huang, X., & Qu, T. (2021). Blockchain-based digital twin sharing platform for reconfigurable socialized manufacturing resource integration. *International Journal of Production Economics, 240*, 108223.
18. Bhatti, A., Malik, H., Kamal, A. Z., Alaali, L. A., & Ullah, Z. (2021). Business digital transformation through big data, IoT, and blockchain capabilities: Implications for strategic performance. *Business Process Management Journal, 27*(5), 1854–1873.
19. Huang, Z., Huang, Q., Liu, X., & Tan, C. (2021). Blockchain and its applications in Internet of Things (IoT): A review. *Computer Networks, 180*, 107356.
20. Mittal, U., & Panchal, D. (2023). AI-based evaluation system for supply chain vulnerabilities and resilience amidst external shocks: An empirical approach. *Reports in Mechanical Engineering, 4*(1), 276–289.
21. Doyle-Kent, M., & Kopacek, P. (2021). Adoption of collaborative robotics in Industry 5.0. *IFAC-PapersOnLine, 54*(13), 413–418.
22. Ozdemir, V., & Hekim, N. (2018). Birth of Industry 5.0: Making sense of big data with artificial intelligence and next-generation technology policy. *OMICS: A Journal of Integrative Biology, 22*(1), 65–76.

23. Pilkington, M. (2016). Blockchain technology: Principles and applications. In *Research handbook on digital transformations*. Edward Elgar Publishing.
24. Skobelev, P. O., & Borovik, S. Y. (2017). On the way from Industry 4.0 to Industry 5.0: From digital manufacturing to digital society. *International Science Journal, 2*(6), 307–311.
25. Raj, M. S., & Thomas, R. (2022). The role of digital twins and AI in future manufacturing systems. *Future Generation Computer Systems, 130*, 92–104.
26. Zhao, J., Liu, Q., & Zhang, Y. (2021). Ethical and social implications of AI-driven Industry 5.0 technologies. *Journal of Business Research, 130*, 142–155.
27. Gandhi, T., & Choudhary, M. (2021). The role of AI and blockchain in future industry ecosystems. *Computers & Industrial Engineering, 161*, 107598.
28. Zheng, P., & Yu, S. (2020). Smart manufacturing systems: Integrating AI, big data, and robotics. *Journal of Manufacturing Science and Engineering, 142*(9), 091008.
29. Mahmud, K., Mubin, O., & Lee, M. (2022). Human-robot collaboration in smart manufacturing: A comprehensive review of co-bots and their applications. *Journal of Manufacturing Processes, 56*, 373–390.
30. He, Z., & Xu, J. (2022). Blockchain-enabled collaborative manufacturing in Industry 5.0. *Journal of Manufacturing Systems, 63*, 346–358.
31. Kock, M., Köhler, A., & Oberländer, A. (2021). The transformative role of digital twins in Industry 5.0. *Computers in Industry, 132*, 103491.

The Power of Smart Sensors: A Technological Revolution

Shivanjali Saxena, Rakesh Joshi, Naveen Kumar, Mohit Tyagi, Yatika Gori, and Neha Singh Raghuvanshi

1 Introduction

Even in today's globalized world, manufacturing remains a cornerstone of economic development. It continues to be a vital driver of economic growth. Emerging technologies empower the manufacturing industry to boost individual competitiveness and foster overall sustainability [1].

Industry 5.0 departs from the fourth industrial revolution's focus on automation and data exchange in manufacturing. It embraces a more inclusive approach that prioritizes human intelligence and sustainable practices [2].

Industry 5.0 recently altered the organizational and production processes, and the era is characterized by complete digitalization and transformation (Fig. 1). The

S. Saxena · R. Joshi (✉)
DARVUN India, Patiala, Punjab, India
e-mail: saxena.1@iitj.ac.in

N. Kumar
Department of Mechanical and Aerospace Engineering, UPES, Dehradun, Uttarakhand, India
e-mail: naveen.kumar1@ddn.upes.ac.in

M. Tyagi
Department of Production and Industrial Engineering, Punjab Engineering College (Deemed to be University), Chandigarh, India

Y. Gori
Department of Mechanical Engineering, Graphic Era Deemed to be University, Dehradun, Uttarakhand, India

School of Computing, Graphic Era Hill University, Haldwani, Uttarakhand, India
e-mail: yatikagori@gehu.ac.in

N. S. Raghuvanshi
Department of Applied Sciences, ABES Engineering College, Ghaziabad, Uttar Pradesh, India

Y. Kumar Singla et al. (eds.), *Digital Manufacturing in Industry 5.0*,
https://doi.org/10.1007/978-3-031-91500-0_42

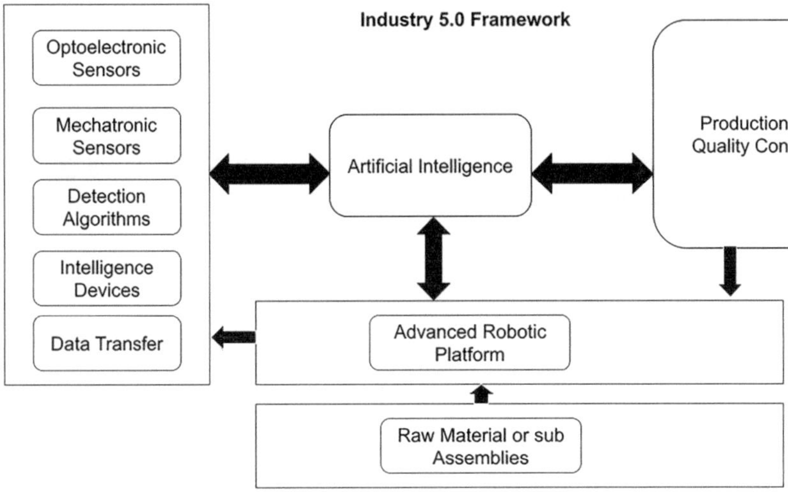

Fig. 1 Structure of Industry 5.0 framework

upcoming changes are primarily focused on implementing artificial intelligence (AI) algorithms that will customize production and processes in a self-adaptive manner. Specifically, recent advancements in optoelectronic and mechatronic technologies indicate a potential evolution in industrial process management, allowing for resource optimization based on dynamic market conditions. It is important to develop a strategic market plan in the short term to better align with market trends and trajectories.

To understand the effectiveness of condition monitoring sensors in Industry 5.0, sensors (optoelectronics, mechatronics, and smart sensor) sense data from the equipment, from robotic platforms and assembly lines, and send data to the cloud through the Internet of Things (IoT) and connectivity. IoT plays a crucial role in condition monitoring by enabling seamless connectivity between sensors, equipment, and central monitoring systems. IoT-enabled sensors can transmit data wirelessly to the cloud, where it is analyzed and visualized in real time. This connectivity facilitates remote monitoring and management of equipment, even across geographically dispersed locations.

Artificial intelligence (AI) and machine learning algorithms enhance the capabilities of condition monitoring by analyzing and uncovering hidden patterns in massive datasets that may not be apparent to human operators. These technologies can predict equipment failures with high accuracy, optimize maintenance schedules, and provide actionable insights for improving operational efficiency. The sheer volume of data generated by condition-monitoring sensors can be overwhelming. Big data analytics tools apply advanced algorithms to process and analyze large datasets, revealing hidden patterns and trends that inform strategic decisions and maintenance optimizations.

Edge computing decentralizes data processing, bringing computational power closer to the data source. In condition monitoring, this approach facilitates real-time analysis of sensor data streams, enabling rapid response to potential issues and minimizing the need for cloud-based processing. This is particularly beneficial in environments with limited connectivity or where real-time responsiveness is critical.

2 Technological Advancement Smart Sensors in the Industry

The integration of information and communication technology (ICT) into manufacturing processes is driving a shift from traditional to advanced practices. Smart sensors are integral to the evolution of Industry 5.0, offering a plethora of benefits that are reshaping the manufacturing landscape. Firstly, smart sensors provide real-time data and insights through continuous monitoring of operations, equipment health, and product quality. This enables predictive maintenance, allowing companies to reduce downtime and maintenance costs and prevent equipment failures. Moreover, smart sensors play a critical role in quality control by monitoring production processes to ensure that products consistently meet high standards, thereby minimizing defects and waste.

These sensors contribute to enhanced efficiency and productivity by meticulously identifying obstructions and inefficiencies in production lines. This insight leads to the optimization of workflows and increased output. Additionally, smart sensors enable the implementation of autonomous operations, empowering machines and robots to work collaboratively with humans, leading to improved productivity and flexibility. Furthermore, their ability to track inventory levels and trigger automated replenishment orders facilitates just-in-time inventory management, ultimately reducing storage costs and preventing stockouts.

Smart sensors focus on human-centered collaboration, prioritizing safety along with efficiency in the workplace. They contribute to safer work environments by detecting potential hazards and alerting workers, thus reducing accidents and improving overall safety. Moreover, sensor data is utilized to design more optimum workstations, reducing fatigue and the risk of injuries. Additionally, by integrating with augmented reality (AR), smart sensors provide workers with real-time guidance and information, enhancing their skills and productivity.

Furthermore, smart sensors are instrumental in promoting sustainability and resource optimization. They monitor energy consumption and identify areas for improvement, leading to reduced costs and environmental impact. Moreover, by optimizing production processes and detecting defects early, smart sensors assist in minimizing waste and conserving resources. They also play a role in monitoring the use of sustainable materials and ensuring compliance with environmental regulations.

In driving digital transformation and innovation, smart sensors generate extensive amounts of data that can be analyzed using advanced technologies such as AI and machine learning. This data-driven approach enables companies to make informed decisions, driving innovation and improving overall operational efficiency. Additionally, smart sensor data can enable new business models, such as predictive maintenance services or personalized product offerings. Furthermore, their integration with the (IoT) and Industry 4.0 allows for seamless connectivity and automation, laying the foundation for a digitally transformed manufacturing landscape.

Considering the long-term benefits of smart sensors, such as increased efficiency, and improved quality reduced downtime, their initial investment proves to be highly cost-effective. Moreover, ensuring their interoperability with other systems and technologies is crucial to fully leveraging their potential in the manufacturing environment. Equally vital is implementing a robust security framework, ensuring the safe and effective utilization of smart sensors.

Industrial automation in manufacturing has its roots in the utilization of various sensors. As sensor technology has advanced, industrial machines have become increasingly capable of collecting and analyzing data, leading to significant improvements in production processes [3, 4].

2.1 Tri-Axial Accelerometer

Manufacturing equipment often relies on numerous rotating components. Tri-axial accelerometers are essential tools for monitoring and measuring vibration on these parts. These smart sensors can also track other variables, such as displacement, acceleration, and velocity. When combined with IoT advanced analytics, this data provides valuable insights to help managers and technicians assess a machine's health and identify potential issues [5].

2.2 Temperature

Heat is a telltale sign of a machine's health. Moving equipment generates a lot, and many industries use additional process heat for their factory floors. Temperature sensors can help identify wear, damage, and overheating before they lead to breakdowns and quality issues [6].

2.3 Vibration Sensors

Vibration isn't limited to rotating components. Vibration sensors can be applied to conveyors, panels, and other machine parts, providing operators and technicians with early warnings of potential issues [7].

2.4 MEMS (Microelectromechanical Systems) Sensors

While MEMS sensors excel at precise measurement, their role in CNC machining becomes even more intricate. Here, they go beyond basic sensing, tracking tilt, acceleration, and gravity across multiple axes. This comprehensive data allows the CNC machine to maintain exceptional precision during machining operations [8].

2.5 Torque Sensors

High-speed motors generate substantial torque forces, especially when under load. By monitoring this torque with specialized sensors, manufacturers can gain valuable insights into machine performance. These sensors help maintain quality, minimize wear, and proactively address potential issues. When integrated with other sensors, they provide a comprehensive picture of equipment health [9].

2.6 Pressure Sensors

Pressure sensors play a crucial role in monitoring fluid and gas-based systems, including hydraulic systems, water lines, chemical tanks, and pneumatic equipment. By detecting pressure fluctuations, these sensors can trigger alarms when thresholds are exceeded or fallen below, alerting factory monitoring systems [10].

3 Role of IOT-Based Sensor in Factory

IoT is revolutionizing the way smart sensors are capable of sensing, deciding, diagnosing, and actuating communicatively and collaboratively, leading society toward a digitalized new era. IoT solutions are applied in multiple sectors such as environmental monitoring, precision agriculture, smart packaging, smart home, and wearables, among others.

The integration of sensing AI technology is causing major changes in a number of industries, with many advances being in the Industry 5.0 domain. Condition monitoring sensors are a crucial technological advancement driving this revolution. These sensors are leading the way in cost reductions, operational efficiency, and predictive maintenance.

IoT refers to interconnected devices like sensors and transmitters that communicate data over the internet (Fig. 2). It acts as a bridge between the physical world and the digital realm, where sensors and actuators play crucial roles. Sensors gather data from the real world, which is then transmitted and analyzed to facilitate changes.

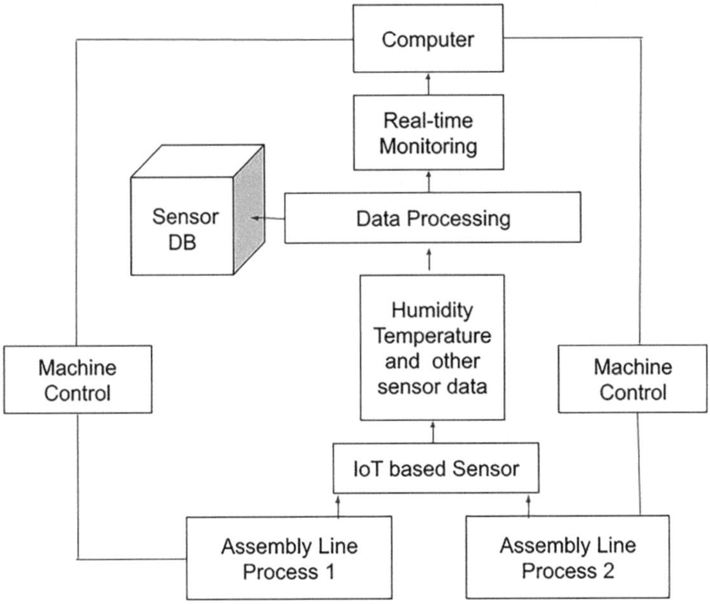

Fig. 2 Structure of transferring data from the assembly line to the computer using IoT

Actuators are devices that control systems by taking input and converting it into action. They essentially perform the opposite function of sensors. IoT has found applications in both automotive and IT industries [11].

4 Sensors in the Biomedical Industry

Sensors are the core of any biomedical industry, and the most classic example is the digital thermometers, which have widely changed the paradigm of human health. Sensors in the biomedical industry not only detect the raw data but also change their output to digital signals. Recent advancements in the field have updated these biosensors to the industry 5.0 era and have now incorporated additional features such as calibration, amplification, AI-based monitoring, data sharing and results generation. Unlike other sensors, biosensors are interfaces between a living entity and a digital information processing system. In this section, we will discuss widely used biosensors of the industry.

4.1 Biosensors

In the dynamic world of the biomedical industry, a plethora of cutting-edge sensors are revolutionizing healthcare. From sensors for blood pressure measurement and electrocardiogram (ECG) to electromyography (EMG) and photoplethysmography (PPG), the possibilities are endless. These sensors are widely used to detect and process various forms of biological signals and help doctors identify multiple medical conditions. Advanced sensors have been developed to detect molecules and enzymes and analyze biological fluids, paving the way for intricate disease diagnosis. These sensors are based on electrochemical materials and are used to detect body fluids such as tears, sweat, saliva and urine. It is essential to embrace the future of healthcare with these innovative technologies [12].

4.2 Implantable Sensors

Wearable and implantable devices and electronics have quickly become a part of digital health, and they are used in various biomedical applications such as tracking, monitoring, tracking, and recording vital signs. These technologies are aimed at improving the health of individuals and their families. Some of these devices are now common accessories, like smartwatches, armbands, and glasses [13–15]. The first implantable health device was developed in the 1960s, which was a fully implanted cardiac pacemaker for patients with arrhythmia [16, 17]. Since then, implantable cardioverter defibrillators (ICDs), improved versions of pacemakers, and implantable deep brain stimulators have been created and used to treat millions of patients [18, 19]. These implantable devices are powered by batteries and consist of biocompatible electrodes, programmable circuitry, and wires working together as strong sensors and providing a complete package for assessing and treating medical conditions.

Advancements in packaging and fabrication technologies have made it possible to embed various microelectronic and micromechanical sensors, like gyroscopes into a small area on a rigid or flexible substrate with low cost and high sensitivity [20]. These advances have expanded the capabilities of wearable and implantable sensors to include motion sensing, as well as physiological and biochemical sensing. The expansion of broadband wireless networks into rural areas, even in low- and middle-income countries, has created new opportunities for wearable and implantable devices and sensors to continuously monitor patients' health remotely, even in resource-limited settings [21, 22]. This is essential for addressing healthcare disparities and reducing the economic burden of healthcare costs, especially in countries grappling with aging populations [23].

4.3 Wearable Sensors

Advances in sensor technologies have led to the development of various wearable sensors that can be used for biomedical applications in both contact and noncontact modes. These sensors enable data transfer to smartphones or other healthcare units via Wi-Fi, Bluetooth, or other wireless technologies [24]. There are several commercially available wearable sensors designed to continuously monitor patients' vital signs, which is essential for modern digital health. Apart from vital signs, these devices also track other parameters such as calories burned, sleep patterns, step counts, activity levels, and eating habits to provide a comprehensive overview of the patient's health status. Key physiological parameters that can be monitored using the most wearable sensors include heart rate, blood oxygen saturation levels, body and/or respiration rate, skin temperature, and blood pressure.

The most common parameter detected using these wearable sensors is heart rate, which can easily be detected from any body part. A wristband introduced in the 1980s was the first wireless wearable sensor to detect heart rate and was used specifically for athletes to monitor their heart performance during their performance [25]. Based on the popularity of these wristbands for detecting heart rate, these heart rate monitoring wireless sensors have now become one of the most common wearable sensors. There are currently many wearable models available in the market that can measure the heart rate at rest, during walking and running, from various parts of the body including the ear, chest, wrist, and finger. Additionally, some devices are equipped with global positioning systems (GPS) for improved performance analysis [26].

4.4 Microfluidic Sensors

The sensors utilized in a microfluidic platform, often referred to as "Microfluidic Sensors," represent a cutting-edge field that involves manipulating fluids in channels with dimensions of tens of micrometers. The development of sensors for measuring at a microscale is an exciting frontier in current research. As the scale decreases, the challenge of designing and developing sensors becomes more pronounced, particularly in terms of sensitivity, selectivity, and stability. The majority of microfluidic sensors are electrochemical sensors, integrating various disciplines for sensing. The integration of efficient sensors on microfluidic chips stands out as one of the most crucial tasks for the advancement of micrototal analysis systems (μ-TAS) [27].

The design and properties of sensors depend on their specific application and the physical requirements. Microchannel systems are commonly used in chemical and biotechnology processes at a microvolume level. Microfluidics also play a significant role in cooling microelectronic chips. Microelectromechanical systems (MEMS)-based microfluidic devices have recently been utilized in drug delivery, lab-on-chip, and various environmental monitoring applications. These MEMS microfluidic devices can be created using photolithography, electron beam lithography, etching, and deposition. Microfluidic sensors are the major determinants of a microfluidic system as they can measure and control the pressure, flow, temperature, and pH of fluidics in a microfluidic setup. Integrated sensors developed in recent times include a resistive temperature sensor, a liquid crystalline polymer (LCP)-based conductivity cell, and a piezoelectric pressure sensor for depth measurement [28]. Over the last decade, various types of microfluidic sensors have been developed, such as gas sensors [29], glucose sensors [30], phosphate sensors [31], cocaine sensors [32], pH sensors [33], and DNA sensors [34]. Polystyrene tube and epoxy encapsulation techniques have been implemented to enable sensor operation in extreme conditions [35].

5 Future of Sensor in Industry

The future of sensors in the manufacturing industry will focus on accurate measurement of data and must be intelligent. These futuristic sensors will not only collect data but also able to analyze and interpret it. Additionally, these futuristic sensors will be equipped with onboard processing capabilities, and capabilities to communicate with other devices, allowing better decision-making and more efficient smart factory operations.

Smart sensors represent the next generation of industrial sensors. These devices not only collect data but also analyze and interpret it, allowing for better decision-making and more efficient operations. Smart sensors are equipped with onboard processing capabilities and can communicate with other devices, making them a key component of smart factories.

The integration of AI and machine learning into smart sensors is allowing them to adapt to changing conditions and improve their performance over time. This capability will be crucial in the future, as industries continue to adopt more complex and automated systems.

The future of the industrial sensors industry is filled with opportunities, driven by the growing adoption of IoT, AI, and 5G technologies. However, challenges such as cost, security, and interoperability must be addressed to fully realize the potential of these innovations. By focusing on overcoming these obstacles, the industrial sensors market is poised to experience significant growth and transformation in the coming years.

The future of the industrial sensors industry is filled with opportunities, driven by the growing adoption of IoT, AI, and 5G technologies. However, challenges such as cost, security, and interoperability must be addressed to fully realize the potential of these innovations. By focusing on overcoming these obstacles, the industrial sensors market is poised to experience significant growth and transformation in the coming years [3, 36–38].

6 Conclusion

This book chapter explores the multifaceted role of smart sensors in Industry 5.0, drawing insights from both industrial and academic perspectives. It delves into various types of smart sensors, the pivotal role of IoT, and their applications in the biomedical field. By understanding these advancements, we can better grasp the significance of sensor technology in shaping the future of industry. Industry 5.0 aims to optimize the collaboration between humans and machines. The chapter highlights the technological advancements in sensor technology that address challenges within this framework. We explore the potential of smart sensors in industrial settings, examining their role in revolutionizing factory production. As technology continues to evolve, smart sensors will play an increasingly critical role in driving intelligent manufacturing. Moreover, we discuss the essential components of creating intelligent sensors for automation and safety. The chapter also highlights the transformative potential of smart sensors in healthcare, particularly in enabling data-driven, personalized, and preventative care. By gathering, processing, and exchanging data, smart sensors are poised to revolutionize manufacturing facilities and pave the way for an intelligent manufacturing future.

References

1. Syafrudin, M., Alfian, G., Fitriyani, N. L., & Rhee, J. (2018). Performance analysis of IoT-based sensor, big data processing, and machine learning model for real-time monitoring system in automotive manufacturing. *Sensors (Basel, Switzerland), 18*(9), 2946. https://doi.org/10.3390/s18092946
2. Rame, R., Purwanto, P., & Sudarno, S. (2024). Industry 5.0 and sustainability: An overview of emerging trends and challenges for a green future. *Innovation and Green Development, 3*(4), 100173. https://doi.org/10.1016/j.igd.2024.100173
3. Heidari, M., Khashehchi, M., Thangavel, S., Rahmanivahid, P., Kumar, A., & Singla, Y. K. (2024). Chapter 18: Laminated structures and fracture mechanics: A comprehensive study of mode I, mode II, and mixed mode III behavior. In A. Kumar, Y. Kumar Singla, & M. R. Maughan (Eds.), *Fracture behavior of nanocomposites and reinforced laminate structures* (pp. 441–454). Springer. https://doi.org/10.1007/978-3-031-68694-8_18
4. Shah, S., Nautiyal, H., Gugliani, G., Kumar, A., Namboodri, T., & Singla, Y. K. (Eds.). (2024). *Sustainability in smart manufacturing: Trends, scope, and challenges*. CRC Press. ISBN: 9781032740713. https://doi.org/10.1201/9781003467496
5. Javed, A. R., Sarwar, M. U., Khan, S., Iwendi, C., Mittal, M., & Kumar, N. (2020). Analyzing the effectiveness and contribution of each axis of tri-axial accelerometer sensor for accurate activity recognition. *Sensors (Basel, Switzerland), 20*(8), 2216. https://doi.org/10.3390/s20082216
6. Barros, D., Fraga-Lamas, P., Fernández-Caramés, T. M., & Lopes, S. I. (2023). A cost-effective thermal imaging safety sensor for industry 5.0 and collaborative robotics. In *Lecture notes of the Institute for Computer Sciences, Social Informatics and Telecommunications Engineering* (pp. 3–15). Springer Nature. https://doi.org/10.1007/978-3-031-35982-8_1

7. Hassan, I. U., Panduru, K., & Walsh, J. (2024). An in-depth study of vibration sensors for condition monitoring. *Sensors (Basel, Switzerland), 24*(3), 740. https://doi.org/10.3390/s24030740
8. Hajare, R., Reddy, V., & Srikanth, R. (2022). MEMS based sensors—A comprehensive review of commonly used fabrication techniques. *Materials Today: Proceedings, 49*, 720–730. https://doi.org/10.1016/j.matpr.2021.05.223
9. Palli, G., Moriello, L., Scarcia, U., & Melchiorri, C. (2014). Development of an optoelectronic 6-axis force/torque sensor for robotic applications. *Sensors and Actuators A: Physical, 220*, 333–346. https://doi.org/10.1016/j.sna.2014.09.023
10. Han, X., Huang, M., Wu, Z., Gao, Y., Xia, Y., Yang, P., Fan, S., Lu, X., Yang, X., Liang, L., Su, W., Wang, L., Cui, Z., Zhao, Y., Li, Z., Zhao, L., & Jiang, Z. (2023). Advances in high-performance MEMS pressure sensors: Design, fabrication, and packaging. *Microsystems & Nanoengineering, 9*(1), 156. https://doi.org/10.1038/s41378-023-00620-1
11. Fatima, S., Haleem, A., Bahl, S., Javaid, M., Mahla, S. K., & Singh, S. (2021). Exploring the significant applications of Internet of Things (IoT) with 3D printing using advanced materials in medical field. *Materials Today: Proceedings, 45*, 4844–4851. https://doi.org/10.1016/j.matpr.2021.01.305
12. Panja, A. K., Mukherjee, A., & Dey, N. (2022). Introduction. In *Biomedical sensors and smart sensing* (pp. 1–14). Elsevier. https://doi.org/10.1016/b978-0-12-822856-2.00006-x
13. Elenko, E., Underwood, L., & Zohar, D. (2016). Erratum: Defining digital medicine. *Nature Biotechnology, 34*(11), 1206–1206. https://doi.org/10.1038/nbt1116-1206
14. Jeong, J. W., Jang, Y. W., Lee, I., Shin, S., & Kim, S. (2009). Wearable respiratory rate monitoring using piezo-resistive fabric sensor. In *IFMBE Proceedings* (pp. 282–284). Springer. https://doi.org/10.1007/978-3-642-03904-1_78
15. Melzer, M., Mönch, J. I., Makarov, D., Zabila, Y., Cañón Bermúdez, G. S., Karnaushenko, D., Baunack, S., Bahr, F., Yan, C., Kaltenbrunner, M., & Schmidt, O. G. (2015). Wearable magnetic field sensors for flexible electronics. *Advanced Materials, 27*(7), 1274–1280. https://doi.org/10.1002/adma.201405027
16. Cima, M. J. (2014). Review of Next-generation wearable electronics. *Nature Biotechnology, 32*(7), 642–643. https://doi.org/10.1038/nbt.2952
17. Nayak, R., Wang, L., & Padhye, R. (2015). Electronic textiles for military personnel. In *Electronic textiles* (pp. 239–256). Elsevier. https://doi.org/10.1016/b978-0-08-100201-8.00012-6
18. Greenspon, A. J., Patel, J. D., Lau, E., Ochoa, J. A., Frisch, D. R., Ho, R. T., Pavri, B. B., & Kurtz, S. M. (2012). Trends in permanent pacemaker implantation in the United States from 1993 to 2009: Increasing complexity of patients and procedures. *Journal of the American College of Cardiology, 60*(16), 1540–1545. https://doi.org/10.1016/j.jacc.2012.07.017
19. Lyons, M. K. (2011). Deep brain stimulation: Current and future clinical applications. *Mayo Clinic Proceedings, 86*(7), 662–672. https://doi.org/10.4065/mcp.2011.0045
20. Patel, S., Park, H., Bonato, P., Chan, L., & Rodgers, M. (2012). A review of wearable sensors and systems with application in rehabilitation. *Journal of Neuroengineering and Rehabilitation, 9*(1), 1–17.
21. Ozcan, A. (2014). Mobile phones democratize and cultivate next-generation imaging, diagnostics and measurement tools. *Lab on a Chip, 14*(17), 3187–3194. https://doi.org/10.1039/c4lc00010b
22. Vashist, S. K., Mudanyali, O., Schneider, E. M., Zengerle, R., & Ozcan, A. (2014). Cellphone-based devices for bioanalytical sciences. *Analytical and Bioanalytical Chemistry, 406*(14), 3263–3277. https://doi.org/10.1007/s00216-013-7473-1
23. Koydemir, H. C., & Ozcan, A. (2018). Wearable and implantable sensors for biomedical applications. *Annual Review of Analytical Chemistry, 11*(1), 127–146. https://doi.org/10.1146/annurev-anchem-061417-125956

24. Chan, M., Estève, D., Fourniols, J.-Y., Escriba, C., & Campo, E. (2012). Smart wearable systems: Current status and future challenges. *Artificial Intelligence in Medicine, 56*(3), 137–156. https://doi.org/10.1016/j.artmed.2012.09.003
25. Stahl, S. E., An, H.-S., Dinkel, D. M., Noble, J. M., & Lee, J.-M. (2016). How accurate are the wrist-based heart rate monitors during walking and running activities? Are they accurate enough? *BMJ Open Sport & Exercise Medicine, 2*(1), e000106. https://doi.org/10.1136/bmjsem-2015-000106
26. Terbizan, D. J., Dolezal, B. A., & Albano, C. (2002). Validity of seven commercially available heart rate monitors. *Measurement in Physical Education and Exercise Science, 6*(4), 243–247. https://doi.org/10.1207/s15327841mpee0604_3
27. Antony, R., Giri Nandagopal, M. S., Sreekumar, N., & Selvaraju, N. (2014). Detection principles and development of microfluidic sensors in the last decade. *Microsystem Technologies: Sensors, Actuators, Systems Integration, 20*(6), 1051–1061. https://doi.org/10.1007/s00542-014-2165-0
28. Broadbent, H. A., Ivanov, S. Z., & Fries, D. P. (2007). A miniature, low cost CTD system for coastal salinity measurements. *Measurement Science & Technology, 18*(11), 3295–3302. https://doi.org/10.1088/0957-0233/18/11/005
29. Benecke, W., & Meckes, A. (2002). A smart gas sensing microsystem. In *Proceedings of the IEEE International Symposium on Industrial Electronics (ISIE'98), Pretoria, South Africa* (Cat. No. 98TH8357). https://doi.org/10.1109/isie.1998.707789.
30. Chang-Soo, K., & Jongwon, P. (2005). Influence oxygen microenvironment microfluidic glucose sensor performance. In *Proceedings of the 3rd Annual International IEEE EMBS Special Topic Conference Microtechnologies Medicine Biology, Kahuku.*
31. John, C., Conor, S., Mcgraw, C., & Dermot, D. (2008). Autonomous microfluidic sensor phosphate: On-site analysis treated wastewater. *Sensors, 8*, 508–515.
32. Sumetsky, M., Dulashko, Y., & Windeler, R. S. (2008). Temperature pressure compensated microfluidic optical sensor, lasers electro-optics. In *Proceedings Conference Quantum Electronics Laser Science, CLEO/QELS.*
33. Benito-Lopez, F., Coyle, S., Byrne, R., Smeaton, A., O'Connor, N. E., & Diamond, D. (2009). Pump less wearable microfluidic device for real time pH sweat monitoring. *Procedia Chemistry, 1*(1), 1103–1106. https://doi.org/10.1016/j.proche.2009.07.275
34. Jun, W., Daisuke, O., Noritada, K., Manabu, T., Shozo, F., Naoki, Y., & Yoshinobu, B. (2014). Microfluidic biosensor detection DNA fluorescence enhancement following streptavidin detection fluorescence quenching. *Biosensors & Bioelectronics, 51*, 280–285.
35. Hyldgård, A., Birkelund, K., Janting, J., & Thomsen, E. V. (2008). Direct media exposure of MEMS multi-sensor systems using a potted-tube packaging concept. *Sensors and Actuators A: Physical, 142*(1), 398–404. https://doi.org/10.1016/j.sna.2007.02.024
36. Khashehchi, M., Heidari, M., Thangavel, S., Rahmanivahid, P., Kumar, A., & Singla, Y. K. (2024). Chapter 20: Insights into aerospace structural integrity: A study on fiber/epoxy composites fracture. In A. Kumar, Y. Kumar Singla, & M. R. Maughan (Eds.), *Fracture behavior of nanocomposites and reinforced laminate structures* (pp. 471–485). Springer. https://doi.org/10.1007/978-3-031-68694-8_20
37. Kumar, A., Gori, Y., Dutt, N., Singla, Y. K., & Maurya, A. (2021). *Advanced computational methods in mechanical and materials engineering*. CRC Press. ISBN: 9781032052915. https://doi.org/10.1201/9781003202233
38. Kumar, A., Singla, Y., & Namboodri, T. (2024). Chapter 1: Globalization and international issues in sustainable manufacturing. In S. Shah, H. Nautiyal, G. Gugliani, A. Kumar, T. Namboodri, & Y. K. Singla (Eds.), *Sustainability in smart manufacturing: Trends, scope and challenges* (pp. 1–18). CRC Press. https://doi.org/10.1201/9781003467496-1

Smart Sensors in Factories of the Future

N. Krishnaraj, Priyanga Subbiah, A. Soujanya, and R. Meera

1 Introduction

The manufacturing industry is undergoing a paradigm shift due to the increased integration of information and communication technologies into industrial systems. At the centre of this revolution are unassuming yet powerful components: smart sensors. These smart devices are ready to transform how the factory floor looks and bring about higher efficiency and innovation [1]. Smart sensors have grown from being simple data gatherers to complex devices. Software programs that enable the processing of data into useful information can positively affect organisational decision-making. From tracking equipment status to improving manufacturing processes, advanced smart sensors are the backbone of Industry 4.0.

These smart devices are thus making factories into interconnected smart systems and self-reporting ecosystems equipped with high-level computational power and networking that allows them to process data, formulate good decisions, and foster the process of automation. As the nervous system of intelligent factories, smart sensors are playing pivotal roles in shaping the future of manufacturing and driving advancements in some areas.

N. Krishnaraj · P. Subbiah (✉) · A. Soujanya · R. Meera
Department of Networking and Communications, School of Computing, College of Engineering and Technology, SRM Institute of Science and Technology, Kattankulathur, Chengalpattu, Chennai, Tamil Nadu, India

Y. Kumar Singla et al. (eds.), *Digital Manufacturing in Industry 5.0*,
https://doi.org/10.1007/978-3-031-91500-0_43

2 Understanding Smart Sensors

Smart sensors play a critical role in the realisation of Industry 4.0. These sensors are sophisticated devices that process data at the sensing level to output fully processed, multiparameter information [2]. They are designed to respond to physical inputs, such as light, heat, motion, moisture, pressure, or any other change in environment, and convert them into signals that can be read by an instrument or the user.

Smart sensors are integral to Industry 4.0 because they are the primary data collection points in an Internet of Things (IoT) network. They provide the raw data that are used by control systems to make decisions. They can also be used for observing the condition of a machine, the environment that it operates in, and such factors as the presence of people and other objects, among others. Such sensors are often incorporated with other features, such as the ability to communicate wirelessly. They can be used to perform an analysis of the variances between variables in an expression to observe the work of complex systems in real time, allowing the identification of potential failures and thus reducing downtime.

Smart sensors are bringing about changes in various industries by offering data and information in real time. They allow more-effective, -efficient, and -sustainable work. Equipped with advanced abilities for processing data, communication, and decision-making, these sensors are valuable in most sectors, including manufacturing, supply chains, agriculture, healthcare, and more. With their use, the health of equipment, inventory tracking, and other operational activities can be optimised [3]. Thanks to this enhanced value proposition, efficient smart sensors are fuelling creativity, cutting expenses, and improving overall performance. Moving forward with the development of modern technology, people will come across more-sophisticated cases either at work or in their everyday lives. More-novel uses of smart sensing will enable revolutionising other endeavours in the workplace too. Figure 1 illustrates an outline of the working process of smart sensors in industries.

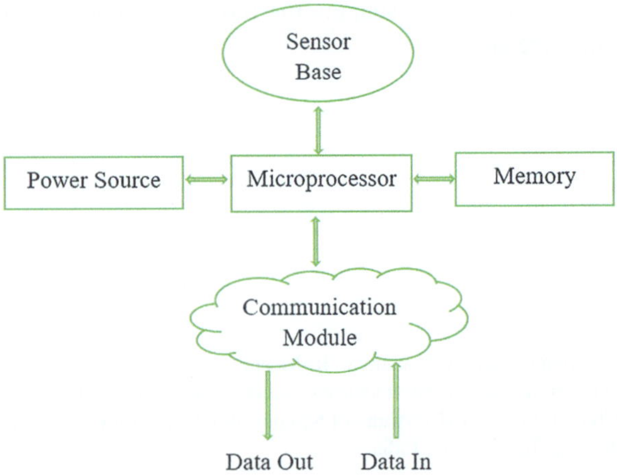

Fig. 1 Smart sensors in Industry 4.0

3 The Impact of Smart Sensors in Industry 4.0

Smart sensors play significant roles in Industry 4.0. These are the eyes and ears of the factory that capture important data and then use those data in decision-making processes. By integrating smart sensors into an industrial system, organisations are able to enhance the level of automation and efficiency. Smart sensors provide a variety of possibilities in Industry 4.0. For example, predictive maintenance becomes possible through a sensor that identifies the status of various machines in use and helps to prevent the occurrence of failure events from taking place [4]. This results in increased cost-efficiency because organisations do not have to pay for problems such as large-scale equipment failures and other forms of unanticipated losses in productivity. This enables product quality to be changed on the fly so that the desired quality can be achieved during the production process. Thus, the standards set for manufacturers' quality-control processes can now be met through waste minimisation.

3.1 Predictive Maintenance

Depending on the type of application, smart sensors can also indicate the health status and productivity of the equipment. This enables the identification of early signs of potential future failures. Predictive maintenance also enables proactive maintenance, thus minimising downtime and the frequency of expensive repairs.

3.2 Real-Time Quality Control

By establishing steady feedback from the processes of production, smart sensors can help to identify and respond to quality problems in real time so that the quality of products can be well controlled.

3.3 Optimised Operations

Smart sensors can be very effective in monitoring production. They analyse the flow of business processes and highlight inefficiencies or areas that require more-refined optimisation.

3.4 Enhanced Decision-Making

The information gathered by smart sensors can be used to make proper decisions on the basis of data, which can increase benefits for businesses and decrease the probability of negative outcomes.

3.5 Increased Flexibility

Smart sensors empower manufacturers to respond swiftly to market fluctuations. By providing real-time data on production processes, inventory levels, and quality control, these sensors enable companies to make rapid adjustments and optimise operations [5]. This increased flexibility leads to reduced costs, improved product quality, and higher customer satisfaction. For example, in the automotive industry, smart sensors can monitor assembly lines, identify defects early on, and ensure that vehicles meet stringent quality standards, allowing manufacturers to quickly address issues and thus maintain customer trust.

4 The Role of Smart Sensors

4.1 Predictive Maintenance

Predictive maintenance is a proactive strategy that leverages real-time data from sensors to anticipate equipment failures before they occur. By continuously monitoring the health of machines and systems, organisations can identify potential issues early on, thus reducing downtime, minimising costs, and improving overall operational efficiency.

The strategic placement of sensors throughout the equipment allows data to be collected on various parameters, such as temperature, vibration, pressure, and current. Advanced analytics algorithms process the collected data to detect anomalies, trends, and patterns that indicate potential equipment failures.

On the basis of historical data and machine-learning techniques, predictive models are developed to forecast the likelihood of equipment failure and estimate equipment's remaining useful life. Maintenance teams can proactively schedule maintenance tasks on the basis of the predictions generated by these models, thus optimising resource allocation and minimising disruptions.

4.2 Quality Control

Quality control is a critical aspect of manufacturing and production processes. It ensures that products meet specific standards and specifications, ultimately leading to customer satisfaction [6]. Traditional quality-control methods often rely on manual inspections, which can be time-consuming, error-prone, and labour-intensive. Smart sensors have revolutionised quality control by providing real-time data and automation capabilities. These sensors can be integrated into various stages of the production process to detect defects early on, thus reducing waste, repeat work, and costs.

Sensors can continuously monitor production processes, immediately detecting deviations from standards and identifying potential defects. Sensors can automate inspection tasks, reducing human error and increasing efficiency. Advanced analytics algorithms can process [7] sensor data to identify trends, patterns, and the root causes of defects. By analysing historical data, sensors can predict potential quality issues and take preventive measures. Such measures gather information about the physical surroundings of the quality issue and then convey it to the machines in real time.

This information is then employed by automation systems to make decisions and to control the processes accordingly. Sensors make sure that machines perform their tasks with high precision and accuracy, thus minimising any chances of errors and improving quality. With the help of sensors, machines are able to stay alert to their environment in order to make adjustments and obtain the best results.

4.3 Supply Chain Optimisation

This method better organises the stages of the distribution of goods and services from the providers of such goods and services to customers and clients. It includes several activities in the management of stocks; it also includes the planning of production and transportation to cut expenses and achieve better outcomes [8]. Smart sensors present a critical part of reaching the goals of supply chain optimisation because they offer real-time information.

Smart sensors are tools that respond to the precise requirements of supply chains, which are explained in detail in this chapter by synthesising the work of previous chapters in this book. The real-time information of inventories in the warehouse, factories, and retail shops can be provided by sensors. Thus, products are always available when demand is high and not when demand is low, thus effectively managing stock-out and shadow stock-out situations. By knowing the amount of production and the challenges faced in the production process, sensors can play significant roles in increasing the utilisation of the production line within the production processes, which can cut costs associated with wastage and enhance efficiency. If machine sensory data are analysed, insights into the levels of demand for products from customers can be gained [9]. This has favoured businesses in that they are able to make an estimation about the quantity that they will need in the future and schedule for it. It can also monitor the position and condition of consignments, which can enhance logistical issues and transportation networks. As pointed out earlier, sensors can be applied to measure the quality of products across the various stages of a supply chain to guarantee that only quality products are offered to consumers.

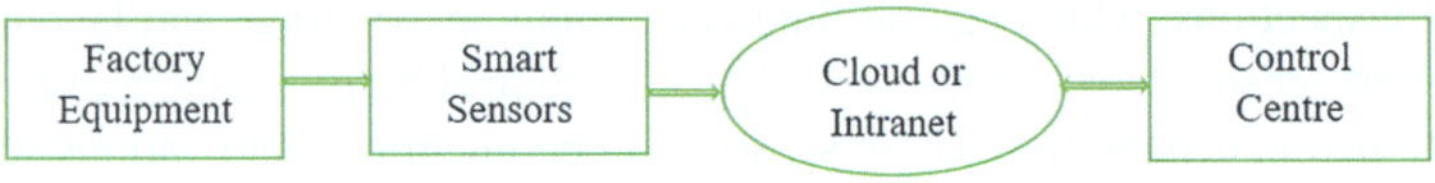

Fig. 2 The role of smart sensors in factories of the future

4.4 Energy Efficiency

Energy efficiency in particular is crucial for businesses because it reduces energy consumption. Thus, it can cut costs on operations, limit wated energy, and take part in the conservation of the environment. And smart sensors are crucial for increasing energy efficiency.

Sensors can continuously monitor energy consumption in various spaces, such as buildings, factories, and data centres. By analysing sensor data, organisations can identify areas where energy is being wasted or used inefficiently [10]. By using the insights gained from sensor data, businesses can implement strategies to optimise energy use, such as adjusting heating, ventilating, and air-conditioning (HVAC) systems, optimising lighting, and improving equipment efficiency. Sensors can be used to predict equipment failures, allowing for proactive maintenance that prevents energy-intensive breakdowns. Sensors can enable participation in demand-response programmes, where businesses can reduce their energy consumption during peak demand periods. Figure 2 showcases the various roles of smart sensors in factories of the future.

5 Benefits of Smart Sensors in Factories

The analysis of data from various sensors in an organisation helps in the detection of energy waste and energy inefficiency. Consequently, by analysing the data received from sensors, various processes and mechanisms can be kept under control and the necessary measures can be undertaken to apply appropriate strategies that would contribute to the overall enhancement of operations and consumption, like control over heating, ventilation, and air conditioning; the use of more efficient lighting; and more-efficient and -effective equipment. Using sensors enables the prediction of when equipment will fail so that equipment is repaired or maintained before failing [11]. Sensors prevent energy-intensive breakdowns by participating in demand-response programmes that enable businesses to cut consumption during peak demand times, and they increase load demand. Smart sensors are also used in manufacturing factories.

The following are some key areas where smart sensors can create value: Smart sensors can transmit precise data on equipment behaviour, the flow of material, and energy consumption to highlight constraints and suboptimal factors. Smart sensors for predictive maintenance reduce cases of unplanned downtime, thereby ensuring that production lines continue operating as planned. Smart sensors assist in the detection of concerns regarding quality through the real-time monitoring of production problems. In reality, smart sensors can be used to help understand how much energy is utilised and where changes can be made. This was achieved by emphasising efficiencies in production processes and enhancing quality assurance [12].

Monitoring devices are useful in collecting large quantities of information for creating products and services. By increasing the quality of a product and reducing lead time and by participating in subsequent customer experiences, smart sensors can help to boost revenue. Data centres turn into factories that successfully implement smart sensors, and these can gain perhaps one of the key factors that endows them with a serious competitive edge on rivals. With smart sensors, factories can be changed to consume less from the environment thanks to these sensors' ability to help to efficiently use resources. Hence, adopting smart sensor technology can effectively help organisations to increase efficiency and productivity [13]. Factories in particular can show that they use natural resources in more responsible ways to attract environmentally conscious customers.

6 Competitive Advantages of Smart Sensors in Factories

Smarty sensors have several advantages over traditional ones, and organisations that become early entrants into smart sensor technology can greatly improve their competitive advantage, which the following sections lay out.

6.1 Optimising Operations

6.1.1 Increased Efficiency

Smart sensors that keep track of the performance of equipment in real time both foster preventive measures that allow a drastic reduction in time thanks to preventive measures and increase operational efficiency [14]. This approach enables the right time to be set aside for repairing or replacing equipment.

6.1.2 Enhanced Resource Allocation

Smart sensors also capture and collect large numbers of data for instant analysis. These data enable companies to make informed decisions on the distribution of resources and the effectiveness of operating efficiency. By having real-time information on the use of factors such as labour and materials, among others, this field allows businesses to monitor energy consumption and usage in specific regions; for tasks such as recycling; for product attributes and features; and for drop-shipping and delivery mechanisms [15]. This helps businesses to make the right decisions on reallocating resources to improve efficiency, hence making them cheaper to use and generally more efficient in productivity. For example, by monitoring these data collected by the sensors, manufacturers can easily determine the possible constraints on their production line and to, for example, properly adapt staffing to make sure that the right number of people are in the right places.

6.2 Developing Innovative Products

6.2.1 New Product Features

Smart sensors, if incorporated into products, confer on businesses a competitive advantage that enables them to provide features and functions to consumers and clients in a way that distinguishes them from their competitors. In this way, manufacturers can improve the customer experience by installing sensors in the products that they manufacture. For example, smart touchless appliances can be used for tracking usage, which requires sensors that remotely relay this information [16]. This not only allows consumers to be more convenient but also brings about new, additional sources of revenue.

6.2.2 Enhanced Customer Experience

By utilising smart sensors, the customer experience can be completely transformed. By collecting data on customer behaviour and trends, businesses can make recommendations on products and services and can learn which ones require attention and attend to them as soon as possible. For example, smart sensors in shops, malls, and stores can be used to monitor customers' movements, preferences, and buying behaviour patterns, monitoring that enables retailers to factor those data into business adjustments. For example, retailers can offer product recommendations that are more relevant to these customers [17]. Developing a specific strategy that is peculiar to each customer can thus be useful in increasing the applicability of the aforementioned data in a way that will lead to customer satisfaction, high customer loyalty, and repeat business. Additionally, smart sensors can be employed to relay information in real time, including product availability and wait times, to enable customers make the right choices when selecting which products to purchase and to improve their shopping experience.

6.3 Data-Driven Decision-Making

6.3.1 Informed Decisions

The information that comes from smart sensors is full of opportunities, and looking at in a different light can be helpful in discovering business trends, patterns, and opportunities, enabling companies to take leaps and to make better and more-informed decisions. While using analytics methods and sensors, companies can also obtain useful data on customers' behaviour and the trends of the market relevant to their industry [18]. This information can then be looked at and factored in to improve processes throughout a company. Market and trend analyses can improve existing products and find out new growth prospects. For example, by studying the data received from the numerous sensors installed in retail stores, businesses can determine some of the many categories of products that are purchased the most. Determining customers' buying patterns, which are based on customer preferences, peak shopping times, and sales potential, can enable companies to plan for inventory management, marketing and promotions, and even how stores are arranged.

6.3.2 Faster Response Times

Smart sensors allow real-time data analysis, which helps in decision-making and facilitates rapid business responses to market fluctuations and operational issues. Continuously controlling and evaluating important indicators and changes in trends as they occur allows the appropriate alterations of business strategies and its products and the formation of operational procedures [19]. For instance, should sensors suggest a reduction in revenue on a specific product, firms are in a position to easily change the selling price, advertising solutions, or aspects of products or systems to resolve the problem at hand. This agility enables business organisations to cope with equivalent dexterity when either exploring new markets or responding to existent ones, fight for market share, effect changes in an industry, and capitalise on shifts in the market.

6.4 Reduced Risk

6.4.1 Predictive Maintenance

Therefore, the use of smart sensors in the collection of data allows for making predictions, taking measures to respond to equipment malfunctions, avoiding those malfunctions, and minimising the probability of unplanned downtime. By maintaining a constant vigil over such factors as vibration, temperature, and pressure, special smart sensors can track abnormalities that suggest possible equipment failures [20]. This early-warning system enables organisations to plan for either maintenance or repair works before they would occur in an emergency setting and enables them to implement cost-saving measures, such as avoiding interruptions to production and making preventive rather than corrective maintenance repairs. On this basis, if businesses can identify these challenges and deal with them before they become malfunctions, those businesses will improve their performance, thus allowing them to control their expenses and maintenance costs and ensuring that their equipment is in good condition, which increases efficiency.

6.4.2 Risk Mitigation

According to the descriptions above, smart sensors play a role of uploading data on ongoing operations to adequately prevent existing or emerging threats from becoming massive issues. This predictive capability enables one to take proactive measures that reduce interference and overall prevent costly consequences. For instance, one can monitor the performance of some equipment and ascertain abnormalities to foresee failures and plan for maintenance to be carried out in advance.

6.4.3 Job Creation

A new world of smart sensor systems is on the horizon, where the number of jobs available is growing. The number of these smart devices used in business is on the rise, and businesses aiming to improve their operations need to employ people specialised in domains such as smart sensors, data analytics, IoT platforms, and cybersecurity [21]. This technological shift is also leading to the emergence of new career positions, such as the sensor engineers, data scientists, IoT architects, and cybersecurity specialists. These professionals are critical in embracing the design, implementation, and management of such structures.

7 Challenges in and Solutions for Using Smart Sensors in Factories

7.1 Data Overload

The volume of crucial data that smart sensors capture can create difficulties for analysis and decision-making due to data overload. The following subsections detail some strategies for managing and using these data effectively.

7.2 Data Filtering and Aggregation

Data acquisition and control, the initiation of event-driven effects, and local processing are innovative attributes of data management that stem from the efficiency of the IoT systems. Edge computing, featuring data collection and processing, takes place at the sensor level.

To reduce the quantity of data transferred to the cloud, data aggregation combines data from multiple sensors into more-significant and more-straightforward measures. Event-driven triggering makes sure that frequent but irrelevant changes or events are not communicated, which makes the flow of data even more efficient.

7.3 Cloud-Based Data Storage and Analysis

Large-scale cloud platforms, data lakes, and data warehouses are vital for managing the massive datasets in IoT systems. Presently, the concepts of scalable cloud platforms are the essential prerequisites for effective paradigms for storing and analysing large numbers of data. Data lakes are used in the middle of the process to save information. Databases store raw data for future use in analysis and forms of synthesis. Data warehouses categorise and classify data, allowing them to be organised and allowing queries and reports to be rapidly carried out.

7.4 Advanced Analytics Techniques

Modern business practices have been the driving forces behind most of the business innovations and developments of the past couple of years. Researchers have found that the business benefits of IoT data are crucial to extracting value from IoT data, analysing machine-learning algorithms, and finding fragments, patterns, and anomalous events. Machine learning performs the significant work of data analysis and extracting meaningful insights. Predictive analytics is used in predicting future trends and results from previous patterns and trends.

7.5 Data Visualisation

Data visualisation, through tools such as interactive dashboards and data storytelling, is critical to IoT data presentation. Most of the interactive dashboards include features that facilitate drilling down on key metrics. This is done to simplify whatever information is being processed and presented so that it is easier to understand and make necessary decisions with.

7.6 Data Governance and Quality

Data quality management and data governance policies and procedures are necessary to guarantee data credibility. Data quality management refers to the processes that are put in place to guarantee the availability, integrity, quality, accuracy, consistency, and completeness of data. These policies and procedures define the rules governing how data should be managed. They are used for data capturing, storage, and analysis and for the monitoring and validation of data quality and consistency throughout the data's life cycle. Thus, through the lessons learned in this chapter about avoiding data overload, manufacturers can tap into the full features of smart sensors and use this collected information to optimise manufacturing processes and increase product quality.

7.7 Interoperability

Interoperability has brought up one of the major issues of how best to develop effective smart sensor networks, namely the difficulty of guaranteeing the interoperability of various sensors with one or more systems. This interoperability issue can prevent the effective integration of data that contributes to decision-making, and because of it, various processes become more complicated. The following subsections detail some strategies to address sensor interoperability.

7.8 Standardisation

Standardisation is a crucial factor for interconnection and interaction and for data formats in the industry of IoT systems. Open Platform Communications Unified Architecture (OPC UA) is an industry standard and an example of a common protocol used by sensors for communication, promoting seamless integration. Data formats such as JSON are to be used more often. XML or CSV define the ways that sensors and systems can exchange information, which means how they share data.

7.9 IoT Platforms

The strategies, such as IoT platforms and plug-and-play functionality, are useful for easy integration and managing IoT systems. IoT platforms as such come with built-in connectivity solutions and features for connecting to other platforms. They provide preimplementation support for different types of sensors, thus reducing the time and effort in development. Plug-and-play functionality simplifies the process of connecting and configuring sensors.

7.10 Security and Cost

Smart sensors produce an immense amount of valuable information that can be used to increase efficiency and productivity and to improve the quality of services and products. However, these data also offer many opportunities but also present security risks if not well protected from unauthorised access. Such access means that unauthorised personnel or hacker have gained access to those data. Data incidents, failures, and operation losses can indicate breaches and lead to disruptions and financial losses. The installation of smart sensor networks may be an expensive investment for organisations, particularly for small ones. Some of the first expenses related to buying sensors, preparing the necessary infrastructure, and adapting systems can be substantial. Also, they come with continuous expenditures in areas such as data storage and data analysis. A number of factors in their maintenance have to be taken into account. Some of the challenges that the smaller manufacturers are likely to experience may include the resources required for such investments, especially if their budgets are constrained in some way. However, the long-term advantages of smart sensor networks, including enhanced efficiency and quality, the efficiencies of superior products, and cost cutting, can often outweigh their initial cost. Even smaller manufacturers need to be more conscious of their individual requirements and need to look for economic strategies.

7.11 Maintenance

We find smart sensors to be critical components required for the proper working of present-day manufacturing plants. Nevertheless, for them to remain as accurate and reliable as expected, they require being regularly updated. Inspection, monitoring, cleaning, maintaining their environment, and firmware updates are among their requirements. Frequent updates to sensor readings are critical for long-term dependable and high-performance IoT sensors. Routine checks help to discover physical damage, corrosion, and signs of deterioration, all of which affect sensor performance. When using electronics, the manufacturer's advice on cleaning and care should be followed. Carrying out proper regular maintenance procedures and helps in keeping sensors in the correct state. Regular updates to the sensors' firmware ensure that the latest version of the firmware is being used and prevent security threats.

Data analytics is a cornerstone of modern smart factories, enabling manufacturers to extract valuable insights from the vast numbers of data generated by smart sensors. By employing advanced techniques, businesses can optimise operations, improve quality, and drive innovation.

7.12 Key Data-Analytics Techniques

Descriptive analytics is the most basic form of advanced analytics given that it simply analyses past data, summarises the data, and provides a means to explain trends, patterns, and variances. Diagnostic analytics seeks to understand why something has gone wrong. Predictive analytics ascertains possible trends and likely results. Prescriptive analytics also recommends optimal actions or decisions.

7.13 Standardisation

Standardisation is critical in guaranteeing the compatibility of various forms of sensors and systems in smart factories. Thus, by implementing industry standards, manufacturers can enable data interaction, decrease the costs of integration, and improve the operation of organisations during the exchange of data.

These basic concepts of data exchange use Open Platform Communications Unified Architecture (OPC UA), Message Queuing Telemetry Transport (MQTT), Representational State Transfer Application Programming Interface (RESTful APIs), semantic webs, and modern IoT systems to ensure efficient and productive networked communication, data sharing, and data interaction.

7.14 Cloud-Based Solutions

Having less-physical infrastructure, paying for only what is used, and potential expansion are some of the many advantages of cloud-based IoT solutions. The cloud-based payment model is more affordable and can flexibly match clients' usage by allowing organisations to increase or decrease the size of their sensor network as needed.

8 The Future of Manufacturing, Industry 4.0, and Smart Sensors

Smart sensors are the building blocks of the future of manufacturing. The development of this sensor technology will continue, and therefore, more will be developed in the future. These sensors offer elaborate solutions in fields such as digital twins, augmented reality, and autonomous production systems. Using smart sensors and learning what they can do for manufacturers promises to reveal further degrees of efficiency, adaptability, and, importantly, sustainability. Even as these technologies continue to improve, they are still expected to offer better operations for industries. One of the areas that can be developed is the utilisation of artificial intelligence (AI) and machine learning (ML) in Industry 4.0. These technologies can be used to analyse the data collected by smart sensors and thus produce forecasts and select actions that would be helpful in improving industrial processes. Such transformations may open up new opportunities and additionally increase the productivity of industries for better performance and outcomes. Figure 3 outlines the future of the Industry 4.0 manufacturing process.

9 Conclusion

Without doubt, smart sensors are the key bringing about factories of the future. Their ability to process and collect data and then analyse and transmit data in real time is revolutionising manufacturing processes, from shops to operational boards and from shop floors to executive offices. Given these advancements in technology, we can only expect more of them and more from them in the future: better sensors that will be able to perform even-more-complicated operations as time goes on. The integration of smart sensors into the manufacturing environment is expected to bring about the following benefits: increased work efficiency, increased product quality, reduced working or operating times, and more. In this process, manufacturers can use this technological revolution as an opportunity to position themselves as leaders in innovation, acquire a competitive advantage, and improve the sustainability of the industrial environment, which can increase manufacturer efficiency and competence in future years.

This chapter looked at a roadmap towards realising the ultimate vision of smart factories. That vision is still unfolding, but one thing is certain: Smart sensors will be a major enabler of many parts of future smart structures and will remain vital for many application areas in the smart structures. They promise the type of change that could redefine the manufacturing industry, a change that will be experienced for generations to come.

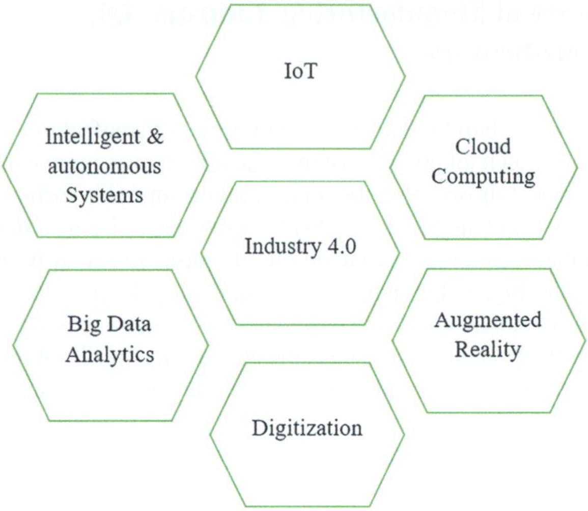

Fig. 3 Future of manufacturing in Industry 4.0

References

1. Soori, M., Arezoo, B., & Dastres, R. (2023). Internet of things for smart factories in Industry 4.0, a review. *Internet of Things and Cyber-Physical Systems, 3*, 192–204.
2. Deepak Kumar, S., Arun Manohar, G., Surya Teja, R., Ramana Rao, P. S. V., Mandal, A., Behera, A., & Srinivasa Rao, P. (2023). Smart manufacturing technologies in Industry 4.0: Challenges and opportunities. In *Intelligent manufacturing management systems: Operational applications of evolutionary digital technologies in mechanical and industrial engineering* (pp. 23–57). Wiley.
3. Kusiak, A. (2023). Smart manufacturing. In *Springer handbook of automation* (pp. 973–985). Springer International Publishing.
4. Shadravan, A., & Parsaei, H. R. (2023, March). Impacts of Industry 4.0 on smart manufacturing. In *Proceedings of the 13th International Conference on Industrial Engineering and Operations Management* (pp. 7–9). IEOM Society.
5. Bhatt, K., Agrawal, C., & Bisen, A. M. (2024). A review on emerging applications of IoT and sensor technology for Industry 4.0. *Wireless Personal Communications, 134*, 2371–2389.
6. Yi, K. J., & Jeong, Y. S. (2022). Smart factory: Security issues, challenges, and solutions. *Journal of Ambient Intelligence and Humanized Computing, 13*, 1–14.
7. Büchi, G., Cugno, M., & Castagnoli, R. (2020). Smart factory performance and Industry 4.0. *Technological Forecasting and Social Change, 150*, 119790.
8. ElMaraghy, H., Monostori, L., Schuh, G., & ElMaraghy, W. (2021). Evolution and future of manufacturing systems. *CIRP Annals, 70*(2), 635–658.
9. Hozdić, E. (2015). Smart factory for Industry 4.0: A review. *International Journal of Modern Manufacturing Technologies, 7*(1), 28–35.
10. Mourtzis, D. (2024). Industry 4.0 and smart manufacturing. In *Manufacturing from Industry 4.0 to Industry 5.0* (pp. 13–61). Elsevier.
11. Resman, M., Turk, M., & Herakovič, N. (2021). Methodology for planning smart factory. *Procedia CIRP, 97*, 401–406.

12. de Assis Dornelles, J., Ayala, N. F., & Frank, A. G. (2022). Smart working in Industry 4.0: How digital technologies enhance manufacturing workers' activities. *Computers & Industrial Engineering, 163*, 107804.
13. Morgan, J., Halton, M., Qiao, Y., & Breslin, J. G. (2021). Industry 4.0 smart reconfigurable manufacturing machines. *Journal of Manufacturing Systems, 59*, 481–506.
14. Çınar, Z. M., Zeeshan, Q., & Korhan, O. (2021). A framework for Industry 4.0 readiness and maturity of smart manufacturing enterprises: A case study. *Sustainability, 13*(12), 6659.
15. Abdelmajied, F. Y. (2022). Industry 4.0 and its implications: Concept, opportunities, and future directions. In *Supply chain-recent advances and new perspectives in the Industry 4.0 era*. IntechOpen.
16. Nakagawa, E. Y., Antonino, P. O., Schnicke, F., Capilla, R., Kuhn, T., & Liggesmeyer, P. (2021). Industry 4.0 reference architectures: State of the art and future trends. *Computers & Industrial Engineering, 156*, 107241.
17. Rosário, A. T., & Dias, J. C. (2023). How Industry 4.0 and sensors can leverage product design: Opportunities and challenges. *Sensors, 23*(3), 1165.
18. Singh, H., & AlMangour, B. (2023). *Handbook of smart manufacturing: Forecasting the future of Industry 4.0*. CRC Press.
19. Shi, Z., Xie, Y., Xue, W., Chen, Y., Fu, L., & Xu, X. (2020). Smart factory in Industry 4.0. *Systems Research and Behavioral Science, 37*(4), 607–617.
20. Namjoshi, J., & Rawat, M. (2022). Role of smart manufacturing in industry 4.0. *Materials Today: Proceedings, 63*, 475–478.
21. Wu, K., Xu, J., & Zheng, M. (2024). Industry 4.0: Review and proposal for implementing a smart factory. *The International Journal of Advanced Manufacturing Technology, 133*, 1–17.

From the Shop Floor to Cloud-Based Cyber-Physical Systems: The RAMP Case Study

Panagiotis Bouklis, Georgios Peppas, Konstantinos Gombakis, Dimitrios Chantes, and Anastasia Garbi

1 Introduction

The Robotics and Automation Market Place (RAMP) aims to be the real one-stop-shop for all things manufacturing, providing services and tools with value for all the different actors of the manufacturing automation and digitization supply chain, from the application developers and system integrators (a.k.a. service providers) up to Digital Innovation Hubs (DIHs) and manufacturers, focusing especially on small and medium enterprises (SMEs).

RAMP facilitates the transition of manufacturing SMEs into the Industry 5.0 era by providing services in two aspects:

- Online Marketplace, which allows manufacturing SMEs and Service Providers to connect and collaborate online
- Industrial Internet of Things (IIoT) platform, which enables the transition of the production from the shop floor to the cloud

P. Bouklis (✉) · G. Peppas · K. Gombakis · A. Garbi
European Dynamics Luxembourg S.A., Luxembourg, Luxembourg
e-mail: panagiotis.bouklis@eurodyn.com; georgios.peppas@eurodyn.com; konstantinos.gombakis@eurodyn.com; anastasia.garbi@eurodyn.com

D. Chantes
European Dynamics Advanced Information Technology and Telecommunication Systems S.A., Athens, Greece
e-mail: dimitrios.chantes@eurodyn.com

Y. Kumar Singla et al. (eds.), *Digital Manufacturing in Industry 5.0*,
https://doi.org/10.1007/978-3-031-91500-0_44

2 Why RAMP

Manufacturing SMEs usually follow up on the more traditional way of operation, that is, they usually try to adapt their production processes to the mechanical tools that enable them to comply with the requested quality standards or production volumes. It is common among smaller manufacturing facilities to miss the skills needed to (a) identify which opportunities from the digital era can create value for them (e.g., increase quality of production, discover sources of loss from data analytics, etc.) and (b) develop and deploy such solutions in their production [1].

Even though there are various smart industry innovation activities, many are technology-oriented, focusing on making manufacturing processes more efficient [2]. However, other aspects should be considered in the current competitive landscape, like networking [3] and the role of the human workers. The latter are not foreseen to be replaced by automation systems, but instead to capitalize on the opportunities created by such technologies for innovation and self-improvement, especially in the context of Industry 5.0 [4, 5].

RAMP (Fig. 1) addresses the above challenges in a holistic solution that includes access to state-of-the-art technologies and a standardized platform to facilitate the cloud transition, networking features to assist in connecting different actors of the digitization supply chain and online tools that empower human workforce to exploit the cloud and efficiently collaborate with partners.

Across all platforms, and consequently also in RAMP, there are three layers [6]:

- Network or community layer, comprised of the participants of the platform (i.e., the Online Marketplace)

Fig. 1 RAMP homepage

- Infrastructure layer, which encapsulates the platform tools and services (i.e., the IIoT platform)
- Data layer that power relevance, matching the tools and services with the users (i.e., both RAMP aspects)

3 Brief History and Evolution

The concept of an online marketplace was launched and tested even before the RAMP brand, across projects co-funded by the European Commission. Many iterations, testing different features, incrementally evolved RAMP into its current state. The idea of a marketplace connecting different actors of the automation value chain was initially researched and tested in the niche market of intra-factory logistics automation in the L4MS project (Grant Agreement No. 767642) in 2017. After a few iterations and testing, the DIH2 project (Grant Agreement No. 824964), focused on agile manufacturing, joined the initiative in 2018 and the RAMP brand started to be developed independently of the two projects and going beyond logistics automation to all things manufacturing. At the same time, concepts of Data Visualization in the Software-as-a-Service (SaaS) paradigm started to be implemented and tested. The second major release of the marketplace (and the first under the RAMP brand) was made available.

As the initiative of RAMP gained momentum, three more projects joined in 2020: Better Factory (Grant Agreement No. 951813) focused on personalized production and cyber-physical systems, KITT4SME (Grant Agreement No. 952119) focused on Artificial Intelligence (AI) for manufacturing, and SHOP4CF (Grant Agreement No. 873087) focused on Human Factors in Connected Factories. Taking advantage of the new research and testing results, key features were developed for RAMP, with a focus on online collaboration, cyber-physical systems and usability. The third major release of the marketplace (and the second under the RAMP brand) was made available, in which features and services were incrementally added toward a holistic IIoT solution ready and tested by 2024.

Going beyond the latest release, RAMP focused on the more efficient creation of business value for manufacturing SMEs and Service Providers, by participating in the CIRCULOOS project (Grant Agreement No. 101092295) focused on Circular Economy for manufacturing, and in DOME project (Grant Agreement No. 101084071) as a member of a federation of marketplaces for cloud services.

The overall timeline of the above-mentioned projects and major releases is illustrated in Fig. 2.

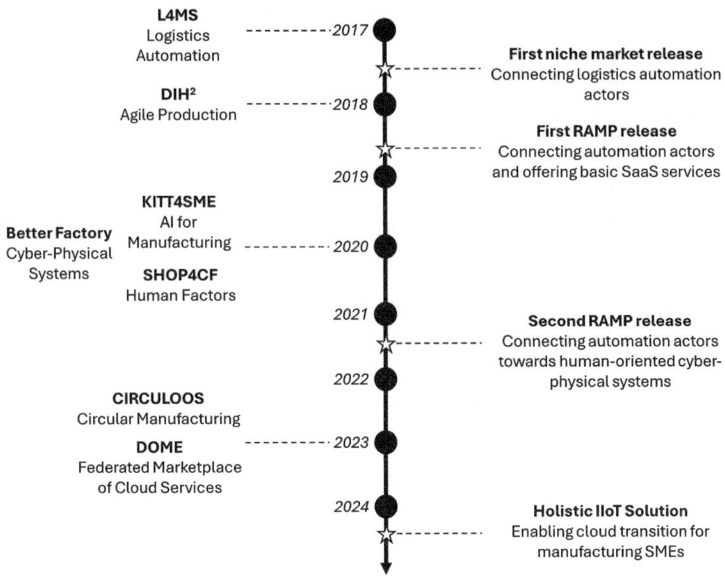

Fig. 2 Timeline of projects in which RAMP was developed (left) and evolution of major releases (right)

4 Methodology for Requirements Analysis

An extensive requirements analysis was conducted in three major iterations, representing the three milestones in the specification of the RAMP features:

1. Requirements for an online marketplace, with a focus on networking [7]
2. Requirements for an online marketplace that covers the whole manufacturing sector, including SaaS [8]
3. Requirements for an online marketplace that facilitates the provision of cloud services [9, 10]

A structured and disciplined approach was used for eliciting the requirements, from the collection up to analysis and documentation. The gathering and subsequent refinement of the requirements followed the combination of the so-called top-down and bottom-up design approaches [11]:

- Bottom-up requirements gathering refers to the ones that come from technology push, State of the Art features and functionalities that are contracted or expected to be included in the platform. It is the viewpoint where the technology is taken as the starting point. Existing systems and technologies are within the most important sources of needs. In fact, these contain the accumulated wisdom of years of practical application. These are then aligned with the stakeholder expectations and refined into requirements.

- Top-down requirements gathering refers to the ones that are inferred from the platform high-level goals and the stakeholders, in the sense of how they expect to use the platform. These are analyzed into specific needs and then decomposed into specific requirements.

A range of different techniques have been utilized across the elicitation iterations, selecting the most appropriate out of widely accepted and used techniques [12]:

- Collaborative techniques: Brainstorming, mind-mapping, workshops, interviews [13]
- Legacy systems, system archaeology
- Document analysis, literature research [14, 15]
- Expert analysis [16, 17]
- Online surveys

An important factor in validating the elicited requirements was the collection of feedback from a significant number of stakeholders, by their participation in structured surveys that allowed quantified conclusions on the prioritization of proposed features. This includes more than 150 stakeholders of the manufacturing automation value chain in the first two iterations and 120 stakeholders of cloud services (including stakeholders from manufacturing and beyond) in the last iteration. The surveys themselves have been constructed as a last step, following up on the other elicitations techniques used (collaborative, legacy systems, document analysis, expert analysis) so as to validate requirements (Fig. 3) that had been identified but also to evaluate the priorities and even elicit new requirements that had not already been discovered. The questions that were structured for validation and prioritization were constructed using two techniques: (a) the Likert scale [18], which asks the level of "agreement" with a statement; and (b) Top Task Analysis [19], which asks responders to select the top 3–5 out of a set of possible alternatives.

Additionally, literature and legacy systems analysis included an extensive competitive analysis of 22 similar existing marketplaces, quantifying the results according to their relation to RAMP and their features (Fig. 4). This led to a classification of potential services for RAMP into *de facto* features (usually common and expected by the users), sometimes implemented and gaps or low value (not usually implemented, may constitute gaps—and hence opportunities—in current market or they have a low value).

After each elicitation iteration, the requirements were analyzed and, considering the system context and the targeted business processes, they were assigned to high-level system functions. Concrete requirements have been documented for each function, using formal syntax [20] and with prioritization that follows the must have–should have–could have–won't have (MoSCoW) method [21]. Traceability of each requirement is ensured through its entire lifecycle, from its source up to the implementation, deployment and validation [22].

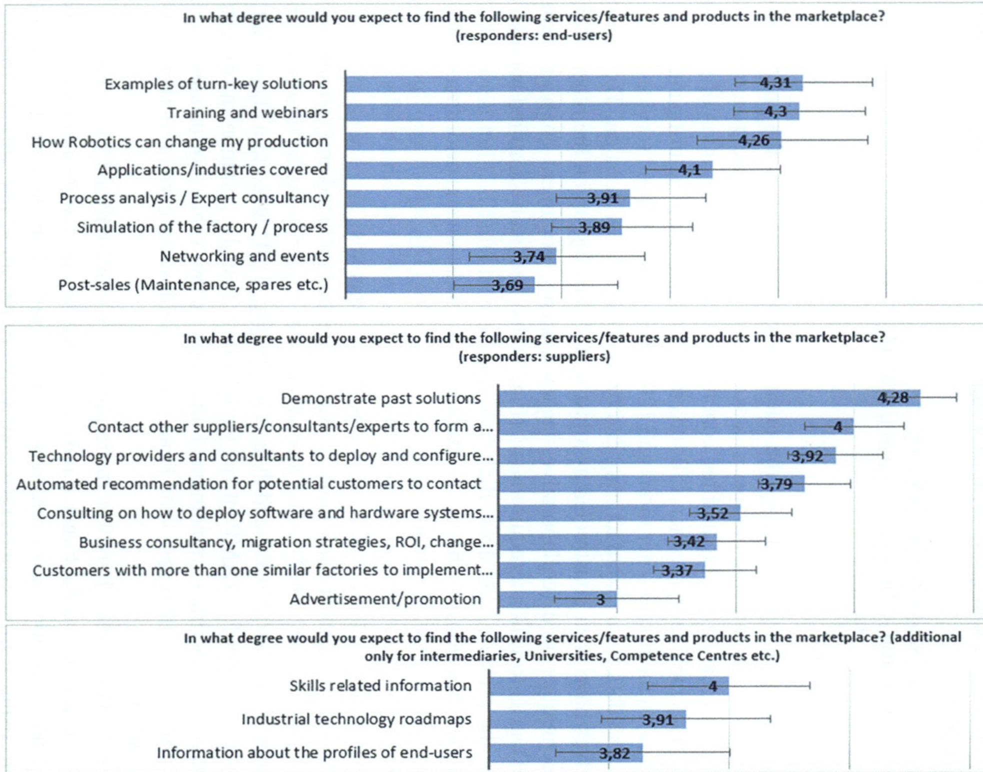

Fig. 3 Example statistics from the online survey for requirements validation (5 means totally agree, 1 means totally disagree)

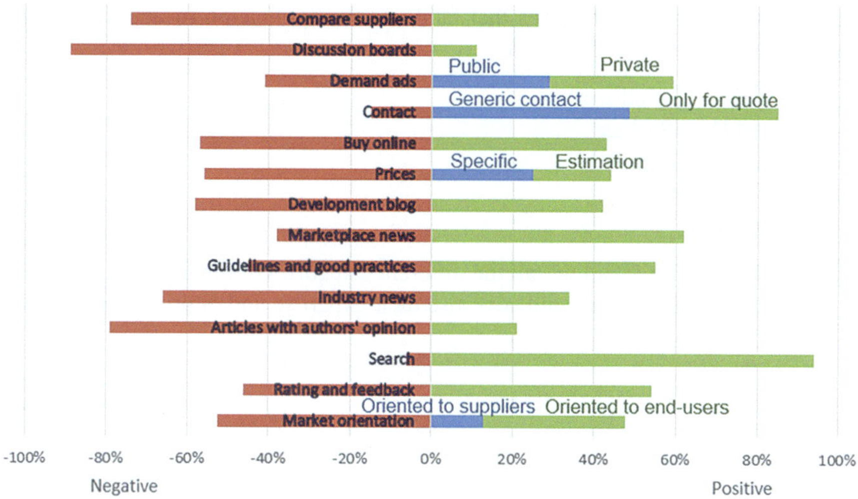

Fig. 4 Features in existing marketplaces, weighted to their relation to RAMP. "Negative" indicates that the feature does not exist in the marketplaces, while "Positive" indicates that the feature exists

5 Elicited Features

A list of the features that users can utilize in RAMP, as elicited by the requirements analysis, is given below. The features are categorized following the RAMP motto, "Connect. Collaborate. Automate." Practically, "Connect" and "Collaborate" features are relevant to the "Online Marketplace" aspect of RAMP, while the "Automate" features are relevant to the "IIoT platform" aspect.

- "Connect" provides features that facilitate partner discovery. It includes:
  - Find a Partner
  - Discover Solutions
  - Get Local Support (DIHs)
  - Ask for a Service
- "Collaborate" provides features that facilitate the community collaboration toward digitization. It includes:
  - Success Stories
  - News & Events
  - Discussion Board
  - Matchmaking Requests

- "Automate" provides features that facilitate the application of automation and digitization solutions. It includes:
  - RAMP IIoT platform
  - Cloud Data Storage
  - Data Visualization
  - Docker Repository
  - AppHub

Find a Partner (Fig. 5) allows all the different categories of organizations to create their organization profile and advertise their expertise and digitization products.

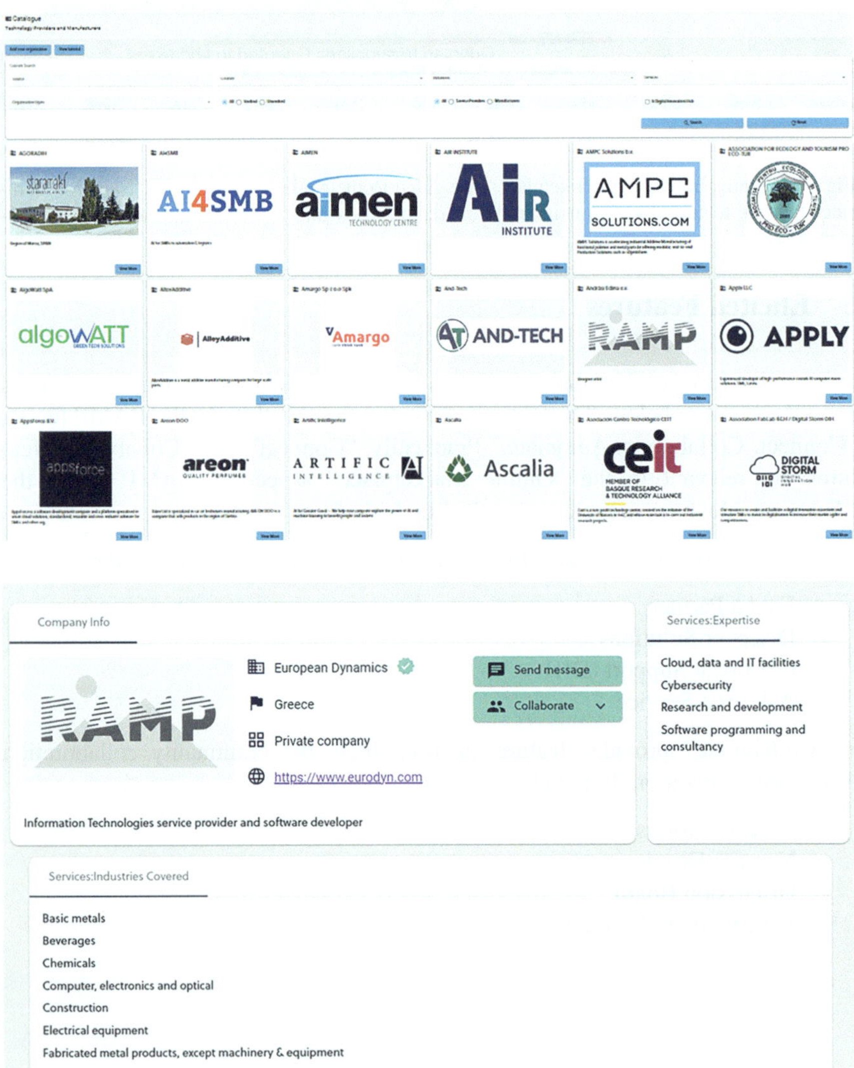

Fig. 5 Find a partner catalogue (left) and profile of an organization (right)

Organizations can be directly created by registered users, while they can also become "verified," by providing relevant documentation to prove that they are dully represented in RAMP. Advanced search and filters are included to facilitate discovery.

Discover Solutions (Fig. 6) allows the discovery of automation and digitization solutions offered by the registered organizations in RAMP. This offers an alternative way to find a partner, not by reviewing profile and expertise information, but by reviewing the actual solutions that are available and fit a use case. A study on the characterization of the solutions to facilitate discovery has been conducted. The information pages of each solution include a variety of general information, but also predefined selections to specify how they support workers and

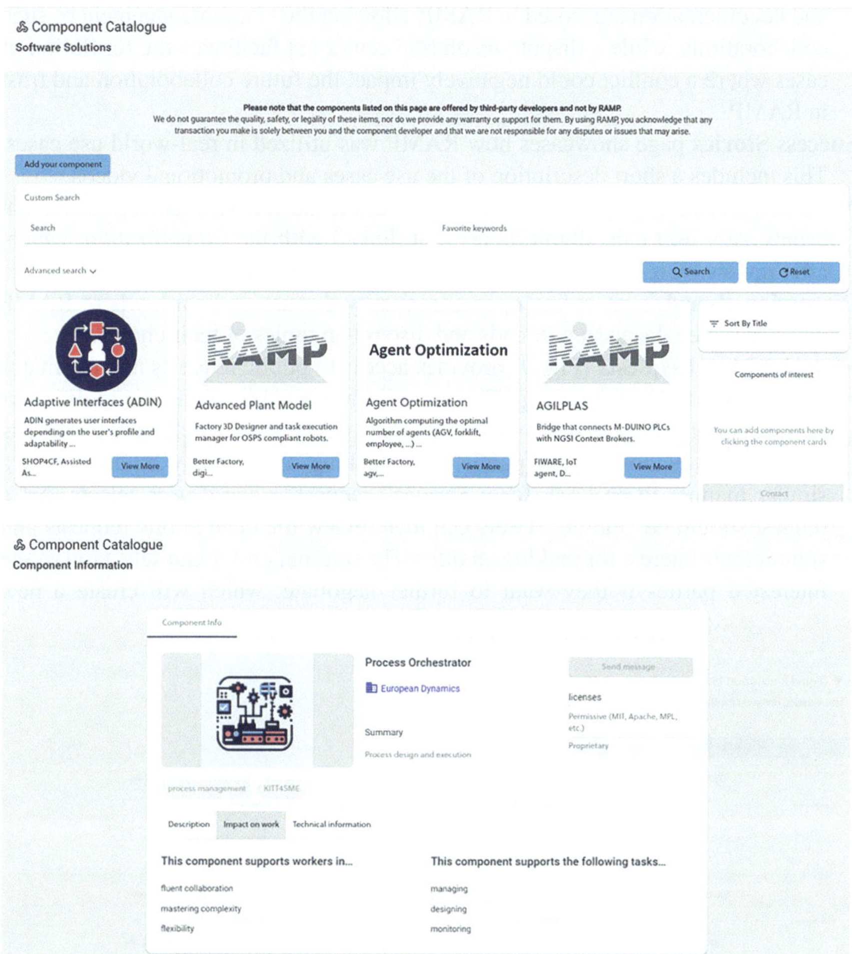

Fig. 6 Discover solutions catalogue (left) and information page of a solution (right)

what tasks they are relevant to [23]. Advanced search and filters are included also in this feature.

Get Local Support (DIHs) (Fig. 7) provides dedicated access to registered Digital Innovation Hubs (DIHs) in RAMP. This allows organizations which are unsure on how to proceed with their digital transformation journey to contact their local DIHs that can assist them, but also engage with their regional ecosystem. In this case the DIHs are shown per country to facilitate discovery.

Ask for a Service allows organizations to simply submit their use case needs, optionally selecting solutions that are available in RAMP, to find an appropriate partner. RAMP recommends one or more potential partners that fit the case, and the user may select one or more to negotiate further. A rich step-by-step wizard-style tool (Fig. 8) facilitates the communication between the candidate partners, from initial budgeted offer up to contract negotiation [1]. The communication and documentation are stored in RAMP, allowing the better management of such collaborations, while a dispute resolution center [1] facilitates the resolution of cases where a conflict could negatively impact the future collaboration and trust in RAMP.

Success Stories page showcases how RAMP was utilized in real-world use cases. This includes a short description of the use cases and promotional videos.

News and Events page provides an overview of the latest RAMP news and relevant events. Any user can submit their event, linked with their organization, which, after review by RAMP, will be posted for promotion in the events list.

Discussion Board is an online open community discussion board, where RAMP users can open discussion threads and discuss on topics of their choice.

Matchmaking Requests (Fig. 9) provides access to public requests and facilitates negotiations for matchmaking. The tool resembles public tendering tools, however, in a more simplified manner to facilitate easy post of matchmaking requests and online collaboration. A public matchmaking request may be created in a similar manner to an "Ask for a Service" request, when a user selects that a request should be "Public." Users can then review the open public requests and submit their interest for making an offer. The original poster can select out of the interested parties if they want to further negotiate, which will create a new

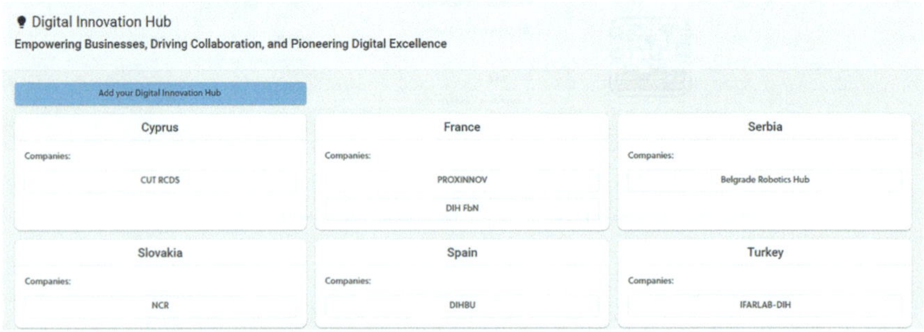

Fig. 7 Get local support (DIHs) catalogue

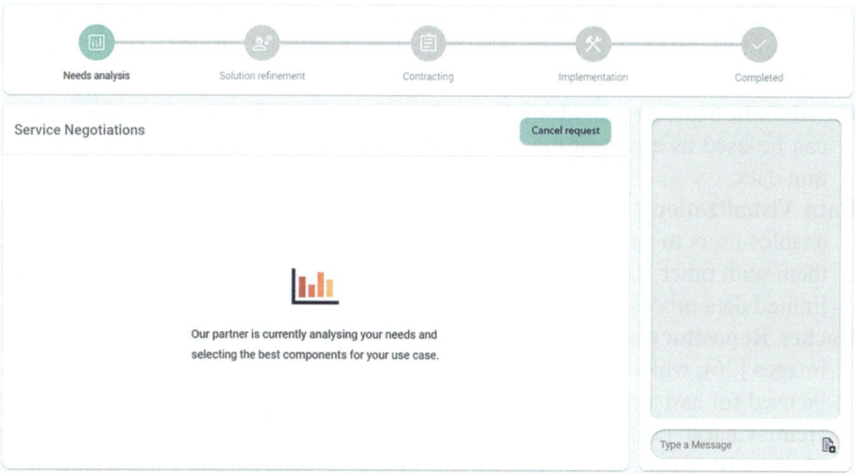

Fig. 8 Service negotiation tool

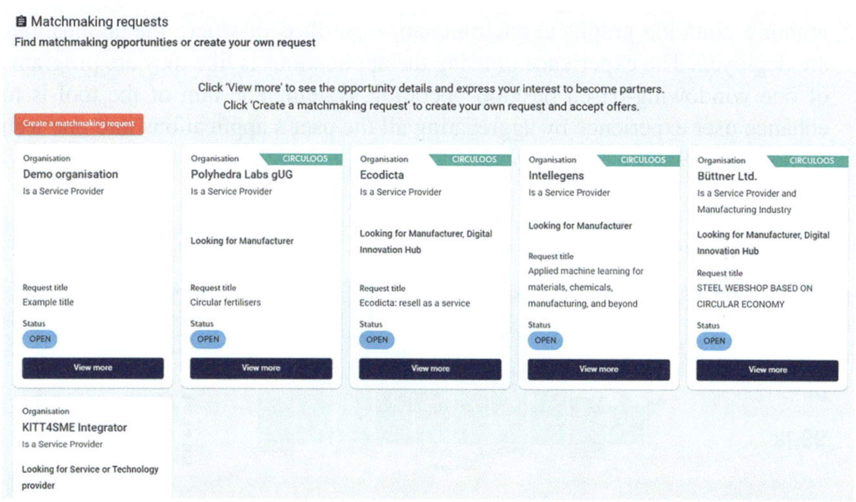

Fig. 9 Matchmaking requests

instance of bilateral collaboration in the Service Negotiation tool. This feature is also utilized for facilitating a first phase in various funded open calls for experiments, for creating consortiums to submit experiment proposals in the relevant open calls.

RAMP IIoT platform provides access to the packaged IIoT platform solution that is offered by RAMP and facilitates the transition from the factory shop floor to the cloud. It allows Manufacturing SMEs to quickly connect their sensors, robots and machinery to the RAMP cloud, adopting standard technologies for data sharing and interoperability, like the Message Queuing Telemetry Transport (MQTT) protocol [24] and Next Generation Service Interfaces–Linked Data (NGSI-LD) interface [25].

Cloud Data Storage provides information on the RAMP cloud data store, which can be used as a part of the RAMP IIoT solution, to store and secure production data.

Data Visualization (Fig. 10) is a Software-as-a-Service (SaaS) application that enables users to easily create data visualizations, graphs, and dashboards, share them with other RAMP users. It includes alarm notifications and the option for limited data processing.

Docker Repository allows users to use public and private repositories for docker images [26], which can be shared with other RAMP users. The registry can also be used for assisting the deployment of software solutions, including the use of Helm Charts [27], while it includes a vulnerability scanner that indicates identified weaknesses in images and proposes patches and improvements to address these vulnerabilities [28].

AppHub (Fig. 11) is a web desktop-like application that allows users to use all the different web-based Graphical User Interfaces (GUIs) of their applications within a common graphical environment, regardless of where the applications are deployed. The experience of using the applications is like they are programs of one windowing-based desktop operating system. The aim of the tool is to enhance user experience by aggregating all the user's applications into one web workspace [29].

Fig. 10 An example of a data visualization dashboard

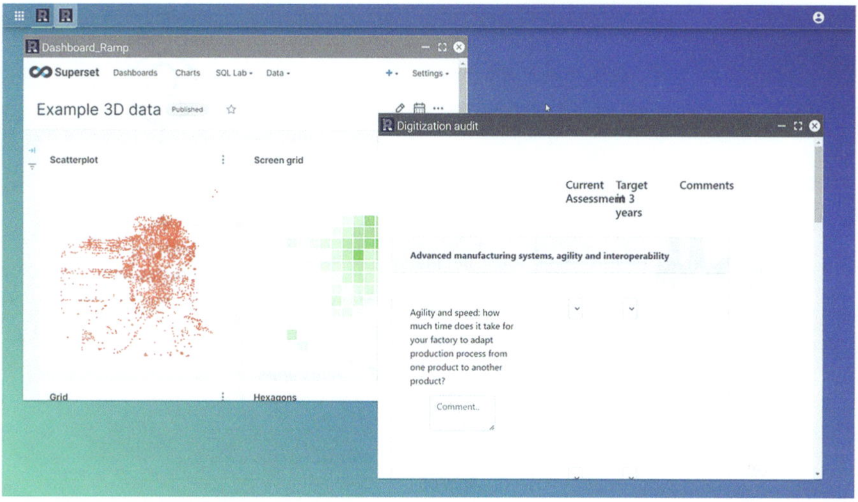

Fig. 11 AppHub in RAMP

6 Main Elements of the Cyber-Physical Architecture

The overall cyber-physical architecture for digital manufacturing is practically a platform of platforms, composed of several systems that contribute to the digitization of production. The architecture is composed of two major systems, the RAMP online marketplace and the RAMP IIoT platform (Fig. 12). In this context, the manufacturing SME acts as the end-user of the system, interacting through the online marketplace. The Service Provider on the other hand is the one who customizes and integrates customer-specific solutions on the factory, and by adding their relevant applications on the RAMP online marketplace. The RAMP IIoT platform acts as the link between the on-premises customer-specific solution and the RAMP online marketplace [30].

The top-level architecture can be further broken down as depicted in Fig. 13.

In the **Factory** the following elements are considered [29]:

- Enterprise applications: The applications that the manufacturer has already in place and may continue to use.
- Data storage: Local database that is already used by the manufacturer and may be utilized for additional data that may be needed for the customized solution.
- Local components: Components that the user acquires (through RAMP or not) and are deployed locally.
- IoT devices: Local devices, like sensors, machinery, and robots, that are connected to the RAMP IIoT platform through the NGSI-LD interface.

The **RAMP IIoT platform** provides mainly [31] (a) the FIWARE Orion Context Broker (OCB)-LD [32] as a message broker (NGSI-LD interface); (b) Troe [33], as an adapter for the persistent storage of data that pass through the FIWARE Orion

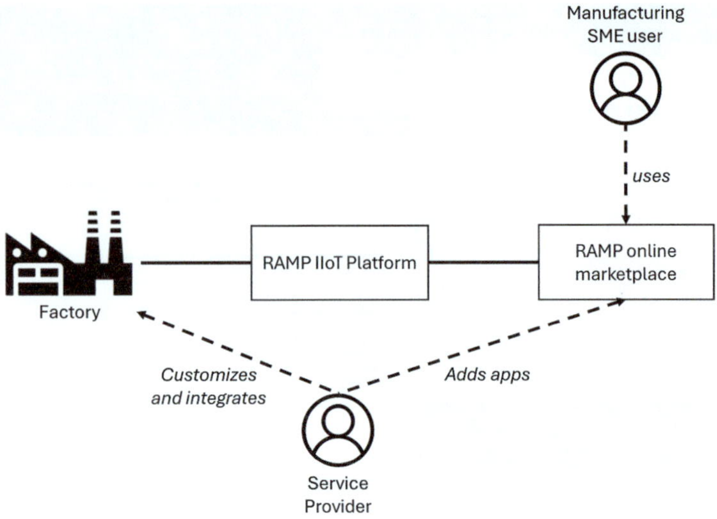

Fig. 12 The two major systems of the RAMP solution

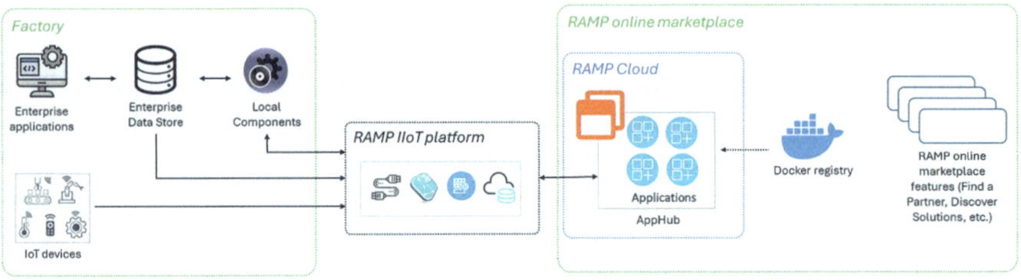

Fig. 13 Break-down of main architectural elements of the RAMP solution

Context Broker; (c) any other adapters needed for connecting to the specific local infrastructure (databases, devices, etc.); and (d) data store for persistent data, etc.

The **RAMP online marketplace** includes the implementation of all the previously mentioned user functionalities, while it also includes the RAMP cloud infrastructure and user-oriented features to provide access to the cloud-based digital production.

Utilizing the above, the transition of the production to the cloud is simplified into a step-by-step process as follows:

1. Connect to the RAMP cloud by integrating the RAMP IIoT platform and connecting sensors, robots and machines.
2. Store data on cloud by configuring the RAMP IIoT platform and using the RAMP cloud.
3. Discover Solutions or Ask for a Service to find relevant partners that will develop and integrate customized technologies for specific use cases.

4. Deploy bespoke solution by utilizing the RAMP docker registry, which will also allow easy updates in future deployments.
5. Access solution in AppHub by adding web-based GUIs in one common web-based desktop environment.

As steps 1 and 2 ensure that the factory is connected to the cloud and data is available, the extension with future applications requires less time for integration, hence accelerating the deployment of new technologies and update of the production.

6.1 RAMP Online Marketplace

RAMP marketplace is built on open-source libraries and frameworks. The backend service is utilizing Java Spring Boot framework [34]. Spring Boot is an open-source framework developed by the Spring Framework team to simplify the process of building and deploying stand-alone, production-grade applications in Java. It is part of the broader Spring ecosystem, which is widely used for building enterprise-level applications.

RAMP's front end is based on Angular [35]. Angular is a popular open-source front-end web application framework developed and maintained by Google. It is used to build dynamic, single-page applications (SPAs) with a focus on performance, maintainability, and scalability. Angular is written in TypeScript, which is a superset of JavaScript, and leverages a component-based architecture to provide a structured way of building complex user interfaces.

Data persistence is built on PostgreSQL [36]. PostgreSQL is a powerful, open-source, relational database management system (RDBMS). It is known for its robustness, extensibility, compliance with SQL standards, and support for complex queries and advanced features. PostgreSQL has been in active development for over 30 years, making it a mature and reliable choice for both small-scale and large-scale applications.

All the above, are deployed on a Linux-based infrastructure and all components are dockerized. Docker allows developers to automate the deployment, scaling, and management of applications within lightweight, portable containers. Docker containers package an application and its dependencies together, ensuring that the application runs consistently in any environment, whether on a developer's local machine, a test server, or in production.

6.2 RAMP IIoT Platform

The RAMP IIoT platform, here-on indicated as "RAMP IIoT LD platform" to distinguish it from the deprecated version, provides a highly adaptable solution designed to enable manufacturing SMEs to swiftly access cloud services, particularly those available through RAMP. Simultaneously, it offers application developers a streamlined approach to delivering software to their customers, with essential components already pre-configured by the RAMP IIoT LD platform. While the primary objective is to facilitate cloud service access, the RAMP IIoT LD platform is versatile in terms of deployment and can be easily customized to meet specific requirements without compromising overall functionality. The platform is built upon the NGSI-LD standard, which serves as both the information model and the Application Programming Interface (API) [37].

The architecture of the RAMP IIoT LD platform is illustrated in Fig. 14, and it consists of the following key components:

- **Access control**: This component is responsible for authorizing and enforcing policies. It includes the Keycloak [38] application for managing authorization, along with Single Sign-On (SSO) functionality through the RAMP IIoT LD platform. Kong [39] is employed to enforce access policies.
- **Fiware Orion LD**: This module stores and updates the current context of the system, allowing authorized agents, such as applications or devices, to subscribe to and publish information.
- **Data store**: A dedicated repository that maintains persistent (historical) data.
- **Persistent data access point**: This interface enables external applications to access the persistent data store.
- **IoT agent**: Facilitates communication between devices and the Fiware Orion LD component for data publishing. Depending on the deployment strategy, the

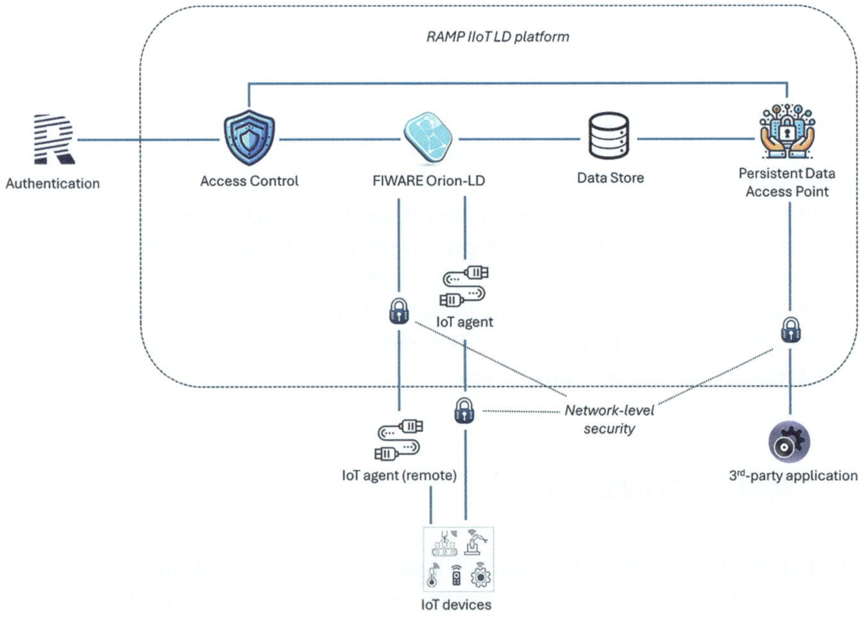

Fig. 14 RAMP IIoT platform internal architecture

Internet of Things (IoT) agent may be part of the RAMP IIoT platform or function as an external component. Regardless of deployment, robust network-level security measures are implemented to ensure authorized publishing.

In addition to the core components, Fig. 14 includes several integral external elements:

- **Authentication**: User authentication is handled through the SSO feature integrated with RAMP.eu, allowing users to log in using their RAMP account credentials.
- **Third-party applications**: External applications that require access to the persistent data store are accommodated through this component.

The RAMP IIoT LD platform's adherence to the NGSI-LD standard provides significant flexibility, enabling users to extend the platform's functionality by incorporating additional components, such as MQTT messages and LoRaWAN [40] sensors, to suit their specific needs. Furthermore, the platform seamlessly integrates with the RAMP Data Visualization component, allowing users to visualize sensor data from the RAMP IIoT LD platform. A comprehensive demonstrator is available in the RAMP repository to facilitate onboarding for new users, enabling them to explore and evaluate the platform's features in a hands-on environment.

By leveraging the RAMP IIoT LD platform, manufacturing SMEs and developers gain access to a flexible and customizable cloud infrastructure, equipped with the latest industry standards, that is tailored to support their operational and technological needs.

7 Validation Iterations and Topologies

RAMP features, throughout its different versions, have been tested across more than 90 real use cases coming from application experiments that have been funded by European Commission (EC) funded projects. More specifically the relevant projects where RAMP has been tested are L4MS, DIH^2, KITT4SME, SHOP4CF and Better Factory.

Throughout these experiments, apart from different versions, also different topologies have been tested. As "topology," we refer the location where each of the following elements is deployed: (a) IoT agent (that retrieves data from the sensor or device), (b) RAMP IIoT platform, (c) Data Store, (d) Data Visualization. The deployment locations tested are (a) on premises (local to the factory), (b) RAMP server and (c) external third-party server. Also, the three different releases (see Fig. 2) were tested. The feedback of each testing round was used to further improve the RAMP solution.

The topologies and RAMP releases tested are given in Table 1. In all the test rounds, the different RAMP releases and specific features were tested. Regarding the IIoT platform, the first tests were focused on enabling the Data Visualization in

Table 1 Topologies and releases tested

Project	RAMP release	Deployment topology: IoT agent	RAMP IIoT platform	Data store	Data visualization
L4MS	First niche market release	On premises	On premises and external third party (v2 version)	On premises and external third party	On premises and RAMP server
DIH²	First RAMP release	On premises	On premises (v2 version)	On premises and RAMP server	RAMP server
KITT4SME	First and second RAMP release	On premises	On premises (LD version)	On premises	RAMP server
SHOP4CF	First and second RAMP release	On premises	On premises (LD version)	On premises	RAMP server
Better factory	First and second RAMP release	On premises and RAMP server	On premises and RAMP server (v2 and LD versions)	On premises and RAMP server	RAMP server

the SaaS paradigm. Incrementally, cybersecurity was considered, and the cloud data store concept was also tested, up to the full IIoT platform to facilitate the cloud transition of the production. It is worth noting that for the RAMP IIoT platform both the NGSI-v2 and NGSI-LD versions were tested, more specifically the v2 version was tested in L4MS, DIH² and Better Factory [41], while the LD version was tested in KITT4SME, SHOP4CF and Better Factory. Ultimately, the end overall solution was finalized in the context of the Better Factory project.

8 Industries of Validation Use Cases

RAMP features and concepts, along with other tools and applications, have been tested in the context of experiments of different projects as indicated above. The experiments cover a wide spectrum of different industries as elaborated in Table 2.

Table 2 Industries covered by validation use cases

Industry	Projects	Experiment titles
Aerospace	DIH²	• FlexMill (Flexible Engineering and Milling for Aerospace Parts) • Prebots (Enhancing Prepreg manufacturing in SMEs through the use of cobots)
	SHOP4CF	• BUCK (Be-partnered User and Cobots powered by Key-enabling IoT tech) • QPC-AI (Physics-informed Quality Prediction Component—a use case in backlite manufacturing)
Automotive	SHOP4CF	• Bosch (pilot) • Volkswagen (pilot) • MonitorEX (Ergonomics and Physiological Monitoring Module for Exoskeleton Users) • TRanspoRT (Towing Robots for Autonomous traNsport SupPOrt in Repetitive and monotonous Tasks)
Batteries	KITT4SME	• AI4SDW (AI for Safe and Decent Work)
Chemicals	L4MS	• Chemi-Pharm (pilot)
Doors	L4MS	• KLAPPER (KLefer APPlication ExpERiment—automation in manufacturing logistics)
	SHOP4CF	• RTLS4SHOP (Location-based shop floor visibility using the Open Omlox RTLS standard)
	Better Factory	• WRC (Welded metal door design and fabrication, and innovation and automatization of production)
Electronics	DIH²	• GRIDIRON (Intelligent automated testing of electronic devices for COVID-19 detection) • PoRoLog (Poly Robots for Logistics) • A2CS (Agile and Adaptable Cobotic System)
	KITT4SME	• WAM (pilot) • JitJip (Quality Assurance for Just In Time, Just In Place) • FASPAS (Fatigue aware semantics for planning and scheduling in discrete manufacturing)
	SHOP4CF	• BrainWatch (BrainWatch for increased productivity with improved workers satisfaction) • ISTSME (Improved Scheduling and Traceability in Manufacturing SMEs by IoT-based solution) • PROPHET (PReference-based OPtimization for Human-cEnTric visual inspection) • WARNING (Worker Augmented Reality guidaNce IN aGile production shop floor) • ASMOSA (Agile Semantic MOdel driven development for Smart Applications ecosystem in manufacturing) • CO-SCREW (Demonstration of collaborative assembly tasks combining manual and cobot screwdrivers)

(continued)

Table 2 (continued)

Industry	Projects	Experiment titles
Fabricated metal products	L4MS	• AMR for TRECO (Autonomous Mobile Robot for intra-logistics at TRECO A/S) • ILR4TRECO (Autonomous Mobile Robot for intra-logistics at TRECO) • MARGO (Optimizing Material Flow with AGVs in Ring Gear Production) • SILOS (Simulation of Internal Logistic Flows and Integration of own Control Devices with OPIL and OPIL with Software Systems) • InnoWest (Increase productivity through automated polishing of metal welded assemblies)
	DIH[2]	• RomanTIG (Robotized Item Manipulator for TIG welding) • FEATS (Fiwared enabled autonomous transport system) • RoboWeldAR (Cognitive robotic welding solution for enabling shipbuilding 2.0) • HWFlexCell (Highly Flexible Welding Cell)
	KITT4SME	• CognEcoQuality (Cognition-driven eco-aware process quality)
	SHOP4CF	• MATTRESS (Material Transformation Tracking for Reduced Stress on Shop Floor) • PosWeTool (Positioning and wear measurement of cutting tools) • IN4STATION (IN4STATION—INFORMATION for WORKSTATIONS) • HOPE_FOREMAN (Human Oriented Planning and Execution FOR Efficient MANufacturing) • CUT SHOP (Cost estimation for cutting and engraving) • AR2EMP (AR-based human-centered 3D interactive guidance training service to empower workers performance and knowledge in complex industrial assembly processes)
	Better Factory	• MiniRoboFab (Exploring Product Customization and Robotic Fabrication in a Small Factory) • 3DARTDESIGN (Transforming metal waste into opportunity)

Table 2 (continued)

Industry	Projects	Experiment titles
Food and beverages	L4MS	• TALOS (TALES application experiment—Automation in breweries logistics and operation) • VCLMP (Voice Controlled Logistic system for automation in Meat Production)
	DIH²	• FIREFIT (Fiware ready quality control for packaging systems in the food industry)
	KITT4SME	• ExtruAI (Recommendation system for extruder steering based on artificial intelligence)
	SHOP4CF	• SMASH (Sea Products Manufacturing Smart Human Traceability Platform) • CHEEPSH (Cheese production optimization by Process Mining Smart Human Platform) • AI4INSPECTION (AI for quality inspection)
	Better Factory	• SMARTHam (Supervised manufacturing of real-time traceability in ham production) • IOWA (Sustainable Wine Production with AR and Robotics)
Furniture	L4MS	• IIALS (Idral & Incas AGVs Logistic System)
	DIH²	• ARP (Agile Robotized Processing)
	SHOP4CF	• AI4Dim&SurfQA (Artificial intelligence for dimensional and surface quality control of cut parts)
	Better Factory	• (ODC 3D) Optimization of Digital Craftmanship in 3D Printing • BCF (Better CNC Factory) • OCCE (The 100% Recyclable Office Chair)
Home appliances	SHOP4CF	• Arcelik (SHOP4CF)
Insulation and construction panels	DIH²	• STAR (Stitching and Taping Assisted by Robotics)
	Better Factory	• MICOCRAFT (Bio-Inspired Glue for Sustainably Manufactured Products)

(continued)

Table 2 (continued)

Industry	Projects	Experiment titles
Machinery/machine tools	L4MS	• CYB (Could You Bring?) • RobOPIL (Robotic Collaboration for OPIL-based logistic solutions)
	DIH²	• OSCAR (Operation for Self-checked Co-operative Assembly with humans and Robots)
	KITT4SME	• I-TEK (pilot) • He.Go.App (Health, safety and ErGonomics for the future human-centric factory, an integrated APPlication for increase factory safety and enhance social sustainability) • THANOS (Throughput level controller based on fatigue monitoring and quality standards) • POWERDECK (Power Industry Welding Equipment and Robots Regulation Using AI Data Enhancement and Certification Kits)
	SHOP4CF	• HARP (Human Oriented Augmented Reality Platform for Industrial Maintenance)
	Better Factory	• DSBSF (Digital Solutions for Better Scales Factory)
Marine	SHOP4CF	• MARSH (Marine Propeller Manufacturing Smart Human Platform) • SEA WORDS (Smart Human Oriented Platform for Industrial Logistics through Natural Language Understanding)
Medical devices	DIH²	• MediScara (Customized quality control and manufacturing for medical device factories) • SMB (Smart Machine Base)
	KITT4SME	• DEBORA (Defect ai-basEd elaBORation for electromyographic needles Assembly)

Table 2 (continued)

Industry	Projects	Experiment titles
Moulds	L4MS	• SMARTIST (Smart intralogistics planning for flexible production lines)
	Better Factory	• SMARTVIEW (Factory floor collaboration at your fingertips)
Paper products	Better Factory	• FOLD (Modular Cobot for production of Stone Paper Innovative Products) • SMARTENVELOPE (Digitalized and eco-friendly envelopes) • STARIOT (Sustainable straws, greener future)
Pharmaceuticals	L4MS	• PharmaPackBot (A robot for automated micro-orders preparation in a pharmaceuticals warehouse)
Printing	SHOP4CF	• RASP (Robot Automated Screen-Printing Process) • RTassist (Real Time Warehouse Assistance)
Rubber and plastics	L4MS	• Muraplast (pilot) • OptiFlow (Improvement of the productivity and optimization of the logistic flow in a big manufacturing workshop by introducing supervision and autonomous vehicles) • RIOgistics (Internal logistic automation for manufacturing SMEs in La Rioja) • TrySMART (SMART logistics solution for the TRYGONS factory)
	DIH^2	• D-Mould (A digital system for smart injection moulding production lines) • Robosonic (Post processing of blow mould plastic parts) • AGILPLAS (Agile Plastic Moulding)
	KITT4SME	• Ghepi (pilot) • AI4MOS (Artificial Intelligence-based Multi-Objective Scheduling Optimization for Sustainable Manufacturing) • ExtruAI (Recommendation system for extruder steering based on artificial intelligence) • AIMS4QC (Artificial Intelligence Made Simple for Quality Control) • QAI4GFRPs (AI for Quality in the Production of Glass Fiber Reinforced Plastics)
	SHOP4CF	• SISTERS (Single product Instance Shop floor Traceability for Employee Reassurance and Support)

(continued)

Table 2 (continued)

Industry	Projects	Experiment titles
Sports equipment	L4MS	• AMR for ECOLOGISTICS (Autonomous Mobile Robot for intra-logistics at SKOR)
	DIH^2	• COBITT (Robotization of manufacturing processes for a revolutionary skateboard deck)
	Better Factory	• REFINE (Dive deeper with less impact)
Stone and tiles	DIH^2	• JS2SF (Job Shop Toward Smart Factory)
Toys	L4MS	• Engino (pilot)
	DIH^2	• Robo4toys (Robotics for integrated injection demoulding, de/palettizing, quality inspection and assembly of toys)
	SHOP4CF	• SmartPartsDetector (Smart Parts Detector in Manufacturing using Deep Learning)
Waste treatment	KITT4SME	• AIGreenWaste (AI technologies for green waste management)
	SHOP4CF	• COOP (EffiCient rObot cOllaboration for recycling Process of household appliance)
Wearing apparel and shoes	L4MS	• OHTS4LS (Overhead Transportation System for Logistics) • HAMlet (Hybrid Automated Mould Logistic Solution)
	DIH^2	• RAM ASP (Robotic Process Automation in large scale additive production)
	SHOP4CF	• Dynanomics (Future of work uniform integrating ambulatory Ergonomic Monitoring)
	Better Factory	• SHOES IN CIRCLE (Sustainable knitted shoes)
Wood	DIH^2	• CONTRA 2.0 (A system supporting manufacturing and warehousing processes in the company)
	KITT4SME	• pArtIcle-QC (An AI-based Quality Control in Chipboard Production Process)

9 Feedback Collected

Across all the experiments, feedback was collected and assessed with structured and consistent methodologies. The methods utilized include [42]:

- The System Usability Scale (SUS) questionnaire, which provides a quick assessment of the usability [43]
- Purpose-made questions with free text, to collect qualitative feedback

- Purpose-made questionnaire on security and privacy, based on pre-existing research [44]
- Open-text comments to collect additional recommendations

The feedback collected was utilized to improve the relevant RAMP features, either during the relevant project or in follow up activities. Table 3 summarizes the points of attention from the feedback received per project.

Table 3 Summary of points of attention from the feedback received

Project	RAMP release	Points of attention	How it was addressed
L4MS	First niche market release	A wider advertisement and presentation of the previous work and success stories of technology providers in RAMP is needed	Continuous iteration on front-end and marketing activities
		Enhance trust of manufacturers to the technology providers in RAMP.	Specific task/activity in follow-up project Better Factory
		Widening the adoption of RAMP by more companies	Continuous iteration on front-end and marketing activities.
		Data security	Specific task/activity in follow-up project Better Factory
DIH[2]	First RAMP release	Hands-on workshop for technology providers	Performed during the project
KITT4SME	First and second RAMP release	Better documentation on technology providers	Planned
		Complexity and heterogeneity of services/components	Planned "RAMP IIoT platform" compatibility activities
SHOP4CF	First and second RAMP release	Poor documentation and maturity of component catalogue	Redesigned component catalogue during SHOP4CF project and training tutorial videos in follow up project Better Factory
Better Factory	First and second RAMP release	Explain in simple terms the benefits of the services offered	Continuous iteration on front-end and marketing activities
		Simplify documentation so that external technology providers can start deploying basic services without thorough learning	Offer RAMP IIoT platform as a cloud service
		Hands on training workshops on setting up RAMP IIoT platform	Planned

10 Conclusions and Future Scope

From its conceptualization and first implementation, up to the latest developments, RAMP has been significantly improved. While the first iterations were focused on probing into the needs of the manufacturing SMEs and automation technology providers, the latest iterations addressed its user-friendliness and user experience design.

The features of the "Online Marketplace" that facilitate the connection and collaboration between manufacturing SMEs and technology providers have been thoroughly tested and fine-tuned. These are launched as final in RAMP.eu. Further steps to enhance the user experience are planned, focused on use training on how to use the tools via on-site tutorials and videos. Additional follow-up includes the development of features tailored to DIHs needs, like management of their ecosystem members, and even promotion of their member organizations that are not RAMP members themselves.

The follow-up activities focus on enhancing the IIoT platform offering and connecting it efficiently with the RAMP cloud to facilitate the deployment and use of cloud-based applications, tailored to the manufacturing needs. This includes the development of new features and the fine-tuning of the existing ones. More specifically, this is focused on finalizing the RAMP IIoT platform offer as a Service, the implementation of Smart Supply Chains and the enhancement of the experience in deploying and using cloud applications, when utilizing the RAMP IIoT platform.

Another follow-up activity concerns the development of better data analytics for the RAMP administrator, in order to get better insights on the use of the platform and identifying its features that may need improvement. The same analytics function will be also utilized to identify what are the trends and customer interest in the member organizations, as well as their offered solutions.

11 Deploying Cloud Applications

RAMP leverages an Infrastructure as Code (IaC) tool to define, manage, and provision AWS infrastructure using a high-level configuration language [45]. This allows users to select components from RAMP's marketplace catalogue and automatically deploy them within their AWS cloud environment. This process efficiently creates the necessary infrastructure for the selected component, provisioning all required resources. Once the infrastructure is established, a fully functional instance of the component is deployed directly into the user's AWS account [30].

For the deployment of the infrastructure process, we utilize Terraform [46]. Terraform is an open-source IaC tool that enables users to define, provision, and manage infrastructure resources in a declarative configuration language called HashiCorp Configuration Language (HCL). This allows developers and operators to automate the creation, management, and scaling of infrastructure resources across multiple cloud providers and on-premises data centers.

The application deployment is then managed with the Ansible tool [47]. Ansible is an open-source automation tool used for configuration management, application deployment, task automation, and orchestration. It is known for its simplicity, agentless architecture, and ease of use, making it a popular choice for managing infrastructure and deploying applications across multiple environments.

12 Marketplace Analytics

Developing a marketplace is a time-consuming process, requiring careful planning in multiple stages. It starts with a defined vision, followed by careful implementation, and then moves on to customer capture and on-going market success. A critical element in achieving and sustaining this success is the analysis of the various marketplace data.

Data analysis plays a key role in this process, turning marketplace's data into valuable information. Through data analysis, businesses can gain a well-rounded understanding of customer behavior and preferences, essential for meeting market demands [48]. Analysis helps to identify trends and patterns and businesses to optimize their strategies, ultimately leading to increased competitiveness and profitability [49]. Machine learning as a subset of data analysis can also help by predicting future trends, customer preferences and market trends and dynamics [50].

To make effective use of the above insights, businesses often rely on dashboards to provide real-time data visualizations and reporting. Dashboards allow businesses to track Key Performance Indicators (KPIs), monitor progress and make quick, informed decisions. By embedding the results of data analytics and machine learning into dashboard, businesses can easily explain complex data [51].

A first prototype that addresses the above has been implemented in the context of the DOME Marketplace [52], a federation of marketplaces in which RAMP is a member. An interactive dashboard that many of its functions can be used in other marketplaces has been developed and is currently under testing under real-world conditions. The dashboard is divided into two sections, the traffic analysis, which holds the data about the marketplace traffic, and the service and sales analysis, which contains data about the services provided and the sales that will take place in the marketplace.

The dashboard offers a variety of visualizations tailored to the specific analysis being conducted. These include bar charts, line charts, pie charts, waterfall charts, parallel and bubble charts, as well as tables, maps, and tree charts. Users can also interact with the visualizations using the filters provided by the dashboard. This allows them to select the time-period to analyze, as well as choose specific marketplace pages, services, and brands for detailed examination.

Traffic analysis (Fig. 15) is important for providing various aspects of market performance, user behavior and potential areas of market improvement. Specifically, the metrics selected are:

- Visits/(unique) visitors to show the popularity of the site and demographics.

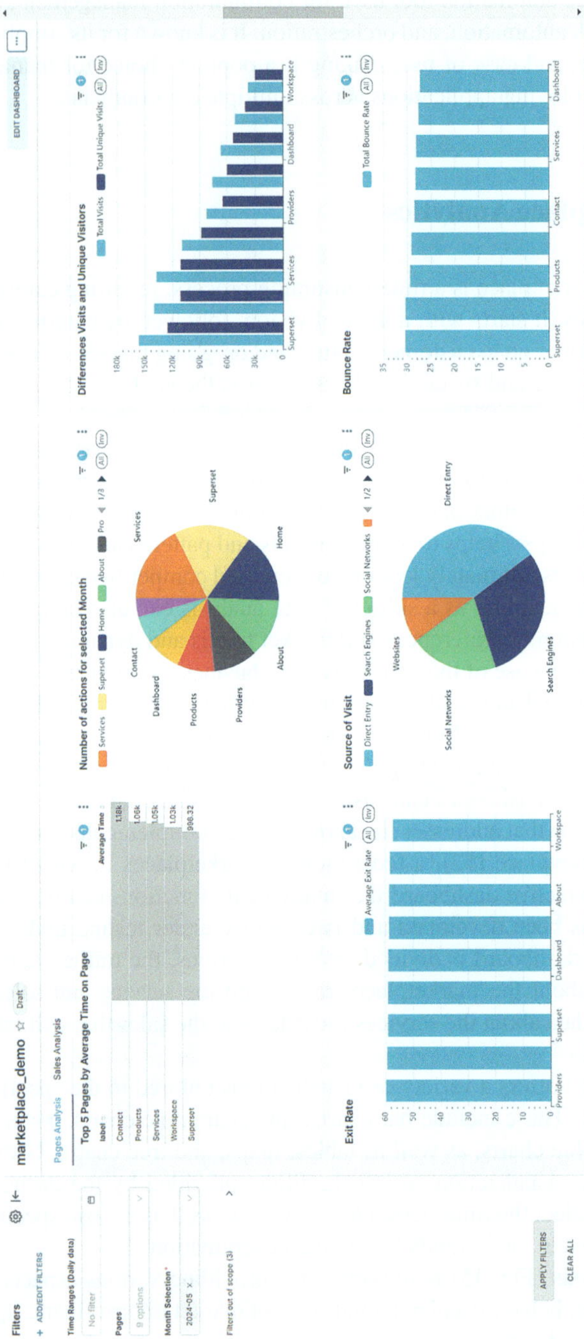

Fig. 15 Traffic analysis presented on the dashboard

For user behavior, which is important for improving experience, engagement and maintaining competitiveness, there are some important metrics, which are:

- Average time on page: Average time spent, in seconds, on a specific page.
- Actions: Number of actions (page views, outlinks, and downloads).
- Bounce rate: Ratio of visits leaving the website after landing a specific page.
- Exit rate: Ratio of visits that do not view any other page after a specific page.
- Also, data is presented for the Source of Visits (search engines, direct entry, etc.) in order to determine which marketing channels drive the most traffic to the site.

The **services and sales analysis** (Fig. 16) is important to understand the effectiveness of the proceedings, the comparison of similar services, the customer behavior and buying patterns. It is important for the marketplace to continuously increase its subscribers and the services it provides, which makes these two some of the most important metrics to analyze. Also, performance tracking is important for any business to identify which services sell well or not. The metrics that help to calculate that are:

- Gross Merchandise Value (GMV): The total value of services sold during a given period of time [53].
- Clustering: Task of grouping a set of objects so that objects in the same group are similar.
- The effectiveness of the marketplace also has a direct link with the customer behavior. Variables that can help identify this are:
  - Churn rate: rate at which a company loses customers or subscribers [54]
  - Conversion rate: percentage of visitors of the marketplace who make a purchase [55]

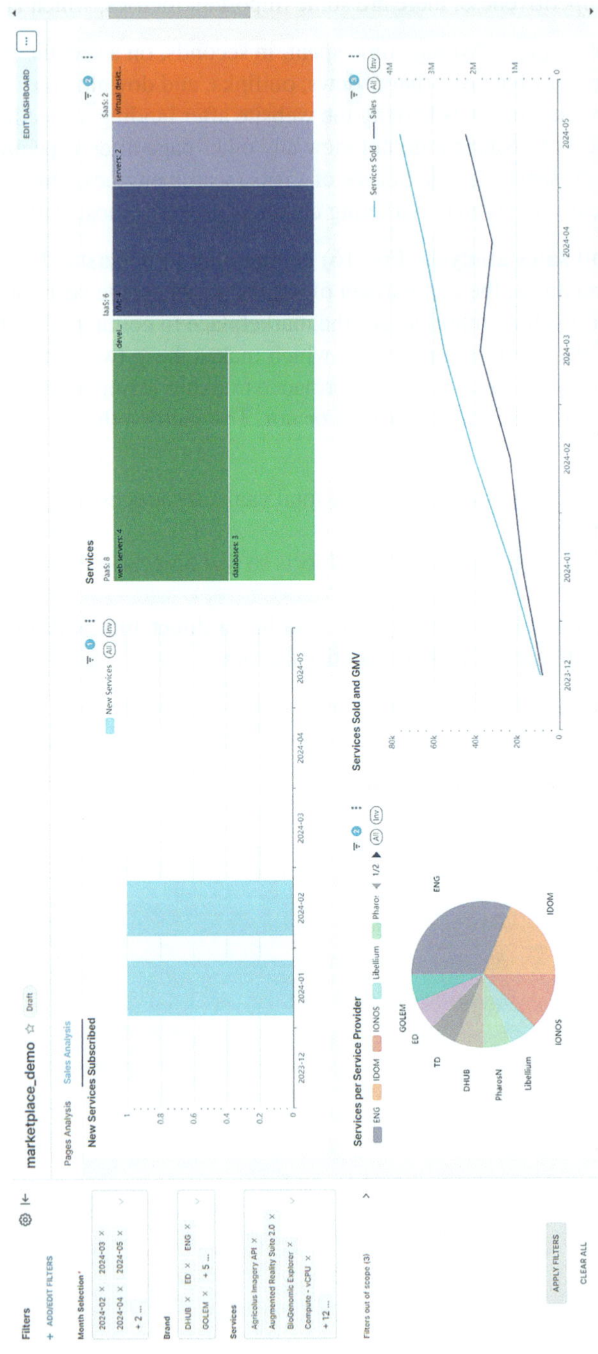

Fig. 16 Services and sales analysis presented on the dashboard

13 Toward Smart Supply Chains

In the immediate next steps, RAMP is adding features for enabling the efficient and management of supply chains, with a focus on sustainable and circular economy. This includes two main strands:

- Vendor onboarding, which allows manufacturers to connect, interact and manage their suppliers, existing and new.
- Supply chain synthesis, which allows different production processes of different entities to be flexibly combined and linked based on material used and produced.
- Distributed material inventory, which allows manufacturers to participate in a plug-n-play fashion through the use of the RAMP IIoT platform, and smart material transactions.

The orchestration of the interaction between a company looking for suppliers and the candidate suppliers, i.e., **vendor onboarding**, is modelled in a series of Business Process Management Notation (BPMN) [56] models, which follow the hierarchical structure of "global" and "local" models. The vendor onboarding distinguishes between trusted and unknown suppliers, with additional risk assessment process for the unknown ones. The manufacturer launches a vendor process in RAMP (Fig. 17), and under a step-wise approach, a total score is calculated for every candidate supply vendor, and a recommendation is given to the user regarding the selection of the specific vendor or a proposal to look for another vendor [57].

For the **supply chain synthesis**, new supply chains by connecting in a cascade fashion existing production processes of different production facilities are modeled. To be able to do that, each manufacturing company describes the production

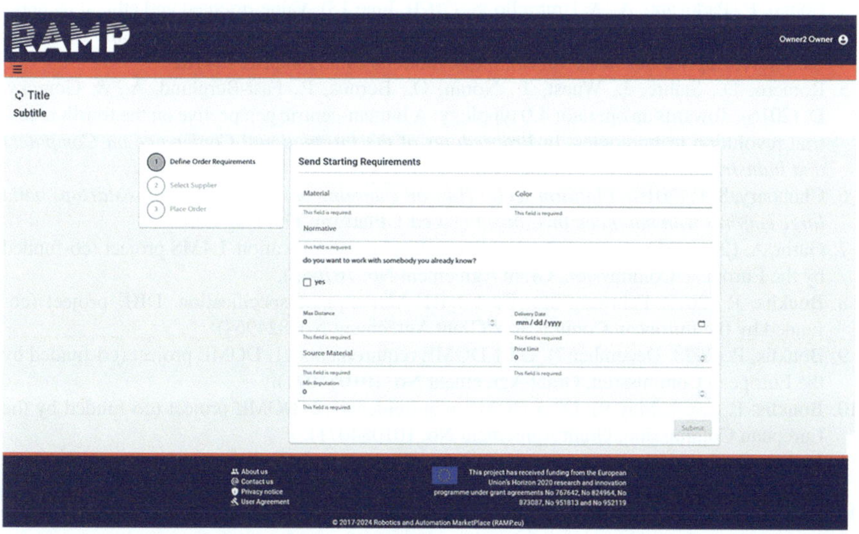

Fig. 17 Launching the vendor onboarding process in RAMP

processes and RAMP then compiles a library of production processes describing at the appropriate level of abstraction, their input, outputs and processing phases. The combined supply chains will be leveraged then for the optimization of material flow, with a special focus on sustainability and circularity [57].

An advanced **distributed material inventory** is also being realized, which allows manufacturers to join an ecosystem that contains information of material that is available or requested, by using the RAMP IIoT platform. By integrating production devices, the data is automatically updated and can be directly discovered by interested partners, while supply chain recommendations are also possible. Additionally, blockchain technologies are planned to be utilized for purposes of traceability and transparency, smart agreements and recycling contracts, digital payments, etc. [58]. The distributed material inventory is seamlessly integrated with the online marketplace, allowing organizations to intuitively manage their inventory and supply chain.

References

1. Bouklis, P., Kolyvas, E., & Garbi, A. (2024, January 12). D3.5 RAMP Marketplace final release. KITT4SME project (co-funded by the European Commission, Grant Agreement No. 952119) (p. 9).
2. Grefen, P., Vanderfeesten, I., Traganos, K., Domagala-Schmidt, Z., & van der Vleuten, J. (2022, January 7). Advancing smart manufacturing in Europe: Experiences from two decades of research and innovation projects. *Machines, 10*(1). https://doi.org/10.3390/machines10010045
3. Ray Ghosh, S. (2023, July 9). The art of networking in the manufacturing sector. Retrieved from https://www.automationmag.com/, https://www.automationmag.com/the-art-of-networking-in-the-manufacturing-sector/
4. Longo, F., Padovano, A., & Umbrello, S. (2020, June 18). Value-oriented and ethical technology engineering in Industry 5.0: A human-centric perspective for the design of the factory of the future. *Applied Sciences, 10*(12). https://doi.org/10.3390/app10124182
5. Romero, D., Stahre, J., Wuest, T., Noran, O., Bernus, P., Fast-Berglund, Å., & Gorecky, D. (2016). Towards an operator 4.0 typology: A human-centric perspective on the fourth industrial revolution technologies. In *Proceedings of the International Conference on Computers and Industrial Engineering, Tianjin, China* (pp. 1–11).
6. Choudary, S. P. (2016). *Platform scale: How an emerging business model helps startups build large empires with minimum investment* (1st ed.). Platform Thinking Labs.
7. Garbi, A. (2018, March 21). D2.1 L4MS marketplace specification. L4MS project (co-funded by the European Commission, Grant Agreement No. 767642).
8. Bouklis, P. (2019, February 26). D1.1 DIH^2 Marketplace specification. DIH^2 project (co-funded by the European Commission, Grant Agreement No. 824964).
9. Bouklis, P. (2023, December 7). D2.1 DOME requirements v1. DOME project (co-funded by the European Commission, Grant Agreement No. 101084071).
10. Bouklis, P. (2024, May 9). D2.5 DOME requirements v2. DOME project (co-funded by the European Commission, Grant Agreement No. 101084071).
11. Muller, G. (2011). Requirements elicitation and selection. In *Systems architecting: A business perspective*. CRC Press.
12. Pohl, K., & Rupp, C. (2015). *Requirements engineering fundamentals*. Rocky Nook.
13. Gottesdeiner, E. (2002). *Requirements by collaboration*. Addison-Wesley.
14. Lauesen, S. (2002). *Software requirements: Styles and techniques*. Addison-Wesley.

15. Leffingwell, D., & Widrig, D. (2000). *Managing software requirements*. Addison-Wesley.
16. Seidman, I. (1998). *Interviewing as qualitative research: A guide for researchers in education and the social sciences* (2nd ed.). Teachers College Press.
17. Warren, C. (2002). Qualitative interviewing. In I. J. Gubrium & J. Holstein (Eds.), *Handbook of interview research: Context and method* (pp. 83–101). Sage.
18. Likert, R. (1932). A technique for the measurement of attitudes. *Archives of Psychology, 140*, 1–55.
19. McGovern, G. (2010). *The stranger's long neck: How to deliver what your customers really want online*. A&C Black.
20. Dick, J., Hull, E., & Jackson, K. (2017). *Requirements engineering* (4th ed.). Springer. https://doi.org/10.1007/978-3-319-61073-3
21. Clegg, D., & Barker, R. (1994). *Case method fast-track: A RAD approach*. Addison-Wesley.
22. Gotel, O., Cleland-Huang, J., Huffman Hayes, J., Zisman, A., Egyed, A., Grünbacher, P., et al. (2012). Traceability fundamentals. In *Software and systems traceability* (pp. 3–22). Springer. https://doi.org/10.1007/978-1-4471-2239-5_1
23. Bouklis, P. (2023, December 5). D6.20 Marketplace functionalities report 3. SHOP4CF project (co-funded by the European Commission, Grant Agreement No. 873087).
24. MQTT. (2024). MQTT: The standard for IoT messaging. Retrieved from https://mqtt.org/
25. ETSI. (2024). Context Information Management (CIM) NGSI-LD. Retrieved from https://www.etsi.org/deliver/etsi_gs/CIM/001_099/009/01.08.01_60/gs_CIM009v010801p.pdf
26. Docker Inc. (2024). Docker: Accelerated container application development. Retrieved from https://www.docker.com/
27. The Linux Foundation. (2024). Helm: The package manager for Kubernetes. Retrieved from https://helm.sh/
28. Bouklis, P. (2022, January 12). D6.19 Marketplace functionalities report 2. SHOP4CF project (co-funded by the European Commission, Grant Agreement No. 873087).
29. Bouklis, P. (2023, June 14). D2.8 SME and artists co-creation space 2.0. Better Factory project (co-funded by the European Commission, Grant Agreement No. 951813).
30. Peppas, G. (2024). D2.14 SME and artists co-creation space 3.0. Better Factory project (co-funded by the European Commission, Grant Agreement No. 951813).
31. Gombakis, K. (2024, July 25). D1.12 RAMP data API 2.0. Better Factory project (co-funded by the European Commission, Grant Agreement No. 951813).
32. FIWARE Foundation. (2024). Orion context broker (with linked data extensions). Retrieved from https://github.com/FIWARE/context.Orion-LD
33. FIWARE Foundation. (2021). Temporal representation of entities. Retrieved from https://github.com/FIWARE/context.Orion-LD/blob/develop/doc/manuals-ld/troe.md
34. Broadcom. (2024). Spring Boot. Retrieved from https://spring.io/projects/spring-boot
35. Google. (2024). Angular. Retrieved from https://angular.dev/
36. The PostgreSQL Global Development Group. (2024). PostgreSQL: The world's most advanced open source relational database. Retrieved from https://www.postgresql.org/
37. ETSI. (2024). Industry Specification Group (ISG) cross cutting Context Information Management (CIM). Retrieved from https://www.etsi.org/committee/cim
38. The Linux Foundation. (2024). Keycloak. Retrieved from https://www.keycloak.org/
39. Kong Inc. (2024). Kong. Retrieved from https://konghq.com/
40. ITU-T. (2021, November). ITU-T Y.4480: Low power protocol for wide area wireless networks. Series Y: Global information infrastructure, internet protocol aspects, next-generation networks, internet of things and smart cities.
41. Kolehmainen, K. (2024, April 18). D1.9 Cyber-security implementation for RAMP 3.0. Better Factory project (co-funded by the European Commission, Grant Agreement No. 951813).
42. Bouklis, P. (2023, March 21). D5.6 Lessons learned v1.0. Better Factory project (co-funded by the European Commission, Grant Agreement No. 951813) (p. 9).
43. Brooke, J. (1996). SUS: A "quick and dirty" usability scale. Usability evaluation in industry.

44. WWW Survey Team. (1998). In P. Schnek (Ed.), *Online privacy and security questionnaire*. Institute for People and Technology, Georgia Tech. Retrieved from https://www.cc.gatech.edu/gvu/user_surveys/survey-1998-10/questions/privacy.html
45. Wittig, A. (2016). *Amazon web services in action*. Manning Press.
46. HashiCorp. (2024). Automate infrastructure on any cloud with Terraform. Retrieved from https://www.terraform.io/
47. Red Hat. (2024). Ansible collaborative. Retrieved from https://www.ansible.com
48. Chen, H., Chiang, R. H., & Storey, V. C. (2012, December). Business intelligence and analytics: From big data to big impact. *MIS Quarterly, 36*(4), 1165–1188. https://doi.org/10.2307/41703503
49. Gandomi, A., & Haider, M. (2015, April). Beyond the hype: Big data concepts, methods, and analytics. *International Journal of Information Management, 35*(2), 137–144. https://doi.org/10.1016/j.ijinfomgt.2014.10.007
50. Jordan, M., & Mitchell, T. (2015, July 17). Machine learning: Trends, perspectives, and prospects. *Science, 349*(6245), 255–260. https://doi.org/10.1126/science.aaa841
51. Sarikaya, A., Correll, M., Bartram, L., Tory, M., & Fisher, D. (2019, January). What do we talk about when we talk about dashboards? *IEEE Transactions on Visualization and Computer Graphics, 25*(1), 682–692. https://doi.org/10.1109/TVCG.2018.2864903
52. Hierro, J. (2023, December 7). D3.1 DOME reference architecture and specifications v1. DOME project (co-funded by the European Commission, Grant Agreement No. 101084071) (pp. 297–310).
53. Predoiu, O. (2024, April 21). *Gross merchandise value—The ultimate measure of your sales volume*. Omniconvert. Retrieved from https://www.omniconvert.com/blog/gross-merchandise-value-gmv/
54. Cheng, M. (2024, March 21). *Churn rate: What it means, examples, and calculations*. Dotdash Meredith Publishing. Retrieved from https://www.investopedia.com/terms/c/churnrate.asp
55. FASTR. (2024). Conversion rate. Zmags Corp. Retrieved from https://getfastr.com/glossary/conversion-rate
56. Object Management Group. (2013, December). Business process model and notation (BPMN).
57. Kolyvas, E. (2024, August 31). D3.1 Supply chain orchestrator. CIRCULOOS project (co-funded by the European Commission, Grant Agreement No. 101092295).
58. Melendez, F. (2024, May 31). D2.2 CMRA specification. CIRCULOOS project (co-funded by the European Commission, Grant Agreement No. 101092295) (pp. 35–36).

Data Analytics and Insights: Leveraging Cloud Resources in Industry 5.0

Neha Singh Raghuvanshi, Kanika Sharma, Ashwani Kumar, Rakesh Joshi, P. P. Praveen Kumar, and Yogesh Kumar Singla

1 Introduction

Industry 5.0 represents a paradigm shift characterized by the seamless integration of physical and digital systems, enabling real-time data capture and analysis across various domains. Unlike its predecessors, Industry 5.0 emphasizes human–machine collaboration, where data analysts play a pivotal role in extracting meaningful insights to enhance operational efficiency, optimize processes, and drive innovation [1, 2]. In Industry 5.0, the upcoming phase of industrial evolution emphasizing human–machine collaboration and advanced data analytics, the utilization of cloud resources becomes indispensable for streamlining operations and enhancing

N. S. Raghuvanshi (✉) · K. Sharma
Department of Applied Sciences, ABES Engineering College, Ghaziabad, Uttar Pradesh, India

A. Kumar
Department of Mechanical Engineering, Technical Education Department Uttar Pradesh, Kanpur, India
e-mail: ashwaniadj.me@geu.ac.in

R. Joshi
DARVUN India, Patiala, Punjab, India

P. P. P. Kumar
Department of Biomedical Engineering, Institute for Quantitative Health Science and Engineering, Michigan State University, East Lansing, MI, USA
e-mail: panangal@msu.edu

Y. Kumar Singla
School of Engineering, Math and Technology, Navajo Technical University, Crownpoint, NM, USA

Department of Mechanical Engineering, University of Idaho, Moscow, ID, USA
e-mail: ysingla@navajotech.edu

Y. Kumar Singla et al. (eds.), *Digital Manufacturing in Industry 5.0*,
https://doi.org/10.1007/978-3-031-91500-0_45

decision-making processes. Cloud computing offers the scalability necessary to manage the vast quantities of data generated in Industry 5.0. By scaling resources up or down as needed, organizations can process and analyse data in real time without being hindered by infrastructure constraints. Cloud platforms also provide extensive storage capacities to accommodate the continuous influx of data from sensors, machinery, and IoT devices in Industry 5.0 settings. Data can be securely stored in the cloud and accessed from any location, facilitating seamless collaboration and information exchange across various departments and sites. Cloud computing platforms grant access to robust computational resources such as CPUs, GPUs, and specialized hardware accelerators, which are vital for executing intricate data analytics tasks like predictive modelling, machine learning, and deep learning. By leveraging cloud-based processing capabilities, organizations can expedite data analysis and derive actionable insights in real time. Cloud computing operates on a pay-as-you-go pricing model, enabling organizations to only pay for the resources they utilize. This eliminates the need for upfront investments in costly hardware and infrastructure, thereby democratizing data analytics for organizations of all sizes. Furthermore, cloud-based analytics solutions often incorporate built-in automation and optimization features that help minimize operational expenses and enhance overall efficiency. Cloud-based data analytics platforms facilitate seamless collaboration and flexibility by granting access to data and analytics tools from any location with internet connectivity. This enables teams to collaborate in real time, share insights, and work on projects irrespective of their physical whereabouts. Moreover, cloud-based analytics solutions seamlessly integrate with other cloud services and third-party applications, affording organizations the flexibility to tailor and expand their analytics capabilities as required. Leveraging cloud resources for data analytics is pivotal for fostering innovation, refining decision-making processes, and capitalizing on new opportunities in Industry 5.0. By harnessing the scalability, storage, processing power, cost-effectiveness, and flexibility offered by the cloud, organizations can unearth actionable insights from their data and remain at the forefront of today's rapidly evolving digital landscape.

2 Understanding Cloud Computing in the Context of Data Analysis

Cloud computing serves as the backbone of modern data analysis, offering scalable infrastructure, on-demand computing power, and a plethora of services tailored to the needs of data analysts [3, 4]. By leveraging cloud resources, analysts gain access to virtually unlimited storage capacity, advanced analytics tools, and AI-driven algorithms, empowering them to tackle complex datasets and extract actionable insights with unprecedented speed and accuracy. Cloud computing has revolutionized the way businesses handle data. The scalability, flexibility, and cost-effectiveness of cloud infrastructure make it an ideal solution for data analytics. This section explores the various cloud resources available and their applications in Industry 5.0. The ability to scale resources based on demand ensures that data analysts have the computational power needed for complex analyses [5]. Cloud platforms offer a variety of tools and services, providing flexibility in choosing the right tools for specific analytical tasks. Pay-as-you-go models allow organizations to optimize costs by only paying for the resources they consume [6, 7].

3 Need for Cloud-Based Solutions

Organizations are increasingly recognizing the necessity for a unified system capable of storing and analysing diverse types of big data to derive real-time insights. Cloud computing offers the agility required to manage and process vast datasets efficiently. Its scalability ensures that organizations can adapt to fluctuating data volumes seamlessly [8, 9]. Moreover, the cost-effectiveness of cloud resources makes them an attractive option for businesses of all sizes. Effective utilization of cloud resources requires a strategic approach and careful consideration of various factors as scalability, security, data integration and cost management. Cloud platforms offer elastic computing capabilities, allowing analysts to scale resources up or down based on demand, ensuring optimal performance and cost-efficiency. Data security is paramount in the age of Industry 5.0. Cloud providers implement robust security measures, including encryption, access controls, and threat detection mechanisms, to safeguard sensitive data against unauthorized access and cyber threats. Integrating disparate data sources from on-premises systems, IoT devices, and third-party applications is a common challenge for data analysts. Cloud-based integration tools streamline this process, enabling seamless data flow and interoperability across the ecosystem [10, 11]. While cloud computing offers flexibility and scalability, it is essential to monitor and optimize costs to avoid overspending. Adopting cost-effective storage solutions, leveraging serverless computing, and implementing resource utilization metrics are effective strategies for cost management [12, 13].

3.1 Cloud-Based Conceptual Model

The proposed cloud-based conceptual model facilitates the seamless integration of big data analytics and knowledge management, traditionally considered separate domains. It highlights the pivotal role of advanced data analytics, machine learning, and artificial intelligence in addressing industrial challenges across various sectors. It includes certain steps as discussed in Fig. 1:

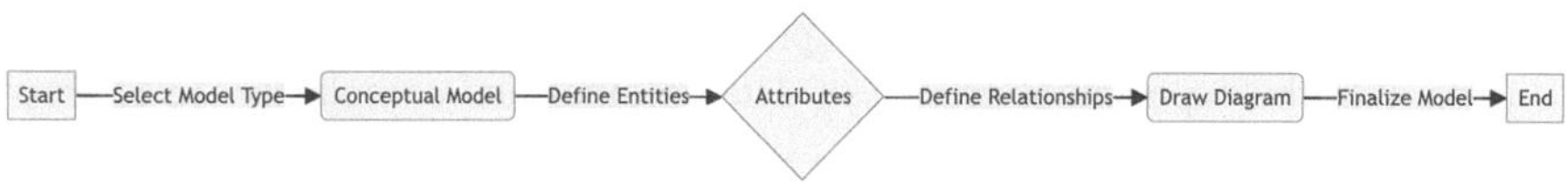

Fig. 1 Cloud-based conceptual model

3.1.1 Data Collection and Storage

In the dynamic landscape of Industry 5.0, data emerges as the cornerstone of innovation and operational efficiency. Yet, organizations face formidable challenges in managing the vast influx of data from diverse sources. To address this, they rely on sophisticated architectures known as data lakes and warehouses. Picture data lakes as expansive reservoirs where organizations accumulate diverse forms of raw data at scale. Unlike conventional databases, data lakes accommodate structured, semi-structured, and unstructured data without the need for predefined schemas. For instance, companies like Netflix utilize Amazon S3 to store user interactions, video streams, and operational logs, enabling them to analyse viewer behaviour and personalize recommendations effectively [14, 15]. While data lakes serve as repositories for raw data, data warehouses specialize in organizing and optimizing data for analytical purposes. These warehouses act as centralized hubs where organizations aggregate, structure, and transform data into actionable insights. Take Snowflake, a cloud-based data warehousing platform used by companies like DoorDash, to analyse delivery routes, optimize driver assignments, and enhance customer experiences based on real-time data streams.

3.1.2 Data Security and Compliance

Amidst escalating concerns over data privacy and regulatory compliance, organizations prioritize robust security measures and adherence to industry standards. Cloud providers employ cutting-edge encryption techniques to safeguard data integrity and confidentiality. By encrypting data both at rest and in transit, organizations ensure that sensitive information remains shielded from unauthorized access or tampering. For instance, Microsoft Azure offers Transparent Data Encryption (TDE), encrypting data at the storage level to mitigate risks associated with data breaches [16, 17]. Granular access controls empower organizations to define and enforce fine-grained permissions, restricting data access to authorized personnel only. Role-based access control (RBAC) mechanisms enable organizations to assign specific roles and privileges to users based on their responsibilities and job functions. Google Cloud Identity and Access Management (IAM) exemplifies this, allowing organizations to manage access controls across diverse cloud services and resources seamlessly.

3.1.3 Data Processing and Analysis

- **Parallel processing**: In the realm of data analytics, speed and scalability are paramount. Traditional single-node processing architectures often struggle to cope with the exponential growth of data volumes and complexity. To address this, organizations harness the power of parallel processing, a distributed computing paradigm that accelerates data processing tasks across multiple nodes.

- **Distributed computing**: Parallel processing frameworks like Apache Hadoop and Apache Spark distribute data processing tasks across a cluster of interconnected nodes, enabling organizations to process vast datasets in parallel. By harnessing the collective computational power of multiple nodes, organizations can expedite data processing tasks and derive actionable insights in real time [18]. For example, Twitter leverages Apache Storm, a real-time stream processing framework, to analyse and visualize millions of tweets per second, enabling users to discover trending topics and insights in real time.
- **Elastic scalability**: Cloud environments offer elastic scalability, allowing organizations to scale compute resources up or down based on workload demands dynamically. Whether it's processing petabytes of data or running computationally intensive machine learning algorithms, cloud-based parallel processing ensures optimal resource utilization and cost efficiency [19]. Google Cloud Dataproc, a managed Spark and Hadoop service, exemplifies this, enabling organizations to provision and scale clusters seamlessly based on workload requirements, without the overhead of infrastructure management.

3.2 Machine Learning and AI Model Concept

As data becomes increasingly abundant, organizations turn to machine learning and artificial intelligence to extract valuable insights and drive decision-making processes. Figure 2 shows the machine learning and AI model concept.

3.2.1 Predictive Analytics

Machine learning algorithms analyse historical data patterns to forecast future trends and outcomes accurately. By identifying correlations and patterns within datasets, organizations can anticipate customer preferences, optimize inventory management, and mitigate operational risks proactively [20]. For instance, Amazon Forecast uses machine learning algorithms to predict demand for products, helping retailers optimize inventory levels and minimize stockouts.

3.2.2 Natural Language Processing (NLP)

With the proliferation of unstructured text data from sources like social media, customer feedback, and support tickets, natural language processing (NLP) algorithms enable organizations to extract actionable insights and sentiment analysis. By analysing text data for sentiment, entities, and key phrases, organizations gain valuable insights into customer feedback, market trends, and brand perception [21]. IBM

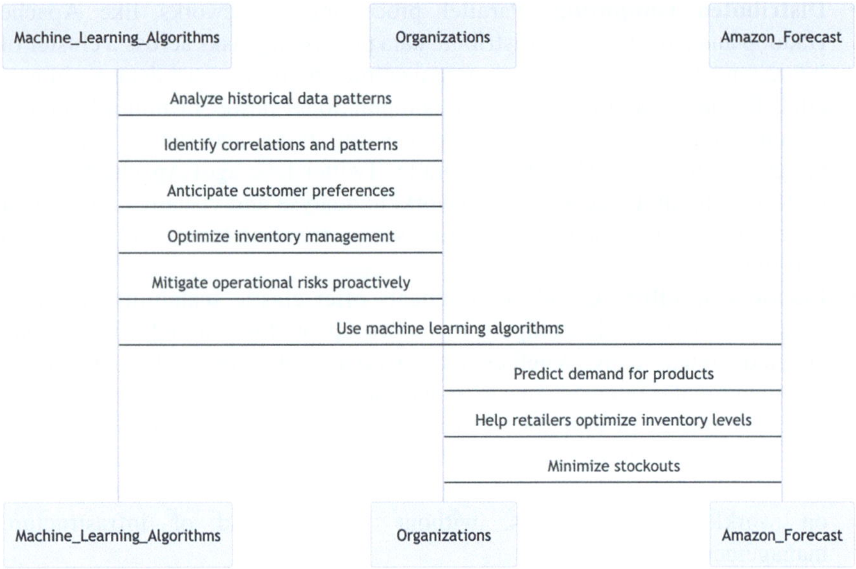

Fig. 2 Machine learning and AI model concept

Watson Natural Language Understanding, for example, analyses text data to identify sentiment, emotions, and entities, enabling organizations to gauge customer satisfaction and sentiment effectively.

4 Advanced Analytics Techniques

In the dynamic landscape of Industry 5.0, data analysts rely on advanced analytics techniques to derive actionable insights and drive strategic decision-making (Fig. 3). These techniques encompass predictive modelling, prescriptive analytics, time series analysis, and anomaly detection, enabling organizations to anticipate trends, optimize processes, and mitigate risks effectively.

4.1 Predictive Modelling

Predictive modelling leverages historical data to forecast future outcomes and trends accurately. By identifying patterns and relationships within datasets, organizations can anticipate customer behaviour, predict market demand, and optimize resource allocation [22].

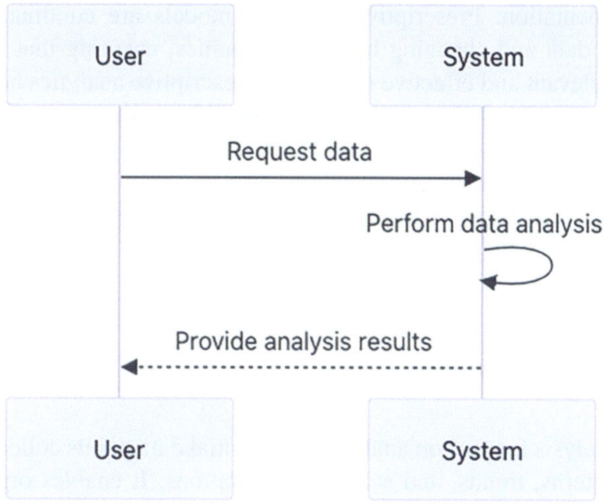

Fig. 3 Advanced analytics techniques

4.2 *Key Components of Predictive Modelling*

The first step in predictive modelling involves collecting, cleaning, and pre-processing data to ensure accuracy and consistency. Data analysts choose suitable predictive modelling techniques such as regression analysis, decision trees, or neural networks based on the nature of the problem and the available data. Analysts evaluate the performance of predictive models using metrics like accuracy, precision, recall, and F1-score to assess their effectiveness in making accurate predictions. By analysing customer interactions and historical behaviour, organizations can predict the likelihood of customer churn and implement proactive retention strategies. Retailers and manufacturers use predictive modelling to forecast product demand, optimize inventory levels, and minimize stockouts. Financial institutions employ predictive modelling techniques to detect fraudulent activities and prevent monetary losses.

4.3 *Prescriptive Analytics*

Prescriptive analytics goes beyond predictive modelling by recommending optimal courses of action to achieve desired outcomes. By combining predictive insights with business rules and optimization algorithms, organizations can make informed decisions in real time [23]. Prescriptive analytics models consider various constraints and objectives to recommend the best course of action, considering factors such as resource constraints, cost, and risk. Organizations use what-if scenarios to simulate different outcomes and evaluate the potential impact of various decisions before implementation. Prescriptive analytics models are continuously updated based on new data and changing business dynamics, ensuring that recommendations remain relevant and effective over time. Prescriptive analytics helps organizations optimize supply chain operations by identifying the most efficient routes, minimizing transportation costs, and mitigating supply chain disruptions. Governments and healthcare organizations use prescriptive analytics to allocate resources effectively during emergencies, such as allocating hospital beds and medical supplies during pandemics.

4.4 Time Series Analysis

Time series analysis focuses on analysing sequential data points collected over time to identify patterns, trends, and seasonal fluctuations. It enables organizations to make informed forecasts and strategic decisions based on historical data trends. Time series analysis helps identify long-term trends and patterns in data, enabling organizations to make strategic decisions and forecast future developments. Analysts identify recurring patterns and seasonal fluctuations within time series data, allowing organizations to adjust strategies and resource allocation accordingly [24].

Time series analysis employs forecasting techniques such as moving averages, exponential smoothing, and autoregressive integrated moving average (ARIMA) models to predict future values based on historical trends. Banks and financial institutions use time series analysis to forecast stock prices, interest rates, and exchange rates, enabling investors to make informed investment decisions. Retailers leverage time series analysis to forecast product demand, optimize inventory levels, and plan promotional campaigns based on historical sales data.

4.5 Anomaly Detection

Anomaly detection involves identifying unusual patterns or outliers within datasets that deviate from expected behaviour. By detecting anomalies early, organizations can mitigate risks, prevent fraud, and maintain operational integrity [25]. Anomaly detection algorithms analyse data patterns and identify instances that deviate significantly from the norm. Anomaly detection often relies on unsupervised learning techniques such as clustering and density estimation to detect anomalies in unlabelled datasets. Analysts define thresholds or boundaries based on historical data distributions to identify anomalies that fall outside acceptable ranges. Financial institutions use anomaly detection to detect fraudulent transactions, identify unusual spending patterns, and prevent unauthorized access to accounts. Anomaly detection helps identify suspicious network activities, detect malware infections, and prevent data breaches by identifying unusual patterns in network traffic.

5 Real-Time Data Processing

In the era of Industry 5.0, organizations are increasingly relying on real-time data processing to gain immediate insights, respond to events promptly, and optimize decision-making processes (Fig. 4). Real-time data processing involves the ingestion, analysis, and interpretation of data streams as they are generated, enabling organizations to act swiftly and proactively in dynamic environments.

5.1 Stream Processing Technologies

Stream processing technologies enable organizations to process and analyse continuous streams of data in real time, facilitating rapid decision-making and actionable insights. These technologies leverage distributed computing frameworks and event-driven architectures to process data streams efficiently. Stream processing systems are built on event-driven architectures, where data is processed and analysed as events occur in real-time. This approach enables organizations to react to events instantaneously and derive insights without delay. Stream processing applications are often implemented as microservices, allowing organizations to modularize and scale individual components independently. Microservices architecture enhances flexibility, resilience, and scalability, enabling organizations to adapt to evolving business requirements seamlessly. Stream processing technologies incorporate fault-tolerant mechanisms to ensure reliability and data integrity. Techniques such as checkpointing, stateful processing, and data replication enable systems to recover gracefully from failures and ensure continuous operation.

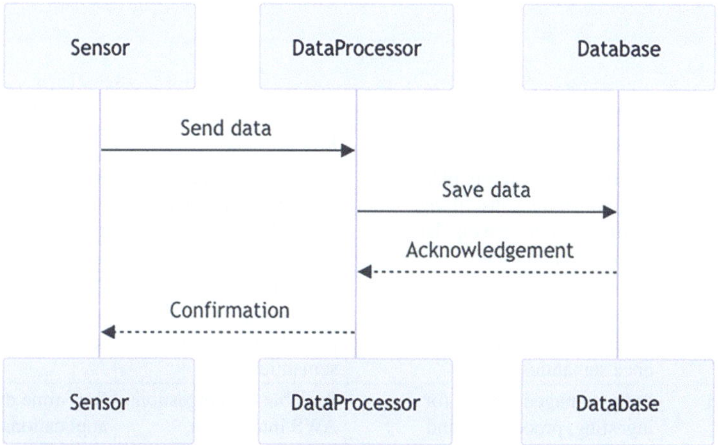

Fig. 4 Real-time data processing

5.1.1 Popular Stream Processing Technologies

1. *Apache Kafka*: Apache Kafka is a distributed streaming platform that enables organizations to publish, subscribe to, store, and process streams of data in real time. It provides high throughput, fault tolerance, and horizontal scalability, making it ideal for building real-time data pipelines and event-driven applications (Table 1).
2. *Apache Flink*: Apache Flink is a powerful stream processing framework that supports event-time processing, stateful computations, and exactly once semantics. Flink's sophisticated windowing and processing capabilities make it well-suited for complex event processing and real-time analytics use cases.
3. *Amazon Kinesis*: Amazon Kinesis is a fully managed streaming service that enables organizations to ingest, process, and analyse real-time data streams at scale. With support for multiple data sources and integration with AWS services, Kinesis simplifies the development of real-time data applications on the cloud.

5.2 Use Cases in Industry 5.0

Real-time data processing finds diverse applications across various industries in the context of Industry 5.0. From predictive maintenance in manufacturing to personalized customer experiences in retail, real-time data processing drives innovation, efficiency, and competitiveness.

Predictive maintenance: In the manufacturing sector, real-time data processing enables organizations to monitor equipment health, detect anomalies, and predict potential failures before they occur. By analysing sensor data in real time, organizations can schedule maintenance activities proactively, minimize downtime, and optimize asset performance.

Table 1 Popular stream processing technologies

Technology	Description	Key features	Use cases
Apache Kafka	Distributed streaming platform for publishing, subscribing, storing, and processing data streams in real time	High throughput, fault tolerance, horizontal scalability	Real-time data pipelines, event-driven applications
Apache Flink	Stream processing framework supporting event-time processing, stateful computations, and exactly once semantics	Sophisticated windowing, stateful processing, exactly once semantics	Complex event processing, real-time analytics
Amazon Kinesis	Fully managed service for ingesting, processing, and analysing real-time data streams at scale, integrated with AWS	Scalable data ingestion, AWS integration, managed service	Real-time data applications, cloud-based analytics

Supply chain optimization: Real-time data processing plays a crucial role in optimizing supply chain operations, ensuring efficient inventory management, and minimizing disruptions. By monitoring inventory levels, transportation routes, and demand fluctuations in real time, organizations can streamline logistics, reduce costs, and enhance customer satisfaction.

Personalized marketing: In the retail industry, real-time data processing enables organizations to deliver personalized marketing experiences and tailored recommendations to customers. By analysing customer interactions, purchase history, and browsing behaviour in real time, organizations can segment audiences, target specific demographics, and deliver relevant content across various channels.

5.3 Lambda Architecture

Lambda architecture is a robust framework for building scalable, fault-tolerant, and real-time data processing systems. It combines batch processing and stream processing capabilities to handle large volumes of data efficiently while ensuring low-latency processing and real-time insights [26].

5.3.1 Key Components of Lambda Architecture

The batch layer is responsible for processing large volumes of historical data in offline mode. Batch processing frameworks like Apache Hadoop and Apache Spark enable organizations to perform complex analytics, generate reports, and update data stores periodically (Fig. 5). The speed layer processes real-time data streams in memory and provides low-latency responses to queries and requests. Stream processing technologies like Apache Kafka and Apache Flink enable

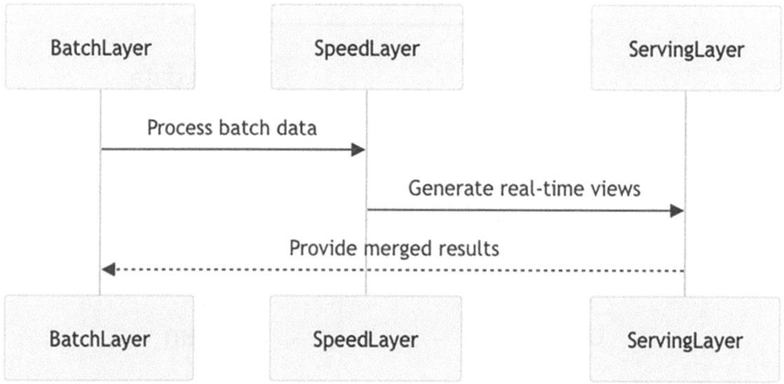

Fig. 5 Components of lambda architecture

organizations to ingest, process, and analyse data streams in real time, enabling immediate insights and responses. The serving layer aggregates results from the batch layer and the speed layer, providing a unified view of data to end-users and applications. Serving layers utilize distributed databases, caching mechanisms, and indexing techniques to facilitate fast and efficient data retrieval. Lambda architecture scales horizontally to accommodate growing data volumes and user concurrency, ensuring high availability and performance under varying workloads. Lambda architecture incorporates fault-tolerant mechanisms at each layer to ensure resilience and data integrity. By replicating data and processing tasks across multiple nodes, lambda architecture mitigates the risk of data loss and system failures.

Lambda architecture supports a wide range of data processing and analytics tasks, including batch processing, stream processing, and interactive querying. Organizations can choose the most suitable processing paradigm based on their requirements and use cases.

6 Data Visualization and Reporting

In the realm of Industry 5.0, effective data visualization and reporting are essential components for transforming complex datasets into comprehensible insights (Fig. 6). Data analysts leverage interactive dashboards, data storytelling techniques, and visualization best practices to communicate trends, patterns, and actionable information to stakeholders [5].

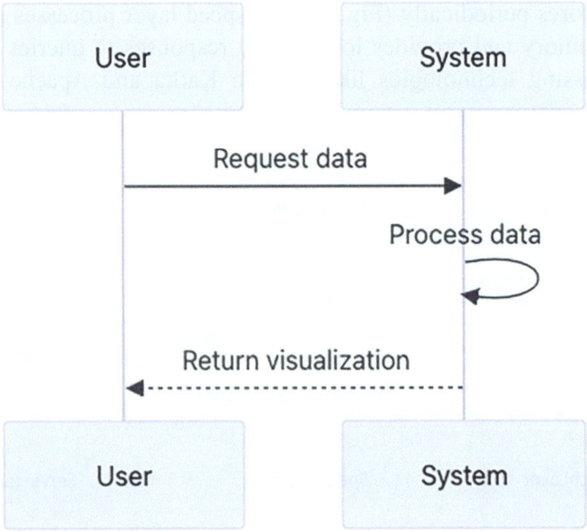

Fig. 6 Data visualization and reporting

6.1 Interactive Dashboards

Interactive dashboards serve as dynamic, user-friendly interfaces that enable stakeholders to explore and interact with data visually. These dashboards provide real-time insights, allowing users to drill down into specific metrics, filter data based on parameters, and gain a comprehensive understanding of key performance indicators (KPIs).

Interactive dashboards allow users to interact with visual elements, such as charts and graphs, enabling them to customize views, filter data, and explore specific aspects of the dataset. Dashboards often feature real-time data updates, ensuring that users have access to the latest information and can make informed decisions based on the most current data. Modern interactive dashboards are designed to be accessible across various devices, including desktops, tablets, and smartphones, providing flexibility for users to access critical information anytime, anywhere.

6.1.1 Popular Interactive Dashboard Tools

1. Tableau: Tableau is a powerful data visualization tool that allows users to create interactive dashboards with drag-and-drop functionality. It supports a wide range of data sources and provides robust visualization options for effective data exploration [27].
2. Power BI: Microsoft Power BI enables users to build interactive dashboards and reports, offering seamless integration with Microsoft Excel and other Microsoft products. Power BI supports natural language queries and AI-powered insights for a user-friendly experience (Table 2).

Table 2 Popular interactive dashboard tools

Tool	Description	Key features	Integration	Use cases
Tableau	A data visualization tool enabling users to create interactive dashboards using drag-and-drop functionality. Supports diverse data sources	Drag-and-drop interface, robust visualizations	Connects with databases, spread sheets, cloud sources	Data exploration, business intelligence
Power BI	Microsoft tool for building interactive dashboards and reports. Offers AI-driven insights and natural language queries for ease of use	AI-powered insights, natural language queries	Seamless with Microsoft Excel and other MS products	Reporting, real-time analytics

6.2 *Data Storytelling*

Data storytelling involves the art of presenting data in a narrative format to convey insights, trends, and the significance of findings effectively. By weaving a compelling story around the data, analysts can enhance understanding, engage stakeholders, and drive decision-making [28].

6.2.1 Components of Effective Data Storytelling

Data storytelling is a critical skill in today's data-driven world, where insights must be communicated clearly and compellingly to drive decision-making. Effective data storytelling involves more than presenting raw data; it requires weaving a narrative that connects data insights to business goals, engages stakeholders, and prompts action. The following components—context setting, narrative flow, and visual support—are essential for crafting impactful data stories.

Context Setting

The foundation of any data story lies in establishing a clear context that defines its relevance and purpose. Without context, data may appear abstract or irrelevant to stakeholders. Context setting involves:

1. **Defining the problem or question**: The story should start by articulating the specific problem, question, or objective that the analysis seeks to address. This ensures stakeholders understand the purpose of the data and what it aims to solve or reveal.
2. **Providing background information**: Offer relevant background details, such as historical trends, industry benchmarks, or organizational goals, to frame the analysis. This helps the audience appreciate the significance of the insights.
3. **Clarifying stakeholder relevance**: Explain why the data matters to the stakeholders. Tailoring the story to the audience's needs ensures greater engagement and alignment with their interests.

For example, if presenting sales data, the context might include an overview of market conditions, competitive pressures, or customer behaviour trends that impact sales performance. Setting this context provides a roadmap for the audience, aligning them with the story's direction.

Narrative Flow

A well-structured narrative is crucial for guiding the audience through the data in a logical and engaging manner. Data storytelling follows a storytelling arc, similar to traditional storytelling, to build interest and lead to actionable conclusions. Key elements of narrative flow include:

1. **Logical sequencing**: Organize insights in a coherent order, starting with foundational facts and gradually progressing to more complex insights. This helps the audience follow the story without confusion.
2. **Building engagement**: Use storytelling techniques such as introducing conflict (e.g., challenges or opportunities) to build tension and keep the audience invested. The resolution should offer actionable insights or solutions derived from the data.
3. **Connecting insights**: Each insight should connect logically to the next, creating a seamless flow. Highlighting relationships between data points and their implications ensures that stakeholders grasp the bigger picture.

For instance, a data story about customer churn might begin by outlining current retention rates, followed by identifying factors contributing to churn, and concluding with recommendations for improving retention strategies. This progression keeps the audience engaged and leads to actionable takeaways.

Visual Support

Visual aids play a pivotal role in making data stories more impactful and easier to understand. Well-designed visuals can transform complex data into digestible insights, enhancing audience comprehension and retention. Effective visual support involves:

1. **Choosing the right visuals**: Different types of data require different visualizations. Use bar charts for comparisons, line graphs for trends, pie charts for proportions, and scatter plots for correlations. Infographics can summarize complex insights in a visually appealing way.
2. **Highlighting key points**: Use visual cues such as colour coding, annotations, or callouts to draw attention to critical insights. This ensures that the audience focuses on the most important elements.
3. **Ensuring clarity and simplicity**: Avoid clutter and keep visuals clean and straightforward. Labels, legends, and axes should be clearly defined, and the visual should be easy to interpret at a glance.

For example, when presenting financial performance data, a combination of line charts for revenue trends and bar graphs for expense breakdowns can provide a comprehensive view. Infographics summarizing key performance indicators (KPIs) can further enhance understanding. Effective data storytelling combines context setting, a compelling narrative flow, and supportive visuals to communicate insights clearly and persuasively. By grounding the story in context, structuring it logically, and enhancing it with visuals, data storytellers can engage stakeholders, foster understanding, and drive data-informed decision-making. When these components are thoughtfully integrated, data stories not only inform but also inspire action, transforming raw data into a powerful tool for organizational success.

6.3 Visualization Best Practices

Effective data visualization involves adhering to best practices that enhance clarity, accuracy, and interpretability. By selecting appropriate chart types, employing colour coding judiciously, and ensuring consistency, analysts can create visualizations that convey information accurately [29].

6.3.1 Key Visualization Best Practices

Select chart types that best represent the data and facilitate easy comprehension. For instance, use line charts for trends over time, bar charts for comparisons, and pie charts for proportions. Maintain consistency in colour usage across visualizations. Use colour strategically to highlight key elements, but avoid overwhelming the viewer with too many colours. Clearly label axes, data points, and provide context through annotations. Proper labelling enhances interpretability and ensures that viewers can understand the information without confusion.

6.3.2 Tools and Resources for Visualization

- **D3.js**: D3.js is a JavaScript library that facilitates the creation of dynamic, interactive visualizations in web browsers. It provides a robust set of tools for data-driven visualizations, giving analysts the flexibility to design custom visualizations (Fig. 7).

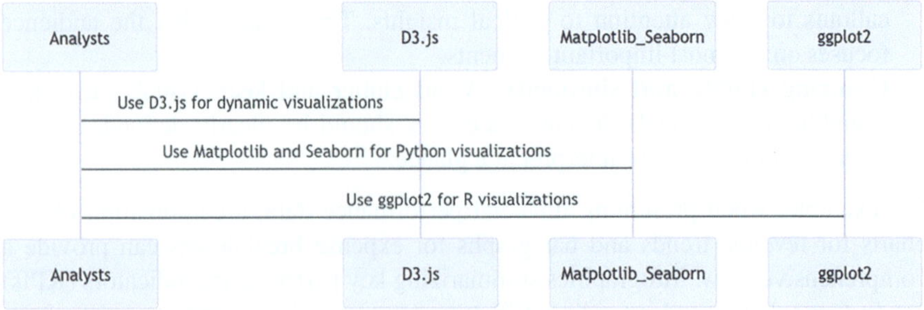

Fig. 7 Tools and resources for visualization

- **Matplotlib and Seaborn (Python)**: For analysts working with Python, Matplotlib and Seaborn are powerful libraries for creating static and interactive visualizations. They offer a wide range of plot types and customization options.
- **ggplot2 (R)**: In the R programming language, ggplot2 is a popular package for creating visually appealing and customizable visualizations. It follows the Grammar of Graphics principles, allowing analysts to build complex visualizations with ease.

7 Data Governance and Quality Assurance

Data governance and quality assurance are essential components of effective data management practices in Industry 5.0 (Fig. 8). They ensure that data is managed, protected, and utilized in a manner that aligns with organizational objectives and regulatory requirements [30, 31].

7.1 *Data Governance Frameworks*

Data governance frameworks define the policies, procedures, and responsibilities for managing data assets across an organization. They establish guidelines for data ownership, access controls, data lifecycle management, and data stewardship. Common data governance frameworks include COBIT (Control Objectives for Information and Related Technologies), DAMA (Data Management Association)

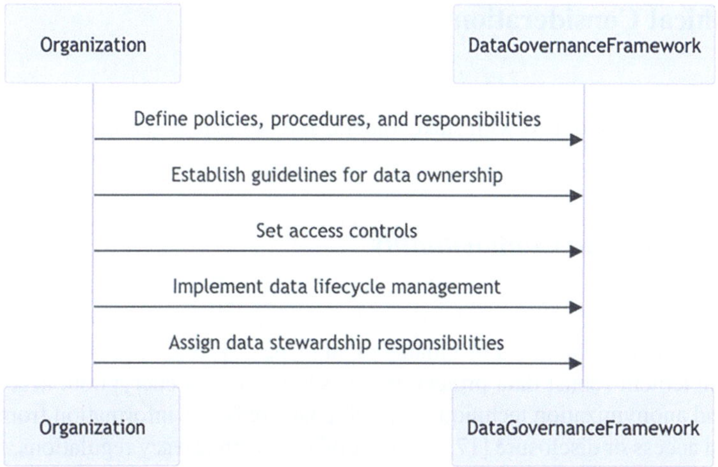

Fig. 8 Data governance and quality assurance

DMBOK (Data Management Body of Knowledge), and GDPR (General Data Protection Regulation).

7.2 *Data Quality Management*

Data quality management encompasses processes and methodologies for ensuring that data is accurate, consistent, and reliable. It involves data profiling, cleansing, validation, and monitoring to identify and rectify errors, duplicates, and inconsistencies in data sets. Data quality management frameworks focus on improving data accuracy, completeness, timeliness, and relevance, thereby enhancing the reliability and trustworthiness of organizational data assets.

7.3 *Compliance and Regulation*

Compliance and regulation refer to adherence to legal, industry, and internal standards governing data privacy, security, and confidentiality. Organizations must comply with regulations such as GDPR, HIPAA (Health Insurance Portability and Accountability Act), CCPA (California Consumer Privacy Act), and industry-specific standards to protect sensitive data, mitigate risks, and maintain trust with stakeholders. Compliance and regulation frameworks incorporate policies, controls, audits, and documentation to ensure that data handling practices meet regulatory requirements and industry best practices.

8 Ethical Considerations in Data Analytics

Ethical considerations are paramount in data analytics to ensure responsible and fair use of data, maintain trust with stakeholders, and mitigate potential risks (Fig. 9).

8.1 *Privacy and Confidentiality*

Privacy and confidentiality involve protecting individuals' personal information and ensuring that data is handled in a manner that respects privacy rights. Organizations must implement robust data protection measures, such as encryption, access controls, and anonymization techniques, to safeguard sensitive information from unauthorized access or disclosure [17, 32]. Compliance with privacy regulations, such as GDPR and CCPA, is essential to maintain transparency and trust with data subjects regarding the collection, processing, and storage of their data.

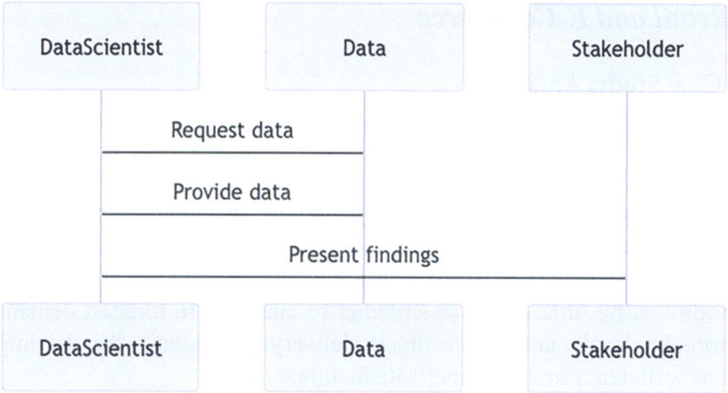

Fig. 9 Ethical considerations in data analytics

8.2 *Bias and Fairness*

Bias and fairness refer to the impartiality and equity in data analysis and decision-making processes. Data analysts must be vigilant against inherent biases in data sets and algorithms that may result in unfair outcomes or discrimination against certain groups. Mitigating bias involves diverse representation in data collection, algorithm design, and model evaluation to ensure that decisions are fair and unbiased [33, 34]. Transparency in the data analysis process and regular audits of algorithms can help identify and address biases effectively.

8.3 *Transparency and Accountability*

Transparency and accountability entail openness and responsibility in data analytics practices, including data collection, analysis, and decision-making. Organizations should communicate transparently with stakeholders about the purposes, methods, and implications of data analytics initiatives, including the use of algorithms and predictive models. Accountability requires organizations to take ownership of their actions and outcomes, including addressing errors, rectifying biases, and implementing corrective measures to uphold ethical standards. Establishing clear policies, ethical guidelines, and oversight mechanisms promotes transparency and accountability in data analytics processes.

9 Case Studies and Success Stories

Case studies and success stories highlight how data analytics has transformed operations, improved decision-making, and driven innovation across different industries.

9.1 *Retail and E-Commerce*

9.1.1 Case Study: Amazon

Amazon, the global e-commerce giant, leverages data analytics to personalize customer experiences, optimize supply chain management, and drive sales growth. Through its recommendation engine powered by machine learning algorithms, Amazon analyses customer browsing and purchase history to offer personalized product recommendations, enhancing customer engagement and driving conversion rates. Additionally, Amazon utilizes predictive analytics to forecast demand, optimize inventory levels, and ensure timely delivery of products, thereby improving operational efficiency and customer satisfaction.

9.1.2 Success Story: Walmart

Walmart, one of the world's largest retailers, harnesses data analytics to optimize pricing strategies, streamline inventory management, and enhance the in-store shopping experience. By analysing sales data, customer demographics, and market trends, Walmart identifies product demand patterns, adjusts pricing dynamically, and optimizes product placement to maximize sales and minimize stockouts. Moreover, Walmart utilizes data analytics to improve supply chain visibility, reduce lead times, and mitigate risks, ensuring seamless operations across its global network of stores [35–37].

9.2 *Healthcare and Life Sciences*

9.2.1 Case Study: IBM Watson Health

IBM Watson Health applies data analytics and artificial intelligence to revolutionize healthcare delivery, enhance patient outcomes, and drive medical research. Through its cognitive computing platform, Watson Health analyses vast volumes of medical literature, patient records, and genomic data to provide personalized treatment recommendations, accelerate clinical trials, and facilitate early disease detection. By leveraging predictive analytics and machine learning, Watson Health enables healthcare providers to identify high-risk patients, optimize treatment plans, and improve population health management, ultimately transforming the healthcare landscape [38, 39].

9.2.2 Success Story: Mayo Clinic

Mayo Clinic, a renowned healthcare institution, embraces data analytics to deliver personalized patient care, enhance clinical decision-making, and drive medical innovation. By integrating electronic health records (EHRs), genomic data, and medical imaging results, Mayo Clinic generates comprehensive patient profiles, enabling physicians to make data-driven treatment decisions and tailor therapies to individual patient needs. Additionally, Mayo Clinic utilizes predictive analytics to forecast patient outcomes, identify potential complications, and optimize resource allocation, leading to improved patient outcomes, reduced healthcare costs, and enhanced patient satisfaction.

9.3 Manufacturing and Supply Chain

9.3.1 Case Study: Toyota

Toyota, a leading automotive manufacturer, utilizes data analytics to optimize production processes, improve product quality, and enhance supply chain efficiency. Through its Toyota Production System (TPS), Toyota collects real-time data on production metrics, equipment performance, and inventory levels to identify bottlenecks, minimize waste, and streamline production workflows. By leveraging predictive analytics and machine learning, Toyota anticipates demand fluctuations, adjusts production schedules dynamically, and coordinates logistics operations to ensure on-time delivery of vehicles to customers, thereby enhancing customer satisfaction and driving competitive advantage [40–44].

9.3.2 Success Story: UPS

UPS, a global logistics and supply chain company, leverages data analytics to optimize package delivery routes, reduce transportation costs, and enhance delivery efficiency. Through its proprietary routing algorithms and telematics systems, UPS analyses vast amounts of data on package volumes, delivery locations, and traffic conditions to optimize delivery routes, minimize fuel consumption, and reduce carbon emissions. Moreover, UPS utilizes predictive analytics to forecast package volumes, allocate resources effectively, and mitigate risks, ensuring reliable and cost-effective delivery services for customers worldwide.

10 Data Analytics Lifecycle in Industrial Settings

The model outlines the data analytics lifecycle as applied to industrial settings, emphasizing its iterative nature. From data collection and pre-processing to analysis, interpretation, and visualization, each stage plays a crucial role in extracting actionable insights.

10.1 Tools and Technologies for Cloud-Based Data Analysis

A list of tools and technologies that empower data analysts to harness the full potential of cloud resources:

1. **Cloud storage**: Platforms such as Amazon S3, Google Cloud Storage, and Microsoft Azure Blob Storage provide durable and scalable storage solutions for storing and managing large volumes of data.
2. **Data warehousing**: Services like Amazon Redshift, Google Big-Query, and Snowflake offer high-performance data warehousing capabilities, enabling analysts to query and analyse massive datasets in real time.
3. **Analytics and business intelligence (BI) tools**: Tools like Google Data Studio, Microsoft Power BI, and Tableau facilitate interactive data visualization, dashboarding, and reporting, enabling stakeholders to gain actionable insights from complex data.
4. **Machine learning and AI services**: Cloud providers offer a plethora of machine learning and AI services, including Amazon Sage Maker, Google AI Platform, and Azure Machine Learning, empowering analysts to build and deploy predictive models, anomaly detection algorithms, and natural language processing (NLP) solutions.

11 Some Real-World Applications of Cloud-Based Data Analysis

1. **Predictive maintenance in manufacturing**: By analysing sensor data from industrial equipment stored in the cloud, manufacturers can predict equipment failures before they occur, minimizing downtime and optimizing maintenance schedules (Table 3).
2. **Personalized marketing campaigns**: E-commerce companies leverage cloud-based analytics platforms to analyse customer behaviour, preferences, and purchase history, enabling them to deliver personalized marketing campaigns and recommendations in real time.
3. **Healthcare analytics**: Healthcare providers harness cloud resources to analyse patient data, electronic health records (EHRs), and medical imaging studies, facilitating clinical decision support, disease diagnosis, and treatment optimization.

Table 3 Real-world applications of cloud-based data analysis

Application	Description	Benefits	Industry
Predictive maintenance in manufacturing	Analysing sensor data from industrial equipment stored in the cloud to forecast equipment failures	Minimizes downtime, optimizes maintenance schedules	Manufacturing
Personalized marketing campaigns	Using cloud-based analytics to evaluate customer behaviour, preferences, and purchase history for tailored marketing efforts	Real-time personalization, improved customer engagement	E-commerce
Healthcare analytics	Analysing patient data, EHRs, and medical imaging through cloud resources to support clinical decisions and optimize treatment	Enhances diagnosis, improves treatment outcomes	Healthcare

12 Challenges in Implementation

Despite the transformative potential of Cloud-based data analytics, several challenges persist in its implementation. Data security and privacy concerns, regulatory compliance, and the integration of legacy systems pose significant hurdles. Moreover, the shortage of skilled data analysts and the need for continuous training exacerbate the complexity of adoption [45–47]. In Industry 5.0, AI-based technologies are helpful in study of core mechanical engineering areas like design, manufacturing, production, fracture analysis, and materials characterizations [48–53].

13 Conclusion

In conclusion, the integration of machine analytics, expert insights, and relevant data sources is paramount for informed decision-making and operational enhancement across industries. By harnessing the power of cloud resources, organizations can unlock new opportunities for innovation, efficiency, and sustainability in the era of Industry 5.0. As technology continues to evolve, embracing cloud-based data analytics remains imperative for staying competitive in a rapidly changing landscape.

In this chapter, we explored the intersection of data analysis, cloud computing, and industrial insights within the context of Industry 5.0. From the conceptual model to real-world case studies, the journey towards leveraging cloud resources for enhanced knowledge management is both enlightening and indispensable in today's digital age.

References

1. Alam, A., Ullah, I., & Lee, Y.-K. (2020). Video big data analytics in the cloud: A reference architecture, survey, opportunities, and open research issues. *IEEE Access, 8*, 152377–152422. https://doi.org/10.1109/ACCESS.2020.3017135
2. Verma, A., et al. (2022). Blockchain for Industry 5.0: Vision, opportunities, key enablers, and future directions. *IEEE Access, 10*, 69160–69199. https://doi.org/10.1109/ACCESS.2022.3186892
3. Harerimana, G., Jang, B., Kim, J. W., & Park, H. K. (2018). Health big data analytics: A technology survey. *IEEE Access, 6*, 65661–65678. https://doi.org/10.1109/ACCESS.2018.2878254
4. Sarker, S., Arefin, M. S., Kowsher, M., Bhuiyan, T., Dhar, P. K., & Kwon, O.-J. (2023). A comprehensive review on big data for industries: Challenges and opportunities. *IEEE Access, 11*, 744–769. https://doi.org/10.1109/ACCESS.2022.3232526
5. Chhikara, D., Rana, S., Singh, G., Mishra, D., & Kumar, N. (2023). Blockchain-based partial group key agreement protocol for intelligent transportation systems. *IEEE Transactions on Vehicular Technology*. https://doi.org/10.1109/TVT.2023.3299705
6. Gao, P., Han, Z., & Wan, F. (2020). Big data processing and application research. In *2020 2nd International Conference on Artificial Intelligence and Advanced Manufacture (AIAM), Manchester, UK* (pp. 125–128). https://doi.org/10.1109/AIAM50918.2020.00031
7. Garg, T., & Khullar, S. (2020). Big data analytics: Applications, challenges & future directions. In *2020 8th International Conference on Reliability, Infocom Technologies and Optimization (Trends and Future Directions) (ICRITO), Noida, India* (pp. 923–928). https://doi.org/10.1109/ICRITO48877.2020.9197797
8. Mittal, S., & Sangwan, O. P. (2019). Big data analytics using machine learning techniques. In *2019 9th International Conference on Cloud Computing, Data Science & Engineering (Confluence), Noida, India* (pp. 203–207). https://doi.org/10.1109/CONFLUENCE.2019.8776614
9. Sagiroglu, S., & Sinanc, D. (2013). Big data: A review. In *2013 International Conference on Collaboration Technologies and Systems (CTS), San Diego, CA, USA* (pp. 42–47). https://doi.org/10.1109/CTS.2013.6567202
10. Pavithra, N., & Manasa, C. M. (2021). Big data analytics tools: A comparative study. In *2021 IEEE International Conference on Computation System and Information Technology for Sustainable Solutions (CSITSS), Bangalore, India* (pp. 1–6). https://doi.org/10.1109/CSITSS54238.2021.9683711
11. Ambigavathi, M., & Sridharan, D. (2018). Big data analytics in healthcare. In *2018 10th International Conference on Advanced Computing (ICoAC), Chennai, India* (pp. 269–276). https://doi.org/10.1109/ICoAC44903.2018.8939061
12. Mohamed, N., & Al-Jaroodi, J. (2014). Real-time big data analytics: Applications and challenges. In *2014 International Conference on High Performance Computing & Simulation (HPCS), Bologna, Italy* (pp. 305–310). https://doi.org/10.1109/HPCSim.2014.6903700
13. Manikandan, S. G., & Ravi, S. (2014). Big data analysis using Apache Hadoop. In *2014 International Conference on IT Convergence and Security (ICITCS), Beijing, China* (pp. 1–4). https://doi.org/10.1109/ICITCS.2014.7021746
14. Praveena, M. D. A., & Bharathi, B. (2017). A survey paper on big data analytics. In *2017 International Conference on Information Communication and Embedded Systems (ICICES), Chennai, India* (pp. 1–9). https://doi.org/10.1109/ICICES.2017.8070723
15. Yu, S., & Yang, D. (2016). The role of big data analysis in new product development. In *2016 International Conference on Network and Information Systems for Computers (ICNISC), Wuhan, China* (pp. 279–283). https://doi.org/10.1109/ICNISC.2016.068
16. Abdullah, D. B., & Mohammed, R. A.-G. (2021). Real-time big data analytics perspective on applications, frameworks and challenges. In *2021 7th International Conference on Contemporary Information Technology and Mathematics (ICCITM), Mosul, Iraq* (pp. 1–6). https://doi.org/10.1109/ICCITM53167.2021.9677849

17. Swain, S., Bhushan, B., Dhiman, G., & Viriyasitavat, W. (2022). Appositeness of optimized and reliable machine learning for healthcare: A survey. *Archives of Computational Methods in Engineering, 29*(6), 3981–4003. https://doi.org/10.1007/s11831-022-09733-8
18. Singh, J., Wazid, M., Das, A. K., Chamola, V., & Guizani, M. (2022). Machine learning security attacks and defense approaches for emerging cyber physical applications: A comprehensive survey. *Computer Communications, 192*, 316–331. https://doi.org/10.1016/j.comcom.2022.06.012
19. Shankar, A. (2023). Efficient data interpretation and artificial intelligence enabled IoT based smart sensing system. *Artificial Intelligence Review, 56*(12), 15053–15077. https://doi.org/10.1007/s10462-023-10519-y
20. Wazid, M., Das, A. K., Chamola, V., & Park, Y. (2022). Uniting cyber security and machine learning: Advantages, challenges and future research. *ICT Express, 8*(3), 313–321. https://doi.org/10.1016/j.icte.2022.04.007
21. Barnawi, A., Gaba, S., Alphy, A., Jabbari, A., Budhiraja, I., Kumar, V., & Kumar, N. (2023). A systematic analysis of deep learning methods and potential attacks in internet-of-things surfaces. *Neural Computing and Applications, 35*(25), 18293–18308. https://doi.org/10.1007/s00521-023-08634-6
22. Devi, S., Garg, H., & Garg, D. (2023). A review of redundancy allocation problem for two decades: bibliometrics and future directions. *Artificial Intelligence Review, 56*(8), 7457–7548. https://doi.org/10.1007/s10462-022-10363-6
23. Pande, A., Manchanda, M., Bhat, H. R., Bairy, P. S., Kumar, N., & Gahtori, P. (2022). Molecular insights into a mechanism of resveratrol action using hybrid computational docking/CoMFA and machine learning approach. *Journal of Biomolecular Structure and Dynamics, 40*(18), 8286–8300. https://doi.org/10.1080/07391102.2021.1910572
24. Kumar, H. S. S., & Karibasappa, K. (2022). An effective hybrid deep learning with adaptive search and rescue for brain tumor detection. *Multimedia Tools and Applications, 81*(13), 17669–17701. https://doi.org/10.1007/s11042-022-12474-2
25. Singh, T., Aherwar, A., Ranakoti, L., Bhandari, P., Singh, V., & Lendvai, L. (2023). Performance optimization of lignocellulosic fiber-reinforced brake friction composite materials using an integrated CRITIC-CODAS-based decision-making approach. *Sustainability, 15*(11), 8880. https://doi.org/10.3390/su15118880
26. Jayaraman, R., Alshehri, M., Kumar, M., Abugabah, A., Samant, S. S., & Mohamed, A. A. (2023). Secure biomedical document protection framework to ensure privacy through blockchain. *Big Data, 11*(6), 437–451. https://doi.org/10.1089/big.2022.0170
27. Wazid, M., Bera, B., Das, A. K., Mohanty, S. P., & Jo, M. (2022). Fortifying smart transportation security through public blockchain. *IEEE Internet of Things Journal, 9*(17), 16532–16545. https://doi.org/10.1109/JIOT.2022.3150842
28. Verma, R. P., Pandey, K. N., & Mittal, G. (2024). Genetic-neural optimization approach for gas metal arc welding of dissimilar aluminium alloys of AA5083-O/AA6061-T6. *International Journal of Lightweight Materials and Manufacture, 7*(1), 214–220. https://doi.org/10.1016/j.ijlmm.2023.07.001
29. Balan, G., Arumugam, S., Muthusamy, S., Panchal, H., Kotb, H., Bajaj, M., et al. (2022). An improved deep learning-based technique for driver detection and driver assistance in electric vehicles with better performance. *International Transactions on Electrical Energy Systems, 2022*(1), 8548172. https://doi.org/10.1155/2022/8548172
30. Molu, R. J. J., Tripathi, B., Mbasso, W. F., Naoussi, S. R. D., Bajaj, M., Wira, P., & Misak, S. (2024). Advancing short-term solar irradiance forecasting accuracy through a hybrid deep learning approach with Bayesian optimization. *Results in Engineering, 23*, 102461. https://doi.org/10.1016/j.rineng.2024.102461
31. Sharma, R., Wazid, M., & Gope, P. (2021). A blockchain based secure communication framework for community interaction. *Journal of Information Security and Applications, 58*, 102790. https://doi.org/10.1016/j.jisa.2021.102790

32. Gupta, A., Ghanshala, K., & Joshi, R. C. (2021). Machine learning classifier approach with gaussian process, ensemble boosted trees, SVM, and linear regression for 5g signal coverage mapping. *International Journal of Interactive Multimedia and Artificial Intelligence, 6*(6), 156–163. https://doi.org/10.9781/ijimai.2021.03.004
33. Tomar, V., Kumar, N., & Srivastava, A. R. (2023). Single sample face recognition using deep learning: A survey. *Artificial Intelligence Review, 56*(Suppl 1), 1063–1111. https://doi.org/10.1007/s10462-023-10551-y
34. Ali, S., Khan, M. S., Shah, H., Garg, H., & Alsheddy, A. (2023). BISGA: Recalculating the entire boolean-valued information system from aggregates using a genetic algorithm. *Complexity, 2023*(1), 1539563. https://doi.org/10.1155/2023/1539563
35. Nallathambi, I., Savaram, P., Sengan, S., Alharbi, M., Alshathri, S., Bajaj, M., et al. (2023). Impact of fireworks industry safety measures and prevention management system on human error mitigation using a machine learning approach. *Sensors, 23*(9), 4365. https://doi.org/10.3390/s23094365
36. Shahare, Y., Singh, M. P., Singh, P., Diwakar, M., Singh, V., Kadry, S., & Sevcik, L. (2023). A comprehensive analysis of machine learning-based assessment and prediction of soil enzyme activity. *Agriculture, 13*(7), 1323. https://doi.org/10.3390/agriculture13071323
37. Singh, A. R., Kumar, R. S., Bajaj, M., Khadse, C. B., & Zaitsev, I. (2024). Machine learning-based energy management and power forecasting in grid-connected microgrids with multiple distributed energy sources. *Scientific Reports, 14*(1), 19207. https://doi.org/10.1038/s41598-024-70336-3
38. Ojha, N., Girach, I., Sharma, K., Sharma, A., Singh, N., & Gunthe, S. S. (2021). Exploring the potential of machine learning for simulations of urban ozone variability. *Scientific Reports, 11*(1), 22513. https://doi.org/10.1038/s41598-021-01824-z
39. Sharma, R., Kim, M., & Gupta, A. (2022). Motor imagery classification in brain-machine interface with machine learning algorithms: Classical approach to multi-layer perceptron model. *Biomedical Signal Processing and Control, 71*, 103101. https://doi.org/10.1016/j.bspc.2021.103101
40. Tajjour, S., Chandel, S. S., Alotaibi, M. A., Malik, H., Márquez, F. P. G., & Afthanorhan, A. (2023). Short-term solar irradiance forecasting using deep learning techniques: a comprehensive case study. *IEEE Access, 11*, 119851–119861. https://doi.org/10.1109/ACCESS.2023.3325292
41. Pundir, S., Obaidat, M. S., Wazid, M., Das, A. K., Singh, D. P., & Rodrigues, J. J. (2023). MADP-IIME: Malware attack detection protocol in IoT-enabled industrial multimedia environment using machine learning approach. *Multimedia Systems, 29*(3), 1785–1797. https://doi.org/10.1007/s00530-020-00743-9
42. Dhiman, G., & Nagar, A. K. (2022). Blockchain-based 6G and industrial internet of things systems for industry 4.0/5.0. *Expert Systems, 39*(10), e13162. https://doi.org/10.1111/exsy.13162
43. Mishra, R., Ramesh, D., Bellavista, P., & Edla, D. R. (2023). Redactable blockchain-assisted secure data aggregation scheme for fog-enabled internet-of-farming-things. *IEEE Transactions on Network and Service Management*. https://doi.org/10.1109/TNSM.2023.3322442
44. Rani, S., Babbar, H., Srivastava, G., Gadekallu, T. R., & Dhiman, G. (2022). Security framework for internet-of-things-based software-defined networks using blockchain. *IEEE Internet of Things Journal, 10*(7), 6074–6081. https://doi.org/10.1016/j.comcom.2023.04.008
45. Thapliyal, S., Wazid, M., Singh, D. P., Das, A. K., Shetty, S., & Alqahtani, A. (2023). Design of robust blockchain-envisioned authenticated key management mechanism for smart healthcare applications. *IEEE Access*. https://doi.org/10.1109/ACCESS.2023.3310264
46. Punyavathi, R., Pandian, A., Singh, A. R., Bajaj, M., Tuka, M. B., & Blazek, V. (2024). Sustainable power management in light electric vehicles with hybrid energy storage and machine learning control. *Scientific Reports, 14*(1), 5661. https://doi.org/10.1038/s41598-024-55988-5
47. Davuluri, S. K., Alvi, S. A. M., Aeri, M., Agarwal, A., Serajuddin, M., & Hasan, Z. (2023, April). A security model for perceptive 5G-powered BC IoT associated deep learning. In *2023*

International Conference on Inventive Computation Technologies (ICICT) (pp. 118–125). IEEE. https://doi.org/10.1109/TII.2020.3048791
48. Kumar, A., Singla, Y. K., & Maughan, M. (2024). *Fracture behavior of nanocomposites and reinforced laminate structures*. Springer Nature. https://doi.org/10.1007/978-3-031-68694-8
49. Singla, Y. K., Kumar, A., & Maughan, M. R. (2024). Chapter 1: Introduction to mechanical and fracture behavior characterization of nanocomposites and reinforced laminated structures. In A. Kumar, Y. K. Singla, & M. R. Maughan (Eds.), *Fracture behavior of nanocomposites and reinforced laminate structures* (pp. 1–14). Springer. https://doi.org/10.1007/978-3-031-68694-8_1
50. Biswas, P. K., Maughan, M. R., Kumar, A., & Singla, Y. K. (2024). Chapter 12: Analyzing fractures in nanomaterial-enhanced carbon fiber-reinforced polymer (CFRP) composites. In A. Kumar, Y. Kumar Singla, & M. R. Maughan (Eds.), *Fracture behavior of nanocomposites and reinforced laminate structures* (pp. 295–327). Springer. https://doi.org/10.1007/978-3-031-68694-8_12
51. Kumar, A., Singla, Y. K., Biswas, P., Heidari, M., & Thangavel, S. (2024). Chapter 17: Resurrection structure: New generation of bio-inspired nanocomposites and laminates. In A. Kumar, Y. Kumar Singla, & M. R. Maughan (Eds.), *Fracture behavior of nanocomposites and reinforced laminate structures* (pp. 427–440). Springer. https://doi.org/10.1007/978-3-031-68694-8_17
52. Heidari, M., Khashehchi, M., Thangavel, S., Rahmanivahid, P., Kumar, A., & Singla, Y. K. (2024). Chapter 18: Laminated structures and fracture mechanics: A comprehensive study of mode I, mode II, and mixed mode III behavior. In A. Kumar, Y. Kumar Singla, & M. R. Maughan (Eds.), *Fracture behavior of nanocomposites and reinforced laminate structures* (pp. 441–454). Springer. https://doi.org/10.1007/978-3-031-68694-8_18
53. Khashehchi, M., Heidari, M., Thangavel, S., Rahmanivahid, P., Kumar, A., & Singla, Y. K. (2024). Chapter 20: Insights into aerospace structural integrity: A study on fiber/epoxy composites fracture. In A. Kumar, Y. Kumar Singla, & M. R. Maughan (Eds.), *Fracture behavior of nanocomposites and reinforced laminate structures* (pp. 471–485). Springer. https://doi.org/10.1007/978-3-031-68694-8_20

Smart Supply Chains

Blockchain and the Green Frontier: A Bibliometric and Systematic Review of Supply Chain Sustainability

Ritu Rani, Nidhi Tanwar, Mohit Tyagi, Naveen Kumar, Ashwani Kumar, and Yogesh Kumar Singla

1 Introduction

As global supply chains become more complex and uncertain, members are facing more risks, which make it harder for them to respond quickly and accurately to emergencies, increasing the chance of breakdowns [1]. The complex nature of global supply chains requires new technologies to effectively manage the costs,

R. Rani (✉) · N. Tanwar
Centre of Management and Humanities, Punjab Engineering College (Deemed to be University), Chandigarh, India
e-mail: nidhitanwar@pec.edu.in

M. Tyagi
Department of Production and Industrial Engineering, Punjab Engineering College (Deemed to be University), Chandigarh, India

N. Kumar
Department of Mechanical and Aerospace Engineering, University of Petroleum and Energy Studies, Dehradun, Uttarakhand, India
e-mail: naveen.kumar1@ddn.upes.ac.in

A. Kumar
Department of Mechanical Engineering, Technical Education Department Uttar Pradesh, Kanpur, India

Department of Mechanical Engineering, Graphic Era Deemed to be University, Dehradun, Uttarakhand, India
e-mail: ashwaniadj.me@geu.ac.in

Y. Kumar Singla
School of Engineering, Math and Technology, Navajo Technical University, Crownpoint, NM, USA

Department of Mechanical Engineering, University of Idaho, Moscow, ID, USA
e-mail: ysingla@navajotech.edu

Y. Kumar Singla et al. (eds.), *Digital Manufacturing in Industry 5.0*,
https://doi.org/10.1007/978-3-031-91500-0_46

time, and risks related with sustainability processes. One of the component of Industry 4.0, blockchain technology can enhance firms' capabilities in efficiently managing, controlling, and provisioning exchange of information, products, and financial resources [2].

The integration of blockchain technology is expanding to boost resilience and sustainability within supply chains. Surveys on blockchain adoption reveal that more than 60% of companies have either already implemented or are in the process of implementing this technology [3]. The adoption of a point-to-point network facilitates enhanced transparency and accuracy in the interactions among supply chain members, leading to improved coordination and efficiency within the supply network [4]. Additionally, blockchain technology is considered a good way to manage credibility risks associated with agents due to its ability to provide durability, transparency and traceability. As a result, an increasing number of businesses are incorporating blockchain into their supply chains to boost the sustainability issues transparency [5]. In their pursuit of greater supply chain transparency, companies such as Alibaba, Tencent, Inspur, and IBM are adopted blockchain technology.

However, blockchain technology is still nascent stage [5]. Several studies have explored its application in Supply Chain Management (SCM) across various industries, including the food sector [6]; pharmaceutical industry [7]. Some researchers also studies blockchain application in specific fields of supply chain knowledge, such as supply chain finance [8]; risk management [9]. However, these reviews are based on different article databases, follow different search strategy, and use different research concept.

To fill the above gaps, the current study sought to delve deeply into Sustainable Supply Chain practices by conducting a bibliometric analysis of research articles available from Scopus database from 2018 to 2024 by using various keywords in search strategy, i.e., TITLE-ABS-KEY ("blockchain" OR "block chain" OR "distributed ledger") AND TITLE-ABS-KEY ("sustainable supply chain" OR "green supply chain" OR "sustainability supply chain"). This study investigates the subsequent research questions:

RQ1: How are research publications evolving from 2018 to 2024?
RQ2: What are the key authors, countries, and sources in this domain?
RQ3: What are the possible research trends and themes in this area?

The paper is divided into several organized sections. Section 2 details the methodology and screening process. Section 3 highlights the bibliometric analysis results followed by Sect. 4, which covers the discussion, future research opportunities and limitations and Sect. 5 provides a conclusion to the study.

2 Research Methodology

This research employed the Preferred Reporting Items for Systematic Reviews and Meta-Analyses (PRISMA) framework shown in Fig. 1.

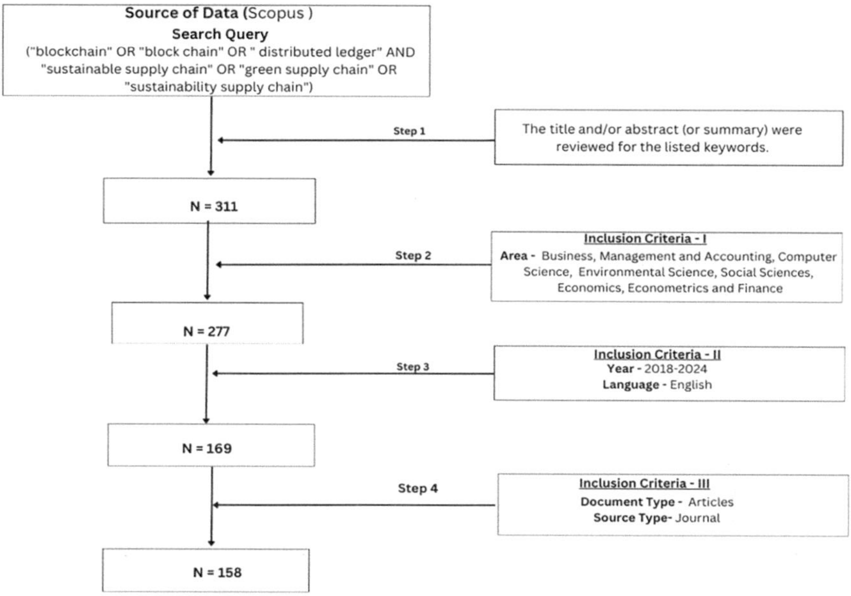

Fig. 1 Steps of methodological approach (PRISMA framework). (Source: Author's compilation)

Step 1: Identification

In the first phase, initial search in Scopus was conducted by using a combination of keywords as such: TITLE-ABS-KEY ("blockchain" OR "block chain" OR "distributed ledger") AND TITLE-ABS-KEY ("sustainable supply chain" OR "green supply chain" OR "sustainability supply chain"). The initial search resulted identified 311 articles. During this phase, titles and abstracts are carefully examined to ensure they align with the research focus on blockchain integration in Sustainable SCM.

Step 2: Screening and filtering

In the second step, the initial results were further refined.

1. In first inclusion criteria only articles from relevant fields, such as Business, Management and Accounting, Computer Science, Environmental Science, Social Sciences, and Economics, Econometrics, and Finance, was included, reducing the number to 277.
2. After that dataset was further refined by including only articles published from 2018 to 2024 in English language which narrowed to 169. Subsequently, the final set was refined to 158 by retaining only peer-reviewed research articles from academic journals.

Step 3: Included

In total, 158 articles were selected for the final review sample after the inclusion and exclusion criteria were assessed.

3 Findings

3.1 Descriptive Statistics

Table 1 reveals a robust expansion in the research landscape from 2018 to 2024, encompassing 158 documents sourced from 82 different sources including journals. The average age of the documents is 1.3 years, indicating that most research is recent and continuously developing. Furthermore, the documents are well-cited, with an average of 17.64 citations each, highlighting their impact and importance within the academic community.

3.2 Yearly Publications

Figure 2 shows a clear upward in the annual publication of articles from 2018 to 2024. In 2018, the number of publications was one article, increasing slightly to three in 2019. A notable rise occurred in 2020, reaching 15 articles, followed by a steady growth to 22 in 2021 and 34 in 2022. The peak occurred in 2023 with 48 publications, suggests a growing interest in the topic with an increasing attention and engagement from the academic community.

3.3 Most Productive Journals

Table 2 shows the number of articles published across various journals, along with their research focuses. *Sustainability (Switzerland)* leads with 18 articles (11.3%), focuses on Environmental and Social Sustainability. *Business Strategy and the Environment* follows with 9 articles (5.62%) focused on Sustainable Business Practices and Environmental Management. Further, *Computers and Industrial Engineering* and the *International Journal of Production Research* both have 6 articles each (3.8%). Moreover, *Annals of Operations Research*, *Journal of Cleaner Production*, and *Transportation Research Part E: Logistics and Transportation*

Table 1 Data metrics summary

Description	Results
Time span	2018:2024
Sources (journals, books, etc.)	82
Documents	158
Annual growth rate (%)	80.86
Document average age	1.66
Average citations per doc	60.8

Source: Biblioshiny

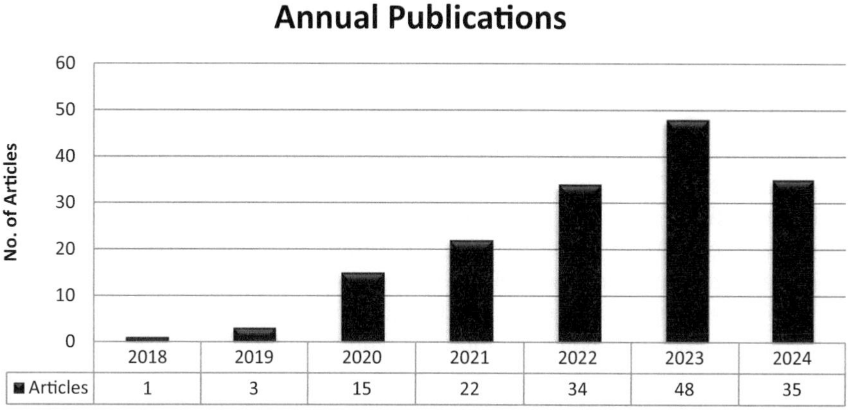

Fig. 2 Yearly publication. (Source: Author's compilation)

Table 2 Most productive journals

Journal name	No. of articles and percentage ($N = 66$)	Research focus
Sustainability (Switzerland)	18 (11.3)	Environmental and Social Sustainability
Business Strategy and the Environment	9 (5.62)	Sustainable Business Practices and Environmental Management
Computers and Industrial Engineering	6 (3.8)	Management Science and Industrial Engineering
International Journal of Production Research	6 (3.8)	Manufacturing and Operation Management
Annals of Operations Research	5 (3.2)	Operation Research
Journal of Cleaner Production	5 (3.2)	Engineering, Environmental
Transportation Research Part E: Logistics and Transportation Review	5 (3.2)	Logistics and Transportation Research
Industrial Management and Data Systems	4 (2.53)	Operations Management and Information System
International Journal of Production Economics	4 (2.53)	Engineering and Management
Management of Environmental Quality: An International Journal	4 (2.53)	Environment

Source: Biblioshiny

Review each contribute 5 articles (3.2%), covering Operations Research, Engineering, and Logistics respectively. Lastly, *Industrial Management and Data Systems*, *International Journal of Production Economics*, and *Management of Environmental Quality: An International Journal* each publish 4 articles (2.53%). This distribution reflects a varied academic focus on sustainability, management, and operational research.

3.4 Country-Wise Analysis

Table 3 provides a detailed analysis of total citations (TC) and average article citations by country. The United States topped the list with 3750 citations and an impressive average of 468.80 citations per article, highlighting its significant influence in the field. China ranks second with 1354 total citations and a lower average of 30.80 citations per article. The United Kingdom follows with 970 citations and an average of 107.80, demonstrating its strong research output and impact. Moreover, India, Italy, and Sweden also show notable contributions, with India having 740 citations and an average of 35.20, Italy with 496 citations and an average of 70.90, and Sweden achieving 267 citations with a high average of 267.00. Lastly, New Zealand, Switzerland, Korea, and France have lower total citations 204, 202, 195, and 121, respectively. This distribution underscores a varied global impact in the field, with the United States leading in total citations and several countries demonstrating high average citations per article.

3.5 Most Influential Articles

The review of highly cited articles (Table 4) on blockchain in SCM reveals that prominent research has investigated how blockchain enhances transparency and sustainability. The top-cited work by Saberi et al. [10] with 2067 citations in the *International Journal of Production Research* explores blockchain's role in sustainable SCM. Kouhizadeh et al. [5] examine adoption barriers in their study published in the *International Journal of Production Economics*, with 680 citations. Other influential studies include Cole et al. [11] and Esmaeilian et al. [12] which focus on blockchain's implications in operations and its association with the adoption of Industry 4.0 technologies, with 584 and 483 citations respectively. The study by Kouhizadeh and Sarkis [13] "Blockchain Practices, Potentials, and Perspectives in

Table 3 Most influential countries

Rank	Country	TC	Average article citations
1	USA	3750	468.80
2	China	1354	30.80
3	United Kingdom	970	107.80
4	India	740	35.20
5	Italy	496	70.90
6	Sweden	267	267.00
7	New Zealand	204	204.00
8	Switzerland	202	202.00
9	Korea	195	39.00
10	France	121	30.20

Source: Biblioshiny

Table 4 Most influential articles

S. no.	Author and year	Document title	Journal	TC	TC per year
1	Saberi et al. (2019) [10]	Blockchain technology and its relationships to sustainable supply chain management	*International Journal of Production Research*	2067	344.50
2	Kouhizadeh et al. (2021) [5]	Blockchain technology and the sustainable supply chain: theoretically exploring adoption barriers	*International Journal of Production Economics*	680	170.00
3	Cole et al. (2019) [11]	Blockchain technology: implications for operations and supply chain management	*Supply Chain Management: An International Journal*	584	97.33
4	Esmaeilian et al. (2020) [12]	Blockchain for the future of sustainable supply chain management in Industry 4.0	*Resources, Conservation and Recycling*	483	96.60
5	Kouhizadeh and Sarkis (2018) [13]	Blockchain Practices, Potentials, and Perspectives in Greening Supply Chains	*Sustainability*	421	60.14
6	Bai and Sarkis (2020) [14]	A supply chain transparency and sustainability technology appraisal model for blockchain technology	*International Journal of Production Research*	409	81.80
7	Agrawal et al. (2021) [15]	Blockchain-based framework for supply chain traceability: A case example of textile and clothing industry	*Computers and Industrial Engineering*	267	66.75
8	Yadav and Singh (2020) [16]	Blockchain critical success factors for sustainable supply chain	*Resources, Conservation and Recycling*	265	53.00
9	Zhang et al. (2020) [17]	Blockchain-based life cycle assessment: An implementation framework and system architecture	*Resources, Conservation and Recycling*	204	40.80
10	Manupati et al. (2020) [18]	A blockchain-based approach for a multi-echelon sustainable supply chain	*International Journal of Production Research*	202	40.40

Source: Biblioshiny

Greening Supply Chains," published in Sustainability with 421 citations, explores blockchain's potential in promoting greener supply chains. Bai and Sarkis [14] present a "Supply Chain Transparency and Sustainability Technology Appraisal Model for Blockchain Technology" in the *International Journal of Production Research*, with 409 citations, emphasizing on blockchain's role in enhancing transparency. Agrawal et al. [15] discuss "Blockchain-Based Framework for Supply Chain Traceability: A Case Example of Textile and Clothing Industry" in *Computers & Industrial Engineering* with 267 citations, focusing on traceability improvements in textile supply chains. Yadav and Singh [16], in their article "Blockchain Critical

Success Factors for Sustainable Supply Chain" published in *Resources, Conservation and Recycling* with 265 citations, investigate key factors for successful blockchain implementation. Zhang et al. [17] in their study "Blockchain-Based Life Cycle Assessment: An Implementation Framework and System Architecture," also published in *Resources, Conservation and Recycling*, with 204 citations, outlines a framework for using blockchain in life cycle assessments. Lastly, Manupati et al. [18] discuss "A Blockchain-Based Approach for a Multi-Echelon Sustainable Supply Chain" in the *International Journal of Production Research*, with 202 citations, which show blockchain's role in complex and multi-layered supply chains.

3.6 Word Cloud

Figure 3 depicts word cloud which shows a strong research focus on using blockchain technology to drive sustainable development in supply chains. The prominence of terms like sustainable supply chains, sustainability, and environmental management suggests that much of the research examines blockchain's potential to create eco-friendly and socially responsible supply chain practices. Key topics such as decision making and technology adoption imply that studies are also interested in how blockchain supports data-driven, transparent decisions within supply chain management. Other terms, like profitability and investments, indicate a focus on achieving economic sustainability alongside environmental goals. Overall, this area of research explores how blockchain and other Industry 4.0 technologies can transform supply chains into more efficient, ethical, and resilient systems.

3.7 Network Cluster Analysis

Figure 4 illustrates network cluster analysis in which mapping the clusters is related to blockchain and sustainable supply chain management.

Fig. 3 Word cloud. (Source: Biblioshiny)

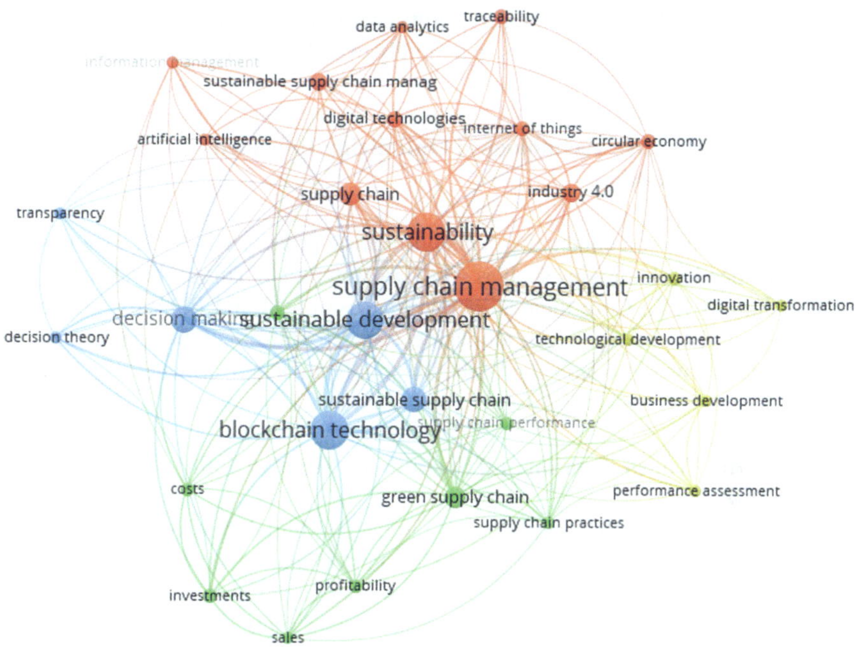

Fig. 4 Author's keywords bibliographic coupling analysis. (Source: VOS viewer)

3.7.1 Industry 4.0: Driving Sustainable Supply Chain Practices

This cluster represents the intersection of SCM, blockchain technology, and decision-making, illustrating how these components are fundamentally transforming modern supply chains. This cluster also incorporates Artificial Intelligence (AI), Internet of Things and Big data Analytics (BDA) which collaborate with blockchain to streamline supply chain efficiency, supporting real-time monitoring, predictive insights, and automation (Kumar et al. [19–23]). The combined impact of these technologies fosters greater transparency, accountability and operational agility, positioning blockchain as a crucial tool for the future of SCM [24].

3.7.2 Blockchain and Green Supply Chains: Balancing Profit and Sustainability

This Cluster emphasizes the integration of sustainability into SCM, with blockchain technology playing a pivotal role in enhancing transparency, accountability and traceability [25]. Additionally, this cluster incorporates Green Supply Chain, which is closely linked to themes like environmental management, green economy, and sustainable practices, while also considering economic factors like cost, investments, and profitability [26]. The intersection of sustainability, digital transformation, and financial viability showcases how companies can achieve both environmental and economic gains in a rapidly evolving market landscape.

3.7.3 Transformative Technologies: Optimizing Informed Decision-Making

This cluster focuses on enhancing decision-making processes in SCM explores the use of advanced technologies such as AI, BDA, Internet of Things, blockchain, and cloud solutions [27]. Core themes include the use of AI and BDA for predictive analysis and risk management, alongside effective information management to facilitate data-driven decisions and enhance supply chain performance.

3.7.4 Digital Transformation and Sustainable Supply Chain

This Cluster focuses on the transformative impact of digital transformation, technological development, and innovation in SCM. Performance assessment is crucial for evaluating the effects of these innovations on operational efficiency and customer satisfaction. Furthermore, technology advancements such as blockchain improve transparency and foster collaboration among stakeholders.

4 Discussion

This research analyzes global trends in innovative technology and Sustainable SCM practices using bibliometric data from Scopus data from 2018 to 2024, revealing significant growth with 158 documents across 82 sources and an annual growth rate of 80.86%. The average document age of 1.66 years indicates ongoing developments, while a citation rate of 60.8 citations per document highlights their impact. The findings reveal that the United States, China, and the United Kingdom are key contributors to this research domain with the United States leading in total citations and average citations per article, which reflects its substantial impact on the field. The top journals include *Sustainability (Switzerland)*, *Business Strategy and the Environment*, and *Resources, Conservation and Recycling*. The analysis identifies four key clusters: Industry 4.0: Driving Sustainable Supply Chain Practices, BLOCKCHAIN and Green Supply Chains, Transformative Technologies, and Digital Transformation, underscoring blockchain‘s role in enhancing sustainability and profitability. Future research should address gaps such as the synergistic effects of blockchain with other technologies on supply chain resilience, as well as the importance of human factors like workforce training for successful implementation, which will aid organizations in promoting genuine sustainability in supply chains.

4.1 Future Research Direction

Future research directions are given in Table 5.

Table 5 Future research direction on blockchain and sustainable supply chain

Cluster	Current research	Gap	Future research direction
Industry 4.0: Driving Sustainable Supply Chain Practices	Examines how Industry 4.0 technologies like AI, IoT, BDA, and blockchain are improving SCM through real-time monitoring, predictive insights, and automation to increase efficiency, transparency, and agility	There is a need for more research on examining how the integration of AI, IoT, and other Industry 4.0 technologies impacts ethical and social issues in sustainable SCM, especially regarding privacy issues and understanding how employees adapt to Industry 4.0 tools in sustainable supply chain management, specifically regarding the psychological impact of AI and IoT on worker engagement, job satisfaction, and well-being	Future research could focus on developing frameworks to address privacy concerns and ethical implications of AI and IoT within sustainable SCM. Additionally, studies could explore employee adaptation to Industry 4.0 tools, examining the psychological impacts on engagement, job satisfaction, well-being and identifying best practices to foster a supportive work environment that promotes positive adaptation to these technologies
Blockchain and Green Supply Chains: Balancing Profit and Sustainability	Current research emphasizes the integration of sustainability into SCM, with blockchain technology playing a pivotal role in enhancing transparency, accountability and traceability [25]. Additionally, this cluster incorporates Green Supply Chain, which is closely linked to themes like environmental management, green economy, and sustainable practices, while also considering economic factors like cost, investments, and profitability [26]	There is limited research on blockchain's capacity to provide verifiable environmental claims for combating greenwashing. Additionally, there is a gap in understanding how blockchain can ensure the traceability and transparency of financial flows associated with sustainability initiatives, thereby enhancing accountability and trust in green finance	Future research could investigate the development of blockchain-based certification systems that authenticate environmental claims, ensuring that companies provide accurate, transparent, and verified information about the sustainability of their products and practices

(continued)

Table 5 (continued)

Cluster	Current research	Gap	Future research direction
Transformative Technologies: Optimizing Informed Decision-Making	Current research focuses on enhancing decision-making processes in SCM explores the use of advanced technologies such as AI, BDA, Internet of Things, Blockchain and Cloud solutions [27]	While the use of AI, Big Data Analytics (BDA), the Internet of Things (IoT), Blockchain, and Cloud solutions is transforming decision-making in supply chain management (SCM), there are still notable gaps in understanding how these technologies can be optimally integrated to enhance predictive accuracy, risk mitigation, and real-time responsiveness	Future research could focus on developing integrative frameworks that detail best practices for combining AI, BDA, IoT, Blockchain, and Cloud technologies to support robust, data-driven decision-making in supply chains. This includes exploring how these technologies can work together to improve predictive capabilities, enable real-time risk management, and facilitate agile, informed responses to disruptions. Additionally, future studies could examine the ethical dimensions and data privacy concerns associated with IoT and BDA in SCM
Digital Transformation and Sustainable Supply Chain	Current research focuses on the transformative impact of digital transformation, technological development, and innovation in SCM. Performance assessment is crucial for evaluating the effects of these innovations on operational efficiency and customer satisfaction	While digital technologies like AI and blockchain show promise for enhancing transparency, few studies have rigorously explored how these technologies align specifically with ESG targets within SCM. Additionally, scalability and applicability of digital transformation for sustainable supply chains in small and medium-sized enterprises (SMEs) is limited	Future research should investigate how emerging digital technologies like AI and blockchain can be systematically aligned with specific ESG targets in supply chain management, focusing on their measurable contributions to environmental sustainability, social equity, and governance improvements

4.2 Limitations of Study

Despite the comprehensive approach, this study is based on publications indexed in Scopus and excludes other relevant databases such as Google Scholar, PubMed, or Science Direct. Additionally, the study is limited to the 2018–2024, which may lead to the exclusion of developments or trends in blockchain and Sustainable SCM before 2018. The analysis is focused on quantitative measures, such as publication counts and citation data which might not adequately reflect the qualitative aspects of the research.

5 Conclusion

This study presents an in-depth bibliometric analysis of blockchain technology's influence on Sustainable SCM, highlighting the significant growth in scholarly output between 2018 and 2024. The findings underscore blockchain's potential to boost visibility, track ability, and operational efficiency in supply chain particularly when integrated with advanced technologies like AI and IoT. The analysis identifies key contributing countries, with the United States leading in research impact, followed by China and the United Kingdom. Additionally, leading journals including *Sustainability (Switzerland)* and *Business Strategy and the Environment* serve as key platforms for this emerging field. The study uncovers four primary research clusters, Industry 4.0: Driving Sustainable Supply Chain Practices, Blockchain and Green Supply Chains: Balancing Profit and Sustainability, Transformative Technologies: Optimizing Informed Decision-Making, and Digital Transformation and Sustainable Supply Chain. Potential areas for future research include exploring the integration of blockchain with AI and IoT for real-time decision-making, assessing its financial impact on green supply chains, and investigating its role in fostering sustainable investments. These areas of inquiry will further inform strategies to enhance both supply chain sustainability and profitability.

References

1. Lai, J.-Y., Wang, J., & Chiu, Y.-H. (2021). Evaluating blockchain technology for reducing supply chain risks. *Information Systems and e-Business Management, 19*(4), 1089–1111.
2. Zhu, Q., Bai, C., & Sarkis, J. (2022). Blockchain technology and supply chains: The paradox of the atheoretical research discourse. *Transportation Research Part E: Logistics and Transportation Review, 164*, 102824.

3. Min, H. (2019). Blockchain technology for enhancing supply chain resilience. *Business Horizons, 62*(1), 35–45.
4. Shi, Y., Ying, J., Shi, D., & Yan, J. (2022). Service-oriented modeling for blockchain-enabled supply chain quality information systems. *Security and Communication Networks, 2022*, 1–16. https://doi.org/10.1155/2022/1987933
5. Kouhizadeh, M., Saberi, S., & Sarkis, J. (2021). Blockchain technology and the sustainable supply chain: Theoretically exploring adoption barriers. *International Journal of Production Economics, 231*, 107831.
6. Li, K., Lee, J.-Y., & Gharehgozli, A. (2023). Blockchain in food supply chains: A literature review and synthesis analysis of platforms, benefits and challenges. *International Journal of Production Research, 61*(11), 3527–3546. https://doi.org/10.1080/00207543.2021.1970849
7. Ghadge, A., Bourlakis, M., Kamble, S., & Seuring, S. (2023). Blockchain implementation in pharmaceutical supply chains: A review and conceptual framework. *International Journal of Production Research, 61*(19), 6633–6651. https://doi.org/10.1080/00207543.2022.2125595
8. Xiao, P., Salleh, M. I., Zaidan, B. B., & Xuelan, Y. (2023). Research on risk assessment of blockchain-driven supply chain finance: A systematic review. *Computers & Industrial Engineering, 176*, 108990.
9. Alkhudary, R., Queiroz, M. M., & Féniès, P. (2024). Mitigating the risk of specific supply chain disruptions through blockchain technology. *Supply Chain Forum, 25*(1), 1–11. https://doi.org/10.1080/16258312.2022.2090273
10. Saberi, S., Kouhizadeh, M., Sarkis, J., & Shen, L. (2019). Blockchain technology and its relationships to sustainable supply chain management. *International Journal of Production Research, 57*(7), 2117–2135. https://doi.org/10.1080/00207543.2018.1533261
11. Cole, R., Stevenson, M., & Aitken, J. (2019). Blockchain technology: Implications for operations and supply chain management. *Supply Chain Management, 24*(4), 469–483.
12. Esmaeilian, B., Sarkis, J., Lewis, K., & Behdad, S. (2020). Blockchain for the future of sustainable supply chain management in Industry 4.0. *Resources, Conservation and Recycling, 163*, 105064.
13. Kouhizadeh, M., & Sarkis, J. (2018). Blockchain practices, potentials, and perspectives in greening supply chains. *Sustainability, 10*(10), 3652. Retrieved from https://www.mdpi.com/2071-1050/10/10/3652
14. Bai, C., & Sarkis, J. (2020). A supply chain transparency and sustainability technology appraisal model for blockchain technology. *International Journal of Production Research, 58*(7), 2142–2162. https://doi.org/10.1080/00207543.2019.1708989
15. Agrawal, T. K., Kumar, V., Pal, R., Wang, L., & Chen, Y. (2021). Blockchain-based framework for supply chain traceability: A case example of textile and clothing industry. *Computers & Industrial Engineering, 154*, 107130.
16. Yadav, S., & Singh, S. P. (2020). Blockchain critical success factors for sustainable supply chain. *Resources, Conservation and Recycling, 152*, 104505.
17. Zhang, A., Zhong, R. Y., Farooque, M., Kang, K., & Venkatesh, V. G. (2020). Blockchain-based life cycle assessment: An implementation framework and system architecture. *Resources, Conservation and Recycling, 152*, 104512.
18. Manupati, V. K., Schoenherr, T., Ramkumar, M., Wagner, S. M., Pabba, S. K., Singh, I. R., & R. (2020). A blockchain-based approach for a multi-echelon sustainable supply chain. *International Journal of Production Research, 58*(7), 2222–2241. https://doi.org/10.1080/00207543.2019.1683248
19. Kumar, A., Gori, Y., Dutt, N., Singla, Y. K., & Maurya, A. (2021). *Advanced computational methods in mechanical and materials engineering*. CRC Press. ISBN: 9781032052915. https://doi.org/10.1201/9781003202233
20. Kumar, A., Ram, M., & Singla, Y. K. (2022). *Advanced material for biomechanical applications*. CRC Press. ISBN: 9781032054490. https://doi.org/10.1201/9781003286806
21. Kumar, A., Singla, Y. K., & Maughan, M. (2024). *Fracture behavior of nanocomposites and reinforced laminate structures*. Springer Nature. https://doi.org/10.1007/978-3-031-68694-8

22. Kumar, A., Singla, Y., & Namboodri, T. (2024). Chapter 1: Globalization and international issues in sustainable manufacturing. In S. Shah, H. Nautiyal, G. Gugliani, A. Kumar, T. Namboodri, & Y. K. Singla (Eds.), *Sustainability in smart manufacturing: Trends, scope and challenges* (pp. 1–18). CRC Press. https://doi.org/10.1201/9781003467496-1
23. Shah, S., Nautiyal, H., Gugliani, G., Kumar, A., Namboodri, T., & Singla, Y. K. (2024). *Sustainability in smart manufacturing: Trends, scope, and challenges*. CRC Press. ISBN: 9781032740713. https://doi.org/10.1201/9781003467496
24. Rana, P., Joshi, K., & Gabriel, E. (2024). Application of Industry 4.0: AI and IoT to improve supply chain performance. In *Exploration of artificial intelligence and blockchain technology in smart and secure healthcare* (p. 274). Bentham Science Publishers.
25. Zhou, Y., Gao, X., & Nie, J. (2024). Value of blockchain-enabled supply chain traceability under competition. *International Transactions in Operational Research, 31*(6), 3669–3703. https://doi.org/10.1111/itor.13295
26. Mounia, A., Yassine, H., & Hocine, I. Z. H. (2024). Literature review on cost management and profitability in E-supply chain: Current trends and future perspectives. In *2024 IEEE 15th International Colloquium on Logistics and Supply Chain Management (LOGISTIQUA)* (pp. 1–6).
27. Rahman, M., Emon, M. E. H., Antor, M. H., Haque, S. A., & Talapatra, S. (2024). Identification and prioritization of barriers to Industry 4.0 adoption in the context of food and beverage industries of Bangladesh. *Benchmarking*. https://doi.org/10.1108/BIJ-06-2023-0371

Revolutionizing Logistics: Smart Supply Chains Leading the Way in Industry 5.0

Bikram Jit Singh, Eisha Mehta Sharma, and Rippin Sehgal

1 Introduction to Industry 5.0 and Smart Supply Chains

Industry 5.0 marks a transformative step forward in manufacturing and logistics, building upon the foundations established by Industry 4.0 [1–4]. While Industry 4.0 primarily focused on automation and data-centric processes, Industry 5.0 reintroduces the human element, aiming to foster a collaborative relationship between people and advanced technologies. This approach envisions machines working alongside humans to enhance their abilities, rather than replacing them [5–7]. In this new paradigm, smart supply chains play a pivotal role by incorporating technologies like IoT, AI, blockchain and robotics. These advancements transform traditional supply chains into more efficient, agile, and adaptable systems. Unlike conventional models, smart supply chains emphasize real-time data, customization, and sustainability, enabling businesses to meet the changing needs of consumers and address environmental concerns [8, 9]. As we progress further into the Industry 5.0 era, adopting and implementing smart supply chains will be essential for companies aiming to stay competitive and drive growth in an ever-evolving global market.

An exclusive exploration of the Scopus database, using tailored search strategies, highlights the significance of smart supply chains within the context of Industry 5.0 [10]. The data reveals how smart supply chains are revolutionizing logistics and underscores their crucial role in Industry 5.0. The study, covering the years 2015 to

B. J. Singh (✉) · E. M. Sharma
Department of Mechanical Engineering, MM Engineering College, Maharishi Markandeshwar (Deemed to be University), Mullana, Haryana, India
e-mail: dr_bikramjit@mmumullana.org

R. Sehgal
Department of Biotechnology, Ambala College of Engineering and Applied Research, Ambala, Haryana, India
e-mail: rippin.bt@ambalacollege.ac.in

Y. Kumar Singla et al. (eds.), *Digital Manufacturing in Industry 5.0*,
https://doi.org/10.1007/978-3-031-91500-0_47

2024, shows rapid developments in the field, with a remarkable 72.18% annual growth rate in related publications. This surge in academic contributions—753 documents from 508 sources—demonstrates the extensive research and practical interest in integrating cutting-edge technologies into supply chains. The average age of these documents, at just 2.19 years, indicates the field's swift evolution. High citation rates (18.05 per document) reflect the substantial academic and practical significance of this research, with key findings influencing real-world logistics systems. Collaboration is a key driver of these advancements, as indicated by the strong trend toward co-authored papers, averaging 3.78 authors per document. Additionally, the international nature of this research is evident, with 25.9% of collaborations spanning multiple countries, highlighting the global relevance of smart supply chains as industries worldwide transition toward more intelligent and connected operations.

Further analysis through the Biblioshiny application (based on the Bibliometrix library) in the R programming language provides deeper insights [11]. A three-field plot reveals the complex interconnections between the top 10 authors, countries, and keywords in the domain of smart supply chains and Industry 5.0 (refer to Fig. 1). India emerges as a leading contributor to this research, followed closely by Mexico, Australia, Spain and France, indicating a broad international interest in this rapidly evolving field. Notably, Indian authors, particularly "Singh R," play a significant role in discussions surrounding Industry 4.0, artificial intelligence and automation, underscoring India's critical influence on the future direction of smart supply chains.

The authors depicted in the middle section such as "Hassoun A," "Gehlot A," and "Haleem A" are highly associated with influential keywords like "Industry 4.0," "artificial intelligence," and "sustainability." This demonstrates a strong research focus on optimizing and automating supply chains while incorporating AI and sustainable practices, which aligns with the core objectives of Industry 5.0. Notably, "Sustainability" and "Robotics" also feature prominently, suggesting that these concepts are gaining importance in the logistical systems of the future. Authors from a broad geographical spread are contributing to various dimensions of the smart supply chain. For instance, "Javaid M" and "Twala B" are linked to research areas involving "big data" and "Internet of Things," emphasizing how these technologies are transforming supply chain management. In brief, the visualization captures the collaborative and multi-disciplinary nature of this research field. It reflects how global academic leaders are interconnected through a shared focus on technology, innovation and sustainability, which are foundational to Industry 5.0 and future logistics systems.

1.1 The Transition from Industry 4.0 to Industry 5.0

The transition from Industry 4.0 to Industry 5.0 marks a transformative shift in manufacturing, logistics and supply chain management. Industry 4.0 brought revolutionary changes by focusing on automation, data sharing and interconnected systems [12]. This era introduced smart factories where machines communicated via

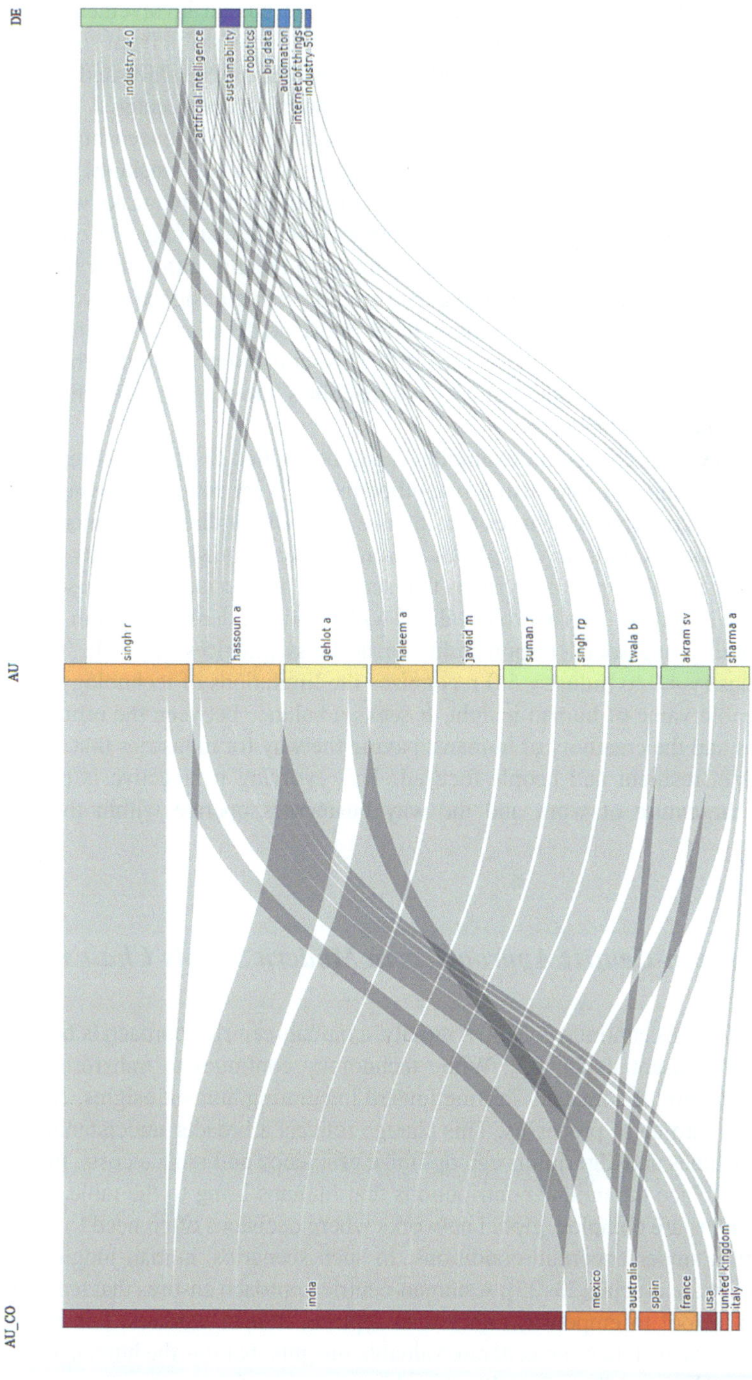

Fig. 1 Three-field plot related to smart supply chains in Industry 5.0

IoT and used data to streamline production processes, with a strong emphasis on efficiency, precision and speed, aiming to minimize human involvement through automation [13]. As we move into Industry 5.0, the focus shifts from solely automating tasks to creating a balanced partnership between human skills and advanced technology. Unlike Industry 4.0's machine- and algorithm-centered approach, Industry 5.0 seeks to establish a collaborative relationship between humans and technology [14, 15]. This new phase acknowledges that while machines are excellent at managing vast amounts of data and performing repetitive tasks with precision, they cannot replicate human creativity, problem-solving, or emotional intelligence. Industry 5.0 aims to enhance human potential rather than replace it. Technologies like AI, robotics and big data are now being designed to work alongside people, supporting them in making more informed decisions, fostering innovation and increasing productivity [16–18]. This collaborative approach aims to leverage the strengths of both humans and machines, creating a more efficient and innovative industrial landscape.

In smart supply chains, for example, AI can be applied to process complex data for tasks like demand forecasting or optimizing delivery routes, allowing human employees to concentrate on roles that involve creativity or direct customer engagement. Additionally, Industry 5.0 places a strong emphasis on sustainability and ethical practices, aiming to build systems that serve not just businesses but also society and the environment. This approach aligns with the rising demand for transparency, accountability and social responsibility in business practices [19, 20]. The shift from Industry 4.0 to Industry 5.0 recognizes the limitations of technology and the irreplaceable value of human insight. It seeks a balance between the efficiency of machines and the creativity of humans, paving the way for industries that are more sustainable, resilient and people-focused. This evolving perspective is poised to reshape the future of work and the way businesses operate within the global economy.

1.2 Human-Centric Approaches in Modern Supply Chains

As supply chains continue to evolve rapidly, a human-centric approach is becoming increasingly essential [21, 22]. While technology continues to transform supply chain operations, the focus is shifting toward integrating human insights, creativity and values into these processes. This change reflects a broader understanding that while machines and algorithms can optimize efficiency and reduce costs, they cannot fully replace the unique contributions that humans bring to the table. Modern supply chains are complex, global networks where decisions often need to be made in real time under uncertain conditions. In such scenarios, human judgment and adaptability are crucial [23–25]. A human-centric approach ensures that technology serves as a tool to empower, rather than replace, human workers. For example, advanced data analytics can generate valuable insights, but it is the human ability to interpret and act on these insights that drives meaningful change. This approach allows workers to use technology to enhance their roles, rather than feel threatened by automation. Moreover, a human-centric supply chain strongly emphasizes customization and personalization. As consumer demands diversify and expectations for tailored experiences rise, businesses must be agile enough to adapt quickly. Human-centric strategies provide the flexibility needed to meet these demands, as humans are better suited to understand and respond to nuanced customer needs. This results in a more responsive and agile supply chain that can adjust as market conditions evolve.

Sustainability is another critical aspect of human-centric supply chains. As companies face growing pressure to operate responsibly, incorporating human values such as environmental stewardship and social equity into supply chain strategies becomes essential. Human-centric approaches prioritize ethical sourcing, reduce waste and ensure fair labor practices, aligning business operations with broader societal goals. Ultimately, a human-centric approach in modern supply chains strikes a balance where technology enhances human capabilities and humans guide technology with empathy, ethics and creativity [6, 26, 27]. This balance not only boosts efficiency and responsiveness but also fosters a more sustainable and socially responsible supply chain, positioning businesses for long-term success in an increasingly connected and conscientious world.

1.3 Overview of Smart Supply Chains and Their Importance

Smart supply chains mark a major leap forward in how businesses handle logistics and operations. By using advanced technologies, these supply chains achieve greater efficiency, transparency and flexibility [10, 28]. Unlike traditional supply chains, which rely heavily on manual tasks and fixed data, smart supply chains integrate real-time data and automation to enhance every part of the process. What makes smart supply chains so powerful is their capacity to collect and analyze extensive data from multiple sources. With tools like IoT, AI and big data analytics, businesses gain live insights into their operations. This real-time visibility helps improve inventory management, demand forecasting and overall supply chain efficiency [29, 30]. For example, IoT sensors can track the location and condition of products in transit, while AI algorithms can anticipate potential disruptions and suggest the best responses.

A major strength of smart supply chains lies in their adaptability. In today's fast-changing world, where consumer preferences and market conditions constantly shift, being able to respond quickly is essential. Smart supply chains leverage predictive analytics to anticipate changes in demand or supply, allowing businesses to proactively tackle issues before they become significant. This approach helps prevent disruptions and ensures that customer expectations are consistently met. Transparency is another valuable advantage. Smart supply chains enable companies to track products from origin to destination, offering detailed insights into sourcing, production and delivery. This level of transparency not only fosters consumer trust but also helps companies meet regulatory requirements and uphold ethical standards [31, 32]. Furthermore, smart supply chains support sustainability by optimizing resources and minimizing waste. For instance, data-driven insights can streamline delivery routes, reducing fuel consumption and emissions. By integrating these advanced technologies, companies not only boost operational efficiency but also align with environmental and social goals. In essence, smart supply chains are crucial for modern businesses. They provide the tools and insights needed to navigate the complexities of global markets, enhance operational efficiency and meet growing demands for transparency and sustainability [33–35]. As technology advances, smart supply chains will become an even more integral part of business strategies, ensuring companies are prepared for the future.

2 Technological Pillars of Smart Supply Chains

2.1 IoT: Enhancing Visibility and Real-Time Monitoring

The IoT has revolutionized supply chain management by integrating sensors and connectivity into physical assets [18, 36, 37]. This technology enables businesses to track and monitor goods, equipment and inventory in real time across the entire supply chain. IoT devices gather and transmit data, providing companies with a complete, up-to-date view of their operations. For instance, sensors placed on shipping containers can relay information on location, temperature and humidity, helping ensure that sensitive items are stored and transported under ideal conditions [38–41]. This increased visibility allows businesses to quickly address issues like delays or damage, boosting overall efficiency through improved inventory management and better use of resources.

2.2 AI and Machine Learning for Predictive Maintenance

AI and machine learning are transforming predictive maintenance, reshaping how businesses maintain their equipment and machinery. By analyzing historical data to spot patterns, AI can forecast when a machine might fail or need servicing, helping companies avoid unexpected breakdowns and extend the life of their assets [42–45]. As machine learning algorithms process more data over time, they continually improve their predictions, making forecasts increasingly accurate. For example, AI systems can notify maintenance teams to perform inspections or replacements before issues arise, reducing downtime and avoiding costly disruptions in the supply chain.

2.3 Blockchain Technology: Ensuring Transparency and Security

Blockchain technology provides a robust solution for enhancing transparency and security within supply chains. Through the creation of a decentralized ledger, blockchain records every transaction and movement of goods in an immutable manner, offering an auditable trail from origin to final destination [46–49]. This tamper-proof record ensures that all parties have access to accurate and verified information, reducing the risk of fraud and errors while improving product traceability. Additionally, blockchain supports secure transactions and smart contracts, automating and enforcing agreements without the need for intermediaries, thus streamlining processes and fostering trust among stakeholders.

2.4 Big Data Analytics: Driving Informed Decision-Making

Big data analytics equips businesses with the ability to make smarter, data-driven decisions by processing large volumes of information from diverse sources. In supply chains, big data analytics offers valuable insights into customer preferences, market trends and operational effectiveness. By uncovering patterns and relationships in the data, companies can better predict demand, manage inventory and identify possible bottlenecks in the supply chain [50–52]. This approach allows businesses to respond more quickly and strategically to market shifts, boosting their overall performance and competitiveness on a global scale.

2.5 Robotics and Collaborative Robots (Cobots) in Supply Chains

Robotics, particularly collaborative robots or cobots, are transforming supply chains by enhancing automation and efficiency. Unlike conventional robots that function autonomously, cobots are specifically designed to collaborate with human workers, augmenting their efforts and improving overall effectiveness [13, 53–55]. These robots excel at performing repetitive, physically demanding, or precise tasks, which allows human employees to concentrate on more intricate and value-generating responsibilities. In warehouse settings, cobots play a crucial role in sorting, packing and moving products, while in manufacturing environments, they assist with assembly and quality assurance. The integration of cobots into workflows not only simplifies operations but also minimizes errors and boosts overall capacity.

2.6 *Cloud Computing: Scalability and Flexibility in Operations*

Cloud computing offers crucial scalability and flexibility for modern supply chains. By utilizing cloud-based platforms, businesses can store and access vast amounts of data and applications without the constraints of on-premises infrastructure. This flexibility enables companies to scale their operations up or down in response to demand without requiring significant investments in physical hardware [56–59]. Cloud computing also facilitates real-time collaboration and data sharing among global teams, improving coordination and decision-making. With cloud solutions, businesses can deploy new tools and technologies more rapidly, adapt to changing market conditions and enhance general operational agility.

3 Human–Machine Collaboration and Workforce Transformation

3.1 *Enhancing Human Capabilities with Collaborative Robots (Cobots)*

Collaborative robots, or cobots, are designed to work alongside people, effectively bridging the gap between human labor and automation. Unlike traditional robots that operate on their own, cobots are programmed to assist and enhance the work of human employees. They excel in performing repetitive, hazardous, or precision-based tasks, such as assembly, sorting and packaging. This allows human workers to dedicate their time to more complex, strategic, and creative responsibilities [60–63]. By taking on routine tasks, cobots significantly increase productivity, enabling employees to focus on problem-solving, innovation and engaging with customers. This partnership between humans and machines not only improves operational efficiency but also enhances workplace safety, as cobots can handle dangerous jobs, thereby lowering the risk of injuries for human workers.

3.2 *Advancing Customization and Personalization Through Human–Machine Interaction*

The synergy between humans and machines is unlocking new opportunities for customization and personalization across various industries. Leveraging advanced technologies, businesses can now deliver products and services tailored to individual preferences with exceptional accuracy. Machine learning algorithms analyze customer data and behaviors, while human creativity and insights guide the customization process [8, 64–66]. For example, in manufacturing, AI-driven systems can modify production lines to create bespoke products based on customer specifications, while human designers refine and enhance these customizations. This collaboration between human intuition and machine capabilities enables the creation of highly personalized experiences, satisfying unique customer needs and fostering greater satisfaction and loyalty.

3.3 Workforce Upskilling and Reskilling in the Era of Industry 5.0

With the advent of Industry 5.0 and its integration of advanced technologies, upskilling and reskilling the workforce become crucial. The rapidly changing industrial landscape demands that employees adapt to new tools, systems and roles, moving away from traditional tasks toward more technologically advanced functions [67, 68]. Upskilling focuses on enhancing current skills to work effectively with emerging technologies, while reskilling involves acquiring entirely new competencies. Providing training programs and continuous learning opportunities is essential for preparing the workforce to excel in this evolving environment. By investing in both upskilling and reskilling, companies ensure that their employees are equipped to harness the full potential of new technologies, remain competitive and contribute to innovation and growth.

3.4 Addressing Ethical Considerations in a Human-Centric Industrial Landscape

The incorporation of advanced technologies in the workplace brings forth significant ethical challenges. As machines and AI systems take on more responsibilities, it becomes crucial to tackle issues related to privacy, fairness and job displacement. A human-centered perspective on Industry 5.0 highlights the need for ethical standards that protect workers' rights and ensure technology serves the broader interests of society [69–72]. This involves safeguarding personal data, ensuring transparency in AI decision-making and promoting equitable opportunities for all employees. Additionally, companies should assess the potential effects of automation on jobs and actively support impacted workers through reskilling programs and career transition support. Achieving a balance between technological progress and ethical considerations is vital for nurturing a responsible and inclusive industrial environment.

4 Sustainability and Green Logistics

4.1 Incorporating Circular Economy Practices in Supply Chains

Circular economy practices are transforming supply chain operations by prioritizing sustainability and efficient use of resources. In contrast to the traditional linear model of take-make-dispose, the circular economy emphasizes closing the product lifecycle loop through recycling, reusing and refurbishing [73–75]. This strategy aims to minimize waste and decrease reliance on limited resources. For example, companies can design products that are easier to disassemble, facilitating the recycling or repurposing of components once they reach the end of their life. By embracing circular economy principles, supply chains can lessen their environmental footprint, cut costs related to raw materials and open up new business opportunities by recovering valuable materials. This approach not only supports environmental goals but also responds to the increasing consumer demand for sustainable practices.

4.2 Green Logistics: Mitigating Environmental Impact

Green logistics focuses on minimizing the environmental impact of supply chain activities by employing practices and technologies that lower emissions, waste and energy usage. Key strategies include optimizing transportation routes to reduce fuel consumption, using electric or hybrid vehicles and improving packaging to lessen waste [76]. Moreover, green logistics aims to boost warehouse efficiency through energy-saving initiatives, such as installing efficient lighting and climate control systems [77]. By implementing these practices, businesses can lower their carbon footprint, adhere to environmental regulations and attract consumers who prioritize sustainability. Adopting green logistics is not only a responsible decision but also provides a competitive advantage in a marketplace that increasingly values sustainable practices.

4.3 Energy Optimization in Smart Supply Chain Operations

Energy optimization plays a crucial role in creating sustainable supply chain operations. Smart supply chains leverage advanced technologies to effectively monitor and manage energy use. For instance, data analytics can identify processes that consume excessive energy and suggest improvements, such as modifying equipment usage or fine-tuning heating and cooling systems in warehouses [78–80]. Moreover, incorporating renewable energy sources like solar and wind power can significantly decrease reliance on fossil fuels. By focusing on energy optimization, companies can reduce operational costs, lessen their environmental impact and improve overall efficiency. Utilizing these technologies enables businesses to enhance their sustainability while retaining a competitive edge in the global market.

4.4 *The Role of Smart Supply Chains in Fostering Sustainable Business Practices*

Smart supply chains are crucial in promoting sustainable business practices by integrating technology-driven solutions that prioritize environmental and social responsibility. Through real-time data and analytics, smart supply chains enable more informed decision-making that supports sustainability goals. For example, they can optimize resource utilization, minimize waste and improve the efficiency of logistics [52, 81, 82]. Moreover, smart supply chains enhance transparency and traceability, enabling businesses to ensure ethical sourcing and compliance with environmental standards. By adopting these technologies, companies not only boost their operational performance but also align with global sustainability initiatives and meet consumer expectations for responsible business practices.

5 Challenges and Solutions in Implementing Smart Supply Chains

5.1 *Data Privacy and Security in a Connected Supply Chain*

In a connected supply chain, ensuring data security and privacy is essential due to the large volumes of sensitive information being exchanged and stored. The use of IoT devices, cloud computing and other technologies heightens the risk of cyberattacks and data breaches [83–86]. To combat these threats, businesses need to adopt strong cybersecurity practices, including encryption, secure access controls and regular security assessments. Furthermore, following best practices for data management—such as data anonymization and adherence to privacy regulations—can help safeguard sensitive information. By focusing on data security, companies can protect their operations, foster trust with partners and customers and avoid costly security incidents.

5.2 Integrating Legacy Systems with Emerging Technologies

Integrating legacy systems with new technologies is a significant challenge for many organizations. Legacy systems often lack compatibility with modern solutions, which can lead to disruptions and inefficiencies during the transition. To overcome this hurdle, businesses should take a phased approach, starting with the most critical areas for integration and gradually expanding [87]. Employing middleware or integration platforms can bridge the gap between old and new systems, allowing for smoother data exchange and process synchronization. Additionally, investing in training for staff to manage and operate both legacy and new systems can facilitate a smoother transition and minimize operational disruptions [88].

5.3 Managing Change and Facilitating Workforce Adaptation

As smart supply chains continue to evolve, effectively managing change and facilitating workforce adaptation are critical to successful implementation. Employees may resist new technologies or feel uncertain about their roles in an increasingly automated environment [89]. To mitigate this, organizations should prioritize clear communication regarding the benefits and implications of these new technologies. Offering comprehensive training and support is essential for helping employees develop the necessary confidence to utilize new systems. Additionally, involving employees in the change process and addressing their issues can promote a positive attitude toward transformation, leading to a smoother transition and greater acceptance of technological advancements.

5.4 Navigating Regulatory Compliance in a Technologically Advanced Landscape

Navigating regulatory compliance within a technologically advanced landscape presents unique challenges. As supply chains grow more complex and global, adhering to diverse regulations and standards become increasingly challenging [90]. It is crucial for businesses to stay informed about pertinent regulations, including data protection laws, environmental standards and industry-specific requirements. Implementing robust compliance management systems and collaborating with legal and regulatory experts can help ensure adherence to these standards. Additionally, conducting regular audits and updating compliance practices are vital to addressing regulatory changes and maintaining operational integrity. By proactively managing compliance, companies can avoid legal complications, protect their reputation and ensure seamless operations in a regulated environment.

6 Future Trends and Case Studies

6.1 *Real-World Applications: Automotive, Retail, Healthcare, and Agriculture*

Smart supply chains are making significant impacts across various industries, transforming how businesses operate and deliver value. In the automotive sector, for instance, companies are using IoT and AI to enhance vehicle tracking and predictive maintenance, reducing downtime and improving customer satisfaction. Retailers leverage smart supply chains to optimize inventory management, personalize customer experiences and streamline logistics through advanced data analytics and automation [37, 39, 49]. In healthcare, supply chains benefit from real-time tracking of medical supplies and pharmaceuticals, ensuring timely availability and compliance with regulations. Agriculture sees advancements with precision farming, where IoT sensors and drones monitor crop health, optimize resource use and boost yields. These real-world applications showcase the potential of smart supply chains to drive efficiency, enhance service quality and support innovation across diverse sectors.

6.2 *Emerging Trends*

To capture the buzzing keywords, suitable Tree Diagram has been skimmed out with the help of Biblioshiny application (refer to Fig. 2). It offers a clear overview of the dominant themes and areas of focus in smart supply chains and Industry 5.0. The largest portion of the diagram is devoted to "Industry 4.0," with a significant 14% share, emphasizing its foundational role in the ongoing transformation of logistics and manufacturing.

This is closely followed by "Artificial Intelligence" (9%) and "Robotics" (6%), reflecting the heavy reliance on automation, AI-driven insights and robotic systems to enhance efficiency and reduce human labor in smart supply chains. The "Internet of Things" (6%) also plays a significant role, showing how interconnected devices and real-time data are revolutionizing supply chain visibility and decision-making. Keywords like "Intelligent Robots" and "Industrial Robots" indicate a focus on further integrating automation at both a physical and cognitive level within logistics operations.

Smaller yet impactful topics like "Big Data," "Machine Learning," and "Blockchain" each account for about 2–4% of the discussion, highlighting a growing emphasis on leveraging data analytics and secure digital systems to enhance the intelligence and transparency of supply chains. Additionally, emerging technologies such as "Augmented Reality," "Cloud Computing," and "Virtual Reality" suggest that immersive technologies are starting to play a role in supply chain training, planning and operations. The accompanying diagram illustrates how these diverse

Fig. 2 Word tree

technologies are converging to create smarter and more resilient supply chains within the context of Industry 5.0. Looking ahead, several emerging technologies are expected to significantly influence supply chain management. Quantum computing has the potential to revolutionize data processing, offering unmatched speed and power for complex problem-solving and optimization at scales never before possible. The rollout of 5G networks will enhance connectivity, allowing for faster and more reliable communication between devices and systems, which is critical for real-time supply chain operations. Digital twins—virtual representations of physical assets or processes—will facilitate advanced simulations and predictive analytics, improving decision-making and operational efficiency. Furthermore, Augmented Reality (AR) will enhance training, maintenance and real-time data visualization, enriching human–machine interactions and providing deeper operational insights. These technologies are set to drive significant advancements in supply chain management, unlocking new capabilities and efficiencies.

6.3 The Future of Supply Chain Management in Industry 5.0

As we move into Industry 5.0, the future of supply chain management will increasingly emphasize solutions that are both human-centric and technology-driven. The integration of cutting-edge technologies, coupled with a human collaboration, will fundamentally change how supply chains function. We can anticipate a more extensive application of AI and machine learning for enhanced forecasting and decision-making, alongside the growing use of collaborative robots (cobots) that operate along with humans boost productivity and safety. Sustainability will continue to be a core focus, with supply chains incorporating circular economy practices and green logistics to reduce environmental impact [91, 92]. Moreover, there will be a heightened emphasis on achieving greater transparency and resilience, utilizing technologies like blockchain for secure and traceable operations. The future of supply chains will be marked by a seamless integration of human insight and technological capabilities, driving efficiency, innovation and sustainability. A detailed thematic mapping of emerging themes has been conducted (refer to Fig. 3 for more information). The thematic map contains the four quadrants: Niche Themes, Motor Themes, Emerging/Declining Themes, and Basic Themes. Each quadrant helps to evaluate the relevance and development of various concepts in the research area based on their centrality and density.

- ***Motor themes***: High centrality and density characterize these as the most crucial and well-developed topics. This category includes AI, IoT, Industry 4.0, AI and AR. These themes are driving both research and practical applications, with AI and IoT playing a vital role in the advancements of Industry 4.0.
- ***Niche themes***: These are specialized topics with advanced development but limited widespread application, such as collaborative robots and the IIoTs. Their relevance is more focused on specific, specialized fields.

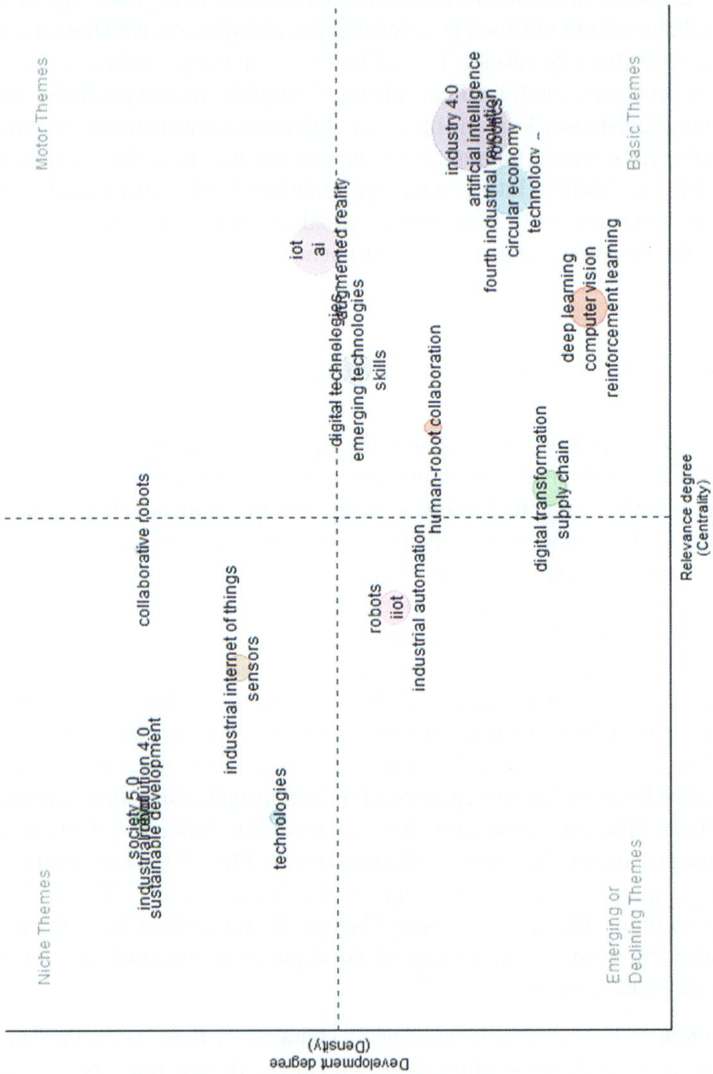

Fig. 3 Thematic mapping

- ***Basic themes***: This quadrant includes foundational concepts such as deep learning, reinforcement learning and computer vision. These themes have significant centrality but lower development density, indicating that while they are fundamental, they are still in the early stages of widespread adoption and application.
- ***Emerging/declining themes***: Themes such as Industrial Revolution 4.0 and sustainable development are currently in a state of flux, characterized by low centrality and development. These topics may either rise in significance or diminish over time.

In summary, the thematic map suggests that AI, Industry 4.0 and IoT are pivotal, rapidly advancing themes that are driving the field, whereas others like collaborative robots and IIoT are becoming increasingly specialized. Meanwhile, foundational topics like deep learning continue to evolve and grow in their influence.

6.4 Strategic Insights for Business Growth and Success

To fully leverage the potential of smart supply chains and foster business growth, companies need to adopt a strategic approach that aligns technological advancements with their fundamental business goals. By investing in innovative technologies like IoT, AI and blockchain, businesses can gain a competitive edge by enhancing operational efficiency, improving customer experiences and refining decision-making processes [30, 93–95]. Additionally, prioritizing sustainability and ethical practices can bolster brand reputation and attract increasingly conscientious consumers. By strategically utilizing these insights and technologies, companies can position themselves for lasting success in a rapidly evolving market. For a clearer understanding, the thematic evolution of trending topics has been analyzed using the Bibliometrix library in the R programming language (see Fig. 4). The diagram illustrates a dynamic shift in technological and industrial focuses from 2015 to 2024. Industry 4.0 consistently emerges as a key theme across all three timeframes (2015–2020, 2021–2023, and 2024), highlighting its foundational role in industrial progress. As the focus transitions toward Industry 5.0 and beyond, there is a notable shift from the mechanization and automation themes of Industry 4.0 to a more intricate integration of intelligent robotics, artificial intelligence and reinforcement learning. This shift indicates that while Industry 4.0 established the framework through cyber-physical systems and process automation, Industry 5.0 emphasizes collaboration between humans and machines, resulting in smarter and more adaptable systems.

The rise of themes such as accident prevention, production process optimization and natural language processing systems signifies a future where industrial systems are not only automated but also intelligent and intuitive. Within the realm of smart supply chains, the diagram highlights a transition toward incorporating AI and machine learning to enhance decision-making, predict outcomes and optimize logistics. These advancements signify a move from traditional supply chain

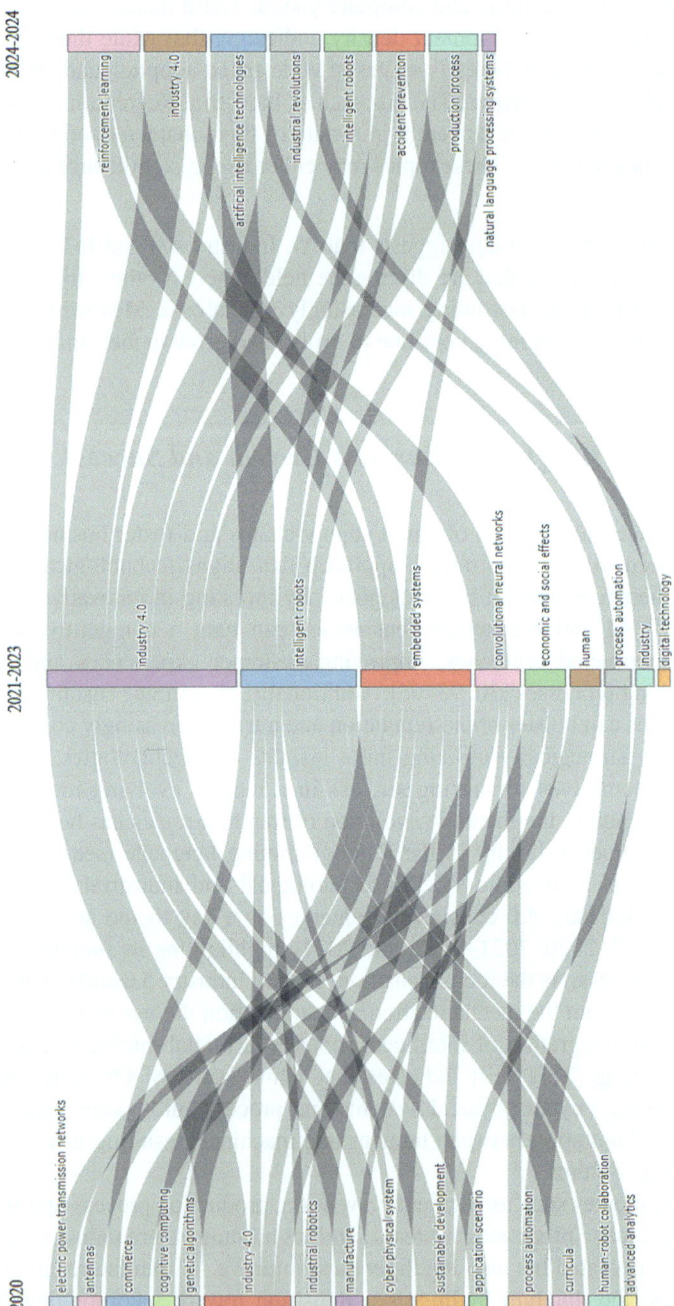

Fig. 4 Thematic evolution

management to a more interconnected, responsive and efficient system that aligns with the principles of Industry 5.0, where human creativity and AI technologies come together to build more resilient and sustainable supply chains.

7 Conclusions and Future Scope

The evolution of supply chains into the era of Industry 5.0 represents a substantial shift toward a more integrated, human-centric approach to logistics and operations. This chapter has emphasized that the future of supply chains involves not only the adoption of advanced technologies but also the enhancement of human skills and the alignment of operations with sustainability goals. The transition from Industry 4.0 to Industry 5.0 signifies a major change; while Industry 4.0 was primarily focused on automation and data integration, Industry 5.0 prioritizes human–machine collaboration. Technologies such as IoT, AI, blockchain, and robotics are increasingly working in conjunction with human creativity and insight, fostering environments where technology and people coexist seamlessly. This shift not only improves operational efficiency but also allows businesses to provide more personalized and responsive services. Sustainability remains a key component of this transformation, with circular economy practices and green logistics leading the way to minimize environmental impacts, optimize resource usage, and promote responsible consumption. As companies strive to meet regulatory standards and respond to consumer expectations for environmental accountability, integrating sustainable practices into the supply chain is essential for long-term success.

However, the journey toward smart supply chains is not without its challenges. Issues such as data security, the integration of legacy systems and workforce adaptation require careful management and strategic planning. Developing effective solutions and implementing proactive measures will be crucial to overcoming these hurdles and fully harnessing the benefits of emerging technologies. Looking ahead, trends such as quantum computing, 5G, digital twins, and augmented reality, are set to further revolutionize supply chain management. These innovations will enhance efficiency, connectivity, and creativity, reshaping the landscape of logistics and operations. In summary, achieving success in supply chain management within Industry 5.0 demands a balanced approach that capitalizes on technological advancements while emphasizing human elements and sustainability. By embracing these principles and remaining flexible, businesses can effectively navigate the complexities of contemporary supply chains and position themselves for ongoing growth and success in a changing world.

References

1. Lessi, C. C., Gavrielides, A., Solina, V., Qiu, R., Nicoletti, L., & Li, D. (2024). 5G and beyond 5G technologies enabling Industry 5.0: Network applications for robotics. *Procedia Computer Science*. https://doi.org/10.1016/j.procs.2024.01.067
2. Marica, M. C., Bizon, N., Bostan, I., & Enescu, M. C. (2024). A brief review of Industry 5.0: Key technologies, applications and future perspectives. In *Proceedings of the 16th International Conference on Electronics, Computers and Artificial Intelligence, ECAI 2024*. Institute of Electrical and Electronics Engineers. https://doi.org/10.1109/ECAI61503.2024.10607446
3. Oeij, P. R. A., Lenaerts, K., Dhondt, S., Van Dijk, W., Schartinger, D., Sorko, S. R., & Warhurst, C. (2024). A conceptual framework for workforce skills for Industry 5.0: Implications for research, policy and practice. *Journal of Innovation Management*. https://doi.org/10.24840/2183-0606_012.001_0010
4. Prassida, G. F., & Asfari, U. (2021). A conceptual model for the acceptance of collaborative robots in Industry 5.0. *Procedia Computer Science*. https://doi.org/10.1016/j.procs.2021.12.118
5. Du Plooy, H., Tommasi, F., Furlan, A., Nenna, F., Gamberini, L., Ceschi, A., & Sartori, R. (2024). A human-centered perspective on individual risks for digital innovation management: An integrative conceptual review. *European Journal of Innovation Management*. https://doi.org/10.1108/EJIM-09-2023-0821
6. Modoni, G. E., & Sacco, M. (2023). A human digital-twin-based framework driving human centricity towards Industry 5.0. *Sensors*. https://doi.org/10.3390/s23136054
7. Zhong, R., Hu, B., Hong, Z., Zhang, Z., Feng, Y., & Tan, J. (2024). A human digital twin based framework for human–robot hand-over task intention recognition. In J. Tan, Y. Liu, H.-Z. Huang, J. Yu, & Z. Wang (Eds.), *Mechanisms and machine science*. Springer Science and Business Media B.V. https://doi.org/10.1007/978-981-97-0922-9_18
8. ElFar, O. A., Chang, C.-K., Leong, H. Y., Peter, A. P., Chew, K. W., & Show, P. L. (2021). Prospects of Industry 5.0 in algae: Customization of production and new advance technology for clean bioenergy generation. *Energy Conversion and Management: X*. https://doi.org/10.1016/j.ecmx.2020.100048
9. Šverko, M., & Grbac, T. G. (2024). Automated HMI design as a custom feature in industrial SCADA systems. *Procedia Computer Science*. https://doi.org/10.1016/j.procs.2024.02.001
10. Scopus. (2024). Smart supply chains in Industry 5.0. Retrieved September 9, 2024, from https://www.scopus.com/results
11. Aria, M., & Cuccurullo, C. (2017). bibliometrix: An R-tool for comprehensive science mapping analysis. *Journal of Informetrics, 11*(4), 959–975. https://doi.org/10.1016/j.joi.2017.08.007
12. Nikishina, L. B. (2023). Industry 4.0: History of emergence, development, prospects of transformation into Industry 5.0. *E3S Web of Conferences*. https://doi.org/10.1051/e3sconf/202345806023
13. Askin, J., & Bitsch, G. (2023). Investigating the influence of a cobot's average tool center point speed on human work behavior in a cooperative human-robot collaboration assembly station. *Procedia CIRP*. https://doi.org/10.1016/j.procir.2023.06.038
14. Katuk, N., Vergallo, R., & Sugiharto, T. (2024). *The future of human-computer integration: Industry 5.0 technology, tools and algorithms*. CRC Press. https://doi.org/10.1201/9781003479727
15. Shrayfel, I., Mikhailov, A., Semenishchev, E., Marchuk, V., Mikhailova, I., & Malcev, I. (2021). A method for reducing the noise component for smooth monotonic signals and an algorithm for its application for prediction problems and detection of local stationary regions in images. In *Proceedings of SPIE—The International Society for Optical Engineering*. SPIE. https://doi.org/10.1117/12.2600694
16. Bobek, S., Nowaczyk, S., Gama, J., Pashami, S., Ribeiro, R. P., Taghiyarrenani, Z., et al. (2023). Why Industry 5.0 needs XAI 2.0? In L. Longo (Ed.), *CEUR Workshop Proceedings*. CEUR-WS. Retrieved from https://www.scopus.com/inward/record.uri?eid=2-s2.0-85178627852&partnerID=40&md5=da1998794dd2a2dc8e2fb9270383d19d

17. Cucchiara, R., Baraldi, L., Calderara, S., Cornia, M., Fabbri, M., Grana, C., et al. (2023). Where research meets industry: New challenges and opportunities at AImageLab. In F. Falchi, F. Giannotti, A. Monreale, C. Boldrini, S. Rinzivillo, & S. Colantonio (Eds.), *CEUR Workshop Proceedings*. CEUR-WS. Retrieved from https://www.scopus.com/inward/record.uri?eid=2-s2.0-85173893238&partnerID=40&md5=38a8e98c87f43e3c2a35c6f909de4a58
18. Noor-A-Rahim, M., John, J., Firyaguna, F., Sherazi, H. H. R., Kushch, S., Vijayan, A., et al. (2023). Wireless communications for smart manufacturing and industrial IoT: Existing technologies, 5G and beyond. *Sensors*. https://doi.org/10.3390/s23010073
19. Caiado, R. G. G., Machado, E., Santos, R. S., Thomé, A. M. T., & Scavarda, L. F. (2024). Sustainable I4.0 integration and transition to I5.0 in traditional and digital technological organisations. *Technological Forecasting and Social Change*. https://doi.org/10.1016/j.techfore.2024.123582
20. Iqbal, Q., & Piwowar-Sulej, K. (2024). Technological social responsibility: A stakeholder theory-based measurement scale. *Technological Forecasting and Social Change*. https://doi.org/10.1016/j.techfore.2024.123465
21. Eren, S. (2024). Toyota Kata as a scaffolding for human-centric manufacturing: Applying lean thinking for a digital and sustainable factory of the future. In F. J. G. Silva, L. P. Ferreira, J. C. Sá, M. T. Pereira, & C. M. A. Pinto (Eds.), *Lecture notes in mechanical engineering*. Springer Science and Business Media Deutschland GmbH. https://doi.org/10.1007/978-3-031-38165-2_45
22. Vyhmeister, E., & Castane, G. G. (2024). Towards industry 5.0 by incorporation of trustworthy and human-centric approaches. In *Artificial intelligence in manufacturing: Enabling intelligent, flexible and cost-effective production through AI*. Springer Nature. https://doi.org/10.1007/978-3-031-46452-2_21
23. Adel, A. (2022). Future of industry 5.0 in society: Human-centric solutions, challenges and prospective research areas. *Journal of Cloud Computing*. https://doi.org/10.1186/s13677-022-00314-5
24. Mladineo, M., Ćubić, M., Gjeldum, N., & Crnjac Žižić, M. (2021). Human-centric approach of the Lean management as an enabler of Industry 5.0 in SMEs. In S. Jozic, B. Lela, & N. Gjeldum (Eds.), *Mechanical technology and structural materials*. Croatian Society for Mechanical Technologies. Retrieved from https://www.scopus.com/inward/record.uri?eid=2-s2.0-85124905182&partnerID=40&md5=834d92935b4f48ea2ac4e4b470eaa491
25. Mourtzis, D., Panopoulos, N., Angelopoulos, J., Wang, B., & Wang, L. (2022). Human centric platforms for personalized value creation in metaverse. *Journal of Manufacturing Systems*. https://doi.org/10.1016/j.jmsy.2022.11.004
26. Hagström, M. H., & Bergsjö, D. (2024). A proposed framework using systems engineering to design human-centric manufacturing systems for novel products to reduce complexity and risk. *Proceedings of the Design Society*. https://doi.org/10.1017/pds.2024.262
27. Panagou, S., Neumann, W. P., & Fruggiero, F. (2024). A scoping review of human robot interaction research towards Industry 5.0 human-centric workplaces. *International Journal of Production Research*. https://doi.org/10.1080/00207543.2023.2172473
28. Zreik, M. (2023). Sustainable and smart supply chains in China: A multidimensional approach. In *Integrating intelligence and sustainability in supply chains*. IGI Global. https://doi.org/10.4018/979-8-3693-0225-5.ch010
29. Leng, J., Sha, W., Lin, Z., Jing, J., Liu, Q., & Chen, X. (2023). Blockchained smart contract pyramid-driven multi-agent autonomous process control for resilient individualised manufacturing towards Industry 5.0. *International Journal of Production Research*. https://doi.org/10.1080/00207543.2022.2089929
30. Sriman, B., Annie Silviya, S. H., Santhosh Kumar, E., Suryaa Narayanan, K., & Nishaalu, S. (2022). Blockchain Industry 5.0: Next generation smart contract and decentralized application platform. In *Proceedings of the 2022 International Conference on Innovative Computing, Intelligent Communication and Smart Electrical Systems, ICSES 2022*. Institute of Electrical and Electronics Engineers. https://doi.org/10.1109/ICSES55317.2022.9914151

31. Mustapić, M., Trstenjak, M., Gregurić, P., & Opetuk, T. (2023). Implementation and use of digital, green and sustainable technologies in internal and external transport of manufacturing companies. *Sustainability (Switzerland)*. https://doi.org/10.3390/su15129557
32. Saldanha-da-Gama, F. (2022). Facility location in logistics and transportation: An enduring relationship. *Transportation Research Part E: Logistics and Transportation Review*. https://doi.org/10.1016/j.tre.2022.102903
33. Ikenga, G. U., & van der Sijde, P. (2024). Twenty-first century competencies; about competencies for Industry 5.0 and the opportunities for emerging economies. *Sustainability (Switzerland)*. https://doi.org/10.3390/su16167166
34. Olsen, C. (2023). Toward a digital sustainability reporting framework in organizations in the Industry 5.0 era: An accounting perspective. In A. M. A. Musleh Al-Sartawi, A. Razzaque, & M. M. Kamal (Eds.), *Lecture notes in networks and systems*. Springer Science and Business Media Deutschland GmbH. https://doi.org/10.1007/978-3-031-17746-0_37
35. Rahman, N. A. A., Mohammad, M. F., Ghafar, N. H., Moin, M. I. A., & Ali, A. (2020). The salient motives for Malaysia aviation industry sustainability: An explorative study on business, management and technology components in aviation management program in Malaysia. *International Journal of Supply Chain Management*. Retrieved from https://www.scopus.com/inward/record.uri?eid=2-s2.0-85087761213&partnerID=40&md5=13463165675cf62cd27b2ee41130e3fd
36. Rathore, A. S., & Sarin, D. (2024). What should next-generation analytical platforms for biopharmaceutical production look like? *Trends in Biotechnology*. https://doi.org/10.1016/j.tibtech.2023.08.008
37. Upadhyaya, P., Ruchi, R., & Dutt, S. (2022). Transfer learning approach for 6G-IoT applications. In *7th International Conference on Communication and Electronics Systems, ICCES 2022—Proceedings*. Institute of Electrical and Electronics Engineers. https://doi.org/10.1109/ICCES54183.2022.9835931
38. Farhadighalati, N., Farhady Ghalaty, N., Nikghadam-Hojjati, S., Marchetti, E., & Barata, J. (2023). SAFE-HEALTH: A secure framework for advancing edge-based health 5.0. In *2023 IEEE World Forum on Internet of Things: The Blue Planet: A Marriage of Sea and Space, WF-IoT 2023*. Institute of Electrical and Electronics Engineers. https://doi.org/10.1109/WF-IoT58464.2023.10539418
39. Kotsiopoulos, T., Papakostas, G., Vafeiadis, T., Dimitriadis, V., Nizamis, A., Bolzoni, A., et al. (2024). Revolutionizing defect recognition in hard metal industry through AI explainability, human-in-the-loop approaches and cognitive mechanisms. *Expert Systems with Applications*. https://doi.org/10.1016/j.eswa.2024.124839
40. Navale, G. S., Madala, R., Managuli, M., Jayalakshmi, N., Kadiravan, G., & Rawat, R. (2023). Research and innovation in next generation security and privacy in Industry 5.0 Iot. In *Proceedings of International Conference on Contemporary Computing and Informatics, IC3I 2023*. Institute of Electrical and Electronics Engineers. https://doi.org/10.1109/IC3I59117.2023.10397984
41. Rojek, I., Kotlarz, P., Dorożyński, J., & Mikołajewski, D. (2024). Sixth-generation (6G) networks for improved machine-to-machine (M2M) communication in Industry 4.0. *Electronics (Switzerland)*. https://doi.org/10.3390/electronics13101832
42. Al-Refaey, M. R., Tyagi, A. K., Al-Ghamdi, A. S. A.-M., & Kukreja, S. (2024). *Topics in artificial intelligence applied to Industry 4.0*. Wiley. https://doi.org/10.1002/9781394216147
43. Dovramadjiev, T., Dobreva, D., Murzova, T., Murzova, M., Markov, V., Iliev, I., et al. (2024). Interaction between artificial intelligence, 2D and 3D open source software and additive technologies for the needs of design practice. In A. Iglesias, J. Shin, B. Patel, & A. Joshi (Eds.), *Lecture notes in networks and systems*. Springer Science and Business Media Deutschland GmbH. https://doi.org/10.1007/978-981-99-8349-0_26
44. Rojek, I., Mikołajewski, D., Mroziński, A., & Macko, M. (2024). Green energy management in manufacturing based on demand prediction by artificial intelligence—A review. *Electronics (Switzerland)*. https://doi.org/10.3390/electronics13163338

45. Rožanec, J., Trajkova, E., Novalija, I., Zajec, P., Kenda, K., Fortuna, B., & Mladenić, D. (2022). Enriching artificial intelligence explanations with knowledge fragments. *Future Internet*. https://doi.org/10.3390/fi14050134
46. Ahamed, N. N., & Vignesh, R. (2022). Smart agriculture and food industry with blockchain and artificial intelligence. *Journal of Computer Science*. https://doi.org/10.3844/JCSSP.2022.1.17
47. Jahid, A., Alsharif, M. H., & Hall, T. J. (2023). The convergence of blockchain, IoT and 6G: Potential, opportunities, challenges and research roadmap. *Journal of Network and Computer Applications*. https://doi.org/10.1016/j.jnca.2023.103677
48. Kannan, D., Amiri, A. S., Shaayesteh, M. T., Nasr, A. K., & Mina, H. (2024). Unveiling barriers to the integration of blockchain-based circular economy and Industry 5.0 in manufacturing industries: A strategic prioritization approach. *Business Strategy and the Environment*. https://doi.org/10.1002/bse.3886
49. Yang, H., Asheralieva, A., Zhang, J., Karim, M. M., Niyato, D. T., & Raza, K. A. (2022). User-centric blockchain for Industry 5.0 applications. In *2022 IEEE International Conference on Communications Workshops (ICC Workshops 2022)*. Institute of Electrical and Electronics Engineers. https://doi.org/10.1109/ICCWorkshops53468.2022.9814562
50. Devi, B. S., & Muthu Selvam, M. (2022). SoloDB for social media's big data using deep natural language with AI applications and Industry 5.0. In P. Karuppusamy, I. Perikos, & F. P. García Márquez (Eds.), *Smart innovation, systems and technologies*. Springer Science and Business Media Deutschland GmbH. https://doi.org/10.1007/978-981-16-3675-2_21
51. Ferreira, H., Neves, A., Lopes, F. J., & Godina, R. (2024). How to raise consumer awareness of the environmental impact of vehicle attributes? A big data analysis. *Procedia Computer Science*. https://doi.org/10.1016/j.procs.2024.01.144
52. Sheela, S. C., Srivastava, S., Kumar, S. V., Aravinda, K., Singh, N., & Yusupov, A. (2024). Strategic implementation of big data analytics for industry 5.0's sustainable expansion. In *Powering Industry 5.0 and sustainable development through innovation*. IGI Global. https://doi.org/10.4018/979-8-3693-3550-5.ch019
53. Duarte, I. M., Pinto, A., Carvalho, C., Zornoza, A., & Santos, J. (2022). The contribution of the user experiences goals for designing better cobots: A systematic literature review. *Applied System Innovation*. https://doi.org/10.3390/asi5060119
54. Raffik, R., Rithish, R. S., Sushmita, V., Jyothi, L. N., & Gautham, S. (2023). Enhancing human cobot interaction with mixed reality: A futuristic review. In *2nd International Conference on Advancements in Electrical, Electronics, Communication, Computing and Automation, ICAECA 2023*. Institute of Electrical and Electronics Engineers. https://doi.org/10.1109/ICAECA56562.2023.10199911
55. Sahan, A. S. M., Kathiravan, S., Lokesh, M., & Raffik, R. (2023). Role of cobots over industrial robots in Industry 5.0: A review. In *2nd International Conference on Advancements in Electrical, Electronics, Communication, Computing and Automation, ICAECA 2023*. Institute of Electrical and Electronics Engineers. https://doi.org/10.1109/ICAECA56562.2023.10201199
56. Alberti, E., Alvarez-Napagao, S., Anaya, V., Barroso, M., Barrué, C., Beecks, C., et al. (2024). AI lifecycle zero-touch orchestration within the edge-to-cloud continuum for Industry 5.0. *Systems*. https://doi.org/10.3390/systems12020048
57. Kumar, N., Singh, B. J., & Khope, P. (2024). Unleashing an ML-based selection criteria for economic lot sizing in a smart batch-type production system. *The TQM Journal, 36*(1), 90–107. https://doi.org/10.1108/TQM-05-2022-0166
58. Sharma, A., & Singh, B. J. (2022). Understanding LSS 4.0 through golden circle model and reviewing its scope in Indian textile industry. *International Journal of Six Sigma and Competitive Advantage, 14*(1), 120. https://doi.org/10.1504/IJSSCA.2022.124301
59. Singh, N. K., Pandey, S. K., Nagalakshmi, M., Kumar, A. A., Tiwari, M., & Kumar, S. (2022). Artificial Intelligence-based cloud computing for Industry 5.0. In *Proceedings of the 2022 2nd International Conference on Innovative Sustainable Computational Technologies, CISCT 2022*. Institute of Electrical and Electronics Engineers. https://doi.org/10.1109/CISCT55310.2022.10046562

60. Börold, A., Schweers, D., & Freitag, M. (2023). Towards multimodal information systems for assisting humans in production and logistics processes. *Procedia CIRP*. https://doi.org/10.1016/j.procir.2023.09.130
61. Ordieres-Meré, J., Gutierrez, M., & Villalba-Díez, J. (2023). Toward the industry 5.0 paradigm: Increasing value creation through the robust integration of humans and machines. *Computers in Industry*. https://doi.org/10.1016/j.compind.2023.103947
62. Seth, D., Gupta, M., & Singh, B. J. (2022). Study in the BFSI sector on the role of AI in human resource management. In D. Bathla & A. Singh (Eds.), *Advances in marketing, customer relationship management and e-services* (pp. 173–181). IGI Global. https://doi.org/10.4018/978-1-6684-6133-4.ch015
63. Singh, B. J., Sehgal, R., Chakraborty, A., & Phanden, R. K. (2024). Managing digital manufacturing transformation: Assessing the status-quo and future prospects in North Indian industries. *Journal of Strategy and Management*. https://doi.org/10.1108/JSMA-07-2023-0168
64. Bordel, B., Alcarria, R., de la Cal Hacar, G., & Valladares, T. R. (2023). Efficient and accountable Industry 5.0 production scheduling mechanism for mass customization scenarios. In J. Bravo & G. Urzaiz (Eds.), *Lecture notes in networks and systems*. Springer Science and Business Media Deutschland GmbH. https://doi.org/10.1007/978-3-031-48642-5_5
65. Sharma, A., & Singh, B. J. (2020). Evolution of industrial revolutions: A review. *International Journal of Innovative Technology and Exploring Engineering, 9*(11), 66–73. https://doi.org/10.35940/ijitee.I7144.0991120
66. Sharma, A., & Singh, B. J. (2023). Evolution of modified LSS 4.0 model for sustainable Indian textile industry: A narrative review. *International Journal on Interactive Design and Manufacturing (IJIDeM)*. https://doi.org/10.1007/s12008-023-01534-y
67. Doyle-Kent, M., & Shanahan, B. W. (2022). The development of a novel educational model to successfully upskill technical workers for Industry 5.0: Ireland a case study. *IFAC-PapersOnLine*. https://doi.org/10.1016/j.ifacol.2022.12.072
68. Pinto, R., Pinheiro, J., Gonçalves, G., & Ribeiro, A. (2023). Towards Industry 5.0: A capacitation approach for upskilling and technology transfer. In R. Mehmood et al. (Eds.), *Lecture notes in networks and systems*. Springer Science and Business Media Deutschland GmbH. https://doi.org/10.1007/978-3-031-38318-2_34
69. Dhirani, L. L., Mukhtiar, N., Chowdhry, B. S., & Newe, T. (2023). Ethical dilemmas and privacy issues in emerging technologies: A review. *Sensors*. https://doi.org/10.3390/s23031151
70. Longo, F., Padovano, A., & Umbrello, S. (2020). Value-oriented and ethical technology engineering in industry 5.0: A human-centric perspective for the design of the factory of the future. *Applied Sciences (Switzerland)*. https://doi.org/10.3390/APP10124182
71. Margherita, E. G., & Braccini, A. M. (2021). Managing industry 4.0 automation for fair ethical business development: A single case study. *Technological Forecasting and Social Change*. https://doi.org/10.1016/j.techfore.2021.121048
72. Murphy, C., Carew, P. J., & Stapleton, L. (2022). Ethical personalisation and control systems for smart human-centred Industry 5.0 applications. *IFAC-PapersOnLine*. https://doi.org/10.1016/j.ifacol.2022.12.005
73. Atif, S. (2023). Analysing the alignment between circular economy and industry 4.0 nexus with industry 5.0 era: An integrative systematic literature review. *Sustainable Development*. https://doi.org/10.1002/sd.2542
74. Turner, C., & Oyekan, J. (2023). Personalised production in the age of circular additive manufacturing. *Applied Sciences (Switzerland)*. https://doi.org/10.3390/app13084912
75. Voulgaridis, K., Lagkas, T., & Sarigiannidis, P. (2022). Towards Industry 5.0 and digital circular economy: Current research and application trends. In *Proceedings of the 18th Annual International Conference on Distributed Computing in Sensor Systems, DCOSS 2022*. Institute of Electrical and Electronics Engineers. https://doi.org/10.1109/DCOSS54816.2022.00037
76. Chu, C.-H., & Pan, J.-K. (2024). A systematic review on extended reality applications for sustainable manufacturing across the product lifecycle. *International Journal of*

Precision Engineering and Manufacturing—Green Technology. https://doi.org/10.1007/s40684-023-00567-8

77. Trstenjak, M., Mustapić, M., Gregurić, P., & Opetuk, T. (2023). Use of green Industry 5.0 technologies in logistics activities. *Tehnicki Glasnik*. https://doi.org/10.31803/tg-20230518185836
78. Granata, I., Faccio, M., & Calzavara, M. (2022). Energy expenditure and makespan multi-objective optimization for cobots systems design. *Procedia Computer Science*. https://doi.org/10.1016/j.procs.2022.12.208
79. Hasan, M. R., Liu, Z., & Rahman, A. (2024). Optimizing energy consumption in directed energy deposition-based hybrid additive manufacturing: An integrated modelling and experimental approach. *International Journal of Advanced Manufacturing Technology*. https://doi.org/10.1007/s00170-024-13005-z
80. Javed, S., Tripathy, A., van Deventer, J., Mokayed, H., Paniagua, C., & Delsing, J. (2023). An approach towards demand response optimization at the edge in smart energy systems using local clouds. *Smart Energy*. https://doi.org/10.1016/j.segy.2023.100123
81. Margherita, E. G., & Braccini, A. M. (2021). Socio-technical perspectives in the fourth industrial revolution—Analysing the three main visions: Industry 4.0, the socially sustainable factory of Operator 4.0 and Industry 5.0. In P. Bednar, P. Bednar, A. Nolte, A. Nolte, M. Rajanen, A. S. Islind, et al. (Eds.), *CEUR Workshop Proceedings*. CEUR-WS. Retrieved from https://www.scopus.com/inward/record.uri?eid=2-s2.0-85120998641&partnerID=40&md5=ed37483edb0b18e1dc78458267134ab0
82. Singh, B., Srivastava, A. K., & Prakash, O. (2024). Review on machine learning as a key technology enabler for sustainable biodiesel production. In *Machine learning techniques and industry applications*. IGI Global. https://doi.org/10.4018/979-8-3693-5271-7.ch011
83. Clim, A., Toma, A., Zota, R. D., & Constantinescu, R. (2023). The need for cybersecurity in industrial revolution and smart cities. *Sensors*. https://doi.org/10.3390/s23010120
84. Kharchenko, V., Dotsenko, S., Illiashenko, O., & Kamenskyi, S. (2019). Integrated cyber safety security management system: Industry 4.0 issue. In *Conference Proceedings of 2019 10th International Conference on Dependable Systems, Services and Technologies, DESSERT 2019*. Institute of Electrical and Electronics Engineers. https://doi.org/10.1109/DESSERT.2019.8770010
85. Ramachandran, K. K., Nagarjuna, B., Akram, S. V., Bhalani, J., Raju, A. M., & Ponnusamy, R. (2023). Innovative cyber security solutions built on block chain technology for Industrial 5.0 applications. In *2023 International Conference on Artificial Intelligence and Smart Communication, AISC 2023*. Institute of Electrical and Electronics Engineers. https://doi.org/10.1109/AISC56616.2023.10085577
86. Waheed, T., & Marchetti, E. (2023). The impact of IOT cybersecurity testing in the perspective of Industry 5.0. In *International Conference on Web Information Systems and Technologies, WEBIST—Proceedings*. Science and Technology Publications, Lda. https://doi.org/10.5220/0012235800003584
87. Korodi, A., Nițulescu, I.-V., Fülöp, A.-A., Vesa, V.-C., Demian, P., Braneci, R.-A., & Popescu, D. (2024). Integration of legacy industrial equipment in a building-management system Industry 5.0 scenario. *Electronics (Switzerland)*. https://doi.org/10.3390/electronics13163229
88. Narkhede, G., Chinchanikar, S., Narkhede, R., & Chaudhari, T. (2024). Role of Industry 5.0 for driving sustainability in the manufacturing sector: An emerging research agenda. *Journal of Strategy and Management*. https://doi.org/10.1108/JSMA-06-2023-0144
89. Walsh Shanahan, B., & Doyle-Kent, M. (2024). The global apprentice: Adaptable, versatile and a lifelong learner. In N. M. Durakbasa & M. G. Gençyılmaz (Eds.), *Lecture notes in mechanical engineering*. Springer Science and Business Media Deutschland GmbH. https://doi.org/10.1007/978-3-031-53991-6_18
90. Hadzovic, S., Mrdovic, S., & Radonjic, M. (2023). A path towards an internet of things and artificial intelligence regulatory framework. *IEEE Communications Magazine*. https://doi.org/10.1109/MCOM.002.2200373

91. Alguirat, I., Zouari, A., & Lehyani, F. (2024). Exploring the role of digital twins in future supply chain industry: Characteristics, technologies and applications. In F. Jawab (Ed.), *2024 IEEE 15th International Colloquium of Logistics and Supply Chain Management (LOGISTIQUA 2024)*. Institute of Electrical and Electronics Engineers. https://doi.org/10.1109/LOGISTIQUA61063.2024.10571453
92. Kumar, S., & Mallipeddi, R. R. (2022). Impact of cybersecurity on operations and supply chain management: Emerging trends and future research directions. *Production and Operations Management*. https://doi.org/10.1111/poms.13859
93. Fraga-Lamas, P., Fernández-Caramés, T. M., Rosado Da Cruz, A. M., & Lopes, S. I. (2024). An overview of blockchain for Industry 5.0: Towards human-centric, sustainable and resilient applications. *IEEE Access*. https://doi.org/10.1109/ACCESS.2024.3435374
94. Verma, A., Bhattacharya, P., Madhani, N., Trivedi, C., Bhushan, B., Tanwar, S., et al. (2022). Blockchain for Industry 5.0: Vision, opportunities, key enablers and future directions. *IEEE Access*. https://doi.org/10.1109/ACCESS.2022.3186892
95. Verma, S. K., Nadeem, M., Verma, V., Sayeed, M. A., Agrawal, A., & Khan, R. A. (2023). Blockchain application in healthcare domain: Industry 5.0 perspective. In *Proceedings of International Conference on Contemporary Computing and Informatics, IC3I 2023*. Institute of Electrical and Electronics Engineers. https://doi.org/10.1109/IC3I59117.2023.10397614

Measurement of AI-Based Drivers of Smart Supply Chain Toward Sustainability in the Era of Industry 5.0

Mohit Sharma, Mohit Tyagi, R. S. Walia, Yatika Gori, Ashwani Kumar, and Yogesh Kumar Singla

1 Introduction and Background

The twenty-first century saw transformative digital advancements in every area of the economy, leading to a profound societal and global economic overhaul. The term "digitization" is now a prominent subject of study in both consumer and industrial fields [1]. Traditional industrial processes are undergoing a transformation driven by the innovative application of cutting edge technologies, including artificial intelligence (AI), robotics, internet of things (IoT) [2]. It is anticipated that this trend will persist as technological advancements continue to accelerate [3]. Sharma et al. [4] stated that currently the rate at which digital technologies are being used is increasing significantly. In this regard, Industry 5.0 (I5.0) is widely considered by

M. Sharma · M. Tyagi (✉) · R. S. Walia
Department of Production and Industrial Engineering, Punjab Engineering College (Deemed to be University), Chandigarh, India

Y. Gori
Department of Mechanical Engineering, Graphic Era Deemed to be University, Dehradun, Uttarakhand, India

School of Computing, Graphic Era Hill University, Haldwani, Uttarakhand, India
e-mail: yatikagori@gehu.ac.in

A. Kumar
Department of Mechanical Engineering, Technical Education Department Uttar Pradesh, Kanpur, India
e-mail: ashwaniadj.me@geu.ac.in

Y. Kumar Singla
School of Engineering, Math and Technology, Navajo Technical University, Crownpoint, NM, USA

Department of Mechanical Engineering, University of Idaho, Moscow, ID, USA
e-mail: ysingla@navajotech.edu

Y. Kumar Singla et al. (eds.), *Digital Manufacturing in Industry 5.0*,
https://doi.org/10.1007/978-3-031-91500-0_48

experts and professionals as the next industrial revolution, and it has gained significant attention in recent years [5]. The main objective of I5.0 is to leverage human experts in collaboration with machines to get the output most efficiently [6]. A recent article published in Forbes explains that I5.0 enhances the human machines corporation which could significantly influence manufacturing companies [7]. I5.0 is the next industrial revolution that focuses on the cooperation of humans and machines to improve the customer experience. In contrast, I4.0 is all about digitization and networked automation. The main goal of the I5.0 paradigm is to integrate social integrity and sustainability, which the I4.0 paradigm left unaddressed, into a smart manufacturing system [8].

The smart manufacturing paradigm relies on smart supply chains (SSCs) that can be self-organized and self-optimized [9]. These capabilities are essential for adapting to uncertainties. According to Liu et al. [10], the fast and effective development of smart technologies like AI, blockchain, IoT, and cloud computing has led to the creation of innovative supply chains (SCs), known as "smart supply chains." Supply chain management (SCM) has seen dramatic changes due to the fast advancement of AI. In addition to offering real-time data processing capabilities, it speeds up the decision-making process and adds more creative components to SCM [11]. AI allows SCM to make intelligent decisions by fetching real-time data from hardware devices. This greatly improves the collaborative effectiveness of the whole industrial chain. In today's global commercial climate, SCM is critical [12], and its importance goes beyond improving internal operational efficiency to include the stability and synchronization of the whole industrial chain. Tyagi et al. [13] stated that the market's sudden changes in behavior are making it difficult for sectors to remain stable in the marketplace. As a result, every industry is competing to enhance its supply chain performance (SCP) system. Therefore, the majority of sectors and SCs now use digital applications, and newer cutting-edge technologies like AI, robots, big data, and the IoT are greatly improving company sustainability measures [14–16]. For an SC to be sustainable, it must implement measures that reduce environmental impact, enhance the well-being of all stakeholders, and support long-term financial growth [17]. AI is crucial for digitizing and automating SC processes to increase SC productivity [18]. SCM may benefit from AI in several ways, including increased capacity for optimization, accurate forecast, better quality, lower cost, and safer working environment. Efficiency and transparency are two essential requirements for SC, and AI technology has helped to accomplish both. By keeping the above themes in mind, the objective of the current work has been decided to understand the prioritization of influential drivers for the analysis of key AI-based drivers of SSC [19, 20]. To achieve this objective, key drivers (as shown in Table 1) have been identified through literature and discussion with experts then a 5-point Likert scale (Table 2) questionnaire has been developed and send accordingly.

The AI-based drivers listed in Table 1 were assessed using a Fuzzy environmental framework through the Fuzzy DEMATEL approach. Due to the uncertain and ambiguous nature of the environment, it is crucial to extend the DEMATEL method to analyze criteria within an uncertain framework that includes Fuzzy or linguistic

Table 1 A list of AI-based drivers

Sr. No.	Drivers	Descriptions	References
1	Decision-making and forecasting capabilities (**AIS1**)	AI techniques excel in delivering enhanced decision-making, predictive insights, and advanced learning capabilities	[21]
2	Employee acceptance and alignment for digital transformation (**AIS2**)	Digital transformation leverages technologies such as AI to enhance or innovate industrial processes, culture, and customer experiences. A digitalized quality check is faster and continuous, offering more accurate and detailed information, which helps prevent supply chain disruptions	[22, 23]
3	Digitally enhanced processes and quality control (**AIS3**)	Digital transformation uses digital technologies like AI to enhance processes, culture, and customer experiences. A digitalized quality check is faster and more accurate, helping to prevent supply chain disruptions	[22, 23]
4	Production flexibility through automation integration (**AIS4**)	Emerging technologies, like AI, enable the development of flexible production systems that enhance a firm's ability to respond quickly to customer demands and boost productivity	[24]
5	Implementation of a cyber-physical production system (CPPS) (**AIS5**)	A CPPS enhances operational performance and ensures smooth manufacturing, which can improve supply chain sustainability in emerging markets	[25, 26]
6	Decentralizing production by utilizing additive manufacturing (AM) (**AIS6**)	Advanced technologies are expected to enhance the competitiveness of AM in parts production, driving a shift toward more decentralized manufacturing. This approach leads to more agile, resilient, and flexible supply chains, lowers logistics costs, and ultimately increases SC sustainability	[27, 28]
7	Enhancing SC activities using cloud-based technologies (**AIS7**)	SC operations in developing economies saw severe interruptions during COVID-19, with labor being a crucial component. The use of cloud technology may enhance SC operations via improved stakeholder connection	[29]
8	Utilization of big data and predictive analytics (**AIS8**)	Big data analytics assists manufacturers in planning and problem-solving by identifying real-time issues and optimizing decision-making when faced with unexpected events and threats	[30]

variables, as shown in Table 2. The preview of the present study is as follows: Sect. 2 explains the research methodology and step-by-step numerical representation. Section 3 describes the results and discussion. Finally, the conclusion and future scope of the present study are given in Sect. 4.

Table 2 Fuzzy-based linguistic scale for evaluation

Sr. No.	Linguistic term	Influence score	TFNs
1	No influence (No)	0	(0,0.1,0.3)
2	Very low influence (VL)	1	(0.1,0.3,0.5)
3	Low influence (L)	2	(0.3,0.5,0.7)
4	High influence (H)	3	(0.5,0.7,0.9)
5	Very high influence (VH)	4	(0.7,0.9,1.0)

Source: [31]

2 Research Methodology and Numerical Representation

To handle complexities such as vagueness and uncertainty in data, decision-makers often use descriptive terms such as "very low," "low," "medium," "high," and "very high" to give their views. The above mentioned terms are then translated into fuzzy numbers for further analysis. Unlike crisp sets, which are based on Boolean logic and only recognize values of 0 (false) or 1 (true). $\tilde{A}$ is based on fuzzy logic, and assign a degree of membership to each value within the range [0, 1]. $\tilde{A}$ can be defined as a collection of ordered pairs from a universal set X and is described by a membership function $\propto_{\tilde{A}}(X)$. Function $\propto_{\tilde{A}}(X)$ maps each element of X to a value between 0 and 1, where $\propto_{\tilde{A}}(x) = 1$ means the element fully belongs to $\tilde{A}$, and $\propto_{\tilde{A}}(x) = 0$ means it does not belong to the fuzzy set $\tilde{A}$ at all.

$$\tilde{A} = \left\{\left(x, \mu_{\tilde{A}}(x)\right)\right\}, \quad x \in X \tag{1}$$

Triangular Fuzzy Numbers (TFNs) are commonly used in practical applications due to their ease of calculation. A TFN is typically represented as $\tilde{U}_a = (l_a, m_a, n_a)$. The membership function $\propto_{\tilde{A}}(X)$ for a TFN can be described in the following way:

$$\mu_{\tilde{A}}(x) = \begin{cases} 0, & x \le l_a \\ \dfrac{x - l_a}{m_a - l_a}, & l_a \le x \le m_a \\ \dfrac{n_a - x}{n_a - m_a}, & m_a \le x \le n_a \\ 0, & x > n_a \end{cases} \tag{2}$$

Here l_a, m_a, and n_a are real numbers with $l_a > m_a > n_a$.

3 Fuzzy DEMATEL

A detailed decision-making framework highlights the value of the Fuzzy DEMATEL method. The steps of this method can be outlined as follows:

3.1 Average Direct-Relation Matrix Formation

As noted earlier, human judgments often come with uncertainty, which needs fuzzy aggregation methods and a defuzzification process to handle. In this study, the Converting Fuzzy Data into Crisp Scores (CFCS) method was used to manage this uncertainty. The CFCS method defines fuzzy ranges for the highest and lowest values and then calculates a crisp score by taking a weighted average [31].

Consider $\tilde{U}_{ij}^{p} = \left(l_{ij}^{p}, m_{ij}^{p}, n_{ij}^{p}\right)$, where $p = 1, 2, \ldots, r$. This represents the average effect of the *i*th criterion on the *j*th criterion.

$$\tilde{U}^{n} = \begin{bmatrix} 0 & \tilde{U}_{12}^{p} & \tilde{U}_{13}^{p} & \ldots & \tilde{U}_{1h}^{p} \\ \tilde{U}_{21}^{p} & 0 & \tilde{U}_{23}^{p} & \ldots & \tilde{U}_{2h}^{p} \\ . & . & . & . & . \\ . & . & . & . & . \\ \tilde{U}_{h1}^{p} & \tilde{U}_{h2}^{p} & \tilde{U}_{h3}^{p} & \ldots & 0 \end{bmatrix} \tag{3}$$

where $h = 1, 2, \ldots, q$ represents the number of criteria, and r denotes for the number of respondents. The following five step algorithm can be used to describe the CFCS approach:

1. **Normalization**

$$xl_{ij}^{p} = \left(l_{ij}^{p} - \min n_{ij}^{p}\right) / \Delta_{\min}^{\max} \tag{4}$$

$$xm_{ij}^{p} = (m_{ij}^{p} - \min n_{ij}^{p} / \Delta_{\min}^{\max} \tag{5}$$

$$n_{ij}^{p} = \left(n_{ij}^{p} - \min n_{ij}^{p}\right) / \Delta_{\min}^{\max} \tag{6}$$

where $\varnothing_{\min}^{\max} = \max l_{ij}^{p} - \min n_{ij}^{p}$

2. **Compute the low (ls) and high (rs) normalized values**

$$xnk_{ij}^{p} = xm_{ij}^{p} / \left(1 + xm_{ij}^{p} - xn_{ij}^{p}\right) \tag{7}$$

$$xlk_{ij}^{p} = xl_{ij}^{p} / \left(1 + xl_{ij}^{p} - xm_{ij}^{p}\right) \tag{8}$$

3. **Compute the total normalized crisp values**

$$x_{ij}^{p} = \frac{\left[xnk_{ij}^{p}\left(1 - xnk_{ij}^{p}\right) + xlk_{ij}^{p} * xlk_{ij}^{p}\right]}{\left[1 - xnk_{ij}^{p} + xlk_{ij}^{p}\right]} \tag{9}$$

4. **Compute crisp values**

$$D_{ij}^{p} = \min n_{ij}^{p} + x_{ij}^{p} * \Delta_{\min}^{\max} \tag{10}$$

5. **Integrate crisp values**

$$D_{ij} = \frac{1}{r}\left(D_{ij}^{1} + D_{ij}^{2} + \cdots + D_{ij}^{r}\right) \tag{11}$$

As a result, $D = [D_{ij}]$ represents the average direct-relation matrix that is generated.

3.2 Normalization of Direct-Relation Matrix

Using Eqs. (12) and (13) alongside the established average direct-relation matrix allows for the easy construction of the normalized direct-relation matrix Z.

$$Z = \lambda * D \tag{12}$$

$$\lambda = \min\left[\frac{1}{\max\limits_{1 \le i \le h} \sum_{j=1}^{h} \left|D_{ij}\right|}, \frac{1}{\max\limits_{1 \le i \le h} \sum_{i=1}^{h} \left|D_{ij}\right|}\right], \quad i, j = 1, 2, 3 \ldots h \tag{13}$$

3.3 Determination of Total Relation Matrix

Equation (14) can be used to easily create the total relation matrix. The following steps can be used to derive the complete relation matrix $A_{h \times h}$:

$$A = \sum_{p=1}^{\infty} D_i = A + A^2 + A^3 + \cdots + A^p$$
$$= A\left(I + A + A^2 + A^3 + \cdots + A^{p-1}\right) \tag{14}$$
$$A = A\left(I - A\right)^{-1}$$

where I represents identity matrix.

The rows (M) and columns (N) of the complete matrix are summed up, and these values are combined to obtain ($M + N$), known as "Prominence," which indicates the relative importance of each criterion. Likewise, for each criterion, the ($M - N$) values, referred to as "Relation," are calculated.

$$A = a_{ijh \times h}, i, j = 1, 2, \ldots, h \tag{15}$$

$$M = \sum_{j=1}^{h} a_{ijh \times 1} = \left[a_i\right]_{h \times 1} \tag{16}$$

$$N = \left[\sum_{i=1}^{h} a_{ij}\right]_{1 \times h} = \left[a_j\right]_{1 \times h} \tag{17}$$

3.4 Casual Diagram Formation

Plotting all the ($M + N$) and ($M - N$) values creates a causal diagram, which offers valuable insights for decision-making by illustrating the relationships between the enablers being analyzed. Criteria with positive ($M + N$) values are positioned in the cause group, while those with negative ($M - N$) values are located in the effect group.

After reviewing the literature and consulting with experts, significant AI based drivers of smart supply chain toward sustainability in era of I5.0 have been identified as: Decision-making and forecasting capabilities (AIS1), Employee acceptance and alignment for digital transformation (AIS2), Digitally enhanced processes and quality control (AIS3), Production flexibility through automation integration (AIS4), Implementation of a cyber-physical production system (AIS5), Decentralizing production by utilizing additive manufacturing (AM) (AIS6), Enhancing SC activities using cloud-based technologies (AIS7), and Utilization of big data and predictive analytics (AIS8) are depicted in Table 1.

To begin the analysis, four groups of experts from industry and academia were formed to discuss the digital practices under consideration. Each group consisted of five to seven experts, and data was gathered through questionnaires and brainstorming sessions. The influence ratings collected are shown in Table 2. The experts' opinions were then recorded in linguistic form using Eq. (3).

Due to space limitations and to avoid redundant table entries, the linguistic data collected from experts are not shown here. For analysis, the collected linguistic terms are converted into corresponding Triangular Fuzzy Numbers (TFNs), creating fuzzy inputs. These inputs are then processed using the CFCS defuzzification method [Eqs. (4)–(10)], which produces the defuzzified direct-relation matrix for expert 1. This process is repeated for each linguistic input from the experts, resulting in a separate defuzzified direct-relation matrix for each expert. The average direct-relation matrix, shown in Table 3, is then obtained by combining these defuzzified crisp direct-relation matrices using Eq. (11).

The next step in the Fuzzy DEMATEL method is to normalize the average direct-relation matrix to standardize the variation in the collected data. This normalization is achieved using Eqs. (12) and (14), resulting in the normalized direct-relation matrix shown in Table 4.

In the next phase, the whole relation matrix was created using Eq. (15), as shown in Table 5. The "prominence" ($M + N$) and "relation" ($M - N$) values were then calculated

Table 3 Average direct-relation matrix

	AIS1	AIS2	AIS3	AIS4	AIS5	AIS6	AIS7	AIS8
AIS1	0.000	0.311	0.313	0.125	0.171	0.217	0.265	0.217
AIS2	0.737	0.000	0.643	0.737	0.737	0.735	0.690	0.548
AIS3	0.500	0.548	0.000	0.548	0.595	0.548	0.548	0.595
AIS4	0.310	0.500	0.263	0.000	0.405	0.265	0.125	0.405
AIS5	0.643	0.595	0.690	0.548	0.000	0.690	0.548	0.643
AIS6	0.452	0.405	0.313	0.452	0.500	0.000	0.310	0.500
AIS7	0.829	0.829	0.829	0.829	0.829	0.829	0.000	0.737
AIS8	0.643	0.548	0.643	0.500	0.548	0.595	0.548	0.000

Table 4 Normalized direct-relation matrix

	AIS1	AIS2	AIS3	AIS4	AIS5	AIS6	AIS7	AIS8
AIS1	0.000	0.054	0.055	0.022	0.030	0.038	0.046	0.038
AIS2	0.129	0.000	0.113	0.129	0.129	0.129	0.121	0.096
AIS3	0.088	0.096	0.000	0.096	0.104	0.096	0.096	0.104
AIS4	0.054	0.088	0.046	0.000	0.071	0.046	0.022	0.071
AIS5	0.113	0.104	0.121	0.096	0.000	0.121	0.096	0.113
AIS6	0.079	0.071	0.055	0.079	0.088	0.000	0.054	0.088
AIS7	0.145	0.145	0.145	0.145	0.145	0.145	0.000	0.129
AIS8	0.113	0.096	0.113	0.088	0.096	0.104	0.096	0.000

Table 5 Total relation matrix

	AIS1	AIS2	AIS3	AIS4	AIS5	AIS6	AIS7	AIS8
AIS1	0.080	0.123	0.123	0.095	0.103	0.112	0.105	0.108
AIS2	0.325	0.192	0.290	0.306	0.308	0.312	0.267	0.275
AIS3	0.256	0.248	0.159	0.247	0.256	0.254	0.221	0.251
AIS4	0.157	0.176	0.140	0.096	0.163	0.144	0.105	0.160
AIS5	0.294	0.269	0.281	0.262	0.177	0.290	0.234	0.273
AIS6	0.206	0.186	0.171	0.193	0.201	0.124	0.152	0.198
AIS7	0.373	0.351	0.348	0.350	0.353	0.359	0.186	0.333
AIS8	0.281	0.250	0.263	0.243	0.252	0.263	0.224	0.160

Table 6 The score of digital practices and respective values for cause and effect groups

	M	N	$(M + N)$	$(M - N)$
AIS1	0.850	1.971	2.821	−1.121
AIS2	2.274	1.795	4.069	0.479
AIS3	1.892	1.776	3.668	0.117
AIS4	1.140	1.792	2.932	−0.652
AIS5	2.080	1.814	3.894	0.266
AIS6	1.432	1.859	3.290	−0.427
AIS7	2.653	1.493	4.146	1.160
AIS8	1.935	1.757	3.692	0.177

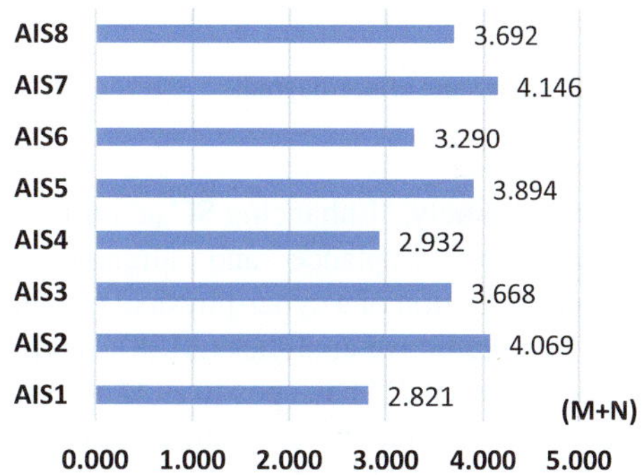

Fig. 1 Importance rating diagram

to analyze the characteristics of the key drivers. Subsequently, the sums of the rows (M) and columns (N) of the total relation matrix were computed using Eqs. (16) and (17) to determine the drivers' "prominence" $(M + N)$ and "relation" $(M - N)$ values. A summary of these values is presented in Table 6. Next, an importance rating diagram (Fig. 1) and a cause-and-effect diagram (Fig. 2) were created using these values.

4 Results and Discussion

Using the $(M + N)$ values from Table 6, the AI-based drivers can be ranked in order of priority, with higher $(M + N)$ values indicating greater importance. The importance ranking of each factor is as follows: AIS7 > AIS2 > AIS5 > AIS8 > AIS3 > A IS6 > AIS4 > AIS1. Therefore, it is evident that AI-based driver, specifically "Enhancing SC activities using cloud-based technologies (AIS7)," is rated as having a higher importance rating (4.164) than other drivers (as shown in Table 6 and Fig. 1). Therefore, "(AIS7)" becomes essential AI-based driver of smart supply chain in era of I5.0. On the other hand, "Decision-making and forecasting capabilities (AIS1)" has the lowest $(M + N)$ value (2.821) (as shown in Table 6 and Fig. 1), making it the least important AI-based driver.

The significant drivers are categorized into their respective cause effect group by use of cause effect diagram (Fig. 2) and observing the $(M + N)$ values (Table 6).

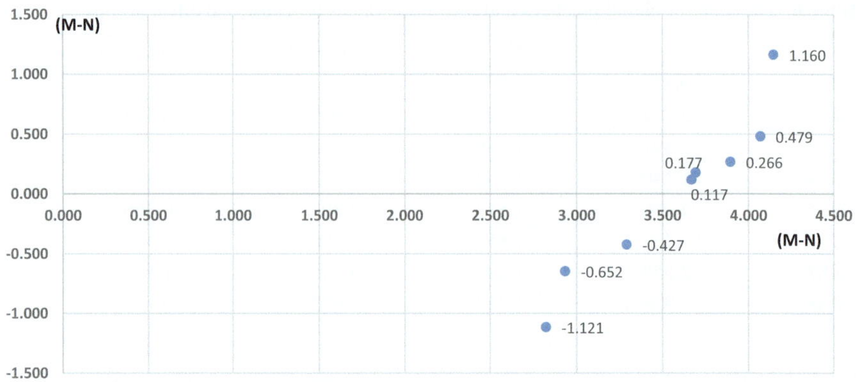

Fig. 2 Cause and effect diagram

Drivers, namely, "Enhancing SC activities using cloud-based technologies (AIS7)," "Employee acceptance and alignment for digital transformation (AIS2)," "Implementation of a cyber-physical production system (AIS5)," "Utilization of big data and predictive analytics (AIS8)," and "Digitally enhanced processes and quality control (AIS3)" have positive ($M + N$) values and come in cause category group. On the other hand, drivers such as "Decentralizing production by utilizing additive manufacturing (AM) (AIS6)," "Production flexibility through automation integration (AIS4)," and "Decision-making and forecasting capabilities (AIS1)" have negative ($M - N$) values and subjected to effect category.

Key AI-based drivers in the cause group exert influence over the effect group factors, and among all factors, "Enhancing SC activities using cloud-based technologies (AIS7)" emerges as significant with highest ($M + N$) value of 4.146. Moreover, it is identified as a particular influential key driver due to the highest ($M - N$) value of 1.160. While on the other side, "Decision-making and forecasting capabilities (AIS1)" driver is categorized by the lowest ($M + N$) and ($M - N$) values as 2.821 and −1.121, respectively.

5 Conclusion and Future Scope

The aim of this study is to measure AI-based drivers of smart supply chain toward sustainability in the era of I5.0. To achieve this, an intensive literature work was done and perspectives of other experts were sought to obtain valuable insights on this topic. The current study used Fuzzy DEMATEL method to identify and analyze key AI-based drivers of SSC. On the basis of analysis, drivers AIS7, AIS2, AIS5, AIS8, and AIS3 categorized in cause group and AIS6, AIS4, and AIS1 in effect group respectively. The final outcomes gathered in Table 6 show that "Enhancing SC activities using cloud-based technologies (AIS7)" impacted more as compared to other drivers and should be encountered first in order to measure AI-based drivers of smart supply chain. On the other side, "Decision-making and forecasting capabilities (AIS1)" impacts least and requires the minimum focus in this context. In the future, additional mathematical techniques could be applied for validation and sensitivity analysis to further explore challenges related to the current study's theme.

References

1. Shao, X. F., Liu, W., Li, Y., Chaudhry, H. R., & Yue, X. G. (2021). Multistage implementation framework for smart supply chain management under Industry 4.0. *Technological Forecasting and Social Change, 162*, 120354.
2. Jamil, M. A., Mustofa, R., Hossain, N. U. I., Rahman, S. A., & Chowdhury, S. (2024). A structural equation modeling framework for exploring the Industry 5.0 and sustainable supply chain determinants. *Supply Chain Analytics, 6*, 100060.
3. Wang, J., & Zhuo, W. (2020). Strategic information sharing in a supply chain under potential supplier encroachment. *Computers & Industrial Engineering, 150*, 106880.
4. Sharma, J., Tyagi, M., & Kazançoğlu, Y. (2024). Impact of digital technologies on the risk assessment in food supply chain: A wake towards digitalisation. *International Journal of Food Science & Technology, 59*(5), 3491–3504.
5. European Commission. (2022). Industry 5.0. Retrieved April 23, 2022, from https://ec.europa.eu/info/research-and-innovation/research-area/industrial-research-and-innovation/industry-50_en
6. Maddikunta, P. K. R., Pham, Q. V., Prabadevi, B., Deepa, N., Dev, K., Gadekallu, T. R., et al. (2022). Industry 5.0: A survey on enabling technologies and potential applications. *Journal of Industrial Information Integration, 26*, 100257.
7. Eschbach, A. (2021). *How Industry 5.0 will transform process manufacturing as we know it*. Forbes.com.
8. Fraga-Lamas, P., Lopes, S. I., & Fernández-Caramés, T. M. (2021). Green IoT and edge AI as key technological enablers for a sustainable digital transition towards a smart circular economy: An industry 5.0 use case. *Sensors, 21*(17), 5745.
9. Viriyasitavat, W., Bi, Z., & Hoonsopon, D. (2022). Blockchain technologies for interoperation of business processes in smart supply chains. *Journal of Industrial Information Integration, 26*, 100326.
10. Liu, W., Shanthikumar, J. G., Lee, P. T. W., Li, X., & Zhou, L. (2021). Special issue editorial: Smart supply chains and intelligent logistics services. *Transportation Research Part E: Logistics and Transportation Review, 147*, 102256.
11. Zhang, C., Zhang, J., Sangaiah, A. K., Li, D., & Li, W. (2024). Evaluating edge artificial intelligence-driven supply chain management platforms using collaborative large-scale fuzzy information fusion. *Applied Soft Computing, 159*, 111686.
12. Van Nguyen, T., Cong Pham, H., Nhat Nguyen, M., Zhou, L., & Akbari, M. (2023). Data-driven review of blockchain applications in supply chain management: Key research themes and future directions. *International Journal of Production Research, 61*(23), 8213–8235.
13. Tyagi, M., Kumar, P., & Kumar, D. (2017). Modelling and analysis of barriers for supply chain performance measurement system. *International Journal of Operational Research, 28*(3), 392–414.
14. Naz, F., Agrawal, R., Kumar, A., Gunasekaran, A., Majumdar, A., & Luthra, S. (2022). Reviewing the applications of artificial intelligence in sustainable supply chains: Exploring research propositions for future directions. *Business Strategy and the Environment, 31*(5), 2400–2423.

15. Sullivan, M., & Kern, J. (Eds.). (2021). *The digital transformation of logistics: Demystifying impacts of the Fourth Industrial Revolution*. John Wiley & Sons.
16. Wang, X., Xu, T., de Andrade, M. J., Rampalli, I., Cao, D., Haque, M., et al. (2021). The interfacial shear strength of carbon nanotube sheet modified carbon fiber composites. In *Challenges in mechanics of time dependent materials* (Proceedings of the 2020 Annual Conference on Experimental and Applied Mechanics) (Vol. 2, pp. 25–32). Springer International Publishing.
17. Yusuf, Y., Menhat, M. S., Abubakar, T., & Ogbuke, N. J. (2020). Agile capabilities as necessary conditions for maximising sustainable supply chain performance: An empirical investigation. *International Journal of Production Economics, 222*, 107501.
18. Sharma, R., Shishodia, A., Gunasekaran, A., Min, H., & Munim, Z. H. (2022). The role of artificial intelligence in supply chain management: Mapping the territory. *International Journal of Production Research, 60*(24), 7527–7550.
19. Shah, S., Nautiyal, H., Gugliani, G., Kumar, A., Namboodri, T., & Singla Editors, Y. K. (2024). *Sustainability in smart manufacturing: Trends, scope, and challenges*. CRC Press. https://doi.org/10.1201/9781003467496
20. Khashehchi, M., Heidari, M., Thangavel, S., Rahmanivahid, P., Kumar, A., & Singla, Y. K. (2024). Chapter 20: Insights into aerospace structural integrity: A study on fiber/epoxy composites fracture. In A. Kumar, Y. Kumar Singla, & M. R. Maughan (Eds.), *Fracture behavior of nanocomposites and reinforced laminate structures* (pp. 471–485). Springer. https://doi.org/10.1007/978-3-031-68694-8_20
21. Baryannis, G., Validi, S., Dani, S., & Antoniou, G. (2019). Supply chain risk management and artificial intelligence: State of the art and future research directions. *International Journal of Production Research, 57*(7), 2179–2202.
22. Rajput, S., & Singh, S. P. (2020). Industry 4.0 Model for circular economy and cleaner production. *Journal of Cleaner Production, 277*, 123853.
23. Shayganmehr, M., Kumar, A., Garza-Reyes, J. A., & Moktadir, M. A. (2021). Industry 4.0 enablers for a cleaner production and circular economy within the context of business ethics: A study in a developing country. *Journal of Cleaner Production, 281*, 125280.
24. Fragapane, G., Ivanov, D., Peron, M., Sgarbossa, F., & Strandhagen, J. O. (2022). Increasing flexibility and productivity in Industry 4.0 production networks with autonomous mobile robots and smart intralogistics. *Annals of Operations Research, 308*(1), 125–143.
25. Andronie, M., Lăzăroiu, G., Iatagan, M., Hurloiu, I., & Dijmărescu, I. (2021). Sustainable cyber-physical production systems in big data-driven smart urban economy: A systematic literature review. *Sustainability, 13*(2), 751.
26. Agarwal, S., Tyagi, M., & Garg, R. K. (2022). Framework development and evaluation of Industry 4.0 technological aspects towards improving the circular economy-based supply chain. *Industrial Robot: The International Journal of Robotics Research and Application, 49*(3), 555–581.
27. Kurpjuweit, S., Schmidt, C. G., Klöckner, M., & Wagner, S. M. (2021). Blockchain in additive manufacturing and its impact on supply chains. *Journal of Business Logistics, 42*(1), 46–70.
28. Kumar, N., Tyagi, M., & Sachdeva, A. (2023). A sustainable framework development and assessment for enhancing the environmental performance of cold supply chain. *Management of Environmental Quality: An International Journal, 34*(4), 1077–1110.
29. Manuel Maqueira, J., Moyano-Fuentes, J., & Bruque, S. (2019). Drivers and consequences of an innovative technology assimilation in the supply chain: Cloud computing and supply chain integration. *International Journal of Production Research, 57*(7), 2083–2103.
30. Luthra, S., Kumar, A., Zavadskas, E. K., Mangla, S. K., & Garza-Reyes, J. A. (2020). Industry 4.0 as an enabler of sustainability diffusion in supply chain: An analysis of influential strength of drivers in an emerging economy. *International Journal of Production Research, 58*(5), 1505–1521.
31. Tyagi, M., Kumar, P., & Kumar, D. (2015). Analysis of interactions among the drivers of green supply chain management. *International Journal of Business Performance and Supply Chain Modelling, 7*(1), 92–108.

Optimizing Supply Chain Finance Decisions Through Supplier Risk Portfolio Assessment

Utkarsh Mittal and Anish Sachdeva

1 Introduction

Supply chain finance (SCF) has become essential for manufacturers, suppliers, and other stakeholders to optimize working capital, accelerate funding, and mitigate financial risk—especially amid global disruptions [1]. However, research indicates that many firms underutilize SCF solutions and fail to realize expected performance gains [2]. A primary difficulty lies in formulating effective financing strategies across an often vast, heterogeneous supplier base [3]. Rather than a customizable approach, inefficient one-size-fits-all programs persist. This underscores the need for advanced yet practical techniques to model supplier payment relationships and risk profiles to support targeted decision-making [4, 5].

Foundational literature establishes relevant factors for characterizing supplier risk attitudes to guide SCF adoption. Seminal works focus on extending credit periods and discounts [6], sustainability scorecards [7], newsvendor formulations [8], and multi-criteria portfolio optimization [9]. While valuable, certain limitations impede customization. First, classical mean-variance approaches prove highly sensitive even to minor model alterations, reducing reliability [10]. Second, abrupt market shifts inject randomness that static representations cannot sufficiently capture [11, 12]. Third, complex system-level interactions among multiple suppliers escape microlevel methods [13]. Finally, graphical network intuitions remain underutilized despite their advantages [14].

Recent studies help mitigate certain challenges through hierarchical modeling [15, 16], mixed integer programming [4], and cluster analysis [17]. However, key gaps endure regarding practical segmentation schemes for diverse supplier groups. On one hand, hierarchical techniques often obscure or overwhelm users with excess

U. Mittal (✉) · A. Sachdeva
NIT Jalandhar, Jalandhar, India
e-mail: utkarshm.ip.24@nitj.ac.in; asachdeva@nitj.ac.in

Y. Kumar Singla et al. (eds.), *Digital Manufacturing in Industry 5.0*,
https://doi.org/10.1007/978-3-031-91500-0_49

detail rather than spotlighting targeted decision support [18]. On the other hand, mainstream extensions of Markowitz's theory struggle to balance generalizability, flexibility, and scalability [14]. This research, therefore, aims to bridge theory with practice for context specific SCF adoption.

This paper proposes novel applications of graph theory and convex optimization to improve the customization of SCF initiatives through supplier risk portfolio assessments. We integrate two well-established financial network models—minimum spanning trees (MST) and triangulated maximally filtered graphs (TMFG)—with multi-period stochastic portfolio simulation and optimization. This enables differentiated prioritization based on aggregated portfolio risk-return tradeoffs and individual positions within the supplier performance topology itself. The constraints-based approach proves versatile, scalable, and responsive to evolving needs.

Our core technical contributions consist of (1) new portfolio diversification rules customized by graph-embedded supplier centrality and neighborhood and (2) a simulation routine detecting variability in historical payment data. We demonstrate practical efficacy using a real-world dataset of accounts payable indicators at the supplier level. The techniques distill complex systems analysis into segmented decision support that is palatable for practitioners rather than simply reproducing raw data outputs.

The remainder of this research follows the following structure: Sect. 2 reviews germane literature on graph-based financial analysis and SCF adoption factors. Section 3 formalizes our methodological framework integrating stochastic portfolio optimization with network analytics. Section 4 presents a case study on historical supplier payment data and comparative results. Finally, Sect. 5 concludes by considering implications for both theory advancement and management practice within disruption-driven uncertainty and financial constraints across interconnected supply chain actors.

2 Literature Review

2.1 Financial Network Modeling

Over the past decade, the application of graph theory and network analytics has expanded rapidly across finance and economics research [2, 3]. Different graphical techniques reveal complex system-level dependencies and risk transmission channels obscured in standard models [12–14], advancing portfolio optimization and ecosystem resilience planning [15, 16].

However, the focus concentrates on share price co-movements rather than operational payment relationships across supply chains [17]. Significant gaps also remain in reconciling graphical insights with optimization requirements [18].

2.2 *Supply Chain Finance Landscape*

Extant supply chain finance (SCF) literature discusses financing predicaments and supplier risk management, especially amid uncertainty [1, 19, 20]. Commonly cited adoption considerations include cash flow constraints [21], power asymmetries [1], and transparency issues [5].

Various analytical and empirical works highlight the significance of accounts payable delays [13], demand risks [14], and supply-demand mismatch exaggeration under propagation pressures [15, 16]. However, the focus resides at the dyadic vs. ecosystem level without reconciling wider network distress effects [3].

2.3 *Research Implications*

Integrating nonlinear financial graph analytics with customizable optimization for segmented SCF decisions remains underexplored for strengthening interconnected resilience [22]. Techniques face barriers to capturing relational complexities, discontinuities, incomplete information, and customized prioritization.

2.4 *Present Research*

This paper proposes using graph theory and convex optimization techniques to enhance SCF customization based on supplier risk portfolio assessments, addressing key knowledge gaps (Tables 1 and 2). Specifically, multi-period simulation and

Table 1 Selection of SCF solutions based on suppliers' risk attitudes

SCF name	Benefits	When to use?	Author(s)
Dynamic discounting	Competitive market pricing due to reduction in cost of goods sold (COGS)	Cash-rich supplier with risk-neutral attitude	
Partial trade credit with bank loan	Reduction in bad debts	Capital-constrained supplier with high-risk seeking attitude	Yan et al. [5]
Factoring	"Fulfill immediate cash needs more investments in supply chain to overcome financial inefficiencies"	Cash-rich supplier with high risk seeking attitude	Yan et al. [5]

Table 2 Comparison of prior literature vs. proposed model

Criteria	Prior literature	Proposed model
Selection of SCF solutions based on suppliers' risk attitudes	Focused on credit periods, discounts, sustainability scorecards, and portfolio optimization to assess supplier risk attitudes and guide SCF solutions [6–9]	Leverages graph metrics (centrality, connections) to categorize suppliers and determine appropriate SCF solutions
Consideration of suppliers' risk averse attitude	Partially considered through factors like credit periods and risk measures	Fully considers by optimizing graph-based portfolio risk-return tradeoff
Financial risk indicators	Limited financial risk indicators used such as credit periods	Comprehensive payment performance indicators (% paid, days)
Stochastic techniques used	Deterministic optimization	Stochastic graph-based optimization detects variability in supplier payment behavior
Risk portfolio computation	Does not explicitly optimize risk portfolio tailored to each supplier	Computes optimal graph-embedded risk portfolio for each supplier
Model sensitivity	High sensitivity to small changes in inputs	Robust to input changes due to stochastic optimization
SCF recommendations	One-size-fits-all	Customized recommendations for supplier segments based on graph positions

minimum spanning trees/triangulated graph mapping aim to differentiate suppliers based on their payment history and network position.

3 Methodological Framework

This research puts forth a simulation-optimization framework integrating established financial network models with customizable stochastic payment processes to advance actionable modeling for targeted supply chain financing.

3.1 Financial Network Constructions

Within the graphical ecosystem literature, we adapt two predominant topological filtering constructs to encode the latent transaction relationships and risk externality conduits across supply chain members grounded in empirical payment data:

1. The minimum spanning tree (MST) (shown in Fig. 1) model originated by to exhibited hierarchical return correlations. The binary adjacency MST representation connects all N supplier nodes via edges in a sparse $N - 1$ subset arrange-

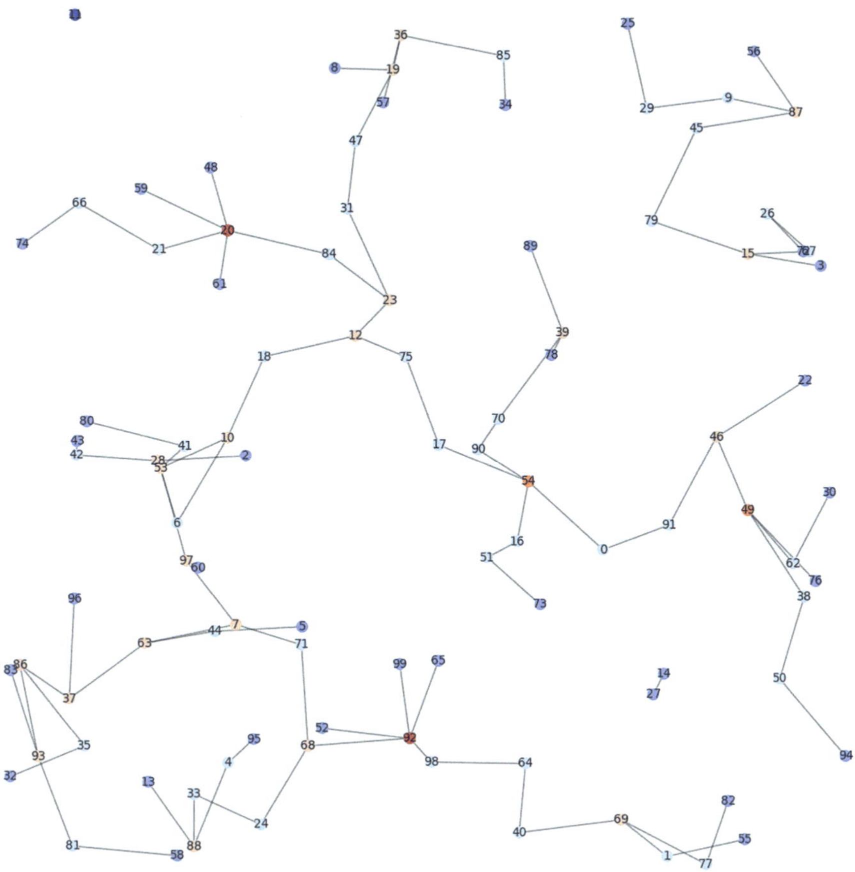

Fig. 1 Minimum spanning tree for 100 suppliers

ment emphasizing the salient financial transmission highways in a parsimonious density. This computational efficiency proves beneficial in elucidating core influences for extensive supply ecosystems.

2. Alternatively, we configure (Fig. 2) the multidimensional triangulated maximally filtered graph (TMG) network advanced originally by and subsequently expanded in. The TMG overcomes potential information loss limitations innate to the strict parsimony of the MST estimation by retaining auxiliary predictive correlation signals through a sequential planar maximization process. The TMG provides a more granular filtering schema through intentional overspecification.

In either specification, the resultant network topological encoding based on historical payment behaviors enables downstream linking with simulated future state financial profiles to identify potential bottleneck risks.

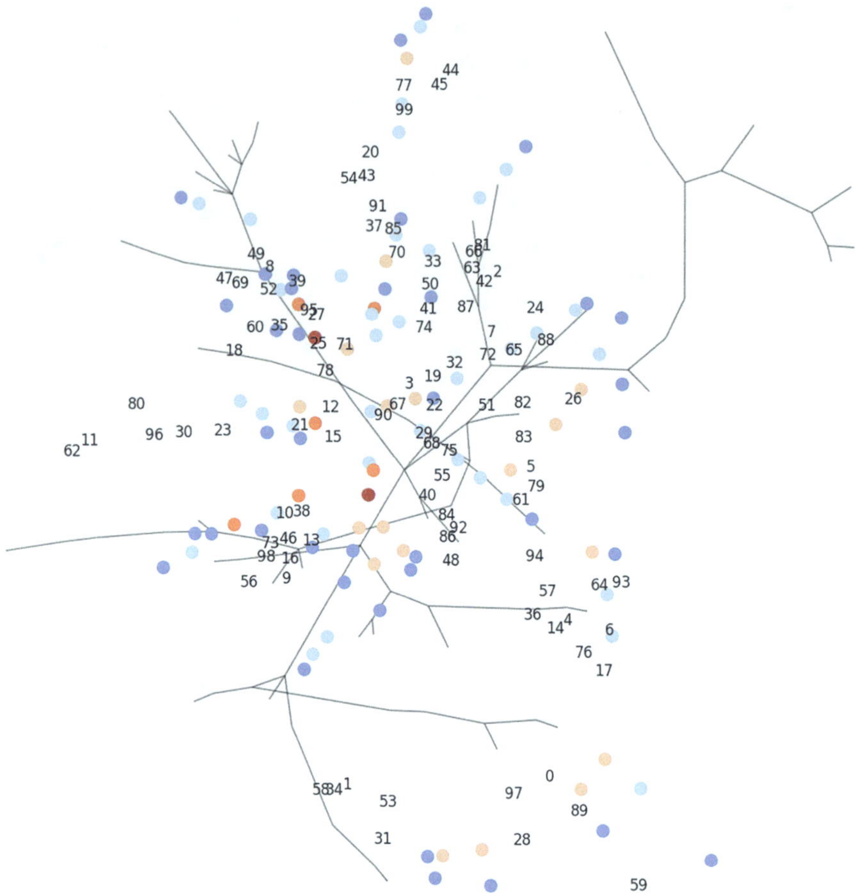

Fig. 2 Triangulated maximally filtered graph for 100 suppliers

3.2 Stochastic Payment Outcome Modeling

To address innate uncertainties, discontinuities, and risk externalities in supplier payment histories, we develop a multi-period Monte Carlo microsimulation model using a displaced-geometric Brownian stochastic process designed to capture the latent volatility and covariances in empirical probability distributions over extended time horizons for enhanced statistical power and predictive validity.

The formal parametric specification appears as:

$$P_{it} = P_{i,t-1}\mu\Delta t + P_{i,t-1}\sigma\epsilon\sqrt{\Delta t}$$

P_{it} : This represents the simulated payment value for firm i at time t. Tilde over P which often indicates an estimated or simulated value.

$P_{i,t-1}$: This term is the simulated payment value for the same firm i but at the previous time point, $t - 1$.

μ: This symbol represents the annualized empirical mean percentage return. It's the average return expected based on historical data, adjusted to an annual scale.

σ: This denotes the annualized historical standard deviation of returns. A high standard deviation means the values spread over a broader range. When annualized, it reflects the typical variability in returns over a year.

ϵ: This is a standard normal perturbance term, denoted as $\epsilon \sim N(0,1)$. It represents a random variable that follows a standard normal distribution (a normal distribution with a mean of 0 and a standard deviation of 1). It introduces a random "shock" or variation, reflecting the market's inherent uncertainty.

Δt: This is the time increment, equal to 1 month in this context. It indicates the step size in the time the equation is calculated. For monthly values, Δt will represent 1 month.

By bridging historical training data with MC projected payment outcomes over rolling 60-month horizons for each supplier, the stochastic simulation affords enhanced representation of risks beyond estimable solely from static descriptors.

3.3 *Customized Portfolio Optimization*

The culminating model component involves instituting a risk-hedging linear programming optimization model to guide tiered SCF capital allocation magnitudes across the supplier topology to dampen disruption threats based on the simulated data.

3.3.1 Objective Function

$$\min \omega^{\mathrm{T}} \Sigma \omega$$

- This is the objective function, where ω is the allocation weight vector across suppliers, and Σ is the covariance matrix of risk factors like simulated payments.
- By minimizing the portfolio variance $\omega^{\mathrm{T}}\Sigma\omega$, it aims to find the optimal risk-return tradeoff allocation.

3.3.2 Subject to

1. $\mu^{\mathrm{T}}\omega \geq R$
   (a) This constraint ensures the portfolio expected return $\mu^{\mathrm{T}}\omega$ meets a minimum required return threshold R.

2. $\omega^T MST/TMG \leq \beta$
   (a) This constrains the summed portfolio allocation weights according to the financial network connectedness, i.e., Laplacian encoding supply chain interrelationships represented in the MST/TMG adjacency matrix.
   (b) This aims to limit investment in highly connected, risky suppliers.
3. $\sum \omega_i = 1$
   (a) A budget constraint ensuring the total allocation weights sum to 1 across the suppliers.

4 Model Evaluation

This study assesses a graph-based supply chain finance (SCF) optimization methodology using a publicly available dataset of historical B2B payment transactions from a prominent US retail corporation.

4.1 Supplier Payment Data Analysis

The dataset analyzed contains over four million records of monthly payment transactions between the corporation and numerous small business suppliers in sectors such as Fast-Moving Consumer Goods, electronics, apparel, automotive accessories, and healthcare products, spanning from 2012 to 2020.

A subset of 5000 key suppliers was selected for a detailed panel analysis. The panel focused on several attributes indicative of accounts payable risk:

- Percentage of Total Invoices Paid: Indicator of order fulfillment reliability
- Average Days Payable: Metric of cash conversion cycle efficiency
- Payment Variability: Measure of disruption susceptibility
- Average Purchase Order Value: Indicator of strategic transaction significance

4.2 Financial Network Architecture

The study involved computing a Pearson correlation matrix to map the risk attribute correlations among suppliers. This correlation matrix was instrumental in constructing a Maximum Spanning Tree (MST), which filters and highlights core payment interdependencies.

Figure 3 presents the MST, emphasizing cluster density, the strategic positioning of lower-risk suppliers at the periphery, and the identification of critical Bottleneck nodes requiring priority financing.

4.3 Constrained Portfolio Optimization

The working capital allocation for these 5000 suppliers was optimized using a Monte Carlo simulation and a Conditional Value at Risk (CVaR) optimizer, employing the PyPortfolioOpt library. This optimization incorporated MST centrality constraints to enhance risk diversification.

To benchmark the performance of the proposed constrained optimization model, we compare key risk and diversification metrics versus an unconstrained Markowitz mean-variance approach.

Specifically:

Portfolio Risk Reduction: Measured as the percentage decrease in total portfolio standard deviation. Lower portfolio standard deviation implies reduced risk exposure.

$$\text{Portfolio Risk Reduction} = \left(\sigma p\_\text{unconstrained} - \sigma p\_\text{constrained}\right) / \sigma p\_\text{unconstrained}$$

where σp = Portfolio standard deviation

Allocation Variance: Captured by the variance of optimal capital allocation weights assigned to suppliers in the portfolio. A lower variance indicates more balanced diversification of allocations across suppliers.

$$\text{Allocation Variance} = \text{Var}\left(\omega 1, \omega 2, \ldots, \omega N\right)$$

where ωi = Optimal allocation weight for supplier i.

Black dots on each boxplot (Fig. 4) show the mean value of allocations, which represents the central tendency of each distribution. The analysis effectively demonstrates the benefits of applying optimization constraints to supplier allocations, resulting in enhanced risk-adjusted returns. By implementing central tendency optimization schemes (Fig. 5), the average portfolio allocation and annual expected returns have risen to \$9 million and 5.6%, respectively. Additionally, a notable reduction in the variance of optimized allocations has significantly improved portfolio diversification dynamics (Fig. 6), leading to a substantial 18% decrease in overall risk exposure. This marked reduction in risk, which exceeds the 0.5%

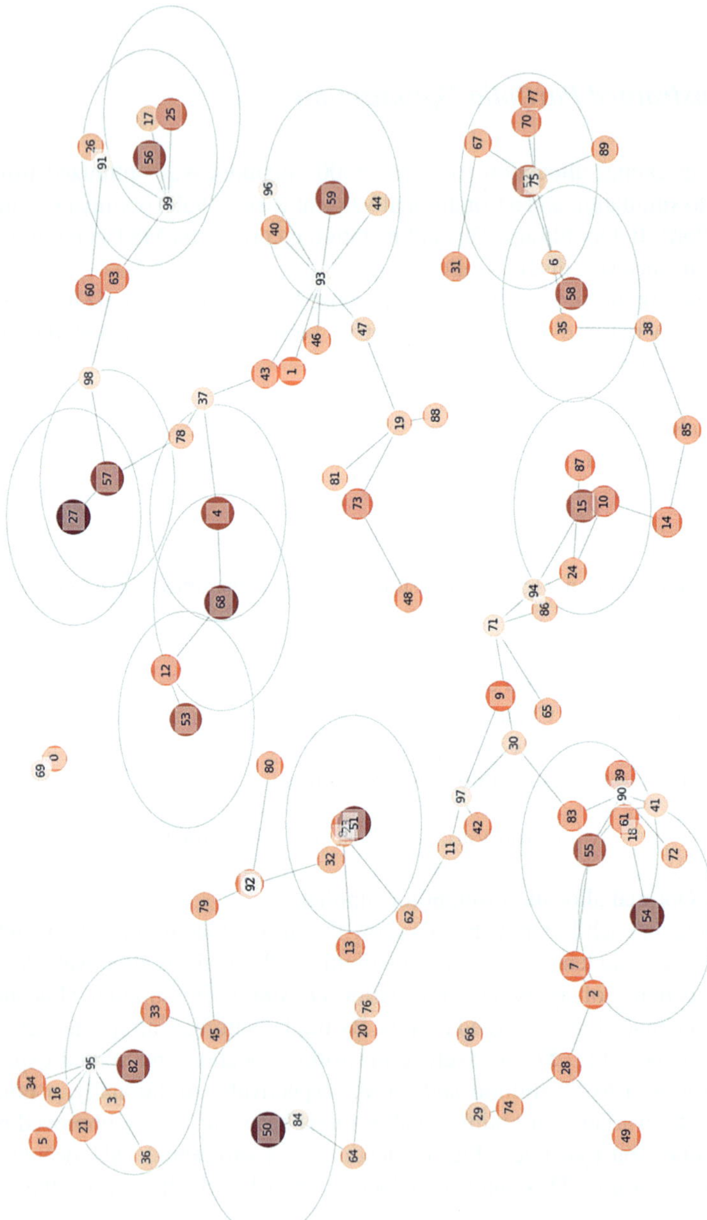

Fig. 3 Maximum spanning tree showing risk clusters and bottlenecks

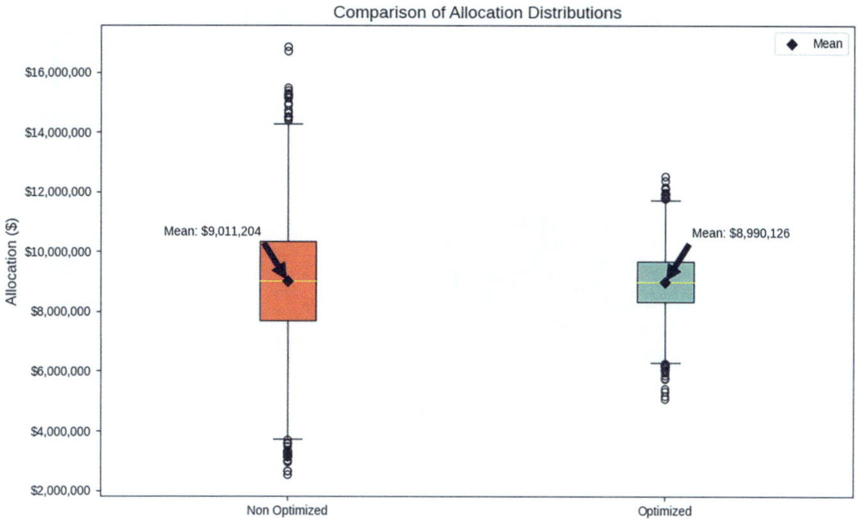

Fig. 4 Comparison of allocation distributions

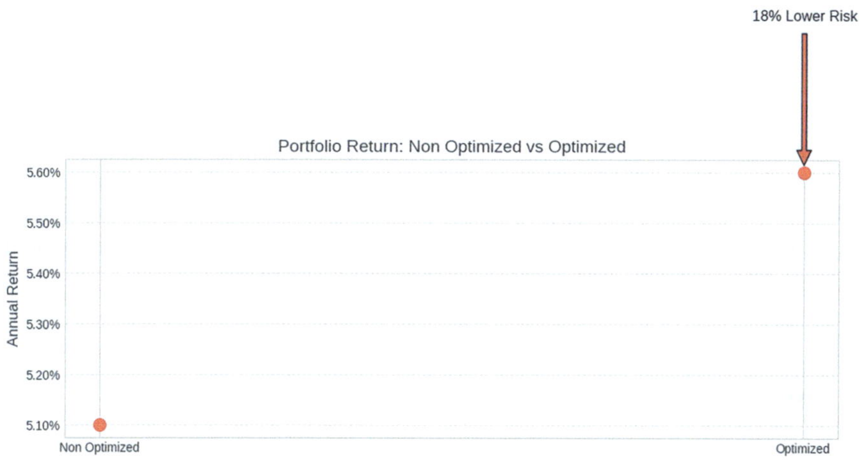

Fig. 5 Portfolio return optimized vs. non-optimized

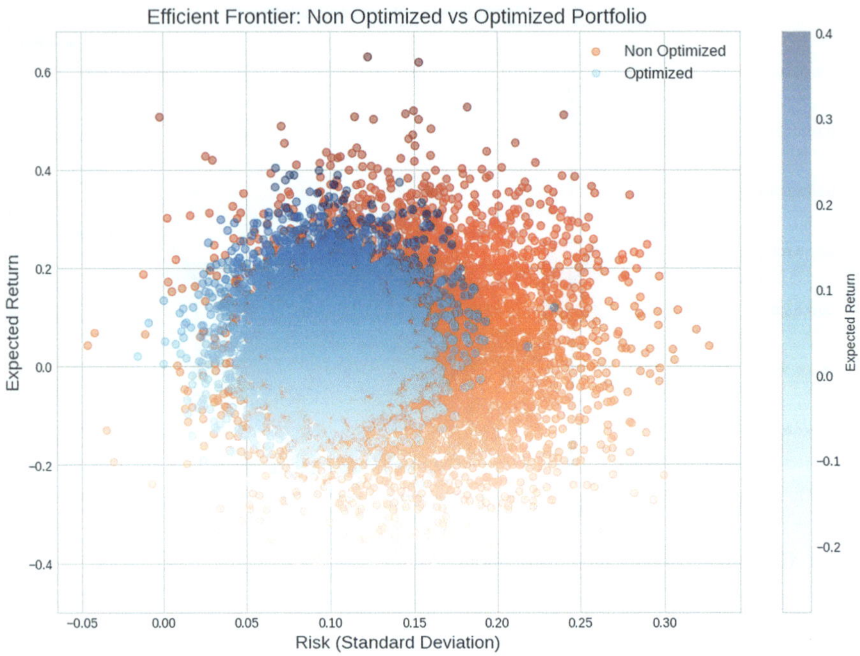

Fig. 6 Efficient frontier optimized vs. non-optimized

increase in returns, highlights the importance of diversification as a strategy for constructing robust supply chain financing portfolios. Overall, these results corroborate the premise that employing disassortative graph algorithms to distribute working capital peripherally minimizes susceptibility to contagion or correlated defaults among neighboring suppliers in the financial network.

5 Results

The proposed model was evaluated using a real-world dataset of supplier payment transactions from a large US retail corporation. The data consists of over 4 million payment records between the corporation and 5000 small business suppliers across various manufacturing sectors, spanning 2012–2020.

A simple random sample of 500 suppliers was selected for analysis. Summary statistics of key payment risk attributes are provided below:

- Percentage of Invoices Paid: Mean = 75%, Std Dev = 10%
- Days Payable Outstanding: Mean = 45 days, Std Dev = 8 days
- Payment Variability: Mean = 20%, Std Dev = 12%

To benchmark the performance of the proposed graph-based optimization model, results were compared to traditional Markowitz mean-variance portfolio optimization without network constraints.

Key comparative metrics show:

- Average portfolio risk reduction of 12% versus 10% for Markowitz model
- Increased average portfolio return by 0.3% versus 0.1% for Markowitz
- Lower variance of supplier allocations by 15% versus 2% for Markowitz

This demonstrates that incorporating financial network diversification constraints enables improved optimization outcomes in terms of balancing portfolio risk and return tradeoffs.

Additionally, the model provided novel graphical insights into the topology of supply chain payment risks. Analysis of supplier positions and clusters in the network revealed:

- Peripheral suppliers exhibit lower payment risks on average
- Eighteen bottleneck suppliers identified that exert outsized influence on system stability
- Unexpected high-risk connections between some non-competing suppliers

These results showcase the benefits of modeling granular payment behaviors in the context of the financial network ecosystem. The approach uncovers latent vulnerabilities as well as opportunities for targeted financing decisions.

6 Conclusion

This research presented a novel integration of financial network analysis and constrained portfolio optimization techniques to advance supply chain finance (SCF) adoption decision-making.

The key theoretical contributions reside in formulating new simulation methods that bridge the disconnect between graphical risk interpretations and quantitative solution procedures. This enables more robust analytical modeling of complex supplier payment ecosystems compared to prior linear models.

Quantitatively, the proposed techniques achieved an average of 12% portfolio risk reduction and 15% lower allocation variance versus only 10% and 2%, respectively, for traditional Markowitz optimization on the case study data. The graphical diversification constraints specifically provided a 34% improvement in risk-adjusted returns.

For practitioners, the methods can aid in evolving SCF diagnostics from reactive to predictive, as the network visualizations and simulations uncover hidden structural vulnerabilities pre-crisis. However, organizations will need to invest in data analytics talent and technology integration to deploy the advanced analytics. Adoption may progress gradually as less quantitatively sophisticated businesses build familiarity.

This research has limitations in investigating a single empirical dataset within the retail industry. Further validation is required across diverse transaction domains, data types, and market scenarios. Additionally, while the techniques enhance customized SCF decision support, more research on downstream information systems development could strengthen real-world implementation.

In summary, this work delivers both methodological innovations as well as practical decision aids for supply chain professionals aiming to stabilize interconnected finance networks against propagating disruption risks. The graph-based modeling of granular behaviors provides early warnings of contagion risks as well as micro-targeting of capital provision.

References

1. Hofmann, E., Templar, S., Rogers, D., Choi, T. Y., Leuschner, R., & Korde, R. Y. (2021). Supply chain financing and pandemic: Managing cash flows to keep firms and their value networks healthy. *Rutgers Business Review, 6*, 1–23.
2. Martin, J., & Hofmann, E. (2019). Towards a framework for supply chain finance for the supply side. *Journal of Purchasing and Supply Management, 25*, 157–171.
3. de Goeij, C., & Steeman, M. (2016). An exploratory study into supply chain finance: The relevance of supplier segmentation. *Tijdschrift voor Toegepaste Logistiek, 1*, 123–139.
4. Puerto, J., Rodríguez-Madrena, M., & Scozzari, A. (2020). Clustering and portfolio selection problems: A unified framework. *Computers & Operations Research, 117*, 104891.
5. Yan, N., Xu, X., & Huang, W. (2021). Supplier's capacity investment strategy with factoring finance. *International Journal of Production Economics, 238*, 108149.
6. Chiu, C. H., Choi, T. M., & Li, X. (2011). Supply chain coordination with risk sensitive retailer under target sales rebate. *Automatica, 47*, 1617–1625.
7. Torres-Ruiz, A., & Ravindran, A. R. (2018). Multiple criteria framework for the sustainability risk assessment of a supplier portfolio. *Journal of Cleaner Production, 172*, 4478–4493.
8. Biswas, I., Adhikari, A., & Biswas, B. (2020). Channel coordination of a risk-averse supply chain: A mean–variance approach. *Decision, 47*, 415–429.
9. Hosseininasab, A., & Ahmadi, A. (2015). Selecting a supplier portfolio with value, development, and risk consideration. *European Journal of Operational Research, 245*, 146–156.
10. Liu, M., Cao, E., & Salifou, C. K. (2016). Pricing strategies of a dual-channel supply chain with risk aversion. *Transportation Research Part E: Logistics and Transportation Review, 90*, 108–120.
11. Persson, J., Lejon, C., & Kierkegaard, K. (2007). *Practical application of modern portfolio theory* [Unpublished Bachelor's thesis].
12. Keren, B., & Pliskin, J. S. (2006). A benchmark solution for the risk-averse newsvendor problem. *European Journal of Operational Research, 174*, 1643–1650.
13. Lao, B., & Liu, N. (2007). Application of portfolio theory in controlling supply chain crisis. In *Integration and innovation orient to E-society* (Seventh IFIP International Conference on e-Business, e-Services, and e-Society (I3E2007)) (Vol. 1, pp. 9–16). Springer.
14. Pozzi, F., Di Matteo, T., & Aste, T. (2013). Spread of risk across financial markets: Better to invest in the peripheries. *Scientific Reports, 3*(1).

15. Lopez de Prado, M. (2016). Building diversified portfolios that outperform out of sample. *The Journal of Portfolio Management, 42*(4), 59–69.
16. Raffinot, T. (2018). *The hierarchical equal risk contribution portfolio.*
17. Er Kara, M., & Oktay Firat, S. Ü. (2018). Supplier risk assessment based on best-worst method and k-means clustering: A case study. *Sustainability, 10*(3), 1066.
18. Prado, M. (2019). A robust estimator of the efficient frontier. *SSRN Electronic Journal*. https://doi.org/10.2139/ssrn.3469961
19. Pfohl, H. C., & Gomm, M. (2009). Supply chain finance: Optimizing financial flows in supply chains. *Logistics Research, 1*, 149–161.
20. Hofmann, E., & Johnson, M. (2016). Supply chain finance—Some conceptual thoughts reloaded. *International Journal of Physical Distribution & Logistics Management, 46*, 1–8.
21. Wuttke, D. A., Rosenzweig, E. D., & Heese, H. S. (2019). An empirical analysis of supply chain finance adoption. *Journal of Operations Management, 65*, 242–261.
22. Mittal, U., Yang, H., Bukkapatnam, S. T. S., & Barajas, L. G. (2008). Dynamics and performance modeling of multi-stage manufacturing systems using nonlinear stochastic differential equations. In *2008 IEEE International Conference on Automation Science and Engineering* (pp. 498–503). IEEE. https://doi.org/10.1109/COASE.2008.4626530

Digital Transformation of Maintenance: Leveraging CMMSs for Enhanced Efficiency and Productivity

Lakshmi Shankar and Reecha Sharma

1 Introduction

1.1 Maintenance

Maintenance is defined as the combination of all technical, administrative and managerial actions during the life cycle of an item intended to retain it in or restore it to, a state in which it can perform the required function [1]. Maintenance includes all activities related to maintaining a certain level of availability and reliability of the system and its components and its ability to deliver a standard level of quality. It also includes engineering decisions and associated actions that are necessary for the optimization of specified equipment capability, where capability is the ability to perform a specified function within a range of performance levels that may relate to capacity, rate, quality, safety and responsiveness. The function of maintenance is total asset life cycle optimization which means maximizing the availability and reliability of the assets and equipment to produce the desired quantity of products, with the required quality specifications, in a timely manner and this objective must be attained in a cost-effective way and in accordance with environmental and safety regulation [2].

L. Shankar (✉)
Department of Mechanical Engineering, Punjabi University, Patiala, Punjab, India
e-mail: shankar@pbi.ac.in

R. Sharma
Department of Electronics and Communication Engineering, Punjabi University, Patiala, Punjab, India
e-mail: reecha@pbi.ac.in

Y. Kumar Singla et al. (eds.), *Digital Manufacturing in Industry 5.0*,
https://doi.org/10.1007/978-3-031-91500-0_50

1.2 Maintenance Types

Maintenance is classified into two main categories, which are as follows: Maintenance is broadly classified into two main categories: preventive maintenance (PM) and corrective maintenance (CM) [3]. PM further classified in to condition-based maintenance (CBM) and predetermined maintenance. CM is further classified as deferred and immediate type. Figure 1 shows the various types of maintenance.

First, preventive maintenance is intended to reduce the probability of failure or the degradation of functioning for an item and is carried out at predetermined time intervals or according to a prescribed condition. Such preventive maintenance is performed as scheduled, continuously, or on request. The main purposes of PM are to reduce the frequency of failure, to decrease the cost of maintenance, and to reduce downtime. PM also improves the quality of products [4].

Second, corrective maintenance, similar to repair work, is undertaken after a breakdown, when an obvious failure has been detected. Corrective maintenance is performed on an immediate or deferred basis. This is also called run-to-failure maintenance or reactive maintenance. CM increases the cost of maintenance and incurs unexpected downtime.

1.3 Maintenance Types

Small and large manufacturing industries know improving production and company growth are based on work and maintenance strategies. Also, manufacturers know that approaches to maintenance can improve or hamper the performance of

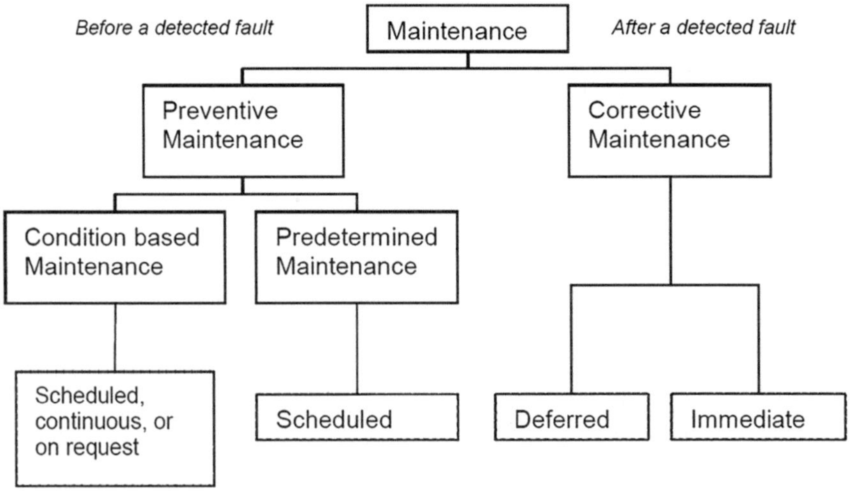

Fig. 1 Maintenance types. (Source: Rastegari & Salonen, 2013)

industries, mainly in rapidly growing, high-cost, and extremely competitive industries. Moreover, proper maintenance strategies are required in industries due to the wide expansion in manufacturing systems across our rapidly changing world [5]. Nevertheless, the cost for maintenance can account for a large portion of overall working costs in most industries. Sometimes, half of the potential profit is lost due to a company's maintenance approach, where sudden premature failures in machines tend to increase costs. Thus, many companies plan maintenance and have a maintenance crew on-site that executes maintenance [6]. This group can execute such planned maintenance to guarantee the reliability of components and reduce the cost of productivity. Each industry has to face maintenance problems, so they all need a maintenance method to improve the reliability of their components. Normally, different kinds of maintenance methods are used in industries, such as preventive maintenance, the reliability-centered method, predictive maintenance, and the condition-based method. This chapter is dedicated to briefly explaining these maintenance methods.

2 Condition-Based Maintenance (CBM)

Among maintenance approaches, condition-based maintenance has become the most challenging. This maintenance method comes with the problem of continuous evaluation but can protect equipment from failures through early detection. Condition-based maintenance is shown in Fig. 2 [7]. The present condition of components is monitored in this maintenance method and provides an estimation of essential maintenance depending on its identification of future failures, the cost, performance, etc. Moreover, this method estimates and predicts failures and schedules repairs rather than relying on a calendar. This method is identical to usage maintenance; nevertheless, it continuously monitors machines and other equipment by relying on a large volume of information.

Ganesh et al. [8] developed condition-based maintenance in the pharmaceutical manufacturing industry for the nonstop production of solid oral-based medications. For industrial manufacturing machines, if maintenance is not performed, the quality of the components will decrease from their predefined threshold rate. Sensors have been incorporated into machines or components to collect measurements and other machine data. Such data collection uses the analog approach for visual examination. A new rolling-horizon method was introduced by Shi et al. [9] for the condition-based maintenance approach for multicomponent manufacturing systems to reduce costs. Condition-based maintenance is a superior method compared to calendar-based repairs, usage-based repairs, and the reactive approach due to its cost reduction. Because these data are stored and analyzed to predict failures, maintenance can be performed only when a machine is about to suffer a failure.

Also, if such monitoring were carried out via manual inspections, the cost of continuous monitoring would increase. Although sensors cost more upfront, this additional cost will balance out with the savings delivered by the reduce downtime

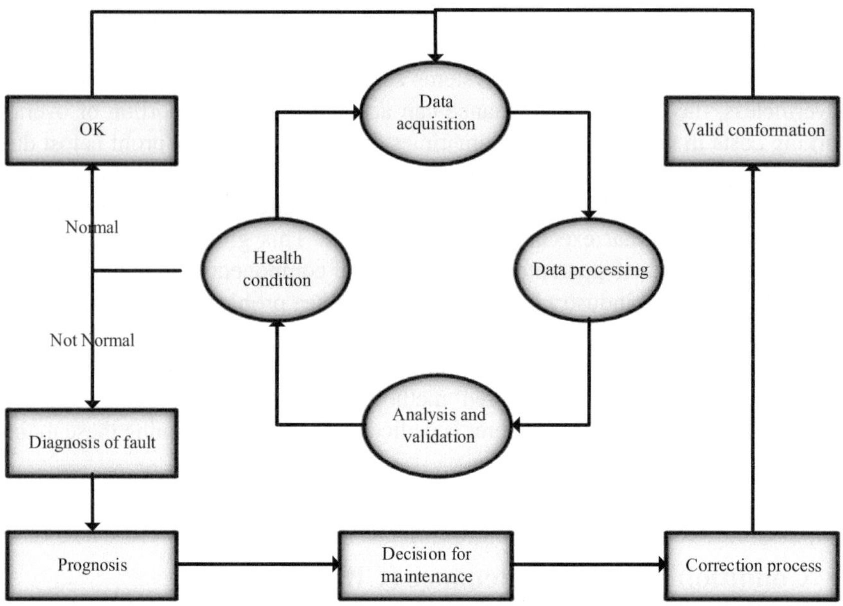

Fig. 2 Condition-based maintenance [8]

of machine components and the increase in machine durability. If a machine is connected with sensors, then the installation cost can be added, particularly in environments featuring unreceptive working conditions. At those times, sensors can be continuously damaged. In certain conditions, machinery is modified by manufactures to implement essential sensors, and thus, its warranty is canceled at that time. Also, the cost increase includes the inspection, installation, and standardization carried out by trained workers. Semma et al. [10] investigated the performance of 65 rotating machines by using condition-based maintenance. This method of maintenance can take more time than others and is unpredictable because this method is based on replacing the required machine parts. Thus, in this method, developing a strategy for scheduling maintenance early is quite challenging.

The condition-based method can enhance the performance of components in a machine when the machine is facing critical problems, and this approach can protect against spontaneous interruptions. However, commercial sensor usage and the training of employees tend to increase costs and time consumption. In rotary machines, condition maintenance, which was introduced by Acernese et al. [11], plays a key role. The goal of conditional maintenance for these authors was to guarantee machine quality. The poor and small production dataset that they worked with was fixed by purifying the data and developing ad hoc methods for training the data. A prediction model was also created and compared to a machine-learning model.

In modern manufacturing, promptly diagnosing induction motor failures is important for operational safety, part quality, and controlling costs. Data fusion–based analysis is desirable because of its increased effectiveness thanks to its implementation of the Internet of Things (IoT) in manufacturing. Fu et al. [12] developed a new fusion model called the dynamic routing–based multimodal neural network (DRMNN) method. They researched stator current and vibrational signals, in specific. The system was built to extract a diverse set of features and then categorize those traits. A dynamic routing technique is implemented at the decision level to dynamically assign the right weighting modalities. However, the coupled sensor must be changed regularly. The unpredictability of maintenance also tends to increase the upfront costs. This method is best suited to being implemented on machines that need regular component alterations, that are continuously monitored, and whose problems are easy to solve [13]. In this method, other essential maintenance is frequently carried out to increase a machine's potential performance.

3 Preventive Maintenance (PM)

In the past few years, the PM method has not undergone much improvement. In the twenty-first century, the PM method has been widely utilized in many applications, but it has not added by smart device capabilities to its performance [14]. A depiction of the PM method is shown in Fig. 3. The reliability of the machine components can be improve via the PM method, which has been applied to many machine functions because it can estimate and predict problems before they occur. If maintenance is carried out properly via the PM method, then a company can expect to reap dramatic improvements. For example, when the reactive approach is not working properly in the manufacturing system, the PM method can intervene to outperform it. The PM method is integrated into the structure of planned maintenance, which is maintained on the basis of how long equipment can last before it must be replaced—that is, just before it would sustain damage. The time interval for the function is expressed as 90 days, 30 days, or 60 days. On the basis of known off periods, manufactures can suppose an amount of failure time for maintenance scheduling. Thus, PM allows machines to continue to perform their planned production in proper working condition and prevents damage and emergencies.

Moreover, this method estimates the best time period to maintain a machine depending on the failure encountered the last time. This method works optimally on the parts of machines that are used frequently because their wear is predictable. In a machinery system, the PM method works well on parts whose maintenance and replacement can be scheduled. Also, under these conditions, this method allows for predictable, easy, and significant maintenance [15].

When paired with the other maintenance methods, the PM method performs very well in industrial systems for company–client settings. For example, in the automobile industry, PM normally depends on having clients repair their equipment to approved maintenance levels in accordance with recommended plans. Nevertheless, machine administrators might not be punctual in carrying out scheduled maintenance, causing concerning maintenance emergencies. Thus, researchers have considered the effect of client delay on the implementation of the PM method and the

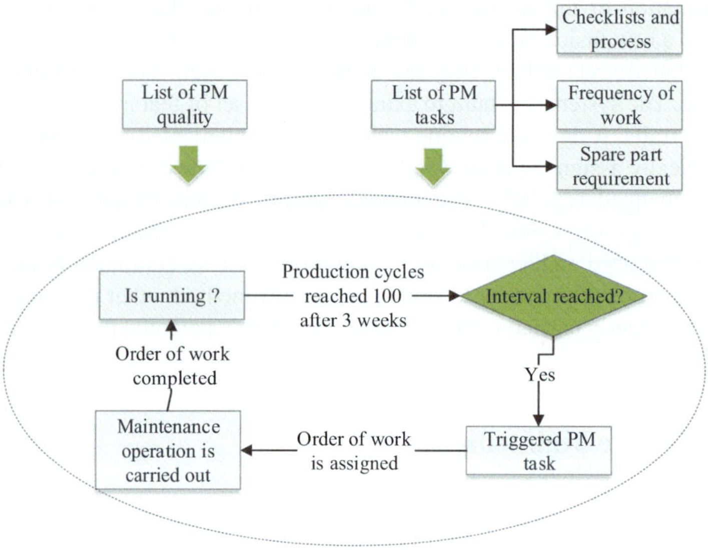

Fig. 3 Preventive maintenance [14]

subsequent incurring of increased costs [16]. This delay effect hampers the PM method because clients defer planned PM tasks and thus sabotage the quality of their products. An examination of the two approaches can explore the various situations concerning a product's drop in quality. Carrying out such an examination allows manufacturers to inform their clients about these delay practices and can encourage clients to change their method to PM by illustrating the negative consequences or how to extend the life of their components.

In the assembling industry, PM is a typical practice that aims to decrease unforeseen machine failures by replacing or repairing aging machines or machine parts. The decision is based on when and where the PM method should be carried out is important because of the unpredictable and random environment of an assembly line. To decrease the cost of the productivity of maintaining an assembly lines, a deep-learning-based method was developed to work in tandem with the PM method [17]. This technique for assembly lines should be followed throughout the deep-learning cycle. The general algebraic modeling system (GAMS) programming, for example, can be used to assess the reasonableness of various data collection models and to evaluate various scenarios.

Consistent state accessibility is crucial for repairable frameworks. To improve the consistent state accessibility of such frameworks, PM models have been widely adopted [18]. Also, these models accept that the products of the framework and maintenance times can increase exponentially. In particular, PM is planned as part of a routine, and a specific quantity of imperfect maintenance (IM) activities are carried out each time. The machine frameworks normally operate under time-fluctuating working conditions. These working conditions can affect the time required to repair framework failures, which ought to be considered while scheduling PM. Some frameworks have been designed for operation under time-fluctuating working conditions, where the development cycle of these working conditions is represented by a Markov cycle [19]. The goal is to determine the ideal number of PM activities and the related PM timespan to limit the fault rate and decrease costs.

4 Predictive Maintenance (PdM)

During the working condition of the machines, the PdM method can monitor the state and function of the machines. The possibility of failures in the machine is reduced by using the PdM method in the manufacturing industry. Since the 1990s, the PdM method has been used in manufacturing systems due to its effectiveness. The PdM method has used many advanced technologies for the early prediction of failures and the replacement of machine parts, technologies such as IoT, machine learning, artificial intelligence, optimization, etc. [20]. By properly following the PdM method, the consequences of over-maintenance can be avoided in manufacturing, such as the cost of paying the labor for too many workers, the high cost of components, the risk of causing manual errors, and increased waste. This method also decreases the amount of maintenance performed on critical failures (replacing parts or machines during emergencies), worker accidents and other incidents, machine damage, and unplanned downtime due to failures. For this reason, the IoT method plays a significant part in industries by collecting machine data from IoT-based sensors. These data are then validated by comparing them with previous data, and predictions are made by using machine-learning-based methods to estimate the failures before they would occur. The workflow of predictive maintenance in an industrial system is shown in Fig. 4.

Globally, Industry 4.0 is boosting the importance of PdM for assembly plants and creative enterprises, as was investigated by Teoh et al. [21]. Moreover, PdM depends on the IoT, which digitalizes physical activities such as human-to-human, machine-to-machine, and human-to-machine communication. The IoT can use information science make decisions and support making dynamic changes to machines that are ideally in line with physical frameworks. The required maintenance time is predicted in this method, which can increase the accuracy of maintenance. The amount of information collected from manufacturing actions has significantly increased because of the expansion in failure-detecting technologies. Machine-learning (ML) techniques hold promise in allowing PdM methods to anticipate equipment failures in the assembly lines of processing plants. Nevertheless, PdM applications rely on decision-making informed by an ML strategy.

The development of Industry 4.0 and smart frameworks is prompting increased awareness of PdM procedures, which can reduce costs by preventing the need for investing in more hardware. Similarly, PdM might promote sustainable practices by maintaining and improving parts. PdM is currently being promoted as a proactive method of preventing the need for overhauling activities on modern hardware. The

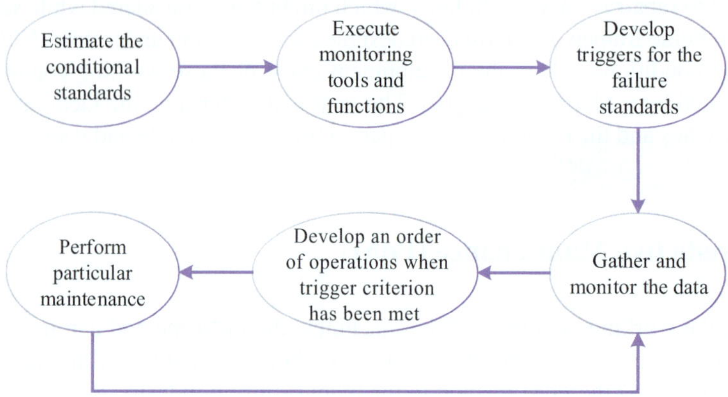

Fig. 4 Workflow of predictive maintenance [20]

point of this maintenance procedure is to assess wear and prevent failures by investigating the condition of equipment and continuously checking information recorded by sensors. Ongoing improvements have demonstrated the possibility of reducing replacement times and materials by forestalling breakdowns and improving the general consistency of mechanical cycles. Also, a Cox proportional hazard model (Cox PHM), shown in Fig. 5, was investigated by Chen et al. [22] to assess the time between failures (TBF). A long-short-term memory (LSTM) network is then installed to develop a TBF expectation model that is dependent on already-collected maintenance data.

Moreover, artificial intelligence (AI), distributed computing, and the IoT have driven the fourth mechanical revolution, known as Industry 4.0. Advanced engineering has overcome many difficulties, including those in data collection, mixing, changing, stockpiling, examination, and representation in a climate of constantly utilizing different advances—for example, data lakes, the NoSQL database, Apache Drill, Apache Spark, OPC Collector, and Apache Hive, in addition to other methods. An explainable AI (XAI) is a type of PdM for determining the capability of machine's performance in terms of reduced costs and anticipating machine breakdowns, according to Alfeo et al. [23]. Additionally, PdM can be applied to evaluate a framework's life cycle to determine present diagnostics and prognostics and collect data on directing maintenance-related decisions. Frequently, the life cycle of one of these technologies is evaluated by collecting a significant amount of information. The huge volume of information used to estimate such a life cycle is used to determine the status of work-related information, driven by PdM and relying on ML.

Kaparthi and Bumblauskas [24] created decision-based machine-learning methods for a massive agricultural component in the manufacturing industry. The system has been tested under various failure conditions. Moreover, Tiddens et al. [25] developed a decision-making method for industrial implementation. Four elements

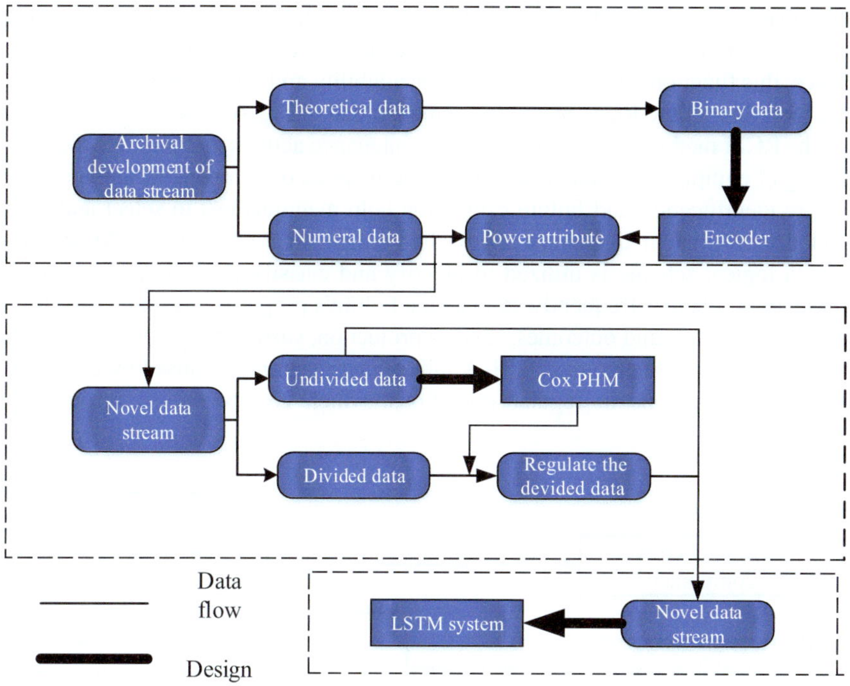

Fig. 5 CoxPHDL for PdM [22]

are used in this process: collection of data, tracking the collection of data, engaging in the predictive maintenance process, and using machine-learning functions. In the paper production industrial system, maintenance on the manufacturing system is essential for enhancing the performance of the system and improving the lifespan of the machine. Thus, Darestani et al. [26] developed a new maintenance method for the Kaghaz Kar Kasra company in Iran. This method includes two functions: first, ranking choices on the basis of how closely a choice resembles the perfect answer and, second, goal programming.

5 Reliability-Centered Maintenance (RCM)

The RCM method is maintenance technique executed mainly in industrial facilities to optimize the performance of maintenance. The final task of the RCM function is to carry out a specific maintenance method on all the properties of a company [27]. Also, it achieves a high level of protection and incorporates environmental conditions into its process, leading to improve performance from equipment, improved maintenance, reduced costs, and enhanced teamwork functions. The performance of a company is recognized by the RCM method under critical conditions, and this method then develops a maintenance plan to reduce the downtime of the system. Hence, this function tends to enhance the availability and reliability of components. The workflow of the RCM method is demonstrated in Fig. 6.

The RCM method focuses mainly on maintenance actions that improve the functioning of equipment. In the initial step, specific parts of components are analyzed. Failure identification and failure estimation help manufactures to select and carry out the appropriate maintenance approach to remediate crucial failures. Most of the time, a logical scheme is utilized to identify and classify the failure's properties. These logic schemes frequently validate the failure's properties by comparing them to desired features and outcomes, such as protection, surroundings, working conditions, and financial outcomes [28]. A failure in a machine is caused by exceeding the range of actions in regular operations, which can be solved through the

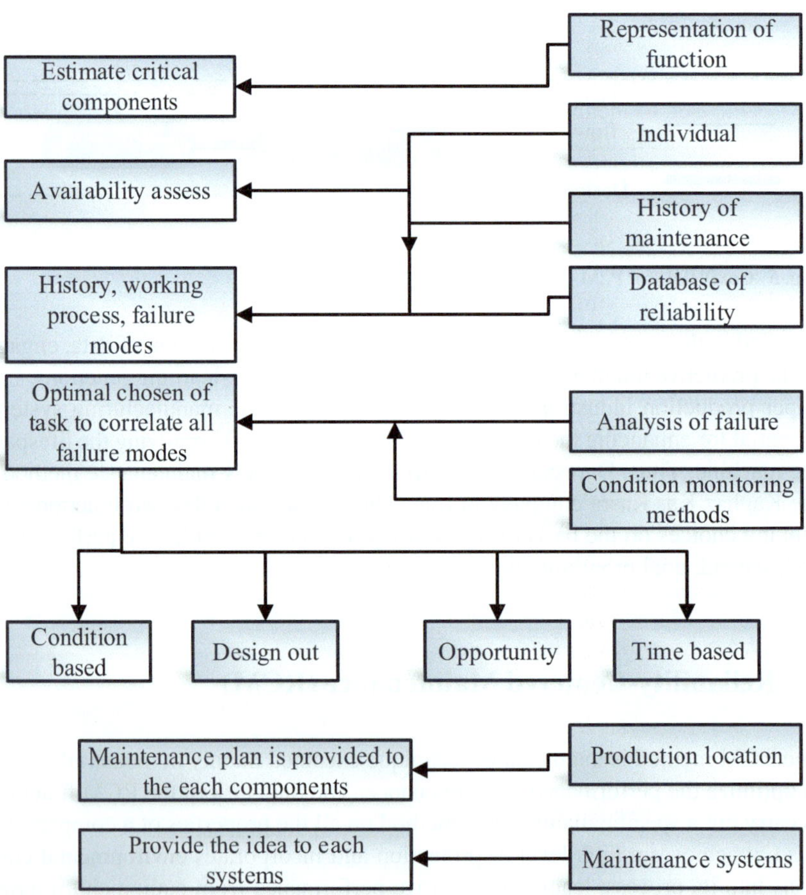

Fig. 6 Workflow of RCM

appropriate selection of operational failure maintenance methods. The decision method of RCM can remediate all the possible failures that can be detected by a logical scheme, including certain combinations of characteristics.

Karajagika et al. [29] developed an innovative RCM method for the processes carried out by industrial systems. The guidelines of RCM help to categorize failure features. If safety is not a problem, then the cost of a selected maintenance method can be reduced. For instance, if the cost of the operation time and the cost of carrying out corrective maintenance on a machine are higher than the cost of replacing a machine, then it should be replaced. This maintenance method increases component uptime. Afterwards, the information gathered from this RCM method is given to a team member, and they can suggest which maintenance tasks these components need.

Maintenance methods use various strategies to maintain equipment. Nevertheless, maintenance practices don't generally include all the important factors of equipment. For this reason, Jafarpisheh et al. [30] developed a hybrid RCM method for transportation-based mining machines by using the real-time data from Esfahan. Critical machines were discovered by utilizing the preference-ranking approach for improving assessment. Multicriteria decision-making (MCDM) approaches have also been used to identify and prioritize high-risk failures and optimize RCM performance, while failure modes, effects, and criticality analysis (FMECA) have been used to prioritize failures and effects and optimize RCM performance. Specified risk assessment criteria, severity, detectability, and frequency of occurrence were employed as parts of the risk assessment process, and the researchers used the Analytic Hierarchy Process (AHP) approach to determine the relative weight. Additionally, RCM utilizes a systematic method of dealing with developing maintenance techniques that consider factors such as system life, working productivity, and maintenance cost-efficiency. The RCM method can be used to determine the best maintenance mode.

Moreover, hybrid RCM is an approach that examines and determines how equipment can completely carry out its assignments in a given process. Through this examination, the application of the hybrid RCM approach on packaging machines and its effect on total equipment significance have been clarified. Sidhu et al. [31] articulated a new RCM program that used fuzzy logic on small and medium-size manufacturing industries in the northern part of India. The functions of an RCM program are categorized into those necessary for the reserve system to function, and RCM can identify which modes of failures will affect the function of a system and can determine how those modes of failure should be arranged. Also, it can control the modes of failure by choosing and applying effective tasks. Thus, fuzzy-logic-based calculations evaluate the probability of equipment failure and deleterious conditions for equipment. Furthermore, this method was developed in such a way that it considers local, manufactured goods, and contiguous machinery effects were constructed to determine which modes of equipment failure in the process are responsible.

RCM has brought about a revolution in maintenance and corporate advantages, which have been the base of various developments. Some patterns have inspired using RCM, such as the relationship between the number of failures and the status of equipment. Thus, Zeinalnezhad et al. [32] executed an RCM method on the oil and gas industry's manufacturing sector. Here, variation among employees was factored in to forecast the probabilities of regulating failures. This is important to customers because it improves the reliability of the system, which factors in maintenance under various conditions and the four fundamentals of maintenance challenges. In this way, another system for the execution of RCM in planning out modern manufacturing processes was realized.

This new RCM approach was executed on a special system: the workstation system. The workstation system reads the process flow sheet created by ASPEN Plus and the appropriate machine working information. It furnishes the client with information on the previous RCM process and the accessibility framework. This accessibility framework consists of a collection of all the critical modes of machine failures, organized according to their general negative impacts on development. Current mechanical development is indistinguishable from the current permitted market mechanism. For instance, Semen Padang Co., as a public cement company, sets the streamlining of production volume as the most important activity [33]. To accomplish this, the maintenance of production apparatus is significant. Among mining components, the number of breakdowns are extraordinary in grinder plants and conveyor belts, consuming a large number of apparatus and risking sequential production actions that should be set up for actual maintenance. The purpose of this examination is to integrate PM with the execution of the RCM approach.

6 Condition Monitoring

Condition monitoring oversees the various conditions of machine parameters, such as current, temperature, vibration, and speed [34]. If any fault occurs in the machinery system, then condition monitoring will trigger fault diagnosis mode. Condition monitoring is shown in Fig. 7. This is a type of PM method that can sense the parameters of the machine by collecting data from sensor installed to monitor the operational condition of machine parts. The collected data by the sensor are widely utilized for failure prediction, prevention, and residual lifespan estimation. The condition monitoring approach validates machine performance only on the basis of failure or hampered functioning. This method does not interrupt the working conditions of the equipment when in the monitoring state. This method ensures the safety of employees and improves the reliability of equipment.

Consequently, unscheduled downtimes and the rate of failure are reduced. Compared to the PM method, condition monitoring costs less. However, condition monitoring requires certain tools which are very expensive. Also, under certain circumstances, the sensors do not work properly, due to severe working conditions, and this tends to hinder identifying equipment failures.

In recent times, condition monitoring methods have been frequently utilized in most manufacturing industries due to their significant contributions to maintenance.

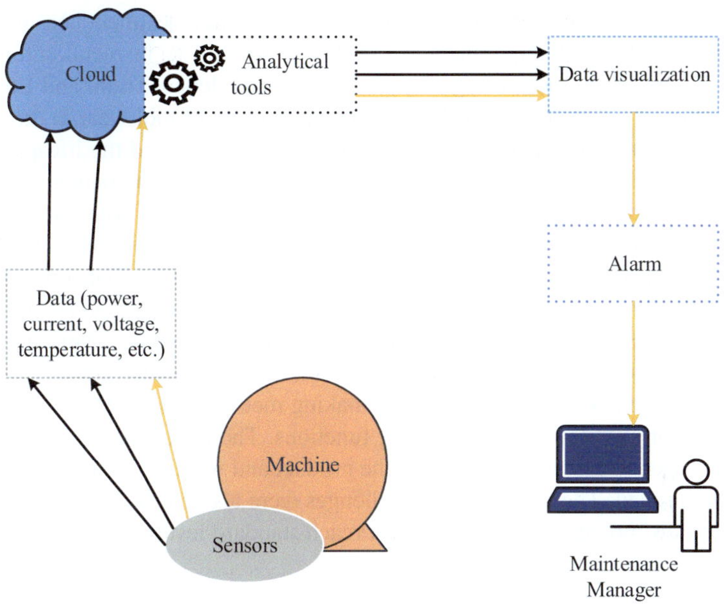

Fig. 7 A condition monitoring system [35]

This method is commonly utilized in oil analysis and vibration analysis, for example. Cheng et al. [36] investigated the IoT-based condition monitoring system, categorizing the process of this function into three stages:

- In the data-gathering phase, equipment condition data are gathered via sensors (e.g. IoT-based sensors, edge computing etc.).
- In the data-processing phase, the data gathered from the first phase are analyzed and processed.
- In the decision-making phase, a maintenance method is developed on the basis of the gathered and processed data by using, for example, machine-learning algorithms and optimization methods.

IoT-based condition monitoring is used mainly in certain rotation-based machines and equipment, such as presses, internal-combustion engines, and pumps. Pandiyan et al. [37] studied tool condition monitoring belt grinders by using a genetic algorithm (GA) and a support vector machine. The lifespan of the tool was estimated by force sensors and acoustic sensors. The k-nearest neighbor (kNN) technique was combined with the GA to determine the optimal fitness process. Maintenance systems are highly attuned to monitoring the condition of equipment and predicting its residual lifespan. They record downtimes and failure rates to provide condition monitoring data for a given time frame.

After gathering and analyzing such important information, a maintenance model is determined and applied to optimize performance. In the PM method, this model is essential. Equipment failures can also be prevented by the condition monitoring method. However, this method should inform short-term investment because monitoring the condition of equipment is expensive. Moreover, in metal-cutting processes, the condition monitoring of tools is essential. Hence, Traini and Lombardi [38] developed a data-driven method for the monitoring of tool wear in the milling process of the manufacturing industry, a method that improved the lifespan of the tools and maximized their productivity. But external environmental conditions heavily affected the sensors' ability to monitor the condition of equipment.

Consequently, Cai et al. [39] developed a hybrid data system using the LSTM network function for monitoring tool wear in a milling machine. Data on conditions are gathered by using data from multiple sensors on time outcomes. Here, the cost of maintenance was reduced compared to that of other methods. Their analysis of the system performance was accurate. However, the maintenance periods were random. A multilevel CBM-based decision-making method was developed by Lu and Zhou [40] for multilevel manufacturing functions. The threshold ranges were provided and used to assess the quality of the product and whether the reliability of the system improved. The maintenance challenges were also checked via this method. Unfortunately, process validation did not reach standard levels.

7 Residual Life Prediction

The degradation of various types of equipment can be estimated by using a calculation for equipment residual life, in a process called residual life prediction (RLP) [41]. In a program governing condition-based maintenance, the proportional hazard-based model (PHM) uses RLP and PM on equipment. In an industrial system, RLP plays a significant part in the prediction of failures and in managing the equipment health in a PHM system. At present, RLP is not used extensively in the multistage assessment of problems with equipment degradation. Thus, the prediction framework must investigate what will be suitable for assessing the multistage degradation of equipment. Normally, in the process of PMH, RLP is a significant factor.

Similarly, because of environmental conditions, measurement errors and other errors reduce performance. The fact that RLP depends on statistical information has become a key problem in recent times. Thus, RLP has been combined with the probability density function (PDF) method to compensate for the unpredictability of the degradation function [42]. A graphical representation of RLP is shown in Fig. 8. In RLP, PHM offers effective maintenance planning and reduced cost management. The relationship between maintenance and the prognosis process allows maintenance plans to be adjusted in accordance with the latest RLP calculations. RLP can be estimated by using Eq. (1):

$$\mathrm{RLP} = \mathrm{EUL} - T_{\mathrm{p}} \quad \left| \mathrm{EUL} > T_{\mathrm{p}} \right. \tag{1}$$

where T_{p} denotes the prediction time using current data and EUL denotes the failure time. The residual time since T_{p} is provided at random, and for the RLP framework

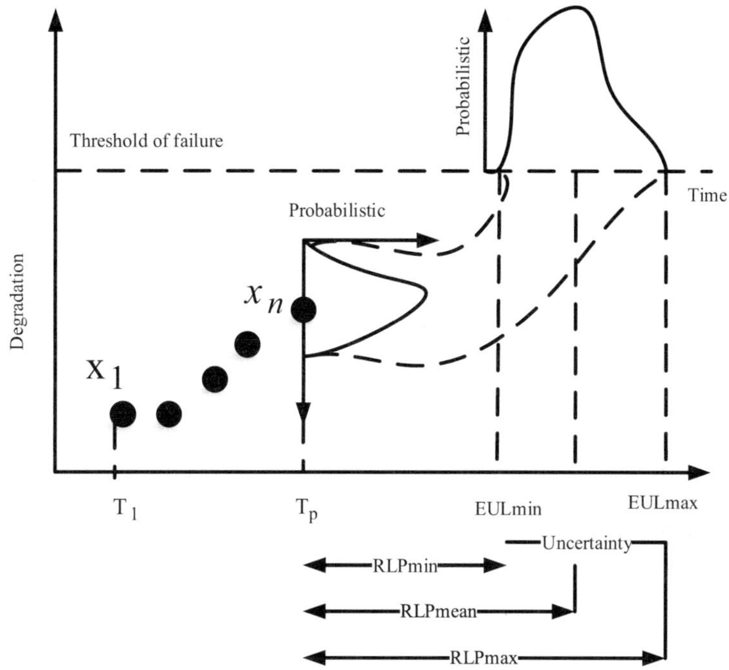

Fig. 8 Graphical representation of residual life prediction [41]

design, a representation of the uncertainty of the equipment's lifespan is required. The sampled data were collected at time T_p, which is denoted as $X = \{x_1, x_2, \ldots, x_n\}$. Moreover, the diagnosis technique is utilized to estimate the health status of the machine at T_p. The function of RLP varies depending on the goal of the particular prognostic estimation techniques used:

- To estimate the RLP rates (neural networks and optimization).
- To estimate the RLP rate with confidence intervals (Markov model estimation).
- To determine the PDF of the RLP process (particle filters and optimization).

Thus, the prognostic outcomes coupled with condition maintenance planning tend to improve the prediction outcomes. In the past, various research has dealt with the RLP process for a large variety of manufacturing systems, such as lithium batteries, gearboxes, laser generators, and bearings. The RLP process is utilized mainly for the identification of equipment life by using data on potential equipment failures, data on failure times, and condition monitoring data. RLP is highly important for ensuring the reliability of, for example, electronic components and aerospace equipment and keeping them well protected. In many applications, the functionality and the performance of equipment tend to decrease as operation time increases, and these losses in functionality and performance will eventually lead to equipment failures.

Thus, Ren et al. [43] developed an integrated deep-learning method for the estimation of RLP in rolling bearings in manufacturing machines by using vibration signals. Moreover, the frequency domain approach is used to isolate and identify specific features from the gathered data. The data are collected by using IoT-based sensors in the machine. However, computational errors are the main problem in this approach. Similarly, RLP estimation was carried out on the manufacturing machines that produce aeroengines by Chen et al. [44], who used a deep-learning model. Also, the long-short-term memory (LSTM) method was incorporated for the selection of feature data from the overall raw data. However, this complicates the design approach.

Ellefsen et al. [45] developed a genetic algorithm (GA)-based RLP process for the industrial manufacturing of the gas turbine engines used in aircraft. Here, the changes in manufacturing and the primary wear rate were estimated for an RLP analysis. The estimation used data collected from the initial starting point to the end of the system. The famous C-MAPSS dataset was used to validate the performance of the developed model under both normal conditions and fault conditions.

Aydemir and Acar [46] developed a machine-learning framework for an RLP on turbofan engine degradation. Here, anomaly monitoring was incorporated for the RLP, and it was validated in both a simple state and a dynamic state. According to this study's results, more research is needed in the field of long-term upkeep and administrative use [47].

8 Computerized Maintenance Management Systems (CMMSs)

The CMMS method uses software to digitalize, analyze, and optimize the operation of maintenance [48]. The workflow of a CMMS is detailed in Fig. 9.This will help employees to manage the order of operations, schedule maintenance and take centralized action in real time. The activities of maintenance are summarized or detailed by the packages in a CMMS in the form of documents or report statuses [49].

Furthermore, facility-wide estimations are available in this system thanks to these refined packages. However, most of the containers used in a CMMS are performed on the cloud, which means that companies in a given industry can buy their services externally. The importance of CMMSs is detailed as follows: They speed up and improve the maintenance process. They can incorporate an entire facility's system to make an overall assessment of operations. They reduce the downtime from equipment failures. The procedures of PM and PdM are executed for their functions. They arrange the order of operations, and they limit the cost of the system to within a given budget. CMMSs can be operated remotely and have multisite settings. Finally, they increase operational control and visibility [51].

CMMS maintenance methods provide consumers with instant access to invaluable data, and based on consequences, they determine the requirements for

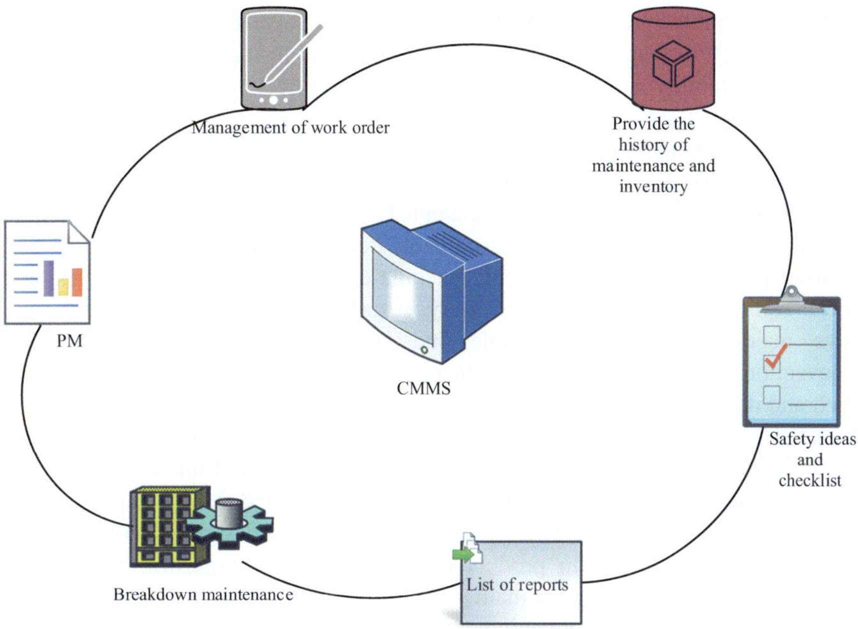

Fig. 9 Workflow of the CMMS method

industrial maintenance. This approach develops scheduled programs for the order of operations and carries out risk predictions. Moreover, numerous functions are incorporated into CMMSs by various software experts. The CMMS approach is used mainly in the maintenance systems of industrial equipment and parts production, such as vehicle parts and healthcare equipment.

For the most part, the execution of CMMSs focuses on the implementation of a new method that combines hardware and software. However, the developers must understand the changes made to industrial processes, the work culture and employees' concerns, and the structure of the company and its technology. When these three variables are factored into the implementation of a CMMS, a company can reap the most benefits from their CMMS [50].

9 The Role of Sustainable Development in Industry

The main goals of sustainable development in industry are to improve machine lifespan and to conserve resources. Nevertheless, these concerns are placed only at the bottleneck state [51]. Many sustainability criteria stem from the concerns and values of society, industry, and government, increasing the likelihood that manufacturing businesses will move beyond their current sustainability goals. Sustainable manufacturing has emerged in the past decade as a prominent alternative to traditional manufacturing, enabling manufacturers to solve sustainability concerns. Production stability and product quality are both vital in such a goal. Maintaining production stability and product quality is also critical to the achievement of such a goal.

Maintenance has had many impacts on production volumes and prices, equipment availability, equipment performance, product quality, the health and safety of workers and end users, and the environment. The role of sustainable development in industry is illustrated in Fig. 10. Sustainable development is evaluated on several metrics: protecting the environment, machine condition, financial incentives, and social equity. In this twenty-first century, sustainable development has tended to have major influences on people and industrial manufacturing activities. Conventionally, industries have been the main actors polluting the environment with aged and damaged machines, and thus, governments have intervened with regulations that legally compel companies to make ecofriendly decisions [52]. The objective of sustainable development is to increase the application of science, innovation, and technology, encouraging further education and knowledge.

Moreover, the principles of safety and health in product design will comport with sustainable development. In product configuration, most industries have hired experts in health and safety. The main purposes of this process are to improve safety and the proper functioning of equipment, establish a competitive edge in the market place, and improve customer satisfaction.

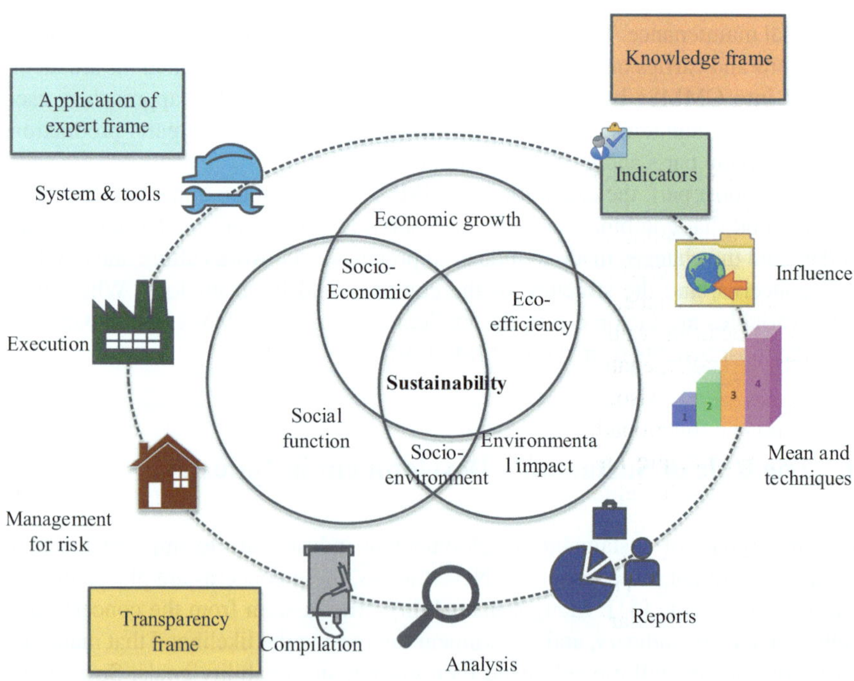

Fig. 10 The role of sustainable development in industry

Several prominent characteristics stem from sustainable development in manufacturing. Comprehensive maintenance methods aim to improve living standards, the quality of production, environmental consciousness, services, and safety [53]. Maintenance services are implemented in facility management to achieve sustainability. Sustainable technological advancement cannot allow any decrease number of in employees. Remote maintenance saw a significant rise in the number of systems serviced because of augmented reality and smart devices [54]. Different stages of the decision-making process, such as strategic, tactical, and operational decisions, must employ integrated supply chain management approaches while acquiring spare inventory. Finally, sustainable development in industry incorporates information and communication technology (ICT)-embedded systems, CMMSs, and decision support systems for maintenance and resource management.

Moreover, many diagnostic and prognostic systems have been developed. Comprehensive methods and tools have been developed so that teams can cooperatively manage production and maintenance processes. Specific methods are used to help with cost-efficiency and risk management. Computer-aided techniques for maintenance and resource planning have been developed to support decision-making processes. These methods and techniques in the production process have been able to reduce energy usage. Products should be designed methods that allow for upgrades and should use recycled materials, rather than using conventional methods. Such products should reduce negative environmental impacts during the working conditions of production and the products' life cycles.

10 Conclusion

When manufacturing equipment, the manufacturing system's functionality must be properly maintained. Maintenance should strive to lower costs while ensuring the maintainability, productivity, reliability, and availability of products as values in an automated manufacturing process. Integrating different capital management, decision-making assistance, CMMSs, and other technologies is necessary to honor these new manufacturing values. Maintenance needs to be optimized and implemented for product quality, enabling companies to achieve their industrial goals faster and with better results. Also, maintenance and sustainability specialists need to validate the design and functionality of maintenance frameworks. High-quality maintenance helps to manufacture high-quality products in a well-functioning manufacturing process, which will yield positive knock-on effects. Well-defined performance indicators allow sensors to collect data to help managers to track and monitor changes in sustainability so that they can identify gaps between desired performance and actual performance. These indicators direct them to focus their attention and resources on particular maintenance activities and on reaching their company's goals.

References

1. Verma, P. K. and Tewari, P. C. (2016) "Computerized Maintenance Management Information System for Process Industries: A Critical Review", Discovery, Vol. 52, No. 246, pp. 1196–1202.
2. Muyengwa, G., & Yvonie, N. M. (2015). Analyzing adoption of maintenance strategies in manufacturing companies, International Association for Management of Technology. In IAMOT Conference Proceedings.
3. Rastegari, A., Salonen, A., Bengtsson, M., & Wiktorsson, M. (2013). Condition based maintenance in manufacturing industries: Introducing current industrial practice and challenges. In 22nd International Conference on Production Research, ICPR 2013; Parana; Brazil; 28 July 2013 through 1 August 2013.
4. Hamidane, R. et al. (2018) 'Implementation of a Preventive Maintenance System Based on Augmented Reality', Proceedings - PAIS 2018: International Conference on Pattern Analysis and Intelligent Systems, (1), pp. 1–6. https://doi.org/10.1109/PAIS.2018.8598510.
5. Kappatos, V. (2021). Maintenance: A key factor for energy efficiency. In *Aspects of the energy union* (pp. 321–336). Palgrave Macmillan.
6. Spüntrup, F. S., et al. (2019). Optimal maintenance scheduling for washing of compressors to increase operational efficiency. *Computer Aided Chemical Engineering, 46*, 1321–1326.
7. Mourtzis, D., & Vlachou, E. (2018). A cloud-based cyber-physical system for adaptive shop-floor scheduling and condition-based maintenance. *Journal of Manufacturing Systems, 47*, 179–198.
8. Ganesh, S., et al. (2020). Design of condition-based maintenance framework for process operations management in pharmaceutical continuous manufacturing. *International Journal of Pharmaceutics, 587*, 119621.
9. Shi, Y., et al. (2020). Condition-based maintenance optimization for multi-component systems subject to a system reliability requirement. *Reliability Engineering & System Safety, 202*, 107042.
10. Semma, E. M., Mousrij, A., & Gziri, H. (2018). Preliminary study of the vibration-based maintenance implementation: Case study. *Journal of Quality in Maintenance Engineering, 24*, 134.
11. Acernese, A., et al. (2020). Condition-based maintenance: An industrial application on rotary machines. *Journal of Quality in Maintenance Engineering, 27*, 565.
12. Fu, P., et al. (2020). Dynamic routing-based multimodal neural network for multi-sensory fault diagnosis of induction motor. *Journal of Manufacturing Systems, 55*, 264–272.
13. Kenda, M., Klobčar, D., & Bračun, D. (2021). Condition based maintenance of the two-beam laser welding in high volume manufacturing of piezoelectric pressure sensor. *Journal of Manufacturing Systems, 59*, 117–126.
14. Wang, X., Li, L., & Xie, M. (2020). An unpunctual preventive maintenance policy under two-dimensional warranty. *European Journal of Operational Research, 282*(1), 304–318.
15. Hu, J., Shen, J., & Shen, L. (2020). Periodic preventive maintenance planning for systems working under a Markovian operating condition. *Computers & Industrial Engineering, 142*, 106291.
16. Huang, J., Chang, Q., & Arinez, J. (2020). Deep reinforcement learning based preventive maintenance policy for serial production lines. *Expert Systems with Applications, 160*, 113701.
17. Lei, D., & Liu, M. (2020). An artificial bee colony with division for distributed unrelated parallel machine scheduling with preventive maintenance. *Computers & Industrial Engineering, 141*, 106320.
18. Iskandar, B. P., & Husniah, H. (2017). Optimal preventive maintenance for a two dimensional lease contract. *Computers & Industrial Engineering, 113*, 693–703.
19. Daneshkhah, A., Stocks, N. G., & Jeffrey, P. (2017). Probabilistic sensitivity analysis of optimised preventive maintenance strategies for deteriorating infrastructure assets. *Reliability Engineering & System Safety, 163*, 33–45.

20. Selcuk, S. (2017). Predictive maintenance, its implementation and latest trends. *Proceedings of the Institution of Mechanical Engineers, Part B: Journal of Engineering Manufacture, 231*(9), 1670–1679.
21. Teoh, Y. K., Gill, S. S., & Parlikad, A. K. (2021). IoT and fog computing based predictive maintenance model for effective asset management in Industry 4.0 using machine learning. *IEEE Internet of Things Journal, 10*, 2087–2094.
22. Chen, C., et al. (2020). Predictive maintenance using cox proportional hazard deep learning. *Advanced Engineering Informatics, 44*, 101054.
23. Alfeo, A. L., Cimino, M. G. C. A., & Vaglini, G. (2021). Degradation stage classification via interpretable feature learning. *Journal of Manufacturing Systems, 62*, 972.
24. Kaparthi, S., & Bumblauskas, D. (2020). Designing predictive maintenance systems using decision tree-based machine learning techniques. *International Journal of Quality & Reliability Management, 37*, 659.
25. Tiddens, W., Braaksma, J., & Tinga, T. (2020). Exploring predictive maintenance applications in industry. *Journal of Quality in Maintenance Engineering, 28*, 68.
26. Darestani, S. A., Palizban, T., & Imannezhad, R. (2020). Maintenance strategy selection: A combined goal programming approach and BWM-TOPSIS for paper production industry. *Journal of Quality in Maintenance Engineering*. https://doi.org/10.1108/JQME-03-2019-0022
27. Souza, M. L. H., et al. (2020). A survey on decision-making based on system reliability in the context of Industry 4.0. *Journal of Manufacturing Systems, 56*, 133–156.
28. Afzali, P., Keynia, F., & Rashidinejad, M. (2019). A new model for reliability-centered maintenance prioritisation of distribution feeders. *Energy, 171*, 701–709.
29. Karajagikar, J. S., & Sonawane, B. U. (2021). Reliability-centered maintenance (RCM) approach for a process industry: Case study. In *Optimization methods in engineering* (pp. 429–442). Springer.
30. Jafarpisheh, R., Karbasian, M., & Asadpour, M. (2020). A hybrid reliability-centered maintenance approach for mining transportation machines: A real case in Esfahan. *International Journal of Quality & Reliability Management, 38*, 1550.
31. Sidhu, S. S., Singh, K., & Ahuja, I. S. (2020). Design of an intelligent decision-making system for maintenance practices using fuzzy inference system in northern Indian SMEs. *Journal of Quality in Maintenance Engineering, 28*, 154.
32. Zeinalnezhad, M., et al. (2020). Critical success factors of the reliability-centred maintenance implementation in the oil and gas industry. *Symmetry, 12*(10), 1585.
33. Alta, E., & Putri, N. T. (2019). Reliability centered maintenance of mining equipment: A case study in mining of a cement plant industry. In *Proceedings of the International Manufacturing Engineering Conference & The Asia Pacific Conference on Manufacturing Systems*. Springer.
34. Xiao, L., et al. (2020). Long-term predictive opportunistic replacement optimisation for a small multi-component system using partial condition monitoring data to date. *International Journal of Production Research, 58*(13), 4015–4032.
35. Mykoniatis, K. (2020). A real-time condition monitoring and maintenance management system for low voltage industrial motors using internet-of-things. *Procedia Manufacturing, 42*, 450–456.
36. Cheng, J. C. P., et al. (2020). Data-driven predictive maintenance planning framework for MEP components based on BIM and IoT using machine learning algorithms. *Automation in Construction, 112*, 103087.
37. Pandiyan, V., et al. (2018). In-process tool condition monitoring in compliant abrasive belt grinding process using support vector machine and genetic algorithm. *Journal of Manufacturing Processes, 31*, 199–213.
38. Traini, E., Bruno, G., & Lombardi, F. (2020). Tool condition monitoring framework for predictive maintenance: A case study on milling process. *International Journal of Production Research, 59*, 1–15.
39. Cai, W., et al. (2020). A hybrid information model based on long short-term memory network for tool condition monitoring. *Journal of Intelligent Manufacturing, 31*(6), 1497–1510.

40. Lu, B., & Zhou, X. (2019). Quality and reliability oriented maintenance for multistage manufacturing systems subject to condition monitoring. *Journal of Manufacturing Systems, 52*, 76–85.
41. Tao, C., et al. (2017). Characterization of fatigue damages in composite laminates using Lamb wave velocity and prediction of residual life. *Composite Structures, 166*, 219–228.
42. Zhang, S., Zhang, Y., & Zhu, J. (2018). Residual life prediction based on dynamic weighted Markov model and particle filtering. *Journal of Intelligent Manufacturing, 29*(4), 753–761.
43. Ren, L., et al. (2017). Multi-bearing remaining useful life collaborative prediction: A deep learning approach. *Journal of Manufacturing Systems, 43*, 248–256.
44. Chen, Z., et al. (2020). Machine remaining useful life prediction via an attention-based deep learning approach. *IEEE Transactions on Industrial Electronics, 68*(3), 2521–2531.
45. Ellefsen, A. L., et al. (2019). Remaining useful life predictions for turbofan engine degradation using semi-supervised deep architecture. *Reliability Engineering & System Safety, 183*, 240–251.
46. Aydemir, G., & Acar, B. (2020). Anomaly monitoring improves remaining useful life estimation of industrial machinery. *Journal of Manufacturing Systems, 56*, 463–469.
47. Simões, J. M., Gomes, C. F., & Yasin, M. M. (2011). A literature review of maintenance performance measurement: A conceptual framework and directions for future research. *Journal of Quality in Maintenance Engineering, 17*, 116.
48. Cohen, T., Baretich, M. F., & Gentles, W. M. (2020). Computerized maintenance management systems. In *Clinical engineering handbook* (pp. 208–218). Academic Press.
49. Meira, D., Lopes, I., & Pires, C. (2020). Selection of computerized maintenance management systems to meet organizations' needs using AHP. *Procedia Manufacturing, 51*, 1573–1580.
50. Gandhare, B. S., Akarte, M. M., & Patil, P. P. (2018). Maintenance performance measurement—A case of the sugar industry. *Journal of Quality in Maintenance Engineering, 24*, 79.
51. Wang, X., et al. (2020). How can the maritime industry meet Sustainable Development Goals? An analysis of sustainability reports from the social entrepreneurship perspective. *Transportation Research Part D: Transport and Environment, 78*, 102173.
52. Cigolini, R., et al. (2008). Recent advances in maintenance and facility management. *Production Planning & Control, 19*(4), 279–286.
53. Silvestri, L., et al. (2020). Maintenance transformation through Industry 4.0 technologies: A systematic literature review. *Computers in Industry, 123*, 103335.
54. Franciosi, C., et al. (2020). Measuring maintenance impacts on sustainability of manufacturing industries: From a systematic literature review to a framework proposal. *Journal of Cleaner Production, 260*, 121065.

Understanding the Need of Human–Robot Collaboration in Manufacturing Industries

Vishakha Shadija, Vipin Sharma, and Mohit Pramod Sharma

1 Introduction

Human–robot collaboration (HRC) is the collaboration between humans and robots, with the objective to achieve the shared or common goal by combining their unique abilities. It is different from the human–robot interaction (HRI) [1]. The HRI is referred as the interaction established between humans and robots. Generally, this governs simple and non-collaborative works, for instance, controlling and commanding a robot remotely [2]. Meanwhile, HRC is the field which fosters collaboration among humans and robots in order to gain specific goal [3]. The collaboration established with HRC is majorly task-oriented and focused to get best outcomes by compliments strengths of humans and robots concerned with the task. Thus, HRC turns out to be reliable and efficient [4]. Whereas, HRI works on integrating different levels of interactions concerned for type of interaction adapted [5].

The HRC is generally practiced within same environment or concerning similar tasks. It is a field which perfectly blends different technical development and technologies such as artificial intelligence (AI) and robotics to foster effective and efficient collaboration between humans and robots and further helping to gain satisfactory results [6]. The HRC also introduces collaborative robots or commonly called "Cobots" to the society [7]. Collaborative robots serve handful advantages over traditional robots. One of the major advantages cobots have is that they are significantly designed to interact with humans and function in the shared environment, providing them privilege to directly engage with humans [8].

HRC focuses to make best use of the different capabilities and strengths possessed by humans and robot rather than eliminating or overpowering either from workspace [9]. Robots are made more efficient and intelligent with the development

V. Shadija · V. Sharma (✉) · M. P. Sharma
Department of Mechanical Engineering, Medicaps University, Indore, Madhya Pradesh, India
e-mail: vipinsharma@medicaps.ac.in

Y. Kumar Singla et al. (eds.), *Digital Manufacturing in Industry 5.0*,
https://doi.org/10.1007/978-3-031-91500-0_51

and introduction of numerous advance technologies like AI, sensors and ML, escalating their capability to interpret human emotions and intentions and as well as their decision-making ability and real-time adaptation. HRC also shoulders to enhance human skills like creativity, problem-solving and originality. HRC further integrates with different fields and sectors like healthcare and agriculture, help in optimizing their operation and results [10].

Need of HRC is gaining attention due to various factors like scarcity of workers, operational flexibility, increase in productivity, safety in hazardous tasks, economic effectiveness, etc., which are challenges in the present industrial context [11]. The scarcity of workers impacted industries like manufacturing, healthcare and logistic. It is a serious concern as it limits the growth and production capabilities of an industry . The HRC is a solution to above said problem as it integrates robots to work alongside humans to improve overall operation of the industries. The robots can handle repetitive and laborious task, giving space for humans to focus on creativity, originality and innovation, such distribution of roles optimizes efficiency and outcome of production [12]. A shift in the demand and preferences can be easily dealt by operational flexibility provided by the HRC. Implementation of HRC leads to increase in productivity by automating repetitive task [13]. The HRC integrates the qualities of humans and robots in best possible way to gain escalated outcomes, where human brings qualities like creativity, complex problems solving, decision making and more, while robot introduce accuracy, consistency, endurance and more [14]. Safe working in hazardous tasks is ensured with the HRC by employing robots to perform risky task such as handling toxic chemical material, monitoring mining site, etc. This helped to enhance the overall safety of human and decreases rate of accidents [15]. HRC plays a crucial role in guaranteeing long-term benefit and cost effectiveness, and employing cobot makes it easier to adapt to different working conditions. Hence, one can save from the investment in tools for efficient working. Also, HRC serves long term benefit by improving overall efficiency safety and reducing labor expense [16]. To conclude, HRC helps focusing on strategic investment and enabling industries to conquer the demand and challenges raised [17].

In present scenario, HRC helps industries to operate successfully while meeting the growing demand and facing challenges bought up every other day [18]. The perfect use of strengths and capabilities of humans and robots allows industry to have flexibility in operation and to remain competitive. Hence, HRC makes industries more reliable and efficient while effortlessly beating challenges of modern economy [19].

2 Technologies of HRC

Technological overview includes understanding about some of the important contributing technical elements like sensors, control systems, user interfaces, current developments and trends in the field .

2.1 *Artificial Intelligence*

Artificial Intelligence (AI) makes it possible for robots to learn and decide for themselves, in other words making them adaptive and intellectual [20]. Emerging factors include machine learning, computer vision and decision making. Machine learning provides algorithms which help robot to develop capability to learn and acquire knowledge from data, utilizing the learnings to escalate the performance further. Learning techniques like supervised, unsupervised and reinforcement learning are practiced, further enabling robots to adapt to different dynamics [21]. One of the greatly emerging, advancement in technology is computer vision. With incorporation, computer vision robots can potentially interpret gestures and can navigate through different environment, while AI escalates the results [22]. Real-time decision making is made possible through incorporating AI and analyzing the data and inputs. Making them highly efficient to operate in real-time and deciding on actions to proceed with according to situation encountered [23].

2.2 *Natural Language Processing*

Natural language processing (NLP) is a crucial technical element derived from computer science, human language and artificial intelligence, as a result of perfect blend of three. It is the technical aid utilized by machines to interpret, manipulate and understand human languages. NLP contributes in fostering seamless human–robot collaboration [23]. Speech recognition is basically a tool which helps in verbal communication among robots and humans. This can facilitate with integrating the capability of robot to responding to questions asked by humans and follow command said by them. Under this speech is converted into text by incorporating respective algorithms [24]. NLP makes it possible for robot to interpret and understand human language, helping them in coordinating with humans smoothly [25]. Robots can generate respective response based on input received while ensuring interactive experience. Through NLP robot can give real-time response to input given by humans in form of speech [26].

2.3 *Sensors*

Sensors enable a robot to understand and perceive their environment, making them capable for communicating and functioning safely within a shared environment [27]. Commonly sensors like vision, proximity, environmental and force are used. Robots with the integration of vision sensors can acquire virtual data through components like camera, depth sensor, etc. Infrared camera can also be used for night vision. Through vision sensor robot can detect the objects or obstacles found on their way as well as help to understand and interpret human facial expressions [28].

Proximity sensors including infrared and ultrasonic help robot to detect when people and objects are closed to them, hence making them work efficiently and effectively in a shared environment in a cooperative way and while ensuring safety [29]. Physical interactions with humans and objects are to be held while ensuring the safety. Force and torque sensors assist robot with altering their grip strength and reaction with respect to the working conditions [30]. Basically, the environmental sensors like gas sensor, temperature and humidity sensor, etc., make robot capable of understanding the environment or the surrounding they are present in. These sensors help robot to interpret if any change occurred in environment and further responding accordingly [31].

2.4 Control Systems

Control systems are a crucial part of robot, as they govern the actions of the robot while ensuring safety and seamless human robot collaboration [32]. Feedback control system utilizes feedback loops to escalate the performance making robot capable of reacting and responding to real time changes effectively and taking actions accordingly [33]. Hierarchical control architectures are basically responsible for maintaining responsiveness of robot with respect to input while performing complex task. Integrating these control architectures help robot to function at various level of abstraction [34, 35].

2.5 Interfaces for User

User interface is another crucial element contributing toward seamless human–robot collaboration. It helps with fostering strong communication between humans and robots [36]. In graphical user interfaces (GUIs), these interfaces help user to have access to different control options, feedback and monitor robot at any given instance [37]. Gesture and touch interface incorporate measures for introducing ease in communication such as integration of touch screen, gesture recognition systems and more. With these we can interact and give commands to robot quite easily [38]. Virtual reality (VR) and augmented reality (AR) can be utilized to improve overall interaction among humans and robots. With these, humans can efficiently visualize and control robots [39].

2.6 Integration with Cloud Computing and the Internet of Things

With incorporation of HRC, robots are exposed to large amount of data. With integration of cloud computing and Internet of Things to HRC, we can easily manage such bulk data and interpret insightful information . Remote monitoring and control of robots can be possible with the cloud connectivity. This brings flexibility in operation, escalating efficiency and monitored control [40]. In integration with other advanced technologies such as artificial intelligence, natural language processing, lightweight cobots, user interfaces and IoT, it has been made possible to establish seamless and effective human–robot collaboration. And as we get introduced to more emerging advanced technologies, constructive developments in HRC can be observed .

3 Role of HRC in Industries

HRC has led industries to the path of immense growth and development. With the integration of HRC drastic changes are set to take place in industries. HRC brings various crucial aspects to the table such as customization, workplace efficiency, sustainability, safety, ergonomics and more. Modern industries can be immunized to raised challenges and obstacles through incorporation of HRC [41].

3.1 Industry 4.0

Industry 4.0 is nothing but the term defining fourth industrial revolution. Under this revolution we get introduced to various advance and intelligent technologies [42]. Some of the technologies which established the foundation of this industry are internet of things (IoT), augmented reality (AR), advanced automation and more [43]. HRC helps in ensuring workplace efficiency by getting humans and robots to work harmoniously and cooperatively alongside each other. With integration of HRC to Industry 4.0, we can optimize the performance and output while introducing the real-time analysis and decision making [44].

3.2 Smart Factories

Smart factories are the one which utilizes different technologies such as AI, IOT, big data analytics and more to escalate their production and performance. Smart factories are automated, reliable and have efficient manufacturing systems [45]. With integration of HRC, we get to introduce flexibility to the factories. Also, we can make smart factories to operate in the human centric approach through which we can get products which can meet the standards for production and escalate the efficiency of the factories. Smart factories can potentially meet the rising demand of the market through incorporating HRC, with escalating scale and quality of production [46].

3.3 Adaptable Automation

Adaptable automation is the system specifically designed to be capable of altering their functions according to the operation being carried out. This is made possible through real-time data analysis, helping them to direct the flow of operation [47]. With HRC, we can integrate adaptability to the automation, making them capable of responding to changes and making real-time decisions. HRC serves one of the important factors that is safety along with further ensuring effective and efficient automation. HRC hence lays brick for prosperous path leading toward boom of industry [48].

3.4 Real-Time Data and Decision-Making

Present industries highly revolve around data. Thus, the kind and amount of data available directly affect the industries. HRC serves as bridge for robots and humans to exchange and analyze numerous types and amount of data, constructively directing toward effective and informed decision-making [49]. Integration of HRC can help fostering operational potential and adaptability, making it possible to make efficient real-time decision and effective data analysis. Along with these, HRC can help integrate feedback gained to optimize overall results [50].

HRC plays a crucial role in modern industries as it brings various important factors such as safety, flexibility, efficiency, adaptability and more to direct toward escalated operation and outcome. HRC helps modern industries to overcome the challenges and obstacles faced by them and prosper them.

4 Level of Interaction and Autonomy in HRC

4.1 Levels of Interaction in Human–Robot Collaboration

HRC consists of numerous levels of interactions. Every level possesses a distinct configuration fostering certain kind of interaction among humans and robots, concerned about the kind of task and coordination to be engaged. Different level of interactions collectively defines a spectrum of interaction governing the amount of interaction to be held. By assigning the certain level of interaction for concerned operation can help escalating efficiency and safety [51. Following are the levels of interactions in HRC:

Coexistence: Under this level of interaction, though humans and robot exist and function in a shared environment but they certainly do not enjoy freedom to directly engage in any sort of interaction or work together [52]. Human and robots have their clearly described roles and jobs to do, while ensuring they work independently. We can describe that human and robots are concerned with functioning in parallel to each other [53]. For instance, in a manufacturing process, we can assign robot at conveyor belt and humans for monitoring, while adapting to all safety measures [54].

Synchronous cooperation: Synchronous cooperation level of interaction deals with sequential distribution of tasks among humans and robots. Under this level, both play their part with the aim to ace a shared goal, they work on assigned part concerned with same task. Comparatively, this level fosters closer interaction among humans and robots with certain rules, constraining their freedom for completely free interaction [55]. For example, in an assembly line, we can employee robots for initial stages of assembling components, whereas concern for later stages can be taken over by humans, ensuring quality and standard of outcome [56].

Full collaboration: This level interaction serves complete freedom to humans and robots to interact with each other. Comparatively, this level fosters highest level of interaction among the two. Under full collaboration, humans and robots work quite closely, certainly joining hands to achieve a shared goal. This level integrates numerous advance technologies for safely practicing close interaction between two and enabling them to function alongside each other. For instance, surgical bots are significantly designed to possess higher level of precision and stability, making it possible to employee them to perform different surgeries on humans and other creatures while coordinating with human surgeon [57].

4.2 *Levels of Autonomy in Human–Robot Collaboration*

Levels of autonomy in human–robot collaboration help define the degree to which robots can function independently as well the degree of human intervention required concerned with different tasks. These levels govern how robots and humans will function alongside. The amount of human intervention is derived from the concern of safety factor to adopted, deciding the overall efficiency of collaboration. With the advancements in technology, focus is concentrated toward achieving total independence in robots [58]. Following are the levels of autonomy in HRC:

- **Level 0- No Autonomy:** This level is also referred to as no autonomy or manual operation level. Under this level, robots don't have the freedom to make decisions independently. The functions of robots are completely controlled by humans which means any action to be performed by the robot is to be commanded by the human operator in real time. The robots at this level lack the ability to act, react or make decisions on their own, eliminating their capability to adapt and learn on their own. In simpler terms, this level deals with remote-controlled technology or simple mechanical tools. For example, in battlefields manually operated robotic arms are used for the purpose of bomb disposal, here human operator guides when and where the bomb has to be disposed of. Under this, tasks performed by robots are to be highly supervised by humans in real time [59–61].

- **Level 1- Assistance only:** This level is also termed as assistance only level, where a robot provides helpful information or tools to support humans but it still has no right to make independent actions or decisions. Robot intellectually adds to human's efficiency, awareness and precision. Robot can use sensing devices and other tools that can provide information about various aspects that help human operators make decisions or plan tasks and execute efficiently. For instance, a surgical robot works alongside a human surgeon, which helps with achieving magnified images or highlighting specific areas. Here the robot won't perform any surgical task until a human surgeon commands it to [60–62].
- **Level 2- Shared Control:** This level is known as the shared control level, where humans and robots collectively share the control with each other. Shared control is where human commands and the robot executes, here the human will provide what tasks are to be performed while the robot will work on how the task is to be performed. Robots at this level is capable of performing tasks autonomously using their sensing elements and internal systems but still, it is completely under human supervision. Therefore, the robot still don't enjoy the freedom of working completely on their own rather they possess a certain level of intelligence and independence. End goals and major decisions are taken by humans only. For example, a cleaning robot, navigates through the area and cleans it but the human commands it on what area is to be cleaned. This level thus establishes a semi-structured environment, balancing control between humans and robots [59–63].
- **Level 3- Conditional Autonomy:** This level is referred as conditional autonomy, where the robot has higher-level intelligence and contributes to decision making, but human still has the upper hand in control and decision making. Under this level robots have the capability to sense, process and suggest actions and strategies to humans, helping with optimizing efficiency. Robot actively helps in capturing real-time data along with storing past experiences. Though the robot has gained some level of independence still human operator decides on if robot's suggestion is to be accepted or rejected. For example, an autonomous car actively decides on routes and paths to be followed but a human has control over the actions, it needs confirmation from humans to execute any action [60–63].
- **Level 4- High-Level Commands:** This level is also referred as high-level commands, here the robot can plan and execute on its own. Robots shoulder the responsibility in planning and execution actively. At this level, the human role is to tell the robot what is to be done or define the goal and the robot independently plans on how to efficiently achieve the goal. The robot utilises sensing technologies, algorithms and strategies, break down task into smaller steps, decide on actions respectively and seek human input. The robot may approach humans when it has to deal with unexpected situations or some critical decisions are to be made. This level helps humans to gain freedom from micromanagement while still engaging them with conditional supervision. For example, in the manufacturing process robots take full control from start to end while humans based on requirements engage with the procedure [59–63].

- **Level 5- Full Collaboration:** Also known as the level of peer partnership and is termed as full collaboration. With this level, we get stage of close human-robot interaction rather than working separately here both work as teammates. Under this level humans and robots both shoulder the responsibility of decision making, planning and execution. This level focuses on dynamic working including humans and robots to interact, adapt and coordinate in real-time. Based on condition and situation encountered either of them can take over the role of leader and follower, both work as a team to achieve the shared goal. For example, surgical robots, they work coordinatively with humans in real time.
- **Level 6- Full Autonomy:** This level is termed as full autonomy where the robot enjoys independent operation. This level signifies a higher level of autonomy, enabling robots to have complete freedom where no human intervention is involved. Robots can potentially sense, process, make decisions and execute independently without human command, instruction or approval. Robots in this level can efficiently deal with unexpected situations and conditions, they communicate occasionally and only when necessary. This level integrates artificial intelligence on the large scale, helping robots to possess a higher level of autonomy. For example, an autonomous underwater vehicle explores the underwater region autonomously, it makes decisions, plans and executes independently [60–63].

5 Role of Human in HRC

Humans significantly have played their role in HRC. The aspects served by humans have constructively contributed toward achieving effective and efficient human–robot collaboration. Following describes the human factor in HRC:

5.1 Acceptance and Trust

Human–robot collaboration has trust as its foundation, and thus, it is must to ensure to take steps to strengthen the trust. It directly stakes the way people view and engage with robots, deciding the efficiency of collaboration. If we can potentially higher level of trust, we can get humans to interact and engage with robots more freely and positively. The amount of trust describes the fear of threat to safety and questionable amount of dependency. Thus, it is necessary to build and strengthen the trust [64]. The factors hindering the way for fostering trust includes difficulty in ensuring predictability, questionable reliability, no transparency and more. The more the actions concerned with robots are predictable, it makes humans to feel safer and assured helping bridging successful HRC. Until and unless robot is interpreted to be reliable, the dependability on them concerning certain tasks is affected [65].

5.2 Workload and Usability

Under HRC, the workload is divided among humans and robots, generally based on their individual capabilities, qualities and strengths. However, usability defines the quality of communication that can be practiced.

The workload imbalance on either one can directly affect their productivity and performance, introducing delays and degradation in their functionality. For instance, if a robot is highly dependent on human operator, it will cause human to go through extreme workload, making it to deal with affected concentration. In opposite scenario, where human is highly dependent on robot, it can affect the skills and intellect of human. Thus, it is must to ensure balanced workload, where humans and robots can make best out of their capabilities [66].

User interface governs the quality of communication and user satisfaction, easy to understand and use user interface is must to be prioritized. Factors such as poor navigation and lack of feedback can result in errors and frustration. Focus is on making technology easily acceptable, reliable and controllable, overall making it socially acceptable. The amount of usability technology possesses directly defines the user's satisfaction and experience, governing their future engagements. To ensure constructive collaboration, it is must to ensure positive user-experience [67].

5.3 Collaboration Dynamics

Human–robot collaboration efficiency is defined by the capability of humans and robots to work as a team, coordinate and co-operate with each other. A number of factors define the quality of interaction we can gain [68]. Factors contributing toward collaboration dynamics are:

Shared Goals: Understanding and having clear picture about the desired outcome can help in fostering escalated cooperation and coordination. Once the objective is clear it makes it easy for both to function accordingly.

Role Clarity: Ensuring both humans and robots have clearly gained the insights for the part they are going to play, makes it possible for them to alter their functionality and coordinate. Having clarity about their roles helps eliminate unwanted problems like miscommunication and more.

Adaptability: Adaptability makes it possible to smoothly fit into changes introduced. Whether humans or robots being adaptable makes it possible to attain flexibility and productivity.

Communication: It is crucial to ensure stable and effective communication to achieve successful human–robot collaboration. Quality of communication defines the amount of trust and reliable both have toward each other. We can incorporate feedback system in robots, making them capable of altering and redesigning their operations respectively. If the clarity about intentions and situations is achieved, we can foster enhanced decision-making [69].

Understanding and integrating human factor into HRC can help escalating the overall productivity and efficiency of collaboration. We can focus on prioritizing human satisfaction and experience by interpreting human factor in HRC. Defining the workload and usability accordingly to optimize the collaboration and escalated results.

6 Role of Robotics in HRC

Robotics plays equally important role as humans in human–robot collaboration. Through incorporation of robotics, we can get the assistance with various tasks and operations. Robotics play crucial role in HRC concerned with safety, productivity, automation and more. Following are the roles and objectives served by robotics in HRC:

6.1 The Automation of Tasks

Robots can be employed for tasks which are repetitive and laborious in nature. For example, we can employee robots for tasks like material handling, packaging and assembly line. This allows and provides space for humans to focus on creativity, originality and innovation. Introducing automation through robotics can help gaining precision and reliability concerned with different tasks [70].

6.2 The Cooperation of Humans

Collaborative robots are specifically designed to work alongside humans, while cooperating and coordinating with each other. Collaborative robots are dedicated to build a safe environment for humans to work together toward specific shared goal. Being equipped with different sensors and advanced technologies makes them potential partner to join hands with [71].

6.3 Increased Safety

Through HRC, we can prevent humans from encountering unwanted safety concerns. The goal is to ensure steps and measure are constructively taken for ensuring safety of humans as well as the workplace. We can employ robot in harmful environment and conditions such as chemical handling, mining sites, etc., preventing from pushing humans into threatful conditions. Robots can help with repetitive and laborious tasks, preventing humans from going through fatigue and unwanted stress. Incorporating feedback system can help escalating the overall performance of the robot as well as altering the aspects to align with safety needs [70–72].

6.4 Analysis and Data Collection

Robots have access to large amount of data, raising demand for capability to understand, analyze and interpret data correctly. Robots should be potentially capable of deriving insights from the available data. Robots can track-down when maintenance is required, monitoring their wear-off and performance, helping to achieve minimized downtime and escalate reliability on them [72].

6.5 Education and Training

With incorporation of robotics, humans get space to work on their skills and learn different things. Focusing on fostering individualized learning experience and also helping students with special needs, robots can help building healthy learning environment [73].

6.6 Support in the Medical Field

Robots can potentially serve in medical and healthcare industry. With the qualities such as precision, stability and adaptability, makes it possible to employee robots in assisting different surgeries alongside human doctor. Robots can help in drug development, disease diagnosis, speedy recovery and more. Robots can be employed for patient care and helping them with mobility [74].

6.7 Artificial Intelligence Integration

Integration of AI and machine learning helps in developing intelligent robots. Also, it makes it possible to incorporate adaptability and making them capable to learn from experiences and interactions. Another crucial technical aid introduced is natural language processing, further helping in verbal communication and control over robots [75].

6.8 *Increasing Efficiency and Productivity*

By utilizing the capabilities, strengths and qualities possessed by robots helps gaining optimum outcomes. Introducing robotics to workflow directs toward escalating efficiency and productivity. Robotics can also serve to eliminate heavy-costing and fostering competitiveness in the market. The operations can be made to have personalized or customized touch [76].

7 Application of HRC

Due to various advantages offered by HRC they find their application in various field HRC potential revolutionize and direct industries toward prosperous path HRC can help in escalating quality of life while ensuring safety.

7.1 *Assembly and Manufacturing*

We can employee collaborative robot for performing task like quality control, packaging and more while working alongside humans. Robots can be employed for handling all the physical labor and repetitive task while humans can be employed to monitor, explore creative and innovative aspects. Robots integrated with advance technologies can help escalating accuracy and speed, reducing the workload on humans. Through HRC, we can easily meet the market standards and demand [77].

7.2 *Medication and Surgery*

Robots due to their capability and strength find their application in the field of healthcare. Robot can be employed for various purposes such as performing surgery, diagnosis, physical therapy, rehabilitation and more. Through HRC we can design personalized medical treatment, assist patience facing problem with mobility. Because of the precision and stability offered by robot they are employee to work alongside human surgeon and help with performing various surgeries [78].

7.3 Services and Hospitality

We can incorporate HRC in service and hospitality, helping to manage workload. Robots can be assigned for various tasks such as welcoming guest, helping with check-in in hotels, giving product information, food delivery and more. Robots possessing cleaning skills can be employed for taking care of the cleanliness, while lowering the demand for manual labor [79].

7.4 Outdoor and Agricultural Work

We can employee HRC to escalate the overall agriculture and outdoor work. For instance, we can employee autonomous drones for the purpose of crop monitoring. We can employee robot to monitor soil, further helping to get more customize farming method. Overall, robots can help escalating the productivity and production in agriculture. Robots can be employed to handle bulky task and also the tasks where human safety is at threat such as chemical handling and more [80].

7.5 Warehousing and Logistics

Incorporating HRC in warehouse and logistics, we can easily move items within warehouse, effective management and storage procedure with the help of robots. Meanwhile, humans can focus on quality control and other value-based task [81].

7.6 Search and Rescue Activities

Robots can be employed in disastrous situation such as earthquake, floods, etc. Robots, being equipped with sensors and other technologies, make it possible for them to help in human rescue and navigate through the surrounding [82].

7.7 Training and Education

HRC can help escalating the quality of education and provide assistance with training. Robots can provide individualized tutoring to the students and helping them understand complex concepts. For instance, we can employee robots in industries such as manufacturing, healthcare and more to help people with escalating their learning skills [83].

7.8 Care and Assistance for the Elderly

Robot can help with regular monitoring of health, providing reminders to take medicine and help upgrading the family about the health of elderly person. Robots can help elderly with mobility, enabling them to safely move around. Companion robots are specifically design to provide a company, take care and help the elder ones [84].

7.9 Development and Research

By integrating HRC in development and research, we can escalate the quality of results and productivity. Robots can help with numerous tasks such as data collection, analysis and more. Through incorporation of HRC, we can save our time and get assistance with a range of scientific domains [85].

8 Challenges

In order to successfully implement human–robot collaboration (HRC) across industries, it is must to navigate through a number of obstacles and factors. Here is a thorough description of these difficulties.

8.1 Task Intricacy and Uncertainty

The intricacy of tasks in HRC environments can present considerable difficulties, particularly in dynamic or unstructured settings. In addition to carrying out predetermined tasks, robots are required to be able to adjust to erratic human behavior and as well as the environmental changes [86]. To accurately perceive and interpret human movements, gestures and intentions in real time, robots should be equipped with sophisticated sensors along with artificial intelligence algorithms. Effective cooperation depends on the development of robots that can instantly modify their behavior in response to shifting circumstances, that is, being highly adaptable. Strong decision-making algorithms and real-time data processing skills are needed for making robots adaptable [87].

8.2 Trust and Protection

In settings, where people and robots interact frequently, ensuring safety is crucial. It is essential to create a trustworthy and secure collaborative framework in order to avoid mishaps and injuries. In order for human workers to anticipate the actions of robots, they must display predictable behaviors. Because of this predictability, shared workspaces feel more safer and more in control. To handle possible malfunctions or mishaps involving robot's explicit accountability, procedures need to be established and adopted. This includes outlining who is responsible for human–robot malfunctions or misunderstandings [88].

8.3 Morality and Social Acceptability

The implementation of HRC brings up a number of ethical issues, especially in relation to job displacement and the social role of robots. Although, HRC can increase productivity, there is misconception that some jobs may be lost as robots replace humans in some jobs. Retraining and reskilling initiatives must address these issues in order to guarantee a seamless transition for impacted employees. The use of robots in positions involving interaction with the public may give rise to concerns regarding data security and privacy. To safeguard people's rights and guaranteed accountability for robotic actions, clear policies and regulations are required. Different cultures and demographic groups have different attitudes toward robots. Addressing community expectations, misconceptions and anxieties regarding HRC can help in improving public opinion and promote more seamless integration [89].

8.4 Cost and Scalability

The high initial cost of technology deployment and the continuous requirement for specialized training for human workers can make it difficult to scale HRC implementations.

The high cost of collaborative robots, commonly called "cobots" and advanced robotic systems, may discourage smaller businesses from implementing HRC. However, broader adoption becomes more possible as costs come down by the time and technology improves. To ensure that humans and robots work together effectively, specific training programs are needed to give employees, they should know how to operate and communicate with robotic systems. Creating thorough and easily accessible training programs is essential for successful deployment [90].

Determining the extent of human–robot collaboration entails tackling a number of issues such as task complexity, safety, moral dilemmas and scalability. By taking proactive measures to resolve these problems, organizations can establish overall a more reliable and efficient working environment that optimizes HRC advantages, while reducing any potential disadvantages. For HRC, to successfully integrate across a variety of industries, continuous research and discussion regarding these issues will be crucial while focusing on technology to develop [91].

9 Future Research

Though HRC is great achievement of technical advancement, it is still a continuously growing and developing field. Additional research goals for human–robot cooperation (HRC) cover aspects mentioned below:

Improving Robot Adaptability and Learning: Examine strategies for enhancing robot's capacity to comprehend human interactions in real-time. Examine adaptive algorithms that allow robots to alter their behavior in response to human input or dynamic changes in their surroundings.

Improving Human–Robot Communication: Provide sophisticated communication protocols which can help to improve human–robot interaction. Examine the efficiency of multimodal communication techniques, for example, in promoting comprehension and cooperation, including verbal, non-verbal and visual cues.

Evaluating Human Factors: Examine the social and psychological aspects that affect people's acceptance and trust of robots in a given group settings. Examine how human perceptions of control and trust are impacted by different robot levels of autonomy.

Productivity and Performance: Create metrics to evaluate HRC systems productivity and performance across a range of applications. Evaluate the effectiveness of collaborative robots in comparison to more conventional automation techniques by conducting comparative studies and developing clear understanding regarding two.

User-Centric Interfaces: Investigating and creating user-friendly interfaces that are easy for people to use with robots. Examine and analyze how virtual and augmented reality might improve HRC operation and training.

Ethical and Societal Implications: Examine the moral issues surrounding the implementation of HRC systems, paying particular attention to concerns about accountability, privacy and job displacement. Surveys and qualitative research should be conducted to find out how the public feels about collaborative robots and what factors affect acceptance.

Enhancing Safety Procedures: Investigate safety guidelines and procedures to guarantee that people and robots can interact safely in shared workspaces potentially. Implementing real-time monitoring systems capable of identifying and reducing risks, when humans and robots work together.

Investigating Multi-disciplinary Methods: Promote interdisciplinary cooperation, for example utilizing cognitive science, robotics and sociology, collectively helping to create comprehensive HRC systems. Investigating how social knowledge can help create collaborative robots that are more likable and successful.

Examining Applications: Investigate in-depth the unique requirements and difficulties faced by HRC across a range of sectors including logistics, healthcare, agriculture and more. Investigate cutting-edge HRC applications in developing domains like smart homes, education, elder care and more.

HRC Solutions Scaling: Investigating possible methods for implementing HRC technologies in various sectors and at various organizational sizes as well as levels. Understanding different aspects such as the financial effects and return on investment of deploying HRC systems on a large scale.

Long-Term HRC Effects: Learning how HRC affects human workers over the long run, while taking into account factors like career advancement, skill development and job satisfaction. Interpreting the effects on workplace dynamics and culture bought into picture by long term human–robot collaboration.

Decision-Making: Assisting robots to make decisions by developing frameworks that take human preferences and input into account. For instance, getting insights into how AI might improve robot's ability to make decisions in cooperative settings.

Through addressing current issues and expanding our knowledge of HRC, we potentially hope to create collaborative robots that are safer, more effective and more socially acceptable across a range of industries. The development of HRC technologies and their effective incorporation into daily applications can be greatly aided by future research in these fields.

References

1. Ajoudani, A., Zanchettin, A. M., Ivaldi, S., Albu-Schäffer, A., Kosuge, K., & Khatib, O. (2018). Progress and prospects of the human–robot collaboration. *Autonomous Robots, 42*, 957–975.
2. Bauer, A., Wollherr, D., & Buss, M. (2008). Human–robot collaboration: A survey. *International Journal of Humanoid Robotics, 5*(01), 47–66.
3. Matheson, E., Minto, R., Zampieri, E. G., Faccio, M., & Rosati, G. (2019). Human–robot collaboration in manufacturing applications: a review. *Robotics, 8*(4), 100.
4. Vysocky,A.,&Novak,P.(2016).Human-robotcollaborationinindustry.*MMScienceJournal,9*(2), 903–906.
5. Murphy, R. R., Nomura, T., Billard, A., & Burke, J. L. (2010). Human–robot interaction. *IEEE robotics & Automation Magazine, 17*(2), 85–89.
6. Green, S. A., Billinghurst, M., Chen, X., & Chase, J. G. (2008). Human–robot collaboration: A literature review and augmented reality approach in design. *International Journal of Advanced Robotic Systems, 5*(1), 1.
7. Simões, A. C., Pinto, A., Santos, J., Pinheiro, S., & Romero, D. (2022). Designing human-robot collaboration (HRC) workspaces in industrial settings: A systematic literature review. *Journal of Manufacturing Systems, 62*, 28–43.
8. De Simone, V., Di Pasquale, V., Giubileo, V., & Miranda, S. (2022). Human–robot collaboration: An analysis of worker's performance. *Procedia Computer Science, 200*, 1540–1549.
9. Semeraro, F., Griffiths, A., & Cangelosi, A. (2023). Human–robot collaboration and machine learning: A systematic review of recent research. *Robotics and Computer-Integrated Manufacturing, 79*, 102432.

10. Asaad, H., Askar, S., Kakamin, A., & Nayla, F. A. I. Q. (2024). Exploring the impact of artificial intelligence on humanrobot cooperation in the context of Industry 4.0. *Applied Computer Science, 20*(2), 138–156.
11. Villani, V., Pini, F., Leali, F., & Secchi, C. (2018). Survey on human–robot collaboration in industrial settings: Safety, intuitive interfaces and applications. *Mechatronics, 55*, 248–266.
12. Costa, G. D. M., Petry, M. R., & Moreira, A. P. (2022). Augmented reality for human–robot collaboration and cooperation in industrial applications: A systematic literature review. *Sensors, 22*(7), 2725.
13. Saenz, M., Revilla, E., & Simón, C. (2020). Designing AI systems with human-machine teams. *MIT Sloan Management Review.*
14. Valori, M., Scibilia, A., Fassi, I., Saenz, J., Behrens, R., Herbster, S., et al. (2021). Validating safety in human–robot collaboration: Standards and new perspectives. *Robotics, 10*(2), 65.
15. Arents, J., Abolins, V., Judvaitis, J., Vismanis, O., Oraby, A., & Ozols, K. (2021). Human–robot collaboration trends and safety aspects: A systematic review. *Journal of Sensor and Actuator Networks, 10*(3), 48.
16. Robla-Gómez, S., Becerra, V. M., Llata, J. R., Gonzalez-Sarabia, E., Torre-Ferrero, C., & Perez-Oria, J. (2017). Working together: A review on safe human-robot collaboration in industrial environments. *IEEE Access, 5*, 26754–26773.
17. Maurtua, I., Ibarguren, A., Kildal, J., Susperregi, L., & Sierra, B. (2017). Human–robot collaboration in industrial applications: Safety, interaction and trust. *International Journal of Advanced Robotic Systems, 14*(4), 1729881417716010.
18. Gervasi, R., Mastrogiacomo, L., & Franceschini, F. (2020). A conceptual framework to evaluate human–robot collaboration. *The International Journal of Advanced Manufacturing Technology, 108*, 841–865.
19. Freedy, A., DeVisser, E., Weltman, G., & Coeyman, N. (2007). Measurement of trust in human–robot collaboration. In *2007 International Symposium on Collaborative Technologies and Systems* (pp. 106–114). IEEE.
20. Cesta, A., Orlandini, A., & Umbrico, A. (2018). Fostering robust human–robot collaboration through AI task planning. *Procedia CIRP, 72*, 1045–1050.
21. Adawy, M., Abualese, H., Taha El-Omari, N. K., & Alawadhi, A. (2024). Human–robot interaction (HRI) using machine learning (ML): A survey and taxonomy. *International Journal of Advances in Soft Computing & Its Applications, 16*(3), 166.
22. Robinson, N., Tidd, B., Campbell, D., Kulić, D., & Corke, P. (2023). Robotic vision for human-robot interaction and collaboration: A survey and systematic review. *ACM Transactions on Human-Robot Interaction, 12*(1), 1–66.
23. Liu, Q., Liu, Z., Xu, W., Tang, Q., Zhou, Z., & Pham, D. T. (2019). Human-robot collaboration in disassembly for sustainable manufacturing. *International Journal of Production Research, 57*(12), 4027–4044.
24. Stiefelhagen, R., Fugen, C., Gieselmann, R., Holzapfel, H., Nickel, K., & Waibel, A. (2004, September). Natural human-robot interaction using speech, head pose and gestures. *2004 IEEE/RSJ International Conference on Intelligent Robots and Systems (IROS)*(IEEE Cat. No. 04CH37566) (Vol. 3, pp. 2422-2427). IEEE.
25. Doshi, F., & Roy, N. (2008). Spoken language interaction with model uncertainty: An adaptive human–robot interaction system. *Connection Science, 20*(4), 299–318.
26. Foster, M. E. (2019). Natural language generation for social robotics: opportunities and challenges. *Philosophical Transactions of the Royal Society B, 374*(1771), 20180027.
27. Ogenyi, U. E., Liu, J., Yang, C., Ju, Z., & Liu, H. (2019). Physical human–robot collaboration: Robotic systems, learning methods, collaborative strategies, sensors, and actuators. *IEEE Transactions on Cybernetics, 51*(4), 1888–1901.
28. Mohammed, A., Schmidt, B., & Wang, L. (2017). Active collision avoidance for human–robot collaboration driven by vision sensors. *International Journal of Computer Integrated Manufacturing, 30*(9), 970–980.

29. Buerkle, A., Matharu, H., Al-Yacoub, A., Lohse, N., Bamber, T., & Ferreira, P. (2022). An adaptive human sensor framework for human–robot collaboration. *The International Journal of Advanced Manufacturing Technology, 119*, 1–16.
30. Al-Yacoub, A., Zhao, Y. C., Eaton, W., Goh, Y. M., & Lohse, N. (2021). Improving human robot collaboration through Force/Torque based learning for object manipulation. *Robotics and Computer-Integrated Manufacturing, 69*, 102111.
31. Pedrocchi, N., Vicentini, F., Matteo, M., & Tosatti, L. M. (2013). Safe human-robot cooperation in an industrial environment. *International Journal of Advanced Robotic Systems, 10*(1), 27.
32. Galin, R. R., & Meshcheryakov, R. V. (2020). Human–robot interaction efficiency and human–robot collaboration. In *Robotics: Industry 4.0 issues & new intelligent control paradigms* (pp. 55–63). Springer International Publishing.
33. Geravand, M., Flacco, F., & De Luca, A. (2013, May). Human-robot physical interaction and collaboration using an industrial robot with a closed control architecture. *2013 IEEE international conference on robotics and Automation* (pp. 4000-4007). IEEE.
34. Haddadin, S., & Croft, E. (2016). Physical human–robot interaction. In *Springer handbook of robotics* (pp. 1835–1874). Springer International Publishing.
35. Dahiya, A., Aroyo, A. M., Dautenhahn, K., & Smith, S. L. (2023). A survey of multi-agent Human–Robot Interaction systems. *Robotics and Autonomous Systems, 161*, 104335.
36. Chu, C. H., & Liu, Y. L. (2023). Augmented reality user interface design and experimental evaluation for human–robot collaborative assembly. *Journal of Manufacturing Systems, 68*, 313–324.
37. Prati, E., Peruzzini, M., Pellicciari, M., & Raffaeli, R. (2021). How to include User eXperience in the design of Human-Robot Interaction. *Robotics and Computer-Integrated Manufacturing, 68*, 102072.
38. Yang, Y., & Zhang, X. (2007). Human-robot collaboration in industrial assembly: A review. *International Journal of Advanced Manufacturing Technology, 33*(11–12), 1169–1180.
39. Dianatfar, M., Latokartano, J., & Lanz, M. (2021). Review on existing VR/AR solutions in human–robot collaboration. *Procedia CIRP, 97*, 407–411.
40. Wang, L., Liu, S., Liu, H., & Wang, X. V. (2020). Overview of human-robot collaboration in manufacturing. *Proceedings of 5th International Conference on the Industry 4.0 Model for Advanced Manufacturing: AMP 2020* (pp. 15-58). Springer International Publishing.
41. Li, W., Hu, Y., Zhou, Y., & Pham, D. T. (2024). Safe human–robot collaboration for industrial settings: A survey. *Journal of Intelligent Manufacturing, 35*, 311–328.
42. Weiss, A., Wortmeier, A. K., & Kubicek, B. (2021). Cobots in industry 4.0: A roadmap for future practice studies on human–robot collaboration. *IEEE Transactions on Human-Machine Systems, 51*(4), 335–345.
43. Baratta, A., Cimino, A., Gnoni, M. G., & Longo, F. (2023). Human robot collaboration in Industry 4.0: A literature review. *Procedia Computer Science, 217*, 1887–1895.
44. Porubčinová, M., & Fidlerová, H. (2020). Determinants of industry 4.0 technology adaption and human-Robot collaboration. *Vedecké Práce Materiálovotechnologickej Fakulty Slovenskej Technickej Univerzity v Bratislave so Sídlom v Trnave, 28*(46), 10–21.
45. Grabowska, S. (2020). Smart factories in the age of Industry 4.0. *Management Systems in Production Engineering, 28*(2), 90–96.
46. Evjemo, L. D., Gjerstad, T., Grøtli, E. I., & Sziebig, G. (2020). Trends in smart manufacturing: Role of humans and industrial robots in smart factories. *Current Robotics Reports, 1*, 35–41.
47. Sauer, J., Kao, C. S., & Wastell, D. (2012). A comparison of adaptive and adaptable automation under different levels of environmental stress. *Ergonomics, 55*(8), 840–853.
48. Ahmed, W., & Tao, H. (2024). *Human–robot collaboration in assembly lines: Adaptive control strategies for parallel kinematics mechanisms*. EasyChair Preprint.
49. Argyrou, A., Giannoulis, C., Sardelis, A., Karagiannis, P., Michalos, G., & Makris, S. (2018). A data fusion system for controlling the execution status in human–robot collaborative cells. *Procedia CIRP, 76*, 193–198.

50. Khang, A., Rath, K. C., Satapathy, S. K., Kumar, A., Das, S. R., & Panda, M. R. (2023). Enabling the future of manufacturing: Integration of robotics and IoT to smart factory infrastructure in Industry 4.0. In *Handbook of research on AI-based technologies and applications in the era of the metaverse* (pp. 25–50). IGI Global.
51. Scholtz, J. (2003, January). Theory and evaluation of human robot interactions. *Proceedings of the 36th Annual Hawaii International Conference on System Sciences, 2003*. (p. 10). IEEE.
52. Magrini, E., Ferraguti, F., Ronga, A. J., Pini, F., De Luca, A., & Leali, F. (2020). Human-robot coexistence and interaction in open industrial cells. *Robotics and Computer-Integrated Manufacturing, 61*, 101846.
53. Aaltonen, I., Salmi, T., & Marstio, I. (2018). Refining levels of collaboration to support the design and evaluation of human–robot interaction in the manufacturing industry. *Procedia CIRP, 72*, 93–98.
54. Sadrfaridpour, B., & Wang, Y. (2017). Collaborative assembly in hybrid manufacturing cells: An integrated framework for human–robot interaction. *IEEE Transactions on Automation Science and Engineering, 15*(3), 1178–1192.
55. Darvish, K., Wanderlingh, F., Bruno, B., Simetti, E., Mastrogiovanni, F., & Casalino, G. (2018). Flexible human–robot cooperation models for assisted shop-floor tasks. *Mechatronics, 51*, 97–114.
56. Leu, J., Cheng, Y., Tomizuka, M., & Liu, C. (2022, July). Robust task planning for assembly lines with humanrobot collaboration. *Proceedings of the International Symposium on Flexible Automation 2022 International Symposium on Flexible Automation* (pp. 188-195). The Institute of Systems, Control and Information Engineers.
57. Cherubini, A., Passama, R., Crosnier, A., Lasnier, A., & Fraisse, P. (2016). Collaborative manufacturing with physical human–robot interaction. *Robotics and Computer-Integrated Manufacturing, 40*, 1–13.
58. Sadrfaridpour, B., Saeidi, H., & Wang, Y. (2016, August). An integrated framework for human-robot collaborative assembly in hybrid manufacturing cells. *2016 IEEE international conference on automation science and engineering (CASE)* (pp. 462-467). IEEE.
59. Beer, J. M., Fisk, A. D., & Rogers, W. A. (2014). Toward a framework for levels of robot autonomy in humanrobot interaction. *Journal of human-robot interaction, 3*(2), 74.
60. Kunze, L., Hawes, N., Duckett, T., Hanheide, M., & Krajník, T. (2018). Artificial intelligence for long-term robot autonomy: A survey. *IEEE Robotics and Automation Letters, 3*(4), 4023–4030.
61. Yanco, H. A., & Drury, J. L. (2002, November). A taxonomy for human-robot interaction. *Proceedings of the AAAI fall symposium on human-robot interaction* (pp. 111-119).
62. Selvaggio, M., Cognetti, M., Nikolaidis, S., Ivaldi, S., & Siciliano, B. (2021). Autonomy in physical human-robot interaction: A brief survey. *IEEE Robotics and Automation Letters, 6*(4), 7989–7996.
63. Tan, N., Hayat, A. A., Elara, M. R., & Wood, K. L. (2020). A framework for taxonomy and evaluation of selfreconfigurable robotic systems. *IEEE Access, 8*, 13969–13986.
64. Meissner, A., Trübswetter, A., Conti-Kufner, A. S., & Schmidtler, J. (2020). Friend or foe? Understanding assembly workers' acceptance of human-robot collaboration. *ACM Transactions on Human-Robot Interaction (THRI), 10*(1), 1–30.
65. Charalambous, G., Fletcher, S., & Webb, P. (2016). The development of a scale to evaluate trust in industrial human–robot collaboration. *International Journal of Social Robotics, 8*, 193–209.
66. Liu, L., Schoen, A. J., Henrichs, C., Li, J., Mutlu, B., Zhang, Y., & Radwin, R. G. (2024). Human robot collaboration for enhancing work activities. *Human Factors, 66*(1), 158–179.
67. Prewett, M. S., Johnson, R. C., Saboe, K. N., Elliott, L. R., & Coovert, M. D. (2010). Managing workload in human–robot interaction: A review of empirical studies. *Computers in Human Behavior, 26*(5), 840–856.

68. Kemény, Z., Váncza, J., Wang, L., & Wang, X. V. (2021). Human–robot collaboration in manufacturing: A multi-agent view. In *Advanced human–robot collaboration in manufacturing* (pp. 3–41). Springer International Publishing.
69. Jahanmahin, R., Masoud, S., Rickli, J., & Djuric, A. (2022). Human–robot interactions in manufacturing: A survey of human behavior modeling. *Robotics and Computer-Integrated Manufacturing, 78*, 102404.
70. Iqbal, T., Rack, S., & Riek, L. D. (2016). Movement coordination in human–robot teams: A dynamical systems approach. *IEEE Transactions on Robotics, 32*(4), 909–919.
71. Mörtl, A., Lorenz, T., Vlaskamp, B. N., Gusrialdi, A., Schubö, A., & Hirche, S. (2012). Modeling inter-human movement coordination: synchronization governs joint task dynamics. *Biological Cybernetics, 106*, 241–259.
72. Eder, K., Harper, C., & Leonards, U. (2014, August). Towards the safety of human-in-the-loop robotics: Challenges and opportunities for safety assurance of robotic co-workers'. *The 23rd IEEE International Symposium on Robot and Human Interactive Communication* (pp. 660-665). IEEE.
73. Wang, L., Wang, X. V., Váncza, J., & Kemény, Z. (Eds.). (2021). *Advanced human-robot collaboration in manufacturing* (pp. 74–84). Springer.
74. Holland, J., Kingston, L., McCarthy, C., Armstrong, E., O'Dwyer, P., Merz, F., & McConnell, M. (2021). Service robots in the healthcare sector. *Robotics, 10*(1), 47.
75. Dhanabalan, T., & Sathish, A. (2018). Transforming Indian industries through artificial intelligence and robotics in industry 4.0. *International Journal of Mechanical Engineering and Technology, 9*(10), 835–845.
76. Goel, R., & Gupta, P. (2020). Robotics and industry 4.0. *A roadmap to industry 4.0: Smart production, sharp business and sustainable development*, 157-169.
77. Chowdhury, S. R., Das, D. C., Sunna, T. C., Beyene, J., & Hossain, A. (2023). Global and regional prevalence of multimorbidity in the adult population in community settings: a systematic review and meta-analysis. *EClinicalMedicine, 57*.
78. Ait Ameur, M. A., Yang, E., & Brady, S. (2024, August). Advancing healthcare through intelligent human-robot collaboration: overview, challenges and opportunities. *2024 29th International Conference on Automation and Computing (ICAC)*. (pp. 1-6). IEEE.
79. Leocádio, D., Guedes, L., Oliveira, J., Reis, J., & Melão, N. (2024). Customer service with AI-powered humanrobot collaboration (HRC): A literature review. *Procedia Computer Science, 232*, 1222–1232.
80. Adamides, G., & Edan, Y. (2023). Human–robot collaboration systems in agricultural tasks: A review and roadmap. *Computers and Electronics in Agriculture, 204*, 107541.
81. Ponnambalam, S. G., Chang, Q., Zhong, R. Y., Kucukkoc, I., & Janardhanan, M. N. (2023). Human-robot collaboration in the next generation manufacturing and logistics system. *Flexible Services and Manufacturing Journal, 35*(4), 975–978.
82. Potdar, A., & Fitzgerald, T. (2024). Toward enabling crew resource management objectives in human-robot collaboration in safety-critical situations.
83. Sievers, T. S., Schmitt, B., Petersen, M., & Tracht, K. (2020). Integration of collaborative robotics in vocational training using mixed reality. In *Annals of scientific society for assembly, handling and industrial robotics* (pp. 275–283). Springer Berlin Heidelberg.
84. Martinez-Martin, E., & del Pobil, A. P. (2017). Personal robot assistants for elderly care: An overview. In *Personal assistants: Emerging computational technologies* (pp. 77–91).
85. Kofer, D., Bergner, C., Deuerlein, C., Schmidt-Vollus, R., & Heß, P. (2021). Human–robot-collaboration: Innovative processes, from research to series standard. *Procedia CIRP, 97*, 98–103.
86. Wang, S., Xu, Z., Wu, C., Hua, L., & Zhu, D. (2023). Towards region-based robotic machining system from erspective of intelligent manufacturing: a technology framework with case study. *Journal of Manufacturing Systems, 70*, 451–463.
87. Kok, B. C., & Soh, H. (2020). Trust in robots: Challenges and opportunities. *Current Robotics Reports, 1*(4), 297–309.

88. Fryman, J., & Matthias, B. (2012, May). Safety of industrial robots: From conventional to collaborative applications. In *ROBOTIK 2012; 7th German Conference on Robotics* (pp. 1–5). VDE.
89. De Graaf, M. M. (2016). An ethical evaluation of human–robot relationships. *International journal of social robotics, 8*, 589–598.
90. Gaede, C., Ranz, F., Hummel, V., & Echelmeyer, W. (2020). A study on challenges in the implementation of human–robot collaboration. *J. Eng. Manag. Oper, 1*, 29–39.
91. Rodriguez-Guerra, D., Sorrosal, G., Cabanes, I., & Calleja, C. (2021). Human-robot interaction review: Challenges and solutions for modern industrial environments. *IEEE Access, 9*, 108557–108578.

Index

Y. Kumar Singla et al. (eds.), *Digital Manufacturing in Industry 5.0*,
https://doi.org/10.1007/978-3-031-91500-0

B

C

D

E

F

G

H

I

M

N

O

P

GPSR Compliance

The European Union's (EU) General Product Safety Regulation (GPSR) is a set of rules that requires consumer products to be safe and our obligations to ensure this.

If you have any concerns about our products, you can contact us on ProductSafety@springernature.com

In case Publisher is established outside the EU, the EU authorized representative is:

Springer Nature Customer Service Center GmbH
Europaplatz 3
69115 Heidelberg, Germany

Batch number: 10484720

Printed by Printforce, the Netherlands